REGISTRATION FORM — FILL THIS FORM IN NOW!

HANDBOOK OF NEURAL COMPUTATION

Thank you for purchasing the *Handbook of Neural Computation*. Please complete the information below and return this registration card so we may promptly notify you of future updates. **It is essential that you return this registration card** in order to receive the first two supplements which are included in the purchase price of the Handbook.

Delivery address

Name: ....

Address: ....
                 ....
                 ....

City: ....

Zip code/Post code: ....

Country: ....   State: ....

E-mail: ....

Invoice address (if different)

Name: ....

Address: ....
                 ....
                 ....

City: ....

Zip code/Post code: ....

Country: ....   State: ....

E-mail: ....

Please return this card to

Handbook of Neural Computation
Electronic Publishing Department
Oxford University Press
198 Madison Avenue
New York, NY 10016-4314
USA

Tel: (212) 726-6246
Fax: (212) 726-6442
E-mail: techsupport@oup-usa.org

or

Handbook of Neural Computation
Customer Services Department
IOP Publishing Ltd
Techno House, Redcliffe Way, Bristol BS1 6NX
UK

Tel: (+44) 117 929 7481
Fax: (+44) 117 929 4318
E-mail: custserv@ioppublishing.co.uk

For the latest information on the *Handbook of Neural Computation* visit one of our websites at:

http://www.oup-usa.org/acadref/honc.html

or

http://www.iop.org/Books/Catalogue/0750303123

T0133077

Handbook of Neural Computation

Handbook of
Neural
Computation

Editors in Chief

Emile Fiesler and Russell Beale

INSTITUTE OF PHYSICS PUBLISHING
Bristol   Philadelphia

and

OXFORD UNIVERSITY PRESS
New York   Oxford
1997

INSTITUTE OF PHYSICS PUBLISHING
Bristol   Philadelphia

and

OXFORD UNIVERSITY PRESS
Oxford   New York
Athens   Auckland   Bangkok   Bogotá
Bombay   Buenos Aires   Calcutta   Cape Town
Dar es Salaam   Delhi   Florence   Hong Kong   Istanbul
Karachi   Kuala Lumpur   Madras   Madrid   Melbourne
Mexico City   Nairobi   Paris   Singapore
Taipei   Tokyo   Toronto

and associated companies in

Berlin   Ibadan

Published by Institute of Physics Publishing,
Techno House, Redcliffe Way, Bristol BS1 6NX, United Kingdom
(US Editorial Office: The Public Ledger Building, Suite 1035, 150 South Independence Mall West,
Philadelphia, PA 19106, USA)
and Oxford University Press, Inc., 198 Madison Avenue, New York, New York 10016, USA

British Library Cataloguing-in-Publication Data and
Library of Congress Cataloging-in-Publication Data are available

ISBN 0 7503 0312 3

This handbook is a joint publication of Institute of Physics Publishing
and Oxford University Press

PROJECT STAFF

INSTITUTE OF PHYSICS PUBLISHING

Publisher: Robin Rees
Project Editor: Sarah Hood
Production Editor: Neil Scriven
Production Manager: Sharon Toop
Assistant Production Manager: Jenny Troyano
Production Assistant: Sarah Plenty
Electronic Production Manager: Tony Cox

OXFORD UNIVERSITY PRESS

Senior Editor: Sean Pidgeon
Project Editor: Matthew Giarratano
Editorial Assistant: Merilee Johnson
Cover Design: Joan Greenfield

Printing (last digit): 9 8 7 6 5 4 3 2 1
Printed in the United Kingdom on acid-free paper

Contents

Preface

The current era of human history has been termed the Information Age. Our new array of information media still includes those relics of a previous era, printed books and journals, but has been expanded immeasurably by the addition of digital modes of information storage and transmission. These media provide a repository for the increasingly distributed and diverse collection of data, theories, models and ideas that constitutes the universe of human knowledge. It might also be argued that the dissemination of information has been one of the successes of this era, although it is important to make the distinction between information volume and effectiveness of distribution. In the academic arena, it seems clear that the quantity of new research materials makes it increasingly difficult to access what is genuinely relevant and useful, as the usual collection mechanisms (libraries, journals, conference proceedings) have become overloaded.

This information explosion has been a particular characteristic of the field of neural computing, which has seen, in the last 10 years, a rapid increase in the number of published papers, together with many new monographs and textbooks. It is this information overload that the *Handbook of Neural Computation* aims to address, by providing a central resource of material that is continually updated and refreshed. It distills the information and expertise of the whole community into a structured set of articles written by leading researchers. Such a reference is of little use if it does not evolve in parallel with the field that it claims to represent; to remain current and useful, therefore, the handbook will be updated by means of regular supplements, allowing it to mirror the continuing development of the field.

Neural computation is at the center of a new kind of multidisciplinary research that adapts natural paradigms and applies them to practical problems. Artificial neural networks are useful tools that have been applied successfully in a broad range of environments (as witnessed by the case studies in Part G of this handbook), and yet they have an intrinsic complexity that provides a continuing stimulus to theoretical investigations. These interesting aspects of the field have attracted a diverse research community. For example, neural networks attract the interest of computer scientists because, as designers of computing systems, they are interested in the possibilities that the technology holds. Engineers, users of the technology, are interested to see how effective the approach can be and therefore want to understand the operational characteristics of networks. Because of their relationship with models of human information processing, neural networks are investigated by psychologists and others interested in human capabilities. Mathematicians and physicists find application for their previously developed tools in modeling complex, dynamic systems, while discovering new challenges that require different techniques. This heterogeneous mix of backgrounds provides the community with a many-pronged attack on the problems posed by the field, with a lively debate available on practically any topic; this collusion, sometimes collision, of cultures has resulted in a spectacularly fast development of the area.

The multidisciplinary character of the field creates some problems for its practitioners, who often have to become familiar with contributions from a number of different disciplines. The diversity of publications and worldwide activity makes it very difficult to develop a feel for the whole field. This problem is partly addressed by conferences and neural network journals, but these present only the leading edge of research. The *Handbook of Neural Computation* aims to bridge this gap, collecting material from across the spectrum of neural network activity and tying it together into a coherent whole. Input from computer scientists, engineers, biologists, psychologists, mathematicians and physicists (and now also those whose background is explicitly in neural networks, a relatively recent phenomenon) has been assembled into a work that forms a central reference repository for the field.

This handbook is not designed to compete with journals or conferences. The latter are well suited to the dissemination of leading-edge research. The handbook provides, instead, an overview of the field, collating and filtering the research findings into a less detailed but broader view of the domain. As well as allowing established practitioners to view the wider context of their work, it is designed to be used by newcomers to the field, who need access to review-style articles. The opening sections of the handbook introduce the basic concepts of neural computation, followed by a comprehensive set of technical descriptions of neural network models. While it is not possible to describe every variant of every model, we have aimed to present the major ones in a structured and self-consistent arrangement. Descriptions of hybrid approaches that couple neural techniques with other methods are followed by details of implementations in hardware. Applications of neural computation to different domains form the next part, followed by more detailed individual case studies, collated under common headings and written in such a style as to facilitate the transfer of applicable techniques between different domains. The handbook finishes with a collection of essays from leading researchers on future directions for research.

We hope that this handbook will become an invaluable reference tool for all those involved in the field of neural computation. It should provide a comprehensive, organized view of the field for many years, supplemented on a regular basis to allow it to remain genuinely up to date. The electronic version of the handbook, comprising both CD-ROM and Internet implementations, will facilitate distributed access to the content and efficient retrieval of information. The handbook should provide a coherent overview of the field, helping to ensure that we are all aware of important developments and thinking in other disciplines that impact our own research activities.

Russell Beale and Emile Fiesler, June 1996

Foreword

James A Anderson

Neural networks are models for computation that take their inspiration from the way the brain is supposed to be constructed and that often try to solve the problems that the brain seems to try to solve. Biological neural networks in mammals are built from neurons (nerve cells) that are themselves remarkably complex biological units. Huge numbers of neurons, connected together and cooperating in poorly understood ways, give rise to the complex behavior of organisms. Artificial neural networks, variants of which are discussed at length in this volume, are smaller, simpler, and more understandable than the biological ones, but are still able to do some remarkably interesting things. Some of the operations that artificial networks are good at—pattern recognition, concept formation, association, generalization, some kinds of inference—seem to be similar to things that brains do well. It is fair to say that artificial neural networks behave a lot more like humans than digital computers do.

There are two related but distinct goals that have driven neural network research since its beginnings:

(i) First, we want to construct and analyze artificial neural networks because that may allow us to begin to understand how the biological neural networks in our brains work. This is the domain of neuroscience, cognitive science, psychology, and perhaps philosophy.

(ii) Second, we want to construct and analyze artificial neural networks because that will allow us to build more intelligent machines. This is the domain of engineering and computer science.

These two goals—understanding the brain and making smart devices—are mixed together in varying proportions throughout this collection though the bias here is toward the careful analysis and application of artificial networks. Although there is a degree of creative tension between these two goals, there is also synergy.

The modern history of artificial neural networks might be said to begin with an often reprinted 1943 paper by Warren McCulloch and Walter Pitts, 'A logical calculus of the ideas immanent in nervous activity'. McCulloch and Pitts were making models for brain function, that is, what does the brain compute and how does it do it? However, only two years after the publication of their paper, in 1945, John von Neumann used their model for neuron behavior and neural computation in an influential discussion of the proper design to be used for future generations of digital computers.

The creative tension arises from the following observation. Consider an engineer who wants to use biology as inspiration for an intelligent adaptive device. Why should engineers be bound by biological solutions? If you are stuck with slow and unreliable biological hardware, perhaps you are also forced to use intrinsically undesirable algorithms.

Ample evidence suggests that our lately evolved species-specific behaviors like language are simply not very well constructed. After only a few tens of thousands of generations of talking ancestors, human language is still no more than an indispensable kludge, grounded in and limited by the circuitry that nature had to work with in the primate brain. Maybe after several million more years of evolution our descendants will finally get it right. Maybe there are better ways to perform the operations of intelligence. Why stick with the second rate?

The synergy between biological neural networks and artificial neural networks arises in several ways.

First, precise analysis of simple, general neural networks is intrinsically interesting and can have unexpected benefits. The McCulloch–Pitts paper developed a primitive model of the brain, but a very good model for many kinds of computation. One of its side effects was to originate the field of finite state automata.

Second, to make intelligent systems usable by humans perhaps we must make artificial systems that are conceptually, though not physically, designed like we are. We would have difficulty communicating with a truly different kind of intelligence. The current emphasis on user-friendly computer interfaces is an example. Large amounts of computer power are spent to provide a translator between a real logic processor and our far less logical selves. For us to acknowledge a system as intelligent perhaps it has to be just like us. As Xenophanes commented 2500 years ago, 'horses would draw the forms of gods like horses, and cattle like cattle, and they would make the gods' bodies the same shape as their own'.

Third, neural networks provide a valuable set of examples of ways that a massively parallel computer could be organized. Current digital computers will soon run up against limitations imposed by the physics of electronic circuitry and the speed of light. One way to keep increasing computer speed is to use multiple CPUs; if one computer computes fast, then two computers should compute twice as fast. Unfortunately, coordinating many CPUs to work fast and effectively on a single problem has proven to be extremely difficult. Neurons have time constants in the millisecond range; present-day silicon devices have time constants in the nanosecond range. Yet somehow the brain has been able to build exceedingly powerful computing systems by summing the abilities of huge numbers of biological neurons, even though each neuron is computing several orders of magnitude more slowly than an electronic device constructed from silicon. The best known example of this design is the mammalian cerebral cortex, where neurons are arranged in parallel arrays in a highly modular structure. Most neural networks described in this collection are abstractions of the architecture of the mammalian cerebral cortex. Knowing, in detail, how this parallel architecture works would be of considerable practical value.

However, the study of human cognitive abilities suggests a price may be paid for using it. The resulting systems, both biological and artificial, may be forced to become very special-purpose and will almost surely lack the universality and flexibility that we are accustomed to in digital computers. The things that make neural networks so interesting as models for human behavior, for example, good generalization, easy formation of associations, and the ability to work with inadequate or degraded data, may appear in less benign form in artificial neural networks as loss of detail and precision, inexplicable prejudice, and erroneous and unmotivated conclusions. Making effective use of artificial neural networks may require a different kind of computing than we are used to, one that solves different problems in different ways but one with great power in its own domain.

All these fascinating, important and very practical issues are discussed in detail in the pages to follow. It is hard to predict what form computers will take in a century. There is a good chance, however, that they will incorporate in some form many of the ideas presented here.

How to Use This Handbook

The *Handbook of Neural Computation* is the first in a series of three updatable reference works known collectively as the Computational Intelligence Library. (The other two volumes are the *Handbook of Evolutionary Computation* and the *Handbook of Fuzzy Computation*.) This handbook has been designed to provide valuable information to a diverse readership. Through regular supplements, the handbook will remain fully up to date and will develop and evolve along with the research field that it represents.

WHERE TO LOOK FOR INFORMATION

An informal categorization of readers and their possible information requirements is given below, together with pointers to appropriate sections of the handbook.

The Research Scientist

This reader has a very good general knowledge of neural computation. She may want to

- develop new neural network models or improve existing ones (Part C: Neural Network Models)
- develop new applications of neural networks (Part F: Applications of Neural Computation; Part G: Neural Networks in Practice: Case Studies)
- improve the underlying theory and/or heuristic principles of neural computation (Part B: Fundamental Concepts of Neural Computation; Part H: The Neural Network Research Community)

The Applications Specialist

This reader is working in a technical environment (such as engineering). He perhaps

- has a problem that may be amenable to a neural network solution (Part F: Applications of Neural Computation; Part C: Neural Network Models)
- wants to compare the cost-effectiveness of the neural network solution with that of other possible solutions (Part F: Applications of Neural Computation)
- is interested in real systems experience as conveyed by case studies (Part G: Neural Networks in Practice: Case Studies)

The Practitioner

This reader is working in a professional discipline that is not closely related to computer science, such as medicine or finance. She may have heard of the potential of neural networks for solving problems in her professional field, but might have little or no knowledge of the principles of neural computation or of how to apply it in practice. She may want to

- find a quick way into the subject (Part A: Introduction; Part B: Fundamental Concepts of Neural Computation)
- look at real case studies to see what neural networks have already achieved in her field of interest (Part G: Neural Networks in Practice: Case Studies; Part F: Applications of Neural Computation)
- find a relatively easy and quick route to implementation of a neural network solution (Part G: Neural Networks in Practice: Case Studies; Part F: Applications of Neural Computation; Part C: Neural Network Models)

The Student (or Teacher)

This reader may be

- looking for an easy way into the subject (Part A: Introduction)
- interested in getting a firm grasp of the fundamentals (Part B: Fundamental Concepts of Neural Computation)
- interested in practical examples for projects (Part G: Neural Networks in Practice: Case Studies)

CROSS-REFERENCES

Most of the articles in the handbook contain cross-references to related articles. A section number in the margin indicates that further information on the concept under discussion may be found in that section of the handbook. The notation in the following example indicates that further information on the multilayer perceptron and the radial basis function network may be found in sections C1.2 and C1.6.2, respectively.

C1.2 Several neural network models have been proposed for applications of this type. The *multilayer*
C1.6.2 *perceptron* and the *radial basis function* network were considered in this case.

In the electronic edition of the handbook, these marginal section numbers become hypertext links to the section in question. (Full details of the functionality of the electronic edition are provided in the application itself.)

NUMBERING OF EQUATIONS, FIGURES, PAGES, AND TABLES

To facilitate incorporation of the regular supplements to the handbook, which will include new material and updates to existing articles, a unique system of numbering of equations, figures, pages and tables has been employed. Each section in the handbook starts at page 1 with the section code preceding the page number. For example, section F1.8 starts on page F1.8:**1** and continues through page F1.8:**6**, and then section F1.9 follows on page F1.9:**1**. Equations, figures, and tables are numbered sequentially throughout each section with the section code preceding the number of the equation, figure, and table. For example, the third equation in section B3.2 is referred to as equation (B3.2.3) or simply (B3.2.3). The third figure or table in the same section would be referred to as figure B3.2.3 or table B3.2.3.

HANDBOOK SUPPLEMENTS

The *Handbook of Neural Computation* will be updated on a regular basis by means of supplements containing new contributions and revisions to existing articles. To receive these supplements it is essential that you complete and return the registration card at the front of the loose-leaf binder and return it to the address indicated on the card. (Purchasers of the electronic edition will receive separate registration information.) **If you have not already completed the registration card, please do so now**. After you have registered, you will receive new supplements as they are published. The first two supplements are free; thereafter, you will be sent subscription renewal notices. If you wish to keep your copy of the handbook fully up to date, it is essential that you renew your subscription promptly.

FURTHER INFORMATION

For the latest information on the *Handbook of Neural Computation*, please visit our website at http://www.oup-usa.org/acadref/hnc.html, or you may contact the editors in chief or the publisher at the contact addresses given below.

Dr Emile Fiesler	Dr Russell Beale	Mr Sean Pidgeon
IDIAP	School of Computer Science	Senior Editor
C.P. 592	University of Birmingham,	Scholarly and Professional Reference
Martigny CH-1920	Edgbaston	Oxford University Press
Switzerland	Birmingham B15 2TT	198 Madison Avenue
e-mail: efiesler@idiap.ch	United Kingdom	New York, NY 10016, USA
	e-mail: r.beale@cs.bham.ac.uk	e-mail: sdp@oup-usa.org

IMPORTANT

Please remember that no part of this handbook may be reproduced
without the prior permission of Institute of Physics Publishing and
Oxford University Press

PART A

INTRODUCTION

PART A

INTRODUCTION

A1

Neural Computation: The Background

Contents

A1.1 The historical background

J G Taylor

Abstract

The brief history of neural network research presented in this section indicates that, although the initial revolution in neural networks lost its early momentum, the second revolution may well avoid the fate of the first. The subject now has strengths that were absent from its earliest version: these are discussed, and especially the fact that the biological origin of the subject is now giving it greater stability. The new avenues opened up by biologically motivated research and by studies in other areas such as statistical mechanics, statistics, functional analysis and machine learning are described, and future directions discussed. The strengths and weaknesses of the subject are compared with those of alternative and competing approaches to information processing.

A1.1.1 Introduction

The discipline of neural networks is presently living through the second of a pair of revolutions, the first having started in 1943 with the publication of a startling result by the American scientists Warren McCulloch and Walter Pitts. They considered the case of a network made up of binary decision units (BDNs) and showed that such a network could perform any logical function on its inputs. This was taken to mean that one could 'mechanize' thought, and it helped to support the development of the digital computer and its use as a paradigm for human thought. The result was made even more intriguing due to the fact that the BDN is a beautifully simple model of the sort of nerve cell used in the human brain to support thinking. This led to the suggestion that here was a good model of human thought.

Before the logical paradigm won the day, another American, Frank Rosenblatt, and several of his colleagues showed how it was possible to train a network of BDNs, called a *perceptron* (appropriate for a device which could apparently perceive), so as to be able to recognize a set of patterns chosen beforehand (Rosenblatt 1962).

C1.1.1

This training used what are called the connection weights. Each of these weights is a number by which one must multiply the activity on a particular input in order to obtain the effect of that input on the BDN. The total activity on the BDN is the sum of such terms over all the inputs. The connection weights are the most important objects in a neural network, and their modification (so-called *training*) is presently under close study. The last word has clearly not yet been said on what is the most effective training algorithm, and there are many proposals for new learning algorithms each year.

B3

The essence of the training rules was very simple: one would present the network with examples and change those connection weights which led to an improvement of the results, so as to be closer to the desired values. This rule worked miracles, at least on a set of rather 'toy' example patterns. This caused a wave of euphoria to sweep through the research community, and Rosenblatt spoke to packed houses when he went to campuses to describe his results.

One of the factors in his success was that he appeared to be building a model duplicating, to some extent, the activity of the human brain. The early result of McCulloch and Pitts indicated that a network of BDNs could solve any logical task; now Rosenblatt had demonstrated that such a network could also be trained to classify any pattern set. Moreover, the network of BDNs used by Rosenblatt, which possessed a more detailed description of the state of the system in terms of the connection weights between the model neurons than did the *McCulloch–Pitts network*, seemed to be a more convincing model of the brain.

B1.2

A1.1.2 Living neurons

A1.2 To justify such a strong claim it is necessary to expand the argument a little. *Living neurons* are, in fact, composed of a cell body and numerous outgrowths. One of these, which may branch into several collaterals, is called the axon. It acts as the output line for the neuron. The other outgrowths are called the dendrites; they are often covered with little 'spines', where the ends of the axons of other cells attach themselves. The interior of the nerve cell is kept at a negative electric potential (usually about -60 mV) by means of active pumps in the cell wall which pump sodium ions outside and keep slightly fewer potassium ions inside. This electrical balance is especially delicately assessed at the exit point of the axon. If the cell electrical potential becomes too positive, usually by about $+10$ to $+15$ mV, then there will be a sudden reversal of the potential to about $+60$ mV, and an almost as sudden return to the usual negative resting value, all in about 2 to 3 ms.

This sequence of potential changes is called an action potential, which moves steadily down the axon and its branches (at about 1 to 10 m s^{-1}). It is this action potential that is the signal sent from one nerve cell to its neighbors. The generation of the signal by the neuron is achieved by the summation of the signals coming to the cell body from the dendrites, which themselves have been affected by action potentials coming to them from nearby cells. The strengths of the action potentials moving along the axons are all the same. It is by means of rescaling the effects of each action potential as it arrives at a synapse or junction from one cell to the next (by means of multiplication of the incoming activity of a nerve impulse by the appropriate connection weight mentioned earlier) that a differential effect is achieved for each cell on its neighbors.

The above description of the actions of the living nerve cells in the brain is highly simplified, but gives a correct overall picture. It is seen that each nerve cell is acting like a BDN, with the decision to respond being that of assessing whether or not the total activity from its neighbors arriving at its axon outgrowth is above the threshold mentioned earlier. This activity is the sum of the incoming action potentials scaled by an appropriate factor, which may be identified with the connection weight of the BDN. The identification of the BDN with the living nerve cell is thus complete. A network of BDNs is, indeed, a simple model of the brain.

A1.1.3 Difficulties to be faced

This, then, was the first neural network revolution. Its attraction to many (although not all) was reduced when Marvin Minsky and Seymour Papert showed in 1969 that perceptrons are very limited. They have an Achilles heel: they cannot solve some very simple pattern classification tasks, such as separating the binary patterns (0, 0), (1, 1) from the patterns (1, 0), (0, 1), known as the parity problem, or XOR. To solve this problem it is necessary to have neurons whose outputs are not available to the outside world. These so-called 'hidden neurons' cannot be trained by causing their outputs to become closer to the desired values given by the training set. Thus, in the XOR case, the input–output training set is (0, 0), 0; (1, 1), 0; (0, 1), 1; (1, 0), 1. The desired outputs of 0 or 1 (in the various cases) for the output neurons are not provided for any hidden neuron. Yet in the case of any linearly inseparable problem, such as XOR, there must be hidden neurons present in the network architecture in order to help turn the problem into a linearly separable one for the outputs.

In addition, there was a further important difficulty which was emphasized by Minsky and Papert, who gave a very thorough mathematical analysis of the time it takes to train such networks, and how this increases with the number of input neurons. It was shown by Minsky and Papert (1969) that training times increase very rapidly for certain problems as the number of input lines increases.

These (and other) difficulties were seized upon by opponents of the burgeoning subject. In particular, this was true of those working in the field of artificial intelligence (AI) who at that time did not want to concern themselves with the underlying 'wetware' of the brain, but only with the functional aspects— regarded by them solely as logical processing. Due to the limitations of funding, competition between the AI and neural network communities could have only one victor.

A1.1.4 Reawakening

Neural networks then went into a relative quietude, with only a few, but very clever, devotees still working on it. Then came new vigor from various sources. One was from the increasing power of computers, allowing simulations of otherwise intractable problems. At the same time, the difficulty of training hidden

neurons was solved by the *backpropagation algorithm*, originally introduced by Paul Werbos (1974), and C1.2.3
independently discovered by Parker (1985) and LeCun (1985); it was highly publicized by the PDP Group
with Rumelhart and McClelland (1986). Backpropagation allowed the error to be transported back from
the output lines to earlier layers in the network so as to give a very precise modification of the weights
on the hidden units. It was possible to simulate ever-larger problems using this training scheme, and so
begin to train neural networks on industrially interesting problems.

Another source of stimulus was the seminal paper of John Hopfield (1982) and related work of
Grossberg and collaborators (Cohen and Grossberg 1983) in analyzing the dynamics of networks by
introducing powerful methods based on Lyapunov functions to describe this development. In all, this
work showed how a network of BDNs, coupled to each other and asynchronously updated, can be seen to
develop in time as if the system were running down an energy hill to find a minimum. Hopfield (1982)
showed, in particular, how it is possible to sculpt the energy landscape so that there are a desired set of
minima. Such a network leads to a content-addressable memory, since a partially correct starting activity
will develop into the complete version quite quickly.

The introduction of an energy function quickly alerted the physics community, ever eager to sharpen
their teeth on a new problem. This led to the spin glass approach, with the global ideas on phase transitions
and temperature entering the field of neural networks for the first time. A spin glass derivation was also
given by Amit (1989) of the capacity limit of $0.14N$ as the limit to the number of patterns which can
usefully be stored in a network of N neurons (and which was originally found experimentally by Hopfield
(1982)). Gardner then introduced the general notion of the 'space' of neural networks (Gardner 1988), an
idea that has been explored more fully by the recent developments of differential geometry by the work of
Amari (1991). It is clear that the statistical mechanical approach is still flourishing, and is leading to many
new insights. For example, it has become clear how the presence of temperature allows the avoidance of
spurious states brought about by the form of the connection weights; these false states are made unstable
if the network is 'hot' enough, and only the correct states are recalled in that case. It has also become
clear as to what was the source of the limit on the storage capacity of these networks, and how this might
be increased by choosing suitable connectivity to obtain the full capacity N (Coombes and Taylor 1993).

Another very important historical development was the creation of the *Boltzmann machine* (Hinton C1.4
and Sejnowski 1983), which may be regarded as the extension of the *Hopfield network* to include hidden B1.3
neurons. The name was assigned since the probability distribution of the states of the network is identical
to the Boltzmann distribution. The Boltzmann machine learning algorithm, based on the Kullback–
Liebler metric as a distance function on the probability distributions of the states, allowed this probability
distribution to move more closely to an external one to be learned. However, the learning algorithm is
slow, and this has prevented many useful applications.

A further network which proved very attractive to those entering the field was the *self-organizing* C2.1.1
map. This had been developed by several workers (Willshaw and von der Malsburg 1976, Grossberg 1976)
and reached a very effective form for applications in terms of the self-organizing feature map (SOFM) of
Kohonen (1982). This allowed the weights of a single-layer network to adapt to an ensemble of inputs so
as to learn the distribution of those inputs in an ordered fashion. Numerous developments have occurred
in this approach more recently (Ritter *et al* 1991).

The other question, of the scaling of training times as the size of the input space increases, which
was raised by Minsky and Papert, is still unsolved. Papert, in a recent paper (Minsky and Papert 1989),
wrote '... the entire structure of recent connectionist theories might be built on quicksand: it is all based
on toy-sized problems with no theoretical analysis to show that performance will be maintained when the
models are scaled up to realistic size. The connectionist authors fail to read our work as a warning that
networks, like brute force, scale very badly'. This is a warning not to be taken lightly. It is being met by
various methods and devices: accelerator cards, ever faster and smaller *hardware devices*, and a deeper E1
understanding of the theory behind neural computation. It is to be noted in this respect that accelerator
cards may offer time saving and tractable training sessions on large databases but still may not help the
convergence to significant solutions. It may be that the second neural network 'revolution' is only just
beginning, but it is very clear that the scaling problem is in the forefront of researchers' minds.

A1.1.5 Forms of networks and their training

In order to understand in more detail the way that greater strength is being brought to the subject of neural
networks, it is important to point out the two extremes that now exist inside the discipline itself. At one end

is the work of those mainly concerned with solving industrial problems. These include engineers, computer scientists, and people in the industrial sector. To them, neural computing is only one of a spectrum of adaptive information processing techniques. At the other extreme are those interested in understanding living systems, such as biologists, psychologists, and philosophers, together with mathematicians and physicists who are interested in the whole range of the subject as throwing up valuable and interesting new problems.

B2.3 The styles of approach of the two extremes are somewhat different. The subject of artificial neural computing is based on networks, some of which have been mentioned earlier, which use the rather simple BDNs defined above. There are two extremes of the architectures of the networks: *feedforward networks* (input streams steadily through the network from a set of input neurons to a set of output ones) and

B2.3 *recurrent networks* (where there is constant feedback from the neurons of the network to each other, as in the Hopfield network mentioned earlier). This is mirrored in the differences between the topologies such networks possess; one is the line, and the other the circle, which cannot be topologically deformed into each other. As is to be expected, there are two extreme styles of computation in these networks. In the feedforward case the input moves through the network to become the output; in the recurrent network the activities in the network develop over time until it settles into some asymptotic value which is used as the output of the network. The network thus relaxes into this asymptotic state.

B3.1, C3 Network training can be classified into three sorts: *supervised, reinforcement* and *unsupervised*. The most popular of the first of these, backpropagation, has been mentioned earlier as the way to train neural networks to solve hard problems like parity, which needs hidden nodes (with no output that might be specified directly by the supervisor or teacher). It uses a set of training data which is assumed to be given, so that the (usually) feedforward network has a set of given inputs and outputs. When a given input is applied to the untrained network, the output is not expected to be the desired one, so that an error is obtained. That is used to assign changes, usually small ones, to the connection weights to all the neurons (including the hidden ones) in the network. This process of change is repeated many times until an acceptably low error level is obtained.

The second training method uses a reward given to the network by the environment on its response to a given input. This reward may also be used to determine modifications to the weights to achieve a maximum reward from the environment. Thus, this form of learning is 'with a critic', to be compared to supervised learning, which is 'with a teacher'. Finally, there is unsupervised learning, which is closer to the style of learning in biological systems (although reinforcement learning also has strong biological roots). In this method correlations between signals are learned by increasing the connection weight between two neurons which are both active together.

At the other end of the subject of neural computation is investigation of nervous systems of the many species of animals, in an attempt to understand them. Since even a single living neuron is very complex, this approach does not aim for application in the marketplace, although simplified versions of mechanisms gleaned from this area of study are turning out to be of great value in commercial applications. This is true, for example, for models of the eye or ear, and also in the area of control, where reinforcement training (related to conditioned learning) has led to some very effective industrial control systems (White and Sofge 1992). The biological neural networks which are of interest are also extremely complex as nonlinear dynamical systems or mappings, although there is steady progress in their unraveling.

The most important lesson to be learned from these studies, besides the detailed network styles being used, is that the brain has developed a very powerful modular scheme for handling the scaling problem mentioned earlier. Exactly how this works is presently under extensive scrutiny, in particular, through the use of noninvasive techniques (EEG, MEG, PET, MRI). The causal chains of activations of various brain regions is being discovered as a subject performs a particular information processing task; the results are allowing more global models of the brain to be constructed.

A1.1.6 Strengths of neural networks

In the face of the difficulties neural networks are still facing, of slow training, incompletely understood complexity and the highly nonlinear neural network system involved, as mentioned earlier, there are several features which will ensure the continued strength of the subject as a viable discipline.

Firstly, increases in computing power that were almost undreamed of several years ago, with gigabytes of memory and giga-interconnection updates per second. That may still be some way from the speed and power of the human brain. But if only specialized devices are to be developed, the total complexity of the human brain need not be a deterrent from attaining a lesser goal.

Secondly, there are developments in the theoretical understanding of neural networks that are impressive. Convergence of training schedules and their speed-up is presently under active investigation. The subject of dynamical systems theory is being brought to bear on these questions, and impressive results are being obtained. The use of concepts like attractor, stability, circle maps and so on are allowing a strong framework to be built for neural networks; in particular, the manner in which the dynamics of learning appears to display the general features of a sequence of phase transitions, as new features of the complexity of the training set are able to be discovered by the network, and new specialized feature detectors in the hidden layers emerge in the training process.

Thirdly, there are several different disciplines which are seen to have a great deal of overlap with neural networks. Thus the branch of statistics associated with regression analysis is now recognized as having been extended in an adaptive manner by the use of neural network representations of time series (Breiman 1994). Computer-intensive techniques, such as bootstrapping, are proving of great value in neural networks for tackling problems with small data sets. *Pattern recognition*, for example, also has important overlaps with the discipline in the areas of classification and *data compression*. Neural networks can extend these areas to give them an adaptability that is proving to be very important, such as in learning the most important features of a scene by means of adaptive principle component analysis (PCA) (Oja 1982). Statistical mechanics (especially spin glasses) has already been noted above as leading to important new insights into the problems of storage and response of neural networks. Machine learning is also of importance for the subject, and under the 'probably approximately correct' (PAC) approach has allowed the study of the complexity of neural networks needed to solve a given problem.

B1.5, B6
F1.5

Fourthly, the field of function approximation has led to the important 'universal approximation theorem' (Hecht-Nielsen 1987, Hornik *et al* 1989). This theorem states that any suitably smooth function can be approximated arbitrarily closely by a neural network with only one hidden layer. The number of nodes required for such an approximation would be expected to increase without bound as the approximation was made increasingly better. The result is of the utmost importance to those who wish to apply neural networks to a particular problem; it states that a suitable network can always be found. This is also true for trajectories of patterns (Funahashi and Nakamura 1993).

There is a similar, but more extended result, for the learning of conditional probability distributions (Allen and Taylor 1994), where now the universal network has to have at least two layers to be able to have a smooth limit when the stochastic series being modeled becomes noise-free. Again, this is very important in the modeling by neural networks of *financial series* which have considerable stochasticity.

G6.3

Fifthly, and already discussed briefly above, is the emerging subject of computational neuroscience. This attempts to create simple models of the neural systems which are important in controlling the response patterns of animals of a given species. This has a vast breadth, encompassing as it does the million or so species of living animals, culminating with man. It is a subject with vast implications for mankind, especially from the medical benefits that better understanding of brain processes would bring, both to those in the field of mental health and in the more general area of understanding of healthy living systems.

The field of computational neuroscience has led to useful devices by the route of 'reverse engineering'. In this, algorithms are developed for information processing based on simple models of the neural processing occurring in the living system. Thus it is not only the single neuron which is proving of value in reverse engineering, as it has already for the development of artificial neural networks (and where also it continues with the incorporation of increasingly complex neurons to achieve more powerful artificial neural networks). It is increasingly occurring in the reverse engineering of the overall architecture of artificial networks from that of living neural networks. This approach has also proved of value at the hardware level, as well as generating new styles of artificial neural computation. Thus, in the first category, is the work of Carver Mead and his colleagues at the California Institute of Technology in the United States (Mead 1989). They have built both a silicon retina and a silicon ear, using VLSI designs based on the known functions of these devices in living systems and their approximate wiring diagrams.

The retina has lateral inhibitory connections between the first (horizontal) layer of cells and the input cells, which leads to a very elegant method of reducing redundancy (say, in patches of constant illumination) of visual inputs. It is also possible to extend this modeling to later layers in the retina, and also to proceed further into the early layers of the visual cortex. The latter appears to use a decomposition of the input into some overcomplete set of functions, such as might arise from differences of Gaussians or similar functions with localized values. This leads into the field of wavelet transforms, another theoretical area proving to be of great value in developing new paradigms for neural networks (Szu and Hopper 1995).

The manner in which more global brain processing can be understood has been developed over the

---

last few years by Teuvo Kohonen in the SOFM mentioned earlier (Kohonen 1982). In more detail, this algorithm is based on the idea of competition between nearby neurons, ending up in one neuron winning and the others being turned off by lateral inhibition from that winner. This winner is then trained by increasing the connection weights to it so that it gives a larger output. This means rotating the weights on the winning neuron so that they are more closely aligned to the input. The same is done for the neurons in a small region round the winner. If this is done repeatedly for a set of training inputs the network ends up representing the inputs in a topographic fashion over its surface (assuming the network is laid out in a two-dimensional fashion). If the inputs have features which are more than two dimensional then the resulting map may have folds in it; such discontinuities are seen, for example, in the map of rotation sensitivity for cells in the visual cortex.

One can search for other tricks that nature may use, and attempt to incorporate them into suitable machines. Thus there are presently attempts to build a 'vision machine' by means of the sensitive response of sets of coupled oscillators to their inputs. Yet again this also leads to some very important mathematical problems in understanding the response patterns of many physical systems.

It also leads to the more general question of whether or not it is possible to use the finer details of the temporal structure of neural activity. An extreme case of this is the use of information by coincidence of a number of nerve impulses impinging on a given cell. Suggestions of this sort have been around for a decade or more, but it is only recently that the improvement in computing power has allowed increasing numbers of simulations to test this idea.

As is well known, chaos and fractals are a key aspect of any physical phenomena. Will they prove to be of importance in improving neural networks? Some, especially Walter Freeman (1995) from Berkeley in connection with olfaction, suggest that such is the case, and that strange attractors may be used to give a very effective method of searching through, or giving access to, a large region of the state space of a neural network. That possibility has not yet been achieved in detail; however, see Quoy *et al* (1995) for an interesting attempt to achieve a useful speed-up by 'living on the edge of chaos' for a neural network. But the question is an important one and again indicates the breadth of possibilities now coming under the banner of neural networks.

A1.1.7 Hybrids and the future

From what has been sketched above about the past and some of the avenues being explored in the present for neural networks, it is clear that the subject now has such breadth and depth that it is unlikely to run out of steam as it did earlier. Indeed, it is becoming increasingly clear that artificial neural networks (ANNs) can be seen to be one of a number of similar tools in the tool-kit of anyone tackling problems
D2, D1   in information processing. Along with *genetic algorithms*, *fuzzy logic*, belief networks, and other areas (such as parallel computing), ANNs are to be used either on their own or in hybrid systems wherever and however is most appropriate. The past divisions, noted above as having existed between different branches of information processing, seem to have been removed by these developments. Moreover, new techniques are being developed to allow the parallel use of these various technologies, or even better, in a manner that allows them to help each other. Thus genetic algorithms are being used to help improve the architecture of a neural network, where the fitness function used to select better descendants at each stage of the generation process is the error on the training set (in the case of a supervised learning problem). Similarly, it has proved of value to obtain help from fuzzy logic to allow for rough initial settings of the weights in a network.

There are some general rules for determining when a neural network is most appropriate for a particular task, compared with one of the other methods mentioned earlier. If the data are noisy, if there are no rules for the decisions or response that are required, or if the training and response must be rapid (something missing from genetic algorithms, for example), then ANNs may be the best bet. It is also necessary to comment finally on the present situation in the relation between ANNs and AI mentioned earlier. As noted above for other adaptive techniques, the move is now to combine an ANN solution for part of a problem with results obtained from a knowledge-based expert system (KBES). That has been done successfully
F1.7.2, G1.4   in *speech recognition*, where the Kohonen network mentioned earlier is good for individual phoneme recognition, but not so good for words (due to difficulty in incorporating context into the ANN). A KBES approach, with about 20 000 expert rules, then allows the total system to be far more effective. Similar
B4.10.2, C1.2.8   greater efficiency can also be obtained using hybrid systems with *time-delayed neural networks* (which involve inputs that are delayed or lagged relatively to each other, so as to cover a spread of input times).

It is clear that a more realistic and effective approach is arising in the relationship between the different branches of information processing. Undoubtedly this use of the best of all possible worlds will increase. But at the same time the neural network approach, in the context of obtaining a better understanding of the human brain, will also give ever increasing powers to the ANN approach. In the end one can only see that as being the most effective (provided there is the computing power) method for many of the deeper problems facing the information industry. Nor is there any serious alternative to the further development of neural network models of ourselves to understand the higher levels of human cognition, including human consciousness.

References

Allen D W and Taylor J G 1994 Learning time series by neural networks *Proc. Int. Conf. on Artificial Neural Networks (Sorrento, Italy, 1994)* ed M Marinaro and P Morasso (Berlin: Springer) pp 529–32

Amari S 1991 Dualistic geometry of the manifold of higher-order neurons *Neural Networks* **4** 443–51

Amit D 1989 *Models of Brain Function* (Cambridge: Cambridge University Press)

Breiman L 1994 Bagging predictors *UCLA Preprint* (unpublished)

Cohen M A and Grossberg S 1983 Absolute stability of global pattern formation and parallel memory storage by competitive neural networks *IEEE Trans. Syst. Man Cybern.* **13** 815–26

Coombes S and Taylor J G 1993 Using generalised principal component analysis to achieve associative memory in a Hopfield net *Network* **5** 75–88

Freeman W 1995 *Society of Brains* (Hillsdale, NJ: Erlbaum)

Funahashi K and Nakamura Y 1993 Approximation of dynamical systems by continuous time recurrent neural networks *Neural Networks* **6** 801–6

Gardner E 1988 The space of interactions in neural network models *J. Phys. A: Math. Gen.* **21** 257–70

Grossberg S 1976 Adaptive pattern classification and universal recoding, I: Parallel development and coding of neural feature detectors *Biol. Cybern.* **23** 121–34

Hecht-Nielsen R 1987 Kolmogorov's mapping neural network existence theorem *Proc. Int. Conf. on Neural Networks III* (New York: IEEE) pp 11–13

Hinton G and Sejnowski T 1983 Optimal perceptual inference *Proc. IEEE Conf. on Computer Vision and Pattern Recognition (Washington)* (New York: IEEE) pp 448–53

Hopfield J 1982 Neural networks and physical systems with emergent collective computational properties *Proc. Natl Acad. Sci., USA* **81** 3088–92

Hornik K, Stinchcombe M and White H 1989 Multi-layer feedforward networks are universal approximators *Neural Networks* **2** 359–66

Kohonen T 1982 Self-organised formation of topologically correct feature maps *Biol. Cybern.* **43** 56–69

LeCun Y 1985 Une procédure d'apprentissage pour réseau á seuil asymetrique *Cognitiva* **85** (Paris: CESTA) pp 599–604

McCulloch W S and Pitts W 1943 A logical calculus of ideas immanent in nervous activity *Bull. Math. Biophys.* **5** 115–33

Mead C 1989 *Analogue VLSI and Neural Systems* (Reading, MA: Addison-Wesley)

Minsky M and Papert S 1969 *Perceptrons* (Boston, MA: MIT Press)

——1989 *Perceptrons* 2nd edn (Boston, MA: MIT Press)

Oja E 1982 A simplified neuron model as a principal component analyser *J. Math. Biol.* **15** 61–8

Parker D B 1985 Learning logic *Technical Report TR-47* Center for Computational Research in Economics and Management Science, Massachusetts Institute of Technology, Cambridge, MA

Quoy M, Doyon B and Samuelides M 1995 Dimension reduction by learning in a discrete time chaotic neural network *Proc. World Congr. on Neural Networks (1995)* (Washington: INNS) pp I-300-303

Ritter H, Martinetz T and Schulten K 1991 Neural computation and self-organising maps (Reading, MA: Addison-Wesley)

Rosenblatt F 1962 *Principles of Neurodynamics* (New York: Spartan)

Rumelhart D E and McClelland J L 1986 *Parallel Distributed Processing* (Boston, MA: MIT Press)

Szu H and Hopper T 1995 Wavelets as preprocessors for neural networks Plenary Talk *Proc. World Congr. on Neural Networks (Washington, DC, 1995)* (Washington: INNS); Kohonen T 1995 Plenary Talk *Proc. World Congr. on Neural Networks (Washington, DC, 1995)* (Washington: INNS)

Werbos P 1974 Beyond regression *PhD Thesis* Harvard University

White D A and Sofge D A (eds) 1992 *Handbook of Intelligent Control* (New York: Van Nostrand Reinhold)

Willshaw D J and von der Malsburg C 1976 How patterned neural connections can be set up by self-organisation *Proc. R. Soc.* B **194** 431–45

A1.2 The biological and psychological background

Michael A Arbib

Abstract

A brief look at how biology and psychology motivate the definitions of artificial neurons presented in other sections of this handbook.

A1.2.1 Biological motivation and neural diversity

In biology, there are radically different types of neurons in the human brain, and further variations in neuron types of other species. In brain theory, the complexities of real neurons are abstracted in many ways to aid an understanding of different aspects of neural development, learning, or function. In neural computation, the artificial neurons are designed as variations on the abstractions of brain theory and implemented in software, VLSI, or other media. Although detailed models of biological neurons are not within the scope of this handbook, it will be useful to provide an informal view of neurons as defined biologically, for it is the biological neurons that inspired the various notions of formal *neuron* used in neural computation B1 (discussed in detail elsewhere in this handbook). The nervous system of animals comprises an intricate network of neurons (a few hundred neurons in some simple creatures; hundreds of billions in a human brain) continually combining signals from receptors with signals encoding past experience to barrage motor neurons with signals which will yield adaptive interactions with the environment. In animals with backbones (vertebrates, including mammals in general and humans in particular) the brain constitutes the most headward part of this *central nervous system* (CNS), linked to the receptors and effectors of the body via the spinal cord. Invertebrate nervous systems (neural networks) provide astounding variations on the vertebrate theme, thanks to eons of divergent evolution. Thus, while the human brain may be the source of rich analogies for technologists in search of 'artificial intelligence', both invertebrates and vertebrates will provide endless ideas for technologists designing neural networks for sensory processing, robot control, and a host of other applications (Arbib 1995).

Although this variety means that there is no such thing as a typical neuron, the 'basic neuron' shown in figure A1.2.1 indicates the main features that carry over into artificial neurons. We divide the neuron into three parts: the *dendrites*, the soma (cell body) and a long fiber called the *axon* whose branches form the *axonal arborization*. The soma and dendrites act as input surface for signals from other neurons and/or receptors. The axon carries signals from the neuron to other neurons and/or effectors (muscle fibers or glands, say). The tips of the branches of the axon are called nerve terminals or boutons. The locus of interaction between a terminal and the cell upon which it impinges is called a *synapse*, and we say that the cell with the terminal *synapses upon* the cell with which the connection is made.

The 'signal' carried along the axon is the potential difference across the cell membrane. For 'short' cells (such as the bipolar cells of the retina) passive propagation of membrane potential carries a signal from one end of the cell to the other, but if the axon is long, this mechanism is completely inadequate since changes at one end will decay away almost completely before reaching the other end. Fortunately, cell membranes have the further property that if the change in potential difference is large enough (we say it exceeds a *threshold*), then in a cylindrical configuration such as the axon, a 'spike' can be generated which will actively propagate at full amplitude instead of fading passively. After a spike has been dispatched to propagate along the axon, there is a *refractory period*, of the order of a millisecond, during which a new spike cannot be started along the axon. The details of axonal propagation can be explained by the

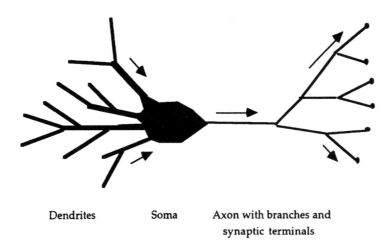

Dendrites   Soma   Axon with branches and
synaptic terminals

Figure A1.2.1. The 'basic' biological neuron. The soma and dendrites act as the input surface; the axon carries the output signals. The tips of the branches of the axon form synapses upon other neurons or upon effectors (though synapses may occur along the branches of an axon as well as at the ends). The arrows indicate the direction of 'typical' information flow from inputs to outputs.

Hodgkin–Huxley equation (Hodgkin and Huxley 1952), which also underlies more complex dynamics that may allow even small patches of neural membrane to act like complex computing elements. At present, most artificial neurons used in applications are much simpler, and it remains for future technology in neural computation to more fully exploit these 'subneural subtleties'.

An impulse traveling along the axon triggers off new impulses in each of its branches, which in turn trigger impulses in their even finer branches. When an impulse arrives at one of the terminals, after a slight delay it yields a change in potential difference across the membrane of the cell upon which it impinges, usually by a chemically mediated process that involves the release of chemical 'transmitters' whereby the presynaptic cell affects the postsynaptic cell. The effect of the 'classical' transmitters is of two basic kinds: either *excitatory*, tending to move the potential difference across the postsynaptic membrane in the direction of the threshold, or conversely, *inhibitory*, tending to move the polarity away from the threshold. Indeed, most neural modeling to date focuses on these excitatory and inhibitory interactions (which occur on a time scale of a millisecond, more or less, in biological neurons). However, neurons may also secrete transmitters which modulate the function of a circuit over some quite extended time-scale. Modeling which takes account of this *neuromodulation* (Dickinson 1995) will become increasingly important in future, since it allows cells to change their function—for example, a cell may change from one which passively responds to stimulation to a pacemaker which spontaneously fires in a rhythmic pattern— enabling a neural network to dramatically switch its overall mode of activity.

The excitatory or inhibitory effect of the transmitter released when an impulse arrives at a terminal generally causes a subthreshold change in the postsynaptic membrane. Nonetheless, the cooperative effect of many such subthreshold changes may yield a potential change at the start of the axon which exceeds the threshold—and if this occurs at a time when the axon has passed the refractory period of its previous firing, then a new impulse will be fired down the axon.

Synapses can differ in shape, size, form and effectiveness. The geometrical relationships between the different synapses impinging upon the cell determine what patterns of synaptic activation will yield the appropriate temporal relationships to excite the cell. A highly simplified example (figure A1.2.2) shows how the properties of nervous tissue just presented would indeed allow a simple neuron, by its very dendritic geometry, to compute some useful function (cf Rall 1964, p 90). Consider a neuron with four dendrites, each receiving a single synapse from a visual receptor, so arranged that synapses a, b, c and d (from left to right) are at increasing distances from the axon hillock (e). We assume that each receptor reacts to the passage of a spot of light above its surface by yielding a generator potential which yields in the postsynaptic membrane the same time course of depolarization. This time course is propagated passively, and the further it is propagated, the later and the lower is its peak. If four inputs reached a, b, c and d simultaneously, their effect might be less than the threshold required to trigger a spike there. However, if an input reaches d before one reaches c, and so on, in such a way that the peaks of the four resultant time courses at the axon hillock coincide, it could well pass the threshold. This then is a cell

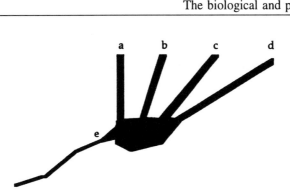

Figure A1.2.2. An example, adapted from Wilfrid Rall, of the subtleties that can be revealed by neural modeling when dendritic properties (in this case, length-dependent conduction time) are taken into account. The effect of simultaneously activating all inputs may be subthreshold, yet the cell may respond when inputs traverse the cell from right to left.

which, although very simple, can detect direction of motion across its input. It responds only if the spot of light is moving from right to left, and if the velocity of that motion falls within certain limits. Our cell will not respond to a stationary object, or one moving from left to right, because the asymmetry of placement of the dendrites on the cell body yields preference of one direction of motion over others. We see, then, that the form (i.e. the geometry) of the cell can have a great impact upon the *function* of the cell and we thus speak of *form–function* relations. Very little work on artificial neurons has taken advantage of subtle properties of this kind, though Mead's (1989) study of *Analog VLSI and Neural Systems*, while inspired E1.3 by biology, does open the door to technological applications in which surprisingly complex computations may be executed by single neurons. Such neurons can compute functions that would require networks of some complexity if one were using the much simpler artificial neurons that are discussed in Chapter B1 B1 of this handbook.

A1.2.2 Psychological motivation and learning rules

Much work in neural computation focuses on the *learning rules* which change the weights of connections B3.3 between neurons to better adapt a network to serve some overall function. Intriguingly, the classic definitions of these learning rules come not from biology, but from the psychological studies of Donald Hebb and Frank Rosenblatt. The work since the early 1980s which has revealed the biological validity of variants of the rules they formulated (Baudry *et al* 1993) is beyond the scope of this handbook. Instead, since the 'line of descent' of neural learning rules may be traced back to this psychological work, we now provide a brief introduction to the ideas of Hebb and Rosenblatt. Hebb (1949) developed a multilevel model of perception and learning, in which the 'units of thought' were encoded by 'cell assemblies', each defined by activity reverberating in a set of closed neural pathways. Hebb introduced a neurophysiological *postulate* (far in advance of physiological evidence): 'When an axon of cell A is near enough to excite a cell B and repeatedly or persistently takes part in firing it, some growth process or metabolic change takes place in one or both cells, such that A's efficiency as one of the cells firing B, is increased.' (Hebb 1949, p 62).

The essence of the Hebb synapse is to increase coupling between coactive cells so that they could be linked in growing assemblies. Hebb developed similar hypotheses at a higher hierarchical level of organization, linking cognitive events and their recall into 'phase sequences'—a temporally organized series of activations of cell assemblies. The simplest formalization of *Hebb's rule* is to increase w_{ij} by B3.3.1

$$\Delta w_{ij} = k y_i x_j \tag{A1.2.1}$$

where synapse w_{ij} connects a presynaptic neuron with firing rate x_j to a postsynaptic neuron with firing rate y_i. Hebb's original learning rule referred exclusively to excitatory synapses, and has the unfortunate property that it can only increase synaptic weights, thus washing out the distinctive performance of different neurons in a network. However, when the Hebbian rule is augmented by a *normalization* rule (e.g. keeping B4.4.1 constant the total strength of synapses upon a given neuron), it tends to 'sharpen' a neuron's predisposition 'without a teacher', causing its firing to become better and better correlated with a cluster of stimulus patterns. This performance is improved when there is some competition between neurons so that if one neuron becomes adept at responding to a pattern, it inhibits other neurons from doing so (competitive learning, see Rumelhart and Zipser 1986).

B1.5, B6   Rosenblatt (1958) explicitly considered the problem of *pattern recognition*, where a 'teacher' is essential—for example, placing 'b' and 'B' in the same category depends on a historico–social convention known to the teacher, rather than on some natural regularity of the environment. He thus introduced

C1.1.1   *perceptrons*, neural networks that change with 'experience', using an *error-correction rule* designed to change the weights of each response unit when it makes erroneous responses to stimuli that are presented to the network. Consider the case in which a set of input lines feeds a single layer of preprocessors whose

B1.2   outputs feed into an output unit which is a *McCulloch–Pitts neuron*. The definition of such a neuron is given in Chapter B1; here we need only note that it has adjustable weights $(w_1, \ldots, w_d)$ and threshold θ and effects a twofold classification: if the preprocessors feed the pattern $x = (x_1, \ldots, x_d)$ to the output unit, then the response of that unit will be 1 if $f(x) = w_1 x_1 + \ldots + w_d x_d - \theta \geqslant 0$, but 0 if $f(x) < 0$. A *simple perceptron* is one in which the preprocessors are not interconnected, *which means that the network has no short-term memory*. (If such connections are present, the perceptron is called *cross-coupled* or

B2.3   *recurrent*. A recurrent perceptron may have multiple layers and loops back from an 'earlier' to a 'later' layer.) Rosenblatt (1958) provided a learning scheme with the property that if the patterns of the training set (i.e. a set of feature vectors, each one classified with a 0 or 1) can be separated by some choice of weights and threshold, then the scheme will eventually yield a satisfactory setting of the weights. The best known perceptron learning rule strengthens an active synapse if the efferent neuron fails to fire when it should have fired, and weakens an active synapse if the neuron fires when it should not have done so:

$$\Delta w_{ij} = k(Y_i - y_i)x_j. \tag{A1.2.2}$$

As before, synapse w_{ij} connects a neuron with firing rate x_j to a neuron with firing rate y_i, but now Y_i is the 'correct' output supplied by the 'teacher.' (This is similar to the Widrow–Hoff (1960) least-mean-squares model of adaptive control.) Notice that the rule does change the response to x_j 'in the right direction'. If the output is correct, $Y_i = y_i$ and there is no change, $\Delta w_{ij} = 0$. If the output is too small, then $Y_i - y_i > 0$, and the change in w_{ij} will add $\Delta w_{ij} x_j = k(Y_i - y_i)x_j x_j > 0$ to the output unit's response to $(x_1, \ldots, x_d)$. Similarly, if the output is too large, Δw_{ij} will decrease the output unit's response. Thus, there is a sense in which $w + \Delta w$ classifies the input pattern x 'more nearly correctly' than w does. Unfortunately, in classifying x 'more correctly' we run the risk of classifying another pattern 'less correctly.' However, the *perceptron convergence theorem* shows that Rosenblatt's procedure does not yield an endless seesaw, but will eventually converge to a correct set of weights, if one exists, albeit perhaps after many iterations through the set of trial patterns.

As Rosenblatt himself noted, extension of these classic ideas to multilayer feedforward networks posed the *structural credit assignment problem:* when an error is made at the output of a network, how is credit (or blame) to be assigned to neurons deep within the network? One of the most popular techniques

C1.2   is called *backpropagation*, whereby the error of output units is propagated back to yield estimates of how much a given 'hidden unit' contributed to the output error. These estimates are used in the adjustment of synaptic weights to these units within the network. In fact, any function $f : X \to Y$ for which X and Y are codeable as input and output patterns of a neural network can be approximated arbitrarily well by a feedforward network with one layer of hidden units. The catch is that very many hidden units may be required for a close fit. It is often an empirical question whether there exists a sufficiently good approximation achievable by a network of a given size—an approximation which a given learning rule may or may not find.

Finally, we note that Hebb's rule (i) *does not depend explicitly* on a teaching signal Y, whereas the perceptron rule (ii) *does depend explicitly* on a teacher. For this reason, Hebb's rule plays an important

B3.1   role in studies of *unsupervised learning* or *self-organization*. However, it should be noted that Hebb's rule

B3.1   can also play a role in *supervised learning* or *learning with a teacher*. This is the case when the neuron being trained has a *teaching input*, separate from the trainable inputs, that can be used to pre-emptively fire

C1.3, F1.4   the neuron. *Supervised Hebbian learning* is often the method of choice in *associative networks*. Moreover, picking up another psychological theme, it is closely related to Pavlovian conditioning: here the response of the cell being trained corresponds to the conditioned and unconditioned response (R), the 'training input' corresponds to the unconditioned stimulus (US), and the 'trainable input' corresponds to the conditioned stimulus (CS). Since the US alone can fire R, while the CS alone may initially be unable to fire R, the conjoint activity of US and CS creates the conditions for Hebb's rule to strengthen the US → R synapse, so that eventually the CS alone is enough to elicit a response.

Acknowledgement

Much of this article is based on the author's article 'Part I–Background' in *The Handbook of Brain Theory and Neural Networks* edited by M A Arbib, Cambridge, MA: A Bradford Book/The MIT Press (1995).

References

Arbib M A (ed) 1995 *The Handbook of Brain Theory and Neural Networks* (Cambridge, MA: Bradford Books/MIT Press)

Baudry M, Thompson R F and Davis J L (eds) 1993 *Synaptic Plasticity: Molecular, Cellular, and Functional Aspects* (Cambridge, MA: Bradford Books/MIT Press)

Dickinson P 1995 Neuromodulation in invertebrate nervous systems *The Handbook of Brain Theory and Neural Networks* ed M A Arbib (Cambridge, MA: Bradford Books/MIT Press)

Hebb D O 1949 *The Organization of Behavior* (New York: Wiley)

Hodgkin A L and Huxley A F 1952 A quantitative description of membrane current and its application to conduction and excitation in nerve *J. Physiol. Lond.* **117** 500–44

Mead C 1989 *Analog VLSI and Neural Systems* (Reading, MA: Addison-Wesley)

Rall W 1964 Theoretical significance of dendritic trees for neuronal input-output relations *Neural Theory and Modeling* ed R Reiss (Stanford, CA: Stanford University Press) pp 73–97

Rosenblatt F 1958 The perceptron: a probabilistic model for information storage and organization in the brain *Psychol. Rev.* **65** 386–408

Rumelhart D E and Zipser D 1986 Feature discovery by competitive learning *Parallel Distributed Processing* ed D E Rumelhart and J L McClelland (Cambridge, MA: MIT Press)

Widrow B and Hoff M E Jr 1960 Adaptive switching circuits *1960 IRE WESCON Convention Record* **4** 96–104

A2

Why Neural Networks?

Paul J Werbos

Abstract

This chapter reviews the general advantages of artificial neural networks (ANNs) which have motivated their use in practical applications. It explains two alternative definitions (computer hardware oriented and brain oriented) of an ANN, and provides an overview of the computational tasks that various classes of ANNs can perform. The advantages include: (i) access to existing sixth-generation computer hardware with huge price–performance advantages; (ii) links to brain-like intelligence; (iii) ease of use; (iv) superior approximation of nonlinear functions; (v) advantages of learning over tweaking, including learning off-line to be adaptive on-line (in control); (vi) availability of many specific designs providing nonlinear generalizations of many familiar algorithms. Among the algorithms and applications are those for image and speech preprocessing, function maximization or minimization, feature extraction, pattern classification, function approximation, identification and control of dynamical systems, data compression, and so on.

Contents

The views presented in this chapter are those of the author and are not necessarily those of the National Science Foundation.

A2.1 Summary

Paul J Werbos

Abstract

See the abstract for Chapter A2.

Artificial neural networks (ANNs) are now being deployed in a growing number of real-world applications across a wide range of industries. There are six major factors which (with varying degrees of emphasis) explain why practical engineers and computer scientists have chosen to use ANNs:

(i) ANN solutions can now be implemented on special-purpose chips and boards which offer considerably more throughput per dollar and more portability than conventional computers or supercomputers.
(ii) Because the brain itself is made up of neural networks, ANN designs seem like a natural way to try to replicate brain-like intelligence in artificial systems.
(iii) ANN designs are often much easier to use than the non-neural equivalents—especially when the conventional alternatives require first-principles models which are not well developed.
(iv) Various universal approximation theorems suggest that ANNs can usually approximate what can be done with other methods anyway and that the approximation can be as good as desired, if one can afford the computational cost of the accuracy required.
(v) ANN designs usually offer solutions based on 'learning' which can be far cheaper and faster than the traditional approach of elaborate prior research followed by tweaking applications until they work.
(vi) The ANN literature includes designs to solve a variety of specific tasks—like function approximation, pattern recognition, clustering, feature extraction, and a variety of novel control-related capabilities— of importance to many applications. In many cases it provides a workable nonlinear generalization of familiar linear methods.

Generally speaking, ANNs tend to have greater advantage when data are plentiful but prior knowledge is limited.

Advantages (i) and (ii) follow directly from the very *definition* of ANNs discussed in Section A2.2. A2.2 Advantages (v) and (vi) are not unique to ANNs; most of the algorithms used to adapt ANNs for specific tasks can *also* be used to adapt other nonlinear structures, such as fuzzy logic systems or physical models based on first principles or econometric models. For example, backpropagation—the most popular ANN algorithm—was originally formulated in 1974 as a *general* algorithm, for use across a wide variety of nonlinear systems, of which ANNs were discussed only as a special case (Werbos 1994). Backpropagation has been used to adapt several different *types* of ANN, but applications to other types of structure are now less common, because it is easier to use off-the-shelf equations or code designed for ANNs. Engineers who wish to achieve neural-like capabilities using *non-neural* designs could benefit substantially by learning about the techniques which have been developed in the neural network field, and subsequently generalized (for example, see White and Sofge 1992, Werbos 1993).

Some ANN advocates have argued that ANNs can perform some tasks which are beyond the reach of 'parametric mathematics'. Some critics have argued that ANNs cannot do anything that cannot be done just as well 'using mathematical methods'. Both of these positions are quite naive insofar as ANNs are simply a *subset* of what can be done with precise mathematics. Nevertheless, they are an interesting and important subset, for the reasons given above.

Many of us believe that the greatest value of ANN research, in the long term, will come when we use it to go back to the brain itself, to develop a more functional, engineering-based understanding of the brain as an engineering device. This belief is shared even by many researchers who believe that 'consciousness' in the largest sense includes more than just an understanding of the brain (Levine and Elsberry 1996, Pribram 1994).

References

Levine D and Elsberry W (ed) 1996 *Optimality in Biological and Artificial Networks* (Hillsdale, NJ: Erlbaum)

Pribram K (ed) 1994 *Origins: Brain and Self-Organization* (Hillsdale, NJ: Erlbaum)

Werbos P 1993 Elastic fuzzy logic: a better fit to neurocontrol and true intelligence *J. Int. Fuzzy Syst.* **1** 365–77

——1994 *The Roots of Backpropagation: From Ordered Derivatives to Neural Networks and Political Forecasting* (New York: Wiley)

White D A and Sofge D A (eds) 1992 *Handbook of Intelligent Control: Neural, Fuzzy and Adaptive Approaches* (New York: Van Nostrand)

A2.2 What is a neural network?

Paul J Werbos

Abstract

See the abstract for Chapter A2.

A2.2.1 Introduction

There are several possible answers to the question, 'What is a neural network?' Years ago, some people would answer the question by simply writing out the equations of *one particular* artificial neural network (ANN) design. However, there are many different ANN designs, oriented towards very different kinds of tasks. Even within the field itself few researchers appreciate how broad the range really is.

A2.2.2 The US National Science Foundation neuroengineering program: a case study

The example of the US National Science Foundation (NSF) neuroengineering program is a useful case study of the varying motivations and concepts behind ANN research. At NSF, the decision to fund a program in neuroengineering was motivated by two very different-looking definitions of what the field is about. Fortunately, in practice, the two definitions ended up including virtually the same set of research efforts. One definition was motivated by computer hardware considerations, and the other by links to the brain.

A2.2.3 Artificial neural networks as sixth-generation computers

The neuroengineering program at NSF started out as an element of the optical technology program. It was intended to support a vision of sixth-generation computing, illustrated in figure A2.2.1.

Most people today are very familiar with fourth-generation computing, illustrated on the left-hand side of the figure. Ordinary personal computers and workstations are examples of fourth-generation computing. In that scheme, there is one CPU chip inside which all the hard-core computing work is done. The CPU processes one instruction at a time. Its capabilities map nicely into familiar computer languages like FORTRAN, BASIC, C or SMALLTALK (in historical order). The key breakthroughs underlying fourth-generation computing were the invention of the microchip (co-invented by Federico Faggin of CalTech) and the development of *VLSI technology*.

E1.3, E1.4.3

A decade or two ago, many computer scientists became excited by the concept of massively parallel processing (MPP) or fifth-generation computing, illustrated in the middle of the figure. In MPP, hundreds or even millions of fully featured CPU chips are inserted into a *single* computer, in the hope of increasing computational throughput a hundred-fold or a million-fold. Unfortunately, MPP computers cannot just run conventional computer programs in FORTRAN or C in a straightforward manner. Therefore, governments in the United States and Japan have funded a large amount of research into high-performance computing, teaching people how to write computer programs within that *subset* of algorithms which can exploit the power of these 'supercomputers'.

In the late 1980s, researchers in optical technology came to NSF and argued that optical computing offers the hope of computational power a thousand or even a million times larger than fifth-generation computing. Since the computing industry is a huge industry, this claim was considered very carefully. NSF consulted with Carver Mead—the father of VLSI—and his colleague, Federico Faggin, among others.

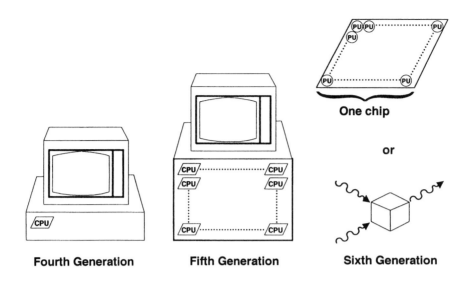

Figure A2.2.1. Three generations of computer hardware.

Mead and Faggin claimed that similar capabilities could be achieved in microchips, if one were willing to put hundreds or millions of *extremely simple processing units* onto a single chip. Thus sixth-generation capability could be implemented either in optical technology or in VLSI. (Michael Conrad of Wayne State University in Detroit has studied a third alternative, using molecular computing.)

The skeptics argued that sixth-generation computers can only run an *extremely* small subset of all possible computer programs. They would not represent a massive improvement in productivity for the computing industry as a whole, because they would be useful only in a few very small niche applications. They would not be suitable for truly generic, general-purpose computing. Carver Mead replied that the human brain itself is based on an extremely massive parallelism, using processors which—like the elements of optical hologram processors—perform the same 'simple' operations over and over again, without running anything at all like FORTRAN code. The human brain appears to demonstrate *very* generic capabilities; it is not just a niche machine. Therefore, he argued, sixth-generation computers should also be able to achieve truly generic capabilities. Mead himself has made a major effort to follow through on these opportunities (Mead 1988).

In evaluating this argument, NSF concluded that Mead's argument was essentially correct, but that extensive research would be needed in order to convert the argument into a working engineering capability. More precisely, they concluded that research would be needed to actually *develop* algorithms or designs, to perform useful generic computational tasks consistent with the constraints of sixth-generation computing. The neuroengineering program was initiated in 1988 to do precisely that. For the purposes of this program, ANNs were *defined* as algorithms or designs of this sort.

The concept of sixth-generation hardware was largely theoretical in 1988. A few years later, there was a great variety of experimental ANN boards and chips available; however, few of these were of direct practical interest, because of limited throughput, reliability or availability. But by 1995, there were a number of practical, reliable high-throughput workstations, boards and chips available on the commercial market—boards available for $5000 or less (retail) and chips available, in some cases, at prices under $10 (wholesale). A few examples follow. Adaptive Solutions Inc, of Beaverton, Oregon, has sold workstations—using digital ANN chips able to implement a variety of ANN designs—which benchmark 100 times as fast as a Cray supercomputer, on the image recognition problems which are currently the main source of funding for the company; they also provide a PC board based on a SIMD architecture. Accurate Automation Corporation of Chattanooga, Tennessee, sells an MIMD board which is slower but more flexible, originally developed for control applications. HNC of San Diego, California, has won a Babbage prize for breakthroughs in price–performance ratios in a neural-oriented array processor workstation. Among the many interesting chips are those designed by Motorola, Adaptive Solutions, Harris Semiconductor (motivated by NASA system identification applications) and a collaboration between Ford Motor Company and the Jet Propulsion Laboratory of Pasadena, California. Some of the chip designers have distributed software simulators of their designs to researchers; such simulators make it possible for engineering researchers, with knowledge of neural networks and applications but *not* of hardware as such, to develop and test designs which *could* be implemented directly in hardware. One should expect even more powerful hardware from a larger set of suppliers to be developed each year; however, the results achieved by 1995 were already enough to make sixth-generation computing a realistic option for practical engineers.

The implications of this are very great. Suppose that you have an existing, conventional algorithm to perform some task like control or pattern recognition—tested on a mainframe or supercomputer. Suppose that your algorithm is not widely used in industry, because of its cost or physical demands. (For example, people do not put mainframes on cars or dedicated supercomputers in every workstation of a factory.) If you develop an equivalent ANN of equal capability and complexity, then these ANN chips and boards would make it far easier for people to actually use your work. In some applications—such as spacecraft— chips could be sent into orbit, and then reprogrammed (virtually rewired) by telemetry, to permit a complete updating of their functions when desired, without the need to replace hardware.

Some researchers believe in the possibility of a seventh-generation style of computing, exploiting quantum effects such as Bell's theorem. Most of the work in true quantum computing today is highly abstract, with little emphasis on useful generic computing tasks; however, H John Caulfield of Alabama A&M University has done preliminary work which might have practical implications involving optical computing and neural networks (Caulfield 1995, Caulfield and Shamir 1992). A few further possibilities along these lines are discussed in the author's chapter in Levine and Elsberry (1996), and in Conrad (1994). In general, we would expect the main computational advantage of quantum computing to involve some exploitation of massive parallelism involving simple operations, as with optical computing; thus ANN approaches may be crucial to practical success in quantum computing.

Most successful projects in neuroengineering do not focus at all on the chips or boards at first. They begin with extensive simulations on PCs or workstations, along with some mathematical analysis and a very aggressive effort to understand and assimilate designs developed elsewhere. After some success in simulations, they proceed to tests on real-world plants or data, which they use to refine their designs and to justify building up a more modular, flexible software system. Then, after there is success on a real-world plant, market forces almost always encourage them to look more intensively at chips and boards.

A2.2.4 Artificial neural networks as brain-like designs or circuits

Figure A2.2.2 represents a different definition of neuroengineering—the definition used at the actual start of the NSF program. The figure emphasizes the link to neuroscience, as well as the difference between neuroscience and neuroengineering. In neuroscience and psychology, one tries to understand what the capabilities of the brain actually *are*. Of special interest to us are the capabilities of the brain in solving difficult computational problems important to engineering. In neuroscience, one also studies how the circuits or architectures in the brain give rise to these capabilities.

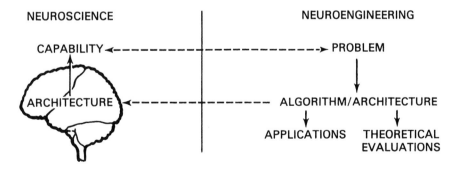

Figure A2.2.2. Neuroscience and neuroengineering. Neuroengineering tries to develop algorithms and architectures, inspired by what is known about brain functioning, to imitate brain capabilities which are not yet achieved by other means. By demonstrating algorithm capabilities and properties, it may raise issues which feed back to questions or hypotheses for neuroscience.

In neuroengineering, we do something different. We try to *replicate* capabilities of the brain, in a practical engineering or computational context. We try to exploit what is known about how the brain achieves these capabilities, in developing designs which are consistent with that knowledge. (We now use the word 'design' rather than 'algorithm' to emphasize the fact that the same equations may be implemented sometimes in software and sometimes as chip architectures.) We then test and improve these designs, based on real-world applications, simulations, and mathematical analysis drawing on a variety of disciplines. Finally, there can be a feedback from what we have learned, allowing us to understand the brain in a new light, hopefully deriving new insights and designs in the process.

Even at this global level, we can see some issues which lead to diversity or even conflict in the neural network community. There are two extreme approaches to developing ANN designs: (i) bottom-up efforts to copy what is *currently known* about biological circuits directly into chips, sometimes without engineering analysis along the way; (ii) totally engineering-based efforts, based on the idea that *today's* knowledge of the brain is very partial, and that 'brain-like circuitry' now requires little more than limiting ourselves to what we could implement on sixth-generation hardware. In informal discussions, people sometimes compare 'paying biologists to teach engineers how to do engineering' versus 'paying engineers to teach biologists how to do biology'.

The NSF program in neuroengineering emphasizes the engineering approach, because it is hard to imagine how a purely bottom-up biological approach, without new engineering-based mathematical paradigms, could replicate or explain something as global as 'intelligence' in the brain (Pribram 1994), let alone 'consciousness' in the broadest sense (Levine and Elsberry 1996). Almost all of the useful basic designs in the ANN field resulted from some sort of biological inspiration, and biology still has a great deal to tell us; however, we have now reached the point where our ability to learn useful new things from biology depends on the participation of people who appreciate how much has already been learned in an engineering context. US government funding is generally available for such collaborations, but it is difficult to locate competent proposals combining both key elements: firstly, engineers with a deep enough understanding to be truly relevant and, secondly, wet, experimental biologists willing to take a novel approach to fundamental issues.

Whatever the limits of today's ANN designs, the brain still provides an existence proof that far more is possible and that research to develop more powerful designs can, in fact, succeed.

References

Caulfield H J 1995 Optical computing benefits from quantum mechanics *Laser Focus World* May 181–4
Caulfield H J and Shamir J 1992 Wave particle duality processors: characteristics, requirements and applications *J. Opt. Soc. Am.* A **7** 1314–23
Conrad M 1994 Speedup of self-organization through quantum mechanical parallelism *On Self-Organization: An Interdisciplinary Search for a Unifying Principle* ed R K Mishra, D Maaz and E Zwierlein (Berlin: Springer)
Levine D and Elsberry W (eds) 1996 *Optimality in Biological and Artificial Networks* (Hillsdale, NJ: Erlbaum)
Mead C 1988 *Analog VLSI and Neural Systems* (Reading, MA: Addison-Wesley)
Pribram K (ed) 1994 *Origins: Brain and Self-Organization* (Hillsdale, NJ: Erlbaum)

A2.3 A traditional roadmap of artificial neural network capabilities

Paul J Werbos

Abstract

See the abstract for Chapter A2.

Practical uses of artificial neural networks (ANNs) all depend on the fact that ANNs can perform specific computational tasks important to engineering or to other economic sectors. Unfortunately, popularized accounts of ANNs often make it sound as though ANNs only perform one or two fundamental tasks, and that the rest is 'mere application'. This is highly misleading.

In 1988, a broad survey of ANNs would have shown the existence of three basic types of design, still in use today:

(i)   *hard-wired designs* to perform highly specific, concrete tasks, such as image preprocessing by a   E1
      'silicon retina';
(ii) designs to perform *static* or *combinatorial optimization*—the minimization or maximization of a   F1.3
      complicated function of many variables;
(iii) designs based on *learning*, where the weights or parameters of an ANN are adjusted or adapted over   B3
      time, so as to permit the system to perform some kind of generic task over a wide range of possible
      applications.

Learning designs now account for the bulk of the field, but the other two categories still merit some discussion

A2.3.1 Hard-wired designs

The hard-wired designs usually try to mimic the details of some brain circuit, complete with all the connections *and* all the parameters as they exist in an adult brain *without further learning*. Major examples would be 'silicon retinas' (used for preprocessing images, as in Mead 1988), 'silicon cochleas' (for preprocessing speech data), and artificial controllers for hexapod robots modeled on studies of the cockroach. Grossberg, like Mead, has put major efforts into developing something like a silicon retina, of great interest to the US Navy, by building on more detailed biological research in his group (Gaudiano 1992).

Even the brain itself uses relatively fixed preprocessors and postprocessors, to simplify the job of the higher centers, based on millions of years of evolution and experience with certain very specific, concrete tasks. Most of the current work on wavelets—which are often used as preprocessors coming before ANNs—could be seen as belonging to this category; however, even wavelet analysis can be made adaptive using neural network methods (Szu *et al* 1992).

A2.3.2 Static optimization

Years ago, *static optimization* based on *Hopfield networks* accounted for perhaps a quarter of all efforts   F1.3, B1.3
towards ANN applications. (Grossberg had discussed the same class of network in earlier years, but Hopfield proposed its use on optimization problems. See the chapter by Hopfield in Lau (1992).) The key idea here was that Hopfield networks always settle down into a (local) minimum of some 'energy'

function, a function which depends on the weights in the network. By choosing the weights and the transfer functions in a clever manner, the user can make the network minimize some desired function of many inputs. This idea was especially natural for people trying to minimize quadratic functions of many variables with constraints. For example, many researchers envisaged using Hopfield networks to maximize very complex likelihood functions taken from image segmentation and image analysis research; they envisaged high-quality segmentation on a chip.

This approach worked very well on toy problems, including toy versions of the traveling salesman problem; however, it encountered great difficulty in scaling up to problems of more realistic scale. With larger problems, there were issues of numerical efficiency and the difficulty of finding a 'good' energy function. Even with smaller problems, these kinds of networks frequently have many, many local minima or 'attractors'. At present, people in industry facing very large static optimization problems still tend to use classical methods; see the chapter by Shanno in Miller *et al* (1990). When there are many local minima, it was popular a few years ago to use simulated annealing or modifications of the Hopfield network (such as Szu's 'Cauchy machine', Scheff and Szu 1987) to provide a kind of random element to help the system escape from local minima. Currently, it is more popular to use genetic algorithms for this purpose.

Unfortunately, genetic algorithms also have difficulties in scaling to larger problems (except when there is a special structure present). There has been a lot of discussion of ANN-genetic hybrids, which could help overcome the scaling problem, but the author is not aware of any large-scale applications to static optimization problems or of any hybrid designs which are truly suitable for this purpose. In any case, it seems very unlikely that neural circuits in the brain would use this particular way of injecting noise. For a credible alternative view of these issues, see the work of Michael Conrad of Wayne State University (Conrad 1993, 1994, Smalz and Conrad 1994).

Many researchers believe that Hopfield networks or Hopfield-like networks could perform much better in optimization, if only the users of these networks could be more 'clever', somehow, in specifying their weights or connections. But from a practical point of view, it is probably not realistic to demand higher levels of 'cleverness' than engineers have displayed in past efforts to use these networks. Fortunately, it is not necessary to rely on cleverness alone when solving large problems. For example, methods which make some use of Kohonen's feature-extraction ANNs have demonstrated accuracy comparable to that of classical methods on a number of large-scale routing and optimization problems; see the chapter by El Ghaziri in Kohonen *et al* (1991). Clearly this approach is worthy of further pursuit.

B3.3.1 More generally, it is possible to use learning methods to derive useful weights in a more reliable manner for Hopfield networks. When Hopfield networks are adapted by use of the well known *Hebbian methods*, they act as associative memories, which are not suitable for solving complex optimization problems. However, it is also possible to adapt them so as to minimize error and solve problems which cannot be solved by more popular feedforward networks. Hopfield networks are a special case of simultaneous recurrent networks (SRNs). See White and Sofge (1992), Chapter 3, and Werbos (1993) for relatively straightforward discussions of how to adapt the weights in such networks so as to minimize error. This is a promising area for future research, but the author is not aware of any working examples as yet in static optimization.

In summary, there are several examples of state-of-the-art performances on large problems by Kohonen-related networks. There is reason to hope for better performance and reliability with Hopfield-like networks in the future, with further research exploiting learning and noise injection.

A2.3.3 Designs based on learning

The vast bulk of the neural network field today is based on designs which *learn* to perform tasks over time. *Learning* can be used to solve extremely complex problems, especially when the human user understands the art of learning in *stages*, using a schedule of related tasks of increasing difficulty.

Many authors have argued that 'intelligence' in the true sense of the word can never be achieved by simply expanding our library of computational algorithms tailored to narrow, application-specific tasks. Instead, 'intelligence' implies the ability of a computational system to learn the algorithms itself, from experience, based on generalized learning principles which can be used in a wide variety of applications. Many authors have argued at length that a deeper understanding of learning must be the foundation of any really scientific explanation of intelligence (Hebb 1949, Pribram 1994, Werbos 1994).

But what kinds of generic tasks can ANNs learn to perform? The ANN field has traditionally used a three-fold taxonomy to describe these tasks:

- Supervised learning
- Unsupervised learning
- Reinforcement learning.

In all three areas, there is a traditional choice between two modes of learning:

- 'off-line learning', where all the observations in a database of 'training data' are analyzed together, simultaneously;
- 'on-line learning', where data are fed into the network one observation at a time. The weights or parameters in the network are changed after each observation, but there is no other record kept of the observation. The system then goes on to the next observation, and so on.

A2.3.3.1 Supervised learning

Intuitively, in on-line mode, *supervised learning* works as follows. Whenever we make an observation, B3.1 we first see a set (or vector) of input values X. We plug in these values as inputs to our ANN and then calculate the outputs of the ANN using the weights or parameters inherited from before. Then, in the training period, we *also* obtain a specification of exactly what the outputs of the ANN *should* have been for that observation. (For example, the inputs might represent the pixels of an image containing a handwritten digit; the desired output might be a coded representation of the correct classification of the digit.) We then adjust the weights of the ANN so as to make its actual output more like the desired output in the future (see figure A2.3.1).

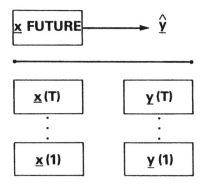

Figure A2.3.1. The supervised learning task.

Many researchers will immediately recognize the similarity between this figure and the well established, well known method called multiple regression or ordinary least squares. As in multiple regression, supervised learning tries to estimate a set of weights which represent the relationship between the input variables X and the dependent or target variables Y, but supervised learning looks for the best *nonlinear* relationship, not just the best linear relationship. It uses ANN forms which are capable of approximating any smooth nonlinear relationship (Barron 1993). Also, it offers numerical techniques which are faster than those generally used in statistics. Conventional statistics normally use the offline mode; however, the on-line mode is more useful in many applications.

Nevertheless, the theoretical issues involved in supervised learning (apart from learning speed) are indeed quite close to those in statistics. The best current research in supervised learning draws heavily on the literature in statistics—including the literature on issues like robustness and multicolinearity, which are neglected all too often in conventional statistical analysis.

Computer tools for supervised learning are now very widespread, though of varying quality. Most of the real-world applications of ANNs today are based at least in part on supervised learning. Supervised learning may be thought of as a tool for function approximation, or as a tool for statistical pattern recognition. Former post office officials have told me that all of the best ZIP-code recognizers today use fairly standard ANNs for digit recognition. This is a remarkable achievement in such a short time, relative to a field (statistical pattern recognition) which had already been highly developed and intensively funded long before ANNs became widely known. Also, this is far from an isolated example; fortunately, there are other sections in this handbook which review some of the many, many applications in this category. There is substantial opportunity to develop even better designs for supervised learning (Werbos 1993), but the tools available today are already quite useful.

A2.3.3.2 Unsupervised learning

On the other hand, supervised learning is clearly absurd as a model of what the human brain does as a whole system. There is no one telling us exactly what to do with every muscle of our body every moment of the day. The term *unsupervised learning* was coined in the 1980s to describe ANN designs which do not require that kind of detailed guidance or feedback.

B3.1

Intuitively, in online mode, unsupervised learning works as follows. Whenever we make an observation, we first see a vector of input values X. We plug these values in as inputs to our ANN, calculate the outputs of our ANN using weights inherited from before, then adapt or adjust the weights *without* using any external information about how 'good' the outputs were.

From an engineering viewpoint, supervised learning is a well defined task—the task of matching or predicting some externally-specified target variables. Unsupervised learning as such is not a well defined task. Some of the designs used in unsupervised learning originated as biological models, models which were formulated well before their value as computational systems was known; fortunately, many of these designs did turn out to have important 'emergent properties', computational capabilities which were discovered only after the models were studied further (see Pribram 1994 for more elaborate discussions of the related concepts of self-organization, chaos and so on).

As a practical matter, unsupervised learning includes useful designs to perform a variety of tasks— most notably, feature extraction, clustering and *associative memory*. In feature extraction, one maps an input vector X into another vector R, which tries to represent the same useful information in a more useful form—usually a more compact form. If the vector R does have fewer components than the original input vector, then this can be used as a data compression scheme. In any event, it can also be used to provide more useful, more tractable input either to a supervised learning scheme or to some other downstream information processor. Clustering offers similar benefits.

C1.3, F1.4

Some of the ANN designs for clustering and feature extraction are based more on experimentation and intuition than on mathematical theory. However, classical methods for clustering, found in standard statistical packages, are usually even more *ad hoc* in nature; they tend to require arbitrary choices of distance measures and sequencing (Duda and Hart 1975). At least some of the ANN designs do provide something like adaptive distance measures to permit a more rational clustering strategy, which is occasionally useful.

Some of the ANN designs for feature extraction are equivalent (in the limit) to conventional principal components analysis (PCA), the most popular classical method for data-based feature extraction. However, PCA itself is a linear design, and it does not represent a true stochastic model (Joreskog and Sorbom 1984). There is another class of ANN design which is truly nonlinear, but approximates PCA in the linear special case; we might say that these 'autoassociator' designs are the nonlinear generalization of PCA (Werbos 1988, Hinton and Beckman 1990, Fleming and Cottrell 1990). These designs have performed reasonably well in moderate-sized applications like diagnostics in aerospace vehicles and chemical plants; however, they have not performed as well in complex data compression applications, and the issue of statistical consistency is a concern. There are other ANN designs—like Kohonen's

C2.1.1 *self-organizing maps* (see Kohonen in Lau 1992) and the stochastic encoder/decoder/predictor (White and Sofge 1992, Chapter 13)—which are firmly rooted in stochastic analysis; they may be viewed as nonlinear generalizations of factor analysis, which is the standard method used by statisticians to model the structure of probability distributions for vectors containing many continuous variables (Joreskog and Sorbom 1984). Both of these have had significant real-world applications, but the details are proprietary in the cases I am most familiar with.

The distinction between supervised and unsupervised systems has been confused at times in the literature, in part because of confusion between systems and subsystems, and in part because of cultural

C2.2.1 differences within the field. For example, there is a design called *ARTMAP* which is used to perform supervised learning tasks, using *components* based on unsupervised learning designs; the system as a whole is worthy of evaluation in the context of supervised learning—because it is a competitor in that market—even though its components are unsupervised (Carpenter *et al* 1992). Heteroassociative memories are similar. On the other hand, the autoassociators mentioned above use a supervised learning approach on the inside in order to solve a problem in unsupervised learning; the design as a whole is unsupervised. The human brain itself clearly has a structure of modules and submodules which is far more complex than anything which has ever been implemented as an ANN; thus it would not be surprising if the brain included supervised *components* as part of a more complex architecture.

A2.3.3.3   *Reinforcement learning*

Many of us believe that the concept of unsupervised learning is just as absurd as the concept of supervised learning, as a description of what the brain does *as a whole system*. Intermediate between supervised learning and unsupervised learning is another classical area called *reinforcement learning*, illustrated in c3 figure A2.3.2.

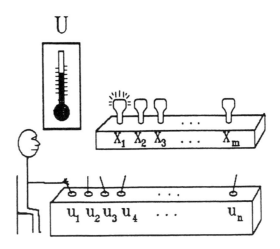

Figure A2.3.2. The reinforcement learning task. (From Miller *et al* 1990 with permission of MIT Press.)

Intuitively, in online mode, reinforcement learning works as follows. When we make an observation, we first see a vector of inputs, X. We plug X into our ANN, calculate the outputs of the ANN, then obtain from the outside a *global* evaluation U of *how good* the outputs were. Instead of obtaining *total* feedback (as in supervised learning) or no feedback (as in unsupervised learning), we obtain a *moderate* degree of feedback. In the modern formulation of reinforcement learning, it is also assumed that $U(t)$ at time t will depend on the observed variables X, which in turn depend on actions taken at an earlier time; the goal is to maximize U over future time, *accounting* for the impact of present actions on future U. An example of such a system might be an ANN which learns how to operate a factory so as to maximize profit over time, or to minimize fuel consumption or pollution or a weighted sum of both.

In figure A2.3.2, we see a cartoon figure representing our ANN system. The cartoon figure has control over certain levers, forming a vector u, and gets to see certain input information X. The cartoon figure starts out with no knowledge about the causal relationships between u, X and U. Its job is to *learn* these relationships, and come up with a strategy of action which will maximize the reward criterion U over time. This is the problem or task of reinforcement learning. Reinforcement learning maps very well into many serious theories and models of human and animal behavior (Levine and Elsberry 1996). It also maps directly into the problem of *optimizing performance over time*, a fundamental task considered in modern control theory and decision analysis. Modern work on reinforcement learning has modified the definition of the problem very slightly, to allow for knowledge of U as a function of X, for reasons beyond the scope of this section. Some of the very largest, socially important applications of ANNs have come precisely in this area.

Reinforcement learning should not be interpreted as an alternative way to perform supervised learning tasks. Rather, it is a large collection of alternative designs aimed at performing a *different task*. These designs typically contain *components* which are supervised, but the designs as a whole are neither supervised nor unsupervised.

Reinforcement learning is only one example—though perhaps the most important example—of neural network designs *for control*. Problems in decision and control can be resolved into a number of specific tasks—including prediction over time or system identification by ANN—which are just as fundamental, in their own way, as the task of supervised learning. In the last few years, there has been a tremendous growth in research, developing new generic designs for use on these generic tasks. Decision and control may itself be seen as a kind of integrating framework—like the human brain itself—which encourages us to combine a wide variety of subtasks and components into a single system, which serves as a unifying framework. This requirement for unification and integration is one of the key factors which distinguishes the ANN approach from earlier styles of research.

References

Barron A R 1993 Universal approximation bounds for superpositions of a sigmoidal function *IEEE Trans. Info. Theory* **39** 930–45

Carpenter G A, Grossberg S, Markuzon N, Reynolds J H and Rosen D B 1992 Fuzzy ARTMAP: a neural network architecture for incremental supervised learning of analog multidimensional maps *IEEE Trans. Neural Networks* **3** 698–713

Conrad M 1993 Emergent computation through self-assembly *Nanobiology* **2** 5–30

Conrad M 1994 Speedup of self-organization through quantum mechanical parallelism *On Self-Organization: An Interdisciplinary Search for a Unifying Principle* ed R K Mishra, D Maaz and E Zwierlein (Berlin: Springer)

Duda R O and Hart P E 1975 *Pattern Classification and Scene Analysis* (New York: Wiley)

Fleming M K and Cottrell G W 1990 Categorization of faces using unsupervised feature extraction *Proc. Int. Joint Conf. on Neural Networks (San Diego, CA)* (New York: IEEE Press) p II-65-70

Gaudiano P 1992 A unified neural network model of spatiotemporal processing in A and Y retinal ganglion cells II: temporal adaptation and simulation of experimental data *Biol. Cybern.* **67** 23–34

Hebb D O 1949 *The Organization of Behavior* (New York: Wiley)

Hinton G E and Beckman S 1990 An unsupervised learning procedure that discovers surfaces in random-dot stereograms *Proc. Int. Joint Conf. on Neural Networks (Washington, DC)* (Hillsdale, NJ: Erlbaum) I-218-222

Joreskog K G and Sorbom D 1984 *Advances in Factor Analysis and Structural Equation Models* (Lanham, MD: University Press of America). See also the classic but out-of-print text by Maxwell and Lawley *Factor Analysis as Maximum Likelihood Method*

Kohonen T, Makisara K, Simula O and Kangas J (eds) 1991 *Artificial Neural Networks* vol 1 (New York: North-Holland)

Lau C G (ed) 1992 *Neural Networks: Theoretical Foundations and Analysis* (New York: IEEE Press)

Levine D and Elsberry W (eds) 1996 *Optimality in Biological and Artificial Networks* (Hillsdale, NJ: Erlbaum)

Mead C 1988 *Analog VLSI and Neural Systems* (Reading, MA: Addison-Wesley)

Miller W T, Sutton R and Werbos P (eds) 1990 *Neural Networks for Control* (Cambridge, MA: MIT Press)

Pribram K (ed) 1994 *Origins: Brain and Self-Organization* (Hillsdale, NJ: Erlbaum)

Scheff K and Szu H 1987 1-D optical Cauchy machine infinite film spectrum search *Proc. IEEE Int. Conf. on Neural Networks* (New York: IEEE Press)

Smalz R and Conrad M 1994 Combining evolution with credit apportionment: a new learning algorithm for neural nets *Neural Networks* **7** 341–51

Szu H H, Telfer B and Kadambe S 1992 Neural network adaptive wavelets for signal representation and classification *Opt. Eng.* **31** 1907–16

Werbos P 1988 Backpropagation: past and future *Proc. Int. Conf. on Neural Networks* (New York: IEEE Press) I-343-353

——1993 Supervised learning: can it escape its local minimum *Proc. WCNN93* (Hillsdale, NJ: Erlbaum)

——1994 *The Roots of Backpropagation: From Ordered Derivatives to Neural Networks and Political Forecasting* (New York: Wiley)

White D A and Sofge D A (eds) 1992 *Handbook of Intelligent Control: Neural, Fuzzy and Adaptive Approaches* (New York: Van Nostrand)

PART B

FUNDAMENTAL CONCEPTS OF NEURAL COMPUTATION

PART B

FUNDAMENTAL CONCEPTS OF NEURAL COMPUTATION

B1

The Artificial Neuron

Michael A Arbib

Abstract

This chapter first describes the basic structure of a single neural unit, briefly relating it to the general notion of a neural network. The interior workings of simple artificial neurons—especially the discrete-time McCulloch–Pitts neuron and continuous-time leaky integrator neuron—are then presented, including the general properties of threshold functions and activation functions. Finally, we briefly note that there are many alternative neuron models available.

Contents

Much of this chapter is based on the author's overview article 'Part I–Background' in *The Handbook of Brain Theory and Neural Networks* edited by M A Arbib, Cambridge, MA: A Bradford Book/The MIT Press (1995).

B1.1 Neurons and neural networks: the most abstract view

Michael A Arbib

Abstract

See the abstract for Chapter B1.

There are many types of artificial neuron, but most of them can be captured as formal objects of the kind shown in figure B1.1.1. There is a set X of signals which can be carried on the multiple input lines x_1, $\ldots$, x_n and single output line y. In addition, the neuron has an internal state s belonging to some state set S.

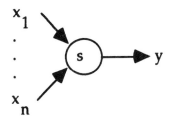

Figure B1.1.1. A 'generic' neuron, with inputs $x_1, \ldots, x_n$, output y, and internal state s.

A neuron may be either discrete-time or continuous-time. In other words, the input values, state and output may be given at discrete times $t \in \mathbb{Z} = \{0, 1, 2, 3, \ldots\}$, say, or may be given at all times t in some interval contained in the real line $\mathbb{R}$. A *discrete-time neuron* is then specified by two functions which specify (i) how the new state is determined by the immediately preceding inputs and (in some neuron models, but by no means all) the previous state, and (ii) how the current output is to be 'read out' from the current state:

The *next-state-function* $f : X^n \times S \to S, s(t) = f(x_1(t-1), \ldots, x_n(t-1), s(t-1))$; and
The *output function* $g : S \to Y, y(t) = g(s(t))$.

As we shall see in later sections, popular choices take the signal-set X to be either a binary set—$\{0, 1\}$ is the 'classical choice', though physicists, inspired by the 'spin-glass' analogy, often use the spin-down, spin-up set denoted by $\{-1, +1\}$—or an interval of the real line, such as $[0, 1]$; while the state-set is often taken to be $\mathbb{R}$ itself. A *continuous-time neuron* is also specified by two functions $f : X^n \times S \to S$, and $g : S \to Y, y(t) = g(s(t))$, but now f serves to define the *rate of change* of the state, that is, it provides the right-hand side of the differential equation which defines the state dynamics:

$$\frac{\mathrm{d}s(t)}{\mathrm{d}t} = f(x_1(t), \ldots, x_n(t), s(t)).$$

Clearly, S at least can no longer be a discrete set. A popular choice is to take the signal-set X to be an interval of the real line, such as $[0, 1]$, and the state-set to be $\mathbb{R}$ itself.

The focus of this chapter will be on motivating and defining some of the best known forms for f and g. But first it is worth noting that the subject of neural computation is not interested in neurons as

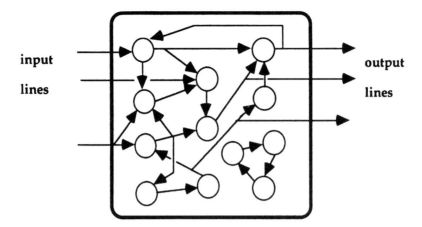

Figure B1.1.2. A neural network viewed as a system (continuous-time case) or automaton (discrete-time case). The input at time t is the pattern on the input lines, the output is the pattern on the output lines; and the internal state is the vector of states of all neurons of the network.

ends in themselves but rather in neurons as units which can be composed into networks. Thus, both as background for later chapters and as a framework for the focused discussion of individual neurons in this chapter, we briefly introduce the idea of a neural network.

We first show how a neural network comprised of continuous-time neurons can also be seen as a continuous-time system in this sense. As typified in figure B1.1.2, we characterize a neural network by selecting N neurons and by taking the output line of each neuron, which may be split into several branches carrying identical output signals, and either connecting each branch to a unique input line of another neuron or feeding it outside the network to provide one of the N_L network output lines. Then every input to a given neuron must be connected either to an output of another neuron or to one of the (possibly split) N_1 input lines of the network. Then the input set X of the entire network is $\mathbb{R}^{N_1}$, the state set $Q = \mathbb{R}^N$, and the output set $Y = \mathbb{R}^{N_L}$. If the ith output line comes from the jth neuron, then the *output function* is determined by the fact that the ith component of the output at time t is the output $g_j(s_j(t))$ of the jth neuron at time t. The *state transition function* for the neural network follows from the state transition functions of each of the N neurons

$$\frac{\mathrm{d}s_j(t)}{\mathrm{d}t} = f_j(x_{1_j}(t), \ldots, x_{n_j j}(t), s_j(t))$$

as soon as we specify whether $x_{i_j}(t)$ is the output of the kth neuron or the value currently being applied on the lth input line of the overall network.

Turning to the discrete-time case, we first note that, in computer science, an *automaton* is a discrete-time system with discrete input, output and state spaces. Formally, we describe an automaton by the sets X, Y and Q of inputs, outputs and states, respectively, together with the *next-state function* $\delta : Q \times X \rightarrow Q$ and the *output function* $\beta : Q \rightarrow Y$. If the automaton is in state q and receives input x at time t, then its next state will be $\delta(q, x)$ and its next output will be $\beta(q)$. It should be clear that a network like that shown in figure B1.1.2, but now a discrete-time network made up solely from discrete-time neurons, functions like a finite automaton, as each neuron changes state synchronously on each tick of the time-scale $t = 0, 1, 2, 3, \ldots$. Conversely, it can be shown (see e.g. Arbib 1987, Chapter 2—that the result was essentially, though inscrutably, due to McCulloch and Pitts 1943) that any finite automaton can be simulated by a suitable network of discrete-time neurons (even those of the 'McCulloch–Pitts type' defined below). Although we can define a neural network for the very general notion of 'neuron' shown in figure B1.1.1, most artificial neurons are of the kind shown in figure B1.1.3 in which the input lines are parametrized by real numbers. The parameter attached to an input line to neuron i that comes from the output of neuron j is often denoted by w_{ij}, and is referred to by such terms as the *strength* or synaptic weight for the *connection* from neuron j to neuron i. Much of the study of neural computation is then devoted to finding settings for these weights which will get a given neural network to approximate some desired behavior. The weights may either be set on the basis of some explicit design principles, or 'discovered' through the use of *learning rules* whereby the weight settings are automatically adjusted 'on the basis of experience'. But all this is meat for later chapters, and we now return to our focal aim:

B3.3

B3.3

introducing a number of the basic models of single neurons which 'fill in the details' in figure B1.1.3. As described in Section A1.2, there are radically different types of neurons in the human brain, and further *A1.2* variations in neuron types of other species.

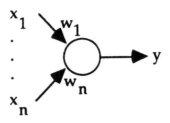

Figure B1.1.3. A neuron in which each input x_i passes through a 'synaptic weight' or 'connection strength' w_i.

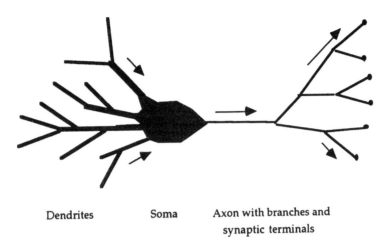

| Dendrites | Soma | Axon with branches and synaptic terminals |

Figure B1.1.4. The 'basic' neuron. The soma and dendrites act as the input surface; the axon carries the output signals. The tips of the branches of the axon form synapses upon other neurons or upon effectors. The arrows indicate the direction of information flow from inputs to outputs.

In neural computation, the artificial neurons are designed as variations on the abstractions of brain theory and implemented in software, *VLSI*, or other media. Figure B1.1.4 indicates the main features *E1.3, E1.4.3* needed to visualize biological neurons. We divide the neuron into three parts: the *dendrites*, the soma (cell body) and a long fiber called the axon whose branches form the *axonal arborization*. The soma and dendrites act as input surface for signals from other neurons and/or input devices (sensors). The axon carries 'spikes' from the neuron to other neurons and/or effectors (motors, etc). Towards a first approximation, we may think of a 'spike' as an all-or-none (binary) event; each neuron has a 'refractory period' such that at most one spike can be triggered per refractory period. The locus of interaction between an axon terminal and the cell upon which it impinges is called a *synapse*, and we say that the cell with the terminal synapses upon the cell with which the connection is made.

References

Arbib M A 1987 *Brains, Machines and Mathematics* 2nd edn (Berlin: Springer)
McCulloch W S and Pitts W H 1943 A logical calculus of the ideas immanent in nervous activity *Bull. Math. Biophys.* **5** 115–33

B1.2 The McCulloch–Pitts neuron

Michael A Arbib

Abstract

See the abstract for Chapter B1.

The work of McCulloch and Pitts (1943) combined neurophysiology and mathematical logic, modeling the neuron as a binary discrete-time element. They showed how excitation, inhibition and threshold might be used to construct a wide variety of 'neurons'. It was the first model to squarely tie the study of neural networks to the idea of computation in its modern sense. The basic idea is to divide time into units comparable to a refractory period (assumed to be the same for each neuron) so that in each time period at most one spike can be initiated in the axon of a given neuron. The McCulloch–Pitts neuron (figure B1.2.1(a)) thus operates on a discrete time-scale, $t = 0, 1, 2, 3, \ldots$. We write $y(t) = 1$ if a spike does appear at time t, $y(t) = 0$ if not. Each connection or *synapse*, from the output of one neuron to the input of another, has an attached *weight*. Let w_i be the weight on the ith connection onto a given neuron. We call the synapse *excitatory* if $w_i > 0$, and *inhibitory* if $w_i < 0$. We also associate a *threshold* θ with each neuron, and assume exactly one unit of delay in the effect of *all* presynaptic inputs on the cell's output, so that a neuron 'fires' (i.e. has value 1 on its output line) at time t only when the weighted values of its inputs at time t are at least θ. Formally, if at time $t - 1$ the value of the ith input is $x_i(t - 1)$ and the output one time step later is $y(t)$, then

$$y(t) = 1 \text{ if and only if } \sum_i w_i x_i(t - 1) \geqslant \theta \,.$$

To place this definition within our general formulation, we note that the *state* of the neuron at time t does *not* depend on the previous state of the neuron itself, but is simply $s(t) = \sum_i w_i x_i(t - 1)$, and that the output may be written as $y(t) = g(s(t))$, where g is now the *threshold function*

$$g(s) = H(s - \theta) \qquad \text{which equals 1 iff} \qquad s \geqslant \theta$$

where H is the Heaviside (unit step) function, with $H(x) = 1$ if $x \geqslant 0$, but $H(x) = 0$ if $x < 0$.

Figures B1.2.1(b)–(d) show how weights and threshold can be set to yield neurons which realize the logical functions AND, OR and NOT. As a result, McCulloch–Pitts neurons are sufficient to build networks which can function as the control circuitry for a computer carrying out computations of arbitrary complexity. This discovery played a crucial role in the development of automata theory and in the study of learning machines (see Arbib 1987 for a detailed account of this relationship). In neural computation, the McCulloch–Pitts neuron is often generalized so that the input and output values can lie anywhere in the range [0, 1] and the function $g(s(t))$ which yields $y(t)$ is a continuously varying function rather than a step function. In this case we call g the *activation function* of the neuron; g is usually taken to be a *sigmoid function*, that is, $g : \mathbb{R} \to [0, 1]$ is continuous and monotonically increasing, with $g(-\infty) = 0$ B3.2.4 and $g(\infty) = 1$ (and, in some studies, with the additional property that it has a single inflection point). Two popular sigmoidal functions are

$$\frac{1}{1 + \exp(-s/\theta)} \quad \text{and} \quad \tfrac{1}{2}(1 + \tanh(s)) \,.$$

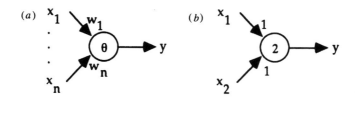

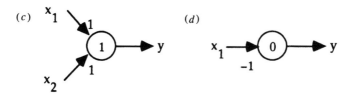

Figure B1.2.1. (a) A McCulloch–Pitts neuron operating on a discrete time-scale. Each input has an attached weight w_i, and the neuron has a *threshold* θ. The neuron 'fires' at time $t + 1$ if the weighted values of its inputs at time t are at least θ. Settings of weights and threshold for neurons that function (b) as an AND gate (the output fires if x_1 and x_2 both fire), (c) an OR gate (the output fires if x_1 or x_2 or both fire), and (d) a NOT gate (the output fires if x_1 does NOT fire).

References

Arbib M A 1987 *Brains, Machines and Mathematics* 2nd edn (Berlin: Springer)

McCulloch W S and Pitts W H 1943 A logical calculus of the ideas immanent in nervous activity *Bull. Math. Biophys.* **5** 115–33

B1.3 Hopfield networks

Michael A Arbib

Abstract

See the abstract for Chapter B1.

Hopfield (1982) contributed much to the resurgence of interest in neural networks in the 1980s by associating an *energy function* with a network, showing that if only one neuron changed state at a time (the so-called asynchronous update), a symmetrically connected network would settle to a local minimum of the energy, and that many optimization problems could be mapped to energy functions for symmetric neural networks. Based on this work, many papers have used neural networks to solve optimization problems (Hopfield and Tank 1985). The basic idea, given a criterion J to be minimized, is to find a Hopfield network whose energy function E approximates J, then let the network settle to an equilibrium and read off a solution from the state of the network. The study of optimization is beyond the scope of this chapter, but it will be worthwhile to understand the notion of network 'energy'.

In a *McCulloch–Pitts network*, every neuron processes its inputs to determine a new output at each time step. By contrast, a *Hopfield network* is a network of such units with (a) *symmetric* weights ($w_{ij} = w_{ji}$) and no self-connections ($w_{ii} = 0$), and (b) *asynchronous* updating. For instance, let s_i denote the state (0 or 1) of the ith unit. At each time step, pick just one unit at random. If unit i is chosen, s_i takes the value 1 if and only if $\sum w_{ij}s_j \geqslant \theta_i$. Otherwise s_i is set to 0. Note that this is an *autonomous* (input-free) network: there are no inputs (although instead of considering θ_i as a threshold we may consider $-\theta_i$ as a constant input, also known as a bias). Hopfield defined a measure called the *energy* for such a network,

$$E = -\tfrac{1}{2}\sum_{ij} s_i s_j w_{ij} + \sum_i s_i \theta_i .$$

This is not the physical energy of the neural network, but a mathematical quantity that, in some ways, does for neural dynamics what the potential energy does for Newtonian mechanics. In general, a mechanical system moves to a state of lower potential energy. Hopfield showed that his symmetrical networks with asynchronous updating had a similar property. For example, if we pick a unit and the foregoing firing rule does not change its s_i, it will not change E. However if s_i initially equals 0, and $\sum w_{ij}s_j \geqslant \theta_i$ then s_i goes from 0 to 1 with all other s_j constant, and the 'energy gap', or change in E, is given by

$$\Delta E = -\tfrac{1}{2}\sum_j (w_{ij}s_j + w_{ji}s_j) + \theta_i$$

$$= -\sum_j w_{ij}s_j s_j + \theta_i, \text{ by symmetry}$$

$$\leqslant 0 \text{ since } \sum w_{ij}s_j \geqslant \theta_i .$$

Similarly, if s_i initially equals 1, and $\sum w_{ij}s_j < \theta_i$ then s_i goes from 1 to 0 with all other s_j constant, and the energy gap is given by

$$\Delta E = \sum w_{ij}s_j - \theta_i < 0 .$$

In other words, with every asynchronous update, we have $\Delta E \leq 0$. Hence the dynamics of the network tends to move E towards a minimum. We stress that there may be different such states—they are local minima—just as, in figure B1.3.1, both D and E are local minima (each of them is lower than

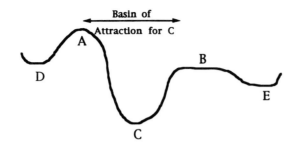

Figure B1.3.1. An energy landscape: For a ball rolling on the 'hillside', point A is an unstable equilibrium, point B lies in a region of neutral equilibrium, and point C is a point of stable equilibrium. Point C is called an *attractor*: the *basin of attraction of* C comprises all states from which the ball's dynamics tend toward C.

any 'nearby' state) but not global minima (since C is lower than either of them). Global minimization is not guaranteed.

The expression just presented for ΔE depends on the symmetry condition, $w_{ij} = w_{ji}$, for, without this condition, the expression would instead be $\Delta E = -\frac{1}{2}\sum_j(w_{ij}s_j + w_{ji}s_j) + \theta_i$. In this case, Hopfield's updating rule would not yield a passage to the energy minimum, but might instead yield a limit cycle, which could be useful in, for example, controlling rhythmic behavior. In a control problem, a link w_{ij} might express the likelihood that the action represented by i would precede that represented by j, in which case $w_{ij} = w_{ji}$ is normally inappropriate.

The condition of asynchronous update is crucial, too. If we consider the simple 'flip-flop' with $w_{12} = w_{21} = 1$ and $\theta_1 = \theta_2 = 0.5$, then the McCulloch–Pitts network will *oscillate* between the states $(0, 1)$ and $(1, 0)$ or will sit in the states $(0, 0)$ or $(1, 1)$; in other words, there is no guarantee that it will converge to an equilibrium. However, with $E = -\frac{1}{2}\sum_{ij} s_i s_j w_{ij} + \sum_i s_i \theta_i$, we have $E(0, 0) = 0, E(0, 1) = E(1, 0) = 0.5$ and $E(1, 1) = 0$, and the Hopfield network will *converge* to the minimum at either $(0, 0)$ or $(1, 1)$.

References

Hopfield J 1982 Neural networks and physical systems with emergent collective computational properties *Proc. Natl Acad. Sci., USA* **79** 2554–8

Hopfield J and Tank D W 1985 Neural computation of decisions in optimization problems *Biol. Cybern.* **52** 141–52

B1.4 The leaky integrator neuron

Michael A Arbib

Abstract

See the abstract for Chapter B1.

The simplest *continuous-time* model of the neuron in frequent use is the *leaky integrator* model, which has become popular in the simpler applications of neural networks which choose *analog VLSI* for their E1.3 implementation. The leaky integrator model uses the 'firing rate' (to mimic the biological measure of the number of spikes traversing the axon in some recent interval: but the artificial neuron need not involve any explicit spike generation) as a continuously varying *output* measure of the cell's activity, in which the *internal state* of the neuron is described by a single variable, the 'membrane potential' (another biological term with no implications for how this value should be stored in digital or analog circuitry). The firing rate is approximated by a simple, sigmoidal function of the membrane potential. That is, for this continuous-time neuron, the state is just the membrane potential, and the *activation function g* converts the membrane B3.2.4 potential m to the firing rate $g(m)$ which increases from 0 to its maximum value, 1 say, as m increases from $-\infty$ to $+\infty$. The biological motivation is this: if the membrane potential is low, the neuron will never reach threshold and so will have 0 as its firing rate; conversely, above a certain membrane potential the neuron will fire at its maximal firing rate, namely once every refractory period.

The time evolution of the cell's membrane potential is given by the differential equation

$$\tau \frac{dm(t)}{dt} = -m(t) + \sum_i w_i X_i(t) + h \tag{B1.4.1}$$

where τ is the time constant, and $X_i(t)$ is the firing rate at the ith input. Thus an excitatory input ($w_i > 0$) will be such that increasing it will increase $dm(t)/dt$, while an inhibitory input ($w_i < 0$) will have the opposite effect. A neuron described by (B1.4.1) is called a *leaky integrator* neuron. This is because the equation

$$\tau \frac{dm(t)}{dt} = \sum_i w_i X_i(t) \tag{B1.4.2}$$

would simply integrate the inputs with scaling constant τ:

$$m(T) = m(0) + \frac{1}{\tau} \int_0^T \sum_i w_i X_i(t) \, dt$$

but the $-m(t)$ term in (B1.4.1) opposes this integration by a 'leakage' of the potential $m(t)$ as it tries to return to its input-free equilibrium h. When all the inputs are zero,

$$\tau \frac{dm(t)}{dt} = -m(t) + h$$

has h as its unique equilibrium, and

$$m(t) = e^{-t/\tau} m(0) + (1 - e^{-t/\tau})h$$

which tends to the *resting level h* with *time constant* τ with increasing t so long as τ is positive.

It should be noted that, even at this simple level of modeling, there are alternatives. In the above model, we have used subtractive inhibition. But one may alternatively use *shunting* inhibition which, applied at a given point on a dendrite, serves to divide, rather than subtract from, the potential change passively propagating from more distal synapses. Again, the 'lumped-frequency' model cannot model relative timing effects corresponding to different delays (corresponding to pathways of different lengths linking neurons). These might be approximated by introducing appropriate delay terms

$$\tau \frac{dm(t)}{dt} = -m(t) + \sum_i w_i x_i (t - \tau_i) + h \,.$$

All this reinforces the observation that there is no modeling approach which is automatically appropriate. Rather, we seek to find the simplest model adequate to address the complexity of a given range of problems.

B1.5   Pattern recognition

Michael A Arbib

Abstract

See the abstract for Chapter B1.

With x_j a 'measure of confidence' that the jth item of a set of features occurs in some input *pattern x*, the *preprocessor* shown in figure B1.5.1 converts x into the *feature vector* $(x_1, x_2, \ldots, x_d)$ in a d-dimensional Euclidean space $\mathbb{R}^d$ called the *pattern space*. The pattern recognizer takes the feature vector and produces a response that has the appropriate one of K distinct values; points in $\mathbb{R}^d$ are thus grouped into at least K different categories. However, a category might be represented in more than one connected region of $\mathbb{R}^d$. To take an example from visual pattern recognition (although the theory of *pattern recognition* B6 *networks* applies to any classification of $\mathbb{R}^d$), 'a' and 'A' are members of the category of the first letter of the English alphabet, but they would be found in different connected regions of a pattern space. In such cases, it may be necessary to establish a hierarchical system involving a separate apparatus to recognize each subset, and a further system that recognizes that the subsets all belong to the same set (see our later discussion of radial basis functions). Here we avoid this problem by concentrating on the case in which the decision space is divided into exactly two connected regions.

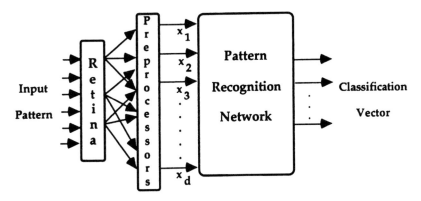

Figure B1.5.1. One strategy in pattern recognition is to precede an adaptive neural network by a layer of 'preprocessors' or 'feature extractors' which replace the image by a finite vector for further processing. In other approaches, the functions defined by the early layers of the network may themselves be subject to training.

We call a function $f : \mathbb{R}^d \rightarrow \mathbb{R}$ a *discriminant function* if the equation $f(x) = 0$ gives the *decision surface* separating two regions of a pattern space. A basic problem of pattern recognition is the specification of such a function. It is virtually impossible for humans to 'read out' the function they use (not to mention *how* they use it) to classify patterns. Thus, a common strategy in pattern recognition is to provide a classification machine with an adjustable function and to 'train' it with a set of patterns of known classification that are typical of those with which the machine must ultimately work. The function may be linear, quadratic, polynomial (see the discussion of polynomial neurons below), or even more subtle yet, depending on the complexity and shape of the pattern space and the necessary discriminations. The experimenter chooses a class of functions with parameters which, it is hoped, will, with proper adjustment, yield a function that will successfully classify any given pattern. For example, the experimenter may decide to use a linear function of the form

$$f(x) = w_1 x_1 + w_2 x_2 + \cdots + w_d x_d + w_{d+1}$$

B1.2 (i.e. a *McCulloch–Pitts neuron*) in a two-category pattern classifier. The equation $f(x) = 0$ gives a hyperplane as the decision surface, and training involves adjusting the coefficients $(w_1, w_2, \ldots, w_d, w_{d+1})$ so that the decision surface produces an acceptable separation of the two classes. We say that two categories are *linearly separable* if an acceptable setting of such linear weights exists. Of course, as will be shown B1.7.3, C1.6.2 in later chapters, many interesting pattern sets are not linearly separable (cf the section on *radial basis functions* below), and so whole networks—rather than single, simple neurons—are needed to categorize most interesting patterns.

B1.6 A note on nonlinearity and continuity

Michael A Arbib

Abstract

See the abstract for Chapter B1.

In both the *McCulloch–Pitts* and *leaky integrator neurons*, the neuron is defined by a linear term followed B1.2, B1.4
by a nonlinearity. Without the nonlinearity, the theory of neural networks reduces to linear systems
theory—an already powerful branch of systems theory. A number of applications of neural networks do
indeed exploit the methods of linear algebra and linear systems. However, with fixed input, a linear system
has only a single equilibrium state whereas a nonlinear system may, depending on its structure, exhibit
multiple equilibrium states, limit cycles, or even chaotic behavior. This rich repertoire takes us far beyond
the range of linear systems, and is exploited in neural network applications. For example, the equilibria
of a network may be considered as 'standard patterns', and the passage of a network from some initial
state (a 'noisy' pattern) to a nearby equilibrium may be considered a means of *pattern recognition*. Since B1.5, B6
stable equilibria are often called 'attractors', this is called 'pattern recognition by attractor networks'. This
complements the style of pattern recognition exemplified in figure B1.5.1 where the 'noisy' pattern is the
input to the network, and the 'classification' of the pattern is the output. In this case, too, nonlinearities
are crucial as, whether by the sharp divide of the Heaviside step function or by the more gentle emphasis
of the sigmoid, they can separate the patterns into, or towards, a vector of binary oppositions. The closest
that a linear system comes to this—and it is a method emulated in some neural network applications
(Oja 1992)—is *principal component analysis* which is a method not of classifying patterns but rather of B4.4.3
reducing them to a low-dimensional representation which contains much of the variance of a given set of
patterns.

Given these reasons for using nonlinear activation functions, are there reasons to choose continuous
ones, rather than the simple step function? There are two main reasons. One is noise resistance: a
step function can amplify noise which a sigmoid function may smooth out, but this may be at the price
of postponing a binary decision until after further statistical analysis has been made. The other is to
allow the use of *training methods* (see Chapter B3) which exploit methods of the differential calculus B3
to adjust synaptic weights to better approximate some desired network behavior. In fact, the classical
Hebbian and *perceptron* training rules do indeed work for binary neurons. However, the widely used B3.3.1, B3.3.2
backpropagation method for training multilayer feedforward networks makes essential use of the fact that C1.2.3
the activation functions are continuous, indeed differentiable. This is not the place to review the details
of backpropagation. Rather, we note the general situation of which it is a special case. If a network
has no loops in it, then the input pattern uniquely determines the output pattern (so long as we hold the
input constant and wait long enough for its effects to propagate through all the layers of the network).
The output y depends, however, not only on the input x itself, but also upon the current setting w of the
weights of the network connections. We write $y = f(x; w)$, where the form of f depends on the actual
structure of the network. The training problem is this: given a set of constraints on the desired values of
input pairs, find a choice w_0 of w such that $y = f(x; w_0)$ 'best' meets these constraints. The definition
of 'best' usually involves some cost function C which measures how well the current $f(-; w)$, at step i
of the training procedure, meets the constraints; call the current cost $C(w, i)$. Training then consists in
adjusting w to try and minimize $C(w, i)$. Since calculus-based methods of minimization rest on the taking
of derivatives, their application to network training requires that C be a differentiable function of w; this,
in turn, requires that $f(x; w)$ be differentiable, and this, in turn, requires that the activation functions be

differentiable. This, then, provides a powerful motivation for using activation functions that are not only continuous but also differentiable. However, minimization can also be conducted by step-wise search and so, as noted before, training methods have been successfully defined for networks employing the Heaviside function as an activation function.

References

Oja E 1992 Principal components, minor components, and linear neural networks *Neural Networks* **5** 927–35

B1.7 Variations on a theme

Michael A Arbib

Abstract

See the abstract for Chapter B1.

There are many variations on the basic definitions given above, and a few are briefly noted here. We first look at integrate-and-fire neurons which add spike generation to the leaky integrator neurons defined above. However, as noted earlier, much of neural computation is devoted to finding settings for the connection weights which will get a given neural network to approximate some desired behavior. This has led authors to define classes of 'neurons' which are defined not because of their similarity to 'real' neurons but simply because of their mathematical utility in an approximation network. We present polynomial neurons and *radial basis functions* as two examples of this kind, before looking at the use of *stochastic neurons* to C1.6.2, C1.4 provide a means of escaping 'local minima'. We close with a brief mention of the use of neurons to form *self-organizing maps*, but can give no details since they depend on ideas about synaptic plasticity that will C2.2.1 not be presented until Chapter B3.

B1.7.1 Integrate-and-fire neurons

Another class of neuron models has continuous-time, continuous state-space $\mathbb{R}$, but discrete signal space $\{0, 1\}$—so that the model approximates spike generation. This model of a spiking cell—the *integrate and fire* model—far antedates the discrete-time model of McCulloch and Pitts: it was introduced by Lapicque (1907). Essentially, it uses the leaky integrator model (1) for the membrane potential, but now an arriving input $X_i(t) = 1$ acts like a delta-function to instantaneously increment the state by w_i. The output instantaneously switches to 1 (a spike is generated) each time the neuron reaches a given threshold value. This model captures the two key aspects of biological neurons: a passive, integrating response for small inputs and a stereotyped impulse once the input exceeds a particular amplitude. Hill (1936) used *two* coupled leaky integrators, one of them representing membrane potential, and the other representing the fluctuating threshold to approximate the effect of the refractory period on neuron dynamics.

B1.7.2 Polynomial neurons

Here the idea is to generalize the input–output power of neurons by replacing the linear next-state function $\sum_i w_i x_i$ by some polynomial combination of the inputs:

$$\sum_{i_1 \ldots i_{j_k} \, w_{i_1} \ldots i_{j_k}} x_{i_1} \ldots x_{i_{j_k}} \, .$$

Here we have some finite set S, say, of tuples of the form $i_1 \ldots i_{j_k}$, where each i_a is the index of one of the inputs to the neuron under consideration. Then, for each such tuple we calculate the monomial $w_{i_1 \ldots i_{j_k}} x_{i_1} \ldots x_{i_{j_k}}$ and then sum them to get the term that drives the activation function of the neuron. We thus regain the usual neuron definition when each tuple is restricted to be of length one, forcing the above sum to be linear. This idea goes back to the work of Gilstrap in the 1960s (see Barron *et al* 1987 for a more recent review). These neurons are also known as *high-order neurons* or 'neurons with high-order connections'; they are also called *sigma-pi* neurons since the above expression is a sum (sigma) of products (pi) of the x_i.

The increased power of polynomial neurons is clear on considering XOR, the simple Boolean operation of addition modulo 2, also known as the exclusive-or. If we imagine the square with vertices $(0, 0)$, $(0, 1)$, $(1, 1)$, and $(1, 0)$ in the Cartesian plane, with (x_1, x_2) being labeled by $x_1 \oplus x_2$, we have 0s at one diagonally opposite pair of vertices and 1s at the other diagonally opposite pair of vertices. It is clear that there is no way of interposing a straight line such that the 1s lie on one side and the 0s lie on the other side; i.e. there is no way of choosing w_1, w_2 and θ such that $w_1 x_1 + w_2 x_2 \geqslant \theta$ iff $x_1 \oplus x_2 = 1$. However, we can realize the exclusive-or with a single *polynomial* neuron with $w_1 = w_2 = 1$, $w_{12} = 2$, since $x_1 + x_2 - 2 x_1 x_2 = x_1 \oplus x_2$.

B1.7.3   Radial basis functions

Suppose that a pattern space can be divided into 'clusters' for each of which there is a single category to which pattern vectors within the cluster are most likely to correspond. We can then address the pattern recognition problem by dividing the pattern space into regions bounded by hyperplanes, where each hyperplane corresponds to a single threshold neuron (figure B1.7.1). By connecting each neuron to an AND gate, we get a network that signals whether or not a pattern falls within the polygonal space approximating the cluster; connecting all these AND gates to an OR gate, we end up with a network that signals whether or not the pattern is (approximately) in any of the clusters belonging to a given category.

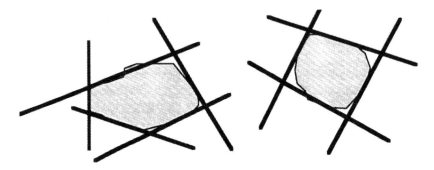

Figure B1.7.1. Here we see two convex 'clusters' approximated by a set of lines ('hyperplanes' in a general d-dimensional set). Each line serves as discriminant function f for a threshold neuron; we choose the sign of f so that most of the points in the cluster satisfy $f(x) > 0$. If we connect these neurons to an AND gate, then the AND gate will fire primarily for x belonging to the cluster. If we can divide the set of instances of patterns in a more complex category into a finite set of convex clusters (two in the above case), and connect AND gates for these clusters to an OR gate, we get a network which will fire primarily for x belonging to any cluster of the pattern.

An alternative to this 'compounding of linear separabilities' (the architecture described above is sometimes referred to as an instance of a *three-layer perceptron*) is the use of *radial basis functions* (RBFs; see Lowe 1995 for a survey). An RBF operates on an input x in $\mathbb{R}^n$ and is characterized by a weight vector w in $\mathbb{R}^n$. However, instead of forming the linear combination $\sum_i w_i x_i$ and passing it through a step or sigmoid activation function, we instead take the norm $||x - w||$ of the difference between x and w, and then pass it through an activation function f which decreases as $||x - w||$ increases (a Gaussian is a typical choice). The 'neuron' thus tests whether or not the current input x is close to w, and can relay the measure of closeness to other units which will use this information about where x lies in the input space to determine how best to process it. Although the details are beyond the scope of this chapter, we briefly discuss the use of RBFs to solve the above 'cluster-based' pattern recognition problem in cases in which it is possible to describe the clusters of data as if they were generated according to an underlying probability density function. The multilayer perceptron method concentrates on class boundaries, while the RBF method focuses upon regions where the data density is highest. In probabilistic classification of patterns, we are primarily interested in the posterior probability $p(c|x)$ that class c is present given the observation x. However, it is easier to model other related aspects of the data such as the unconditional distribution of the data $p(x)$, or the probability $p(x|c)$ that the data were generated given that they came from a specific class c—the Bayes theorem then tells us that $p(c_i|x) = p(c_i)p(x|c_i)p(x)$. Of interest here is the case where the distribution of the data is modeled as if it were generated by a mixture distribution, that is, a linear combination of parameterized states, or basis functions such as Gaussians. Since individual

C1.2, C1.6.2

© 1997 IOP Publishing Ltd and Oxford University Press

data clusters for each class are not likely to be approximated by a single Gaussian distribution, we need several basis functions per cluster. (Think of each Gaussian as defining an elliptical 'hill' resting on the ocean floor. Then we may need to superimpose a set of such hills to cover a given area which rises above 'sea level' to form an island.) We assume that the likelihood and the unconditional distribution can both be modeled by the same set of distributions, $q(x|s)$ but with different coefficients (e.g. Gaussians with different means, variances and orientations of the axes of the ellipsoid), that is,

$$p(x) = \sum_s \hat{p}(s)q(x|s).$$

This gives a radial basis function architecture (see Lowe 1955 for further details).

B1.7.4 Stochastic neurons

Finally, we note that there are many cases in which a noise term is added to the next-state function or the *activation function*, allowing neural networks (such as the *Boltzmann machine* of Ackley *et al* 1985, see Aarts and Korst 1995 for a recent review) to perform a kind of stochastic approximation. We have earlier spoken of deterministic discrete-time neurons in which the quantity $s(t) = \sum_i w_i x_i(t-1)$ is passed through a sigmoidal function to determine the output   B3.2.4, C1.4

$$y(t) = \frac{1}{1 + \exp(-s(t)/\theta)}.$$

The twist in Boltzmann machines is to use a noisy binary neuron; it has two states, 0 and 1, and the formula

$$p(t) = \frac{1}{1 + \exp(-s(t)/T)}$$

is now interpreted as the *probability* that the state of the neuron will be 1 at time t. When T is very large, the neuron's behavior is highly random; when $T \to 0$, the next state will be 1 only when $s(t) > 0$. T is thus a noise term, often referred to as 'temperature' on the basis of an analogy with the Boltzmann distribution used in statistical mechanics. In most cases, the response of a Boltzmann machine to given inputs starts with a large value of T. Subsequently, the value of T is decreased to eventually become 0. This is an example of the strategy of *simulated annealing* which uses controlled noise to escape from local minima during a minimization process (recall our discussion of figure B1.7.1 in relation to *Hopfield networks*) to almost surely find the global minimum for the function being minimized. The idea is to use   B1.3 noise to 'shake' a system out of a local minimum and let it settle into a global minimum. Returning to figure B1.3.1, consider, for example, shaking strong enough to shake the ball from D to A, and thus into the basin of attraction of C, but not strong enough to shake the ball back from C towards D.

B1.7.5 Learning vector quantization and Kohonen maps

The input patterns to a neural network define a continuous vector space. Vector quantization provides a means to 'quantize' this space by forming a 'code book' of significant vectors linked to useful information— we can then analyze a novel vector by looking for the vector in the code book to which it is most similar. *Learning vector quantization* provides a means whereby a neural network can self-organize, both to provide   C1.1.5 the code book (one neuron per entry) and to find (by a winner-take-all technique) the code associated with a novel input vector. If this methodology is augmented by constraints which force nearby neurons to become associated with similar codes, the result is a *self-organizing feature map* (also known as a Kohonen map),   C2.1.1 whereby a high-dimensional feature space is mapped quasi-continuously onto the neural manifold (Kohonen 1990). These methods of self-organization are extensions of the *Hebbian learning* mechanisms described   B3.3.1 in Chapter B3, and thus further description lies beyond the scope of this introduction.

References

Aarts E H L and Korst J H M 1995 Boltzmann machines *The Handbook of Brain Theory and Neural Networks* ed M A Arbib (Cambridge, MA: Bradford Books/MIT Press) pp 162–5
Ackley D H, Hinton G E and Sejnowski T J 1985 A learning algorithm for Boltzmann machines *Cog. Sci.* **9** 147–69

Barron R L, Gilstrap L O and Shrier S 1987 Polynomial and neural networks: analogies and engineering applications *Proc. Int. Conf. on Neural Networks* (New York: IEEE Press) II 431–93

Hill A V 1936 Excitation and accommodation in nerve *Proc. R. Soc.* B **119** 305–55

Kohonen T 1990 The self-organizing map *Proc. IEEE* **78** 1464–80

Lapicque L 1907 Recherches quantitatifs sur l'excitation électrique des nerfs traitée comme une polarisation *J. Physiol. Paris* **9** 620–35

Lowe D 1995 Radial basis function networks *The Handbook of Brain Theory and Neural Networks* ed M A Arbib (Cambridge, MA: Bradford Books/MIT Press) pp 779–82

© 1997 IOP Publishing Ltd and Oxford University Press

B2

Neural Network Topologies

Abstract

An artificial neural network consists of a topology and a set of rules that govern
the dynamic aspects of the network. This section contains a detailed treatment of
the topology of a neural network, that is, the combined structure of its neurons and
connections. It starts with the basic concepts including neurons, connections, and layers,
followed by symmetry and high-order aspects. Next, fully and partially connected
topologies are discussed, which is complemented by an overview of special topologies
like modular, composite, and ontogenic ones. The next section discusses aspects of a
formal framework, which is an underlying theme that unites this section in which a
balance is sought between clarity and mathematical rigor in the hope of providing a
useful basis and reference for the other chapters of this handbook. This section proceeds
with a discussion on modular topologies and concludes with theoretical considerations
for choosing a neural network topology.

Contents

B2.1 Introduction

Emile Fiesler

Abstract

See the abstract for Chapter B2.

A neural network is a network of neurons. This high-level definition applies to both biological neural networks and artificial neural networks (ANNs). This chapter is mainly concerned with the various ways in which neurons can be interconnected to form the networks or network topologies used in ANNs, even though some underlying principles are also applicable to their biological counterparts. The term 'neural network' is therefore used to stand for 'artificial neural network' in the remainder of this chapter, unless explicitly stated otherwise. The main purpose of this chapter is to provide a base for the rest of the Handbook and in particular for the next chapter, in which the training of ANNs is discussed.

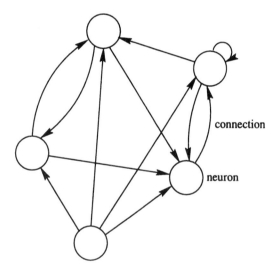

Figure B2.1.1. An unstructured neural network topology with five neurons.

Figure B2.1.1 shows an example neural network topology. A node in such a network is usually called an *artificial neuron*, or simply *neuron*, a tradition that is continued in this handbook (see Chapter B1). B1 The widely accepted term 'artificial neuron' is specific to the field of ANNs and therefore preferred over its alternatives. Nevertheless, given the length of this term and the need to frequently use it, it is not surprising that its abbreviated form, 'neuron', is often used as a substitute instead. However, given that the primary meaning of the word 'neuron' is a biological cell from the central nervous system of animals, it is good practice to clearly specify the meaning of the term 'neuron' when using it. Instead of '(artificial) neuron', other terms are also used:

- *Node.* This is a generic term, related to the word 'knot' and used in a variety of contexts, one of them being graph theory, which offers a mathematical framework to describe neural network topologies (see Section B2.8.4). B2.8.4

- *Cell.* An even more generic term, that is more naturally associated with the building blocks of organisms.
- *Unit.* A very general term used in numerous contexts.
- *Neurode.* A nice short term coined by Caudill and Butler (1990), which contains elements of both the words 'neuron' and 'node', giving a cybernetic flavor to the word 'neuron'.

The first three words are generic terms, borrowed from other fields, which can serve as alternative terminology as long as their meaning is well defined when used in a neural network context. The neologism 'neurode' is specifically created for ANNs, but unfortunately not widely known and accepted.

A connectionist system, better known as artificial neural network, is in principle an abstract entity. It can be described mathematically and can be manifested in various ways, for example in hardware and software implementations. An artificial neural network comprises a collection of artificial neurons connected by a set of links, which function as communication channels. Such a link is called an *interconnection* or *connection* for short.

References

Caudill M and Butler C 1990 *Naturally Intelligent Systems* (Cambridge, MA: MIT Press)

B2.2 Topology

Emile Fiesler

Abstract

See the abstract for Chapter B2.

A *neural network topology* represents the way in which neurons are connected to form a network. In other words, the neural network topology can be seen as the relationship between the neurons by means of their connections. The topology of a neural network plays a fundamental role in its functionality and performance, as illustrated throughout the handbook.

The generic terms *structure* and *architecture* are used as synonyms for network topology. However, caution should be taken when using these terms since their meaning is not well defined as they are also often used in contexts where they encompass more than the neural network topology alone or refer to something different altogether. They are for example often used in the context of hardware implementations (*computer architectures*) or their meaning includes, besides the network topology, also the learning rule (see for example the book by Zurada (1992)).

More precisely, the topology of a neural network consists of its *frame* or *framework* of neurons, together with its *interconnection structure* or *connectivity*:

$$\text{neural network topology} \begin{cases} \text{neural framework} \\ \text{interconnection structure} \end{cases}$$

The next two subsections are devoted to these two constituents respectively.

B2.2.1 Neural framework

Most neural networks, including many biological ones, have a layered topology. There are a few exceptions where the network is not explicitly layered, but those can usually be interpreted as having a layered topology, for example in some *associative memory networks*, which can be seen as a one-layer neural network where all neurons function both as input and output units. C1.3

At the framework level, neurons are considered as abstract entities, thereby not considering possible differences between them. The framework of a neural network can therefore be described by the number of neuron layers, denoted by L, and the number of neurons in each of the layers, denoted by N_l, where l is the index indicating the layer number:

$$\text{neural framework} \begin{cases} \text{number of neuron layers} & L \\ \text{number of neurons per layer} & N_l \text{ where } 1 \leq l \leq L. \end{cases}$$

The number of neurons in a layer (N_l) is also called the *layer size*.

The following neuron types can be distinguished.

- *Input neuron.* A neuron that receives external inputs from outside the network.
- *Output neuron.* A neuron that produces some of the outputs of the network.
- *Hidden neuron.* A neuron that has no direct interaction with the 'outside world', only with other neurons within the network.

Similar terminology is used at the layer level for *multilayer neural networks*.

- *Input layer.* A layer consisting of input neurons.
- *Hidden layer.* A layer consisting of hidden neurons.
- *Output layer.* A layer consisting of output neurons.

In multilayer and most other neural networks the neuron layers are ordered and can be numbered: the input layer having index one, the first hidden layer index two, the second hidden layer index three, and so forth until the output layer, which is given the highest index L, equal to the total number of layers in the network. The number of neurons in the input layer can thus be denoted as N_1, the number of neurons in the first hidden layer as N_2, in the second hidden layer as N_3 and so on, until the output layer, whose size would be N_L. In figure B2.2.1 a four-layer neural network topology is shown, together with the layer sizes.

Layer name	l	N_l
output layer	$4 = L$	$N_4 = N_L = 1$
second hidden layer	3	$N_3 = 2$
first hidden layer	2	$N_2 = 4$
input layer	1	$N_1 = 2$

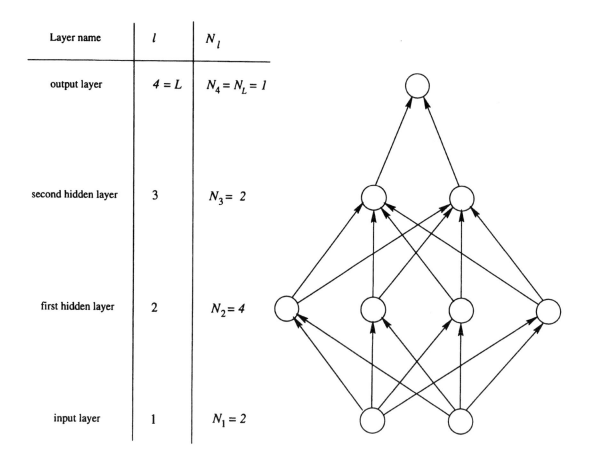

Figure B2.2.1. A fully interlayer connected topology with four layers.

Combining all layer sizes yields

$$N = \sum_{l=1}^{L} N_l \tag{B2.2.1}$$

where N is the total number of neurons in the network. Besides being clearer, the indexed notation for layer sizes is preferred since the number of layers in neural networks varies from one model to another and there are even some models that adapt their topology dynamically during the training process, thereby varying the number of layers (see Section C1.7). Also, if one assigns a different variable to each layer (for example l, m, n, ...), one soon runs out of variables and into notational conflicts; this is especially the case for generic descriptions of multilayer neural networks and *deep networks*, which are networks with many layers.

In some neural networks, neurons are grouped together, as in layered topologies, but there is no well-defined way to order these groups. The groups of neurons in networks without an ordered structure

C1.7

are called *clusters*, *slabs*, or *assemblies*, which are therefore generic terms which include the layer concept as a special case.

The neurons within a layer, or cluster, are usually not ordered, all neurons being equally important. However, the neurons within a cluster are sometimes numbered for convenience to be able to uniquely address them, for example in computer simulations. Layers are likewise shapeless and can be represented in various ways. Exceptions are the input and output layers, which are special since the application constraints can suggest a specific shape, which can be one, two, or higher dimensional. Note however, that this structural shape is usually only present in pictorial representations of the neural network, since the individual neurons are still equally important and 'unaware' of each other's presence with respect to relative orientation. An exception could be an application specific partial connectivity where only certain neurons are connected to each other, thereby embedding positional information, such as the feature detectors of LeCun *et al* (1989).

Likewise, there is also no fixed way of representing neural networks in pictorial form. Neural networks are most often drawn bottom up, with the input layer at the bottom and the output layer at the top, as in figure B2.2.1. Besides this, a left-to-right representation is also used, especially for *optical neural networks* E1.5 since the direction of the passing light in optical diagrams is by default assumed to be from left to right. Besides these, other pictorial orientations are also conceivable. This representational flexibility is also present in graph theory (see Section B2.8.4).

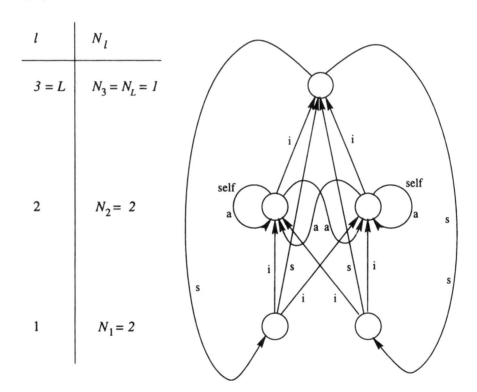

Figure B2.2.2. A three-layer neural network topology with six interlayer connections (i), four supralayer connections (s) between the input and output layer, and four intralayer connections (a) including two self-connections (self) in the hidden layer.

B2.2.2 Interconnection structure

The interconnection structure of a neural network determines the way in which the neurons are linked. Based on a layered structure, several different kinds of connection can be distinguished (see figure B2.2.2 for an illustration):

- *Interlayer connection.* This connects neurons in adjacent layers whose layer indices differ by one.
- *Intralayer connection.* This is a connection between neurons in the same layer.

- *Self-connection.* This is a connection that connects a neuron to itself. It is a special kind of intralayer connection.
- *Supralayer connection.* This is a connection between neurons that are in distinct layers that are not adjacent; in other words these connections 'cross' or 'jump' at least one hidden layer.

With each connection an *(interconnection) strength* or *weight* is associated which is a weighting factor that reflects its importance. This weight is a scalar value (a number), which can be positive (*excitatory*) or negative (*inhibitory*). If a connection has a zero weight is it considered to be nonexistent at that point in time.

Note that the basic concept of layeredness is based on the presence of interlayer connections. In other words, every layered neural network has at least one interlayer connection between adjacent layers. If interlayer connections are absent between any two adjacent clusters in the network, a spatial reordering can be applied to the topology, after which certain connections become the interlayer connections of the transformed, layered, network.

References

Le Cun Y, Boser B, Denker J S, Henderson D, Howard R E, Hubbard W and Jackel L D 1989 Backpropagation applied to handwritten zip code recognition *Neural Comput.* **1** 541–51
Zurada J M 1992 *Introduction to Artificial Neural Systems* (St Paul, MN: West)

B2.3 Symmetry and asymmetry

Emile Fiesler

Abstract

See the abstract for Chapter B2.

The information flow through a connection can be symmetric or asymmetric. Before elaborating on this, it should be stated that 'information transfer' or 'flow', in the following discussion, refers to the *forward propagation*, where network outputs are produced in reaction to external inputs or stimuli given to the neural network. This in contrast to the information used to update the network parameters as determined by the neural network *learning rule*.

B3.3

A connection in a neural network is either *unidirectional* when it is only used for information transfer in one direction at all times, or *multidirectional* where it can be used in more than one direction (the term *multidirectional* is used here instead of bidirectional to include the case of high-order connections (see Section B2.4)). A multidirectional connection can either have one weight value that is used for information flow in all directions, which is the symmetric case (see figure B2.3.1), or separate weight values for information flow in specific directions, which is the asymmetric case (see figure B2.3.2).

B2.4

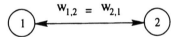

Figure B2.3.1. A symmetric connection between two neurons.

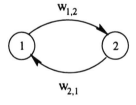

Figure B2.3.2. Two asymmetric connections between two neurons.

Hence, a *symmetric connection* is a multidirectional connection which has one weight value associated with it that is the same when used in any of the possible directions. All other connections are *asymmetric connections*, which can be either unidirectional connections (see figure B2.3.3) or multidirectional connections with more than one weight value per connection. Note that a multidirectional connection can be represented by a set of unidirectional connections (see figure B2.3.2), which is closer to biological reality where synapses are also unidirectional. In a unidirectional connection the information flows from its *source neuron* to its *sink neuron* (see figure B2.3.3).

The definitions regarding symmetry can be extended to the network level: a *symmetric neural network* is a network with only symmetric connections, whereas an *asymmetric neural network* has at least one

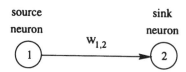

Figure B2.3.3. A unidirectional connection between a source and a sink neuron.

asymmetric connection. Most neural networks are asymmetric, having a unidirectional information flow or a multidirectional one with distinct weight values.

An important class of neural networks is the so called *feedforward neural networks* with unidirectional information flow from input to output layer. The name *feedforward* is somewhat confusing since the best-known algorithm for training a feedforward neural network is the *backpropagation learning rule*, whose name indicates the backward propagation of (error gradient) information from the output layer, via the hidden layers, back to the input layer, which is used to update the network parameters. The opposite of feedforward is 'feedback'; a term used for those networks that contain loops where information is fed back to neurons in previous layers. This terminology is not recommended since it is most often used for networks which have unidirectional supralayer connections from the output to the input layer, thereby excluding all other possible topologies with loops from the definition. Preferred is the term *recurrent neural network* for networks that contain at least one loop. Some common examples of recurrent neural networks are symmetric neural networks with bidirectional information flow, networks with self-connections, and networks with unidirectional connections from output back to input neurons.

C1.2.3

B2.4 High-order topologies

Emile Fiesler

Abstract

See the abstract for Chapter B2.

Most neural networks have only *first-order connections* which link one source neuron to one sink neuron. However, it is also possible to connect more than two neurons by a *high-order connection* (the term *higher order* is sometimes used instead of 'high order') (see figure B2.4.1).

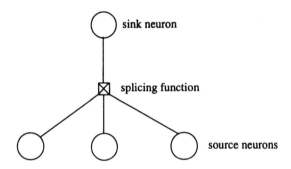

Figure B2.4.1. A third-order connection.

High-order connections are typically asymmetric, linking a set of source neurons to a sink neuron. The *connection order* (ω) is defined as the cardinality of the set of its source neurons, which is the number of elements in that set. As an example, figure B2.4.1 shows a third-order connection. The information produced by the source neurons is combined by a *splicing function* which has ω inputs and one output. The most commonly used splicing function for high-order neural networks is multiplication, where the connection outputs the product of the values produced by its source neurons. The set of source neurons of a high-order connection is usually located in one layer. The connectivity definitions of Section B2.2.2 apply therefore also to high-order connections.

The concept of higher orders can also be extended to the network level. A *high-order neural network* has at least one high-order connection and the *neural network order* (Ω) is determined by the highest-order connection in the network:

$$\Omega = \max_{w} \; \omega_w \tag{B2.4.1}$$

where w ranges over all weights in the network.

Having high-order connections gives the network the ability to extract higher-order information from the input data set, which is a powerful feature.

Layered high-order neural networks with multiplication as splicing function are also called *sigma–pi* ($\Sigma\Pi$) *neural networks*, since a summation (Σ) of products (Π) is used in the forward propagation:

$$a_j = \sum_{\{s_j\}} w_{\{s_j\}j} \prod_{i \in \{s_j\}} a_i \tag{B2.4.2}$$

where a_j is the activation value of the sink neuron, $\{s_j\}$ is the set of source neurons, $w_{\{s_i\}j}$ the associated weight, and a_i the activation values of the source neurons. The layer indices are omitted from this formula for notational simplicity. In Section B2.8.8 notational issues concerning weights are discussed. For more information on sigma–pi neural networks, see Rumelhart *et al* (1986), which is based on the work of Williams (1983).

The history of high-order neural networks includes the work of Poggio (1975) where the term 'high order' is used, and Feldman and Ballard (1982) where multiplication is used as splicing function and the connections are named *conjunctive connections*. An important and fundamental contribution to the area of high-order neural networks, which has given rise to their wider dissemination, is the work by Lee *et al* (1986).

For completeness *functional link networks* (Pao 1989) and *product unit neural networks* (Durbin and Rumelhart 1989) are mentioned here since they can be considered as special cases of high-order neural networks. In these types of network there is no combining of information from several source neurons taking place, but incoming information from a single source is transformed by means of a nonlinear splicing function.

References

Durbin R and Rumelhart D E 1989 Product units: a computationally powerful and biologically plausible extension to backpropagation networks *Neural Comput.* **1** 133–42

Feldman J A and Ballard D H 1982 Connectionist models and their properties *Cogn. Sci.* **6** 205–54

Lee Y C, Doolen G, Chen H, Sun G, Maxwell T, Lee H and Giles C L 1986 Machine learning using a higher order correlation network *Physica* D **22** 276–306

Pao Yoh-Han 1989 *Adaptive Pattern Recognition and Neural Networks* (Reading, MA: Addison-Wesley)

Poggio T 1975 On optimal nonlinear associative recall *Biol. Cybernet.* **19** 201–9

Rumelhart D E, McClelland J L and the PDP Research Group 1986 *Parallel Distributed Processing: Explorations in the Microstructure of Cognition. vol 1: Foundations* (Cambridge, MA: MIT Press)

Williams R J 1983 *Unit Activation Rules for Cognitive Network Models* ICS Technical Report 8303, Institue for Cognitive Science, University of California, San Diego

B2.5   Fully connected topologies

Emile Fiesler

Abstract

See the abstract for Chapter B2.

The simplest topologies are the fully connected ones, where all possible connections are present. However, depending on the neural framework and learning rule, the term *fully connected neural network* is used for several different interconnection schemes, and it is therefore important to distinguish between these.

The most commonly used topology is the *fully interlayer-connected* one, where all possible interlayer connections are present but no intra- or supralayer ones. This is the default interconnectivity scheme for most nonrecurrent multilayer neural networks.

A truly fully connected or *plenary neural network* has all possible inter-, supra-, and intralayer connections including self-connections. However, only a few neural networks have a plenary topology. A slightly more popular 'fully connected' topology is a plenary neural network without self-connections, as used for example for some *associative memories*.

C1.3

B2.5.1   Connection counting

In order to compare different neural network topologies, and more specifically their complexities, it is useful to know how many connections a certain topology comprises. The connection counting is based on *fully connected* topologies since they are the most commonly used and since they enable a fair and yet simple comparison. Fully interlayer-connected topologies are considered as well as the various combinations of interlayer connections together with intra- and supralayer connections (see Section B2.2.2); and *fully connected* means here that all possible connections of each of those kinds are present in the topology. Before starting the counting of the connections, a few related issues need to be discussed and defined.

The total number of weights in a network can be denoted by W. For most neural networks this number is equal to the number of connections, since one weight is associated with one connection. In neural networks with *weight sharing* (Rumelhart *et al* 1986), where a group of connections shares the same weight, the number of weights can be smaller than the number of connections. However, even in this case it is common practice to assign a separate weight to each connection and to update shared weights together and in an identical way. Given this, the number of connections is again equal to the number of weights and the same notation (W) can be used for both.

When counting the number of weights, it has to be decided whether to also count the neuron biases. The bias of a neuron, which determines its threshold level, can also be regarded as a special weight and its value is often modified in the same way as normal weights. This can be explained in the following way. The weighted sum of inputs to a neuron n, which has $W_{l,n}$ input providing connections, can be denoted as

$$\sum_{i=1}^{W_{l,n}} w_{i,n} a_i - \Theta_n = \sum_{i=1}^{W_{l,n}} w_{i,n} a_i + \Theta_n(-1) \qquad (B2.5.1)$$

where a_i is the activation value of the neuron providing the ith input, and $w_{i,n}$ is the weight between that neuron providing the ith input to neuron n and neuron n itself (see Section B2.8.2 for a discussion on notational issues concerning weights). Renaming Θ_n as $w_{0,n}$ and assuming a_0 to be a virtual activation with a constant value of -1, equation B2.5.1 becomes equal to:

$$\sum_{i=0}^{W_n} w_{i,n} \, a_i. \tag{B2.5.2}$$

Hence, the bias of a neuron can be seen as the weight of a virtual connection that receives its input from a virtual or dummy neuron that has a constant activation value of -1. In this section biases are not counted as weights. They can be included in the connection counting by initializing the appropriate summation indices with zero instead of one.

For networks where intralayer connections are present, two cases need to be distinguished: with and without self-connections. Both cases can be conveniently combined in one formula by using the $\pm$ symbol, as utilized in the following section. If self-connections are present, the addition has to be used, else the subtraction has to be used.

The maximum number of connections in asymmetric neural networks is twice that of their symmetric counterparts, except for self-connections, which are intrinsically directed. Asymmetric topologies are therefore not elaborated upon in this context. The most common neural networks have symmetric first-order topologies, which will be discussed first, followed by symmetric high-order ones.

B2.5.1.1 Counting symmetric first-order connections

The simplest and most widely used topologies have interlayer connections only. The total number of possible interlayer connections can be obtained by multiplying the layer sizes of each pair of adjacent layers and summing these over the whole network:

$$W = \sum_{l=1}^{L-1} W_l = \sum_{l=1}^{L-1} N_l \, N_{l+1} \tag{B2.5.3}$$

where W_l represents the number of connections between layer l and $l+1$.

When intralayer connections are also present, a number equal to the number of possible connections within a layer $((N_l/2)(N_l \pm 1))$ has to be added for each layer in the network, and the total becomes

$$\sum_{l=1}^{L} \frac{N_l}{2} (N_l \pm 1) + \sum_{l=1}^{L-1} N_l \, N_{l+1} = \frac{(N_L)^2 \pm N}{2} + \sum_{l=1}^{L-1} N_l \left(N_l + \frac{N_{l+1}}{2} \right). \tag{B2.5.4}$$

The number of connections in networks with both interlayer and supralayer connections can be calculated by summing over all the layer sizes, multiplied by the sizes of all the layers of a higher index:

$$\sum_{l=1}^{L-1} N_l \sum_{m=l}^{L-1} N_{m+1} = \sum_{m=1}^{L-1} N_{m+1} \sum_{l=1}^{m} N_l. \tag{B2.5.5}$$

Plenary neural networks have all possible connections and are equivalent to a fully connected undirected graph with N nodes (see Section B2.8.4), which has

$$\frac{N}{2} (N \pm 1) \tag{B2.5.6}$$

connections.

In summary, the number of connections in (fully connected) first-order topologies is quadratic in the number of neurons:

$$W = O(N^2) \tag{B2.5.7}$$

where $O()$ is the 'order' notation as used in complexity theory (see for example Aho et al (1974)).

B2.5.1.2 Counting high-order connections

In this subsection the counting of connections is extended to high-order topologies. In order to focus the high-order connection counting on the most common case, all the source neurons of a high-order connection are assumed here to share the same layer and the possibility of having multiple instances of the same source neuron providing input to one high-order connection is excluded.

It is illustrative to first examine the case of one single sink neuron in a high-order network. The total number of possible connections of order ω that can provide information for one specific sink neuron is equal to the number of possibilities of combining the corresponding source neurons. This number is equal to

$$\left(\begin{array}{c} n \\ \omega \end{array} \right) := \frac{n!}{\omega!(n-\omega)!} \tag{B2.5.8}$$

where n is the number of potential source neurons. Note that ω can be maximally n.

Adding up these numbers over all possible orders, the maximum number of connections associated with a high-order neuron† then becomes

$$\sum_{i=1}^{\Omega} \left(\begin{array}{c} n \\ i \end{array} \right). \tag{B2.5.9}$$

Since Ω is bounded by n, the total number of high-order connections is bounded by

$$\sum_{i=1}^{n} \left(\begin{array}{c} n \\ i \end{array} \right) = 2^n - 1. \tag{B2.5.10}$$

The virtual bias connection of the neuron can be added to this sum to obtain the crisp maximum of 2^n.

To obtain the connectivity count of a high-order topology, these high-order neurons need to be combined into a network. Given the scope of this handbook, only the most prevalent case, that of asymmetric fully interlayer connected high-order networks is presented here (high-order connections are usually unidirectional and counting multidirectional high-order connections is complicated since the set of source neurons can no longer be assumed to share the same layer). For a more elaborate treatment of this subject the reader is referred to the article by Fiesler *et al* (1996), which also contains a comparison between the various topologies based on these connection counts.

The number of connections in a fully interlayer-connected neural network of order Ω is

$$\sum_{l=1}^{L-1} N_{l+1} \sum_{i=1}^{\Omega} \left(\begin{array}{c} N_l \\ i \end{array} \right). \tag{B2.5.11}$$

In general, the number of connections in (fully connected) high-order topologies is exponential in the number of neurons:

$$W = O(2^N). \tag{B2.5.12}$$

References

Aho A V, Hopcroft J E and Ullman J D 1974 *The Design and Analysis of Computer Algorithms (Computer Science and Information Processing)* (Reading, MA: Addison-Wesley)

Fiesler E, Caulfield H J, Choudry A and Ryan J P 1996 Maximal interconnection topologies for neural networks, in preparation

Rumelhart D E, McClelland J L and the PDP Research Group 1986 *Parallel Distributed Processing: Explorations in the Microstructure of Cognition. vol 1: Foundations* (Cambridge, MA: MIT Press)

---

† Note that the concept of 'order' can be seen from the connection point of view as well as from the neuron point of view.

B2.6 Partially connected topologies

Emile Fiesler

Abstract

See the abstract for Chapter B2.

Even though most neural network topologies are fully connected according to any of the definitions given in Section B2.5, this choice is usually an arbitrary one and based on simplicity. Partially connected topologies offer an interesting alternative with a reduced degree of redundancy and hence a potential for increased efficiency. As shown in Sections B2.5.1.1 and B2.5.1.2, the number of connections in fully connected neural networks is quadratic in the number of neurons for first-order networks and exponential for high-order networks. Although it is outside the scope of this chapter to discuss the amount of redundancy desired in neural networks, one can imagine that so many connections are in many cases an overkill with a serious overhead in training and using the network. On the other hand, partial connectedness brings along the difficult question of which connections to use and which not. Before giving an overview of the different strategies followed in creating partially connected topologies, a number of metrics are presented, providing a base for studying them.

B2.6.1 Connectivity metrics

Some basic neural network connectivity metrics are presented in this section. They can be used for the analysis and comparison of partially connected topologies, but are also applicable to the various kinds of fully connected topology discussed in Section B2.5.

The *degree* of a neuron is equal to the number of connections linked to it. More specifically, the degree of a neuron can be subdivided into an *in degree* (d^{in}) or *fan-in*, which is the number of connections that can provide information for the neuron, and an *out degree* (d^{out}) or *fan-out*, which is the number of connections that can receive information from the neuron. It therefore holds that

$$d_n = d_n^{in} + d_n^{out} \tag{B2.6.1}$$

where d_n is the degree of neuron n. For the network as a whole, the *average degree* $(\overline{d})$ can be defined as

$$\overline{d} = \frac{\sum_{l=1}^{L} \sum_{i=1}^{N_l} (d_{l,i}^{in} + d_{l,i}^{out})}{N} \tag{B2.6.2}$$

where $d_{l,i}$ denotes the degree of neuron i in layer l. Another useful metric is the *connectivity density* of a topology, which is defined as

$$\frac{W}{W_{max}} \tag{B2.6.3}$$

where W is the number of connections in the network and W_{max} the total number of possible connections for that interconnection scheme; these are given in Sections B2.5.1.1 and B2.5.1.2.

The last metric given here is the *connectivity level*, which provides a ratio of the number of connections with respect to the number of neurons in the network:

$$\frac{W}{N}. \tag{B2.6.4}$$

B2.6.2 A classification of partially connected neural networks

As mentioned earlier, choosing a suitable partially connected topology is not a trivial task. This task is most difficult if one strives to find a scheme for choosing such a topology *a priori*, that is, independent of the application. Most approaches leading to partially connected topologies are therefore assuming a number of constraints, which can aid in the topology choice. Based on this, the methods for constructing partially connected networks can be classified as follows:

- *Methods based on theoretical and experimental studies.* These methods usually assume a fixed, possibly random, connectivity distribution with either a constant degree or connectivity level. The created networks are typically used for theoretical studies to determine fundamental aspects of these networks, as for example their storage capacity.

- *Methods derived from biological neural networks.* The goal of these methods is to mimic biological neural networks as well as possible, or at least to use certain criteria from biology as constraints to aid the network building.

- *Application dependent methods.* This is an important class of methods where the choice of topology is directly based on information obtained from a given application domain.

- *Methods based on modularity.* Modular neural networks, which are discussed in a later section, are a special kind of partially connected neural networks that can be seen as a subclass of the application-dependent models. They consist of sets of modules, which can each be either fully or partially connected internally. The modules themselves are typically sparsely connected to each other, again often based on application-dependent knowledge. (See also Sections B2.7 and B2.9.)

- *Methods developed for hardware implementation.* These methods are based on constraints that arise from hardware limitations in analog or digital electronic, optical, or other hardware implementations. An important subclass are the *locally connected* neural networks, such as *cellular neural networks* (see Section E1.2.4), that minimize the amount of wiring needed for the network, which is of fundamental importance for electronic implementations.

E1.2.4

- *Ontogenic methods.* An important class of methods, where the topology is dynamically adapted during the training process by adding and/or deleting connections and/or neurons, are the *ontogenic methods*. The ontogenic methods that include the removal and/or addition of individual connections provide an automatic way to create partially connected neural networks. The various kinds of ontogenic neural network are discussed in Sections C1.7 and C2.4

C1.7, C2.4

An extensive review of partially connected neural networks, based on this classification, can be found in the atricle by Elizondo *et al* (1996). A short summary of this work, restricted to nonontogenic methods, is the article by Elizondo *et al* (1995).

Besides these purely neural-network-based methods, other artificial intelligence techniques, such as evolutionary computation and *inductive knowledge*, have been used to aid the construction of partially connected networks.

D2

For completeness, a technique that does not necessarily reduce the number of connections but reduces the number of modifiable parameters by reducing the number of weights needs to be mentioned here, which is weight sharing (see also Section B2.5.1). Using this technique, groups of connections are assigned only one updatable weight. These groups of connections can for example act as feature detectors in pattern recognition applications.

References

Elizondo D, Fiesler E and Korczak J 1995 Non-ontogenic sparse neural networks *Proc. Int. Conf. on Neural Networks (Perth)* (Piscatawat, NJ: IEEE) pp 290–5
—— 1996 A survey of partially connected neural networks, in preparation

B2.7 Special topologies

Emile Fiesler

Abstract

See the abstract for Chapter B2.

Besides the common layered topologies, which are usually at least fully interlayer connected, there exists a variety of other topologies that are not necessarily layered, or at least not homogeneously layered. In this section a number of these are discussed.

Modular neural networks are composed of a set of smaller subnetworks (the modules), each performing a subtask of the complete problem. The topology design of modular neural networks is typically based on knowledge obtained from a specific application or application domain. Based on this knowledge, the problem is split up into subproblems, each assigned to a neural module. These individual modules do not have to belong to the same category and their topologies can therefore differ considerably. The global interconnectivity of the modular network, that links the modules, is often irregular as it is usually tuned to the application. The overall topology of modular neural networks is therefore often irregular and without a uniform layered structure.

Somewhat related to modular neural networks are *composite neural networks*. A composite neural network consists of a concatenation of two or more neural network models, each with its associated topology, thereby forming a new neural network model. A layered structure can therefore be observed at the component level, since they are stacked, but the internal topologies of the components themselves can differ from each other, yielding an inhomogeneous global topology. Composite neural networks are often called *hybrid neural networks*, a context-dependent term that is even more popular for describing combinations of neural networks with other artificial intelligence techniques such as expert systems and evolutionary systems. In this handbook, the term 'hybrid neural network' is therefore reserved for these latter systems (see part D of this handbook).

Another kind of topology that is sometimes used in the context of neural computation is the *tree*, which refers to the graph theoretical definition of a connected acyclic graph (see Section B2.8.4 for the relationship between graph theory and neural network topologies). The typical tree topology used is a *rooted* one, where connections branch off from one point or a set of points. These points are usually the output neurons of the network. Tree-based topologies are usually deep and sparse, and the neurons have a restricted fan-in and fan-out. If these networks are trees according to the definition, that is, without cross-connections between the branches of the tree, it can be argued whether they should be classified as neural networks or as *decision trees* (Kanal 1979, Breiman *et al* 1984). In this context it should be mentioned that it is in some cases possible to convert the tree-based topology into a conventional layered neural network topology (see for example Frean 1990).

An important class of networks which can have a nonstandard topology are the *ontogenic neural networks*, as discussed in the previous section, where the topology can change over time during the training process. Even though their topology is dynamic, it is usually homogeneous at each point in time during the training; this in contrast with modular neural networks, which are usually inhomogeneous.

One of the fundamental motivations behind ontogenic neural networks is to overcome the notorious problem of finding a suitable topology for solving a given problem. The ultimate goal is to find the optimal topology, which is usually the *minimal topology* that allows a successful solution of the problem. For this reason, but also for establishing a base for comparing the resulting topologies of different ontogenic training methods, it is important to define the minimal topology (Fiesler 1993).

B2.9

C1.6, C2.3

C1.7, C2.4

Definition. A *minimal neural network topology* for a given problem is a topology with a minimal computational complexity that enables the problem to be solved adequately.

In practice, the topological complexity of neural networks can be estimated by the number of high-complexity operations, like multiplications, to be performed during one recall phase. In the case where the splicing function is either the multiplication operation or a low-complexity operation, the count can be restricted to the number of multiplications only. For first-order networks, where the number of multiplications to be performed in the recall process is almost equal to the number of weighted connections, this can be further simplified as:

Definition. A *minimal first-order neural network topology* for a given problem is a neural network topology with a minimal number of weighted connections that solves the problem adequately.

To illustrate the concept of minimal topology, the well-known *exclusive OR* (XOR) problem can be used. The exclusive OR function has two Boolean inputs and one Boolean output which yields FALSE either when both inputs are TRUE or when both inputs are FALSE, and yields TRUE otherwise. This function is the simplest example of a *nonlinearly separable* problem. Since nonlinearly separable problems cannot be solved by first-order perceptrons without hidden layers (Minsky and Papert 1969), the minimal topology of a perceptron that can solve the XOR problem has either hidden layers or high-order connections.

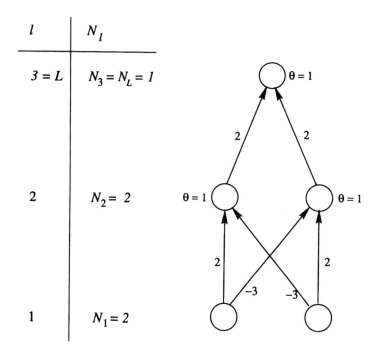

Figure B2.7.1. A first-order neural network with a minimal interlayer-connected topology that can solve the XOR problem. It has three layers and six interlayer connections.

In the following three examples, binary (0, 1) inputs, outputs, and activation values are assumed, as well as a *hard-limiting threshold* or *Heaviside function* ($\mathcal{H}$) as activation function:

$$\mathcal{H}(x) = \begin{cases} 0 & \text{if } x \leq \Theta \\ 1 & \text{if } x > \Theta \end{cases} \tag{B2.7.1}$$

and the activation value of a neuron in layer $l + 1$ is calculated by the following forward propagation formula:

$$a_{l+1_j} = \mathcal{H}\left(\sum_i W_{l_{i,j}} a_{l_i}\right) \tag{B2.7.2}$$

where a_{l_i} is the activation value of neuron i in layer l, and $W_{l_{i,j}}$ the weight of the connection between this neuron and neuron j in layer $l + 1$, in accordance with the abbreviated notation of Section B2.8.2.

 Handbook of Neural Computation release 97/1

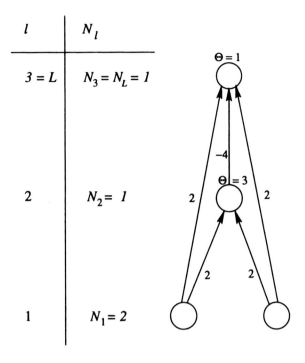

l	N_l
$3 = L$	$N_3 = N_L = 1$
2	$N_2 = 1$
1	$N_1 = 2$

Figure B2.7.2. A first-order neural network with a minimal topology that can solve the XOR problem. It has three layers, three interlayer connections, and two supralayer connections.

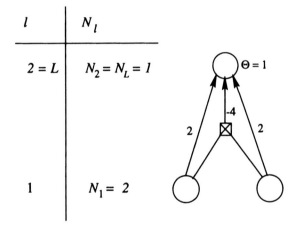

l	N_l
$2 = L$	$N_2 = N_L = 1$
1	$N_1 = 2$

Figure B2.7.3. A high-order neural network with a minimal topology that can solve the XOR problem. It has two layers, two first-order connections, and one second-order connection.

Figure B2.7.1 shows the minimal topology of an interlayer-connected first-order neural network able to solve the XOR problem, and figure B2.7.2 the smallest first-order solution which uses supralayer connections.

Figure B2.7.3 shows the smallest high-order solution with two first-order connections and one second-order connection.

References

Breiman L, Friedman J H, Olsen R A and Stone C J 1984 *Classification and Regression Trees* (Belmont, CA: Wadsworth)

Fiesler E 1993 Minimal and high order network topologies *Proc. 5th Workshop on Neural Networks: Academic/Industrial/NASA/Defense; Int. Conf. on Computational Intelligence: Neural Networks, Fuzzy Systems, Evolutionary Programming and Virtual Reality (WNN93/FNN93) (San Francisco, CA); SPIE Proc.* **2204** 173–8

Frean M 1990 The upstart algorithm: a method for constructing and training feedforward neural networks *Neural Comput.* **2** 198–209

Kanal L N 1979 Problem solving models and search strategies for pattern recognition *IEEE Trans. Pattern Anal. Machine Intell.* **1** 194–201

Minsky M L and Papert S A 1969 *Perceptrons* (Cambridge, MA: MIT Press)

B2.8   A formal framework

Emile Fiesler

Abstract

See the abstract for Chapter B2.

Even though ANNs have been studied for several decades, a unifying formal theory is still missing. An important reason for this is the nonlinear nature of neural networks, which makes them difficult to study analytically, since most of our mathematical knowledge relates to linear mathematics. This lack of formalization is further illustrated by the upsurge in progress in neurocomputing during the period when computers became popular and widespread, since they enable the study of neural networks by simulating their nonlinear dynamics. It is therefore important to strive for a formal theoretical framework that will aid the development of formal theories and analytical studies of neural networks. A first step towards this goal is the standardization of terminology, notations, and several higher-level neural network concepts to enable smooth information dissemination within the neural network community, including users, that consists of people with a wide variety of backgrounds and interests. The IEEE Neural Network Council Standardization Committee is aiming at this goal. A further step towards this goal is a formal definition of a neural network that is broad enough to encompass virtually all existing neural network models, yet detailed enough to be useful. Such a topology-based definition, supported by a consistent terminology and notation, can be found in the article by Fiesler (1994); other examples of formal definitions can be found in the artices by Valiant (1988), Hong (1988), Farmer (1990), and Smith (1992).

A deep-rooted nomenclature issue, that of the definition of a layer, will be addressed in the next section. Further, in order to illustrate the concept of a consistent and mnemonic notation, the notational issue of weights, the most important neural network parameters, is discussed in the subsequent section, which is followed by a structured method to visualize and study weights and network connectivity. Lastly, the relationship between neural network topologies and graph theory is outlined; this offers a mathematical base for neural network formalization from the topology point of view.

B2.8.1   Layer counting

A fundamental terminology issue which gives rise to much confusion throughout the neural network literature is that of the definition of a *layer* and, related to this, how to count layers in a network. The problem is rooted in the generic nature of the word 'layer', since it can refer to at least three network elements:

- A layer of neurons
- A layer of connections and their weights
- A combination of a layer of neurons plus their connections and weights.

Some of these interpretations need further explanation. The second meaning, that of the connections and associated weights, is difficult to use if there are other connections present besides interlayer connections only, for example intralayer connections, which are inherently intertwined with a layer of neurons. Defining a layer as a set of connections plus weights is therefore very limited in scope and its use should be discouraged. For both the second and the third meaning, the relationship between the neurons and 'their' connections needs to be defined. In this context of layers, all incoming connections, that is, those that are capable of providing information to a layer of neurons, are usually the ones that are associated with that ayer. Nevertheless, independent of which meaning is used, an important part of this terminology issue can be solved by simply defining what one means by a layer.

C1.1   An early neural network in history with a layered topology was the perceptron (Rosenblatt 1958), which is sometimes called the *single-layer perceptron*. It has a layer of input units that duplicate and fan-out incoming information, and a layer of output units that perform the (nonlinear) weighted sum operation. The name *single-layer perceptron* reflects the third meaning of the word 'layer' as given above, and is based on not counting the input layer as a layer, which is explained below. Since the conception of the perceptron, many other neural network models have been introduced. The topology of some of these models does not match with the layer concept given by the third interpretation. This is for example the case for networks which have intralayer connections in the input (neuronal) layer or where a certain

C1.4 amount of processing takes place in the input layer, such as the *Boltzmann machine* and related *stochastic*

B2.3 neural network models and such as *recurrent* neural networks that feed information from the output layer back to the input layer.

     Currently, the most popular neural network models belong to the family of multilayer neural networks. The terminology associated with these models includes the terms *input layer*, *hidden layer*, and *output layer* (see Section B2.2.1), which corresponds to the first interpretation of the word 'layer' as a layer of neurons.

     The issue of defining a layer also gives rise to the problem of counting the number of layers, which is mainly caused by the dilemma of whether one should count the input layer as a layer. The argument against counting the input layer is that in many neural network models the input layer is used for duplicating and fanning out information and does not perform any further information processing. However, since there are neural network models where the input neurons are also processing units, as explained above, the best solution is to include the input layer in the counting. This policy has therefore been adopted by this handbook.

     The layer counting problem manifests itself mainly when one wants to label or classify a neural network as having a certain number of layers. An easy way to circumvent the layer counting problem is therefore to count the number of hidden layers instead of the total number of layers. This approach avoids the issue of whether to count the input layer.

     In can be concluded that the concept of a layer should be based on a layer of neurons. For a number of popular neural network models it would be possible to also include the incoming interlayer connections into the layer concept, but this should be discouraged given its limited scope of validity. In general it is best to clearly define what is understood by a layer, and in order to avoid the layer counting problem one can count the number of hidden layers instead.

B2.8.2   Weight notation

To underline the importance and to illustrate the use of a consistent and mnemonic notation, the notation of the most fundamental and abundant neural network parameters, that of the weights, is discussed in this section.

     A suitable and commonly used notation for a connection weight is the letter w, which is also mnemonic, using the first letter of the word 'weight'. Depending on the topology, there are several ways to uniquely address a specific weight in the network.

     The best and most general way is to specify the position of both the source and the sink neuron that are linked by the connection associated with a weight, by specifying the layer and neuron indices of both: $w_{l_i m_j}$, where l and m are the indices of the source and sink layers respectively and i and j the neuron indices within these layers. This notation specifies a weight in a unique way for all the different kinds of first-order connection as defined in Section B2.2.2.

     For neural networks with only interlayer connections, the notation can be simplified if necessary. Since the difference between the layer indices (l and m) is always one for these networks, one of the two indices could be omitted: $w_{l_{ij}}$. In cases where this abbreviated notation is used, it is important to clearly specify which layer the index l represents: whether it represents the layer containing the source or the sink neuron.

     A further notational simplification is possible for first-order networks with one neuronal layer or networks without any cluster structure, where all neurons in the network are equally important. The weights in these networks can be simply addressed by w_{ij}, where the i and j indices point to the two neurons linked by the connection 'carrying' this weight.

High-order connections require a more elaborate notation since they combine the information of several source neurons. Hence, the set of source neurons ($\{s_i\}$) needs to be included in the notation and the weight of a high-order connection can be denoted as $w_{\{s_i\}m_j}$. When desired, this notation can be abbreviated for certain kinds of networks, analogous to first-order connections as described above.

Similarly to the weight notation, mnemonic notations for other network parameters are also recommended and used in this handbook.

B2.8.3 Connectivity matrices

A compact way to represent the connections and/or weights in a neural network is by means of a *connectivity matrix*. For first-order neural networks this is a two-dimensional array where each element represents a connection or its associated weight. A global connectivity matrix describes the complete network topology with all neuron indices enumerated along each of its two axes. Note that a symmetric neural network has a symmetric connectivity matrix and an asymmetric neural network an asymmetric one. Feedforward neural networks can be represented by a triangular matrix without diagonal elements. Figure B2.8.1 shows an example for the fully interlayer connected topology of figure B2.2.1.

	1,1	1,2	2,1	2,2	2,3	2,4	3,1	3,2	4,1
1,1			•	•	•	•			
1,2			•	•	•	•			
2,1							•	•	
2,2							•	•	
2,3							•	•	
2,4							•	•	
3,1									•
3,2									•
4,1									

Figure B2.8.1. Connectivity matrix for the four-layer fully interlayer-connected neural network topology as depicted in figure B2.2.1. On the vertical axis the source neurons are listed by a tuple consisting of the layer number followed by the neuron number in that layer. On the horizonal axis the sink neurons are listed using the same notation. A '•' symbol marks the presence of a connection in the topology.

For layered networks, the order of the neuron indices should reflect the sequential order of the layers, starting with the input layer neurons at one end of the matrix and ending with the output neurons at the other end of the matrix. The matrix can be subdivided into blocks based on the layer boundaries (see figure B2.8.1). In such a matrix, subdivided into blocks, the diagonal elements, which are the matrix elements with identical indices, represent the self-connections and the diagonal blocks containing these diagonal elements contain the intralayer connections. The interlayer connections are found in the blocks that are horizontally or vertically adjacent to the diagonal blocks. All other blocks represent supralayer connections. Figure B2.8.2 shows the global connectivity matrix for the network depicted in figure B2.2.2.

	1,1	1,2	2,1	2,2	3,1
1,1			•	•	•
1,2			•	•	•
2,1			•	•	•
2,2			•	•	•
3,1	•	•			

Figure B2.8.2. Global connectivity matrix for the layered neural network topology with various kinds of connection as depicted in figure B2.2.2. The notation is the same as in figure B2.8.1.

For layered neural networks with only interlayer connections, individual connectivity matrices can be constructed for each of the connection sets between adjacent layers.

The connectivity matrices for high-order neural networks need to have a dimensionality of $\Omega + 1$, corresponding to the maximum number of source neurons (Ω) plus one sink neuron.

Based on the definitions of Section B2.2.2, the *span* of a connection, measured in number of layers, can be defined as the difference between the indices of the layers in which the neurons that are linked by that connection are located. That is, the span of a connection which connects layer l with layer m is $|l - m|$. For example, interlayer connections have a span of one, intralayer connections a zero span, and supralayer connections a span of two or more. Different kinds of supralayer connection can be distinguished based on their span. The span of a connection can be easily visualized with the aid of a global connectivity matrix, since it is equal to the horizontal or vertical distance, in blocks, from the matrix element corresponding to that connection to the closest diagonal element of the connectivity matrix. The span of a high-order connection, which is equal to the maximum difference between any of the indices of the layers it connects, is more difficult to visualize given the increased dimensionality of the connectivity matrix.

B2.8.4 Neural networks as graphs

Graph theory (see for example Harary 1969) provides an excellent framework for studying and interpreting neural network topologies. A neural network topology is in principle a graph $(\mathcal{N}, \mathcal{W})$, where $\mathcal{N}$ is the set of neurons and $\mathcal{W}$ the set of connections, and when the network has a layered structure it becomes a *layered graph* (Fiesler 1993). More specifically, neural networks are directed layered graphs, specifying the direction of the information flow. In the case where the information between neurons can flow in more than one direction, there are two possibilities:

- if distinct weight values are used for the information flow (between some neurons) in more than one direction, the topology remains a directed graph but with multiple connections between those neurons that can have a multidirectional information flow;

- if the same weight value is used in all directions, the topology becomes symmetric (see Section B2.3) and corresponds to the topology of an undirected graph.

Figure B2.1.1 shows a neural network topology without a layered structure, which is a directed graph. If all possible connections are present, as in a plenary neural network, its topology is equivalent to a *fully connected graph*.

References

Farmer J D 1990 A Rosetta stone for connectionism *Physica* D **42** 153–87

Fiesler E 1993 Layered graphs with a maximum number of edges *Circuit Theory and Design 93: Proc. 11th Eur. Conf. on Circuit Theory and Design (Davos, 1993)* part I, ed H Dedieu (Amsterdam: Elsevier) pp 403–8

—— 1994 Neural network classification and formalization *Comput. Standards Interfaces* **16** 231–9

Harary F 1969 *Graph Theory* (Reading, MA: Addison-Wesley)

Hong Jiawei 1988 On connectionist models *Commun. Pure Appl. Math.* **41** 1039–50

Rosenblatt F 1958 The perceptron: a probabilistic model for information storage and organization in the brain *Psychol. Rev.* **65** 386–408

Smith L S 1992 A framework for neural net specification *IEEE Trans. Software Eng.* **18** 601–12

Valiant L G 1988 Functionality in neural nets *Proc. 7th Natl Conf. Am. Assoc. Artificial Intell. (AAAI)-88 (St Paul, MN, 1988)* vol 2 (San Mateo, CA: Morgan Kaufmann) pp 629–34

B2.9 Modular topologies

Massimo de Francesco

Abstract

See the abstract for Chapter B2.

B2.9.1 Introduction

The beauty of neural network programming, and certainly one of the reasons why early models were found so appealing by computer science researchers, is the idea of a distributed, uniform method of computation, where a few decisions concerning simple topologies of fully connected layers of neurons are enough to define a complete system able to carry out any assigned task. Indeed, the dream of a self-programming system, coupled with the mathematical purity of a regular structure, has been the primary focus of research in neural networks.

This uniformity, however, can be the major shortcoming when trying to cope with real-world problems. The brain itself, the most perfected biological neural system, is far from being a regular and uniform structure: millions of years of evolution and genetic selection ended up in a highly organized, hierarchical system, which can be better described by the expression *network of networks*. From nature's point of view, uniformity is a waste of resources.

B2.9.2 The complexity problem

As a matter of fact, uniform architectures such as *multilayer perceptrons* have proved to be able to tackle problems in an effective way, and approximation theorems show that these networks are able under certain conditions to represent virtually any mapping. However, the computational costs associated with training a uniformly connected network can be unacceptably high, and the learning rules commonly used are not guaranteed to converge to the global optimum. C1.2

Scaling properties of uniform multilayer perceptrons are a matter of concern, because the number of weights usually grows more than linearly with the size of the problem. Since an interesting result of computational learning theory tells us that we need proportionally as many examples as weights to achieve a given accuracy (Baum and Haussler 1989), the actual number of examples and the time needed to train the system can become prohibitively large as the problem size increases.

Furthermore, uniform feedforward architectures are subject to interference effects from uncorrelated features in the input space. By trying to exploit all the information a given unit receives, it becomes much more sensible to apparent relationships between unrelated features, which arise especially with high input dimensionality and insufficient training data.

Problems such as *image or speech recognition* convey such an amount of information that their F1.6, F1.7
treatment by a uniform architecture is not conceivable without relying on heavy preprocessing of the data in order to extract the most relevant information.

Modular architectures try to cope with these problems by restricting the search for a good approximation function to a smaller but potentially more interesting set of candidates. The idea that led to the investigation of more modular architectures came from the observation that class boundaries in large, real-world problems are usually much smoother and more regular than those found in such toys but extremely difficult problems as *n*-parity or the double spiral, and do not require the excessively powerful approximation capability of uniform architectures. For instance, we do not expect a face classification system to completely change its output as one single bit in the input space is altered.

Modularity is also the natural outcome of divide and conquer strategies, where *a priori* knowledge about the problem can be exploited to shape the network architecture.

B2.9.3 Modular topologies

Although any simple categorization could not account for all the types of architecture commonly called modular, published work seems to focus on three main levels of modularity related to neural computation: modular multinetwork systems, modular topologies, and (biological) modular models. We will essentially discuss the former, with special emphasis on modular topologies, although we will give a definition of and pointers to the latter.

B2.9.3.1 Modular systems

Modular systems usually decompose a difficult problem into easier subproblems, so that each subproblem can be successfully solved by an eventually uniform neural network. Different options have been investigated regarding the way input data is fed into the different modules, how the different results are finally combined, and whether the subnetworks are trained independently or in the context of the global system.

Some of these modular systems rely on the decomposition of the training data itself, by specializing different networks on different subsets of the input space. Sabourin and Mitiche (1992) for instance describe a character recognition system where high-level features in the input data, such as presence or absence of loops, are used to select a specifically trained subnetwork. Others rely on the fact that different instantiations of the same network trained on the same data (or on different representations of the same data) usually converge to different global minima (because of the randomized starting conditions), so that a simple voting procedure can be implemented (see for instance the article by Lincoln and Skrzypek (1990). Others again add specific neural circuitry to provide more sophisticated combination of the partial results (see for instance the article by Waibel (1989)).

Among modular systems, the multiexpert model (Jacobs *et al* 1991) deserves special consideration, since no *a priori* knowledge regarding the task decomposition is required: the system itself learns the correct allocation of training cases by performing gradient descent on an altered error function enforcing the competition between the expert networks and thus inducing their specialization to local regions of the input space.

Most of the modular systems described here claim better generalization than a comparable uniform architecture, although some of them achieve this at the expense of increased computation.

B2.9.3.2 Modular models

CALM networks (Murre *et al* 1992) or cortical column models (Alexandra *et al* 1991) are original neural network models which are intrinsically modular. The basic computing structures of CALM and cortical column models are small modules composed by neuron-like elements, and the models describe the interaction, learning, and computing properties of assemblies of these modules. The main focus here is on biological resemblance, rather than computational efficiency.

B2.9.3.3 Modular topologies

The final category of modular architectures includes simple topological variations of otherwise well known and widely used neural models such as multilayer perceptrons. Units of the hidden and possibly output layers in these networks are further organized into several clusters which have only local connectivity to units in the previous layer. Modules are thus composed by one or more units having connections limited to a local field (or a union of local fields) in the previous layer, and several modules operating in parallel are needed to completely cover the input space. This eventually overlapping tiling can be repeated for the subsequent layers, but is especially useful between the input and the first hidden layer. These architectures do not require modification of the standard learning rules, so that standard backpropagation can be applied. They are therefore very easy to implement, yet achieve very good results by diminishing the total number of weights, by partially avoiding interference effects, and by enforcing a divide and conquer strategy. If it is possible to load the training set in a modular topology, then we will obtain a network which is faster and which generalizes better than a corresponding uniform network.

The study of a printed optical character recognition task from De Francesco (1994) will help illustrate these points with some numbers. Suppose we are processing a 16×16 binary image with a feedforward neural network. With 50 hidden units, the first layer in a fully connected topology would contain $50 \times 16 \times 16 = 12\,800$ weights. If we define a modular architecture using nine modules with an 8×8 local input field overlapping the whole image, and if each of these modules contains six units (for a total of $9 \times 6 = 54$ hidden units), the combined first layer would have $9 \times 6 \times 8 \times 8 = 3456$ weights, roughly a quarter of the uniform architecture. The results reported in table B2.9.1 show that the modular architecture is much more accurate than the uniform one. Furthermore, since the modular architecture has much fewer weights, it is tighter and executes faster, so that it can be more easily deployed in an industrial application where speed and space constraints are an important factor.

Table B2.9.1. A comparison of modular and uniform topologies.

Topology	No of modules	No of weights	No of hidden layers	No of outputs	Accuracy (%)
Uniform	2 (2 layers)	$\sim W$	~ 25	~ 100	$< 85^*$
Uniform	2	$\sim 2W$	~ 50	~ 100	98.2
Modular	10	W	~ 50	~ 100	99.5

* The uniform architecture with the same number of weights as the modular network was most of the time unable to converge on the training set; 85% represents the accuracy on the test set of the most converged network in the batch. Accuracy values of the two other architectures are averaged over ten runs.

Similar results have been reported by Le Cun (1989) on a smaller problem, with a topology combining local fields with additional constraints of equality between weights in different clusters. This is known as the weight sharing technique, described by Rumelhart *et al* (1986). Today, weight sharing is especially used in *time delay neural networks*, which have been extensively applied to speech recognition tasks.   C1.2.8, F1.7

Recent theoretical results on sample size bounds for shared weight networks (Taylor 1995) indicate that the generalization power of these networks depends on the number of classes of weights (shared weights are counted only once), rather than on the total number of connections, which explains their improved performance over uniform architectures.

B2.9.4 A need for further research

It must be noted that many modular architectures are in fact subsets of uniform topologies, in the sense that they are equivalent to a uniform architecture with some of the connections fixed with zero-valued weights. It can thus be objected that these modular networks are intrinsically less powerful than uniform ones, and this is certainly true in the general case. The point is that modular architectures can and must be adapted to the particular problem or class of problems to be effective, where uniform ones only depend on the problem dimensions. This raises the issue of determining whether and how a given architecture is suited to the particular task. Local receptive fields for instance can be easily justified in image processing, but much less so in financial forecasting or medical diagnosis, where the input is composed of complex variables with no evident topological relationship. Which knowledge is useful and how it can be translated into the network architecture is still an open question from a theoretical point of view.

Some *ontogenic networks* attempt to cope with the architectural dilemma by modifying themselves   C1.7, C2.4
during training, usually pruning apparently unused connections, trying in this way to prevent some of the problems associated with fully connected networks. They however fail to produce any intelligible modularity in the final architecture, and their global performance is usually not as good as successfully trained networks with a fixed modular topology.

Although important experimental evidence supporting the superiority of modular architectures has been cumulated over the last few years, and even if large-scale problems such as speech recognition have shown to be tractable only by modular topologies, the lack of important theoretical results and the additional efforts needed to choose and specify a modular architecture have certainly diminished their interest among researchers in neural networks. Therefore, before hoping to find a more widespread use of modular neural networks, some fundamental and related questions will have to be answered more precisely:

- How can problems be categorized in order to establish which ones benefit the most from modularity?
- How can we exploit topological data in the theoretical determination of optimal bounds for the size of the training set?
- Conversely, given a problem, is there any computationally effective way to determine a good topology to solve it?

References

Alexandre D, Guyot F, Haton J-P and Burnod Y 1991 The cortical column: a new processing unit for multilayered networks *Neural Networks* **4** 15–25

Baum E B and Haussler D 1989 What size net gives valid generalization? *Neural Comput.* **1** 151–60

De Francesco M 1994 Functional networks: a new computational framework for the specification, simulation and algebraic manipulation of modular neural systems *PhD Thesis* University of Geneva

Jacobs R A, Jordan M I, Nowlan S J and Hinton G E 1991 Adaptive mixtures of local experts *Neural Comput.* **3** 79–87

Murre J M J, Phaf R H and Wolters G 1992 CALM: a building block for learning neural networks *Neural Networks* **5** 52–82

Le Cun Y 1989 Generalization and network design strategies *Technical report* CRG-TR-89-4, University of Toronto Connectionist Research Group

Lincoln W and Skrzypek J 1990 Synergy of clustering multiple back propagation networks *Advances in Neural Information Processing Systems 2* (Denver, CO, 1989) ed D S Touretzky (San Mateo, CA: Morgan Kaufmann) pp 650–9

Rumelhart D E, Hinton G E and Williams R G 1986 Learning internal representation by error propagation *Parallel Distributed Processing* vol 1, ed D E Rumelhart and J L McClelland (Cambridge, MA: MIT Press) pp 318–62

Sabourin M and Mitiche A 1992 Optical character recognition by a neural network *Neural networks* **5** 843–52

Taylor J S 1995 Sample sizes for threshold networks with equivalences *Information Comput.* **118** 65–72

Waibel A 1989 Modular construction of time delay neural networks for speech recognition *Neural Comput.* **1** 39–46

B2.10 Theoretical considerations for choosing a network topology

Maxwell B Stinchcombe

Abstract

A minimal criterion for choosing a network topology is 'denseness'. A network topology is dense if it contains networks that can come arbitrarily close to any functional relation between inputs x and outputs y. Within a chosen dense class of networks, the question is how large a network to choose. Here a minimal criterion is consistency. A method of choosing the size of the the network is consistent if, as the number of data or training examples grows large, all avoidable errors disappear. This means that the choices cannot overfit. The most widespread consistent methods of choice are variants of a statistical technique known as cross-validation.

B2.10.1 Introduction

Neural networks provide an attractive set of models of the unknown relation between a set of input variables $x \in \mathbb{R}^k$ and output variables $y \in \mathbb{R}^m$. The different topologies or architectures provide different classes of nonlinear functions to estimate the unknown relation. The questions to be answered are as follows:

(i)   What class of relations is, at least potentially, representable?
(ii) What parts of the potential are actually realizable?
(iii) How might we actually learn (or estimate) the unknown relation?
(iv) How well does the estimated relation do when presented with new inputs?

The formal answers to the first question have taken the form of denseness (or universal approximation) theorems—if some aspect of the architecture goes to infinity, then, up to any $\epsilon > 0$, all relations in some class $\mathcal{X}$ of functions from $\mathbb{R}^k$ to $\mathbb{R}^m$ can be ϵ-captured. If an architecture does not have this property, then there are relations between x and y that will not be captured.

The formal answers to the second question have taken the form of consistency theorems—if the number of data (read number of training examples) becomes large, then, up to any $\epsilon > 0$, all relations in $\mathcal{X}$ between x and y can be ϵ-learned (read estimated). The previous denseness results are a crucial ingredient here.

Imbedded in the consistency theorems are two kinds of answer to the third question. The first class of consistency theorems delivers asymptotic learning if the complexity of the architecture (measured by the number of parameters) goes to infinity at a rate sufficiently slow relative to the the amount of data. These results provide little practical guidance—multiplication of the complexity by any positive constant maintains the asymptotic relation. The second, more satisfactory class of consistency theorems delivers asymptotic learning if the complexity of the architecture is chosen by *cross-validation* (CV). The focus B3.5.2, C1.2.6 here will be CV and related procedures.

The essential CV idea is to divide the N data points into two disjoint sets of N_1 and N_2 points, $N_1 + N_2 = N$, estimate the relation between x and y using the N_1 points, and (providing an answer to the fourth question) evaluate the generalization capacity using the N_2 points. This simple idea has many variants. Related procedures include complexity regularization (loss-minimization procedures that include penalties for overparametrization), and nonconvergent methods (N_1-estimated gradient descent on overparametrized models with an N_2 deterioration-of-fit stopping rule).

B2.10.2 Measures of fit and generalization

The aim is to use artificial neural network (ANN) models to estimate an unknown relationship between x and y and to estimate the quality of estimate's fit to the data, and its capacity for generalization. The starting point is a representative sample of N data points, $(x_i, y_i)_{i=1}^N$. The most widely used measures of generalization of an estimated relation, φ, are of the form $\langle \ell_\varphi, \mu \rangle$ where $\ell_\varphi : \mathbb{R}^{k+m} \to \mathbb{R}_+$ is the (measurable) 'loss function', μ is the (countably additive Borel) probability on $\mathbb{R}^{k+m}$ from which the generalization points (x, y) will be drawn, and $\langle f, \nu \rangle := \int f \, d\nu$ for any nonnegative function f and probability ν. By far the most common loss function is $\ell_\varphi^2(x, y) = (y - \varphi(x))^2$, but any $\ell_\varphi^p = (y - \varphi(x))^p$, $p \in [1, \infty]$ (with the usual L^p convention for $p = \infty$) is feasible. Extremely useful for theoretical purposes are the Sobolev loss functions that depend on $f(x)$, the true conditional mean of y given x, and the distance between the derivatives of f and φ, for example, $\ell_\varphi^{2,\text{Sob}}(x, y) = \sum_{|\alpha| \leq M} (D^\alpha f(x) - D^\alpha \varphi(x))^2$. (In these last two sentences and from here onwards, we will assume that $y \in \mathbb{R}^1$. This is for notational convenience only, the results and discussion apply to higher output dimensions.)

This loss function approach covers both the case of noisy and noiseless observations. If $f(x)$ denotes (a version of) $E(y|x)$, then a complete description of μ is given by $y = f(x) + \epsilon$ where x is distributed according to P, the marginal of μ on $\mathbb{R}^k$, and ϵ is a mean-zero random variable with distribution $Q(x)$ on $\mathbb{R}^m$. If ϵ is independent of x and $Q(x) \equiv Q$, we have the standard additive noise model. If $Q(x)$ is a point mass on 0, i.e. if the conditional variance of ϵ is a.e. 0, we have noiseless observations (the additive noise model with zero variance).

When the data are a random sample drawn from μ, and both N_1 and N_2 are moderately large, the Glivenko–Cantelli theorem tells us that the empirical distributions μ_N, μ_{N_1}, and μ_{N_2} are good approximations to μ. If we pick a model, $\hat{\varphi}$, to minimize $\langle \ell_\varphi, \mu_{N_1} \rangle$, then $\langle \ell_{\hat{\varphi}}, \mu_{N_1} \rangle$ is an underestimate of $\langle \ell_{\hat{\varphi}}, \mu \rangle$. However, $\langle \ell_{\hat{\varphi}}, \mu_{N_2} \rangle$ is unbiased, and this is the basis of CV. We can not expect good generalization of our estimated models if the empirical distribution of the $(x_i, y_i)_{i=1}^N$ is very far from μ.

B2.10.3 Denseness

C1.1   *Single-layer feedforward* (SLFF) networks are (for present purposes) functions of the form $f(x, \theta, J) = \beta_0 + \sum_{j=1}^J \beta_j G(\gamma_j' x + \gamma_{j,0})$ where $\gamma_j' x$ is the inner product of the k-vectors γ_j' and x, $\gamma_{j,0}$ is a scalar, $G : \mathbb{R} \to \mathbb{R}$, and θ is the vector of the β and γ. The first formal denseness results were proved for SLFF networks in Funahashi (1989), followed nearly immediately (and independently) by Cybenko (1989) and Hornik *et al* (1989). All three of these showed that, if G is a sigmoid, then for any continuous g defined on any compact set $K \subset \mathbb{R}^k$, and for any $\epsilon > 0$, if J is sufficiently large, then there exists a θ such that $\sup_{x \in K} |f(x, \theta, J) - g(x)| < \epsilon$. (This is 'denseness in $C(\mathbb{R}^k)$ in the compact-open topology'.) Note carefully that this is a statement about the existence of a network with this type of architecture, not a guarantee that the network can be found, something that the consistency results deliver.

C1.2   In the article by Hornik *et al* (1989) there is an inductive proof that the same result is true for *multilayer feedforward* (MLFF) networks (feedforward networks applied to the outputs of other feedforward networks). An immediate consequence of denseness in the compact open topology is the result that for the ℓ_φ^p loss functions with compactly supported P, for large J, there exist θ such that the loss associated with $f(x, \theta, J)$ is within any $\epsilon > 0$ of the theoretical minimum loss (which is zero in the noiseless case, and is the expected value of the conditional variance in the ℓ_φ^2 case). Using some of the techniques in Funahashi (1989) and Cybenko (1989), Hornik *et al* (1990) show that the same results are true using the various $\ell_\varphi^{\text{Sob}}$ loss functions; Stinchcombe and White (1989, 1990) and Hornik (1991, 1993) have expanded these results in various directions, loosening the restrictions on G and allowing for different restrictions on the θ.

C1.6.2   *Radial basis function* (RBF) networks are (for present purposes) functions of the form $h(x, \theta, J) = \beta_0 + \sum_{j=1}^J \beta_j G((x - c_j)' \mathbf{M}(x - c_j))$ where the β are scalars, the c_j are k-vectors, $\mathbf{M}$ is a positive definite matrix, and $G : \mathbb{R} \to \mathbb{R}$. Park and Sandberg (1991, 1993a, 1993b) show that for large J, the ℓ_φ^p loss function is within any $\epsilon > 0$ of its theoretical minimum.

The sum of dense networks is again dense, meaning that combination networks will also have denseness properties. One expects that architectures more complicated than SLFF, MLFF, and RBF networks will also have denseness properties, and the techniques used in the literature just cited are well-suited to delivering such results.

Denseness is a minimal property, and, unfortunately rather too crude to usefully compare different dense network architectures—given two different architectures, there are typically two corresponding disjoint, dense sets $\mathcal{X}_1, \mathcal{X}_2 \subset \mathcal{X}$ of possible relations for which the two architectures are better suited. Further, the known rates at which the loss can can be driven to its theoretical minimum as a function of the number of parameters is the same for both RBF and SLFF networks (Stinchcombe *et al* 1993). The empirical process techniques used in Stinchcombe *et al* (1993) (and previously for a class of SLFF networks in Barron 1993) seem broadly applicable (see also Hornik *et al* 1993).

B2.10.4 Consistency

Let $\hat{\varphi}(N)$ be an estimator of the relationship between x and y based on the data set N. A consistency result for $\hat{\varphi}(N)$ is a statement of the form, 'as $N \uparrow \infty$, $\langle \ell_{\hat{\varphi}(N)}, \mu \rangle$ converges to its theoretical minimum'. The methods of Grenander (1981), Gallant (1987), White and Wooldridge (1991) allow denseness results to be turned into consistency results (White 1990, Gallant and White 1992, also Hart and Wehrly 1993).

For SLFF networks, the two consistency results in White (1990) concern the ℓ_φ^2 loss function and have very different flavors. The first gives conditions on the rates at which different aspects of SLFF architecture can go to infinity, the second concerning leave-one-out cross-validation (see below). By contrast, the article by Gallant and White (1992) concerns the $\ell_\varphi^{p,\mathrm{Sob}}$ loss functions, $p < \infty$, imposes a prior compactness condition on the set of possible relations between x and y, and requires only that the complexity of the network become infinite in the limit. In particular, this allows for the many variants of CV.

B2.10.5 Cross-validation

Cross-validation (CV) refers to the simple idea of splitting the data into two parts, using one part to find the estimated relation, and then judging the quality of the fit using the other part of the data. There are many variants of this simple idea.

Let $\mathcal{M} = \cup_J \mathcal{M}_J$ be the union of different classes of models of the relation between x and y. (The classical example has $\mathcal{M}_J$ as the class of linear models in which regressors $1, \ldots, J$ are included. In fitting either an SLFF or an RBF, $\mathcal{M}_J$ is the class of functions where J nonlinear terms are included in the summation. If the choice is to be between architectures that vary in more than the number of nonlinear terms to be added, the appropriate choice of $\mathcal{M}_J$ should be clear.) Let $\hat{\varphi}_J(S) \in \mathcal{M}_J$ denote the loss minimizing estimate of the relation between x and y based on the data in $S \subset N$, that is, $\hat{\varphi}_J(S)$ minimizes $\langle \ell_\varphi, \mu_S \rangle$ over $\varphi \in \mathcal{M}_J$.

Originally (Stone 1974), CV meant 'leave-one-out CV' or 'delete-one CV,' picking that $\hat{\varphi}_J$ that minimizes the average $\mathrm{Ave}\langle \ell_{\hat{\varphi}_J(N\setminus\{i\})}^2, \mu_i \rangle$ where the average is taken over all $i \in N$ and μ_i is a point mass on the ith data point. Intuitively, this works because 'overfitting' the data leads on $N\setminus\{i\}$ to larger errors in predicting y_i from x_i. The variants in the statistics literature (Zhang 1993) include delete-d CV (the obvious variant of classical delete-one CV), r-fold CV, picking $\hat{\varphi}_J$ to minimize $\mathrm{Ave}\langle \ell_{\hat{\varphi}_J(N\setminus N_r)}^2, \mu_{N_r} \rangle$ where the average is taken over a random division of the data into r equally sized parts, and repeated learning-testing, a bootstrap method which consists of picking $\hat{\varphi}_J$ to minimize $\mathrm{Ave}\langle \ell_{\hat{\varphi}_J(N_1)}^2, \mu_{N_2} \rangle$ where the average is taken over random independent selections of size d subsets N_2 of N and $N_1 = N\setminus N_2$. Note that this list includes sample-splitting CV, which is just twofold CV, splitting the data in half, fitting on one half, and picking the model from the predicted loss estimated with the second half.

Delete-d CV requires fitting the model N choose d times, and is, computationally, the most expensive of the procedures. The least expensive is r-fold CV with $r = 2$. Generally, in the classical case (described above), the computationally more intensive procedures have a better chance of picking the correct model (Zhang 1993, 1992). Even though there is a tendency to overfit in the classical case, provided $\mathcal{M}$ is dense, the CV procedure will deliver a consistent estimate of the functional relationship between x and y. That is, as $N \uparrow \infty$, the loss approximates its theoretical minimum (Hart and Wehrly 1993). Thus, when data (training examples) are cheap relative to the computational problems of picking the $\hat{\varphi}$, 2-fold CV recommends itself.

B2.10.6 Related procedures

Complexity regularization and noncovergent methods either are or can be understood as variants of cross-validation.

B2.10.6.1 Complexity regularization

Complexity regularization picks that model $\hat{\varphi} \in \mathcal{M}$ that minimizes $\langle \ell_\varphi, \mu_N \rangle + \lambda P(\varphi)$ where $P(\varphi)$ is a penalty term for the complexity of φ, and λ is a scalar. This is an idea that goes back (at least) to ridge regression (Hoerl and Kennard 1970). For example, $P(\varphi)$ could be the minimal J such that $\varphi \in \mathcal{M}_J$ when $\mathcal{M}_J \subset \mathcal{M}_{J+1}$. Intuitively, the tendency to overfit by picking too complex a φ is countered by the penalty.

Akaike's information criterion (AIC; Akaike 1973) works for the independent additive noise model. It has ℓ_φ being the sample log likelihood, $\lambda = 1$, and $P(\varphi)$ being the number of parameters used in specifying φ (in the case that the additive noise is i.i.d. Gaussian, this is the same as the ℓ_φ^2 loss function). Stone (1977) showed that delete-one CV is equivalent to maximizing the sample log likelihood plus $e_J \geq 0$. He also showed that if one of the classes of models, say $\mathcal{M}_{J^*}$, is exactly correctly specified, then e_{J^*} is equal to the number of parameters used in specifying $\mathcal{M}_{J^*}$. There is a tendency to overinterpret this result; e_J may not be equal to the number of parameters for $J \neq J^*$, and there is no guarantee that the two criteria make the same choice. The Kullback–Leibler (1951) information criterion can provide a (slight) generalization of the AIC.

The general difficulty in applying complexity regularization procedures is correctly choosing λ. This can be done by CV (though it seems rather indirect)—simply let $\varphi(\lambda)$ be the choice as a function of λ based on the subset N_1 of the data, and pick λ to minimize $\langle \ell_{\varphi(\lambda)}, \mu_{N_2} \rangle$ (see Lukas 1993 for the asymptotic optimality of this procedure).

B2.10.6.2 Nonconvergent methods

The nonconvergent methods of model selection (Finnoff *et al* 1993) is a form of twofold CV. One starts with a model that is tremendously overparametrized (e.g. the number of nonlinear terms in an ANN might be set at $N/2$). By gradient descent (or its backpropagation variant), the parameters in the model are moved in a direction chosen to improve $\langle \ell_\varphi, \mu_{N_1} \rangle$, continuing until $\langle \ell_\varphi, \mu_{N_2} \rangle$ begins to increase. This is a model selection procedure in two separate senses. First, if the starting point of the parameters is zero, then gradient descent will not have pushed very many of the parameters away from zero by the time the N_2 fit has begun to deteriorate. Parameters close to zero identify nonlinear units that can be ignored and so an $\mathcal{M}_J$ has been chosen. The second point arises from a shift away from the statistical viewpoint of nested sets of models. The aim is a model (or estimate) of the relation between x and y. The fact that our model has 'too many' parameters is not, in principle, an objection if the model itself has not been overfit.

References

Akaike H 1973 Information theory and an extension of the maximum likelihood principle *Second Int. Symp. on Information Theory* ed B N Petrov and F Csaki (Budapest: Akademiai Kiado) pp 267–81

Barron A 1993 Universal approximation bounds for superpositions of a sigmoidal function *IEEE Trans. Info. Theory* **39** 930–45

Billingsley P 1968 *Convergence of Probability Measures* (New York: Wiley)

Cybenko G 1989 Approximation by superpositions of a sigmoidal function *Math. Control Signals Syst.* **2** 303–14

Finnoff W, Hergert F and Zimmermann H G 1993 Improving model selection by nonconvergent methods *Neural Networks* **6** 771–83

Funahashi K 1989 On the approximate realization of continuous mappings by neural networks *Neural Networks* **2** 183–92

Gallant R 1987 Identification and Consistency in Seminonparametric Regression ed T F Bewley *Fifth World Conf. on Advances in Econometrics* vol 1 (New York: Cambridge University Press) pp 145–170

Gallant R and White H 1992 On learning the derivatives of an unknown mapping with neural networks *Neural Networks* **5** 129–138

Grenander U 1981 *Abstract Inference* (New York: Wiley)

Hart J D and Wehrly T E 1993 Consistency of cross-validation when the data are curves *Stochastic Processes and their Applications* **45** 351–61

Hoerl A and Kennard R 1970 Ridge regression: biased estimation for non-orthogonal problems *Technometrics* **12** 55

Hornik K 1991 Approximation capabilities of multilayer feedforward networks *Neural Networks* **4** 251–7

——1993 Some new results on neural network approximation *Neural Networks* **6** 1069–72

Hornik K, Stinchcombe M and White H 1989 Multilayer feedforward networks are universal approximators *Neural Networks* **2** 359–66 (Reprinted in White H (ed) 1992 *Artificial Neural Networks: Approximation & Learning Theory* (Oxford: Blackwell) and in Rao Vemuri V (ed) *Artificial Neural Networks: Concepts and Control Applications* (IEEE Computer Society))

——1990 Universal approximation of an unknown mapping and its derivatives using multilayer feedforward networks *Neural Networks* **3** 551–560 (Reprinted in White H (ed) 1992 *Artificial Neural Networks: Approximation & Learning Theory* (Oxford: Blackwell))

Hornik K, Stinchcombe M, White H and Auer P 1994 Degree of approximation results for feedforward networks approximating unknown mappings and their derivatives *Neural Comput.* **6** 1262–75

Kullback L and Leibler R A 1951 On information and sufficiency *Ann. Math. Stat.* **22** 79–86

Lukas M A 1993 Asymptotic optimality of generalized cross-validation for choosing the regularization parameter *Numerische Mathematik* **66** 41–66

Park J and Sandberg I W 1991 Universal approximation using radial basis-function networks *Neural Comput.* **3** 246–57

——1993a Approximation and radial-basis function networks *Neural Comput.* **5** 305–16

——1993b Nonlinear approximations using elliptic basis function networks *Circuits, Syst. Signal Processing* **13** 99–113

Stinchcombe M and White H 1989 Universal approximation using feedforward networks with non-sigmoid hidden layer activation functions *Proc. Int. Joint Conf. on Neural Networks (Washington, DC)* vol I (San Diego: SOS Printing) pp 613–7 (Reprinted in White H (ed) 1992 *Artificial Neural Networks: Approximation & Learning Theory* (Oxford: Blackwell))

——1990 Approximating and learning unknown mappings using multilayer feedforward networks with bounded weights *Proc. Int. Joint Conf. on Neural Networks (Washington, DC)* vol III (San Diego: SOS Printing) pp 7–16 (Reprinted in White H (ed) 1992 *Artificial Neural Networks: Approximation & Learning Theory* (Oxford: Blackwell))

Stinchcombe M, White H and Yukich J 1995 Sup-norm approximation bounds for networks through probabilistic methods *IEEE Trans. Info. Theory* **41** 1021–7

Stone M 1974 Cross-validitory choice and assessment of statistical predictions *J. R. Stat. Soc.* B **35** 111–33

——1977 An asypmtotic equivalence of choice of model by cross validation and Akaike's criterion *J. R. Stat. Soc.* B **39** 44–47

White H 1990 Connectionist nonparametric regression: multilayer feedforward networks can learn arbitrary mappings *Neural Networks* **3** 535–50

White H and Wooldridge J 1991 Some results for sieve estimation with dependent obserations *Nonparametric and Semiparametric Methods in Econometrics and Statistics* ed W Barnett, J Powell and G Tauchen (New York: Cambridge University Press)

Zhang P 1992 On the distributional properties of model selection criteria *J. Am. Stat. Assoc.* **87** 732–7

——1993 Model selection via multifold cross validation *Ann. Stat.* **21** 299–313

B3

Neural Network Training

James L Noyes

Abstract

The characteristics of neural network models are discussed, including a four-parameter generic activation function and an associated generic output function. Both supervised and unsupervised learning rules are described, including the Hebbian rule (in various forms), the perceptron rule, the delta and generalized delta rules, competitive rules, and the Klopf drive reinforcement rule. Methods of accelerating neural network training are described within the context of a multilayer feedforward network model, including some implementation details. These methods are primarily based upon an unconstrained optimization framework which utilizes gradient, conjugate gradient, and quasi-Newton methods (to determine the improvement directions), combined with adaptive steplength computation (to determine the learning rates). Bounded weight and bias methods are also discussed. The importance of properly selecting and preprocessing neural network training data is addressed. Some techniques for measuring and improving network generalization are presented, including cross validation, training set selection, adding noise to the training data, and the pruning of weights.

Contents

B3.1 Introduction

James L Noyes

Abstract

See the abstract for Chapter B3.

Neural networks do not learn by being programmed; they learn by being trained. Sometimes the words *training* and *learning* are used interchangeably within the context of neural networks, but here a distinction will be made between them. Learning, in a neural network, is the adjustment of the network in response to external stimuli; this adjustment can be permanent. In biological neural networks, both memory and the formation of thoughts involve neuronal synaptic changes. An artificial neural network models the synaptic states of its artificial neurons by means of numerical weights. A successful neural network learning process causes these weights to change and eventually to stabilize.

Learning may be supervised or unsupervised. *Supervised learning* is a process in which the external network input data and the corresponding target data for network output are provided and the network adjusts itself in some fashion so that a given input will produce the desired target. This can be done by determining the network output for a given input, comparing this output with the corresponding target, computing any error (difference) between the output and target, and using this error to provide the *external feedback*, based upon external target data, that is necessary to adjust the network. In *unsupervised learning*, the network adjusts itself by using the inputs only. It has no target data, and hence cannot determine errors upon which to base external feedback for learning. An unsupervised network can, however, group similar sets of input patterns into clusters predicated upon a predetermined set of criteria relating the components of the data. Based upon one or more of these criteria, the network discovers any existing regularities, patterns, classifications or separating properties. The network adjusts itself so that similar inputs produce the same representative output.

Training, in a neural network, refers to the presentation of the inputs, and possibly targets, to the network. This is done during the *training phase*. Training, and hence learning, is just the means to an end. This end is effective recall, generalization, or some combination of the two during the *application phase*, when the network is used to solve a problem. *Recall* is based upon the decoding and output of information that has previously been encoded and learned. *Generalization* is the ability of the network to produce reasonable outputs associated with new inputs. This is usually an important property for a neural network to possess. Recall and generalization take place during the use of a neural network for a particular application. In general, these are quite fast, whereas learning is commonly much slower because the network weights must typically be readjusted many times during the learning process. These weight adjustments, which are based upon the particular learning rule employed, are the main characteristics of training. Once a neural network has been trained and tested, it is used in an application mode until it no longer performs to the satisfaction of the user. When this point is reached, the training data set may be modified by adding or removing data, and the training and testing process repeated (Rumelhart and McClelland 1986, Noyes 1992, Fausett 1994).

References

Fausett L 1994 *Fundamentals of Neural Networks* (Englewood Cliffs, NJ: Prentice-Hall)

Noyes J L 1992 *Artificial Intelligence with Common Lisp: Fundamentals of Symbolic and Numeric Processing* (Lexington, MA: D C Heath)

Rumelhart D E and McClelland J L 1986 *Parallel Distributed Processing* vol 1 (Cambridge, MA: MIT)

B3.2 Characteristics of neural network models

James L Noyes

Abstract

See the abstract for Chapter B3.

Before discussing the concepts of neural network training, a brief discussion outlining the characteristics of general neural network models is necessary.

B3.2.1 Biological and applications-oriented modeling

A neural network model may be developed to simulate various features of the human or animal brain (for example, to study the effectiveness of different neural connection schemes, or how the absence of myelin affects response times, or how the loss of a collection of neurons degrades memory). This type of modeling can be characterized as *biologically oriented* (McClelland and Rumelhart 1986, Klopf 1988, Hertz *et al* 1991, Kandel 1991).

On the other hand, a neural network model may be developed to help solve a problem that has nothing in common with biology or neurophysiology. The network model is designed or chosen with a specific application in mind, such as the identification of handwritten letters, face recognition, function approximation, robotic control, or prediction of credit risk. This type of model can be characterized as *application oriented*. The majority of neural network models are of this type. In this type of model one need not concern oneself with developing constructs that have any biological counterpart at all. If the network performs well on a certain class of problem, then it is deemed adequate.

B3.2.2 The neuron

The purpose of the neuron is to receive information from other neurons, perform some relatively simple processing on this combined information and send the results on to other neurons. For neural network models it is convenient to classify these neurons into one of three types: (i) An *input neuron* is one that has only *one* input, no weight adjustment, and the input is from an external source (i.e. the input values used for training or in applications). (ii) An *output neuron* is one whose output is used externally as a network result. For example, the values from all of the output neurons are used during a supervised training session. (iii) A *hidden neuron* receives its inputs *only* from other neurons and sends its output *only* to other neurons. Neural network topologies are discussed in detail in Chapter B2 of this handbook. B2

The following general notational conventions will be followed in the remainder of this chapter. A scalar variable will be written with one or more italicized lower-case letters, such as *net*, w, or w_{ij}. A vector is written as a lower-case letter in italicized boldface. For example, an input vector is written as $\boldsymbol{x}$ and an output vector is written as $\boldsymbol{y}$. All vectors are assumed to be column vectors. A matrix is written as an upper-case letter in bold sans serif. For example, a weight matrix could be denoted by **W**. A transpose of a vector or matrix is indicated with a small upper-case T as a superscript, such as $\boldsymbol{x}^{\mathrm{T}}$ (a row vector) and $\mathbf{W}^{\mathrm{T}}$. Since there are typically many of these scalars, vectors, and matrices needed to describe neural network processing, subscripts will be used frequently.

B3.2.3 Neuron signal propagation

For a given neuron to fire, the incoming signals from other neurons must be combined in some fashion. One early solution was to use a simple weighted sum as a *firing rule*. When this weighted sum reaches a given *threshold* value θ, the neuron will fire. For neuron i this is written as:

$$\sum_{j=1}^{m} w_{ij} x_{ij} \geq \theta .$$

This approach was adopted by Warren McCulloch and Walter Pitts in one of the first neural network models ever devised (McCulloch and Pitts 1943). Here a signal of 1 was output when its weighted sum reached or exceeded the threshold and a 0 was output when it did not. Even though these signals were limited to binary values, they were able to demonstrate that *any* arbitrary logical function could be constructed by an appropriate combination of such 'logical threshold elements'. The *learning* issue was not actually addressed.

In general, a *propagation rule* describes how the signal information coming into a hidden or output neuron is combined to achieve a net input into that neuron. The *weighted-sum rule* is the most common way to do this and for neuron i is given by:

$$net_i = w_{i0} + \boldsymbol{w}_i^{\mathrm{T}} \boldsymbol{x}_i = w_{i0} + \sum_{j=1}^{m} w_{ij} x_{ij} . \tag{B3.2.1}$$

Here w_{i0} is an optional *bias* value for this neuron, $\boldsymbol{x}_i$ is the vector of input values (signals) from other neurons, and $\boldsymbol{w}_i$ is the vector of the associated connection weights. Sometimes the bias is incorporated into the vector $\boldsymbol{w}_i$, in which case the vector $\boldsymbol{x}_i$ is given an extra first-component value of unity. It should be noted that the above m-term *inner product* is very computationally intensive. In general, the number of inputs to a neuron will depend on the connection topology, so it is sometimes more accurate to say that m_i inputs are used, instead of just m.

One could use this bias to implement the above threshold value θ and cause the neuron to output a value if the above inner-product value meets or exceeds this threshold. This type of firing scheme could be incorporated into the weighted-sum rule by setting $w_{i0} = -\theta$ and then producing an output only when $net_i \geq 0$. This is equivalent to the previous firing rule.

B3.2.4 Neuron inputs and outputs

The output of input neurons is usually identical to their input (i.e. $y_i = x_i$). For hidden and output neurons, the inputs into one neuron come from the output of the other neurons, so it is sufficient to discuss output signals only. The neuron outputs can be of different types. The simplest type of output is *binary* output, where y_i takes the value 0 or 1. A similar type of output with slightly different properties, is *bipolar* output, where each y_i takes on the value -1 or $+1$. While the binary output is simpler and more natural to use, it is frequently more computationally advantageous to use bipolar output. Alternatively, the output may be continuous: this is sometimes called an *analog* output. Here y_i takes on real-number values, often within some predefined range. This range depends upon the choice of the activation function and its parameters (described below).

An *activation rule* describes how the neuron simulates the firing process that sends the signal onward. This rule is normally described by a mathematical function called an *activation function* which has certain desired properties. Here is a useful *generic sigmoid activation function* associated with a hidden or output neuron:

$$f(z) = a/(1 + \mathrm{e}^{-bz+c}) + d . \tag{B3.2.2}$$

This function has one variable (z) and four controlling parameters ($a, b, c,$ and d) which typically remain constant during the network training process. This activation function performs the mapping $f : \mathbb{R} \rightarrow (d, a+d)$, is monotonically increasing, and has the shape of the s-curve for learning. This type of curve is often called a *sigmoid curve*. The parameter b has the most significant effect on the slope of this curve: a small value of b corresponds to a gradual curve increase, while a large value corresponds to a steep increase. The case $b = \infty$ corresponds to a hard-limiting step function. (One can define the *steepness* by the product ab.) The parameter c causes a shifting along the horizontal axis (and is usually zero). The parameters a and d define the range limits for scaling purposes. Here are some specific examples:

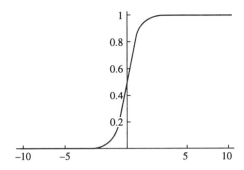

Figure B3.2.1. Logistic function with $b = 2$.

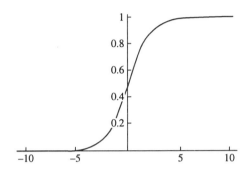

Figure B3.2.2. Simple logistic function.

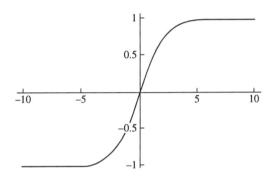

Figure B3.2.3. Bipolar function with $b = 1$.

$a = 1, b > 0, c = 0, d = 0$	gives the *logistic function* $1/(1 + e^{-bz})$ with a range of $(0, 1)$ as shown in figure B3.2.1.
$a = 1, b = 1, c = 0, d = 0$	gives the *simple logistic function* with a range of $(0, 1)$ as shown in figure B3.2.2.
$a = 2, b > 0, c = 0, d = -1$	gives the *bipolar function* $2/(1 + e^{-bz}) - 1$ with a range of $(-1, 1)$ as shown in figure B3.2.3.
$a = 2, b = 2, c = 0, d = -1$	gives the *simple hyperbolic tangent function* $\tanh(z)$ with a range of $(-1, 1)$ as shown in figure B3.2.4.

All four of these functions are frequently used in neural network learning models. Once the activation function has been selected, the output of neuron i is typically given by

$$y_i = f(net_i). \tag{B3.2.3}$$

Notice that the generic sigmoid activation function is also *differentiable*, which is a requirement for many of the training methods to be discussed later in this chapter. In particular, its derivative is given by

$$f'(z) = ab\, e^{-bz+c}/(1 + e^{-bz+c})^2 = (b/a)[f(z) - d][(a + d) - f(z)] \tag{B3.2.4}$$

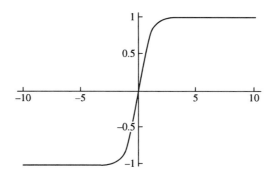

Figure B3.2.4. Simple hyperbolic tangent.

which performs the mapping $f' : \mathbb{R} \rightarrow (0, ab/4)$, where the derivative maximum of $ab/4$ occurs when $z = c/b$.

Many other activation functions may be used in neural network models. A common discontinuous function is the *step function*. However, because it is discontinuous, it cannot be used for training methods that require differentiability.

In addition to the activation function, it is sometimes useful to define an *output function* that is applied to the activation function for each *output neuron* in order to modify its result (it is *not* normally used to modify the result computed by input neurons or hidden neurons). One common modification is to convert continuous output into discrete output (e.g. real output into binary or bipolar output). One can define a *generic output function*, which is compatible with the generic sigmoid activation function previously described, when one sets $d = y_L$ and $a = y_U - y_L$, where y_L and y_U are given problem-dependent lower and upper limits:

$$F(z) = \begin{cases} y_L & \text{if } z \leq y_L + ae \\ z & \text{if } y_L + ae < z < y_U - ae \\ y_U & \text{if } z \geq y_U - ae . \end{cases} \quad \text{(B3.2.5)}$$

This function performs the mapping: $F : (d, a+d) \rightarrow [d, a+d]$. The parameter e is a measure of closeness and must lie within the interval $[0, 1/2)$. This function is not differentiable and hence is typically used only in conjunction with the display of the results produced by the output neurons and in a supervised training algorithm that has a termination condition that stops the iteration when all of the y_i values produced by the output neurons are within e of the corresponding target values t_i. When continuous target values are being matched, a sum of squared errors is frequently used in a termination condition, stopping when the sum of all of the $[t_{iL} - y_{iL}]^2$ values are small enough, where L is the output layer. When something like binary or bipolar target values are to be matched, one can compute an *auxiliary* sum of squares by using $[t_{iL} - F(y_{iL})]^2$ as an *additional* termination condition, stopping when this sum is exactly zero—which can often happen before the regular sum of squares is small and thereby save additional training iterations. This can also help prevent overtraining.

For example, suppose one requires a bipolar range with $y_L = -1$ and $y_U = 1$. One then sets $d = y_L = -1$ and $a = y_U - y_L = 2$. One choice is to set $e = 0.4$. This leads to what is sometimes called the *40–20–40 rule* (Fahlman 1988). The generic sigmoid activation and output functions become:

$$f(z) = 2/(1 + e^{-bz}) - 1$$

for $c = 0$ and

$$F(z) = \begin{cases} -1 & \text{if } z \leq -0.2 \text{ (lower 40\% of the range)} \\ z & \text{if } -0.2 < z < 0.2 \text{ (middle 20\% of the range)} \\ 1 & \text{if } z \geq 0.2 \text{ (upper 40\% of the range).} \end{cases}$$

The smaller the value of e, the more stringent the matching requirement. Another choice is $e = 0.1$, which yields a more stringent *10–80–10 rule*.

B3.2.5 Neuron connections

The way in which neurons communicate information is determined by the types of connections that are allowed. For the purposes of this chapter, some basic definitions will be given. For further information

the reader should consult Chapter B2 of this handbook, which provides a detailed discussion of *neural* B2
network topology.

A *feedforward network* is one for which the signal only flows in a forward direction from the input B2.3
neurons through possible intermediate (hidden) neurons to the output neurons during their use, without
any connections back to previous neurons. On the other hand, a *recurrent network* contains one or more B2.3
cycles and hence allows a neuron to have a closed-loop signal path back to itself either directly or through
other neurons.

Neural networks only work properly if they have a suitable connection structure for the given
application. One common structure groups the neurons into layers. Neurons within these layers usually
have the same characteristics and are typically not connected at all or else are fully interlayer connected.
Multiple layers are common and are called *multilayer networks*. The input neurons are all in the first layer, C1.2
known as the *input layer*, the output neurons are all in the last layer, known as the *output layer*, and any
hidden neurons are contained in *hidden layers* between the input and output layers. The input layer is
unique in that *no* weights affect the input into it so it is not considered to be a *computational layer* that
has weights to compute.

A *single-layer network* is a neural network that has only one computational layer (i.e. it really has C1.1
two layers, an input layer that is not computational and an output layer that is). A *multilayer feedforward
network* (MLFF) is one in which the neuron outputs of one layer feed into the neuron inputs of the
subsequent layer.

References

Fahlman S E 1988 An empirical study of learning speed in back-propagation networks *Carnegie Mellon Computer Science Report* CMU-CS-88-162

Hertz J, Krogh A and Palmer R G 1991 *Introduction to the Theory of Neural Computation* Santa Fe Institute Lecture Notes vol 1 (Redwood City, CA: Addison-Wesley)

Kandel E R (ed) 1991 *Principles of Neural Science* 3rd edn (New York: Elsevier)

Klopf A H 1988 A neuronal model of classical conditioning *Psychobiology* **16** 85–125

McClelland J L and Rumelhart D E 1986 *Parallel Distributed Processing* vol 2 (Cambridge, MA: MIT Press)

McCulloch W S and Pitts W 1943 A logical calculus of the ideas immanent in nervous activity *Bull. Math. Biophys.* **5** 115–33

B3.3 Learning rules

James L Noyes

Abstract

See the abstract for Chapter B3.

This section describes some of the more important learning rules that have been used in neural network training. It is not intended to present the complete training algorithms themselves (one training rule could be incorporated in many algorithmic variations; specific algorithmic implementations are discussed in Part C. Each of these rules describes a learning process that modifies a specified neural network to incorporate new information. There are two standard ways to do this: (i) The *on-line training* approach, sometimes called *case* or *exemplar* updating, updates the appropriate weights after each single input (and target) vector. (ii) The *off-line training* approach, sometimes called *batch* or *epoch* updating, updates the appropriate weights after each complete pass through the entire sequence of training data.

As indicated above, the term 'learning' applied to neural networks usually refers to learning the *weights*, and that is what is discussed in this section. This definition excludes other information about the network that might be learned, such as the way in which the neurons are connected, the activation function and parameters that it uses, the propagation rule, and even the learning rules themselves.

B3.3.1 Hebbian rule

Donald O Hebb, a psychologist at McGill University, developed the first commonly used learning rule for neural networks in his classic book *Organization of Behavior* (Hebb 1949). His rule was a very general one which was based upon synaptic changes. It stated that when an axon of neuron A repeatedly stimulates neuron B while neuron B is firing, a metabolic change takes place such that the weight w between A and B is increased in magnitude. The simplest versions of Hebbian learning are unsupervised. Denoting these neurons by n_j and n_i, if neuron n_i receives positive input x_j while producing a positive output y_i, this rule states that for some learning rate $\eta > 0$:

$$w_{ij} := w_{ij} + \Delta w_{ij} \tag{B3.3.1}$$

where the increase in the weight connecting n_j and n_i can be given by

$$\Delta w_{ij} := \eta y_i x_j \tag{B3.3.2}$$

where on-line training is normally used. Of all the learning rules, Hebbian learning is probably the best known. It established the foundation upon which many other learning rules are based.

Hebb proposed a *principle*, not an algorithm, so there are some additional details that must be provided in order to make this computable. (i) It is implicitly assumed that all weights w_{ij} have been initialized (e.g. to some small random values) prior to the start of the learning process. (ii) The parameter η must be specified precisely (it is typically given as a constant, but it could be a variable). (iii) There must be some type of normalization associated with this increase or else w_{ij} can become infinite. (iv) Positive inputs tend to excite the neuron while negative inputs tend to inhibit the neuron.

Example: Suppose one wishes to train a single neuron, n_1, which has $m = 4$ inputs from other neurons and has a bipolar activation function of $f(z) = \text{sgn}(z)$. Layer notation will be used. Assume a fixed learning rate is used with $\eta = 1/4$, an initial random weight vector of $w = (0.1, -0.4, -0.1, 0.3)^{\text{T}}$

is given with a bias value of $w_{10} = 0.5$, and that $k = 2$ training input vectors are to be used; these are given as: $x_1 = (0, 1, 0, -1)^T$, $x_2 = (1, 0, 0, 1)^T$. The computation is performed as follows, starting with x_1:

$$net_1 = 0.5 + (0.1)(0) + (-0.4)(1) + (-0.1)(0) + (0.3)(-1) = -0.2$$
$$y_1 = f(net_1) = \text{sgn}(-0.2) = -1$$
$$\Delta w_{11} = \tfrac{1}{4}(-1)(0) = 0 \qquad \Delta w_{12} = \tfrac{1}{4}(-1)(1) = -\tfrac{1}{4}$$
$$\Delta w_{13} = \tfrac{1}{4}(-1)(0) = 0 \qquad \Delta w_{14} = \tfrac{1}{4}(-1)(-1) = \tfrac{1}{4}.$$

The updated weight vector becomes $w = (0.1, -0.65, -0.1, 0.55)^T$. Continuing this computation for x_2:

$$net_1 = 0.5 + (0.1)(1) + (-0.65)(0) + (-0.1)(0) + (0.55)(1) = 1.15$$
$$y_1 = f(net_1) = \text{sgn}(1.15) = 1$$
$$\Delta w_{11} = \tfrac{1}{4}(1)(1) = \tfrac{1}{4} \qquad \Delta w_{12} = \tfrac{1}{4}(1)(0) = 0$$
$$\Delta w_{13} = \tfrac{1}{4}(1)(0) = 0 \qquad \Delta w_{14} = \tfrac{1}{4}(1)(1) = \tfrac{1}{4}.$$

The updated weight vector now becomes $w = (0.35, -0.65, -0.1, 0.8)^T$.

In the example above, the Hebbian rule was used in an unsupervised fashion. Notice that the appropriate weight was also increased when the input and output were both 'off' (negative) at the same time. That is a common *modification* to what the Hebbian rule originally stated and it leads to a stronger form of learning sometimes called the *extended Hebbian rule*.

Suppose now that the Hebbian rule is used in another way, namely in a supervised learning situation. In this situation the weight improvement is given by:

$$\Delta w_{ij} := \eta t_i x_j \qquad (B3.3.3)$$

where t_i is a given target value. In this form it is sometimes called the *correlation rule* (Zurada 1992).

Example. Suppose one wishes to train a single neuron, n_1, which has $m = 4$ inputs and an identity activation (and output) function of $f(z) = z$. Assume a fixed learning rate is used with $\eta = 1$, an initial weight vector of $w = 0$ is given with a bias value of $w_0 = 0$ and that $k = 4$ *orthogonal unit vectors* and corresponding targets are to be used for training. These training pairs are given as: $x_1 = (1, 0, 0, 0)^T$, $t_1 = 0.73$; $x_2 = (0, 1, 0, 0)^T$, $t_2 = -0.32$; $x_3 = (0, 0, 1, 0)^T$, $t_3 = 1.24$; $x_4 = (0, 0, 0, 1)^T$, $t_4 = -0.09$. Now consider how well the weights can be determined with just *one pass* through the training set. The training computation can now be simplified to:

$$w_{ij} := w_{ij} + t_i x_j .$$

The *training* phase proceeds as follows:

$$w_{11} = 0 + (0.73)(1) = 0.73 \qquad w_{12} = 0 + (-0.32)(1) = -0.32$$
$$w_{13} = 0 + (1.24)(1) = 1.24 \qquad w_{14} = 0 + (-0.09)(1) = -0.09 .$$

Using equation (B3.2.1), the propagation rule is given by

$$f(net_1) = net_1 = 0.73 x_{11} - 0.32_{12} + 1.24 x_{13} - 0.09 x_{14} .$$

Hence, by inspection, it may be seen that the training input vectors produce their target values exactly with just one pass through the training set. This network has been trained as an associative memory.

The previous example worked well because of the particular selection of input vectors. The suitability of this rule depends upon the orthogonality (correlation) of the input training vectors. When the input vectors are not orthogonal, the output will include a portion of each of their target values. However, if the training input vectors are linearly independent, then they can be orthogonalized by the Gram–Schmidt process (Anderson and Hinton 1981). Unfortunately, the Gram–Schmidt process can be unstable, so other techniques such as *Householder transformations* may be used (Tucker 1993). The advantage is that the $m \times m$ weight matrix $\mathbf{W}$ may be readily determined to satisfy

$$\mathbf{W}x_1' = y_1, \mathbf{W}x_2' = y_2, \ldots, \mathbf{W}x_m' = y_m \text{ or simply } \mathbf{W}\mathbf{X}' = \mathbf{Y}$$

where x_i' are the orthogonalized input training vectors and the $\mathbf{X}'$ and $\mathbf{Y}$ matrices are constructed from these respective column vectors. Since $\mathbf{X}'$ is orthogonal, its inverse is equal to its transpose so that the weight matrix is simply computed by:

$$\mathbf{W} = \mathbf{Y}(\mathbf{X}')^{\mathrm{T}}. \tag{B3.3.4}$$

There have been several variations of the Hebbian learning rule that offer certain improvements (Hertz *et al* 1991). One simple variation has already been illustrated, that of extended Hebbian learning. A second simple variation is to normalize the weights that are found by a factor of $1/N$ where N is the number of neurons in the system. Another more substantial variation, called by some *neo-Hebbian learning*, utilizes a component that incorporates forgetting, together with learning (Kosko 1992). Still another variation, called *differential Hebbian learning*, computes the weight increase based upon the product of the rates of change (i.e. the derivatives with respect to time) of the input and output signals instead of the x_i and y_i values themselves (Wasserman 1989, Kosko 1992). Only when both of these signals increase or decrease at the same time is their product positive, causing a weight increase.

B3.3.2 Perceptron rule

The psychologist Frank Rosenblatt invented a device known as the perceptron during the late 1950s (Rosenblatt 1962, McCorduck 1979). The *perceptron* used layers of neurons with a binary step activation function. Most perceptrons were trained, but some were self-organizing. Rosenblatt's original perceptron device was designed to simulate the retina. His idea was to be able to classify patterns appearing on the retina (the input layer) into categories. A common type of perceptron model is a neural network using linear threshold neurons with m neurons in the input layer and one neuron in the output layer. The outputs could be binary or bipolar. This is a supervised scheme that updates the weights by using equation (B3.2.1) where the weight change for the learning rate $\eta > 0$ is given by

$$\Delta w_{ij} := \eta(t_i - y_i)x_j. \tag{B3.3.5}$$

Here $y_i = f(net_i)$ where $f(z)$ is now defined by the discontinuous *threshold activation function*

$$f(z) = \begin{cases} 1 & \text{for } z \geq \theta \\ 0 & \text{for } z < \theta \end{cases}$$

where θ is a given threshold. This type of neuron is called a *linear threshold neuron*. As stated in section B3.2.1, this can be accomplished by setting $w_{i0} = -\theta$ in the weighted-sum rule that determines net_i.

Here, as in the Hebbian rule, $\eta > 0$, but now the *error* is multiplied instead of just the output alone. Because of the incorporation of the target value, it is easy to see that this is a supervised learning method. It is also more powerful than the Hebbian rule. Notice that whenever the output of neuron i is equal to the desired target value, the weight change is zero. As with Hebbian learning, on-line training is normally used.

There is a theorem called the *perceptron convergence theorem* (Rosenblatt 1962) which states the following: if a set of weights exists that allow the perceptron to respond correctly to all of the training patterns, then the rule's learning method will find a set of weights to do this and it will do it in a finite number of iterations.

Perceptrons became very successful at solving certain types of pattern recognition problem. This led to exaggerated claims about their applicability to a broad range of problems. Marvin Minsky and Seymour Papert spent some time studying these types of model and their limitations. They authored a text in 1969 (reprinted with additional notes in Minsky and Papert 1988) which presented a detailed analysis of the capabilities and limitations of perceptrons. The best-known example of a very simple limitation was the impossibility of modeling an XOR gate. This is called the XOR problem (exclusive OR). To solve this problem a model has to learn two weights so that the following XOR table can be reproduced:

x_1	x_2	t_1
0	0	0
0	1	1
1	0	1
1	1	0.

These four input points can easily be plotted on the x_1–x_2 axis as the corners of a unit square. Dropping the neuron i index for simplicity, the output is then defined by:

$$f(net) = \begin{cases} 1 & \text{for } w_1 x_1 + w_2 x_2 \geq \theta \\ 0 & \text{for } w_1 x_1 + w_2 x_2 < \theta. \end{cases}$$

Hence, to match the target values, the following four inequalities would have to be satisfied:

$$w_1(0) + w_2(0) < \theta \text{ or } 0 < \theta$$
$$w_1(0) + w_2(1) \geq \theta \text{ or } w_2 \geq \theta$$
$$w_1(1) + w_2(0) \geq \theta \text{ or } w_1 \geq \theta$$
$$w_1(1) + w_2(1) < \theta \text{ or } w_1 + w_2 < \theta.$$

This is a contradiction, because it is impossible for each individual weight to be greater than or equal to θ while their sum is less than θ.

This was a two-dimensional example of a general *inability* of a single-layer network to map functions (solve problems) that are *not* linearly separable. A *linearly separable function* is a function for which there exists a *hyperplane* of the form

$$\boldsymbol{w}^{\mathrm{T}} \boldsymbol{x} = \sum_{j=1}^{m} w_j x_j = \theta$$

for which all points on one side of this hyperplane have one function value and all points on the other side of this plane have a different function value. For example, if $m = 2$ the AND gate function and OR gate function are linearly separable on the plane since a straight line can be shown to separate their points with the same function values, but this is not the case with the XOR gate function. However, as will be seen later, a multilayer network can solve such a problem.

B3.3.3 Delta rule

Bernard Widrow and Marcian E (Ted) Hoff developed an important learning rule to solve problems in adaptive signal processing. It may be considered to be more general than the perceptron rule because their rule could handle continuous as well as discrete inputs and outputs for problems. This rule, which they called the least-mean-square (LMS) rule, could be used to solve a variety of problems without using hidden neurons (Widrow and Hoff 1960). Because it uses the 'delta' correction difference, it is often called the delta rule.

The delta rule is a supervised scheme that updates the weights by using equation (B3.3.1) where the weight change is given for a fixed learning rate $\eta > 0$ by

$$\Delta w_{ij} := \eta(t_i - net_i)x_j \tag{B3.3.6}$$

with *no* activation function needed. (An alternative view of this is to use the delta as $(t_i - y_i)$, as was the case in the perceptron rule, where the activation function is the simple linear identity function $f(z) = z$.)

The LMS name derives from the idea of training until the weights have been adjusted so that the total least-mean-square error of a single neuron in the output layer, namely

$$E(\boldsymbol{w}) = \tfrac{1}{2} \sum_{j=1}^{k} (t_{1j} - net_{1j})^2 = \tfrac{1}{2} \sum_{j=1}^{k} (t_j - net_j)^2 \tag{B3.3.7}$$

is minimized, summing over all $j = 1, 2, \ldots, k$ training cases (where the index 1 is dropped since there is only one output). It is important to remember that E is a function of all the weight and bias variables, since the input and target data are all known.

Using equation (B3.2.1) for this single output neuron, equation (B3.3.7) becomes

$$E(\boldsymbol{w}) = \tfrac{1}{2} \sum_{j=1}^{k} (t_j - w_0 - w_1 x_{1j} - \cdots - w_m x_{mj})^2.$$

 Handbook of Neural Computation release 97/1

The delta rule may be viewed as an adaptive way of solving the least-squares minimization problem where the parameters $w_0, w_1, \ldots, w_m$ of a multiple linear regression function are to be determined. This method has been used successfully in conjunction with both on-line and off-line training.

Widrow and Hoff called the single output model an adaptive linear element or *adaline*. They showed   C1.1.3
that the training algorithm for this network would converge for any function that the network is capable of representing. This single neuron in the output layer was later extended to a multiple-neuron model called *madaline* (many adalines).   C1.1.4

B3.3.4 Generalized delta rule

This rule (sometimes also just called the delta rule) was proposed by several researchers including Werbos, Parker, Le Cun, and Rumelhart (Rumelhart and McClelland 1986). It is also related to an early method presented by Bryson for solving optimal control problems (Dreyfus 1990). David Rumelhart and the PDP Research Group helped popularize this learning rule in conjunction with a complete training method known as *backpropagation*. This training method is one of the most important techniques in neural network   C1.2
training. As will be shown later, this is a gradient descent method which moves a positive distance along the negative gradient in 'weight space'. The associated learning rule requires that the activation function $f(z)$ be semilinear. A *semilinear* activation function is one in which the output of a neuron is a *nondecreasing* and *differentiable* function of the *net* total input. Note that the generic sigmoid activation function given by equation (B3.2.2) is semilinear.

The generalized delta rule again uses equation (B3.3.1). Here the weight changes for the *output layer* are given for a fixed learning rate $\eta > 0$ by

$$\Delta w_{ij} := \eta[t_i - f(net_i)]f'(net_i)x_j .\tag{B3.3.8}$$

Note that the term in braces is the same as $(t_i - y_i)$, which was used in the perceptron rule (see equation (B3.3.5)) so the weight changes will be small when these values are close together. However, now the weight changes will also be small whenever the derivative of the activation function is close to zero (i.e. the function is nearly flat at the net_i point). Examination of the derivative of the generic sigmoid activation function shows that $f'(net_i)$ is always positive and it approaches zero as net_i becomes large. This helps ensure the stability of the weight changes so that they do not oscillate. Backpropagation has been shown to be very effective for a variety of problems, and the added hidden layers can overcome the separability problem. However, there are three difficulties with this method. If some of the weights become too large during the training cycle, the corresponding derivatives will approach zero and the weight improvements also approach zero (even though the output is not close to the target). This can cause what is sometimes called *network paralysis* (Wasserman 1989). It can lead to a termination of the training even though a solution has not yet been found. A second difficulty is that, like all gradient methods, it may stop at a local minimum instead of a global one. A third difficulty, also common with unmodified gradient methods, is that of slow convergence (i.e. a lengthy learning process). Using a smaller learning rate η may help some of these situations, or it may just increase the training time. This indicates the value of a *variable* learning rate, as will be seen later.

The weight changes for the *hidden layers* are more involved since this derivative is multiplied by the inner product of a weight vector and an error vector. For each prior layer l, summing over j, it has the form:

$$\Delta w_{l,ij} := \left\{ \sum_j [w_{l+1,ij} \cdot \Delta w_{l+1,j}] \right\} f'(net_{l,i})x_{lj} .\tag{B3.3.9}$$

The basic idea behind both of these weight correction formulas is to determine a way to make the appropriate correction to a weight in proportion to the error that it causes. The importance of this method is that it makes it possible to make these weight corrections in all of the computational layers. The details of the backprojection method are described more fully by Rumelhart and McClelland (1986).

B3.3.5 Kohonen rule

This rule is typically used in an *unsupervised* learning network to bring about what is called *competitive learning*. A competitive learning network is a neural network in which each group (cluster) of neurons competes for the right to become active. This is accomplished by specifying an additional *criterion* for

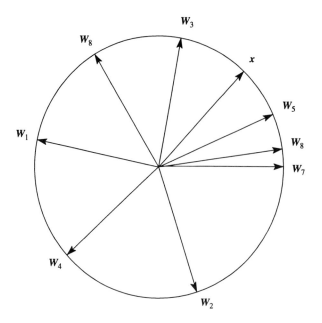

Figure B3.3.1. Two-dimensional unit vectors in the unit circle.

the network so that it is forced to make a choice as to which neurons will respond. The simplest network of this kind consists of a single layer of computational neurons, each fully connected to the inputs. A common type of layer may be viewed as a two-dimensional self-organizing topographic feature map. Here the location of the most strongly excited neurons is correlated with certain input signals. Neighboring excited neurons correspond to inputs with similar features. Teuvo Kohonen is the person most often associated with the *self-organizing network*, which is one in which the network updates the connection weights based only upon the characteristics of the input patterns presented. Kohonen devised a learning rule that can be used in various types of competitive learning situation to cause the neural network to organize itself by selecting representative neurons.

C2.1.1

The most extreme competitive learning strategy is the *winner take all* criterion where the activation of the neuron with the largest *net* input is the one to have its weights updated.

This type of competitive learning assumes that the weights in the network are typically initialized to random values. Their weight vectors and input vectors are normalized by using their corresponding Euclidean norms. If the current normalized m-dimensional input vector is x, and there are q neurons in the group, then one computes

$$w_p^\mathrm{T} x = \max\{w_1^\mathrm{T} x, w_2^\mathrm{T} x, \cdots, w_q^\mathrm{T} x\}. \tag{B3.3.10}$$

This represents a collection of q m-dimensional weight vectors and one input vector all emanating from the origin of a unit hypersphere (in two dimensions this is a circle). See figure B3.3.1, where $q = 8$ and $p = 5$. This means that neuron p is the winning neuron in this group if its weight vector w_p makes a smaller angle with x than the weight vector associated with any other neuron.

The weight improvement is given for a *decreasing* learning rate $\alpha > 0$ by

$$w_{pj} := w_{pj} + \alpha\,\Delta w_{pj} \tag{B3.3.11}$$

where the weight changes associated with neuron p are given as:

$$\Delta w_{pj} := x_j - w_{pj}. \tag{B3.3.12}$$

For the winner take all criterion, this corresponds to modifying the corresponding w_p vector (only) by a fraction of the difference between the current input vector and the current weight vector. (Notice that no activation function is needed in order to do this.) After this improvement, the weights associated with neuron p tend to better estimate this input. Unfortunately, neurons which have weight vectors that are far from any input vector may never win and hence never learn; these are like 'dead neurons'. Solutions to this difficulty and other variations of this learning rule are given by Hertz *et al* (1991).

Other less extreme variations of this strategy allow the *neighboring neurons* to have their weights updated also. Here a 'geometry' is chosen that can be used to define these neighbors. For example, suppose the group of neurons is considered to be arranged in a two-dimensional array. A *linear neighborhood* would be all neurons within a certain distance away in either the same row or the same column (e.g. if the distance were 2, then two neurons on each side would also have their weights updated). A *hexagonal neighborhood* is one in which the neighbors are within a certain distance in all directions in this plane (e.g. two hexagons away from a neuron in a plane would correspond to 17 neighbors that would also have their weights updated). Other choices are possible (Caudill and Butler 1992). Kohonen also proposed a modification of his rule called the 'Mexican hat' variation, which is described by Hertz *et al* (1991). In this variation, a *neighborhood function* is defined and used as a multiplier.

This type of learning can be used for determining the statistical properties of the network inputs (it generates a model of the distribution of the input vectors around the unit hypersphere). Competitive learning, in general, is well suited as a regularity detector in pattern recognition.

B3.3.6 Outstar rule

Steven Grossberg coined the terms *instar* and *outstar* to characterize the way in which actual neurons behave. Here instar refers to a neuron that receives (dendrite) inputs from many other neurons in the network. Outstar refers to a neuron that sends (axon) outputs to many other neurons in the network, and again the connecting synapses modify this output.

Instar training, which is unsupervised, is accomplished by adjusting the connecting weights to match   C1.1.6
the input vector. This can be achieved by using the Kohonen rule defined in the last section. The instar neuron fires whenever a specific *input* vector is used. On the other hand, the outstar produces a desired pattern to be sent to other neurons when it fires, and hence it is a supervised training method. One way to accomplish *outstar training* is to adjust its weights to be like the desired *target* vector. The weight   C1.1.6
improvement here is given for a *decreasing* learning rate $\beta > 0$ by

$$w_{ji} := w_{ji} + \beta \, \Delta w_{ji} \qquad (B3.3.13)$$

where the weight changes associated with the neurons $j = 1, 2, \ldots$ to which neuron i sends output are given as

$$\Delta w_{ji} := t_j - w_{ji} . \qquad (B3.3.14)$$

Here the outstar weights are iteratively trained, based upon the distribution of the target vectors (Wasserman 1989). Outstar training is distinctive in that the neuron weight adjustments are not applied to the neuron's own input weights, but rather applied to the weights of receiving neurons. *Counterpropagation networks*,   C2.3.2
such as those proposed by Hecht-Nielsen (1990), can utilize a combination of Kohonen learning and Grossberg outstar learning.

B3.3.7 Drive reinforcement rule

Drive reinforcement learning was developed by Harry Klopf of the Air Force Wright Laboratories. This name arises from the fact that the signal levels, called the *drives*, are used together with the changes in signal levels, which are considered as *reinforcements*. This approach is a discrete variation of differential Hebbian learning and does well at modeling several different types of *classical conditioning* phenomenon. Classical conditioning involves the following components: an unconditional stimulus, an unconditional response, a conditioned stimulus, and a conditioned response. One important feature of this type of model is the *time* between stimulus and response.

Klopf suggested the following changes to the original Hebbian model (Klopf 1988):

(i) Instead of correlating presynaptic levels of activity with postsynaptic activity levels, *changes* in these levels are correlated. Specifically, only positive changes in the first derivatives of these input levels are correlated with changes in output levels.

(ii) A time interval is incorporated into the learning model by correlating *earlier* changes in presynaptic levels with *later* changes in postsynaptic levels.

(iii) The change in synapse efficacy should be proportional to its current efficacy in order to account for experimental s-shaped learning curves.

This model predicts a learning acquisition curve that has a positive initial acceleration and a subsequent negative acceleration (like the s-curve) and which is not terminated by conditioned inhibition. First one defines a new net_i as

$$net_i(t) := \sum_{j=1}^{n} w_{ij}(t)x_{ij}(t) - \theta \qquad (B3.3.15)$$

where n is the number of synaptic weights.

The output, or drive, for neuron i may then be defined as

$$y_i(t) = \begin{cases} 0 & \text{for } net_i(t) \leq 0 \\ net_i(t) & \text{for } 0 < net_i(t) < \lambda \\ \lambda & \text{for } net_i(t) \geq \lambda. \end{cases} \qquad (B3.3.16)$$

Here each $y_i(t)$ is nonnegative and bounded. (Negative values have no meaning because they would correspond to negative firing frequencies.) A common range is from 0 to $\lambda = 1$. The time value t is computed by adding a discrete time step for each iteration. The weight update has the form:

$$w_{ij}(t+1) := w_{ij}(t) + \Delta w_{ij}(t). \qquad (B3.3.17)$$

Here the weight change is given by

$$\Delta w_{ij}(t) := \Delta y_i(t) \left\{ \sum_{k=1}^{\tau} \eta_k |w_{ij}(t-k)| \, \Delta x_{ij}(t-k) \right\} \qquad (B3.3.18)$$

where the sum is from $k = 1$ to $k = \tau$ (the upper time interval limit) and absolute weight values are used.

The change in the input presynaptic signal at time $t - k$ is given by

$$\Delta x_{ij}(t-k) := x_{ij}(t-k) - x_{ij}(t-k-1). \qquad (B3.3.19)$$

If $\Delta x_{ij}(t-k) < 0$, then it is reset to zero before computing the above weight change.

The change in the output postsynaptic signal, the reinforcement, at time t is

$$\Delta y_i(t) := y_i(t) - y_i(t-1). \qquad (B3.3.20)$$

For this learning rule there are τ constants $\eta_1 > \eta_2 > \ldots > \eta_\tau \geq 0$. These are ordered to indicate that the most recent stimuli have the most influence. For example, if $\Delta t = 1/2$ second, then one might choose $\tau = 6$ so that $t - 1, t - 2, \ldots, t - 6$ would correspond to half-second time intervals back 3 seconds from the present time, and η_6 could be zero. For example, $\eta_1 = 5$, $\eta_2 = 3$, $\eta_3 = 1.5$, $\eta_4 = 0.75$, $\eta_5 = 0.25$, $\eta_6 = 0$ can be used to model an exponential recency effect (Kosko 1992).

A lower bound is set on the absolute values of the weights, which means that positive (excitatory) weights remain positive and negative (inhibitory) weights remain negative (e.g. $|w_{ij}(t)| \geq 0.1$). These weights are typically initialized to small positive and negative values such as $+0.1$ and -0.1. Finally, the change in $\Delta y_i(t)$ is usually restricted to positive changes only. Learning does not occur if this signal is decreasing in strength.

This type of learning allows the corresponding neural network to perceive *causal* relationships based upon temporal events. That is, by keeping track of past events, these may be associated with present events. None of the other learning rules presented in this chapter can do this. The drive reinforcement method has also been used to develop adaptive control systems. As an example, this method has been used to solve the pole balancing problem with a self-supervised control model (Morgan *et al* 1990). In this problem the object is to balance a pole that is standing up on a movable cart by moving it back and forth. This learning rule can also be used to help train hierarchical control systems (Klopf *et al* 1993).

B3.3.8 Comparison of learning rules

The following is a general summary of the main features of these rules and how they compare with one another.

The *Hebbian rule* is the earliest and simplest of the learning rules. Learning occurs by modifying the connecting weight between each pair of neurons that are 'on' (fire) at the same time, and weights are usually updated after each example (on-line training). The concept of how to connect a collection of such neurons into a *network* was not explicitly defined. The Hebbian rule can be used in either an unsupervised or a supervised training mode. It is still a common learning rule for a neural network designed to act as an associative memory. It can be used with training patterns that are either binary or bipolar. The original Hebbian rule only referred to neurons firing at the same time and did not address neurons that do not fire at the same time (see the discussion on asynchronous updating in section B3.4.3). A stronger form of learning arises if the weights are increased when both neurons are 'off' at the same time as well as 'on' at the same time.

The *perceptron rule* is a more powerful learning rule than the Hebbian rule. Here a layered network of neurons is defined explicitly. Single-computational-layer perceptrons are the simplest types of network. The perceptron rule is normally used in a supervised training mode. The convergence theorem states that if a set of weights exist that will permit the network to associate correctly all input–target training patterns, then its training algorithm will learn a set of weights that will perform this association in a finite number of training cycles. Weights are updated after each example is presented (on-line training). The original perceptron with a binary-valued output served as a classifier. It essentially forms two decision regions separated by a hyperplane.

The *delta rule* is also known as the Widrow–Hoff or least-mean-square (LMS) learning rule. It is also a supervised rule which may be viewed as an extension of the single-computational-layer perceptron rule since this rule can handle both discrete and continuous (analog) inputs. The 'delta' in this rule is the difference between the target and the net input with the weight improvement proportional to this difference. The weights are typically adjusted after each example is presented (on-line training), so the method is adaptive in nature just as the two previous learning methods. The LMS name refers to the fact that the sum of squares of these deltas is minimized. It can be used when the data are not linearly separable. A commonly employed special case of this network is the adaline that only uses one (bipolar or binary) output unit.

The *generalized delta rule* can be viewed as an extension of the delta rule (or the perceptron rule). Specifically, it extends the previous delta rule in two important ways that significantly increase the power of the learning process. First, it generalizes the delta difference of the previous rule by replacing the net input by a function of the net input and then multiplying this difference by the function's rate of change (derivative). This activation function, providing a neuron's output, is required to be both nondecreasing and differentiable. Typically this is some type of s-shaped sigmoid function. In the previous learning rules, the neuron outputs were typically quite simple (such as step functions and identity functions) and not always differentiable. Second, by requiring differentiability of the activation function, it permits learning methods (e.g. backpropagation) to be developed that can train weights in *multiple*-layer networks. This supervised learning rule can be used with discrete or continuous inputs and can update the weights through either on-line or off-line training. Off-line training is equivalent to a gradient descent method. With only *three* layers (one hidden layer) and continuous data, these networks can form any decision region and can learn any continuous mapping to an arbitrary accuracy (Kolmogorov 1957, Sprecher 1965, Hecht-Nielsen 1987).

The *Kohonen rule* also utilizes a network of layered neurons, but the layer can be of a different type than the layers associated with the previous three learning rules. In those rules the neurons were in one-dimensional layers (i.e. each is considered as a column or row of neurons). The Kohonen rule uses either a one- or two-dimensional layer of neurons, the latter being somewhat more common. The neurons in a layer can form cluster units. This is a self-organizing unsupervised network in which the neurons *compete* with one another to become active. Different competition criteria have been used. For example, during the training process, the neuron whose weight vector most closely matches the input training pattern becomes the winner. Only this neuron and its neighbors update their weights. A more extreme *winner take all* criterion only allows the winning neuron to update its weights. This type of network can be used to determine the statistical properties of the network inputs.

The *outstar rule* utilizes the ability of a neuron to send its output to many other neurons. It is a supervised training method that directly adjusts its weights to be just like a given target vector. It is distinctive from the other learning rules in that the weight adjustments are applied to the weights of the receiving neurons, not its own input weights.

The *drive reinforcement rule* allows a neural network to identify causal relationships and solve certain adaptive control problems. Klopf modified the original Hebbian rule to incorporate changes in neuron input levels, time intervals, and current weight values in order to determine how weights should be modified.

Overall, it is seen that the Hebbian rule, perceptron rule, delta rule, and sometimes the generalized delta rule are typically employed when one has an on-line training situation. The generalized delta rule and the others can be used in the off-line mode. The generalized delta rule is very flexible and can also be used as a general function approximator. The Hebbian rule and Kohonen rule may be considered as operating in an unsupervised mode, while the others are typically supervised (the Hebbian rule has a supervised form also). The drive reinforcement rule is the only one of these that incorporates rates of change over time and is designed to deal with cause and effect learning.

References

Anderson J A and Hinton G E 1981 Models of information processing in the brain *Parallel Models of Associative Memory* ed G E Hinton and J A Anderson (Hillsdale, NJ: Lawrence Erlbaum Associates) pp 9–48

Caudill M and Butler C 1992 *Naturally Intelligent Systems* (Cambridge, MA: MIT Press)

Dreyfus S E 1990 Artificial neural networks, backpropagation, and the Kelley-Bryson gradient procedure *J. Guidance, Control Dynamics* **13** 926–8

Hebb D O 1949 *The Organization of Behavior* (New York: Wiley)

Hecht-Nielsen R 1987 Kolmogorov's mapping neural network existence theorem *IEEE Int. Conf. on Neural Networks* vol III (New York: IEEE Press) pp 11–4

——1990 *Neurocomputing* (Reading, MA: Addison-Wesley)

Hertz J, Krogh A and Palmer R G 1991 *Introduction to the Theory of Neural Computation* Santa Fe Institute Lecture Notes vol 1 (Redwood City, CA: Addison-Wesley)

Klopf A H 1988 A neuronal model of classical conditioning *Psychobiology* **16** 85–125

Klopf A H, Morgan J S and Weaver S E 1993 A hierarchical network of control systems that learn: modeling nervous system function during classical and instrumental conditions *Adaptive Behavior* **1** 263–319

Kolmogorov A N 1957 On the representation of continuous functions of many variables by superposition of continuous functions of one variable and addition *Dokl. Akad. Nauk USSR* **114** 953–6

Kosko B 1992 *Neural Networks and Fuzzy Systems: a Dynamical Systems Approach to Machine Intelligence* (Englewood Cliffs, NJ: Prentice Hall)

McCorduck P 1979 *Machines Who Think* (San Francisco, CA: Freeman)

Minsky M and Papert S 1988 *Perceptrons: an Introduction to Computational Geometry* expanded edition reprinted from the 1969 edition (Cambridge, MA: MIT Press)

Morgan J S, Patterson E C and Klopf A H 1990 Drive-reinforcement learning: a self-supervised model for adaptive control *Network* **1** 439–48

Rosenblatt F 1962 *Principles of Neurodynamics* (Washington, DC: Spartan Books)

Rumelhart D E and McClelland J L 1986 *Parallel Distributed Processing* vol 1 (Cambridge, MA: MIT)

Sprecher D 1965 On the structure of continuous functions of several variables *Trans. Am. Math. Soc.* **115** 340–55

Tucker A 1993 *Linear Algebra: an Introduction to the Theory and Use of Vectors and Matrices* (New York: Macmillan)

Wasserman P D 1989 *Neural Computing: Theory and Practice* (New York: Van Nostrand Reinhold)

Widrow B and Hoff M E 1960 Adaptive switching circuits *Wescon Convention Record* part 4 (New York: Institute of Radio Engineers) pp 96–104

Zurada J M 1992 *Introduction to Artificial Neural Systems* (St Paul, MN: West Publishing)

B3.4 Acceleration of training

James L Noyes

Abstract

See the abstract for Chapter B3.

Early neural network training methods, such as *backpropagation*, often took quite a long time to train. The time that it takes to train a network has long been an issue when different types of applications have been considered. The length of training time depends upon the number of iterations (passes through the training data). The number of iterations required to train a network depends on several interrelated factors including data preconditioning, choice of *activation function*, the size and topology of the network, initialization of weights and biases, *learning rules* (weight updating schemes), the way in which the training data are presented (on-line or off-line), and the type and number of training data used.

C1.2

B3.2.4
B3.3

In this section, some of these factors will be addressed and suggestions will be made to accelerate network training in the context of multilayer feedforward networks.

B3.4.1 Data preprocessing

Of all the quantities that one can set or modify prior to a neural network training phase, the single modification that can have the greatest effect on the convergence (training time) is data preprocessing. The training data that a network uses can have a significant effect on the values computed during the learning process. Data preprocessing can help condition these computations so they are not as susceptible to roundoff error, overflow, and underflow. Preprocessing of the training data typically refers to some simple type of data transformation achieved by some combination of scaling, translation, and rotation. Sometimes a less sophisticated algorithm can work as well with preconditioned data as a more sophisticated algorithm can work with unconditioned data.

It has generally been found that problems with discrete $\{0, 1\}$ binary values should be transformed into equivalent problems with corresponding bipolar values (or their equivalent), unless one has a good reason to do otherwise. This is because training problems are often exacerbated by zero (0) input values. Not only do these values cause the corresponding net_i *not* to contain (add) any w_{ij} components because the corresponding $x_j = 0$, but the zero values also prevent the same w_{ij} values from being efficiently corrected because the term $x_j error_i = 0$ for that value (it behaves just as though $error_i = 0$).

The simple linear transformation $T(z) = 2z - 1$ will transform binary $\{0, 1\}$ values into bipolar $\{-1, 1\}$ values. To employ these bipolar training values requires that the generic sigmoid activation function (equation (B3.2.2)) use $a = 2$ and $d = -1$ as parameters. Another common mapping range, as an alternative to the bipolar range, is $\{-0.5, +0.5\}$ with $T(z) = z - 1/2$. As always, when the training data are transformed and the network is trained with these transformed data, the problem data must be transformed in the same manner. Simple symmetric scaling can sometimes make a significant difference in the training time.

If continuous (analog) data, rather than discrete data, are to be used for network training, then other scaling techniques can be used, such as normalizing each input data value by using the transformation $z_i = (x_i - \mu)/\sigma$, where μ is the mean and σ is the standard deviation of the underlying distribution. In practice, the sample mean and standard deviation are used. This is a statistically based data scaling technique and can be used to help compensate for networks that have variables with widely differing magnitudes (Bevington 1969). In general, all of the standard deterministic and statistically based scaling techniques are candidates for use in the preprocessing of neural network data.

B3.4.2 Initialization of weights

Initialization of the network weights (and biases) can also have a significant influence upon both the solution (the final trained weights) and the training time. it is important to avoid choices of these weights that would make either the activation function values or the corresponding derivatives near zero. The most common type of initialization is that of uniformly distributed 'random' numbers. Here a pseudorandom number (PRN) generator is used (Park and Miller 1988). Usually the initial weights are generated as small positive and negative weights distributed around zero in some manner. It is *not* generally a good idea to use large initial weights since this can lead to small error derivatives which produce small weight improvements and slow learning.

It is common to use a PRN generator to compute initial weights within the interval $[-\rho, \rho]$ where ρ is typically set to a constant value within some range, say $1/4 \leq \rho \leq 5$. In general, the choice of ρ depends upon the gain of the activation function (as specified by its parameters), the training data set, the learning method, and learning rate used during training (Thimm *et al* 1996).

For the standard backpropagation method using the simple logistic function, the most commonly used intervals are probably $[-1, 1]$ and $\left[-1/2, 1/2\right]$. For example, Fahlman (1988) conducted a detailed investigation of the learning speed for backpropagation and backprop-like algorithms (e.g. Quickprop). These were applied to a benchmark set of encoder and decoder problems of various sizes, mostly of size 8 or 10; for example, a 10–5–10 multilayer feedforward (MLFF) network was common. In this empirical study he found that even though PRNs in the interval $[-1, 1]$ worked well, there were good results for ρ as large as 4.

Success has also been achieved with other schemes whereby the hidden layer weights are initialized in a different manner than the output layer weights. For example, one might initialize the hidden layer weights with small PRNs distributed around zero and initialize the weights associated with the output layer with an equal distribution of $+1$ and -1 values (Smith 1993). Here the idea is to keep hidden layer outputs at a mid-range value and to try achieve output layer values that do not make the derivatives too small.

If one choice of initial weights does not lead to a solution, then another set is tried. Even if a solution is reached, it is sometimes a good strategy to generate two or three other sets of initial weights in order to see if the corresponding solution is the same or at least equally as good. Other useful weight initialization schemes have also been developed and studied, such as by Fausett (1994). Thimm and Fiesler (1994) present a detailed comparison of neural network initialization techniques. They conclude that all methods are equally or less effective compared with a simple initialization scheme with a fixed range of random numbers. The range $[-0.77, 0.77]$ is found to be most suitable for multilayer neural networks.

B3.4.3 Updating schemes

Synchronous updating of a neural network means that the activation function is applied simultaneously for all neurons. *Asynchronous updating* means that each neuron computes its activation function independently (e.g. randomly) which corresponds to independent neuron firings. The corresponding output is then propagated to other neurons before another neuron is selected to fire. This type of updating can add stability to a neural network by preventing oscillatory behavior sometimes associated with synchronous updating (Rumelhart and McClelland 1986).

B3.4.4 Adaptive learning rate methods

Adaptive learning rates have been shown to provide a substantial improvement in neural network training times. This can be especially important in real-time training problems. A significant class of adaptive learning rate methods is based upon solving the *unconstrained minimization problem* (UMP). In the following, this problem and the methods for its solution will be given, they will then be placed within the framework of neural network training.

B3.4.4.1 *The unconstrained minimization problem*

The general unconstrained minimization problem (UMP) consists of finding a real vector such that a given scalar objective function of that vector is maximized or minimized. In the following, the minimization problem will be addressed in the context of minimizing the errors associated with an MLFF network.

Handbook of Neural Computation   release 97/1   © 1997 IOP Publishing Ltd and Oxford University Press

However, it is possible to formulate other supervised neural network models as optimization problems also. The vector to be determined is the n-dimensional vector $w = (w_1, w_2, \ldots, w_n)^T$ of network weights and biases, which is typically called the weight vector. The UMP may then be formulated as

$$\text{minimize:} \quad E(w) \tag{B3.4.1}$$

where w is unconstrained (not restricted in its n-dimensional real domain). $E(w)$ is the neural network objective function and it is possible that many local minima exist.

There are many well-known methods for solving the general UMP. Most of these methods are extremely effective and have been perfected over the years for the solution of scientific and engineering problems. Once the neural network problem has been formulated as a UMP, all of the theory of unconstrained optimization, such as that relating to the existence of solutions, problem conditioning, and solution convergence rates, may be applied to neural network problems. In addition, all of the practical knowledge such as efficient optimization algorithms, scaling techniques, and standard UMP software may be applied to help facilitate neural network learning (Noyes 1991).

The optimization methods are broadly classified by the type of information that they use. These are:

(i)   Search methods. These use evaluations of the objective function $E(w)$ only and do not utilize any partial derivative information of the objective function with respect to the weights. These methods are usually very slow and are seldom used in practice unless no derivative information is available. Sometimes, however, n-dimensional search methods can be used to augment derivative methods.

(ii) First-derivative (gradient) methods. These use both objective function evaluations and evaluations of the first partial derivatives of $E(w)$. The *gradient* $\nabla E(w)$ is an n-dimensional real vector consisting of the first partial derivatives of $E(x)$ with respect to each weight w_i for $i = 1, 2, \ldots, n$. These gradient methods are the optimization methods that are typically used for neural network training. Most are relatively fast and require only a moderate amount of information. These methods include: (a) steepest descent, (b) conjugate gradient descent, and (c) quasi-Newton descent. These are called descent methods because they guarantee a decrease in $E(w)$ at each iteration (e.g. training epoch).

(iii) Second-derivative (Hessian) methods. These use function evaluations and both first- and second-partial-derivative evaluations. The *Hessian* $\nabla^2 E(w)$ is an $n \times n$ real matrix consisting of the second-partial derivatives of $E(w)$ with respect to both w_i and w_j for $i = 1, 2, \ldots, n$ and $j = 1, 2, \ldots, n$. These methods are used less often than the first-derivative methods, because they require more information and often more computation. These methods typically require the fewest number of iterations, especially when they are close to the solution. Even though these methods may often be the fastest, they are typically not that much faster than the modified gradient methods (i.e. conjugate gradient and quasi-Newton). Hence these modified gradient methods are usually the methods of choice.

In general, all of these classes of methods for solving the UMP find a *local minimum* point w^* such that $E(w^*) \leq E(w)$ for all weight vectors w in a neighborhood of w^*. (If w^* is a local minimum of $E(w)$ then the norm of $\nabla E(w^*)$ is zero and $\nabla^2 E(w^*)$ is positive semidefinite.) Only additional conditions on $E(w)$, such as convexity, will guarantee that this local minimum is also global. In practice, several 'widely scattered' initial weight vectors w^0 can be employed, each yielding a solution w^*. The w^* associated with the smallest $E(w^*)$ is then selected as the best choice for the global minimum weight vector.

B3.4.4.2 The neural network optimization framework

Suppose one chooses the multilayer feedforward (MLFF) network as the neural network model. The objective function is then typically a least-squares function so the neural network optimization model can be given by:

$$E(w) = \tfrac{1}{2} \sum_{p=1}^{P} \sum_{q=1}^{N_L} [t_{pq} - y_{pq}]^2 . \tag{B3.4.2}$$

Here P is the total number of presentations (input–target cases) in the training set given by $\{(x_p, t_p); p = 1, 2, \ldots, P\}$. N_L is the number of components in t_p, t_{pq} is the qth component of the pth target vector and y_{pq} is the corresponding computed output from the output layer that depends upon w. The multiplier of $1/2$ is simply used for normalization purposes.

Even a moderately sized neural network problem can lead to a large, high-dimensional optimization problem and hence the storage required by certain algorithms can be a major issue. This is easily seen since the number of weights and biases needed for an L-layer MLFF network of the form $N_1-N_2-N_3-\ldots-N_L$ is given by

$$n = (N_1 + 1)N_2 + (N_2 + 1)N_3 + \ldots + (N_{L-1} + 1)N_L \tag{B3.4.3}$$

where N_i is the number of units in the ith layer. Note that the added constant '1' indicates the inclusion of the bias term with the other weight terms.

Example: Consider the previously discussed XOR gate problem modeled as a 2–2–1 network with bipolar training data given by

x_1	x_2	t_1
-1	-1	-1
-1	$+1$	$+1$
$+1$	-1	$+1$
$+1$	$+1$	-1.

The corresponding activation function of $f(z) = 2/(1 + e^{-bz}) - 1$ could then be used with the parameter $b > 0$ controlling the slope of this s-curve. The number of weights and biases is $n = (2 + 1)2 + (2 + 1)1 = 9$. There are $P = 4$ input–target cases, with $N_L = 1$ component in the target vector (in this case it is a scalar). Fortunately, $E(w)$ seldom needs to be explicitly formulated in practice. Here it will be done in order to show the presence of the weights and biases which are to be chosen optimally so that $E(w)$ is minimized:

$$
\begin{aligned}
E(w) = {}& \tfrac{1}{2}\{[t_{11} - y_{11}]^2 + [t_{21} - y_{21}]^2 + [t_{31} - y_{31}]^2 + [t_{41} - y_{41}]^2\} \\
= {}& \tfrac{1}{2}\{[-1 - f(w_{74} + w_{75}f(w_{51} - w_{52} - w_{53}) + w_{76}f(w_{61} - w_{62} - w_{63}))]^2 \\
& + [1 - f(w_{74} + w_{75}f(w_{51} - w_{52} + w_{53}) + w_{76}f(w_{61} - w_{62} + w_{63}))]^2 \\
& + [1 - f(w_{74} + w_{75}f(w_{51} + w_{52} - w_{53}) + w_{76}f(w_{61} + w_{62} - w_{63}))]^2 \\
& + [-1 - f(w_{74} + w_{75}f(w_{51} + w_{52} + w_{53}) + w_{76}f(w_{61} + w_{62} + w_{63}))]^2\}.
\end{aligned}
$$

The nine-element vector w is defined by

$$w = (w_{51}, w_{52}, w_{53}, w_{61}, w_{62}, w_{63}, w_{74}, w_{75}, w_{76})^{\mathrm{T}}$$

where the first index is the index of the receiving neuron and the second index is that of the transmitting neuron in the previous layer.

Even without making the final substitution of $2/(1 + e^{-bz}) - 1$ for the activation function $f(z)$, one can see the complexity of this objective function $E(w)$. Fortunately, however, this problem together with many much larger problems can often be solved easily with the right optimization method.

In the above example, the elements w_{51}, w_{52}, w_{53}, respectively, represent the bias and the two weights associated with the first neuron in the second (hidden) layer. The elements w_{61}, w_{62}, w_{63}, respectively, represent the bias and the two weights associated with the second neuron in the hidden layer. The elements w_{74}, w_{75}, w_{76}, respectively, represent the bias and the two weights associated with the first (and only) neuron in the output layer.

Based upon the objective function, it is relatively easy to write the computer code for a function and procedure that will evaluate the function $E(w)$ and gradient $\nabla E(w)$ respectively. To evaluate $E(w)$ requires P forward passes through the network (no backward passes are needed). A training *epoch* consists of one pass through all of the input–target vectors in the training set. To evaluate the gradient $\nabla E(w)$ requires P forward and backward passes (just like the backpropagation method). With a little extra computation, $E(w)$ can also be computed in the gradient procedure.

The reason for making this last statement is that, by using the best-known optimization methods for solving the neural network training problem, not only is a weight improvement *direction* recomputed during each training epoch, but an *adaptive learning rate* can be computed as well (Gill *et al* 1981). None of the well known optimization methods would use a *fixed* learning rate, because it would be extremely inefficient to do so. The standard backpropagation method typically uses a 'small' fixed learning rate and this is why it is typically quite slow. The reason this is done is because a small enough learning rate is guaranteed to produce a decrease in the objective function as long as the gradient $\nabla E(w)$ is not zero. However, adaptive learning rates can be chosen to guarantee such a decrease also and they are usually much faster.

In addition, most optimization methods *modify* $-\nabla E(w)$, the negative gradient at the current point, in order to compute a new direction. This is because other information, such as gradients at nearby points, can frequently yield a better direction of decrease. Only one method, steepest descent, uses just the negative gradient for the direction to move at each iteration, but even this method does not use a fixed step. This method is typically slow also, but not nearly as slow as a fixed-step gradient algorithm (e.g. backpropagation). Within a neural network context, a judicious computation of both the direction and learning rate can guarantee a *sufficient decrease* in the objective function during each training epoch. Specifically, this means that the computed learning rate must be large enough to reduce the magnitude of the directional derivative by a prescribed amount and must also reduce the objective function by a given amount. On the other hand, the learning rate cannot be too large or a functional increase may result. The equations to test these conditions are standard and are given below. The variable ν is the counter for the training epochs—it is *not* an exponent. It is typically used as a subscript for scalars and as a superscript for vectors (so that the counter is not confused with the indices).

$$|\nabla E(w^\nu + \eta_\nu d^\nu)^{\mathrm{T}} d^\nu| \leq -\alpha \, \nabla E(w^\nu)^{\mathrm{T}} d^\nu \qquad \text{where} \quad 0 \leq \alpha < 1 \qquad \text{(B3.4.4)}$$

$$E(w^\nu) - E(w^\nu + \eta_\nu d^\nu) \geq -\beta \eta_\nu \, \nabla E(w^\nu)^{\mathrm{T}} d^\nu \qquad \text{where} \quad 0 < \beta \leq \tfrac{1}{2}. \qquad \text{(B3.4.5)}$$

The value of the constant α determines the accuracy with which the learning rate approximates a *stationary point* of $E(w)$ along a direction d^ν. If $\alpha = 0$, the learning rate procedure is normally associated with an 'exact line search'. If α is 'small', the procedure is usually associated with an 'accurate line search'. However, the objective function $E(w)$ must also be sufficiently reduced at the same time, using the constant value β as a multiplier. If $\beta \leq \alpha$, then there is at least one solution (at least one value for η_ν) that satisfies these two conditions (Gill *et al* 1981). This sufficient decrease at each iteration, in turn, *guarantees* convergence to a local minimum since the least-squares objective function is bounded below by zero. In addition, most of these methods usually have a superlinear convergence rate (Fletcher 1987). In neural network terminology, this means that the learning will be much faster than backpropagation, which has a linear rate.

B3.4.4.3 Adaptive learning rate algorithm

Before presenting a generic minimization algorithm, a simple adaptive learning rate algorithm will be given (Dennis and Schnabel 1983).

Given ϵ in $(0, 1/2)$, e.g. $\epsilon = 10^{-4}$, $0 < \beta < \alpha < 1$ as chosen constants along with w^ν and d^ν, the current weight and direction, start with a learning rate of $\eta_\nu = 1$:

> While $E(w^\nu + \eta_\nu d^\nu) > E(w^\nu) + \epsilon \eta_\nu \nabla E(w^\nu)^{\mathrm{T}} d^\nu$
> adjust $\eta_\nu := \lambda \eta_\nu$ for some λ in $[\beta, \alpha]$
> Then set $w^{\nu+1} := w^\nu + \eta_\nu d^\nu$.

In this implementation, if $\lambda < \beta$, a search failure is indicated and $w^{\nu+1}$ is automatically reset to a new *random* value which restarts the process. This modification makes the adaptive learning rate algorithm more robust.

B3.4.4.4 Neural network minimization algorithm

A generic neural network minimization algorithm that encompasses *all* of the classes of methods mentioned in this chapter is now presented. This represents a framework for neural network training. The geometrical interpretation of this algorithm is that for each current weight vector w^ν a direction d^ν is chosen which makes a strictly acute angle with the negative of the gradient vector $-\nabla E(w^\nu)$. The new weight vector $w^{\nu+1}$ is obtained by using a positive learning rate of size η_ν with a direction d^ν that will sufficiently decrease $E(w)$. The extreme case is to choose a value η_ν that *minimizes* $E(w)$ along this direction line (instead of just reducing $E(w)$), but this is a time-consuming process and is not usually implemented in practice. As with most algorithms of this nature, it is only guaranteed to approximate a stationary point (i.e. a point where the gradient is zero).

0. Set $\nu := 0$, select an initial weight vector w^0 and choose *numax*, the maximum number of iterations to use.

1. Solve the *direction subproblem* by finding a search direction d^ν from the current weight vector w^ν that guarantees a function decrease. This can be achieved if the gradient $\nabla E(w^\nu)$ is not zero. If the norm of the gradient $\|\nabla E(w^\nu)\|$ is suitably small, the algorithm terminates successfully.

2. Solve the *learning rate subproblem* by finding a positive learning rate η_ν so that a sufficient decrease is obtained. (In particular, this means that $E(w^\nu + \eta_\nu d^\nu)$ is sufficiently smaller than $E(w^\nu)$.) Set the improvement $p^\nu := \eta_\nu d^\nu$.

3. Update $w^{\nu+1} := w^\nu + p^\nu$ and $\nu := \nu + 1$. If $\nu > numax$, the algorithm terminates unsuccessfully, otherwise return to step 1.

Table B3.4.1. Weight and bias improvement vectors.

Simple gradient (SBP):	$p^\nu := \eta d^\nu = -\eta \nabla E(w^\nu)$
Modified gradient (MBP):	$p^\nu := \eta d^\nu = -\eta[\nabla E(w^\nu) + \gamma p^{\nu-1}]$
Steepest descent:	$p^\nu := \eta_\nu d^\nu = -\eta_\nu \nabla E(w^\nu)$
Conjugate gradient (CG):	$p^\nu := \eta_\nu d^\nu = -\eta_\nu[\nabla E(w^\nu) + \gamma_\nu p^{\nu-1}]$
Quasi-Newton (QN):	$p^\nu := \eta_\nu d^\nu = -\eta_\nu S(w^\nu) \nabla E(w^\nu)$
Newton:	$p^\nu := \eta_\nu d^\nu = -\eta_\nu \{\nabla^2 E(w^\nu)\}^{-1} \nabla E(w^\nu)$

In table B3.4.1, η is a fixed learning rate, while η_ν is an adaptive learning rate which depends upon the current training epoch, d^ν is the current direction vector, γ is a fixed scalar multiplier, γ_ν is a variable scalar multiplier involving two inner product calculations, $S(w^\nu)$ is an $n \times n$ matrix built up from the differences in successive gradients and improvement vectors, $\nabla E(w^\nu)$ is the current n-component gradient vector, and finally $\nabla^2 E(w^\nu)$ is the current $n \times n$ Hessian matrix. In practice, since both of these matrices are symmetric, only the upper-triangular part of $S(w^\nu)$ and $\nabla^2 E(w^\nu)$ are usually stored (requiring $n(n+1)/2$ locations instead of n^2 locations). For the Newton method, a linear system of equations is solved instead of finding a matrix inverse for $\nabla^2 E(w^\nu)$ and multiplying the inverse by $-\nabla E(w^\nu)$. That is, one solves the linear system $\nabla^2 E(w^\nu) d^\nu = -\nabla E(w^\nu)$ for the current direction d^ν.

The specific algorithm classes are usually based upon how the direction subproblem is solved. Table B3.4.1 shows the improvement vector p^ν for some of these classes. Notice that the first two of these methods are the standard backpropagation method (SBP) and the backpropagation method with a momentum term added (MBP). Notice also that these are the only methods that use *fixed* learning rates (steplengths). This helps explain why SBP and MBP often take a great many training epochs to converge, when they do.

B3.4.4.5 Algorithm efficiency

The following example demonstrates that the choice of learning rate can significantly affect convergence.

Example: This example uses the standard backpropagation method (SBP) to solve the XOR gate problem with the training set shown using layers containing 2–2–1 neurons and a logistic activation function with $b = 1$. The training data are as follows:

x_1	x_2	t_1
0	0	0
0	1	1
1	0	1
1	1	0.

Using the *same* randomly chosen starting point, one can use SBP with several *fixed* learning rates and count the number of training epochs (iterations) needed. Note the differences in training efficiency.

Learning rate (η)	Training epochs (ν)
0.9	932
1.7	494
3.0	280
5.0	160
10.0	121
$\eta > 10$	(convergence failure)

Convergence is also affected by the initial weight vector and the fact that these *same* fixed learning rates will produce a *different* number of training epochs when different initial weight vectors are used. The only efficient way to perform this minimization is to have the algorithm adjust the learning rate as it goes. That adjustment requires additional computation (more forward passes through the training set), but the overall training computations will normally be greatly reduced.

Of course, measuring efficiency by simple iteration (epoch) counts is not the whole story. The computation of the improvement p^ν can require many floating point operations. Even though the actual implementation of these 'formulas' is typically more efficient than that shown here, the adaptive learning rate methods usually require a lot more operations *per iteration* than SBP or MBP. However, they frequently require a lot fewer operations *per problem*, and this is the real measure of algorithm efficiency. The number of operations required for various optimization schemes is calculated and described by Moreira and Fiesler (1995).

B3.4.4.6 Quasi-Newton and conjugate gradient methods

In unconstrained optimization practice, quasi-Newton (QN) methods and conjugate gradient (CG) methods are the methods of choice, because of their superlinear convergence rates. Both of these methods are based upon minimizing a quadratic approximation to a given objective function. However, there are significant differences between these two methods. CG uses a simpler updating method that is easier to code and requires fewer floating point operations and much less memory (see table B3.4.1). The coefficient γ_ν is the quotient of two inner products, and there are three formulas that have been used in practice to compute this coefficient: Fletcher–Reeves, Polak–Ribiere, and Hestenes–Stiefel. (These formulas are fully described by Gill *et al* 1981.) The CG method requires $O(n)$ memory locations, while QN requires $O(n^2)$ memory locations; this is the most significant factor for neural network models because of their potentially large size of n. This can be seen by examining equation (B3.4.3) and is illustrated in table B3.4.2. However, the QN method is typically less sensitive to the accuracy in computing the learning rate in order to produce a sufficient decrease in the objective function and directional derivative. The earliest method of this type was called the DFP (Davidon–Fletcher–Powell) variable-metric method. Because the QN method is similar to the Newton method, a learning rate of unity is often satisfactory and eliminates the need for an adaptive learning rate determination. The contemporary method for computing the matrix $\mathbf{S}(w^\nu)$ is typically the BFGS (Broyden–Fletcher–Goldfarb–Shanno) method and has been found to work well in practice (Fletcher 1987). For these reasons, QN is usually faster than CG and is usually the preferred method for small-to-moderate-size optimization problems. Unfortunately, while some neural networks are small, others can be quite large, as shown by the MLFF examples in table B3.4.2. The value of n is obtained from equation (B3.4.3).

Table B3.4.2. Multilayer feedforward storage size examples.

N_1–N_2–N_3 Network	n	n^2	$n(n+1)/2$	$10n$
2–2–1	9	81	45	90
10–5–10	115	13 225	6 670	1 150
25–10–8	348	121 104	60 726	3 480
81–40–8	3608	13 017 664	6 510 636	36 080

B3.4.4.7 Low-storage methods

Because of these sizes, several practitioners have chosen the CG method over the QN method as a means of speeding up neural network learning (Barnard and Cole 1989, Johansson *et al* 1990). However, there is still another class of methods called *low-storage methods* which have the advantages of the QN speed, but require not much more memory than CG, taking O(n) memory locations. For example, one low-storage version of the quasi-Newton method requires approximately $10m$ additional memory locations (see table B3.4.2).

One such technique that has successfully been used for neural network training is Nocedal's low-storage L-BFGS method (Nocedal 1980). L-BFGS employs a low storage approximation to the standard BFGS direction transformation matrix, combined with an efficient adaptive learning rate determination. The matrix used approximates the inverse Hessian, so this method is of the quasi-Newton variety, but it is not explicitly stored. Instead, it uses a rotational vector storage algorithm where only the most recent gradient differences are stored (the oldest are overwritten by the newest). The learning rate $\eta_\nu = 1$ is always tried first. If this fails to produce a sufficient decrease, a safeguarded and efficient cubic/quadratic polynomial fitting algorithm is used to find an appropriate value of η_ν. L-BFGS has both reduced memory requirements and improved convergence speed (Liu and Nocedal 1989). It has been employed to solve a variety of MLFF neural network problems (Noyes 1991).

Low-storage optimization techniques belong to a relatively recent class of methods. Other methods of this class have been proposed by Griewank and Toint (1982), Buckley and Lenir (1983), and Fletcher (1990). Fletcher's method is described as using less storage than L-BFGS at the expense of more calculations.

B3.4.4.8 Other optimization methods

Many other optimization strategies could be tried. The best-known methods for solving the UMP are the *line search* methods which are the one-dimensional search methods used to solve the learning rate subproblem discussed earlier in this chapter. A newer class of methods is based upon *trust regions*, which could be used to restrict the size of the learning rate at any iteration, based upon the validity of the Taylor series approximation (Fletcher 1987). Another optimization strategy that can be used to limit the weight and bias values is that of *constrained optimization* where the weight values are constrained in some fashion (discussed in section B3.4.5).

There are other ways to compute adaptive learning rates for the solution of optimization problems. One such method, developed by Jacobs and Sutton, has been used in conjunction with accelerating the backpropagation method. It is called the *delta bar delta* method and was designed to compute a *different* learning rate for each weight in the network based upon a weighted average of the weight's current and past partial derivative values (Jacobs 1988, Smith 1993).

No matter what adaptive learning rate method is used, it is clear that adaptive learning rate methods have the potential of significantly accelerating the network learning process over that of a fixed learning rate for gradient-based methods. They tend to be very robust and free the user from the often difficult decision of what learning rate to use for a given application.

B3.4.5 Weight constraints

A general neural network training problem is frequently modeled through the use of an *unconstrained* objective function $E(w)$ that depends upon the training data as well as the n-vector (n-dimensional vector) w of weights and biases. Another type of optimization is called *constrained optimization* in which some or all of the variables are constrained in some way, often by algebraic equalities or inequalities. For the neural network problem, the simplest types of constraint are upper and lower bounds upon each of the weights and biases. These simple bounds could be enforced for each. More computation per iteration would typically be necessary, but convergence could be faster overall if reasonable bounds were known (because these values could not be overadjusted).

Any least-squares function to be minimized, such as that resulting from training an MLFF network, possesses the special property that its minimum objective function value is bounded below by zero. In the usual problem statement, the w vector is not constrained and hence not bounded at all. However, there are certain problems such as those with physical parameters (such as scientific models) in which it is useful to consider the employment of *simple bounds* of the form

$$w_\mathrm{L} \leq w^\nu \leq w_\mathrm{U}$$

where $w_\mathrm{L} = w_\mathrm{L} e$, $w_\mathrm{U} = w_\mathrm{U} e$ for given scalars w_L, w_U and the n-vector $e = (1, 1, 1, \ldots, 1)^\mathrm{T}$. Note that this is a special case in which the *same* simple bounds are used for all weights and biases.

There can be advantages in bounding these weights. As the network is trained, unconstrained weights can occasionally become very large, which can force the neuron units to produce excessive net_i values (especially if a fixed learning rate is used which is too large). Small derivatives with proportionally small propagation error corrections can result, and little improvement will be made to the weights and biases. This brings the training process to a standstill, which is called *network paralysis*. Bounding the weights and biases will prevent them from becoming too large. Such bounds can also limit the weights and biases from ever being overcorrected and producing floating point overflow during the iteration process. If any *a priori* information is known about realistic limits for a given problem, this information can be easily and naturally incorporated. Finally, because well-chosen bounds w_L and w_U can be employed to restrict the sequence w^ν from going too far in a given direction, convergence can be improved in some cases. Notice, however, that poorly chosen bounds can actually prevent the sequence w^ν from converging to an optimum point.

There are different ways of implementing such bound limits in an algorithm. Here is the simplest method that adjusts each component w_i *after* the vector $w^{\nu+1}$ has been computed. Sometimes this method is called 'clipping':

$$\text{if } w_i^{\nu+1} < w_\mathrm{L} \text{ then } w_i^{\nu+1} := w_\mathrm{L} \qquad \{\text{lower-limit check}\}$$
$$\text{else if } w_i^{\nu+1} > w_\mathrm{U} \text{ then } w_i^{\nu+1} := w_\mathrm{U} \qquad \{\text{upper-limit check}\}.$$

This has the advantage of being very easy to code, being relatively fast, and requiring no additional storage. Its disadvantage is that the adjusted $w^{\nu+1}$ point may not lie in the same direction as the improvement vector, and hence may slow down the convergence process.

With a small amount of additional work, the aforementioned disadvantage may be corrected by computing a modified learning rate which is the minimum of the previously computed adaptive learning rate and the learning rate which would place $w^{\nu+1}$ on the nearest constraint bound. Here both w^ν and $r^\nu = -\nabla E(w^\nu)$ are used, with their respective components denoted by w_i and r_i:

$$\text{if } r_i < 0 \text{ then } s_\nu := \min\{s_\nu, (w_\mathrm{L} - w_i)/r_i\} \qquad \{\text{lower-limit check}\}$$
$$\text{else if } r_i > 0 \text{ then } s_\nu := \min\{s_\nu, (w_\mathrm{U} - w_i)/r_i\} \qquad \{\text{upper-limit check}\}.$$

This may be derived from a more general set of standard linear constraint conditions (Gill *et al* 1981). This is done *before* the vector $w^{\nu+1}$ is computed. These conditions check each component r_i in the direction vector r^ν. The constraints to be checked are the potentially binding ones having normal vectors which make an acute angle with the direction vector (otherwise a decrease in $E(w)$ cannot be guaranteed). The most binding limit is the nearest bound, which corresponds to the minimum s_ν. No learning rate, fixed or adaptive, is allowed to exceed this limit.

B3.4.6 Implementation issues

This section briefly describes two important implementation issues that may be used to further enhance *all* neural network training methods. *Extended precision computation* can help ensure that gradient directions and improvements are computed accurately. Neural network models can be very ill conditioned in that a small perturbation in the modeling expressions or training data can produce a large perturbation in the final weights and biases. Consequently, it is usually important to code the necessary expressions so as to reduce roundoff error and the possibility of floating point overflow. One simple technique is to test the argument of any exponential or hyperbolic activation function in order to ensure that the function evaluation will not produce overflow. Another more general technique to employ whenever possible is to perform all floating point computations, or at least the critical ones such as inner products, weight updates, and function evaluations, in extended precision (e.g. double precision). While using a higher precision will always take more storage and a little more execution time per iteration, it usually results in fewer iterations per problem and can often make the difference between convergence and failure to solve a neural network problem.

Dynamic data structures can permit even larger problems to be modeled. Neural network models are natural candidates for such an approach because of their potentially large size and inherent dynamic character. Several high-level computer programming languages such as Ada, C, C++, Modula-2, and Pascal contain the capability of accessing additional primary memory known as dynamic memory. This allows the algorithm implementor to utilize *both* regular static memory *and* dynamic memory to solve much larger problems. Usually this is accomplished by using pointers and dynamic variables to create some type of linked structure in dynamic memory. Since several data structures such as linked scalars, linked vectors, and linked matrices are possible, it is important to choose a dynamic data structure suitable for the type of neural network model at hand (Freeman and Skapura 1991). Here 'suitable' means a structure that supports efficient floating point computation *and* makes efficient use of memory.

References

Barnard E and Cole R A 1989 A neural-net training program based on conjugate-gradient optimization *Technical Report CSE 89-014* July Oregon Graduate Center

Bevington P R 1969 *Data Reduction and Error Analysis for the Physical Sciences* (New York: McGraw-Hill)

Buckley A and Lenir A 1983 QN-like variable storage conjugate gradients *Mathematical Programming* **27** 155–75

Dennis J E Jr and Schnabel R B 1983 *Numerical Methods for Unconstrained Optimization and Non-linear Equations* (Englewood Cliffs, NJ: Prentice-Hall)

Fahlman S E 1988 An empirical study of learning speed in back-propagation networks *Carnegie Mellon Computer Science Report* CMU-CS-88-162

Fausett L 1994 *Fundamentals of Neural Networks* (Englewood Cliffs, NJ: Prentice-Hall)

Fletcher R 1987 *Practical Methods of Optimization* 2nd edn (New York: Wiley)

——1990 Low storage methods for unconstrained optimization *Computational Solution of Non-linear Systems of Equations* (*Lectures in Applied Mathematics* **26**) ed E L Allgower *et al* (Providence, RI: American Mathematical Society) pp 165–79

Freeman J A and Skapura D M 1991 *Neural Networks: Algorithms, Applications and Programming Techniques* (Reading, MA: Addison-Wesley)

Gill P E, Murray W and Wright M H 1981 *Practical Optimization* (San Diego, CA: Academic)

Griewank A and Toint P L 1982 Partitioned variable metric updates for large structured optimization problems *Numerische Mathematik* **39** 119–37

Jacobs R A 1988 Increased rates of convergence through learning rate adaptation *Neural Networks* **1** 295–307

Johansson E M, Dowla F U and Goodman D M 1990 *Backpropagation Learning for Multi-Layer Feed-Forward Neural Networks using the Conjugate Gradient Method* Lawrence Livermore National Laboratory, UCRL-JC-104850 *Preprint* September 26

Liu D C and Nocedal J 1989 On the limited memory BFGS method for large scale optimization *Math. Programming* B **45** 503–28

Moreira M and Fiesler E 1995 Neural networks with adaptive learning rates and momentum terms *IDIAP Technical Report No 95-04*

Nocedal J 1980 Updating quasi-Newton matrices with limited storage *Math. Comput.* **35** 773–82

Noyes J L 1991 Neural network optimization methods *Proc. 4th Conf. Neural Networks and Parallel Distributed Processing* (Fort Wayne, IN: Indiana-Purdue University) pp 1–12

Park S K and Miller K W 1988 Random number generators: good ones are hard to find *Communications of the ACM* **31** 1192–203

Rumelhart D E and McClelland J L 1986 *Parallel Distributed Processing* vol 1 (Cambridge, MA: MIT)

Smith M 1993 *Neural Networks for Statistical Modeling* (New York, NK: Van Nostrand Reinhold)

Thimm G and Fiesler E 1994 High Order and Multilayer Perceptron Initialization *IDIAP Technical Report 94-07* 1994 (Institut Dalle Molle D'Intelligence Artificielle Perceptive, Case Postale 609 1920 Martigny Valais Suisse)

Thimm G, Moerland P and Fiesler E 1996 The interchangeability of learning rate and gain in backpropagation neural networks *Neural Comput.* **8**

B3.5 Training and generalization

James L Noyes

Abstract

See the abstract for Chapter B3.

In a neural network, the number, dimension, and type of training data have a substantial effect upon the network's training phase as well as its subsequent performance during the application phase. In particular, training affects generalization performance. The connection topology chosen and the activation function used are usually influenced by the available training data. Different neural network models and their associated solution methods may have different training data requirements. If a particular model is to be employed, then the user should determine whether there are any special training approaches recommended. This section addresses some general approaches to training and generalization, often within the context of a multilayer feedforward (MLFF) network baseline model.

Some basic terminology must first be established. A set of *training data* is the data set that is used to train a given network (i.e. determine all weights and biases). A *validation data* set can be used to determine when the network has been satisfactorily trained. A set of *test data* is used to determine the quality of this trained network. Typically, the neural network modeler is familiar with the characteristics of both training data and validation data. The test data are the data associated with the problem that the neural network is designed to solve. In some cases, the characteristics of the data associated with the problem may not be completely known before it is used in the network. The real goal of the network is to perform well on these actual *problem data* because of the network's ability to generalize. Typically, some balance between recall and generalization is desired. A lengthy training phase tends to improve recall at the expense of generalization. It is possible to quantify the notion of generalization, but some of these quantification methods can be rather complex (Hertz *et al* 1991).

To many, the generalization ability is the most valuable feature of neural networks. This leads to further questions relating to the size of the training set (the size of the potential application set may not even be known), the amount of training employed, the order in which the training data are presented, and the degree to which the training data are representative of the problem data.

B3.5.1 Importance of appropriate training data

When discussing the problem of selecting appropriate *training data*, one can consider the neural network B4 to be a *mapping* from an N_1-dimensional space into an N_L-dimensional space, where these dimensions are the number of neurons in the input and output layers, respectively. In a supervised network, the number of input and output neurons is dictated by the problem. However, when layers or clusters are to be used, the modeler is able to choose other topology defining characteristics. There are many similarities between designing and training a neural network and that of approximating a function (with a statistical emphasis). To start, one first picks the underlying network topology (with the form of the approximating function) so that it will adequately be able to model the anticipated data. Having selected the topology, one then attempts to determine the weights and biases (parameters of the approximation function) so that the training error is small. However, as will be seen, this does not guarantee that the error associated with the actual problem data will also be small.

The set of training data should be representative of the anticipated problem data. A polynomial fitting analogy may be used to illustrate why this is true. If only a very small sample of data is used where none of the data used has an ordinate value larger than a given number, then the corresponding polynomial is not guaranteed to give a close approximation for any abscissa that does have a large ordinate, even if the data are error free. Put another way, the statistical characteristics of the training data (the sample) should be close to the statistics of the actual problem data (the underlying population) for a network to be properly trained. In addition, the statistics of the validation data (a different sample) should also be close to the statistics of the actual problem data.

In the following it will be assumed that the chosen network topology can adequately model the application data and that the training data, validation data, and actual problem data all come from the same underlying distribution.

The size of the network model, as well as the type of model used, should depend upon the number of data to be used to train it. These two sizes are interrelated. A model with a lot of weights and biases to determine generally requires a lot of training data or else it will memorize well, but not generalize well. That is, it may train faster and do quite well reproducing desired training results, but it may give a very unsatisfactory performance when any kind of nontraining data is used. On the other hand, a model with too few weights compared with the size of the training data set may train very slowly or not train at all. (The training speed depends upon the difficulty of the problem itself as well as the size of the training data set.) These data set sizes must often be determined empirically, after a lot of experimentation. Normally one chooses the smallest network that trains well and performs satisfactorily on the test data.

C1.2.4 Another consideration is the *robustness* of the network—its sensitivity to perturbations in its parameters and weights. For example, it has been shown that the probability of error increases with the number of layers in an MLFF network (Stevenson *et al* 1990).

During the application period when the network is used to solve actual problems, it may be found that there are new types of data case for which the network is not producing the anticipated or required output. This could result from obtaining new problem data having different characteristics than the data used to train the network. This could also result from trying to solve a problem containing data from a different underlying distribution than that of the training data. Assuming that these new problem data are valid for the intended application, some or all of the data from these new cases can be added to the training (and validation) data sets and the network can be retrained.

B3.5.2 Measuring and improving network generalization

Network generalization may be addressed in two stages: how to detect and measure the generalization error, and how to reduce this error by improving generalization.

B3.5.2.1 Measures of generalization

Quantitative measures of generalization try to predict how well a network will perform on the actual problem data. If a network's generalization ability cannot be bounded or estimated, then it may not reliably be used for unseen problem data. Given a test data set of m examples from some arbitrary probability distribution, what size of MLFF network will provide a valid generalization? Alternatively, given a network, what is the minimum and maximum number of samples needed to train it adequately?

A method of quantifying the number of training data needed for an L-layer MLFF network was given by Mehrotra *et al* (1991) and a perceptron-based example of this was given by Wasserman (1993). Consider an MLFF network with N_1 inputs. For this type of network, assume there are W weight and bias values to be determined. Each input corresponds to a single point in N_1-dimensional space. If one were to partition each dimension into K intervals, then there are K^{N_1} uniformly distributed hypercubes in this N_1-space. As the number of input components increases, the number of hypercubes increases exponentially. If it is desired to have a training point in each hypercube in order to have the set of training data uniformly distributed, then the number of training examples needed is also K^{N_1}. For example, suppose one had to design a 5–N_2–3 network (so $N_1 = 5$) and wanted $K = 2$ intervals. This would mean that $2^5 = 32$ input examples would be needed in the training set. The number of weights and biases would then be $W = (5 + 1)N_2 + (N_2 + 1)3 = 9N_2 + 3$. So an N_2 of 2 or 3 should be reasonable to try for a good generalization capability, but an N_2 of 5 or higher would probably be too large. One can work this in the other direction, choosing N_2 first, then picking a K value to determine the number of training cases needed.

B3.5.2.2 The Vapnik–Chervonenkis dimension

An even more theoretical way to try to determine the number of training data needed to achieve good generalization is by using the Vapnik–Chervonenkis dimension or *VC dimension* (Vapnik and Chervonenkis 1971, Baum and Haussler 1989, Sakurai 1993, Wasserman 1993). The VC dimension can be used to relate a neural network's memorization capacity to its generalization capacity. The VC dimension is closely related to the number of weights and biases in a network, in analogy with the number of degrees of freedom (coefficients) in polynomial least-squares data fitting problems. Roughly speaking, for a fixed number of training cases, the smaller the network, the better the generalization since it is more likely to behave similarly on another training set of the same size with the same characteristics.

If $\mathcal{F}$ is a class of $\{-1, +1\}$-valued functions on $\mathbb{R}^{N_1}$ (where N_1 is the number of input neurons), and S is a set of m points in $\mathbb{R}^{N_1}$, then VCdim($\mathcal{F}$) is the cardinality of the largest set $S \subset \mathbb{R}^{N_1}$ that is shattered (i.e. all partitions S^+ and S^- of S can be induced by functions in $\mathcal{F}$). The VC dimension for a network of this type with only one computational layer can be shown to be just n, the number of unknown weights and biases.

There is no closed-form solution for the VC dimension for a general MLFF network, but it is closely related to the number of weights and biases in the network. Even though no closed-form solution has been found, a theoretical bound has been obtained. Baum and Hausler (1989) define an accuracy parameter ϵ and try to predict correctly at least a fraction $1 - \epsilon$ of examples from a test data set with the same distribution. Assuming $0 < \epsilon \le 1/8$, theoretical order of magnitude bounds for m are given by $\Omega(n/\epsilon)$ and $O((n/\epsilon)\log_2(N/\epsilon))$, where N is the number of neurons in a single-hidden-layer network and n is the total number of weights and biases. For example, this means that one needs on the order of n/ϵ training examples in order to have a generalization error under ϵ.

Yamasaki (1993) has given a precise expression for the number of test examples that can be memorized in an MLFF network that employs a logistic activation function (see section B3.2.4) and a single unit in the output layer L. This expression is given by

$$N_1 \cdot \lceil N_2/2 \rceil + \lfloor N_2/2 \rfloor \cdot \lceil N_3/2 - 1 \rceil + \ldots + \lfloor N_{L-2}/2 \rfloor \cdot \lceil N_{L-1}/2 - 1 \rceil$$

where the ceiling (least-integer) and floor (greatest-integer) functions are used.

Although upper and lower bounds have been defined for certain network types, these bounds often tend to be quite conservative about the number of training examples required.

B3.5.2.3 The generalized prediction error

Other approaches to the measurement of a network's generalization have been tried. Moody (1992) proposed a measure called the generalized prediction error (GPE) to estimate how well a given network would perform on nontest data. The GPE is based upon the weights and biases, the number of examples in the training set, and the amount of error in the training data. It works by appending an additional term to the objective function to be minimized during the training process.

B3.5.2.4 Cross validation

A more empirical method of measuring generalization error is that of *cross validation* (Stone 1959, 1974, White 1989, Smith 1993, Liu 1995). The idea here is to use additional examples from test data sets that were not used in training the network. The network is trained with the training data set (only) to determine the weights and biases, and a test data set is selected. Each input pattern from the test set is presented to the trained network and the corresponding output is computed. That output is then compared with the corresponding target data in the test set to determine each error. These errors can be combined to produce an overall error for the given test set by using the same error measure as was used when the network was trained (e.g. a least-squares error). This is done for all the test data sets. If each of these overall errors is small enough, then the neural network model generalizes well and is said to be validated. If not, then some adjustments are made either in the training or in the model itself to improve generalization, and the entire process is repeated.

C1.2.6

B3.5.2.5 The 'leave one out' approach

In some cases there are not enough data to make more than one test data set. In some cases there may only be enough data to place in the training set and train the network, but none for the test set to validate the network. In this situation a typical strategy is the 'leave one out' approach. That is, one trains the network with $m - 1$ examples in the training set, then evaluates the network with the unused example. This can then be done m times and a determination made as to whether the results are satisfactory. This approach can be extended to 'leave some out' with more combinations to be tried. A different type of approach is to effectively synthesize new data from the old by adding random errors to the training data (see below).

B3.5.2.6 Reducing the number of weights

Perhaps the simplest methods to improve generalization are to simply *increase the training set* or *decrease the number of weights and biases* in the model (e.g. by reducing the size of the hidden layers). Both of these methods tend to reduce the effects of any errors in the training data. If the ability to generalize is important, then one wants to be sure that there are not too many hidden neurons for the amount of training data used. Extra neurons can cause overfitting. This situation is analogous to the task of fitting a polynomial to a given set of data. If the polynomial has too high a degree, then extra coefficients must be determined. So even though the polynomial fits the data points well (perhaps even exactly), it can be highly oscillatory between the given data points so that it does not accurately represent the data trend, even at nearby data points.

B3.5.2.7 Early training termination

C1.2.6 Another relatively simple method to improve generalization is that of *early training termination* used by Smith (1993) and others. The training algorithm determines weights and biases based upon training data that often include errors. If the network models this type of training data too closely, then it is not likely to perform well on the actual problem data, even if both are from the same distribution. This tends to happen when one overfits the data by training with the goal of making the overall training error as small as possible (this is the normal goal of any minimization algorithm). The resulting network then models too much of the training data error. To prevent this from happening one pauses periodically in the training process to compute an overall (cross validation) test case error for one or more test sets using the current weight and bias values. These values, together with the corresponding overall test case error, are then saved. The training is then resumed. As the training continues, the overall training error usually gets smaller. However, at some stage of the training process, the overall testing error gets larger. When this happens, one terminates the training and uses the previous weights and biases that were saved. An alternative method of early training termination is even simpler and can be employed when binary or bipolar training data is used. This method uses a generic sigmoid output function (equation (B3.2.5)) to compute an auxiliary sum of squares and stops when this sum is exactly zero instead of stopping when the regular sum of squares (equation (B3.4.2)) is small (see section B3.2.4)).

B3.5.2.8 Adding noise to the data

Another method of using the available training data in such a way as to improve generalization without using exceptionally large training sets involves *adding noise to the data*, effectively augmenting the original training data with generated training data. This is done by applying a small, say 1–5%, random error to each component of each training example each time the network processes it. This does two things: it has the effect of adding more training data, and it prevents memorization. Here the training examples actually used are different for every presentation (the original training data are unchanged), and it is impossible for any of the weights to adjust themselves so that any single input is memorized. In addition, the trained network tends to be more robust when there is a relatively smooth mapping from the input space into the output space (Matsuoka 1992).

B3.5.2.9 Weight decay and weight pruning

There are several methods of improving generalization by causing the weights and biases to be computed
in a different manner. *Weight decay* methods try to force some of the weights toward zero. *Weight* C1.2.6
pruning methods actually seek to eliminate small weights entirely. One way to implement weight decay is
by adding a nonnegative penalty term to the objective function to be minimized (Krogh and Hertz 1992,
Smith 1993). This could take the form

$$A(w) = E(w) + \rho C(w)$$

where $E(w)$ is the original objective function (e.g. a least-squares function), $\rho > 0$ is a scaling multiplier,
and $C(w)$ is a 'complexity' measure that frequently includes some or all of the weights and biases directly.
For example, $C(w) = (\Sigma w_i^2)/2$ helps keep the weights small since small weights help minimize $A(w)$.

The multiplier ρ should be chosen so that it is neither too small (allowing a close fit with possible
overfitting) nor too large (allowing an excessive error influence). It can either be fixed or it can be adjusted
successively by using the previous test validation methods.

Often the penalty term is differentiable, where the partial derivatives are easily formulated and
incorporated into any gradient-based or Hessian-based descent methods. Other penalties can be based
upon Taylor series expansions (Le Cun *et al* 1990) or weight smoothing methods (Jean and Wang 1994).

After the initial training of a neural network, one may decide to prune the weights, and perhaps
neurons (when all input weights are zero). It is possible effectively to remove any weights and biases that
are too small, and will therefore have the least effect on the training error, by setting the weights to zero
and retraining the network. When the network is fully or partially retrained, the zero weights and biases
are treated as constants so that they are not altered. This can be accomplished with or without the aid of
automation since the pruning algorithm to do this can be directly followed by the network modeler when
the model is small or implemented on the computer when the model is large and many weights and biases
must be checked (Ying *et al* 1993). The use of this type of method is an alternative to methods that limit
the number of hidden neurons. This method can also be used in conjunction with weight decay methods.

One may combine some of the above methods to help further improve a neural network's
generalization capability.

References

Baum E B and Haussler D 1989 What size net gives valid generalization? *Neural Information Processing Systems*
vol 1, ed D S Touretzky (San Mateo, CA: Morgan Kaufmann) pp 81–90

Hertz J, Krogh A and Palmer R G 1991 *Introduction to the Theory of Neural Computation* Santa Fe Institute Lecture
Notes vol 1 (Redwood City, CA: Addison-Wesley)

Jean J S N and Wang J 1994 Weight smoothing to improve network generalization *IEEE Trans. Neural Networks* **5**
752–63

Krogh A and Hertz J A 1992 A simple weight decay can improve generalization *Advances in Neural Information
Processing Systems* vol 4 ed J Moody, S J Hanson and R P Lippman (San Mateo, CA: Morgan Kaufmann)
pp 950–7

Le Cun Y L, Denker J S and Solla S A 1990 Optimal brain damage *Advances in Neural Information Processing
Systems* vol 2 ed D S Touretsky (San Mateo, CA: Morgan Kaufmann) pp 598–605

Liu Y 1995 Unbiased estimate of generalization error and model selection in neural networks *Neural Networks* **8**
215–9

Matsuoka J 1992 Noise injection into inputs in back-propagation learning *IEEE Trans. Systems, Man, Cybern.* **22**
436–40

Mehrotra K G, Mohan C K and Ranka S 1991 Bounds on the number of samples needed for neural learning *IEEE
Trans. Neural Networks* **2** 548–58

Moody J E 1992 The effective number of parameters: an analysis of generalization and regularization in nonlinear
learning systems *Advances in Neural Information Processing Systems* vol 4, ed J Moody, S J Hanson and
R P Lippman (San Mateo, CA: Morgan Kaufmann) pp 847–54

Sakurai A 1993 Tighter bounds of the VC-dimension of three-layer networks *World Congress on Neural Networks*
vol III (International Neural Network Society) 540–3

Smith M 1993 *Neural Networks for Statistical Modeling* (New York, NK: Van Nostrand Reinhold)

Stevenson M, Winter R and Widrow B 1990 Sensitivity of feedforward neural networks to weight errors *IEEE Trans.
Neural Networks* **1** 71–80

Stone M 1959 Application of a measure of information to the design and comparison of regression experiments *Ann. Math. Statistics* **30** 55–69

——1974 Cross-validatory choice and assessment of statistical predictions *J. R. Statistical Soc.* B **36** 111–47

Vapnik V N and Chervonenkis A 1971 On the uniform convergence of relative frequencies of events to their probabilities *Theory Probab. Appl.* **16** 264–80

Wasserman P D 1993 *Advanced Methods in Neural Computing* (New York: Van Nostrand Reinhold)

White H 1989 Learning in artificial neural networks: a statistical perspective *Neural Comput.* **1** 425–64

Yamasaki M 1993 The lower bound of the capacity for a neural network with multiple hidden layers *World Congress on Neural Networks* vol III (International Neural Network Society) 544–7

Ying X, Surkan A J and Guan Q 1993 Simplifying neural networks by pruning alternated with backpropagation training *World Congress on Neural Networks* vol III (International Neural Network Society) July 364–7

B4

Data Input and Output Representations

Thomas O Jackson

Abstract

Neural networks are adaptive systems that have 'automatic' learning properties, that is, they adapt their internal parameters in order to satisfy constraints imposed by a training algorithm and the input and output training data. In order to extract the maximum potential from the training algorithms very careful consideration must be given to the form and characteristics of the data that are presented to the network at the input and output stages. In this chapter we discuss the requirements for data preparation and data representation. We consider the issue of feature extraction from the data sample to enhance the information content of the data used for training, and give examples of data preprocessing techniques. We consider the issue of data separability and discuss the mechanisms by which neural networks can partition and categorize data. We compare and contrast the different means by which real-world variables can be represented at the input and output of neural networks, looking in detail at the properties of local and distributed schemes and discrete and continuous methods. Finally, we consider the representation of more complex or abstract properties such as time and symbolic information. The objective in this chapter is to highlight the fundamental role that data preparation plays in developing successful neural network systems, and to provide developers with the necessary methods and understanding to approach this task.

Contents

B4.1 Introduction

Thomas O Jackson

Abstract

See the abstract for Chapter B4.

The past decade has seen a meteoric rise in the popularity of neural network techniques. One reason for this increase may be that neural computing can offer relatively simple solutions to complex pattern classification problems. In simple terms, the neural computing approach can be described by the following algorithm.

(i)   Gather the data sample.
(ii) Choose and prepare the training set from the sample.
(iii) Select an appropriate network topology.
(iv) Train the network until it displays the desired properties.

It has been described as a 'black box' solution (even 'statistics for amateurs' (Anderson 1995)) because the internal representations or mechanics of the network need not be known, or understood, in order to find a solution to the problem in hand. Neural networks have been, and perhaps continue to be, applied in this 'simplistic' manner. However, this approach obscures a realm of complexities which contribute to the successful performance of neural computing methods. One major issue, which is the focus of this chapter, is the manner in which data are presented to a neural network. That is, the mechanisms by which the data set is transformed into input vectors such that the salient information is presented in a 'meaningful' manner to a network. It is true to say that the familiar maxim applied to conventional computing systems—'garbage in, garbage out'—is equally valid in the neural computing paradigm.

The theme of data representation receives minimal attention in many neural texts. This is a major oversight. The structures used to represent data at the input to a neural network contribute as much to the successful solution of any given problem as the choice of network topology. It could be argued that the data representations are more critical than the network *topology*; the flexibility inherent in neural learning   B2 algorithms can accommodate nonoptimal selection of topological parameters such as weights or the number of nodes. However, if a network is trained with inappropriately structured data then it is unlikely that the network will learn a mapping function that has any useful correlation with the training data. Similarly, the representations used at the *output* of a neural network play a crucial role in the training process.

The aim of this chapter is to illustrate the techniques and data structures that ensure appropriate representation of the input and output data. There are two issues: (i) enhancement of feature information from the data set, and (ii) how to represent features (as variables) at the network input and output layers. We will discuss these two problems from a number of different viewpoints. In Section B4.2 we start with fundamental principles and consider data complexity and data separability. In the course of this discussion we shall examine the mechanisms by which neural networks are able to partition and categorize data. The motivation for this discussion is simple—in order to understand the constraints that determine satisfactory data representations it is first necessary to understand how a network 'processes' data. Section B4.3 considers data preprocessing. Sections B4.4 to B4.10 deal with the specifics of data representation, considering discrete versus continuous data formats, local and distributed schemes and data encoding techniques.

It is worth emphasizing that this chapter does not address the issue of *internal* data representations but rather the means by which data are represented at the *input* and *output* stages of a network. The subject of internal representations is discussed within Chapter B5.   B5

References

Anderson J A 1995 *An Introduction to Neural Networks* (MIT Bradford Press)

Anderson J A 1995 *An Introduction to Neural Networks* (MIT Bradford Press)

B4.2 Data complexity and separability

Thomas O Jackson

Abstract

See the abstract for Chapter B4.

There are a number of different mathematical frameworks which might be used to illustrate the point that data representation is a fundamental issue in neural computing. The approach adopted here is to consider the problem in terms of *pattern space partitioning*. To identify the properties that distinguish 'good' data representations we must first review how a neural network performs pattern classification within a given pattern space. To do this a hypothetical and somewhat trivial pattern classification problem will be discussed. Consider the data set shown in figure B4.2.1; it describes two data classes distributed across a two-dimensional feature space. The data points are representative samples taken from each class. The pattern classification task is defined as follows: given any random vector, A, taken from the same feature space, which class should it be assigned to?

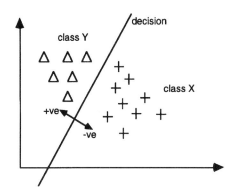

Figure B4.2.1. Class separation using a linear decision boundary.

One traditional pattern classification technique which is commonly used to solve this categorization problem is pattern space partitioning using *decision boundaries*. A decision boundary is a hyperplane partition in the pattern space which segregates pattern classes. The simplest example of a decision boundary is the *linear decision boundary* shown in figure B4.2.1. Any vector that falls on the (arbitrarily assigned) positive side of the boundary is attributed to class Y, similarly, any vector that falls on the negative side of the boundary is attributed to class X. The field of *statistical pattern recognition* has given rise B6.2.3 to many forms of decision boundary (two good reference texts on this subject are Duda and Hart (1973) and Fu (1980)). However, the challenge of decision boundary methods is not in defining the form of the hyperplane boundaries, but in *positioning* the planes in the pattern space.

In the trivial example shown in figure B4.2.1, a simple visual inspection is sufficient to identify where a linear partition may be positioned. Clearly, however, the problem becomes nontrivial when we move to data sets with three or more dimensions, and complex analytical methods are required in these cases. The compelling attraction of neural computing techniques is that they provide adaptive learning algorithms which can position decision boundaries 'automatically' through repetitive exposure to representative samples of the data.

C1.1.1, F1.2.3 The *perceptron* (Rosenblatt 1958) is the simplest *neural classifier* and it can be easily demonstrated that the network functions as a linear discriminator. The analysis is straightforward and is worth considering briefly here. The definition of the perceptron classifier is given by

$$y = Hv\left[\sum_{i=1}^{n} w_i x_i - \theta\right].$$
(B4.2.1)

where w_i are the weight vectors, x_i are the input vector components, θ is a constant *bias* input and Hv is the Heaviside function.

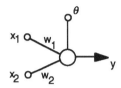

Figure B4.2.2. The perceptron classifier.

The output, y, will take on a positive or negative value dependent upon the input data and weight vector values. A positive response indicates class Y, a negative response indicates class X.

We can rearrange (B4.2.1) and express it in the inner product form

$$y = (|W \| X| \cos \phi) - \theta.$$
(B4.2.2)

The $\cos \phi$ term (where ϕ is the angle between the weight vector, W, and the input vector X) has a range between ± 1. Any value of ϕ greater than $\pm 90°$ will reverse the value of the output, y. This produces a linear decision boundary because the crossover point is at $\pm 90°$. The weight parameters and the bias value determine the position of the decision boundary in the pattern space. If we consider the crossover region where $y = 0$, we can demonstrate this point

$$0 = \sum_{i=1}^{n} w_i x_i - \theta.$$
(B4.2.3)

Expanding this for the perceptron two weight network:

$$0 = w_1 \times x_1 + w_2 \times x_2 - \theta.$$
(B4.2.4)

Rearranging this for x,

$$x_2 = -\frac{w_1}{w_2} x_1 + \frac{\theta}{w_2}.$$
(B4.2.5)

Comparing (B4.2.5) to the equation for a straight line, $y = mx + c$, we can see that the slope of the decision boundary, m, is controlled by the ratio of w_1/w_2, and the axis intercept, c, is controlled by the bias term, θ.

During the learning cycle the weight values are modified iteratively, in order to arrive at a satisfactory position of the decision plane. Satisfactory in this context means minimizing the number of classification errors to a predefined acceptable level across the training set (which of course should converge to zero in B3 the optimal case). Details of the training algorithms are discussed in Chapter B3.

The brief analysis of the perceptron has demonstrated that it can partition a pattern space by placing a linear decision boundary within it. Identifying representative data samples is clearly a key issue. Placement of the boundary is made on the assumption that the samples taken from classes X and Y are fully representative of the class types. Inadequate training data can lead to the boundary being positioned incorrectly. For example, in figure B4.2.3 exclusion of the samples X_1 and X_2 from the training data could result in classification errors.

In 'real world' classification tasks the data sets are rarely separated or partitioned as easily as the trivial example we have discussed, and, in practice, the range of problems that can be solved with simple linear decision boundaries is extremely limited. For most nontrivial pattern classification problems we must

© 1997 IOP Publishing Ltd and Oxford University Press

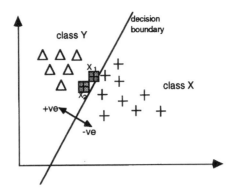

Figure B4.2.3. Misclassification due to incorrectly positioned decision boundary.

contend with data sets which have complex class boundaries. Examples are shown in figure B4.2.4(a) and (b).

The data spread shown in figure B4.2.4(b) is an example of the XOR classification problem. This classification task was used by Minsky and Papert (1969) to highlight the limitations of the single-layer perceptron classifier.

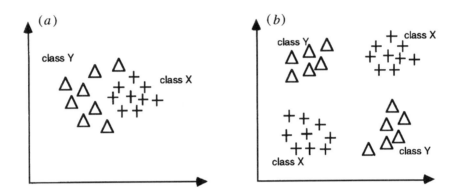

Figure B4.2.4. (a) Meshed classes. (b) XOR problem.

A simple visual inspection shows that neither of these data sets can be separated using a single linear classification boundary. In such cases, a perceptron could not converge to a satisfactory solution. Complex data sets, as typified in the examples of figure B4.2.4, must be partitioned by combining multiple decision boundaries. For example, the XOR problem shown in figure B4.2.4(b) can be resolved in the following manner.

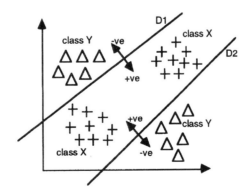

Figure B4.2.5. Piece-wise linear classification achieved by combining decision planes.

By placing two decision boundaries it is possible to logically combine the classification decisions of each and partition the data satisfactorily. This technique is known as *piece-wise linear* classification. A truth table illustrating the combination of the decision boundaries is shown in table B4.2.1.

Table B4.2.1. Truth table for piece-wise linear classification scheme.

Classification	Sign of decision line	
	D1	D2
Class X	+	+
Class Y	−	+
Class Y	+	−

Partitioned regions of this type are known as *convex regions* or alternatively *convex hulls*. A convex region is one in which any point in the space can be connected by a straight line to any other without crossing the boundary of that region. Convex regions may be *open* or *closed*—examples of each type are shown in figure B4.2.6.

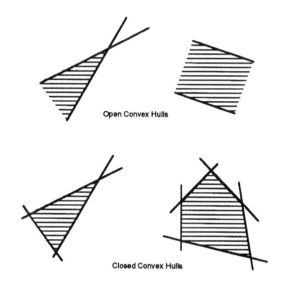

Open Convex Hulls

Closed Convex Hulls

Figure B4.2.6. Examples of open and closed convex hulls.

In a perceptron classifier convex hulls are created by combining the output of two parallel perceptron units into a third unit, figure B4.2.7. The third unit, which forms a second layer in the network, is configured to perform the logical AND function (i.e. it becomes active when both its inputs are active) so that it implements the condition for class X in table B4.2.1. There are, however, many classes of problems which cannot be partitioned by convex regions. The meshed class example shown in figure B4.2.4(*a*) is one example. The solution to this class of problems is to combine perceptrons into a network of three C1.2 or more layers. This class of networks are generally termed *multilayer perceptrons*. The third layer of units receives *regions* as inputs and is able to combine these regions into areas of arbitrary complexity. Examples are shown in figure B4.2.8.

The number of units in the first layer of the network controls the number of linear planes. The complexity of the regions that can be created in the pattern space is defined by the number of linear planes that are combined. There is a mathematical proof, the Kolmogorov theorem (Kolmogorov 1957), which states that regions of arbitrary complexity can be generated with just three layers. The proof will not be explored here, but a useful analysis can be found in (Hecht-Nielsen 1987).

To summarize, we have seen that the class of networks based upon perceptron classifiers are able to partition a pattern space using decision boundaries. We have also seen that the position of the boundaries in the pattern space is determined by the weight constants in the network and the bias terms. At this point the fundamental link between the classification performance and the quality of the training data becomes

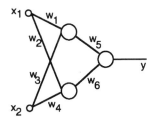

Figure B4.2.7. Two-layer perceptron network for partitioning convex hulls.

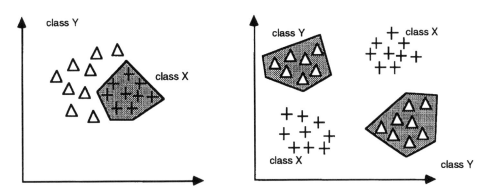

Figure B4.2.8. Arbitrary complex regions partitioned by perceptron networks of three or more layers.

apparent; the weights of the network are modified in response to the training data. Clearly, for a network to generate meaningful internal representations that adequately partition the pattern space, we must present the network with data that accurately define that pattern space.

References

Duda R O and Hart P E 1973 *Pattern Classification and Scene Analysis* (New York: Wiley)

Fu K S 1980 *Digital Pattern Recognition* (Berlin: Springer)

Hecht-Neilsen R 1987 Kolmogorov's mapping neural network existence theorem *1st IEEE Int. Conference on Neural Networks* **3** San Diego 11–14

Kolmogorov A N 1957 On the representation of continuous functions of many variables by superposition of continuous functions of one variable and addition *Dokl. Akad. Nauk USSR* **114** 953–6

Minsky M and Papert S 1969 *Perceptrons: An Introduction to Computational Geometry* (Cambridge, MA: MIT Press)

Rosenblatt F 1958 The Perceptron: a probabilistic model for information storage and retrieval in the brain *Psych. Rev.* **65** 386–408

B4.3 The necessity of preserving feature information

Thomas O Jackson

Abstract

See the abstract for Chapter B4.

The preceding discussion provides us with an important insight into neural network classification techniques; the clustering of the data has a large impact upon the complexity of the neural network classifier. From this we conclude that the data presentations should preserve the clustering inherent in the data set. This implies that the properties which determine the class distribution must be understood. Neural computing offers no 'short cuts' here; data analysis is a prerequisite, and we need to draw from established statistical and numerical analysis techniques (again, Duda and Hart (1973) and Fu (1980) are useful references).

As an example of how we might approach this task, consider the following *character recognition* G1.2 problem: a neural network will be used to map the five bitmaps, figure B4.3.1, onto their respective vowel classes.

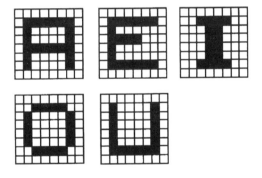

Figure B4.3.1. Five 'character' bitmaps.

The 'raw' data is the set of five 64-bit binary vectors representing the bitmaps. One simple approach to this problem might be to use the 64-bit vector as the input to the network. Another option is to assign each bitmap an arbitrary code, for example 110011 to represent the bitmap for character 'A'. However, a more productive approach might be to recognize that it is the information contained in the *shape* of the characters which uniquely defines them. This information can be used to derive representations that explicitly define the shape. For example, we might consider counting the number of the horizontal and vertical spars, the relative positions of the spars, and the ratio of vertical to horizontal spars. This approach allows contextual or *a priori* knowledge to be captured in the data presented to a network. One advantage of this approach is that similar shape characters, such as 'O' and 'U', would have similar representations (that is, there would be many common features in the two feature vectors). In many applications this is a desirable property as it can lead to more robust generalization.

Wasserman (1993) has suggested that in some circumstances it may be desirable to use the 'raw' data as the input to the network. Many classification problems are difficult to solve using traditional pattern recognition partially because the task of identifying and extracting appropriate feature information is so complex and ill-defined. In such cases a neural network *may* prove more adept at identifying underlying features or data trends than a human analyst. Consequently, there may be an advantage gained from presenting a network with large, unprocessed data vectors and expecting that the adaptive training procedure will be able to identify the underlying information. There is clearly a compromise which must be reached between these two approaches. Unfortunately there are few analytical methods available to assist in the decision process.

To demonstrate that a data representation is capable of destroying the clustering properties we will consider an example using binary coding. Binary codings map a discrete valued number from a single dimension into a much higher, complex dimension space. For example, if a feature with a range of values 0–32 is mapped into a binary representation, the set of values is mapped onto a six-dimension feature space. However, this transform is not an appropriate mapping because the binary representation has many discontinuities between neighboring states. For example, consider the transition of values from 29–32 in binary form.

Value	Binary
29	011101
30	011110
31	011111
32	100000

We can see that there is a common pattern in bits 3–5 of the vectors for the values 29–31. However, there is no corresponding pattern in the binary vector for value 32. In terms of pattern vectors this would suggest that the two feature values, 31 and 32, are quite separate in pattern space. These discontinuities destroy the inherent clustering of the data set and fragment the data. In general, the fragmentation leads to more complex pattern spaces and a more demanding partitioning task.

This simple example leads us to an important general principle: the metric we use to gauge similarity in the pattern domain should be preserved in the data representation. In the example above, we are using a Euclidean metric to determine the similarity of the discrete representation, but the similarity of the binary patterns is determined by the Hamming metric, and, as we have argued, these are not equivalent.

B1.3   This is not to say that binary codings are universally inappropriate. The discrete *Hopfield network*, for example, makes good use of binary representations. However, it is important to note that the inputs to a Hopfield network generally encode states or events rather than feature values. For example, one application of the Hopfield network is in optimization problems such as the traveling salesman. In this problem the binary input vectors record the event that a particular salesman has visited a certain city (represented by a discrete node).

In conclusion, the primary objective for any data representation is to capture the appropriate information from the data set in order to adequately constrain the classification problem. Careful consideration of the problem characteristics and suitable preprocessing will, in general, lead to more predictable classification performance.

References

Duda R O and Hart P E 1973 *Pattern Classification and Scene Analysis* (New York: Wiley)
Fu K S 1980 *Digital Pattern Recognition* (Berlin: Springer)
Wasserman P D 1993 *Advanced methods in neural computing* (New York: Van Nostrand Reinhold)

B4.4 Data preprocessing techniques

Thomas O Jackson

Abstract

See the abstract for Chapter B4.

Data sets are often plagued by problems of noise, bias, large variations in the dynamic range or sampling range, to highlight a few. These problems may obscure the major information content or at least make it far more problematic to extract. There are a number of general data processing algorithms available which can remove these unwanted variances, and enhance the information content in the data. We will discuss these in the following sections.

B4.4.1 Normalization

Data sets can exhibit large dynamic variances over one or more dimensions in the data. These large variances can often dominate more important but smaller trends in the data. One technique for removing these variations is *normalization*. Normalization removes redundant information from a data set, typically by compacting it or making it invariant over one or more features. For example, when building a *pattern recognition* system to recognize surface textures in gray-scale images it is often desirable to F1.2 make the system invariant to changes in light conditions (i.e. contrast and brightness) within the image. Normalization techniques allow the variations in the contrast and brightness to be removed such that the images have a consistent gray-scale range.

Similarly when processing speech signals, for example in a *voice recognition* system, it is advantageous F1.7 to make the system invariant to changes in the absolute volume level of the signal. This is described in figure B4.4.1.

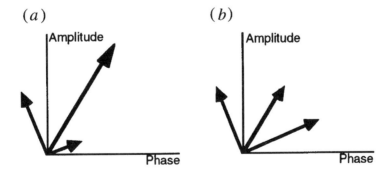

Figure B4.4.1. (*a*) Varying magnitudes; (*b*) normalized amplitudes.

The vectors represent the phase and amplitude of the signal. In figure B4.4.1(*a*), the three vectors are shown with varying amplitudes and phases, however, it may only be the phase information that is of relevance to the classification problem. In figure B4.4.1(*b*) the vectors have been normalized to unit length, such that all amplitude variations have been removed, whilst leaving the phase information intact.

We may also want to normalize data with respect to its position. For example, in a character recognition system it is typical that the input data are normalized with respect to position and size. In classification systems which use template matching schemes this preprocessing step can substantially reduce the number of templates required. A simple example is shown in figure B4.4.2.

One point of caution should be noted from this example. Normalization procedures can remove important feature information as well as redundant information. For example, consider the case of a character 'C'. If it is normalized to remove scale variations then it is possible to normalize upper case 'C' and lower case 'c' to the same representation. This may or may not be a desirable transform, depending upon the application. This example stresses the importance of understanding the context of the normalization with respect to the classification task in hand.

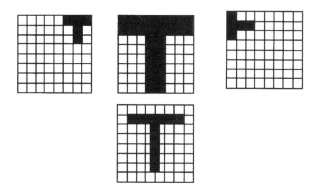

Figure B4.4.2. Scale and position normalization. The three 'T' characters in the top of the diagram can be normalized and reduced to a single representation shown below.

B4.4.2 Normalization algorithms

The principle of normalization is to reduce a vector (or data set) to a standard unit length; usually 1, for convenience. To do this we compute the length of the vector and divide each vector component by its length. The length, l, of a vector, Y, is given by

$$l = \left(\sum_{1}^{m} y_i^2 \right)^{1/2} \tag{B4.4.1}$$

where l is the length, and m is the dimensionality of Y. Hence, a normalized, unit length vector Y' is given by

$$Y' = \frac{Y}{l}. \tag{B4.4.2}$$

A vector (or data set) can be normalized across many different dimensions, and with respect to many different statistical measures such as the mean or variance. We shall describe three approaches which Wasserman (1993) has termed *total normalization, vertical normalization* and *horizontal normalization*.

Total normalization. This is the most widely applied normalization method. The normalization is performed globally across the whole data set. For example, to remove unnecessary offsets from a data set we can normalize with respect to the mean. This is described in equation (B4.4.3).

Evaluate the mean of the data vectors, $\overline{y}$, across the full data set (1 to p vectors):

$$\overline{y} = \frac{\sum_{j=1}^{p} \sum_{i=1}^{m} y_{ji}}{p \times m} \tag{B4.4.3}$$

where m is the number of components in a vector.

For each vector, divide by the mean:

$$Y' = \frac{Y}{\overline{y}}. \tag{B4.4.4}$$

Vertical normalization. In some applications normalizing over the total data set is not appropriate, for example when the components of a feature vector represent different data types. In these circumstances it is more appropriate to evaluate the mean or variance measure of the *individual* vector components. An algorithm to normalize by removing the mean is described in equation (B4.4.5).

Determine the mean $\bar{y}_i$ of each component, i, over each vector in the data set (1 to p):

$$\bar{y}_i = \frac{\sum_{j=1}^{p} y_{ji}}{p} \qquad \text{for } i = 1 \text{ to } m. \tag{B4.4.5}$$

For all vectors, divide each component by the corresponding component mean:

$$Y' = \frac{y_i}{\bar{y}_i} \qquad \text{for } i = 1 \text{ to } m \tag{B4.4.6}$$

Horizontal normalization. When handling vectors that incorporate temporal properties, for example, a vector that represents an ordered time series, we must normalize the vectors individually. Hence, to normalize with respect to the mean, we can perform the following equation.

For each vector, $j = 1$ to p, establish the mean, $\bar{y}_j$:

$$\bar{y}_j = \frac{\sum_{i=i}^{m} y_{ji}}{m}. \tag{B4.4.7}$$

For each vector, $j = 1$ to p, divide by the mean:

$$Y'_j = \frac{Y_j}{\bar{y}_j}. \tag{B4.4.8}$$

The algorithms described above describe techniques to remove offsets from a data set. The same methods can be used to remove unwanted variations in vector magnitude by dividing by the vector length.

These descriptions present details of three possible approaches to normalization. They are not a definitive set of algorithms. However, they highlight the fact that caution must be exercised when normalizing vectors to ensure that only the redundant information is removed. Normalization is a powerful technique when applied correctly and can significantly enhance the information content within a data set.

B4.4.3 Principal component analysis

Normalization is one scheme by which pertinent feature information can be enhanced in a data set. Another scheme which is often linked to neural networks, largely due to the work of Oja (1982, 1992) and Linsker (1988), is principal component analysis (PCA) (also known as the Karhunen–Loeve transform, (Papoulis 1965)). It is a data compression technique that extracts characteristic features from the data whilst minimizing the information loss. It is typically used in statistical analysis for high-dimensional data sets, where the features with the greatest significance are obscured by the size and complexity of the data.

The basic principle of PCA is the representation of the data by a reduced set of unit vectors (eigenvectors). The eigenvectors are positioned along the directions of greatest data variance. They are positioned so that the projections from the data points onto the axis of the vector are minimized across the full data set. A simple example is shown in figure B4.4.3. The vector, Y, is positioned along the direction of the greatest data spread in the two-dimensional space. Any point in the data sample can now be described in terms of its projection along the axis of Y, with only a small reduction in positional accuracy. As a consequence, a two-dimensional position vector has been reduced to a single-dimensional description. In high-dimensional spaces the objective is to find the minimum set of eigenvectors that can describe the data spread whilst ensuring a tolerably low loss in accuracy.

Having discussed the approach in general terms, we can now provide a mathematical framework for PCA. The eigenvectors that are required are members of the covariance matrix, R, for the data set. This matrix is generated from the outer product equation:

$$R = \frac{1}{N} \sum_{k=1}^{n} (x_k - \bar{x})(x_k - \bar{x})^T \tag{B4.4.9}$$

where $\bar{x}$ is the mean vector of the data sample and N is the number of vectors.

Once the eigenvectors of this matrix are found, $(\lambda_1, \lambda_2, K, \lambda_n)$, they can be ordered in terms of their eigenvalues. The principal components are those which minimize the mean squared error between the data

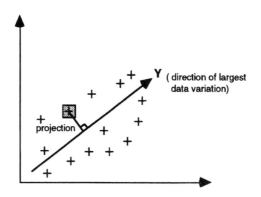

Figure B4.4.3. Determining the direction of greatest variation in a data set.

and its projection onto the new axis. The smaller eigenvectors are discarded (i.e. those with the smallest variance) and the data vectors are approximated by a linear sum of the remaining m eigenvectors:

$$\tilde{x} = \sum_{i=1}^{m} (\lambda_i x) \lambda_i. \tag{B4.4.10}$$

$\tilde{x}$ will be close to x if the appropriate eigenvectors were chosen. Note that the dimensionality of $\tilde{x}$ is less than that of the original vector. Proof that the information loss in this reduction is minimal will not be discussed here, however, a detailed analysis can be found in Haykin (1994), and a formal analysis of eigenvectors and eigenvalues is presented in Rumelhart and McClelland (1986). Principal component analysis is a useful statistical technique in a data preprocessing 'toolkit' for neural networks.

References

Haykin S 1994 *Neural Networks: A comprehensive foundation* (New York: Macmillan College Publishing Company)
Linsker R 1988 Self-organisation in a perceptual network *Computer* **21** 105–17
Oja E 1982 A simplified neural model as a principal component analyzer *J. Math. Biol.* **15** 267–73
——1992 Principal components, minor components and linear neural networks *Neural Networks* **5** 927–36
Papoulis A 1965 *Probability, random variables and stochastic processes* (New York: McGraw-Hill)
Rumelhart D E and McClelland J L 1986 *Parallel Distributed Processing: Explorations in the Microstructure of Cognition* (Cambridge, MA: MIT Press)
Wasserman P D 1993 *Advanced methods in neural computing* (New York: Van Nostrand Reinhold)

B4.5   A 'case study' review

Thomas O Jackson

Abstract

See the abstract for Chapter B4.

To consolidate the ideas discussed so far, we will review a neural network application as a small case study. The application is a *face-recognition* system using gray-scale camera images. The neural system F1.6.5 was developed at Rutgers University and reported in Wilder (1993). The recognition system was required to identify individual faces captured by a CCD camera, under controlled and constant lighting conditions. The neural network used was the Mammone–Sankar neural tree network (NTN) (the details of this are not important for our discussion).

The CCD camera produces a gray-scale image that is 416×320 pixels in size. A 'holistic' analysis approach was used, whereby the facial image is processed as a whole, rather than being partitioned into regions of high interest features (such as eyes, ears, mouth etc). The question is, given the 416×320 pixel image, where do we start on the task of generating data suitable for developing a neural network solution? Clearly, we would not wish to take the 'easy' option and treat the image as a pixel map; this would generate a 133, 120 component vector. This approach would quickly leave us bereft of computer resources and sufficient hours (or patience) to complete the training task! Obviously some form of data reduction is required.

The method selected was gray-scale projections. This involves generating a 'gray-scale' profile of an image by summing the gray-scales along predetermined paths in the image (e.g. along pixel rows or columns). If a number of projections are made, along several high interest planes, then a two-dimensional image can be represented by a one-dimensional gray-scale profile vector. The images were partitioned into 16 horizontal and vertical planes, and the gray-scale data were integrated over these planes. These profiles provided strong delineation of the facial features in each orientation. A schematic representation is provided in figure B4.5.1.

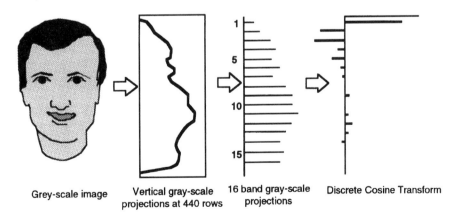

Grey-scale image   Vertical gray-scale projections at 440 rows   16 band gray-scale projections   Discrete Cosine Transform

Figure B4.5.1. Feature extraction processing stages.

This step reduces the 133, 120 pixel image into two one-dimensional vectors, each with 16 components describing the vertical and horizontal gray-scale profiles. One could potentially consider using these vectors as the basis for the network training data. However, a further data transform was applied to these vectors, mapping them into a spatial frequency domain using a unitary orthogonal transform. The authors cite several reasons for this step:

- unitary transforms are energy and entropy preserving;
- they decorrelate highly correlated vectors, and;
- the major percentage of the vector information is mapped onto the low frequency components, allowing the high frequency components to be discarded with minimum information loss.

Three transforms were tested: the discrete cosine transform (DCT), the Karhunen–Loeve (PCA, described in section B4.4.3) and the Hadamard. All three gave similar recognition performance. However, the DCT was chosen due to the fact that it has an efficient and fast hardware implementation. The feature decorrelation provided by the transform also creates some invariance to small localized changes in the input image (caused, for example, by the subject changing a facial expression or removing spectacles). The final step in the preprocessing phase was to discard some of the high frequency components (which had minimal information content) of the DCT. This resulted in a final training vector with 23 feature components.

A number of important principles for data preprocessing are demonstrated in this example. Firstly, there is a solid grasp of the underlying characteristics of the classification problem. As a result efficient techniques for extracting the high interest features within the images were derived. Secondly, a clear method for data reduction with minimal information loss was applied (that is, gray-scale projections). Thirdly, transforms were applied to the 'reduced' vector descriptions which enhanced the information content and allowed further redundant information to be discarded. These transforms provided some invariance to small changes in the images and increased the separability between individual images. These principles should be uppermost in our thinking when developing a pattern recognition system (neural or otherwise).

References

Wilder J 1993 Face recognition using transform codings of gray scale projections and the neural tree network *Artificial Neural Networks for Speech and Vision* ed R J Mammone (London: Chapman and Hall) pp 520–36

B4.6 Data representation properties

Thomas O Jackson

Abstract

See the abstract for Chapter B4.

Having looked at data preparation techniques in broad terms we can now focus on the details of data representations. Anderson (1995) has suggested that there are five general rules to consider when adopting data representations. Summarizing, these are broadly as follows:

- similar events should give rise to similar representations;
- things that should be separated should be given different representations (ideally separate categories should have orthogonal representations);
- if an input feature is important (in the context of the recognition task) then it should have a large number of elements associated with it;
- carrying out adequate preprocessing will reduce the computational task in the adaptive parts of the network;
- the representation should be easy to program and flexible.

Wasserman (1993) has also proposed a list of properties for data representation schemes. He suggests that there are four principal characteristics of a good representation:

- Compactness
- Information preservation
- Decorrelation
- Separability.

We shall discuss each of these properties in turn.

Compactness. Large networks require longer training times. For example, it has been shown that the training times for the simple perceptron network increase exponentially with the number of inputs, within the range $2^M < t < M^M$, where M is the number of inputs. Also it has been proposed that learning times for MLPs increase at a rate proportional to the number of connections cubed. Hence, it is advantageous to keep input vectors short.

Information preservation. The need for compact representations must be balanced against the need to preserve information in the data vector. Consequently, we need to utilize data transforms which allow a reduction in dimensionality without a reduction in the amount of information represented. Also, the transform should be reversible—such that when the reduced vector is expanded all of the original information is recovered. Data transforms of this nature are in use in the analog domain, for example techniques such as fast Fourier transforms, which represent complex frequency modulated signals in terms of a number of sinusoid components. Similarly, in the digital domain there are numerous encoding techniques, such as Manchester encoding, which also reduce the dimensionality of a digital signal without a reduction in the information content.

Decorrelation. This supports Anderson's suggestion that objects which belong to different classes should be given different representations.

Separability. Ideally the data transforms should increase the separation between disparate classes but enhance the grouping of similar classes. This is complementary to the requirement for decorrelation.

These lists outline the broad objectives that need to be satisfied by a data representation scheme. In the following sections, we discuss appropriate coding schemes which meet some or all of these constraints.

References

Anderson J A 1995 *An Introduction to Neural Networks* (MIT Bradford Press)

Wasserman P D 1993 *Advanced methods in neural computing* (New York: Van Nostrand Reinhold)

B4.7   Coding schemes

Thomas O Jackson

Abstract

See the abstract for Chapter B4.

In the following section we consider the pragmatic issue of how to present features or variables to a neural network using discrete or continuous values input nodes. Discrete codings typically refer to binary $(0,1)$ or bipolar $(-1, +1)$ activation functions but can also include nodes with graded output levels. Continuous valued variables can take any value in the set of real numbers. There are many alternative coding schemes, so to structure the discussion we categorize them in terms of local or distributed schemes, and discrete hence, continuous representations. There has been only marginal effort expended to date on comparing the quantitative and qualitative benefits of the various representation schemes, although the work of Hancock (1988) is one useful reference. Walters (1987) has also suggested a mathematical framework within which the various schemes may be compared.

B4.7.1   Local versus distributed schemes

One of the first issues that needs to be resolved when considering schemes to present data to a neural network is the choice of *distributed* or *local representations*. A local representation is one in which the feature space is divided into a fixed number of intervals or categories, and a single node (or a cluster of nodes) is used to represent each category. For example, a local input representation for a neural network to classify the range of colors in the visible spectrum would use a seven node input, in which each node is assigned one of the colors, figure B4.7.1.

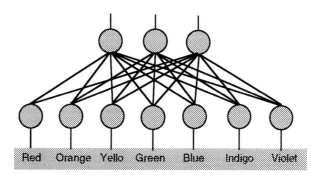

Figure B4.7.1. A local representation scheme.

Each node has a unique interpretation and they are nonoverlapping. A color is represented by activating the appropriate node. Local representations typically use binary (or bipolar) activation levels. However, it is possible to use continuous valued nodes and introduce the concept of *fuzzy* or probabilistic   D1.2 representation. The representation usually operates in a one-of-n mode, but it is also possible to indicate the presence of two or more features by turning on each of the relevant nodes simultaneously.

A distributed representation is one in which a concept or feature is represented by a pattern of activity over a large set of units. The units are not specific to any individual feature but each unit contributes

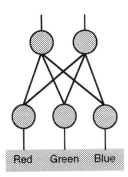

Figure B4.7.2. A distributed coding scheme.

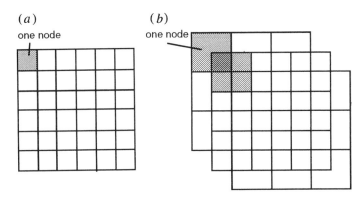

Figure B4.7.3. (*a*) A local representation. (*b*) Coarse distributed representation.

to the representation of many features. For example, a distributed representation to encode the spectrum described above could employ just three nodes to represent the primary colors (red, blue, green) and describe the full color spectrum in terms of the combinations of the primary colors, figure B4.7.2.

Table B4.7.1. Characteristics of local representation schemes.

Advantages	Disadvantages
It is a simple representation scheme which allows direct visibility of variables.	Local schemes do not scale well—a node is required for each input feature.
More than one concept can be represented at any time by activating units simultaneously.	A new node has to be added in order to encode a new feature.
If continuous valued units are used then probabilistic representations can be implemented.	They are sensitive to node failures and are consequently less robust than distributed schemes.

One example of a distributed scheme is Hinton's *coarse coding* (Rumelhart and McClelland 1986). In coarse coding each node has an overlapping receptive field, and a feature or value is represented by the simultaneous activation of several fields. Hinton (1989) has contrasted the two schemes in the following manner.

In figure B4.7.3(*a*), a local representation scheme is depicted. The state space is divided into 36 states, and a neuron is assigned to each state. Figure B4.7.3(*b*) shows how the state space could be mapped onto a coarse coding scheme using neurons with wider, and overlapping, receptive fields. In this example each neuron in the coarse coding scheme has a receptive field four times the size of that in the local representation. The feature space is represented with only 27 nodes in the coarse coding, but requires 36 nodes in the local representation scheme. The economy offered by coarse coding can be improved by increasing the size of the receptive field. The accuracy of the coarse coding scheme is also

Table B4.7.2. Characteristics of distributed representation schemes.

Advantages	Disadvantages
Distributed schemes are efficient (in the ideal case they require $\log n$ nodes, where n is the number of features).	Distributed schemes are more complex than local schemes.
Similar inputs give rise to similar representations.	Variables are not directly accessible but must be 'decoded' first.
They are robust to noise or faulty units because the representation is spread across many nodes.	Distributed schemes can only represent a single variable at any one time.
Addition of a new concept does not require the addition of a new unit.	

improved by increasing the size of the receptive fields. This is possibly counterintuitive, but the increased field size ensures that the overlapping field zones become increasingly more specific. Hence, accuracy is proportional to nr where n is the number of nodes and r is the receptive field (or radius).

Hinton suggests that coarse coding is only effective when the features to be represented are relatively sparsely distributed. If many features co-occur within a receptive field, then the patterns of activity become ambiguous and individual features cannot be distinguished. As a rule of thumb, Hinton suggests that the size of the receptive fields should be similar to the spacing of the feature set.

In tables B4.7.1 and B4.7.2 the properties of local and distributed coding schemes are described.

References

Hancock P 1988 Data representation in neural nets: an empirical study *Proc. 1988 Connectionist Models Summer School (Carnegie Mellon University)* ed D Touretzky, G Hinton and T Sejnowski (San Mateo, CA: Morgan Kauffman)

Hinton G 1989 Neural networks *1st Sun Annual Lecture in Computer Science* (University of Manchester, UK)

Rumelhart D E and McClelland J L 1986 *Parallel Distributed Processing: Explorations in the Microstructure of Cognition* (Cambridge, MA: MIT Press)

Walters D K W 1987 Response mapping functions: classification and analysis of connectionist representations. *IEEE 1st Int. Conf. on Neural Networks* ed M Caudill and C Butler (New York: IEEE Press)

B4.8 Discrete codings

Thomas O Jackson

Abstract

See the abstract for Chapter B4.

In general continuous codings provide better performance than discrete. This point will not be justified here, but a detailed investigation is reported in Hancock (1988). However, in some circumstances we may have to use discrete codings and discrete nodes. For example, if we are using an off-the-shelf *VLSI*   E1.3, E1.4.3 neural network; many commercial neural network chips use discrete implementations. Hence, despite the performance advantage of continuous codings we shall look at both discrete and continuous schemes for representing numbers. We will start with a discussion of discrete schemes.

B4.8.1 Simple sum scheme

The most basic coding scheme for representing real values using a layer of discrete input nodes is the *simple sum scheme*. This scheme represents a number, N, by setting an equivalent number of nodes to an active state. For example, the number 5 could be represented by the binary patterns 0000111111, or 110000111 or 111110000. This scheme offers simplicity as well as some inherent fault tolerance (the loss of an individual node does not result in large error in the value of the variable represented). For small numeric ranges this approach is practical. However, it does not scale well; representing a large range of numbers (e.g. 1–1000) soon becomes prohibitive.

B4.8.2 Value unit encoding

An encoding closely related to the sum scheme is *value unit encoding* (also known as *point approximation* Gallant (1993)). In this method each node is assigned a unique interval within the input range $[u, v]$. A node becomes active if the input value lies within its interval. The intervals do not overlap, so only one unit is active during the representation of a number (i.e. it is a local representation scheme). The precision of the representation is bounded by the interval width, which in turn is defined by the number of units used. The scheme can be represented in the following manner:

$$\left. \begin{array}{c} a_1 \\ a_2 \\ \vdots \\ a_k \end{array} \right\} = (+1)\,\text{iff} \left\{ \begin{array}{cccc} u < x \le u + \partial \\ u < x \le u + 2\partial \\ \vdots & \vdots & \vdots & \vdots \\ u < x \le u + n\partial \end{array} \right. \tag{B4.8.1}$$

where n is the number of nodes, a_n is the output activation of unit n, and ∂ is the interval size given by $(v - u)/n$. Note that the lower limit of the range, u, is represented by an all zero representation.

As an example, to represent a range of values $[0,15]$ using five input nodes, an interval width of 3 is required. Representations for the values 2 and 10 would be as in figure B4.8.1.

The efficiency of the value unit encoding scheme is clearly dependent upon the degree of precision required; higher precision requires the use of more units and a reduction in the economy of representation. Unlike the sum scheme, this technique does not offer fault tolerance because the failure of a single node can lead to a loss of representation.

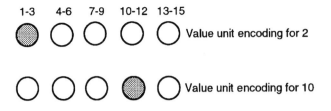

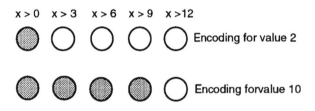

Figure B4.8.1. Example of value unit encoding.

B4.8.3 Discrete thermometer

Discrete thermometer encoding is an extension to *value unit encoding*; the units are coded to respond over some interval of the input range $[u, v]$. However, thermometer coding is a distributed scheme and a unit is always active if the input value is equal to, or greater than, its interval threshold. To represent a value in the range $[0,15]$ the following representations would be used, figure B4.8.2.

Figure B4.8.2. Example of a discrete thermometer encoding.

For an input range of $[u, v]$ the thermometer code can be expressed in the following manner:

$$\left.\begin{array}{c} a_1 \\ a_2 \\ \vdots \\ a_k \end{array}\right\} = (+1)\,\text{iff} \begin{cases} x \geq u + \partial \\ x \geq u + 2\partial \\ \vdots \quad \vdots \quad \vdots \\ x \geq u + n\partial \end{cases} \tag{B4.8.2}$$

where n is the number of nodes, a_n is the output activation of unit n, and ∂ is the interval size given by $(v - u)/n + 1$.

The thermometer scheme has some inherent fault tolerance, due to the fact that the failure of a node does not result in a large error in the value represented. The maximum error introduced by the failure of a single node is equivalent to the value of the interval width.

One of the benefits of the thermometer scheme is that variable precision can be controlled in a simple manner: the precision can be improved by reducing the size of the intervals. The cost of this improved resolution is the need to use more units for any given range of input values. Where economy of representation is required (for example in hardware implementations) precision can be traded for larger interval widths and fewer nodes. In situations where both precision and compactness are required, the *group and weight* scheme may be more appropriate.

B4.8.4 Group and weight scheme

Takeda and Goodman (1986) have proposed a discrete representation which combines the economy of binary representations with the strengths of the simple sum scheme. A number is represented as a bit pattern, using N bits. The bit pattern is split into K groups, each of which has M bits (hence $N = KM$). The bits in each group are summed and multiplied by a base number given by $M + 1$. The algorithm to transform a number using this group and weight approach is as follows:

$$\sum_{k=1}^{K}\left[(M + 1)^{k-1}\sum_{i=1}^{M} x_{ki}\right] \tag{B4.8.3}$$

where x_{ki} is bit i of group k.

For example, to represent the number 5 using a 6-bit pattern, with two groups of three bits (i.e. $M = 3$, $k = 2$). This can be represented by 100×100. Expanding this using equation (B4.8.3) gives us

$$[4^1 \times (1 + 0 + 0) + 4^0 \times (1 + 0 + 0)] = 5.$$

The binary and simple sum scheme are special cases for equation (B4.8.3). If $M = 1$ and $K = N$, then it reduces to the binary case. If $M = N$ and $K = 1$, then we have the simple sum scheme. One difficulty with this scheme is that there are many possible permutations for representing any number. In the above example (010 100), (001 010) (001 001) (etc) are all valid bit patterns for the number 5. This can make generating a training set problematic.

B4.8.5 Bar coding

A simple variation on the thermometer scheme has been employed by Anderson (1995), which can be loosely described as 'bar coding'. This scheme incorporates elements of linear thermometer coding with aspects of topographical map representation (see Section C2.1), and is modeled on neurobiological mechanisms observed in the cerebral cortex regions. A continuous parameter is represented by a state vector with two fields. The first field is a 'symbolic' field which provides a unique code for the value (e.g. Anderson has used binary ASCII codes to represent characters). The second field is an analog code represented by a 'sliding bar' of activity on a 'topographical scale'. The activity bar is represented by activating consecutive nodes in the input layer. This is described in figure B4.8.3.

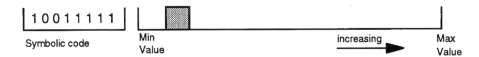

Figure B4.8.3. Two-field state vector with 'symbolic' field and sliding analog field (after Anderson (1995)).

Vectors in this representation scheme can be concatenated together to represent multiple parameters. A further variant on the theme is the use of an activity bar that can increase or decrease in width in order to represent the degree of similarity between two states, figure B4.8.4.

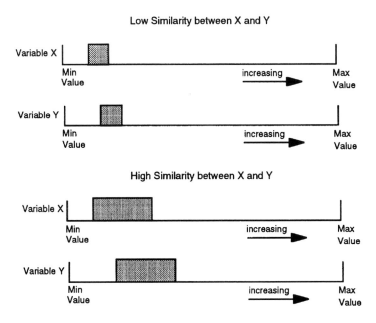

Figure B4.8.4. The use of an activity bar of increasing or decreasing width is used to represent the degree of similarity between two vectors (after Anderson (1995)).

Anderson has used this scheme in a neural classification system to represent multiparameter continuous valued signals from a radar. A typical input vector was composed of five signal parameters and had the following form:

azimuth	elevation	frequency	pulse-width	pseudo-spectra
[0000111100]	[0111000000]	[0000011110]	[0011110000]	[0001010101000]

The variables (e.g. azimuth, elevation) are represented by an 'activity bar' consisting of three or four active nodes. The position within the frame represents the magnitude. The 'pseudo-spectra' field is used to encode category information about the type of the radar signal. There were three signal types used in the training example: a monochromatic pulse, a phase modulated signal or a continuous frequency sweep signal. A single active node was used to represent a monochromatic pulse, an alternating sequence (as shown in the example) was used to represent a phase modulated frequency. A continuous block of active nodes was used to represent a signal with a continuous frequency sweep. The patterns used are 'caricature' representations of the spectrum produced by Fourier analysis of each signal type. The signal codes are positioned within the pseudo-spectra data field relative to the center frequency of the signal.

The approach used here by Anderson raises an interesting issue, namely mixing data types within any single or output vector. In practice many data sets will be composed of diverse data types, for example, continuous, discrete, binary, symbolic. There is no reason, other than hardware constraints, why these diverse types cannot be represented simultaneously within a network input or output layer. For example, to generate a feature vector to capture information for trading on a financial market, we may need to represent each of the following: share-price, share-price-index, share-price-rising, month, company. This could map onto a feature vector with the following data types: continuous value, continuous value, bipolar (Y,N), discrete, symbolic. An example of a vector to represent this data may be: (4.59, 101.3, +1, 10, 111000).

B4.8.6 Nonlinear thermometer scales

The discrete thermometer and bar coding schemes we have discussed so far have used linear scales and constant width intervals. However, these schemes can also be adapted to use nonlinear numeric scales, to accommodate nonlinear trends in data. For example, if the data have a large range we may wish to make the intervals logarithmic in order to enhance the regions of interest. Wasserman (1993) suggests that Tukey's (1977) transformational ladder lists a useful set of methods to consider for monotonically increasing or decreasing nonlinear representations. The list is as follows:

- $\exp(\exp(y))$
- $\exp(y)$
- y^4
- y^2
- $y^{0.5}$
- $y^{0.25}$
- $\log(y)$
- $\log(\log(y))$

Monotonically increasing data sets would use the transforms in the upper half of the list, decreasing distributions would use the transforms in the bottom of the list. Other methods such as normal and Gaussian distributions would also clearly be applicable. These methods can also be applied in the continuous valued variants for thermometer coding.

B4.8.7 *N*-tupling preprocessing

The representation schemes we have considered so far are biased towards multilayer networks derived from the perceptron model. However, there is a class of neural network schemes which do not use nodes and weights architectures. The class of networks in question are binary *associative networks* such as the binary associative memory (Anderson 1995), *WISARD* (Aleksander and Morton 1990), and the *advanced distributed associative memory* (Austin 1987). These networks rely on binary input representations, and

C1.3, F1.4
C1.5.4, C1.5.8

place quite different demands upon the form of representations that can be employed. In particular these networks rely upon the use of sparsely distributed binary input vectors.

One representation technique that is applicable in this domain is N-tuple preprocessing (Browning and Bledsoe 1959). N-tupling is a one-step mapping process that semi-orthogonalizes the input data by greatly increasing the dimensionality of the input vector. The input is sampled by an arbitrary number of N-tuple units. The function of a tuple unit is to map an N-bit binary vector onto a discrete location in a 2^N address space (i.e. a tuple unit is a one-of-N decoder), this is shown in figure B4.8.5. The N-tuple sampling produces a high-dimensional but *sparse* coded binary representation of the input vector.

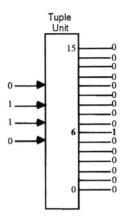

Figure B4.8.5. A 4-tuple unit, showing the 4 to 16-bit vector expansion.

The increase in dimensionality is defined by

$$\dim(\overset{r}{x}) \rightarrow 2^N \left[\frac{\dim(\bar{x})}{N} \right] \tag{B4.8.4}$$

where N is the dimensionality of the tuple units, and $\bar{x}$ is the input vector. From (B4.8.4) it can be seen that N-tuple sampling increases the dimensionality of the input vector x_n and reduces the density x_s/x_n of the vector. For binary networks N-tupling is an effective preprocessing method.

References

Aleksander I and Morton H 1990 *An Introduction to Neural Computing* (London: Chapman and Hall)

Anderson J A 1995 *An Introduction to Neural Networks* (MIT Bradford Press)

Austin J 1987 ADAM: A distributed associative memory for scene analysis *1st IEEE Int. Conf. on Neural Networks* ed M Caudill and C Butler (San Diego, CA: IEEE)

Browning W and Bledsoe W 1959 Pattern recognition and reading by machine *Proc. astern J. Comp. Conf.* pp 225–232

Gallant S I 1993 *Neural Network Learning and Expert Systems* (MIT Bradford Press)

Hancock P 1988 Data representation in neural nets: an empirical study *Proc. 1988 Connectionist Models Summer School (Carnegie Mellon University)* ed D Touretzky, G Hinton and T Sejnowski (San Mateo, CA: Morgan Kauffman)

Takeda M and Goodman J W 1986 Neural networks for computation: number representations and programming complexity *Appl. Opt.* **25** 3033–47

Tukey J W 1977 *Exploratory data analysis* (Reading, MA: Addison-Wesley)

Wasserman P D 1993 *Advanced methods in neural computing* (New York: Van Nostrand Reinhold)

B4.9 Continuous codings

Thomas O Jackson

Abstract

See the abstract for Chapter B4.

Continuous codings provide more robust and flexible means for coding numbers, both real valued and integer. There are several popular forms for continuous coding of inputs, all of which rely on the use of units with a continuous graded output response. These schemes will now be discussed.

B4.9.1 Simple analog

The simplest continuous valued representation scheme is the use of direct analog coding, whereby the activation level of a node is directly proportional to the input value. It would be a reasonable approximation to suggest that this method is probably used in 60–70% of neural network applications. Neuron models typically use an activation range of $[0, 1]$ or $[-1, +1]$. In order to use the analog coding scheme over any given number range, $[u, v]$, we simply linearly scale the representation. If the number range is offset from zero then we can use a simple transform:

$$\text{value in range } (u, v) = (v - u)[a_i] + u \tag{B4.9.1}$$

where a_i is the activation of the node.

The simple analog scheme is robust and economical. The most significant weakness in this technique is the potential loss of precision when scaling the input over a large range. For example, given an input range of $[0, 1000]$, the difference in representation between two input values such as 810 and 890 can be masked by the precision of the neuron transfer function. This effect is more pronounced at the extremes of the range due to the nonlinearity of the *sigmoid transfer function*. Some of these difficulties can be avoided   B3.2.4 by careful preprocessing of the data, using methods such as *normalization* (see section B4.4.1). Also, a   B4.4.1 data set that has a large dynamic range can be preprocessed using a logarithmic representation. This will allow the large range of the data to be compressed, but will emphasize small percentage deviations which may be of greatest relevance to the classification problem.

The effect of the nonlinearity in the sigmoid transfer function is of greater concern when the scheme is used for representing variables at the output stage of a *multilayer perceptron* network. Care must be taken   C1.2 to avoid using output values which place the nodes in their saturation mode (i.e. outside of the nonlinear region of the sigmoid function); failure to do so can lead to excessively long training times. This is due to the fact that the output error value propagated through the network during the backpropagation training phase is proportional to the derivative of the sigmoid function. At the points of saturation the rate of change in output with respect to input activation tends to zero. As a consequence the rate of change of weights also tends to zero, and training rates crawl along at a prohibitively slow pace. To combat this problem, the outputs should be offset from the limits by some scaling factor. Guyon (1991) has demonstrated that the multilayer perceptron algorithm training performance is improved by biasing the sigmoid function such that it is asymmetric, figure B4.9.1. He proposed the following modifications to the sigmoid function to make it asymmetric about the origin:

$$f(x) = \frac{2a}{1 + e^{-bx}} - a. \tag{B4.9.2}$$

Suggested values of a and b (which are scaling and bias terms) are:

$$a = 1.716$$

and

$$b = 0.66666.$$

For convenience it is useful to set the target output range for the MLP between the limits of ± 1. These bias values allow an adequate offset of ± 0.716.

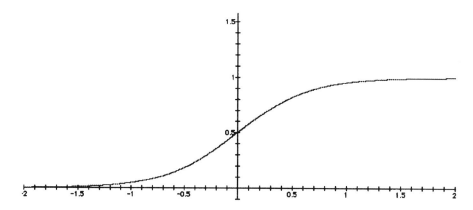

Figure B4.9.1. Offset, asymmetric, transfer function.

A typical example of this encoding technique can be found in Gorman and Sejnowski's (1988) neural sonar recognition system. Here a neural network is trained to classify sonar returns, distinguishing between mines and similarly shaped natural objects. The sonar signal is a power/frequency spectrum, as shown in figure B4.9.2. The spectral envelope is sampled at sixty points by sixty analog neuron nodes. Each node records a single value in the envelope. This example illustrates the inherent simplicity of analog codings. However, one downside to this simplicity is that the scheme offers no fault tolerance; if a node fails then the representation is lost.

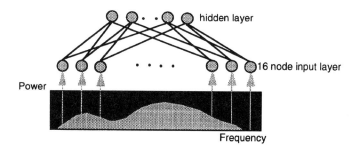

Figure B4.9.2. Sampling of the spectral envelope by the analog coding scheme.

B4.9.2 Continuous thermometer

The continuous thermometer coding is a mix of the discrete thermometer and simple analog methods. The advantage of the continuous scheme over the discrete scheme is that higher precision can be achieved using fewer nodes. This is due to the fact that each node can represent a continuous range of values within its interval. It offers similar fault tolerant properties to the discrete scheme. An example is shown in figure B4.9.3.

B4.9.3 Interpolation coding

Interpolation coding, proposed by Ballard (1987), is a multiunit extension of simple analog coding. In the simplest case, a single analog unit is replaced by two units, with the output activation functions mapped in

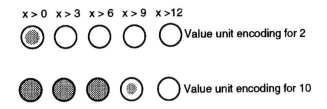

Figure B4.9.3. Continuous thermometer scheme.

opposition to each other. The outputs of the units always sum to a total of one, but one unit's activation decreases linearly with the increase in the other. The scheme can also be used in thermometer type codings, with pairs of units being assigned to each interval. For example, using a thermometer range of 0–12, the output for the value 2, and the output for the value 10 can be encoded as shown in figure B4.9.4.

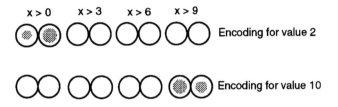

Figure B4.9.4. Two-unit interpolation encoding.

This method can also be extended across multiple units. This scheme has been found to have good resilience to noise (Hancock 1988).

The output is decoded using the following algorithm.

- Determine the value of the node with maximum response, o_1 and the value of the highest neighbor, o_2. The peak responses (or center response), p, for the selected nodes are then weighted by the actual response, and the output value is given by

$$\text{output} = \frac{(p_1 o_1 + p_2 o_2)}{(o_1 + o_2)}. \tag{B4.9.3}$$

B4.9.4 Proportional coarse coding

In section B4.7.1 we described how a coarse distributed scheme can represent a feature space using the simultaneous activation of many discrete units. Coarse coding can also be implemented with nonlinear activation functions. The contribution to the output value from each node is not linear but is proportional, the relative contributions being controlled by the activation function. Saund (1986) has developed a scheme which uses the derivative of the sigmoid function as the proportionality function

$$f(x) = \frac{1}{1 + e^{-x}}$$

$$f'(x) = \frac{e^{-x}}{(1 + e^{-x})^2}. \tag{B4.9.4}$$

Examples of the derivative are shown in figure B4.9.5. The width of the function can be controlled by

a gain parameter. The width of the function controls the degree of distribution across the nodes (i.e. the coarseness of the representation). Saund calls this a *smearing function*.

The layer of units is configured in the same manner as a thermometer coding: each unit is assigned a response interval. However, the scheme differs from thermometer coding in that intervals overlap. To represent a variable the smearing function is centered at the value of the variable, x, and the units within the range of the function are activated to the level determined by the smearing function.

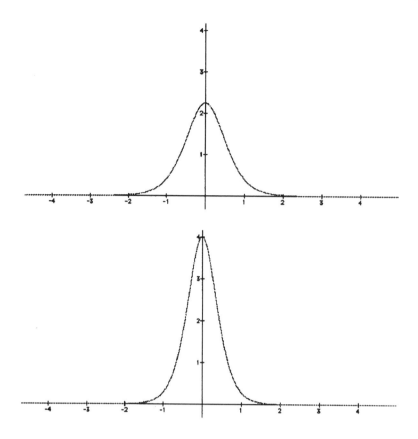

Figure B4.9.5. Proportionality functions based on the derivative of the sigmoid function.

To determine the value of a number represented by a pattern of activity, the smearing function is 'slid' across the outputs until a best-fit is found. The best-fit is determined by the placement which minimizes the least square difference

$$Q(x) = \sum_i (s_{x-i} - a_i)^2 \tag{B4.9.5}$$

where a is the activation value of the node at interval i and s_{x-i} is the value of the smearing function at point x within the interval. The placement of the function at the best-fit point indicates the value of the variable. An example is shown in figure B4.9.6. Saund reports that variable precision of better than 2% can be achieved using eight units.

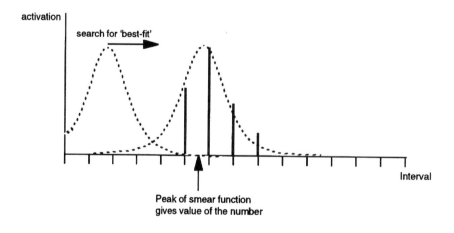

Figure B4.9.6. The smearing function determines the point of maximum response (after Saund (1986)).

B4.9.5 Computational complexity of distributed encoding schemes

The advantage of distributed schemes is their compactness and robustness to damage or noise. The penalty paid for this compactness is complexity. For example, in Hancock (1988) a proportional coarse coding scheme is described which is based upon a Gaussian distribution:

$$\text{output} = \exp(-0.5(\Delta/\sigma)) \tag{B4.9.6}$$

where Δ is the distance of the input from the node's center value, and σ is the standard deviation of the Gaussian curve.

Hancock describes a one-pass algorithm which is used to 'decode' the representation. The example is based upon a four-node representation. Each of the units, a_1–a_4 has a value at which it gives peak response, p_1–p_4. The purpose of the algorithm is to establish the distance of the actual response from the peak response, and subsequently determine the value represented by the nodes. The algorithm is as follows:

- find the unit, a_1, with the highest output, o_1;
- find the neighboring unit a_2 with the next highest output, o_2;
- calculate the offset Δ_2 from the peak response p_2, using

$$\Delta = [-2\ln(o_2)]^{1/2} \, (|p_2 - p_1|)/\sigma;$$

- calculate an initial estimate x_2 of the output value:

$$\text{if } p_1 > p_2 \text{ then } x_2 = p_2 + \Delta_2 \text{ else } x_2 = p_2 - \Delta_2;$$

- form an estimate x_i for each of the other units, i:

$$\text{if } x_2 > p_2 \text{ then } x_i = p_i + \Delta_i \text{ else } x_i = p_i - \Delta_i;$$

- calculate the output value by weighting the individual estimates according to the actual outputs of each unit:

$$\text{output} = \frac{x_1 o_1 + x_2 o_2 + x_3 o_3 + x_4 o_4}{o_1 + o_2 + o_3 + o_4}.$$

This example highlights the computational overhead that is associated with some of the more complex distributed encoding schemes. It is worth highlighting this issue because this decoding must be performed as a postprocessing activity, and hence requires additional computer resource. In software implementations of neural systems this may not present a problem; however, it is more problematic (or costly) in systems that use dedicated hardware. In some circumstances the computational overhead associated with these coding methods may be too high, and simpler schemes may prove more pragmatic.

References

Ballard D H 1987 Interpolation coding: a representation for numbers in neural models *Biol. Cybern.* **57** 389–402

Gorman R P and Sejnowski T J 1988 Analysis of hidden units in a layered network trained to classify sonar targets *Neural Networks* **1** 75–89

Guyon I P 1991 Application of neural networks to character recognition *Int. J. Patt. Recog. Artif. Intell.* **5** 353–82

Hancock P 1988 Data representation in neural nets: an empirical study *Proc. 1988 Connectionist Models Summer School (Carnegie Mellon University)* ed D Touretzky, G Hinton and T Sejnowski (San Mateo, CA: Morgan Kauffman)

Saund E 1986 Abstraction and representation of continuous variables in connectionist networks *Proc. A.A.A.I-86: Fifth National Conference on Artificial Intelligence (Philadelphia, PA: Los Altos, Kaufmann)* 638–43

B4.10 Complex representation issues

Thomas O Jackson

Abstract

See the abstract for Chapter B4.

B4.10.1 Introduction

In our review of data representations we have so far restricted the discussion to the representation of real-valued variables. However, in some application domains we may wish to represent more complex variables and concepts, such as time or symbolic information. There are many diverse methods being developed to facilitate the representation of these complex parameters, but an in-depth review of these methods is outside the scope of this chapter. However, we shall highlight a number of techniques which are broadly representative of developments in this area. Firstly, we shall consider how to represent time in neural networks. Secondly, we shall review the work of Pollack and discuss symbolic representation. It will become apparent that the network topology and the form of data representation become highly interdependent in these domains.

B4.10.2 Representing time in neural systems

The question of representing time in neural systems raises many interesting issues. We shall discuss three fundamental approaches to the problem, and illustrate them with examples of their use in typical applications. These approaches broadly split into the following methods:

- representing time by transforming it into a spatial domain;
- making the representation of data to a network time-dependent through the use of delays or filters in *time delay networks*;
- making a network time-dependent by the use of recursion.

B4.10.2.1 Transforming between time and spatial domains

Many *signal processing* domains produce data that have important temporal properties, for example, in F1.8, G3.3 *speech processing* applications. In general, neural network topologies are configured to handle static data, F1.7, G1.4 and are not able to process time-varying data. One method to resolve this problem is to transform time varying signals into a spatial domain. The simplest way to do this is to sample a time-varying signal, using n samples, and represent it as a time ordered series of measurements in a static feature vector: $[t_1, t_2, \ldots, t_n]$. Alternatively, the signal can be sampled and transformed into a spatial domain using mathematical techniques such as fast Fourier transforms (FFTs) or spectrograms.

Examples of this approach can be seen in many neural network applications, for example in Kohonen's phonetic typewriter, and in the NETtalk system, both of which are speech processing systems.

Kohonen (1988) has developed a neural based system for real-time speech-to-text translation (for phonetic languages). The key to Kohonen's system is the transformation of a time-varying speech signal into a spatial representation using FFTs. The speech signal is sampled at 9.83 millisecond intervals. This is achieved using a D/A converter, the output of which is analyzed using a 256 point fast Fourier transform. The Fourier transform extracts 15 spectral components, which, after normalization, form the features of the

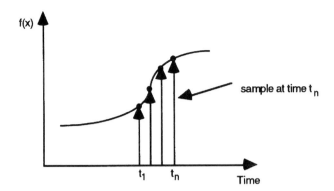

Figure B4.10.1. Sampling a time-varying signal, into *n* discrete measurements.

input vector. This is a static vector, representing the spatial relationships between the instantaneous values of 15 frequency components. The sampling interval of 9.83 milliseconds is much shorter than the duration of a typical speech phoneme (which vary in duration from 40 to 400 milliseconds) and as a consequence the classification of a phoneme is made on the basis of several consecutive samples (typically seven). A rule-based system is used to analyze the transitions between the samples and subsequently classify the speech phonemes. Hence, the neural network is used to identify and classify the static, spectral signals, but rule-based postprocessing is used to capture the temporal properties.

A similar approach can be seen in the NETtalk system, although in this application the spatial relationships in the data are of more specific concern than the temporal properties. The NETtalk system was developed by Sejnowski and Rosenberg (1987). It is a neural system which produces synthesized speech from written English text. The neural network generates a string of phonemes from a string of input text; the phonemes are used as the input to a traditional speech synthesis system. Pronouncing English words from written text is a nontrivial task because the rules of English pronunciation are idiosyncratic and the sound of an individual character is dependent upon the context provided by the surrounding characters contained in a word. As a consequence the neural network uses a 'sliding' window that is able to 'view' characters behind and ahead of any individual input character. The NETtalk system uses a seven character window, which slides over a string of input text. This is described in figure B4.10.2. Each of the characters within the frame is fed to one of seven groups within the input layer. Each input cluster is composed of 29 input units. The clusters use local representation; a character is represented by activating one of the nodes (26 alphabet characters plus three special characters including a 'space' character). Using this approach, and a supervised training algorithm, the network is able to learn the phonetic translation of each central character input, whilst accounting for the context of the surrounding characters. Although this application is not strictly a problem with temporal properties, it can be appreciated that this type of approach could be usefully applied to time-varying signals.

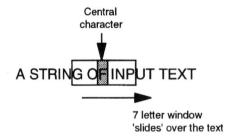

Figure B4.10.2. The text 'window' used in the NETtalk system.

These two examples demonstrate how it is possible, using appropriate preprocessing and postprocessing, to generate data representations in time-dependent domains that are devoid of explicit temporal properties, and which make use of spatial relationships that standard neural network topologies can readily process.

B4.10.2.2 Time-delay neural networks

In the preceding section we described methods for representing time-varying signals using spatial representations. However, in some applications we are not concerned with analyzing a signal at a specific point in time, but in predicting the state of a signal at a future point in time. In these circumstances, we need to encapsulate the notion of time dependency within the neural network solution. This can be achieved using time delays or filters to control the effect, with time, of the network inputs on the internal representations. One network incorporating this approach is the time-delay neural network (TDNN) developed by Lang and Hinton (1988) for phoneme classification.

The operation of the TDNN relies on two key modifications to the standard multilayer network topology; the introduction of time delays on inter-layer connections and duplication of the internal layers of the network. The hidden layer and the output layer are replicated (in Lang and Hinton's example there are ten duplicate copies of the hidden layer and five duplicate copies of the output layer) with identical sets of weights and nodes. The input vector is time sliced with a moving window (in a similar fashion to the NETtalk system), and a sampled section, at time t_n, is presented to one copy of the hidden layer via time delays of t_n, t_{n+1}, t_{n+2}, and so on. In a similar manner, the activity represented at the hidden layer is passed to one copy of the output layer via five time delays. At time t_{n+1}, the input is moved to the next time slice, and this is presented to the next copy of the hidden layer and the next copy of the output layer. Using this approach the variation of the input signal over time has a direct impact on the internal representations formed by the network during training. The detailed mechanics of the network will not be discussed here, but are presented in Section C1.2. For the purposes of our discussion we wish to highlight the fact that there are no specific constraints on the data representation to capture the time series. The temporal properties are captured, via the time delays, in the network topology itself.

B4.10.2.3 Time sensitivity through recursion

The two methods described above both suffer from the same limitation that all temporal sequences must be of the same (predetermined) length or sampled on a fixed time base. This may be acceptable in some applications but clearly not in all. Elman (1990) has addressed this issue by developing networks that incorporate the concept of 'memory' through the use of recursion. Memory allows time to be represented in a network by its impact upon the current input state. In figure B4.10.3 a schematic diagram is shown which describes Elman's feedback mechanisms that create a short-term memory module to modify the internal network state parameters on a time-dependent basis.

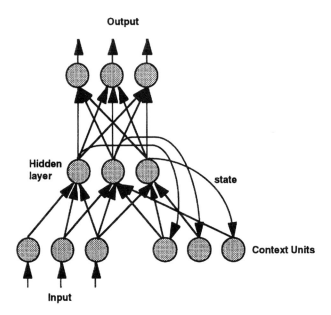

Figure B4.10.3. A simple recurrent network used by Elman to represent time. (Note the feedback connections from the hidden layer to the context layer.) Not all connections are shown (after Elman 1990).

The network shown in the diagram has a memory component; the *context units*. The context units have a one-to-one mapping with the hidden layer, so that any activation at the hidden layer is directly mirrored at the context layer. The context units also have feedforward connections to the hidden layer; each context unit activates all of the hidden units. At time t, the first input is presented to the network. The activation at the hidden layer is replicated at the hidden layer via the feedback connections. At time $t + 1$ the next input is presented and propagated through the network. However, both the input and the context units activate the hidden units. Consequently, the total input to the hidden layer is a function of the present input plus the previous input activation at time t. The context units therefore provide the network with a dynamic 'memory' which is time sensitive.

To demonstrate the principles involved we shall discuss Elman's use of the network for learning sentence structure. In the test application, a set of sentences was randomly generated, using a lexical dictionary of 29 items (with 13 classes of noun and verb) containing 10 000 two- and three-word sentences. Each lexical item was represented by a randomly assigned sparse coded vector (one-bit set in 31, so that each vector was orthogonal to the others). The training process consisted in presenting a total of 27 534 31-bit binary vectors to the network, which were formed from the stream of the 10 000 sentences. The training was supervised, such that the first input word-vector was trained to map onto the next word in the sentence sequence. For example, the sentence 'man eats food' meant that the first input would be the binary representation for 'man'. The associated target vector would be the vector for 'eats'. Similarly, the next input would be 'eats' which would be associated with 'food' as the output target.

Elman discovered that the network had many highly interesting emergent properties when trained on this test set. The prediction task is nondeterministic, sentence sequences cannot be learned 'rote' fashion. However, it was found that the network functioned in a predictive manner and suggested probable conclusions for incomplete sentence inputs.

B4.10.3 Representation of symbolic information

One area of neural computing where the issue of data representation acquires a very different perspective is the domain of cognitive science or artificial intelligence. A wide range of neural networks are being developed which form the basis for cognitive models. The issues in this domain are far reaching and the range of methods that have been developed are highly diverse. However, to draw attention to some of the issues in this novel area of neural computing we shall highlight the work of Jordan Pollack who has developed neural network models for high-level symbolic data representation. This work focuses on the issues of recursion, and the need for flexible data structures when representing symbolic information. The primary reason for discussing this work rather than any of the other major efforts in this area is that Pollack's approach places emphasis on the data representation issues. By way of introduction we shall first define the concept of a 'symbol' and 'symbolic reasoning'.

The most widely accepted model for cognitive reasoning is currently the 'symbolic processing' paradigm. This paradigm hypothesizes that reasoning ability is derived from our mental capacity to manipulate symbols and structures of symbols. A symbol is a token which represents an object or a concept. The formal definition of the symbolic paradigm has been credited to Newell and Simon (1976) and reads as follows: 'a physical symbol system consists of a set of entities, called symbols, which are physical patterns that can occur as components of another type of entity called an expression (or symbol structure)'. One important issue to highlight in this definition is that the symbol representations must display *compositionality*, that is, that they can be combined, systematically, to form new or higher-level concepts.

The challenge facing the neural computing community is to derive neural architectures that are capable of manipulating symbols and symbol structures, whilst adhering to the formalisms defined by the symbol paradigm. Alternatively, the challenge is to propose new, viable models to replace the symbol model of reasoning. To date the bulk of the effort in neural network cognitive research has been focused towards symbolic models. However, there are also a number of researchers calling for a paradigm shift and developing models based at the 'sub-symbolic' level (e.g. Hinton 1991, Smolensky 1988). As we have already stated, these issues are largely outside the scope of our current discussions, but we shall consider some of the data structure issues raised in Pollack's work.

Pollack (1991) has argued that a major failing of connectionism in addressing high-level cognition is the inadequacy of its representations, especially in addressing the problem of how to represent *variable length* data structures (as typified by trees and lists). He has proposed a neural network solution to this problem which draws extensively on the properties of reduced descriptions and recursion. A reduced description is a compact, symbol representation for a larger concept or object. In principle, reduced descriptions support the notion of *compositionality*. The system is called a recursive autoassociative memory (RAAM). He suggests that the RAAM demonstrates that neural systems can learn rules for compositionality if they use appropriate internal representations. The RAAM principle is best described by way of a diagram, see figure B4.10.4.

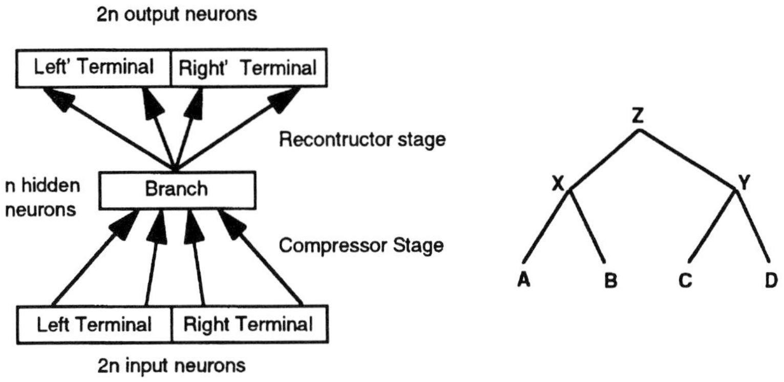

Figure B4.10.4. RAAM network, with typical ternary tree structure which the network can encode.

The RAAM is a two-stage encoding network with a *compressor* stage and a *reconstructor* stage. The input layer to hidden layer is the compressor stage—this combines two n-bit inputs (i.e. two nodes in the tree) into a single n-bit vector. The hidden layer to output layer is the reconstructor, which maps the compressed vector back into its two constituent parts. For example, considering the tree structure in figure B4.10.4, the compressor stage of the network maps the terminals A and B onto a compressed vector representation for terminal X. Similarly C and D are mapped onto a representation for Y. Applying this mechanism recursively X and Y are reapplied to the input layer and are mapped onto a reduced vector representation for the node Z. The reconstructor layer learns the reciprocal mappings, hence Z would be mapped back onto nodes X and Y, and X back to A and B etc. The representation for Z can consequently be considered a reduced representation for the complete tree. These mappings are trained using standard autoassociative backpropagation learning algorithms. A tree of any depth can be represented by this recursive approach. To support the recursion the network uses an external stack (not shown in figure B4.10.4) to store intermediary representations.

The RAAM system can be also used to represent sequences, for example, $(X \rightarrow Y \rightarrow Z)$ by exploiting the fact that they map onto left-branching binary trees, that is, $(((NIL\ X)\ Y)\ Z)$. Pollack suggests that, using these principles, the RAAM can represent complex syntactic and semantic trees (such as required in natural language processing) and represent propositions of the type 'Pat loved John', 'Pat knew John loved Mary'. Given that the propositional sentences can be parsed into ternary trees of type (action agent object), the network can represent a proposition of arbitrary depth. For example, the sentence 'Pat knew John loved Mary' can be broken into the triple sequence (KNEW PAT (LOVED JOHN MARY)). Pollack demonstrated the properties of the network using a training set of 13 propositional sentences, with recursion varying from 1 to 4 levels.

The constituent parts of the propositions were encoded using binary codings (e.g. the human agent set—John, Man, Mary, Pat—was encoded using the binary patterns 100, 101, 110, 111 respectively). Once trained, the system was shown to perform productive generalization. For example, given the triple (LOVED X Y) the network is able to represent all sixteen possible instantiations of the triple even though only four were present in the training set. Pollack argues that this demonstrates that the RAAM is not simply *memorizing* the training set but is learning the high-level principles of compositionality.

Although we do not have time to explore the implications of the network performance in the cognitive domain, it highlights an important issue with respect to data representation. The RAAM network provides mechanisms for representing *arbitrary* length data structures within a fixed topology network. These types of mechanisms are a prerequisite if neural networks are to make any future impact in the domain of symbolic processing. The following references are recommended to readers who may wish to pursue this topic further: Shastri and Ajjanggade (1989), Hinton (1991), Smolensky (1988).

The discussion of the time-dependent networks and Pollack's work demonstrate that in these complex domains the data representations do not differ greatly from the techniques we have discussed in the context of neural networks for pattern recognition. However, it is evident that the structure of the networks play a much more significant role than the input or output representations in determining how the data are interpreted.

References

Elman J L 1990 Finding structure in time *Cognitive Sci.* **14** 179–211

Hinton G E 1991 *Connectionist symbol processing* (Cambridge, MA: MIT/Elsevier)

Kohonen T 1988 The Neural Phonetic Typewriter *IEEE Computer* **21** 25–40

Lang K J and Hinton G E 1988 The development of time-delay neural network architecture for speech recognition *Technical Report CMU-CS-88-152* Carnegie-Mellon University, Pittsburgh, PA

Newell A and Simon H A 1976 Computer science as empirical enquiry: symbols and search *Commun. ACM* **19**

Pollack J B 1991 Recursive distributed representations *Connectionist Symbol Processing* (Cambridge, MA: MIT/Elsevier) ed G E Hinton pp 77–106

Sejnowski T J and Rosenberg C R 1987 Parallel networks that learn to pronounce English text *Complex Systems* **5** 145–68

Shastri L and Ajjanggade V 1989 A connectionist system for rule based reasoning with multi-place predicates and variables *Technical report MS-CIS-89-06* University of Pennsylvania

Smolensky P 1988 Connectionism, constituency and the language of thought *Fodor and his Critics* ed B L G Rey (Oxford: Blackwell)

B4.11 Conclusions

Thomas O Jackson

Abstract

See the abstract for Chapter B4.

The successful design and implementation of a pattern classification system hinges on one central principle—'know your data'. This cannot be overstated. A thorough understanding of the characteristics of the data—its properties, trends, biases and distribution—is a prerequisite to generating training data for neural networks. Poor training data will confound even the most sophisticated neural network training algorithm.

In this chapter we have drawn attention to this issue, and provided a broad overview of techniques for data preparation and variable representation that will contribute to developing efficient neural network classification systems. Neural networks are being applied extensively in many diverse application domains. It would be a mammoth task to try to provide a set of definitive techniques that would cater for all cases, and clearly we have not taken this approach. Instead, we have emphasized the approach to data preparation and analysis which should be adopted, stressing that traditional data analysis techniques, appropriate to the domain in question, should be exploited to the full. Attention to detail in data preparation will reap major benefits in the ease with which a neural solution to a classification task will be found.

We will close with a quote from Saund (1986):

'A key theme in artificial intelligence is to discover good representations for the problem at hand. A good representation makes explicit information useful to the computation, it strips away obscuring clutter, it reduces information to its essentials.'

References

Saund E 1986 Abstraction and representation of continuous variables in connectionist networks *Proc. A.A.A.I-86: Fifth National Conference on Artificial Intelligence (Philadelphia, PA: Los Altos, Kaufmann)* pp 638–43

Further reading

1. Rumelhart D E and McClelland J L 1986 *Parallel Distributed Processing* vol 1 and 2 (Cambridge, MA: MIT Press)

   The PDP volumes provide broad coverage of representation issues. The appendix of volume 1 also contains useful tutorial material on linear algebra.

2. Anderson J A 1995 *An Introduction to Neural Networks* (Cambridge, MA: MIT Press)

   Anderson's book provides a very thorough and interesting discussion of data representation, taking on board developments within the field of neuroscience.

3. Wasserman P D 1993 *Advanced Methods in Neural Computing* (New York: Van Nostrand Reinhold)

   Wasserman has a lengthy section on 'neural engineering' in this book which covers many issues relating to data representation and the application of neural computing methods.

4. Haykin S 1994 *Neural Networks: A Comprehensive Foundation* (New York: MacMillan)

   This book provides a very mathematical treatise of neural computing methods, including discussions of theorems for pattern separability. Not for the mathematically faint-hearted.

B5

Network Analysis Techniques

Contents

B5.1 Introduction

Russell Beale

One of the oft-quoted advantages of neural systems is that they can be used as a black box, able to learn a task without the user having a detailed understanding of the internal processes. While this is undoubtedly true, it is also the case that many errors and cases of poor performance are created by users who use inappropriate networks, architectures or learning paradigms for their problems, and that having a grasp of what the network is trying to do and how it is going about it will inevitably result in the more appropriate and effective use of neural systems.

It is natural to want to extend this understanding to a deeper level, and to ask what exactly is happening inside the network—it is often not sufficient to know that a network appears to be doing something; we want to know how and why it is doing it. Analyzing networks in order to understand their internal dynamics is not an easy task, however. In general, networks learn a complex nonlinear mapping between inputs and outputs, parametrized by the weights, and sometimes the architecture, of the network. This mapping may be distributed over the whole of the network, and it can be difficult or impossible to disentangle the different contributions that make up the overall picture. Any connectist system that has learned a representation is unlikely to have developed a highly localized one in which individual nodes represent specific, atomic concepts, though these do occur in some systems that are specifically designed for a more symbolic approach. Equally, truly distributed representations, in which the contribution of any one element of the network only marginally affects the overall output, are hard to point to. There are visualization tools that allow, for example, the weight values to be pictured, but these do not give the whole story, and the representation of often huge numbers of weights in a two- or three-dimensional space is restrictive at best, useless at worst.

The two sections that follow present different approaches to understanding the behavior of networks and their internal representations. Stephen Luttrell discusses the creation of analyzable networks, in which the network is constructed in such a manner that it is immediately amenable to analysis. While this has the advantage of being comprehensible in terms of its behavior, it results in a network structure that is unfamiliar to most neural network researchers. Alexander Linden presents a different angle on the problem. He discusses the use of iterative inversion techniques on previously trained networks, which helps in finding, for example, false-positive and false-negative cases, and answering 'what if' questions. This approach, in comparison to Luttrell's, can be applied to any pretrained network.

It is likely that future supplements to this handbook will contain descriptions of other approaches to network analysis, and that ongoing research will bring this aspect of neural computation to full maturity.

B5.2   Iterative inversion of neural networks and its applications

Alexander Linden

Abstract

In this section we survey the iterative inversion of neural networks and its applications, and we discuss its implementation using gradient descent optimization. Inversion is useful for analyzing already trained neural networks, for example, finding false positive and false negative cases and answering related 'what-if' questions. Another group of applications addresses the reformulation of knowledge stored in neural networks, for example, compiling transition knowledge into control knowledge (model-based predictive control). Among the applications that will be discussed are inverse kinematics, active learning and reinforcement learning. At the end of this section, the more general case of constrained solution spaces is discussed.

B5.2.1   Introduction

Many problems can be formulated as inverse problems, where events or inputs have to be determined, that cause desired or observed effects in some given system or environment. The corresponding forward formulation models the causal direction, that is, it takes causal factors as input and predicts the outcome due to the system's reaction. Examples of inverse problems are briefly presented here, jointly with their forward formulation.

- For a robot manipulator, the forward model maps its joint angle configuration to the coordinates of the end-effector. The *inverse kinematics* takes a specified desired position of the end-effector as input and determines the configurations that cause it. Usually there will be infinitely many configurations in the solution space (DeMers 1996) for a robot manipulator with excess degrees of freedom.
- In process control, the forward model predicts the next state of some dynamic system, based on its current state and the control signals applied to it. The *inverse dynamics* determines the control signals that would cause a given desired state given the current state (Jordan and Rumelhart 1992).
- In remote sensing (e.g. medical imaging, astronomy, geophysical sensing with satellites) the forward model maps known or speculated characteristics of objects (e.g. geo- and biophysical parameters like nature of soil and vegetation) to sensed measurements (e.g. electromagnetic or acoustic waves). The inverse task is to infer the characteristics of the remote objects given their measurements (Davis *et al* 1995)—see also *Inverse Problems* **10** 1994 for more applications.

It will be assumed, unless otherwise stated, that the problems considered here are such that causes and effects can be adequately described by vectors of physical measurements. Under this assumption, forward models are usually many-to-one functions, since many causes may have the same effects. The inverse does only exist as a set-valued function and learning this with neural networks will cause problems. It can be shown (Bishop 1995) that if specific inputs of a neural network are trained onto many targets, the output will converge to their weighted average, which is usually not an inverse solution.

To avoid this problem, the methodology discussed here will consider inversion as an optimization problem. Inverse solutions will be calculated iteratively based on a given forward model (Williams 1986).

B5.2.2 Introduction to inversion as an optimization problem

Assume a feedforward neural network has already been trained (e.g. by supervised learning) to implement a forward mapping for a given problem. In other words, it implements a differentiable function f, that maps real-valued inputs $x = (x_1, \ldots, x_L)$ to real-valued outputs $y = (y_1, \ldots, y_M)$. Since only the differentiability of f is assumed, the method described here applies to *statistical regression* and *fuzzy systems* as well.

D1

The problem of inversion can now be stated as follows: *for which input vectors x does $f(x)$ approximate a desired y^*?* This question can be translated into an optimization problem: find the x that minimize

$$E = \|y^* - f(x)\|^2 . \tag{B5.2.1}$$

Since f is differentiable, gradient optimization is applicable, whereby the input components of x are considered as free parameters, while the weights of the neural network are held constant. The procedure requires the calculation of the partial derivatives δ_i for each of the input components $x_1, \ldots, x_L$:

$$\delta_i = \frac{\partial E}{\partial x_i} \tag{B5.2.2}$$

$$= \frac{\partial \|y^* - f(x)\|^2}{\partial x_i} = 2(y^* - f(x)) \cdot \frac{\partial f(x)}{\partial x_i} . \tag{B5.2.3}$$

The procedure of computing the δ_i is very similar to the error backpropagation procedure for training the weights of a neural network. The only difference is that error signals are now also computed for the input units and that the partial derivatives for the weights $\partial E / \partial w_{ij}$ need not be computed, since the weights are held constant.

Starting with an initial point $x^{(0)}$ in input space, the gradient-descent step rule for the nth iteration is

$$x_i^{(n)} = x_i^{(n-1)} - \eta \delta_i^{(n-1)} \tag{B5.2.4}$$

where $\eta > 0$ is the step-width. Its iteration over n yields a sequence of inputs $x^{(1)}, x^{(2)}, \ldots, x^{(n)}$, which subsequently minimizes $\|y^* - f(x^{(n)})\|^2$. As is common for gradient-descent techniques, this procedure can get trapped into local minima, that is, if $\|y^* - f(x^{(n)})\|^2$ converges to some $c \gg 0$.

D2.1, C1.4.2 In these cases more global techniques like *genetic algorithms* or *simulated annealing* could be used. Furthermore, gradient descent techniques are sometimes a little slow for real-time applications. Faster gradient optimization methods have already been developed for the purpose of training the weights and are hence applicable to iterative inversion as well. The techniques discussed here are also applicable to

B2.3, C1.2.8 other types of structures, for example, *recurrent neural networks, time-delay neural networks* (Thrun and Linden 1990) and *Hidden Markov Models*. The key idea is to transform these structures into feedforward neural network representation (unfolding from time to space). Therefore, without loss of generality, the following discussion can be focused on feedforward neural networks.

B5.2.3 An example: iterative inversion for network analysis

Although classification† is usually treated as a forward problem, we consider it here as a first demonstration on iterative inversion. Furthermore, it will be illustrated how it can be applied to the analysis of already trained neural networks. The domain of numerical character recognition was chosen for demonstration purposes only.

G1.3 Consider a feedforward neural network (Linden and Kindermann 1989) that has already been trained on *classifying handwritten numerals‡*. Inputs to the network are 8×11 gray-level pixel maps and its ten output units specify the corresponding categories. In figure B5.2.1 the task is to find an input, without looking at the training set, that gets classified as a '3'. Consequently, the output of the network must come close to the vector $(0, 0, 0, 1, 0, 0, 0, 0, 0, 0)$. The process starts in figure B5.2.1(*a*) with the null matrix (hence all pixels are white). A modification to equation (B5.2.4) ensures that input activations do not leave the interval $[0, 1]$:

† The task of classification is to assign categorical symbols to given patterns.
‡ The details will be ignored, because iterative inversion is independent of the structure and the training of the neural network. It should be noted however, that the training set contained 49 different versions of the ten numerals.

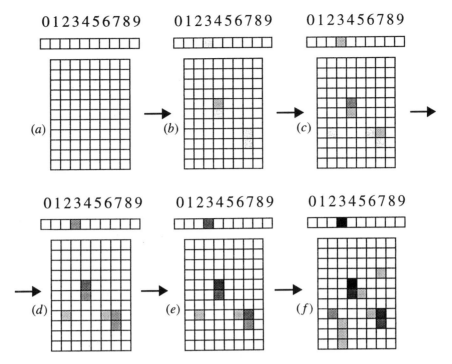

Figure B5.2.1. Example of iterative inversion in a numerical character recognition domain. The snapshots from initial input (*a*) to the final result (*f*) have ten iterations in between. White pixels indicate input activations near zero and black indicates a one.

$$x_i^{(n)} = \min\left[1, \max\left[0, x_i^{(n-1)} + \eta\delta_i^{(n-1)}\right]\right]. \tag{B5.2.5}$$

After a number of iterations, the classification of the input pattern in figure B5.2.1 comes gradually closer to a '3'. Inverse solutions as in figure B5.2.1(*f*) are quite sensitive to the particular choice of initial starting points. Often, domain knowledge can help in choosing good starting points, especially if an expectation about the solution already exists. If no good domain knowledge exists, a neutral or a selection of parallel initial starting points (possibly combined with genetic algorithms) can be chosen.

Sometimes it is required to integrate additional constraints to restrict the number of possible inverse solutions, which is is also called *regularization*. For example, minimizing the extended objective function

$$E = \|y^* - f(x)\|^2 + \lambda\|x^* - x\|^2 \tag{B5.2.6}$$

will favor inverse solutions x that are in the neighborhood of x^* (Kindermann and Linden 1992). The weighting factor $\lambda > 0$ sets a priority between the different objectives. A choice of $\lambda < 0$ favors solutions that are distant from x^*.

This method can also be used to improve the training technique considerably. It is possible, for example, to detect false positive input patterns which are very close to the null matrix, but still get classified as a '7' (figure B5.2.2(*a*)). Augmenting the training set with this and similar derived input patterns and training with the correcting output (Hwang *et al* 1990) leads to improved behavior. For example, figure B5.2.2(*b*) is derived using the same conditions as for figure B5.2.2(*a*), but is less of a false positive. This technique of augmenting a training set can be considered as a kind of *knowledge acquisition* or *selective querying*: a human is put into the loop in order to correct the outputs of the neural network by analyzing its input/output behavior.

The same principle can also be applied to spot false negatives. Figure B5.2.2(*c*) shows an input pattern not classified as a '7' but still close to a typical '7' (x^* has been set to a '7' used during training). This example shows that having access to the classifier can be abused for camouflaging fraud, such that it is not detected. Iterative inversion provides a way to proactively detect possible fraudulent situations.

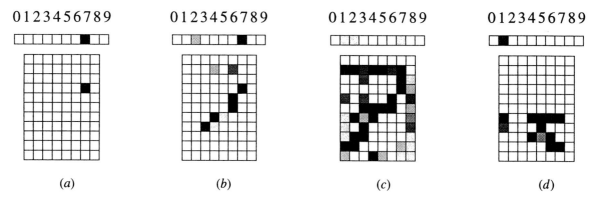

Figure B5.2.2. Interesting input/output relationships can be found with the iterative inversion technique: (*a*) depicts an input pattern that is as 'white' as possible; (*b*) same as (*a*), but with an improved classification network; (*c*) depicts an input pattern that looks like a '7' but does explicitly not get classified as such; (*d*) depicts an input pattern that is 'white' in its upper half, but still gets classified as a '1'.

It is also useful, as will be pointed out in the next section, to hold specific parts of the input vector constant. In figure B5.2.2(*d*), only the lower half of the pixel map was allowed to vary while searching for an input pattern that would be classified as a '1'.

B5.2.4 Applications of knowledge reformulation by inverting forward models

B5.2.4.1 From transition knowledge to control knowledge

Control problems have a natural inverse formulation: given a current state description x_t of a process and a description of a desired state d, what control input u_t should be applied to the dynamic process to yield a given desired state? The corresponding forward formulation is a mapping g which predicts the next state $\hat{x}_{t+1}$ given a current state x_t and a current control u_t as input:

$$\hat{x}_{t+1} = g(x_t, u_t).$$
(B5.2.7)

The following assumes that a forward model g has been identified† for a given process. Iterative inversion can be now applied to calculate a control vector $\tilde{u}_t$ in order to get the dynamic process closer to a desired state $d = g(x_t, u_t)$ given a current state x_t. Inputs to g which represent x_t are held constant during the gradient descent optimization.

This procedure actually implements a technique called *model-based predictive control* (Bryson and Ho 1975) with lookahead 1. The generalization to k-step lookahead can be achieved by k-times concatenating g (see the left part of figure B5.2.3 for an example of $k = 3$). In the general case, the objective function is

$$E = \|d - g(\hat{x}_{t+k-1}, \tilde{u}_{t+k-1})\|^2$$
(B5.2.8)

where $\hat{x}_{t+i}$ is the result of repeatedly applying g to $\hat{x}_{t+i-1}$ and $\tilde{u}_{t+i-1}$ until x_t and $\tilde{u}_t$ are reached. The control signal vectors $\{\tilde{u}_{t+i}\}_{i=1}^{k-1}$ are considered the free variables of the optimization. Only the control vector $\tilde{u}_t$ is sent to the process to be controlled. After the state transition into x_{t+1} is observed, the other control signal vectors $\{\tilde{u}_{t+i}\}_{i=2}^{k-1}$ can be used as starting points for the next iterative inversion.

This neurocontrol method is very flexible and has the potential to deal with even discontinuous control laws, since the control action is computed as the result of gradient descent. It has been applied for dynamic robot manipulator control (Kawato *et al* 1990, Thrun *et al* 1991). Its main drawback is that for real-time purposes the method might be slow, especially if the lookahead k is large. There have been a couple of techniques developed to speed this process up (Thrun *et al* 1991, Nguyen and Widrow 1989). Their basic idea is to use a second neural network trained on the results of iterative inversion in order to quickly compute u_t given x_t and d. This second neural network can either provide good initial starting points or can be used as the controller.

† The field of system identification deals with obtaining approximations of g.

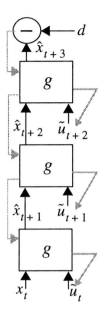

Figure B5.2.3. A cascaded neural network architecture for performing three-step look ahead model-based predictive control. The gray arcs represent the flow of error signals. The gray arcs running into the control variables denote the fact that their corresponding partial derivatives (i.e. error signals) have to be computed for the gradient descent search.

B5.2.4.2 Inverse kinematics

Consider a simple planar robot arm with three joints. The forward kinematics takes the joint angles $\theta = (\theta_1, \theta_2, \theta_3)^\mathsf{T}$ as input and calculates the (x, y)-position of the arm's fingertip. In this simple example, the forward kinematics can be represented by a differentiable trigonometric mapping $K(\theta_1, \theta_2, \theta_3) = (x, y)$. It is again straightforward to derive inverse solutions by iterative inversion (Thrun *et al* 1991, Hoskins *et al* 1992). Figure B5.2.4 illustrates this process by showing the robot arm in each of the joint positions that gradient descent steps through from the initial starting point (i.e. the current position of the robot manipulator) to the final configuration, where its fingertips are at a specified (x^*, y^*) position. Even in this simple case, the inverse mapping is not a function, since many joint angles yield the same fingertip position. Regularization constraints can be included to relax the joints as much as possible or to have minimum joint movement. In analogy to the human planning process, this kind of search can be considered as mental planning, because the robot arm is moved 'mentally' through the workspace (Thrun *et al* 1991) until it coincides with the 'goal'.

B5.2.5 Other applications of search in the input space of neural networks

B5.2.5.1 Function optimization

Optimization of a univariate function f with respect to its input x can be achieved by either performing gradient ascent (for maximization) or descent (for minimization):

$$x_i^{(n)} = x_i^{(n-1)} \pm \frac{\partial f(x^{(n-1)})}{\partial x_i^{(n-1)}} . \tag{B5.2.9}$$

This is a special case of iterative inversion, because the application of equation B5.2.9 is equivalent to iteratively assigning $y^* = f(x) \pm 1$ as desired target and using equation B5.2.3. The following two applications will briefly illustrate the use of function extremization.

B5.2.5.2 Active learning

In active learning (Cohn 1996) the objective is to learn forward models with minimum data collection efforts. Usually one starts with an incomplete or nonexistent forward model. The idea is to derive

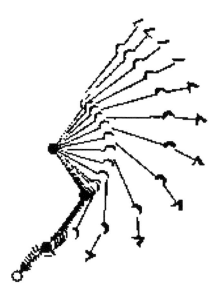

Figure B5.2.4. A planar robot manipulator in each of the calculated points in joint space during an iterative inversion.

points in input space, such that maximal information can be gained for the forward model by querying the environment for the corresponding outputs at these input points. Consider a committee of neural networks†, where a large disagreement between individual neural networks on the same input can be interpreted as something 'interesting' in terms of information gain (Krogh and Vedelsby 1995). The measure of disagreement is a function $A(x)$ based on some kind of variance calculation of the outputs $y_i = f_i(x)$. Query points x are then calculated by maximizing $A(x)$ by equation (B5.2.9). A query on x yields a target y^* which once integrated into the training set will reduce the disagreement of the committee (at least on x and its neighborhood). Other methods in active learning use other heuristics to specify the 'interestingness' or 'novelty' of input points to derive new useful queries (Cohn 1996).

B5.2.5.3 Converting evaluation knowledge into actionable knowledge

Evaluation models estimate the utility or value of being in a particular state or performing a certain control action while being in a state, that is, they calculate functions like $Q(x)$ or $Q(x, u)$. As iterative inversion was applied to infer control knowledge from transition knowledge, it can in the same way calculate actions
C3 from evaluation models. *Reinforcement learning* is one of the most prominent ways of obtaining evaluation models, for example, Q-learning. Control actions can be directly calculated by maximizing $Q(x, u)$ with respect to u for any given x (Werbos 1992). If only state evaluations $Q(x)$ are available, the existence of a transition model $g(x, u)$ is needed to calculate control actions by maximizing $Q(g(x, u))$ with respect to u. Both techniques assume differentiable evaluation models. Unfortunately, some applications have the property that the evaluation models make sudden jumps in the state space (Linden 1993), that is, are not differentiable.

B5.2.6 The problem of unconstrained search in input space

When searching in input space some input configurations may be impossible by the nature of the domain. The information about the validity of inputs is not captured by the structure and parameters of the model f. For example, consider that the variables x_1 and x_2 describe the position of an object on a circle. Hence, x_1 and x_2 have to obey $x_1^2 + x_2^2 = 1$. But gradient descent on x_1 and x_2 in order to minimize $E(d, x) = \|d - f(x_1, x_2)\|^2$ would yield values x_1 and x_2 for which $x_1^2 + x_2^2 \neq 1$. The idea is to find a way of restricting the search space. In this example one would minimize $E(d, \theta) = \|d - f(\sin\theta, \cos\theta)\|^2$ with respect to θ and obtain provable valid solutions.

† A committee of neural networks is a set of neural networks which all try to model the same function. The resulting output of the committee is usually the mean of the individual neural networks: $f(x) = (\sum f_i(x))/n$.

The key idea is to know (or to learn to know) where the input data are actually coming from. If all input data lie on a lower-dimensional manifold $\mathcal{X}' \subset \mathcal{X}$ and it is possible to describe $\mathcal{X}'$ by an auxiliary space $\mathcal{A}$ and a mapping $h : \mathcal{A} \mapsto \mathcal{X}'$ such that

- for each point $a \in \mathcal{A}$ the image $h(a) \in \mathcal{X}'$
- for each point $x' \in \mathcal{X}'$ the inverse image $a \in \mathcal{A}$ exists such that $h(a) = x'$
- and h is differentiable

then, instead of minimizing $E(d, x) = \|d - f(x)\|^2$ with respect to x, one can now minimize $E(d, a) = \|d - f(h(a))\|^2$ in an unconstrained way with respect to $\mathcal{A}$-space, but still conforming to the constraints defined by h. An example for this is the case where all inputs $x_1, \ldots, x_L$ describe a discrete probability distribution, that is, they satisfy $\sum x_i = 1$ and $x_i \geq 0$. In this example, the function h should be the softmax function

$$x_i = \frac{e^{a_i}}{\sum_{j=1}^{L} e^{a_j}} \qquad (B5.2.10)$$

whereby $\mathcal{A}$ is the whole $\Re^L$. Another frequent constraint is the positivity of input variables (e.g. if they describe distances). Here h is simply the component-wise application of the exp function, that is, $x_i = \exp a_i$.

The real challenge is how to acquire h when little is known about the domain. In this context, methods used for dimensionality reduction such as nonlinear principal component analysis might turn out to be useful. The idea is to train autoassociative networks with a bottle-neck hidden layer (Oja 1991) on all input data. The bottle-neck hidden layer here represents the auxiliary search space $\mathcal{A}$. The part of the network that maps the bottle-neck layer representation to the output would represent the function h.

B5.2.7 Alternative approaches

Indirect approaches for obtaining an inverse. Jordan and Rumelhart (1992) presented an approach of learning exactly *one* inverse function by training a second neural network g such that the composite function $f \circ g$ accomplishes an autoassociation task. The only way for g to achieve $x = (f \circ g)(x)$ for all relevant cases x is that g approximates *one* inverse of f. A nice application of this approach is a lookahead controller for a truck backer-upper (Nguyen and Widrow 1989). A drawback of this method is that only one of the many inverse solutions is compiled into g.

Density estimation. Ghahramani (1994) and Bishop (1995) propose a probability density framework to deal with inverse problems. Here, the joint probability distribution of the inputs and outputs $p(x; y^*)$ is learned from data. Inputs x are determined by maximizing the conditional probability $p(x|y)$. Although this framework results only in valid inputs that have actually been used in the training process, high-dimensional input or output spaces make estimating joint probabilities much more data-intensive than simple function estimation. It is also not obvious how to include domain knowledge, for example in the form of fuzzy rules, into a joint density estimation framework.

Mathematical programming. Lu (1993) addresses the question of inverting neural networks with mathematical programming techniques. The advantage of this technique is that there is no need to choose initial starting points. On the other hand, it seems difficult to extend this framework to other neural network architectures, for example, radial basis functions or mixtures of experts, because it assumes that the activation functions are monotone.

Acknowledgements

Most of this work originates from my time at the GMD (German National Research Center for Information Technology) in Sankt Augustin, Germany, and ICSI (International Computer Science Institute) in Berkeley, California. I am very grateful for all the joint work at these places, in particular with Jörg Kindermann, Frank Weber, Heinz Mühlenbein, Gerd Paass, Sebastian Thrun, and Christoph Tietz (during my time at GMD) and Ben Gomes and Steven Omohundro (during my time at ICSI). Many thanks go also to my colleagues in the Information Technology Lab at the General Electric Corporate Research and Development Center (New York) for commenting on earlier versions of this paper: Bill Cheetham, Özden Gur Ali, and in particular Pratap Khedkar.

References

Bishop C M 1995 *Neural Networks for Pattern Recognition* (Oxford: Oxford University Press) pp 202–4

Bryson A E and Ho Y C 1975 *Applied Optimal Control* (Chichester: Wiley) (revised version of 1969 edition) pp 15ff

Cohn D A 1996 Neural network exploration using optimal experiment design *Neural Networks* (at press) also appeared as Technical Report, AI MEMO no 1491, MIT, Cambridge (ftp to publications.ai.mit.edu)

Davis D T *et al* 1995 Solving inverse problems by Bayesian iterative inversion of a forward model with applications to parameter mapping using SMMR remote sensing data *IEEE Trans. Geoscience and Remote Sensing* **33** 1182–93

DeMers D E 1996 Canonical parameterization of excess motor degrees of freedom with self-organizing maps *IEEE Trans. Neural Networks* **7** (to appear)

Ghahramani Z 1994 Solving inverse problems using an EM approach to density estimation *Proc. 1993 Connectionist Models Summer School* ed Mozer M *et al* (Hillsdale, NJ: Erlbaum) pp 316–23

Hoskins D A, Hwang J N and Vagners J 1992 Iterative inversion of neural networks and its application to adaptive control *IEEE Trans. Neural Networks* **3** 292–301

Hwang J N, Choi J J, Oh S and Marks R J 1990 Query learning based on boundary search and gradient computation of trained multilayer perceptrons *Proc. Int. Joint Conf. on Neural Networks (San Diego, 1990)*

Jordan M I and Rumelhart D E 1992 Forward models: supervised learning with a distal teacher *Cognitive Science* **16** 307–54

Kawato M, Maeda Y, Uno Y and Suzuki R 1990 Trajectory formation of arm movement by cascade neural network model based on minimum torque-change criterion *Biol. Cybern.* **62** 275–88

Kindermann J and Linden A 1992 Inversion of neural networks by gradient descent *Artificial Neural Networks: Concepts and Control Applications* ed R Vemuri (Washington, DC: IEEE Computer Society Press) also appeared 1990 *J. Parallel Comput.* **14** 3 277–86

Krogh A and Vedelsby J 1995 Neural network ensembles, cross validation and active learning *Advances in Neural Information Processing Systems 7* (Cambridge, MA: MIT Press) p 231

Linden A 1993 On discontinuous Q-functions in reinforcement learning *Proc. German Workshop on Artificial Intelligence (Lecture Notes in Artificial Intelligence)* (Berlin: Springer)

Linden A and Kindermann J 1989 Inversion of multilayer nets *Proc. 1st Int. Joint Conf. on Neural Networks (Washington DC)* (San Diego, CA: IEEE)

Lu B L 1993 Inversion of feed-forward neural networks by a separable programming *Proc. World Congress on Neural Networks, (Portland, OR)* pp IV-415–420

Nguyen D and Widrow B 1989 The truck backer-upper: an example of self-learning in neural networks *Proc. First Int. Joint Conf. on Neural Networks* (Washington, DC: IEEE)

Oja E 1991 Data compression, feature extraction, and autoassociation in feed-forward networks *Artificial Neural Networks* (North-Holland: Elsevier) pp 737–45

Thrun S and Linden A 1990 Inversion in time *Proc. EURASIP Workshop on Neural Networks* (Sesimbra, Portugal)

Thrun S, Möller K, and Linden A 1991 Planning with an adaptive world model *Advances in Neural Information Processing Systems 3: Proc. 1990 Conf.* ed R P Lippmann, J E Moody and D S Touretzky (San Mateo, CA: Morgan Kaufmann Publishers) pp 450ff

Werbos P 1992 Neurocontrol and fuzzy logic: connections and designs *Int. J. Approximate Reasoning* **6** 185–219

Williams R J 1986 Inverting a connectionist network mapping by backpropagation of error *8th Annual Conf. of the Cognitive Science Society* (Hillsdale, NJ: Lawrence Erlbaum) pp 859ff

Further reading

1. Lee S and Kil R M 1994 Inverse Mapping of continuous functions using local and global information *IEEE Trans. Neural Networks* **5** 409–23

   Discusses an approach to deal with local minima while doing gradient descent in input space.

2. Weigend A S, Zimmermann H G and Neuneier R 1995 The observer–observation dilemma in neuro-forecasting: reliable models from unreliable data through learning *AI Applications on Wall Street* ed R Freedman (New York) pp 308–17

   Uses gradient descent in input space to modify the training data. The word 'clearning' is a contraction of the two words 'cleaning' and 'learning'. The authors consider this technique as a cleaning procedure for noisy training data based on the belief in the structure and generalization of the model.

B5.3   Designing analyzable networks

Stephen P Luttrell

Abstract

In this section a unified theoretical model of unsupervised neural networks is presented. The analysis starts with a probabilistic model of the discrete neuron firing events that occur when a set of neurons is exposed to an input vector, and then uses Bayes' theorem to build a probabilistic description of the input vector from knowledge of the firing events. This sets the scene for unsupervised training of the network, by minimization of the expected value of a distortion measure between the true input vector and the input vector inferred from the firing events. Various models of this type are investigated. For instance, if the model of the neurons permits firing to occur only within a defined cluster of neurons, and further, if only one firing event is observed, then the theory approximates the well known topographic mapping network of Kohonen.

B5.3.1   Introduction

The purpose of this article is to present an analysis of an unsupervised neural network whose behavior closely approximates the well known *topographic mapping network* (Kohonen 1984) in which the neural C2.1.1 network was tailored in a purely algorithmic fashion to have topographically ordered neuron properties, some of which were derived by considering the convergence properties of the training algorithm (for instance, see Ritter and Schulten 1988). An alternative approach will be described which is based on optimization (e.g. by gradient ascent/descent) of an objective function. This approach allows some of the properties of the neural network to be derived directly from the objective function, which is not possible in the original topographic mapping network because it does not have an explicit objective function. The main novel feature of the new approach is that it uses a neuron model in which each neuron fires discretely in response to the presentation of an input vector. If these firing events are assumed to be the only information about the input vector that is preserved by the neural network, then it is possible to define an objective function that satisfies two constraints: (i) it seeks to maximize a suitably chosen measure of the information preserved about the input vector and (ii) it yields network properties that are as close to those of the original topographic mapping network as possible. Subject to these two constraints there is very little freedom of choice in the form of the chosen objective function, which may then be used to derive many interesting and useful properties.

In section B5.3.2 the neural network model is presented together with its probabilistic description. In section B5.3.3 the network optimization criterion (i.e. an objective function) is presented and analyzed, and in section B5.3.4 a useful upper bound to the objective function is derived that is much easier to optimize than the full objective function. In section B5.3.5 a very simple neural network model is discussed in which only one neuron is permitted to fire in response to the input vector; this is equivalent to a *vector quantizer* C1.1.5 (Linde *et al* 1980). In section B5.3.6 a related neural network model is discussed in which neurons in a single cluster fire in response to the input vector; this is equivalent to the well known topographic mapping network (Kohonen 1984), as was shown in Luttrell (1990, 1994). The theory provides a natural interpretation of the topographic neighborhood function. In section B5.3.7 a neural network model is discussed in which a single neuron in each of many clusters of neurons fires in response to the input; this is equivalent to the 'self-supervised' network that was discussed in Luttrell (1992, 1994). In section B5.3.8 various pieces of research that are related to the theory presented in this section are briefly mentioned.

B5.3.2 Probabilistic neural network model

The basic neural network model will describe the behavior of a pair of layers of neurons, called the 'input' and 'output' layer. The locations of the neurons that 'fire' in the output layer will be described probabilistically. Denote the rates of firing of the neurons in the input layer by the vector x, where $\dim x$ is equal to the number of neurons in the input layer. Denote the location of a neuron that fires in the output layer by the vector y, which is assumed to sit on a d-dimensional rectangular lattice of size m (where $\dim m = d$), so $\dim m = 2$ for a two-dimensional sheet of output neurons.

The answer to the question 'Which output neuron will fire next?' is then $\Pr(y|x)$, which is the probability distribution over possible locations y of the next neuron that fires, given that the input x is known. More generally, the answer to the question 'Which n neurons will fire next?' is then $\Pr(y_1, y_2, \ldots, y_n|x)$ which is a joint probability distribution over the possible locations $(y_1, y_2, \ldots, y_n)$ of the next n neurons that fire. Note that the y_i are not restricted to being different from each other, so a given neuron might fire more than once. Marginal probabilities may be derived from $\Pr(y_1, y_2, \ldots, y_n|x)$ to give the probability of occurrence of a subset of the events in $(y_1, y_2, \ldots, y_n)$. Thus, to obtain a marginal probability, the locations of the unobserved firing events must be summed over. Care has to be taken when forming marginal probabilities. For instance, in the $n = 3$ case the marginal probabilities for $(?, y_1, y_2)$, $(y_1, ?, y_2)$ and $(y_1, y_2, ?)$ are all *different* (where the ? denotes the unobserved event). However, if the *order* in which the neurons fire is *not* observed, then $\Pr(y_1, y_2, \ldots, y_n|x)$ is the sum of the probabilities for all $n!$ permutations of the sequence of firings, in which case $\Pr(y_1, y_2, \ldots, y_n|x)$ is a symmetric function of $(y_1, y_2, \ldots, y_n)$, and in the $n = 3$ case the marginal probabilities for $(?, y_1, y_2)$, $(y_1, ?, y_2)$ and $(y_1, y_2, ?)$ are all the *same*. If the number of firings is itself known only probabilistically (i.e. as $\Pr(n)$) then an appropriate average $\sum_{n=0}^{\infty} \Pr(n)(\cdots)$ must be formed.

It is important to distinguish between the neural network *itself*, whose input–output state after n neurons have fired is described by the vector $(y_1, y_2, \ldots, y_n; x)$, and the *knowledge of* the network input–output relationship, which is written as $\Pr(y_1, y_2, \ldots, y_n|x)$. For instance, a piece of software that is written to compute quantities like $\Pr(y_1, y_2, \ldots, y_n|x)$ is not really a 'neural network' program; rather, it is a program that makes probabilistic statements about how a neural network behaves. The utility of $\Pr(y_1, y_2, \ldots, y_n|x)$ it that it allows average properties of the neural network to be computed. One particular property that is of great interest is the network objective function; this is the quantity that measures the network's average performance. This is the subject of the next section.

B5.3.3 Optimization criterion

B3.4.4 A neural network is trained by minimizing a suitably defined *objective function*, which will be chosen to be the average Euclidean distortion D defined as (Luttrell 1994)

$$D \equiv \sum_{y_1, y_2, \ldots, y_n=1}^{m} \int dx\, dx'\, \Pr(x)\, \Pr(y_1, y_2, \ldots, y_n|x)\, \Pr(x'|y_1, y_2, \ldots, y_n) \|x - x'\|^2 \qquad (\text{B5.3.1})$$

where x and x' are both vectors in input space, the y_i are vectors in output space, $\|x - x'\|^2$ is the square of the Euclidean distance between x and x', $\int dx\, \Pr(x)(\cdots)$ is the average over input space using probability density $\Pr(x)$. It will be assumed that $\int dx\, \Pr(x)(\cdots)$ is accurately approximated by an average over a suitable training set. Thus, if samples x are drawn from the training set and plotted in input space, then after a large number of samples has been drawn the density of plotted points approximates $\Pr(x)$. $\sum_{y_1, y_2, \ldots, y_n=1}^{m} \Pr(y_1, y_2, \ldots, y_n|x)(\cdots)$ is the average over output space as specified by the probabilistic neural network model, and $\int dx'\, \Pr(x'|y_1, y_2, \ldots, y_n)(\cdots)$ is the average over input space as specified by the *inverse* of the probabilistic neural network, i.e. the probability density of input vectors given that the location of the firing neurons is known. This is determined entirely by the other probabilities already defined, and may be written as

$$\Pr(x'|y_1, y_2, \ldots, y_n) = \frac{\Pr(y_1, y_2, \ldots, y_n|x')\, \Pr(x')}{\int dx''\, \Pr(y_1, y_2, \ldots, y_n|x'')\, \Pr(x'')}$$

which is an application of Bayes' theorem. This may be used to eliminate $\Pr(x'|y_1, y_2, \ldots, y_n)$ from the expression for D in (B5.3.1) to obtain

$$D = 2 \int dx \, \Pr(x) \sum_{y_1, y_2, \ldots, y_n=1}^{m} \Pr(y_1, y_2, \ldots, y_n|x) \|x - x'(y_1, y_2, \ldots, y_n)\|^2 \qquad \text{(B5.3.2)}$$

where the $x'(y_1, y_2, \ldots, y_n)$ are defined as $x'(y_1, y_2, \ldots, y_n) \equiv \int dx \, \Pr(x|y_1, y_2, \ldots, y_n)\,x$. The $x'(y_1, y_2, \ldots, y_n)$ will be called 'reference vectors'. This means that there is a separate reference vector for each possible set of locations for the n neurons that fire. Thus the total number of reference vectors increases exponentially with n, which soon leads to an unacceptably large number of reference vectors. The next section introduces a theoretical trick for circumventing this difficulty.

B5.3.4 Least upper bound trick

The exponential increase with n of the number of reference vectors $x'(y_1, y_2, \ldots, y_n)$ in (B5.3.2) can be avoided (Luttrell 1994) by minimizing not D, but a suitably defined upper bound to D that depends on simplified reference vectors with the functional form $x'(y)$, rather than $x'(y_1, y_2, \ldots, y_n)$. When this upper bound is minimized it yields a *least* upper bound on D, rather than its ideal lower bound. This is the price that has to be paid for not using the full reference vectors $x'(y_1, y_2, \ldots, y_n)$. The upper bound is derived as follows. Use the following identity, which holds for all $x'(y_i)$

$$x - x'(y_1, y_2, \ldots, y_n) \equiv \frac{1}{n} \sum_{i=1}^{n} (x - x'(y_i)) + \frac{1}{n} \sum_{i=1}^{n} (x'(y_i) - x'(y_1, y_2, \ldots, y_n))$$

to separate x from $x'(y_1, y_2, \ldots, y_n)$ and assume that $\Pr(y_1, y_2, \ldots, y_n|x)$ is a symmetric function of $(y_1, y_2, \ldots, y_n)$, to write D in (B5.3.2) in the form $D = D_1 + D_2 - D_3$, where

$$D_1 \equiv \frac{2}{n} \int dx \, \Pr(x) \sum_{y=1}^{m} \Pr(y|x) \|x - x'(y)\|^2$$

$$D_2 \equiv \frac{2(n-1)}{n} \int dx \, \Pr(x) \sum_{y_1, y_2=1}^{m} \Pr(y_1, y_2|x)(x - x'(y_1)) \cdot (x - x'(y_2)) \qquad \text{(B5.3.3)}$$

$$D_3 \equiv 2 \sum_{y_1, y_2, \ldots, y_n=1}^{m} \Pr(y_1, y_2, \ldots, y_n) \left\| x'(y_1, y_2, \ldots, y_n) - \frac{1}{n} \sum_{i=1}^{n} x'(y_i) \right\|^2.$$

D_1 is $1/n$ times the average Euclidean distortion that would occur if only 1 out of the n neuron firing events is observed (assuming that $x'(y)$ is chosen to be $\int dx \, \Pr(x|y)\,x$). D_2 is a new type of term that cannot be interpreted as a simple Euclidean distortion. Suppose that the locations y_1 and y_2 of two out of the n neuron firing events are observed (which two does not matter, because it is assumed that the order in which the events occur is not observed), and an attempt is made to reconstruct the input vector *independently* from each of these firing events. This produces two vectors $x'(y_1)$ and $x'(y_2)$, and two error vectors $(x - x'(y_1))$ and $(x - x'(y_2))$. The covariance of these error vectors is the average of their outer product $\int dx \, \Pr(x) \sum_{y_1, y_2=1}^{m} \Pr(y_1, y_2|x)(x - x'(y_1))(x - x'(y_2))^{\mathsf{T}}$, and D_2 is $2(n-1)/n$ times the trace of this covariance matrix (i.e. the sum of its eigenvalues). Because $D_3 \geq 0$, it follows that $D \leq D_1 + D_2$, so minimization of $D_1 + D_2$ yields a least upper bound to D, as required. Note that D_2 and D_3 contribute only for $n \geq 2$. In the $n \to \infty$ limit the contribution of D_1 vanishes, and then D_2 is the value that D would take if $x'(y_1, y_2, \ldots, y_n)$ were approximated by the expression $\frac{1}{n} \sum_{i=1}^{n} x'(y_i)$ and the error term D_3 were ignored. Many useful results can be obtained by minimizing $D_1 + D_2$ as defined in (B5.3.3) when $n \geq 2$ (or minimizing D itself when $n = 1$) and some of these will be discussed in the following sections.

B5.3.5 Vector quantizer model: single neuron approximation

In the expression for D in (B5.3.2) assume that only a single neuron fires n times, so that $\Pr(y_1, y_2, \ldots, y_n|x)$ is given by $\Pr(y_1, y_2, \ldots, y_n|x) = \delta_{y_1, y(x)} \delta_{y_2, y(x)} \cdots \delta_{y_n, y(x)}$, where $\delta_{y, y(x)} = 1$ if

$y = y(x)$, and 0 otherwise. The role of the 'encoding function' $y(x)$ is to convert the input vector x into the index of the 'winning' neuron (i.e. the one that fires). This allows D to be simplified to the form

$$D = 2 \int dx \ \Pr(x) \|x - x'(y(x))\|^2 \qquad (B5.3.4)$$

where the n argument reference vector $x'(y(x), y(x), \ldots, y(x))$ has been written using an abbreviated notation $x'(y(x))$. In (B5.3.4) D can be minimized with respect to $y(x)$ to give

$$y(x) = \arg \min_y \|x - x'(y)\|^2 \qquad (B5.3.5)$$

where 'arg $\min_y \ldots$' means 'the value of y that minimizes $\ldots$'. This is a 'nearest-neighbor' encoding rule because the winning neuron y has the reference vector that is closest to the input vector, in the Euclidean distance sense. In (B5.3.4) D can be minimized with respect to $x'(y)$ to give

$$x'(y) = \frac{\int dx \ \Pr(x) \, \delta_{y,y(x)} \, x}{\int dx \ \Pr(x) \, \delta_{y,y(x)}}$$

$$= \int dx \ \Pr(x|y) \, x \qquad (B5.3.6)$$

where the second line has been obtained by using Bayes' theorem. The term $x'(y)$ is the centroid of the input vectors x that are permitted given that the location y of the firing neuron is known. In effect, $x'(y)$ is the decoder corresponding to the encoder $y(x)$. Because the optimizations of $y(x)$ and $x'(y)$ are mutually coupled, these two results (i.e. (B5.3.5) and (B5.3.6)) must be iterated in order to obtain a consistent solution. This is essentially the LBG algorithm (Linde *et al* 1980) for training a vector quantizer, which may be summarized as follows.

(i)   Initialize the reference vectors $x'(y)$, for example, set them to different randomly selected vectors chosen from the training set.
(ii) Encode each vector x in the training set using the nearest-neighbor rule $y(x)$ in (B5.3.5).
(iii) Compute the centroids on the right-hand side of (B5.3.6).
(iv) Update the reference vectors $x'(y)$ as in (B5.3.6).
(v)   Test if the reference vectors $x'(y)$ have converged, and if not then go to step (ii), otherwise stop. There are many possible convergence tests. For instance, have all the reference vectors moved by less than some predefined fraction of the diameter of the volume of input space that they live in? Another possibility is: has D decreased by less than some predefined fraction of its value on the previous iteration? There is no method that is guaranteed to avoid premature termination.

The LBG algorithm is a 'batch' training algorithm. An 'online' training algorithm can be obtained by updating the $x'(y)$ in the direction of $-\partial D/\partial x'(y)$ (i.e. gradient descent), which yields the update prescription

$$\Delta x'(y(x)) = \varepsilon \, (x - x'(y(x))) \qquad (B5.3.7)$$

which operates as follows.

(i)   Initialize the reference vectors $x'(y)$, for example, set them to different randomly selected vectors chosen from the training set.
(ii) Encode a vector x from the training set using the nearest-neighbor rule $y(x)$ in (B5.3.5).
(iii) Move the corresponding reference vector $x'(y(x))$ a small amount towards the input vector x as in (B5.3.7).
(iv) Test whether the reference vectors $x'(y)$ have converged, and if not then go to step (ii), otherwise stop.

Neither the batch nor the online training algorithms can avoid the problem of becoming trapped in a local minimum. It is prudent to run these algorithms several times on each training set, but starting from a different initial configuration of reference vectors on each run.

B5.3.6 Topographic mapping model: single cluster approximation

Generalize the vector quantizer case studied in section B5.3.5 so that the neurons that fire are not all forced to be the same neuron. Thus, in the expression for D in (B5.3.2) assume that the neurons that fire are located in a single cluster and fire independently, so that $\Pr(y_1, y_2, \ldots, y_n|x)$ is given by $\Pr(y_1, y_2, \ldots, y_n|x) = \Pr(y_1|y(x)) \Pr(y_2|y(x)) \cdots \Pr(y_n|y(x))$, where the 'shape' of the cluster is modeled by $\Pr(y|y(x))$. The results for D_1 and D_2 in (B5.3.3) then permit an upper bound on D to be obtained as

$$D \le \frac{2}{n} \int dx \ \Pr(x) \sum_{y=1}^{m} \Pr(y|y(x)) \|x - x'(y)\|^2$$
$$+ \frac{2(n-1)}{n} \int dx \ \Pr(x) \left\| x - \sum_{y=1}^{m} \Pr(y|y(x)) \, x'(y) \right\|^2 . \qquad \text{(B5.3.8)}$$

In the special case $n = 1$, this inequality reduces to an equality, and the second term on the right-hand side of (B5.3.8) vanishes. The first term of (B5.3.8) is $1/n$ times the average Euclidean error that occurs when only one neuron firing event is observed. The second term of (B5.3.8) is $2(n-1)/n$ times the average Euclidean error that occurs when an attempt is made to reconstruct the input vector from the weighted average $\sum_{y=1}^{m} \Pr(y|y(x)) \, x'(y)$ of the reference vectors. This term dominates when $n \gg 1$.

It is possible to interpret the second term of (B5.3.8) in terms of a *radial basis function* network. The $\Pr(y|y(x))$ are a set of nonlinear functions that connect the input layer to a hidden layer, $x'(y)$ is the set of weights connecting the yth hidden neuron to the output layer, and $x - \sum_{y=1}^{m} \Pr(y|y(x)) \, x'(y)$ is the error vector between the input and output layers. This use of a nonlinear input-to-hidden transformation plus a linear hidden-to-output transformation is the same as is used in a radial basis function network, except that here the nonlinear basis functions add up to 1, and the error is measured between the input and output, rather than between a target and the output.

B1.7.3

B5.3.6.1 Optimization of the n = 1 case

D itself in (B5.3.2) (and not merely its upper bound in (B5.3.8)) may be minimized with respect to $y(x)$ and $x'(y)$ to give (Luttrell 1990, 1994)

$$y(x) = \arg\min_y \sum_{y'=1}^{m} \Pr(y'|y) \|x - x'(y')\|^2$$
$$x'(y) = \frac{\int dx \ \Pr(x) \ \Pr(y|y(x)) \, x}{\int dx \ \Pr(x) \ \Pr(y|y(x))} \qquad \text{(B5.3.9)}$$
$$\Delta x'(y) = \varepsilon \ \Pr(y|y(x)) \, (x - x'(y)) .$$

The term $y(x)$ is no longer a nearest-neighbor encoding rule as it was in the vector quantizer case in (B5.3.5). It is a 'minimum distortion' encoding rule where the winning neuron is the one that leads to the minimum expected Euclidean error. Note that the phrase 'winning neuron' is used loosely in this context; it is actually the neuron that determines where the cluster of firing neurons is located. When $n = 1$ the neuron that actually fires is somewhere in the cluster located around the winning neuron. $x'(y)$ is a straightforward generalization of the vector quantizer case in (B5.3.6). Both the batch and online versions of the training algorithm are implemented as a straightforward generalization of the batch and online vector quantizer training algorithms, so they will not be repeated here. In the online training algorithm, an important change is that each training vector x causes each reference vector $x'(y)$ to be updated by an amount that is proportional to $\Pr(y|y(x))$. In the vector quantizer case in (B5.3.7) only the winning reference vector $x'(y(x))$ was updated.

It is useful to approximate $y(x)$ (Luttrell 1990) by doing a Taylor expansion of $\|x - x'(y')\|^2$ in (B5.3.9) in powers of $(y'-y)$ to obtain

$$y(x) = \arg\min_y \left(\|x - x'(y)\|^2 + \sum_{i=1}^{d} \frac{\partial \|x - x'(y)\|^2}{\partial y_i} \sum_{y'=1}^{m} \Pr(y'|y) \, (y'_i - y_i) + \cdots \right)$$

where the derivatives are evaluated as finite-difference expressions on the lattice of points on which y sits. If the 'arg min' operation is applied to the first term in isolation, then it returns a y that guarantees that $\partial \|x - x'(y)\|^2 / \partial y = 0$, which ensures that the first-order term in the Taylor series vanishes. So $y(x)$ reduces to $y(x) = \arg \min_y (\|x - x'(y)\|^2) +$ second-order terms, which is a nearest-neighbor encoding rule. Using this approximation, the online training algorithm is the same as the well known topographic mapping training algorithm (Kohonen 1984) and $\Pr(y'|y)$ plays the role of the 'neighborhood function' around the yth neuron.

B5.3.6.2 Optimization of the n ≫ 1 case

If $n \gg 1$ in (B5.3.6) then $D_1 \ll D_2$, so D_1 can be ignored, in which case the upper bound for D in (B5.3.8) can be approximately minimized with respect to $y(x)$ and $x'(y)$ to give

$$y(x) \approx \arg \min_y (\|x - \bar{x}'(y)\|^2)$$

$$\Delta x'(y) \approx \varepsilon \, \Pr(y|y(x)) \, (x - \bar{x}'(y(x)))$$

where $\bar{x}'(y)$ is a weighted average of the reference vectors $x'(y)$ defined as $\bar{x}'(y) \equiv \sum_{y'} = \sum_{y'=1}^m \Pr(y'|y) \, x'(y')$. These results may also be obtained directly from the original definition of D in (B5.3.2) for $n \gg 1$ by making the approximation $x'(y_1, y_2, \ldots, y_n) \approx \frac{1}{n} \sum_{i=1}^n x'(y_i)$ (i.e. ignoring D_3) and noting that $\frac{1}{n} \sum_{i=1}^n x'(y_i) \approx \sum_{y'=1}^m \Pr(y'|y(x)) x'(y')$ (i.e. the n neurons that fire allow a good estimate of the cluster shape $\Pr(y'|y(x))$ to be made).

B5.3.7 Topographic mapping model: multiple cluster approximation

In the expression for D in (B5.3.2) assume that one neuron located in each of c clusters fires, so that $\Pr(y|x)$ has the form $\Pr(y|x) = \Pr(y^1, y^2, \ldots, y^c | y^1(x), y^2(x), \ldots, y^c(x))$, where superscripts have been used for cluster indices, and the encoding function $y(x)$ has been partitioned as $y(x) \equiv (y^1(x), y^2(x), \ldots, y^c(x))$ to separate the pieces that locate each cluster. This allows D to be written as

$$D = 2 \int dx \, \Pr(x) \sum_{y^1, y^2, \ldots, y^c = 1}^m \Pr(y^1, y^2, \ldots, y^c | y^1(x), y^2(x), \ldots, y^c(x))$$
$$\times \|x - x'(y^1, y^2, \ldots, y^c)\|^2.$$

Partition the input space into c nonoverlapping subspaces, so that the input vector x is written as $x = (x^1, x^2, \ldots, x^c)$, and use the following identity, which holds for all $x'^i(y^i)$

$$x - x'(y^1, y^2, \ldots, y^c) \equiv ((x^1, x^2, \ldots, x^c) - (x'^1(y^1), x'^2(y^2), \ldots, x'^c(y^c)))$$
$$- (x'(y^1, y^2, \ldots, y^c) - (x'^1(y^1), x'^2(y^2), \ldots, x'^c(y^c)))$$

where $x'^i(y^i)$ lies in input subspace i, to write D in the form $D = D_1 - D_3$, where

$$D_1 \equiv 2 \int dx \, \Pr(x) \sum_{i=1}^c \sum_{y^i=1}^m \Pr(y^i | y^1(x), y^2(x), \ldots, y^c(x)) \|x^i - x'^i(y^i)\|^2$$

$$D_3 \equiv 2 \sum_{y^1, y^2, \ldots, y^c = 1}^m \Pr(y^1, y^2, \ldots, y^c) \left\| \begin{array}{c} x'(y^1, y^2, \ldots, y^c) \\ -(x'^1(y^1), x'^2(y^2), \ldots, x'^c(y^c)) \end{array} \right\|^2$$

which should be compared with the results in (B5.3.3). Note that in D_1 the ith cluster contributes only to the average Euclidean error in the ith input subspace; this was enforced by the assumed functional dependence in $(x'^1(y^1), x'^2(y^2), \ldots, x'^c(y^c))$. Because $D_3 \geq 0$ it follows that $D \leq D_1$, so minimization of D_1 leads to a least upper bound on D. Minimization of D_1 with respect to $y^i(x)$ and $x'^i(y^i)$ then gives

$$(y^1(x), y^2(x), \ldots, y^c(x)) = \arg \min_{y^1, y^2, \ldots, y^c} \sum_{i=1}^c \sum_{y'^i=1}^m \Pr(y'^i | y^1, y^2, \ldots, y^c) \|x^i - x'^i(y'^i)\|^2$$

$$x'^i(y^i) = \frac{\int dx \, \Pr(x) \Pr(y^i | y^1(x), y^2(x), \dots, y^c(x)) \, x^i}{\int dx \, \Pr(x) \Pr(y^i | y^1(x), y^2(x), \dots, y^c(x))}$$

$$\Delta x'^i(y^i) = \varepsilon \Pr(y^i | y^1(x), y^2(x), \dots, y^c(x)) (x^i - x'^i(y^i)) \qquad \text{(B5.3.10)}$$

which is equivalent to the 'self-supervised' network training algorithm that was discussed in Luttrell (1992, 1994). If the c subspaces were treated completely separately, then in (B5.3.10) the results for the ith subspace would read the same as the $n = 1$ topographic mapping case in (B5.3.9), with a superscript i inserted where appropriate. Now examine (B5.3.10) in detail. When there is more than one cluster of firing neurons, the effective shape of each cluster is modified by the locations of the other clusters, i.e. $\Pr(y'^i | y^i(x)) \rightarrow \Pr(y'^i | y^1(x), y^2(x), \dots, y^c(x))$. So, the cluster shapes determine the winning neurons, which, in turn, determine the cluster shapes. Note, as in the single cluster case in section B5.3.6, that the phrase 'winning neurons' refers to the neurons that determine the cluster locations $(y^1(x), y^2(x), \dots, y^c(x))$. This feedback makes the determination of which neurons are the winners a nontrivial coupled optimization problem, in which the $y^i(x)$ affect each other, so they must be *jointly* optimized. In particular, the optimal $y^i(x)$ is a function of the whole input vector x, and not merely a function of the part of x that lies in the ith subspace (i.e. x^i), as it would be if the subspaces were considered separately. In practice, the problem of optimizing the $y^i(x)$ could be solved by iterating the following set of equations

$$y^i(x) = \arg \min_{y^i} \sum_{y'^i = 1}^{m} \Pr(y'^i | y^i, \{y^j(x) : j \neq i\}) \|x^i - x'^i(y'^i)\|^2$$

where the $\{y^j(x) : j \neq i\}$ on the right-hand side is obtained from the previous iteration of the equation. If this converges, then it solves the coupled optimization problem. Although only one neuron was permitted to fire in each of the c clusters, it is straightforward to generalize these results to the case where any number of neurons may fire in each cluster. It is also possible to generalize to the more realistic case where the input subspaces overlap each other.

B5.3.8 Related research

In section B5.3.6 the density of reference vectors can be derived for an optimized network (Luttrell 1991) and the result obtained is *independent* of the topographic neighborhood function. This contrasts with the result obtained for a standard topographic network in Ritter (1991) where the density is *dependent* on the topographic neighborhood function. This difference arises from the choice of encoding prescriptions used in the two approaches; minimum distortion in Luttrell (1991), and nearest neighbor in Ritter (1991). The results of section B5.3.6 may also be used to derive a hierarchical vector quantizer (Luttrell 1989a) for encoding high-dimensional vectors in easy-to-implement stages. An example of the use of this approach in *image compression* can be found in Luttrell (1989b). The results of section B5.3.6 may also be interpreted F1.5.2 as vector quantization for communication along a noisy channel (Luttrell 1992). This type of coding problem was analyzed in Kumazawa *et al* (1984) and Farvardin (1990), but the connection with neural networks was not made.

References

Farvardin N 1990 A study of vector quantization for noisy channels *IEEE Trans. Info. Theory* **36** 799–809

Kohonen T 1984 *Self Organization and Associative Memory* (Berlin: Springer)

Kumazawa H, Kasahara M and Namekawa T 1984 A construction of vector quantizers for noisy channels *Electron. Eng. Japan* B **67** 39–47

Linde Y, Buzo A and Gray R M 1980 An algorithm for vector quantizer design *IEEE Trans. Commun.* **28** 84–95

Luttrell S P 1989a Hierarchical vector quantization *Proc. IEE* I **136** 405–13

——1989b Image compression using a multilayer neural network *Patt. Recog. Lett.* **10** 1–7

——1990 Derivation of a class of training algorithms *IEEE Trans. Neural Networks* **1** 229–32

——1991 Code vector density in topographic mappings: scalar case *IEEE Trans. Neural Networks* **2** 427–36

——1992 Self-supervised adaptive networks *Proc. IEE* F **139** 371–7

——1994 A Bayesian analysis of self-organizing maps *Neural Comput.* **6** 767–94

Ritter H 1991 Asymptotic level density for a class of vector quantization processes *IEEE Trans. Neural Networks* **2** 173–5

Ritter H and Schulten K 1988 Convergence properties of Kohonen's topology conserving maps: fluctuations, stability and dimension selection *Biol. Cybern.* **60** 59–71

B6

Neural Networks: A Pattern Recognition Perspective

Christopher M Bishop

Abstract

The majority of current applications of neural networks are concerned with problems in pattern recognition. In this chapter we show how neural networks can be placed on a principled, statistical foundation, and we discuss some of the practical benefits which this brings.

Contents

B6.1 Introduction

Christopher M Bishop

Abstract

See the abstract for Chapter B6.

Neural networks have been exploited in a wide variety of applications, the majority of which are concerned with pattern recognition in one form or another. However, it has become widely acknowledged that the effective solution of all but the simplest of such problems requires a *principled* treatment, in other words one based on a sound theoretical framework.

From the perspective of *pattern recognition*, neural networks can be regarded as an extension of the many conventional techniques which have been developed over several decades. Lack of understanding of the basic principles of statistical pattern recognition lies at the heart of many of the common mistakes in the application of neural networks. In this chapter we aim to show that the 'black box' stigma of neural networks is largely unjustified, and that there is actually considerable insight available into the way in which neural networks operate, and how to use them effectively. *B1.5*

Some of the key points which are discussed in this chapter are as follows:

(i) Neural networks can be viewed as a general framework for representing nonlinear mappings between multidimensional spaces in which the form of the mapping is governed by a number of adjustable parameters. They therefore belong to a much larger class of such mappings, many of which have been studied extensively in other fields.

(ii) Simple techniques for representing multivariate nonlinear mappings in one or two dimensions (e.g. polynomials) rely on linear combinations of *fixed* basis functions (or 'hidden functions'). Such methods have severe limitations when extended to spaces of many dimensions; a phenomenon known as the *curse of dimensionality*. The key contribution of neural networks in this respect is that they employ basis functions which are themselves adapted to the data, leading to efficient techniques for multidimensional problems.

(iii) The formalism of statistical pattern recognition, introduced briefly in section B6.2.3, lies at the heart of a principled treatment of neural networks. Many of these topics are treated in standard texts on *statistical pattern recognition*, including those by Duda and Hart (1973), Hand (1981), Devijver and Kittler (1982), and Fukunaga (1990). *B6.2.3*

(iv) Network training is usually based on the minimization of an *error function*. We show how error functions arise naturally from the principle of maximum likelihood, and how different choices of error function correspond to different assumptions about the statistical properties of the data. This allows the appropriate error function to be selected for a particular application. *B6.3*

(v) The statistical view of neural networks motivates specific forms for the *activation functions* which arise in network models. In particular we see that the logistic sigmoid, often introduced by analogy with the mean firing rate of a biological neuron, is precisely the function which allows the activation of a unit to be given a particular probabilistic interpretation. *B3.2.4*

(vi) Provided the error function and activation functions are correctly chosen, the outputs of a trained network can be given precise interpretations. For regression problems they approximate the conditional averages of the distribution of target data, while for classification problems they approximate the posterior probabilities of class membership. This demonstrates why neural networks can approximate the optimal solution to a regression or classification problem.

B6.3 (vii) *Error backpropagation* is introduced as a general framework for evaluating derivatives for feedforward networks. The key feature of backpropagation is that it is computationally very efficient compared with a simple direct evaluation of derivatives. For network training algorithms, this efficiency is crucial.

(viii) The original learning algorithm for multilayer feedforward networks (Rumelhart *et al* 1986) was based on gradient descent. In fact the problem of optimizing the weights in a network corresponds to unconstrained nonlinear optimization for which many substantially more powerful algorithms have been developed.

(ix) Network complexity, governed for example by the number of hidden units, plays a central role in
B6.4 determining the *generalization* performance of a trained network. This is illustrated using a simple curve-fitting example in one dimension.

These and many related issues are discussed at greater length by Bishop (1995).

References

Anderson A and Rosenfeld E (eds) 1988 *Neurocomputing: Foundations of Research* (Cambridge, MA: MIT)

Bishop C M 1995 *Neural Networks for Pattern Recognition* (Oxford: Oxford University Press)

Devijver P A and Kittler 1982 *Pattern Recognition: A Statistical Approach* (Englewood Cliffs, NJ: Prentice-Hall)

Duda R O and P E Hart 1973 *Pattern Classication and Scene Analysis* (New York: Wiley)

Fukunaga K 1990 *Introduction to Statistical Pattern Recognition* (2nd edn) (San Diego, CA: Academic)

Hand D J 1981 *Discrimination and Classification* (New York: Wiley)

Rumelhart D E, Hinton G E and Williams R J 1986 Learning internal representations by error propagation *Parallel Distributed Processing: Explorations in the Microstructure of Cognition Volume 1: Foundations* ed D E Rumelhart, J L McClelland, and the PDP Research Group (Cambridge, MA: MIT) pp 318–62 (reprinted in Anderson and Rosenfeld (1988).)

B6.2   Classification and regression

Christopher M Bishop

Abstract

See the abstract for Chapter B6.

In this section we concentrate on the two most common kinds of pattern recognition problem. The first of these we shall refer to as *regression*, and is concerned with predicting the values of one or more continuous output variables, given the values of a number of input variables. Examples include prediction of the temperature of a plasma given values for the intensity of light emitted at various wavelengths, or the estimation of the fraction of oil in a multiphase pipeline given measurements of the absorption of gamma beams along various cross-sectional paths through the pipe. If we denote the input variables by a vector x with components x_i where $i = 1, \ldots, d$ and the output variables by a vector y with components y_k where $k = 1, \ldots, c$ then the goal of the regression problem is to find a suitable set of functions which map the x_i to the y_k.

The second kind of task we shall consider is called *classification* and involves assigning input patterns to one of a set of discrete classes C_k where $k = 1, \ldots, c$. An important example involves the automatic interpretation of handwritten digits (Le Cun 1989). Again, we can formulate a classification problem in terms of a set of functions which map inputs x_i to outputs y_k where now the outputs specify which of the classes the input pattern belongs to. For instance, the input may be assigned to the class whose output value y_k is largest.

In general, it will not be possible to determine a suitable form for the required mapping, except with the help of a data set of examples. The mapping is therefore modeled in terms of some mathematical function which contains a number of adjustable parameters, whose values are determined with the help of the data. We can write such functions in the form

$$y_k = y_k(x; w) \tag{B6.2.1}$$

where w denotes the vector of parameters $w_1, \ldots, w_W$. A neural network model can be regarded simply as a particular choice for the set of functions $y_k(x; w)$. In this case, the parameters comprising w are often called *weights*.

The importance of neural networks in this context is that they offer a very powerful and very general framework for representing nonlinear mappings from several input variables to several output variables. The process of determining the values for these parameters on the basis of the data set is called *learning* B3 *or training*, and for this reason the data set of examples is generally referred to as a *training set*. Neural network models, as well as many conventional approaches to statistical pattern recognition, can be viewed as specific choices for the functional forms used to represent the mapping (B6.2.1), together with particular procedures for optimizing the parameters in the mapping. In fact, neural network models often contain conventional approaches (such as linear or logistic regression) as special cases.

B6.2.1   Polynomial curve fitting

Many of the important issues concerning the application of neural networks can be introduced in the simpler context of curve fitting using polynomial functions. Here, the problem is to fit a polynomial to a set of N data points by minimizing an error function. Consider the Mth-order polynomial given by

$$y(x) = w_0 + w_1 x + \ldots + w_M x^M = \sum_{j=0}^{M} w_j x^j . \qquad (B6.2.2)$$

This can be regarded as a nonlinear mapping which takes x as input and produces y as output. The precise form of the function $y(x)$ is determined by the values of the parameters $w_0, \ldots, w_M$, which are analogous to the weights in a neural network. It is convenient to denote the set of parameters $(w_0, \ldots, w_M)$ by the vector w in which case the polynomial can be written as a functional mapping in the form (B6.2.1). Values for the coefficients can be found by minimization of an error function, as will be discussed in detail in Section B6.3. Examples of polynomial curve fitting are given in Section B6.4.

B6.2.2 Why neural networks?

Pattern recognition problems, as we have already indicated, can be represented in terms of general parametrized nonlinear mappings between a set of input variables and a set of output variables. A polynomial represents a particular class of mapping for the case of one input and one output. Provided we have a sufficiently large number of terms in the polynomial, we can approximate a wide class of functions to arbitrary accuracy. This suggests that we could simply extend the concept of a polynomial to higher dimensions. Thus, for d input variables, and again one output variable, we could, for instance, consider a third-order polynomial of the form

$$y = w_0 + \sum_{i_1=1}^{d} w_{i_1} x_{i_1} + \sum_{i_1=1}^{d} \sum_{i_2=1}^{d} w_{i_1 i_2} x_{i_1} x_{i_2} + \sum_{i_1=1}^{d} \sum_{i_2=1}^{d} \sum_{i_3=1}^{d} w_{i_1 i_2 i_3} x_{i_1} x_{i_2} x_{i_3} . \qquad (B6.2.3)$$

For an Mth-order polynomial of this kind, the number of independent adjustable parameters would grow like d^M, which represents a dramatic growth in the number of degrees of freedom in the model as the dimensionality of the input space increases. This is an example of the *curse of dimensionality* (Bellman 1961). The presence of a large number of adaptive parameters in a model can cause major problems as discussed in Section B6.4. In order that the model make good predictions for new inputs it is necessary that the number of data points in the training set be much greater than the number of adaptive parameters. For medium to large applications, such a model would need huge numbers of training data in order to ensure that the parameters (in this case the coefficients in the polynomial) were well determined.

There are, in fact, many different ways in which to represent general nonlinear mappings between multidimensional spaces. The importance of neural networks, and similar techniques, lies in the way in which they deal with the problem of scaling with dimensionality. In order to motivate neural network models it is convenient to represent the nonlinear mapping function (B6.2.1) in terms of a linear combination of *basis* functions, sometimes also called 'hidden functions' or *hidden units*, $z_j(\boldsymbol{x})$, so that

$$y_k(\boldsymbol{x}) = \sum_{j=0}^{M} w_{kj} z_j(\boldsymbol{x}) . \qquad (B6.2.4)$$

Here the basis function z_0 takes the fixed value 1 and allows a constant term in the expansion. The corresponding weight parameter w_{k0} is generally called a *bias*. Both the one-dimensional polynomial (B6.2.2) and the multidimensional polynomial (B6.2.3) can be cast in this form, in which basis functions are fixed functions of the input variables.

We have seen from the example of the higher-order polynomial that to represent general functions of many input variables we have to consider a large number of basis functions, which in turn implies a large number of adaptive parameters. In most practical applications there will be significant correlations between the input variables so that the effective dimensionality of the space occupied by the data (known as the *intrinsic dimensionality*) is significantly less than the number of inputs. The key to constructing a model which can take advantage of this phenomenon is to allow the basis functions themselves to be adapted to the data as part of the training process. In this case the number of such functions only needs to grow as the complexity of the problem itself grows, and not simply as the number of input variables grows. The number of free parameters in such models, for a given number of hidden functions, typically only grows linearly (or quadratically) with the dimensionality of the input space, as compared with the d^M growth for a general Mth-order polynomial.

One of the simplest, and most commonly encountered, models with adaptive basis functions is given by the two-layer feedforward network, sometimes called a *multilayer perceptron*, which can be expressed C1.2 in the form of (B6.2.4) in which the basis functions themselves contain adaptive parameters and are given by

$$z_j(x) = g\left(\sum_{i=0}^{d} w_{ji} x_i\right) \qquad \text{(B6.2.5)}$$

where w_{j0} are bias parameters, and we have introduced an extra 'input variable' $x_0 = 1$ in order to allow the biases to be treated on the same footing as the other parameters and hence be absorbed into the summation in (B6.2.5). The function $g(\cdot)$ is called an *activation function* and must be a nonlinear B3.2.4 function of its argument in order that the network model can have general approximation capabilities. If $g(\cdot)$ were linear, then (B6.2.4) would reduce to the composition of two linear mappings which would itself be linear. The activation function is also chosen to be a differentiable function of its argument in order that the network parameters can be optimized using gradient-based methods as discussed in section B6.3.3. Many different forms of activation function can be considered. However, the most common are sigmoidal (meaning 'S shaped') and include the logistic sigmoid

$$g(a) = \frac{1}{1 + \exp(-a)} \qquad \text{(B6.2.6)}$$

which is plotted in figure B6.2.1. The motivation for this form of activation function is considered in section B6.3.2. We can combine (B6.2.4) and (B6.2.5) to obtain a complete expression for the function represented by a two-layer feedforward network in the form

$$y_k(x) = \sum_{j=0}^{M} w_{kj} g\left(\sum_{i=0}^{d} w_{ji} x_i\right). \qquad \text{(B6.2.7)}$$

The form of network mapping given by (B6.2.7) is appropriate for regression problems, but needs some modification for classification applications as will also be discussed in section B6.3.2. It should be noted that models of this kind, with basis functions which are adapted to the data, are not unique to neural networks. Such models have been considered for many years in the statistics literature and include, for example, *projection pursuit regression* (Friedman and Stuetzle 1981, Huber 1985) which has a form remarkably similar to that of the feedforward network discussed above. The procedures for determining the parameters in projection pursuit regression are, however, quite different from those generally used for feedforward networks.

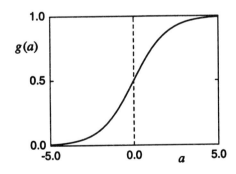

Figure B6.2.1. Plot of the logistic sigmoid activation function given by (B6.2.6).

It is often useful to represent the network mapping function in terms of a network diagram, as shown in figure B6.2.2. Each element of the diagram represents one of the terms of the corresponding mathematical expression. The bias parameters in the first layer are shown as weights from an extra input having a fixed value of $x_0 = 1$. Similarly, the bias parameters in the second layer are shown as weights from an extra hidden unit, with activation again fixed at $z_0 = 1$.

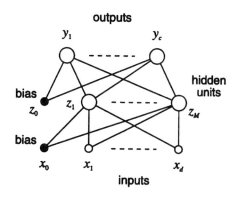

Figure B6.2.2. An example of a feedforward network having two layers of adaptive weights.

More complex forms of feedforward network function can be considered, corresponding to more complex topologies of network diagram. However, the simple structure of figure B6.2.2 has the property that it can approximate any continuous mapping to arbitrary accuracy provided the number M of hidden units is sufficiently large. This property has been discussed by many authors including Funahashi (1989), Hecht-Nielsen (1989), Cybenko (1989), Hornik *et al* (1989), Stinchcombe and White (1989), Cotter (1990), Ito (1991), Hornik (1991), and Kreinovich (1991). A proof that two-layer networks having sigmoidal hidden units can simultaneously approximate both a function and its derivatives was given by Hornik *et al* (1990).

The other major class of network model, which also possesses universal approximation capabilities, is B1.7.3 the *radial basis function network* (Broomhead and Lowe 1988, Moody and Darken 1989). Such networks again take the form of (B6.2.4), but the basis functions now depend on some measure of *distance* between the input vector x and a prototype vector μ_j. A typical example would be a Gaussian basis function of the form

$$z_j(x) = \exp\left(-\frac{\|x - \mu_j\|^2}{2\sigma_j^2}\right) \tag{B6.2.8}$$

where the parameter σ_j controls the width of the basis function. Training of radial basis function networks usually involves a two-stage procedure in which the basis functions are first optimized using input data alone, and then the parameters w_{kj} in (B6.2.4) are optimized by error function minimization. Such procedures are described in detail by Bishop (1995).

B6.2.3 Statistical pattern recognition

We turn now to some of the formalism of statistical pattern recognition, which we regard as essential for a clear understanding of neural networks. For convenience we introduce many of the central concepts in the context of classification problems, although much the same ideas also apply to regression. The goal is to assign an input pattern x to one of c classes C_k where $k = 1, \ldots, c$. In the case of handwritten digit recognition, for example, we might have ten classes corresponding to the ten digits $0, \ldots, 9$. One of the powerful results of the theory of statistical pattern recognition is a formalism which describes the theoretically best achievable performance, corresponding to the smallest probability of misclassifying a new input pattern. This provides a principled context within which we can develop neural networks, and other techniques, for classification.

For any but the simplest of classification problems it will not be possible to devise a system which is able to give perfect classification of all possible input patterns. The problem arises because many input patterns cannot be assigned unambiguously to one particular class. Instead the most general description we can give is in terms of the probabilities of belonging to each of the classes C_k *given* an input vector x. These probabilities are written as $P(C_k|x)$, and are called the *posterior* probabilities of class membership, since they correspond to the probabilities after we have observed the input pattern x. If we consider a large set of patterns all from a particular class C_k then we can consider the probability distribution of the corresponding input patterns, which we write as $p(x|C_k)$. These are called the class conditional distributions and, since the vector x is a continuous variable, they correspond to probability density functions rather than probabilities. The distribution of input vectors, irrespective of their class labels, is written as $p(x)$ and is called the *unconditional* distribution of inputs. Finally, we can consider the probabilities of occurrence of the different classes irrespective of the input pattern, which we write as $P(C_k)$. These correspond to the relative frequencies of patterns within the complete data set, and are called *prior* probabilities since they correspond to the probabilities of membership of each of the classes before we observe a particular input vector.

These various probabilities can be related using two standard results from probability theory. The first is the *product rule* which takes the form

$$P(C_k, x) = P(C_k|x)p(x) \tag{B6.2.9}$$

and the second is the *sum rule* given by

$$\sum_k P(C_k, x) = p(x). \tag{B6.2.10}$$

From these rules we obtain the following relation

$$P(C_k|x) = \frac{p(x|C_k)P(C_k)}{p(x)} \tag{B6.2.11}$$

which is known as *Bayes' theorem*. The denominator in (B6.2.11) is given by

$$p(x) = \sum_k p(x|C_k)P(C_k) \tag{B6.2.12}$$

and plays the role of a normalizing factor, ensuring that the posterior probabilities in (B6.2.11) sum to one, $\sum_k P(C_k|x) = 1$. As we shall see shortly, knowledge of the posterior probabilities allows us to find the optimal solution to a classification problem. A key result, discussed in section B6.3.2, is that under suitable circumstances the outputs of a correctly trained neural network can be interpreted as (approximations to) the posterior probabilities $P(C_k|x)$ when the vector x is presented to the inputs of the network.

As we have already noted, perfect classification of all possible input vectors will, in general, be impossible. The best we can do is to minimize the probability that an input will be misclassified. This is achieved by assigning each new input vector x to that class for which the posterior probability $P(C_k|x)$ is largest. Thus an input vector x is assigned to class C_k if

$$P(C_k|x) > P(C_j|x) \qquad \text{for all } j \neq k. \tag{B6.2.13}$$

We shall see the justification for this rule shortly. Since the denominator in Bayes' theorem (B6.2.11) is independent of the class, we see that this is equivalent to assigning input patterns to class C_k provided

$$p(x|C_k)P(C_k) > p(x|C_j)P(C_j) \qquad \text{for all } j \neq k. \tag{B6.2.14}$$

A pattern classifier provides a rule for assigning each point of feature space to one of c classes. We can therefore regard the feature space as being divided up into c *decision regions* $\mathcal{R}_1, \ldots, \mathcal{R}_c$ such that a point falling in region $\mathcal{R}_k$ is assigned to class C_k. Note that each of these regions need not be contiguous, but may itself be divided into several disjoint regions all of which are associated with the same class. The boundaries between these regions are known as *decision surfaces* or *decision boundaries*.

In order to find the optimal criterion for placement of decision boundaries, consider the case of a one-dimensional feature space x and two classes C_1 and C_2. We seek a decision boundary which minimizes the probability of misclassification, as illustrated in figure B6.2.3. A misclassification error will occur if we assign a new pattern to class C_1 when in fact it belongs to class C_2, or vice versa. We can calculate the total probability of an error of either kind by writing (Duda and Hart 1973)

$$
\begin{aligned}
P(\text{error}) &= P(x \in \mathcal{R}_2, C_1) + P(x \in \mathcal{R}_1, C_2) \\
&= P(x \in \mathcal{R}_2|C_1)P(C_1) + P(x \in \mathcal{R}_1|C_2)P(C_2) \\
&= \int_{\mathcal{R}_2} p(x|C_1)P(C_1)\,dx + \int_{\mathcal{R}_1} p(x|C_2)P(C_2)\,dx
\end{aligned}
\tag{B6.2.15}
$$

where $P(x \in \mathcal{R}_1, C_2)$ is the joint probability of x being assigned to class C_1 and the true class being C_2. From (B6.2.15) we see that, if $p(x|C_1)P(C_1) > p(x|C_2)P(C_2)$ for a given x, we should choose the regions

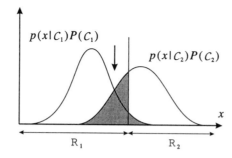

Figure B6.2.3. Schematic illustration of the joint probability densities, given by $p(x, C_k) = p(x|C_k)P(C_k)$, as a function of a feature value x, for two classes C_1 and C_2. If the vertical line is used as the decision boundary then the classification errors arise from the shaded region. By placing the decision boundary at the point where the two probability density curves cross (shown by the arrow), the probability of misclassification is minimized.

$\mathcal{R}_1$ and $\mathcal{R}_2$ such that x is in $\mathcal{R}_1$, since this gives a smaller contribution to the error. We recognize this as the decision rule given by (B6.2.14) for minimizing the probability of misclassification. The same result can be seen graphically in figure B6.2.3, in which misclassification errors arise from the shaded region. By choosing the decision boundary to coincide with the value of x at which the two distributions cross (shown by the arrow) we minimize the area of the shaded region and hence minimize the probability of misclassification. This corresponds to classifying each new pattern x using (B6.2.14), which is equivalent to assigning each pattern to the class having the largest posterior probability. A similar justification for this decision rule may be given for the general case of c classes and d-dimensional feature vectors (Duda and Hart 1973).

It is important to distinguish between two separate stages in the classification process. The first is *inference* whereby data are used to determine values for the posterior probabilities. These are then used in the second stage which is *decision making* in which those probabilities are used to make decisions such as assigning a new data point to one of the possible classes. So far we have based classification decisions on the goal of minimizing the probability of misclassification. In many applications this may not be the most appropriate criterion. Consider, for instance, the task of classifying images used in medical screening into two classes corresponding to 'normal' and 'tumor'. There may be much more serious consequences if we classify an image of a tumor as normal than if we classify a normal image as that of a tumor. Such effects may easily be taken into account by the introduction of a *loss matrix* with elements L_{kj} specifying the penalty associated with assigning a pattern to class C_j when in fact it belongs to class C_k. The overall expected loss is minimized if, for each input x, the decision regions $\mathcal{R}_j$ are chosen such that $x \in \mathcal{R}_j$ when

$$\sum_{k=1}^{c} L_{kj} p(x|C_k)P(C_k) < \sum_{k=1}^{c} L_{ki} p(x|C_k)P(C_k) \qquad \text{for all } i \neq j \qquad \text{(B6.2.16)}$$

which represents a generalization of the usual decision rule for minimizing the probability of misclassification. Note that, if we assign a loss of 1 if the pattern is placed in the wrong class, and a loss of 0 if it is placed in the correct class, so that $L_{kj} = 1 - \delta_{kj}$ (where δ_{kj} is the Kronecker delta symbol), then (B6.2.16) reduces to the decision rule for minimizing the probability of misclassification, given by (B6.2.14).

Another powerful consequence of knowing posterior probabilities is that it becomes possible to introduce a *reject criterion*. In general, we expect most of the misclassification errors to occur in those regions of x-space where the largest of the posterior probabilities is relatively low, since there is then a strong overlap between different classes. In some applications it may be better not to make a classification decision in such cases. This leads to the following procedure

$$\text{if} \quad \max_{k} P(C_k|x) \quad \begin{cases} \geq \theta & \text{then classify } x \\ < \theta & \text{then reject } x \end{cases} \qquad \text{(B6.2.17)}$$

where θ is a threshold in the range $(0, 1)$. The larger the value of θ, the fewer points will be classified. For the medical classification problem, for example, it may be better not to rely on an automatic classification system in doubtful cases, but to have these classified instead by a human expert.

Yet another application for the posterior probabilities arises when the distributions of patterns between the classes, corresponding to the prior probabilities $P(\mathcal{C}_k)$, are strongly mismatched. If we know the posterior probabilities corresponding to the data in the training set, it is then a simple matter to use Bayes' theorem (B6.2.11) to make the necessary corrections. This is achieved by dividing the posterior probabilities by the prior probabilities corresponding to the training set, multiplying them by the new prior probabilities, and then *normalizing* the results. Changes in the prior probabilities can therefore B4.4.1 be accommodated without retraining the network. The prior probabilities for the training set may be estimated simply by evaluating the fraction of the training set data points in each class. Prior probabilities corresponding to the operating environment can often be obtained very straightforwardly since only the class labels are needed and no input data are required. As an example, consider again the problem of classifying medical images into 'normal' and 'tumor'. When used for screening purposes, we would expect a very small prior probability of 'tumor'. To obtain a good variety of tumor images in the training set would therefore require huge numbers of training examples. An alternative is to increase artificially the proportion of tumor images in the training set, and then to compensate for the different priors on the test data as described above. The prior probabilities for tumors in the general population can be obtained from medical statistics, without having to collect the corresponding images. Correction of the network outputs is then a simple matter of multiplication and division.

The most common approach to the use of neural networks for classification involves having the network itself directly produce the classification decision. As we have seen, knowledge of the posterior probabilities is substantially more powerful.

References

Bellman R 1961 *Adaptive Control Processes: A Guided Tour* (New Jersey: Princeton University Press)

Bishop C M 1995 *Neural Networks for Pattern Recognition* (Oxford: Oxford University Press)

Broomhead D S and Lowe D 1988 Multivariable functional interpolation and adaptive networks *Complex Syst.* **2** 321–55

Cotter N E 1990 The Stone–Weierstrass theorem and its application to neural networks *IEEE Trans. Neural Networks* **1** 290–5

Cybenko G 1989 Approximation by superpositions of a sigmoidal function *Math. Control, Signals Syst.* **2** 304–14

Duda R O and P E Hart 1973 *Pattern Classication and Scene Analysis* (New York: Wiley)

Friedman J H and W Stuetzle 1981 Projection pursuit regression *J. Am. Stat. Assoc.* **76** 817–23

Funahashi K 1989 On the approximate realization of continuous mappings by neural networks *Neural Networks* **2** 183–92

Hecht-Nielsen R 1989 Theory of the back-propagation neural network *Proc. Int. Joint Conf. on Neural Networks* vol 1 pp 593–605 (San Diego, CA: IEEE)

Hornik K 1991 Approximation capabilities of multilayer feedforward networks *Neural Networks* **4** 251–7

Hornik K, Stinchcombe M and White H 1989 Multilayer feedforward networks are universal approximators *Neural Networks* **2** 359–66

——1990 Universal approximation of an unknown mapping and its derivatives using multilayer feedforward networks *Neural Networks* **3** 551–60

Huber P J 1985 Projection pursuit *Ann. Stat.* **13** 435–75

Ito Y 1991 Representation of functions by superpositions of a step or sigmoid function and their applications to neural network theory *Neural Networks* **4** 385–94

Kreinovich V Y 1991 Arbitrary nonlinearity is sufficient to represent all functions by neural networks: a theorem *Neural Networks* **4** 381–3

Le Cun Y, Boser B, Denker J S, Henderson D, Howard R E, Hubbard W and Jackel L D 1989 Backpropagation applied to handwritten zip code recognition *Neural Comput.* **1** 541–51

Moody J and Darken C J 1989 Fast learning in networks of locally-tuned processing units *Neural Comput.* **1** 281–94

Stinchcombe M and White H 1989 Universal approximation using feed-forward networks with non-sigmoid hidden layer activation functions. *Proc. Int. Joint Conf. on Neural Networks* (San Diego, CA: IEEE) vol 1 pp 613–8

B6.3 Error functions

Christopher M Bishop

Abstract

See the abstract for Chapter B6.

We turn next to the problem of determining suitable values for the weight parameters w in a network.

Training data are provided in the form of N pairs of input vectors x^n and corresponding desired output vectors t^n where $n = 1, \ldots, N$ labels the patterns. These desired outputs are called *target* values in the neural network context, and the components t_k^n of t^n represent the targets for the corresponding network outputs y_k. For associative prediction problems of the kind we are considering, the most general and complete description of the statistical properties of the data is given in terms of the conditional density of the target data $p(t|x)$ conditioned on the input data.

A principled way to devise an error function is to use the concept of *maximum likelihood*. For a set of training data $\{x^n, t^n\}$, the likelihood can be written as

$$\mathcal{L} = \prod_n p(t^n|x^n) \tag{B6.3.1}$$

where we have assumed that each data point (x^n, t^n) is drawn independently from the same distribution, so that the likelihood for the complete data set is given by the product of the probabilities for each data point separately. Instead of maximizing the likelihood, it is generally more convenient to minimize the negative logarithm of the likelihood. These are equivalent procedures, since the negative logarithm is a monotonic function. We therefore minimize

$$E = -\ln \mathcal{L} = -\sum_n \ln p(t^n|x^n) \tag{B6.3.2}$$

where E is called an *error function*. We shall further assume that the distribution of the individual target variables t_k, where $k = 1, \ldots, c$, are independent, so that we can write

$$p(t|x) = \prod_{k=1}^{c} p(t_k|x). \tag{B6.3.3}$$

As we shall see, a feedforward neural network can be regarded as a framework for modeling the conditional probability density $p(t|x)$. Different choices of error function then arise from different assumptions about the form of the conditional distribution $p(t|x)$. It is convenient to discuss error functions for regression and classification problems separately.

B6.3.1 Error functions for regression

For regression problems, the output variables are continuous. To define a specific error function we must make some choice for the model of the distribution of target data. The simplest assumption is to take this distribution to be Gaussian. More specifically, we assume that the target variable t_k is given by some deterministic function of x with added Gaussian noise ϵ, so that

$$t_k = h_k(x) + \epsilon_k. \tag{B6.3.4}$$

We then assume that the errors ϵ_k have a normal distribution with zero mean, and a standard deviation σ which does not depend on x or k. Thus, the distribution of ϵ_k is given by

$$p(\epsilon_k) = \frac{1}{(2\pi\sigma^2)^{1/2}} \exp\left(-\frac{\epsilon_k^2}{2\sigma^2}\right). \tag{B6.3.5}$$

We now model the functions $h_k(x)$ by a neural network with outputs $y_k(x; w)$ where w is the set of weight parameters governing the neural network mapping. Using (B6.3.4) and (B6.3.5) we see that the probability distribution of target variables is given by

$$p(t_k|x) = \frac{1}{(2\pi\sigma^2)^{1/2}} \exp\left(-\frac{\{y_k(x; w) - t_k\}^2}{2\sigma^2}\right) \tag{B6.3.6}$$

where we have replaced the unknown function $h_k(x)$ by our model $y_k(x; w)$. Together with (B6.3.2) and (B6.3.3) this leads to the following expression for the error function

$$E = \frac{1}{2\sigma^2} \sum_{n=1}^{N} \sum_{k=1}^{c} \{y_k(x^n; w) - t_k^n\}^2 + Nc \ln \sigma + \frac{Nc}{2} \ln(2\pi). \tag{B6.3.7}$$

We note that, for the purposes of error minimization, the second and third terms on the right-hand side of (B6.3.7) are independent of the weights w and so can be omitted. Similarly, the overall factor of $1/\sigma^2$ in the first term can also be omitted. We then finally obtain the familiar expression for the sum of squares error function

$$E = \frac{1}{2} \sum_{n=1}^{N} \left\| y(x^n; w) - t^n \right\|^2. \tag{B6.3.8}$$

Note that models of the form (B6.2.4), with fixed basis functions, are linear functions of the parameters w and so (B6.3.8) is a quadratic function of w. This means that the minimum of E can be found in terms of the solution of a set of linear algebraic equations. For this reason, the process of determining the parameters in such models is extremely fast. Functions which depend linearly on the adaptive parameters are called *linear* models, even though they may be nonlinear functions of the input variables. If the basis functions themselves contain adaptive parameters, we have to address the problem of minimizing an error function which is generally highly nonlinear.

The sum of squares error function was derived from the requirement that the network output vector should represent the conditional mean of the target data, as a function of the input vector. It is easily shown (Bishop 1995) that minimization of this error, for an infinitely large data set and a highly flexible network model, does indeed lead to a network satisfying this property.

We have derived the sum-of-squares error function on the assumption that the distribution of the target data is Gaussian. For some applications, such an assumption may be far from valid (if the distribution is multimodal for instance) in which case the use of a sum-of-squares error function can lead to extremely poor results. Examples of such distributions arise frequently in inverse problems such as robot kinematics, the determination of spectral line parameters from the spectrum itself, or the reconstruction of spatial data from line of sight information. One general approach in such cases is to combine a feedforward network with a *Gaussian mixture model* (i.e. a linear combination of Gaussian functions) thereby allowing general conditional distributions $p(t|x)$ to be modeled (Bishop 1994).

B6.3.2 Error functions for classification

In the case of classification problems, the goal, as we have seen, is to approximate the posterior probabilities of class membership $P(C_k|x)$ given the input pattern x. We now show how to arrange for the outputs of a network to approximate these probabilities.

First we consider the case of two classes C_1 and C_2. In this case we can consider a network having a single output y which should represent the posterior probability $P(C_1|x)$ for class C_1. The posterior probability of class C_2 will then be given by $P(C_2|x) = 1 - y$. To achieve this we consider a target coding scheme for which $t = 1$ if the input vector belongs to class C_1 and $t = 0$ if it belongs to class C_2. We can combine these into a single expression, so that the probability of observing either target value is

$$p(t|x) = y^t(1 - y)^{1-t} \tag{B6.3.9}$$

Handbook of Neural Computation release 97/1   © 1997 IOP Publishing Ltd and Oxford University Press

which is a particular case of the binomial distribution called the Bernoulli distribution. With this interpretation of the output unit activations, the likelihood of observing the training data set, assuming the data points are drawn independently from this distribution, is then given by

$$\prod_n (y^n)^{t^n} (1 - y^n)^{1-t^n}. \tag{B6.3.10}$$

As usual, it is more convenient to minimize the negative logarithm of the likelihood. This leads to the *cross-entropy* error function (Hopfield 1987, Baum and Wilczek 1988, Solla *et al* 1988, Hinton 1989, Hampshire and Pearlmutter 1990) in the form

$$E = -\sum_n \left\{ t^n \ln y^n + (1 - t^n) \ln(1 - y^n) \right\}. \tag{B6.3.11}$$

For the network model introduced in (B6.2.4) the outputs were linear functions of the activations of the hidden units. While this is appropriate for regression problems, we need to consider the correct choice of output unit activation function for the case of classification problems. We shall assume (Rumelhart *et al* 1995) that the class conditional distributions of the outputs of the hidden units, represented here by the vector z, are described by

$$p(z|\mathcal{C}_k) = \exp \left\{ A(\theta_k) + B(z, \phi) + \theta_k^{\mathrm{T}} z \right\} \tag{B6.3.12}$$

which is a member of the *exponential family* of distributions (that includes many of the common distributions as special cases such as Gaussian, binomial, Bernoulli, Poisson, and so on). The parameters θ_k and ϕ control the form of the distribution. In writing (B6.3.12) we are implicitly assuming that the distributions differ only in the parameters θ_k and not in ϕ. An example would be two Gaussian distributions with different means, but with common covariance matrices. (Note that the decision boundaries will then be linear functions of z but will of course be nonlinear functions of the input variables as a consequence of the nonlinear transformation by the hidden units.)

Using Bayes' theorem, we can write the posterior probability for class $\mathcal{C}_1$ in the form

$$
\begin{aligned}
P(\mathcal{C}_1|z) &= \frac{p(z|\mathcal{C}_1)P(\mathcal{C}_1)}{p(z|\mathcal{C}_1)P(\mathcal{C}_1) + p(z|\mathcal{C}_2)P(\mathcal{C}_2)} \\
&= \frac{1}{1 + \exp(-a)}
\end{aligned}
\tag{B6.3.13}
$$

which is a logistic sigmoid function, in which

$$a = \ln \frac{p(z|\mathcal{C}_1)P(\mathcal{C}_1)}{p(z|\mathcal{C}_2)P(\mathcal{C}_2)}. \tag{B6.3.14}$$

Using (B6.3.12) we can write this in the form

$$a = w^{\mathrm{T}} z + w_0 \tag{B6.3.15}$$

where we have defined

$$w = \theta_1 - \theta_2 \tag{B6.3.16}$$

$$w_0 = A(\theta_1) - A(\theta_2) + \ln \frac{P(\mathcal{C}_1)}{P(\mathcal{C}_2)}. \tag{B6.3.17}$$

Thus the network output is given by a logistic sigmoid activation function acting on a weighted linear combination of the outputs of those hidden units which send connections to the output unit.

Incidentally, it is clear that we can also apply the above arguments to the activations of hidden units in a network. Provided such units use logistic sigmoid activation functions, we can interpret their outputs as probabilities of the presence of corresponding 'features' conditioned on the inputs to the units.

As a simple illustration of the interpretation of network outputs as probabilities, we consider a two-class problem with one input variable in which the class conditional densities are given by the Gaussian mixture functions shown in figure B6.3.1. A feedforward network, with five hidden units having sigmoidal activation functions, and one output unit having a logistic sigmoid activation function, was trained by minimizing a cross-entropy error using 100 cycles of the BFGS quasi-Newton algorithm (section B6.3.3). B6.3.3

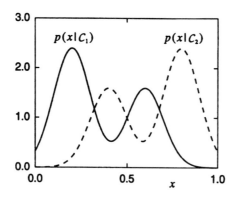

Figure B6.3.1. Plots of the class conditional densities used to generate a data set to demonstrate the interpretation of network outputs as posterior probabilities. The training data set was generated from these densities, using equal prior probabilities.

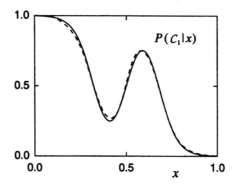

Figure B6.3.2. The result of training a multilayer perceptron on data generated from the density functions in figure B6.3.1. The full curve shows the output of the trained network as a function of the input variable x, while the broken curve shows the true posterior probability $P(C_1|x)$ calculated from the class-conditional densities using Bayes' theorem.

The resulting network mapping function is shown, along with the true posterior probability calculated using Bayes' theorem, in figure B6.3.2.

For the case of more than two classes, we consider a network with one output for each class so that each output represents the corresponding posterior probability. First of all we choose the target values for network training according to a 1-of-c coding scheme, so that $t_k^n = \delta_{kl}$ for a pattern n from class C_l. We wish to arrange for the probability of observing the set of target values t_k^n, given an input vector x^n, to be given by the corresponding network output so that $p(C_l|x) = y_l$. The value of the conditional distribution for this pattern can therefore be written as

$$p(t^n|x^n) = \prod_{k=1}^{c}(y_k^n)^{t_k^n} . \tag{B6.3.18}$$

If we form the likelihood function, and take the negative logarithm as before, we obtain an error function of the form

$$E = -\sum_n \sum_{k=1}^{c} t_k^n \ln y_k^n . \tag{B6.3.19}$$

Again we must seek the appropriate output unit activation function to match this choice of error function. As before, we shall assume that the activations of the hidden units are distributed according to (B6.3.12). From Bayes' theorem, the posterior probability of class C_k is given by

$$p(C_k|z) = \frac{p(z|C_k)P(C_k)}{\sum_{k'} p(z|C_{k'})P(C_{k'})} . \tag{B6.3.20}$$

Substituting (B6.3.12) into (B6.3.20) and rearranging we obtain

$$p(\mathcal{C}_k|z) = y_k = \frac{\exp(a_k)}{\sum_{k'} \exp(a_{k'})} \tag{B6.3.21}$$

where

$$a_k = w_k^{\mathrm{T}} z + w_{k0} \tag{B6.3.22}$$

and we have defined

$$w_k = \theta_k \tag{B6.3.23}$$

$$w_{k0} = A(\theta_k) + \ln P(\mathcal{C}_k). \tag{B6.3.24}$$

The activation function (B6.3.21) is called a *softmax* function or *normalized exponential*. It has the properties that $0 \leq y_k \leq 1$ and $\sum_k y_k = 1$ as required for probabilities.

It is easily verified (Bishop 1995) that the minimization of the error function (B6.3.19), for an infinite data set and a highly flexible network function, indeed leads to network outputs which represent the posterior probabilities for any input vector x.

Note that the network outputs of the trained network need not be close to 0 or 1 if the class conditional density functions are overlapping. Heuristic procedures, such as applying extra training using those patterns which fail to generate outputs close to the target values, will be counterproductive, since this alters the distributions and makes it *less* likely that the network will generate the correct Bayesian probabilities!

B6.3.3 Error backpropagation

Using the principle of maximum likelihood, we have formulated the problem of learning in neural networks in terms of the minimization of an error function $E(w)$. This error depends on the vector w of weight and bias parameters in the network, and the goal is therefore to find a weight vector w^* which minimizes E. For models of the form (B6.2.4) in which the basis functions are fixed, and for an error function given by the sum-of-squares form (B6.3.8), the error is a quadratic function of the weights. Its minimization then corresponds to the solution of a set of coupled linear equations and can be performed rapidly. We have seen, however, that models with fixed basis functions suffer from very poor scaling with input dimensionality. In order to avoid this difficulty we need to consider models with adaptive basis functions. The error function now becomes a highly nonlinear function of the weight vector, and its minimization requires sophisticated optimization techniques.

We have considered error functions of the form (B6.3.8), (B6.3.11) and (B6.3.19) which are differentiable functions of the network outputs. Similarly, we have considered network mappings which are differentiable functions of the weights. It therefore follows that the error function itself will be a differentiable function of the weights and so we can use gradient-based methods to find its minima. We now show that there is a computationally efficient procedure, called *backpropagation*, which allows the required derivatives to be evaluated for arbitrary feedforward network topologies.   C1.2.3

In a general feedforward network, each unit computes a weighted sum of its inputs of the form

$$z_j = g(a_j) \qquad a_j = \sum_i w_{ji} z_i \tag{B6.3.25}$$

where z_i is the activation of a unit, or input, which sends a connection to unit j, and w_{ji} is the weight associated with that connection. The summation runs over all units which send connections to unit j. Biases can be included in this sum by introducing an extra unit, or input, with activation fixed at $+1$. We therefore do not need to deal with biases explicitly. The error functions which we are considering can be written as a sum over patterns of the error for each pattern separately so that $E = \sum_n E^n$. This follows from the assumed independence of the data points under the given distribution. We can therefore consider one pattern at a time, and then find the derivatives of E by summing over patterns.

For each pattern we shall suppose that we have supplied the corresponding input vector to the network and calculated the activations of all of the hidden and output units in the network by successive application of (B6.3.25). This process is often called *forward propagation* since it can be regarded as a forward flow of information through the network.

Now consider the evaluation of the derivative of E^n with respect to some weight w_{ji}. First we note that E^n depends on the weight w_{ji} only via the summed input a_j to unit j. We can therefore apply the chain rule for partial derivatives to give

$$\frac{\partial E^n}{\partial w_{ji}} = \frac{\partial E^n}{\partial a_j}\frac{\partial a_j}{\partial w_{ji}}. \tag{B6.3.26}$$

We now introduce a useful notation

$$\delta_j \equiv \frac{\partial E^n}{\partial a_j} \tag{B6.3.27}$$

where the δ are often referred to as *errors* for reasons which will become clear shortly. Using (B6.3.25) we can write

$$\frac{\partial a_j}{\partial w_{ji}} = z_i. \tag{B6.3.28}$$

Substituting (B6.3.27) and (B6.3.28) into (B6.3.26) we then obtain

$$\frac{\partial E^n}{\partial w_{ji}} = \delta_j z_i. \tag{B6.3.29}$$

Equation (B6.3.29) tells us that the required derivative is obtained simply by multiplying the value of δ for the unit at the output end of the weight by the value of z for the unit at the input end of the weight (where $z = 1$ in the case of a bias). Thus, in order to evaluate the derivatives, we need only to calculate the value of δ_j for each hidden and output unit in the network, and then apply (B6.3.29).

For the output units the evaluation of δ_k is straightforward. From the definition (B6.3.27) we have

$$\delta_k \equiv \frac{\partial E^n}{\partial a_k} = g'(a_k)\frac{\partial E^n}{\partial y_k} \tag{B6.3.30}$$

where we have used (B6.3.25) with z_k denoted by y_k. In order to evaluate (B6.3.30) we substitute appropriate expressions for $g'(a)$ and $\partial E^n/\partial y$. If, for example, we consider the sum-of-squares error function (B6.3.8) together with a network having linear outputs, as in (B6.2.7) for instance, we obtain

$$\delta_k = y_k^n - t_k^n \tag{B6.3.31}$$

and so δ_k represents the error between the actual and the desired values for output k. The same form (B6.3.31) is also obtained if we consider the cross-entropy error function (B6.3.11) together with a network with a logistic sigmoid output, or if we consider the error function (B6.3.19) together with the softmax activation function (B6.3.21).

To evaluate the δ for hidden units we again make use of the chain rule for partial derivatives, to give

$$\delta_j \equiv \frac{\partial E^n}{\partial a_j} = \sum_k \frac{\partial E^n}{\partial a_k}\frac{\partial a_k}{\partial a_j} \tag{B6.3.32}$$

where the sum runs over all units k to which unit j sends connections. The arrangement of units and weights is illustrated in figure B6.3.3. Note that the units labeled k could include other hidden units and/or output units. In writing down (B6.3.32) we are making use of the fact that variations in a_j give rise to variations in the error function only through variations in the variables a_k. If we now substitute the definition of δ given by (B6.3.27) into (B6.3.32), and make use of (B6.3.25), we obtain the following *backpropagation* formula

$$\delta_j = g'(a_j)\sum_k w_{kj}\delta_k \tag{B6.3.33}$$

which tells us that the value of δ for a particular hidden unit can be obtained by propagating the δ backwards from units higher up in the network, as illustrated in figure B6.3.3. Since we already know the values of the δ for the output units, it follows that by recursively applying (B6.3.33) we can evaluate the δ for all of the hidden units in a feedforward network, regardless of its topology. Having found the gradient of the error function for this particular pattern, the process of forward and backward propagation is repeated for each pattern in the data set, and the resulting derivatives summed to give the gradient $\nabla E(w)$ of the total error function.

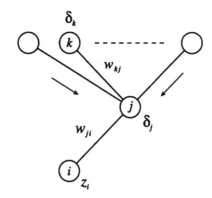

Figure B6.3.3. Illustration of the calculation of δ_j for hidden unit j by backpropagation of the δ from those units k to which unit j sends connections.

The backpropagation algorithm allows the error function gradient $\nabla E(w)$ to be evaluated efficiently. We now seek a way of using this gradient information to find a weight vector which minimizes the error. This is a standard problem in unconstrained nonlinear optimization and has been widely studied, and a number of powerful algorithms have been developed. Such algorithms begin by choosing an initial weight vector $w^{(0)}$ (which might be selected at random) and then making a series of steps through weight space of the form

$$w^{(\tau+1)} = w^{(\tau)} + \Delta w^{(\tau)} \qquad (B6.3.34)$$

where τ labels the iteration step. The simplest choice for the weight update is given by the gradient descent expression

$$\Delta w^{(\tau)} = -\eta\, \nabla E|_{w^{(\tau)}} \qquad (B6.3.35)$$

where the gradient vector ∇E must be reevaluated at each step. It should be noted that gradient descent is a very inefficient algorithm for highly nonlinear problems such as neural network optimization. Numerous *ad hoc* modifications have been proposed to try to improve its efficiency. One of the most common is the addition of a *momentum* term in (B6.3.35) to give

$$\Delta w^{(\tau)} = -\eta\, \nabla E|_{w^{(\tau)}} + \mu\, \Delta w^{(\tau-1)} \qquad (B6.3.36)$$

C1.2.4

where μ is called the momentum parameter. While this can often lead to improvements in the performance of gradient descent, there are now two arbitrary parameters η and μ whose values must be adjusted to give best performance. Furthermore, the optimal values for these parameters will often vary during the optimization process. In fact, much more powerful techniques have been developed for solving nonlinear optimization problems (Polak 1971, Gill *et al* 1981, Dennis and Schnabel 1983, Luenberger 1984, Fletcher 1987, Bishop 1995). These include conjugate gradient methods, quasi-Newton algorithms, and the Levenberg–Marquardt technique.

It should be noted that the term backpropagation is used in the neural computing literature to mean a variety of different things. For instance, the multilayer perceptron architecture is sometimes called a backpropagation network. The term backpropagation is also used to describe the training of a multilayer perceptron using gradient descent applied to a sum-of-squares error function. In order to clarify the terminology it is useful to consider the nature of the training process more carefully. Most training algorithms involve an iterative procedure for minimization of an error function, with adjustments to the weights being made in a sequence of steps. At each such step we can distinguish between two distinct stages. In the first stage, the derivatives of the error function with respect to the weights must be evaluated. As we shall see, the important contribution of the backpropagation technique is in providing a computationally efficient method for evaluating such derivatives. Since it is at this stage that errors are propagated backwards through the network, we use the term backpropagation specifically to describe the evaluation of derivatives. In the second stage, the derivatives are then used to compute the adjustments to be made to the weights. The simplest such technique, and the one originally considered by Rumelhart *et al* (1986), involves gradient descent. It is important to recognize that the two stages are distinct. Thus, the first-stage process, namely the propagation of errors backwards through the network in order to evaluate derivatives, can be applied to many other kinds of network and not just the multilayer perceptron. It can also be applied to error functions other than the simple sum-of-squares, and to the evaluation of other quantities such as the Hessian matrix whose elements comprise the second derivatives of the error function with respect to the weights (Bishop 1992). Similarly, the second stage of weight adjustment using the calculated derivatives can be tackled using a variety of optimization schemes (discussed above), many of which are substantially more effective than simple gradient descent.

One of the most important aspects of backpropagation is its computational efficiency. To understand this, let us examine how the number of computer operations required to evaluate the derivatives of the error function scales with the size of the network. A single evaluation of the error function (for a given input pattern) would require $O(W)$ operations, where W is the total number of weights in the network. For W weights in total there are W such derivatives to evaluate. A direct evaluation of these derivatives individually would therefore require $O(W^2)$ operations. By comparison, backpropagation allows all of the derivatives to be evaluated using a single forward propagation and a singlebackward propagation together with the use of (B6.3.29). Since each of these requires $O(W)$ steps, the overall computational cost is reduced from $O(W^2)$ to $O(W)$. The training of multilayer perceptron networks, even using backpropagation coupled with efficient optimization algorithms, can be very time consuming, and so this gain in efficiency is crucial.

References

Anderson A and Rosenfeld E (eds) 1988 *Neurocomputing: Foundations of Research* (Cambridge, MA: MIT)

Baum E B and Wilczek F 1988 Supervised learning of probability distributions by neural networks *Neural Information Processing Systems* ed D Z Anderson pp 52–61 (New York: American Institute of Physics)

Bishop C M 1992 Exact calculation of the Hessian matrix for the multilayer perceptron *Neural Comput.* **4** 494–501

——1994 Mixture density networks *Technical Report NGRG/94/001* Neural Computing Research Group, Aston University, Birmingham, UK

——1995 *Neural Networks for Pattern Recognition* (Oxford: Oxford University Press)

Dennis J E and R B Schnabel 1983 *Numerical Methods for Unconstrained Optimization and Nonlinear Equations* (Englewood Cliffs, NJ: Prentice-Hall)

Fletcher R 1987 *Practical Methods of Optimization* (2nd edn) (New York: Wiley)

Gill P E, Murray W and Wright M H 1981 *Practical Optimization* (London: Academic)

Hampshire J B and Pearlmutter B 1990 Equivalence proofs for multi-layer perceptron classifiers and the Bayesian discriminant function *Proc. 1990 Connectionist Models Summer School* ed D S Touretzky, J L Elman, T J Sejnowski and G E Hinton (San Mateo, CA: Morgan Kaufmann) pp 159–72

Hinton G E 1989 Connectionist learning procedures *Artif. Intell.* **40** 185–234

Hopfield J J 1987 Learning algorithms and probability distributions in feed-forward and feed-back networks *Proc. Natl Acad. Sci.* **84** 8429–33

Luenberger D G 1984 *Linear and Nonlinear Programming* (2nd edn) (Reading, MA: Addison-Wesley)

Polak E 1971 *Computational Methods in Optimization: A Unified Approach* (New York: Academic)

Rumelhart D E, Durbin R, Golden R and Chauvin Y 1995 Backpropagation: the basic theory *Backpropagation: Theory, Architectures, and Applications* ed Y Chauvin and D E Rumelhart (Hillsdale, NJ: Lawrence Erlbaum) pp 1–34

Rumelhart D E, Hinton G E and Williams R J 1986 Learning internal representations by error propagation *Parallel Distributed Processing: Explorations in the Microstructure of Cognition Volume 1: Foundations* ed D E Rumelhart, J L McClelland, and the PDP Research Group (Cambridge, MA: MIT) pp 318–62 (reprinted in Anderson and Rosenfeld (1988).)

Solla S A, Levin E and Fleisher M 1988 Accelerated learning in layered neural networks *Complex Syst.* **2** 625–40

B6.4 Generalization

Christopher M Bishop

Abstract

See the abstract for Chapter B6.

The goal of network training is not to learn an exact representation of the training data itself, but rather to build a statistical model of the process which generates the data. This is important if the network is to exhibit good *generalization*, that is, to make good predictions for new inputs.   B3.5.2

In order for the network to provide a good representation of the generator of the data it is important that the effective complexity of the model be matched to the data set. This is most easily illustrated by returning to the analogy with polynomial curve fitting introduced in section B6.2.1. In this case the model complexity is governed by the order of the polynomial which in turn governs the number of adjustable coefficients. Consider a data set of 11 points generated by sampling the function

$$h(x) = 0.5 + 0.4 \sin(2\pi x) \tag{B6.4.1}$$

at equal intervals of x and then adding random noise with a Gaussian distribution having standard deviation $\sigma = 0.05$. This reflects a basic property of most data sets of interest in pattern recognition in that the data exhibit an underlying systematic component, represented in this case by the function $h(x)$, but are corrupted with random noise. Figure B6.4.1 shows the training data, as well as the function $h(x)$ from (B6.4.1), together with the result of fitting a linear polynomial, given by (B6.2.2) with $M = 1$. As can be seen, this polynomial gives a poor representation of $h(x)$, as a consequence of its limited flexibility. We can obtain a better fit by increasing the order of the polynomial, since this increases the number of *degrees of freedom* (i.e. the number of free parameters) in the function, which gives it greater flexibility.

Figure B6.4.2 shows the result of fitting a cubic polynomial ($M = 3$) which gives a much better approximation to $h(x)$. If, however, we increase the order of the polynomial too far, then the approximation to the underlying function actually gets worse. Figure B6.4.3 shows the result of fitting a ten-order polynomial ($M = 10$). This is now able to achieve a perfect fit to the training data, since a ten-order polynomial has 11 free parameters, and there are 11 data points. However, the polynomial has fitted the data by developing some dramatic oscillations and consequently gives a poor representation of $h(x)$. Functions of this kind are said to be *overfitted* to the data.

In order to determine the generalization performance of the different polynomials, we generate a second independent *test* set, and measure the root mean square error E^{RMS} with respect to both training and test sets. Figure B6.4.4 shows a plot of E^{RMS} for both the training data set and the test data set, as a function of the order M of the polynomial. We see that the training set error decreases steadily as the order of the polynomial increases. However, the test set error reaches a minimum at $M = 3$, and thereafter increases as the order of the polynomial is increased. The smallest error is achieved by that polynomial ($M = 3$) which most closely matches the function $h(x)$ from which the data were generated.

In the case of neural networks the weights and biases are analogous to the polynomial coefficients. These parameters can be optimized by minimization of an error function defined with respect to a training data set. The model complexity is governed by the number of such parameters and so is determined by the network architecture and in particular by the number of hidden units. We have seen that the complexity cannot be optimized by minimization of training set error since the smallest training error corresponds to an overfitted model which has poor generalization. Instead, we see that the optimum complexity can be

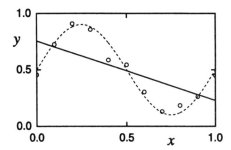

Figure B6.4.1. An example of a set of 11 data points obtained by sampling the function $h(x)$, defined by (B6.4.1), at equal intervals of x and adding random noise. The broken curve shows the function $h(x)$, while the full curve shows the rather poor approximation obtained with a linear polynomial, corresponding to $M = 1$ in (B6.2.2).

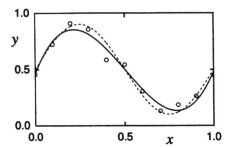

Figure B6.4.2. This shows the same data set as in figure B6.4.1, but this time fitted by a cubic ($M = 3$) polynomial, showing the significantly improved approximation to $h(x)$ achieved by this more flexible function.

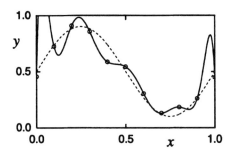

Figure B6.4.3. The result of fitting the same data set as in figure B6.4.1 using a ten-order ($M = 10$) polynomial. This gives a perfect fit to the training data, but at the expense of a function which has large oscillations, and which therefore gives a poorer representation of the generator function $h(x)$ than did the cubic polynomial of figure B6.4.2.

chosen by comparing the performance of a range of trained models using an independent test set. A more elaborate version of this procedure is *cross-validation* (Stone 1974, 1978, Wahba and Wold 1975).

Instead of directly varying the number of adaptive parameters in a network, the effective complexity of the model may be controlled through the technique of *regularization*. This involves the use of a model with a relatively large number of parameters, together with the addition of a penalty term Ω to the usual error function E to give a total error function of the form

$$\widetilde{E} = E + \nu\Omega \tag{B6.4.2}$$

where ν is called a regularization coefficient. The penalty term Ω is chosen so as to encourage smoother network mapping functions since, by analogy with the polynomial results shown in figures B6.4.1–B6.4.3, we expect that good generalization is achieved when the rapid variations in the mapping associated with overfitting are smoothed out. There will be an optimum value for ν which can again be found by comparing the performance of models trained using different values of ν on an independent test set. Regularization is usually the preferred choice for model complexity control for a number of reasons: it

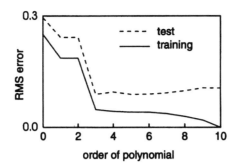

Figure B6.4.4. Plots of the RMS error E^{RMS} as a function of the order of the polynomial for both training and test sets, for the example problem considered in the previous three figures. The error with respect to the training set decreases monotonically with M, while the error in making predictions for new data (as measured by the test set) shows a minimum at $M = 3$.

allows prior knowledge to be incorporated into network training; it has a natural interpretation in the Bayesian framework (discussed in Section B6.5); and it can be extended to provide more complex forms of regularization involving several different regularization parameters which can be used, for example, to determine the relative importance of different inputs.

References

Stone M 1974 Cross-validatory choice and assessment of statistical predictions *J. R. Stat. Soc.* B **36** 111–47
—— 1978 Cross-validation: a review *Math. Operationsforsch. Statist., Ser. Statistics* **9** 127–39
Wahba G and Wold S 1975 A completely automatic French curve: fitting spline functions by cross-validation *Commun. Stat.* A **4** 1–17

B6.5   Discussion

Christopher M Bishop

Abstract

See the abstract for Chapter B6.

In this chapter we have presented a brief overview of neural networks from the viewpoint of statistical pattern recognition. Due to lack of space, there are many important issues which we have not discussed or have only touched upon. Here we mention two further topics of considerable significance for neural computing.

In practical applications of neural networks, one of the most important factors determining the overall performance of the final system is that of data preprocessing. Since a neural network mapping has universal approximation capabilities, as discussed in section B6.2.2, it would in principle be possible to use the original data directly as the input to a network. In practice, however, there is generally considerable advantage in processing the data in various ways before they are used for network training. One important reason why preprocessing can lead to improved performance is that it can offset some of the effects of the 'curse of dimensionality' discussed in section B6.2.2 by reducing the number of input variables. Input can be combined in linear or nonlinear ways to give a smaller number of new inputs which are then presented to the network. This is sometimes called *feature extraction*. Although information is often lost in the process, this can be more than compensated for by the benefits of a lower input dimensionality. Another significant aspect of preprocessing is that it allows the use of *prior knowledge*, in other words information which is relevant to the solution of a problem which is additional to that contained in the training data. A simple example would be the prior knowledge that the classification of a handwritten digit should not depend on the location of the digit within the input image. By extracting features which are independent of position, this translation invariance can be incorporated into the network structure, and this will generally give substantially improved performance compared with using the original image directly as the input to the network. Another use for preprocessing is to clean up deficiencies in the data. For example, real data sets often suffer from the problem of missing values in many of the patterns, and these must be accounted for before network training can proceed.

The discussion of learning in neural networks given above was based on the principle of maximum likelihood, which itself stems from the *frequentist* school of statistics. A more fundamental, and potentially more powerful, approach is given by the *Bayesian* viewpoint (Jaynes 1986). Instead of describing a trained network by a single weight vector w^*, the Bayesian approach expresses our uncertainty in the values of the weights through a probability distribution $p(w)$. The effect of observing the training data is to cause this distribution to become much more concentrated in particular regions of weight space, reflecting the fact that some weight vectors are more consistent with the data than others. Predictions for new data points require the evaluation of integrals over weight space, weighted by the distribution $p(w)$. The maximum-likelihood approach considered in Section B6.3 is related to a particular approximation in which we consider only the most probable weight vector, corresponding to a peak in the distribution. Aside from offering a more fundamental view of learning in neural networks, the Bayesian approach allows error bars to be assigned to network predictions, and regularization arises in a natural way in the Bayesian setting. Furthermore, a Bayesian treatment allows the model complexity (as determined by regularization coefficients, for instance) to be treated without the need for independent data as in cross-validation.

Although the Bayesian approach is very appealing, a full implementation is intractable for neural networks. Two principal approximation schemes have therefore been considered. In the first of these

(MacKay 1992a, b, c) the distribution over weights is approximated by a Gaussian centered on the most probable weight vector. Integrations over weight space can then be performed analytically, and this leads to a practical scheme which involves relatively small modifications to conventional algorithms. An alternative approach to the Bayesian treatment of neural networks is to use Monte Carlo techniques (Neal 1994) to perform the required integrations numerically without making analytical approximations. Again, this leads to a practical scheme which has been applied to some real-world problems.

An interesting aspect of the Bayesian viewpoint is that it is not, in principle, necessary to limit network complexity (Neal 1994), and that overfitting should not arise if the Bayesian approach is implemented correctly. A more comprehensive discussion of these and other topics can be found in the book by Bishop (1995).

References

Bishop C M 1995 *Neural Networks for Pattern Recognition* (Oxford: Oxford University Press)
Jaynes E T 1986 Bayesian methods: general background *Maximum Entropy and Bayesian Methods in Applied Statistics* ed J H Justice (Cambridge: Cambridge University Press) pp 1–25
MacKay D J C 1992a Bayesian interpolation *Neural Comput.* **4** 415–47
—— 1992b The evidence framework applied to classification networks *Neural Comput.* **4** 720–36
—— 1992c A practical Bayesian framework for back-propagation networks *Neural Comput.* **4** 448–72
Neal R M 1994 Bayesian learning for neural networks *PhD Thesis* University of Toronto, Canada

PART C

NEURAL NETWORK MODELS

PART C

NEURAL NETWORK MODELS

C1

Supervised Models

Contents

C1.1   Single-layer networks

George M Georgiou

Abstract

In this section single-layer neural network models are considered. Some of these models are simply single neurons, which, however, are used as the building blocks of larger networks. We discuss the perceptron which was developed in the late 1950s, and played a pivotal role in the history of neural networks. Nowadays, it is rarely used in real-life applications as more versatile and powerful models are available. Nevertheless, the perceptron remains an important model due to its simplicity and the influence it had in the development of the field. Today most neural networks consist of a large number of neurons, each largely resembling the perceptron. The adaline, also a single neuron model, was developed contemporaneously with the perceptron and is trained by the widely applied least mean square (LMS) algorithm. Both adaline and its extension known as madaline found many real applications, especially in signal processing. Notable is that the backpropagation algorithm is a generalization of LMS. A powerful technique, called learning vector quantization (LVQ) is also presented. This technique is used often in data compression and data classification applications. Another model discussed is the CMAC (cerebellar model articulation controller), which has many applications especially in robotics. All of these models are trained in a supervised manner: for each input, there is a target output, based on which an error signal is generated, based on which the weights are adapted. Also discussed are the instar and outstar models, single neurons which are closer to biology, and are primarily of theoretical interest.

C1.1.1   The perceptron

C1.1.1.1   Introduction

The perceptron was developed by Frank Rosenblatt in the late 1950s (Rosenblatt 1957, 1958) and the proof of convergence of the perceptron algorithm, also known as the perceptron theorem, was first outlined in Rosenblatt (1960). This result was enthusiastically received, and stimulated research in the area of neural networks, which was at the time called machine learning. The hope was that since the perceptron can eventually learn all mappings it can represent, then it might be possible that the same is true for networks of perceptrons arranged in multiple layers, to enable them to perform more complex mapping tasks. By the mid-1960s, in absence of a major breakthrough, enthusiasm in the area subsided. The landmark book *Perceptrons* by Minsky and Papert (1969, 1988) scrutinized the learning ability of single-layer perceptrons (i.e. perceptrons arranged on a single layer with no interconnections) to learn different functions. While mathematically accurate, the book was highly critical and pessimistic of the ultimate utility of perceptrons. It showed that such networks cannot learn to perform certain simple pattern recognition tasks, either within a reasonable amount of time or with reasonable weight magnitudes, or perform the task at all. The heart of the problem is that this type of neural network cannot represent nonlinearly separable functions, and thus cannot possibly learn such functions. What the book did not consider was multilayer networks of perceptrons, which *can* represent arbitrary functions. Yet, until now, we did not have algorithms for such networks that were equivalent to the elegant perceptron theorem, which guarantees learning without classification errors, if possible, in finite time. The renewed interest in neural networks in the 1980s was

C1.2.3 largely due to the development of *backpropagation*, which is used to train multilayer neural networks. Learning in these networks is neither exact nor guaranteed, but in practice it gives good solutions. The B3.2.4 activation function of the neurons is not the *Heaviside function*, as in the case of the perceptron, but instead B3.2.4 the *sigmoid function*.

C1.1.1.2 Purpose

The perceptron is used as a two-class classifier. The input patterns belong to one of two classes. The perceptron adjusts its weights so that all input patterns are correctly classified. This can only happen when they are linearly separable. Geometrically the algorithm finds a hyperplane that separates the two classes. After training, other input patterns of unknown class can be classified by observing on which side of the hyperplane each of them lies.

C1.1.1.3 Topology

The perceptron is a single-neuron model shown in figure C1.1.1. Each of the input vector components x_i is multiplied with the corresponding weight w_i, and these products are summed up yielding the net linear output, upon which the Heaviside function is applied to obtain the activation, which is either 1 or -1:

$$\text{net} = \sum_{i=1}^{n+1} w_i x_i \tag{C1.1.1}$$

$$a = f(\text{net}) = \begin{cases} 1 & \text{if net} \geq 0 \\ -1 & \text{if net} < 0. \end{cases} \tag{C1.1.2}$$

The input vector is $X = (x_1, x_2, \ldots, x_n, 1)$. The extra component 1 corresponds to the extra weight component w_{n+1}, which accounts for the threshold of the perceptron.

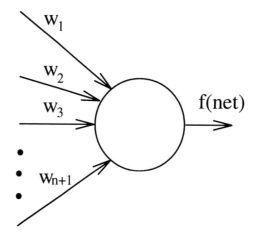

Figure C1.1.1. The perceptron.

C1.1.1.4 Learning

Learning is done in a supervised manner. The input patterns are cyclically presented to the perceptron. The order of presentation is not important. The error for input pattern X is calculated as the difference between the target output and the activation value. The weights are updated according to this formula:

$$w_i(k+1) = w_i(k) + \alpha \varepsilon(k) x_i(k) \tag{C1.1.3}$$

where k is the iteration counter, $\alpha > 0$ is the learning rate, a positive constant, and $\varepsilon(k)$ is the error produced by the input vector at iteration k:

$$\varepsilon(k) = t(k) - a(k) \tag{C1.1.4}$$

where $t(k)$ is the target value and $a(k)$ the activation of the perceptron, both at step k. The exact value of the learning rate α does affect the speed of learning, but regardless of its exact value, as long as it is positive, the algorithm will eventually converge. The algorithm can be described as follows.

(i)   Compute activation for input pattern X.
(ii) Compute the output error ε.
(iii) Modify the connection weight by adding to it the factor $\alpha \varepsilon X$.
(iv) Repeat steps (i), (ii) and (iii) for each input pattern.
(v)   Repeat step (iv) until error is zero for all input patterns.

C1.1.2   The perceptron theorem and its proof

In this section we formally state the perceptron theorem and present its proof.

Theorem: (Rosenblatt) It is given that the input pattern vectors X belong to two classes C_1 and C_2, and that there exists a weight vector W_0 that linearly separates them. In other words, the two classes are linearly separable. The weight vector W is randomly initialized at step 0 to $W(0)$. The input pattern vectors are repeatedly presented to the perceptron in finite intervals, and the weight vector W at step k is modified according to this rule (which is the vector form of (C1.1.3)):

$$W(k+1) = W(k) + \alpha \varepsilon(k) X(k) \qquad (C1.1.5)$$

where α is a real positive constant, $\varepsilon(k)$ is the error as defined in (C1.1.4), and $X(k)$ the input vector. Then there exists an integer N such that for all $k \geq N$, the error $\varepsilon(k) = 0$, and therefore $W(k+1) = W(k)$. In words, in a finite number of steps the algorithm will find a weight vector W that will correctly classify all input vectors.

Proof: Without loss of generality, it is assumed that $\alpha = 1$ and that $W(0) = \mathbf{0}$. It is also assumed that the iteration counter k counts only the steps at which the weight vector is corrected, that is the error ε is nonzero. Thus, the weight vector at step $k+1$ can be written as

$$W(k+1) = \varepsilon(1) X(1) + \varepsilon(2) X(2) + \cdots + \varepsilon(k) X(k). \qquad (C1.1.6)$$

We multiply both sides of (C1.1.6) by the row vector W_0^{T}:

$$W_0^{\mathrm{T}} W(k+1) = \sum_{j=1}^{k} \varepsilon(j) W_0^{\mathrm{T}} X(j). \qquad (C1.1.7)$$

Since all input vectors $X(j)$ are missclassified, $\varepsilon(j) W_0^{\mathrm{T}} X(j)$ is strictly positive. To see this, consider the case when $W_0^{\mathrm{T}} X(j)$ is positive. Since W_0 correctly classifies all input vectors, then the target value of $X(j)$ is $t(j) = 1$ and $\varepsilon(j) = 1 - (-1) > 0$, and therefore $\varepsilon(j) W_0^{\mathrm{T}} X(j)$ is positive. Following similar reasoning for the case when $\varepsilon(j) W_0^{\mathrm{T}} X(j)$ is negative, we conclude that $\varepsilon(j) W_0^{\mathrm{T}} X(j)$ is always positive. We define the strictly positive number a as

$$a = \min_{j}(\varepsilon(j) W_0^{\mathrm{T}} X(j)). \qquad (C1.1.8)$$

Then, from (C1.1.7),
$$W_0^{\mathrm{T}} W(k+1) \geq ka. \qquad (C1.1.9)$$

The Cauchy–Schwartz inequality for two vectors A and B in finite-dimensional real space, is $\|A\|^2 \|B\|^2 \geq |A^{\mathrm{T}} B|^2$, and when applied to W_0 and $W(k+1)$ we get

$$\|W(k+1)\|^2 \geq \frac{|W_0^{\mathrm{T}} W(k+1)|^2}{\|W_0\|^2} \qquad (C1.1.10)$$

where $\|\cdot\|$ is the Euclidean distance metric, or length, of its vector argument, and $|\cdot|$ indicates the absolute value of its real-valued argument. Combining equations (C1.1.9) and (C1.1.10), we arrive at the following inequality:

$$\|W(k+1)\|^2 \geq \frac{(ka)^2}{\|W_0\|^2}. \qquad (C1.1.11)$$

This last inequality will be combined with another one (C1.1.15), to be derived now, and it will be concluded that k must be finite.

We take the square of the Euclidean distance metric of both sides of the update rule (C1.1.5):

$$\|W(j+1)\|^2 = \|W(j)\|^2 + \|\varepsilon(j)X(j))\|^2 + 2\varepsilon(j)W^{\mathrm{T}}(j)X(j). \qquad \text{(C1.1.12)}$$

Defining Q as

$$Q = \max_j \|\varepsilon(j)X(j)\|^2 \qquad \text{(C1.1.13)}$$

and using the fact that $\varepsilon(j)W^{\mathrm{T}}(j)X(j) \le 0$ (recall that $X(j)$ is missclassified), we can write

$$\|W(j+1)\|^2 \le \|W(j)\|^2 + Q. \qquad \text{(C1.1.14)}$$

Adding the inequalities that are generated by the last inequality for $j = 1, 2, \ldots, k$, we obtain

$$\|W(k+1)\|^2 \le Qk. \qquad \text{(C1.1.15)}$$

Now we combine (C1.1.15) with (C1.1.10) to obtain

$$Qk \ge \|W(k+1)\|^2 \ge \frac{(ka)^2}{\|W_0\|^2}. \qquad \text{(C1.1.16)}$$

Dividing all sides by Qk, we finally arrive at this inequality

$$1 \ge \frac{ka^2}{Q\|W_0\|^2} \qquad \text{(C1.1.17)}$$

from which it is clear that k cannot grow without bound, as it would violate the inequality, and therefore k must be finite. This concludes the proof of the perceptron theorem.

Equation (C1.1.17) defines a bound on k, which can be computed by converting the inequality to equality and rounding up to the next integer:

$$k = \left\lceil \frac{Q\|W_0\|^2}{a^2} \right\rceil. \qquad \text{(C1.1.18)}$$

This upper bound for the number of (nonzero) corrections to the weight vector is of little practical use, since it depends on knowledge of a solution weight vector W_0, which normally would not be known beforehand.

C1.1.2.1 Pseudocode representation of the perceptron algorithm

The learning process will stop when either the weight vector causes all input vectors to be classified, or when the number of iterations has exceeded a maximum number ITER_MAX.

```
program perceptron;
{The perceptron algorithm}


    type
    pattern = record              {input pattern data structure}
        inputs : array[ ] of float;   {array of input values}
        targetout : integer;          {the target output}
        end; {record}


    var
    patterns : ^pattern[ ];        {array of input patterns}
    weights : ^float[ ];           {array of weights}
    input : ^float[ ];             {array of input values}
    alpha : float;                 {learning rate}
    target : integer;              {the target output}
```

```
    net: float;                    {the net (linear) output}
    i, j, k : integer;             {iteration indices}
    iter : integer;                {iteration count}
    finished : boolean;            {finish flag}

begin
alpha = 1;                         {initialize alpha}
for i = 1 to length(weights) do
    weights[i] = 0.0;              {initialize weights to zero}
    end do;

repeat                             {loop until done}
    finished = true;               {assume finished}
    for i = 1 to length(patterns) do
        net = 0.0;                 {initialize net output}
        end do;

        input = patterns[i].inputs;
                                   {find inputs}
        target = patterns[i].targetout;
                                   {find target output}
        for j = 1 to length(weights) do
                                   {calculate net output}
            net = net + weights[j] * input[j];
            end do;

        if sgn(net) < > target[i]  {if input pattern not correctly classified}
            begin
            finished = false;      {at least one input pattern is not correctly classified.}
            for k = 1 to length(weights) do
                                   {update weight vector}
                weights[k] = weight[k] + alpha * (targetout − sign(net))
                    * input[k];
                end do;
            end;
        end do;
    end do;
until finished or (iter > ITER_MAX)) {loop until done}
end do;

end. {Program}
```

C1.1.2.2 Advantages

The perceptron guarantees that it will learn to correctly classify two classes of input patterns, provided that the classes are linearly separable. The *adaline* (LMS algorithm) cannot guarantee that it will learn to separate two linearly separable classes. C1.1.3

C1.1.2.3 Disadvantages

If the two classes are not linearly separable, then the perceptron algorithm becomes unstable and fails to converge at all. In many such cases the weight vector appears to wander in a random-like fashion in space. Determining whether two classes are linearly separable beforehand is not easy.

The adaline, on the other hand, ordinarily converges to a good solution regardless of linear separability, but it does not guarantee separation of the two classes even if it is possible. Another disadvantage of the perceptron is that the target output must be binary, unlike adaline which can take any real value.

C1.1.2.4 Hardware implementations

Rosenblatt, with the help of others, built in hardware the Mark I Perceptron (1968), which operated as a
G1.3 *character recognizer*. It is considered to be the first successful neurocomputer (Hecht-Nielsen 1990).

C1.1.2.5 Variations and improvements

In Gallant (1986) the perceptron algorithm was modified to the pocket perceptron algorithm, which can
handle nonlinearly separable data. The idea is quite simple: have an extra set of weights which are kept 'in
your pocket'. Whenever the perceptron weights have a longest run of consecutive correct classifications,
they replace the pocket weights. The training input vectors are randomly selected. It is guaranteed that
changes in the pocket weights will become less and less frequent. Most of the changes will replace one
set of optimal weights with another. Occasionally, nonoptimal weights will replace the pocket weights,
but this will happen less and less frequently as training continues. The pocket algorithm, as well as other
related variations, are discussed in Gallant (1990).

Another extension of the perceptron is the complex perceptron (Georgiou 1993), where the input
vectors and the weights are complex-valued, and the output is multivalued.

C1.1.3 Adaline

Adaline (*ada*ptive *lin*ear *el*ement) is a simple single-neuron model that is trained using the LMS (least
B3.3.3 mean square) algorithm, otherwise known as the *delta rule* and also as the *Widrow–Hoff algorithm*. The
input patterns of the adaline, like those of the perceptron, are multidimensional real vectors, and its output
is the inner product of the input pattern and the weight vector. Training is supervised: for each input
pattern, there is a desired output. For each input pattern, the weights are corrected based on the difference
between the activation value, that is the actual output value, and the target value.

In general, it converges quite fast to a small mean square error, which is defined in terms of the
difference between the target output and the actual output.

It differs from the perceptron in that its output is not discrete (-1 or 1) but is instead continuous
and its value can be anywhere on the real line. It has been widely used in filtering and signal processing.
Being a simple linear model, the range of problems it can solve is limited. Being an early success in neural
computation, it bears historical significance. Also note that the widely used backpropagation algorithm is
a generalization of the LMS algorithm.

Unlike the perceptron, it cannot guarantee separation of two linearly separable classes, but it has the
advantage that it converges fast and training in general is stable even in classification problems where the
two classes are not linearly separable.

C1.1.3.1 Introduction

The adaline was introduced by Widrow and Hoff (1960) a few months after the publication of the perceptron
theorem (Rosenblatt 1960). Adaline and the perceptron are considered to be landmark developments in
the history of neural computation. Widrow and his students generalized adaline to the madaline, many
F1.2, F1.8, F1.9 adalines, network (Widrow 1962). Adaline found many applications in areas such as *pattern recognition,
signal processing, adaptive antennas, adaptive control* and others.

Like the perceptron, the adaline is a single-neuron model and is shown in figure C1.1.2. The output
is calculated as the inner product of the weight vector and the input vector:

$$a = f(\text{net}) = \sum_{i=1}^{n+1} w_i x_i . \tag{C1.1.19}$$

The extra component w_{n+1} accounts for the threshold of the neuron. The input at w_{n+1} is set to 1
for all input vectors, and is called the bias.

The LMS (least mean square) algorithm minimizes the mean square error function E, hence its name,
using the numerical analysis method of steepest descent.

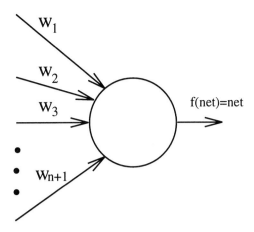

Figure C1.1.2. The adaline.

C1.1.3.2 Purpose

The adaline is used as a pattern classifier, and also as an approximator of input–output relations. Both the inputs and the target values can take real values.

C1.1.3.3 Topology

Adaline, like the perceptron, is a single-neuron model (figure C1.1.2). The difference is that the output is not discrete, like for the perceptron where the output is binary (0 or 1) or bivalent (-1 or 1), but is instead continuous (C1.1.19).

C1.1.3.4 Learning

The objective of the LMS algorithm is to minimize the mean square error (MSE) function, which is a measure of the difference between the target outputs and the corresponding actual outputs. Thus, LMS tries to find a weight vector W that would cause the actual outputs to be as close to the the target outputs as possible.

The training process is a statistical one, and the MSE function J for the weight vector $W = W(k)$ is defined as

$$J = \tfrac{1}{2} E[\varepsilon(k)^2] \tag{C1.1.20}$$

where k is the step and $E[\cdot]$ is the statistical expectation operator. The error $\varepsilon(k)$ is the difference between the target output and the actual output:

$$\varepsilon(k) = t(k) - W^{\mathrm{T}}(k)X. \tag{C1.1.21}$$

The MSE J is expanded to the following:

$$J = \tfrac{1}{2} E[t^2(k)] - E[tX^{\mathrm{T}}]W(k) + \tfrac{1}{2} W^{\mathrm{T}}(k)E[XX^{\mathrm{T}}]W(k). \tag{C1.1.22}$$

The cross-correlation P, a vector, between the target output and the corresponding input vector is defined as

$$P^{\mathrm{T}} = E[tX^{\mathrm{T}}]. \tag{C1.1.23}$$

Also, the input correlation matrix $\mathbf{R}$ is defined as

$$\mathbf{R} = E[XX^{\mathrm{T}}]. \tag{C1.1.24}$$

Thus, the mean square error function (C1.1.22) is simplified to

$$J = \tfrac{1}{2} E[t^2(k)] - P^{\mathrm{T}}W(k) + \tfrac{1}{2} W^{\mathrm{T}}(k)\mathbf{R}W(k). \tag{C1.1.25}$$

Considering that $\mathbf{R}$ is a real, semi-definite (in most practical cases) and symmetric matrix, we conclude that J is a non-negative quadratic function of the weights. Thus, in most cases, J can be viewed as a bowl-shaped surface with a unique minimum. The optimal weight vector $W^\star$, which is called the Wiener weight vector, that minimizes J, can be found by taking the gradient of J with respect to $W(k)$, and setting it to $\mathbf{0}$:

$$\nabla_{W(k)} J = -P + \mathbf{R} W(k) \tag{C1.1.26}$$

which yields

$$W^\star = \mathbf{R}^{-1} P. \tag{C1.1.27}$$

LMS approximates the gradient of the MSE function (C1.1.26), which is difficult to compute in the neural networks context, by using the gradient of the square of the instantaneous error:

$$\nabla_{W(k)} \left(\frac{1}{2} \varepsilon^2(k) \right) = \varepsilon(k) \frac{\partial \varepsilon(k)}{\partial W(k)} = \varepsilon(k) \frac{\partial (t(k) - W^{\mathrm{T}}(k) X(k))}{\partial W(k)} = -\varepsilon(k) X(k). \tag{C1.1.28}$$

The steepest descent method requires that the weight vector be updated by adding to it a quantity that is proportional to the negative gradient. Thus, the LMS learning rule is derived to be this equation:

$$W(k+1) = W(k) + \alpha \varepsilon(k) X(k). \tag{C1.1.29}$$

Note that the LMS learning rule (C1.1.29) is identical to that of the perceptron (C1.1.3). The difference lies in the fact that in the perceptron the error $\varepsilon(k)$ is computed using discrete values for the target and actual outputs. In LMS, those values are real (continuous-valued).

Learning is supervised and it resembles that of the perceptron: the input patterns are cyclically presented to the adaline. Ordinarily the order of presentation is not important. The error for input pattern $X = (x_1, x_2, \ldots, x_n, 1)$ is calculated as the difference between the target output and the activation value (C1.1.21). The weights are updated according to this formula:

$$w_i(k+1) = w_i(k) + \alpha \varepsilon(k) x_i(k) \tag{C1.1.30}$$

where k is the iteration counter, $\alpha > 0$ is the learning rate, a positive constant. The algorithm can be described as follows.

(i)   Initialize total error E to zero.
(ii) Compute activation for input pattern X.
(iii) Compute the output error ε.
(iv) Modify the connection weight by adding to it the factor $\alpha \varepsilon X$.
(v)   Add output error ε to total error E.
(vi) Repeat steps (ii), (iii), (iv) and (v) for each input pattern.
(vii) Repeat steps (i)–(vi) until total error E at the end of step (vi) is small.

The LMS algorithm *converges in the mean* if the mean value of the weight vector $W(k)$ approaches the optimum weight vector $W^\star$ as k grows large. The learning rate α determines the convergence properties of the algorithm, and, for most practical purposes, convergence in the mean is obtained when

$$0 < \alpha < 2/\lambda_{\max} \tag{C1.1.31}$$

where $\lambda_{\max}$ is the maximum eigenvalue of the correlation matrix $\mathbf{R}$ (C1.1.24).

C1.1.3.5   Pseudocode representation of the LMS algorithm

The learning process will stop either when the total error is smaller than MIN_ERROR, or when the number of iterations has exceeded a maximum number ITER_MAX.

program adaline;
{The LMS algorithm for the adaline}

```
    type
    pattern = record              {input pattern data structure}
        inputs : array[ ] of float;   {array of input values}
        targetout : integer;      {the target output}
        end; {record}
```

```
var
    patterns : ^pattern[ ];          {array of input patterns}
    weights : ^float[ ];             {array of weights}
    input : ^float[ ];               {array of input values}
    alpha : float;                   {learning rate}
    target : integer;                {the target output}
    net: float;                      {the net (linear) output}
    i, j, k : integer;               {iteration indices}
    iter : integer;                  {iteration count}
    error : float;                   {total error}

begin
alpha = 0.2;                         {initialize alpha}
for i = 1 to length(weights) do
    weights[i] = random(−0.5, 0.5);  {initialize weights to small values}
    end do;

repeat                               {loop until done}
    error = 0.0;                     {initialize error}
    for i = 1 to length(patterns) do
        net = 0.0;                   {initialize net output}
        end do;

        input = patterns[i].inputs;
                                     {find inputs}
        target = patterns[i].targetout;
                                     {find target output}
        for j = 1 to length(weights) do
                                     {calculate net output}
            net = net + weights[j] * input[j];
            end do;
        error = error + (target − net);
            for k = to length(weights) do
                                     {update weight vector}
                weights[k] = weight[k] + alpha * (target − net)
                    * input[k];
                end do;
            end do;
        end do;
until (error < MIN_ERROR) or (iter > ITER_MAX)
                                     {loop until done}
    end do;

end. {Program}
```

C1.1.3.6 Advantages

The adaline ordinarily converges to a good solution quite fast, even in the case where the two classes are not linearly separable. It can handle datasets where the target output is real-valued (nonbinary).

C1.1.3.7 Disadvantages

Unlike the perceptron, it cannot guarantee separation of two linearly separable classes.

C1.1.4 Madaline

C1.1.4.1 Introduction

Madaline is an early example of a trainable network having more than one layer of neurons. It consists of a layer of trainable adalines that feed a second layer, the output layer, which consists of neurons that function as logic gates, such as AND, OR and MAJ (majority-vote-taker) gates. The weights of the output neurons, however, are not trainable but fixed. Therefore, we classify madaline as a single-layer network. Widrow and Lehr (1990) provide an excellent first-hand account of the history of madalines, as well as for the adaline.

Madaline was developed by Bernard Widrow (Stanford University) (Widrow 1962) and Marcian Hoff in his PhD thesis (Hoff 1962). It is noteworthy that a 1000-weight Madaline I was built in hardware in the early 1960s (Widrow 1987). In its early beginning Madaline I was used in applications such as speech and pattern recognition (Talbert *et al* 1963), weather prediction (Hu 1964) and adaptive controls (Widrow 1987), and later to adaptive signal processing (Widrow and Stearns 1985), where it was used quite successfully in many applications.

The more powerful backpropagation algorithm superseded Madaline I, as this algorithm handles the training of networks with multiple layers, each having adjustable weights.

C1.1.4.2 Purpose

Madaline I, as well as its variants, are commonly used as classifiers.

C1.1.4.3 Topology

The Madaline I network consists of two layers of neurons (figure C1.1.3). The first layer consists of adalines, each of which receives input directly from the input pattern. The output from the adalines is then passed through a hard-limiter, that is the Heaviside function, which in turn feeds the the second layer, which consists of one or more neurons. The neurons of this layer are logical function gates, such as AND gates, OR gates or majority-vote-taker (MAJ) gates. The MAJ gate gives output 1 if at least half of its inputs are 1, and output −1 otherwise. The weights of the logic gate neurons are fixed, whereas those of the adalines in the first layer are adjustable.

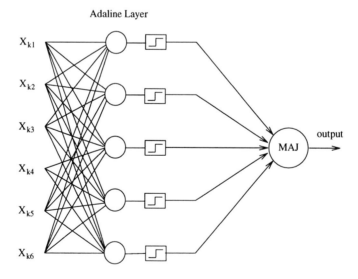

Figure C1.1.3. The madaline.

C1.1.4.4 Learning

Learning is supervised—each input pattern in the training set has a target pattern, usually either 1 or −1. The input patterns are presented to the network. A random order of presentation is preferable over a cyclical one, since the latter may cause cyclic repetition of values of the weights, and thus convergence is not possible (Ridgeway 1962). The Heaviside (hard-threshold) function is applied to each of the outputs of the adalines in the first layer, and the result (1 or -1) is fed as input to the output neuron(s) (the logic gate(s)). Then, the output of the network is compared with the target output for the particular input. If the two agree, no correction is made to the weights of any adaline; if they disagree, then the weights of one or more adalines are adjusted.

The question now becomes 'which adalines should be chosen to have their weights adjusted?' This is answered by the following procedure: start from the adaline whose (net) linear output is closest to zero. (The idea here is to start from the adaline whose output can most easily take the reverse sign, thus changing from positive to negative, or vice versa.) Then, reversing the sign of the corresponding hard-limiter (Heaviside function) of the chosen adaline, check the output to see if it agrees with the target output. If yes, then no other adaline is chosen to have its weights adjusted. If not, repeat the process by choosing the adaline with the next closest value to zero. Thus, this procedure chooses the minimum number of adalines—whose linear output is closest to zero—that when reversing the sign of their linear outputs, the correct target output is obtained.

The next question is 'how to adjust the weights of the chosen adalines?' This adjustment of the weights can be done in two ways: the first way is by changing the weights by a sufficient amount in the LMS direction (see previous section) so that the linear output of the adaline changes sign. In other words, choose a large enough learning rate α in (C1.1.29) so that the output of the adaline, for the same input vector, reverses its sign. This type of learning is called 'fast'. It is possible, and quite often it is the case, that by changing the weights to achieve the correct output for a specific input, the wrong output is obtained for previously learned input–output pairs.

The second way of adjusting the weights is by changing them by a small amount in the LMS direction, without considering whether the change would be large enough to cause the sign of the linear output to be reversed.

In both cases, it is expected (but not guaranteed) that after many iterations, the weights will assume values that will correctly classify all, or at least most, input vectors.

The intuitive idea behind the choice of adalines to adjust their weights, and the way of adjusting their weights, is known as the 'least disturbance principle' (Widrow and Lehr 1990): *adapt to reduce the output error for the current input pattern with minimal disturbance to the responses already learned.* This principle is adhered to by the madaline learning algorithm in various ways: the least number of adalines that can cause the output to change is chosen (minimal disturbance); the adalines with outputs closest to zero are chosen (disturbance is minimal); and the weights are changed in the direction of the negative gradient, which is the direction toward the input vector (error correction with minimal weight change). This heuristic principle is applicable to LMS, madaline, backpropagation and other neural network learning algorithms.

As an example consider the case where there are three adalines in the first layer and a MAJ gate at the output, and that an input pattern X, with desired output $+1$, causes only one out of three adalines to have positive linear output, thus the hard-thresholded output of the madaline is -1. Thus, only one adaline, that has negative linear output at present, will have its weights adjusted, since a single reversal of the output of an adaline will cause the correct output. The general algorithm can be described as follows.

(i)   Initialize the weights of the adalines with small random numbers.

(ii) Consider first input pattern.

(iii) Compute the linear output of the adalines.

(iv) Compute the outputs of the Heaviside functions.

(v)   Compute the value of output logic gate(s).

(vi) Compute error = (target output) − (actual output).

(vii) If the error is different than zero, determine the adalines to be adjusted.

(viii) Adjust the weights of the adaline.

(ix) Repeat step (viii) for each adaline to be adjusted.

(x)   Repeat steps (iii) through (ix) for each input pattern.

(xi) Repeat step (x) until error is zero for all input patterns.

C1.1.4.5 Pseudocode representation of Madaline I

program Madaline_I;
{The Madaline_I algorithm. The output unit is a single AND gate.}

```
type
pattern = record                    {input pattern data structure}
    inputs : array[ ] of float;     {array of input values}
    targetout : integer;            {the target output}
    end; {record}
unit = record
    weight : array[ ] of float;     {The weights of the adaline}
    net : float;                    {The linear output of unit}
    end; {record}

var
patterns : ^pattern[ ];             {array of input patterns}
weights : ^float[ ];                {array of weights}
input : ^float[ ];                  {array of input values}
units : ^unit[ ];                   {array of adaline units}
alpha : float;                      {learning rate}
target : integer;                   {the target output}
net: float;                         {the net (linear) output}
i, j, k : integer;                  {iteration indices}
iter : integer;                     {iteration count}
error: integer;                     {output error}
finished : boolean;                 {finish flag}
sum : integer;                      {the number of adalines with positive output}
output : integer;                   {value of output (AND gate)}
iter : integer;                     {iteration counter}

begin
alpha = 0.2;                        {initialize alpha}
for j = 1 to length(units) do       {initialize weights to small values}
    weights = units[j].weight;
    for i = 1 to length(weights) do
        weights[i] = random(−0.5, 0.5);
        end do;
    end do;
iter = 0;                           {initialize iteration counter}
repeat                              {loop until done}
    iter = iter +1;                 {update iteration counter}
    finished = true;                {assumed finished}

    for k = 1 to length(units) do
                                    {initialize net output of adalines}
        units[k].net = 0.0;
        for i = 1 to length(patterns) do
        units[k].net = units[k].weights[i] ∗ units[k].net
        end do;
    end do;

    for k = 1 to length(patterns) do

        sum = 0;                    {initialize sum}
        for i = 1 length(units) do
```

```
                                          {calculate number of adalines with positive output}
        if sgn(units[i].net) = 1 then
            sum = sum +1;
            end if;
        end do;
    if sum = length(units) then
                                          {If all outputs of adalines are positive, AND output is 1}
        output = 1;
    else                                  {else 0}
        output = 0;
    end if;


    error = patterns[k].targetout − output;
                                          {calculate error}


    if error < > 0 then
        finished = false;                 {at least one correction made}

                                          {update weights of units with wrong output}
        for i = 1 to length(units) do
            if sgn(units[i].net) < > patterns[k].targetout then
                for j = 1 to length(weights) do
                                          {update using adaline rule}
                    units[i].weights[j] = units[i].weights[j]
                        + alpha ∗ (patterns[k].targetout − units[i].net)
                        ∗ patterns[k].input[j];
                end do;
            end if;
        end for;
    end if;
    end do;
until finished or (iter > ITER_MAX)

end. {Program}
```

C1.1.4.6 Advantages

Obviously, madaline is more powerful than adaline. It is one of the earliest, if not the earliest, feasible schemes of training multilayer neural networks. It can learn to separate two nonlinearly separable classes.

C1.1.4.7 Disadvantages

It is not as flexible or powerful as backpropagation, where the weights of the output units are adjustable as well.

C1.1.4.8 Hardware implementations

A 1000-weight madaline was built in hardware in the early 1960s (Widrow 1987).

C1.1.5 Learning vector quantization

C1.1.5.1 Introduction

Learning vector quantization (LVQ) was first studied in the neural network context by Teuvo Kohonen (Kohonen 1986). It is related to Kohonen's *self-organizing maps* (SOM) (Kohonen 1984), with the main C2.1.1 difference being that LVQ is a supervised method, which takes advantage of the class information of the

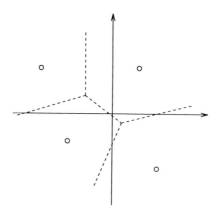

Figure C1.1.4. Voronoi tessellation in two dimensions. The circles represent the prototype vectors of each region.

F1.6, F1.7 input patterns in the training set. It is also related to the well known K-means clustering algorithm (Lloyd 1957, 1982, MacQueen 1967). Traditional LVQ algorithms, primarily used for *speech and image data compression*, are reviewed in Gray (1984) and Nasrabadi and King (1988).

In LVQ, input pattern space is divided into disjoint regions. Each region is represented by a *prototype* vector. Thus, each prototype vector represents a cluster of input vectors. The collection of prototype vectors is called the *codebook*. Learning vector quantization as a classifier can be used in the following manner. The input vector to be classified is compared with all prototypes in the codebook. The prototype that is closest, using the Euclidean distance metric, to the input vector is chosen, and the input vector is classified to the same class as the prototype. It is assumed that each prototype is tagged with the label of the class it belongs to.

The other major use of LVQ is in data compression. When used for this purpose, the input space is again divided into regions and prototype vectors are chosen. Each input vector is compared with all prototypes, and is replaced with the index of the prototype in the codebook that it is closest to, using Euclidean distance. Thus the original vectors are replaced with indices, which point to prototype vectors in the codebook. (The term *vector quantization* refers to the act of replacing an input vector with its corresponding prototype.) Replacing vectors with indices can potentially achieve high compression ratios. Decompression is achieved by looking-up in the codebook the prototypes that correspond to the indices. When the compressed data are transmitted over a channel, substantial bandwidth savings can be achieved. However, it is necessary for the receiver to have the codebook to be able to decompress. Of course, LVQ is a lossy compression technique, as the original vectors cannot be exactly reconstructed—unless, of course, there are as many prototype vectors as there are input vectors. To achieve higher resolution, it is necessary to have a finer subdivision of space, and thus more prototypes.

The question now becomes 'how are the prototypes arrived at?' This is exactly what the LVQ algorithm does. Note that division of space into regions is implicit. All that is needed is the prototypes, since each prototype defines a region. The regions are defined using the *nearest-neighbor rule*. That is, a vector X_j belongs to the region of the prototype vector W_i that is closest to it:

$$\text{Region}(W_i) = \{X_j : \|W_i - X_j\| \le \|W_k - X_j\|, \text{ for all } k\} \qquad (C1.1.32)$$

where $\| \cdot \|$ is the Euclidean distance metric. This partition of space into distinct regions, using prototype vectors and the nearest-neighbor rule, is called *Voronoi* tessellation. A two-dimensional example of such tessellation appears in figure C1.1.4. Notice that the boundaries of the regions are perpendicular bisector lines (planes in three dimensions and hyperplanes in higher dimensions) of the lines joining neighboring prototypes.

The weight vectors of the neurons in an LVQ neural network are the prototypes, the number of which is usually fixed before training begins. Training the network means adjusting the weights with the objective of finding the best prototypes, that is, prototypes that would give best classification or best image compression. The LVQ training algorithm is a case of competitive learning. That is, during training, when an input vector is presented, only a small group of winner neurons (usually one or two) are allowed to adjust their weight vectors. The winner neuron or neurons are the ones closest to the input vector. At the end of training, the weight vectors are frozen, and the network operates in its normal mode: when an input vector is presented, only one neuron becomes active; that is, the one whose weight vector best matches the input vector.

C1.1.5.2 Purpose

Learning vector quantization can be used both as a classifier and as a data compression technique.

C1.1.5.3 Topology

The network consists of a single layer of neurons, each of which receives the same input, which is the input pattern currently presented to the network (figure C1.1.5). The weight vectors of the neurons correspond to the prototype vectors.

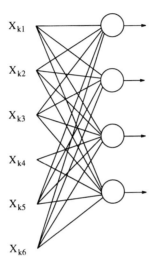

Figure C1.1.5. The learning vector quantization (LVQ) network. It is a single layer of neurons that all receive the same inputs.

C1.1.5.4 Learning

This is a description of the basic LVQ algorithm (LVQ1) (Kohonen 1990c). The training set consists of n input patterns. Each of these vectors is labeled as being one of k classes. The next step is to decide how many prototype vectors there should be, or equivalently, how many neurons the network should have. Quite often one neuron per class is used, but having more neurons per class may be more appropriate in some cases, since a class may be comprised of more than one cluster. It is common to initialize the weight vectors of the neurons to the first input pattern vectors that have the corresponding class. Then, the input vectors are presented to the network either cyclically or randomly. Being a competitive learning process, for each presentation of input vector X_j, a winner neuron W_i is chosen to adjust its weight vector:

$$\|X_j - W_i\| \leq \|X_j - X_k\|, \qquad \text{for all } k. \tag{C1.1.33}$$

Updating of $W_i(t)$ to the next time step $t + 1$ is done as follows:

$$W_i(t + 1) = W_i(t) + \alpha(X_j - W_i(t)) \qquad \text{if } X_j \text{ and } W_i \text{ belong to the same class} \tag{C1.1.34}$$

and

$$W_i(t + 1) = W_i(t) - \alpha(X_j - W_i(t)) \qquad \text{if } X_j \text{ and } W_i \text{ belong to different classes.} \tag{C1.1.35}$$

The idea is to move W_i towards X_j if the class of W_i is the same as that of X_j, else move it away from X_j. The learning rate $0 < \alpha < 1$ may be kept constant during training, or may be decreasing monotonically with time for better convergence. It is suggested that the initial value of α is less than 0.1 (Kohonen *et al* 1995).

The algorithm should stop when some optimum is reached, after which the generalization ability of the network degrades, a condition known as overtraining. The optimal number of iterations depends on many factors, including the number of neurons, the learning rate, the number of input patterns and their distribution, amongst others, and can only be determined by experimentation. It was found that the optimum number of iterations is roughly between 50 and 200 times the number of neurons (Kohonen *et al* 1995).

C1.1.5.5 Pseudocode representation of the LVQ algorithm

```
program lvq1;   {The LVQ1 algorithm.}
    type
    pattern = record                    {input pattern data structure}
        inputs : array[ ] of float;     {array of input values}
        class : integer;                {the target output}
        end; {record}
    unit = record
        weight : array[ ] of float;     {The weights of the unit}
        class : integer;                {The class of the unit}
end; {record}


    var
    patterns : ^pattern[ ];             {array of input patterns}
    units : ^unit[ ];                   {array of units}
    alpha : float;                      {learning rate}
    i, j, l, m : integer;               {iteration indices}
    dis, distance: float;               {Euclidean distance}
    winner: integer;                    {The winning neuron}
begin
alpha = 0.05;                           {initialize alpha}
```

It is assumed that the weights of the neurons (units) are initialized

```
for i = 1 to MAX_ITER do
    for j = 1 to length(patterns) do
    distance = 100000;                  {a large number (plus infinity)}
        {find the closest neuron to the input pattern}
    for l = to length(units) do
        {find the Euclidean distance between the two vectors}
    dis = DISTANCE(patterns[j].inputs,units[l].weight);
    if (dis < distance) then
    begin
        winner = l;
distance = dis;
end;
    end do;
{Modify weight vector of neuron closest to input pattern}
    If (patterns[j].class = units[winner].class) then
{If they belong to the same class}
        for m = 1 to length(weights) do
    units[winner].weight[m] = units[winner].weight[m] +
            alpha * (patterns[j].weight[m] −units[winner].weight[m])
        else
{They belong to different class}
        for m = 1 to length(weights) do
```

```
units[winner].weight[m] = units[winner].weight[m] −
    alpha * (patterns[j].weight[m] −units[winner].weight[m])
end if;
end do;

end. {Program}
```

C1.1.5.6 Variations and improvements

Several improvements and variations of the basic algorithm (LVQ1) (Kohonen 1990c) have also been proposed by Kohonen (1990a, b, c), as well as others.

In LVQ2 not only the weights of the winning neuron (nearest neighbor of input vector X) are updated, but also so are the weights of the next-nearest neighbor, but only under these conditions:

(i)   The nearest neighbor W_i must be of different class than input vector X.
(ii) The next to the nearest neighbor W_j must be of the same class as input vector X.
(iii) The input vector X must be within a window defined about the bisector plane of the line segment that connects W_i and W_j.

Mathematically, 'X falls in a "window" of width w' if it satisfies

$$\min\left(\frac{\|X - W_i\|}{\|X - W_j\|}, \frac{\|X - W_j\|}{\|X - W_i\|}\right) > \frac{1 - w}{1 + w} \tag{C1.1.36}$$

where w is recommended to take values in the interval from 0.2 to 0.3. Thus, if X falls within the window, the weight vectors W_i and W_j are updated according to these equations:

$$W_i(t + 1) = W_i(t) - \alpha(t)(X(t) - W_i(t)) \tag{C1.1.37}$$

$$W_j(t + 1) = W_j(t) + \alpha(t)(X(t) - W_j(t)). \tag{C1.1.38}$$

The idea behind the LVQ2 algorithm is to try to shift the bisector plane closer to the Bayes decision surface. There is no mechanism to ensure that in the long run the weight vectors of the neurons will reflect the class distributions.

The LVQ3 algorithm improves on LVQ2 by trying to make the weight vectors roughly follow the class distributions, by adding an extra case where updating takes place: if the two nearest neighbors W_i and W_j of input vector X belong to same class as X, then update them according to this equation:

$$W_k(t + 1) = W_k(t) + \epsilon\alpha(t)(X(t) - W_k(t)) \tag{C1.1.39}$$

where k is in $\{i, j\}$. Recommended values of ϵ range between 0.1 and 0.5 (Kohonen et al 1995).

C1.1.6 Instar and outstar

C1.1.6.1 Introduction

These two neuron models—or concepts of a neuron—were introduced by Stephen Grossberg of Boston University in Grossberg (1968) in the context of modeling various biological and psychological phenomena. In that paper and in others that followed (Grossberg 1982), he demonstrated that variations of the outstar model can account for many cognitive phenomena such as Pavlovian learning and others that can be informally described as *practice makes perfect, overt practice unnecessary, self-improving memory*, and so on.

A neuron when viewed as the center of activity, receiving input signals from other neurons, is called an *instar* (figure C1.1.6). When the the neuron is viewed as distributing its activation signal to other neurons it is called an *outstar* (figure C1.1.7). Thus, a neural network can be considered as a tapestry of interwoven instars and outstars. By having various ways of learning, i.e. adjusting the weights and obtaining the activation signal of a neuron, one obtains a rich mathematical structure in such networks, the analysis of which quickly becomes difficult. A contributing factor to the difficulty is the fact that time delays are accounted for in Grossberg's formulation. There is little work done on the instar and

B3.3.6

outstar concepts beyond what has been done by Grossberg and his associates. However, in artificial neural networks the outstar model, though not used by itself, is used as a building block of larger networks, most notably in all versions of *adaptive resonance theory* (ART) (Carpenter and Grossberg 1987a, b, 1990) and the counterpropagation network (Hecht-Nielsen 1987, 1988). In these networks, part of the training is done using variations of the outstar learning. A characteristic of outstar learning, unlike other neuron models, is that the weights to be adjusted are outgoing from the neuron under consideration, as opposed to being incoming.

C2.2.1

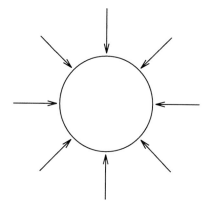

Figure C1.1.6. The instar

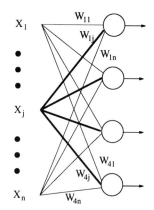

Figure C1.1.7. The outstar.

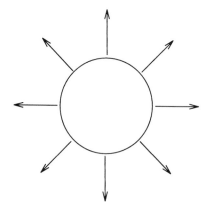

Figure C1.1.8. The outstar network. The *j*th outstar supplies input to a layer of neurons.

C1.1.6.2 Purpose

Originally, instar and outstar were developed as mathematical models of various biological and psychological mechanisms.

In artificial neural networks the outstar model, though not used by itself, is used as a building block of larger neural network models, most notably in all ART networks (Art 1, Art 2, Art 3) and for the *counterpropagation network.*

C2.3.2

C1.1.6.3 Topology

The instar appears in figure C1.1.6 and the outstar in figure C1.1.7. The outstar model in a network is shown in figure C1.1.8. The jth outstar supplies input to a layer of neurons, and the corresponding weights, which appear as thicker lines, are to be adjusted.

C1.1.6.4 Learning

A rare readable tutorial discussion of Grossberg's ideas on instar and outstar appears in Caudill (1989a), from which the following discussion draws. This is a collection of eight papers which originally appeared in the magazine *AI Expert*. In particular, these two (Caudill 1988, 1989b) are relevant to the present discussion.

The activation function a_j of an instar j is not explicit, but instead is given as a time-evolving differential equation, a variant of which, not the most general, is the following:

$$\frac{\mathrm{d}a_j(t)}{\mathrm{d}t} = -Aa_j(t) + I_j(t) + \sum_{i=1}^{n} w_{ij}[a_i(t - t_0) - T]^+ \tag{C1.1.40}$$

where A is a positive constant which accounts for forgetting (exponential decay); $I_j(t)$ is the external input to instar j, which is known as the conditioning stimulus (which corresponds to the bell in the well-known Pavlovian experiment with a salivating dog); $a_i(t - t_0)$ the activation function of neuron i from which neuron j receives input; and w_{ij} the corresponding weight. The time delay t_0 is included to account for the time it takes for signal a_i to arrive at neuron j. The constant T is a threshold value, and the function $[\cdot]^+$ takes the value of its argument, if the argument is positive. If is negative, the quantity is zero:

$$[x]^+ = \begin{cases} x & \text{if } x \geq 0 \\ 0 & \text{if } x < 0. \end{cases} \tag{C1.1.41}$$

This is a noise suppression mechanism, as any activation signal less than the threshold T does not contribute to the computation of a_j. Small fluctuations in the levels of activity in surrounding neurons are ignored, just as happens in biological neurons in the brain.

Now we will proceed with more explanation of the three terms on the right-hand side of (C1.1.40). The first term accounts for the decay of the neuron activation level with the passage of time—a well-known characteristic of biological neurons. This can be clearly seen when the external input $I_j(t)$ is zero and the inputs from other neurons are all less than the threshold, and thus are noncontributing. In such a case, (C1.1.40) simplifies to

$$\frac{\mathrm{d}a_j(t)}{\mathrm{d}t} = -Aa_j(t) \tag{C1.1.42}$$

whose solution, has the form of a decaying exponential, and in simplified form is $a_j(t) = \mathrm{e}^{-At}$. Thus, the larger the positive constant A is, the faster the decay.

Considering only the external input $I_j(t)$, (C1.1.40) becomes:

$$\frac{\mathrm{d}a_j(t)}{\mathrm{d}t} = I_j(t) \tag{C1.1.43}$$

which implies that as long as $I_j(t)$ is greater than zero, the activation $a_j(t)$ increases. Finally, considering the effect of the activity values of other neurons (without precluding the possibility that neuron j receives input from itself), (C1.1.40) is simplified to

$$\frac{\mathrm{d}a_j(t)}{\mathrm{d}t} = \sum_{i=1}^{n} w_{ij}[a_i(t - t_0) - T]^+ \tag{C1.1.44}$$

which accounts for the cumulative effect of the inputs received by neuron j from other neurons. If weight w_{ij} has a negative value, it represents an inhibitive connection.

The other important aspect of the instar–outstar view of neurons is the instar (or outstar, depending on how neurons are viewed during application) learning equation, which specifies how the weights are updated, and again is a time-dependent differential equation. Consider outstar j giving input to neuron i with connection (weight) w_{ij}. Then, w_{ij} is changing according to

$$\frac{dw_{ij}(t)}{dt} = -Fw_{ij}(t) + Ga_j[a_i(t - t_0) - T]^+ \tag{C1.1.45}$$

where the positive constant F accounts for weight decay, otherwise known as forgetting. It is very similar in function to A in (C1.1.40), but it should be noted that A is considerably larger than F since neuron activation level decay happens a lot faster than forgetting learned memories, i.e. the erasing of old weight values. The factor $a_j[a_i(t - t_0) - T]^+$ accounts for Hebbian learning: when the input a_j to a synapse (weight) and activation a_i of a neuron are both high, then the weight is to be strengthened. The constant G is called gain, and it corresponds to the usual learning rate coefficient in neural networks: the larger it is, the faster the learning.

In artificial neural networks the outstar learning equations are substantially simpler, one reason being that updating happens at discrete intervals and thus time delays are easier to handle. As was mentioned earlier, two well-known networks use outstar learning: counterpropagation and ART. In counterpropagation there are two layers of neurons: one which is trained using Kohonen learning and the other using the outstar type of learning equation:

$$w_{ij}(k + 1) = w_{ij}(k) + \alpha(b_j(k) - w_{ij}(k))a_i(k) \tag{C1.1.46}$$

where k is the step, a_i is the output of the Kohonen neuron i (note that only one Kohonen neuron has nonzero activation) and b_j is the desired output.

The basic outstar learning algorithm in ART networks, for outstar j, is given by this equation:

$$w_{mj}(k + 1) = w_{mj}(k) + \alpha(t_m(k) - w_{mj}) \tag{C1.1.47}$$

where k is the step parameter; w_{mj} is the weight being modified, which emanates from outstar j and feeds neuron i; and α is the learning rate; t_m is the target output of neuron m. The subscript m runs through all neurons that receive input from outstar j.

C1.1.7 CMAC

C1.1.7.1 Introduction

The CMAC (cerebellar model articulation controller) model was invented and developed by James Albus in a number of papers in the 1970s (Albus 1971, 1972, 1975a, b). Originally, it was formulated as a model of the cerebellar cortex of mammals (Albus 1971) and was subsequently applied to the control of a robotic arm manipulators. Albus applied CMAC to the control of a three-axis master–slave arm in Albus (1972), and in Albus (1975a) to a seven-degrees-of-freedom manipulator arm. In the latter reference, he gave a detailed description of CMAC and it is considered to be a standard reference. The robotic arms were to learn certain trajectories. After many years of relative obscurity, CMAC was re-examined and shown to be a viable model for complicated control tasks, where the popular backpropagation algorithm could be used (Ersü and Militzer 1984, Ersü and Tolle 1987, Miller 1986, 1987, Miller *et al* 1990a, Moody 1989).

In Parks and Militzer (1989) the convergence of Albus' learning algorithm was proven. In Parks and Militzer (1992) it is discussed that the algorithm is identical to the Kaczmarz technique (Kaczmarz 1937) which is for finding approximate solutions of systems of linear equations.

CMAC is a neural network that generalizes locally; that is, inputs that are close to each other in the input space will yield similar outputs, whereas distant inputs will yield uncorrelated outputs. In the latter case, different parts of the network will be active. Thus, CMAC will likely not discover higher-order correlations in the input space. It has been shown, however, to yield good results for a variety of problems, with the added advantage that training is exceptionally fast. Unlike most common neural network models, CMAC is not merely an ensemble of neurons that produce the output for a given input. Instead, it can be viewed as a single neuron (when the output is one-dimensional) of which a small subset of weights are

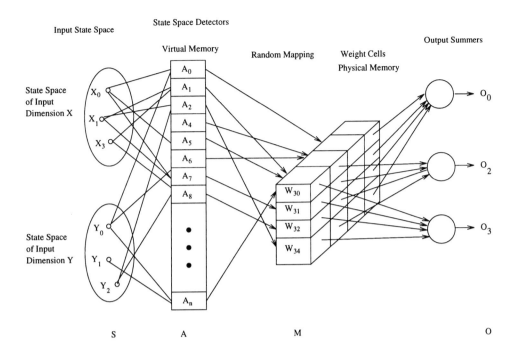

Figure C1.1.9. The CMAC network.

summed to obtain the output and are subsequently modified using the LMS algorithm, considering their input to be 1. The rest of the weights are ignored. The specification of this subset of weights for a given input constitutes that heart of CMAC.

C1.1.7.2 Purpose

CMAC is used as a classifier or as an *associative memory*. It has also been used extensively in robotic F1.4 control.

C1.1.7.3 Topology

A schematic diagram of CMAC appears in figure C1.1.9. Differing from other neural networks, its description includes the invocation of memory cells, both in virtual and in physical memory. The only conventional neurons are the ones that give the output, which are labeled 'output summers'. A detailed explanation of the diagram is included in the next section.

C1.1.7.4 Learning

The operation of CMAC is perhaps not as simple to describe as other neural network models. This is due to the fact that the nonlinearity in the network is not the result of activation functions used, as usual, but instead it is the result of some peculiar mappings.

CMAC can be thought of as a series of mappings (see figure C1.1.9) (Burgin 1992):

$$S \to A \to M \to O \qquad (C1.1.48)$$

where S is the input vector, notated as such for 'stimulus'; A is a large binary array, often impractical, due to its size, to be saved in memory; M is a multidimensional table in memory which holds the weights of the output summers; and O is the output vector.

An input vector S, causes a fixed number C, called the generalization parameter, of elements of array A to be set to 1, while the rest are set to 0. Then, the array A is mapped using random hashing on M. The 1s in A 'activate' the corresponding weights in M. The output is obtained by summing the activated weights of each summer.

Training is done by cyclically presenting the input vectors to CMAC. For each input the output is obtained, and then activated weights in M are adjusted using the usual LMS algorithm (C1.1.29), using input $x_i = 1$. The weights that have not been activated are not modified, which is equivalent to considering their input to be 0 in the LMS algorithm.

It remains to be explained exactly how S is mapped to A; this mapping is called the input mapping. Each of the input dimensions is quantized, and thus the input space becomes discrete. Figure C1.1.9 shows a case where the input is two-dimensional, with dimensions X and Y.

The value that each element of A gets is the output of an AND gate (not shown in figure C1.1.9). The AND gates are called state-space detectors. Each AND gate receives inputs from the input sensors, one input per input dimension. The input sensors are excited whenever the input falls within their receptive fields. If all input sensors that are inputs to an AND gate are excited, then the output of the AND gate is 1, and 0 otherwise.

Each point on the one-dimensional grid in an input dimension excites exactly C input sensors. The input sensors have overlapping receptive fields. If, for example, $C = 3$ and sensor a is excited by the consecutive points $\{4, 5, 6\}$ on a hypothetical grid in the X-dimension, then sensor b is excited by points $\{5, 6, 7\}$, sensor c by $\{6, 7, 8\}$, and so on. Thus, two neighboring points will excite some input sensors in common, whereas two distant points will not.

The input sensors feed the AND gates in such a way that exactly C AND gates have output 1 for each input vector S. One can visualize the effect of the input smoothly traveling in the input space on the output of the AND gates, by imagining the AND gates as bulbs: the number of bulbs that are ON is a constant C, and whenever there is a change, only a small number of bulbs turn OFF and a like number of OFF bulbs turn ON at the same time.

C1.1.7.5 Advantages

In general, learning in CMAC, both in software and in hardware, is substantially faster than in other neural networks such as backpropagation (Miller *et al* 1990b). The speed-up can sometimes be measured in orders of magnitude. This speed advantage makes it feasible to have large CMAC networks, with weights present into the hundreds of thousands, that solve large problems.

The local generalization property of CMAC can be considered an advantage in certain cases. For example, it is possible to add input patterns in a remote area of the input space incrementally, without affecting the already learned input/output relations.

C1.1.7.6 Disadvantages

The local generalization property prevents CMAC from discovering global relations in the input space, which other neural networks, such as backpropagation, are capable of.

Collisions that can occur in the hashing scheme that maps the virtual memory into the real memory, cause interference, or noise, during learning. However, this can be avoided with proper design.

References

Albus J S 1971 A theory of cerebellar functions *Math. Biosciences* **10** 25–61
——1972 Theoretical and experimental aspects of a cerebellar model *PhD Thesis* University of Maryland, USA
——1975a Data storage in the cerebellar model articulation controller CMAC *Trans. ASME, J. Dynamic Systems, Measurement, and Control* 228–33
——1975b A new approach to manipulator control: the cerebellar model articulation controller (CMAC) *Trans. ASME, J. Dynamic Systems, Measurement, and Control* **97** 220–7
Burgin G 1992 Using cerebellar arithmetic computers *AI Expert* **7** 32–41
Carpenter G A and Grossberg S 1987a ART 2: Self-organization of stable category recognition codes for analog input patterns *Appl. Opt.* **26** 4919–30
——1987b A massively parallel architecture for a self-organizing neural pattern recognition machine *Computer Vision, Graphics and Image Processing* **37** 54–115
——1990 ART 3: Hierarchical search using chemical transmitters in self-organizing pattern recognition architectures *Neural Networks* **3** 129–52
Caudill M 1988 Neural networks primer part v *AI Expert* 57–65
——1989a *Neural Networks Primer* (San Francisco, CA: Miller Freeman)
——1989b Neural networks primer part vi *AI Expert* 61–7

Ersü E and Militzer J 1984 Real-time implementation of an associative memory-based learning control scheme for non-linear multivariable processes *Manuscript, Symposium 'Application of Multivariate System Technique' (Plymouth, UK)*

Ersü E and Tolle H 1987 Hierarchical learning control—an approach with neuron-like associative memories ed D Anderson *Proc. IEEE Conf. on Neural Information Processing (Denver)* (AIP, Denver, CO: IEEE)

Gallant S I 1986 Optimal linear discriminants *Eighth Int. Conf. on Pattern Recognition* (New York: IEEE) 849–52

——1990 Perceptron-based learning algorithms *IEEE Trans. Neural Networks* **1** 179

Georgiou G M 1993 The multivalued and continuous perceptrons, *World Congress on Neural Networks* (Portland, OR) vol IV 679–83

Gray R M 1984 Vector quantization *IEEE ASSP Magazine* 4–29

Grossberg S 1968 Some nonlinear networks capable of learning a spatial pattern of arbitrary complexity *Proc. Natl Acad. Sci. USA* **59** 368–2

Grossberg S (ed) 1982 *Studies of Mind and Brain: Neural Principles of Learning, Perception, Development, Cognition and Motor Control* (Boston: Reidel)

Hecht-Nielsen R 1987 Counterprogagation networks *Appl. Opt.* **26** 4979–84

——1988 Applications of counterpropagation networks *Neural Networks* **1** 131–9

Hoff M 1962 Learning phenomena in networks of adaptive switching circuits *Technical Report 1554-1* Stanford Electron. Labs, Stanford, CA

Hu M 1964 Application of the adaline system to weather forecasting *Thesis, Technical Report 6775-1* Stanford University

Kaczmarz S 1937 Angenäherte Auflösung von Systemen Linearer Gleichungen *Bull. Int. Acad. Polon. Sci. Cl. Math. Nat. Ser. A.*

Kohonen T 1984 *Self-Organization and Associative Memory* (Berlin: Springer) 3rd edn 1989

——1986 Learning vector quantization for pattern recognition, *Report TKK-F-A601*, Helsinki University of Technology, Espoo, Finland.

——1990a Internal representations and associative memory, *Parallel Processing in Neural Systems and Computers* ed R Eckmiller, G Hartman and G Hauske (Amsterdam: Elsevier) pp 177–82

——1990b The self-organizing map *Proc. IEEE* **78** 1464–80

——1990c Statistical pattern recognition revisited *Advanced Neural Networks* ed R Eckmiller (Amsterdam: Elsevier) pp 137–44

Kohonen T, Hynninen J, Kangas J, Laaksonen J and Torkkola K 1995 LVQ_PAK: The learning vector quantization program package, *Technical report*, Helsinki University of Technology, Espoo, Finland

Lloyd S P 1957 Least squares quantization in PCMs *Technical report* Bell Telephone Laboratories, Murray Hill, NJ

——1982 Least-squares quantization in PCM *IEEE Trans. Information Theory* **28** 129–37

MacQueen J 1967 Some methods for classification and analysis of multivariate observations *Proc. Fifth Berkeley Symposium on Mathematics, Statistics and Probability* vol 1 pp 281–96

Miller W T 1986 *A nonlinear learning controller for roboting manipulators* vol 726 *Intelligent Robots and Computer Vision* SPIE 416–23

——1987 Sensor-based control of robotic manipulators using a general learning algorithm *IEEE J. Robotics and Automation* **3** 157–65

Miller W T and Glanz F H and Kraft L G 1990a CMAC: an associative neural network alternative to backpropagation *Proc. IEEE* **78** 1561–7

Miller W T, Hewes R P, Glanz F H and Kraft G 1990b Real-time dynamic control of an industrial manipulator using a neural-network-based learning controller *IEEE Trans. Robotics and Automation* **6** 1–9

Minsky M L and Papert S A 1969 *Perceptrons* (Cambridge, MA: MIT Press)

——1988 Epilogue: the new connectionism *Perceptrons* ed M L Minsky and S A Papert expanded edition (Cambridge, MA: MIT Press)

Moody J 1989 *Fast learning in multi-resolution hierarchies* (San Mateo, CA: Morgan Kaufmann)

Nasrabadi N M and King R A 1988 Image coding using vector quantization: a review *IEEE Trans. Communications* **36** 957–71

Parks P C and Militzer J 1989 Convergence properties of associative memory storage for learning control systems *Automation and Remote Control* **50** part 2 254–86

——1992 A comparison of five algorithms for the training of CMAC memories for learning control systems *Automatica* **28** 1027–35

Ridgeway W C III 1962 An adaptive logic system with generalizing properties *Phd Thesis, Technical Report 1556-1* Electron. Labs, Stanford, CA

Rosenblatt F 1957 The perceptron: a perceiving and recognizing automaton *Technical Report 85-460-1* Cornell Aeronautical Laboratory

——1958 The perceptron: a probabilistic model for information storage in the brain *Psych. Rev.* **65** 386–408

——1960 On the convergence of reinforcement procedures in simple perceptrons *Cornell Aeronautical Laboratory Report VG-1196-G-4* Buffalo, NY

Talbert L R *et al* 1963 A real-time adaptive speech recognition system *Technical Report* Stanford University

Widrow B 1962 Generalisation and information storage in networks of adaline *Self-organizing systems* ed Yovits *et al* (Washinton, DC: Wiley)

——1987a Adaline and madaline—1963 *Proc. IEEE 1st Int. Conf. on Neural Networks* vol 1 143–57 Plenary speech

——1987b The original adaptive neural net broom-balancer *Proc. IEEE Int. Symp. Circuits and Systems* pp 351–7

Widrow B and Hoff M 1960 Adaptive switching circuits *Western Electronic Show and Convention, Convention Record* vol 4 Institute of Radio Engineers (now IEEE) 96–104

Widrow B and Lehr M A 1990 30 years of adaptive neural networks: perceptron, madaline, and backpropagation *Proc. IEEE* **78** 1415–42

Widrow B and Stearns S 1985 *Adaptive Signal Processing* (Englewood Cliffs, NJ: Prentice-Hall)

C1.2 Multilayer perceptrons

Luis B Almeida

Abstract

This section introduces multilayer perceptrons, which are the most commonly used type of neural network. The popular backpropagation training algorithm is studied in detail. The momentum and adaptive step size techniques, which are used for accelerated training, are discussed. Other acceleration techniques are briefly referenced. Several implementation issues are then examined. The issue of generalization is studied next. Several measures to improve network generalization are discussed, including cross validation, choice of network size, network pruning, constructive algorithms and regularization. Recurrent networks are then studied, both in the fixed point mode, with the recurrent backpropagation algorithm, and in the sequential mode, with the unfolding in time algorithm. A reference is also made to time-delay neural networks. The section also includes brief mention of a large number of applications of multilayer perceptrons, with pointers to the bibliography.

C1.2.1 Introduction

Multilayer perceptrons (MLPs) are the best known and most widely used kind of neural network. They are formed by units of the type shown in figure C1.2.1. Each of these units forms a weighted sum of its inputs, to which a constant term is added. This sum is then passed through a nonlinearity, which is often called its *activation function*. Most often, units are interconnected in a *feedforward manner*, that is, with B3.2.4 interconnections that do not form any loops, as shown in figure C1.2.2. For some kinds of applications, recurrent (i.e. nonfeedforward) networks, in which some of the interconnections form loops, are also used.

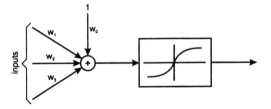

Figure C1.2.1. A unit of a multilayer perceptron.

Training of these networks is normally performed in a supervised manner. One assumes that a *training set* is available, which contains both input patterns and the corresponding desired output patterns (also called *target patterns*). As we shall see, the training is normally based on the minimization of some error measure between the network's outputs and the desired outputs. It involves a backward propagation through a network similar to the one being trained. For this reason the training algorithm is normally called *backpropagation*.

In this chapter we will study multilayer perceptrons and the backpropagation training algorithm. We will review some of the most important variants of this algorithm, designed both for improving the training speed and for dealing with different kinds of networks (feedforward and recurrent). We will also briefly

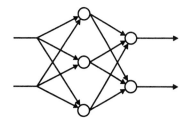

Figure C1.2.2. Example of a feedforward network. Each circle represents a unit of the type indicated in figure C1.2.1. Each connection between units has a weight. Each unit also has a bias input, not depicted in this figure.

mention some theoretical and practical issues related to the use of multilayer perceptrons and other kinds of supervised networks.

C1.2.2 Network architectures

We saw in figure C1.2.2 an example of a feedforward network, of the type that we will consider in this chapter. As we noted above, the interconnections of the units of this network do not form any loops, and hence the network is said to be *feedforward*. Networks in which there are one or more loops of interconnections, such as the one in figure C1.2.3, are called *recurrent*.

B2.3

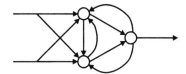

Figure C1.2.3. A recurrent network.

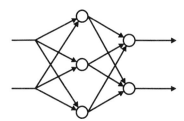

Figure C1.2.4. A layered network.

In feedforward networks, units are often arranged in layers, as in figure C1.2.4, but other topologies can also be used. Figure C1.2.5 shows a network type that is useful in some applications, in which direct links between inputs and output units are used. Figure C1.2.6 shows a three-unit network that is fully connected, i.e. that has all the interconnections that are allowed by the feedforward restriction.

The nonlinearities in the network's units can be any differentiable functions, as we shall see below. The kind of nonlinearity that is most commonly used has the general form shown in figure C1.2.7. It has two horizontal asymptotes, and is monotonically increasing, with a single point where the curvature changes sign. Curves with this general shape are usually called *sigmoids*. Some of the most common expressions of sigmoids are

B3.2.4

$$S(s) = \frac{1}{1 + e^{-s}} = \frac{1 + \tanh(s/2)}{2} \qquad (\text{C1.2.1})$$

$$S(s) = \tanh(s) \qquad (\text{C1.2.2})$$

$$S(s) = \arctan(s). \qquad (\text{C1.2.3})$$

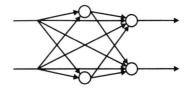

Figure C1.2.5. A network with direct links between input and output units.

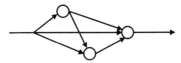

Figure C1.2.6. A fully connected feedforward network.

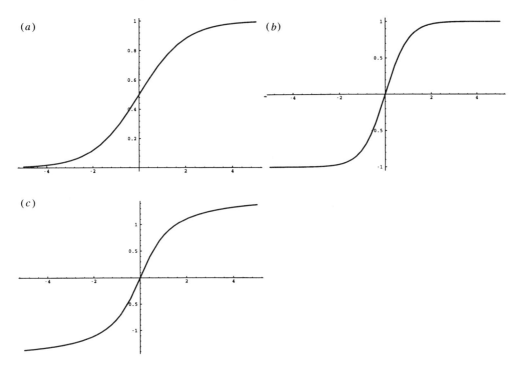

Figure C1.2.7. Sigmoids corresponding to: (*a*) equation (C1.2.1), (*b*) equation (C1.2.2) and (*c*) equation (C1.2.3).

Sigmoid (C1.2.3) is sometimes scaled to vary between -1 and $+1$. Sigmoid (C1.2.1) is often designated as the *logistic function*. As we said above, interconnections between units have *weights*, that multiply the values which go through them. Besides the variable inputs that come through weighted links, units normally also have a fixed input, which is often called *bias*.

It is through the variation of the weights and biases that networks are trained to perform the operations that are desired from them. As an example of how weight changes can affect the behavior of networks, figure C1.2.8 shows three one-unit networks that differ in their weights and that perform different logical operations. Figure C1.2.9 shows two networks with different topologies, that both perform the logical XOR operation. These two networks were trained by the backpropagation algorithm, to be described below. Note that since these networks have analog outputs, the output values are often not exactly 0 or 1. A usual convention, for binary applications, is that output values above the middle of the range of the sigmoid are taken as *true* or 1, and output values below that are taken as *false* or 0. This is the convention adopted here.

As we shall see below, it is sometimes convenient to consider input nodes as units of a special kind, which simply copy the input components to their outputs. These units are then normally designated as *input units*. The number of units and the number of layers that a given network is said to have may depend on whether this convention is taken or not. Another convention that is normally made is to designate as *hidden units* the units that are internal to the network, i.e. those units that are neither input nor output units. The two networks of figure C1.2.9 have, respectively, two and one hidden units.

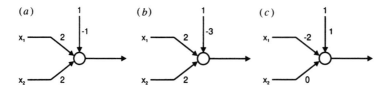

Figure C1.2.8. Single-unit networks implementing simple Boolean functions. (*a*) OR. (*b*) AND. (*c*) NOT. The units are assumed to have logistic nonlinearities.

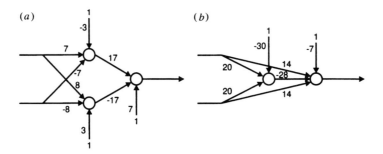

Figure C1.2.9. Two networks that have been trained to perform the XOR operation. The units are assumed to have logistic nonlinearities. The weight values have been rounded, for convenience.

C1.2.3 The backpropagation algorithm for feedforward networks

Let us represent the input pattern of a network by an m-dimensional vector x (italic bold characters shall represent vectors) and the outputs of the units of the network by an N-dimensional vector y. To keep the notation compact, we will represent the input nodes of the network as units (numbered from 1 to m). These units simply copy the components of the input pattern, i.e.

$$y_i = x_i \qquad i = 1, \ldots, m.$$

We will also assume that there is a unit number 0, whose output is fixed at 1, i.e. $y_0 = 1$. The weights from this unit to other units of the network will represent the bias terms of those units. The remaining units, $m + 1$ to N, are the operative units, that have the form shown in figure C1.2.1. In this way, all the parameters of the network appear as weights in interconnections among units, and can therefore be treated jointly, in a common manner. Denoting by w_{ji} the weight in the branch that links unit j to unit i, we can write the weighted sum performed by unit i as

$$s_i = \sum_{j=0}^{N} w_{ji} y_j \qquad i = m + 1, \ldots, N. \tag{C1.2.4}$$

Note that w_{0i} represents the unit's bias term and w_{ji}, with $j = 1, \ldots, m$, are the weights linking the inputs to unit i. We will make the convention that if a branch from one unit to another does not exist in the network, the corresponding weight is set to zero. The unit's output will be

$$y_i = S(s_i) \qquad i = m + 1, \ldots, N \tag{C1.2.5}$$

where S represents the unit's nonlinearity. For the sake of simplicity, we shall assume that the same nonlinearity is used in all units of the network (it would be straightforward to extend the reasoning in this chapter to situations in which nonlinearities differ from one unit to another). As we shall see, the only restriction on the nonlinearities is that they must be differentiable. The output pattern of the network is formed by the outputs of one or more of its units. We will collect these outputs into the output vector o.

Let us denote by x^k the kth pattern of the training set. We assume the training set to have K patterns (the training sets that are most often used are of finite size; infinite-sized training sets are sometimes used, and this would imply slight modifications in what follows, essentially amounting to changing the sums over training patterns into series or integrals, as appropriate). If we assume that we are presenting x^k at the input of the network, we can define an error vector e^k between the actual outputs o^k and the desired outputs d^k for the current input pattern:

$$e^k = o^k - d^k . \tag{C1.2.6}$$

The squared norm of the error vector, $E^k = \|e^k\|^2$ can be seen as a scalar measure of the deviation of the network from its ideal behavior, for the input pattern x^k. In fact, E^k is zero if $o^k = d^k$. Otherwise it is positive, progressively increasing as the network outputs deviate from the desired ones. We can define a measure of the network's deviation from the ideal, in the whole training set, as

$$E = \sum_{k=1}^{K} E^k \tag{C1.2.7}$$

where K is the number of patterns of the training set. If the training set and the network architecture are fixed, E is only a function of the weights of the network, that is, $E = E(w)$ (when convenient, we will assume that we have collected all the weights as components of a single vector w). We can think of the task of training the network on the given training set as the task of finding the weights that minimize E. If there is a set of weights that yields $E = 0$, then a successful minimization will result in a network that performs without error in the whole training set. Otherwise, the weights that minimize E will correspond to the network that performs best in the quadratic error sense.

The quadratic error may not be the best measure of the deviation from ideal in all situations, though it is by far the most commonly used one. If convenient, however, some other cost function $C(e)$ can be used, with $E^k = C(e^k)$. The total cost to be minimized is still given by (C1.2.7). The cost function C should be chosen so as to represent, as closely as possible, the relative importances of different errors in the situation where the network is to be applied. In general, $C(e)$ has an absolute minimum for $e = 0$, and in what follows the only restriction on C is that it be differentiable relative to all components of e.

C1.2.3.1 The basic algorithm

There are, in the mathematical literature, several different methods for minimizing a function such as $E(w)$. Among these, one that results in a particularly simple procedure is the gradient method. Essentially, this method consists of iteratively taking steps, in weight space, proportional to the negative gradient of the function to be minimized, that is, of iteratively updating the weights according to

$$w^{n+1} = w^n - \eta \nabla E \tag{C1.2.8}$$

where ∇E represents the gradient of E relative to w. This iteration is repeated until some appropriate stopping criterion is met. If $E(w)$ obeys some mild regularity conditions and η is small enough, this iteration will converge to a local minimum of E. The parameter η is normally designated as the *learning rate parameter* or *step size parameter*.

The main issue in applying this algorithm is the computation of the gradient components, $\partial E / \partial w_{ji}$. For feedforward networks, this computation takes a very simple form (Bryson and Ho 1969, Werbos 1974, Parker 1985, Le Cun 1985, Rumelhart *et al* 1986). This is best described by means of an example. Consider the network of figure C1.2.10(a). From this network we obtain another one (figure C1.2.10(b)) as follows: we first linearize all nonlinear elements of the original network, replacing them by linear branches with gains $g_i = S'(s_i)$. We then *transpose* it (Oppenheim and Schafer 1975) that is, we reverse the direction of flow of all branches, replacing summing nodes by divergence nodes and vice-versa, and changing outputs into inputs and vice-versa. This new network is often called the *backpropagation network*, or *error propagation network*, for reasons that will soon become clear. As indicated in the figure, we denote the variables in this network by the same letters as the corresponding ones in the MLP, with an overbar.

For feedforward networks, the *backpropagation rule* for computing the gradient components, which we shall describe next, can be easily derived by repeated application of the chain rule of differentiation;

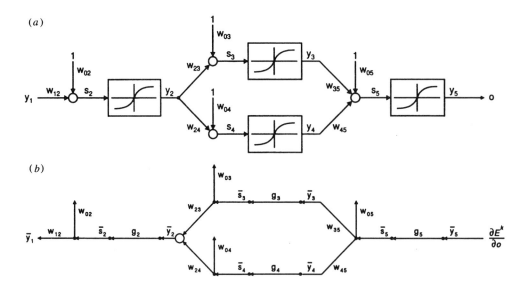

Figure C1.2.10. Example of a multilayer perceptron and of the corresponding backpropagation network. (*a*) Multilayer perceptron. (*b*) Backpropagation network, also called error propagation network.

see for example (Rumelhart *et al* 1986). We will not make that derivation here, however, because in section C1.2.8.1 we will make the derivation for a certain class of recurrent networks that includes feedforward networks as a special case. Here, we will therefore simply describe the rule. First of all, note that, from (C1.2.7)

$$\frac{\partial E}{\partial w_{ji}} = \sum_k \frac{\partial E^k}{\partial w_{ji}}.$$

We place the pattern x^k at the inputs of the MLP, we compute the output error according to (C1.2.6) and we place at the inputs of the error propagation network the values $\partial E^k/\partial o_i$ as shown in figure C1.2.10. The backpropagation rule states that the partial derivatives can then be obtained as

$$\frac{\partial E^k}{\partial w_{ji}} = y_j \bar{s}_i \tag{C1.2.9}$$

i.e. the partial derivative relative to a weight is the product of the inputs of the branches corresponding to that weight in the MLP and in the backpropagation network. As we said, the proof of this fact will be given in section C1.2.8.1.

If the quadratic error is used as a cost function, then $\partial E^k/\partial o_i = 2e_i^k$. Since the backpropagation network is linear, we can place at its inputs e_i^k, instead of $2e_i^k$, and compute the derivatives according to

$$\frac{\partial E^k}{\partial w_{ji}} = 2y_j \bar{s}_i. \tag{C1.2.10}$$

In this case the backpropagation network is propagating errors. This justifies the name of *error propagation network* that is commonly given to the backpropagation network. The variables $\bar{s}_i$ are often called *propagated errors*.

To apply this training procedure, we must have a training set, containing a collection of input patterns and the corresponding target outputs, and we must select a network architecture to be trained (number of units, arranged or not in layers, interconnections among units, activation functions). We must also choose an initial weight vector, w^1 (weights are normally initialized in a random manner, usually with a uniform distribution in some symmetric interval $[-a, a]$—see section C1.2.5.3 below), a step size parameter η and an appropriate stopping criterion.

The backpropagation algorithm can be summarized as follows, where we denote by K the number of patterns in the training set.

(i) Set $n = 1$. Repeat steps (a) through (c) below until the stopping criterion is met.

  (a) Set the variables g_{ji} to zero. These variables will be used to accumulate the gradient components.

(b) For $k = 1, \ldots, K$ perform steps (1) through (4).

    (1) Propagate forward: apply the training pattern x^k to the perceptron and compute its internal variables y_i and outputs o^k.

    (2) Compute the cost function derivatives: compute $\partial E^k / \partial o_i^k$.

    (3) Propagate backwards: apply $\partial E^k / \partial o_i^k$ to the inputs of the backpropagation network and compute its internal variables $\bar{s}_i$.

    (4) Compute and accumulate the gradient components: compute the values $\partial E^k / \partial w_{ji} = y_j \bar{s}_i$ and accumulate each of them in the corresponding variable, i.e. $g_{ji} = g_{ji} + y_j \bar{s}_i$.

(c) Update the weights: set $w_{ji}^{n+1} = w_{ji}^n - \eta g_{ji}$. Increment n.

This algorithm can be used with any differentiable cost function. When the quadratic error is used as a cost function, the factor 2 that appears in (C1.2.10) is usually incorporated into the learning rate constant η, and steps (2) to (4) are replaced by the following.

    (2) Compute the output errors: compute $e^k = o^k - d^k$.

    (3) Propagate backwards: apply e_i^k to the inputs of the backpropagation network and compute its internal variables $\bar{s}_i$.

    (4) Compute and accumulate the gradient components: compute the values $y_j \bar{s}_i$ and accumulate each of them in the corresponding variable, $g_{ji} = g_{ji} + y_j \bar{s}_i$.

For finite minima, i.e. for minima that are not situated at infinity, the above algorithm is guaranteed to converge for η below a certain value $\eta_{\max}$, if the activation functions and the cost function are continuous and differentiable. However, the upper bound $\eta_{\max}$ depends on the network, on the training set and on the cost function, and cannot be specified in advance. On the other hand, the fastest convergence is normally obtained for an optimal value of η that is somewhat below this upper bound. For η below the optimal value, the convergence speed can decrease considerably. This makes the choice of the learning rate parameter η a critical aspect of the training procedure. Often, preliminary tests have to be made with different learning rates, in order to try to find a good value of η for the problem to be solved. In section C1.2.4.2 we will describe a modification of the algorithm, involving adaptive step sizes, which solves this difficulty almost completely, and also yields faster training.

    The stopping criterion to be used depends on the problem being addressed. In some situations, the training is stopped when the cost function E becomes lower than some prescribed value. In other situations, the algorithm is stopped when the maximum absolute value of the error components e_i^k becomes lower than some given limit. In other situations still, training is stopped when the variation of E or of the weights becomes too slow. Often, an upper bound on the number of iterations n is also incorporated, to prevent the algorithm from running forever if the chosen conditions are never met.

C1.2.3.2 Stochastic backpropagation

When the training set is large, each weight update (which involves a sweep through the whole training set) may become very time-consuming, making learning very slow. In such cases, another version of the algorithm, performing a weight update per pattern presentation, can be used.

(i) Set $n = 1$. Repeat step (a) below until the stopping criterion is met.

    (a) For $k = 1, \ldots, K$, perform steps (1) through (5).

        (1) Propagate forward: apply the training pattern x^k to the perceptron, and compute its internal variables y_i and outputs o^k.

        (2) Compute the cost function derivatives: compute $\partial E^k / \partial o_i^k$.

        (3) Propagate backwards: apply $\partial E^k / \partial o_i^k$ to the inputs of the backpropagation network, and compute its internal variables $\bar{s}_i$.

        (4) Compute the gradient components: compute the values $\partial E^k / \partial w_{ji} = y_j \bar{s}_i$.

        (5) Update the weights: set $w_{ji}^{n+1} = w_{ji}^n - \eta y_j \bar{s}_i$. Increment n.

To differentiate between the two forms of the algorithm, the former is often qualified as *batch*, *off-line* or *deterministic*, while the latter is called *real-time*, *on-line* or *stochastic*. This last designation stems from the fact that, under certain conditions, the latter form of the algorithm implements a *stochastic gradient descent*. Its convergence can then be guaranteed if η is varied with n, in such a way that (i) $\eta(n) \to 0$ and

(ii) $\sum_{n=1}^{\infty} \eta(n) = \infty$. In fact, the algorithm can then be shown to satisfy the conditions for convergence introduced by Ljung (1978). In practice, since any training is in fact finite, it is not always clear how best to decrease η. A solution that is sometimes used is to train first in real-time mode, until convergence becomes slow, and then switch to batch mode. Frequently, the largest speed advantage of real-time training occurs in the first part of the training process, and the later switch to batch mode does not bring about any significant increase in training time.

C1.1.3   Backpropagation is a generalization of the delta rule for training single linear units: *adalines*. In fact, it is easy to see that, when applied to a single linear unit (i.e. a unit without nonlinearity), backpropagation

B2.3.4 coincides with the delta rule. For this reason, backpropagation is sometimes designated the *generalized delta rule*.

C1.2.3.3 Local minima

An issue that may have already come to the reader's mind is that gradient descent, like any other local optimization algorithm, converges to local minima of the function being minimized. Only by chance will it converge to the global minimum. A solution that can be used to try to alleviate this problem is to perform several independent trainings, with different random initializations of the weights. Even this, however, does not guarantee that the global minimum will be found, although it increases the probability of finding lower local minima. On the other hand, this solution cannot be used for large problems, where training times of days or even weeks can be involved. When the function $E(w)$ is very complex, with many local minima, one must essentially abandon the hope of finding the optimum, and accept local minima as the best that can be found. If these are good enough, the problem is solved. Otherwise, the only viable solution normally involves using a more complex architecture (e.g. with more hidden units, and/or with more layers) that will normally have lower local minima. It must be said, however, that although local minima are a drawback in the training of multilayer perceptrons, they do not usually cause too many difficulties in practice.

C1.2.3.4 Universal approximation property

An important property of feedforward multilayer perceptrons is their universality, that is, their capacity to approximate, to any desired accuracy, any desired function. The main result in this respect was first obtained by Cybenko (1989), and later, independently, by Funahashi (1989) and by Hornik *et al* (1989). It shows that a perceptron with a single hidden layer of sigmoidal units and with a linear output unit can uniformly approximate any continuous function in any hypercube (and therefore also in any closed, bounded set). More specifically, it states that, if a function f, continuous in a closed hypercube $H \subset \mathbb{R}^k$, and an error bound $\varepsilon > 0$ are given, then a number h, weight vectors w_i and output weights a_i ($i = 1, \ldots, h$) exist such that the output of the single hidden layer perceptron

$$o(x) = \sum_{i=1}^{h} a_i S(w_i \cdot x)$$

approximates f in H with an error smaller than ε, that is, $|f(x) - o(x)| < \varepsilon$ for all $x \in H$, if the nonlinearity S is continuous, monotonically increasing and bounded. Here, for compactness of notation, we have assumed that the input vector x has been extended with a component $x_0 = 1$ and that the weight vectors w_i have components from 0 to k, so that the inner product $(w_i \cdot x)$ incorporates a bias term.

      This result is rather reassuring, since it guarantees that even perceptrons with a single hidden layer can approximate essentially all useful functions. However, the limitations of this result should also be understood. First of all, the theorem only guarantees the existence of a network, but does not provide any constructive method to find it. Second, it does not give any bounds on the number of hidden units h needed for approximating a given function to a desired level of accuracy. It may well turn out that, for some specific problems, while a single hidden layer perceptron must exist which gives a good enough approximation to the desired result, either it is too hard to find, or it has too large a number of hidden units (or both). A large number of units, and therefore of weights, may be a strong drawback, meaning that a very large number of training patterns is required for adequately training the network (see the discussion on generalization in section C1.2.6). On the other hand, it may happen that networks with more than one hidden layer can yield the desired approximation with a much smaller number of weights. The situation is somewhat similar to what happens with combinatorial digital circuits. Although any digital function can be implemented in two layers (e.g. by expressing it as a sum of products), a complex function, such as an output of a binary adder for a large word size, can require an intractable number of product terms, and therefore of gates in the first layer. However, by using more layers, the implementation may become easily tractable.

C1.2.4 Accelerated training

The training of multilayer perceptrons by the backpropagation algorithm is often rather slow, and may require thousands or tens of thousands of *epochs*, in complex problems (the name *epoch* is normally given to a training sweep through the whole training set, either in batch or in real-time mode). The essential reason for this is that the error surface, as a function of the weights, normally has narrow ravines (regions where the curvature along one direction is rather strong, while it is very weak in an orthogonal direction, the gradient component along the latter direction being very small). In these regions, the use of a large learning rate parameter η will lead to a divergent oscillation across the ravine. A small η will lead the weight vector to the 'bottom' of the ravine, and convergence to the minimum will then proceed along this bottom, but at a very low speed, because the gradient and η are both small. In the next sections we will describe two methods of *improving the training speed* of multilayer perceptrons, especially in situations B3.4 where narrow ravines exist.

C1.2.4.1 Momentum technique

Let us rewrite the weight update equation C1.2.8 as

$$w^{n+1} = w^n + \Delta w^n$$

with

$$\Delta w^n = -\eta \nabla E .$$

The momentum technique (Rumelhart *et al* 1986) replaces the latter equation with

$$\Delta w^n = -\eta \nabla E + \alpha w^n$$

in which $0 \leq \alpha < 1$. The second term in the equation, called the *momentum term*, introduces a kind of B6.3.3 'inertia' in the movement of the weight vector, since it makes successive weight updates similar to one another, and has an accumulation effect, if successive gradients are in similar directions. This increases the movement speed along the ravine, and helps to prevent oscillations across it. This effect can also be seen as a linear low-pass filtering of the gradient ∇E. The effect becomes more pronounced as α approaches 1, but normally one has to be conservative in the choice of α because of an adverse effect of the momentum term: the ravines are normally curved, and in a bend the weight movement may be up a ravine wall, if too much momentum has been previously acquired. Like the learning rate parameter η, the momentum parameter α has to be appropriately selected for each problem. Typical values of α are in the range 0.5 to 0.95. Values below 0.5 normally introduce little improvement relative to backpropagation without momentum, while values above 0.95 often tend to cause divergence at bends. The momentum technique may be used both in batch and real-time training modes. In the latter case, the low-pass filtering action also tends to smooth the randomness of the gradients computed for individual patterns.

With momentum, the batch-mode backpropagation algorithm becomes the following.

(i) Set $n = 1$ and $\Delta w_{ji}^0 = 0$. Repeat steps (a) through (d) below until the stopping criterion is met.

    (a) Set the variables g_{ji} to zero. These variables will be used to accumulate the gradient components.

    (b) For $k = 1, \dots, K$ (where K is the number of training patterns), perform steps (1) through (4).

        (1) Propagate forward: apply the training pattern x^k to the perceptron and compute its internal variables y_j and outputs o^k.

        (2) Compute the cost function derivatives: compute $\partial E^k / \partial o_i^k$.

        (3) Propagate backwards: apply $\partial E^k / \partial o_i^k$ to the inputs of the backpropagation network and compute its internal variables $\bar{s}_i$.

        (4) Compute and accumulate the gradient components: compute the values $\partial E^k / \partial w_{ji} = y_j \bar{s}_i$ and accumulate each of them in the corresponding variable, i.e. $g_{ji} = g_{ji} + y_j \bar{s}_i$.

(c) Apply momentum: set $\Delta w_{ji}^n = -\eta g_{ji} + \alpha \Delta w_{ji}^{n-1}$

(d) Update the weights: set $w_{ji}^{n+1} = w_{ji}^n + \Delta w_{ji}^n$. Increment n.

The real-time backpropagation algorithm with momentum is

(i) Set $n = 1$ and $\Delta w_{ji}^0 = 0$. Repeat step (a) below until the stopping criterion is met.

    (a) For $k = 1, \ldots, K$, perform steps (1) through (6).

        (1) Propagate forward: apply the training pattern x^k to the perceptron and compute its internal variables y_j and outputs o^k.

        (2) Compute the cost function derivatives: compute $\partial E^k / \partial o_i^k$.

        (3) Propagate backwards: apply $\partial E^k / \partial o_i^k$ to the inputs of the backpropagation network and compute its internal variables $\bar{s}_i$.

        (4) Compute the gradient components: compute the values $\partial E^k / \partial w_{ji} = y_j \bar{s}_i$.

        (5) Apply momentum: set $\Delta w_{ji}^n = -\eta y_j \bar{s}_i + \alpha \Delta w_{ji}^{n-1}$.

        (6) Update the weights: set $w_{ji}^{n+1} = w_{ji}^n + \Delta w_{ji}^n$. Increment n.

C1.2.4.2 Adaptive step sizes

The adaptive step size method is a simple acceleration technique, proposed in Silva and Almeida (1990a, b) for dealing with ravines. For related techniques see Jacobs (1988) and Tollenaere (1990). It consists of using an individual step size parameter η_{ji} for each weight, and adapting these parameters in each iteration, depending on the successive signs of the gradient components:

$$\eta_{ji}^n = \begin{cases} \eta_{ji}^{n-1} u & \text{if } \left(\frac{\partial E}{\partial w_{ji}}\right)^n \text{ and } \left(\frac{\partial E}{\partial w_{ji}}\right)^{n-1} \text{ have the same sign} \\ \eta_{ji}^{n-1} d & \text{if } \left(\frac{\partial E}{\partial w_{ji}}\right)^n \text{ and } \left(\frac{\partial E}{\partial w_{ji}}\right)^{n-1} \text{ have different signs} \end{cases} \quad \text{(C1.2.11)}$$

$$\Delta w_{ji}^n = -\eta_{ji}^n \frac{\partial E}{\partial w_{ji}} \quad \text{(C1.2.12)}$$

where $u > 1$ and $d < 1$. There are two basic ideas behind this procedure. The first is that, in ravines that are parallel to some axis, use of appropriate individual step sizes is equivalent to eliminating the ravine, as discussed in Silva and Almeida (1990b). Ravines that are not parallel to any axis but are not too diagonal either, are not completely eliminated, but are made much less pronounced. The second idea is that quasi-optimal step sizes can be found by a simple strategy: if two successive updates of a given weight were performed in the same direction, then its step size should be increased. On the other hand, if two successive updates were in opposite directions, then the step size should be decreased.

As is apparent from the explanation above, the adaptive step size technique is especially useful for ravines that are parallel, or almost parallel, to some axis. Since the technique is less effective for ravines that are oblique to all axes, use of a combination of adaptive step sizes and the momentum term technique is justified. This combination is normally done by replacing (C1.2.12) with

$$z_{ji}^n = \frac{\partial E}{\partial w_{ji}} + \alpha z_{ji}^{n-1}$$

$$\Delta w_{ji}^n = -\eta_{ji}^n z_{ji}^n$$

that is, we first filter the gradient with the momentum technique, and then multiply the filtered momentum by the adaptive step sizes.

For applying the backpropagation algorithm with adaptive step sizes and momentum, one must choose the following parameters:

η_0 initial value of the step size parameters

u 'up' step size multiplier

d 'down' step size multiplier

α momentum parameter.

Typical values, which will work well in most situations, are $u = 1.2$, $d = 0.8$ and $\alpha = 0.9$. The initial value of the step size parameters is not critical, but is normally chosen small to prevent the algorithm from diverging in the initial epochs, while the step size adaptation still did not have enough time to act. The step size parameters will then be increased by the step size adaptation algorithm, if necessary. If the robustness measures indicated in section C1.2.4.3 are incorporated in the algorithm, even large initial step size parameters will not cause divergence, and essentially any value can be chosen for η_0.

The batch-mode training algorithm with adaptive step sizes and momentum is as follows.

(i)   Set $n = 1$, $\eta_{ji}^1 = \eta_0$ and $z_{ji}^0 = 0$. Repeat steps (a) through (d) below until the stopping criterion is met.

    (a)   Set the variables g_{ji}^n to zero. These variables will be used to accumulate the gradient components.

    (b)   For $k = 1, \dots, K$ (where K is the number of training patterns), perform steps (1) through (4).

        (1)   Propagate forward: apply the training pattern x^k to the perceptron and compute its internal variables y_j and outputs o^k.

        (2)   Compute the cost function derivatives: compute $\partial E^k / \partial o_i^k$.

        (3)   Propagate backwards: apply $\partial E^k / \partial o_i^k$ to the inputs of the backpropagation network and compute its internal variables $\bar{s}_i$.

        (4)   Compute and accumulate the gradient components: compute the values $\partial E^k / \partial o_i^k$ and accumulate each of them in the corresponding variable, i.e. $g_{ji}^n = g_{ji}^n + y_j \bar{s}_i$.

    (c)   Apply momentum: set $z_{ji}^n = g_{ji}^n + \alpha z_{ji}^{n-1}$.

    (d)   Adapt the step sizes: if $n \geq 2$ set

$$\eta_{ji}^n = \begin{cases} u \eta_{ji}^{n-1} & \text{if } g_{ji}^n \text{ and } g_{ji}^{n-1} \text{ have the same sign} \\ d \eta_{ji}^{n-1} & \text{if } g_{ji}^n \text{ and } g_{ji}^{n-1} \text{ have opposite signs}. \end{cases}$$

    (e)   Update the weights: set $w_{ji}^{n+1} = w_{ji}^n - \eta_{ji}^n z_{ji}^n$. Increment n.

The adaptive step size technique was designed, in principle, for batch training. It has, however, been used with success in real-time training, with the following modifications: (i) while weights are adapted after every pattern presentation, step sizes are adapted only at the end of each epoch, and (ii) instead of comparing the signs of the derivatives, in the step size adaptation (C1.2.11), we compare the signs of the total changes of the weight in the last and next to last epochs.

C1.2.4.3 Robustness

As was said in section C1.2.3.1, the step size parameter η has to be small enough for the backpropagation algorithm to converge. During the course of training, either with or without adaptive step sizes, one may come to a region of weight space for which the current step size parameters are too large, causing an increase in the cost function from one epoch to the next. A similar increase can also occur in a curved ravine if too much momentum has previously been acquired, as noted in section C1.2.4.1. To prevent the cost function from increasing, one must then go back to the step with lowest cost function, reduce the step size parameters and set the momentum memory to zero. To do this, after each epoch we must compare the current value of the cost function with the lowest that was ever found in the current training, and take the above-mentioned measures if the current value is higher than that lowest one (a small tolerance for cost function increases is allowed, as we will see below). To be more specific, these measures are as follows.

(i)   Return to the set of weights that produced the lowest value of the cost function.

(ii)   Reduce all the step size parameters (or the single step size parameter, if adaptive step sizes are not being used) by multiplying by a fixed factor $r < 1$.

(iii) Set the momentum memories z_{ji}^{n-1} (or Δw_{ji}^{n-1} if adaptive step sizes are not being used) to zero.

After this, an epoch is again executed. If the error still increases, the same measures are repeated: returning to the previous point, reducing step sizes and setting momentum memories to zero. This repetition continues until an error decrease is observed. The normal learning procedure is then resumed. A value that is often used for the reduction factor is $r = 0.5$. A tolerance is normally used in the comparison of values of the cost function, that is, a small increase is allowed without taking the measures indicated above. In batch mode, the allowed increase is very small (e.g. 0.1%) just to allow for small numerical errors in the computation of the cost function. In real-time mode, a larger increase (e.g. 20%) has to be allowed, because the exact cost function is normally never computed. Instead, the cost function contributions from the different patterns are added during a whole epoch, while the weights are also being updated. This sum of cost function contributions is only an estimate of the actual cost function at the end of the epoch, and this is why a larger tolerance is needed. If desired, the actual cost function could be computed at the end of each epoch, by presenting all the patterns while keeping the weights frozen, but this would increase computation significantly.

The procedure described in this section is rather effective in making the training robust, irrespective of whether it is combined with adaptive step sizes and/or momentum or not. When combined with adaptive step sizes and momentum, it yields a very effective MLP training algorithm.

C1.2.4.4 *Other acceleration techniques*

B3.4 In this section we will summarize other existing techniques for *fast MLP training*. Most of them are based on a local second-order approximation to the cost function, attempting to reach the minimum of that approximation in each step (for a review of a number of variants see Battiti (1992)). These techniques make use of the Hessian matrix, that is, of the matrix of second derivatives of the cost function relative to the weights. Some methods compute the full Hessian matrix. Since the number of elements of the Hessian is the square of the number of weights, these methods have the important drawback that their amount of computation per epoch is proportional to that square. These methods reduce the number of training epochs but, for large networks, they involve a very large amount of computation per epoch. Other methods assume that the Hessian is diagonal, thereby achieving a linear growth of the computation per epoch with the number of weights. Among these, a variant (Becker and Le Cun 1989) estimates the diagonal elements of the Hessian through a backward propagation, similar to the one described in section C1.2.3.1 for computing the gradient. Another variant, called *quickprop* (Fahlman 1989) estimates the second derivatives based on the variation of the first derivatives from one epoch to the next. It should be noted that the adaptive step size algorithm described in section C1.2.4.2, and the related algorithms referenced in that section, can also be viewed as indirect ways to estimate diagonal Hessian elements.

Another class of second-order techniques is based on the method of conjugate gradients (Press *et al* 1986). This is a method which, when employed with a second-order function, can find its minimum in a number of steps equal to the number of arguments of the function. The various conjugate gradient techniques that are in use differ from one another, essentially, in the approximations they make to deal with non-second-order functions. Among these techniques, one of the most effective appears to be the one of Moller (1990).

B1.7.3, C1.6.2 We should not conclude this section without mentioning that, when the input patterns have few components (up to about 5–10), networks of local units (e.g. *radial basis function networks*) are normally much faster to train than multilayer perceptrons. However, as the dimensionality of the input grows, networks of local units tend to require an exponentially large number of units, making their training very long, and requiring very large training sets to be able to generalize well (cf section C1.2.6).

C1.2.5 Implementation

In this section we discuss some issues that are related to the practical implementation of multilayer perceptrons and of the backpropagation algorithm.

C1.2.5.1 *Sigmoids*

As we said above, the activation functions that are most commonly used in units of multilayer perceptrons are of the sigmoidal type. Other kinds of nonlinearities have sometimes been tried, but their behavior generally seems to be inferior to that of sigmoids. Within the class of sigmoids there still is, however, a wide room for choice. The characteristic of sigmoids that appears to have the strongest influence on the performance of the training algorithm is symmetry relative to the origin. Functions like the hyperbolic tangent and the arctangent are symmetric relative to the origin, while the logistic function, for example, is symmetric relative to a point of coordinates $(0, 0.5)$. Symmetry relative to the origin gives sigmoids a bipolar character that normally tends to yield better conditioned error surfaces. Sigmoids like the logistic tend to originate narrow ravines in the error function, which impair the speed of the training procedure (Le Cun *et al* 1991).

C1.2.5.2 Output units and target values

Most practical applications of multilayer perceptrons can be divided, in a relatively clear way, into two different classes. In one of the classes, the target outputs take a continuous range of values, and the task of the network is to perform a nonlinear regression operation. Normally, in this case, it is convenient not to place nonlinearities in the outputs of the network. In fact, we normally wish the outputs to be able to span the whole range of possible target values, which is often wider than the range of values of the sigmoids. We could, of course, scale the amplitudes of the output sigmoids appropriately, but this rarely has any advantage relative to the simple use of units without nonlinearities at the outputs. Output units are then said to be linear. They simply output the weighted sum of their inputs plus their bias term.

In the other class, which includes most classification and pattern recognition applications, the target outputs are binary, that is, they take only two values. In this case it is common to use output units with sigmoid nonlinearities, similar to other units in the network. The binary target values that are most appropriate depend on the sigmoids that are used. Often, target values are chosen equal to the two asymptotic values of the sigmoids (e.g. 0 and 1 for the logistic function, and ± 1 for the tanh and the scaled arctan functions). In this case, to achieve zero error, the output units would have to achieve full saturation, i.e. their input sums would have to become infinite. This fact would tend to drive the weights linking to these units to grow indefinitely in absolute value, and would slow down the training process. To improve training speed, it is therefore common to use target values that are close, but not equal, to the asymptotic values of the sigmoids (e.g. 0.05 and 0.95 for the logistic function, and ± 0.9 for the tanh and the scaled arctan functions).

C1.2.5.3 Weight initialization

Before the backpropagation algorithm can be started, it is necessary to set the weights of the network to some initial values. A natural choice would be to initialize them all with a value of zero, so as not to bias the result of training in any special direction. However, it can easily be seen, by applying the backpropagation rule, that if initial weights are zero, all gradient components are zero (except for those that concern weights on direct links between input and output units, if such links exist in the network). Moreover, those gradient components will always remain at zero during training, even if direct links do exist. Therefore, it is normally necessary to initialize the weights to nonzero values. The most common procedure is to initialize them to random values, drawn from a uniform distribution in some symmetric interval $[-a, a]$. As we mentioned above, several independent trainings with independent random initializations may be used, to try to find better minima of the cost function.

It is easy to understand that large weights (resulting from large values of a) will tend to saturate the respective units. In saturation the derivative of the sigmoidal nonlinearity is very small. Since this derivative acts as a multiplying factor in the backpropagation, derivatives relative to the unit's input weights will be very small. The unit will be almost 'stuck', making learning very slow.

If the inputs to a given unit i in the network all have similar root mean square (rms) values and are all independent from one another, and if the weights are initialized in some given, fixed interval, the rms value of the unit's input sum will be proportional to $(f_i)^{1/2}$, where f_i is the number of inputs of unit i (often called the unit's *fan-in*). To keep the rms values of the input sums similar to one another, and to avoid saturating the units with largest fan-ins, the parameter a, controlling the width of the initialization interval, is sometimes varied from unit to unit, by making $a_i = k/(f_i)^{1/2}$. There are different preferences for the choice of k. Some people prefer to initialize the weights very close to the origin, making k very small (e.g. 0.01 to 0.1), and therefore keeping the units in their central linear regions in the beginning of the training process. Other people prefer larger values of k (e.g. 1 or larger), that lead the units into their nonlinear regions right from the start of training.

C1.2.5.4 Input normalization and decorrelation

Let us consider the simplest network that one can design, formed by a single linear unit. Single-unit linear networks (adalines) have been in use for a long time, in the area of discrete-time signal processing.

Finite-impulse response (FIR) filters (Oppenheim and Schafer 1975) can actually be viewed as single linear units with no bias. The inputs are consecutive samples of the input signal, and the weights are the filter coefficients. Therefore, adaptive filtering with FIR filters is essentially a form of real-time training of linear-unit networks. It is therefore no surprise that the first adaptive filtering algorithms were derived from the delta rule (Widrow and Stearns 1985).

It is a well-known fact from adaptive filter theory that training is fastest, because the error function is best conditioned (without any ravines) if the inputs to the linear unit are uncorrelated among themselves, that is, $\langle x_i x_j \rangle = 0$ for $i \neq j$, and have equal mean-squared values, that is, $\langle x_i^2 \rangle = \langle x_j^2 \rangle$ for all i, j. Here $\langle \cdot \rangle$ represents the expected value (most often, when training perceptrons, the expected value can be estimated simply by averaging in the training set).

If a bias term is also used in the linear unit, it acts as an extra input that is constantly equal to 1. Its mean squared value is 1, and therefore the mean squared values of all other inputs should also be equal to 1. On the other hand, cross-correlations of other inputs with this new input are simply the expected values of those other inputs, which should be equal to zero, as all cross-correlations between inputs: $\langle x_i 1 \rangle = \langle x_i \rangle = 0$. In summary, for fastest training of a single linear unit with bias one should preprocess the data so that the average of each input component is zero,

$$\langle x_i \rangle = 0$$

and the components are decorrelated and normalized:

$$\langle x_i x_j \rangle = \delta_{ij}$$

where δ_{ij} is the Kronecker symbol. It has been found by experience that this kind of preprocessing also tends to accelerate the training in the case of multilayer perceptrons. Setting the averages of input components to zero can simply be performed by adding an appropriate constant to each of them. Decorrelation can then be performed by any orthogonalization procedure, for example, the Gram–Schmidt technique (Golub and Van Loan 1983). Finally, normalization can be performed by an appropriate scaling of each component. The most cumbersome of these steps is the orthogonalization, and people sometimes skip it, simply setting means to zero and mean-squared values to one. This simplified preprocessing is usually designated *input normalization*, and is often quite effective at increasing the training speed of networks. A more elaborate acceleration technique, involving the adaptive decorrelation and normalization of the inputs of all layers of the network, is described in (Silva and Almeida 1991).

C1.2.5.5 Shared weights

In some cases one would wish to constrain some weights of a network to be equal to one another. This situation may arise, for example, if we wish to perform the same kind of processing in various parts of the input pattern. It is a common situation in image processing, where one may want to detect the same feature in different parts of the input image. An example, in a handwritten digit application, is given in (Le Cun *et al* 1990a). Two examples of shared weight situations will also be found below, in the discussion of recurrent networks.

The difficulty in handling shared weights comes from the fact that even if these weights are initialized with the same value, the derivatives of the cost function relative to each of them will usually be different from one another. The solution is rather simple. Assume that we have collected all weights in a weight vector $w = (w_1, w_2, \ldots)^{\mathrm{T}}$ (where T denotes transposition), and that the first m weights are to be kept equal to one another. These weights are not, in fact, free arguments of the cost function E. To keep all of the arguments of E free, one should replace all of these weights by a single argument a, to which all of them will be equal. Then, the partial derivative of E should be computed relative to a, and not relative to each of these weights individually. But

$$\frac{\partial E}{\partial a} = \sum_{i=1}^{m} \frac{\partial E}{\partial w_i} \frac{\partial w_i}{\partial a}$$

$$= \sum_{i=1}^{m} \frac{\partial E}{\partial w_i} \, .$$

The derivatives that appear in the last line can be computed by the normal backpropagation procedure. In summary, one should compute the derivatives relative to each of the individual weights in the normal way, and then use their sum to update *a* and therefore to update all the shared weights. One should also remember that shared weights should be initialized to the same value.

C1.2.6 Generalization

Until now we have been discussing the training of multilayer perceptrons based on the assumption that we wish to optimize their performance (measured by the cost function) in the training set. However, this is a simplification of the situation that we normally find in practice. Consider, for example, a network being trained to perform a classification task. We assume that we are given a training set, which is usually finite, containing examples of the desired classification. This set is usually only a minute fraction of the universe in which the network will be used after training. After training, the network will be used to classify patterns that were not in the training set.

We see that ideally we would like to minimize the cost function computed in the whole universe. That is normally either impossible or impractical, however, because the universe is infinite, because we do not know it all in advance, or simply because that would be too costly in computational terms. Until now we have been using the cost function evaluated in the training set as an estimate of its value in the whole universe. Whenever possible, precautions should be taken to ensure that the training set is as representative of the whole universe as possible. This may be achieved, for example, by randomly drawing patterns from the universe, to form the training set. Even if this is done, however, the statistical distribution of the training set will only be an approximation to the distribution of the universe. A consequence of this is that, since we optimize the performance of the network in the training set, its performance in that set will normally be better than in the whole universe. A network whose performance in the universe is similar to the performance in the training set is said to *generalize* well, while a network whose performance degrades B3.5 significantly from the training set to the universe is said to generalize poorly.

These facts have two main implications. The first is that if we wish to have an unbiased estimate of the network's performance in the universe, we should not use the performance in the training set, but rather in a *test set* that is independent from the training set. The second implication is that we should try to design networks and training algorithms in order to ensure good generalization, and not only good performance in the training set.

C1.2.6.1 Network size

An important issue in what concerns generalization is the size of the network. Intuitively, it is clear that one cannot effectively train a large network with a training set containing only a few patterns. Consider a network with a single output. When we present at the input a given training pattern, we can idealize writing an expression of the output of the network as a function of the weights. If we wish to make the output equal to the desired output, we can set that expression equal to the desired output, and we will obtain an equation whose unknowns are the weights. The whole training set will therefore yield a set of equations. If the network has more than one output, the situation is similar, and the number of equations will be the number of training patterns times the number of outputs. These equations are usually nonlinear and very complex, and therefore not solvable by conventional means. They may even have no exact solution. Training algorithms are methods to find exact or approximate solutions for such sets of equations.

By making an analogy with the well-known case of the systems of linear equations, we can gain some insight into the issue of generalization. If the number of unknowns (i.e. weights) is larger than the number of equations, there will generally be an infinite number of solutions. Since each of these solutions corresponds to a different set of weights, it is clear that they will generalize differently from one another, and only by chance will the specific solution that we find generalize well. If the number of weights is equal to the number of equations, a linear system will usually have a single solution. A nonlinear system will usually have no solutions, a single solution or a finite number of solutions. Since these are optimal for the training set, which is different from the universe, they will still often not generalize well. The interesting situation is the one in which there are fewer weights than equations. In this case, there will be no solution, unless the set of equations is redundant. Even the existence of an approximate solution implies that there must be some kind of redundancy, or regularity, in the training set (e.g. in a digit-recognition problem, regularities are the facts that all zeros have a round shape, all ones are approximately vertical bars, and so on). With fewer weights than training patterns, the only way for the network to approximately satisfy the training equations is to exploit the regularities of the problem, and the fewer weights the network has, the more it will have to rely on the training set's regularities to be able to perform well on that set. But these regularities are exactly what we expect to be maintained, from the training set to the universe. *Therefore, small networks, with fewer weights than the number of equations, are the ones that can be expected to generalize best, if they can be trained to perform well on the training set.* Note that the latter condition means that network topology is a very important factor. A network with the appropriate number of weights but with an inappropriate topology will not be able to perform well in the training set, and therefore cannot also be expected to perform well in the universe. On the other hand, a network with an appropriately small number of weights and with the appropriate topology will be able to perform well in the training set, and also to generalize well. As a rule of thumb, we would say that the number of weights should be around or below one tenth of the product of the number of training patterns by the number of outputs. In some situations, however, it may go up to about one half of that product.

There are other methods to try to improve generalization. The methods that we will mention are *stopped training, network pruning, constructive techniques* and the use of a *regularization term.*

C1.2.6.2 Stopped training and cross-validation

B3.5.2 In *stopped training*, one considers all the successive weight vectors found during the course of the training process, and tries to find the vector that corresponds to the best generalization. This is normally done by
B3.5.2 *cross-validation.* Another set of patterns, independent from the training and test sets, is used to evaluate the network's performance during the training (this set of patterns is often designated the *validation set*). At the end of training, instead of selecting the weights that perform best in the training set, we select the weights that performed best in the validation set. This is equivalent, in fact, to performing an early stop of the training process, before convergence in the training set, which justifies the designation of 'stopped training'. Since the performance in the validation set tends to oscillate significantly during the training process, it is advisable to continue training even after the first local minimum in the validation performance is observed, because better validation performance may still arise later in the process. Note that, since the validation set is used to select the set of weights to be kept, it effectively becomes part of the training data, i.e. the performance of the final network in the validation set is not an unbiased estimate of its performance on the universe. Therefore, an independent test set is still required, to evaluate the network's performance after training is complete.

C1.2.6.3 Pruning and constructive techniques

B3.5.2 *Network pruning* techniques start from a large network, and try to successively eliminate the least important interconnections, thereby arriving at a smaller network whose topology is appropriate for the problem at hand, and which has a good probability of generalizing well. Among the pruning techniques we mention the skeletonization method of Mozer and Smolensky (1989), *optimal brain damage* (Le Cun *et al* 1990b) and *optimal brain surgeon* (Hassibi *et al* 1993). Network pruning, while effective, tends to be rather time-consuming, since after each pruning some retraining of the network has to be performed (an interesting and efficient technique, which is a blend of pruning and regularization, is mentioned below in section C1.2.6.4). Constructive techniques work in the opposite way to pruning: they start with a small network and add units until the performance is good enough. Several constructive techniques have appeared in the literature, the best known of which is probably cascade-correlation (Fahlman and Lebiere 1990). Other constructive techniques can be found in Frean (1990) and Mézard and Nadal (1989).

C1.2.6.4 Regularization

Regularization is a class of techniques that comes from the field of statistics (MacKay 1992a, b). In its simplest form, it consists of adding a *regularization term* to the cost function to be optimized:

$$E_{\text{total}} = E + \lambda E_{\text{reg}}$$

where E is the cost function that we defined in the previous sections, E_{reg} is the regularization term, λ is a parameter controlling the amount of regularization and E_{total} is the total cost function that will be minimized. The regularization term is chosen so that it tends to smooth the function that is generated by the network at its outputs. This term should have small values for weight vectors that generate smooth outputs, and large values for weight vectors that generate unsmooth outputs. An intuitive justification for the use of such a term can be given by considering a simple example (figure C1.2.11). Assume that a number of training data points are given (in the figure these are represented by dark circles). There is an infinite number of functions that pass through these points, two of which are represented in the figure. Of these, clearly the most reasonable are the ones that are smoothest. If the function to be approximated is smooth, then the approximator's output should be smooth also. On the other hand, if the function to be approximated is unsmooth, then only by chance would an unsmooth function generated by a network approximate the desired one, in the regions between the given data points, since unsmooth functions have a very large variability. Therefore, only by chance would the network generalize well, in such a case. Only a larger number of training points would allow us to expect to be able to successfully approximate such a function. Therefore, one should bias the training algorithm towards producing smooth output functions. This can be done through the use of a regularization term (in the theory of statistics, supervised learning can be viewed as a form of maximum-likelihood estimation, and in this context the use of a regularization term can be justified in a more elaborate way, by taking into consideration a prior distribution of weight vectors (MacKay 1992a, b)).

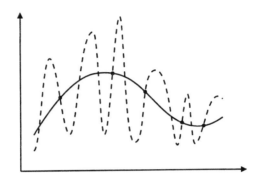

Figure C1.2.11. An illustration of generalization. Given the data points denoted by full circles, there is an infinite number of functions that pass through them. Only the smooth ones can be expected to generalize well.

One of the simplest regularization terms, which is often used in practice (Krogh and Hertz 1992), is the squared norm of the weight vector

$$E_{\text{reg}} = \sum_{j,i} w_{ji}^2 \, .$$

Use of such a regularization term is justified since smaller weights tend to produce slower-changing (and therefore smoother) functions. The use of this term leads to gradient components that are given by

$$\frac{\partial E_{\text{total}}}{\partial w_{ji}} = \frac{\partial E}{\partial w_{ji}} + \lambda w_{ji} \, .$$

The first term on the right-hand side of this equation is still computed by the backpropagation rule. Since the derivative of E_{total} is to be subtracted (after multiplication by the step size parameter) from the weight itself, we see that if the derivative of E is zero, the weight will decay exponentially to zero. For this reason, this technique is often called *exponential decay*. Other forms of regularization terms have been proposed in the literature, which are based e.g. on minimizing derivatives of the function generated by the network (Bishop 1990), or on placing a smooth cost on the individual weights, in an attempt to reduce their number (Weigend *et al* 1991).

A type of regularization term that appears to be particularly promising has been recently introduced (Williams 1994). Instead of the sum of the squares of the weights, it uses the sum of their absolute values:

$$E_{\text{reg}} = \sum_{j,i} |w_{ji}| \, .$$

Use of this term leads to

$$\frac{\partial E_{\text{total}}}{\partial w_{ji}} = \frac{\partial E}{\partial w_{ji}} + \lambda \text{sgn}(w_{ji})$$

where 'sgn' denotes the sign function. If the derivative of E is zero, the weight will decay linearly to zero, reaching that value in a finite time. Only if the derivative of E relative to a weight has absolute value larger than λ will this weight be able to escape the zero value. Therefore, this E_{reg} term acts simultaneously as a regularizer, tending to keep the weights small, and as a pruner, since it automatically sets the least important weights to zero. Experience with this technique is still limited, but its ability to perform both regularization and pruning during the normal training of the network gives it a potential that should not be overlooked. We will designate this form of regularization as *linear decay*, for the reasons given above, or *Laplacian regularization*, since it can be justified, in a statistical framework, by assuming a Laplacian prior on the weights. One word of caution regarding the use of this form of regularization concerns the fact that the regularizer term E_{reg} is not differentiable relative to the weights when these have a value of zero. A way to deal with this problem is discussed in Williams (1994). A simpler way, which this author has used with success, is to check, in every training step, whether each weight has changed sign, and to set the weight to zero if it did. The weight is allowed to leave the zero value in later training steps, if $|\partial E / \partial w_{ji}| > \lambda$.

In finalizing this section, we should point out that there are several other approaches to the issue of trying to find a network with good generalization ability, and also to other related issues, such as trying to estimate the generalization ability of a given network. One of the best known of these approaches is based on the concept of *Vapnik–Chervonenkis* dimension (often designated simply *VC dimension*) (Guyon *et al* 1992).

B3.5.2.2

C1.2.7 Application examples

We have already seen, in figure C1.2.9, two examples of networks trained to perform the logical XOR operation. Another artificial problem that is often used to test network training is the so-called *encoder problem*. A network with m inputs and m outputs is trained to perform an identity mapping (i.e. to yield output patterns that are equal to the respective input patterns) in a universe consisting of m patterns: those obtained by setting one of the components to 1 and all other ones to 0. The difficulty lies in the fact that the network topology that is adopted has a hidden layer with fewer than m units, forming a bottleneck. The network has to learn to encode the m patterns into different combinations of values of the hidden units, and to decode these combinations to yield the correct outputs. An example of a 4–2–4 encoder is shown in figure C1.2.12. Table C1.2.1 shows the encoding learned by a network with the topology of figure C1.2.12, trained by backpropagation. In this case target values were 0.05 and 0.95 instead of 0 and 1, respectively, as explained in section C1.2.5.2. It should be noted that, with the given architecture, the network cannot reproduce the target values exactly. This is why it sometimes outputs 0.02 and sometimes 0.06, instead of 0.05.

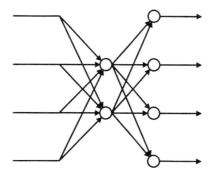

Figure C1.2.12. A 4–2–4 encoder.

Multilayer perceptrons have a rather widespread use, in very diverse application areas. We cannot give a full description of any of these applications here. We shall only give brief accounts of some of them, with references to publications where the reader can find more details.

Often, perceptrons are used as classifiers. A well-known example is the application to the *recognition of handwritten digits* (Le Cun *et al* 1990a). Normally, digit images are segmented, normalized in size and de-skewed. After this, their resolution is lowered to a manageable level (e.g. 16 × 16 pixels), before they are fed to a recognizer MLP. Recognition error rates of only a few percent can be achieved. A significant

G1.3

Table C1.2.1. Encoding learned by the network of figure C1.2.12.

inputs				hidden units		outputs			
1.0	0.0	0.0	0.0	0.95	0.94	0.95	0.06	0.02	0.06
0.0	1.0	0.0	0.0	0.07	0.95	0.06	0.95	0.06	0.02
0.0	0.0	1.0	0.0	0.10	0.03	0.02	0.06	0.95	0.06
0.0	0.0	0.0	1.0	0.95	0.08	0.06	0.02	0.06	0.95

percentage of errors normally comes from the segmentation, which is not performed by neural means. In the author's group (unpublished work), an error rate of 3.8% on zipcode digits was achieved, with automatic segmentation followed by manual elimination of the few gross segmentation errors (segments with no digit at all, or with two or more complete digits). For digits that are pre-segmented, e.g. by being written in forms with boxes for individual digits, it is now possible to achieve recognition errors below 1%, a performance that is already suitable for replacing manual data entry. Several such systems are probably in use these days. The author knows of one designed and being used in Spain (López 1994). However, the problems of automatic digit segmentation and, more generally, of segmentation of cursive handwriting are still hard to deal with (Matan *et al* 1992).

Another important example of a classification application is *speech recognition*. Here, perceptrons can be used *per se* (Waibel 1989) or in hybrid systems, combined with hidden Markov models. See Robinson *et al* (1993) for an example of a state-of-the-art hybrid recognizer for large vocabulary, speaker independent, continuous speech. In hybrid systems, MLPs are actually used as probability estimators, based on an important property of supervised systems: when they are trained for classification tasks, using as cost function the quadratic error (or certain other cost functions), they essentially become estimators of the probabilities of the classes given the input vectors. This property is discussed in Richard and Lippmann (1991). In another example of a classification application, MLPs have been used to validate sensor readings in an industrial plant (Ramos *et al* 1994). F1.7.2, G1.4

In nonclassification, analog tasks, an important class is formed by *control applications*. An interesting example is that of a neural network system that is used to drive a van, controlling the steering based on an image of the road supplied by a forward-looking video camera (Pomerleau 1991). This kind of system has already been used to drive the vehicle on a highway at speeds up to 30 mph. It can also be used, with appropriately trained networks, to drive the vehicle on various other kinds of roads, including some that are hard to deal with by classical means (e.g. dirt roads covered with tree shadows) (Pomerleau 1993). F1.9

Another example of a control application is the control of fast movements of a robot arm, a problem that is hard to handle by more formal, theoretical means (Goldberg and Pearlmutter 1989). For further examples of applications to control, see White and Sage (1992). There have already been in the market, for a few years, industrial control modules that incorporate multilayer perceptrons.

Another important area of application is prediction. Multilayer perceptrons (and also other kinds of networks, namely those based on radial basis functions) have been used in the academic problem of predicting chaotic time series (Lapedes and Farber 1987), but also to predict consumptions of commodities (Yuan and Fine 1993), crucial variables in *industrial plants* (Cruz *et al* 1993) and so on. A very appealing, but also somewhat controversial area is prediction of *financial time series* (Trippi and Turban 1993). G2.8
G6.3

The practical applications of neural networks are constantly increasing in number. Given the impossibility of making an exhaustive listing here, we shall content ourselves with the above examples.

C1.2.8 Recurrent networks

Recurrent networks are networks with unit interconnections that form loops. They can be employed in two very different modes. One is nonsequential, that is, it involves no memory, the desired output for each input pattern depending only on that pattern and not on past ones. The other mode is sequential, that is, desired outputs depend not only on the current input pattern, but also on previous ones. We shall deal with them separately.

C1.2.8.1 Nonsequential networks

In this mode, as said above, desired outputs depend only on the current input pattern. Furthermore, it is assumed that whenever a pattern is presented at the network's input, it is kept fixed long enough to allow the network to reach equilibrium. As is well known from the theory of nonlinear dynamic systems (Thompson and Stewart 1986), a network with a fixed input pattern can exhibit three different kinds of behavior: it can converge to a fixed point, it can oscillate (either periodically or quasi-periodically) and it can have chaotic behavior. In what follows, we shall assume that for each input pattern the network will have stable behavior, with a single fixed point. The conditions under which this will happen are discussed later in this section.

Recurrent backpropagation. In this nonsequential situation, the gradient of the cost function E can still be computed by backward propagation of derivatives through a backpropagation network, in a natural extension of the backpropagation rule of feedforward networks (this extension is usually designated *recurrent backpropagation*). The proof of this fact was first given by Almeida (1987), and soon thereafter independently by Pineda (1987). Here we shall give a version of the proof based on graphs, which is more intuitive than the ones given in those references.

Consider first a recurrent nonlinear network N (not necessarily a multilayer perceptron), which has a single output, any number of inputs, and an internal branch which is linear with a gain w. Such a network, with the notation that we will adopt for its variables, is depicted in figure C1.2.13(a). A single input is shown, for simplicity, but multiple inputs would be treated in exactly the same manner, as we shall see. We assume that this network, as well as all other networks used in this proof, are in equilibrium at fixed points. We wish to compute the derivative of the network's output relative to w, and therefore we shall give an infinitesimal increment dw to w. This can be done by changing w to $w + dw$, but it can also be achieved by adding an extra branch with gain dw, as shown in figure C1.2.13(b). Of course, all internal variables, as well as the output, will suffer increments, as indicated in the figure.

The state of the network will not change if we replace the new branch by an input branch, as long as its contribution to its sink node is unchanged. This could be achieved by keeping the gain dw and the input $y + dy$ of this branch unchanged. We can, however, change the input to y, since the contribution dy dw is a higher order infinitesimum, and can therefore be disregarded (figure C1.2.13(c)).

We shall now linearize the network around its fixed point, obtaining a linear network NL that takes into account only increments (figure C1.2.13(d)). Note that the original input branch disappears, since its contribution has suffered no increment. If we had multiple inputs, the same would have happened to all of them.

We will now divide the contribution of the input branch by dw, by changing its gain to unity. Since this network is linear, its node variables and its output will change to derivatives relative to w, which we will represent by means of upper dots, for compactness (i.e. for example, $\dot{o} = \partial o / \partial w$; see figure C1.2.13(e)).

Finally, we will transpose the network, obtaining network NLT, shown in figure C1.2.13(f) (recall that transposition of a linear network consists in changing the direction of flow of all branches, keeping their gains; inputs become outputs, and vice-versa; summation points become divergence points, and vice-versa). From the *transposition theorem* (Oppenheim and Schafer 1975) we know that the input–output relationship of the network is not changed by transposition, i.e. if we place y at its input we will still obtain $\dot{o}$ at its output. Therefore, we can write

$$\dot{o} = t y$$

where t is the total gain from the input to the output node of the NLT network.

Now consider a recurrent perceptron P (figure C1.2.14(a)) with several outputs, and assume that we wish to compute the derivative of an output o_p relative to a weight w_{ji}. By the same reasoning, we can write

$$\dot{o}_p = t_{ip} y_j$$

where we now use the upper dot to designate the derivative relative to w_{ji}. The factor t_{ip} is the total gain of the linearized and transposed network, PLT, from input p to node i (cf figure C1.2.14(b)). Finally, let us consider the derivative of a cost function term E^k (corresponding to a given input pattern x^k) relative to w_{ji}. Using the chain rule, we can write

$$\frac{\partial E^k}{\partial w_{ji}} = \sum_{p \in P} \frac{\partial E^k}{\partial o_p} \dot{o}_p$$

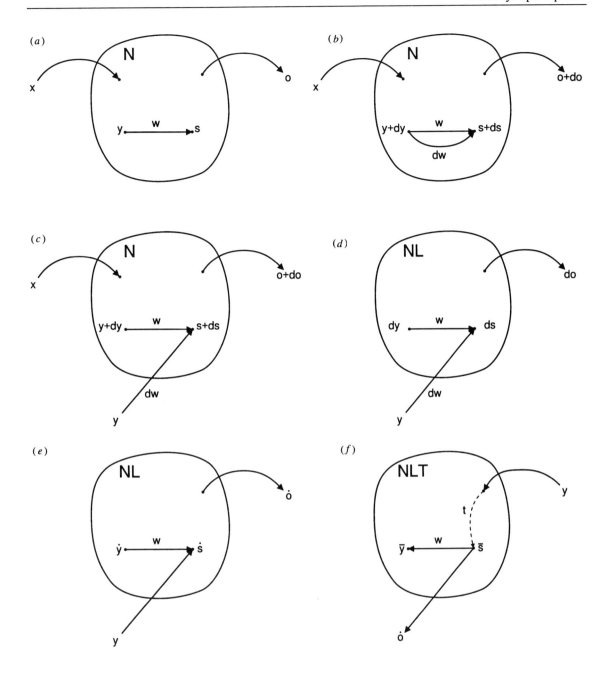

Figure C1.2.13. Illustration of the proof of validity of the backpropagation rule for recurrent networks. Case of a general network. See text for explanation.

and therefore

$$\frac{\partial E^k}{\partial w_{ji}} = \sum_{p \in P} \frac{\partial E^k}{\partial o_p} t_{ip} y_j$$

$$= y_j \sum_{p \in P} \frac{\partial E^k}{\partial o_p} t_{ip} t_{ip}$$

where P is the set of indices of units that produce outputs. Noting that network PLT is linear, we can write

$$\frac{\partial E^k}{\partial w_{ji}} = y_j \bar{s}_i \tag{C1.2.13}$$

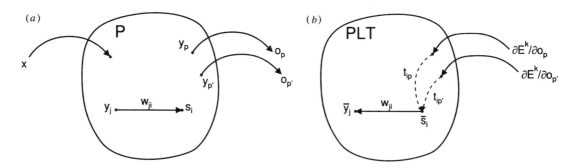

Figure C1.2.14. Illustration of the proof of validity of the backpropagation rule for recurrent networks. Case of a recurrent perceptron. See text for explanation.

where, as depicted in figure C1.2.14(*b*), $\bar{s}_i$ is obtained in the corresponding node of network PLT when the values $\partial E^k / \partial o_p$ are applied at its inputs.

If we assume that the original perceptron was feedforward, we recognize network PLT as the backpropagation network. Equation (C1.2.13) is the same as (C1.2.9), proving the validity of the backpropagation rule for feedforward networks, described in section C1.2.3.1. We will keep the designation of *backpropagation network* for network PLT in the case of recurrent networks. As we saw, this network is still obtained from the original perceptron by linearization followed by transposition. The recurrent backpropagation rule states that, if we apply the values $\partial E^k / \partial o_p$ to the corresponding inputs of the backpropagation network, the partial derivative of the cost function relative to a weight will be given by the product of the inputs of that weight's branches in the perceptron network and in the backpropagation network. Of course, the special case of the quadratic error, described in section C1.2.3.1, where one places the errors at the inputs of the backpropagation network, and then uses (C1.2.10), is also still valid in the recurrent case. For this reason, the backpropagation network is still often called the *error propagation network*, in the recurrent case.

Training a recurrent network by backpropagation takes essentially the same steps as for a feedforward network. The difference is that, when a pattern is applied to the perceptron network, this network must be allowed to stabilize before its outputs and node values are observed. The error propagation network must also be allowed to stabilize, when the derivatives $\partial E^k / \partial o_p$ are applied to its inputs. In digital implementations (including computer simulations) this involves an iteration in the propagation through the perceptron, until a stable state is found, and a similar loop in the propagation through the backpropagation network. In analog implementations the networks will evolve, through their own dynamics, to their stable states.

An important practical remark is that, in recurrent networks, the gradient's components can easily have a much larger dynamic range than in feedforward networks. The use of a technique such as adaptive step sizes, and of the robustness measures described in section C1.2.4.3, is therefore even more important here than for feedforward networks. Note that the gradient can even become infinite, at some points in weight space. This, however, does not cause any significant practical problem: gradient components can simply be limited to some convenient large value, with the proper sign.

Network stability. We assumed above that, with any fixed pattern at its input, the perceptron network was stable and had a single fixed point. It is this author's experience that often, when training recurrent networks with recurrent backpropagation, the networks that are obtained during the training process are all stable and all have single fixed points. There are exceptions, however, and it would be desirable to be able to guarantee that networks will in fact always be stable, and will always have a single fixed point. The issue of stability can be dealt with by means of a sufficient condition for stability, which we shall discuss next. The discussion of the number of fixed points will be deferred to the end of this section.

To derive a sufficient condition for stability, we first note that, while the static equations (C1.2.4) and (C1.2.5) suffice to describe the static behavior of a network, and therefore to find its fixed points, the dynamic behavior of the network is only defined if we specify the dynamic behavior of its units. Therefore, a discussion of network stability will always involve the units' dynamic behavior.

C1.3.4   If some restrictions are imposed on it, a recurrent perceptron is formally equivalent to a *Hopfield network* with graded units (Hopfield 1984). These restrictions are that the units' dynamic behavior is as schematized in figure C1.2.15(*a*), that weights between units are symmetrical, i.e. $w_{ji} = w_{ij}$ for

$i, j = m + 1, \ldots, N$, and that the units' nonlinearities are all increasing, bounded functions. The stability of such networks has been proved in Hopfield (1984) (we have assumed that the network variables are voltages; if currents were considered instead, then the resistor and capacitor should both be connected from the input to ground, as in Hopfield (1984)).

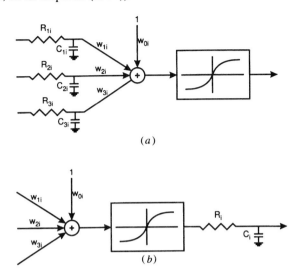

Figure C1.2.15. Typical dynamic behaviors assumed for units of continuous-time recurrent networks.

The behavior of figure C1.2.15(a) normally arises from attempting to model the dynamic behavior of biological neurons. When considering network realizations based on analog electronic systems, it is more natural to consider the dynamic behavior of figure C1.2.15(b). This is because, unless special measures are taken, an analog electronic circuit will have a lowpass behavior that can be modeled, to a first approximation, by a first-order lowpass system. The two behaviors are equivalent if all RC time constants are equal, but otherwise they are not. Here we shall give the proof of stability for the behavior of figure C1.2.15(b). This proof was first given in Almeida (1987), and is very similar to the proof given in Hopfield (1984) for the dynamic behavior of figure C1.2.15(a).

Using the notation given in figure C1.2.15(b), we can write

$$s_i = \sum_{j=0}^{N} w_{ji} y_j$$

$$u_i = S(s_i)$$

$$\frac{dy_i}{dt} = \frac{1}{\tau_i}(u_i - y_i) \qquad (C1.2.14)$$

where $\tau_i = R_i C_i$ is the time constant of the RC circuit of the ith unit. Here we assume that the index i varies from $m + 1$ to N, as in (C1.2.4) and (C1.2.5). We shall prove the network's stability by showing that it has a Lyapunov function (Willems 1970) that always decreases with time. The Lyapunov function that we will consider is

$$W = -\frac{1}{2} \sum_{j,i}^{N} w_{ji} y_i y_j + \sum_{i=m+1}^{N} U(y_i)$$

where U is a primitive of S^{-1}, the inverse of S (see figure C1.2.16). We are still assuming, as in section C1.2.3, that y_0 has a fixed value of 1, and that $y_1, \ldots, y_m$ represent the input components. We are also still assuming that the nonlinearities of all units are equal (it would again be straightforward to extend this proof to the situation in which the nonlinearities differ from one unit to another, but are all increasing and bounded; the proof could still be easily extended to the case in which all nonlinearities are decreasing and bounded; in this case the function W would increase with time, instead of decreasing).

Since we assumed that the inputs do not change, the time derivative of W is given by

$$\frac{dW}{dt} = \sum_{i=m+1}^{N} \frac{dW}{dy_i} \frac{dy_i}{dt}. \qquad (C1.2.15)$$

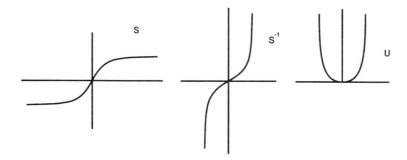

Figure C1.2.16. The functions S, S^{-1} and U. See text for explanation.

For $i = m + 1, \ldots, N$, we have

$$\frac{\partial W}{\partial y_i} = -\sum_{j=0}^{N} w_{ji} y_j + U'(y_i)$$
$$= -[s_i - S^{-1}(y_i)]$$
$$= -[S^{-1}(u_i) - S^{-1}(y_i)].$$

Since S is an increasing function, S^{-1} also is, and therefore either the difference in the last equation has the same sign as the difference in (C1.2.14), or they are simultaneously zero. Therefore, the products in (C1.2.15) are all negative or zero, and dW/dt must be negative or zero. It is zero if and only if all the $\partial W/\partial y_i$ and the dy_i/dt are simultaneously zero. In that case the network is in a fixed point, and W is at a point of stationarity. Since W always decreases in time during the network's evolution, the network's state cannot oscillate or have chaotic behavior. It can only move towards a fixed point, or to infinity. But since the y_i are bounded (because S is bounded), movement towards infinity is not possible, and the state must converge towards some fixed point. As we saw, these fixed points occur at the points of stationarity of W.

A useful remark (Almeida 1987) is that, except for marginally stable states, whenever the perceptron network is stable, the backpropagation network will also be stable, if the same RC-type dynamics are used in it. In fact, if the perceptron is in a nonmarginal stable state, the linearized perceptron network will also be stable. If we write its equations in the standard state space form (Willems 1970)

$$\frac{du}{dt} = \mathbf{A}u$$

where u is the vector of state variables and $\mathbf{A}$ is the system matrix, then it will be stable if and only if all the eigenvalues of $\mathbf{A}$ have negative real parts. The backpropagation network, being the transpose of this system, has state equations

$$\frac{du}{dt} = \mathbf{A}^{\mathrm{T}}u$$

where $\bar{u}$ is the state vector of the backpropagation network and $\mathbf{A}^{\mathrm{T}}$ is the transpose of $\mathbf{A}$. But the eigenvalues of a matrix and of its transpose are equal. Therefore, if the linearized perceptron was stable, the backpropagation network will also be stable. Here, *transpose* is taken in the dynamic system sense. In practice this means that the RC dynamics have to be kept in the backpropagation network too.

The above remark is always true, except for marginally stable states, which are those stable states for which the linearized network is not stable. They lie at the boundary between stability and instability, and can normally be disregarded in practice, since the probability of their occurrence is essentially zero. To train a network with the guarantee that it will always be stable, we therefore have to obey three conditions.

(i)   To use nonlinearities which are increasing and bounded. Networks with sigmoidal units always satisfy this condition.

(ii) To keep the weights symmetrical. For this purpose, we have first to initialize them in a symmetrical way, and then to keep them symmetrical during training. This is an example of a situation of shared weights, and is dealt with in the manner we described in section C1.2.5.5: the two derivatives

$\partial E^k / \partial w_{ij}$ and $\partial E^k / \partial w_{ji}$ are both computed using recurrent backpropagation, and their sum is used for updating both w_{ji} and w_{ij}.

(iii) To implement the RC dynamics both in the perceptron and in the backpropagation network. In digital implementations this means performing a numerical simulation of the continuous-time dynamics. If stability is not achieved, the numerical simulation is too coarse, and its time resolution should be increased. In analog implementations, RC circuits can actually be placed both in the perceptron and in the backpropagation network, to ensure that they have the appropriate dynamics.

Clearly, weight symmetry is a sufficient, but not necessary condition for stability. For example, feedforward networks are always stable, but do not obey the symmetry condition. Weight symmetry is a restriction on the network's adaptability, and it can be argued that it will reduce the network's capabilities. This is a price to be paid for being sure to obtain a network that will always be stable. But as we said at the beginning of this section, training without enforcing symmetry often yields stable networks, and in many situations it may be worth trying first, before resorting to symmetrical networks.

We come now to the discussion of the requirement that there be a single fixed point for each input pattern. Unfortunately, we do not know of any sufficient condition for guaranteeing that this will be true. The discussion of this issue can therefore only be made in qualitative terms. In practice, we have observed situations with multiple stable states only very seldom, and we never needed to take any special measures to cope with them—multiple stable states normally merged by themselves, during training. This can be explained by noting that, when training a recurrent network, we are in fact trying to move its stable states to given areas that are determined by the desired values of the outputs. If two different stable states exist for the same input pattern, and if the network stabilizes sometimes in one and sometimes in the other, then we will be trying to move them both to the same region. It is therefore not too surprising that they will merge. On the other hand, if there are multiple stable states but the network always stabilizes in the same one, then the other ones can be disregarded, as if they did not exist, since they do not influence the network's behavior in any way.

C1.2.8.2 Sequential networks

Besides the nonsequential mode described in section C1.2.8.1, recurrent networks can also be used in a sequential, or dynamic mode. In this case, network outputs depend not only on the current input, but also on previous inputs. There are several variants of the sequential mode, and we will concentrate here on the one that is most commonly used: discrete-time recurrent networks.

In this mode, it is assumed that the network's inputs only change at discrete times $t = 1, 2, \ldots$, and that there are units in the network whose outputs are also only updated at these discrete times, synchronously with the inputs. We shall designate these units *discrete-time units*. The other units, whose outputs immediately follow any variations of their inputs, will be called *instantaneous units*. Wherever interconnections between units form loops, there must be at least one discrete-time unit in the loop. There may, however, be more than one of these units per loop. Often, people build networks in which all units are discrete-time ones, as in figure C1.2.17(a). However, nothing prevents us from using discrete-time and instantaneous units in the same network, as long as there is at least one discrete-time unit per loop. A simple example of a network with one instantaneous and two discrete-time units is given in figure C1.2.17(b). We will use this second network as an example, to better specify the operation of networks of this kind. To be consistent with the conventions used above, we will identify unit 1 with the input, that is, $y_1^n = x^n$. The input has some initial value x^0 (here, we will denote by an upper index the time step that variables refer to). Units 2 and 3, which are the discrete-time ones, have initial states y_2^0 and y_3^0. Unit 4, which is instantaneous, immediately reflects at its output whatever is present at its input. Therefore, its output is always given by

$$y_4^n = S(w_{24} y_2^n)$$

(here n denotes the discrete time, and not the iteration number as in previous sections). Whenever a new discrete-time step arises, the input changes from x^n to x^{n+1}, and the outputs of the discrete-time units change to new values that are computed using the values of variables before that time step:

$$y_2^{n+1} = S(w_{12} x^n + w_{32}^n y_3^n)$$
$$y_3^{n+1} = S(w_{33} y_3^n + w_{43} y_4^n).$$

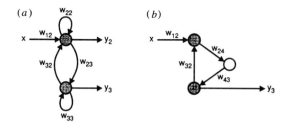

Figure C1.2.17. Examples of sequential networks. Shaded units are discrete time ones, unshaded units are instantaneous ones. (*a*) A network that has only discrete time units. (*b*) A network with both discrete time and instantaneous units.

The output of unit 4 instantaneously changes to reflect the changes of the other units and of the input:

$$y_4^{n+1} = S(w_{24} y_2^{n+1}).$$

We see that, given the initial state of the network, for each input sequence $x^0, x^1, x^2, \ldots, x^T$ the network's outputs will yield a sequence of values. The network's operation is sequential because each output value will depend on previous values of the input.

It is now easy to see why it is required that in every loop of interconnections there be at least one discrete-time unit. In a loop formed only by instantaneous units, there would be a never-ending sequence of updates, always going around the loop.

Training of this kind of recurrent network consists in finding weights so that, for given input sequences, the network approximates, as closely as possible, desired output sequences. The desired output sequences may specify target values for all time steps, or only for some of them. For example, in some situations only the desired final value of the outputs is specified. Different input sequences may be of different lengths, in which case the corresponding output sequences will also have different lengths. Naturally, training, test and validation sets will be formed by pairs of input and desired output sequences.

A great advantage of discrete-time recurrent networks is that, as we shall see, they can be reduced to feedforward networks, and can therefore be trained with ordinary backpropagation. This had already been noted in the well known book by Minsky and Papert (1969). To see how it can be done, consider again the network of figure C1.2.17(*a*). Assume that we construct a new network (figure C1.2.18(*a*)) where each unit of the recurrent network is unfolded into a sequence of units, one for each time step. Clearly, this network will always be feedforward since, in the original network, information could only flow forward in time. The input pattern of this unfolded network will be formed by the sequence of input values $x^0, x^1, x^2, \ldots, x^T$, presented all at once to the respective input nodes. The output sequence can also be obtained all at once, from the respective output nodes. The outputs can be compared with target values (for those times for which target values do exist), and errors (or, more generally, cost function derivatives) can be fed into a backpropagation network, obtained from the feedforward network in the usual way. The only remark that needs to be made, regarding the training procedure, concerns the fact that each weight from the recurrent network appears unfolded, in the feedforward network (and also in the backpropagation network) T times. All instances of the same weight must be kept equal, since they actually correspond to a single weight in the recurrent network. This is again a situation of shared weights, that we have already seen how to handle: the derivatives relative to each of the instances of the same weight are all added together, and the sum is used to update the weight (in all its instances). Networks involving both discrete-time and instantaneous units can also be easily handled. Figure C1.2.18(*b*) shows the unfolding of the network of figure C1.2.17(*b*).

The training method that we have described is normally called *unfolding in time*, or *backpropagation through time*. It requires an amount of storage that is proportional to the number of units and to the length of the sequence being trained, since the outputs of the units at intermediate time steps must be stored until the backward propagation is completed and the cross-products of (C1.2.9) are computed. The total amount of computation per presentation of an input sequence is $O(WT)$, where W is the number of weights in the network, and T is, as above, the length of the input sequence.

Unfolding in time can clearly be used in the batch and real-time modes, if real-time is understood to mean that weights are updated once per presentation of an input sequence. In some situations, instead of having a number of input sequences with the corresponding desired output sequences, one has a single very long (or even indefinitely long) input sequence, with the corresponding desired output sequence. It

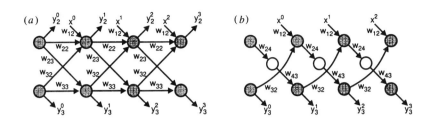

Figure C1.2.18. The unfolded networks corresponding to the sequential networks of figure C1.2.17.

would then be desirable to be able to make a weight update per time step, without having to wait for the end of the sequence to update weights. In such cases, unfolding-in-time may become rather inefficient (or even unusable, if the sequence is indefinitely long). Even in cases where there are several sequences in the training set, it might be more efficient to perform one update per time step. On the other hand, if training sequences are long, it may also be desirable not to have to store the values corresponding to all time steps, as required by the unfolding in time procedure, since these values may consume a large amount of memory. A few algorithms exist which do not need to wait for the end of the sequence to compute contributions to gradients, and which require only a limited amount of memory, irrespective of the length of the input sequence. We will mention only the best known one, often designated *real-time recurrent learning* (RTRL), which was originally proposed by Robinson and Fallside (1987) under the name of *infinite impulse response algorithm*, and is best known from later publications of Williams and Zipser (1989). This algorithm carries forward, in time, the information that is necessary to compute the derivatives of the cost function, and therefore does not need to store previous network states, and also does not need to perform backward propagations in time. There are two prices to be paid for this. One is computational complexity. While, for a fully interconnected network with N units (and therefore $W = N^2$ weights) unfolding in time requires $O(N^2T)$ operations per sequence presentation, RTRL requires $O(N^4T)$ operations. This quickly makes it impractical for large networks. The other price to be paid is that, if weight updates are performed at every time step, what is computed is only an approximation to the actual gradient of the cost function. Depending on the situation, this approximation may be good or bad. For some problems this is of little importance, but for others it may affect convergence, and even lead the training process to converge to wrong solutions. A variant of RTRL that deserves mentioning is called the *Green's function algorithm* (Sun *et al* 1992). It has the advantage of reducing the number of operations to $O(N^3T)$. However, in numerical implementations it involves an approximation that may affect its validity for long sequences.

Several examples of the application of unfolding in time to the training of recurrent networks have appeared in the literature. A very interesting one is described in Nguyen and Widrow (1990), where a controller is trained to park a truck with a trailer in backward motion. A very early example of an application to speech was given in Watrous (1987). Examples of the use of RTRL have also appeared in the literature; for example, for the learning of grammars (Giles *et al* 1992).

Besides the discrete-time mode, recurrent networks are also sometimes used in a continuous-time mode. In this case, the outputs of units change continuously in time according to given dynamics. Inputs and target outputs of the network are then both functions of continuous time, instead of being sequences. A training algorithm for this kind of network, which is an extension of unfolding in time to the continuous time situation, was presented in Pearlmutter (1989).

C1.2.8.3 Time-delay neural networks

An architecture that is often used for sequential applications is shown in figure C1.2.19. It consists of a feedforward neural network that is fed by a delay line which stores past values of the input. In this case the sequential capabilities of the system do not come from the neural network itself, which is a plain feedforward one. They come, instead, from the delay line. An advantage of this structure is that it can be trained with standard backpropagation, since the neural network is feedforward. The disadvantages come from the facts that the architecture is not recursive and that its memory capabilities are fixed and cannot be adapted by training. For several kinds of problems, like those involving a long-time memory, this architecture may need many more weights (and therefore many more training patterns) than a recurrent one. Systems of this kind are often designated *time-delay neural networks* (TDNN). They have been applied to several kinds of problems. See Waibel (1989) for an example of an application to *speech recognition*, in which this architecture is extended by using delay lines at multiple levels, with multiple time resolutions.

F1.7.2

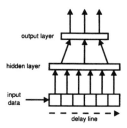

Figure C1.2.19. A time-delay neural network.

Acknowledgement

We wish to acknowledge the use of the 'United States Postal Service Office of Advanced Technology Handwritten ZIP Code Data Base (1987)', made available by the Office of Advanced Technology, United States Postal Service.

References

Almeida L B 1987 A learning rule for asynchronous perceptrons with feedback in a combinatorial environment *Proc. IEEE First Int. Conf. on Neural Networks* (New York: IEEE Press) pp 609–18

Battiti R 1992 First- and second-order methods for learning: between steepest descent and Newton's method *Neural Comput.* **4** 141–66

Becker S and Le Cun Y 1989 Improving the convergence of back-propagation learning with second order methods *Proc. 1988 Connectionist Models Summer School* ed D Touretzky, G Hinton and T Sejnowski (San Mateo, CA: Morgan Kaufmann) pp 29–37

Bishop C M 1990 Curvature-driven smoothing in backpropagation neural networks *Technical Report CLM-P-880* (Abingdon, UK: AEA Technology, Culham Laboratory)

Bryson A E and Ho Y C 1969 *Applied Optimal Control* (New York: Blaisdell)

Cruz C S, Rodriguez F, Dorronsoro J R and López V 1993 Nonlinear dynamical system modelling and its integration in intelligent control *Proc. Workshop on Integration in Real-Time Intelligent Control Systems* (Miraflores de la Sierra) pp 30-1 to 30-9

Cybenko G 1989 Approximation by superpositions of a sigmoidal function *Math. Control, Signal Syst.* **2** 303–14

Fahlman S E 1989 Fast-learning variations on back-propagation: an empirical study *Proc. 1988 Connectionist Models Summer School* ed D Touretzky, G Hinton and T Sejnowski (San Mateo, CA: Morgan Kaufmann) pp 38–51

Fahlman S E and Lebiere C 1990 The cascade-correlation learning architecture *Advances in Neural Information Processing Systems 2* ed D S Touretzky (San Mateo, CA: Morgan Kaufmann) pp 524–32

Frean M 1990 The upstart algorithm: a method for constructing and training feedforward neural networks *Neural Comput.* **2** 198–209

Funahashi K 1989 On the approximate realization of continuous mappings by neural networks *Neural Networks* **2** 183–92

Giles C L, Miller C B, Chen D, Sun G Z, Chen H H and Lee Y C 1992 Extracting and learning an unknown grammar with recurrent neural networks *Advances in Neural Information Processing Systems 4* ed J E Moody, S J Hanson and R P Lippmann (San Mateo, CA: Morgan Kaufmann) pp 317–24

Goldberg K Y and Pearlmutter B A 1989 Using backpropagation with temporal windows to learn the dynamics of the CMU direct-drive arm II *Advances in Neural Information Processing Systems 1* ed D S Touretzky (San Mateo, CA: Morgan Kaufmann) pp 356–65

Golub G H and Van Loan C F 1983 *Matrix Computations* (Baltimore, MD: Johns Hopkins University Press)

Guyon I, Vapnik V, Boser B, Bottou L and Solla S A 1992 Structural risk minimization for character recognition *Advances in Neural Information Processing Systems 4* ed J Moody, S J Hanson and Lippmann R P (San Mateo, CA: Morgan Kaufmann) pp 471–9

Hassibi B, Stork D G and Wolff G J 1993 Optimal brain surgeon and general network pruning *Proc. IEEE Int. Conf. on Neural Networks* (San Francisco, CA) pp 293–9

Hopfield J J 1984 Neurons with graded response have collective computational properties like those of two-state neurons *Proc. Natl Acad. Sci. USA 81* 3088–92

Hornik K, Sithcombe M and White H 1989 Multilayer feedforward networks are universal approximators *Neural Networks* **2** 359–66

Jacobs R 1988 Increased rates of convergence through learning rate adaptation *Neural Networks* **1** 295–307

Krogh A and Hertz J A 1992 A simple weight decay can improve generalization *Advances in Neural Information Processing Systems 4* ed J E Moody, S J Hanson and R P Lippmann (San Mateo, CA: Morgan Kaufmann) pp 950–7

Lapedes A S and Farber R 1987 Nonlinear signal processing using neural networks: prediction and system modelling *Technical Report LA-UR-87-2662* (Los Alamos, NM: Los Alamos National Laboratory)

Le Cun Y 1985 Une procédure d'apprentissage pour réseau à seuil assymétrique *Cognitiva 85* 599–604

Le Cun Y, Boser B, Denker J S, Henderson D, Howard R E, Hubbard W and Jackel L D 1990a Handwritten digit recognition with a backpropagation network *Advances in Neural Information Processing Systems 2* ed D S Touretzky (San Mateo, CA: Morgan Kaufmann) pp 396–409

Le Cun Y, Denker J S and Solla S 1990b Optimal brain damage *Advances in Neural Information Processing Systems 2* ed D S Touretzky (San Mateo, CA: Morgan Kaufmann) pp 598–605

Le Cun Y, Kanter I and Solla S 1991 Second order properties of error surfaces: learning time and generalization *Advances in Neural Information Processing Systems 3* ed R P Lippmann, J E Moody and D S Touretzky (San Mateo, CA: Morgan Kaufmann) pp 918–24

Ljung L 1978 Strong convergence of a stochastic approximation algorithm *Ann. Statistics* **6** 680–96

López V 1994 Private communication

MacKay D J 1992a Bayesian interpolation *Neural Comput.* **4** 415–47

MacKay D J 1992b A practical bayesian framework for backprop networks *Neural Comput.* **4** 448–72

Matan O, Burges C J, Le Cun Y and Denker J S 1992 Multi-digit recognition using a space displacement neural network *Advances in Neural Information Processing Systems 4* ed J E Moody, S J Hanson and R P Lippmann (San Mateo, CA: Morgan Kaufmann) pp 488–95

Mézard M and Nadal J P 1989 Learning in feedforward layered networks: the tiling algorithm *J. Phys. A: Math. Gen.* **22** 2191–204

Minsky M L and Papert S A 1969 *Perceptrons* (Cambridge, MA: MIT Press)

Moller M F 1990 A scaled conjugated gradient algorithm for fast supervised learning *Preprint PB-339* (Aarhus, Denmark: Computer Science Department, University of Aarhus)

Mozer M C and Smolensky P 1989 Skeletonization: a technique for trimming the fat from a network via relevance assignment *Report CU-CS-421-89* (Boulder, CO: Department of Computer Science, University of Colorado)

Nguyen D and Widrow B 1990 The truck backer-upper: an example of self-learning in neural networks *Advanced Neural Computers* ed R Eckmiller (Amsterdam: Elsevier) pp 11–20

Oppenheim A V and Schafer R W 1975 *Digital Signal Processing* (Englewood Cliffs, NJ: Prentice-Hall)

Parker D B 1985 Learning logic *Technical Report TR-47* (Cambridge, MA: Center for Computational Research in Economics and Management Science, MIT)

Pineda F J 1987 Generalization of backpropagation to recurrent neural networks *Phys. Rev. Lett.* **59** 2229–32

Pearlmutter B A 1989 Learning state space trajectories in recurrent neural networks *Neural Comput.* **1** 263–9

Pomerleau D A 1991 Efficient training of artificial neural networks for autonomous navigation *Neural Comput.* **3** 89–97

Pomerleau D A 1993 Input reconstruction reliability estimation *Advances in Neural Information Processing Systems 5* ed S J Hanson, J D Cowan and C L Giles (San Mateo, CA: Morgan Kaufmann) pp 279–86

Press W H, Flannery B P, Teukolsky S A and Vetterling W T 1986 *Numerical Recipes* (Cambridge: Cambridge University Press)

Ramos H S, Langlois T, Xufre G, Amaral J D, Almeida L B and Silva F M 1994 Neural networks in industrial modeling and fault detection Proc. *Workshop on Artificial Intelligence in Real-Time Control (Valencia)*

Richard M D and Lippmann R P 1991 Neural network classifiers estimate Bayesian *a posteriori* probabilities *Neural Comput.* **3** 461–83

Robinson A J and Fallside F 1987 The utility driven dynamic error propagation network *Technical Report CUED/F-INFENG/TR.1* (Cambridge, UK: Cambridge University Engineering Department)

Robinson A J *et al* 1993 A neural network based, speaker independent, large vocabulary, continuous speech recognition system: the Wernicke project *Proc. Eurospeech'93 Conf. (Berlin)* pp 1941–4

Rumelhart D E, Hinton G E and Williams R J 1986 Learning internal representations by error propagation *Parallel Distributed Processing: Explorations in the Microstructure of Cognition* vol 1 ed D E Rumelhart, J L McClelland and the PDP research group (Cambridge, MA: MIT Press) pp 318–62

Silva F M and Almeida L B 1990a Acceleration techniques for the backpropagation algorithm *Neural Networks* ed L B Almeida and C J Wellekens (Berlin: Springer) pp 110–19

Silva F M and Almeida L B 1990b Speeding up backpropagation *Advanced Neural Computers* ed R Eckmiller (Amsterdam: Elsevier) pp 151–60

Silva F M and Almeida L B 1991 Speeding-Up backpropagation by data orthonormalization *Artificial Neural Networks* vol 2, ed T Kohonen, K Mäkisara, O Simula and J Kangas (Amsterdam: Elsevier) pp 149–56

Sun G Z, Chen H H and Lee Y C 1992 Green's function method for fast on-line learning algorithm of recurrent neural networks *Advances in Neural Information Processing Systems 4* ed J E Moody, S J Hanson and R P Lippmann (San Mateo, CA: Morgan Kaufmann) pp 333–40

Thompson J M and Stewart H B 1986 *Nonlinear Dynamics and Chaos* (Chichester: Wiley)

Tollenaere T 1990 SuperSAB: fast adaptive back propagation with good scaling properties *Neural Networks* **3** 561–74

Trippi R R and Turban E (eds) 1993 *Neural Networks in Finance and Investing* (Chicago, IL: Probus)

Waibel A 1989 Modular construction of time-delay neural networks for speech recognition *Neural Comput.* **1** 39–46

Watrous R L 1987 Learning phonetic features using connectionist networks: an experiment in speech recognition *Proc. IEEE 1st International Conf. on Neural Networks* (New York: IEEE Press) pp 381–7

Weigend A S, Rumelhart D E and Huberman B A 1991 Generalization by weight-elimination with application to forecasting *Advances in Neural Information Processing Systems 3* ed R P Lippmann, J E Moody and D S Touretzky (San Mateo, CA: Morgan Kaufmann) pp 875–82

Werbos P J 1974 Beyond regression: new tools for prediction and analysis in the behavioral sciences *PhD Thesis* (Cambridge, MA: Harvard University)

White D A and Sage D A (eds) 1992 *Handbook of Intelligent Control: Neural, Fuzzy and Adaptive Approaches* (New York: Van Nostrand Reinhold)

Widrow B and Stearns S D 1985 *Adaptive Signal Processing* (Englewood Cliffs, NJ: Prentice-Hall)

Willems J L 1970 *Stability Theory of Dynamical Systems* (London: Thomas Nelson)

Williams P M 1994 Bayesian regularization and pruning using a Laplace prior *Cognitive Science Research Paper CSRP-312* (Brighton: School of Cognitive and Computing Sciences, University of Sussex)

Williams R J and Zipser D 1989 A learning algorithm for continually running fully recurrent neural networks *Neural Comput.* **1** 270–80

Yuan J L and Fine T L 1993 Forecasting demand for electric power *Advances in Neural Information Processing Systems 5* ed S J Hanson, J D Cowan and C L Giles (San Mateo, CA: Morgan Kaufmann) pp 739–46

C1.3 Associative memory networks

Mohamad H Hassoun and Paul B Watta

Abstract

One of the most extensively analyzed classes of artificial neural networks is the class of associative networks or associative neural memories. These memory models can be classified in various ways depending on their architecture (static versus recurrent), their retrieval mode (synchronous versus asynchronous), the nature of the stored associations (autoassociative versus heteroassociative), the complexity and capability of the memory storage/recording algorithm, and so on. This section discusses various architectures and recording algorithms for the storage and retrieval of information in neural memories with emphasis on dynamic (recurrent) associative memory (DAM) architectures. The Hopfield model and the bidirectional associative memory are discussed in detail, and criteria for high-performance dynamic memories are outlined for the purpose of comparing the various models.

C1.3.1 Feedback models: associative memory networks

C1.3.1.1 Introduction

One of the most extensively analyzed classes of artificial neural networks is the class of associative networks or *associative neural memories* (ANMs). In fact, the neural network literature over the last two decades abounds with papers of proposed associative neural memory models (e.g. Amari 1972a, b, Anderson 1972, Nakano 1972, Kohonen 1972 and 1974, Kohonen and Ruohonen 1973, Hopfield 1982, Kosko 1987, Okajima *et al* 1987, Kanerva 1988, Chiueh and Goodman 1988, Baird 1990). For an accessible reference on various associative neural memory models the reader is referred to the edited volume by Hassoun (1993). These memory models can be classified in various ways depending on their *architecture* (static versus recurrent), their retrieval mode (synchronous versus asynchronous), the nature of the stored *associations* (autoassociative versus heteroassociative), the complexity and capability of the memory storage/recording algorithm, and so on.

This section discusses various architectures and learning algorithms for the storage and retrieval of information in neural memories with emphasis on dynamic (recurrent) associative memory (DAM) architectures. These dynamic, or feedback, models arise when *recurrent connections* are made between the input and output lines of the network. Analytically, feedback models are treated as nonlinear dynamical systems. From this perspective, information retrieval is viewed as a process whereby the state of the system evolves from an initial state representing a noisy or partial input pattern (key) to a stationary state which represents the stored or retrieved information. With this dynamic model of associative memory, it is crucial that the system exhibit asymptotically stable behavior.

The remainder of this section is organized as follows. First, some fundamental concepts, definitions and terminology of associative memories are introduced. Then, it is shown how artificial neural networks may be used to act as associative memories by constructing both *feedforward* (static) and *feedback* (dynamic) neural architectures. Criteria for high-performance dynamic memories are outlined for the purpose of comparing the various models. Static models are discussed in order to introduce some of the commonly used recording recipes. Finally, dynamic models, including the Hopfield model and the bidirectional associative memory, are discussed in detail.

C1.3.2 Fundamental concepts and definitions

C1.3.2.1 Statement of the associative memory problem

Associative memory may be formulated as an input–output system, as shown schematically in figure C1.3.1. Here, the input to the system is an n-dimensional vector $x \in R^n$ called the *memory key*, and the output is an L-dimensional vector $y \in R^L$ called the *retrieved pattern*. The relation between the memory key and the retrieved pattern is given by $y = G(x)$, where $G : R^n \to R^L$ is the *associative mapping* of the memory. Each input–output pair or *memory association* (x, y) is said to be *stored* or *recorded* in the memory.

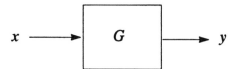

Figure C1.3.1. A block diagram representation of the operation of an associative memory.

The associative memory design problem may be formulated mathematically as follows. Given a finite set of desired memory associations $\{(x^k, y^k) : k = 1, 2, \ldots, m\}$, the first task is to determine an associative mapping which captures these associations as input–output pairs; that is, we are required to determine a function G which satisfies

$$y^k = G(x^k) \qquad \text{for all } k = 1, 2, \ldots, m. \tag{C1.3.1}$$

Recalling that G is a function of the form $G : R^n \to R^L$, equation (C1.3.1) is not the end of the story because it only specifies the value of G at k points in R^n; the question is: where does G map all the remaining vectors? This leads to the second task of associative memory design: here, we require G to not only store the given associations, but also provide *noise tolerance* and *error correction* capabilities. In this case, for each noisy† version $\tilde{x}^k$ of x^k, we require the memory to retrieve the uncorrupted output, that is, we require $y^k = G(\tilde{x}^k)$.

The given set of associations $\{(x^k, y^k)\}$ is called the *fundamental memory set* and each association (x^k, y^k) in the fundamental set is called a *fundamental memory*. A special case of the above problem arises when the fundamental memory set is of the form $\{(x^k, x^k) : k = 1, 2, \ldots, m\}$. In this case, the memory is required to store the *autoassociations* $\{(x^k, x^k)\}$ and is said to be an *autoassociative memory*. In general, though, when the output y^k is different from the input x^k, the memory is said to be *heteroassociative*.

The process of designing an associative memory is called the *recording phase*. As discussed above, the recording phase consists of determining or *synthesizing* an associative mapping G which provides for (i) storage of the fundamental memory set and (ii) error correction. Given a fundamental memory set, an algorithm that specifies how G is to be synthesized is called a *recording recipe*. It is usually the case that the complexity of a recording recipe is related to the quality of the resulting associative mapping. In particular, simple recording recipes tend to produce associative memories which exhibit poor performance in the sense that the memory fails to fully capture the fundamental memory set and/or provides very limited error correction. One of the most common performance problems associated with simple recording algorithms is the creation of a large number of spurious or false memories. A *spurious memory* is a memory association that is unintentionally stored in the memory, that is, a memory association which was not part of the fundamental memory set.

Once recording is complete, the memory is ready for operation, which is called the *retrieval phase*. Here, the memory may be tested to verify that the fundamental memories are properly stored, and the error correction capability of the memory may be measured by corrupting each fundamental memory key with various amounts of noise and observing the resulting output.

C1.3.2.2 Neural network architectures for associative memories

In the neural network approach to associative memory design, a network of artificial neurons is used to realize the desired associative mapping G. Figure C1.3.2(*a*) shows the architecture for a *static* or

---

† The type of noise depends on the application. For example, if the x^k are binary patterns, noise could be measured in terms of bit errors. On the other hand, if the x^k are real-valued, then the noise may appear as additive Gaussian noise.

feedforward associative neural memory. This network consists of L noninteracting neurons. The output of the lth neuron y_l is given by

$$y_l = f_l\left(\sum_{i=1}^{n} w_{li} x_i\right)$$

where $f_l : R \to R$ is the *activation function* and $w_l = (w_{l1}, w_{l2}, \ldots, w_{ln})$ are the *weights* associated with the lth neuron. Usually, each neuron in the network uses an identical activation, which is typically a linear, sigmoidal, or threshold function. Figure C1.3.2(*b*) shows a block diagram description of the network. Here, the weight vectors are collected in an $L \times n$ *weight* or *interconnection matrix* $\mathbf{W} = (w_{li})$, where w_{li} is the synaptic weight connecting the ith input to the lth neuron. Similarly, the activation functions are collected as a vector mapping $F(\bullet) = (f_1(\bullet), f_2(\bullet), \ldots, f_L(\bullet))$. The associative mapping implemented by this feedforward network may be expressed as $y = G(x) = F(\mathbf{W}x)$. Note that in the autoassociative case, there are n inputs and n output units, hence the weight matrix is a square $n \times n$ matrix.

B3.2.4

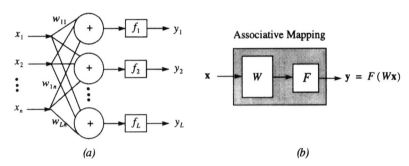

(a)   *(b)*

Figure C1.3.2. (*a*) The architecture of a static neural network for heteroassociative memory. (*b*) A block diagram representation of the neural network.

Although simple, the feedforward architecture can usually provide only limited error correction capability. More powerful architectures can be constructed by including feedback or recurrent connections. To see why feedback improves error correction, consider an autoassociative version of the single-layer associative memory employing units with the sign-activation function. Now assume that this memory is capable of associative retrieval of a set of m bipolar binary memories $\{x^k\}$. Upon the presentation of a key $\tilde{x}^k$, which is a noisy version of one of the stored memory vectors x^k, the associative memory retrieves (in a single pass) an output y which is closer to the stored memory x^k than $\tilde{x}^k$. In general, only a fraction of the noise (error) in the input vector is corrected in the first pass (presentation). Intuitively, we may proceed by taking the output y and feeding it back as an input to the associative memory, hoping that a second pass would eliminate more of the input noise. This process could continue with more passes until we eliminate all errors and arrive at a final output y equal to x^k.

Note that with feedback connections, care must be taken to distinguish between autoassociative and heteroassociative operation. Block diagrams for both the autoassociative and heteroassociative architectures are shown in figures C1.3.3(*a*) and (*b*), respectively. In both cases, memory retrieval may be viewed as a temporal process and described by a system of difference (assuming a discrete-time system) or differential (assuming a continuous-time system) equations.

The dynamics of a (discrete-time) *dynamic autoassociative memory* (DAM) corresponding to figure C1.3.3(*a*) may be described by the system equation

$$x(t + 1) = F(\mathbf{W}[x(t)]) \qquad t = 0, 1, 2, 3, \ldots. \tag{C1.3.2}$$

The actual interpretation of equation (C1.3.2) depends on the type of updating chosen. The two most common updating modes for such a system are called synchronous and sequential. In *synchronous updating*, all states are updated simultaneously at each time instant. In *sequential updating*, only one (randomly chosen) state is updated at each time instant.

B3.4.3

B3.4.3

The dynamic autoassociative memory operates as follows: given a memory key x, the dynamical system of equation (C1.3.2) is iterated starting from the initial state $x(0) = x$, until the dynamics converge to some stationary state which is then taken to be the retrieved pattern, that is,

$$y = G(x) = \lim_{t \to \infty} x(t).$$

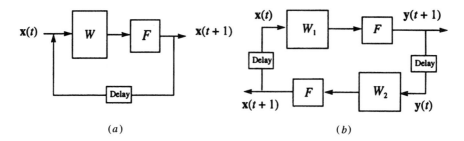

Figure C1.3.3. (*a*) Architecture for a dynamic autoassociative memory and (*b*) dynamic heteroassociative memory.

The above description of the associative mapping of the DAM makes sense only in the case when equation (C1.3.2) represents a stable dynamical system. In the case of an unstable, oscillatory or chaotic system, the limit $\lim_{t \to \infty} x(t)$ may not exist, and hence for certain memory keys (initial states) the memory may not produce a retrieval. This type of open-ended† behavior can be avoided by insisting that the dynamic memory represents a stable dynamical system. The most optimal DAM consists of a state space with n attractors, corresponding to the fundamental memories to be stored.

The architecture for a heteroassociative dynamic associative neural memory (HDAM) is shown in figure C1.3.3(*b*). This system operates similarly to the autoassociative memory, but is described by two sets of equations

$$y(t + 1) = F[\mathbf{W}_1 x(t)] \tag{C1.3.3}$$

$$x(t + 1) = F[\mathbf{W}_2 y(t)]. \tag{C1.3.4}$$

Here, F is usually the sgn‡ activation operator. Similarly to the autoassociative case, it can be operated in the parallel (synchronous) or serial (asynchronous) version, where one and only one unit updates its state at a given time. The stability analysis of this type of network is generally more difficult than for the single-layer feedback network.

C1.3.2.3 Characteristics of high-performance DAMs

In Hassoun (1993), a set of desirable performance characteristics for the class of dynamic associative neural memories is given. Figures C1.3.4(*a*) and (*b*) present conceptual diagrams of the state space for high- and low-performance DAMs, respectively (Hassoun 1993, 1995).

The high-performance DAM in figure C1.3.4(*a*) has large basins of attraction around all fundamental memories. It has a relatively small number of spurious memories, and each spurious memory has a very small basin of attraction. This DAM is stable in the sense that it exhibits no oscillations. The shaded background in this figure represents the region of state space for which the DAM converges to a unique ground state (e.g. zero state). This ground state acts as a default 'no decision' attractor state where unfamiliar or highly corrupted initial states converge.

A low-performance DAM has one or more of the characteristics depicted conceptually in figure C1.3.4(*b*). It is characterized by its inability to store all desired memories as fixed points; those memories which are stored successfully end up having small basins of attraction. The number of spurious memories is very high for such a DAM, and they have relatively large basins of attraction. This low-performance DAM may also exhibit oscillations. Here, an initial state close to one of the stored memories has a significant chance of converging to a spurious memory or to a limit cycle.

To summarize, high-performance DAMs must have the following characteristics: (1) high capacity, (2) tolerance to noisy and partial inputs (this implies that fundamental memories have large basins of attraction); (3) the existence of relatively few spurious memories and few or no limit cycles with negligible basin of attraction; (4) provision for a 'no decision' default memory/state (inputs with very low 'signal-to-noise' ratios are mapped, with high probability, to this default memory), and (5) fast memory retrievals.

† As an analogy, consider the frustrating scenario of asking someone a question and patiently listening to a long-winded response, only to find out that the person cannot answer your question after all! On the other hand, some researchers have advocated the notion that oscillatory and chaotic neural systems are more closely related to the processing of natural biological systems; see Hirsch (1989) for a concise summary of this discussion.
‡ The sgn activation is defined as $\mathrm{sgn}(x) = -1$ for all $x < 0$, and $\mathrm{sgn}(x) = 1$ for all $x \geq 0$.

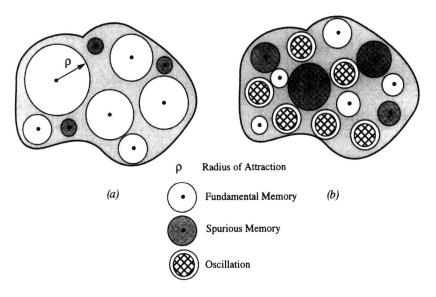

Figure C1.3.4. A conceptual diagram comparing the state space of (*a*) high-performance and (*b*) low-performance autoassociative DAMs.

This list of high-performance DAM characteristics can act as performance criteria for comparing various DAM architectures and/or DAM recording recipes.

C1.3.3 Static models and simple recording recipes

C1.3.3.1 The LAM model and correlation recording

One of the earliest associative neural memory models is the *linear associative memory* (LAM), also called *correlation memory* (Anderson 1972, Kohonen 1972, Nakano 1972). For this memory, given an input key vector $x \in R^n$, the retrieved or output pattern $y \in R^L$ is computed by the simple linear relation

$$y = \mathbf{W}x \tag{C1.3.5}$$

where $\mathbf{W}$ is the $L \times n$ weight or interconnection matrix. The architecture for this network is given in figure C1.3.2(*a*) with linear (identity mapping) activation functions for each neuron. Note the simplicity of this associative mapping—it is characterized by a simple matrix–vector multiplication. Hence, it is referred to as a linear associative memory (LAM).

Having constructed an architecture for a simple neural memory, the question now is: how does one record the memory set $\{x^k, y^k\}$ into this LAM architecture? More specifically, how do we determine or synthesize an appropriate weight matrix $\mathbf{W}$ such that $y^k = \mathbf{W}x^k$ for all $k = 1, 2, \ldots, m$? The correlation memory is a simple recording/storage recipe whereby $\mathbf{W}$ is given by the following outer product rule:

$$\mathbf{W} = \sum_{k=1}^{m} y^k (x^k)^{\mathrm{T}} . \tag{C1.3.6}$$

In other words, the interconnection matrix $\mathbf{W}$ is simply the correlation matrix of m association pairs. Another way of expressing equation (C1.3.6) is

$$\mathbf{W} = \mathbf{Y}\mathbf{X}^{\mathrm{T}} \tag{C1.3.7}$$

where $\mathbf{Y} = [y^1, y^2, \ldots, y^m]$ and $\mathbf{X} = [x^1, x^2, \ldots, x^m]$. Note that for the autoassociative case where the set of association pairs (x^k, x^k) is to be stored, one may still employ equation (C1.3.6) or (C1.3.7) with y^k replaced by x^k.

This recording recipe is simple enough, but how well does it work? That is, what are the requirements on the $\{x^k, y^k\}$ associations which will guarantee the successful retrieval of all recorded vectors (memories) from their associated 'perfect key' x^k? Substituting equation (C1.3.6) into (C1.3.5) and assuming that the key x^h is one of the x^k vectors, we get an expression for the retrieved pattern as

$$\tilde{y}^h = \left[\sum_{k=1}^{m} y^k (x^k)^\mathrm{T} \right] x^h = \|x^h\| y^h + \sum_{k \neq h}^{m} y^k (x^k)^\mathrm{T} x^h \,. \tag{C1.3.8}$$

The second term on the right-hand side of equation (C1.3.8) represents the *cross-talk* between the key x^h and the remaining $m - 1$ patterns x^k. This term can be reduced to zero if the x^k vectors are orthogonal†. The first term on the right-hand side of equation (C1.3.8) is proportional to the desired memory y^h, with a proportionality constant equal to the square of the norm of the key vector x^h. Hence, a sufficient condition for the retrieved memory to be the desired perfect recollection is to have orthonormal‡ vectors x^k, independent of the encoding of the y^k (note, though, how the y^k affects the cross-talk term if the x^k are not orthogonal).

An appealing feature of correlation recording is the relative ease with which memory associations may be added or deleted. For example, if after recording the m associations (x^1, y^1) through (x^m, y^m) it is desired to record one additional association (x^{m+1}, y^{m+1}), then one simply updates the current $\mathbf{W}$ by adding to it the matrix $y^{m+1} (x^{m+1})^\mathrm{T}$. Similarly, an already recorded association (x^i, y^i) may be 'erased' by simply subtracting $y^i (x^i)^\mathrm{T}$ from $\mathbf{W}$.

C1.3.3.2 A simple nonlinear associative memory model

In the case of binary-valued associations $x^k \in \{-1, 1\}^n$ and $y^k \in \{-1, 1\}^L$, a simple nonlinear memory may be constructed by using threshold activations. In this case, F is a clipping nonlinearity operating componentwise on the vector $\mathbf{W}x$ (i.e. each unit now employs a sgn or sign-activation function) according to

$$y = F(\mathbf{W}x) \,. \tag{C1.3.9}$$

The advantage of this nonlinear memory is that some of the constraints imposed by correlation recording of a LAM for perfect retrieval can be relaxed. That is, we require only that the sign of the corresponding components of y^k and $\mathbf{W}x^k$ agree. For this nonlinear memory, it is more convenient to use the *normalized correlation recording* given by

$$\mathbf{W} = \frac{1}{n} \sum_{k=1}^{m} y^k (x^k)^\mathrm{T} \tag{C1.3.10}$$

which automatically normalizes the x^k vectors (note that the square of the norm of an n-dimensional bipolar binary vector is n). Now, suppose that one of the recorded key patterns x^h is presented as input, then the retrieved pattern $\hat{y}^h$ can be written as

$$\hat{y}^h = F\left[y^h + \frac{1}{n} \sum_{k \neq h}^{m} y^k (x^k)^\mathrm{T} x^h \right] = F\left[y^h + \Delta^h \right] \tag{C1.3.11}$$

where Δ^h represents the cross-talk term. For the ith component of $\hat{y}^h$, equation (C1.3.11) gives

$$\hat{y}_i^h = \mathrm{sgn}\left[y_i^h + \frac{1}{n} \sum_{j=1}^{n} \sum_{k \neq h}^{m} y_i^k x_j^k x_j^h \right] = \mathrm{sgn}\left[y_i^h + \Delta_i^h \right]$$

from which it can be seen that the condition for perfect recall is given by the requirements

$$\Delta_i^h > -1 \qquad \text{for } y_i^h = 1$$

and

$$\Delta_i^h < 1 \qquad \text{for } y_i^h = -1$$

for $i = 1, 2, \ldots, L$. These requirements are less restrictive than the orthonormality requirement of the x^k in a LAM. The error correction capability of the above nonlinear correlation associative memory has been analyzed by Uesaka and Ozeki (1972) and later Amari (1977, 1990) (see also Amari and Yanai 1993).

† A set of vectors $\{q_1, \ldots, q_p\}$ is said to be *orthogonal* if $q_i^\mathrm{T} q_j = 0$ for each $i \neq j = 1, 2, \ldots, p$.
‡ A set of vectors $\{q_1, \ldots, q_p\}$ is said to be *orthonormal* if it is orthogonal and if $q_i^\mathrm{T} q_i = 1$ for all $i = 1, 2, \ldots, p$.

C1.3.3.3 The OLAM model and projection recording

It is possible to derive another recording technique which guarantees perfect retrieval of stored memories as long as the set $\{x^k : k = 1, 2, \ldots, m\}$ is linearly independent. Such a learning rule is desirable since linear independence is a less stringent requirement than orthonormality. This recording technique used in conjunction with the LAM architecture (linear network of neurons) is called the *optimal linear associative memory* (OLAM) (Kohonen and Ruohonen 1973).

For perfect storage of *m fundamental associations* $\{x^k, y^k\}$, a LAMs interconnection matrix $\mathbf{W}$ must satisfy the matrix equation

$$\mathbf{Y} = \mathbf{W}\mathbf{X} \tag{C1.3.12}$$

where $\mathbf{X}$ and $\mathbf{Y}$ are as defined earlier in this section. This equation always has at least one solution if all *m* vectors x^k (columns of $\mathbf{X}$) are linearly independent, which necessitates that *m* must be less than or equal to *n*. For the case $m = n$, the matrix $\mathbf{X}$ is square and a unique solution for $\mathbf{W}$ in equation (C1.3.12) may be computed:

$$\mathbf{W}^* = \mathbf{Y}\mathbf{X}^{-1}. \tag{C1.3.13}$$

Here, we require that the matrix inverse $\mathbf{X}^{-1}$ exists, which can be guaranteed when the set $\{x^k\}$ is linearly independent. Thus, this solution guarantees the perfect recall of any y^k upon the presentation of its associated key x^k.

Returning to equation (C1.3.12) with the assumption that $m < n$ and the x^k are linearly independent, it can be seen that an exact solution $\mathbf{W}^*$ is not unique. In this case, we are free to choose any of the $\mathbf{W}^*$ solutions satisfying equation (C1.3.12). In particular, the minimum Euclidean norm solution (Rao and Mitra 1971):

$$\mathbf{W}^* = \mathbf{Y}(\mathbf{X}^{\mathrm{T}}\mathbf{X})^{-1}\mathbf{X}^{\mathrm{T}} \tag{C1.3.14}$$

is desirable since it leads to the best error-tolerant (optimal) LAM (Kohonen 1984). Equation (C1.3.14) will be referred to as the *projection recording recipe* since the matrix–vector product $(\mathbf{X}^{\mathrm{T}}\mathbf{X})^{-1}\mathbf{X}^{\mathrm{T}}x^k$ transforms the *k*th stored vector x^k into the *k*th column of the $m \times m$ identity matrix. Note that if the set $\{x^k\}$ is orthonormal, then $\mathbf{X}^{\mathrm{T}}\mathbf{X} = \mathbf{I}$ and equation (C1.3.14) reduces to the correlation recording recipe of equation (C1.3.7). An iterative version of the projection recording recipe exists (Kohonen 1984). This iterative method is convenient since a new association can be learned (or an old association can be deleted) in a single update step without involving other earlier-learned memories. Other adaptive versions of equation (C1.3.14) can be found in Hassoun (1993, 1995).

The error correcting capabilities of OLAMs have been analyzed by Kohonen (1984) and Casasent and Telfer (1987), among others, for the case of real-valued associations, and by Amari (1977) and Stiles and Denq (1987) for the case of bipolar binary key/recollection vectors.

C1.3.4 Dynamic models: the autoassociative case

C1.3.4.1 The Hopfield model

Consider the nonlinear active electronic circuit shown in figure C1.3.5. In this circuit, each ideal amplifier provides an output voltage given by $x_i = f(u_i)$, where u_i is the input voltage and f is a nonlinear activation function. Each amplifier is also assumed to provide an inverting terminal for producing the output $-x_i$. The resistor R_{ij} connects the output voltage x_j (or $-x_j$) of the *j*th amplifier to the input of the *i*th amplifier. Since, as will be seen later, the conductances R_{ij}^{-1} play the role of interconnection weights, positive as well as 'negative' resistors are required. Connecting a resistor R_{ij} to $-x_i$ helps avoid the complication of actually realizing negative resistive elements in the circuit. The R and C are positive quantities and are assumed equal for all *n* amplifiers. Finally, the current I_i represents an external input signal (or bias) to the *i*th amplifier.

The circuit in figure C1.3.5 is known as the *Hopfield network*, and can be thought of as a single-layer, continuous-time feedback network. The dynamical equations describing the evolution of the *i*th state x_i, $i = 1, 2, \ldots, n$, in the Hopfield network can be derived by applying Kirchhoff's current law to the input node of the *i*th amplifier. After rearranging terms, the *i*th nodal equation can be written as   B1.3

$$C\frac{\mathrm{d}u_i}{\mathrm{d}t} = -\alpha_i u_i + \sum_{j=1}^{n} f(u_j) - \frac{u_i}{R} + I_i \tag{C1.3.15}$$

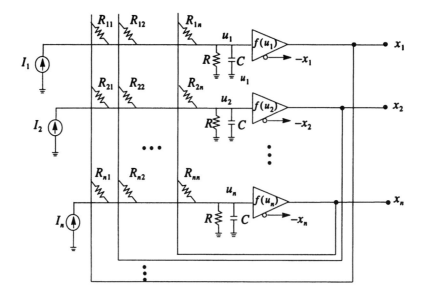

Figure C1.3.5. Circuit diagram for an electronic dynamic associative memory.

where $\alpha_i = \sum_{j=1}^{n} \frac{1}{R_{ij}} + \frac{1}{R}$ and $w_{ij} = \frac{1}{R_{ij}}$ (or $w_{ij} = -\frac{1}{R_{ij}}$ if the inverting output of unit j is connected to unit i). The above Hopfield network can be considered as a special case of a more general dynamical network developed and studied by Cohen and Grossberg (1983) which has ith state dynamics expressed by

$$\tau_i \frac{du_i}{dt} = -\alpha_i u_i + (\gamma_i - \beta_i u_i)\left[\sum_{j=1}^{n} f(u_j) + I_i\right]. \tag{C1.3.16}$$

Using vector notation, the dynamics of the Hopfield network can be described in compact form as

$$C\frac{du}{dt} = -\alpha u + \mathbf{W}x + \theta \tag{C1.3.17}$$

where $\mathbf{C} = C\mathbf{I}$ ($\mathbf{I}$ is the $n \times n$ identity matrix), $\alpha = \text{diag}(\alpha_1, \alpha_2, \ldots, \alpha_n)$, $x = F(u) = [f(u_1), f(u_2), \ldots, f(u_n)]^{\mathrm{T}}$, $\theta = [I_1, I_2, \ldots, I_n]^{\mathrm{T}}$ and $\mathbf{W}$ is an interconnection matrix defined as

$$\mathbf{W} = \begin{bmatrix} w_{11} & w_{12} & \cdots & w_{1n} \\ w_{21} & w_{22} & \cdots & w_{2n} \\ \vdots & \vdots & \ddots & \vdots \\ w_{n1} & w_{n2} & \cdots & w_{nn} \end{bmatrix}.$$

The equilibria of the dynamics in equation (C1.3.17) are determined by setting $\frac{du}{dt} = 0$, giving

$$\alpha u = \mathbf{W}x + \theta = \mathbf{W}F(u) + \theta. \tag{C1.3.18}$$

It can be shown (Hopfield 1984) that the Hopfield network is stable if (i) the interconnection matrix $\mathbf{W}$ is symmetric, and (ii) the activation function f is smooth and monotonically increasing. Furthermore, Hopfield showed that the stable states of the network are the local minima of the bounded computational energy function (Lyapunov function)

$$E(x) = -\tfrac{1}{2}x^{\mathrm{T}}\mathbf{W}x - x^{\mathrm{T}}\theta + \sum_{j=1}^{n} \int_{0}^{x_j} f^{-1}(\xi)\, d\xi \tag{C1.3.19}$$

where $x = [x_1, x_2, \ldots, x_n]^{\mathrm{T}}$ is the network's output state, and $f^{-1}(x_j)$ is the inverse of the activation function $x_j = f(u_j)$. Note that the value of the right-most term in equation (C1.3.19) depends on the specific shape of the nonlinear activation function f. For high gain approaching infinity, $f(u_j)$ approaches the sign function, that is, the amplifiers in the Hopfield network become threshold elements. In this case, the computational energy function becomes approximately the quadratic function

$$E(x) = -\tfrac{1}{2}x^{\mathrm{T}}\mathbf{W}x - x^{\mathrm{T}}\theta. \tag{C1.3.20}$$

It has been shown (Hopfield 1984) that the only stable states of the high-gain, continuous-time, continuous-state system in equation (C1.3.17) are the corners of the hypercube, i.e. the local minima of equation (C1.3.20) are states $x^* \in \{-1, 1\}^n$. For large but finite amplifier gains, the third term in equation (C1.3.19) begins to contribute. The sigmoidal nature of $f(u)$ leads to a large positive contribution near hypercube boundaries, but a negligible contribution far from the boundaries. This causes a slight drift of the stable states toward the interior of the hypercube.

Another way of looking at the Hopfield network is as a gradient system which searches for local minima of the energy function $E(x)$ defined in equation (C1.3.19). To see this, simply take the gradient of E with respect to the state x and compare with equation (C1.3.17). Hence, by equating terms, we have the following gradient system:

$$\frac{\mathrm{d}u}{\mathrm{d}t} = -\rho\nabla E(x) \tag{C1.3.21}$$

where $\rho = \mathrm{diag}(1/C, 1/C, \ldots, 1/C)$. The gradient system in equation (C1.3.21) converges asymptotically to an equilibrium state which is a local minimum or a saddlepoint of the energy E (Hirsch and Smale 1974) (fortunately, the unavoidable noise in practical applications prevents the system from staying at the saddlepoints and convergence to a local minimum is achieved). To see this, we first note that the equilibria of the system described by equation (C1.3.21) correspond to local minima (or maxima or points of inflection) of $E(x)$, since $\mathrm{d}u/\mathrm{d}t = 0$ means that $\nabla E(x) = 0$. For each isolated local minimum x^*, there exists an open neighborhood over which the candidate function $V(x) = E(x) - E(x^*)$ has continuous first partial derivatives and is strictly positive except at x^* where $V(x) = 0$. Additionally,

$$\frac{\mathrm{d}V}{\mathrm{d}t} = \frac{\mathrm{d}E}{\mathrm{d}t} = \nabla E(x)^{\mathrm{T}}\dot{x}(t) = \sum_{j=1}^{n}\frac{\partial E}{\partial x_j}\frac{\mathrm{d}x_j}{\mathrm{d}t} = -C\sum_{j=1}^{n}\frac{\mathrm{d}u_j}{\mathrm{d}t}\frac{\mathrm{d}x_j}{\mathrm{d}t} = -C\sum_{j=1}^{n}\frac{\mathrm{d}x_j}{\mathrm{d}u_j}\left(\frac{\mathrm{d}u_j}{\mathrm{d}t}\right)^2 \tag{C1.3.22}$$

is always negative since $\mathrm{d}x_j/\mathrm{d}u_j$ is always positive, because of the monotonically nondecreasing nature of the relation $x_j = f(u_j)$, or zero at x^*. Hence V is a Lyapunov function, and x^* is asymptotically stable.

The operation of the Hopfield network as an autoassociative memory is straightforward; given a set of memories $\{x^k\}$, the interconnection matrix $\mathbf{W}$ is encoded such that the states x^k become local minima of the Hopfield network's energy function $E(x)$. Then, when the network is initialized with a noisy key $\hat{x}$, its output state evolves along the negative gradient of $E(x)$ until it reaches the closest local minimum which, hopefully, is one of the fundamental memories x^k. In general, however, $E(x)$ will have additional local minima other than the desired ones encoded in $\mathbf{W}$. These additional undesirable stable states represent spurious memories.

When used as a DAM, the Hopfield network is usually operated with very high activation function gain. In this case, the Hopfield memory stores binary-valued associations. The synthesis of $\mathbf{W}$ can be done according to the correlation recording recipe or the more optimal projection recipe. These recording recipes lead to symmetrical $\mathbf{W}$ (since autoassociative operation is assumed, that is, $y^k = x^k$ for all k) which guarantees the stability of retrievals. Note that the external bias may be eliminated in such DAMs. The elimination of bias, the symmetric $\mathbf{W}$, and the use of high-gain amplifiers in such DAMs lead to the truncated energy function

$$E(x) = -\tfrac{1}{2}x^{\mathrm{T}}\mathbf{W}x. \tag{C1.3.23}$$

This discrete-time discrete-state Hopfield model (Hopfield 1982) may be derived by starting with the dynamical system in equation (C1.3.15) and replacing the continuous activation function by the sign function

$$x_i(k+1) = \mathrm{sgn}\left[\sum_{j=1}^{n}w_{ij}x_j(k) + I_i\right]. \tag{C1.3.24}$$

It can be shown that the discrete Hopfield network with a symmetric interconnection matrix ($w_{ij} = w_{ji}$) and with non-negative diagonal elements ($w_{ii} \geq 0$) is stable with the same Lyapunov function as that of a continuous-time Hopfield network in the limit of high amplifier gain, that is, it has the Lyapunov function in equation (C1.3.20). Hopfield (1984) showed that both networks (discrete and continuous networks with the above assumptions) have identical energy maxima and minima. This implies that there is a one-to-one correspondence between the memories of the two models. Also, since the two models may be viewed as minimizing the same energy function E, one would expect that the macroscopic behaviors of the two models are very similar; that is, both models will perform similar memory retrievals.

C1.3.4.2 Capacity of the Hopfield DAM

DAM capacity is a measure of the ability of a DAM to store a set of m unbiased random binary patterns $x^k \in \{-1, 1\}^n$ (that is, the vector components x_i^k are independent random variables taking values 1 or -1 with probability 0.5) and at the same time be capable of associative recall (error correction). One common capacity measure is known as *absolute capacity* and is defined as an upper bound on the pattern ratio m/n such that (with probability approaching 1) all fundamental memories are stored as equilibrium points. This capacity measure, though, does not say anything about error correction behavior, that is, it does not require that the fundamental memories x^k be attractors with associated basins of attraction. Another capacity measure, known as *relative capacity*, has been proposed which is an upper bound on m/n such that the fundamental memories or their 'approximate' versions are attractors (stable equilibria).

It has been shown (Amari 1977, Hopfield 1982, Amit *et al* 1985) that if most of the memories in a correlation-recorded discrete Hopfield DAM, with $w_{ii} = 0$, are to be remembered approximately (i.e. nonperfect retrieval is allowed), then m/n must not exceed 0.15. This value is the relative capacity of the DAM. Another result on the capacity of this DAM for the case of error-free memory recall by one-pass parallel convergence is (in probability) given by the absolute capacity (Weisbuch and Fogelman-Soulié 1985, McEliece *et al* 1987, Amari and Maginu 1988, Newman 1988), expressed as the limit

$$\max\left(\frac{m}{n}\right) \to \frac{1}{4 \ln n} \qquad \text{as } n \to \infty. \tag{C1.3.25}$$

Equation (C1.3.25) indicates that the absolute capacity approaches zero as n approaches infinity! Thus, the correlation-recorded discrete Hopfield network is an inefficient DAM model. Another, more useful DAM capacity measure gives a bound on m/n in terms of error correction and memory size (Weisbuch and Fogelman-Soulié 1985, McEliece *et al* 1987). According to this capacity measure, a correlation-recorded discrete Hopfield DAM must have its pattern ratio m/n satisfy

$$\frac{m}{n} < \frac{(1 - 2\rho)^2}{4 \ln n} \qquad n \to \infty \tag{C1.3.26}$$

in order that error-free one-pass retrieval of a fundamental memory (say x^k) from random key patterns lying inside the Hamming hypersphere (centered at x^k) of radius ρn ($\rho < \frac{1}{2}$) is achieved with a probability approaching 1. Here, ρ defines the radius of attraction of a fundamental memory. In other words, ρ is the largest normalized Hamming distance from a fundamental memory within which almost all the initial states reach this fundamental memory in one pass.

In general, projection-recorded autoassociative DAMs outperform correlation recorded DAMs in terms of capacity and overall performance. Recall that with projection recording, any linearly independent set of memories can be memorized error-free (note that linear independence restricts m to be less than or equal to n). In particular, projection DAMs are well suited for memorizing unbiased random vectors $x^k \in \{-1, 1\}^n$, since it can be shown that the probability of m ($m < n$) of these vectors to be linearly independent approaches 1 in the limit of large n (Komlós 1967). The relation between the radius of attraction of fundamental memories ρ and the pattern ratio m/n is a desirable measure of DAM retrieval/error-correction characteristics. For correlation-recorded binary DAMs, such a relation has been derived analytically for single-pass retrieval and is given by equation (C1.3.26). On the other hand, deriving similar relations for multiple-pass retrievals and/or more complex recording recipes (such as projection recording) is a much more difficult problem. In such cases, numerical simulations with large n values (typically equal to several hundred) are a viable tool (e.g. see Kanter and Sompolinsky 1987, Amari and Maginu 1988).

C1.3.4.3 The brain-state-in-a-box DAM

The brain-state-in-a-box (BSB) model (Anderson *et al* 1977) is one of the earliest DAM models. It is a discrete-time continuous-state parallel-updated DAM whose dynamics are given by

$$x(t + 1) = F[\gamma x(t) + \alpha \mathbf{W} x(t) + \delta \theta] \tag{C1.3.27}$$

where the input key is presented as the initial state $x(0)$ of the DAM. Here, $\gamma x(t)$, with $0 \leq \gamma \leq 1$, is a decay term of the state $x(t)$ and α is a positive constant which represents feedback gain. The vector $\theta = [I_1, I_2, \ldots, I_n]^T$ represents a scaled external input (bias) to the system, which persists for all time t. Some particular choices for δ are $\delta = 0$ (i.e. no external bias) or $\delta = \alpha$. The operation $F(\xi)$ is a piecewise linear operator which maps the ith component ξ_i of its argument vector ξ according to

$$f(\xi_i) = \begin{cases} 1 & \text{if } \xi_i \geq 1 \\ \xi_i & \text{if } -1 < \xi_i < 1 \\ -1 & \text{if } \xi_i \leq -1 . \end{cases} \qquad (C1.3.28)$$

The BSB model gets its name from the fact that the state of the system is continuous and constrained to be in the hypercube $[-1, 1]^n$. When operated as a DAM, the BSB model typically employs an interconnection matrix W given by the correlation recording recipe to store a set of m n-dimensional bipolar binary vectors as attractors (located at corners of the hypercube $[-1, 1]^n$). Here, one normally sets $\delta = 0$ and assumes the input to the DAM (i.e. $x(0)$) to be a noisy vector which may be anywhere in the hypercube $[-1, 1]^n$. The performance of this DAM with random stored vectors, large n and $m \ll n$, has been studied through numerical simulations by Anderson (1993). These simulations particularly address the effects of model parameters on memory retrieval.

The stability of the BSB model in equation (C1.3.27) with symmetric W, $\delta = 0$ and $\gamma = 1$ has been analyzed by several researchers including Golden (1986), Greenberg (1988), Hui and Žak (1992) and Anderson (1993). In this case, the model reduces to

$$x(t + 1) = F[x(t) - \alpha W x(t)] . \qquad (C1.3.29)$$

Golden (1986, 1993) analyzed the dynamics of the system in equation (C1.3.29) and found that it behaves as a *gradient-descent system* that minimizes the energy

$$E(x) = -\tfrac{1}{2} x^T W x . \qquad (C1.3.30)$$

B5.2.2

He also proved that the dynamics in equation (C1.3.29) always converge to a local minimum of $E(x)$ if W is symmetric and $\lambda_{\min} \geq 0$ (i.e. W is positive semidefinite) or $\alpha < 2/|\lambda_{\min}|$, where $\lambda_{\min}$ is the smallest eigenvalue of W. With these conditions, the stable equilibria of this model are restricted to the surface and/or vertices of the hypercube. It is interesting to note here that when this BSB DAM employs correlation recording (with a preserved diagonal of W), it always converges to a minimum of $E(x)$ because of the positive-semidefinite symmetric nature of the autocorrelation matrix.

C1.3.5 Dynamic models: the heteroassociative case

C1.3.5.1 The heteroassociative DAM

The heteroassociative DAM (HDAM) architecture is shown in figure C1.3.3(b) (Okajima *et al* 1987). It consists of two processing paths which form a closed loop. The first processing path computes a vector $y \in \{-1, 1\}^L$ from an input $x \in \{-1, 1\}^n$ according to the parallel update rule

$$y(t + 1) = F(W_1 x(t)) \qquad (C1.3.31)$$

or its serial (asynchronous) version, where one and only one unit updates its state at a given time. Here, F is usually the sgn activation operator. Similarly, the second processing path computes a vector x according to

$$x(t + 1) = F(W_2 y(t)) \qquad (C1.3.32)$$

or its serial version. The HDAM can be operated in either parallel or serial retrieval modes. In the parallel mode, the HDAM starts from an initial state $x(0)$, computes its state y according to equation (C1.3.31), and then updates state x according to equation (C1.3.32). This process is iterated until convergence, i.e. until state x (or equivalently y) ceases to change. On the other hand, in the serial update mode, only one randomly chosen component of the state x or y is updated at a given time.

Various methods have been proposed for storing a set of heteroassociations $\{x^k, y^k\}$, $k = 1, 2, \ldots, m$ in the HDAM. In most of these methods, the interconnection matrices W_1 and W_2 are computed independently by requiring that all one-pass associations $x^k \rightarrow y^k$ and $y^k \rightarrow x^k$, respectively, are stored perfectly. Here, it is assumed that the set of associations to be stored forms a one-to-one mapping; otherwise, perfect storage becomes impossible. Examples of such HDAM recording methods include the use of projection recording (Hassoun 1989a, b) and Householder transformation-based recording (Leung and Cheung 1991). These methods require the linear independence of the vectors x^k (also y^k) for which a capacity of $m = \min(n, L)$ is achievable. One drawback of these techniques, however, is that they do not guarantee the stability of the HDAM, i.e. convergence to spurious cycles is possible. Empirical results show (Hassoun 1989b) that parallel updated projection-recorded HDAMs exhibit significant oscillatory behavior only at memory loading levels close to the HDAM capacity.

Kosko (1987, 1988) proposed a heteroassociative memory with the architecture of the HDAM but with the restriction $\mathbf{W}_2^T = \mathbf{W}_1 = \mathbf{W}$. This memory is known as a *bidirectional associative memory* (BAM). The interesting feature of a BAM is that it is stable for any choice of the real-valued interconnection matrix $\mathbf{W}$ and for both serial and parallel retrieval modes. This can be shown by starting from the bounded Lyapunov (energy) function

$$E(x, y) = -\tfrac{1}{2}x^T\mathbf{W}y - \tfrac{1}{2}y^T\mathbf{W}x = -x^T\mathbf{W}y \geq -\sum_i \sum_j |w_{ij}| \qquad (C1.3.33)$$

of the BAM and showing that each serial or parallel state update decreases E. One can also prove BAM stability by noting that a BAM can be converted to a discrete autoassociative DAM (discrete Hopfield DAM) with state vector $x' = [x^T, y^T]^T$ and interconnection matrix $\mathbf{W}'$ given by

$$\mathbf{W}' = \begin{bmatrix} \mathbf{0} & \mathbf{W} \\ \mathbf{W}^T & \mathbf{0} \end{bmatrix}. \qquad (C1.3.34)$$

Now, since $\mathbf{W}'$ is a symmetric zero-diagonal matrix, the autoassociative DAM is stable if serial update is assumed. Therefore, the serially updated BAM is stable. One may also use this equivalence property to show the stability of the parallel-updated BAM (note that a parallel-updated BAM is not equivalent to the (nonstable) parallel-updated discrete Hopfield DAM; this is because either states x or y, but not both, are updated in parallel at each step).

From above, it can be concluded that the BAM always converges to a local minimum of its energy function defined in equation (C1.3.33). It can be shown (Wang *et al* 1991) that these local minima include all those that correspond to associations $\{x^k, y^k\}$ which are successfully loaded into the BAM (i.e. associations which are equilibria of the BAM dynamics).

The most simple storage recipe for storing the associations as BAM equilibrium points is the correlation recording recipe. This recipe guarantees the BAM requirement that the forward-path and backward-path interconnection matrices $\mathbf{W}_1$ and $\mathbf{W}_2$ are the transpose of each other, since

$$\mathbf{W}_1 = \sum_{k=1}^m y^k (x^k)^T \qquad (C1.3.35)$$

and

$$\mathbf{W}_2 = \sum_{k=1}^m x^k (y^k)^T. \qquad (C1.3.36)$$

However, some serious drawbacks of using the correlation recording recipe are low capacity and poor associative retrievals; when m random associations are stored in a correlation-recorded BAM, the condition $m \ll \min(n, L)$ must be satisfied if good associative performance is desired (Hassoun 1989b, Simpson 1990). Heuristics for improving the performance of correlation-recorded BAMs can be found in Wang *et al* (1990).

C1.3.6 Other models

It should be noted that the above models of associative memories are by no means exhaustive. A number of other interesting models have been reported in the literature (interested readers may find the volume edited by Hassoun (1993) useful in this regard). For example, in terms of overall performance, the Ho–Kashyap model (Hassoun and Youssef 1989) has been shown to outperform both correlation and projection recorded DAMs. Other models are interesting because of their connection to biological memories (e.g. see Kanerva 1988, 1993 and Alkon *et al* 1993).

References

Alkon D L, Blackwell K T, Vogl T P and Werness S A 1993 Biological plausibility of artificial neural networks: learning by non-Hebbian synapses *Associative Neural Memories: Theory and Implementation* ed M H Hassoun (New York: Oxford University Press) pp 31–49

Amari S-I 1972a Learning patterns and pattern sequences by self-organizing nets of threshold elements *IEEE Trans. Comput.* **21** 1197–206

——1972b Characteristics of random nets of analog neuron-like elements *IEEE Trans. Syst. Man Cybern.* **2** 643–57

——1977 Neural theory of association and concept-formation *Biol. Cybern.* **26** 175–85

——1990 Mathematical foundations of neurocomputing *Proc. IEEE* **78** 1443–63

Amari S-I and Maginu K 1988 Statistical neurodynamics of associative memory *Neural Networks* **1** 63–73

Amari S-I and Yanai H-F 1993 Statistical neurodynamics of various types of associative nets *Associative Neural Memories: Theory and Implementation* ed M H Hassoun (New York: Oxford University Press) pp 169–83

Amit D J, Gutfreund H and Sompolinsky H 1985 Storing infinite numbers of patterns in a spin–glass model of neural networks *Phys. Rev. Lett.* **55** 1530–3

Anderson J A 1972 A simple neural network generating interactive memory *Math. Biosci.* **14** 197–220

——1993 The BSB model: a simple nonlinear autoassociative neural network *Associative Neural Memories: Theory and Implementation* ed M H Hassoun (New York: Oxford University Press) pp 77–103

Anderson J A, Silverstien J W, Ritz S A and Jones R S 1977 Distinctive features, categorical perception, and probability learning: some applications of neural model *Psychol. Rev.* **84** 413–51

Baird, B 1990 Associative memory in a simple model of oscillating cortex *Advances in Neural Information Processing Systems 2 (Denver, 1989)* ed D S Touretzky (San Mateo, CA: Morgan Kaufmann) pp 68–75

Casasent D and Telfer B 1987 Associative memory synthesis, performance, storage capacity, and updating: new heteroassociative memory results *SPIE, Int. Robots Comput. Vision* **848** 313–33

Chiueh T D and Goodman R M 1988 High capacity exponential associative memory *Proc. IEEE Int. Conf. on Neural Networks (San Diego, CA)* vol I (New York: IEEE Press) pp 153–60

Cohen M A and Grossberg S 1983 Absolute stability of global pattern formation and parallel memory storage by competitive neural networks *IEEE Trans. Syst. Man Cybern.* **13** 815–26

Golden R M 1986 The brain-state-in-a-box neural model is a gradient descent algorithm *J. Math. Psychol.* **30** 73–80

——1993 Stability and optimization analysis of the generalized brain-state-in-a-box neural network model *J. Math. Psychol.* **37** 282–98

Greenberg H J 1988 Equilibria of the brain-state-in-a-box (BSB) neural model *Neural Networks* **1** 323–4

Hassoun M H 1989a Adaptive dynamic heteroassociative neural memories for pattern classification *Proc. SPIE, Optical Pattern Recognition* vol 1053 ed H-K Liu pp 75–83

——1989b Dynamic heteroassociative neural memories *Neural Networks* **2** 275–87

——(ed) 1993 *Associative Neural Memories: Theory and Implementation* (New York: Oxford University Press)

——1995 *Fundamentals of Artificial Neural Networks* (Cambridge, MA: MIT)

Hassoun M H and Youssef A M 1989 A high-performance recording algorithm for Hopfield model associative memories *Opt. Eng.* **27** 46–54

Hirsch M 1989 Convergent activation dynamics in continuous time networks *Neural Networks* **2** 331–49

Hirsch M and Smale S 1974 *Differential Equations, Dynamical Syst., and Linear Algebra* (New York: Academic)

Hopfield J J 1982 Neural networks and physical systems with emergent collective computational abilities *Proc. Natl Acad. Sci., USA* **79** 2445–558

——1984 Neurons with graded response have collective computational properties like those of two-state neurons *Proc. Natl Acad. Sci., USA* **81** 3088–92

Hui S and Żak S H 1992 Dynamical analysis of the brain-state-in-a-box neural models *IEEE Trans. Neural Networks* **3** 86–9

Kanerva P 1988 *Sparse Distributed Memory* (Cambridge, MA: Bradford/MIT)

——1993 Sparse distributed memory and other models *Associative Neural Memories: Theory and Implementation* ed M H Hassoun (New York: Oxford University Press) pp 50–76

Kanter I and Sompolinsky H 1987 Associative recall of memory without errors *Phys. Rev. A* **35** 380–92

Kohonen T 1972 Correlation matrix memories *IEEE Trans. Comput.* **21** 353–9

——1974 An adaptive associative memory principle *IEEE Trans. Comput.* **23** 444–5

——1984 *Self-Organization and Associative Memory* (Berlin: Springer)

Kohonen T and Ruohonen M 1973 Representation of associated data by matrix operators *IEEE Trans. Comput.* **22** 701–2

Komlós J 1967 On the determinant of (0, 1) matricies *Stud. Sci. Math. Hung.* **2** 7–21

Kosko B 1987 Adaptive bidirectional associative memories *Appl. Opt.* **26** 4947–60

——1988 Bidirectional associative memories *IEEE Trans. Syst. Man Cybern.* **18** 49–60

Leung C S and Cheung K F 1991 Householder encoding for discrete bidirectional associative memory *Proc. Int. Conf. on Neural Networks (Singapore 1991)* pp 237–41

McEliece R J, Posner E C, Rodemich E R, and Venkatesh S S 1987 The capacity of the Hopfield associative memory *IEEE Trans. Info. Theory* **33** 461–82

Nakano K 1972 Associatron: a model of associative memory *IEEE Trans. Syst. Man Cybern.* **2** 380–8

Newman C 1988 Memory capacity in neural network models: rigorous lower bounds *Neural Networks* **3** 223–39

Okajima K, Tanaka S and Fujiwara S 1987 A heteroassociative memory network with feedback connection *Proc. IEEE First Int. Conf. on Neural Networks (San Diego, CA)* vol II, ed M Caudill and C Butler pp 711–8

Rao C R and Mitra S K 1971 *Generalized Inverse of Matrices and its Applications* (New York: Wiley)

Simpson P K 1990 Higher-ordered and intraconnected bidirectional associative memory *IEEE Trans. Syst. Man Cybern.* **20** 637–53

Stiles G S and Denq D-L 1987 A quantitative comparison of three discrete distributed associative memory models *IEEE Trans. Comput.* **36** 257–63

Uesaka G and Ozeki K 1972 Some properties of associative type memories *J. Inst. Elec. Commun. Eng. Japan* **D-55** 323–30

Wang Y-F, Cruz J B Jr and Mulligan J H Jr 1990 Two coding strategies for bidirectional associative memory *IEEE Trans. Neural Networks* **1** 81–92

——1991 guaranteed recall of all training pairs for bidirectional associative memory *IEEE Trans. Neural Networks* **2** 559–67

Weisbuch G and Fogelman-Soulié F 1985 Scaling laws for the attractors of Hopfield networks *J. Physique Lett.* **46** L623–30

C1.4   Stochastic neural networks

Harold Szu and Masud Cader

Abstract

Deterministic neural networks such as backpropagation of error, multilayer perceptrons, and locally based radial basis methods have been a major focus of the neural network community in recent years. However, there has been a distinct, albeit less pronounced, interest in stochastic neural networks. In this review we provide the reader with a sense of the defining components of a stochastic neural network, as well as some of the issues arising from working with stochastic neural networks. In particular, issues revolving around hardware implementation, software simulation, and innovation are developed.

C1.4.1   Introduction

The term stochastic neural network refers to a model of computation whose output is a stochastic function of its inputs and interactions among its neurons. It primarily differs from the more popular deterministic gradient descent algorithms (e.g., *backpropagation*) in that a unit activation is not a deterministic sigmoid function of the inputs, but rather a stochastic function. In addition, the learning algorithm for a stochastic machine usually implements a procedure for finding a minimum on the energy surface as well as entropy maximization (Szu 1986). Although the stochastic component increases the complexity of understanding and implementation, the reward follows from the fact that a training algorithm based on simulated annealing is, theoretically, assured to converge to the global minimum, albeit slowly.

C1.2.3

Recent developments in stochastic neural network modeling have attempted to improve computational performance either by parallel implementation or by replacing the computation of stochastic dynamics with simpler deterministic mean field approximations (Peterson 1987, Hertz 1991, Zerubia and Rama 1993, Yuille 1994, Kappen 1995a, b), that is, estimating stochastic transitions by the mean of the transitions. The performance of such annealed estimates is addressed by Tishby (1995).

Primarily, this line of accelerating the search algorithms has been based on a deterministic Boltzmann learning procedure proposed by Hinton (1989); the 'Cauchy Machine', invented by Szu (Szu and Messner 1986, Szu 1987), which uses a Cauchy distribution to generate random flights as well as walks to new states; 'adaptive simulated annealing' models (Ingber 1995) which permit fast learning via the use of differing annealing schedules across parameter dimensions; and Markov chain Monte Carlo sampling methods for state generation (Geyer 1993). These approaches offer a faster learning procedure than the original Boltzmann machine (Hinton *et al* 1984); however, they are not without their drawbacks (Wasserman 1989a, b, Galland 1993, Ingber 1995).

Other approaches, based on the fact that sufficiently simple architectures of Boltzmann machines can learn by gradient descent on the objective function (Hopfield 1987), utilize hierarchical configurations of simple Boltzmann machines so that training may proceed by gradient descent rather than involving simulated annealing. More complex interaction is enabled through the use of configurations or Boltzmann trees (Saul and Michael 1994). Still other approaches consider the problem of training a Boltzmann machine from an information geometrical view. The alternating minimization algorithm (Byrne 1992) proposes that the learning problem be addressed by minimizing the information divergence of repeated projections of the machine states and shows the equivalence of the algorithm to gradient descent and the expectation maximization technique under specific conditions.

Further, the stochastic Helmholtz machine (Dayan *et al* 1995) illustrates an innovative statistical learning algorithm (the wake-sleep algorithm of Hinton *et al* 1995) where the stochastic neural network architecture is unsupervised. That is, a multilayer network of stochastic binary neurons is augmented by two groups of weights, a top-down generative set in addition to the bottom-up recognition set (resembling very much the biweight connectivity of the *ART* model of Carpenter and Grossberg).

C2.2.1

Parberry and Schnitger (1989) augment the 'classical' Boltzmann machine model, and show that in some cases Boltzmann machines may not be much more powerful than combinatorial circuits built from Boolean threshold gates. They make a number of useful comments about the practical implementation of Boltzmann machines.

E1.3, E1.4

An *electronic chip implementation* of a Boltzmann machine has been developed by Alspector *et al* (1989) at Bellcore, and Skubiszewski (1992), with an *optical version* by Farhat (Farhat and Psaltis 1987, Farhat 1987). Similarly, an electronic Cauchy machine has been designed by Takefuji and Szu (1989), and its optical version realized by Scheff and Szu (1987). Recently, a Gaussian machine based on both the minimization of Helmholtz's free energy and the maximization of entropy has been studied and implemented in a chip by Akiyama *et al* (1990) at Keio University.

E1.5

C1.4.2 Simulated annealing

Since the major ingredient in stochastic machines is the simulated annealing algorithm, we compare the Boltzmann machine and Cauchy machine in terms of different algorithms: *Boltzmann annealing* (BA) and *Cauchy annealing* (CA) in section C1.4.2. Then, we review two benchmark applications; one for finding the global minimum solution of the Traveling Salesman Problem (TSP) and a second which searches for the mini-max feature in an image processing problem in section C1.4.3.

We shall discuss the sequential algorithms used in the above parallel machine implementations as follows. In BA, a Gaussian random process is used to generate new states in the sequential algorithm. Geman and Geman (1984) have proved that the cooling schedule $T(t)$ must be inversely proportional to the logarithm of time t, in order to guarantee convergence to the global minimum. This relatively slow convergence is due to the bounded variance of the Gaussian process which constrains the neighborhood of successive samples. This bounded-variance random walk is called a local search strategy. On the other hand, if one uses an infinite-variance Cauchy random process, a faster cooling schedule that is inversely proportional to time t has been deduced by Szu (1987) in one dimension and Szu and Hartley (Szu 1987, Szu and Hartley 1987) in arbitrary higher dimensions (as applied to solving the bearing fix problem with multiple sensors and multiple targets). This new class of algorithms, implementing a semilocal search strategy, permits occasionally long steps (the so-called Lévy–Doob diffusion) far from the neighborhood of the previous sample. These random flights are indicative of the divergence of the second moment of the Cauchy probability distribution.

In a convex optimization problem, one can start at any point in the function space, measure the local gradient, and take a step in any direction which is lower in altitude than the current position. Repetition of this process will assure asymptotic convergence to the minimum (i.e., optimum) solution. In a nonconvex problem, the optimization function has multiple local minima, each with different depths, for which the optimum is defined to be the global minimum. The application of local gradient techniques to nonconvex optimization creates a problem where one becomes caught in a local minimum with no way of determining whether the local minimum is also the desired global minimum. One solution to this dilemma is to permit steps whose magnitude and direction are dependent on the local gradient and to add random noise in an annealing process Wasserman (1989a, b).

Further, for the algorithm to converge, the magnitude of the random component of the step size must decrease in a statistically monotonic fashion. In the physical annealing process these steps can be equated with Brownian motion of a particle, traveling at statistical velocity V, over an intersample time Δt. The expectation of V^2 is linearly related to the temperature of the particle. The simulated annealing community (Kirkpatrick *et al* 1983) therefore refers to the 'temperature' of the random process and uses the term 'cooling schedule' to refer to the algorithm for monotonically reducing the temperature.

An annealing methodology requires three major steps: (i) the generation of a new search state by means of a random process covering all phase space without the barrier of an energy landscape (section C1.4.2.1); (ii) the acceptance criterion of the new state, based on the energy landscape property at the new and the old states (section C1.4.2.2); and (iii) the cooling schedule for quenching the random noise used to generate a new state together with an appropriate change in the new-state acceptance criterion (section C1.4.2.3).

C1.4.2.1 State-generating probability density

The Boltzmann machine uses a Gaussian probability density to generate the incremental displacement X between the old state x and the new state x' as follows:

$$G_T(x'|x' = x + X) = (1/2\pi T^{1/2} \exp(-X^2/T). \tag{C1.4.1}$$

Based on the central limit theorem (CLT), any random variable with a bounded variance approaches the Gaussian distribution in the large-sampling limit.

The Cauchy state generating probability density is:

$$G_T(x'|x' = x + X) = [T/\pi(T^2 + |X|^2)]. \tag{C1.4.2}$$

Both distributions can be expanded in Taylor series and become identically quadratic for small displacements. This means that locally they are both identical to random walks. However, when the second moment is taken, the Cauchy density produces an infinite divergence while the Gaussian density gives the value of the temperature. This illustrates that the Cauchy distribution will generate random flights in long steps (Lévy flights), and that the CLT does not apply.

For an optical implementation, the random displacement X can be easily generated by a uniform angle distribution between $\pm\pi/2$ by a light beam deflected from a suspended mirror on a flat screen as demonstrated previously for an optical Cauchy machine (Scheff and Szu 1987). The displacement X is measured from the center and is given by

$$X = T\tan(\theta) \tag{C1.4.3}$$

since with $d\tan(\theta)/d\theta = 1/(1 + \tan(\theta)^2)$, we can replace $\tan(\theta)$ with X/T yielding equation (C1.4.3).

C1.4.2.2 Local and distributed acceptance criteria

The primary difference between sequential simulations and parallel implementations of simulated annealing is that the former relies on a centralized acceptance criterion (an *uphill* energy concept), while parallel versions require a distributed criterion (an against peer pressure concept).

The total system energy is convenient for a top-down design, but is not suited for parallel implementations. Any criterion based on the total system energy requires a central processor to tally the contribution from all distributed processors. If each processor is waiting for a centralized decision, the speed of parallel execution will be slowed down.

A natural choice for a distributed acceptance criterion is one based on the interaction forces carried by local communication links. These interactions can be related to the entire energy landscape.

For example, the natural phenomenon occurring in a water–ice phase transition is a parallel and collective computation without central control where a slow cooling or annealing schedule insures the low-energy crystalline state of ice. In other words, during the occasional uphill climb of the energy landscape to detrapping (or a metastable crystalline state), there is an occasional thermal fluctuation against peer pressure. This fluctuation manifests itself via the interacting Coulomb forces which communicate instantaneously among all processors or molecules, rather than through the posterior energy landscape. A neural network is similar to this liquid–solid phase transition which promises the minimum-energy crystal state if it is cooled down properly.

If the energy change $\Delta E = E_{\mathrm{new}} - E_{\mathrm{old}}$ is less than zero, the new state is accepted. On the other hand, if the energy change ΔE is greater than zero, then the following acceptance probability is computed:

$$P_T = 1/[1 + \exp(-\Delta E/T)] \tag{C1.4.4}$$

(which is larger than a uniformly generated random number) and the uphill state is accepted, otherwise the state is rejected. Such an energy landscape formula can be thought of as a two state normalized transition probability: $\exp(-E_{\mathrm{new}})/[\exp(-E_{\mathrm{new}}) + \exp(-E_{\mathrm{old}})]$ and therefore works well on a conventional serial machine for one neuron decision at a time.

For a Gaussian noise model, the appropriate Metropolis acceptance criterion (Metropolis *et al* 1953) cannot be integrated into an elementary quadrature, which yields, by the steepest-descent approximation, the energy landscape concept ΔE. Hinton and Sejnowski have interpreted the acceptance criterion, equation (C1.4.4), as the energy change for each neuron, ΔE_i, which is used to derive a specific hidden layer weight, in order to derive a local acceptance criterion (cf see appendix of Hinton and Sejnowski (1986)).

A one-dimensional optically implemented neural network utilizing CA has been developed as the Cauchy Machine (Scheff and Szu 1987). However, a local distributed VLSI design could not be implemented until a distributed acceptance criterion was derived for the Cauchy density (Takefuji and Szu 1989).

If the total input u_i to the McCulloch–Pitts model of a binary neuron is defined as

$$u_i = \sum_j T_{ij} v_j$$

then, as consistent with the Metropolis acceptance criterion, the output v_i is locally set to be one only if random numbers generated within the interval [0, 1] are less than the acceptance function—which is integrated exactly for each total input as follows:

$$(1/\pi T) \int_0^\infty dx/[1 + ((x - u_i)/T)^2] = (1/2) + \tan^{-1}(u_i/T(t))/\pi. \tag{C1.4.5}$$

In the case of annealing, the inverse of the cooling schedule is defined to be the piecewise constant gain coefficient, G_n, at a positive integer time point t_n:

$$G(t_n) = 1/T(t_n) = G_n. \tag{C1.4.6}$$

Then, the output v_i also fluctuates within a finite bound described as both firing rate transfer functions:

$$v_i = \sigma_{1n}(u_i) = (1/2) + \tan^{-1}(u_i G_n)/\pi. \tag{C1.4.7}$$

Note that equation (C1.4.7) is almost identical to the standard sigmoidal/logistic function of $1/[1 + \exp(-u_i G_n)]$, except that the arctangent function becomes slightly rounded near the central region. In the case of the sigmoidal function, the slope σ_n' is proportional to the gain coefficient G_n:

$$\sigma_n' = dv_i/du_i = G_n v_i (1 - v_i). \tag{C1.4.8}$$

When $T = 0$, the infinite gain G implies an infinite slope. In this limit, both firing rate transfer functions become a binary step function $v_i = \text{step}(u_i)$ describing a binary neuron model. Thus, the annealing process gradually changes a sigmoidal neuron toward a binary neuron.

C1.4.2.3 Annealing cooling schedules

The cooling schedule is critical to the performance of the learning algorithm. For a given random process, cooling at too fast a rate will probably 'freeze' the system in a nonglobal minimum. Cooling at too slow a rate, while reaching the desired global minimum, is a waste of computational resources. The technical problem is to derive the fastest cooling schedule that will guarantee convergence to the global minimum. With this understanding, the term 'cooling schedule' is synonymous with 'permissible fastest cooling schedule' during which the complete phase space is guaranteed to be available for searching at all time.

Without any knowledge of energy landscapes, one can only hope to derive an appropriate cooling schedule for a specific stochastic process. The necessary condition is that at any temperature the phase space is always accessible infinitely often in time (IOT). In other words, an inappropriately fast cooling schedule may quench the IOT availability of some remote states, and hence, not find the global minimum. The specific energy landscape and an appropriate acceptance criterion must be taken into consideration to determine whether the minimum will be actually be found. Ingber (1993), has shown that exponential cooling schedules may be used but only with specific distributional forms used as the state generating function.

For a Gaussian random process, Geman and Geman (1984) have proved that the simulated annealing cooling schedule of the temperature $T(t)$ must be decreased (from a given sufficiently high temperature T_0 down to zero 'degrees') according to the inverse logarithmic formula:

$$T = T_0/\log(1 + t). \tag{C1.4.9}$$

Thus, in the interest of speeding up the annealing process and yet maintaining the capability of finding the global minimum, Szu *et al* applied Cauchy colored noise to the problem, instead of a Gaussian random process. The resultant cooling schedule for an arbitrary initial temperature is derived:

$$T = T_0/(1 + t) \tag{C1.4.10}$$

which is indeed faster, and was shown to insure that the complete search space is available at all temperatures.

The mathematical truth in both proofs is based on the fact that the infinite series of the inverse time steps is divergent from an arbitrary initial time point t_0

$$\sum_{t=t_0}^{\infty} \frac{1}{t} = \infty. \tag{C1.4.11}$$

The complete proofs for both are provided in appendix A.

It is useful to note that CA is $t/\log(t)$ faster than a Gaussian (white noise) simulated annealing algorithm which in turn is superior to the conventional Monte Carlo method in which the temperature is held constant.

C1.4.3 Applications

A stochastic neural network model, the Boltzmann machine, has been used in demonstrating the celebrated Net-Talk (Sejnowski and Rosenberg 1987). Similarly, the problem of obtaining rapid and accurate estimations of the locations of moving emitters from samples of imprecise bearing only data has been addressed with the Cauchy machine (Szu 1987). Another innovative application has been the use a class of BM (in which visible units are connected only to hidden units) to repair a dataset with missing values (Kappen 1995a, b).

In the remainder of this section, we illustrate two applications of the Cauchy machine. The first, a benchmark problem in this area, is that of determining the shortest tour length of a traveling salesman through a set of cities only taking into account the constraint of distance between cities. The second is a problem related to optical character recognition, where the idea is to automatically extract features from the character pattern sets.

C1.4.3.1 Constraint specifications

A traveling salesman problem (TSP), which attempts to find the shortest possible tour through a given number of cities, can be stochastically solved by generating noise via the leptokurtic Cauchy probability density, $T/\pi(T^2 + X^2)$ (Szu 1990). The noise must be quenched with the inversely linear cooling schedule: $T = T_0/(1 + t)$ as described earlier. Moreover, the schedule must be followed consistently for every time step, both in generating new states and in visiting some of the states whenever the acceptance criterion is met.

The performance of CA was calibrated by comparing against the results obtained by an exhaustive search through all possible TSP solutions. This is possible due to a novel factorial number representation for each TSP configuration by an integer n described as follows.

We require a one-dimensional coding scheme for the TSP search space that is one-to-one unique. Due to the combinatorial nature of the TSP, a good guess at a number representation might be a factorial number base system. We adopt, in the following manner, a coding scheme as follows:

(i)   The real line x is sampled by the set of real integers x, using the function Int( ).
(ii) Then, integers are made periodically in the modulus base set of $(N - 1)!$, using the Mod(,) function.
(iii) Such an integer number represents a state of a valid tour since a factorial base set is related to the tour order permutations. Thus, one represents the integer in terms of the factorial number base system by calculating the most significant numbers denoted by the index tuple,

$$X_{\text{new}} = \sum_n \text{index}_n \times n! \tag{C1.4.12a}$$

$$X_{\text{new}} \leftrightarrow (\text{index}_{N-1}, \text{index}_{N-2}, \ldots, \text{index}_1, \text{index}_0) \tag{C1.4.12b}$$

sequentially for all n beginning with $N - 1$ down to 0.

To produce the set of indices, one considers an example for five cities, $N = 5$, denoted by city: Nos 1–5. Given $X_{old} = 0 = $ (No 1, No 2, No 3, No 4, No 5) as a reference (the diagonal matrix element of Hopfield–Tank), where the arbitrary tour order is that city No 1 is visited first, and so on. One finds,

$$X_{new} = 15 = 0 \times 0! + 1 \times 1! + 1 \times 2! + 2 \times 3! + 0 \times 4! \leftrightarrow (\text{No 1, No 4, No 3, No 5, No 2})$$

where the representation index $= (0, 1, 1, 2, 0)$ is obtained with respect to the base set $(0!, 1!, 2!, 3!, 4!)$.

The energy corresponding to each of the possible round-trip routes through n cities, $4 \leq n \leq 10$, has been reported (Szu and Scheff 1990), so, while the exhaustive search through hundreds of thousand of possible cases used several hours of computer time on a Mac II (with factorial scaling implying that five hours for ten cities would require 50 hours for 11 cities). In contrast, CA took about 10 minutes or less to find the global minima for the ten-city problem. As the shortest tours agreed with those found by CA, it is clear that CA is superior because the search required a sampling of less than 1% of the states, with another 2% sampling to verify the stability. Thus, it is evident that traditional Monte Carlo random sampling should be replaced with the CA algorithm.

F1.2, F1.6 ## C1.4.3.2 *Image processing and pattern recognition*

Geman and Geman (1984) have applied Boltzmann annealing to the problem of noisy image restoration. Smith *et al* (1983) have also applied BA to radiology image reconstructions. Szu and Scheff (1990) have shown that CA can also be useful in pattern recognition. In particular, they have used a minimax cost function to investigate the self-extraction of unkown features, previously accomplished using self-reference matched filters (Szu *et al* 1980, Szu and Blodgett 1982, Szu and Messner 1986).

Let the critical feature of the template class c be denoted as $f_c(x, y)$. Then, a space-filling curve, Peano N-curve, is employed to replace the traditional line-by-line scan sampling, in order to preserve the neighborhood proximity relationship.

The performance criterion seeks to minimize the distance between the image template I_c of the c-class ($c = 1, 2$), to minimize the inner product between classes $\langle f_c | f_{c'} \rangle$, and to maximize the distance $|f_c - f_{c'}|^2$ between two feature vectors. Thus, the *minimax* energy for the determination of the global minimum associated with the unknown feature f_c is given by,

$$E(f_c) = a \sum_{c \neq c'} (\langle f_c | f_{c'} \rangle) + b \sum_{c=1,2} |f_c - I_c|^2 + d \sum_{c \neq c'} 1/|f_c - f_{c'}|^2. \tag{C1.4.13}$$

Note how the representation permits parametrization of relative feature importance. For example, the Lagrangian multipliers $a = 10$ and $d = 10$ are set higher than $b = 1$ to reflect the less important fact that feature f_c should resemble image I_c. The results using the CA algorithm are provided by Szu (1990). A sample listing in a variant of Basic is provided in appendix B.

In these examples, we have focused on representation issues which clearly have significant impact on the performance of the algorithms. There is nothing significantly unique about the need for representation encodings in neural network applications; however, in digital simulations of stochastic neural models any time improvement afforded by clever representation greatly facilitates the application.

C1.4.4 Summary

We have illustrated that Cauchy annealing is superior to Boltzmann annealing, which in turn is superior to conventional Markov Monte Carlo methods. We have illustrated or referenced how digital implementations of stochastic neural networks are inefficient, except, perhaps, when coupled with mitigating factors such as clever problem representation, deterministic annealing, adaptive simulated annealing, or composite hierarchical architectural topologies (trees for example). It is also clear that for hard problems of large scale, analog (especially optical) implementations hold great promise for extending the applicability of stochastic neural networks.

Appendix A. Proofs of both cooling schedules

There are a number of similarities in the proofs of the cooling schedules for the CA and BA algorithms in D-dimensional vector spaces. For the convenience of comparison, the proofs will be demonstrated in parallel. In locating the minimum, one must start at some position or state in a D-dimensional space, evaluate the function at that state, and generate the next state vector.

The CA and BA algorithms are different in that CA uses a Cauchy distribution and BA uses a Gaussian distribution in their respective state generating functions. Both the BA and CA algorithms will use as their next state either the current state vector or the next state vector provided its incremental cost increase is less than the time-dependent noise bound, which is temperature (and therefore time) dependent.

The CA algorithm requires that *state generating* be *infinitely often in time (IOT)* (in the sense of accumulation in time defined by the negation below) whereas the BA requires the *state visiting* be IOT. At some cooling temperature $T_c(t)$ at time t, let the state generating probability of being within a specific neighborhood be lower bounded by g_t. Then the probability of not generating a state in that neighborhood is upper bounded by $(1 - g_t)$. To insure a globally optimum solution for all temperatures, a state in an arbitrary neighborhood must be able to be generated IOT, which however does not imply ergodicity, the latter requiring actual visits IOT. To prove that a specific cooling schedule maintains the state generation IOT, it is easier to prove the *negation* of the *converse*, namely the *impossibility of never generating a state in the neighborhood after an arbitrary time* t_0. Mathematically this is equivalent to stating that the infinite product of $|1 - g_t|$ terms is zero. Taking the Taylor series expansion of the logarithm of the product, one can alternatively prove that the sum of the g_t terms is infinite. One can now verify cooling schedules in a D-dimensional neighborhood $|\Delta x_0|$ and arbitrary time t_0. Among the various Lévy–Doob distributions (including Cauchy, Holtzmach, and Gaussian) there are two different classes, local (as in CA) and semilocal (as in CA). There exists an initial temperature T_0 and for $t > 0$, such that

$$\text{BA} : T_a(t) = T_0 / \log(t) \tag{C1.4.A1a}$$

$$\text{CA} : T_c(t) = T_0 / t \tag{C1.4.A1b}$$

$$\text{BA} : g_t \approx \exp\left(\frac{|\Delta x_0^2|}{-T_a(t)}\right) T_a(t)^{D/2} \tag{C1.4.A2a}$$

$$\text{CA} : g_t \approx \frac{T_c(t)}{\left(T_c^2 - |\Delta x_0|^2\right)^{(D+1)/2}} \approx \frac{T_0}{t|\Delta x_0|^{D+1}} \tag{C1.4.A2b}$$

$$\text{BA} : \sum_{t=t_0}^{\infty} g_t \geq \sum_{t=t_0}^{\infty} \exp(-\log(t)) = \sum_{t=t_0}^{\infty} \frac{1}{t} = \infty \tag{C1.4.A3a}$$

$$\text{CA} : \sum_{t=t_0}^{\infty} g_t \approx \frac{T_0}{|\Delta x_0|^{D+1}} \sum_{t=t_0}^{\infty} \frac{1}{t} = \infty. \tag{C1.4.A3b}$$

Appendix B.   Cauchy annealing algorithm (Macintosh QuickBasic version)

```
! input two known images and known feature
DATA 4,5,8,9,11,14,15,16,17,38,41,44,46,47,50,51,52,53,56,57 !input 81 Peano-scanning pixel#
          for the black value=1
DATA 58,59,67,69,70,71,72,78,79        !1= black gun barrel, track belt
DATA 4,5,8,9,12,13,14,15,16,17,30,31,37,42,43,46,47,50,51,52
DATA 53,56,57,58,59,62,63,69,70
DIM f1(81),f2(81),ave1(81),ave2(81),ft1(81),ft2(81)
MAT ave2 =0        ! True_Basic Matrix Equating
FOR n=1 to 29        ! read the tank into ave1, namely I1
    READ k
    LET ave1(k)=1
NEXT n
FOR m = 30 to 58        ! read the carrier into ave2, namely I2
READ J
LET ave2(J)=1
NEXT m
! set up Cauchy annealing to determine the unknown feature by mini-max
RANDOM        !random number rnd generated [0,1]
FOR t=1 to tmax        !after initialize the display-
```

```
        LET temp=To/(1+t)        !Cauchy annealing cooling schedule
        LET theta=(rnd-.5)*Pi        !uniform theta using the radian angle option
        LET dx=int(temp*tan(theta))        ! new pixel by T tan(theta)
        LET xnew=mod(x+dx,82)        ! modulo for 81 scan pixels
        IF xnew=0 then LET xnew=81
        IF f2(xnew)=0 THEN
            LET ft2(xnew)=ave2(xnew)
            LET ft1(xnew)=0
        ELSE
            LET ft2(xnew)=0
            LET ft1(xnew)=ave1(xnew)
        END IF
        LET enew= 0
        LET denominator=0
        LET ef1=0
        LET ef2=0
        FOR n=1 to 81
            LET ef1=ef1+(ft1(n)-ave1(n))*(ft1(n)-ave1(n))
            LET ef2=ef2+(ft2(n)-ave2(n))*(ft2(n)-ave2(n))
            LET denominator=denominator+(ft1(n)-ft2(n))*(ft1(n)-ft2(n))
            LET enew = enew + ft1(n)*ft2(n)
        NEXT n
        LET enew= a*enew + b*ef1 + c*ef2 + (d/denominator)
        IF enew<eold then
            MAT f2=ft2
            MAT f1=ft1
            LET eold=enew
            LET x=xnew
        END IF
        IF enew>=eold then
            IF(rnd*0.5)< (1/(1 + exp((enew-eold)/temp))) then        !up-hill climbing
            MAT f2=ft2
            MAT f1=ft1
            LET eold=enew
            LET x=xnew
            END IF
        END IF
! plotting search states, accepted states, and its minimax energy value
        PLOT POINTS :t,xnew+200
        PLOT POINTS :t,x+100
        PLOT POINTS :t,eold/2
```

References

Akiyama Y Y, Anzai Y and Aiso H 1990 The Gaussian machine: a stochastic, continuous neural network model *J. Neural Network Comput.* **2** (3) 43–51

Alspector J, Guputa B and Allen R 1989 Performance of stochastic learning microchip *Neural Information Processing Systems I* (Morgan Kaufmann)

Byrne W 1992 Alternating minimization and Boltzmann machine learning *IEEE Trans. Neural Networks* **3** 612–20

Dayan P, Hinton G E and Neal R M 1995 The Helmholtz machine *Neural Comput.* **7** 889–904

Farhat N H 1987 Optoelectronic analogs of self-programming neural nets: Architecture and methodologies for implementing fast stochastic learning by simulated annealing *Appl. Opt.* **26** 5093–103

Farhat N H and Psaltis D 1987 Optical implementation of associative memory based on models of neural networks *Optical Signal Processing* ed J L Horner (New York: Academic)

Galland C C 1993 The limitations of deterministic Boltzmann machine learning *Network: Computational Neural Syst.* **4** 355–79

Geman S and Geman D 1984 Stochastic relaxation, Gibbs distributions and the Bayesian restoration of images *IEEE Trans. Pattern Anal. Machine Intell.* **6** 614–34

Geyer C J 1993 *Annealing Markov Chain Monte Carlo with Applications to Ancestral Inference* University of Minnesota

Hertz J A P, Krogh R G and Anders S 1991 *Introduction to the Theory of Neural Computation* (Reding, MA: Addison-Wesley)

Hinton G E 1989 Deterministic Boltzmann learning performs most steep descent is weight space. *Neural Comput.* **1** 143–50

Hinton G E, Dayan P, Frey B J and Neal R N 1995 *The Wake–Sleep Algorithm for Unsupervised Neural Networks* University of Toronto

Hinton G E and Sejnowski T J 1986 Learning and Relearning in Boltzmann Machines *Parallel Distributed Processing* ed J Clelland and D Rumelhart (Cambridge, MA: MIT Press) pp 282–317

Hinton G E, Sejnowski T J and Ackley D H 1984 *Boltzmann Machines: Constrained Satisfaction Networks that Learn* Carnegie Mellon University

Hopfield J J 1987 Learning algorithms and probability distributions in feed-forward and feed-back networks *Proc. Natl Acad. Sci. USA* **84** 8429–33

Ingber L 1993 Simulated annealing: Practice versus theory *Math. Comp. Modeling* **18** 29–57

——1995 Adaptive simulated annealing (ASA): lessons learned *Control and Cybernetics Preprint*

Kappen H J 1995a Deterministic learning rules for Boltzmann Machines. *Neural Networks* **8** 537–548

——1995b *Radial basis Boltzmann machines and learning with missing values* University of Nijmegen

Kirkpatrick S, Gelatt C Jr and Vecchi M P 1983 Optimization by simulated annealing *Science* **220** 671–80

Metropolis N, Rosenbluth A W, Rosenbluth M N and Teller A H 1953 Equations of state calculations for fast computing machines *J. Chem. Phys.* **21** 1087–91

Parberry I and Schnitger G 1989 Relating Boltzmann machines to conventional models of computation. *Neural Networks* **2** 29–67

Peterson C 1987 A mean field theory learning algorithm for neural networks *Complex Syst.* **1** 995–1019

Saul L J and Michael I 1994 Learning in Boltzmann trees *Neural Comput.* **6** 1174–84

Scheff K and Szu H 1987 1-D optical Cauchy machine infinite film spectrum search *IEEE Int. Conf. on Neural Networks (San Diego, 1987)*

Sejnowski T J and Rosenberg C R 1987 Parallel networks that learn to pronounce English text *Complex Syst.* **1** 145–68

Skubiszewski M 1992 *An Exact Hardware Implementation of the Boltzmann Machine* Digital Equipment Corporation

Smith W E, Barrett H H and Paxman R G 1983 Reconstruction of objects from coded images by simulated annealing *Opt. Lett.* **8** 199–201

Szu H H 1986 Non-convex optimization. Real time signal processing IX *SPIE* vol 698 (Bellingham, WA: SPIE) pp 59–65

——1987 *Fast simulated annealing* Neural Networks for Computing, Snow Bird, Utah (New York: AIP)

Szu H and Blodgett J 1982 Self-reference spatiotemporal image–restoration technique. *J. Opt. Soc. Am.* **72** 1666–9

Szu H, Blodgett J and Sica L 1980 Local instances of good seeing *Opt. Commun.* **35** 317–22

Szu H and Hartley R 1987 Nonconvex optimization by fast simulated annealing *Proc. IEEE* **75** 1538–40

Szu H and Messner R 1986 Adaptive invariant novelty filters *Proc. IEEE* **74** 519

Szu H and Scheff K 1990 Simulated annealing feature extraction from occluded and cluttered objects *Int. Joint Conf. on Neural Networks (Washington, DC, 1990)*

Takefuji Y and Szu H 1989 Parallel distributed cauchy machine *Int. Joint Conf. on Neural Networks (Washington, DC, 1990)*

Tishby N 1995 Statistical physics models of supervised learning *The Mathematics of Generalization. Proc. SFI/CNLS Workshop on Formal Approaches to Supervised Learning* (Santa Fe, NM: Addison-Wesley)

Wasserman P D 1989a A combined back-propagation/Cauchy machine network *J. Neural Network Comput.* **1** (3) 34–40

——1989b *Neural Computing Theory and Practice* (New York: Van Nostrand Reinhold)

Yuille A L 1994 Statistical Physics Algorithms that Converge. *Neural Comput.* **6** 341–56

Zerubia J and Rama C 1993 Mean field annealing using compound Gauss–Markov random fields for edge detection and image estimation *IEEE Trans. Neural Networks* **4** 703–9

C1.5 Weightless and other memory-based networks

Igor Aleksander and Helen B Morton

Abstract

Several models are described which stem from the notion that an artificial neuron is a variable-logic decision device and may be regarded as an intelligent lookup table. Feedforward and feedback systems are discussed and related to standard networks. The systems have a history which stretches back to 1965 and have led to machines which have been used to some effect in industrial applications. One such machine is the WISARD which is described in some detail. The key features of such systems are that they may be easily implemented using conventional digital technology and that they provide optimal results with one-shot learning. The systems are unashamedly binary but may be used in a probabilistic mode and may be adjusted for different levels of generalization. The relationship of weightless systems to similar schemes is discussed: specifically, Kanerva's sparse memory methods and the ADAM system developed by Austin.

C1.5.1 Introduction

The concept described first in this section is 'weightlessness'—the use of memory nodes whose function is altered not by the changing of weights but by altering the contents of a memory device. In the world of neural computing, this approach is seen as being somewhat unorthodox. It starts by using a conventional random access memory as a neural node. It is shown that a single-layer group of such nodes (called a 'discriminator') acts very much like a *single-layer perceptron*. The advantages of the discriminator C1.1 approach are that its behavior is amenable to simple analysis and that it is easily implemented with conventional computer hardware. The article looks at a multidiscriminator system used in industrial settings—the WISARD. A way of calculating and optimizing the behavior of this system is given. On the whole, it turns out that usable neural systems with a clear, predictable performance may be obtained through the multidiscriminator approach. The WISARD is a product of the early 1980s. More recently there have been other variants of this technique which are probabilistic in kind and in which generalization can be selected and controlled. These are described, as are other binary systems related to weightless notions, notably the work on sparse memory by Kanerva and Austin's ADAM system.

WISARD is an acronym for WIlkie, Stonham and Aleksander's Recognition Device. It is an adaptive *pattern recognition* machine which is based on neural principles. The prototype was completed in 1981 at B6 Brunel University in London by a team under the direction of one of the authors (IA). Bruce Wilkie was the design engineer and John Stonham his faculty supervisor. The machine was subsequently patented and produced commercially in 1984. Here the object is to examine the principles on which these machines are based. The history of these principles goes back to 1965, when it was suggested that a simple memory device (which nowadays would be called a ROM, read-only memory) has neural-like properties (Aleksander 1965). The ROM is a once-only learning device. Over the years, it has been shown that the same neural-like properties are held by RAM (random-access memory). RAM devices can learn and relearn. Here we examine these learning properties of RAM systems and develop simple ways of analyzing networks of such devices.

Broadly, the difference between a conventional *McCulloch and Pitts* (MCP) weighted node and a B1.2
RAM node is that the RAM can achieve any of the functions of its inputs but cannot *generalize*. This B3.5
makes it appear to be a less interesting device than the MCP. However, we shall show that *networks* of
RAM nodes *do* generalize and that this makes them as interesting as networks of MCP nodes (if not more
so). In the latter parts of the article we show how even single RAMs can be made to generalize. The
main advantage of the RAM node is that systems may be built using conventional digital circuitry, without
the need to develop special VLSI devices. It will also be shown that learning in RAM networks is much
faster than in MCP networks.

C1.5.2 The RAM neuron

The random-access memory device is the silicon building brick of the local memory of any modern
computer. Designing neurocomputers which exploit silicon RAM leads more directly to usable machines
of significant capacity than relying on node designs which still require VLSI development. The principal
components of a RAM device are an address decoder, a group of memory registers, a data-in register
and a data-out register. To store information, an N-input binary *address* is supplied to the input of the
decoder. The output of this decoder has 2^N lines, one for each possible address, i.e. a combination of 0s
and 1s on the N input terminals. The presence of one such pattern at the input of the decoder energizes
the corresponding line and makes the memory register connected to the line active. The active memory
register absorbs the data held in the data-in register and stores it. For a typical, commercially available
RAM chip, M (the number of bits of information that can be stored in each memory register) is 8. This
type of chip is said to be 'byte-oriented'—a byte being eight bits. Typically, N could be 18 and this would
be called a 256 K RAM (2^{18} is 262 144, but it is conventional to name these devices with the first three
digits of the nearest power of 2 which, in this case, is 256).

In order to relate the RAM to the neuron, two further points need to be understood. First, it is noted
that taking one bit of the data input and the corresponding bit of the output, the values of this bit can
be set independently, and represent precisely the truth table of a logic device with one output. As whole
words are always written into the memory one row at a time, this independent setting of a particular bit
in a column is done by selecting the selected bit in the word, leaving the other bits unaltered. Therefore,
a RAM with M bits in the memory register can be thought of as M RAMs each with one bit per memory
register, and each connected to the same N input variables.

The second step is to concentrate only on one of these one-bit-per-word RAMs. The sense in which
such a RAM is *like a neuron* is that, given an input at X_1 to X_N at the address terminals and a desired
output to be held at the data-in terminal, setting the RAM into the reading mode will cause it to 'learn'
this desired response, and this can be overwritten by a subsequent training step. The sense in which this
is *not like a neuron* is that there is no need for a sophisticated training algorithm—the setting for one
input does not affect another and, therefore, the description in the last paragraph *is* the training algorithm.
Admittedly, there is no generalization in the RAM itself. While this could be seen as a disadvantage, it is
shown in the subsequent sections that *networks* of RAMs generalize in a way which is similar to networks
of neurons. It will also be shown later that the RAM itself can be made to generalize.

C1.5.3 A discriminator of RAM neurons

The simplest RAM network with properties of generalization is called a *discriminator* and is shown in
figure C1.5.1. This consists of a layer of K RAMs with N inputs and thus each RAM stores 2^N one-bit
words, and the single layer receives a binary pattern of KN bits. It is assumed that, before any training
takes place, all the memory cells in the RAMs are set to 0. Training consists of applying an 'input pattern'
of 0s and 1s at the input terminals shown in figure C1.5.3. This is called a *training pattern* and is an
example of the class of patterns to be learnt by the discriminator. To record this pattern, a 1 is stored in
that memory location of each RAM which is addressed by this input pattern. Effectively, this causes each
RAM to record the occurrence of part of the input pattern—the part 'sampled' by that RAM. This is done
for other input patterns leading to further 1s being stored in the RAMs.

The RAMs can be switched from a 'write' mode for training, as described above, to a 'read' mode
during which what has been learnt can be used. In this latter phase, when another, previously unseen
pattern is later presented at the input, the summing device (denoted $\sum$) produces a number which is equal
to the number of RAMs that output a 1. This number is said to be the *response* of the discriminator and

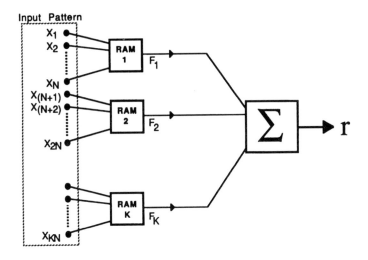

Figure C1.5.1. A RAM discriminator.

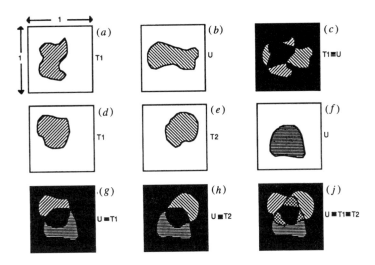

Figure C1.5.2. Discriminator analysis based on overlapping areas.

given the symbol r. Clearly, if one of the patterns used in the training set were to be entered later at the input of the network, it would find storage locations that contain a 1 in each of the RAMs and therefore r would have its maximum value of K, whereas, if the unknown input pattern were to be totally different from any of the training patterns (in the sense that no RAM would receive an individual input on which it had been trained) then the value of r would be 0. Therefore, r is, in some way, a measure of the similarity of an unknown pattern to each of the patterns in the training set. It is worth looking at this idea in greater depth. Figure C1.5.2 is an aid in visualizing the nature of this analysis.

We assume that a RAM with N inputs is connected quite arbitrarily somewhere within an image area whose dimensions are chosen to be one unit by one unit as shown in figure C1.5.2(a). Also in figure C1.5.2(a) is the first training pattern for this network, T_1. Before going further, we shall show how the network responds to an unknown test pattern U shown in figure C1.5.2(b). For a RAM to output a 1 for U, all its N inputs must receive exactly the same pattern for T_1 and for U. In figure C1.5.2(c) the overlap area that is the same for T_1 and U is shown in black. Let us say that it measures A_1 area units. (As the total area of the image is unity, A_1 must be less than 1.) If any point within the image can be selected with equal probability, the probability of such a point being in A_1 is $A_1/1$, that is, precisely A_1. As this is the probability for receiving the same input in T_1 and U for any of the inputs of the RAM in question, the probability for all N RAMs being so connected is $A_1 \times A_1 \times \cdots \times A_1$ (N times), i.e. $(A_1)^N$. That is, the probability of an arbitrarily connected RAM outputting a 1 is $(A_1)^N$.

Assume now that there is a large number of such arbitrarily connected RAMs. By the law of large numbers, a proportion $(A_1)^N$ of this number would fire with a 1 and the rest with a 0. This is called the *relative response* of the network, R. So the absolute response r for a network with K RAMs is

$$r = KR = K(A_1)^N.$$

The same form of reasoning can now be extended to a system trained on two patterns T_1 and T_2 as shown in figures C1.5.2(d) and (e), respectively. The unknown pattern, U, is shown in figure C1.5.2(f). This time, the arbitrarily connected RAM will output a 1 if all its inputs are in the overlap area between T_1 and U (shown in figure C1.5.2(g) as a black area and dubbed A_1) or in the overlap area between T_2 and U (shown in figure C1.5.2(h) as a black area and dubbed A_2). Again the probability of connecting a RAM to the first of these areas is $(A_1)^N$ and the second is $(A_2)^N$. To get the total probability of the RAM firing with a 1, the probability of these two events can be added, provided that one subtracts the event of connecting to an area *common to the two events* which would otherwise be counted twice. This area is shown in black in figure C1.5.2(j) and is the overlap of U, T_1 and T_2. We call this area A_{12} and note that the probability of connecting to it is $(A_{12})^N$. So, the relative response for the system trained on two patterns is:

$$R = (A_1)^N + (A_2)^N - (A_{12})^N. \tag{C1.5.1}$$

The final step of this analysis is to extend equation (C1.5.1) to any number (say E) of training patterns $T_1, T_2, T_3, \ldots, T_E$. The form of such an expression is the same as (C1.5.1). That is, first U is overlapped with all the training patterns to calculate the contribution to the response, then overlap with pairs has to be removed to correct for double counting: but this takes away too much as it also removes the overlap of three training patterns once too many and this has to be put back, and so on. Formally, this is written as:

$$R = (A_1)^N + (A_2)^N + \cdots + (A_E)^N - (A_{12})^N - (A_{13})^N$$
$$- \cdots - (A_{1E})^N - \cdots - (A_{[E-1]E})^N + (A_{123})^N + (A_{124})^N + \cdots - (A_{1234\ldots E})^N. \tag{C1.5.2}$$

(The last term in an equation such as this is negative if E is even and positive if E is odd.)

This formidable looking formula is really no different from (C1.5.1), there is just more of it. The main characteristic of the system (and equation (C1.5.2)) is that R is 1 if U is one of the training patterns. This can be understood either from the description of the system in the earlier parts of this section or from realizing that if, in equation (C1.5.2), $U = T_1$, say, then $A_1 = 1$ and the rest of the equation adds up to 0. Also if U is close to any one of the training patterns, this makes R closer to 1 by an amount which we shall find depends on N.

A legitimate question that can be asked at this point is why is it worth going to all the trouble of inventing discriminators, since the value of R as given by (C1.5.2) could clearly be calculated on any computer. In other words, U could be compared to all the stored training patterns and their combinations, thereby generating all the overlap counts needed by equation (C1.5.2). The value of R could then be computed as a result of this exhaustive search. On the other hand, a specially built hardware discriminator delivers R in just one computation (one pass through the network) thus avoiding long searches and overlap calculations as would be carried out in a simulation on a conventional serial machine.

C1.5.4 The WISARD

A multidiscriminator system has each of its discriminators trained to a different class of object. If the task is one of recognizing the hand-printed letters of the alphabet, say, then the scheme would contain 26 discriminators, one for each letter. The notion of a multidiscriminator system is quite general and takes the form shown in figure C1.5.7. The WISARD is a *hardware implementation* of this scheme directed towards the *recognition of images*. The hardware will be discussed later—here we concentrate on the principles of the arrangement.

We assume that there are 26 discriminators, each of which covers a binary 'image' with K RAMs of N address inputs each while, in theory (such as led to the calculations and predictions in the examples of the last section), it is assumed that the K RAMs are randomly connected to the image with no constraints; in practice a constraint is added—each image input is connected to precisely one RAM input. In other words the size of the image is KN binary picture points. This is done to ensure that the image is evenly covered with a minimum number of RAMs. Several points should be noted here.

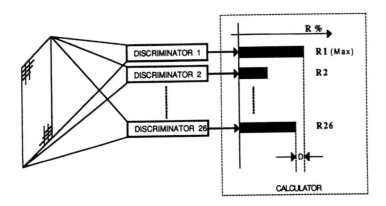

Figure C1.5.3. A multidiscriminator system (the WISARD).

(i)   Although one talks of 'images', these ideas can be applied to any binary data (e.g. sampled *speech* F1.7 *signals*).
(ii) Although one refers to binary picture points, an image could use more than one binary input per picture point. For example, if 16 gray levels are used there are at least 4 binary inputs for each such point. In this case, it is assumed that all these binary points are input randomly to the learning network.
(iii) Whether the discriminators have identical input connections or not hardly matters. Each discriminator is trained to its own class of pattern, and therefore whether it is wired similarly to or differently from other discriminators is not of much consequence.

It is assumed that at the start of any training regime, all the RAMs of all the discriminators are set to 0. The training consists of setting to 1 the outputs of all the RAMs in the discriminator which is appropriate to the desired class. Say that a 26-discriminator system is being trained to recognize hand-printed characters, that the system is currently being trained to recognize a hand-printed letter A, and that discriminator 1 is designated to recognize As. Then discriminator 1 is trained to respond to a version of A with a 1 at all the RAMs it possesses. This is repeated for many other slightly different versions of As. The entire process is repeated for many examples of each of the other letters, taking care that only the appropriate discriminator for each letter is trained.

After training is complete, a response R_j (% of K RAMs that output a 1) will occur at the jth discriminator (indeed, this is true for all values of j from 1 to 26) for the presentation of an unknown pattern to the entire system. The system *recognizes* the unknown pattern as belonging to the class for which R_j is highest. This comparison and selection is performed by the calculator section of the system shown in figure C1.5.3. The key mechanism at work in determining the response of each discriminator is that described by equation (C1.5.2). Some examples will be used to illustrate this, but first, two more tasks performed by the calculator need to be described. The first is a measure of *absolute confidence*. This is merely the actual value of the highest R_j. Should this be close to 100%, the system is saying 'not only is this a member of class j, but also it is very much like one of the training patterns in that class'. Should the highest R_j be low, however, this can be interpreted as the system saying 'this pattern is not much like any that have been used in training, but, if pushed, I will say that it is a member of class j'.

The second additional task done by the calculator is to provide a measure of *relative confidence C*. This is calculated by looking at the difference D between the highest R_j and the second highest. C is then given by the simple formula:

$$C = \frac{D}{R_j}. \tag{C1.5.3}$$

To illustrate the operation of this system we look at an example which involves the prediction of its performance when faced with having to recognize patterns in the presence of noise. One of the very important areas of application of neural networks is in the monitoring of premises for the presence of intruders. Figure C1.5.8 shows a highly stylized version of what this entails.

The task centers on training one discriminator on a particular scene (a room, or an airfield, say) under normal conditions, that is, with no intruders present. A second discriminator is trained with intruders in many possible positions. This has sometimes to be done under very poor lighting conditions, and so the

Figure C1.5.4. Intruder detection in a noisy image of a room.

machine is required to operate correctly even with very poor images. Noise (which looks like 'snow' on a television set) occurs when TV cameras are made to operate in very poor lighting conditions. On a binary image we represent this as the alteration of some picture points from black to white and from white to black. Noise is measured in percentage terms, as a percentage probability of a picture point being affected by it.

Figure C1.5.8 shows the normal image (T_1) of a room on which discriminator 1 is trained. I_1, I_2 and I_3 are images containing an intruder on which disriminator 2 is trained. U_1 is a test pattern which is intruder-free, but contains roughly 40% noise, while U_2 is a test image containing an intruder and 40% noise. Using the theory developed earlier, the responses of the two discriminators to the two test patterns may be calculated once the effect of noise has been formulated. This is done as follows. Taking the overlap between any training pattern X and some unknown test pattern V as being A, then, given $s\%$ noise, the A area will lose $s\%$ (i.e. it will be $A(1 - s/100)$), while the $1 - A$ area will gain an overlap of $(1 - A)s/100$. Hence we have an expression for the noisy overlap A' as a function of the non-noisy one:

$$A' = A - \frac{As}{100} + \frac{(1 - A)s}{100}.$$

This may be simplified a little to

$$A' = A\left(1 - \frac{2s}{100}\right) + \frac{s}{100}.$$

So, using equation (C1.5.2), assuming that the intruder shadow covers 1/6 of the image and that the shadows in I_1, I_2 and I_3 do not overlap, then using the overlap values modified by noise as shown above, the *relative confidence* of the two-discriminator system may be calculated from (C1.5.3) and tabulated as shown in table C1.5.1 (rounded) for $N = 8$.

Table C1.5.1. Relative confidence of the two-discriminator system. (In these calculations it is assumed that the intruder has overlapped with one of the intruder training patterns.)

Noise	Intruder absent			Intruder present		
$s\%$	$R_1\%$	$R_2\%$	$C\%$	$R_1\%$	$R2\%$	$C\%$
0	100	58	42	23	100	77
10	43	28	34	12	43	72
20	17	13	25	5.7	16	65
30	5.7	4.9	14	2.5	5.7	55
40	1.7	1.6	3.7	1.1	1.7	37
50	0.39	0.39	0	0.39	0.39	0

Where only one discriminator is used (say, discriminator number 1) putting a threshold on the response that discerns the presence of an intruder could be done only if the exact amount of noise were known

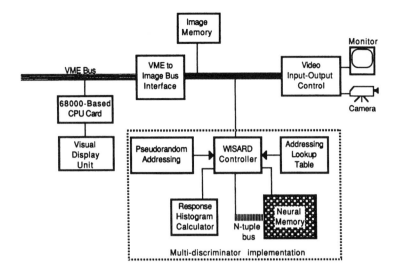

Figure C1.5.5. Details of the WISARD architecture.

beforehand. That is, were it known that the noise was precisely 20%, the presence of an intruder of the size assumed would cause the response of the discriminator to fall from 17% to 5.7%, which would be easily discernible with a threshold of, say, 10%. But were the noise to change from 20% to 30% this would be interpreted as the presence of an intruder and would raise a false alarm.

However, in these calculations it is the relative values of the two discriminators that are indicative of the presence of the intruder, R_1 being greater than R_2 without the intruder and the reverse when the intruder is there. This is true of any noise value up to 50%. Of course, 50% noise obliterates any meaningful pattern, as it is no longer possible to know whether any bit has its true value or a value due to noise. The confidence too is an indication of the level of noise. Perhaps it is worth noting that the confidence is greater in the presence of the intruder—this is due to the fact that the 'intruder-detecting' discriminators have had more training and generally give a stronger response to anything. In practice, this imbalance could be corrected by training the non-intruder discriminator on noisy images.

In a more general sense, it is the fact that there is no need to select a threshold that gives strength to the multidiscriminator method. Put simply, it allows the system to say 'the image before me is nothing like the images that I have been trained on, but if pressed, I will say that it is more like X than any other training pattern'. The WISARD is currently being used in security applications such as described above and in *quality control tasks* where it is used to identify and classify faults in products and to measure the alignment of parts on production lines. G2.8, G2.12

Details are given below of the hardware that has been engineered by Computer Recognition Systems in the United Kingdom to produce a commercial version of the WISARD idea. Figure C1.5.5 shows a block diagram of the system.

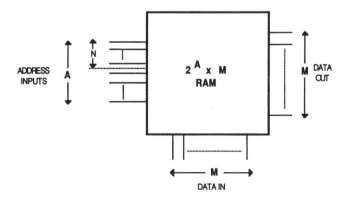

Figure C1.5.6. Partitioning a large memory to make the neural memory.

Most of this equipment is a general purpose image processing system. A video input/output controller digitizes the image picked up by the camera and transfers the resulting bits to an image memory or 'framestore'. Typically this picks up an image of 512×512 picture points (or pixels), which in the case of WISARD are only on or off. The digitized image may be output, again via the image bus and input/output controller, and displayed on a monitor over the original image. The controller can also allow the user to select the size and position of the digitized image.

The part of the system described so far is interfaced to a standard microprocessor system which carries the control software for the entire system and allows the user to select parameters displayed in 'menu' fashion on the visual display unit. The not-so-conventional part of the system is shown within the dotted frame and largely consists of a large memory which can be partitioned to act as the notional system shown in figure C1.5.3. The way in which this partitioning may be achieved is shown in figure C1.5.6.

Under control of the 68000 microprocessor chip, the user can select a value of N and the size of the window which he wishes to use (say $X \times Y$). This determines the number of notional RAMs per discriminator (XY/N). The number of discriminators available is M, that is, each bit of a stored word contributes to the output of a different discriminator. If the total amount of memory available is 2^A words of M bits it requires A address terminals. N of these are used as RAM inputs while the other $A - N$ are used to index the individual RAMs. Therefore the number of RAMs per discriminator cannot exceed 2^{A-N}.

The random connection to the input image is arranged by a pseudorandom generator or a predefined lookup table, which builds up the N address values for the memory after 'picking off' specific picture points from the image memory. The other $(A - N)$ address terminals of the memory are addressed in a systematic manner each time a complete group of N (N-tuple) has been brought together. During training only one of the M terminals at one time (corresponding to the discriminator being trained) is energized and set to 1. The rest of the M terminals are left in the 'non-writing' state. During the 'use' phase, counters carry out a tally of the number of 1s that are generated by each of the M data-out lines, providing the histogram of R_j responses on which the output of the system can rapidly be calculated. The user can set values of confidence and response that are required to drive overall output lines (e.g. robot controls or relays that operate gates on a conveyor belt).

Clearly, the operation of this system is serial, but the access time of the image memory can be made high enough for all the RAMs in the neural memory to be addressed within a short time. Some typical figures are given below.

- Size of neural memory: 2 megabytes
- Value of M: 8
- Value of A : 20
- Value of N: 4
- Number of RAMs per discriminator: $2^{20-4} = 64\,536$
- Number of input image points that can be covered by a discriminator: $4 \times 64\,536 = 2^{18}$
- This means that a 512×512 image (2^{18}) can be completely covered
- Time for a training or testing operation of the entire network (independent of N and window size): 0.08 seconds.

Provision is made in the WISARD for partial coverage if large windows are being used with large values of N. The parameters (N, M, window size and position, coverage) are selected by the user from the menu on the VDU. A warning is issued if the memory requirement is exceeded.

Further details on the architecture of the commercial version of the WISARD are in Aleksander *et al* (1984).

C1.5.5 Probabilistic and generalizing weightless neurons

Recently, the B bits at each storage location of a RAM have been considered to store a number in the interval from 0 to 1 which represents the 'firing probability', $P(1)$, of the neuron. Gorse and Taylor (1989) have indeed assumed that B is so large as to allow them to analyze their node (which they called a p-RAM) as storing continuous values of firing probability. Myers (1990) has investigated RAM systems with $M = 2^B$ well-defined probabilistic states calling them M-PLNs: M-valued probabilistic logic nodes.

It is assumed that a RAM node (the discourse is restricted here to M-PLNs) receives global training

C3 signals of a *reinforcement* kind. That is, a reward or punish signal is distributed globally to a prescribed section of the system, and every node in that section receives the same signal. In this paper it is assumed that the set of M values has at least three elements:

$$P(1) = 0 \qquad P(1) = 1 \quad \text{and} \quad P(1) = 0.5.$$

At the commencement of training all nodes for all inputs store $P(1) = 0.5$. As a final assumption, a clocked timing system is required. At the arrival of every clock signal each node will either fire or not fire (i.e. output a 1 or a 0, respectively). Over a stretch of many clock periods the node will have fired with a frequency approaching $P(1)$. At the arrival of a reinforcement signal, each node 'knows' whether at the last clock point it fired or not. If the reinforcement is positive (i.e. a reward) the last firing value (0 or 1) is stored. If the reinforcement is negative and the stored value is 0 or 1 the value is returned to 0.5.

A modified RAM—the G-RAM—which generalizes internally, has been suggested by Aleksander (1990). This device operates in three phases. Two of them are the usual learning phase and operating phase during which the device records the addresses with their required response and uses the stored information, respectively. The third phase is unusual: it is called the 'spreading' phase. Spreading refers to a process of affecting the content of storage locations, not addressed during training, by the use of a suitable algorithm which may be implemented on-chip or through appropriate actions in the control machinery. Whatever the implementation, spreading is something that can be done 'off-line', that is, between the time that training information has been captured (which may have to be done in some kind of 'real time') and the time that the nodes have to use what has to be learnt (which may also have to be done at speed).

We assume that spreading has taken place and use a simple model of its effect. The training set sets some of the addresses to 0 and others to 1. Full generalization means that any other address will be set to 0 if it is closest in Hamming distance to one of the 0 training patterns or 1 if it is closest to one of the 1 training patterns. So if the training patterns were 00000000 set to 0 and 11110000 set to 1, only addresses that are equidistant from these two patterns (such as 00110011) would be left with $P(1) = 0.5$, while others such as 00001110 and 10000000 would both be set to 0 as they are distinctly nearer to the 00000000 pattern. The simple model is this:

> If a node is sampling N points of a pattern, given an unknown pattern u and that a majority of N is in a pattern area that distinguishes between a required 0 output and a required 1 output, the appropriate pattern will be generated for u.

C1.5.6 The general neural unit

The structure of a general neural unit (GNU) is shown in figure C1.5.7. The circuit parameters of the GNU are: K, the number of neurons in the unit; W, the width of a binary input interface; N, the number of connections that each neuron receives from the input and Q, the proportion of K that each neuron receives as input. So the number of inputs to each neuron is $N + QK$. We generally let $F = QK$ as this is the number of inputs that a node receives from other nodes in the GNU. A further parameter that needs specification is the degree of generalization G of each neuron. It is of some interest that the properties of the unit can be discussed without detailed reference to the function or structure of the node. In fact, the most direct thing that can be done is to assume that the node is a G-RAM with maximum generalization as specified earlier.

With this set of parameters, particularly with the variation of Q, the GNU can be varied from a single layer of a *feedforward* structure ($Q = 0$) to a Hopfield-like *autoassociator* ($N = 0$, $Q = 1$). But what is more interesting are the modes of behavior that can be obtained between these extremes. Some examples now follow. The *Hopfield model* provides an explanation for useful autoassociation properties in systems which by our formulation have $Q = 1$, $N = 0$ and the node generalization of linearly separable functions associated with weighted neurons. Additionally, much of the analysis depends on the existence of reciprocal connections between neurons. From the point of view of biological modeling this seems wrong, because as far as the author is aware, neural clusters with $Q = 1$ have not been found in living brains. More typical are areas such as C_{A3} in the hippocampus where Q is of the order of 5% . These have been highlighted by Rolls (1989) and have led others to use statistical mechanics methods to analyze low-connectivity autoassociators. From a technological standpoint too, it is important to understand the effect that low Q has on the memory properties of the GNU, as Q determines the cost growth of the unit with K.

B2.3, C1.3.2

B1.3

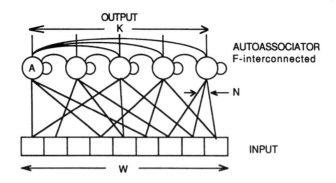

Figure C1.5.7. A general neural unit.

Here, systems with $N = 0$ are analyzed and it is shown that a constant number of inputs per neuron leads to constant performance independently of K. Technologically this is good news because cost increases linearly with the size of the GNU, the performance being set by the number of feedback inputs F per neuron. So Q becomes a dependent parameter of the system with $Q = F/K$. The training of a GNU with $N = 0$ consists of assuming that, one by one, the patterns of a training set $T = \{t_1, t_2, \ldots, t_n\}$ are forced onto the output terminals of the neurons so determining their inputs as well as their target outputs. The adaptation in the neurons is such as to cause the inputs to generate the target outputs, keeping each of t_j 'stable' in the network with time. Given an unknown pattern u it is expected that, after a number of transient states, the GNU will stabilize in the element of T, which in some way (say, in Hamming distance) is most similar to u. The primary performance parameters, therefore, are, first, this retrieval ability and second, the storage capacity (i.e. the number of patterns S that T can contain and that are absorbed by the GNU without interference). A secondary performance parameter is the speed (i.e. number of transient steps) within which the retrieval is achieved. It will first be shown that storage capacity is independent of K and heavily dependent on F.

Consider a binary pattern u currently present at the K binary state variables of what, after all, is an autonomous finite-state machine. The next value (assuming the presence of a clock) of a particular state variable will be solely determined by the computation that can be achieved on the basis of its F input terminals. Assuming that they are randomly connected to the K state variables, and that the computation is related to some measure of similarity with the F-tuples seen during training, the statistical distribution of the next value of the K state variables can only be related to the amount of information contained in K independent F-tuples. This depends on the relative similarities of the patterns in T and pattern u, which can be specified in terms of proportional pattern areas rather than being a function of K. To be clearer about this we shall look at some specific performance parameters in simplified circuit conditions.

Storage capacity needs to be defined probabilistically, to which end the concept of a *contradiction* is used. A contradiction occurs if, after training on t_j, training on t_k causes t_j no longer to be a stable pattern in the GNU. For two patterns t_j and t_k that are the same for a_{jk} of the area of the entire pattern, a contradiction occurs if for any one of the K neurons different targets are required for the same input. The storage capacity of the GNU is then determined by the number of training patterns (and assumptions about their similarity) that yield some tolerable level of contradiction probability. More precisely, consider two patterns t_j and t_k that have a proportional overlap area a_{jk}. The probability that any neuron will have the same input for the two patterns is $(a_{jk})^F$ while the probability that such neurons require different outputs is $(1 - a_{jk})$. So the probability of a contradiction for patterns t_j and t_k, $P(C_{jk})$, is

$$P(C_{jk}) = (1 - a_{jk})(a_{jk})^F . \tag{C1.5.4}$$

It can easily be shown that if S patterns are to be stored in such a system, and if there is choice over the coding of the patterns, such that the overlap between any two of them is the same and minimal, this minimal overlap $a_{jk}(\min)$ is

$$a_{jk}(\min) = 1 - \frac{2}{S} . \tag{C1.5.5}$$

This gives a lowest bound on the probability of contradiction for any two training patterns:

$$P(C_{jk}) \min = \frac{2}{S}\left(1 - \frac{2}{S}\right)^F . \tag{C1.5.6}$$

This needs to be extended to take into account the cumulative effect of $S-1$ patterns on the disruption of any one pattern. The principle involved in doing this is to account for new disruptions threatened as new training patterns are added to the GNU. As seen, the disruptive effect of t_2 on t_1 is $(2/S)[(S-2)/S]^F$. If one now considers the additional effect of t_3 on t_1, a new group of $1/S$ differing neuron outputs becomes threatened again by an area of $[(S-2)/S]^F$ equal inputs. Also $1/S$ differing neuron outputs that have already been accounted for in t_2, are now threatened by a new group of inputs that is identified by the expression

$$\left(\frac{S-2}{S}\right)^F - \left(\frac{S-3}{S}\right)^F.$$

(C1.5.7)

Repeating this to account for all the $S-1$ patterns leads to the overall probability of disrupting a trained pattern $Pd(S)$:

$$Pd(S) = \frac{2(S-1)[(S-2)/S]^F - (S-2)[(S-3)/S]^F}{S}.$$

(C1.5.8)

There are several characteristics of this somewhat bizarre expression that are worth noting.

- It confirms that which is obvious—any GNU can store two orthogonal patterns as $Pd(S)$ evaluates to 0 for $S = 2$
- It also confirms that for large S the probability of disruption tends to 1
- More interestingly, it provides a relationship between F and S for a given limit of acceptability for the value of $Pd(S)$. For example, the following list of values has been computed empirically for $Pd(S)$ between 10% and 15%:

$$
\begin{array}{llllllll}
F = & 2 & 4 & 8 & 16 & 32 & \cdots & \text{Large } F \\
S = & 3 & 4 & 7 & 14 & 27 & \cdots & 0.8 \text{ (Large } F)
\end{array}
$$

- Also, it shows that if S is held to the same value as F, the probability of disruption tends to 22.08%.
- The last observation can be generalized as one would expect the limiting value of $Pd(S)$ for large S to fall logarithmically with F/S. An empirical relationship is:

$$\log_{10}[Pd(S)] = 0.3 - 0.85\frac{F}{S}.$$

(C1.5.9)

A *major conclusion* can be drawn from this analysis. The storage capacity depends primarily on the *fan-in* B2.6.1
F of each node as a result of being able to model the K-node output pattern as a field of signals which the node inputs sample. This has implications not only for the design of artificial perceptual systems but also for the analysis of biological systems as F is a measurable parameter.

Work on weightless systems has recently developed into areas of combined weighted and weightless algorithms (Aleksander *et al* 1994) and neural state machines which are aimed at representing symbols and their perceptual meaning (Aleksander *et al* 1984).

C1.5.7 Kanerva's sparse distributed memory methods

Pennti Kanerva is a Finnish scientist who, while working in the United States, developed a method of associative storage which is both physiologically plausible and pragmatically attractive for implementation in hardware. It is impossible to do justice in a few paragraphs to the well argued case that Kanerva himself has made (Kanerva 1988) and the depth of understanding that this represents of the similarity distributions of randomly selected binary vectors. The interested reader should refer to the original text, while here we merely illustrate the principles on which the method is based.

The method starts with the observation that, in common with the weightless methods discussed in the above paragraphs, a pattern recognition memory should be addressed by the patterns to be recognized with the contents of the addresses being the results of the recognition. Kanerva observes, as we have done in the case of the WISARD system, that the number of meaningful addresses of, say, an n-bit pattern vector, is very small with respect to the total of 2^n. A conventional computer memory would require 2^n storage locations most of which would not be used. In WISARD this redundancy is removed by sampling of the n-bit space, while in Kanerva's method a special memory with a reduced number of locations is constructed, retaining the entire n-bit width as the address vector. The locations have arbitrary addresses, and the method is based on the way that patterns are mapped into these arbitrary addresses. The system is best understood by referring to a physical implementation as shown in figure C1.5.8.

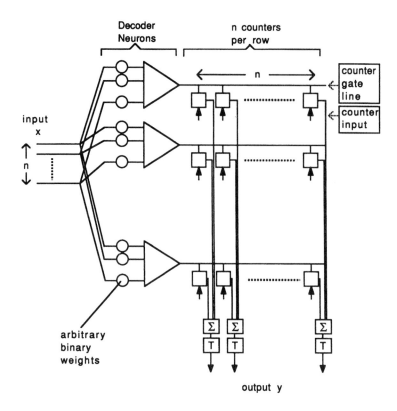

Figure C1.5.8. Kanerva's sparse, distributed memory.

The key to this arrangement lies in the 'decoder neuron'. Each of these is primarily tuned to one arbitrary pattern by means of its binary weights. So if the weights were set to, say, 011101 ($n = 6$), then the neuron activation is a maximum of 6, for $x = 011101$ (counting the contribution of one synapse as unity if its weight matches its input). Were the threshold of the neuron set at ≥ 6 then the neuron would act as a memory address decoder only for the x address 011101. But the threshold is set and left fixed at a lower value in which case the neuron decodes more than one address. For example, were the threshold set at ≥ 5 then the neuron would decode not only 011101, but also all patterns differing from 011101 by one bit (i.e. at a Hamming distance of 1). So in this case a total of seven patterns would be decoded by this neuron. The distance which each decoder is prepared to accept is called the *radius* of the decoder. The word 'sparse' is used to describe this method because the number of decoders is very much less than 2^n. In the above example the number of decoders might be, say 3, the other two being tuned arbitrarily to, say, 100001 and 010011.

With the decoding neurons set as suggested above (i.e. with the weights determined by the toss of a coin) say that we wish to store three patterns in the system:

$$\alpha = 110000$$
$$\beta = 001100$$
$$\gamma = 100011 \, .$$

Let the three decoders be A (for 011101), B (for 100001), and C (for 010011), then we can show the distance of each pattern that needs to be stored from the decoder weight patterns:

α from A is 4 bits; from B is 2 bits; from C is 3 bits;
β from A is 2 bits; from B is 4 bits; from C is 5 bits;
γ from A is 5 bits; from B is 1 bit; from C is 1 bit.

Say that the radius of the decoders is set at ≤ 2 bits. Ignoring for a moment the counter circuitry of figure C1.5.8, α would energize only decoder B. The object of Kanerva's memory is to retrieve the stored patterns and therefore α itself would *somehow* be stored at location B. Similarly β would be stored at A, while γ would be stored at B *and* C. This indicates that, in contrast with conventional memory, the sparse memory can activate more than one address decoder. This means that a space of 2^n, being sparsely occupied, is mapped into a space of as many dimensions as decoders, and this is less than n. The fact that a pattern can now be stored in several locations explains why the term 'distributed' is used in this method. The only remaining question is how does one store more than one word in the addressed location. This is where the counters come in. In the above example, each address, when activated, gates n counters which are initially set to 0. When writing to the memory, the 1 bits in the patterns to be stored increment the counters. So when α is presented for storage, only B is addressed and the contents of its counters go to 110000. That is, after the first training step the content of the counters is

$$A : 000000$$
$$B : 110000$$
$$C : 000000 \, .$$

When β is presented, only A is addressed and the contents of its counters go to 001100. That is, after the second training step the content of the counters is

$$A : 001100$$
$$B : 110000$$
$$C : 000000 \, .$$

When γ is presented, both B and C are addressed and the contents of their counters are incremented by 100011. So after the third training step the contents of the counters are

$$A : 001100$$
$$B : 210011$$
$$C : 100011 \, .$$

To read from the memory, the contents of the addressed counters are added and thresholded bit by bit as shown in the figure (how the threshold is set is discussed below).

So when α is presented, the bitwise sums are 210011 (from the B counters only) . Pattern β gives bitwise sums of 001100 (from A counters only). However, γ behaves a little differently as the bitwise sums are 310022 as the contents of counters B and C are summed bit by bit. Kanerva's point is that this pooling of memory in a distributed way retrieves a distinctive representation of an input if not the input itself. The success of this depends on the setting of the threshold. A threshold of ≥ 1 gives the representations

$$\alpha: \ 110011$$
$$\beta: \ 001100$$
$$\gamma: \ 110011 \, .$$

This leaves α and γ undistinguished. So a threshold of ≥ 2 is attempted. This gives the result

$$\alpha: \ 100000$$
$$\beta: \ 000000$$
$$\gamma: \ 110011 \, .$$

This achieves Kanerva's predicted result that individual internal representations are created for the training patterns, provided that the judicious choice of threshold is made. Kanerva recommends that a value of about half the possible maximum of the summed bits be used.

In fact, it is slightly misleading to judge the behavior of sparse distributed memory on an example with very low dimensions as chosen above. This merely illustrates the mechanism. The strength of the method comes to the fore when systems are large enough for the law of large numbers to become effective and the behavior to benefit from the advantageous statistical properties that the method embodies. Kanerva showed that a system with $n = 1000$ and 20 decoders can store 10 000 patterns which may be retrieved by inputs up to a distance of 420 bits from the centers of the decoder retrieval centers. He also showed that such systems can retrieve sequences and converge on prototypes if organized as state machines in a manner similar to the methods discussed in earlier parts of this article.

C1.5.8 Correlation matrices and ADAM

Another family of binary neural systems owes its existence to an early suggestion by Willshaw *et al* (1969) for an adaptive correlation matrix and its later modification by Austin (1987) which combines the correlation matrix with N-tuple processing similar to that described in section C1.5.2 above. The latter is called ADAM as it is a distributed associative memory.

A correlation matrix such as suggested by Willshaw can be thought of as a set of p horizontal wires and q vertical ones. Binary input patterns are applied at horizontal wires $h_1, h_2, \ldots, h_p$ and, for training, the desired output is placed on vertical wires $v_1, v_2, \ldots, v_q$. A binary 'weight' is placed at the crosspoint of two wires if, at any point during training, there is a logical 1 both at the input wire and the output wire. After training, an unknown input is said to activate those weights for which the input is 1. The output wires simply sum all the activated weights and produce a raw response which is precisely this sum. In the original version, a threshold had to be applied to the output wires so as to decide whether they would output a 1 or not. This makes these outputs precise equivalents to McCulloch and Pitts neurons with binary weights.

The designers of ADAM made a series of modifications to this scheme. First they introduced n-point operation at the output wires. This means that only codes containing n 1s are allowed at the output vector of the matrix. This enables the automatic adjustment of the output threshold which is effected by, say, increasing the threshold from 0 until the output code contains exactly n 1s. A further addition is that of N-tuple processing at the input wires. Say that the total number of bits in an input vector is x, N-tuples require that this group be broken down into x/N groups of N bits each. Then the input vector to each N-tuple is decoded into a single 1 on 2^N wires. This is one way of doing things; other versions use a tighter coding, that is, between N and 2^N wires. The effect in each case is to make more redundant the encoding of the possible codes at the input of the correlation matrix which helps to prevent saturation in the matrix itself.

In the ADAM system itself two correlation matrices have been used, one to turn the input vector into a set of prototype n-point codes, and the second to turn these codes back into the prototypes of the input of the training set. The result of this is that an autoassociator is formed which has good resistance to noise. Austin (1989) has also shown that this scheme may be used for the recognition of two-dimensional shapes which can be made independently of their orientation.

C1.5.9 Conclusions on memory-based networks

In what has become the classical paradigm of neural networks the convention has been to think of the variable element in the neuron as an analog memory—a weight. It has been shown in this article that digital memory has a major role to play in neural systems not merely as a way of implementing weights but also as variable-logic, 'weightless' processors. Both feedforward and recursive applications benefit from this design philosophy. A further principle which comes into view is the concept of sparse codes which through systems with binary weights can perform useful neural functions. Kanerva networks, Willshaw matrices and the ADAM concept have been described as examples of the application of this principle.

References

Aleksander I 1965 Fused logic element which learns by example *Electron. Lett.* **1** 173–7

—— 1990 Ideal neurons for neural computers *Proc. ICNC (Dusseldorf)* (Berlin: Springer)

—— 1994 Developments in artificial neural systems: towards intentional computers *Sci. Prog.* **77** 43–55

Aleksander I, Clarke T J W, Braga A P 1994 Binary neural systems: combining weighted and weightless properties *IEE J. Intell. Syst. Eng.* **3** 211–20

Aleksander I, Thomas W and Bowden P 1984 Wisard, a radical new step forward in image recognition *Sensor Rev.* **2** 120–4

Aleksander I and Wilson M J D 1985 Adaptive windows for image processing *Proc. IEE* E **132** 233–45

Austin J 1987 ADAM: a distributed associative memory for scene analysis *Proc. 1st Int. Conf. on Neural Networks IV* (San Diego, CA: IEEE) pp 285–89

—— 1989 Application of the ADAM system to rotation invariant pattern recognition *Proc. IEE Conf. on Artificial Neural Networks* (London: Plenum)

Gorse D and Taylor J G 1989 An analysis of noisy RAM and neural nets *Physica* **34D** 90–114

Kanerva P 1988 *Sparse Distributed Memory* (Boston, MA: MIT Press)

© 1997 IOP Publishing Ltd and Oxford University Press

Myers C 1990 Learning with delayed reinforcement in an exploratory, probabilistic logic neural network *PhD Thesis* University of London

Rolls E T 1989 *The Computing Neuron* ed R Durbin, C Miall and G Mitchison (Reading, MA: Addison-Wesley) pp 125–59

Willshaw D J, Buneman O P and Longuet-Higgins H C 1969 Non-holographic associative memory *Nature* **222** 960–62

C1.6 Supervised composite networks

Christian Jutten

Abstract

Composite neural networks consist of multilayer networks, in which each layer may use different models of neurons: the classical sigmoidal neuron, the kernel neuron (like radial basis function neurons), the logical neuron, and so on. This section is devoted to supervised composite neural networks and contains three main parts. The first is focused on radial basis function (RBF) networks, as introduced by Poggio and Girosi. The second presents a special class of neural Bayesian classifier based on the kernel density estimator. In the third part, we briefly explain neural tree architectures, and the architecture of the well-known restricted Coulomb energy (RCE) algorithm, stressing their limitations.

C1.6.1 Introduction

Most of the models and algorithms described in this section are constituted of three or four layers (including the input layer). Each layer may use different neuron models, may have different topology, and may be associated with a specialized task.

In neural models described in the previous sections of this chapter, neurons are basically computing units whose output is a sigmoidal function, a Heaviside function, or some other function of its activation. The activation is the inner product between the input vector and the weight vector of the neuron. Other neural models have been proposed, based on another neural model whose output is a nonlinear decreasing function of the distance between the input vector and the weight vector: such a neuron will be called a *kernel neuron* or a *radial basis function (RBF) neuron* in this section. Neurons used in RBF neural networks, in kernel neural networks (KNNs)—also called probabilistic neural networks (PNNs)—and in the famous *self-organizing feature maps* (SOFMs) belong to this family.

B1.7.3

C2.1.1

Note that, in both cases, from a statistical point of view, the neuron is nothing other than a particular nonlinear regressor, whose assemblies have the interesting property of being able to model any nonlinear function.

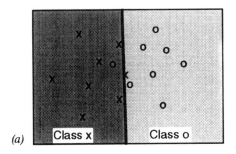
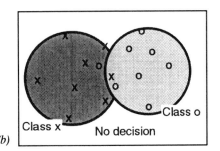

Figure C1.6.1. Boundaries defined by (*a*) hard-limiter neuron, (*b*) RBF neuron.

The difference and the interest of the two neuron models can be easily explained within the framework of classification. For the sake of simplicity, consider a simple binary classification task. In both cases, each neuron is a basic discriminant function, which divides the observation (features or patterns) space into two parts. However, with the hard-limiter neuron the boundary is a hyperplane, while with the RBF neuron the boundary is the circumference of a hypervolume (hypersphere with Euclidean distance) (see figure C1.6.1) centered around class samples. The RBF neuron gives a local decision, and regions of the feature located too far from samples are not classified. This neuron property, which allows rejection and avoids misclassification, is preserved at network level.

This section is devoted to *supervised* composite networks and is divided into three subsections. The first addresses RBF neural networks. It is followed by a section on KNNs. Finally, we present two other approaches leading to composite neural networks: restricted Coulomb energy (RCE) and related models, and neural trees (NTs).

C1.6.2 Radial basis function neural networks

C1.6.2.1 Introduction

In the neural network world, the paradigm of RBF was first introduced by Broomhead and Lowe (1988) and Moody and Darken (1989). Another major contribution was the paper by Poggio and Girosi (1990) who explained the design and interest in RBF networks with regularization theory. RBF networks can perform both classification and function approximation. For classification, the interest in RBF can be explained by the concept of ϕ-separable patterns proposed by Cover (1965). Concerning function approximation, theoretical results on multivariate approximation constitute the basic framework. For more details see Powell (1985), Poggio and Girosi (1990), Haykin (1994 ch 7, pp 237–44).

C1.6.2.2 Purpose of the model

RBF neural networks are general purpose approximators. In the literature, numerous applications involving function approximation as well as classification properties are encountered: time series analysis (Saha and Keeler 1990, Kadirkamanathan *et al* 1991), equalization (Cheng *et al* 1992), classification of seismic events (Chang and Lippmann 1992), handwritten digit recognition (Lee 1991), speech recognition (Lee and Lippmann 1990), adaptive control (Sanner and Slotine 1992), spectral estimation (Nedir *et al* 1993), and so on.

C1.6.2.3 Topology

An RBF network consists of three layers:

(i)   the first contains simple neurons which transmit input without distortion,
(ii) the second (hidden) layer contains the RBF neurons,
(iii) neurons in the output layer are simple linear units.

Each layer is fully-connected to the next one with simple first-order connections (figure C1.6.2).

Basically, the number of input units (output units, respectively) is equal to the dimension n of the input vectors (of the output space, respectively). However, the number of output units can vary according to the coding of the outputs. For instance, for binary classification, we can choose:

(i)   1 output unit which is close to 0 for class 0, and close to 1 for class 1,
(ii) 2 output units: unit 0 is close to 1 and unit 1 is close to 0 if class 0 is decided, and vice versa.

Finally, the number N_2 of RBF units is equal to the number of samples, N, in the learning database. The weight vector between the input vector and the jth RBF unit is simply equal to the input vector of the jth samples of the database: $w_j = x^j$.

The output of the ith neuron of the output layer is then

$$y_i(x) = \sum_{j=1}^{N} w_{ij}\phi(\|x - x^j\|) \tag{C1.6.1}$$

where $\phi(\cdot)$ is a function from $\mathcal{R}^+$ to $\mathcal{R}$, generally decreasing, x is the input vector, and x^j are input examples of the learning database. In equation (C1.6.1), the weights w_{ij} (between RBF units and output units) are tuned during the training, as we will explain in section C1.6.2.5. In what follows, for the sake

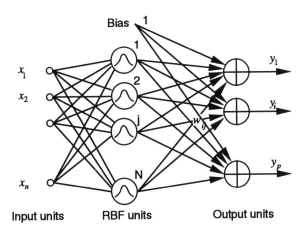

Figure C1.6.2. Topology of an RBF network.

of simplicity, we always consider 1-output networks, and we omit the index i. Equation (C1.6.1) then becomes:

$$y(x) = \sum_{j=1}^{N} w_j \phi(\|x - x^j\|) \,. \tag{C1.6.2}$$

Finally, we remark that the number of RBF units becomes very large with a huge learning database: practical methods to reduce the number will also appear in section C1.6.2.5.

C1.6.2.4 Choice of function

There now remains an essential question: what radial basis function must we use? Poggio and Girosi (1990) addressed this question in the framework of multivariate interpolation with regularization, for function f from $\mathcal{R}^n$ to $\mathcal{R}$. In fact, learning consists of designing a mapping f from N empirical input/output examples (x^j, d^j), $1 \le j \le N$, which are currently noisy examples. It is thus an ill-posed problem in the Hadamard sense, especially since the same input can produce various outputs, in which case we must exploit other information in order to transform the problem to a well-posed problem. This can be done by looking for the function f minimizing a functional consisting of two terms:

$$
\begin{aligned}
\mathcal{E}[f] &= \sum_{j} (\|y(x^j) - d^j\|)^2 + \lambda (\|Pf\|)^2 \\
&= \sum_{j} (\|f(x^j) - d^j\|)^2 + \lambda (\|Pf\|)^2
\end{aligned}
\tag{C1.6.3}
$$

where d^j is the (noisy) target output in response to input x^j, λ is a scalar parameter (regularization parameter), and P is usually a differential operator. The first term of the functional measures the fitting on data, while the second term imposes smoothing on f. It can be shown that equation (C1.6.3) leads to a Euler–Lagrange partial differential equation, solutions of which involve Green's functions $G(x, x^j)$:

$$f(x) = \frac{1}{\lambda} \sum_{j=1}^{N} (d^j - f(x^j)) G(x, x^j) \,. \tag{C1.6.4}$$

The optimal choice of the Green function depends on the operator P. For instance, for one-dimensional data, P can be defined such that

$$(\|Pf\|)^2 = \int_{\mathcal{R}} \left[\frac{\mathrm{d}^2 f(x)}{\mathrm{d}x^2} \right]^2 \mathrm{d}x \,. \tag{C1.6.5}$$

In that case, the Green function is a cubic spline (Haykin 1994, pp 249–50). Furthermore, if we constrain the operator P to be invariant under rotations and translations, a solution of the Green function leads to the Gaussian RBF (Poggio and Girosi 1990):

$$G(x, x^j) = \exp\left(-\frac{1}{2\sigma_j^2}\|x - x^j\|^2\right).$$ (C1.6.6)

Finally, the RBF network approximation becomes

$$f(x) = \sum_{j=1}^{N} w_j \exp\left(-\frac{1}{2\sigma_j^2}\|x - x^j\|^2\right)$$ (C1.6.7)

that is, a linear superposition of Gaussian RBF, whose centers are samples x^j and variances are σ_j^2.

C1.6.2.5 Learning

Three types of parameter are adjusted by training. Weights w_j and variances σ_j^2 are trained by supervised learning. Finally, to avoid too large a complexity, the number of RBF units may be reduced by selecting, usually unsupervised, a small but representative number of samples in the database. RBF networks being universal approximators, there is no restriction on either inputs or outputs, which may be integer as well as real.

Weight computation without center selection. In the simplest case, all the RBFs have the same width. The location of the RBF and the weight w_j must be computed. If the number of samples, N, in the learning database is not too large, one chooses N RBF units, each one being centered on each sample. The N weights w_j, $1 \le j \le N$, are solutions of the set of N linear equation:

$$y(x^k) = d^k = \sum_{j=1}^{N} w_j \phi(\|x^k - x^j\|) \qquad 1 \le k \le N$$ (C1.6.8)

which can be written:

$$\Phi w = y$$ (C1.6.9)

where $\Phi = (\phi_{kj})$ is an $N \times N$ matrix such that $\phi_{kj} = \phi(\|x^k - x^j\|)$, w is the unknown vector $(w_1, w_2, \ldots, w_N)^{\mathrm{T}}$ and y is the target vector $(d^1, d^2, \ldots, d^N)^{\mathrm{T}}$.

According to Light's theorem (Light 1992), the matrix Φ is positive definite if the input vectors x^1, x^2, ..., x^N are distinct. If so, the above set of equations has a unique solution. Note that such a set of N equations must be solved for each output unit.

Weight computation with center selection. If N is large, to avoid computation and memory being too large, one selects $N_2 \ll N$ samples in the learning database, which will be associated with N_2 RBF units. Such RBF neural networks are usually called *generalized RBF* (GRBF) neural networks. The selection of representative samples is usually done by simple *vector quantization* (VQ) algorithms or *Kohonen's algorithm*, which are unsupervised algorithms. We denote these N_2 new samples, usually different from database samples, by c_j. Then, for each output unit, we have a set of N equations with N_2 unknowns:

C1.1.5, B3.3.5

$$y(x^k) = d^k = \sum_{j=1}^{N_2} w_j \phi(\|x^k - c_j\|) \qquad 1 \le k \le N$$ (C1.6.10)

or again:

$$\Phi w = y.$$ (C1.6.11)

The matrix Φ is now rectangular, $N \times N_2$. An optimal solution, in the mean-square error sense, is then given by

$$w = \Phi^+ y$$ (C1.6.12)

where Φ^+ denotes the pseudo-inverse matrix of Φ. This pseudo-inverse matrix can be computed by direct computation, using the relation $\Phi^+ = (\Phi^{\mathrm{T}}\Phi)^{-1}\Phi^{\mathrm{T}}$, iterative algorithm or adaptive (least-square) algorithms.

Supervised selection of centers. Supervised selection of centers was first proposed by Poggio and Girosi (1990), and is more efficient than unsupervised selection (Wettschereck and Dietterich 1992). The idea is based on gradient descent of the cost function:

$$\mathcal{E} = \tfrac{1}{2} \sum_{k=1}^{N} \left(d^k - \sum_{j=1}^{N_2} w_j \phi(\|x^k - c_j\|)^2 \right). \tag{C1.6.13}$$

Gradients of $\mathcal{E}$ with respect to w_j, c_j are easy to compute (see Haykin 1994 for detailed gradient computations) and adaptive algorithms are simply

$$w_i(t+1) = w_i(t) - \mu_w \frac{\partial \mathcal{E}(t)}{\partial w_i(t)} \tag{C1.6.14}$$

$$c_j(t+1) = c_j(t) - \mu_c \frac{\partial \mathcal{E}(t)}{\partial c_j(t)}. \tag{C1.6.15}$$

Adaptation of the radial basis function width. In the most general case, it is interesting to have non-radial basis functions (Poggio and Girosi 1990). This is equivalent to having a weighted norm and is simply obtained by replacing $\|x^k - c_j\|^2$ by $(x^k - c_j)^{\mathrm{T}} \Sigma^{-1} (x^k - c_j)$, where Σ^{-1} is a positive definite matrix. The weighting matrix can also be adapted by a gradient procedure on the cost $\mathcal{E}$ (see Haykin 1994 for details) and the learning rule is:

$$\Sigma^{-1}(t+1) = \Sigma^{-1}(t) - \mu_s \frac{\partial \mathcal{E}(t)}{\partial \Sigma^{-1}(t)}. \tag{C1.6.16}$$

With radial functions, the covariance matrix reduces to $\Sigma = \sigma^2 \mathbf{I}$, and (C1.6.16) adapts only the parameter σ^2.

For RBF classifiers, Musavi *et al* (1992) proposed another approach to adjust the matrix Σ^{-1} of an RBF centered on a point c_i. We briefly explain the procedure for a 2-class problem. We first assume that the cluster i centered on c_i corresponds to class i. The idea is to define the largest cluster possible using a Gram–Schmidt procedure. First, one looks for the nearest input of c_i, for instance x_1^i belonging to the opposite class. The vector $e_1 = x_1^i - c_i$ determines the least principal axis. Then, one looks for the nearest input, for instance x_2^i, with respect to e_1, whose projection on e_1 is less than $\|e_1\|$. The second principal axis is then $e_2 = (x_2^i - c_i) - [e_1^{\mathrm{T}}(x_2^i - c_i)e_1]/\|e_1\|^2$, and so on. Finally, eigenvalues of Σ are defined from $\|e_i\|$, with a correction factor taking into account the empty space phenomenon for high dimensions.

Orthogonal least-square learning. Chen *et al* (1991) proposed an orthogonal least-square (OLS)—supervised—algorithm to select, one by one, the best centers c_i within database samples x^j. Assume the best approximation with q RBF units involves the input samples x^i, $1 \le i \le q$, as centers:

$$y(x) = \sum_{j=1}^{q} w_j \phi(\|x - x^j\|). \tag{C1.6.17}$$

To improve the approximation, we choose, within the remaining $N - q$ samples of the database, the vector x^k which constitutes the best $(q+1)$th regressor, that is, minimizing the square error on the whole database. Note that the criterion must be computed for every remaining point! The algorithm is still a variation of the Gram–Schmidt orthonormalization procedure. Its main drawback is computational cost, the main attractions are incrementality and the small size of the network with respect to a random selection.

C1.6.2.6 Related neural network models

Many kernels can be used for multivariate interpolation. In the RBF approach, all RBF units have the same shape with different width (for instance Gaussian shape with different variances). However, similar approaches suggest approximation based on a family of functions.

- Baldi (1991) has shown theoretical results based on Bernstein polynomials of degree n:

$$B_n(k, x) = C_n^k x^k (1-x)^{n-k}. \tag{C1.6.18}$$

- The kernel basis function (KBF), introduced by Hlaváčková (1995), may include various types of classical kernels: Féjer kernel, Dirichlet kernel, Jackson kernel, and so on.
- Mukherjee and Nayar (1995) designed an RBF network based on wavelets.

Other authors develop spline networks (Friedman 1991, Williamson and Bartlett 1992) or networks using hyperbolic kernels (Jones 1994).

C1.6.3 Kernel density estimators

C1.6.3.1 Introduction

These networks were first proposed by Comon (1990), and Specht (1990) who called them probabilistic neural networks (PNNs). In fact, they are very close to Parzen's windows (Parzen 1962). This approach is very interesting for a Bayesian approach to neural classifiers, as proved by the results of the European Esprit project ELENA (Comon *et al* 1993, 1994, 1995, Comon 1995), but has only been developed by a few other researchers.

C1.6.3.2 Purpose

Kernel density estimators are special RBF neural networks devoted to the estimation of probability density functions (PDFs). They constitute the first step to computing a Bayesian decision. In fact, associated with a winner-take-all (WTA) network, we obtain a complete neural Bayesian classifier.

C1.6.3.3 Topology

The complete network consists of three *networks* in cascade (figure C1.6.3) (Jutten and Comon 1993):

(i) the first is the kernel network, very close to the RBF network, but devoted to density estimation,
(ii) the second is purely linear, and computes terms $-I_k$ (see *Bayesian classification* below); the sums of these are densities weighted by prior probabilities and decision costs,
(iii) the last is a winner-take-all network which computes the largest term $-I_k$ (that is, the smallest I_k) and produces an estimate of the Bayesian decision.

All the connections are first-order and direct.

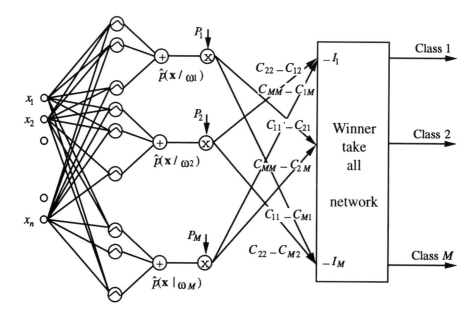

Figure C1.6.3. Topology of a neural Bayesian classifier.

Bayesian classification. The classification problem consists of designing a mapping f, from a set of patterns $\mathcal{X}$ to a set of classes $\mathcal{C}$. In supervised classification, the mapping is known for N examples $(\boldsymbol{x}^i, \omega_{j(i)})$, where $\boldsymbol{x}^i$ $(1 \le i \le N)$ are the patterns, and $\omega_{j(i)}$, $(1 \le j(i) \le M)$ are the associated classes.

We have assumed that every feature vector belonging to the same class ω^j is drawn independently from the same PDF $p(\boldsymbol{x}|\omega_j)$. The optimal Bayesian classifier is the mapping which minimizes the Bayesian risk:

$$R = \sum_{i=1}^{M} \sum_{j=1}^{M} C_{ij} P_j \int \cdots \int_{\boldsymbol{x} \in D_i} p(\boldsymbol{x}|\omega_j)\, \mathrm{d}\boldsymbol{x} \tag{C1.6.19}$$

where P_j is the prior probability of class ω_j, C_{ij} is the cost associated with the decision: 'say ω_i while ω_j is true', and D_i is the region, in the feature (pattern) space, in which each point is assigned to the class ω_i. If $C_{ij} = 1 - \delta_{ij}$, where δ_{ij} is the Kronecker symbol, the Bayesian risk R reduces to the average error probability of the decision process. Arranging equation (C1.6.19) leads to

$$R = \sum_{i=1}^{M} P_i C_{ii} + \sum_{i=1}^{M} \int \cdots \int_{\boldsymbol{x} \in D_i} \sum_{j \neq i} P_j (C_{ij} - C_{jj}) p(\boldsymbol{x}|\omega_j)\, \mathrm{d}\boldsymbol{x} \tag{C1.6.20}$$

where the first term is a constant cost. Then, minimization of equation (C1.6.20) only depends on the second term. Assuming that $C_{jj} < C_{ij}$, each integrand I_i is positive. The risk R will be minimized if and only if we assign to any pattern $\boldsymbol{x}$ the class $\omega_{k(\boldsymbol{x})}$ satisfying

$$k(\boldsymbol{x}) = \operatorname*{argmin}_{1 \le j \le M} \sum_{j \neq i} P_j (C_{ij} - C_{jj}) p(\boldsymbol{x}|\omega_j)\, \mathrm{d}\boldsymbol{x} . \tag{C1.6.21}$$

Then the neural Bayesian classifier (figure C1.6.3) is simply computed using the Bayesian decision (C1.6.21).

Kernel density estimators. Kernel density estimators are well-known tools in statistics, and are long-established (Rosenblatt 1956, Parzen 1962, Cacoulos 1966, Silverman 1986, Härdle 1990).

Let us consider the set $\mathcal{X}_j = \{\boldsymbol{x}(t) \in \mathcal{R}^n, 1 \le t \le N_j\}$ of patterns belonging to class ω_j. Kernel estimation of the conditional PDF $p(\boldsymbol{x}|\omega_j)$ is then

$$\hat{p}(\boldsymbol{x}|\omega_j) = \frac{1}{N_j} \sum_{\boldsymbol{x}(t) \in \mathcal{X}_j} \frac{1}{h(t,j)^n} K\left(\frac{\boldsymbol{x} - \boldsymbol{x}(t)}{h(t,j)}\right) \tag{C1.6.22}$$

where $K(\cdot)$ is a kernel function and $h(t, j)$ is the width of the kernel. In the simplest case, kernels have fixed width: $h(t, j)$ only depends on N_j. Then, comparing equations (C1.6.22) and (C1.6.2), a basic kernel estimator can be viewed as a special RBF network whose connections are all equal to $1/N_j$.

Usually, a kernel $K(\boldsymbol{x})$ is a decreasing function of $\|\boldsymbol{x}\|$. More precisely, it has been proved that $\hat{p}(\boldsymbol{x}|\omega_j)$ is asymptotically an unbiased PDF estimator if $K(\boldsymbol{x})$ is positive and bounded and satisfies

$$\int K(\boldsymbol{x})\, \mathrm{d}\boldsymbol{x} = 1 \tag{C1.6.23}$$

$$\lim_{\|\boldsymbol{x}\| \to \infty} \|\boldsymbol{x}\|^n K(\boldsymbol{x}) = 0 \tag{C1.6.24}$$

$$\lim_{N_j \to \infty} h = 0 \tag{C1.6.25}$$

$$\lim_{N_j \to \infty} N_j h^n \to \infty. \tag{C1.6.26}$$

the convergence is in quadratic mean. However, as h increases, the bias decreases and the variance increases. Usually, one can then choose the width h which minimizes the integrated mean-square error (MISE). Unfortunately, MISE requires knowledge of the Laplacian of the unknown density! This point is essential, because h determines the smoothness of the estimator and avoids overfitting. This parameter should be related to the regularization term in RBF.

For $n = 1$, according to the above conditions, it is easy to define many candidates:

$$\text{uniform: } K(x) = \frac{1}{2}\operatorname{Rect}\left(\frac{x}{2}\right) \tag{C1.6.27}$$

$$\text{triangle: } K(x) = (1 - |x|)\operatorname{Rect}\left(\frac{x}{2}\right) \tag{C1.6.28}$$

$$\text{Epanechnikov: } K(x) = \frac{3}{4}(1 - |x|^2)\operatorname{Rect}\left(\frac{x}{2}\right) \tag{C1.6.29}$$

$$\text{Gaussian: } K(x) = \frac{1}{(2\pi)^{1/2}\exp(-x^2/2)} \tag{C1.6.30}$$

where $\operatorname{Rect}(u) = 1$ if $|u| < 1/2$, and 0 otherwise. In higher dimensions, one usually used radial kernels with similar shapes. Practical algorithms for choosing a good but suboptimal value of h have been proposed by Comon (1995), Voz *et al* (1995).

C1.6.3.4 Learning

Two parameters must be adapted in kernel networks:

(i) as in RBF networks, the number of kernels must often be reduced, usually using unsupervised procedures,
(ii) the width of the kernels.

Reducing the kernel number. The kernel number is basically equal to the number of samples in the database. With large databases, it is essential to select a small subset of samples, leaving the underlying distribution unchanged. This can be roughly done by unsupervised vector quantization (adaptive k-means algorithm) or self-organizing feature maps, as already suggested for reducing the center number in RBF networks.

Another constructive approach (suboptimal) (Comon and Cheneval 1995), based on non-radial kernels, suggests an algorithm able to design the best network with restricted complexity (number of kernels) and directly maximizing the classification rate.

Finally, if a small number of samples must be canceled, vector quantization gives poor results. In that case, for fixed kernels, Fambon and Jutten (1995) have proposed an efficient method, inspired by optimal brain surgery (OBS) (Hassibi *et al* 1993), to prune a few kernels in KNN. The method computes location modifications of remaining kernels, which minimizes the integrated mean-square error between the pruned approximation and the initial approximation. The method can be considered unsupervised, because it uses as a reference the current network approximation.

Computing the kernel width. In the simplest case, all the kernels have the same width. It is easy to show that $h_{\text{opt}} = \text{O}(N^{-1/(d+4)})$, but, as we previously said, the optimal width h_{opt} explicitly contains the unknown density and its second derivative! The problem is well-known in statistics, and Härdle (1990) proposed three approaches:

(i)   if the unknown density is close to a reference distribution, we can compute these unknown terms and h_{opt};
(ii) another idea is to directly estimate the second derivative, but the problem of width choice is still encountered;
(iii) finally, two cross-validation methods.

For a small sample size variable kernel estimation, first proposed by Silverman (1986), seems a more reliable method. However, other problems appear. For instance, variable kernel estimation may not integrate to 1, contrary to fixed kernel estimation (with weights equal to $1/N_j$). Silverman (1986) simply chooses the width of the Gaussian kernel, proportional to the kth nearest neighbor. Recently, Lowe (1995) suggested a refinement and instead used the average distance of the first k neighbors. Optimization is done again using a cross-validation method, and experimental results point out an excellent performance/complexity ratio.

After a center selection by vector quantization, Voz *et al* (1995) computed the average square coding error (called 'inertia' and denoted i_m) inside each Voronoi region coded by a center c_m. Assuming the PDF estimation is almost constant between two neighboring kernels (if the number of kernels is large), they show for Gaussian kernels that $h_{mopt} = \gamma^2 3 i_m / (2n \ln 2)$, where n is the pattern dimension. The factor γ varies from about 0.7 to 2, according to n and the number of initial samples in each cluster. Experimental results show excellent performance. Curves for the choice of γ are given in their paper.

For variable kernels, Comon (1995) suggests estimating the density iteratively. Initially, a rough estimate of the density is obtained using a rough estimate $\hat{h}_R(t, j)$ by a k-nearest-neighbor estimator. Then, a new value $\hat{h}(t, j)$ can be computed and used again to refine the density estimator. Experimentally, it seems a two-pass iteration is enough. Details and complete equations are given by Comon (1995).

C1.6.3.5 Advantages

The kernel classifier method is able to give a very good estimate of conditional density functions provided that we have enough samples (for each class) in the database. The learning is simple, and only one parameter, for the whole network or for each cluster, is awkward to compute. It is then possible to converge toward the optimal Bayesian classifier, and to compute ultimate bounds for a given problem. This is very useful in evaluating the efficiency of any suboptimal algorithm. Moreover, suboptimal algorithms, simple enough and very efficient, may be designed from this approach.

Finally, with good PDF estimates, the neural Bayesian classifier (figure C1.6.3) always remains optimal without retraining because prior probability and cost modifications are canceled by adjusting parameters in the second and third layers, but the kernel estimators are not influenced.

However, a good density estimate is not necessary to achieve a good classifier. In fact, it is enough to have precise estimates near boundaries. Consequently, optimization directly based on classification performance may be very efficient (Comon and Cheneval 1995).

C1.6.4 Other composite networks for classification

C1.6.4.1 Restricted Coulomb energy neural network

Introduction. This algorithm was proposed by Reilly *et al* (1982) for classification. It is a very simple constructive procedure, very popular, although the efficiency strongly reduces with pattern overlapping.

Topology. The network consists of three layers: an input layer F which transmits patterns, a prototype layer G which codes classes, and a decision layer H (see figure C1.6.4). Connections are direct from one layer to the next. First and second layers are fully connected, but only one connection starts from each neuron of the second layer.

Neurons G_k, $1 \leq k \leq N_2$ of the prototype layer are special RBF neurons. For the input pattern $\boldsymbol{x}$, the output of neuron G_k is

$$y_{G_k}(\boldsymbol{x}) = \mathcal{H}[\lambda_k - \|\boldsymbol{x} - \boldsymbol{w}_k\|] \tag{C1.6.31}$$

where $\mathcal{H}[\cdot]$ denotes the Heaviside function ($\mathcal{H}[u] = 0$ if $u < 0$, and 1 otherwise) and λ_k is the radius of influence of the neuron G_k.

The aim of the neurons of the prototype layer is to approximate the classes ω_j, $1 \leq j \leq M$ by a superposition of hypervolumes. In fact, if $\boldsymbol{w}_k$ corresponds to a pattern of class ω_j, (C1.6.31) defines around the pattern $\boldsymbol{w}_k$ a region (hypersphere with Euclidean distance), with a radius λ_k, assigned to class ω_j in the pattern space.

Then, the output of neuron G_k is only connected to the neuron H_j, with a weight equal to 1. The output of neuron H_j is

$$y_{H_j}(\boldsymbol{x}) = \mathcal{H}\left[\sum_{G_k \in \omega_j} y_{G_k}(\boldsymbol{x}) - 1\right]. \tag{C1.6.32}$$

The unit H_j is then active if and only if at least one of cells G_k connected to it is equal to 1. Assuming the input pattern $\boldsymbol{x}$:

- if only one output cell H_j is active, the pattern is assigned to class ω_j,
- if no output cell is active, the pattern is not classified,
- if two output cells are simultaneously active, for instance H_j and H_k, there is a misclassification.

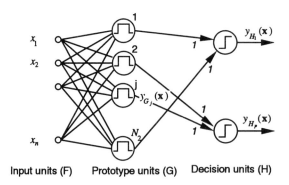

Input units (F)   Prototype units (G)   Decision units (H)

Figure C1.6.4. Topology of the RCE neural network.

Learning. The learning is a very simple one-shot supervised learning. At the beginning, the second and third layers are empty. Assume the first learning sample is the pair pattern/class (x^1, ω_j). Because the network is empty, the pattern cannot be classified. One then adds a prototype cell G_1, with a radius $\lambda_1 = \lambda_0$, and with input connections $w_1 = x^1$. The output of G_1 is connected, with a weight equal to 1, to a new neuron H_j.

Assume the second sample is (x^2, ω_k). Four cases may occur:

(i) $y_{G_1}(x^2) = 1$, consequently $y_{H_j}(x^2) = 1$ and $\omega_k = \omega_j$. There is nothing to do.

(ii) $y_{G_1}(x^2) = 0$, consequently $y_{H_j}(x^2) = 0$ and $\omega_k \neq \omega_j$. We must create a new prototype neuron G_2, with a radius $\lambda_2 = \lambda_0$, coding the class ω_k, whose input weight is $w_2 = x^2$. Another decision neuron H_k, devoted to class ω_k, is created. It is connected to the output of neuron G_2 with a weight equal to 1.

(iii) $y_{G_1}(x^2) = 0$, consequently $y_{H_j}(x^2) = 0$ and $\omega_k = \omega_j$. There is no classification. We improve the coding of class ω_j by adding one neuron G_2, whose input weight is $w_2 = x^2$, and whose output is connected to the neuron H_j.

(iv) $y_{G_1}(x^2) = 1$, consequently $y_{H_j}(x^2) = 1$ but $\omega_k \neq \omega_j$. There is a misclassification. We must code the new class by adding a new prototype neuron G_2 coding the class ω_k, whose input weight is $w_2 = x^2$. Another decision neuron H_k, devoted to class ω_k, is also created and connected to the output of neuron G_2 with a weight equal to 1. To avoid future misclassification, radii λ_1 and λ_2 must decrease and satisfy $\lambda_1 = \lambda_2 < \|x^1 - x^2\|$.

And so on.

Advantages. The algorithm is very simple, but also presents two main drawbacks. First, if the initial radius λ_0 is too small, the number of prototype neurons becomes equal to the number of samples in the database! Secondly, if patterns of different classes overlap, the learning leads to overfitting, because the learning imposes that each pattern must be correctly learned. In that case, this algorithm is not recommended because the generalization error rate becomes large. Figure C1.6.5 explains this overfitting effect for a simple two-class problem by comparing the RCE algorithm boundary with the k-nearest-neighbor boundary, that asymptotically tends towards the Bayesian boundary.

C1.6.4.2 Neural trees

Introduction. Trees, especially binary trees, are very well-known tools in supervised classification. They are easy to use and are efficient. Moreover, a neural implementation of classification binary trees is straightforward. First attempts for neural trees seem to be due to Koutsougeras and Papachristou (1988), then Sethi (1990) and Sirat and Nadal (1990). In all these approaches, the network is automatically constructed: at each step, one adds a new cell corresponding to the best hyperplane, which is defined using an entropy measure, and other simple 'AND' neurons to define new regions corresponding to new leaves of the tree, and connections or simple 'OR' neurons to define the class.

Topology. Neural trees are networks with four layers (figure C1.6.6):

(i)   the first layer transmits input patterns;

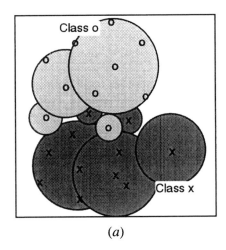 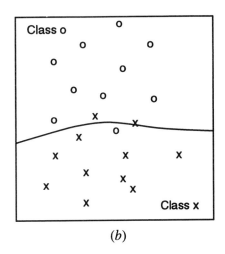

(a)   (b)

Figure C1.6.5. Comparison of boundaries obtained with (a) RCE and (b) k-NN algorithms with two overlapping classes.

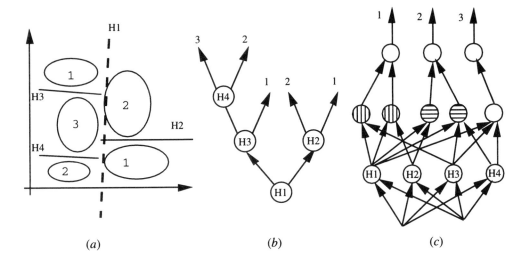

(a)   (b)   (c)

Figure C1.6.6. Classification of patterns (a) leads to the tree (b) which can be easily implemented by a four-layer neural network (c).

(ii) neurons of the second layer are binary neurons which define hyperplanes in the pattern space—the number of neurons is then equal to the number of hyperplanes (nodes in the tree);

(iii) units of the third layer are logical AND neurons which share the space in regions by combining outputs of hyperplane layers—the number of AND neurons is equal to the number of leaves of the tree;

(iv) units of the last layer are logical OR neurons which sum together regions belonging to the same classes and gives the final decision—the number of OR neurons is then equal to the number of classes.

Neural trees have direct, first-order connections. Input layer and hyperplane layer are fully-connected with adaptive connections, while connections between other layers are sparse and fixed.

Learning. Learning is supervised and mainly concerns connections between the input layer and the second layer, which determine the hyperplane equations. Neuron weights can be adapted using the backpropagation algorithm, the pocket algorithm (Gallant 1986) and so on, in order to minimize the error number. An information-based criterion is usually used (Koutsougeras and Papachristou 1988, Sethi 1990, Sirat and Nadal 1990, Omohundro 1987, Willshaw *et al* 1969) to choose the best partition. Sirat and Nadal (1990) have also proposed a class dichotomy based on principal component analysis.

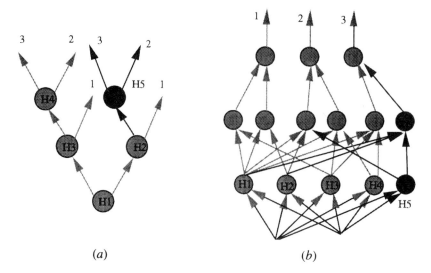

(a) (b)

Figure C1.6.7. Improvement of the tree: black node in (a) implies very simple modification on the associated neural tree: black neurons and connections in (b).

Advantages. This method allows one to find very simply a good initial network architecture which can then be refined. As suggested by Sethi, an appropriate architecture can be obtained by training hyperplane neurons on a reduced learning database. Then, to improve generalization, one may replace hard nonlinearities of neurons (logical AND and OR neurons of third and fourth layers) by soft nonlinearities (sigmoid function) and then continue the training on all the connections using a larger database. To avoid overfitting by adding more and more neurons, a stopping criterion based for instance on entropy measure can control the network expansion (Sethi 1990). Moreover, it is very simple to improve the network: addition of new neurons does not change existing connections' as shown in figure C1.6.7. Note also that addition of new classes can be easily obtained by only adding new neurons and connections, but without teaching again existing neurons and connections.

References

Baldi P 1991 Computing with arrays of bell-shaped and sigmoid functions *Advances in Neural Information Processing Systems 3* ed R P Lippmann, J Moody and D S Touretzky (San Mateo, CA: Morgan Kaufmann) pp 735–42

Broomhead D S and Lowe D 1988 Multivariable functional interpolation and adaptive networks *Complex Syst.* **2** 321–55

Cacoulos T 1966 Estimation of multivariate density *Ann. Inst. Stat. Math.* **18** 178–89

Chang E I and Lippmann R P 1992 A boundary hunting radial basis function classifier which allocates centers constructively *Advances in Neural Information Processing Systems 4* ed J Moody, S J Hanson and R P Lippmann (San Mateo, CA: Morgan Kaufmann) pp 139–46

Chen S, Cowan C F and Grant P M 1991 Orthogonal least squares learning algorithm for radial basis function networks *IEEE Trans. Neural Networks* **2** 302–9

Comon P 1990 Classification Bayésienne distribuée *Revue Technique Thomson* **22** 543–61

—— 1995 Supervised classification: a probabilistic aproach *Invited paper in European Symp. on Artificial Neural Networks (Brussels)* ed M Verleysen (Brussels: D facto) pp 111–28

Comon P and Cheneval Y 1995 Bayesian supervised classification: an approach with variable kernel estimators *Int. Workshop on Artificial Neural Networks (Malaga)*

Comon P, Jutten C, Blayo F, Cheneval Y, Fambon O, Thissen Ph and Verleysen M 1993 Deliverable R1-A-P Axis A: Theory *Esprit Basic Research Project 6891 ELENA*

Comon P *et al* 1994 Deliverable R2-A-P Axis A: Theory *Esprit Basic Research Project 6891 ELENA*

Comon P, Jutten C, Cheneval Y, Chentouf R and Fambon O 1995 Deliverable R3-A-P Axis A: Theory *Esprit Basic Research Project 6891 ELENA*

Cover T M 1965 Geometrical and statistical properties of systems of linear inequalities with application in pattern recognition *IEEE Trans. Electronic Computers* **14** 326–34

Fambon O and Jutten C 1995 Pruning kernel density estimators *European Symp. on Artificial Neural Networks (Brussels)* ed M Verleysen (Brussels: D facto) pp 147–52

Friedman J 1991 Adaptive spline networks *Advances in Neural Information Processing Systems 3* ed R P Lippmann, J Moody and D S Touretzky (San Mateo, CA: Morgan Kaufmann) pp 675–83

Gallant S I 1986 Optimal linear discriminants *Proc. 8th Conf. on Pattern Recognition (Paris)* (Washington, DC: IEEE Computer Society Press) pp 849–52

Härdle W 1990 *Smoothing techniques, with implementations in S* (Berlin: Springer)

Hassibi B, Stork D G and Wolff G J 1993 Optimal brain surgeon and general network pruning *IEEE Int. Conf. on Neural Networks (San Francisco, CA)* vol 1, pp 293–9

Haykin S 1994 *Neural networks, a comprehensive foundation* (Macmillan College Publishing Company)

Hlaváčková K 1995 An upper estimate of the error of approximation of continuous multivariable functions by KBF *European Symp. on Artificial Neural Networks (Brussels)* ed M Verleysen (Brussels: D facto) pp 333–4

Koutsougeras C and Papachristou C A 1988 Training of a neural network model for pattern classification based on an entropy measure *Proc. IEEE Int. Conf. on Neural Networks (San Diego, CA)* pp 247–54

Jones L K 1994 Goodweights and hyperbolic kernels for neural networks, projection pursuit, and pattern classification: Fourier strategies for extracting information from high-dimentional data *IEEE Trans. Information Theory* **40** 439–54

Jutten C and Comon P 1993 Neural Bayesian classifier *New Trends in Neural Computation (Lecture notes in Computer Sciences)* ed J Mira, J Cabestany and A Prieto (Berlin: Springer) 686 pp 119–24

Kadirkamanathan V, Niranjan M and Fallside F 1991 Sequential adaptation of radial basis function neural networks and its application to time-series analysis *Advances in Neural Information Processing Systems 3* ed R P Lippmann, J Moody and D S Touretzky (San Mateo, CA: Morgan Kaufmann) pp 721–7

Lee Y 1991 Handwritten digit recognition using K nearest-neighbor, radial basis function, and backpropagation neural networks *Neural Comput.* **3** 440–9

Lee Y and Lippmann R P 1990 Practical characteristics of neural network and conventional pattern classifiers on artificial and speech problems *Advances in Neural Information Processing Systems 2* ed D S Touretzky (San Mateo, CA: Morgan Kaufmann) pp 168–77

Light W A 1992 Some aspects on radial basis function approximation *Approximation Theory, Spline Functions and Applications* ed D S Singh NATO ASI series 256 (Dordrecht: Kluwer) pp 163–90

Lowe D G 1995 Similarity metric learning for a variable-kernel classifier *Neural Comput.* **7** 72–85

Moody J E and Darken C J 1989 Fast learning in networks of locally-tuned processing units *Neural Comput.* **1** 281–94

Mukherjee S and Nayar S K 1995 Automatic generation of GRBF networks using the integral wavelet transform *SPIE'95*

Musavi M T, Ahmed W, Chan K H, Faris K B and Hummels D M 1992 On the training of radial basis function classifiers *Neural Networks* **5** 595–603

Nedir K, Vesin J-M and Eyer L 1993 Spectral estimation of inequally-spaced data via radial basis function interpolation *XIVth Colloque sur le traitement du signal et ses applications (GRETSI) (Juan-Les-Pins)* 217–20

Omohundro S M 1988 Efficient algorithm with neural network behaviour *Complex Syst.* **1** 273–347

Parzen E 1962 On estimation of probability density function and mode *Ann. Math. Stat.* **33** 1065–76

Poggio T and Girosi F 1990 Networks for approximation and learning *Proc. IEEE* **78** 1481–97

Powell M J D 1985 Radial basis function for multivariable interpolation: a review *IMA Conf. on Algorithms for the Approximation of Functions and Data* pp 143–67

Reilly D L, Cooper L N and Erlbaum C 1982 A neural model for category learning *Biol. Cybern.* **45** 35–41

Rosenblatt M 1956 Remarks on some non parametric estimates of a density *Ann. Math. Stat.* **27** 832–7

Saha A and Keeler J D 1990 Algorithms for better representation and faster learning in radial basis function networks *Advances in Neural Information Processing Systems 2* ed D S Touretzky (San Mateo, CA: Morgan Kaufmann) pp 482–9

Sanner R M and Slotine J-J E 1992 Gaussian networks for direct adaptive control *IEEE Trans. Neural Networks* **3** 837–63

Sethi I K 1990 Entropy nets: from decision trees to neural networks *Proc. IEEE* **78** 1605–13

Silverman B W 1986 *Density estimation for statistics and data analysis* (London: Chapman and Hall)

Sirat J A and Nadal J-P 1990 Neural trees: a new tool for classification *Network* **1** 423–38

Specht D 1990 Probabilistic neural networks *Neural Networks* **3** 109–18

Voz J-L, Verleysen M, Thissen Ph and Legat J-D A practical view of suboptimal Bayesian classification with radial Gaussian kernels *From Natural to Artificial Neural Computation (Lecture Notes in Computer Sciences)* ed J Mira and F Sandoval (Berlin: Springer) vol 930 404–11

Williamson R C and Bartlett P L 1992 Splines, rational functions and neural networks *Advances in Neural Information Processing Systems 4* ed J Moody, S J Hanson and R P Lippmann (San Mateo, CA: Morgan Kaufmann) pp 1040–7

Wettschereck D and Dietterich T 1992 Improving the performance of radial basis function networks by learning center locations *Advances in Neural Information Processing Systems 4* ed J Moody, S J Hanson and R P Lippmann (San Mateo, CA: Morgan Kaufmann) pp 1133–40

Willshaw D J, Buneman O P and Longuet–Higgins H P 1969 Non-holographic associative memory *Nature* **222** 960–2

C1.7 Supervised ontogenic networks

Emile Fiesler and Krzysztof Cios

Abstract

One of the most powerful aspects of neural networks is their ability to adapt to problems
by changing their interconnection strengths according to a given *learning rule*. On the
other hand, one of the main drawbacks of neural networks is the lack of knowledge
for determining the topology of the network, that is, the number of layers and number
of neurons per layer. A relatively new class of neural networks tries to overcome this
problem by letting the network also automatically adapt its topology to the problem.
These are the so-called *ontogenic neural networks*. This section provides an extensive
survey and comparison of these methods.

C1.7.1 Introduction

One of the main strengths of artificial neural networks is their adaptivity, or, more precisely, their ability to
adapt their interconnection weights to solve a given problem. One of their drawbacks, on the other hand,
is the lack of methodology for determining the topology, or architecture, of the network, which affects
important characteristics of the network's learning, such as training time and generalization capability. As
for the network's *generalization*, larger networks tend to overfit (memorize) the training data, while the B3.5
small ones may not be able to learn the training data at all. Ontogenic neural networks overcome this
problem by allowing the network to adapt, besides their weights, also their topology during the training
process.

This section provides an overview and comparison of the various ontogenic training algorithms, and
consists of the following parts. In this section ontogenic neural networks are defined. Section C1.7.2
presents a classification scheme for ontogenic neural networks. Section C1.7.3 describes methods for
layered ontogenic networks. In section C1.7.4 tree-based ontogenic networks are described, and in
section C1.7.5 the methods discussed in C1.7.2.1 and C1.7.2.2 are summarized. The material presented in
this section is partly based on work by Klotz and Fiesler (1996).

What distinguishes ontogenic algorithms from other neural network training algorithms is that they
are used not only to train the weights, but also to learn the topology needed to correctly learn the training
data. With a conventional training algorithm for multilayer networks, such as *backpropagation*, the typical C1.2.3
topology design procedure involves several steps. The only modifiable parameters are usually the weight
values of the interconnections and the gain of the activation function. The first design step is to choose
the number of hidden layers and the number of neurons in each hidden layer, together with a connectivity
scheme which is usually fully interlayer connected. Next, the weights of this network are adapted using
the training algorithm. Then, if the training does not result in an acceptable solution, that is, inputs do not
yield the expected outputs, a new topology is guessed and the process is repeated until a suitable solution
is found. Ontogenic neural networks greatly simplify this process by learning the topology needed for
solving a given task with no or minimal intervention of a user. Supervised ontogenic networks, which
need a tutor that provides target outputs during the training phase, are described in this section, whereas
unsupervised ontogenic networks are described in Section C2.4. C2.4

A *growing*, or *constructive*, ontogenic neural network typically starts with a very small topology, for
example a single hidden neuron, and keeps adding new neurons and connections, until the task at hand

is solved. *Pruning*, or *destructive*, neural networks start with a large topology and reduce its size by eliminating neurons, connections, or both. Growing–pruning methods performs both operations, either in separate stages or in an interleaved manner.

The concept of automatically changing the topology of a neural network is not new (Cameron 1964, Hopcroft and Mattson 1965, Ivakhnenko 1968), but a widespread interest in ontogenic neural networks did not start until the latter part of the 1980s (Gallant 1986a, Mjolsness and Sharp 1986, Lansner and Ekeberg 1987). Since then, various ontogenic methods have been proposed by research teams all over the world. In this section an overview and comparison of these methods is given.

In order to give the reader good insight into how they work, the neural tree algorithm (Sirat and Nadal 1990), which is an extension of the earlier tiling algorithm of Nadal (1989), is explained in greater detail. The aim of this section is to present a broad overview of the area of ontogenic neural networks. It should be noted, however, that this area is a relatively new one and some of the presented methods may not yet be sufficiently evaluated or widely applied. The focus of this section is therefore on the not-so-recent models which are usually more established and tested. Furthermore, since it is impossible to be complete, with new methods and refinements of existing methods being introduced frequently, no claim can be made that all 'important' models are represented.

C1.7.2 Classification of ontogenic neural networks

Many ontogenic training procedures are based on layered neural networks. The neurons in such networks are usually (fully) interlayer connected, whereas some also have supralayer connections which link to neurons in further layers, not only the adjacent ones (see Section B2.2). Another topology used for ontogenic neural networks is the *tree* (see Section B2.7), which is sometimes used in growing methods. Trees can theoretically be regarded as special layered topologies, but will be treated separately in this section. It should be noted that in some cases a tree topology can be converted into an equivalent layered topology, and sometimes a layered topology corresponds to a virtual tree structure.

In order to define a classification of the various supervised ontogenic methods, a first distinction is made between methods based on layered topologies and those based on tree topologies. A further subdivision is made based on whether the method grows and/or prunes the network. This classification is used for subdividing this section into sections, where the following aspects are studied:

- the type of learning rule used, and whether the ontogenic method uses any heuristics,
- the effectiveness, applicability, and flexibility of a method,
- its generalization capability,
- its learning time complexity,
- the existence of a convergence proof,
- the need to retrain the network after its topology is modified, and
- whether the ontogenic method is *local*, where local means that the method is only based on information available at each individual neuron.

C1.7.3 Methods for layered neural networks

Growing methods. These ontogenic methods usually start with a small topology, which is often an input layer of neurons without nonlinearity and an output layer, to which hidden neurons are added until the problem is solved. The size of the input and output layer for supervised learning method s is determined by the application.

In Mezard and Nadal (1989), the authors present the tiling algorithm, which builds several hidden layers. In this method, a layer is composed of two kinds of neuron: the master neuron, which delivers a partial output, meaning that it usually does not deliver the right answer for all training patterns, and so-called *ancillary neurons*. The input patterns are randomly presented to the network, and new sets of weights are generated. Using the *pocket algorithm* (Gallant 1986b), the set of weights which correspond to the smallest error value (that is, the largest number of correctly classified input patterns) is kept. When the total error is zero, the generation of the network is completed. Otherwise, ancillary neurons are created such that they are correctly learning a subset of prototypes. The convergence of this method is guaranteed by the fact that the addition of each new layer gives better results than the previous one: the output error of the master neuron in layer $l + 1$ is smaller than the output error of the master neuron in layer l.

Nadal (1989) study a simpler architecture. Here, each new hidden neuron is connected both to the most recently added hidden neuron and to the input layer. All the neurons in the output layer are connected to every hidden neuron as well as to all the neurons in the input layer. The addition of a hidden layer always decreases the error. The methods described by Mezard and Nadal (1989) and Nadal (1989) give similar performance and have a good generalization ability. The method of Nadal (1989) builds a simpler topology than the tiling algorithm. However, the tiling algorithm uses a smaller number of weights and is less constrained (several topologies with different numbers of hidden layer neurons or with a different number of hidden layers can be obtained for the same problem). The main difference between the two methods concerns the ancillary neurons, which only exist in the tiling algorithm.

Ash (1989) proposed the dynamic node creation (DNC) method which adds new hidden neurons, whose weights are initialized to zero or random values, to a three-layer network when the average error curve begins to flatten rapidly, that is, the error is below a given small value during a given number of epochs, where an epoch is a complete pass through the training set. Standard backpropagation (BP) is used and no retraining is required. The author reports that DNC is usually more efficient than standard BP. In his experiments, DNC always converged, even when standard BP did not. Although more training iterations are needed for the DNC, less computational time is required for the total computation.

A variation and extension of the tiling algorithm is the so-called *cascade correlation* method (CC) (Fahlman and Lebiere 1990). It has been designed in an attempt to overcome certain limitations of BP, especially the slow training time caused by the fixed learning step size, since it is difficult to choose a correct BP step such that the method converges and is not too slow. The second limitation of BP is the moving-target problem, where each neuron is trying to be a feature detector, and all other neurons are adjusting their weights at the same time. It may therefore occur that the feature a neuron is detecting changes during the training. The *quickprop* learning rule (Fahlman 1988) solves the first problem. As it B3.4.2 is not specifically designed for ontogenic neural networks it will not be described here. In order to avoid the second problem, in CC, the hidden layer neurons are added one at a time and their incoming weights are *frozen*. The learning rule attempts to maximize the correlation between the output of the new neuron and the residual error of the active network. The decision to create a new neuron is taken when the error has not significantly changed after a certain number of epochs. This number of epochs must be tuned by a human designer. To create a hidden neuron, a candidate neuron (or even a pool of candidate neurons) is connected to all existing neurons. After a few epochs the correlation is calculated and a gradient descent is performed to maximize it. In the case of a pool of candidates (each with a different set of initial weights, or different thresholds or sigmoid functions), only the neuron whose correlation score is the highest is kept. The weights of the new neurons are then frozen and a new one can be created. An advantage of the method is that no retraining is required since the existing weights do not change. A recurrent version of the algorithm (RCC) was developed later (Fahlman 1991) with the purpose of being able to map a sequence of inputs onto a desired sequence of outputs. The main problem is how to adapt the CC method, where the weights are frozen, to a recurrent topology where self-recurrent links are aimed to modify the weights. Hence, a hybrid architecture has been developed where only recurrent links are allowed to change. The remainder of the method is similar to ordinary CC.

To eliminate a disadvantage of RCC, namely that it cannot learn all finite-state automata, Dong Chen *et al* (1995) propose a simple scheme to dynamically generate a recurrent network topology. In recurrent neural networks, as opposed to feedforward networks where the information flows from the input to the output and no residual information is left in the network after each input is processed, the weights of the so-called recurrent neurons are dependent on their previous values. This characteristic enables recurrent networks to deal with temporal information. The proposed network structure, like that of the RCC, consists of inputs, a hidden layer of fully intralayer connected neurons, and outputs. Both inputs and outputs are connected to all hidden layer recurrent neurons. The main difference between RCC and the method of Dong Chen *et al* (1995) is that the recurrent layer is fully intralayer connected and is allowed to expand when needed. The criterion the authors use for expansion is as follows. A new neuron is added to the recurrent layer after every 50 or so epochs, until all training data are recognized correctly. One of the method's disadvantages is the same as for all recurrent neural networks: their training is much more computationally complex than for non-recurrent networks. Another disadvantage is that the resulting network size is often larger than a minimal one found by using the trial and error method.

The method described by Baum and Lang (1991) uses an *oracle* to obtain additional information. Assume one has two training patterns with opposite desired outputs. A hyperplane must intersect the hypersegment bounded by the two patterns. To increase precision, one can ask the oracle the output value of the point which is in the middle of the segment. Each query produces a new equation which can be solved to find the weights of the hidden units. This method can build a neural network which is as accurate as desired.

Ruján and Marchand (1989) follow the same philosophy: they divide the input space into several classes with as few hyperplanes as possible. The input space is represented in a hypercube whose corners represent the possible input patterns. According to the desired output, each corner can be colored, for example black is assigned the -1 value and white the $+1$ value. Next, the colored points can be separated by hyperplanes such that each region contains points of the same color and the hyperplanes do not intersect inside the hypercube. This operation is called *regular partitioning*. The interesting result is that each partitioning is linearly separable and a single hidden neuron can do the work. The partitioning is performed by a greedy algorithm.

Zollner *et al* (1992) describe another ontogenic method based on a three-layer network. Starting from the observation that a simple perceptron can correctly learn a linearly separable subset of patterns, the method divides the training set into linearly separable subsets which all have the same output value. Each neuron in the hidden layer is responsible for a certain set of patterns. In order to find these subsets, the input patterns are normalized such that they form a hypersphere. Next, a hyperplane cutting the hypersphere is defined such that only patterns inside the resulting hypercone have final output $+1$ (here, *inside* means the smaller part of the cut sphere), and the patterns outside the hypercone should have intermediate output -1 (although they have a final output $+1$). This distinction is made because the patterns inside and outside the hypersphere are not linearly separable and thus must be distinguished, even if their final outputs are the same. In this way, intermediate hidden neurons are created one by one. Next, the output layer performs a logical OR function on the intermediate neurons; at least one $+1$ means the output must be $+1$, and lack of $+1$ means the output must be -1. The method is reported to be more efficient than the tiling algorithm. It creates fewer neurons and the number of created neurons versus the number of training patterns follows a smooth curve where the tiling algorithm produces a more erratic (less predictable) curve.

Reilly *et al* (1982) focus on data classification. Three kinds of neuron are defined: input neurons, coding neurons, and decision neurons. An *influence radius* is defined for each pair of coding/decision neurons. This model can be compared to a three-layer network where the coding neurons are in the hidden layer and the decision neurons are in the output layer. Initially, the neural network consists of only input neurons. Then, when the first pattern is presented, a coding neuron is connected to the input layer and a decision neuron is attached to the coding neuron. The weights are updated in such a fashion that the decision neuron fires when the first pattern is presented. When another training pattern is presented which belongs to the same class, the network is not modified. Otherwise, a new coding neuron and a new decision neuron are created such that the decision neuron fires only when the second pattern is presented. The influence radius of the first decision neuron is decreased such that the first decision neuron does not fire. This process is repeated for each pattern. Since some points may belong to several influence spheres, the networks obtained by this method may have a bad generalization rate, although a proper initialization of the influence radius can overcome the problem.

Murphy (1990) was the first to employ Voronoi diagrams, or Dirichlet tesselations, for the design of feedforward ontogenic neural networks for pattern classification problems. Bose and Garga (1993) provide an improved and systematic way of designing such networks. Their method is based on the construction of a Voronoi diagram over the set of points representing patterns in a feature space, using recent advances in computational geometry and graph theory. They take advantage of the algorithm of Dwyer (1991) which enables quick construction of Voronoi diagrams. A Voronoi cell is a multidimensional polytope which can be defined as an intersection of a number of closed half-spaces (hyperplanes); it is this particular feature that is used in their algorithm. Once the Voronoi diagram is constructed from the training data, the number of separating hyperplanes is also known, and thus their weights, which translates into the first hidden layer of their network. These hyperplanes, in turn, define a number of regions, interiors of Voronoi cells, which correspond to the number of neurons in the second hidden layer. The number of neurons in the output layer is equal to the number of classes one is trying to learn. The advantage of the algorithm is that several alternate network topologies can be generated for a given problem. Its biggest disadvantage is the fact that it is only well suited for classification problems having a distinct clustering structure.

Another interesting growing method is described by Ring (1993). In this method, high-order connections (see Section B2.4) are created by adding new units. A hierarchical organization allows modification of the weights with information from any past period, without the need for a recurrent network. The network starts with a two-layer topology, with single-order connections, and the learning

B2.4

is performed by gradient descent. High-order units are added when a weight is forced to increase and decrease at the same time. This method gives good results for solving temporal tasks (such as grammar recognition) but human intervention is still needed to tune the many parameters. It creates a many new units, and the generalization capability is therefore poor. On the other hand, the network learns very fast.

Tan and Vandewalle (1994) propose a design technique where a neural network is constructed from an arbitrary set of binary associations. Their key idea is to treat these associations as a binary matrix mapping, and to decompose it into a sequence of so-called primitive operations. They define only three such operations: deflation, augmentation, and entry flipping, which they prove to form a complete set of operations for binary input-output vectors. Their algorithm can be summarized in three steps:

- Decompose a given input-output mapping (treated as matrix to matrix mapping) into a sequence of primitive operations
- Construct a sequence of simple feedforward subnetworks realizing each of the primitive operations
- Cascade the subnetworks to construct the overall architecture realizing the input-output mapping.

The algorithm has an algebraic flavor. Its computation time grows linearly with $N_1 + N_L$, the dimension of input and output vectors, for a fixed number of training pairs, which is very efficient. Another advantage of the algorithm is that the output of any neuron in a so-constructed network is connected to at most two other neurons. This feature makes the algorithm very attractive for very large-scale integration (VLSI) implementations. Its disadvantage is that it works only with binary vectors.

Moody and Antsaklis (1995) proposed a *dependence identification* (DI) algorithm which is similar to a method of Marchand *et al* (1990) for constructing a Boolean network. The difference is that it works with continuous training data and uses the concept of linear dependence for grouping similar patterns. The DI algorithm first transforms the training problem into a set of quadratic optimization problems and then constructs the appropriate network. In other words it grows from a small network to a large one. The network is constructed one layer at a time. At first, an attempt is made to train a two-layer network without hidden neurons. If successful, the training is complete. Otherwise, hidden layers are created by selecting subsets of the pattern space, until each pattern is classified correctly by at least one hidden layer neuron. Outputs of the hidden layer neurons are then used as inputs for the next layer, until the maximum number of layers has been generated or an error criterion has been met. The latter two are given as parameters to the network. Another parameter required by the network is used to determine whether the output of the network for an individual pattern is within specified error bounds. The DI algorithm can be used not only with sigmoidal activation functions, required for gradient-based techniques, but also with discontinuous switching functions. The advantage of the algorithm over standard BP is its training speed. A disadvantage is that it requires batch training which makes it unsuitable for on-line learning.

The *Hopfield* (1982) model is usually used for associative memory applications. In the Hopfield model the neurons are fully intralayer connected, without self-connections. An unusual topology derived from the Hopfield network is an *associative network* used by Lansner and Ekeberg (1987) and Ekeberg and Lansner (1989). In this method, one layer of input/output nodes is connected to a set of symmetrically intralayer connected internal nodes. When two input patterns are very similar, the network makes mistakes, and the authors propose the creation of new nodes to avoid this by replacing two strongly dependent neurons by three more selective ones. Suppose both neurons A and B are active at the same time. They will be replaced by three neurons, performing each of the functions A AND B, (NOT A) AND B, and A AND (NOT B). The memorization capacity is thus improved, but the main drawback of the method remains: these networks are aimed at memorizing instead of learning.

In order to teach a trained neural network new samples, Diederich (1988) proposes a system which can extend its topology. The network is composed of two kinds of neuron: committed neurons and free neurons. The committed neurons are partially connected to each other and already represent some information. They are also partially connected to the free neurons. The ontogenic method recruits free neurons, that is, it strengthens the connections between a cluster of committed neurons and some free neurons. After recruitment, these free neurons become committed neurons. The network consists of four parts.

- *The concept space.* Each concept (feature) is represented by a three-unit subnetwork. These units are connected to each other by bidirectional links. The affirmative unit is on only when the corresponding concept is present in the input pattern, and the negative unit is on only when it is not present. When the three units are off, the concept is neither affirmative nor negative. Each affirmative neuron is connected to the affirmative neuron of a substructure representing a more general concept and each

C1.3.4

C1.3

negative neuron is connected to a negative neuron representing a more specific concept. Neurons of the concept space are also connected to neurons in the attribute space (see below), so that concepts are associated with their properties.

- *The attribute space.* Each attribute is materialized by a subnetwork in which each neuron represents a possible attribute value. Exactly one attribute value can be on in an attribute subnetwork.
- *The instance space*, which is single neurons connected to the concept neurons. This corresponds to the input layer.
- *The free space*, which has the same topology as the concept space. This structure is partially connected to the attribute space.

In order to avoid a combinatorial explosion of the number of connections, only about 50% of all the possible connections between the concept space and the free space can be initially created. The learning process is based on the winner-take-all method. When a new training example, which is a set of (attribute, value) pairs, is presented to the network, a high activity in the attribute space will lead to an activation of some neurons in the free space. The neuron with the strongest links will be the only one whose weights are reinforced and will become a committed neuron. It is interesting to note that the total number of neurons does not increase, but the total number of neurons playing an active role in the network does increase. This kind of network is especially designed to build semantic networks and comparing it to more classical neural networks is therefore not very meaningful.

Hanson (1990) starts from the principle that the division of an input space into linearly separable regions is a problem whose complexity dramatically increases with the number of elements in the input space. He also outlines the stochastic (nondeterministic) nature of neurons' output signals. Each weight can therefore be represented as a random variable, with mean zero and standard deviation one. Hidden neurons with a great standard deviation are considered having a low ability in reducing errors and will be split into two neurons, each copying half the information of their parent.

B3.5.2.9   *Pruning methods.* Destructive, or pruning, methods generally start with a large network and then eliminate connections and/or neurons. A popular method used here is known as *weight decay*, where all the weights in the network are slightly decreased during each training step. The weights which are rarely changed during training will diminish to almost zero in magnitude and become candidates to be eliminated. If all the weights associated with a given neuron go down to zero, then this particular neuron also becomes a candidate for elimination. Pruning methods still require the user to guess an initial topology and may thus be seen as less 'ontogenic' than growing methods. A good paper on pruning methods is that of Reed (1993).

Thodberg (1990) proposes to start with a large network trained with standard backpropagation, which has many decision hyperplanes, which usually means a poor generalization capability. Next, a connection is removed, and the network is retrained. If the error is not significantly increased, the pruning is made permanent; otherwise the connection is restored. All the connections are examined in this manner. For a network with a large number of hidden units, the pruning scheme is very slow. During the initial training process the author proposes to use a method similar to that of Chauvin (1989) in order to minimize the number of weights after the initial training.

LeCun *et al* (1990) (optimal brain damage, OBD) first train a network with a backpropagation-like learning rule. Next, some neurons are removed according to a *saliency* order (that is, the neurons whose deletion will have the least effect on the training error will be removed first). To calculate this saliency, the effect of a small perturbation of the value of a weight on the error function can be analytically computed using an approximation of the second derivative of the error function. The disadvantage is that it needs to relearn the entire training set after the removal of each neuron.

In a work by Hassibi (1993), in the network called optimal brain surgeon (OBS), he has shown that OBD sometimes removes the wrong weights (that is, weights with a small magnitude that play an important role in keeping the error low). Without these important weights the neural network is sometimes unable to find any good solution, and no further training of the network can overcome this mistake. In comparison to OBD, OBS removes more weights, and never the wrong ones. It is based on the computation of all second-order derivatives of the error function (where OBD computes only approximations) and generalizes better than OBD and other magnitude-based methods. An advantage is that no relearning of the training patterns is needed because, in addition to deleting the weights, it calculates and changes the other weights affected by the pruning. Drawbacks are the large cost of calculation of all second derivatives and the nonlocality of the method.

Lozowski *et al* (1996) followed up the work of Ishikawa (1994), who proposed to use the sum of the absolute values of the weights as the penalty term, scaled by a constant referred to as the *forgetting rate*. The new method provides a theoretical framework for investigating the problem of selecting the forgetting rate by considering learning as a dynamic system and solutions to the learning process as fixed points in the system. The authors observed that the forgetting rate controls the maximum eigenvalue of these fixed points and thus their stability properties. As the forgetting rate is increased, starting from a small value, fixed points corresponding to poorly pruned networks are removed from the learning dynamics. Their method may provide the basis for an improved means of selecting the forgetting rate.

Growing–pruning methods. These methods are a combination of growing and pruning algorithms.

An attempt to escape from local minima is described by Hirose *et al* (1991). New neurons are added one by one to the hidden layer when the backpropagation error step does not decrease by more than 1% after 100 iterations. The weights of the new neurons are initialized to zero or random values. Once the network converges, the most recently added neurons are removed one by one and the last network that converged is kept as the final network. Some variations in the removal order have been tried and it does not seem to be a good heuristic to remove the first-added neurons first.

An ontogenic method particularly designed for shape recognition has been developed by Honavar and Uhr (1988, 1989). The network contains a pyramid of hidden layers (each layer is smaller and has a more specialized function than the previous one) where adjacent neurons are grouped to form a so-called recognition cone. Each cone is trained to recognize features, where higher layers represent more complex features. The weights between transforms are updated according to a rule similar to the BP learning rule, but when some condition is reached (computed by a minimal complexity heuristic), the method can create new transforms or discard useless ones.

Alpaydin (1990a, b) presents the grow and learn (GAL) method that allows incremental category learning. It is partially based on the work of Reilly *et al* (1982). The topology used is a three-layer network where only the size of the input layer is fixed. Each neuron in the output layer represents a category and a winner-take-all mechanism in the hidden layer allows exactly one neuron to fire when the appropriate pattern is presented to the network, and only the winner neuron is allowed to adjust its weights. During learning, new classes (output neurons) are added each time a training patern does not belong to any of the existing classes. The role of the hidden layer is to build *domination regions* for each class. Domination regions are separated by hyperplanes which equally divide the space between two hidden neurons. Since some classes may belong to more general classes, it is necessary to remove some of the hidden neurons in order to remove domination regions that are included in others'. After learning, the action of such neurons is disabled and the whole training set is presented to test whether the classification has not changed. The disabled neuron is removed from the network if the response of the network is unchanged. The GAL method can be combined with the unsupervised grow and represent method, described in Alpaydin (1990b), to perform both classification and feature detection.

In order to solve time-series prediction problems, Kadirkamanathan and Niranjan (1992a,, b) propose an ontogenic method based on *radial basis functions*. Each time a new training example is presented C1.6.2 to the network, the method has the possibility of adding a new basis function. The criterion for adding a new unit or not is a comparison between the complexity of introducing a new basis function and the approximation error that would incur otherwise. The same principle has been extended to multiple outputs and successfully applied to *speech pattern recognition*. However, pruning schemes (or a modified grow F1.7 criterion) would improve the network behavior, since the authors show that some redundancy has been found in the network.

The first descriptions of the group method of data handling (GMDH) can be found in the articles by Ivakhnenko (1968, 1971). This method is typically used to approximate a continuous function. Each neuron (except the input neurons) has exactly two inputs and one real-valued output. Each neuron of layer l is connected to one or more neurons of layer $l + 1$. The final network is thus a partially interlayer-connected feedforward network. Each neuron computes a quadratic polynomial combination of its two inputs. The output signal of a neuron is the value of its polynomial. The growing process starts with an input layer only. The next layer is configured in such a way that each pair of outputs from the previous layer is associated with its own neuron. Each neuron has its own polynomial and will try to produce an output equal to the overall desired output, in contrast to BP-based methods where neurons are specialized to recognize a particular feature. The coefficients of the polynomials are adjusted while training patterns are presented. The association of these patterns with the desired output values gives a set of linear equations whose solution is approximated in order to minimize the mean squared error. If the ratio of the smallest mean squared error versus the mean squared error of one neuron is greater than a given threshold, this neuron is eliminated from its layer. Layers are added one by one until the neuron of layer l with the smallest mean squared error has a greater mean squared error than the best neuron of layer $l - 1$. If there are still several neurons in this last layer, only the neuron which best approximates the desired output is kept and all useless neurons which do not belong to the same graph as the final output neuron are removed from the network, that is, neurons that are not, directly or indirectly, connected to the final output neuron. Criteria other than the mean squared error have been proposed in Farlow (1984) and Barron (1984). They construct networks which are less sensitive to noise and have a better generalization ability. Barron (1975) describes a method very similar to the GMDH method which is based on generalized McLaurin series. Tenorio and Lee (1989) propose a variation of the GMDH method which is combined with simulated annealing Kirkpatrick *et al* (1983). The GMDH method has been widely used in various domains such as

F1.6, F1.8, G6.3   *target recognition*, *image analysis*, *signal processing*, and *economic forecasting*.

C1.4.6   The *Cauchy machine* is a stochastic neural network related to the *Boltzmann machine* that uses fast simulated annealing instead of classical simulated annealing. Both the methods have the advantage of enabling a global search for the minimum of the error function. Their main drawback is the slow training process and, as with other multilayer neural networks, the lack of knowledge for choosing the number of hidden neurons. If higher-order connections are employed, a smaller number of training cycles can be expected and hidden neurons can be avoided. Hence, a high-order generalization of the Cauchy machine has been developed (Cuche and Fiesler 1996). However, the problem with fully interlayer-connected higher-order neural networks is that the number of possible connections increases exponentially with the

B2.5.1.2   order (see Section B2.5.1.2). To overcome this problem, three ontogenic methods, a pruning, a growing, and a hybrid growing–pruning method have been developed to produce a sparsely connected network topology during the training process (Cuche and Fiesler 1996).

C1.7.4   Tree-based methods

In this subsection, both tree-based ontogenic methods and those based on virtual trees that result in a layered topology are described.

In Deffuant (1990), the author describes his NEURAL (neural units recruitment algorithms) method and provides a convergence proof. The NEURAL method builds a tree structure. Initially, a pool of candidate neurons is randomly generated, and the best one (the neuron producing the lowest number of errors) is kept. The training set is repeatedly divided into two parts until the error on the training set is low enough. The parent neuron is tested on one of the subsets. The test is simply the execution of a specified number of training epochs and the computation of the corresponding error value. If it fails, two new neurons are recruited (again from a pool of candidates) and each one is trained with one of the subsets of patterns. Otherwise, only one new neuron is created. Generalization is achieved by looking for the neuron which is more likely to give the correct output. Two versions of the method have been implemented: one using the perceptron and one using the BP learning rule. Note that further learning is possible after training since new branches and leaves can be added to the tree when needed.

The upstart algorithm, proposed in Frean (1990), starts from another point of view: instead of building layers from the input layer outward until convergence, new neurons are interpolated between the input layer and the output layer. A convergence proof is given which is based on the principle that child neurons will always make fewer errors than their parent, and the addition of a child always reduces the number of errors. When a given neuron makes a mistake (being ON or OFF) its answer can be corrected by a new neuron which would be active at the right time. These neurons are called daughter neurons, and the flow of information proceeds from daughter to parent. Two different kinds of neuron are used for the daughter neurons. These neurons correct the mistakes made by the output neurons. When a given neuron answers 'ON' when an 'OFF' is expected, a new neuron (called X) is connected. When this same neuron answers 'OFF' when 'ON' was expected, another neuron (called Y) is connected to its parent. The resulting network has a hierarchical tree structure and its root is the output unit. The author shows that the final neural network is equivalent to a single-hidden-layer network, and notes that the network size is smaller than the one produced by the tiling algorithm (Mezard and Nadal 1989).

As an illustration, the details of topology determination in the neural tree algorithm are described here.

While the method of Nadal (1989) method can be seen as a tiling algorithm without ancillary units, the method of neural trees (NT) proposed by Sirat and Nadal (1990) can be considered as a tiling algorithm without master units.

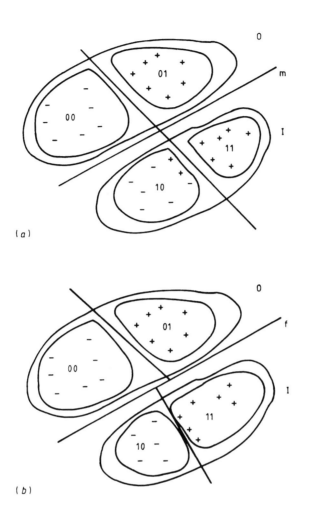

Figure C1.7.1. Tiling and neural tree. (*a*) Tiling algorithm: successive dichotomies of entire input space until convergence. (*b*) Neural tree algorithm: successive dichotomies only of inhomogeneous subspaces.

The idea behind the tiling algorithm is that, given a pattern space, a root node is trained, and a dichotomy is obtained that splits the pattern space into two subspaces. If the dichotomy successfully separates the space such that no classification errors occur, the pattern space is linearly separable, and the process stops. If, however, classification errors still exist, another node is added and trained. This process is repeated until all patterns are classified correctly. The drawback of the algorithm is that when considering the entire pattern space for each additional node, it is possible to achieve a good dichotomization in one of the subspaces, but disrupt correct classification in other subspaces. The advantage of the NT algorithm is that after each dichotomy the two resulting subspaces are treated independently. The difference between the two algorithms is demonstrated in figure C1.7.1. The NT algorithm is very similar to the one described before. A dichotomy is found to separate the pattern space. Subsequently, each of the subspaces is tested for homogeneity. If one subspace is inhomogeneous (contains classification errors), a node is added and trained to dichotomize that subspace only. The same operation is performed on the other subspace. In the same manner, the newly created subspaces are tested for homogeneity, and nodes are added accordingly. This recursive process continues until all patterns are correctly classified. Figure C1.7.2 shows several steps of this process. The upper part depicts the initial pattern space, the next two are two possible dichotomization steps, and the bottom part shows the completed process where all subspaces are homogeneous. It is thus possible to visualize the NT forming, as nodes are added onto other nodes. Actually, the tree that is being

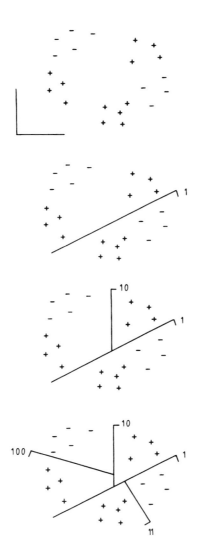

Figure C1.7.2. Successive dichotomies of pattern space.

discussed is only a virtual one. As is shown next, the physical formation of the nodes is quite different. To understand the NT topology, one needs to depart from the traditional way of thinking about feedforward networks. Physically, the NT consists of only a single layer of neurons. Every node receives the networks inputs, that is, every node gets the same information. The outputs of the nodes are considered the network outputs and there are no feedback paths. This is not a tree structure; in fact considering this structure alone it is impossible to determine which node represents the output of which subspace of inputs. When constructing the tree, a *virtual tree*, or a map, of the topology is placed in memory. This map explicitly shows which nodes are the children or parents of other nodes. Hence, during testing this map must be referenced in order to determine which node contains the true classification. This is done by testing the root nodes' output. If the output is +1, the right child, the node number which is specified by the virtual tree map, is subsequently tested. It the output is −1, the left child is tested. This process is then repeated where the previously selected child node becomes the root node for the next cycle. The tree traversal is repeated until a leaf node is reached. This leaf will then hold the correct value for the classification. For the network implementation it is important to distinguish between the actual topology and the virtual tree. A typical NT is shown in figure C1.7.3. The NT algorithm is a two-phase process. The first phase is the training phase, where the network is designed and the map is built:

(i) present training set to a root node
(ii) dichotomize the training set

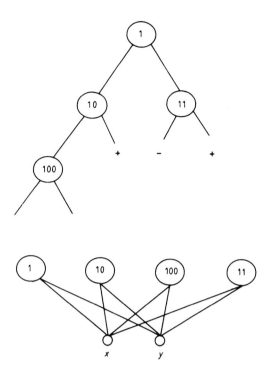

Figure C1.7.3. Typical neural tree architecture. This departs from traditional feedforward networks and consists of (i) a virtual tree architecture (top) and (ii) a physical architecture (bottom).

      a try to minimize epoch error
      b separate into two categories ($+1$ and -1)

(iii) test if either side is homogeneous

      a if yes, label it as a leaf
      b if no, repeat the process with this subset only.

The second phase is the classification of a test vector:

(i)   apply test vector
(ii)   determine output of the node
(iii) test output

      a if output is -1, test left child node
      b if output is $+1$, test right child node

(iv) repeat until a leaf is reached.

The NT training algorithm is able to recognize several classes and includes an epoch weight adjustment, not the usual adjustment after each pattern presentation. An emphasis is placed on the generation of an evenly balanced tree, which is crucial for reducing the worst-case classification time.

    The NT training algorithm can be broken down into two subalgorithms: class reduction and tree-optimized training. Combining features of both enables the algorithm to construct well-balanced trees. The class reduction algorithm essentially converts a multiclass subspace into a two-class subspace. This is beneficial since a neuron, which represents a decision hyperplane, can only determine whether a pattern evaluates to a $+1$ or a -1. Furthermore, it makes this conversion in a way that greatly enhances the possibility of splitting the subspace in half, thus helping in generating a well-balanced tree. The algorithm is described by the following five steps:

(i)   determine the number of classes
(ii)   divide the number of classes in half
(iii) randomly choose half of the classes present and

a assign all patterns in the subspace of those classes a temporary classification of $+1$

b assign all patterns belonging to the other half a temporary classification of -1

(iv) if there is an odd number of classes, randomly decide whether the odd-numbered class will be represented by a $+1$, or a -1

(v) run the tree-optimized training algorithm.

By choosing half of the classes as belonging to $+1$ and the other belonging to -1, the probability of dichotomizing the space into two equal halves is greatly enhanced. In the tree-optimized training algorithm all patterns are presented before any weight adjustments are made. To make this possible, and to emphasize a balanced tree, several pieces of information are collected during each epoch presentation. These include the number of incorrectly classified $+1$ patterns and the number of incorrectly classified -1 patterns, and the exemplars of all incorrectly classified $+1$ patterns and -1 patterns are found. If, after all patterns have been presented, there are more $+1$ errors than -1 errors, the positive error exemplar pattern is used in the traditional updating equation. This has the effect of moving along the gradient in the positive direction. If the -1 errors were greater, the negative error exemplar is used in the updating equation thus moving in the negative gradient direction. This process continues until either of the two following situations occurs. The first is the event that the total error, $+1$ errors plus -1 errors, reaches a local minimum. The other is the event that an even split has been reached, $+1$ errors $= -1$ errors. This criterion helps in the construction of a well-balanced tree. The algorithm consists of the following steps:

(i) choose learning constant, c

(ii) generate random weights

(iii) **repeat while** olderr $\geq$ newerr

(iv) clear poserr, negerr, and exemplar patterns

(v) **for** each pattern vector in subspace **do**

(vi) sum $=$ sum $+$ (weight(j) $*$ pattern vector(i)(j))

(vii) $s =$ sgn(sum)

(viii) **if** $s *$ output vector $= -1$ **then** increase negerr and accumulate negexemplar **else** increase poserr and accumulate posexemplar

(ix) **end for**

(x) newerr $=$ negerr $+$ poserr

(xi) **if** negerr $>$ poserr **then** $s = -1$ $w(j) = w(j) + c * s *$(negexemplar/negerr)

(xii) **if** poserr $>$ negerr **then** $s = +1$ $w(j) = w(j) + c * s *$(posexernplar/poserr)

(xiii) **if** poserr $=$ negerr **then** quit

(xiv) **end repeat while**

The NT algorithm guarantees a satisfactory solution in one iteration, thus saving a lot of time needed for repeated training, and finally, since the *multiclass tree-optimized training algorithm*, used in the NT algorithm, is much quicker than the error BP, the required training time is greatly reduced. These benefits, coupled with the ease of implementing the algorithm (all that is needed to design a network is training data and a learning rate constant) make the NT algorithm an attractive ontogenic network. The only drawback of the algorithm is the fact that some postprocessing must be done before a test vector can be classified.

The ideas of the ID3 machine learning algorithm (Quinlan 1987) and Nadal's algorithms (Nadal 1989, Nadal 1990) inspired Cios and Liu (1992) to design an ontogenic algorithm called *continuous ID3* (CID3). The main differences between the NT algorithm and the CID3 algorithm are as follows. The CID3 algorithm starts with random weights and uses the entropy function to guide the placement of a new hyperplane. By doing this, the algorithm finds a suboptimal number of hidden layer neurons where the entropy is reduced to zero for the first time. Shannon's entropy function is used as a measure of separation of training examples. This is not to say that the generated topology is optimal. According to the CID3 algorithm new neurons (hyperplanes) are added in such a way that each new neuron decreases the entropy function. The entropy is calculated by counting the numbers of patterns belonging to each category on each side (positive–negative) of all hyperplanes. When it has reached zero it means that all training examples are correctly recognized. In addition, the concept of Cauchy training (changing a weight by a random number) was incorporated into the algorithm to help in escaping from local minima. The algorithm will always reach a conclusion with a single output, since each additional layer will have fewer neurons than the one before, thus giving a correct response for all patterns. For details the reader is referred to Cios and Liu (1992) and Sections D1.4 and G2.12 of this handbook.

D1.4, G2.12

In what appears to be the first application of fuzzy logic to ontogenic neural networks, Cios and Sztandera (1992, 1996) greatly simplified the architecture of the original CID3 algorithm by the use of fuzzy sets. The key feature of the ontogenic fuzzy CID3 algorithm, or F-CID3 in short, is its definition of fuzzy sets for each neuron of the hidden layer at which the entropy is reduced to zero for the first time. The fuzzy sets are defined based on the numbers of examples from both categories positioned on the positive and negative sides of the hyperplanes defined by the hidden layer, much in the same way as the Shannon entropy is calculated in the CID3 algorithm. Then, these fuzzy sets are ranked in order to decide a category the pattern belongs to. For many problems, such as distinguishing two spirals, application of the F-CID3 algorithm results in much smaller, in terms of fewer layers, topologies than the CID3 algorithm, and thus with a much smaller number of connections. For more details of the F-CID3 algorithm the reader is referred to papers by Cios and Sztandera (1992, 1996) as well as to Section D1.4 of this handbook.

Golea and Marchand (1990), following up Ruján and Marchand (1989), a tree architecture is proposed: each hidden neuron generates zero, one, or two neurons and is connected to them such that neurons are organized in a hierarchical way. Each child node tries to classify one of the two regions its parent has to classify. Typically a child node needs to classify half of its parent's patterns. Thanks to this, convergence is always guaranteed. The principle has been improved by Marchand *et al* (1990). Some conditions for choosing the hyperplanes have been relaxed, allowing more freedom. The tree structure has been replaced by a more classical three-layer network. Each neuron in the hidden layer is built such that they are correctly classifying a subset of the input patterns. New hidden neurons are sequentially added to classify the remaining patterns.

C1.7.5 Summary

We now summarize the characteristics of the different methods discussed in sections C1.7.3 and C1.7.4 in the form of tables which list the most crucial information about each method. It highlights those aspects that bring about the advantages of ontogenic neural networks, which are

- no trial-and-error process is needed to find a suitable network topology
- the design time can be substantially reduced
- the memory required to implement the network is smaller
- the total training time can be reduced
- the recall time can be shorter if an optimal topology is generated
- they may generalize better
- they may avoid local minima
- flexible topology can enable the absorption of new data through retraining.

The following notations and abbreviations are used in the tables. The 'model' column indicates the neural network model and the learning rule used, if known. 'FF' stands for feedforward, 'BP' for backpropagation, 'RBF' for radial basis functions, 'SDR' for stochastic delta rule, 'HONN' for higher-order neural networks, and 'GMDH' for group method of data handling. The 'descriptor' column gives the name or the most important characteristic of a method. 'DNC' stands for dynamic node creation, 'NEURAL' for neural units recruitment algorithms, 'OBD' for optimal brain damage, 'OBS' for optimal brain surgeon, 'GAL' for grow and learn, 'CID3' for Continuous ID3, 'F-CID3' for fuzzy CID3, 'DI' for dependence identification, and 'OCM' for orthogonal complement method. The 'L' column indicates the number of layers (including the input and output layer); $l + x$ means that the number of layers is initially l and can grow. The column marked '$\nearrow$' indicates that the method increases the network size by adding layers (L), neurons (N), or weights (W), where adding a layer implies addition of neurons, and adding a neuron implies addition of weights; 'N' is also used in the case of an initial two-layer network growing to a three-layer network, although a new layer is created with the first added neuron. The column marked '$\searrow$' indicates that the method reduces the network size, and whether layers (L), neurons (N), or weights (W) are removed. The column 'Loc' indicates whether the method is local (Loc) or nonlocal. 'IC' indicates the initial conditions of the method, which are encoded using up to three characters: the first digit indicates the number of layers (0 for an empty topology) in the initial topology, the first letter indicates the initial connectivity ('F' for fully interlayer connected, 'T' for a tree structure, 'O' for other topologies), and the second letter indicates the values of the initial weights ('R' when they are initialized with (small) random values, 'T' if the ontogenic method starts with an already trained network). Column 'Prf' (proof) indicates whether a convergence proof exists ('M', mathematical proof, 'I', informal proof). In general, the symbol

'—' indicates that the condition is not satisfied, while '?' indicates that the information was not available.

Table C1.7.1. Perceptron-based growing methods.

Reference	Model	Descriptor	L	↗	↘	Loc	IC	Prf
Ash (1989)	MLP[a] + BP[b]	DNC[c]	3	N	—	—	3FR	—
Fahlman and Lebiere (1990)	MLP	cascade correlation	$2+x$	L		X	2FR	—
Fahlman (1991)	MLP	recurrent cascade correlation	$2+x$	L	—	—	2OR	—
Mézard and Nadal (1989)	MLP	tiling	$2+x$	L	—	?	2FR	M
Nadal (1989)	MLP	—	$2+x$	L	—	?	xOR	M
Ruján and Marchand (1989)	MLP	regular partitioning	3	N	—	—	2FR	M
Zollner et al (1992)	MLP	—	3	N	—	—	2FR	I
Sirat and Nadal (1990)	MLP	—	3	N	—	X	1	I
	MLP + BP	—	3	N	—	X	2FR	—
Hanson (1990)	SDR[d]	meiosis networks	3	N	—	X	2FR	—
Deffuant (1990)	MLP + BP	NEURAL[e]	tree	N	—	X	0	M
Golea and Marchand (1990)	MLP + BP	decision trees	tree	N	—	X	2	M
Frean (1990)	MLP + BP	Upstart algorithm	tree $\equiv 3$	N	—	?	2OR	I

[a] Multilayer perceptron.
[b] Backpropagation.
[c] Dynamic node creation.
[d] Stochastic delta rule.
[e] Neural units recruitment algorithms.

Table C1.7.2. Perceptron-based pruning methods.

Reference	Model	Descriptor	L	↗	↘	Loc	IC	Prf
Chauvin (1989)	MLP + BP	energy functions	3	—	W	—	3FR	—
Ji et al (1990)	MLP + BP	—	3	—	W	—	3FR	—
Thodberg (1990)	MLP + BP	Ockham's razor	3	—	W	—	3FT	—
Rumelhart et al (1991)	MLP + BP	weight elimination	3	—	W	—	3FR	—
Hanson and Pratt (1989)	MLP + BP	penalty function	3	—	W	—	3FR	—
LeCun et al (1990)	MLP	OBD[a]	3	—	W	—	3FR	—
Hassibi (1993)	MLP	OBS[b]	3	—	W	—	3FR	—

[a] Optimal brain damage.
[b] Optimal brain surgeon.

Table C1.7.3. Perceptron-based growing and pruning methods.

Reference	Model	Descriptor	L	↗	↘	Loc	IC	Prf
Hagiwara (1990)	MLP + BP	—	3	N	W	—	3FR	—
Hirose et al (1991)	MLP + BP	—	3	N	N	—	?	?
Honavar and Uhr (1988)	cone structure	units recruitment	x[a]	N[b]	N	X	x O	—
Honavar and Uhr (1989)	MLP + BP				W			

[a] The method can handle any number of layers.
[b] Transforms = clusters of neurons.

© 1997 IOP Publishing Ltd and Oxford University Press

Table C1.7.4. Hopfield-model-based growing method.

Reference	Model	Descriptor	L	$\nearrow$	$\searrow$	Loc	IC	Prf
Lansner and Ekeberg (1987)	Hopfield-like	—	2	N	—	X	2OR	—

Table C1.7.5. Radial-basis-function-based ontogenic methods.

Reference	Model	Descriptor	L	$\nearrow$	$\searrow$	Loc	IC	Prf
Kadirkamanathan and Niranjan (1992a) Kadirkamanathan and Niranjan (1992b)	RBF[a]	—	3	N	—	—	2FR	—
Bonnlander and Mozer (1993)	RBF	latticed networks	2	X	—	—	2OR	—

[a] Radial basis functions.

Table C1.7.6. Other ontogenic methods.

Reference	Model	Descriptor	L	$\nearrow$	$\searrow$	Loc	IC	Prf
Reilly *et al* (1982)	—	category learning	3	N	—	X	0	I
Alpaydin (1990a	—	GAL[a]	3	N	N	—	1O	I
Diederich (1988)	—	neuron recruitment	4	N	—	—	—	—
Ivakhnenko (1968)	GMDH	—	$1+x$	L	N	X	1	U
Barron (1975)	GMDH	McLaurin series	$1+x$	L	N	X	1	—
Tenorio and Lee (1989)	GMDH	simulated annealing	$1+x$	L	N	X	1	—
Fujita (1990)	?	OCM[b]	?	?	?	?	2FR	?
Fujita (1992)	MLP	optimization of an objective function	$3+x$	L	?	?	?	?
Baum and Lang (1991)	MLP	query learning	3	X	—	—	2FR	—
Ring (1993)	HONN[c]		$2+x$	L	—	—	2FR	—

[a] Grow and learn.

[b] Orthogonal complement method.

[c] High-order neural networks.

C1.7.5.1 Methods based on layered networks

Growing methods. An important reason for using ontogenic methods is to automatically determine the (optimal) topology of a neural network. Growing methods offer such a possibility since they do not require the use to select an initial topology. The question of how to find the optimal topology, however, is still open because the optimal topology is only known for very small artificial problems (Fiesler 1993), such as the XOR problem, where the minimal topology is the optimal topology since the problem is completely defined and generalization does not apply.

Another reason for using growing methods is to avoid the training becoming stuck in local minima, since the addition of new units changes the dimensionality of the problem. The goal here is to establish that the error surface does not have the same local minima in the higher-dimensional space as in the lower-dimensional space. Growing methods usually start with a small size and then grow the topology. By doing so they require much smaller computational effort than pruning methods, even if the final network size is the same, since they deal with smaller networks most of the time. In addition, a user does not have to guess a 'large enough' initial network topology. The disadvantage of growing methods can be, in some cases, their relatively large final topology. The latter can be alleviated by either eliminating some connections and neurons, for example by using fuzzy logic. It becomes then a growing–pruning method. It is the opinion of the authors that growing or growing–pruning methods are most promising and can be considered as truly ontogenic ones. Table C1.7.1 shows an overview of growing methods for feedforward networks.

Pruning methods. Since small networks have a better generalization capability, several researchers have developed methods for pruning networks. In the simplest scheme the weights are removed one by one after training until the performance of the network is satisfactory. Some other methods are based on the reduction of the weights' magnitude during learning (a neuron connected to other neurons by only small weights is often considered not important and might be removed) by experimenting with the addition of penalty terms to the error function. These penalty terms often suppose a global knowledge of the network's complexity. Since the magnitude-based methods do not seem to be optimal and sometimes encourage the removal of important neurons, more complex methods based on second-derivative computation have been developed. Table C1.7.2 summarizes the pruning methods.

Growing–pruning methods. Most methods in this category are a combination of existing growing and pruning methods. Usually, they first increase the size of the network when a local minimum has been encountered or when the network seems to be too small to learn correctly. Later, redundant or useless neurons or connections are removed. Table C1.7.3 summarizes the growing and pruning methods.

Other methods are summarized in tables C1.7.4–6.

References

Alpaydin E 1990a Grow-and learn: an incremental method for category learning *Proc. In. Neural Network Conf. (INNC) 90 (INNS–IEEE) (Paris, 1990)* vol 2 (dordrecht: Kluwer) pp 761–4

—— 1990b Neural models of incremental supervised and unsupervised learning *PhD Thesis* Ecole Polytechnique Fédérale, Lausanne

Ash T 1989 Dynamic node creation in backpropagation networks *Connection Sci.* **1** 365–75

Barron A R 1984 Predicted squared error: a criterion for automatic model selection *Self-Organizing Methods in Modeling: GMDH Type Algorithms* ed S J Farlow (New York: Dekker)

Barron R L 1975 Learning networks improve computer-aided prediction and control *Comput. Design* 65–70

Baum E B and Lang K J 1991 Constructing hidden units using examples and queries *Advances in Neural Information Processing Systems (NIPS) 3 (IEEE; Denver, CO 1990* ed R P Lippmann, J E Moody and D S Touretzky (San Mateo, CA: Morgam Kaufmann)

Bonnlander B and Mozer M C 1993 Latticed RBF networks: an alternative to constructive methods *Advances in Neural Information Processing Systems (NIPS)—Natural and Synthetic 5 (1992)*

Cameron S H 1964 The generation of minimal threshold nets by an integer program *IEEE Trans. Electron. Comput.* **13** 299–302

Chauvin Y 1989 A back-propagation algorithm with optimal use of hidden units *Advances in Neural Information Processing Systems (NIPS) 1 (IEEE; Denver, CO 1988)* ed D S Touretzky (San Mateo, CA: Morgan Kaufmann)

Cuche S and Fiesler E 1996 Generalized Cauchy machines *Neurocomputing* Submitted to special issue on recurrent networks

Deffuant G 1990 Neural units recruitment algorithm for generation of decision trees *Proc. Int. Joint Conf. on Neural Networks (IJCNN) (IEEE; San Diego, CA 1990* vol I (Ann Arbor, MI: Edward) pp 637–42

Diederich J 1988 Connectionist recruitment learning *Proc. 8th Eur. Conf. on Artificial Intelligence (Munich, 1988)* pp 351–6

Ekeberg Ö and Lansner A 1989 Automatic generation of internal representations in a probabilistic artificial neural network *Neural Networks from Models to Application; Proc. nEuro-88, 1st Eur. Conf. on Neural Networks (ESPCI; Paris 1988)* ed L Personnaz and G Dryfus (Paris: IDSET) pp 178–86

Fahlman S E 1988 Faster-learning variations on back-propagation: an empirical study *Proc. 1988 Connectionist Models Summer School* (San Mateo, CA: Morgan Kaufmann)

—— 1991 The recurrent cascade–correlation architecture *Advances in Neural Information Processing Systems (NIPS)—Natural and Synthetic 3 (Denver, CO, 1990)* ed R P Lippmann, J E Moody and D S Touretzky (San Mateo, CA: Morgan Kaufmann)

Fahlman S E and Lebiere C 1990 The cascade–correlation learning architecture *Advances in Neural Information Processing Systems (NIPS) (IEEE; Denver, CO, 1989)* vol 2, ed D S Touretzky (San Mateo, CA: Morgan Kaufmann) pp 524–32

Farlow S J (ed) 1984 *Self-Organizing Methods in Modeling: GMDH Type Algorithms* (New York: Dekker)

Minimal and high order neural network topologies *Proc. 5th Workshop on Neural Networks; an Int. Conf. on Computational Intelligence: Neural Networks, Fuzzy Systems, Evolutionary Programming and Virtual Reality (San Francisco, CA, 1993). Proc. SPIE* 173–8

Frean M 1990 The upstart algorithm: a method for constructing and training feedforward neural networks *Neural Comput.* **2** 198–209

Fujita O 1990 A method for designing the internal representations of neural networks *Proc. Int. Joint Conf. on Neural Networks (IJCNN) (IEEE; San Diego, CA, 1990)* vol III (Ann Arbor, MI: Edward) pp 149–54

—— 1992 Optimization of the hidden unit function in feedforward neural networks *Neural Networks* **5** 755–64

Gallant S I 1986a Three constructive algorithms for network learning *Proc. 8th Ann. Conf. Cognitive Sci. Soc. (Amherst, 1986* (Hillsdale, NJ: Erlbaum) pp 652–60

—— 1986b Optimal linear discriminants *Proc. 8th Int. Conf. on Pattern Recognition*

Golea M and Marchand M 1990 A groeth algorithm for neural network decision trees *Europhys. Lett.* **12** 205–10

Hagiwara M 1990 Novel back propagation algorithm for reduction of hidden units and acceleration of convergence using artificial selection *Proc. Int. Joint Conf. on Neural Networks (IJCNN) (IEEE; San Diego, CA, 1990)* vol I (Ann Arbor, MI: Edward) pp 625–30

Hanson S J 1990 Meiosis networks *Advances in Neural Information Processing Systems (NIPS) (IEEE; Denver, CO, 1989)* vol 2, ed D S Touretzky (San Mateo, CA: Morgan Kaufmann) pp 533–41

Hanson S J and Pratt L Y 1989 Comparing biases for minimal network construction with back-propagation *Advances in Neural Information Processing Systems 1 (IEEE; Denver, CO, 1988)* ed D S Touretzky (San Mateo, CA: Morgan Kaufmann) pp 177–85

Hirose Y, Yamashita K and Hijaya S 1991 Back-propagation algorithm which varies the number of hidden units *Neural Networks* **4** 61–6

Honavar V and Uhr L 1988 A network of neuron-like units that learn to perceive by generation as well as reweighting of its links *Proc. 1988 Connectionist Models Summer School (1988)* ed D Touretzky, G Hinton and T Sejnowski (San Mateo, CA: Morgan Kaufmann) pp 472–84

—— 1989 Generative learning structures and processe for generalized connectionist networks *Connection Sci.* **1** 139–59

Hopcroft J E and Mattson R L 1965 Synthesis of minimal threshold logic networks *IEEE Trans. Electron. Comput.* **14** 552–60

Hopfield J J 1982 Neural networks and physical systems with emergent collective computational abilities *Proc. Int. Acad. Sci. USA* **79** 2554

Ivakhnenko A G 1968 The group method of data handling—a rival of stochastic approximation *Sov. Automat. Control* **13** (3) 43–55

—— 1971 Polynomial theory of complex systems *IEEE Trans. Ststems Man Cybernet.* **1** 364–78

Ji Chuanyi, Snapp R R and Psaltis D 1990 Generalizing smoothness constraints from discrete samples *Neural Comput.* **2** 188–97

Kadirkamanathan V and Niranjan M 1992a Application of an architecturally dynamic network for speech pattern classification *Proc. Inst. Acoust.* **14**

—— 1992b A function estimation approach to sequential learning with neural networks *Technical Report CUED/F-INFENG/TR.111* Cambridge University Engineering Department

Kirkpatrick S, Gelatt C D Jr and Vecchi M P 1983 Optimization by simulated annealing *Science* **220** 671–80

Klotz J and Fiesler E 1996 Ontogenic neural networks *Expert Syst.* in preparation

Lansner A and Ekeberg Ö 1987 An associative network solving the '4-bitADDER problem' *Proc. IEEE 1st Int. Conf.on Neural Networks (ICNN) (IEEE; San Diego, CA, 19870* vol II, ed M Caudill and C Butler (San Diego, CA: SOS)

Le Cun Y, Denker J S and Solla S A 1990 Optimal brain damage *Advances in Neural Information Processing Systems (NIPS) 2 (IEEE, Denver, CO, 1989)* ed D S Touretzky (San Mateo, CA: Morgan Kaufmann) pp 598–605

Marchand M, Golea M and Ruján P 1990 A convergence theorem for sequential learning in two-layer perceptrons *Europhys. Lett.* **11** 487–92

Mézard M and Nadal J-P 1989 Learning in feedforward layered networks: the tiling algorithm *J. Phys. A: Math. Gen.* **22** 2191–203

Mjolsness E and Sharp D H 1986 A preliminary analysis of recursively generated networks *Neural Networks for Computing (Snowbird, UT, 1986) (AIP Conf. Proc. 151)* ed J S Denker, pp 309–14

Moody J O and Antsaklis P J 1995 The dependence identification neural network construction algorithm *IEEE Trans. Neural Networks* **7**

Nadal J-P 1989 Study of a growth algorithm for a feedforward neural network *Int. J. Neural Syst.* **1** 55–9

Reed R 1993 Pruning algorithms—a survey *IEEE Trans. Neural Networks* **4** 740–7

Reilly D L, Cooper L N and Elbaum C 1982 A neural model for category learning *Biol. Cybernet.* **45** 35–41

Ring M 1993 Sequence learning with incremental high-order neural networks *CSE Technical Report AI 93-193* Department of Computer Sciences, University of Texas at Austin

Ruján P and Marchand M 1989 Learning by minimizing resources in neural networks *Complex Syst.* **3** 229–41

Rumelhart D E, Weigend A S and Huberman B A 1991 Generalization by weight-elimination with application to forecasting *Advances in Neural Information Processing Systems (NIPS)–Natural and Synthetic 3 (Denver, CO, 1990)* ed R P Lippmann, J E Moody and D S Touretzky (San Mateo, CA: Morgan Kaufmann) pp 875–82

Sirat J A and Nadal J-P 1990 Neural trees: a new tool for classification *Network; Comput. Neural Syst.* **1** 423–38

Tenorio M F and Lee Wei-Tsih 1989 Self organizing neural networks for the identification problem *Advances in Neural Information Processing Systems 1 (IEEE; Dencer, CO, 1988)* ed D S Touretzky (San Mateo, CA: Morgan Kaufmann) pp 57–64

Thodberg H H 1990 Improving generalization of neural networks through pruning *Int. J. Neural Syst.* **1** 317–26

Zollner R, Schmitz H J, Wünch F and Krey U 1992 Fast generating algorithm for a general three-layer perceptron *Neural Networks* **5** 771–7

C1.8 Adaptive logic networks

William W Armstrong and Monroe M Thomas

Abstract

An adaptive logic network (ALN) is a multilayer perceptron that accepts vectors of real (or floating point) values as inputs and produces a logic 0 or 1 as output. The ALN has a number of linear threshold units (perceptrons) acting on the network inputs, and their (Boolean) outputs feed into a tree of logic gates of types AND and OR. An ALN represents a real-valued function of real variables by giving a logic 1 response to points on and under the graph of the function, and a logic 0 otherwise. It cannot compute a real-valued function directly, but it can provide information about how to perform that computation in a separate decision-tree-based program. If a function is invertible, then the same ALN can be used to derive a second decision tree to compute an inverse. Another way to look at function synthesis is that linear functions are combined by a tree expression of MAXIMUM and MINIMUM operations. In this way, ALNs can approximate any continuous function defined on a compact set to any degree of precision. The logic tree structure can control qualitative properties of learned functions, for example convexity. Constraints can be imposed on monotonicities and partial derivatives. ALNs can be used for prediction, data analysis, pattern recognition and control applications. They may be particularly useful for extremely large systems, where lazy evaluation allows large parts of a computation to be omitted. A second, earlier type of ALN is also discussed where the inputs are fixed thresholds on variables and the nodes adapt by changing their logical functions.

C1.8.1 Introduction

For the purposes of this section, adaptive logic networks (ALNs) are defined as feedforward neural networks formed from *linear threshold units (LTUs)*, also called *perceptrons*, which send their Boolean outputs to   C1.1.1
a tree of units that realize logic gates AND and OR. A four-layer ALN is shown in figure C1.8.1. The LTUs can be attached at any level of the logic tree. They produce a logic 1 whenever a given inequality is satisfied.

Training of an ALN can involve changing the weights in the LTUs, the functions at the nodes of the logic tree and/or the structure of the tree itself. Different training algorithms and what is called the *relational approach to neural networks* for representing real-valued functions with ALNs distinguish the present work from early work on perceptrons (Minsky and Papert 1969) which was mainly concerned with classification of inputs into two classes.

Architecturally, the ALN is a special case of the early *multilayer perceptron*, which consisted of a   C1.2
loop-free interconnection of LTUs. This is because one can set the weights of LTUs in all layers but the first, so they compute AND and OR functions. For example an AND outputs a 1 if and only if the sum of the Boolean inputs is greater than or equal to the *fan-in* of the unit. There is no loss of generality in   B2.6.1
using trees of logic gates instead of loop-free networks, since one can always transform such a network into an equivalent tree by replication of those units having *fan-out* greater than one.   B2.6.1

When, in the late 1960s, the (single-unit) perceptron was shown to be unable to synthesize nonlinearly-separable classifications (Minsky and Papert 1969), multilayer networks of LTUs were tested.

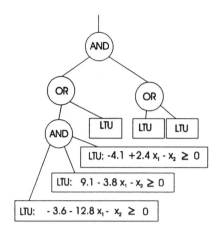

Figure C1.8.1. A four-layer ALN.

Unfortunately, the training algorithms of the day appeared inadequate to train them. In the mid-1970s, a solution to the problem of training multilayer perceptrons was found by Werbos. Instead of LTUs, Werbos used units that formed a combination of their input signals $u = w_0 + w_1x_1 + \cdots + w_nx_n$ and passed this through a differentiable nonlinearity such as $\varphi(u) = 1/(1 + e^{-u})$, the 'logistic sigmoid'. The error backpropagation training algorithm based on *gradient descent* in the weight space was proposed as a method for training the network (Werbos 1974).

B6.3.3, B5.2.2

Shortly before that, a patent was applied for on a system for training multilayer logic networks using a different approach that did not involve differentiable nonlinearities (Armstrong 1974). Generalization properties of purely logical networks were found to depend on the choice of gate functions and the unit fan-ins (Bochmann and Armstrong 1974). In the early 1990s, ALN software became available in source form (Armstrong and Thomas 1992) but until 1993, it was still not convenient to use ALNs for problems requiring real outputs. In 1993, LTUs were combined with logic trees, giving rise to the newer type of ALN described in this article which was first used in the problem of controlling a *vehicle active suspension* system (Armstrong and Thomas 1994). ALNs are now being applied to problems in diverse areas including cardiology (Polak *et al* 1995), predictive maintenance (Armstrong *et al* 1995), rehabilitation (Stein *et al* 1992, Kostov *et al* 1994, 1995, 1996), nondestructive testing and high-energy physics (Kremer and Melax 1994). Educational and commercial software is becoming available (Armstrong and Thomas 1995).

G2.1

Even earlier than that work, Aleksander was using RAM elements, called SLAMs, beginning the study of what he has termed *weightless neurocomputing* (Aleksander 1991). Recently, other variants of adaptive logic have been used which generate pulse trains of 0 and 1 according to a probability stored in a RAM (Gorse *et al* 1994).

C1.5

C1.8.2 Uses of adaptive logic networks

The ALN has the role of a data analysis tool. ALNs compute Boolean outputs in a feedforward fashion, and hence are classifiers. If the class is the set of points on or under the graph of a real-valued function of real variables, an ALN can *represent* that function, though it cannot compute the real-valued outputs directly. Evaluation can be done very rapidly by a decision-tree-based program derived from the ALN, that computes a piecewise linear function. A decision tree partitions the input space so that, for each input vector, only a few linear expressions have to be evaluated. Because of the way ALNs represent functions, it is possible to impose qualitative and quantitative constraints on the functions produced by ALN training. ALNs are of particular interest where speed of evaluation or conformity to a specification are important. We shall now look at aspects of ALN use in detail.

C1.8.2.1 Logic networks as classifiers

A *linear threshold unit (LTU)*, or perceptron, accepts input vectors $(x_1, x_2, \ldots, x_n)$ having n real values as components, and outputs the Boolean value 1 or TRUE if $w_0 + w_1x_1 + \cdots + w_nx_n \geq 0$. Otherwise, the LTU outputs the Boolean value 0 or FALSE. The sequence of weights $w_0, \ldots, w_n$ is called the *weight vector* of the LTU. The function computed by the LTU, which maps vectors of n real components to Booleans, is determined by its weight vector. An LTU thus classifies points in Euclidean n-space into two classes: a 1-class where the LTU outputs a 1, and a 0-class where it outputs a 0. Each class is a *half-space* on one side of an $(n-1)$-dimensional *hyperplane*, the hyperplane being defined by the equation $w_0+w_1x_1+\cdots+w_nx_n = 0$. The points of the hyperplane itself are placed in the 1-class by our convention. In the case where $n = 2$, the hyperplane becomes a line L, and the two classes are half-planes, as shown in figure C1.8.2.

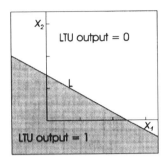

Figure C1.8.2. The half-space (shaded) on and under the line L, where the LTU outputs 1.

An ALN is a classifier that builds upon the classifications of LTUs. Its 1-class is formed by taking a finite sequence of unions and intersections, starting from the 1-classes of the LTUs, according to the structure of the ALN (OR becomes union and AND becomes intersection in constructing the 1-class). This means that ALNs are very flexible classifiers since the 1-class can approximate any open set in n-space†. An example of an ALN classification is shown in figure C1.8.3. The shaded set of figure C1.8.2 is intersected with the union of two half-spaces on and under the dotted lines A1 and A2. Then the union is taken of that result with the intersection of two half-spaces as suggested by the solid lines B1 and B2. (Note that the order of these two operations cannot be interchanged.)

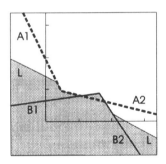

Figure C1.8.3. The 1-class of a small ALN (shaded).

The ALN which performs the classification of figure C1.8.3 is shown in figure C1.8.4.

C1.8.2.2 The relational approach to function representation

For computing functions with real-valued outputs one cannot use the ALN directly because it can output only 0 and 1. Instead, the ALN is used to *represent* the function. This is done by having the set on and under the graph of a function be the 1-class of the ALN. Figure C1.8.5 shows the graph of the function which maps values on the horizontal axis into values on the vertical axis.

Since each LTU must have the 1-class below the 0-class in representing a function, we must have a negative weight on w_n in the formula of the LTU. Then we can divide all weights by the absolute value of that weight without changing the 1-set of the LTU. Thus we can normalize all LTUs used in function approximation to have weight on w_n, the output variable, equal to minus 1.0.

---

† An ALN can be created with a 1-class arbitrarily close to any open set in Euclidean n-space. This is because any open set of n-space is a countable union of parallelepipeds, and we can choose an ALN based on a finite number of those parallelepipeds.

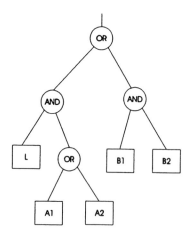

Figure C1.8.4. An ALN which performs the classification of figure C1.8.3.

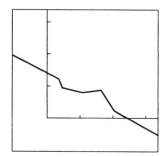

Figure C1.8.5. The graph of the function represented by the ALN in figure C1.8.5.

We shall refer to this method of using the set under the graph of a function to represent the function as the *relational* approach to neural networks. A very large class of functions can be represented in this way. A function mapping a rectangle in two-dimensional space to the real values is illustrated in figure C1.8.6. The function is formed by taking two sinc functions ($y = (\sin x)/x$), rotating them about the y-axis, scaling, translating and adding them. The approximation shown uses 576 LTUs.

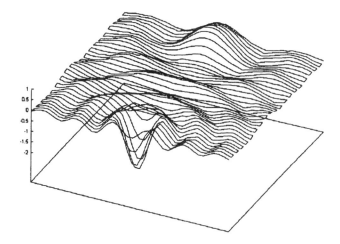

Figure C1.8.6. Graph of a function mapping a rectangle (horizontal) to an interval (vertical).

Instead of talking about the 1-sets defined by LTUs, one can use functions directly, in which case, instead of the LTU defined by $w_0 + w_1 x_1 + \cdots + w_n x_n \geq 0$, we have the real-valued function given by $x_n = w_0 + w_1 x_1 + \cdots + w_{n-1} x_{n-1}$. The operations AND and OR in the ALN become operations

MINIMUM and MAXIMUM when operating on these functions. This function language is sometimes more convenient. The disadvantage is that one loses the strong connection to the ordinary multilayer perceptron when describing functions using MAXIMUM and MINIMUM.

Under the relational approach, an ALN output can also be used directly. In many cases, a real-valued output from the usual kind of neural network is compared to a threshold in order to make a binary decision. By inserting the threshold value as one of the inputs to the ALN, the user can obtain the decision directly by simply evaluating the ALN's logical output.

C1.8.2.3 Functions that may be approximated using adaptive logic networks

The class of functions that may be approximated using ALNs is very large. We can state and prove the following.

Theorem. An ALN with a layer of LTUs and two layers of gates with AND in one and OR in the other can represent an approximation, to any degree of precision, of any continuous function defined on an $(n-1)$-dimensional compact set.

Proof. The proof is based on simple calculus and topology. We recall from integral calculus that Riemann sums, which are made up of thin parallelepipedal pillars reaching up to the function surface, can approximate the n-dimensional volume under the graph of a continuous function. The only modification which has to be made in a Riemann sum to get an ALN is to slant the edges of each pillar slightly so that it represents a function corresponding to the AND of several LTUs. This is suggested in figure C1.8.7 where only three pillars are shown. The OR of the output of the ANDs of LTUs approximates the set under the function graph†.

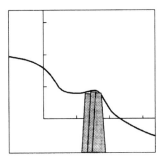

Figure C1.8.7. A Riemann sum is converted to an ALN representation of a function.

C1.8.2.4 Qualitative properties of functions versus logic tree shape

If it is known that a function $y = g(x)$ is convex-up (that is to say, the function values are always greater than or equal to the value given by linear interpolation between any two points of the domain) then it is possible to force the ALN to approximate it with a convex-up function. The graph of any convex-up function is an intersection of (usually infinitely many) half-spaces. We can choose a finite number of half-spaces to approximate the function, and the corresponding LTUs must be combined by a single AND

---

† There are a few more details that may be helpful. We need a fixed base height for the construction, and we take that as the minimum on the compact set of the function to be approximated, minus the required precision ε. We have to choose the width of the base of each pillar such that the function does not vary within the corresponding part of its domain by more than $\varepsilon/2$ from its value at the center of the pillar. We take the top of the pillar to be $\varepsilon/2$ below that value. Now we slope the sides of the pillar inward so that it has half the size at the top as at the bottom (so the sides become functions). Now we cover the compact set which is the domain of the function with open sets, namely the interiors of the sets representing the tops of the pillars. To show that it is possible to select the pillars to form the ALN that approximates the function to within ε, we use the theorem from toplogy that any covering of a compact set with open sets has a finite subcover (Kelley 1955). This completes the proof. The informed reader will note that this is the ALN counterpart of Kolmogorov's theorem, which is much harder to prove. In fact the ALN also gives a lower bound on the function, and a similar construction gives an upper bound.

gate. This is shown in figure C1.8.8, where we suppose that ln N is convex-up (this example will be used below in a predictive maintenance illustration).

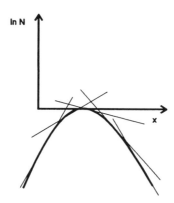

Figure C1.8.8. A convex-up function is represented by an ALN using one AND gate.

Similarly, a convex-down function can be represented by an ALN with one OR gate. Clearly, if we know in advance something about the structure of the function surface, we can force it to be defined by a certain logic tree. For example, a saddle-shaped function surface could be computed by an AND of ORs of LTUs (with some restrictions on the weights—otherwise we can approximate any continuous function). The details of the relationship between types of surfaces, logic trees and weights remain to be worked out. It is clear that this aspect of ALNs represents a departure from the usual idea that random connections of units is the substrate out of which the desired function emerges by training. When there is inadequate *a priori* knowledge, one can take a tree that is large enough to contain an appropriate tree for the given surface type.

C1.8.2.5 Using adaptive logic networks to compute fuzzy sets

D1.2 ALNs synthesize piecewise linear functions. Such functions are often used in applications of *fuzzy sets*. In one application, the goal was to compute a *fuzzy set* N of measurements (pressures, temperatures, etc) representing 'normal' operation of a compressor. Instead of stating which vectors of measurements on a compressor are 'normal', a fuzzy set allows one to express *to what extent* a measurement vector x is a member of the normal set by assigning it a number $N(x)$ between 0 and 1. A value of N near 1 represents normal operation of the compressor, while a value near 0 represents a state close to breakdown. If it is assumed that the function N has qualitative properties like a multivariate normal density, then one way to use ALNs on the problem is to try to approximate the natural logarithm of N. The logarithm of a multivariate normal density of mean m is, up to an additive constant

$$-\tfrac{1}{2}(x - m)'\Sigma^{-1}(x - m)$$

where the prime indicates transposition. The covariance matrix Σ is symmetric and positive definite. Since this function is convex-up, an ALN can be set up using a single AND gate to learn it based on samples of measurements on the compressor, each together with an estimate of ln N. Figure C1.8.8 shows an ALN approximant of this type.

Even if a decision requires a Boolean output, it may be desirable to formulate it as a problem with fuzzy, that is, real-valued, outputs. In this way, the real output can express how close to 0 or 1 the result is. A value too close to 0.5 could be interpreted as 'uncertain'.

C1.8.3 Adaptive logic network training

The fundamental operation behind the adaptation of an LTU for an ALN is least-squares fitting of a collection of data points.. Given a function, the 1-class will be defined as the set on and under the graph of the function. This is illustrated in figure C1.8.9. The data points of the training set are shown, where each point has some error in the value of the output variable compared to the correct function value.

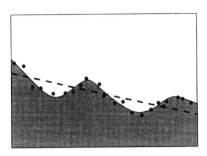

Figure C1.8.9. Fitting nonlinear data with a single LTU.

To train a single LTU, a least-squares fitting algorithm is used (linear regression, in statistical terminology). In figure C1.8.9, the broken line shows the fit that might be determined by linear regression. This part of the training algorithm solves the same problem as backpropagation training of a one-unit network.

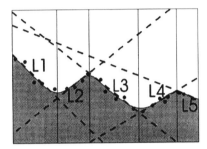

Figure C1.8.10. Fitting nonlinear data with an ALN containing five LTUs.

We consider in figure C1.8.10 the situation when an ALN with five LTUs is used to approximate the function in figure C1.8.9. The logical expression for the 1-set according to these five lines is OR(L1, AND(L2, L3), AND(L4, L5)). The vertical lines in figure C1.8.10 indicate the parts of the horizontal axis where each of the five lines is *active*. For example, L1 is active on the left end of the horizontal axis. The line L1 could be determined by linear regression on the leftmost five data points. Then L2 could be determined by the four points in the next interval, and so on. Performing a piecewise linear regression is well understood, and the only complication here is that as a linear piece moves to fit its data points, some data points that make that piece active may become inactive for this piece and active for a different one. ALN training is an iterative procedure which adjusts the linear piece computed by each LTU, taking into account the data points that make that LTU active. Activity of an LTU for a given data point is computed recursively from the root of the ALN tree based on the AND and OR gates at the nodes of the tree. For example, if two functions are represented by subtrees entering an AND gate, then the one which has the minimum value is the one which is responsible for the given data point. This is essentially the training method used in current software implementations.

C1.8.4 Adaptive logic network decision trees

If we look at the linear pieces as functions of the variable on the horizontal axis of figure C1.8.10, we can write the function represented by the above logical combination of LTUs as MAX(L1, MIN(L2, L3), MIN(L4, L5)). This expression computes the function represented by the ALN on the whole space. To evaluate it, the system could first compute the values of all the linear pieces at the given point and then apply the MIN and MAX operations. A much faster way will now be suggested. As shown in figure C1.8.11, we can partition the horizontal axis as follows:

(i)   divide the axis into two parts so that about half of the active line segments into which the curve is divided lie to the left, and half to the right;

(ii) keep repeating the procedure for the resulting intervals until at most two linear pieces are active in any interval of the partition.

Figure C1.8.11 shows the horizontal axis divided into four intervals. Using this division scheme, at most two linear pieces are involved in the computation of the function in any block of the partition.

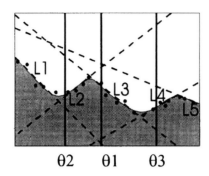

Figure C1.8.11. Partitioning the input space to reduce the number of LTUs that have to be evaluated.

In the leftmost interval, the expression MAX(L1, L2) has to be evaluated, in the next, MIN(L2, L3), in the next, MAX(L3, L4), and finally MIN(L4, L5). The correct interval is determined by a decision tree.

C1.8.4.1 Adaptive logic network decision tree computation

(1) If $x < \theta_1$ then go to step 2, otherwise go to step 3.
(2) If $x < \theta_2$ then go to step 4, otherwise go to step 5.
(3) If $x < \theta_3$ then go to step 6, otherwise go to step 7.
(4) Compute MAX(L1, L2).
(5) Compute MIN(L2, L3).
(6) Compute MAX(L3, L4).
(7) Compute MIN(L4, L5).

The number of pieces in the partition ultimately depends on the complexity of the function being represented. At each comparison of the decision tree, the number of linear pieces that have to be evaluated drops by about half. In general, some pieces will cross a dichotomy, which makes the reduction less than half.

Such an ALN decision tree can be found for problems of any dimension, and the number of linear pieces active in a block of the partition can be made at most equal to the dimension of the space. A minor adjustment may be required. If more pieces than the dimension of the space intersect at a point (giving an overdetermined solution to a system of linear equations) some pieces can be slightly perturbed to correct the situation. With this kind of upper bound on the number of linear pieces that have to be evaluated, it becomes possible to *guarantee hard real-time bounds* on the computation time using an ALN decision tree.

C1.8.4.2 Function inversion using adaptive logic network decision trees

The result of ALN training can be converted into an ALN decision tree computation with a different output variable from the one used in training provided the function is monotonic in the new output variable. A function inverse is obtained in this way, as shown in figure C1.8.12. Note that the ALN as a classifier is increasing in the new output variable, whereas it was decreasing in the original one.

Function inversion is particularly useful in control. ALNs have been used in controlling a mechanical model of a vehicle active suspension system (Armstrong and Thomas 1995, 1996). During training, ALNs learned what the future vertical displacement and velocity of the cab would be, based on results of random strut commands. After training, the desired displacement and velocity were used to compute two strut commands that were averaged to provide a satisfactory control signal.

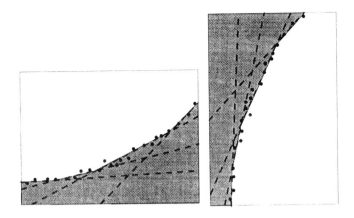

Figure C1.8.12. A monotonic function $y = f(x)$ can be inverted to get $x = g(y)$ using the same ALN.

C1.8.5 Adaptive logic networks with adaptive nodes

To avoid additions and multiplications, and thereby accelerate the computation, one can use fixed thresholds on individual variables, instead of LTUs, as inputs to a logic tree of ANDs and ORs (Armstrong *et al* 1991). Such a network is shown in figure C1.8.13.

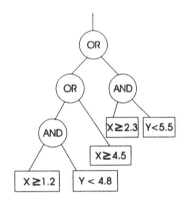

Figure C1.8.13. Structure of an ALN network with fixed thresholds at the leaves.

The synthesis of functions with real-valued outputs may be achieved with a collection of such networks producing one bit each. A bidirectional encoding scheme is then required to convert between the bit vectors output by the trees and the real number values of the function output (training needs one direction, evaluation the other). The hardware required to compute a fixed threshold given the bits of a number x is very simple: two layers of transistors computing NAND (or NOR). A tree of ANDs and ORs can also be turned into several layers of transistors. For applications requiring extremely high speed decisions, trees of transistors may be very attractive.

Training may be done by modifying the functions in the nodes of the logic tree (Armstrong and Thomas 1992). We shall now turn our attention to ALNs with adaptive nodes. ALNs with adaptive nodes have been applied to control (Supynuk and Armstrong 1992), rehabilitation (Kostov *et al* 1992, Popovic *et al* 1993, Armstrong *et al* 1993), high-energy particle physics (Kremer and Melax 1994) and nondestructive grading of fat in beef (McCauley *et al* 1994). The case of binary inputs is straightforward, where an input bit, sometimes complemented, is sent to the inputs of the adaptive logic tree. An optical character recognition (OCR) system has been developed based on this approach, as well as an image processor (Armstrong and Gecsei 1978).

C1.8.5.1 Lazy evaluation

The original adaptive Boolean logic elements were thought of as being realized in hardware (Armstrong 1974). These networks learn Boolean functions by choosing functions AND, OR, LEFT or RIGHT at nodes of a binary tree of flexible logic gates. The LEFT and RIGHT functions are defined by $\text{LEFT}(x, y) = x$, $\text{RIGHT}(x, y) = y$. These functions, in effect, help to change the connections to the network inputs as well as within the logic tree. When simulated serially on a computer, this kind of system can take advantage of the property of lazy evaluation of Boolean logic. 'Lazy' in this context refers to the omission of computations that are *unnecessary*.

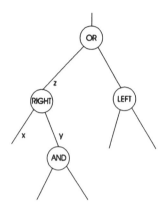

Figure C1.8.14. A network of adaptive Boolean logic elements.

Consider the value on the z connection in figure C1.8.14. If it is a 1, then the output of the OR above it will be a 1 no matter what the result of evaluation is of the entire subtree at the right input to the OR. Hence the right subtree does not need to be evaluated at all! The same situation holds with a 0 signal entering an AND gate. Using lazy evaluation, a very large part of a network can be ignored during evaluation of the response to a particular input. Averaged over all possible node functions and inputs, asymptotically only a fraction $2 (3/4)^L$ of the nodes of an L-layer balanced binary tree has to be evaluated, and this approaches zero as the tree becomes infinitely large.

C1.8.5.2 The role of monotonicity

The four functions used in the adaptive Boolean logic element are exactly all the nonconstant, weakly monotonic increasing Boolean functions of two variables. A function $z = f(x, y)$ with real variables is said to be weakly monotonic increasing (respectively decreasing) in the variable x if, whenever x is increased, the output z either increases (respectively decreases) or stays the same. For example, $z = 5x - 3y$ is monotonic increasing in x and monotonic decreasing in y. In the case of Boolean functions, which map vectors of 0s and 1s to output values 0 and 1, the same definitions apply, where $1 > 0$ is the convention for Boolean values. If x and y are Boolean vectors, then $x \geq y$ if every component of x is greater than or equal to the corresponding component of y.

Monotonicity can be very useful in a learning system with many layers. If all the units in a network compute monotonic increasing functions, then the overall function is monotonic increasing. This forces a positive correlation, however weak, to exist between the output of a unit and the output of the network. This correlation can be exploited in a statistical way by having each adaptive Boolean logic element count the 1s and 0s desired of the network during training, and choose its output for a given input it receives according to the *more frequently desired network response* when the given input to that unit occurs. Statistics to this effect can be accumulated by up–down counters that count up when the desired network output is 1 and down when the desired network output is 0. The counters are bounded, which prevents counts in some cases. In the particular case of functions AND, OR, LEFT and RIGHT, the unit must always output a 0 when the two inputs are 0, and a 1 if the two inputs are 1. For this reason, only two counters are required in the adaptive Boolean logic element, one for the input pair (0,1) and one for the input pair (1,0) (Armstrong and Gecsei 1979).

In ALNs with LTUs, increasing monotonicity internal to the logic tree is assured by using only AND and OR at the nodes. This internal monotonicity of the network opens up the possibility of controlling the monotonicity of the output of a real-valued function represented by the network as a function of each of its real inputs by controlling the signs of weights in the LTUs. Monotonicity is a powerful tool when using ALNs. For backpropagation-type networks with sigmoids, the specification of monotonicities on parts of the input space is difficult in the presence of weights of both signs, since all weights influence each output value.

C1.8.5.3 Insensitivity to perturbations of inputs

The natural progression from two-input logic elements to n-input logic elements was not very successful from the point of view of generalization. Analysis (Bochmann and Armstrong 1974, Armstrong and Godbout 1975) showed that a problem lay in having networks that tended to change output value with minor perturbations of the inputs if n exceeded two. It was also shown that restricting the allowable node functions gave networks with the greatest insensitivity to perturbations of their inputs, i.e. networks which tended to generalize well in the sense of Hamming distance between Boolean vectors. Insensitivity was another reason, in addition to monotonicity and lazy evaluation, for choosing AND, OR, LEFT and RIGHT as the set of functions that can be realized in an adaptive Boolean logic element. An arbitrary binary tree network of the above node functions can always be turned into an equivalent tree having alternating layers of AND and OR gates of arbitrary fan-in. This is the form of the logic tree for ALNs with LTUs.

One of the problems with the usual multilayer perceptrons having sigmoids is that weights can grow large and cause large effects on the output for only small changes of input. With ALNs, only weights in the LTUs have to be examined; there are no sums and products of weights and derivatives of sigmoids to be considered. If one imposes bounds on the weights during training, these become bounds on the partial derivatives of the represented function (or bounds on difference quotients where the derivatives do not exist.).

C1.8.5.4 All-or-nothing credit assignment

The credit assignment problem for learning systems made up of many units is the problem of informing a unit to what extent it has been involved with some good or bad result of network action. In the backpropagation training algorithm, credit assignment is expressed by a real quantity, the backpropagated error. All the units in the network receive some blame for the error. This is disadvantageous in the sense that much computation may be involved, proportional to the complexity of the problem.

In the case of logic networks, there is an obvious approach to credit assignment: make a unit responsible for its action (in response to a given input pattern to the network) if and only if that unit has had an effect on the network output. This will cause a large part of a logic network to be left unchanged. An example will make this clear. In figure C1.8.14, if the value on z is 1, then the output of the network is 1 because of the OR. The LEFT function in the node at the right below the OR cannot possibly influence the output of the network, i.e. the subtree gets no credit or blame. No change of state occurs in the right subtree for this pattern presentation. In a similar fashion, an AND with a 0 entering one input lead cuts off the entire effect of the opposite subtree. A concept of *heuristic responsibility* is thus defined which selects, for each pattern presented to the network, the subset of units which adapt. If the credit assignment computation makes few units change, then learning will be much faster than if adaptation of all units is required.

The particular algorithm just mentioned for assigning heuristic responsibility for a given input to the tree has been called *true responsibility* (Armstrong 1974).

True responsibility.

(i) The only heuristically responsible nodes are the ones selected by the following rules.

(ii) The root unit of the network is always heuristically responsible.

(iii) A child of a heuristically responsible AND node (respectively OR node) is heuristically responsible if the opposite child (unit or input variable) has value 1 (respectively 0).

(iv) The left (respectively right) child of a heuristically responsible node is heuristically responsible if the function in the node is LEFT (respectively RIGHT).

An important lesson learned from the adaptive Boolean logic element was that the best heuristics for assigning credit and blame to particular units of the network were more complex than true responsibility above. The flaw in true responsibility is that the function in the node may change, hence the training of all descendants may be done based on the wrong node function. The following is a heuristic that worked much better.

Heuristic responsibility. Rules (i), (ii) and (iii) are the same as above; rule (iv) is modified to (iv)(a) below, and a new rule (v) is added.

(iv) (a) Both children of a heuristically responsible node are heuristically responsible if the function in the node is LEFT or RIGHT.

(v) If one input of a heuristically responsible node is 'in error', meaning its value is different from the desired network output, then the child at the *opposite* input is heuristically responsible.

The above form of heuristic responsibility goes beyond causing a node to adapt if and only if it has an effect on the network output. It adapts a child node that maybe only potentially has an effect if the function in its parent were to change. It also makes a subtree adapt if the other subtree appears not to be contributing.

If there is one clear lesson from many years of ALN experimentation, it is this: finding a subset of units to adapt, i.e. solving the credit assignment problem, is not easy. In general gradient descent is too weak.

C1.8.6 Advantages of adaptive logic networks

C1.8.6.1 Speedup based on lazy evaluation

As networks get larger, lazy evaluation becomes essential to keep the computation time from growing in proportion to the complexity of the problem (or worse). The usual reasoning that massive hardware is a remedy for long computation times is only true as long as one can afford the hardware. Once the same hardware has to be used iteratively for evaluating different parts of a network, it is useful to have an algorithm which allows some computations to be omitted. The speedup offered by lazy evaluation of ALNs or ALN decision trees is unbounded as problems become larger.

C1.8.6.2 Bounds on rates of change

ALNs with LTUs allow arbitrary bounds to be imposed, during training, on the rate of change of the output variable with respect to any of the input variables (i.e. partial derivatives). This is done simply by imposing the bounds on weights in the normalized LTUs (output weight $= -1.0$). In particular, the learned function may be constrained to be (weakly) monotonic increasing (or decreasing) in any variable by maintaining the desired sign of the weight.

C1.8.6.3 Easy invertibility

If a function $y = f(x_1, \ldots, x_n)$ represented by an ALN with LTUs is strongly monotonic increasing or decreasing in some variable x_i, then the function has an inverse of the form $x_i = g(x_1, \ldots, y, \ldots, x_n)$. By extracting an ALN decision tree from the ALN using x_i as the output variable, one can compute the inverse function directly with no further training. This is important in applications like *control*.

F1.9

C1.8.6.4 Usefulness of expert knowledge

If one chooses an ALN with LTUs consisting of a single AND (OR) node, then the function represented must be convex-up (-down). This simple form occurs very frequently in practice. More complex functions are made up of convex pieces, e.g. an S-curve is made up of two convex parts. There is a strong relationship among the architecture of the logic tree, the constraints on the weights of the LTUs, and the qualitative properties of the functions synthesized. This can be used to force the result of training to have properties dictated by physical or economic laws.

C1.8.6.5 Narrowly targeted credit assignment

In an ALN with LTUs, an LTU may change its weights if its linear piece is active for the given input vector (or if it should *become* active). It is critical to reducing *training time* that only a few LTUs are adjusted at B3.4 each pattern presentation. In ALNs with adaptive nodes, credit assignment is limited through the heuristic responsibility mechanism. Both methods of ALN credit assignment contrast with credit assignment through the backpropagation algorithm, which allows a small bit of blame or credit to be assigned to almost every node in the network. We feel that narrowly targeted credit assignment is a powerful tool for achieving machine learning in very large systems. ALN systems with over one hundred thousand logic nodes have been trained successfully.

C1.8.6.6 Availability of educational software

Educational software illustrating the ALNs discussed in this article is available free on the Internet (Armstrong and Thomas 1992, 1994). For further information on software availability, the reader should contact one of the authors.

C1.8.7 Disadvantages of adaptive logic networks

ALNs offer rich possibilities for further research and experimentation, in particular to determine their advantages and limitations.

C1.8.7.1 Relatively little has been published

The ALN is nowhere near as thoroughly investigated as the backpropagation-type networks. Because the number of persons working on ALNs is much smaller than the number working on other types of neural networks, the results appear more slowly. The main sources of information are a few dozen published research papers and documented software (with online help) available electronically.

C1.8.7.2 Reinforcement learning needs work

The concept of *reinforcement learning* using backpropagation through time was demonstrated long ago by C3 Paul Werbos (Miller *et al* 1990). ALNs have been used on a pendulum problem to learn a cost function of the state and the applied torque. This result, though tentative, suggests reinforcement learning with ALNs is a possibility, but much more work will be required to demonstrate it convincingly. This question and many other questions about ALNs constitute an open invitation to neural network researchers.

References

Aleksander I 1991 Connectionism or weightless neurocomputing? *Proc. 1991 Int. Conf. Artificial Neural Networks (Espoo)* ed T Kohonen, K Makisara, O Simula, J Kangas vol 2 (New York: Elsevier) pp 991–1000
Armstrong W W 1974 *Adaptive Boolean Logic Element* US Patent # 3934231, Feb. 28
Armstrong W W, Chu C and Thomas M M 1995 Using adaptive logic networks to predict machine failure *World Congress on Neural Networks (Washington, DC)* pp II-80–II-83
Armstrong W W, Dwelly A, Liang J-D, Lin D and Reynolds S 1991 Learning and generalization in adaptive logic networks *Proc. 1991 Int. Conf. Artificial Neural Networks (Espoo)* vol 2 ed T Kohonen, K Makisara, O Simula, J Kangas (New York: Elsevier) pp 1173–6
Armstrong W W and Gecsei J 1978 Architecture of a tree-based image processor *12th Asilomar Conf. on Circuits, Systems and Computers (Pacific Grove, CA)* (New York: IEEE) pp 345–9
——1979 Adaptation algorithms for binary tree networks *IEEE Trans. Syst., Man Cybern.* **9** 276–85
Armstrong W W and Godbout G 1975 Properties of binary trees of flexible elements useful in pattern recognition *IEEE Int. Conf. on Cybernetics and Society (San Francisco, CA) IEEE Cat. No. 75* CHO 997-7 SMC pp 447–9
Armstrong W W, Stein R B, Kostov A, Thomas M, Baudin P, Gervais P and Popovic D 1993 Application of adaptive logic networks and dynamics to study and control of human movement *2nd Int. Symp. on 3D Analysis of Human Movement (Poitiers)* (International Society of Biomechanics) pp 81–4
Armstrong W W and Thomas M M 1992 Atree 2.7 ALN Development System for Windows including C++ source code and on-line help. Available from ftp.cs.ualberta.ca in pub/atree/atree2/atre27.exe Unix version in pub/atree/atree2/atree2.tar.Z (compressed tar file with C source code).

—— 1994 Control of a vehicle active suspension system using adaptive logic networks, in the on-site published *Program Addendum of the World Congress on Neural Networks (San Diego, CA)* pp 9–14, also from ftp.cs.ualberta.ca in pub/atree/docs/wcnnpub.ps.Z.

—— 1995 Atree 3.0 ALN Educational Kit for Windows ftp.cs.ualberta.ca in pub/atree/atree3/atree3ek.exe (binary mode ~900 Kilobytes).

—— 1996 Feasibility of control of a vehicle active suspension system using adaptive logic networks *Handbook of Neural Computation* (Bristol and New York: IOP Publishing and Oxford University Press)

Bochmann G V and Armstrong W W 1974 *Properties of Boolean Functions with a Tree Decomposition (Reprint BIT-14)* pp 1–13

Gorse D, Taylor J G and Clarkson T G 1994 A pulse-based reinforcement algorithm for learning continuous functions *Proc. World Congress on Neural Networks (San Diego, CA)* vol 2 pp 73–8

Kelley J L 1955 *General Topology* (Van Nostrand)

Kostov A, Andrews B J, Popovic D B, Stein R B and Armstrong W W 1995 Machine learning in control of functional electrical stimulation systems for locomotion *IEEE Trans. Biomed. Eng.* **42** 541–51

Kostov A, Armstrong W W, Thomas M M and Stein R B 1996 Adaptive logic networks in rehabilitation of persons with incomplete spinal cord injury *Handbook of Neural Computation* (Bristol and New York: IOP Publishing and Oxford University Press)

Kostov A, Stein R B, Armstrong W W and Thomas M M 1992 Evaluation of adaptive logic networks for control of walking in paralyzed patients *14th Ann. Int. Conf. IEEE Engineering in Medicine and Biology Society (Paris)* vol 4 (New York: IEEE) pp 1332–4

Kostov A, Stein R B, Popovic D B and Armstrong W W 1994 Improved methods for control of FES for locomotion *Proc. Int. Federation of Automatic Control (IFAC) Symp. on Modeling and Control in Biomedical Systems (Galveston, TX)* pp 422–7

Kremer S C and Melax S 1994 Using adaptive logic networks for quick recognition of particles *IEEE Int. Conf. on Neural Networks (Orlando, FL)* vol 5 (Piscataway, NJ: IEEE Service Center) pp 3015–19

McCauley J D, Thane B R and Whittaker A D 1994 Fat estimation in beef ultrasound images using texture and adaptive logic networks *Trans. Am. Soc. Agricult. Eng.* **37** 997–1002

Miller W T III, Sutton R S and Werbos P 1990 *Neural Networks for Control* (Cambridge, MA: MIT Press)

Minsky M L and Papert S A 1969 *Perceptrons* (Cambridge, MA: MIT Press)

—— 1988 *Perceptrons* Revised edition (Cambridge, MA: MIT Press)

Polak *et al* 1995 Using ALNs to detect ischemia *Congress on Cardiology (Vienna)* pp 217–20

Popovic D B, Stein R B, Jovanovic K L, Dai R, Kostov A and Armstrong W W 1993 Sensory nerve recording for closed-loop control to restore motor functions *IEEE Trans. Biomed. Eng.* **40** 1024–31

Stein R B, Kostov A, Belanger M, Armstrong W W and Popovic D B 1992 Methods to control functional electrical stimulation in walking *1st Int. Functional Electrical Stimulation Symp. (Sendai)* pp 135–40

Supynuk A G and Armstrong W W 1992 Adaptive logic networks and robot control *Proc. Vision Interface Conf. '92 (Vancouver)* pp 181–6

Werbos P 1974 Beyond Regression: New tools for prediction and analysis in the behavioral sciences *Doctoral Dissertation* Applied Mathematics, Harvard University

C2

Unsupervised Models

Contents

C2.1 Feedforward models

Michel Verleysen

Abstract

Feedforward unsupervised models cover a wide range of neural networks with various applications. In this section, we discuss three widely used models. (i) Kohonen's self-organizing map, also called the *Kohonen network*, the *self-organizing feature map*, or the *topological map*, is intended to map a high-dimensional space into a one- or two-dimensional space, preserving the topology of the input space; it has a strong biological plausibility and is basically intended to be used in applications where preserving the topology between input and output spaces is important (e.g. control, inverse mapping, image compression). It is an unsupervised model, but can be extended to a supervised one by adding a supplementary layer. In addition to the topology-conserving property, the Kohonen model also acts as a vector quantizer. (ii) The neural gas is another vector quantization algorithm that may be considered as a neural network method because it relies on the same principle of adaptation, may be represented in the form of a feedforward graph, and may be described by the same formalism as used in many other neural models. It is different from the Kohonen map in the sense that it does not have the topology preserving property but it generally performs better giving a smaller final distortion error. (iii) The neocognitron is a complex feedforward model formed by several layers each containing a large number of neurons. Its goal is to automatically detect features in two-dimensional arrays of points through self-organization and reinforcement principles. The network is built to be insensitive to shifts in position of the patterns or of small parts of them, thus also allowing for distorted patterns. The network is primarily intended to be used in feature extraction and pattern recognition tasks, for example in OCR (optical character recognition).

C2.1.1 Kohonen's self-organizing map

C2.1.1.1 Introduction

The self-organizing feature map (SOFM) has been developed by Teuvo Kohonen (Helsinki University of Technology, Laboratory of Computer and Information Sciences, FIN-02150 Espoo 15, Finland). While it has been described in several research papers, an interesting and self-contained description of the model, its biological background, its implementation, and possible applications can be found in one of the three editions of Kohonen's seminal book *Self-Organization and Associative Memory* (Kohonen 1989).

The SOFM is presented as a biologically plausible network, which takes its inspiration from the fact that some regions of the cortex 'map' either physical locations of sensory neurons, or the ordering of some physical properties like the acoustic frequencies in the auditory cortex.

C2.1.1.2 Purpose of the model

The purpose of the self-organizing feature map is basically to map a continuous high-dimensional space into a discrete space of lower dimension (usually 1 or 2). This goes through two more or less independent properties obtained through the topology and the learning of the network.

C1.1.5   First, as in adaptive vector quantization methods like the *LVQ algorithms*, the input space is quantized by assigning each neuron to a defined region in the input space. The number of (external) inputs to each neuron being equal to the dimension of the input space, the weight vector of each neuron can be interpreted as a location in this space. The region then assigned to a neuron is the set of locations nearer to the corresponding neuron than to any other one; this is called the 'Voronoi tessellation' of the input space (see figure C2.1.1 for a Voronoi tessellation of a two-dimensional space).

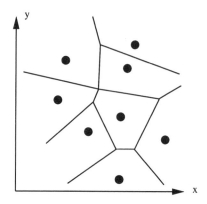

Figure C2.1.1. Voronoi tessellation of a two-dimensional space.

In figure C2.1.1, the eight neurons are represented by using their weight vector (of two components, the input space being of two dimensions) as coordinates in a two-dimensional space; this is the common representation standard for Kohonen maps.

The second property concerns the preservation of the topology. The principle is that the network organizes its topology in the following way. Inherently, all neurons of the network, regardless of their weight vectors, are arranged according to a defined topology. For example, they can be arranged on a one-dimensional string or a two-dimensional grid, as shown in figure C2.1.2.

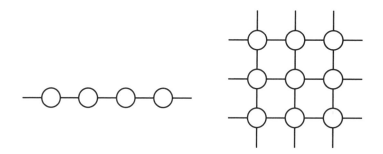

Figure C2.1.2. One-dimensional string and two-dimensional grid.

The 'dimension' d_{out} of the string (one) or of the grid (two) is usually referred to as the output dimension; it is absolutely independent of the dimension of the input space d_{in}, which is determined by the number of external inputs to each neuron.

Having defined these two dimensions, the second property of self-organizing feature maps can now be explained as follows. After learning, that is, after the network has converged to an 'ordered' topology, two input vectors (of dimension d_{in}), one close to the other (according to the definition of a distance measure in the input space), will be projected on two close neurons in the output space; close neurons means here that they are close on the string or on the grid, depending on the output dimension.

Of course, a good topology preservation is not always possible if the input and output dimensions are different; for example, if the dimension of the input space is three, and the neurons are arranged on a two-dimensional grid, and if the input space is uniformly filled, it is not possible to find a projection from the three-dimensional space to the two-dimensional one that will respect the topology (the neighborhood property) at all locations.

Nevertheless, the property of the Kohonen map is that the learning is targeted to the 'best' weight vector for each neuron in order to best match the neighborhood property; this will be detailed in a later section, mainly through examples.

C2.1.1.3 Biological origin

It is well known now that the organization of cells in the brain, and more especially in the cortex, is not a result of chance. Without doubt, a large part of this organization is determined genetically, while it is also proven that learning plays an important role. For example, persons having had a sensory organ amputated will develop different sensory sites than other people, through a rearrangement of dedicated sites in the brain.

Moreover, the locations of nervous cells assigned to sensory or motor tasks in the brain are, at least in several known parts of it, arranged in a way such that they map either the physical locations of the sensory or motor organs themselves, or some of their physical properties: receptive fields are arranged in the cortex according to the sensory organs, neighboring locations in the visual cortex correspond to neighboring locations in the retina, close neurons in the auditory cortex are activated by close frequencies, and so on.

Of course, because of the complex physical structure of the brain and cortex, these mappings are highly nonlinear; their main property is that 'close' elements before the projection will correspond to 'close' elements after the projection. It is this specificity that researchers such as Cristoph Von der Malsburg (1973) and Teuvo Kohonen (1989) tried to model through topological maps.

C2.1.1.4 Topology

A self-organizing feature map contains one layer of neurons, but two layers of connections, as illustrated in figure C2.1.3. For the purpose of simplicity, it will be assumed in the following that the output dimension d_{out} is two, i.e. that neurons are arranged on a two-dimensional grid. Each neuron has d_{in} external connections, to the d_{in} inputs. In addition, each neuron is laterally connected to its neighbors (on the grid), up to a certain distance; for example, neuron n_{43} in the grid can be connected to its four nearest neighbors n_{33}, n_{42}, n_{44}, and n_{53}, or to its eight nearest neighbors (the four preceding ones plus n_{32}, n_{34}, n_{52}, and n_{54}), and so on.

Computations are feedforward in the first layer of connections: the network computes the scalar product between the input vector x and each of the neuron weight vectors w. The self-organizing feature map usually supposes that input and weight vectors are normalized; the scalar product can thus be considered as a distance measure between the two vectors x and w (it is the angle formed by these two vectors on a unit circle).

The second layer of connections acts as a *recurrent* excitatory/inhibitory network, whose aim is to B2.3 reinforce the activation values of 'strong' neurons and to decrease the activation of the 'weak' ones. The values of the lateral connections are fixed (they are not changed during the learning), and are only dependent on the physical distance between neurons on the grid. A typical function representing the values

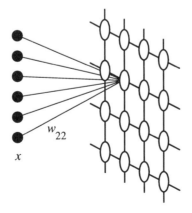

Figure C2.1.3. Self-organizing feature map.

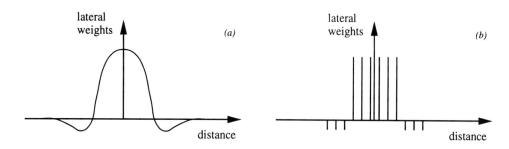

Figure C2.1.4. Lateral connections between neurons.

of the lateral weights versus the distance between neurons on the grid is plotted in figure C2.1.4(a), while practically used values (at discretized locations since there is a discontinuous set of distances between neurons) are shown in figure C2.1.4(b).

It can be seen from figure C2.1.4 that connections for close neurons are positive (excitatory synapses), while they are negative for more distant ones (inhibitory synapses); they tend to zero for larger values of the distance between neurons.

While the connections in the first layer (connections to input vectors) are strictly feedforward, the lateral ones between neurons are bidirectional: there is obviously the same connection between neurons n_{ij} and n_{kl} as between neurons n_{kl} and n_{ij}, since the value of the weight only depends on the distance between the neurons (measured as $|k - i| + |l - j|$).

C2.1.1.5 Operation of the network

The behavior of the network is formally described by

$$a_{ij} = \sigma \left[\sum_{k=1}^{d_{\text{in}}} w_{ijk} x_k + \sum_{n_{mn} \in N(n_{ij})} v_{mnij} a_{mn} \right] \tag{C2.1.1}$$

where a_{ij} is the activation of neuron n_{ij}, w_{ijk} the weight between neuron n_{ij} and the kth component x_k of the input vector x, $N(n_{ij})$ the neighborhood of neuron n_{ij} as defined above, v_{mnij} the lateral weight between neurons n_{mn} and n_{ij}, and $\sigma[\cdot]$ a standard sigmoid-type nonlinearity. In this equation and in the following equations, neurons (and related values as activations) have been numbered by two indices, corresponding to their location on the Kohonen grid ($d_{\text{out}} = 2$), and layer indices have been omitted for simplicity; lateral fixed weights between neurons are denoted by v to avoid the confusion with adaptable weights w.

Equation (C2.1.1) forms a complex system of coupled nonlinear equations that must be solved to find activation values a_{ij}. However, a simple interpretation of the behavior of the lateral connection layer is that it reinforces the activity of neurons in the areas of the map where strong activations already exist through the first summing term of equation (C2.1.1), while it decreases the activations of neurons in other areas. This usually leads to the formation of an 'activity bubble' in the map, as shown in figure C2.1.5, where the neuron activities in a 15×15 neuron map are represented by their gray level.

With this property, Kohonen showed that it is possible to calculate the activations in a simpler and much more straightforward way. Since the activity bubble in the map will be located where the activations due to the first summing term of equation (C2.1.1) are the most important ones, Kohonen proposes to determine the center of this bubble by choosing the 'best match' between the input vector x and the weight vectors w_{ij}, under a defined distance measure. To avoid the necessity of normalization with the scalar product, the Euclidean distance can be chosen. The best-match neuron n_{ab} will thus be chosen as:

$$\|x - w_{ab}\| \leq \|x - w_{ij}\| \qquad \forall 1 \leq i \leq N_i, \forall 1 \leq j \leq N_j \tag{C2.1.2}$$

where N_i and N_j are, respectively, the number of neurons in the x and y directions of the map ($N = N_i N_j$ is the total number of neurons). The selection of the best-match neuron will be used in the learning process of the network.

In this simplified model, much more convenient for practical computations, the activities of the neurons are computed by:

$$a_{ij} = \|x - w_{ij}\|. \tag{C2.1.3}$$

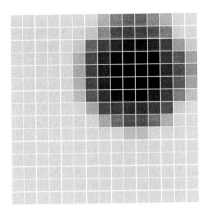

Figure C2.1.5. Activities and 'bubble' in a 15×15 neuron Kohonen map.

While the values of the activities will obviously not be the same in the original and simplified models, the same phenomena of bubble formation and of topology preservation will be encountered in both models, through appropriate learning.

C2.1.1.6 Learning

In order to ensure the above described property of topology conservation between the input space and the locations of 'winning' (best-match) neurons on the grid, appropriate values of the weights w and v in the two layers of connections must be chosen.

Values of lateral connection weights v are chosen according to some Mexican-hat-like function as illustrated in figure C2.1.4. The width of the function (i.e. the distance between neurons above which connection weights are null) is initially chosen as 'reasonable' according to the size of the map, and then decreased during learning.

Values of weights w between the neurons and the input vector x are the result of an adaptive process. Learning in a Kohonen map is a typical example of *unsupervised learning*: input vectors are presented to the network, and the weights are adapted without any knowledge of a 'desired output value' to the network. During learning, the values of the weights are adapted according to

$$w_{ij}(t+1) = w_{ij}(t) + \alpha(t)(x(t) - w_{ij}(t)) \qquad \text{if} \quad n_{ij} \in N(n_{ab}) \qquad \text{(C2.1.4)}$$

$$w_{ij}(t+1) = w_{ij}(t) \qquad \text{if} \quad n_{ij} \notin N(n_{ab}) \qquad \text{(C2.1.5)}$$

where $w_{ij}(t)$ is the weight vector between neuron n_{ij} and input vector $x(t)$ at time step t of the learning, $x(t)$ is the input vector presented to the network at time step t, $\alpha(t)$ is a learning factor that decreases with time to ensure convergence to fixed states, n_{ab} is the best-match neuron according to equation (C2.1.2), and $N(n_{ab})$ is a neighborhood of the best-match neuron n_{ab}, that also decreases with time.

In some other versions of this simplified model, a weaker adaptation can be chosen for neurons 'far' from the best-match one n_{ab}, by adding a multiplying term in the equation

$$w_{ij}(t+1) = w_{ij}(t) + \alpha(t)\beta(|i-a| + |j-b|)(x(t) - w_{ij})(t) \qquad \text{if} \quad n_{ij} \in N(n_{ab})$$

$$w_{ij}(t+1) = w_{ij}(t) \qquad \text{if} \quad n_{ij} \notin N(n_{ab}) \qquad \text{(C2.1.6)}$$

where $\beta(\cdot)$ is a monotonically decreasing function of the distance between neurons n_{ij} and n_{ab} in the map.

The choice of the adaptation factor $\alpha(t)$ and the neighborhood domain $N(n_{ab})$ as a function of time is important, but not critical. Usually, $\alpha(t)$ is chosen according to the Robbins–Monro conditions (Robbins and Monro 1951), that is,

$$\lim_{t \to \infty} \alpha(t) = 0$$

$$\sum_{t=0}^{\infty} \alpha(t) = \infty$$ (C2.1.7)

$$\sum_{t=0}^{\infty} \alpha^2(t) = \infty.$$

The learning algorithm of Kohonen maps in natural language representation is:

(i) compute the Euclidean distance between the input vector and the weight vector associated with the first neuron in the map;

(ii) repeat step (i) for all neurons in the map;

(iii) determine the best-match neuron, the neuron whose distance computed at step (i) is minimum;

(iv) determine a topological neighborhood of the best-match neuron in the Kohonen map;

(v) update the first weight associated with the first neuron in that topological neighborhood by adding a fraction of the difference between the input vector and the weight of this neuron;

(vi) repeat step (v) for all neurons in the neighborhood determined at step (iv).

The learning rule is basically local: the operation at each neuron only requires the knowledge of the weight associated with that neuron and of the input vector. However, a nonlocal function (but implementable as a tree structure) is needed to choose the best-match neuron. Adaptation is then again local, as soon as a neighborhood of the best-match neuron is chosen.

C2.1.1.7 Convergence of the algorithm

As explained by Marie Cottrell (Cottrell *et al* 1994) in her review paper about the different proofs of convergence of the Kohonen algorithm in specific situations,

'Despite the large use and the different implementations in multi-dimensional settings,
the Kohonen algorithm is surprisingly resistant to a complete mathematical study'.

In high dimensions, two obstacles prevent a thorough study of the convergence of the algorithm. First, while ordering is an obvious concept in one dimension, it is difficult to know what is a correctly ordered situation in a dimension greater than one; secondly, it is proven that it is impossible to associate a global decreasing potential function with the algorithm in the general case (Erwin *et al* 1992). In that situation, convergence results of the algorithm are only partial or apply in specific situations. Details of the actual state of the research in that domain may be found in the article by Cottrell *et al* (1994); here are, very briefly, some of these results:

- The self-organization property and the convergence are proved in the one-dimensional case for a Kohonen string, for a large class of input distributions and neighborhood functions.

- Self-organization in dimension 1 is proven for a large class of neighborhood functions when the input and the weights are quantized.

B5.2.2 - A potential function (from which the Kohonen algorithm is a stochastic *gradient descent* function) can be found when the input values belong to a finite discrete set; this potential function is not differentiable, but the convergence to a stationary point can be proven under some conditions on the adaptation parameter and the neighborhood function.

- The '0-neighbor' algorithm (Kohonen procedure with neighborhood restricted to the best-match vector only) always corresponds to a gradient descent procedure.

- Results in higher dimensions are only partial.

Details and references on actual results can be found in the article by Cottrell *et al* (1994).

C2.1.1.8 Examples of results

The Kohonen map is basically aimed to project a continuous high-dimensional space onto a discrete one- or two-dimensional one. For illustration purposes, we will show by a few examples the projection of a two-dimensional space onto a one- or two-dimensional one. Of course, the real interest of Kohonen maps

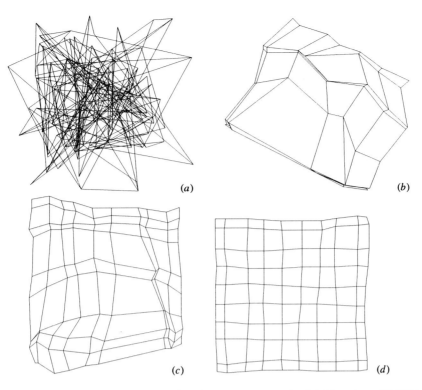

Figure C2.1.6. Locations and neighboring relations between neurons after (*a*) 0 iterations, (*b*) 100 iterations, (*c*) 1000 iterations, and (*d*) 10 000 iterations, for a uniform distribution of inputs.

is rarely found in such low-input-dimension examples; the idea of organization principles is, however, easily expandable to higher dimensions.

The way to easily represent a Kohonen map (in two dimensions) is to locate in the input space each neuron by the coordinates formed by its weight vector. Moreover, to represent the neighborhood relations in the output space, neighboring neurons are linked in the representation. Before learning, there is no ordering since the weights are randomly initialized, and the network resembles a dish of spaghetti. Progressively, however, a local ordering relationship develops in the map, so that two neurons which are neighbors in the map will be close by their locations in the input space. Figure C2.1.6 shows the locations and neighboring relations of a 10×10 Kohonen map, respectively, after 0, 100, 1000, and 10 000 learning iterations; the simulation was carried out with a uniform distribution of input patterns in a square area. The final distribution of neurons is effectively quite uniform, except on the sides of the map where a border effect can be noticed.

Of course, mapping a two-dimensional uniform distribution onto a two-dimensional square map is not the most useful application of a Kohonen map. Figure C2.1.7 shows the mapping of a uniform triangular distribution onto a square Kohonen map; the mapping cannot be perfect because of the border and corner effects, but one can see that the *vector quantization* property is well achieved, while the topological C1.1.5 neighborhood property is maintained at least locally.

Figure C2.1.8 shows the mapping of a nonuniform distribution (left-hand side of the figure) by a 10×10 Kohonen map. As expected, the density of neurons approximates the density of points in the distribution, which proves the vector quantization effect.

Finally, figure C2.1.9 shows how a three-dimensional distribution of points which are effectively arranged on a two-dimensional surface maps onto a two-dimensional Kohonen grid. From such an example, it can be concluded that the nonlinear projection realized by a Kohonen map will be optimal when the dimension of the map is approximately equal to the intrinsic dimension of the data.

C2.1.1.9 Kohonen map and principal component analysis

The last example above clearly shows the projection property of Kohonen maps: points in a high-dimensional space are projected onto a low-dimensional (usually one- or two-dimensional) space. If

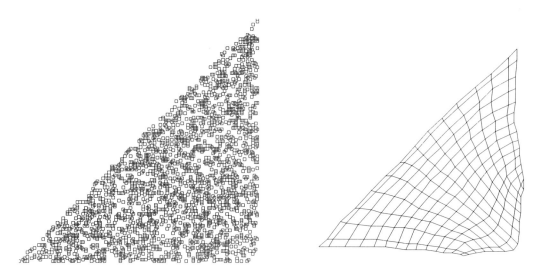

Figure C2.1.7. Final locations and neighboring relations between neurons for a uniform distribution of inputs in a triangle.

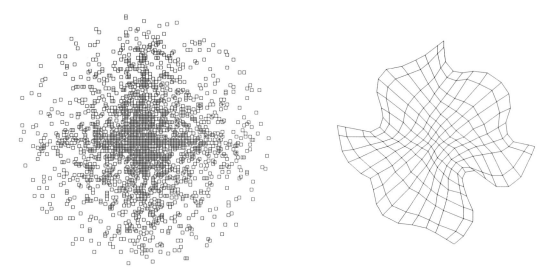

Figure C2.1.8. Final locations and neighboring relations between neurons for a nonuniform distribution of inputs.

we do not take into account the quantization property of Kohonen maps (which can be 'bypassed' by some linear or nonlinear interpolation between adjacent nodes in the map if necessary) the projection is similar to the one obtained with principal component analysis (PCA) according to the fact that in both cases the projected space is chosen to best fit the initial distribution.

The main difference, of course, relies on the nonlinear projection in the case of the Kohonen map, while PCA is purely linear; this can be a strong advantage in many situations.

C2.1.1.10 Related neural network models

The principle of auto-organization, which is the key aspect of the self-organizing feature map, is at the basis of many neural network models, more or less derived from Kohonen's one. Among these extensions, Fritzke's *growing cell structures* are of primary interest. The organization principles of Fritzke's network are similar to those of the Kohonen map; the main difference relies, however, on the fact that the map in Fritzke's network is not fixed *a priori* as in Kohonen's model, but is built during learning. The dimension of the map d_{out}, its number of neurons, and the connectivity between them are progressively adapted to form a structure more adapted to the input distribution. Other differences with respect to Kohonen's model

C2.4

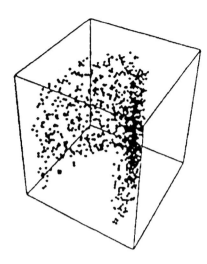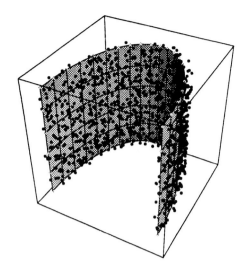

Figure C2.1.9. Final locations and neighboring relations between neurons for three- to two-dimensional mapping of inputs arranged on a two-dimensional surface.

are a fixed neighborhood limited to the direct neighbors, and fixed adaptation parameters. The principle of the network growth is to add cells in regions where neurons are most frequently chosen as the best-match ones, in order to obtain similar firing rates for all neurons, which is a characteristic of good quantization.

The Kohonen self-organizing map is also related to many adaptive models of vector quantization, such as K-means algorithm, maximum-entropy clustering, and neural gas, and the like. The principle of any vector quantization method is to approximate a continuous generally nonuniform distribution of vectors (or a discrete distribution with a large number of vectors) by another distribution being formed by a much smaller number of units of the same dimension. If we consider the weight vector of each neuron as a unit in the input space, a Kohonen network implements this vector quantization scheme. The above-mentioned methods of vector quantization find their place in the field of artificial neural networks (ANNs): they are adaptive, may be represented in the form of a feedforward graph, and may be described by the same formalism as used in many other neural models, including Kohonen's. For this reason, in the next section we will describe the 'neural-gas' method, a vector quantization method known in the field of neural networks as a powerful algorithm that quickly converges to low distortion errors, generally reaches a final distortion under that obtained by other algorithms, and obeys a gradient descent on an energy function surface, like the K-means clustering, but unlike Kohonen's algorithm.

C2.1.2 Neural gas

C2.1.2.1 Introduction

It is a tricky question to know whether adaptive vector quantization techniques find their place in the frame of ANNs or not. We do not pretend to give an answer to this question, but would like to point out some common points between methods which are usually classified as 'artificial neural networks' and vector quantization techniques. Neural networks are adaptive techniques, which can usually be represented in the form of a graph of computational units, linked by synaptic weights. Neural networks have a biological origin, although this is not acknowledged in all models of learning. Neural networks usually realize some kind of function or distribution approximation, or classification.

In that order of ideas, many adaptive methods of vector quantization resemble neural networks. Algorithms like K-means clustering or any derived method such as maximum-entropy clustering and frequency-sensitive learning are also adaptive, can be represented as a similar graph of computational units, and realize some kind of distribution (or probability density) estimation. It is not our intention here, however, to describe all vector quantization methods, since it would greatly exceed the scope of a handbook of neural networks. However, it would not be fair to fully omit and completely dissociate them from the field of neural networks; their common characteristics, and the advantages that both vector quantization and neural networks can take from a common approach, fully justify their insertion here.

Vector quantizers are feedforward unsupervised models: they have no 'teacher', and there is no feedback between the positioning of the neurons and the distribution of input vectors. Rather than going into the details of many similar (while different) vector quantization algorithms, we will focus our discussion on one model, the neural gas, which seems to have three advantages with respect to most other methods (Martinetz *et al* 1993):

- it converges quickly
- it reaches a lower distortion error after convergence than with other methods
- it obeys a gradient descent on an energy function surface, which is not the case for example with the Kohonen maps.

The neural-gas algorithm has been developed by Thomas Martinetz and Klaus Schulten and a good description of the model and its applications can be found in the article by Martinetz *et al* (1993).

C2.1.2.2   Purpose of the model

The purpose of the neural-gas model is to quantize multidimensional vectors, that is, to transform an initial large set of d_{in}-dimensional vectors into a reduced set of N vectors (neurons) of the same dimension, with a minimal mean distortion between each of the vectors in the initial set and its best-match neuron, the best-match neuron being defined as the vectors in the final set closest to the input vector. Each neuron will thus define in the input space a so-called Voronoi region, being the set of locations closer to this neuron than to any other one; this property is identical to the vector quantization property of Kohonen's algorithm, while the topology conservation property does not hold in the neural-gas model. In comparison with other adaptive vector quantization techniques, the neural-gas algorithm is fast, converges to low average distortions, and obeys a gradient descent on a known energy function surface.

C2.1.2.3   Topology

A neural-gas network has one layer of neurons and one layer of connections. As in the Kohonen model, each d_{in}-dimensional neuron is connected to the input vector x of the same dimension, through a weight vector w. There is no connection between units in the neuron layer.

C2.1.2.4   Learning

The neuron activations in the neural-gas model may be defined as the distance between the input vector x and weight w_i associated with neuron n_i:

$$a_i = \|w_i - x\| \tag{C2.1.8}$$

where a_i is the activation of neuron n_i.

Learning, i.e. adaptation of weights w_i to a distribution of input vectors, goes through the selection of the best-match neuron n_b (of weight w_b) at each presentation of an input vector $x(t)$ to the network:

$$\|x(t) - w_b\| \leq \|x(t) - w_i\| \qquad \forall 1 \leq i \leq N. \tag{C2.1.9}$$

After selection of the winner, the weights of several neurons are adapted, as in the Kohonen map. However, the neurons to adapt are not selected according to a topological relation with the best-match neuron n_b, but are selected according to the rank they have in the ordered list of distances between their weights and the input vector.

Each time an input vector $x(t)$ is presented to the network, all neurons are 'ranked' according to their distance from the input; w_{b_0} is the weight of the closest neuron to x (according to the Euclidean distance), w_{b_1} is the second-closest neuron to x, and so on. In other words, the ranks b_i, $1 \leq i \leq N - 1$, are determined according to

$$\|x - w_{b_0}\| \leq \|x - w_{b_1}\| \leq \ldots \leq \|x - w_{b_{N-1}}\|. \tag{C2.1.10}$$

After ranking, neuron weights are adapted according to

$$w_i(t + 1) - w_i(t) = \alpha\beta(j)(x - w_i(t)) \tag{C2.1.11}$$

where α is an adaptation factor, j is chosen so that $b_j = i$, and $\beta(j)$ is a monotonic decreasing function. Usually, function $\beta(j)$ is chosen according to

$$\beta(j) = e^{-j/\lambda} \tag{C2.1.12}$$

where λ is a decay constant. For $\lambda \to 0$, the algorithm becomes equivalent to the K-means method.

It can be proven that the neural-gas algorithm obeys a gradient descent on the surface defined by the energy function

$$E(w, \lambda) = \frac{1}{2C(\lambda)} \sum_{i=1}^{N} \int P(x(t)) \beta j (x - w_i)^2 \, dx \tag{C2.1.13}$$

where $P(x)$ is the distribution of input vectors $x(t)$, j is chosen so that $b_j = i$, the integral is taken over the whole input space, and

$$C(\lambda) = \sum_{j=0}^{N-1} \beta(j) \,. \tag{C2.1.14}$$

A proof of this can be found in the article by Martinetz *et al* (1993). As a stochastic gradient descent algorithm on a bounded function, the neural-gas algorithm will converge, but may be trapped in one of the local minima of the energy function (C2.1.13). It must also be noticed that the parameter λ usually decreases with time, a large parameter corresponding to a smooth energy function, while when $\lambda \to 0$ the energy function (C2.1.13) becomes equivalent to the energy function of the K-means algorithm. Reducing the value of λ with time thus reduces the risk of being trapped in a 'bad' local minimum of function (C2.1.13).

The neural-gas learning algorithm in natural language representation is:

(i) compute the Euclidean distance between the input vector and the weight vector associated with the first neuron in the map;
(ii) repeat step (i) for all neurons in the map;
(iii) rank the neurons by their respective Euclidean distances to the input neuron;
(iv) update the first neuron in the network by adding a fraction of the difference between the input vector and the weight of the neuron, multiplied by a factor depending on its rank;
(v) repeat step (iv) for all neurons in the network.

The learning rule is basically local: the operation at each neuron only requires the knowledge of the weight associated with that neuron and of the input vector. However, a nonlocal function (but implementable as a tree structure) is needed to rank the neurons according to their distance from the input vector.

C2.1.2.5 Examples of results

An illustration of the neural-gas algorithm behavior in two dimensions is given in figure C2.1.10. The input distribution (left-hand part of the figure) is formed by three Gaussian distributions, of which two overlap. The right-hand side of the figure illustrates the positions of the neurons after convergence. It can be seen that more neurons are concentrated in the center of the Gaussian functions than in the tails; this was expected as more points from the input distribution are concentrated in these regions too.

The neurons seem also to be concentrated in the tails of the distribution which is in contradiction to the vector quantization principle. In fact, this visual effect is due to the fact that the Voronoi regions of the neurons in the tails of the Gaussian functions are very narrow but long. The (large) size of the Voronoi regions associated with these neurons, compared to the regions associated with neurons in the centers of the distribution, is thus in accordance with the (low) density of points in the tails.

C2.1.2.6 Related neural network models

As already mentioned above, the neural-gas algorithm is closely linked to Kohonen's self-organizing feature maps, because of its vector quantization property. While the neural gas does not have the topology-preservation property of Kohonen maps, it achieves a similar quantization of input data. In that case, the neural-gas algorithm is also usually faster and more accurate (lower final mean-square error) than other vector quantization methods.

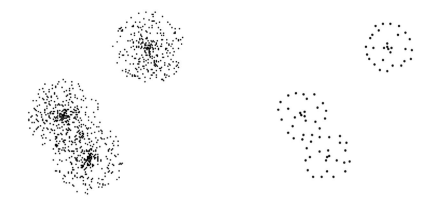

Figure C2.1.10. Input distribution formed by three Gaussian functions in two dimensions, and the corresponding neuron locations after convergence of the neural-gas algorithm.

Many other vector quantization methods exist which are more or less similar to the neural gas; including them in the neural network field or not is a question of personal feeling. We can, however, mention that most of them are derived from or have links to the K-means Lloyd and MacQueen algorithm, which uses the same principle as the above-described neural-gas method, except that only the best-match neuron is moved at each presentation of an input vector; the ranking procedure is thus not used. Several methods also try to avoid confinement to local minima of the distortion function, by adding some probabilistic term in the best-match neuron choice, or by moving several neurons at each presentation of an input vector, depending on the distance between the corresponding weights and this input, and so on.

C2.1.3 Neocognitron

C2.1.3.1 Introduction

The neocognitron has been developed by Kunihiko Fukushima (Osaka University, Department of Biophysical Engineering, Toyonaka, Osaka 560, Japan). It has been described in many research papers and conference communications since 1980; a good description of the network and of the learning process can be found in the article by Fukushima (1988). In this section, we will follow Fukushima's description from that reference.

The neocognitron is presented as a biologically plausible network. It can indeed be proved that in the visual cortex and in the higher areas there are cells which respond to very simple patterns like segments of lines or curves at an early stage, and then respond to more complex figures (squares, circles, and the like) at a higher stage, and so on. The neocognitron is based on the same hierarchical structure; cells in the first layers extract simple patterns, while they become progressively more complex up to the last layers of the network.

The neocognitron is devoted to pattern extraction, in applications where this property of extracting features of increasing complexity may be exploited, for example in optical character recognition, where simple features are combined to recognize more complex characters.

C2.1.3.2 Purpose of the model

As will be described in the next section the neocognitron is formed by an even number of layers, odd layers having adaptable input connections and even layers having fixed input ones. Each odd layer is intended to detect features in the plane formed by the preceding layer; in the first stages of the neocognitron, the detected features are very simple, such as line segments, circles, and angles. Each neuron in an odd layer will be adapted to detect a particular feature in a particular location of the preceding layer. In the even layers, fixed connections between each neuron in that layer and a set of close neurons in the previous one are intended to cancel the effect of small shifts in position of the features detected in the previous layer.

The neocognitron is thus intended to detect complex patterns in a plane of cells, through the progressive detection of simple to elaborated features from the first to the last layers. One of its applications, largely described by the author of the model, is *optical character recognition* (OCR), where simple features (strokes) may be combined to detect more complex patterns (the characters).

G1.2, G1.3

C2.1.3.3 Topology

The topology of a neocognitron is described in figure C2.1.11. It contains an even number of layers (the input plane U_0, having only the task of redistributing the inputs on the different neurons of the first layer U_{S1}, is not counted here as a 'layer'). Layers U_{Si} and U_{Ci} alternate in the network, their functionality being different. Neurons used in the S-layers (odd layers) are typical nonlinear neurons with excitatory and inhibitory inputs, and with positive outputs. Neurons in C-layers (even layers) act as OR functions (they fire if one of their inputs fires).

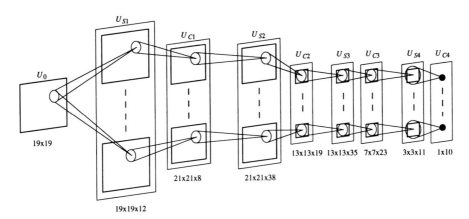

Figure C2.1.11. Layer structure of the neocognitron (according to Fukushima 1988).

All neuron layers are divided into 'cell planes', being represented as squares in figure C2.1.11; each cell plane contains neurons which detect similar features at different locations in the previous layer; similarly, neurons at corresponding locations in the different cell planes of a layer respond to different features at the same location of the previous layer. The numbers of cells indicated at the bottom of figure C2.1.11 are given as an example (these values are those given in the application example described by Fukushima 1988).

Each S-layer contains S-cells and inhibitory V-cells. The S-cells are feature extracting cells; after learning, they respond to specific features at specific locations in the preceding layer. In general, features extracted at lower stages of the network are elementary (line segments, strokes, branchings, and the like), while those extracted at upper stages are more complex (e.g. characters).

C-cells in the C-layers are inserted in the network to allow for shifts in the feature detection in the preceding S-layers. Because of their (local) OR function, C-cells fire when one of the S-cells in a local neighborhood of the location corresponding to the C-cell, but in the previous layer, is activated; this functionality makes the network less sensitive to shifts in the position of the detected features. Since C-layers are each inserted between two S-layers, this property applies to the features detected at all stages of the network; simple features in one layer of the network being elements of more complex features in the subsequent layers, this property also allows for some insensitivity to distortion of complex features.

Subsidiary V-cells in the S-layers have another functionality. These cells have fixed connections to the same neurons in the preceding layer as their associated S-cells (there are as many V-cells as S-cells in S-layers) have variable connections to. The outputs of the V-cells are inhibitory connections to their associated S-cells. Their objective is mainly a winner-take-all functionality, to allow only one neuron firing at a time in one cell plane.

C2.1.3.4 Learning

Learning in a neocognitron is primarily unsupervised: neurons in S-layers find themselves the patterns they have to extract, according to a competitive adaptive scheme. Some supervised learning can, however, be added in the unsupervised scheme to have more control over the kind of feature that will be extracted by each S-layer.

Two principles are at the basis of learning in the neocognitron. The first one concerns the reinforcement of maximum-output cells, the second one the development of iterative connections.

The first principle consists in reinforcing the variable connections (to the S-layers) under two conditions:

(i) the cell receiving the connection has a stronger activation than other cells in its neighborhood;

(ii) the cell sending out the connection has a nonzero activation value.

The weight change at this connection is then proportional to the activation of the neuron sending out the connection.

Connections to the S-cells are thus adapted so that the S-cells will match the template presented at the previous layer. Inhibitory V-cells whose outputs are also connected to the S-cells fire when the pattern presented in the previous layer at the location scanned by the S- and V-cells does not match the feature already learned by the cell; the V-cells thus watch for irrelevant features, and increase the ability of the network to differentiate between different features. This differentiation property is also due to the first condition mentioned above, that a single cell in a small area will have its connections reinforced when a pattern is presented: the connections leading to this cell already match more or less the presented feature (since it is the neuron with the largest activation) and are even reinforced in that direction because of the proportionality between the reinforcement and the feature.

The other principle leading to self-organization consists in selecting seed cells. In each hypercolumn, defined as the group of S-cells in a layer detecting features at approximately the same location in the previous layer, the cell that fires most is chosen as a candidate for seed cells. However, only one seed cell can be selected in each cell plane; if two S-cells are selected as candidates in the same cell plane, only the one having the greatest activation will be selected.

Having selected (maximum) one seed cell in each cell plane, all other S-cells in that plane grow in order to have the spatial distribution of their inputs identical to that of the seed cell. This ensures that all cells in one cell plane will respond to the same feature, but of course at different locations. In contrast, all cells in the same hypercolumn will respond to different features at the same locations.

In order to guarantee that the auto-organization principle of the network can take place, small random values must be attributed as initial conditions for the adaptable weights; if all weights were null at the beginning of the process, no selection of winner could occur, and the competitive process could not be initiated.

The cooperative and simultaneous action of these two principles is at the basis of the auto-organization of the network, where the S-cells will progressively adapt their connections to extract the most frequent features detected when input patterns are presented.

Let us finally mention that supervised learning can easily be inserted into the network, by replacing the automatic selection of seed cells by a teacher's choice. This is possible at any stage of the network, and can greatly help in order to add available knowledge in the network (such as conventional similarity between visually different characters in OCR).

C2.1.3.5 *Related neural network models*

The neocognitron is not 'similar' to other widely used models in the neural network field. Basically, it is much more complex, has more layers, more weights, and more neurons than most other networks. The learning process is also complex, and not obvious to implement. In fact, an implementation of the neocognitron requires a lot of 'fine-tuning' effort; the size of the network is obviously dependent on the application, and some tricks can be given to guide the choice of the number of layers, neurons, cell planes, and so on. There are also parameters in the equations of the network (both use and learning) which are not obvious to choose, such as the exact form of the activation functions, the neuron thresholds, and so on. This is why we did not go into the details of the equations, since this would have largely exceeded the scope of this description.

However, in the opinion of the author, the network has remarkable properties of adaptation to the examples given to the network. Mixing unsupervised and supervised learning can also be a determining advantage, to benefit from the inherent properties of the network when no supplementary information is available, but also to allow for the insertion of knowledge into the network when this knowledge is available.

References

Cottrell M, Fort J C and Pages G 1994 Two or three things that we know about the Kohonen algorithm *Proc. Eur. Symp. on Artificial Neural Networks* (Brussels: D facto) pp 235–44

Erwin E, Obermayer K and Shulten K 1992 Self-organizing maps: ordering, convergence properties and energy functions *Biol. Cyber.* **67** 47–55

Fukushima K 1988 Neocognitron: a hierarchical neural network capable of visual pattern recognition *Neural Networks* **1** 119–30

Kohonen T 1989 *Self-Organization and Associative Memory* (Berlin: Springer)

Martinetz T M, Berkovich S G and Schulten R J 1993 'Neural-Gas' network for vector quantization and its application to time-series prediction *IEEE Trans. Neural Networks* **4** 558–69

Robbins H and Monro S 1951 A stochastic approximation method *Ann. Math. Stat.* **22** 400–7

von der Malsburg C 1973 Self-organization of orientation sensitive cells in the striate cortex *Kybernetik* **14** 85–100

C2.2   Feedback models

Gail A Carpenter (C2.2.1), *Stephen Grossberg* (C2.2.1, C2.2.3), *and*
Peggy Israel Doerschuk (C2.2.2)

Abstract

See the individual abstracts for sections C2.2.1, C2.2.2, and C2.2.3.

C2.2.1   Adaptive resonance theory: self-organizing networks for stable learning, recognition, and prediction

Gail A Carpenter† and *Stephen Grossberg‡*

Abstract

Adaptive resonance theory (ART) is a neural theory of human and primate information processing and of adaptive pattern recognition and prediction for technology. Biological applications to attentive learning of visual recognition categories by inferotemporal cortex and hippocampal system, medial temporal amnesia, corticogeniculate synchronization, auditory streaming, speech recognition, and eye movement control are noted. ARTMAP systems for technology integrate neural networks, fuzzy logic, and expert production systems to carry out both unsupervised and supervised learning. Fast and slow learning are both stable responses to large nonstationary databases. Match tracking search conjointly maximizes learned compression while minimizing predictive error. Spatial and temporal evidence accumulation improve accuracy in three-dimensional object recognition. Other applications are summarized.

C2.2.1.1   Introduction

The problem whereby the brain learns quickly and stably without catastrophically forgetting its past knowledge has been called the *stability–plasticity dilemma* (Grossberg 1980). The stability–plasticity dilemma must be solved by every brain system that needs to rapidly and adaptively respond to the flood of signals that subserves even the most ordinary experiences. Design principles that show how brain systems can stably learn an accumulating knowledge base in response to changing conditions throughout life should clarify how the brain unifies diverse sources of information into coherent moments of conscious experience.

This section summarizes neural models that realize and develop a theory called adaptive resonance theory, or ART, that was introduced 20 years ago (Grossberg 1976a, b). ART principles have been used to explain challenging behavioral and brain data in the areas of visual perception, visual object recognition, auditory source identification, variable-rate speech and word recognition, and aspects of adaptive sensory-motor control, among others. Some of these analyses are reviewed below. In addition, ART concepts have been developed into precise mathematical systems that have been used in a wide variety of technological

† Supported in part by the National Science Foundation (NSF IRI-94-1659) and the Office of Naval Research (ONR N00014-95-1-0409 and ONR N00014-95-1-0657).
‡ Supported in part by the Advanced Research Projects Agency (ONR N00014-92-J-4015) and the Office of Naval Research (ONR N00014-95-1-0657, ONR N00014-95-1-0409, and ONR N00014-92-J-1309).

applications, including control of mobile robots, learning and search of airplane part inventories, control of nuclear reactors, medical diagnosis, three-dimensional visual object recognition, music analysis, and recognition of seismic, sonar, lasar radar, and Landsat imagery. These biological and technological applications exploit the key properties of ART systems that are summarized below. Note that, despite the broad explanatory scope of these systems, fundamentally different types of learning seem to govern processes such as spatial navigation and certain aspects of sensory-motor control. In these latter task domains, it *is* adaptive to forget old coordinate transformations as the brain's control systems adjust to a growing body and to other changes in the body's sensory-motor endowment throughout life.

In addition to the development of ART as a cognitive and neural theory, families of ART neural network architectures have been progressively developed at Boston University. These models include ART 1, ART 2, ART 2-A, ART 3, Fuzzy ART, ARTMAP, Fuzzy ARTMAP, and Fusion ARTMAP (Asfour *et al* 1993, Carpenter and Grossberg 1987a, b, 1990, 1991, Carpenter *et al* 1991a, b, c, 1992, 1993, 1995). Other ART models have also been developed and applied by a number of investigators.

C2.2.1.2 Some key ART properties

ART systems can autonomously learn, recognize, and make predictions with the following properties.

Fast learning of rare events. A successful autonomous agent must be able to learn about rare events that have important consequences, even if these rare events are similar to a surrounding cloud of frequent events that have different consequences. *Fast learning* is needed to categorize a rare event before it is supplanted by more frequent subsequent events.

Stable learning of large nonstationary databases. Individual events may also occur with variable probabilities and durations, and arbitrarily large numbers of events may need to be processed. ART systems contain a *self-stabilizing memory* that permits accumulating knowledge to be stably stored in response to arbitrarily many events in a nonstationary environment under incremental learning conditions, until the algorithm's full memory capacity (which can be chosen to be arbitrarily large) is exhausted.

Efficient learning of morphologically variable events. Multiple scales of generalization, from fine to coarse, need to be employed on an as-needed basis. Supervised ART systems can automatically adjust their scale of generalization to match the morphological variability of the data using a minimax learning rule that conjointly *minimizes* predictive error and *maximizes* generalization using only information that is locally available under incremental learning conditions.

MANY-TO-ONE MAP

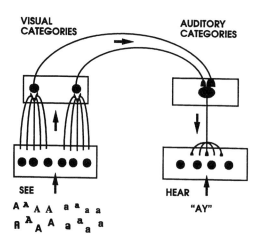

Figure C2.2.1. Many-to-one learning combines categorization of many exemplars into one category and labelling of many categories with the same name. (Reprinted with permission from Carpenter and Grossberg 1994.)

Associative learning of many-to-one and one-to-many maps. Many-to-one learning includes both categorization and associative prediction (figure C2.2.1). For example, during categorization of printed letter fonts, many similar instances of the same printed letter may establish a single recognition category, or compressed representation. During prediction, all of the categories that represent a given letter may be associatively mapped into the letter name, or prediction. This is a second, distinct, type of many-to-one map, since there need be no relationship between the visual features that define a printed letter A and a written letter A, yet both categories have the same name.

One-to-many learning is used to discover and accumulate expert knowledge about an object or event (figure C2.2.2). In many learning algorithms, including backpropagation, the attempt to learn more than one prediction about an event leads to unselective forgetting of previously learned predictions, for the same reason that these algorithms may become unstable in response to nonstationary data.

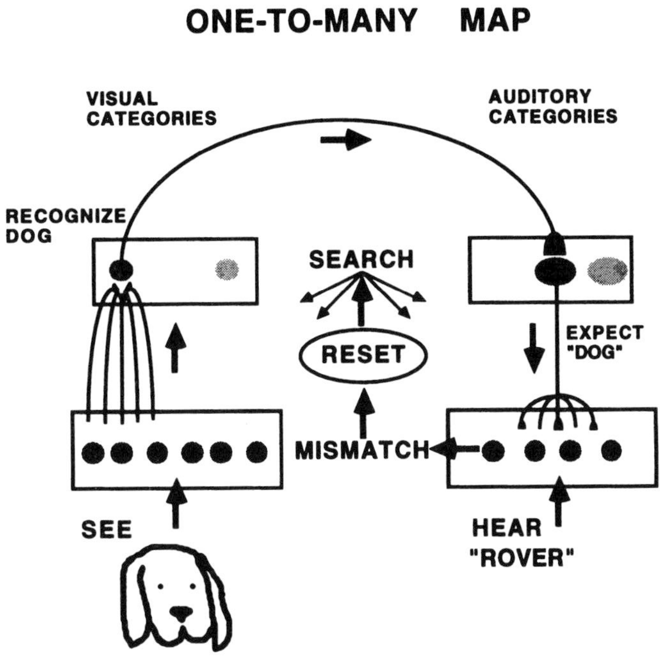

ONE-TO-MANY MAP

Figure C2.2.2. One-to-many learning enables one input vector to be associated with many output vectors. If the system predicts an output that is disconfirmed at a given stage of learning the predictive error drives a memory search for a new category to associate with the new prediction without degrading its previous knowledge about the input vector. (Reprinted with permission from Carpenter and Grossberg 1994.)

Paying attention and top-down priming. ART systems learn top-down expectations (also called primes, or queries) that can bias the system to ignore masses of irrelevant data. A large mismatch between a bottom-up input vector and a top-down expectation can suppress features in the input pattern that are not confirmed by the top-down prime and can thereby drive an adaptive memory search that carries out a bout of hypothesis testing.

Hypothesis testing and match learning. The system hereby selectively searches for recognition categories, or hypotheses, whose top-down expectations provide an acceptable match to bottom-up data. Each top-down expectation begins to focus attention upon, and bind, that cluster of input features that are part of the prototype which it has already learned, while suppressing features that are not.

Choosing the globally best answer without recursive search. After learning has stabilized, an input pattern first selects the category whose top-down expectation provides the globally best match. In addition, the top-down expectation read out by the selected category acts as a *prototype* for the class of all the input patterns that the category represents.

Learning both prototypes and exemplars. The brain sometimes appears to learn prototypes, or abstract types of knowledge, such as being able to recognize that a particular object is a face or an animal and at other times to learn individual exemplars, or concrete types of knowledge, such as being able to recognize a particular face or a particular animal. Supervised ART systems can learn both types of knowledge.

Controlling vigilance to calibrate confidence. A confidence measure, called *vigilance*, calibrates how well an exemplar needs to match the prototype that it reads out in order for the corresponding category to resonate with it and be chosen. The minimax learning rule is realized by *match tracking*, a process that raises the vigilance parameter in response to a predictive error just enough to initiate hypothesis testing to discover a better category.

Rule extraction and fuzzy reasoning. The IF–THEN rules of supervised ART systems can be read off from the learned adaptive weights of the system at any stage of the learning process. This property is particularly important in applications such as medical diagnosis from a large database of patient records, where doctors may want to study the rules by which the system reaches its diagnostic decisions. Tables C2.2.1–3 summarize some medical and other benchmark studies that compare the performance of fuzzy ARTMAP with alternative recognition and prediction models. These and other benchmarks are described elsewhere in greater detail (Carpenter *et al* 1991a, 1992, Carpenter and Tan 1995).

Properties scale to arbitrarily large databases. All of the desirable properties of ART systems scale to arbitrarily large problems. On the other hand, ART helps to solve only learned categorization and prediction problems. These problems are, however, core problems in many intelligent systems, and have been technology bottlenecks for many alternative approaches.

C2.2.1.3 Adaptive resonance theory topology and learning

Since its introduction as a theory of human cognitive information processing (Grossberg 1976b, 1980), theoretical developments of adaptive resonance theory have continued to explain and predict cognitive and neural databases; see Carpenter and Grossberg (1991, 1993), Grossberg (1987a, b, 1988, 1994, 1995), Grossberg and Merrill (1992, 1996) and Grossberg *et al* (1994) for illustrative contributions. In addition, an evolving series of self-organizing neural network models have been developed for applications to adaptive pattern recognition and prediction. These self-organizing models can operate in either an unsupervised or a supervised mode. Unsupervised learning occurs when network predictions do not receive environmental feedback. Supervised learning occurs when prediction contingent feedback is available. This option does not occur in many supervised learning algorithms, such as backpropagation, which can learn only when feedback is available. Unsupervised ART models learn stable recognition categories in response to arbitrary input sequences with either fast or slow learning. These model families include ART 1 (Carpenter and Grossberg 1987a), which can stably learn to categorize binary input patterns presented in an arbitrary order; ART 2, ART2-A, and fuzzy ART (Carpenter and Grossberg 1987b, Carpenter *et al* 1991b, c), which can stably learn to categorize either analog or binary input patterns presented in an arbitrary order, and ART 3 (Carpenter and Grossberg 1990), which can carry out parallel search, or hypothesis testing, of distributed recognition codes in a multilevel network hierarchy. Variations of these models adapted to the demands of individual applications have been developed by a number of authors.

Figure C2.2.3 illustrates one example from the family of ART 1 models, and figure C2.2.4 illustrates a typical ART search cycle. Level F_1 in figure C2.2.3 contains a network of nodes, each of which represents a particular combination of input components, such as sensory features. Level F_2 contains a network of nodes that represent recognition codes which are selectively activated by patterns of activation across F_1. The activities of nodes in F_1 and F_2 are also called short-term memory (STM) traces. STM is the type of memory that can be rapidly reset without leaving an enduring trace. For example, it is easy to reset the STM of a list of numbers that a person has just heard once by distracting the person with an unexpected event. STM is distinct from LTM, or long-term memory, which is the type of memory that we usually ascribe to learning. For example, we do not forget our parents' names when we are distracted by an unexpected event.

As shown in figure C2.2.4(*a*), an input vector I registers itself as a pattern X of activity across level F_1. The F_1 output vector S is then transmitted through the multiple converging and diverging adaptive filter pathways emanating from F_1. This transmission event multiplies the vector S by a matrix of adaptive weights, or LTM traces, to generate a net input vector T to level F_2. The internal competitive dynamics of F_2 contrast-enhance vector T. Whereas many F_2 nodes may receive inputs from F_1, competition or

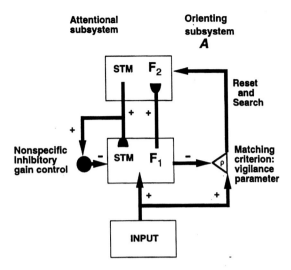

Figure C2.2.3. An example of a model ART circuit in which attentional and orienting circuits interact. Level F_1 encodes a distributed representation of an event by a short term memory (STM) activation pattern across a network of feature detectors. Level F_2 encodes the event using a compressed STM representation of the F_1 pattern. Learning of these recognition codes occurs at the long term memory (LTM) traces within the bottom-up and top-down pathways between levels F_1 and F_2. The top-down pathways read-out learned expectations whose prototypes are matched against bottom-up input patterns at F_1. The size of mismatches in response to novel events are evaluated relative to the vigilance parameter ρ of the orienting subsystem A. A large enough mismatch resets the recognition code that is active in STM at F_2 and initiates a memory search for a more appropriate recognition code. Output from subsystem A can also trigger an orienting response. (Adapted with permission from Carpenter and Grossberg 1987a.)

lateral inhibition between F_2 nodes allows only a much smaller set of F_2 nodes to store their activation in STM. A compressed activity vector Y is thereby generated across F_2. In ART 1, the competition is tuned so that the F_2 node that receives the maximal $F_1 \rightarrow F_2$ input is selected. Only one component of Y is nonzero after this choice takes place. Activation of such a winner-take-all node defines the category, or symbol, of the input pattern I. Such a category represents all the inputs I that maximally activate the corresponding node. So far, these are the rules of a *self-organizing feature map* (SOFM), also called competitive learning, or learned vector quantization (Grossberg 1972, 1976a, 1978, von der Malsburg 1973, Kohonen 1984/1989).

C2.1.1

C2.2.1.4 The link between matching, hypothesis testing and attention

The ART scheme for self-stabilizing its embedded SOFM model incorporates heuristics that are also used in expert production systems and fuzzy systems. In particular, ART systems carry out a form of hypothesis testing to discover new recognition categories and to stabilize learning. Thus in an ART model (Carpenter and Grossberg 1987a, 1991), learning does not occur whenever some winning F_2 activities are stored in STM. Instead activation of F_2 nodes may be interpreted as 'making a hypothesis' about an input I. When Y is activated (figure C2.2.4(*a*)), it generates an output vector U that is sent top-down through the second adaptive filter. After multiplication by the adaptive weight matrix of the top-down filter, a net vector V inputs to F_1 (figure C2.2.4(*b*)). Vector V plays the role of a learned top-down expectation. Activation of V by Y may be interpreted as 'testing the hypothesis' Y, or 'reading out the category prototype' V. When the category representation Y makes a choice (winner-take-all), ART networks employ *outstar learning* (Grossberg 1968) to train the top-down ($F_2 \rightarrow F_1$) adaptive filter. The distributed outstar (Carpenter 1994) allows activity Y in an outstar source field F_2 to be arbitrarily distributed.

C1.1.6

The ART 1 network is designed to match the 'expected prototype' V of the category against the active input pattern, or exemplar, I. Nodes that are activated by I are suppressed if they do not correspond to large LTM traces in the prototype pattern V (figure C2.2.4(*c*)). Thus F_1 features that are not 'expected' by V are suppressed. Expressed in a different way, the matching process may change the F_1 activity pattern

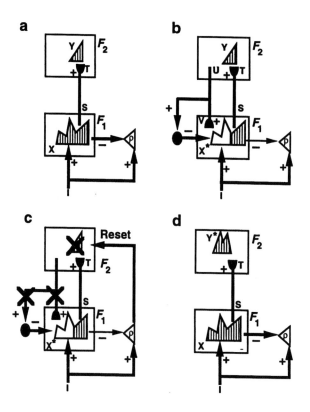

Figure C2.2.4. ART search for a recognition code: (*a*) The input pattern I is instated across the feature detectors at level F_1 as a short term memory (STM) activity pattern X. Input I also nonspecifically activates the orienting subsystem A; see figure C2.2.1 STM pattern X is represented by the hatched pattern across F_1. Pattern X both inhibits A and generates the output pattern S. Pattern S is multiplied by long term memory (LTM) traces and added at F_2 nodes to form the input pattern T, which activates the STM pattern Y across the recognition categories coded at level F_2. (*b*) Pattern Y generates the top-down output pattern U which is multiplied by top-down LTM traces and added at F_1 nodes to form the prototype pattern V that encodes the learned expectation of the active F_2 nodes. If V mismatches I at F_1, then a new STM activity pattern X^* is generated at F_1. X^* is represented by the hatched pattern. It includes the features of I that are confirmed by V. Inactivated nodes corresponding to unconfirmed features of X are unhatched. The reduction in total STM activity which occurs when X is transformed into X^* causes a decrease in the total inhibition from F_1 to A. (*c*) If inhibition decreases sufficiently A releases a nonspecific arousal wave to F_2, which resets the STM pattern Y at F_2. (*d*) After Y is inhibited its top-down prototype signal is eliminated and X can be reinstated at F_1. Enduring traces of the prior reset lead X to activate a different STM pattern Y^* at F_2. If the top-down prototype due to Y^* also mismatches I at F_1, then the search for an appropriate F_2 code continues until a more appropriate F_2 representation is selected. Then an attentive resonance develops and learning of the attended data is initiated. (Adapted with permission from Carpenter and Grossberg 1987a.)

X by suppressing activation of all the feature detectors in I that are not 'confirmed' by hypothesis Y. The resultant pattern X* encodes the cluster of features in I that the network deems relevant to the hypothesis Y based upon its past experience. Pattern X^* encodes the pattern of features to which the network 'pays attention'. This type of attentional focusing prevents irrelevant features from being incorporated into the prototype through learning.

C2.2.1.5 *The link between attention, resonance, and learning*

If the expectation V is close enough to the input I, then a state of *resonance* develops as the attentional focus takes hold. The pattern X* of attended features reactivates hypothesis Y which, in turn, reactivates X*. The network locks into a resonant state through the mutual positive feedback that dynamically links X* with Y. In ART, the resonant state, rather than bottom-up activation, drives the learning process. The resonant state persists long enough, at a high enough activity level, to activate the slower learning process;

hence the term *adaptive resonance* theory. ART systems learn prototypes, rather than exemplars, because the attended feature vector X^*, rather than the input I itself, is learned. These prototypes may, however, also be used to encode individual exemplars, as described below.

C2.2.1.6 The link between intentionality and the stability of learning

The ART attentive matching process may be realized in several ways (Carpenter and Grossberg 1987a). In one instantiation, three different types of inputs are combined at level F_1 (figure C2.2.3): bottom-up inputs, top-down expectations, and attentional gain control signals. The attentional gain control channel sends the same top-down inhibitory signal to all F_1 nodes; it is a 'nonspecific', or modulatory, channel.

The ART matching rule allows F_1 nodes to generate suprathreshold outputs in response to bottom-up inputs, since an input directly activates its target F_1 features (figure C2.2.4(a)). After the input instates itself at F_1, leading to selection of a hypothesis Y and a top-down prototype V, the matching rule ensures that only those active F_1 nodes that are confirmed by the top-down prototype can remain active and be attended at F_1 after an F_2 category is selected, since top-down nonspecific inhibitory feedback shuts off the F_1 nodes that do not receive large learned top-down excitatory signals.

The ART matching rule enables an ART network to realize a self-stabilizing learning process. Carpenter and Grossberg (1987a) proved that ART learning and memory are stable in arbitrary environments, but become unstable when the ART matching rule is eliminated. They also defined several circuits that generate the desired matching properties. Thus a type of matching that guarantees stable learning also enables the network to selectively pay attention to feature combinations that are confirmed by a top-down expectation.

C2.2.1.7 Vigilance control of category generalization

The criterion of an acceptable match between bottom-up inputs and top-down prototypes is defined by a parameter ρ called *vigilance* (Carpenter and Grossberg 1987a, 1991). The vigilance parameter is computed in the orienting subsystem A. Vigilance weighs how similar an input exemplar must be to a top-down prototype in order for resonance to occur. Resonance occurs if $\rho|I| - |X^*| \leq 0$, where $0 \leq \rho \leq 1$. This inequality says that the F_1 attentional focus X^* inhibits A more than the input I excites it. If A remains quiet, then an $F_1 \leftrightarrow F_2$ resonance can develop.

Vigilance calibrates how much novelty the system can tolerate before activating A and searching for a different category. If the top-down expectation and the bottom-up input are too different to satisfy the resonance criterion, then hypothesis testing, or memory search, is triggered. Memory search leads to selection of a better category at level F_2 with which to represent the input features at level F_1. During search, the orienting subsystem interacts with the attentional subsystem, as in figures C2.2.4(c) and (d); to rapidly reset mismatched categories and to select other F_2 representations with which to learn about novel events, without risking unselective forgetting of previous knowledge. Search may select a familiar category if its prototype is similar enough to the input to satisfy the vigilance criterion. The prototype may then be refined by top-down attentional focusing. If the input is too different from any previously learned prototype, then an uncommitted population of F_2 cells is selected and learning of a new category is initiated.

Because vigilance can vary across learning trials, recognition categories capable of encoding widely differing degrees of generalization or abstraction can be learned by a single ART system. Low vigilance leads to broad generalization and abstract prototypes since then $\rho|I| - |X^*| \leq 0$ for all but the poorest matches. High vigilance leads to narrow generalization and to prototypes that represent fewer input exemplars, even a single exemplar. Thus a single ART system may be used, say, to recognize abstract categories of faces and dogs, as well as individual faces and dogs. A single system can learn both, during supervised learning, by increasing vigilance just enough to activate A if a previous categorization leads to a predictive error (Carpenter and Grossberg 1987a, 1992, Carpenter *et al* 1991a, 1992). ART systems hereby provide a new answer to whether the brain learns prototypes or exemplars. Various authors have realized that neither one nor the other alternative is satisfactory, and that a hybrid system is needed (Smith 1990). ART systems can perform this hybrid function in a manner that is sensitive to environmental demands.

C2.2.1.8 Memory consolidation and direct access to the globally best category

As inputs are practiced over learning trials, the search process eventually converges upon stable categories. The process whereby search is automatically disengaged may be interpreted as a form of memory consolidation. Inputs familiar to the network access their correct category directly, without the need for search. The category selected is the one whose prototype provides the globally best match to the input pattern. If both familiar and unfamiliar events are experienced, familiar inputs can directly activate their learned categories, while unfamiliar inputs continue to trigger adaptive memory searches for better categories, until the network's memory capacity, which can be chosen arbitrarily large, is fully utilized (Carpenter and Grossberg 1987a).

C2.2.1.9 Some biological applications

These ART properties have been used to explain and predict various cognitive and brain data that have, as yet, received no other theoretical explanation (Carpenter and Grossberg 1991, Grossberg 1987a, b). For example, a formal lesion of the orienting subsystem creates a memory disturbance that mimics properties of medial temporal amnesia (Carpenter and Grossberg 1987c, 1993, Grossberg and Merrill 1992). These and related data correspondences to orienting properties (Grossberg and Merrill 1992, 1996) have led to a neurobiological interpretation of the orienting subsystem in terms of the hippocampal formation of the brain. In applications to visual object recognition, the interactions within the F_1 and F_2 levels of the attentional subsystem are interpreted in terms of data concerning the prestriate visual cortex and the inferotemporal cortex (Desimone 1992), with the attentional gain control pathway interpreted in terms of the pulvinar region of the brain.

ART processing properties have also helped to explain behavioral and neural data from several other sensory, cognitive, and motor systems. The following sections briefly review some recent contributions. In all these models, top-down priming effects are due to a top-down nonspecific inhibitory gain control signal that is released in parallel with specific excitatory signals.

C2.2.1.10 Neural dynamics of multisource audition

How does the brain's auditory system construct coherent representations of acoustic objects from the jumble of noise and harmonics that relentlessly bombards our ears throughout life? Bregman (1990) has distinguished at least two levels of auditory organization, called primitive streaming and schema-based segregation, at which such representations are formed in order to accomplish auditory scene analysis. The work summarized here models data about both levels of organization, and suggests that ART mechanisms of matching and resonance play a key role in achieving the selectivity and coherence that are characteristic of our auditory experience. In environments with multiple sound sources, the auditory system is capable of teasing apart the impinging jumbled signal into different mental objects, or streams, as in its ability to solve the cocktail party problem.

Govindarajan *et al* (1994) have developed a neural network model of this primitive streaming process, called the ARTSTREAM model. This model groups different frequency components based on pitch and spatial location cues, and selectively allocates the components to different streams. The grouping is accomplished through a resonance that develops between a given object's pitch, its harmonic spectral components, and (to a lesser extent) its spatial location. Those spectral components that are not reinforced by being matched with the top-down prototype read-out by the selected object's pitch representation are suppressed, thereby allowing another stream to capture these components, as in the 'old-plus-new heuristic' of Bregman (1990). These resonance and matching mechanisms are specialized versions of ART mechanisms.

C2.2.1.11 Neural dynamics of variable-rate speech categorization

F1.7 What is the neural representation of a *speech code* as it evolves in real time? Grossberg *et al* (1995) have developed a neural model of this schema-based segregation process, called the ARTPHONE model. It is used to quantitatively simulate data concerning segregation and integration of phonetic percepts, as exemplified by the problem of distinguishing 'topic' from 'top pick' in natural discourse. Psychoacoustic data (Repp 1980) concerning categorization of stop consonant pairs indicate that the closure time between syllable final vowel–consonant (VC) and syllable initial consonant–vowel (CV) transitions determines whether consonants are segregated (perceived as distinct) or integrated (fused into a single percept). Hearing two stops in a VC–CV pair that are phonetically the same, as in 'top pick', requires about 150 ms more closure time than hearing two stops in a VC_1–C_2V pair that are phonetically different, as in 'odd ball'. The ARTPHONE model traces these properties to dynamical interactions between a working memory for short-term storage of phonetic items and a list categorization network that groups, or chunks, sequences of the phonetic items in working memory. The speech code in the model is a resonant wave that emerges after bottom-up signals from the working memory select list chunks which, in turn, read out top-down expectations that amplify consistent working memory items.

C2.2.1.12 Neural dynamics of boundary and surface representation

In the area of visual perception, Gove *et al* (1995) have developed a neural network model, called a FACADE theory model, to explain how visual thalamocortical interactions give rise to boundary percepts such as illusory contours and surface percepts such as filled-in brightnesses. Top-down feedback interactions are needed in addition to bottom-up feedforward interactions to simulate these data. One feedback loop is modeled between lateral geniculate nucleus (LGN) and cortical area V1, and another within cortical areas V1 and V2. The first feedback loop realizes a resonant matching process, as in ART, which enhances LGN cell activities that are consistent with those of of active cortical cells, and suppresses LGN activities that are not.

C2.2.1.13 Neural dynamics for multimodal control of saccadic eye movements

Saccades are eye movements by which an animal can scan a rapidly changing enviroment. While the saccadic system plans where to move the eyes, it also retains reflexive responsiveness to fluctuating light sources. These two types of saccade ultimately result in control of the same set of eye muscles. Visually reactive cells encode gaze error in a *retinotopically* activated motor map. Planned targets are coded in *head-centered* coordinates. When two conflicting commands attempt to share control of the saccadic eye movement system, the system must resolve the conflict and coordinate command of one set of eye muscles.

The superior colliculus is a brainstem region that plays a prominent role in both planned and reactive saccades. This region coordinates information to adjust movements of the head and eyes to a stimulus. In order to combine these visual, somatic, and auditory saccade targets in the superior colliculus, the targets in head-centered coordinates are mapped to a gaze motor error in retinotopic coordinates.

How does the saccadic movement system select a target when visual and planned movement commands differ? How do retinal, head-centered, and motor error coordinates learn to interact during the selection process? ART matching and resonance are proposed to control the stability of this learning and the attentive selection of saccadic target locations. Targets in retinotopic and head-centered coordinates are rendered dimensionally consistent so that they can compete for attention to generate a movement command in motor error coordinates.

These results illustrate the scope of ART processing in the brain. The remaining discussion focuses upon ART applications, notably applications wherein *fuzzy logic* computations are incorporated into ART D1 algorithms.

C2.2.1.14 Fuzzy adaptive resonance theory

Fuzzy ART is a generalization of ART 1 that incorporates operations from fuzzy logic (Carpenter *et al* 1991c). While ART 1 can learn to classify only binary input patterns, fuzzy ART can learn to classify both analog and binary input patterns. Moreover, fuzzy ART reduces to ART 1 in response to binary input patterns. As shown in figure C2.2.5, the generalization to learning both analog and binary input patterns is achieved simply by replacing appearances of the binary intersection operator ($\cap$) in ART 1 by the analog MIN operator ($\wedge$) of fuzzy set theory. The MIN operator reduces to the intersection operator in the binary case. In particular, as parameter α approaches 0, the function T_j which controls category choice through the bottom-up filter (figure C2.2.4(*a*)) then measures the degree to which the adaptive weight vector w_j is a fuzzy subset (Kosko 1986) of the input vector I. The network first chooses the category j that maximizes T_j.

In fuzzy ART, input vectors are L^1 (city block) normalized at a preprocessing stage (figure C2.2.6). This normalization procedure, called complement coding, leads to a symmetric theory in which the MIN

<table>
<tr><th>ART 1
(BINARY)</th><th>FUZZY ART
(ANALOG)</th></tr>
</table>

CATEGORY CHOICE

$$T_j = \frac{|I \cap w_j|}{\alpha + |w_j|} \qquad T_j = \frac{|I \wedge w_j|}{\alpha + |w_j|}$$

MATCH CRITERION

$$\frac{|I \cap w_J|}{|I|} \geq \rho \qquad \frac{|I \wedge w_J|}{|I|} \geq \rho$$

FAST LEARNING

$$w_J^{(new)} = I \cap w_J^{(old)} \qquad w_J^{(new)} = I \wedge w_J^{(old)}$$

$\cap$ = logical AND $\wedge$ = fuzzy AND
Intersection   minimum

Figure C2.2.5. Comparison of ART 1 and fuzzy ART (Reprinted with permission from Carpenter Grossberg and Rosen 1991b.)

operator ($\wedge$) and the MAX operator ($\vee$) of fuzzy set theory (Zadeh 1965) play complementary roles. Geometrically, the categories formed by fuzzy ART are then hyper-rectangles. Figure C2.2.7 illustrates how MIN and MAX define these rectangles in the two-dimensional case, with the MIN and MAX values defining the acceptable range of feature variation in each dimension. Complement coding uses on-cell (with activity a in figure C2.2.6) and off-cell (with activity a^c in figure C2.2.6) opponent processes to represent the input pattern. This representation preserves individual feature amplitudes while normalizing the total on-cell/off-cell vector. The on-cell portion of a prototype encodes features that are critically present in category exemplars, while the off-cell portion encodes features that are critically absent (figure C2.2.6). The on-cell components of a category weight vector define the lower left-hand corner of the category rectangle in figure C2.2.7, and the complements of the off-cell components define the upper right-hand corner. Each category is then defined by an interval of expected values for each input feature. Thus for the category 'man', fuzzy ART would encode the feature of 'hair on head' by a wide interval ($[A, 1]$) and the feature 'hat on head' by a wide interval ($[0, B]$). For the category 'dog', two narrow intervals, $[C, 1]$ for hair and $[0, D]$ for hat correspond to narrower ranges of expectations for these two features.

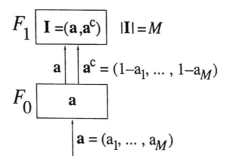

Figure C2.2.6. Complement coding uses on-cell and off-cell pairs to normalize input vectors. (Reprinted with permission from Carpenter Grossberg and Rosen 1991b.)

Learning in fuzzy ART converges because all adaptive weights can only decrease in time. Decreasing weights correspond to increasing sizes of category 'boxes'. A box can grow to a maximum size of $M(1 - \rho)$, so smaller vigilance values permit larger category boxes. Learning stops when the input space is covered by boxes. Input complement coding thus works with the property of increasing box size to prevent a proliferation of categories. With fast learning, constant vigilance, and a finite input set of arbitrary size and composition, learning stabilizes after just one presentation of each input pattern. A fast-commit slow-recode option combines fast learning with a forgetting rule that buffers system memory against noise. Using this option, rare events can be rapidly learned, yet previously learned memories are not rapidly erased in response to statistically unreliable input fluctuations. The equations that define the fuzzy ART and fuzzy ARTMAP algorithms are given in Carpenter *et al* (1991c, 1992).

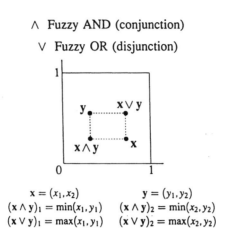

$$\wedge \ \text{Fuzzy AND (conjunction)}$$
$$\vee \ \text{Fuzzy OR (disjunction)}$$

$$x = (x_1, x_2) \qquad y = (y_1, y_2)$$
$$(x \wedge y)_1 = \min(x_1, y_1) \qquad (x \wedge y)_2 = \min(x_2, y_2)$$
$$(x \vee y)_1 = \max(x_1, y_1) \qquad (x \vee y)_2 = \max(x_2, y_2)$$

Figure C2.2.7. Fuzzy AND (MAX) and OR (MIN) operations generate category hyper-rectangles. (Reprinted with permission from Carpenter *et al* 1991b.)

Since fuzzy ARTMAP match tracking allows vigilance ρ_a to vary, a predictive error can create new categories that could not be learned if vigilance were constant. Supervised learning permits the creation of complex categorical structures without a loss of stability.

C2.2.1.15 Fuzzy ARTMAP

Each fuzzy ARTMAP system includes a pair of fuzzy ART modules ART_a and ART_b (figure C2.2.8). During supervised learning, ART_a receives a stream $\{a^{(p)}\}$ of input patterns and ART_b receives a stream $\{b^{(p)}\}$ of input patterns, where $b^{(p)}$ is the correct prediction given $a^{(p)}$. These modules are linked by an associative learning network and an internal controller that ensures autonomous system operation in real time. The controller is designed to create the minimal number of ART_a recognition categories, or 'hidden units,' needed to meet accuracy criteria. As noted above, this is accomplished by realizing a minimax learning rule that conjointly minimizes predictive error and maximizes category generalization. This scheme automatically links predictive success to category size on a trial-by-trial basis using only local operations. It works by increasing the vigilance parameter ρ_a of ART_a by the minimal amount needed to correct a predictive error at ART_b (figure C2.2.9).

Parameter ρ_a calibrates the minimum confidence that ART_a must have in a recognition category, or hypothesis, that is activated by an input $a^{(p)}$ in order for ART_a to accept that category, rather than search for a better one through an automatically controlled process of hypothesis testing. As in ART 1, lower values of ρ_a enable larger categories to form. These lower ρ_a values lead to broader generalization and higher code compression. A predictive failure at ART_b increases the minimal confidence ρ_a by the least amount needed to trigger hypothesis testing at ART_a, using a mechanism called *match tracking* (Carpenter *et al* 1991a). Match tracking sacrifices the minimum amount of generalization necessary to correct the predictive error. Speaking intuitively, match tracking embodies the idea that the criterion confidence level that permitted selection of the active hypothesis needs to be raised to satisfy the demands of the current environment. Match tracking increases the criterion confidence just enough to trigger hypothesis testing. Hypothesis testing leads to the selection of a new ART_a category, which focuses attention on a new cluster of $a^{(p)}$ input features that is better able to predict $b^{(p)}$. The combination of match tracking and fast learning allows a single ARTMAP system to learn a prediction for a rare event different from that for a cloud of similar frequent events in which it is embedded.

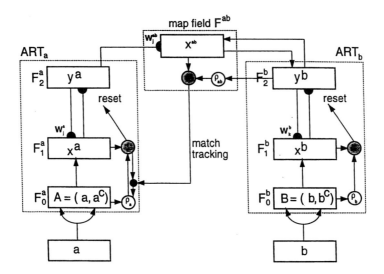

Figure C2.2.8. Fuzzy ARTMAP architecture. The ART_a complement coding preprocessor transforms the M_a-vector a into the $2M_a$-vector $A = (a, a^c)$ at the ART_a field F_0^a. A is the input vector to the ART_a field F_1^a. Similarly the input to F_1^b is the $2M_b$-vector (b, b^c). When a prediction by ART_a is disconfirmed at ART_b, inhibition of map field activation induces the match tracking process. Match tracking raises the ART_a vigilance ρ_a to just above the F_1^a to F_0^a match ratio $|x^a|/|A|$. This triggers an ART_a search which leads to activation of either an ART_a category that correctly predicts b or to a previously uncommitted ART_a category node. (Reprinted with permission from Carpenter *et al* 1992.)

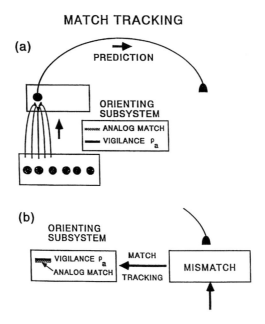

Figure C2.2.9. Match tracking: (a) A prediction is made by ART_a when the baseline vigilance ρ_a is less than the analog match value. (b) A predictive error at ART_b increases the baseline vigilance value of ART_a until it just exceeds the analog match value and thereby triggers hypothesis testing that searches for a more predictive bundle of features to which to attend. (Reprinted with permission from Carpenter and Grossberg 1994.)

C2.2.1.16 Some technological applications

ART models, ranging from ART 1 to fuzzy ARTMAP, outperform many expert systems, genetic algorithms, and other neural networks in benchmark studies (tables C2.2.1–3) and have been used to help solve outstanding technological problems (Bachelder *et al* 1993, Baraldi and Parmiggiani 1995, Caudell *et al* 1994, Christodoulou *et al* 1995, Dubrawski and Crowley 1994, Gan and Lua 1992, Gopal *et al* 1994, Ham and Han 1993, Harvey 1993, Johnson 1993, Kasperkiewicz *et al* 1994, Keyvan *et al* 1993, Kumara *et al* 1994, Mehta *et al* 1993, Moya *et al* 1993, Murshed *et al* 1995, Suzuki *et al* 1993, Tarng *et al* 1994, Wienke 1994, Wienke and Kateman 1994, Wienke *et al* 1994). Recent unsupervised ART models have been used to explain behavioral and brain data in auditory source analysis, variable-rate speech perception, visual perception, and visual object recognition, as described above. A self-organizing neural architecture, called VIEWNET (Bradski and Grossberg 1995), that can learn to recognize noisy three-dimensional objects from sequences of their two-dimensional views is next reviewed to show how fuzzy ARTMAP can be embedded into larger systems. Another three-dimensional object recognition application illustrates how the ART-EMAP architecture (Carpenter and Ross 1993, 1995) uses distributed network activity to improve noise tolerance while retaining the speed advantage of fast learning. Both architectures illustrate how temporal evidence accumulation can augment ARTMAP capabilities.

Table C2.2.1. ARTMAP benchmark studies. (Reproduced with permission from Carpenter and Grossberg 1993.)

---

1. Medical database
   Mortality following coronary bypass grafting (CABG) surgery
      Fuzzy ARTMAP significantly outperforms:
         Logistic regression
         Additive model
         Bayesian assignment
         Cluster analysis
         Classification and regression trees
         Expert-panel-derived sickness scores
         Principal component analysis

2. Mushroom database
Decision trees (90–95% correct)
ARTMAP (100% correct)
   Training set an order of magnitude smaller

3. Letter recognition database
   Genetic algoritm (82% correct)
   Fuzzy ARTMAP (96% correct)

4. Circle-in-the-square task
   Backpropagation (90% correct)
   Fuzzy ARTMAP (99.5% correct)

5. Two-spiral task
   Backpropagation (10 000–20 000 training epochs)
   Fuzzy ARTMAP (1–5 training epochs)

C2.2.1.17 Two Applications of fuzzy ARTMAP

VIEWNET: neural architectures for learning to recognize three-dimensional objects from sequences of two-dimensional views. VIEWNET (View Information Encoded With NETworks) accumulates evidence across sequences of possibly noisy or incomplete two-dimensional views of a three-dimensional object in order to generate more accurate object identifications than would otherwise be possible (Bradski and Grossberg 1995). The simplest VIEWNET architecture, VIEWNET 1, incorporates a preprocessor that generates a compressed but two-dimensional invariant representation of an image, a supervised incremental learning system that classifies the preprocessed representations into two-dimensional view categories whose outputs are combined into three-dimensional invariant object categories, and a working memory that makes a three-dimensional object prediction by accumulating evidence from three-dimensional object category nodes as multiple two-dimensional views are experienced. Evidence accumulation has also been successfully used in neural network machine vision applications that are based on aspect networks (Baloch and Waxman

Table C2.2.2. Fuzzy ARTMAP applied to the Landsat image database (Feng *et al* 1993). With the exception of K-N-N, fuzzy ARTMAP test set performance exceeded that of other neural network and machine learning algorithms. Compared to K-N-N, fuzzy ARTMAP showed a 6:1 code compression ratio.

Algorithm	Accuracy (%)
K-N-N	91
Fuzzy ARTMAP	89
RBF	88
Alloc80	87
INDCART	86
CART	86
Backprop	86
C4.5	85
NewID	85
CN2	85
Quadra	85
SMART	84
LogReg	83
Discrim	83
CASTLE	81

	Test (%)	Compression
K-N-N	91	1:1
Fuzzy ARTMAP	89	6:1

Table C2.2.3. On the Pima Indian Diabetes (PID) database fuzzy ARTMAP test set performance was similar to that of the ADAP algorithm (Smith *et al* 1988) but with far fewer rules and faster training. An ARTMAP pruning algorithm (Carpenter and Tan 1995) further reduces the number of rules by an order of magnitude and also boosts test set accuracy to 79%. An instance counting algorithm ARTMAP-IC (Carpenter and Markuzon 1996) boosts accuracy to 81%.

Supervised learning	
Training	576
Test	192

ADAP
(Smith et al *1988)*

- 100 000 rules
- 76% correct on test set
- Slow learning

Fuzzy ARTMAP
(Carpenter et al *1992)*

- 50–80 rules
- 76% correct on test set
- Fast learning (6–15 epochs)

1991, Seibert and Waxman 1990). Recognition was studied with noisy and clean images using slow and fast learning. Slow learning at the fuzzy ARTMAP map field was designed to learn the conditional probability of the three-dimensional object given the selected two-dimensional view category. VIEWNET 1 was demonstrated on an MIT Lincoln Laboratory database of 4000 128 × 128 two-dimensional views of aircraft with and without additive noise. A recognition rate of up to 90% was achieved with one two-dimensional view and a rate of up to 98.5% correct with three two-dimensional views.

ART-EMAP: object recognition by spatial and temporal evidence accumulation. ART-EMAP also incorporates fuzzy ARTMAP into a larger architecture for three-dimensional recognition. It uses spatial and temporal evidence accumulation to recognize target objects and pattern classes in noisy or ambiguous input environments (Carpenter and Ross 1993, 1995). During performance, ART-EMAP integrates spatial evidence distributed across recognition categories to predict a pattern class. During training, ART-EMAP is equivalent to fuzzy ARTMAP and so inherits the advantages of fast, on-line, incremental learning, such as speed, stability, and the ability to encode rare cases. Distributed activation during performance also endows the network with the advantages of slow learning, including noise tolerance and error correction. When a decision criterion determines the pattern class choice to be ambiguous, additional input from the same unknown class may be sought. Evidence from multiple inputs accumulates until the decision criterion is satisfied and the system makes a high-confidence prediction. Accumulated evidence can also fine tune performance during unsupervised rehearsal learning. Thus, in four incremental stages, ART-EMAP improves predictive accuracy of fuzzy ARTMAP and extends its domain to include spatiotemporal recognition and prediction.

C2.2.1.18 Concluding remarks

The above examples illustrate an emerging picture of how the adaptive brain works wherein issues of stability and plasticity are joined with properties of attention, intention, hypothesis testing, and consciousness. The mediating events are adaptive resonances that achieve a dynamic balance between the complementary demands of stability and plasticity, and of expectation and novelty, in response to rapidly changing environments. Similar issues arise in technological problems wherein intelligent agents or controllers are desired that can support a significant level of autonomous performance. This is why ART systems are finding their way into solutions to a rapidly expanding set of applied problems.

Acknowledgement

The authors wish to thank Robin Locke for her valuable assistance in the preparation of this manuscript.

C2.2.2 Resonance correlation network

Peggy Israel Doerschuck

Abstract

The resonance correlation network (RCN) is a self-organizing feedback network which classifies either binary- or continuous-valued patterns. Its architecture is based on adaptive resonance theory (ART), but it eliminates the need for a sequential search through previously learned categories which is a drawback of ART. The RCN uses a dot product activation function and uses normalized correlation as its similarity measure. This guarantees that the most highly activated recognition layer node is the one which is most similar to the pattern currently under consideration, thus obviating the need for search.

C2.2.2.1 Introduction

The resonance correlation network (RCN) was developed by Ryan (1988). The RCN was inspired by *adaptive resonance theory* (ART) (Carpenter and Grossberg 1987a). Its main advantage is that it eliminates   C2.2.1
the need for sequential search which is characteristic of ART.

C2.2.2.2 Purpose

The RCN is a self-organizing network which is used to classify binary- or continuous-valued patterns. The network automatically forms classes of data by clustering similar patterns together. The user supplies a vigilance parameter which specifies how similar patterns must be in order to be clustered together. The RCN is inspired by ART, but uses a different similarity measure which eliminates the need for sequential search for a matching prototype. This is particularly advantageous where patterns are to be classified in real time.

C2.2.2.3 Topology

The architecture of the RCN is illustrated in figure C2.2.10. It consists of an input layer I, a comparison layer C, a recognition layer R, and a reset assembly Re. A vector of bottom-up weights $B_j = (b_{1j}, b_{2j} \ldots)$ goes from each node in C to each node j in R, and a vector of top-down weights $T_j = (t_{j1}, t_{j2}, \ldots)$ goes from each node j in R to each node in C. The input layer merely stores the input pattern. Each node in C corresponds to a component of the input pattern, and each node in R corresponds to a learned pattern prototype. The prototype associated with node R_j is stored in the vector of top-down weights T_j.

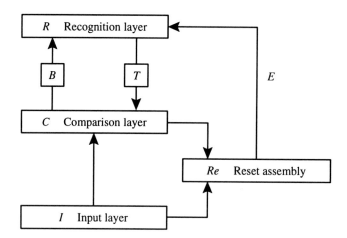

Figure C2.2.10. High-level architecture of the resonance correlation network (copyright 1988 IEEE).

Each node j in R receives signals from C which are modulated by B_j. The nodes in R compete with each other, and the one with the highest activation is called the winner. Layer C receives input from I and also receives weighted signals from the winning node R_j which are proportional to its top-down weights T_j, so that the pattern of activation in C is a combination of the input pattern and the winning prototype pattern. The reset assembly receives as inputs vectors I and C and produces a signal E which will inhibit the winning node R_j if I and C are not similar enough.

The architecture described thus far is the same as in ART. However, the RCN includes an additional set of nodes U associated with the recognition layer. These nodes are used to distinguish those nodes in R which correspond to previously learned prototypes from unused R nodes. In the case of a mismatch, the reset assembly uses this information to inhibit not only the winning node in R but also any other R node which corresponds to a previously learned prototype. This prevents the firing of each of the next highest responders in turn and thereby eliminates the sequential search for a matching prototype which is characteristic in ART. In addition, U controls whether the input pattern I or the winning node's prototype T_j is registered on C, as is described further in section C2.2.2.5.

C2.2.2.4 Learning

Learning in the RCN is unsupervised. The user supplies a vigilance parameter v which specifies how similar patterns must be in order to be clustered together. Learning is accomplished by changing the values of the bottom-up and top-down weights and the U nodes. These values are governed by first-order nonlinear ordinary differential equations. Like ART, the RCN operates in either fast learning mode or slow learning mode. In fast learning, it is assumed that these values are allowed to reach their asymptotic limits for each presentation of an input pattern. More than one pass through the training data may be needed before the classes stabilize. The RCN works on either binary- or continuous-valued patterns.

C2.2.2.5 Learning rule

Bottom-up weights are initialized such that $b_{ij} < d^{-1/2}$, where d is the dimensionality of the input pattern. This ensures that the activation of a previously encoded R node will always be greater than that of an unused node (Ryan 1988). Top-down weights are initialized to 0. The node activations and weights of the network are governed by the following differential equations (Ryan 1988).

Recognition layer nodes:

$$\tau \dot{R}_j = -R_j + (1 - R_j)[C \cdot B_j + f(R_j)] - (A + R_j)\left[\sum_{k \neq j} f(R_k) + E_j\right]. \tag{C2.2.1}$$

Comparison layer nodes:

$$\tau \dot{C}_i = -C_i + I_i\left[1 - \alpha \sum_j u_j f(R_j)\right] + \alpha \sum_j t_{ji} f(R_j). \tag{C2.2.2}$$

Reset mechanism:

$$r = v\|I\|\|C\| - I \cdot C \tag{C2.2.3}$$

$$\tau \dot{E}_j = f(r)[f(R_j) + u_j] - E_j. \tag{C2.2.4}$$

Bottom-up weights:

$$\dot{b}_{ij} = f(R_j)(C_i - \|C\|b_{ij}). \tag{C2.2.5}$$

Top-down weights:

$$\dot{t}_{ji} = f(R_j)(C_i - t_{ji}) \tag{C2.2.6}$$

$$U : \dot{u}_j = f(R_j)(1 - u_j). \tag{C2.2.7}$$

The primary forcing function for R_j is the dot product:

$$F_j = C \cdot B_j. \tag{C2.2.8}$$

Here, $f(R_j)$ is 1 if R_j is the winning node and 0 otherwise:

$$f(R_j) = \begin{cases} 1 & \text{if } C \cdot B_j > C \cdot B_k \; \forall k \neq j \\ 0 & \text{otherwise}. \end{cases} \tag{C2.2.9}$$

The parameter τ is a time constant $\ll 1$, and A is a constant $> \|I\|_{max}$ which keeps losing nodes from becoming active. $\|X\| = (\sum x_i^2)^{1/2}$ denotes the norm of X.

The vector of nodes U is initialized to $\mathbf{0}$, and u_j converges to 1 if R_j is used to encode a pattern, as shown in (C2.2.7). The presence of u_j in (C2.2.2) will cause the pattern T_j to be registered on C if the winning node R_j has been used and will cause I to be registered on C if R_j is unused.

The variable r in (C2.2.3) is a measure of the correlation between I and C. The value $f(r)$ in (C2.2.4) is 1 if the similarity is too low. This makes E_j in (C2.2.4) respond, so that it inhibits the winning node in (C2.2.1). The presence of u_j in (C2.2.4) causes previously used recognition nodes to be inhibited in (C2.2.1) also if there is a reset. The parameter α, which is a value between 0 and 1, controls the stability of the prototypes, as is described in more detail later in this section.

In fast learning mode, the weights are assumed to reach their asymptotic values upon each presentation of an input pattern. Under certain conditions, the fast learning algorithm is as follows (assuming K prototypes have already been learned).

Repeat until convergence:

(i)   present the next input pattern P at layer I;
(ii) copy I into C;
(iii) calculate the activation level of each node in R and find the winning node R_i for which $F_i > F_k$ for all $k \neq j$;
(iv) send R_i's activation down via T_i, to C, so that C becomes some linear combination of I and T_i;
(v)   if $C \cdot T/\|C\|\|T_i\| < v$ then the pattern is novel so select an unused node $j = K + 1$; otherwise $j = i$;
(vi) update the weights as follows: $T_j = C$ and $B_j = C/\|C\|$.

In step (iv), the pattern of activation in the comparison layer becomes

$$C = I(1 - \alpha u_i) + \alpha T_i \qquad \text{where } R_i \text{ is the winning node}. \tag{C2.2.10}$$

With $\alpha = 1$, C takes on the value of T_i, so that the ith prototype remains stable at the location of the first pattern in cluster i. With $\alpha = 0$, $C = I = P$ at step (iv), so the ith prototype is moved to the position of the current pattern. The parameter α is thus used to control the stability of the prototypes.

Figure C2.2.11 shows how the receptive domains of a prototype can move where $\alpha \neq 1$. In this figure, it is assumed that the patterns are of three dimensions. Normalization projects the patterns onto the surface of a unit sphere. The receptive domains are shown plotted in rectangular coordinates for ease of illustration. Each circle represents the intersection of a cone of fixed angle with the surface of the sphere, where the angle of the cone is determined by v.

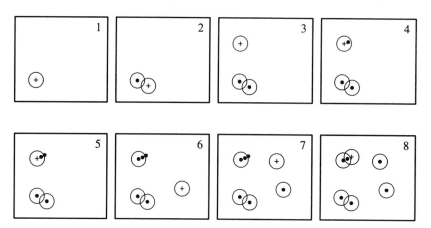

Figure C2.2.11. Fast-learn receptive domains. '+' represents the current input pattern and '•' represents previously coded patterns. Addition and adjustment of receptive domain as training patterns are presented to the network. In step 5, the receptive domain initiated at step 3 has shifted and no longer includes the original prototype. In step 8, the pattern previously seen at step 3 is presented again and generates a new receptive domain (copyright 1988 IEEE).

It has been shown (Ryan 1988) that where a dot product forcing function is used, the use of normalized correlation as a similarity measure provides direct access upon presentation of a pattern P to a node which encodes P. It is easy to see that there is no need for a sequential search through all prototypes since that R node with the greatest activation also has the best normalized correlation similarity measure:

If R_j is the winner, then

$$C \cdot B_j > C \cdot B_k \qquad \text{for all } k \neq j. \tag{C2.2.11}$$

Substituting $B_j = T_j/\|T_j\|$ (from step (vi)) into (C2.2.11) yields:

$$C \cdot T_j/\|T_j\| > C \cdot T_k/\|T_k\| \qquad \text{for all } k \neq j. \tag{C2.2.12}$$

Dividing each side by $\|C\|$ yields:

$$C \cdot T_j/\|C\|\|T_j\| > C \cdot T_k/\|C\|\|T_k\| \qquad \text{for all } k \neq j. \tag{C2.2.13}$$

If $C \cdot T_j/\|C\|\|T_j\| < v$ then (C2.2.13) yields:

$$C \cdot T_k/\|C\|\|T_k\| < v \qquad \text{for all } k \neq j. \tag{C2.2.14}$$

From (C2.2.14), if the winning node's prototype does not pass the vigilance test, no other node's prototype will. Therefore, it is not necessary to compute the similarity of C with any prototype other than the winner.

C2.2.2.6 Related neural network models

The RCN has a control structure like that of ART1 but, since it works for both binary- and continuous-valued patterns, is functionally more similar to ART2 (Carpenter and Grossberg 1987b). The main difference between the RCN and ART2 is that (i) RCN uses normalized bottom-up weights, while ART2 normalizes the input pattern; (ii) RCN uses normalized correlation as its similarity measure; and (iii) RCN eliminates the need for sequential search for a matching prototype, at the cost of an extra set of nodes U which keep track of encoded R nodes.

Simplified ART1 (Israel *et al* 1992) is another variation of ART1 which eliminates search. It is simpler than RCN, and uses only one set of bidirectional weights, but works only on binary patterns.

Fuzzy ART (Carpenter *et al* 1991c) eliminates search in ART by using complement coding, which   C2.2.1.14 doubles the size of the input and weights. Fuzzy ART works only on (0, 1) continuous patterns.

C2.2.2.7 Advantages

The RCN is faster than ART because there is no need for sequential search. It is more general-purpose than simplified ART1 and fuzzy ART because it works on both binary- and continuous-valued patterns not restricted to (0, 1). It also has a simpler architecture and learning rule than fuzzy ART.

C2.2.2.8 Disadvantages

For classification of binary-valued patterns, the RCN is more complex than simplified ART1, which uses only one set of bidirectional weights and does not require normalization. Also, the RCN uses normalized prototypes. This approach may not be appropriate where scale is an important factor in classification.

C2.2.2.9 Typical applications

The RCN is used for pattern clustering and classification of binary- or continuous-valued patterns where the class of the sample patterns is not *a priori* known. It is functionally equivalent to ART2, but eliminates the need for sequential search for a matching category. The RCN can be used in the same types of applications as ART1 or ART2.

C2.2.2.10 Variations and improvements

In the RCN, cluster centers are placed at either (i) the location of the first seen pattern in the cluster (if $\alpha = 1$), (ii) the most recently seen pattern in the cluster (if $\alpha = 0$); or (iii) some linear combination of the two (for α between 0 and 1). Case (i) results in stationary cluster centers, but these 'centers' may actually be patterns which are outliers. Cases (ii) and (iii) can lead to cluster drift, as illustrated in figure C2.2.11. Both of these situations can cause additional categories to be formed if the 'centers' are positioned at outlying patterns. While using a value of α between 0 and 1 can control this to some extent, results depend on the value selected for α and the order of presentation of the patterns.

Fuzzy clustering techniques have recently been incorporated into a fast-learning RCN to improve cluster formation. The adaptive fuzzy leader clustering resonance correlation network (Cleary and Israel 1994) incorporates fuzzy leader clustering into the RCN structure to produce better-formed categories. Here each cluster center's position is determined by taking a weighted average of all the patterns in a category. Each pattern's contribution to the average is weighted by its degree of membership in the category, which is a function of its relative proximity to the center. Outlying patterns therefore have less influence on movement of the cluster center. This tends to produce fewer categories and reduces the order dependence of category formation, but at the price of additional storage required to keep track of all sample patterns.

C2.2.3 Boundary and feature contour systems

Stephen Grossberg

Abstract

When humans gaze upon a scene, their brains rapidly combine several different types of locally ambiguous visual information to generate a globally consistent and unambiguous representation of form-and-color-and-depth, or FACADE. This state of affairs raises the question: what new computational principles and mechanisms are needed to understand how multiple sources of visual information cooperate automatically to generate a percept of three-dimensional form? This section reviews some modeling work aimed at developing such a general-purpose vision architecture. This architecture clarifies how scenic data about boundaries, textures, shading, depth, multiple spatial scales and motion can be cooperatively synthesized in real-time into a coherent representation of three-dimensional form. It embodies a new vision theory that attempts to clarify the functional organization of the visual brain from the lateral geniculate nucleus (LGN) to the extrastriate cortical regions V4 and MT. Moreover, the same processes which are useful towards explaining how the visual cortex processes retinal signals are equally valuable for processing noisy multidimensional data from artificial sensors, such as synthetic aperture radar, laser radar, multispectral infrared, magnetic resonance and high-altitude photographs. These processes generate three-dimensional boundary and surface representations of a scene.

C2.2.3.1 Introduction

The difficulties inherent in computationally understanding biological vision can be appreciated by considering a few examples. Figure C2.2.12 depicts an Ehrenstein figure and an offset grating. When we view the offset grating, we see and recognize horizontal black lines on white paper, but we also *recognize* a vertical boundary between the lines that we do not *see*. The vertical boundary does not generate brightnesses or colors that differ significantly from the background. Such a boundary is often said to be an *amodal* percept. Thus there is a profound difference between seeing and recognizing, and we can sometimes recognize groupings that we cannot see. This state of affairs raises the central question: if we can recognize things that we cannot see, then why do we bother to see?

The other side of the coin is equally perplexing; namely, we can sometimes see things that are not in the image. Thus, in viewing the Ehrenstein figure shown in figure C2.2.12, we can see a bright disk within the perpendicular lines, although the luminance across all white parts of the figure is the same.

C2.2.3.2 The hierarchical resolution of uncertainty

In order to computationally understand such labile relationships between recognized boundary segmentations and seen surface brightnesses, a qualitatively different type of vision theory, called FACADE theory, is being developed to clarify how representations of form-and-color-and-depth are generated (Francis *et al* 1994, Gove *et al* 1995, Grossberg 1994, Grossberg and Mingolla 1985a, b, Grossberg and Todorović 1988, Pessoa *et al* 1995). FACADE theory holds that the paradoxes of figure C2.2.12 can be understood as manifestations of adaptive neural mechanisms. Specifically, our visual systems are designed to detect, complete and regularize relatively invariant object boundary structures amid noise caused by the eyes' own optics or occluding objects; to fill in relatively invariant surface colors under variable illumination conditions; and to learn to recognize familiar objects or events in the environment. These three principal functions are performed by the three main subsystems of the theory: the boundary contour system (BCS), the feature contour system (FCS) and the object recognition system (ORS), respectively, as indicated in the macrocircuit of figure C2.2.13.

A unifying theme constraining the design of the theory's mechanisms is that there exist fundamental computational limitations at each stage of the visual measurement process—that is, uncertainty principles are just as important in vision as in physics. For example, the computational demands on a system

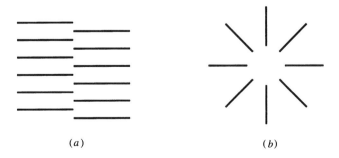

(a)   *(b)*

Figure C2.2.12. (*a*) an offset grating and (*b*) an Ehrenstein figure.

that computes invariant boundary structures are, in many respects, complementary to those on a system that computes invariant surface colors. For example, boundaries are completed in an *oriented* fashion *inwardly* between pairs or larger numbers of boundary inducers. The output of the boundary system is *insensitive* to contrast polarity because it pools outputs from cells that are sensitive to opposite contrast polarities in order to build object boundaries despite contrast reversals against a textured background. 'All boundaries are invisible' within the boundary system because they add up signals from opposite contrast polarities. Surfaces, on the other hand, fill-in in an *unoriented* fashion *outwardly* using a diffusive process whose output is *sensitive* to contrast polarity. Surface representations can therefore support visible percepts.

The complementary computations of the BCS and FCS clarify why they process the signals from each monocular preprocessing (MP) stage in parallel (figure C2.2.13). This is not to say that the BCS and FCS are independent modules. Figure C2.2.14 depicts in greater detail how levels of the BCS and FCS interact through multiple feedforward and feedback pathways to generate a three-dimensional surface representation at the final level of the FCS, which is called the binocular filling-in domain, or FIDO.

In addition to the complementary relationship between the FCS and the BCS, there also exist informational uncertainties at processing levels within each of these systems. As indicated below, the computations within the FCS which reduce uncertainty due to variable illumination conditions create new uncertainties about surface brightnesses and colors that are resolved at a higher FCS level by the process that fills-in surface properties such as brightness, color and depth. Likewise, the computations within the BCS which reduce uncertainty about boundary orientation create new uncertainties about boundary position that are resolved at a higher BCS level by the process of boundary completion.

C2.2.3.3 Model architecture

Preprocessing by a model lateral geniculate nucleus. The BCS consists of multiple copies, each with cells whose receptive fields are sensitive to a different range of image sizes. Each BCS copy consists of a filter followed by a grouping, or boundary completion, network. There are two parallel BCS architectures. One models the formation of static boundary segmentations by the LGN parvo → interblob → interstripe → V4 processing stream in figure C2.2.15. The other models boundary segmentations that are derived from

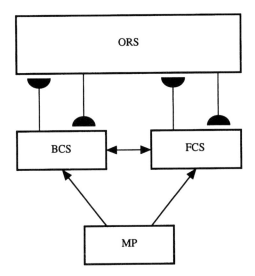

Figure C2.2.13. A macrocircuit of processing stages: Monocular preprocessed (MP) signals are sent independently to both the boundary contour system (BCS) and the feature contour system (FCS). The BCS preattentively generates coherent boundary structures from these MP signals. These structures send outputs to both the FCS and the object recognition system (ORS). The ORS, in turn, rapidly sends top-down learned template signals, or expectations, to the BCS. These template signals can modify the preattentively completed boundary structures using learned, attentive information. The BCS passes these modifications along to the FCS. The signals from the BCS organize the FCS into perceptual regions wherein filling-in of visible brightnesses and colors can occur. This filling-in process is activated by signals from the MP stage. The completed FCS representation, in turn, also sends signals to the BCS and the ORS.

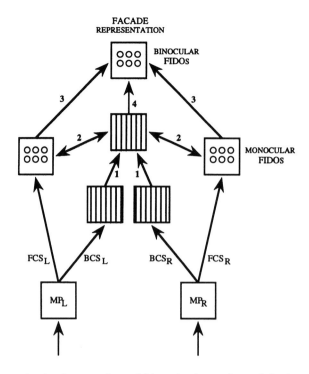

Figure C2.2.14. Macrocircuit of monocular and binocular interactions of the boundary contour system (BCS) and the feature contour system (FCS): left eye and right eye monocular preprocessing stages (MP_L and MP_R) send parallel pathways to the BCS (boxes with vertical lines, designating oriented responses) and the FCS (boxes with three pairs of circles, designating opponent colors). The monocular signals BCS_L and BCS_R activate simple cells which, in turn, activate bottom-up pathways, labeled 1, to generate a binocular boundary segmentation using the complex, hypercomplex, and bipole cell interactions of figure C2.2.16. The binocular segmentation generates output signals to the monocular filling-in domains, or FIDOs, of the FCS via pathways labeled 2. This interaction selects binocularly consistent FCS signals and suppresses the binocularly inconsistent FCS signals. Reciprocal FCS → BCS interactions enhance consistent boundaries and suppress boundaries corresponding to further surfaces. The surviving FCS signals activate the binocular FIDOs via pathways 3, where they interact with an augmented binocular BCS segmentation to fill-in a multiple-scale surface representation of form-and-color-and-depth, or FACADE. Processing stages MP_L and MP_R are compared with LGN data; the simple-complex cell interaction with V1 data; the hypercomplex-bipole interaction with V2 and (possibly) V4 data, notably about inter-stripes; the monocular FCS interaction with blob and thin stripe data; and the FACADE representation with V4 data. (Reprinted with permission from Grossberg (1994).)

moving forms by the LGN magno → 4B → thick stripe → MT processing stream in figure C2.2.15. The summary herein will consider only the static BCS, and only a single scale of its monocular processing properties, as summarized in figure C2.2.16. For extensions to binocular processing and three-dimensional figure–ground separation, see Grossberg (1994). For summaries of the motion BCS, see Francis and Grossberg (1996), Grossberg and Mingolla (1993) and Grossberg and Rudd (1992).

The model LGN ON and OFF cells receive input from retinal ON and OFF cells. ON cells are turned on by increments in image contrasts, whereas OFF cells are turned off. (See Schiller (1992) for a review.) Because these ON and OFF cells have antagonistic surrounds and obey shunting, or membrane, equations, they help to discount the illuminant, normalize image activities, and extract ratio contrasts from an image (Grossberg 1983). These image preprocessing properties are needed to simulate even the most basic brightness percepts (Grossberg and Todorović 1988).

The LGN model also receives feedback from model cortical cells, and this feedback can cause the resultant LGN activity to differ under certain circumstances from that caused solely by its retinal input. Grossberg (1980) suggested that the feedback pathway realizes a top-down pattern-matching process that helps to select activities of monocular LGN cells that support the activities of binocular cortical cells, and to suppress the activities of LGN cells that do not, via positive corticogeniculate feedback linked to internal LGN opponent processes. Topographic correspondence is necessary to carry out such a

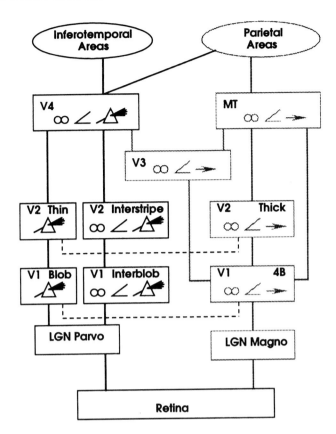

Figure C2.2.15. Schematic diagram of anatomical connections and neuronal selectivities of early visual areas in the macaque monkey. LGN = lateral geniculate nucleus (parvocellular and magnocellular divisions). Divisions of V1 and V2: blob = cytochrome oxidase blob regions; interblob = cytochrome oxidase-poor regions surrounding the blobs; 4B = lamina 4B; thin = thin (narrow) cytochrome oxidase strips; interstripe = cytochrome oxidase-poor regions between the thin and thick stripes; thick = thick (wide) cytochrome oxidase strips; V3 = visual area 3; V4 = visual area(s) 4; MT = middle temporal area. Areas V2, V3, V4, MT have connections to other areas not explicitly represented here. Area V3 may also receive projections from V2 interstripes or thin stripes. Heavy lines indicate robust primary connections, and thin lines indicate weaker, more variable connections. Broken lines represent observed connections that require additional verification. Icons: rainbow = tuned and/or opponent wavelength selectivity (incidence at least 40%); angle symbol = orientation selectivity (incidence at least 20%); spectacles = binocular disparity selectivity and/or strong binocular interactions (V2) (incidence at least 20%); pointing arrow = direction of motion selectivity (incidence at least 20%). (Adapted with permission from DeYoe and van Essen (1988).)

matching process. A similar modulatory role for top-down feedback is assumed to be active during monocular viewing. Experimental support for this ART prediction has been reported by Sillito *et al* (1994).

Corticogeniculate feedback was hypothesized to be part of a more general and ubiquitous model of top-down feedback in stabilizing adaptive synapses in thalamocortical and corticocortical circuits, while also regulating the gain of these circuits. In this more general *adaptive resonance theory*, or ART, modeling C2.2.1 framework, bottom-up processing in the absence of top-down processing can activate its target circuits, top-down processing represents a form of hypothesis testing that can subliminally prime these circuits, and a combination of bottom-up and top-down processing can select those bottom-up activations that are consistent with top-down feedback and suppress those that are not.

A boundary contour system model of cortical boundary segmentation. The LGN cell outputs activate the first stage of cortical BCS processing, the simple cells (see figures C2.2.16 and C2.2.17) whose oriented receptive fields respond to a prescribed contrast polarity, or direction-of-contrast. The model LGN cells input to pairs of like-oriented simple cells that are sensitive to opposite directions-of-contrast. The simple cell pairs, in turn, send their rectified output signals to like-oriented complex cells. By pooling rectified

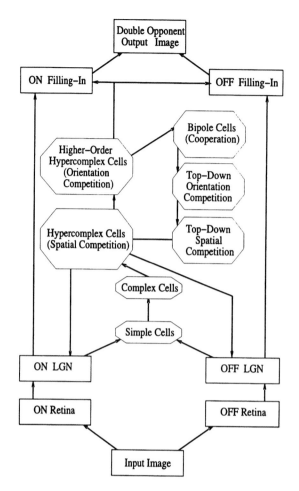

Figure C2.2.16. A monocular boundary contour system circuit: BCS stages are designated by octagonal boxes, FCS stages by rectangular boxes. Model BCS stages may be divided into a static oriented contrast-sensitive filter (SOC filter) and a static oriented cooperative-competitive grouping network (SOCC loop). The simple-complex-hypercomplex cells form the filter. The feedback network between hypercomplex and bipole cells forms the grouping network. (Reprinted from Gove *et al* (1995) with the permission of Cambridge University Press.)

outputs from oppositely polarized simple cells, complex cells realize a full-wave rectified filter that responds to both directions-of-contrast, as do all subsequent BCS cell types in the model.

Complex cells activate hypercomplex cells through an on-center off-surround network, or spatial competition, whose off-surround carries out an endstopping operation (see figure C2.2.17). In this way, complex cells excite hypercomplex cells of the same orientation and position, while inhibiting hypercomplex cells of the same orientation at nearby positions. One role of this spatial competition is to spatially sharpen the neural responses to oriented luminance edges. Another role is to initiate the process, called *end cutting*, whereby boundaries are formed that abut a line end at orientation perpendicular or oblique to the orientation of the line itself, as in figure C2.2.18(*c*).

The hypercomplex cells input to a competition across orientations at each position among higher-order hypercomplex cells. This competition acts to sharpen up orientational responses at each position. Outputs from the higher-order hypercomplex cells feed into bipole cells that initiate long-range boundary grouping and completion (figure C2.2.17). Bipole cells have two oriented receptive fields. Their cell bodies fire if both of their receptive fields are sufficiently activated by appropriately oriented hypercomplex cell inputs. Bipole cells act like a type of statistical and-gate that controls long-range cooperation among the outputs of active higher-order hypercomplex cells. For example, a horizontal bipole cell is excited by activation of horizontal hypercomplex cells that input to its horizontally oriented receptive fields. A horizontal bipole cell is also inhibited by activation of vertical hypercomplex cells. In this way, groupings among horizontal contrasts may be blocked by intervening contrasts of different orientation.

Output signals from bipole cells feed back to the hypercomplex cells after undergoing competitive processing. First, bipole cell outputs compete across orientation to determine which orientation is receiving the largest amount of cooperative support (see figure C2.2.16). Competition also takes place across nearby locations to select the best spatial location of the emerging boundary. These competitive interactions are needed to select and sharpen the best boundary grouping because the bipole cell receptive fields are themselves rather broad. Broad bipole receptive fields are needed because, in many situations, neither the image contrasts to be grouped nor the cortical cells that group them are precisely aligned across space. Broad receptive fields allow the grouping to get started and the competitive interactions sharpen and deform it. Hypercomplex cells that receive the most cooperative support from bipole grouping after cooperative-competitive feedback takes hold further excite the corresponding bipole cells.

This cycle of bottom-up and top-down interaction between hypercomplex cells and bipole cells rapidly converges to a final boundary segmentation (see figure C2.2.18(c)). Feedback among bipole cells and hypercomplex cells hereby drives a resonant cooperative-competitive decision process that completes the statistically most favored boundaries, suppresses less favored boundaries, and coherently binds together appropriate feature combinations in the image.

C2.2.3.4 Filling-in of surface representations within the FCS

Each BCS boundary segmentation generates topographic output signals to the ON and OFF filling-in domains, or FIDOs (see figure C2.2.16). These FIDOs also receive inputs from the ON and OFF LGN cells, respectively. The LGN inputs activate their target cells, which allow activation to diffuse rapidly across gap junctions to neighboring FIDO cells. This diffusive filling-in process is restricted to the compartments derived from the BCS boundaries, which create barriers to filling-in by decreasing the permeability of their target gap junctions. The filled-in OFF activities are subtracted from the ON activities at double-opponent cells, whose activities represent the surface brightness of each percept (see figure C2.2.16). This double-opponent representation is illustrated in figure C2.2.18(d).

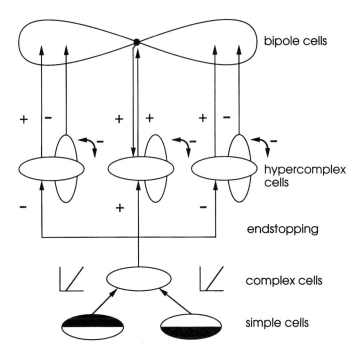

Figure C2.2.17. A simplified monocular model of the SOC filter interactions that convert simple cells into complex cells and then into two successive levels of hypercomplex cells. The interactions (simple cell) → (complex cell) and (complex cell) → (hypercomplex cell) describe two successive spatial filters. Simple cells form one filter. Their rectified outputs combine as inputs to complex cells. A second filter is created by the on-center off-surround, or endstopping, network that generates hypercomplex cell receptive fields from combinations of complex cell outputs. Higher-order hypercomplex cells further transform hypercomplex cell outputs via a push–pull competition across orientations. These hypercomplex cells interact with cooperative bipole cells to complete boundary groupings.

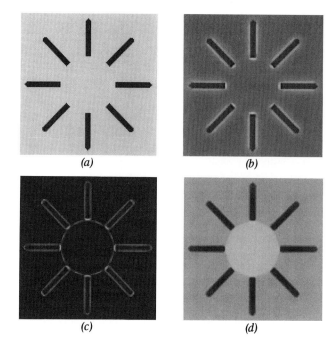

Figure C2.2.18. (*a*) The Ehrenstein figure. (*b*) The LGN stage response. Both ON and OFF activities are coded as rectified deflections from a neutral gray. Note the brightness buttons at the line ends. (*c*) The equilibrium BCS boundaries. (*d*) In the filled-in result, the central circle contains stronger FCS signals than the background, corresponding to the perception of increased brightness. In this figure, the representations of boundaries at multiple orientations are superimposed. (Reprinted from Gove *et al* (1995) with the permission of Cambridge University Press.)

Figure C2.2.19. (*a*) Top left: unprocessed SAR image of upstate New York scene consisting of highway with bridge overpass. (*b*) Top right: logarithm-transformed SAR image. (*c*) Bottom left: ON-minus-OFF cell responses averaged across spatial scales. (*d*) Bottom right: Multiple-scale FCS surface representation derived by averaging the filled-in responses within each scale of figure C2.2.20.)

C2.2.3.5 Typical application

The BCS and FCS have been used successfully to process images from artificial sensors, including synthetic aperture radar (SAR) sensors (Grossberg *et al* 1995). SAR sensors are used to produce range imagery of high spatial resolution under difficult weather conditions (Munsen *et al* 1983, Munsen and Visentin 1989). This application uses several key properties of BCS/FCS circuits.

First, the model ON and OFF cells normalize, and thereby compress, the large dynamic range (five orders of magnitude) of the signal. Later boundary segmentation and filling-in stages compensate for image speckle that has characteristics of random multiplicative noise. They do so by detecting, regularizing and completing boundary structures wherein diffusion of normalized signals can complete surface representations that compensate for speckle in a form-sensitive way.

Figure C2.2.19 shows an SAR image and the result of multiple-scale BCS/FCS processing applied to the image. Figure C2.2.19(*a*) shows the original SAR image of a highway with bridge overpass in upstate New York. Figure C2.2.19(*b*) shows the logarithmically transformed ($\log_{10}$) version of the original image for comparison. Figure C2.2.19(*c*) displays the result of center-surround processing by model LGN ON and OFF cells that normalize the image and detect its local ratio contrast. Figure C2.2.19(*d*) displays the FCS surface representation.

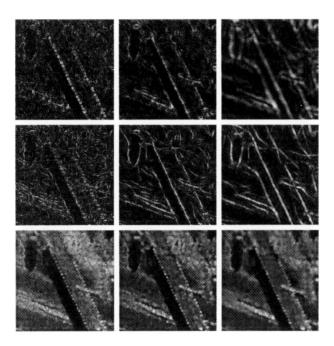

Figure C2.2.20. Top row: complex cell processing at three scales. Intensity of each pixel depicts the total activity of the oriented complex cells at that position. Middle row: higher-order hypercomplex cell processing at three spatial scales. Intensity of each pixel depicts the total activity of the cells at that position. Bottom row: result of surface filling-in processing result at three different scales on the example image. A linear combination of these images is used to obtain the final multiple-scale output in figure C2.2.19(*d*).

Figure C2.2.20 displays, in its first row, the outputs of complex cells using three different receptive field sizes: small, medium and large. Row two shows the corresponding boundaries after sharpening and completion by hypercomplex-bipole cell feedback. The third row shows the filled-in surface that forms when the LGN ON and OFF cell outputs diffuse within the boundary compartments defined by each scale. The final surface representation in figure C2.2.19(*d*) is a weighted sum of the three images in row three of figure C2.2.20.

Acknowledgements

This work was supported in part by the Office of Naval Research (ONR N00014-95-1-0409 and ONR N00014-95-1-0657). The author wishes to thank Cynthia E Bradford for her valuable assistance in the preparation of the manuscript.

References

Asfour Y R, Carpenter G A, Grossberg S and Lesher G W 1993 Fusion ARTMAP: A neural network architecture for multi-channel data fusion and classification *Proc. World Congr. on Neural Networks (WCNN-93)* vol II (Hillsdale, NJ: Lawrence Erlbaum Associates) pp 210–5

Bachelder I A, Waxman A M and Seibert M 1993 A neural system for mobile robot visual place learning and recognition *Proc. World Congr. on Neural Networks (WCNN-93)* vol I (Hillsdale NJ: Lawrence Erlbaum Associates) pp 512–7

Baloch A A and Waxman A M 1991 Visual learning adaptive expectations and behavioral conditioning of the mobile robot MAVIN *Neural Networks* **4** 271–302

Baraldi A and Parmiggiani F 1995 A neural network for unsupervised categorization of multivalued input patterns: An application to satellite image clustering *IEEE Trans. Geosci. Remote Sensing* **33** 305–16

Bradski G and Grossberg S 1995 Fast learning VIEWNET architectures for recognizing 3-D objects from multiple 2-D views *Neural Networks* **8** 1053–80

Bregman A S 1990 *Auditory scene analysis* (Cambridge, MA: MIT Press)

Carpenter G A 1994 A distributed outstar network for spatial pattern learning *Neural Networks* **7** 159–68

Carpenter G A and Grossberg S 1987a A massively parallel architecture for a self-organizing neural pattern recognition machine. *Computer Vision Graphics and Image Processing* **37** 54–115

——1987b ART 2: Stable self-organization of pattern recognition codes for analog input patterns. *Appl. Opt.* **26** 4919–30

——1987c Neural dynamics of category learning and recognition: Attention memory consolidation and amnesia *The Adaptive Brain I: Cognition Learning Reinforcement and Rhythm* ed S Grossberg (Amsterdam: Elsevier/North Holland) pp 238–86

——1990 ART 3: Hierarchical search using chemical transmitters in self-organizing pattern recognition architectures. *Neural Networks* **3** 129–52

——(eds) 1991 *Pattern Recognition by Self-Organizing Neural Networks* (Cambridge MA: MIT Press)

——1992 Fuzzy ARTMAP: Supervised learning recognition and prediction by a self-organizing neural network. *IEEE Commun. Mag.* **30** 38–49

——1993 Normal and amnesic learning recognition and memory by a neural model of cortico-hippocampal interactions *Trends in Neurosci.* **16** 131–7

——1994 Fuzzy ARTMAP: A synthesis of neural networks and fuzzy logic for supervised categorization and nonstationary prediction *Fuzzy Sets, Neural Networks and Soft Computing* ed R R Yager and L A Zadeh (New York: Van Nostrand Reinhold) pp 126–65

Carpenter G A, Grossberg S and Iizuka K 1993 Comparative performance measures of Fuzzy ARTMAP, learned vector quantization and back propagation for handwritten character recognition *Proc. Int. Joint Conf. on Neural Networks (WCNN-93)* vol I (Piscataway NJ: IEEE Service Center) pp 794–9

Carpenter G A, Grossberg S, Markuzon N, Reynolds J H and Rosen D B 1992 Fuzzy ARTMAP: A neural network architecture for incremental supervised learning of analog multidimensional maps *IEEE Trans. Neural Networks* **3** 698–713

Carpenter G A, Grossberg S and Reynolds J H 1991a ARTMAP: Supervised real-time learning and classification of nonstationary data by a self-organizing neural network *Neural Networks* **4** 565–88

Carpenter G A, Grossberg S and Rosen D B 1991b ART2-A: An adaptive resonance algorithm for rapid category learning and recognition *Neural Networks* **4** 493–504

——1991c Fuzzy ART: Fast stable learning and categorization of analog patterns by an adaptive resonance system *Neural Networks* **4** 759–71

Carpenter G A and Markuzon N 1996 ARTMAP-IC and medical diagnosis: Instance counting and inconsistent cases *CAS/CNS Technical Report TR-96-017* Boston University

Carpenter G A and Ross W D 1993 ART-EMAP: A neural network architecture for learning and prediction by evidence accumulation *Proc. World Congr. on Neural Networks (WCNN'93)* vol III (Hillsdale, NJ: Lawrence Erlbaum) pp 649–56

——1995 ART-EMAP: A neural network architecture for object recognition by evidence accumulation *IEEE Trans. Neural Networks* **6** 805–18

Carpenter G A and Tan A-H 1995 Rule extraction: From neural architecture to symbolic representation *Connection Sci.* **7** 3–27

Caudell T P, Smith S D G, Escobedo R and Anderson M 1994 NIRS: Large scale ART-1 neural architectures for engineering design retrieval *Neural Networks* **7** 1339–50

Christodoulou C G, Huang J, Georgiopoulos M and Liou J J 1995 Design of gratings and frequency selective surfaces using Fuzzy ARTMAP neural networks *J. Electromag. Waves and Applications* **9** 17–36

Cleary R and Israel P 1994 A resonance correlation network with adaptive fuzzy leader clustering *Proc. Sixth Int. Conf. on Tools with Artificial Intelligence 1994* (Los Alamitos, CA: IEEE Computer Society Press) pp 168–74

Desimone R 1992 Neural circuits for visual attention in the primate brain eds G A Carpenter and S Grossberg *Neural Networks for Vision and Image Processing* (Cambridge, MA: MIT Press) pp 343–64

DeYoe E A and van Essen D C 1988 Concurrent processing streams in monkey visual cortex *Trends in Neurosciences* **11** 219–26

Dubrawski A and Crowley J L 1994 Learning locomotion reflexes: A self-supervised neural system for a mobile robot *Robotics and Autonomous Syst.* **12** 133–42

Feng C, Sutherland A, King S, Muggleton S and Henery R 1993 Comparison of machine learning classifiers to statistics and neural networks *Proc. Fourth Int. Workshop on Artificial Intelligence and Statistics* pp 363–8

Francis G and Grossberg S 1996 Cortical dynamics of form and motion integration: persistence, apparent motion, and illusory contours. *Vision Research* **36** 149–173

Francis G, Grossberg S and Mingolla E 1994 Cortical dynamics of feature binding and reset: control of visual persistence *Vision Research* **34** 1089–104

Gan K W and Lua K T 1992 Chinese character classification using an adaptive resonance network *Patt. Recog.* **25** 877–82

Gopal S, Sklarew D M and Lambin E 1994 Fuzzy neural networks in multi-temporal classification of landcover change in the Sahel *Proc. DOSES Workshop on New Tools for Spatial Analysis (Lisbon Portugal)* (Brussells: ECSC-EC-EAEC) pp 55–68

Gove A, Grossberg S and Mingolla E 1995 Brightness perception illusory contours and corticogeniculate feedback *Visual Neurosci.* **12** 1027–52

Govindarajan K K, Grossberg S, Wyse L L and Cohen M A 1994 A neural network model of auditory scene analysis and source segregation *Technical Report CAS/CNS-TR-94-039* Boston University

Grossberg S 1968 Some nonlinear networks capable of learning a spatial pattern of arbitrary complexity *Proc. Nat. Acad. Sci. USA* **59** 368–72

——1972 Neural expectation: Cerebellar and retinal analogs of cells fired by learnable or unlearned pattern classes *Kybernetik* **10** 49–57

——1976a Adaptive pattern classification and universal recoding I: Parallel development and coding of neural feature detectors *Biol. Cybern.* **23** 121–34

——1976b Adaptive pattern classification and universal recoding II: Feedback expectation olfaction and illusions *Biol. Cybern.* **23** 187–202

——1978 A theory of human memory: Self-organization and performance of sensory-motor codes maps and plan *Progress in Theoretical Biology* ed R Rosen and F Snell vol 5 (New York: Academic) pp 233–374 (Reprinted in Grossberg S 1982 *Studies of Mind and Brain: Neural Principles of Learning Perception Development Cognition and Motor Control* (Boston, MA: Reidel))

——1980 How does a brain build a cognitive code? *Psychol. Rev.* **87** 1–51

——1983 The quantized geometry of visual space: the coherent computation of depth, form, and lightness. *Behavioral and Brain Sciences* **6** 625–57

——(ed) 1987a *The Adaptive Brain I: Cognition Learning Reinforcement and Rhythm* (Amsterdam: Elsevier/North-Holland)

——(ed) 1987b *The Adaptive Brain II: Vision Speech Language and Motor Control* (Amsterdam: Elsevier/North-Holland)

——(ed) 1988 *Neural Networks and Natural Intelligence* (Cambridge, MA: MIT Press)

——1994 3-D vision and figure-ground separation *Perception and Psychophys.* **55** 48–120

——1995 The attentive brain *Am. Sci.* **83** 438–49

Grossberg S, Boardman I and Cohen M A 1995 Neural dynamics of variable-rate speech categorization *Technical Report CAS/CNS-TR-94-038* Boston University

Grossberg S and Merrill J W L 1992 A neural network model of adaptively timed reinforcement learning and hippocampal dynamics *Cognitive Brain Research* **1** 3–38

——1996 The hippocampus and cerebellum in adaptively timed learning recognition and movement *Technical Report CAS/CNS-TR-93-065* Boston University (*J. Cognitive Neuroscience* in press)

Grossberg S and Mingolla E 1985a Neural dynamics of form perception: boundary completion, illusory figures, and neon color spreading *Psychological Review* **92** 173–211

——1985b Neural dynamics of perceptual grouping: textures, boundaries, and emergent segmentations *Perception and Psychophysics* **38** 141–171

——1993 Neural dynamics of motion perception: direction fields, apertures, and resonant grouping *Perception and Psychophysics* **53** 243–78

Grossberg S, Mingolla E and Ross W D 1994 A neural theory of attentive visual search: Interactions of visual spatial and object representations *Psychol. Rev.* **101** 470–789

Grossberg S, Mingolla E, and Williamson J 1995 Synthetic aperture radar processing by a multiple scale neural system for boundary and surface representation *Neural Networks* **8** 1005–28

Grossberg S and Rudd M E 1992 Cortical dynamics of visual motion perception: short-range and long-range apparent motion *Psychological Review* **99** 78–121

Grossberg S and Todorović D 1988 Neural dynamics of 1-D and 2-D brightness perception: a unified model of classical and recent phenomena *Perception and Psychophysics* **43** 241–277

Ham F M and Han S W 1993 Quantitative study of the QRS complex using fuzzy ARTMAP and the MIT/BIH arrhythmia database *Proc. World Congr. on Neural Networks (WCNN-93)* vol I (Hillsdale NJ: Lawrence Erlbaum Associates) pp 207–11

Harvey R M 1993 Nursing diagnosis by computers: an application of neural networks *Nursing Diagnosis* **4** 26–34

Israel P, Yu S and Ryan P 1992 Simplified ART1 *Proc. SPIE Conf. on Science of Neural Networks entitled Applications of Artificial Neural Networks III* (Chicago, IL: SPIE) vol 1709, part 1, pp 476–85

Johnson C 1993 Agent learns user's behavior *Electrical Engineering Times* June 28 pp 43, 46

Kasperkiewicz J, Racz J and Dubrawski A 1994 HPC strength prediction using an artificial neural network *ASCE J. Comput. Civil Eng.* submitted

Keyvan S, Durg A and Rabelo L C 1993 Application of artificial neural networks for development of diagnostic monitoring system in nuclear plants *Am. Nucl. Soc. Conf. Proc. (April 18–21)*

Kohonen T 1984/1989 *Self-organization and Associative Memory* 3rd edn (New York: Springer)

Kosko B 1986 Fuzzy entropy and conditioning *Info. Sci.* **40** 165–74

Kumara S R T, Merchawi N S, Karmarthi S V and Thazhutaveetil M 1994 *Neural Networks in Design and Manufacturing* (London: Chapman and Hall)

Mehta B V, Vij L and Rabelo L C 1993 Prediction of secondary structures of proteins using fuzzy ARTMAP *Proc. World Congr. on Neural Networks (WCNN-93)* vol I (Hillsdale, NJ: Lawrence Erlbaum Associates) pp 228–32

Moya M M, Koch M W and Hostetler L D 1993 One-class classifier networks for target recognition applications *Proc. World Congr. on Neural Networks (WCNN-93)* vol III (Hillsdale, NJ: Lawrence Erlbaum Associates) pp 797–801

Munsen D Jr, O'Brien J, and Jenkins W 1983 A tomographic formulation of spotlight-mode synthetic aperture radar *Proc. IEEE* **72** 917–925

Munsen D Jr and Visentin R L 1989 A signal processing view of strip-mapping synthetic aperture radar *IEEE Trans. Acoustics, Speech, and Signal Processing* **37** 2131–47

Murshed N A, Bortolozzi F and Sabourin R 1995 Off-line signature verification without a priori knowledge of class ω_2. A new approach *Proc. ICDAR 95: Third Int. Conf. on Document Analysis and Recognition* (Los Alamitos, CA: IEEE Computer Society Press) pp 191–6

Pessoa L, Mingolla E and Neumann H 1995 A contrast- and luminance-driven multiscale network model of brightness perception *Vision Research* **35** 2201–23

Repp B H 1980 A range-frequency effect on perception of silence in speech *Haskins Laboratories Status Report on Speech Research* **SR-61** 151–65

Ryan T W 1988 The resonance correlation network *Proc. IEEE Int. Conf. on Neural Networks* vol 1 (New York: IEEE) pp 673–80

Schiller P 1992 The ON and OFF channels of the visual system *Trends in Neurosciences* **15** 86–92

Seibert M and Waxman A 1990 Learning aspect graph representations from view sequences ed Touretzky D *Advances in Neural Information Processing Systems 2* (San Mateo, CA: Morgan Kaufmann) pp 258–65

Sillito A M, Jones H E, Gerstein G L and West D C 1994 Feature-linked synchronization of thalamic relay cell firing induced by feedback from the visual cortex *Nature* **369** 479–82

Smith E E 1990 *An Invitation to Cognitive Science* ed D O Osherson and E E Smith (Cambridge, MA: MIT Press)

Smith J W, Everhart J E, Dickson W C, Knowler W C and Johannes R S 1988 Using the ADAP learning algorithm to forecast the onset of diabetes mellitus *Proc. Symp. on Computer Applications and Medical Care* (Piscataway, NJ: IEEE Computer Society Press) pp 261–5

Suzuki Y, Abe Y and Ono K 1993 Self-organizing QRS wave recognition system in ECG using ART 2 *Proc. World Congr. on Neural Networks (WCNN-93)* vol IV (Hillsdale, NJ: Lawrence Erlbaum Associates) pp 39–42

Tarng Y S, Li T C and Chen M C 1994 Tool failure monitoring for drilling processes *Proc. Third Int. Conf. on Fuzzy Logic, Neural Nets and Soft Computing (Iizuka, Japan)* (Fukuoka, Japan: Fuzzy Logic Systems Institute) pp 109–11

von der Malsburg C 1973 Self-organization of orientation sensitive cells in the striate cortex *Kybernetik* **14** 85–100

Wienke D 1994 Neural resonance and adaptation—towards nature's principles in artificial pattern recognition *Chemometrics: Exploring and Exploiting Chemical Information* ed L Buydens and W Melssen (Nijmegen, The Netherlands: University Press)

Wienke D and Kateman G 1994 Adaptive resonance theory based artificial neural networks for treatment of open-category problems in chemical pattern recognition—Application to UV-Vis and IR spectroscopy *Chemometrics and Intelligent Laboratory Systems* ed L Buydens and W Melssen (Nijmegen, The Netherlands: University Press)

Wienke D, Xie Y and Hopke P K 1994 An adaptive resonance theory based artificial neural network (ART 2-A) for rapid identification of airborne particle shapes from their scanning electron microscopy images *Chemometrics and Intelligent Laboratory Systems* ed L Buydens and W Melssen (Nijmegen, The Netherlands: University Press)

Zadeh L 1965 Fuzzy sets *Info. Control* **8** 338–53

C2.3 Unsupervised composite networks

Cris Koutsougeras

Abstract

This section concerns neural networks which are hybrid either in terms of structure or in terms of training algorithms. The counterpropagation network is one that incorporates structural characteristics of the Kohonen and Grossberg networks and it is trained by composite supervised–unsupervised methods. The adaptive critic concept concerns neural network implementations of reinforcement learning where teacher information is available, a supervised learning characteristic, but target outputs are not specified, an unsupervised learning characteristic. The counterpropagation network as well as a number of adaptive critic implementations are taken up in this section.

C2.3.1 Introduction

The use of *feedforward networks* for the purposes of learning is based on the proven fact (Funahashi 1989, Kolmogorov 1963) that such structures with at least two layers of neurons (at least one hidden layer) can be universal function approximators. An example of a feedforward network is shown in figure C2.3.1. In producing the input-to-output mapping, the network maps the input space in an intermediate space (output of the hidden layer) and subsequently through the output layer, the intermediate space is mapped to the network's output space. The intermediate space can be viewed as an alternative representation of the input space retaining those properties at the original (input) space which pertain to the target learning task. The intermediate space may thus emphasize statistical, topological or other properties of the input space which make the target task describable or expressible, while suppressing other properties or information contained in the original space which do not have an apparent correlation to the target task.

In pure supervised learning methods and algorithms the intermediate mapping emerges in a rather spontaneous way as a result of the overall network adaptation. No specific mechanism is dedicated to the intermediate mapping because in the overall objectives of the method there are no specific or explicit quality criteria or requirements imposed or referring to this intermediate mapping. Such requirements are implicit to the overall adaptation objectives and the intermediate mapping emerges through the adaptation process as one that happens to aid the overall input–output mapping. In other words it can be viewed as a 'fortunate' side effect but otherwise no specialized attention is paid to it. Incidentally, this is the reason for difficulties relating to the generalization capabilities of feedforward networks and for the requirements of extensive trials of various combinations of numbers of hidden layer neurons, initial weights and learning rates (Koutsougeras *et al* 1992a).

The combination of supervised and unsupervised techniques allows the exercise of some control over the evaluation of the intermediate mapping, so that the input-to-intermediate and the intermediate-to-output mappings are developed by essentially different methods with objectives tuned to each one's role, while the development of the two individual mappings happens in a synergistic way. Two representative paradigms combining supervised and unsupervised techniques are the subject of this section. First we examine the counterpropagation network and then the concept of adaptive critic networks.

C2.3.2 The counterpropagation network

C2.1.1 The counterpropagation network was introduced in 1986 by Robert Hecht-Nielsen (1988, 1987a, 1990, 1987b) who was trying to utilize Kohonen's *self-organizing maps* for the purposes of function approximation. The resulting counterpropagation network behaves as a statistically optimal self-adaptive lookup table. If we were to classify it in one of the broader categories then it would belong to the general function approximators. It would work best with continuous mappings from $R^n \rightarrow R^m$ (or at least piecewise continuous). The way the network functions can be illustrated intuitively as follows: a hidden layer effectively clusters the input space in a collection of regions by a clustering method based on unsupervised techniques. The output layer, based on supervised techniques, effectively learns the average of the output values associated with each region of the input space. This average value associated with a region then becomes the designated output for all inputs falling within the corresponding input space region.

C2.3.2.1 Topology

There are two functional layers as shown in figure C2.3.1. The first functional layer (hidden layer) is a typical Kohonen layer. This layer operates according to the classic Kohonen unsupervised scheme (Kohonen 1982a, 1982b, 1988). Each neuron in this layer effectively functions as a receptor for a certain cluster of inputs, responding when the network's input belongs to this cluster. Typically, each neuron in this layer 'sees' the entire input vector which is fed to the network, so there are connections from every input component to every neuron of this layer. Usually there are no physical connections among neurons in the Kohonen layer; however, as we will see in the following, there is a competition among these neurons and it involves their corresponding network excitation values. So interactions among them effectively exist by means of which a scalar value from each neuron is broadcast to all other neurons in this layer.

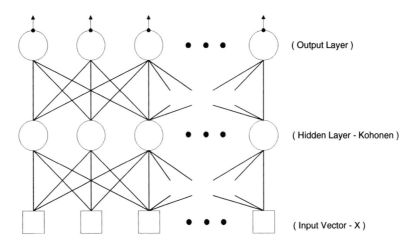

(Output Layer)

(Hidden Layer - Kohonen)

(Input Vector - X)

Figure C2.3.1. Example of a feedforward structure. The use of a Kohonen type of hidden layer is specific to the counterpropagation network model.

The output layer consists of m neurons where m is the dimensionality of the output space. Thus, if the target input–output mapping for the overall network is from R^n to R^m, the output layer consists of m neurons, that is, one neuron dedicated to producing one component of the output vector. This layer produces an output vector for each one of the regions in which the input space has been effectively partitioned by the intermediate layer. Incidentally, the vector which gets to be associated with each region is the average of the outputs associated with those training set samples which fall within that same region. Thus, the only information needed at each neuron of the output layer is which region the input vector falls within. This information is provided by the receptor neuron associated with that region (these receptor neurons are the ones in the hidden layer). Since each neuron of the output layer needs to know which neuron of the hidden layer is active at any given time, there are feedforward (unidirectional) connections from every neuron of the hidden layer to every neuron at the output layer. Since only the region identification is needed at each output neuron, there is no exchange among output neurons and thus no intralayer (lateral) connections.

In summary, the topology of the network is as follows. There are two layers: one hidden and one output layer. All pairs from input lines to neurons of the hidden layer are connected. All pairs of neurons within the hidden layer are laterally *effectively* connected. All pairs of neurons from the hidden layer to the output layer are connected. No lateral connections within the output layer neurons exist. All connections are unidirectional carrying single scalar values. The direction of computation is feedforward. Connections are unidirectional. Since there is no feedback, for every one connection the inverse one does not exist.

C2.3.2.2 Learning

Different learning paradigms are used for each of the two layers since the hidden layer is adjusted using unsupervised techniques while the output layer is adjusted using supervised ones. As mentioned before, the hidden layer implements a Kohonen self-organizing map while the output layer implements a Grossberg *instar* (Grossberg 1982, 1971, 1969, Hecht-Nielsen 1990).

C1.1.6

The hidden layer. The input to the network is a vector of real values and is fed to all neurons in the hidden layer. The weights associated with the connections feeding the kth hidden layer neuron form a weight vector W_k associated with that neuron. This neuron simply sums up all the weight-scaled inputs feeding to it thereby effectively computing the scalar product $W_k^T X$ of its associated weight vector W_k and the current input vector X. The value computed by each neuron is broadcast to all other neurons of the hidden layer and a competition takes place. The neuron which has produced the largest value wins and stays active while all others are shut off thereby setting their output to 0. Thus, one hidden-layer neuron responds to a given input vector, and it is this same neuron which will provide input to the output layer for determining the final output mapping. So far, the input vector X has been mapped to an intermediate binary vector which has one component set to 1 (corresponding to the active hidden layer neuron) and all others set to 0.

The learning rule (hidden layer). While the network is still in the adaptation phase, the weight vectors change after the competition step is decided. Only the weight vector of the winning neuron changes by moving closer to the current input X. This is done by means of the rule:

$$W_k^{\text{new}} = W_k^{\text{old}} + a(t)(X - W_k^{\text{old}})$$

where $a(t)$ is a time-decreasing learning rate which guarantees convergence (stability). Thus the W_k is changed by adding to it a fraction of its difference from the current input X. After a number of iterations, each weight vector W_k will identify a set S_k of input vectors for which the kth neuron will always be winning the competition and W_k will approximate their mean. It is also obvious that the kth neuron will still win the competition for any new input vector falling in the region which is bounded by the envelope of the S_k vectors.

In this way the input space is partitioned in a number of regions equal to the number of hidden-layer neurons. One hidden-layer neuron is associated with each region and this neuron wins the competition for any future input that falls within this region. Thus the neuron effectively becomes a 'receptor' for its corresponding region (receptive field). This clustering evolves in an autonomous way without teacher feedback. The weight vectors arrange themselves around the input space thereby identifying their receptive fields automatically. The topological distribution of the weight vectors relates to that of the input space samples which constitute the training set.

The output layer. This layer learns to produce the mean output vector value for each of the regions in which the input space has been clustered so far. The rationale is as follows. Let us assume that S_k is the set of input vectors which falls within the region which has become the receptive field of the kth hidden layer neuron. Then we may ask what is the output for a new input vector X which falls within the same region (for which the same kth neuron responds). All that is known about the behavior of the overall input–output function which is to be approximated is the set of outputs corresponding to the input vectors in the set S_k. Thus, if these values are taken as the basis for a guess, a simple reasonable choice of statistical approximation is to use the average of these values. During the learning phase the weight between the kth neuron of the hidden layer and the ith neuron of the output layer is set to the average of the ith component of the output vectors produced by the inputs which fall within the receptive field of the kth neuron of the hidden layer. Whenever this kth neuron is responding to an input, the weight value associated with the connection from the kth neuron to the ith neuron becomes the output of the ith neuron. Given that the outputs of the hidden layer neurons are binary and that only one of them is nonzero, the activation function of the ith neuron of the output layer can be expressed as $y_i = W_i^T B$, where B is the intermediate vector produced by the hidden layer and W_i is the weight vector associated with the ith neuron.

The learning rule (output layer). Since learning in the output layer is supervised, each output neuron 'sees' the target output which corresponds to each input vector of the training set during training. Specifically, the ith neuron sees the ith component of each such target output vector. The binary vector produced by the hidden layer is input to every output neuron. When the output of the hidden layer is asserted (competition settled) all of the weights associated with all output neurons are updated according to the local rule

$$W_i^{\text{new}} = W_i^{\text{old}} + a(Y_i - W_i^{\text{old}})^T B$$

where $Y_i = y_i I$; I is the unit vector and y_i is the target value of the ith output neuron corresponding to the current input. In other words, Y_i is a vector with all its components equal to the y_i target value. This updating rule (originally used by Grossberg) causes each connection weight to approximate the average values for the corresponding receptive fields as mentioned earlier.

C2.3.2.3 Related neural network models

C1.2 The *backpropagation network* (Rumelhart *et al* 1986) is an alternative to the possible uses of the counterpropagation network. A related model was developed by Koutsougeras and Papadourakis in 1991 using a two-layer feedforward structure, sigmoid nonlinear neurons and blended supervised–unsupervised learning (Koutsougeras and Papadourakis 1992b). Feedforward networks trained by other interesting alternative algorithms have been presented by Psaltis and Neifeld (1988), Kollias and Anastassiou (1988), deFigueiredo (1992), Karayiannis (1992), and Klassen and Pao (Klassen *et al* 1988). These networks are B5.3.5 also competitors for the uses of the counterpropagation network. Finally, a program package for *vector quantization* called LVQ_PAK was introduced in 1992 by Kohonen *et al* (1992).

C2.3.2.4 Advantages

The advantages of the counterpropagation network lie in its simplicity. The learning rules are simple and easily computable thus the computing requirements are not greedy. The advantages also lie in the fact that it provides a somewhat direct control on the internal representations which are developed for the target function. It does not try to implement a mapping or transformation mechanism that applies uniformly in the entire input space; instead, it clusters the input space in regions and tries to approximate the input–output function independently within each region. An unexpected post-training behavior is less likely than with backpropagation or the model proposed by Koutsougeras. While an optimal number of hidden-layer neurons for a target input–output function cannot be determined *a priori*, it is not so sensitive to this choice and as a matter of fact with this model the more hidden-layer neurons the better it behaves. In the corresponding case of backpropagation there is a certain unknown number of hidden-layer neurons beyond which the excessive inherent nonlinearity produces unexpected results.

C2.3.2.5 Disadvantages

The disadvantages of this network are those carried over by the Kohonen layer, and also the fact that within each region of the clustered input space (receptive field) the target input–output function is approximated by a single constant. Let us take these up one by one. First, the Kohonen layer works well with certain initial weight distributions relating to the topology of the intended mapping. In other words, the resulting clustering is sensitive to the choice of initial weights. Second, the output of the entire network depends on the hidden-layer neuron which responds to the input, thus the output over the entire receptive field of each neuron will be constant. Thus, a continuous input–output function will be approximated with piecewise constant segments. This is not the case with the aforementioned competitor networks.

C2.3.2.6 Typical applications

An attempt to visualize the function of this network would reveal constant output values over segments of a continuous input space. The network implies a segmentation of its continuous multidimensional real space into convex segments. The output values of all known samples within each segment are averaged and the resulting mean value is stored and used as output over the entire segment. So there is a constant value over each segment although the corresponding values for two different segments may be different. In view of this fact, this network is useful in applications requiring a function approximation with piecewise constant segments. Typical examples of this category of applications are *pattern recognition/classification* F1.2 problems which are such that the pattern space (input space) can be divided into clusters (segments) so that each cluster contains patterns of a single class. This network can also be used in any application requiring function approximation if piecewise linear approximation with a relatively small number of segments can be acceptable. Since backpropagation is a competitor in these same cases, counterpropagation has not usually been a preference in significant applications.

C2.3.2.7 Hardware implementations

It is rather straightforward to implement simulators for this network, so there are no known hardware implementations of it.

C2.3.2.8 Variations and improvements

Variations of this network are related to variants of the Kohonen self-organizing maps. These reflect ways to deal with the problems associated with the choice of initial weights. There are the methods of radial sprouting of the input and weight vectors (Hecht-Nielsen 1990), the addition of uniformly distributed noise vectors to the input data, or the conscience method of Desieno (1988).

   Another interesting variant is to allow more than one neuron in the hidden layer to respond to a given input. In this case the receptive fields of the various hidden-layer neurons are overlapping rather than being disjoint. In this case the output produced by the network should be a weighted sum of the average output values corresponding to all the intersecting receptive fields. However, a good way for assessing the weights that should be used in computing the above weighted sum is not known.

C2.3.2.9 State of the art

An open problem is how to determine the weights to be used for computing the output as the weighted sum of the average outputs of a relevant set of receptive fields, as described in the extension above, where more than one hidden-layer neuron responds to an input. A method using barycentric coordinates is described by Hecht-Nielsen (1990) but such coordinates are not unique and presently no good solution that works better than the original network is known.

C2.3.3 Adaptive critic networks

C2.3.3.1 Purposes

The adaptive critic is a concept referring to an automated agent capable of producing a critique for the 'goodness' or the 'utility' of its input. In most interesting cases the input is the state of an observed system or a sequence of such states or even actions applied to that system. A typical example where this concept applies has been the automatic control of a system known as a 'plant' in the control systems jargon. Such a typical control structure is shown in figure C2.3.2.

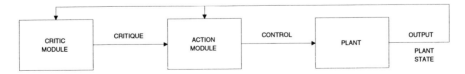

Figure C2.3.2. A typical control structure utilizing a critic module and an action module.

According to the arrangement in this figure the control mechanism consists of a module which actually produces the control inputs (actions) to the plant and a critic module which provides a critique to the action module as to how well the latter is doing. The task charged to the critic module is to track the evolution of the plant states or outputs and to pass an evaluation of how likely it is that a global goal (the purpose of the controlling) will be reached with the observed trajectory of states or outputs. In the literature the term 'utility' is used to refer to the likelihood or to a measure of expectation that a future global goal will be reached based on the information at the current state or of a sequence of states within a window extending from the past to the present (Bryson and Ho 1969, Raiffa 1968). Thus the purpose of the adaptive critic is to approximate the utility as a function of the current state or the recent history of states. If the exact value of this utility corresponding to an input is known then the adaptive critic would reduce to a pure function approximator and any supervised scheme for function approximation can be used to implement it. However, in real-world applications this utility function is not known or it may not even be unique. In such cases target output values are not available for the learning or adaptation and thus pure supervised schemes cannot be applied. Unsupervised schemes seem to be better candidates because they do not assume the availability of target outputs, and instead of *a priori* targets, it is the rationale behind the design of the scheme that determines its goal and function. However, it is usually the case that the critic receives some sort of feedback which consists of the occasional information that the global goal has been reached or that the control process has led to definite failure. An example is the problem of pole balancing, where the control scheme of figure C2.3.2 is used to balance a pole (the pole is the plant in this case). The critic occasionally receives the information that the pole has fallen (control strategy unrecoverably failed) or that it is still up. This feedback does not provide explicit targets for the critic but rather a reinforcement which C3 can be further used for punishment/reward used in *reinforcement learning* schemes. Thus the adaptive critic designs can be classified as reinforcement learning methods which places them in between the pure supervised and the pure unsupervised schemes.

The above discussion leads to the credit assignment task which is yet another use of the adaptive critic designs and is necessary to their overall performance goal. Out of the knowledge that a (usually long) sequence of states or control actions has failed or has not failed, the critic has to figure out how to assign credit to the individual states or actions (or to any specific subsequence for that matter). This is a requirement implicit to the goal of determining the utility of an element of a sequence on the basis that the only information is the judgment (evaluation) about the sequence as a whole (as opposed to individual elements). If it is known that a control sequence led to failure, it is challenging to determine which elements in the sequence (or which part of the sequence) led to the unrecoverable downfall.

A number of adaptive critic designs with the above properties have been implemented with neural networks. The most widely known implementations of adaptive critics through neural networks are those proposed by Barto *et al* (1983), Anderson (1989, 1987) and Werbos (1990, 1989). Two versions of adaptive critics by Werbos based on backpropagation networks are summarized in the following. The first concerns a general purpose stand-alone critic network and the second assumes a specific structure of the action module that contains a predictive model of the plant to be controlled. Both implement a mechanism that is functionally equivalent to a heuristic dynamic programming (HDP) (Howard 1960, Bryson and Ho 1969).

C2.3.4 Dynamic programming adaptive critics

Two models of dynamic programming adaptive critics are discussed in this section. These are the heuristic dynamic programming (HDP) and the dual heuristic programming (DHP) critic models which are due to Werbos. The HDP model is basically an extension of the DHP model but it requires a slightly different system architecture.

C2.3.4.1 HDP adaptive critic

Purpose. A crucial assumption in these implementations is that the utility function $U(X)$ is known. This assumption essentially reduces the problem to a function approximation for which the classic supervised method of backpropagation (using a feedforward network) can be directly applied. The twist, however, is that the critic is not called to learn the supplied $U(X)$ function but another function $J(X)$ which Werbos calls the 'strategic utility' function and is supposed to be an approximation of $U(X)$. The rationale of this choice is twofold. First, the real significance of $U(X)$ is in the relative values which it produces for two different events X_1 and X_2. Any other function which fluctuates in the same way as $U(X)$ would in principle be just as good as $U(X)$ since it would be consistent with $U(X)$ in telling which of any two events X_1 and X_2 has better utility. Second, the function U is defined on a vector X which represents the perceived reality of the state of the plant. However, there may be additional parameters involved in the whole decision-making process and so the reality viewed by the controller may be described as another vector R which contains X (that is, it is an expanded version of X). So J reflects the thesis that the utility function which is actually output by the critic should be good for the vector of the state of reality R which is actually chosen and should condense the information content of U, being equivalent to U for practical purposes without having to be exactly the same as U. This causes a deviation from the principles of the pure supervised schemes in that teacher information is provided for adaptation but the target values are flexible.

Thus the HDP adaptive critic is an approximator for the strategic utility function. It is implemented using backpropagation as a basis. A general utility function $U(X(t))$ is assumed given. The strategic utility function $J(X(t))$ is supposed to be an estimate of $\sum_{k=t}^{\infty} U(t)$.

Topology. Since this implementation is based on the backpropagation network, the topology is the same. There must be at least two neuron layers (that is, at least one hidden layer). The reason for requiring at least two layers is the proven fact (Funahashi 1989) that a feedforward network with two layers and sigmoid neuron activation functions is a universal analytic function approximator. Connection types are interlayer and supralayer connections only. No intralayer connection is allowed. No high-order connections are allowed. Within the constraint that no intralayer connections are allowed, all other feedforward connections are allowable between pairs of neurons. The usual choice is a two-layer topology with sparse interlayer connectivity. The direction of computation is strictly feedforward and thus unidirectional. Since there is no feedback, for every connection the inverse one does not exist. The basic topology remains unaltered during training.

Learning. Basic backpropagation is used for adaptation. However, the originality occurs in determining the targets for the supervised learning scheme which the backpropagation implements. For input $X(t)$ the target is set to $J(X(t+1)) + U(X(t))$. In intuitive terms this says that the strategic utility of a state X is the utility $U(X)$ asserted by the teacher plus the strategic utility of the next state to which X leads. Each adaptation pass is carried out in two steps. In the first step the target for input $X(t)$ is computed and in the second step the actual weight update takes place. These steps are as follows:

(i)   The $X(t+1)$ is used as input and the network produces the value $J(X(t+1))$.

(ii) Then $X(t)$ is used as input and $J(X(t+1)) + U(X(t))$ is used as target and the weights are updated using a backpropagation pass.

Detailed descriptions of the algorithm and the handling of boundary cases can be found in Werbos (1990, 1989).

Learning rule. The learning rule is the standard local *Delta rule* used in the backpropagation network (Rumelhart *et al* 1986).   B3.3.3

C2.3.4.2   DHP adaptive critic

Purpose. The rate of change of the strategic utility function can be used as an alternative measure of goodness for the input being evaluated by the critic. Instead of producing a single scalar value as evaluation of the whole input vector, the critic produces a vector of values which reflect evaluations for each component of the input vector. Thus the output of the DHP critic network is a vector of the same dimension as that of the input vector. The ith component of the output vector is an approximation of the partial derivative of the reference strategic utility function with respect to the ith component of the input vector. The strategic utility function is the same J function used in the HDP critic network.

In addition, the DHP critic network assumes that the control structure in which it is contained has a particular architecture which is, in fact, the one shown in figure C2.3.3. There are three backpropagation-type networks in that structure. The A network produces the actions. The C network is the critic. The M network is a model of the plant and is supposed to simulate that plant. Given an input $X(t)$, the network M is supposed to produce an approximation of the expected next input $X(t+1)$; for example, if $X(t)$ is the plant's current state, M is supposed to predict the next state.

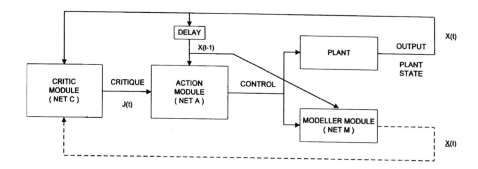

Figure C2.3.3. The control structure of figure C2.3.2 augmented to include a predictive process module.

Topology. The topology of the C network is the same as that of the HDP critic network, except that this network has as many outputs as inputs. The topologies of the A and M networks are also standard backpropagation networks.

Learning. As noted earlier, the ith output of C is the derivative of the strategic utility function J with respect to the ith input X_i. The function $J(X)$ is supposed to have the same form as in the case of the HDP critic approximating $J(X(t))$ by $J(X(t+1)) + U(X(t))$. If L_i is the ith output of the DHP critic network, then L_i should be:

$$L_i(t) = \sum_m \frac{\partial J(X(t))}{\partial X_m(t+1)} \frac{\partial X_m(t+1)}{\partial X_i(t)} + \frac{\partial U}{\partial X_i(t)}$$

or

$$L_i(t) = \sum_m L_m(t+1) \frac{\partial M}{\partial X_i(t)} + V_i(t).$$

The $V_i(t)$ values are assumed to be available. The L_i values are computed by the module M. A pass of the learning algorithm consists of two parts. The first part concerns the computation of the L_i values. The second concerns the update of the critic network's weights.

A single pass of the learning algorithm proceeds as follows:

(i) First step: use $X(t+1)$ as input to the critic network C and set $L(t+1)$ to the values produced by the critic network C. Use these $L(t+1)$ values as targets of the M network and extend the use of the standard backpropagation delta rule to the inputs to obtain the derivatives $\partial M / \partial X_i(t)$.

(ii) Second step: set $L_i(t)$ to $V_i(t)$ plus the values of the derivatives computed in the first step. Use these new values $L_i(t)$ as targets of the critic network C when $X(t)$ is input. Update the weights of the network according to standard backpropagation.

In the first step above, the delta rule is used to compute the derivatives $\partial J(X(t))/\partial X_m(t+1)$ since in backpropagation the delta rule provides a standardized way to express the effect of a variable on the network's output. Therefore, it should be assumed that the modeler network M has independently been trained beforehand to simulate the plant. The action network A is also adjusted concurrently with the critic network C but as this network does not relate directly to the adjustment of the C network it is outside the scope of this section. Details can be found in Werbos (1990, 1989).

Related neural network models. In place of the backpropagation network, one can employ any other scheme which implements the supervised learning mechanism required in the above mechanisms. In terms of functionality, related models are the GDHP adaptive critics (see below) (Werbos 1990), the adaptive critic element (Barto *et al* 1983) and Anderson's adaptive critic network (Anderson 1989, 1987).

Advantages. The HDP is the simplest of the general-purpose adaptive critics implemented by multilayer networks. Other models (like the DHP and GDHP) require the use of an additional module which acts as a model of the plant to be controlled. The HDP model does not require such an extra modeler and thus it is simpler.

The advantage of the DHP critic over the HDP critic is that it relies on the derivatives of the utility function and it is all about computing such derivatives. This alleviates the 'drift to infinity' problem (see below) which is possible with HDP.

Disadvantages. The disadvantages of the HDP model are those carried over by the backpropagation network which serves as the means for its implementation. Another disadvantage is the 'drift to infinity' problem: the target values for the function $J(X)$ may become too large (remember that these values are not provided *a priori* nor are they fixed so they could grow large).

The DHP critic assumes the existence of the modeler M network and this introduces a rather large number of cascaded layers in the whole design. As a result, adaptation is slow and the structure amplifies the standard problems involved with standard backpropagation such as sensitivity to initial weight, learning rate, number of neurons in the hidden layers, etc.

Typical applications. As mentioned already, the HDP and DHP critic networks are to be used in combination with action-producing modules as general control mechanisms. Thus automatic *control* F1.9 *problems* are the typical applications. Also, other applications mostly relating to forecasting can be found in Werbos (1977, 1979, 1987, 1988, 1989).

Variations. Werbos has presented an extension of the DHP critic which he calls Globalized DHP (Werbos 1990, 1989). This assumes the same structure as DHP and functions similarly except that it reverts to the HDP choice of a single output evaluating the entire input X.

C2.3.5 The adaptive critic element

The earliest work relating to the adaptive critic element (ACE) was done by Klopf (1982). More recent refinements were introduced through the work of Barto *et al* (1983). Through this work the ACE gained more visibility. Other work bearing similarities is the work of Widrow and his colleagues (Widrow *et al* 1973, Widrow and Smith 1964).

C2.3.5.1 Purposes

In its original presentation, the ACE is supposed to work together with an associative search element (ASE) for the purposes of controlling an external plant. The structure of this controller is the same as that of figure C2.3.2 where the ACE is the adaptive critic and the ASE is the component producing actions. Thus the purpose that the ACE must fulfill is the same as that of any general type of adaptive critic according to the analysis and description of section C2.3.3.

C2.3.5.2 Topology

Only one neuron constitutes this adaptive critic as seen in figure C2.3.4. Input to the ACE is a vector of values which come from a 'decoder' module. Only one component of the input vector is 1 at any given time and all others are 0. Each input line of the ACE corresponds to a region of the plant's state space; whenever the plant's state is in that particular region, this same input will be active (set to 1) and all others will be inactive (set to 0). This sort of mapping is performed by a decoder, the details of which are outside the scope of this section, but they are available in Barto *et al* (1983).

An external reinforcement value $r(t)$ is also provided as input to the ACE. This input is positive when rewarding is intended and negative when punishment is intended. The output of the ACE is a time-dependent scalar value $y(t)$. Let $X(t)$ be the input vector and W be the weight vector associated with the inputs of the ACE. If we denote the weighted sum of inputs $W^T X(t)$ by $p(t)$, then the output of ACE is $y(t) = r(t) + \gamma p(t) - p(t-1)$, where $0 \leq \gamma \leq 1$ is a discount factor.

C2.3.5.3 Learning

Learning is supervised since the reinforcement essentially corresponds to teacher information guiding the adaptation but no specific targets are provided for the actual output.

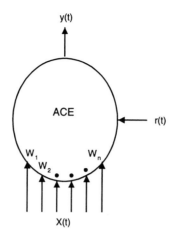

Figure C2.3.4. The adaptive critic element of Barto *et al.*

C2.3.5.4 Learning rule

A local learning rule updates each component W_i of the weight vector in successive steps as follows:

$$W_i(t + 1) = W_i(t) + \beta y(t) X'_i(t)$$

where β is a learning rate and $X'_i(t)$ is called the trace of X_i computed by $X'_i(t + 1) = \lambda X'_i(t) + (1 - \lambda) X_i(t)$, with $0 \leq \lambda \leq 1$ being a trace decay rate. In intuitive terms, the ACE learns the expected value of a discounted sum of future failure reinforcement signals. At each step, predictions are adjusted proportionally to the network input and the difference between the new prediction at the current state and the previous prediction at the previous state. This reflects the method of *temporal differences* developed by Sutton (1984, 1988) who has also proved convergence theorems for the temporal differences class of algorithms.

C2.3.5.5 Related neural network models

Other related models are the HDP and DHP adaptive critics (Werbos 1990, 1989) and Anderson's two-layer adaptive critic network (Anderson 1989, 1987).

C2.3.5.6 Advantages

The ACE has the advantage of simplicity. With only one neuron it is very versatile and easy to use. However, its significance lies more in the paradigm which it establishes. The ACE has been compared to the BOXES method developed by Michie and Chambers (1968) and has been found superior. The comparison involved the inverted pendulum application in which both systems would be able to achieve incrementally longer balancing times with more experience. The ACE did better because it learns continuously rather than just at failures.

C2.3.5.7 Disadvantages

The ACE assumes the existence of a decoder module which produces a (binary) vector input to the ACE depending on what the plant's state is. A lot depends on this clustering of the plant's state space and its properties. This clustering is extraneous to the ACE and can be done by other methods (e.g. the BOXES method (Michie and Chambers 1968)). The limited functional capacity of the ACE causes all the burden of a real complex problem to be transferred to the decoder part of the design. Another problem arises from the fact that the ACE learns certain behaviors which apply to the regions in which the plant's state space is partitioned by the decoder. This means that ultimately it approximates a global behavior over all the state space by means of constant segments or segments of very small variability. For very complex (highly nonlinear) target behaviors required for controlling the plant, the state space partition must be very fine so that the behavior required within each region can be approximated by the linear function performed by the ACE. This problem led to the replacement of the ACE element by a two-layer feedforward network in the work of Anderson (1989, 1987).

C2.3.5.8 Typical applications

The pole balancing problem (Barto *et al* 1983) has been the classic application related to the ACE.

C2.3.5.9 Hardware implementations

No hardware implementations are known by the author.

C2.3.5.10 Variations and improvements

Lin and Kim (1991) describe an extension of Barto's original ACE-ASE control structure where they integrate the cerebellar model articulation controller (CMAC). Also, Anderson's adaptive critic network implements the function of the ACE by a two-layer network (Anderson 1989, 1987). In this work the decoder which partitions the plant's state space and provides input to the ACE has been replaced by a layer of trainable neurons. Essentially, the decoder-ACE structure has been replaced by a two-layer feedforward backpropagation type of network which learns to perform the decoder-ACE composite function. Being a universal function approximator, the two-layer network can learn (using a backpropagation type of algorithm (Anderson 1987)) the function that needs to be performed by the system's segment which extends from the raw state variables (former inputs to the decoder) all the way to the output of the former ACE. This also allows the learning of rather complex and nonlinear control behaviors alleviating the disadvantage of the ACE as explained in section C2.3.5.7. Anderson (1989) reports that this scheme works much better than the original ACE.

References

Anderson C W 1987 Strategy learning with multilayer connectionist representations *Technical Report TR87-509.3* GTE Laboratories, Waltham, MA (corrected version of report in *Proc. Fourth Int. Workshop on Machine Learning (Irvine, CA)* pp 103–14)
——1989 Learning to control an inverted pendulum using neural networks *IEEE Control Syst. Mag.* **9** 31–7
Barto A G, Sutton R S and Anderson C W 1983 Neuronlike elements that can solve difficult learning control problems *IEEE Trans. Syst. Man Cybern.* **13** 835–46
Bryson A E Jr and Ho Y C 1969 *Applied Optimal Control* (Waltham, MA: Ginn)
deFigueiredo R J P 1992 An optimal multilayer neural interpolating (OMNI) net in generalized fock space setting *Proc. IEEE Int. Joint Conf. on Neural Networks* vol I (New York: IEEE Press) pp 111–20
Desieno D 1988 Adding a conscience to competitive learning *Proc. Int. Conf. on Neural Networks* vol I (New York: IEEE Press) pp 117–24
Funahashi K 1989 On the approximate realization of continuous mappings by neural networks *Neural Networks 2* **3** 183–92
Grossberg S 1969 Embedding fields: a theory of learning with physiological implications *J. Math. Psychol.* **6** 209–39
——1971 Embedding fields: underlying philosophy, mathematics, and applications to psychology, physiology and anatomy *J. Cybern.* **1** 28–50
——1982 *Studies of Mind and Brain: Neural Principles of Learning, Perception, Development, Cognition, and Motor Control* (Boston, MA: Reidel)
Hecht-Nielsen R 1987a counterpropagation networks *Proc. Int. Conf. on Neural Networks* vol II (New York: IEEE Press) pp 19–32
——1987b Counterpropagation networks *Appl. Opt.* **26** 4979–84
——1988 Applications of counterpropagation networks *Neural Networks* **1** 131–9
——1990 *Neurocomputing* (Reading, MA: Addison-Wesley) © 1990 by Addison-Wesley Publishing Company, Inc.
Howard R 1960 *Dynamic Programming and Markov Processes* (Cambridge, MA: MIT Press)
Karayiannis N B 1992 ALADIN: algorithms for learning and architecture determination *Proc. IEEE Int. Joint Conf. on Neural Networks* vol I (New York: IEEE Press) pp 601–6
Klassen M, Pao Y H and Chen A D C 1988 Characteristics of the functional link net: a higher order delta rule net *Proc. IEEE Int. Joint Conf. on Neural Networks* vol I (New York: IEEE Press) pp 507–14
Klopf A H 1982 *The Hedonistic Neuron: A Theory of Memory, Learning and Intelligence* (Washington, DC: Hemisphere)
Kohonen T 1982a Self-organized formation of topologically correct feature maps *Biol. Cybern.* **43** 59–69

——1982b A simple paradigm for the self-organized formation of structured feature maps *Competition and Cooperation Neural Networks (Lecture Notes in Biomathematics 45)* ed S Amari and M Arbib (Berlin: Springer)

——1988 *Self-Organization and Associative Memory* 2nd edn (Berlin: Springer)

Kohonen T, Kangas J, Laaksonen J and Torkkola K 1992 LVQ_PAK: a program package for the correct application of learning vector quantization algorithms *Proc. IEEE Int. Joint Conf. on Neural Networks* vol I (New York: IEEE Press) pp 725–30

Kollias S and Anastasiou D 1988 Adaptive training of multilayer neural networks using least squares estimation technique *Proc. IEEE Int. Joint Conf. Neural Networks* vol I (New York: IEEE Press) pp 383–90

Kolmogorov A N 1963 On the representation of continuous functions of many variables by superposition of continuous functions of one variable and addition *Am. Math. Soc. Trans.* **28** 55–9

Koutsougeras C, Georgiou G and Papachristou C A 1992 A feed forward classifier model: multiple classes, confidence output values, and implementation *Int. J. Patt. Recog. Artif. Intell.* **6** 539–69

Koutsougeras C and Papadourakis G 1992 Coupling supervised and unsupervised techniques in training feedforward networks *Int. J. Artif. Intell. Tools* **1** 37–55

Lin C S and Kim H 1991 CMAC-based adaptive critic self-learning control *IEEE Trans. Neural Networks* **2** 530–3

Michie D and Chambers R A 1968 BOXES: an experiment in adaptive control *Machine Intelligence* vol 2 ed E Dale and D Michie (Edinburgh: Oliver and Boyd) pp 137–52

Psaltis D and Neifeld M 1988 The emergence of generalization in networks with constrained representations *Proc. IEEE Int. Joint Conf. on Neural Networks* vol I (New York: IEEE Press) pp 371–82

Raiffa H 1968 *Decision Analysis: Introductory Lectures on Making Choices Under Uncertainty* (Reading, MA: Addison-Wesley)

Rumelhart D E, Hinton G E and Williams R J 1986 Learning internal representations by error propagation *Parallel Distributed Processing: Explorations in the Microstructure of Cognition. vol 1: Foundations* ed D E Rumelhart and J L McClelland (Cambridge, MA: MIT Press)

Sutton R S 1984 Temporal aspects of credit assignment in reinforcement learning *PhD Dissertation* University of Massachusetts

——1988 Learning to predict by the methods of temporal differences *Machine Learning* **3** 9–44

Werbos P J 1977 Advanced forecasting methods for global crisis warning and models of intelligence *General Systems Yearbook* **22** 25–38

——1979 Changes in global policy analysis procedures suggested by new methods of optimization *Policy Anal. Info. Syst.* **3** 1

——1986 Generalized information requirement of intelligent decision-making systems *SUGI-11 Proc.* (Cary, NC: SAS Institute)

——1987a Building and understanding adaptive systems: a statistical/numerical approach to factory automation and brain research *IEEE Trans. Syst. Man Cybern.* **17** 7–19

——1987b Learning how the world works *Proc. 1987 IEEE Int. Conf. on Systems, Man and Cybernetics* vol I (New York: IEEE Press) pp 302–10

——1988a Backpropagation: past and future *Proc. IEEE Int. Conf. on Neural Networks* vol I (New York: IEEE Press) pp 343–353

——1988b Generalization of backpropagation with application to a recurrent gas market model *Neural Networks* **1** 339–56

——1989a Backpropagation and neurocontrol: a review and prospectus *Proc. IEEE Int. Joint Conf. on Neural Networks* vol I (New York: IEEE Press) pp 209–16

——1989b Maximizing long-term gas industry profits in two minutes in Lotus using neural network methods *IEEE Trans. Syst. Man Cybern.* **1** 315–33

——1990 *Neural Networks for Control* ed W T Miller, R S Sutton and P J Werbos (Cambridge, MA: MIT Press)

Widrow B and Smith F W 1964 Pattern-recognizing control systems *Computer and Information Sciences* ed J T Tou and R H Wilcox (Washington, DC: Spartan) pp 288–317

Widrow B Gupta N K and Maitra S 1973 Punish/reward: learning with a critic in adaptive threshold systems *IEEE Trans. Syst. Man Cybern.* **3** 455–65

C2.4 Unsupervised ontogenic networks

Bernd Fritzke

Abstract

In this section, network models are described which learn unsupervised and generate their topology during learning. Among the models described, one can distinguish those having a specific dimensionality (e.g. two or three) from models whose dimensionality varies locally with the input data. Furthermore, models with a fixed number of units but variable connectivity can be distinguished from models which also change their number of units through insertion and/or deletion. Application areas of the described models include vector quantization, data visualization and clustering.

C2.4.1 Introduction

First, an incremental self-organizing network model from the author is described which is called 'growing cell structures'. The model is able to generate dimensionality reducing mappings which may be used, for example, for visualization of high-dimensional data or for clustering. The basic building blocks of the generated topology are hypertetrahedrons of a certain dimensionality k chosen in advance. In contrast to Kohonen's self-organizing feature map, which serves similar purposes, neither the number of units nor the exact topology has to be predefined in this model. Instead, a growth process successively inserts units and connections. All parameters in the model are constant over time rather than following a decay schedule. This makes it possible to continue the growth process until a specific network size is reached or until an application-dependent performance criterion is fulfilled. There is no need to define *a priori* the number of adaptation steps (in contrast, for example, to the self-organizing feature map). The input data directly guide the insertion of new units. Generally, this leads to network structures reflecting the given input distribution better than a predefined topology could.

Next, several methods are described which have been proposed by Martinetz and Schulten. *Competitive Hebbian learning* uses a given data distribution to generate topological neighborhoods among B3.3.1 a fixed set of centers. 'Neural gas' is a particular vector quantization method. If used in combination— referred to then as 'topology-representing networks'—these methods can generate optimally topology-preserving mappings for a network of predefined size.

Finally, a variation of the growing cell structures is described which is able to do 'topology learning'. Rather than mapping the given input data onto a structure of specific dimensionality this model, called 'growing neural gas', incrementally generates a topology which closely reflects the topology of the input data. In particular, the resulting graph may have different dimensionalities in different areas of the input space. Applications include clustering, vector quantization and function approximation. Again all parameters are constant, which leads to substantial advantages over other approaches such as, for example, the topology-representing networks of Martinetz and Schulten (1994).

C2.4.2 Growing cell structures

C2.4.2.1 Introduction

The model described here has been proposed earlier (Fritzke 1991) in order to overcome some problems of the *self-organizing feature map* (Kohonen 1982), mainly the difficult choice of the network structure C2.1.1

and the need to define a decay schedule for various parameters. Kohonen's model, which builds upon ideas of Willshaw and von der Malsburg (1976), aims at mapping high-dimensional input signals onto an (often two-dimensional) neural sheet of fixed size and structure in such a way that neighborhood relations among the input signals are preserved as well as possible. An essential prerequisite for this is a network structure matching the structure of the distribution. If this is approximately the case, Kohonen's model is able to find appropriate mappings (figure C2.4.1). In many cases, however, missing knowledge of the probability distribution of the data prevents the choice of a matching topology. The often chosen standard topology (a square grid) can then lead to poor mappings with rather arbitrary neighborhood relations (figure C2.4.2).

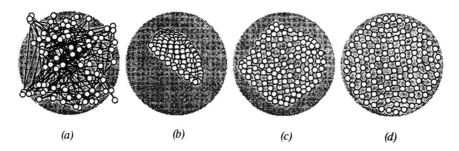

(a)　　　　*(b)*　　　　*(c)*　　　　*(d)*

Figure C2.4.1. A Kohonen self-organizing map finds an ordered mapping from the 2D input space onto a two-dimensional grid. The figure shows a 10×10 map which orders according to a uniform distribution in a circle. Shown here are the initial stage with randomly initialized reference vectors (a), two intermediate stages (b), (c), and and the final stage (d). In this case the resulting mapping reflects very well the topology of the input manifold: small changes in the input space lead to small (or no) changes of the corresponding neuron on the grid. Moreover, neighboring units on the grid have similar reference vectors. (a) 0 signals; (b) 30 signals; (c) 2000 signals; (d) 10 000 signals.

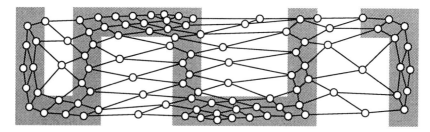

Figure C2.4.2. Kohonen map result after 10 000 signals for a clustered data distribution which is uniform in the shaded areas. Due to the mismatch between data distribution and network topology there are a large number of connections between cells with very different reference vectors. The given data distribution is not represented well.

C2.4.2.2 Purpose

The purpose of the growing cell structures model is the generation of a topology-preserving mapping from the input space $\mathbf{R}^n$ onto a topological structure A of equal or lower dimensionality k. By *topology preservation* we (informally) mean the following.

- Input vectors which are close in $\mathbf{R}^n$ should be mapped onto neighboring (or identical) nodes in A.
- Neighboring nodes in A should have similar input vectors mapped onto them.

In many situations this property is not achievable in a strict sense since a reversible mapping from high-dimensional spaces onto low-dimensional structures does not exist in general. In any case we intend to
B4.4.3　preserve at least the more prominent similarity relations by the mapping. There is some relation to *principal component analysis* (PCA). When doing a PCA on the data followed by a projection onto the subspace spanned by the eigenvectors corresponding to the k-largest eigenvalues, the *linear* dimensionality reduction is achieved which maximally preserves information. A growing cell structures model which generates a k-dimensional topological structure can be seen as a projection onto a *nonlinear*, discretely sampled

submanifold. The same holds for the self-organizing feature map (Kohonen 1982) which, however, relies on a predefined structure.

C2.4.2.3 Topology

The model comprises a set A of nodes. Each node $c \in A$ has an associated n-dimensional *reference vector* w_c which indicates the center of its receptive field in input space $\mathbf{R}^n$. A given set of nodes with their reference vectors defines a particular partition of the input space, the so-called Voronoi tessellation (see figure C2.4.9). The receptive field of each node c is the Voronoi field $V(c)$ of its reference vector w_c and can be characterized by

$$V(c) = \{p \in \mathbf{R}^n \mid (\|p - w_c\| < \|p - w_d\|)\, \forall d \in A, d \neq c\}.$$

The Voronoi field of c thus consists of those points in $\mathbf{R}^n$ for which w_c is the nearest of all currently existing reference vectors.

Between certain pairs of nodes there are edges indicating neighborhood. The resulting topology is strictly k-dimensional whereby k is some positive integer chosen in advance. The basic building block and also the initial configuration of each network is a k-dimensional simplex. This is, for example, a line for $k = 1$, a triangle for $k = 2$, and a tetrahedron for $k = 3$. Figure C2.4.3 shows some topologies for different values of k.

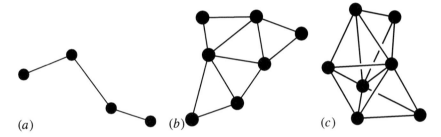

(a) (b) (c)

Figure C2.4.3. Example topologies for the growing cell structures model for different network dimensionalities k. The initial topology is always a k-dimensional simplex. Topologies for $k = 1$, $k = 2$ and $k = 3$, each consisting of several k-dimensional simplices, are shown.

For a given network configuration a number of adaptation steps are used to update the reference vectors of the nodes and to gather local error information at each node (see figure C2.4.4).

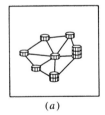 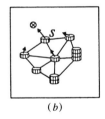 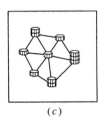

(a) (b) (c)

Figure C2.4.4. One adaptation step for a two-dimensional growing cell structures model. Only the best-matching unit s and its direct neighbors are adapted. The columns represent the local error variables. The error variable of the best-matching unit s is increased.

This error information is used to decide where to insert new nodes. A new node is always inserted by splitting the longest edge emanating from the node q with maximum accumulated error. In doing this, additional edges are inserted such that the resulting structure consists exclusively of k-dimensional simplices again (see figure C2.4.5).

C2.4.2.4 Learning rule

Learning in the described model comprises adaptation of the reference vectors *and* insertion of nodes and connections. This terminology makes sense since inserting a new node and interpolating its reference

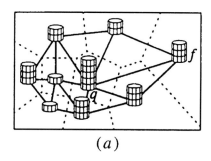 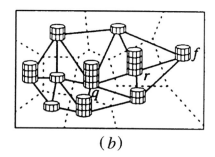

(a) (b)

Figure C2.4.5. Insertion of a new node in the growing cell structures model. (a) Situation before an insertion. Unit q has accumulated the most error during the preceding adaptation steps. The broken lines indicate the Voronoi regions of the units which can be interpreted as receptive fields. (b) A new unit r has been inserted in between the unit q and its direct neighbor which is furthest in input space (among all direct neighbors). The error variables of the neighbors of r have been reduced and r is given a nonzero initial value.

vector from the neighbors is in essence equivalent to performing a number of adaptation steps for a so-far unused unit. Insertions, however, can be done much faster than positioning by stepwise adaptation. Moreover, if adaptation is not required to move units over large distances, the parameters chosen can be small and constant.

The growing cell structures learning procedure is described in the following.

(i)   Start with a k-dimensional simplex. The $(k+1)$ vertices are initialized to random vectors in $\mathbf{R}^n$.

(ii) Choose an input signal ξ according to the input distribution $P(\xi)$.

(iii) Determine the best-matching unit s (the unit with the nearest reference vector):

$$\|w_s - \xi\| \leq \|w_c - \xi\| \qquad (\forall c \in A). \tag{C2.4.1}$$

(iv) Add the squared distance† between the input signal and the best-matching unit s to a local error variable E_s:

$$\Delta E_s = \|w_s - \xi\|^2.$$

(v)   Move s and its direct topological neighbors‡ towards ξ by fractions ϵ_b and ϵ_n, respectively, of the total distance:

$$\Delta w_s = \epsilon_b(\xi - w_s)$$
$$\Delta w_i = \epsilon_n(\xi - w_i) \qquad (\forall i \in N_s).$$

(With N_c we denote the set of topological neighbors of a unit c, i.e. those units which are connected to c by an edge.)

(vi) If the number of input signals generated so far is an integer multiple of a parameter λ, insert a new unit as follows.

   • Determine the unit q with the maximum accumulated error:

$$E_q \geq E_c \qquad (\forall c \in A).$$

   • Insert a new unit r by splitting the longest edge emanating from q, say an edge leading to a unit f. Insert the connections (q, r) and (r, f) and remove the original connection (q, f). To rebuild the structure such that it again consists only of k-dimensional simplices, the new unit r is also connected with all common neighbors of q and f, i.e. with all units in the set $N_q \cap N_f$.

---

† Depending on the problem at hand other local measures are also possible, for example, the *number* of input signals for which a particular unit is the winner or even the positioning error of a robot arm controlled by the network. The local measure should generally be something which one is interested in reducing and which is likely to be reduced by the insertion of new units.

‡ Throughout this paper the term *neighbors* denotes units which are topological neighbors in the graph (as opposed to units within a small Euclidean distance of each other in input space).

- Interpolate the reference vector of r from the vectors of q and f:

$$w_r = 0.5(w_q + w_f).$$

- Decrease† the error variables of all neighbors of r:

$$\Delta E_i = -\frac{\alpha}{|N_i|} E_i \qquad (\forall i \in N_r). \qquad (C2.4.2)$$

- Set the error variable of the new unit r to the mean value of its neighbors:

$$E_r = \frac{1}{|N_r|} \sum_{i \in N_r} E_i$$

(vii) Decrease the error variables of all units:

$$\Delta E_c = -\beta E_c \qquad (\forall c \in A). \qquad (C2.4.3)$$

(viii) If a stopping criterion (e.g. net size or some performance measure) is not yet fulfilled, continue with step (ii).

How does this method work? The accumulation of distortion error (iv) enables the identification of units generating high distortion error. Inserting a new unit at the same position, however, as the unit q with maximum accumulated error would not decrease the expected distortion error‡. Therefore, new units are always inserted between a unit and one of its neighbors (v). Instead of erasing all error variables after an insertion, only the variables of cells in the vicinity of the new cell are decreased. This preserves the accumulated error in other regions of the input space and makes it possible to insert new units always after a constant number λ of adaptation steps. Erasing the error variables would force us to choose λ proportional to the current network size. The exponential decay of all error variables (vii) stresses the impact of recently accumulated error, which makes sense since units are moving around slightly.

C2.4.2.5 Examples

Figure C2.4.6 shows some stages of the resulting growth process for a simple distribution. Figure C2.4.7 depicts the result of the self-organizing process for a more complicated distribution. The parameters used in both simulations were: $\alpha = 0.2$, $\varepsilon_b = 0.02$, $\varepsilon_n = 0.006$, $\beta = 0.0005$ and $\lambda = 200$.

C2.4.2.6 Related neural network models

In addition to the self-organizing feature map (Kohonen 1982) the described model is related to the elastic net model (Durbin and Willshaw 1987). The supervised variant of the growing cell structures (see, for example, Fritzke 1994b) is linked to the *radial basis function* model (Poggio and Girosi 1990, Moody and Darken 1989). If one replaces the strict k-dimensional structure with an unconstrained topology one arrives at the growing neural gas model described in section C2.4.3.

C1.6.2

† In principle, one could also erase all error variables each time a new unit is inserted. In this case one had to perform a number of adaptation steps per insertion which is proportional to the current network size since all units need a sufficient chance to be hit by some input signals and to accumulate error information. By keeping the error information, however, and only adjusting the values near newly inserted units, we only need a constant number of adaptation steps per insertion. The amount of adjustment actually used is a heuristic value. The correct determination of the appropriate error values after an insertion would require detailed knowledge of the probability distribution $P(\xi)$ of the input signals and the inherent dimensionality of the data. Assuming this knowledge for a unit i the expected summed square error would be the following integral in the Voronoi polyhedron $V(i)$:

$$E_i = \frac{1}{\beta} \int_{V(i)} P(\xi) \|w_i - \xi\|^2 \, d\xi$$

with β being the decay parameter from equation C2.4.3. One could now approximate the Voronoi polyhedron (which is hard to compute in high dimensions) and make assumptions on the unknown probability density $P(\xi)$ to improve (C2.4.2). From our experience, however, this is not necessary. One reason might be that the adaptation steps provide a constant flow of valid error information.

‡ One can consider the decrease in expected distortion error as a function $\Delta E(w_r)$ of the position w_r of the new unit r. This function is non-negative and has minima (where it takes the value 0) exactly at the positions of the existing units.

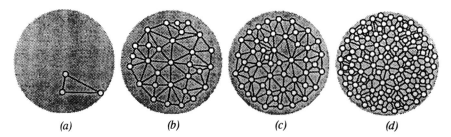

Figure C2.4.6. Growing cell structures simulation for the circular data distribution. Starting with one triangle of units, more units are inserted by splitting existing edges. The connectivity is updated such that the structure consists exclusively of triangles at every moment. The final structure covers the relevant area rather evenly. (*a*) 0 signals; (*b*) 6000 signals; (*c*) 10 000 signals; (*d*) 20 000 signals.

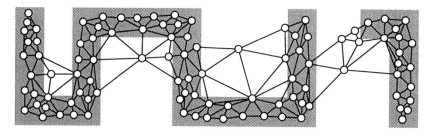

Figure C2.4.7. Growing cell structures result after 20 000 signals for the same data distribution as used previously for the Kohonen feature map. The growth process has led to a two-dimensional structure with a topology rather well adapted to the given data distribution.

C2.4.2.7 Advantages

In contrast to other approaches there is no need to specify the network size in advance. Moreover, all parameters are constant, which eliminates the need to define a decay (or annealing) schedule. Generally, parameter setting is unproblematic.

C2.4.2.8 Disadvantages

Due to the irregular and dynamic graph structure employed, the model is more difficult to implement than models with a predefined regular structure.

C2.4.2.9 Typical applications

F1.3 • *Combinatorial optimization.* A one-dimensional growing cell structures network which is closed to a ring can be used to generate approximate solutions to Euclidean traveling salesman problems (Fritzke and Wilke 1991).

 • *Data visualization.* By choosing a small network dimension k for high-dimensional input data it is possible to visualize the data with an embedding of the network in the k-dimensional space. Such an embedding can be constructed during the growth process (Fritzke 1994a). It always exists due to the dimensionality k of the network. See figure C2.4.8 for a simple example.

C2.4.2.10 Variations and improvements

Closed networks. Networks which have no 'outer' nodes, (e.g. a ring in the one-dimensional case or a closed two-dimensional mesh of triangles) and which are useful, for example, to generate approximate solutions for the traveling salesman problem (Fritzke 1991).

Supervised growing cell structures. A combination of the proposed networks with the radial basis function approach. New nodes are inserted on the basis of accumulated classification error which leads to near-minimal nets with strong generalization abilities (Fritzke 1994b).

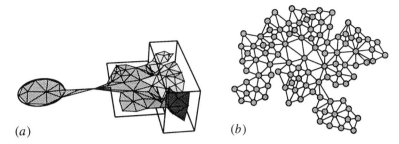

Figure C2.4.8. A two-dimensional growing cell structures network has adapted to a signal distribution which has different dimensionalities in different areas of the input space. The distribution, which has been used earlier by Martinetz and Schulten, consists of a (one-dimensional) ring, a line segment, a rectangle and a parallelepiped. (*a*) shows the network in the input space with the triangles in gray shades. (*b*) shows an embedding of the network in the plane. Since the network is two-dimensional, such an embedding exists independently of the input data dimensionality (three in this case). Thus, the growing cell structures network realizes a mapping from the input space to the plane while trying to preserve the topology of the input data distribution. This works best for the rectangular part. The 1D part is mapped onto two dimensions (see lower end of (*b*)). In the 3D region of the input space the network is folded, which is the only way to preserve the topology at least to some degree (like a space-filling curve in two dimensions).

Removal of units. By introducing the removal of units it is possible to get networks consisting of different unconnected graphs representing clusters in the input data. See Fritzke (1994a) for a removal criterion based on estimated probability density. This criterion is applicable if the *number of input signals* is locally accumulated. For the error-minimizing networks presented in this section, however, we are still in the process of formulating a suitable removal criterion. This criterion will be based on the quantization error which would occur if a specific unit would be removed.

C2.4.3 Competitive Hebbian learning, neural gas and topology-representing networks

C2.4.3.1 Introduction

The models described in this section were proposed at the same time (even in the same session of the 1991 International Conference on Artificial Neural Networks in Helsinki) as the growing cell structures. They represent another kind of extension of the original self-organizing map method with its predefined topology.

C2.4.3.2 Purpose

The purpose of the methods described here is to distribute a number of centers according to some probability distribution (neural gas) and to generate a topology among these centers which has a dimensionality which is equal to the *local* dimensionality of the data and may be different in different parts of the input space (competitive Hebbian learning). The generation of such a topology can be denoted as 'topology learning'.

C2.4.3.3 Topologies

Competitive Hebbian learning (Martinetz 1993) assumes a number of centers in $\mathbf{R}^n$ and successively inserts topological connections among them by evaluating input signals drawn from a data distribution $P(\xi)$. The principle of this method is:

For each input signal ξ connect the two closest centers (measured by Euclidean distance) by an edge.

The resulting graph is a subgraph of the Delaunay triangulation (figure C2.4.9(*a*)) corresponding to the set of centers. This subgraph (figure C2.4.9(*b*)), which is called the induced Delaunay triangulation, is limited to those areas of the input space $\mathbf{R}^n$ where $P(\xi) > 0$. The induced Delaunay triangulation has been shown to optimally preserve topology in a very general sense (Martinetz *et al* 1993).

The neural gas method does not use a particular topology. A 'network' simply consists of a number of disconnected centers in $\mathbf{R}^n$.

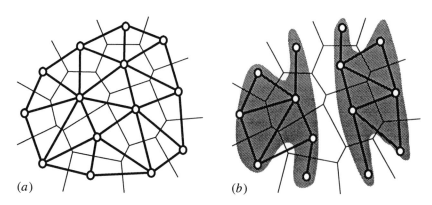

Figure C2.4.9. Two ways of defining closeness among a set of points. (*a*) The *Delaunay triangulation* (thick lines) connects points having neighboring *Voronoi polygons* (thin lines). Basically, this reduces to points having small Euclidean distance with respect to the given set of points. (*b*) The *induced Delaunay triangulation* (thick lines) is obtained by masking the original Delaunay triangulation with a data distribution $P(\xi)$ (shaded). Two centers are only connected if the common border of their Voronoi polygons lies at least partially in a region where $P(\xi) > 0$ (adapted from Martinetz and Schulten (1994)).

C2.4.3.4 Learning rules and examples

Formally, competitive Hebbian learning can be described as follows.

(i) Initialize the set A to contain N units c_i, at random positions $w_{c_i} \in \mathbf{R}^n$, $i = 1, 2 \ldots, N$:

$$A = \{c_1, c_2, \ldots, c_N\}.$$

Initialize the connection set C, $C \subset A \times A$, with the empty set (start with no connections):

$$C = \{\}.$$

(ii) Generate at random an input signal ξ according to $P(\xi)$.
(iii) Determine units s_1 and s_2 ($s_1, s_2 \in A$) such that

$$\|w_{s_1} - \xi\| \leq \|w_c - \xi\| \qquad (\forall c \in A)$$

and

$$\|w_{s_2} - \xi\| \leq \|w_c - \xi\| \qquad (\forall c \in A \setminus s_1).$$

(iv) If it does not exist already, insert a connection between s_1 and s_2 to C:

$$C = C \cup \{(s_1, s_2)\}.$$

(v) Continue with step (ii) unless the maximum number of signals is reached.

Only centers lying on the input data submanifold or in its vicinity actually develop any edges (see figures C2.4.10 and C2.4.11). The others are useless for the purpose of topology learning and are often called *dead units*. To make use of all centers they have to be placed in those regions of $\mathbf{R}^n$ where $P(\xi)$ differs from zero. This could be done by any vector quantization procedure. Martinetz and Schulten (1991) have proposed the neural gas method for this purpose (which is a vector quantization method). The main principle of neural gas is the following:

For each input signal ξ adapt all centers according to their *rank order* (nearest, second-nearest, etc) with respect to ξ. Decrease the number of significantly moved centers over time until only the winner is moved.

The important idea here is to use the distance in input space only for determining a rank order and then do adaptations on the basis of this rank order. This makes the method rather invariant to initialization. To achieve convergence, a decay schedule is needed for the parameters which again demands to define the total number of adaptation steps in advance.

Formally neural gas can be described as follows.

Handbook of Neural Computation release 97/1 © 1997 IOP Publishing Ltd and Oxford University Press

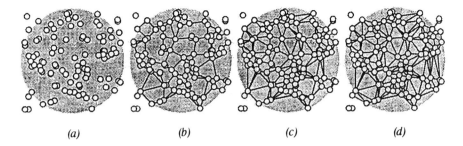

(a)   (b)   (c)   (d)

Figure C2.4.10. Competitive Hebbian learning simulation for the circular data distribution. Starting with a number of units at random positions, edges are inserted between best-matching and second-best-matching unit for each input signal. The positions of the units, however, remain unchanged which possibly leads to 'dead units' not contributing to the network. (a) 0 signals; (b) 500 signals; (c) 3000 signals; (d) 80 000 signals.

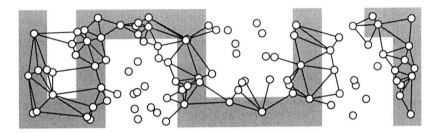

Figure C2.4.11. Competitive Hebbian learning result for the clustered data distribution after 80 000 signals. A considerable number of 'dead units' without any connections can be seen. The random initialization of center positions leads to unevenly well represented regions of the input data submanifold. The two clusters, however, could be detected by inspecting the graph structure.

(i)   Initialize the set A to contain N units c_i, at random positions: $w_{c_i} \in \mathbf{R}^n$ ($i = 1, 2, \ldots, N$).

$$A = \{c_1, c_2, \ldots, c_N\}.$$

Initialize the time parameter t:
$$t = 0.$$

(ii)   Generate at random an input signal ξ according to $P(\xi)$.

(iii)   Order all elements of A according to their distance to ξ, i.e. find the sequence of indices $(i_0, i_1, \ldots, i_{N-1})$ such that w_{i_0} is the reference vector closest to ξ, w_{i_1} is the reference vector second-closest to ξ and w_{i_k}, $k = 0, \ldots, N - 1$ is the reference vector such that k vectors w_j exist with $\|w_j - \xi\| < \|w_k - \xi\|$. Following Martinetz *et al* (1993) we denote with $k_i(\xi, A)$ the number k associated with w_i.

(iv)   Adjust the reference vectors according to

$$\Delta w_i = \epsilon \cdot h_\lambda(k_i(\xi, A)) \cdot (\xi - w_i)$$

with the following time dependencies:

$$\lambda(t) = \lambda_i(\lambda_f/\lambda_i)^{t/t_{max}} \qquad \epsilon(t) = \epsilon_i(\epsilon_f/\epsilon_i)^{t/t_{max}} \qquad \text{and} \quad h_\lambda(k) = \exp(-k/\lambda).$$

(v)   Increase the time parameter t:
$$t = t + 1.$$

(vi)   If $t < t_{max}$ continue with step (ii).

For the time-dependent parameters suitable initial values (λ_i, ϵ_i) and final values (λ_f, ϵ_f) have to be chosen. In figures C2.4.12 and C2.4.13 simulation results for our example distributions are shown. Following Martinetz *et al* (1993) we used the following parameters: $\lambda_i = 10$, $\lambda_f = 0.01$, $\epsilon_i = 0.5$, $\epsilon_f = 0.005$, $t_{max} = 40\,000$.

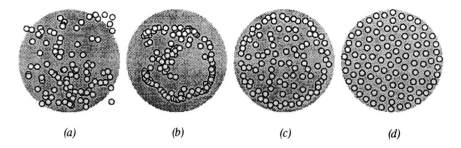

Figure C2.4.12. Neural gas simulation for the circular data distribution. Strong neighborhood interaction at the beginning leads to a rapid clustering of the units which loosens up towards the end of the simulation where a very uniform distribution of centers is achieved. (*a*) 0 signals; (*b*) 250 signals; (*c*) 6000 signals; (*d*) 40 000 signals.

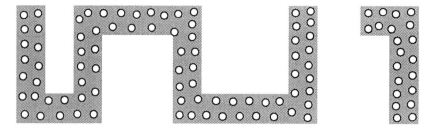

Figure C2.4.13. Neural gas result after 40 000 input signals for the clustered data distribution. The distribution of the centers reflects well the underlying data distribution. There is no topological information involved in neural gas, i.e. there are no neighborhood connections.

For a given data distribution one could now first run the neural gas algorithm to distribute a certain number of centers and then use competitive Hebbian learning to generate the topology. It is, however, also possible to apply both techniques concurrently (Martinetz and Schulten 1991). In this case a method for removing obsolete edges is required, since the motion of the centers may make edges invalid which have been generated earlier. Martinetz and Schulten use an *edge aging* scheme for this purpose. One should note that the competitive Hebbian learning algorithm does not influence the outcome of the neural gas method in any way since the adaptations in neural gas are based only on distance in input space and not on the network topology. On the other hand neural gas does influence the topology generated by competitive Hebbian learning, since it moves the centers around. The straightforward combination of competitive Hebbian learning and neural gas has been called 'topology-representing networks'. We do not describe this method separately due to lack of space (Martinetz and Schulten (1994) give a comprehensive description). Simulation results for the topology-representing networks for our example distributions are shown in figures C2.4.14 and C2.4.15.

C2.4.3.5 Related neural network models

C1.1.5 The neural gas method is related to other *vector quantization* methods. Competitive Hebbian learning and the topology-representing networks model are related to the growing neural gas model.

C2.4.3.6 Advantages

The combination of neural gas and competitive Hebbian learning described above, which is sometimes referred to as topology-representing networks (Martinetz and Schulten 1994), is an effective method for topology learning. If a sufficient number of units is used a topological structure develops which characterizes very well the topology of the underlying data distribution. The nature of competitive Hebbian learning prevents the method forming 'topological defects' as have been observed for the self-organizing feature map.

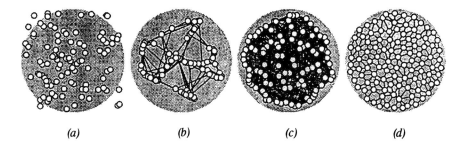

Figure C2.4.14. Neural gas plus competitive Hebbian learning simulation for the circular data distribution. Starting with a number of units at random positions, edges are inserted between best-matching and second-best-matching unit for one input signal (competitive Hebbian learning). At the same time the unit positions are adapted with the neural gas method. Strong neighborhood interaction at the beginning leads to a rapid clustering of the units which loosens up towards the end of the simulation. In intermediate stages a very high connectivity may occur which, however, vanishes towards the end. (*a*) 0 signals; (*b*) 250 signals; (*c*) 6000 signals; (*d*) 40 000 signals.

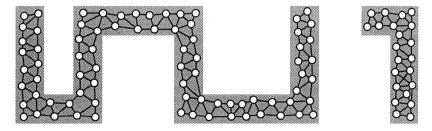

Figure C2.4.15. Neural gas plus competitive Hebbian learning result after 40 000 signals for the clustered data distribution. The final structure reflects very well the structure of the underlying distribution. Since a parameter decay is needed for neural gas, the number of adaptation steps has to be fixed in advance. The same holds for the number of units which must be fixed beforehand. If no knowledge of the distribution is present, this is a difficult step.

C2.4.3.7 Disadvantages

A problem in practical applications may be to determine *a priori* a suitable number of centers. Depending on the complexity of the data distribution which one wants to model, different numbers of centers may be appropriate. The underlying neural gas algorithm requires a decision in advance and, if the result is not satisfying, one or more new simulations have to be performed from scratch.

The parameter λ determining the degree of neighborhood adaptation must be set large enough to 'capture' all units in early stages. Otherwise dead units might occur which do not contribute to the network. On the other hand λ must decay fast enough to allow the units to be distributed according to the underlying data distribution (as opposed to the position dictated by neigborhood interaction as can be observed in intermediate stages of the simulation (see figure C2.4.14(*b*)).

C2.4.3.8 Typical applications

The methods described can be used for vector quantization and clustering. Also, a supervised version has been proposed which associates a local linear mapping with every unit and may be used, for example, for time-series prediction (Martinetz *et al* 1993).

C2.4.4 Growing neural gas

C2.4.4.1 Introduction

The model described in this section can be seen as a variant of the growing cell structures without the strict topological constraints of the former model. One could also interpret it as an incremental variant of the topology-representing networks described in the previous section.

C2.4.4.2 Purpose

The purpose of the growing neural gas model is to generate a graph structure which reflects the topology of the input data manifold (topology learning). This graph has a dimensionality which varies with the dimensionality of the input data. The resulting structure can be used to identify clusters in the input data. The nodes by themselves can be used as a codebook for vector quantization.

Moreover, node positions and the neighborhood information contained in the edges can be used to set up an interpolation scheme where function values for arbitrary position in $\mathbf{R}^n$ are computed from the values stored at the nearest node and its neighboring nodes in the graph. The growing neural gas model shares some properties with the topology-representing networks (Martinetz *et al* 1993). In particular, competitive Hebbian learning (Martinetz *et al* 1993) is used to generate the topology. The incremental structure of our method, however, leads to some inherent advantages of the approach presented. The growing neural gas model relates to the topology-representing networks in a similar way as the growing cell structures model relates to the self-organizing feature map.

C2.4.4.3 Topology

In the growing neural gas model there are no such restrictions on the topology as in the growing cell structures model. In principle, arbitrary edges are allowed and, in particular, the topology may have different dimensionalities in different parts of the input space.

C2.4.4.4 Learning rule

In the following we consider networks consisting of

- a set A of units (or nodes). Each unit $c \in A$ has an associated *reference vector* $w_c \in \mathbf{R}^n$. The reference vectors can be regarded as positions in input space of the corresponding units.
- a set C of connections (or edges) among pairs of units. These connections are not weighted. Their sole purpose is the definition of topological structure.

Moreover, there is a (possibly infinite) number of n-dimensional input signals obeying some unknown probability density function $P(\xi)$.

The main idea of the method is successively to add new units to an initially small network by evaluating local statistical measures gathered during previous adaptation steps. This is the same approach as used in the growing cell structures model (Fritzke 1994a) which, in contrast, has a topology with a fixed dimensionality.

In the approach described here, the network topology is generated incrementally by competitive Hebbian learning and has a dimensionality which depends on the input data and may vary locally. The complete algorithm for our model which we call growing neural gas is given by the following:

(i) Start with a set A of two units a and b at random positions w_a and w_b in $\mathbf{R}^n$:

$$A = \{a, b\}.$$

Initialize the connection set C to contain an edge between a and b and set the age of this connection to zero:

$$C = \{(a, b)\} \qquad \text{age}_{(a,b)} = 0.$$

(ii) Generate an input signal ξ according to $P(\xi)$.

(iii) Determine units s_1 and s_2 ($s_1, s_2 \in A$) such that

$$\|w_{s_1} - \xi\| \le \|w_c - \xi\| \qquad (\forall c \in A)$$

and

$$\|w_{s_2} - \xi\| \le \|w_c - \xi\| \qquad (\forall c \in A \setminus s_1).$$

(iv) If it does not already exist, insert a connection between s_1 and s_2 to C:

$$C = C \cup \{(s_1, s_2)\}.$$

In any case: set the age of the connection between s_1 and s_2 to zero ('refresh' the edge):

$$\text{age}_{(s_1, s_2)} = 0.$$

(v) Add the squared distance between the input signal and the nearest unit in input space to a local error variable:

$$\Delta E_{s_1} = \|w_{s_1} - \xi\|^2 .$$

(vi) Move s_1 and its direct topological neighbors towards ξ by fractions ϵ_b and ϵ_n, respectively, of the total distance:

$$\Delta w_{s_1} = \epsilon_b(\xi - w_{s_1})$$
$$\Delta w_i = \epsilon_n(\xi - w_i) \qquad (\forall i \in N_{s_1}) .$$

(vii) Increment the age of all edges emanating from s_1:

$$\text{age}_{(s_1,i)} = \text{age}_{(s_1,i)} + 1 \qquad (\forall i \in N_{s_1})$$

whereby N_c is the set of direct topological neighbors† of c.

(viii) Remove edges with an age larger than $a_{\max}$. If this results in units having no emanating edges, remove them as well.

(ix) If the number of input signals generated so far is an integer multiple of a parameter λ, insert a new unit as follows.

- Determine the unit q with the maximum accumulated error.

$$E_q \geq E_c \qquad (\forall c \in A)$$

- Interpolate a new unit r from q and its neighbor f with the largest error variable:

$$A = A \cup \{r\} \qquad w_r = 0.5(w_q + w_f) .$$

- Insert edges connecting the new unit r with units q and f, and remove the original edge between q and f:

$$C = C \cup \{(r, q), (r, f)\} \qquad C = C \setminus (q, f) .$$

- Decrease‡ the error variables of q and f.

$$\Delta E_q = -\alpha E_q \qquad \Delta E_f = -\alpha E_f .$$

Interpolate the error variable of r from q and f:

$$E_r = 0.5(E_q + E_f) .$$

(x) Decrease the error variables of all units:

$$\Delta E_c = -\beta E_c \qquad (\forall c \in A) .$$

(xi) If a stopping criterion (e.g. net size or some performance measure) is not yet fulfilled, continue with step (ii).

How does the method described work? The adaptation steps towards the input signals (vi) lead to a general movement of all units towards those areas of the input space where signals come from ($P(\xi) > 0$). The insertion of edges (vii) between the nearest and the second-nearest unit with respect to an input signal generates a single connection of the induced Delaunay triangulation (see figure C2.4.9(b)) *with respect to the current position of all units*. The removal of edges (viii) is necessary to get rid of those edges which are no longer part of the induced Delaunay triangulation because their ending points have moved. This is achieved by *local* edge aging (vii) around the nearest unit combined with age resetting of those edges (iv) which already exist between nearest and second-nearest units. With insertion and removal of edges the model tries to construct and then track the induced Delaunay triangulation which is a slowly moving target due to the adaptation of the reference vectors. The accumulation of squared distances (v) during the adaptation helps to identify units lying in areas of the input space where the mapping from signals to units causes much error. To reduce this error, new units are inserted in such regions.

† From the definition of N_c and C it follows that for each unit c the following holds: $i \in N_c$ iff $(c, i) \in C$. We do not generally distinguish between edges (c, i) and (i, c), i.e. edges are always undirected.
‡ This step is heuristic and makes it possible to use only a constant number λ of adaptation steps per insertion. If, on the other hand, one were to use a number of adaptation steps proportional to the current network size, the error values could simply be erased after each insertion. See also the corresponding remarks for the growing cell structures model.

C2.4.4.5 Examples

In figures C2.4.16 and C2.4.17 the results of our two standard examples can be seen. The parameters for this simulation were: $\lambda = 200$, $\epsilon_b = 0.05$, $\epsilon_n = 0.0006$, $\alpha = 0.5$, $a_{max} = 88$ and $\beta = 0.0005$. For the circle the growth process looks very similar to that of the growing cell structures. One should note, however, that the dimensionality of the growing neural gas network is not strictly fixed but depends only on the dimensionality of the input data which in this example happens to be two everywhere.

This property can be seen very well for the probability distribution in figure C2.4.18 which has been proposed by Martinetz and Schulten (1991) to demonstrate their (nonincremental) neural gas/competitive Hebbian learning model. We have used this distribution for the growing neural gas model and it is obvious that the model quickly identifies the important topological relations in this rather complicated distribution by forming structures of different dimensionalities. See figure C2.4.8 for the very different result of the growing cell structures model for the same distribution.

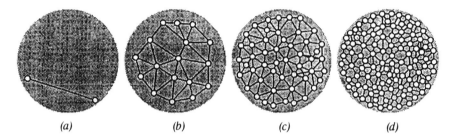

(a)   (b)   (c)   (d)

Figure C2.4.16. Growing neural gas evolves a topology for the circular distribution. Starting with two units (which may or may not be connected) more and more units are inserted. Connections are generated by competitive Hebbian learning and die away if they are not refreshed for a while. (*a*) 0 signals; (*b*) 2800 signals; (*c*) 9400 signals; (*d*) 20 000 signals.

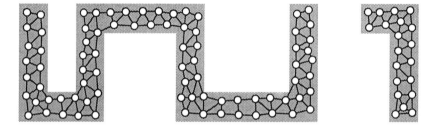

Figure C2.4.17. Growing neural gas result after 20 000 signals for the data distribution used for the other models. The growth process (stopped after 100 units were present) has led to a structure with a topology very well adapted to the given data distribution. Note that the structure consists of two clusters reflecting the clustered data distribution and that all reference vectors lie in those regions where data actually come from.

C2.4.4.6 Related neural network models

This model is related in particular to the growing cell structures model from which it differs through its less rigid topology definition. Moreover, there is a relation to topology-representing networks since also competitive Hebbian learning and edge aging are also used to generate the topology.

C2.4.4.7 Advantages

In contrast to the neural gas/competitive Hebbian learning combination of Martinetz and Schulten, the network size need not be predefined in this model. All parameters are constant and also the total number of adaptation steps need not be defined *a priori*. The growth process can be interrupted when a user-defined performance criterion has been fulfilled. Due to the fractal-like growth all intermediate stages of the network are rather good descriptions of the underlying distribution with a resolution depending on the current network size. The model of Martinetz and Schulten, however, has intermediate stages with very

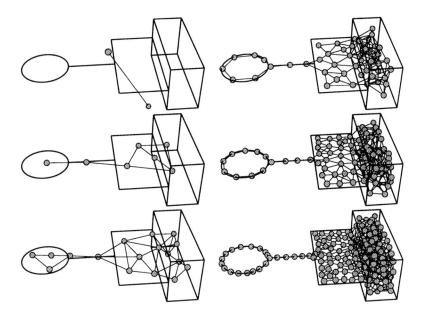

Figure C2.4.18. The growing neural gas network adapts to a signal distribution which has different dimensionalities in different areas of the input space. The initial network consisting of two randomly placed units and the networks of size 7, 17, 50, 100 and 200 after 1000, 3000, 9600, 19 600 and 39 600 input signals, respectively, are shown. The last network shown is not necessarily the 'final' one since the growth process could, in principle, be continued indefinitely. The parameters for this simulation were: $\lambda = 200$, $\epsilon_b = 0.05$, $\epsilon_n = 0.0006$, $\alpha = 0.5$, $a_{max} = 88$ and $\beta = 0.0005$.

unevenly distributed centers and extremely high connectivity (see figure C2.4.14) which provide only poor descriptions of the underlying data.

C2.4.4.8 Disadvantages

An inherent disadvantage of this type of model (which it shares with the topology-representing networks of the previous section) is the fact that a visualization of the network is only (directly) possible if the data are low-dimensional. The reason is, of course, that no dimensionality reduction is intended by topology-learning networks so that they have the same dimensionality as the data.

C2.4.4.9 Typical applications

Applications of this unsupervised model are clustering or vector quantization. If one combines this model with local linear mappings or the radial basis function model, incremental supervised models are also possible.

References

Durbin R and Willshaw D 1987 An analogue approach to the travelling salesman problem using an elastic net method *Nature* **326** 689–91

Fritzke B 1991 Unsupervised clustering with growing cell structures *Proc. Int. Joint Conf. on Neural Networks 1991 (Seattle, WA)* vol II pp 531–6

——1994a Growing cell structures—a self-organizing network for unsupervised and supervised learning *Neural Networks* **7** 1441–60

——1994b Supervised learning with growing cell structures *Advances in Neural Information Processing Systems 6* ed J Cowan, G Tesauro and J Alspector (San Mateo, CA: Morgan Kaufmann) pp 255–62

Fritzke B and Wilke P 1991 FLEXMAP—A neural network with linear time and space complexity for the traveling salesman problem *Proc. Int. Joint Conf. on Neural Networks 1991 (Singapore)* pp 929–34

Kohonen T 1982 Self-organized formation of topologically correct feature maps *Biol. Cybern.* **43** 59–69

Martinetz T M 1993 Competitive Hebbian learning rule forms perfectly topology preserving maps *Int. Conf. Artificial Neural Networks* (Amsterdam: Springer) pp 427–34

Martinetz T M, Berkovich S G and Schulten K J 1993 Neural-gas network for vector quantization and its application to time-series prediction *IEEE Trans. Neural Networks* **4** 558–69

Martinetz T M and Schulten K J 1991 A 'neural-gas' network learns topologies *Artificial Neural Networks* ed T Kohonen, K Mäkisara, O Simula and J Kangas (Amsterdam: North-Holland) pp 397–402

——1994 Topology representing networks *Neural Networks* **7** 507–22

Moody J E and Darken C 1989 Fast learning in networks of locally-tuned processing units *Neural Comput.* **1** 281–94

Poggio T and Girosi F 1990 Networks for approximation and learning *Proc. IEEE* **78** 1481–97

Willshaw D J and von der Malsburg C 1976 How patterned neural connections can be set up by self-organization *Proc. R. Soc.* B **194** 431–45

C3

Reinforcement Learning

S Sathiya Keerthi and B Ravindran

Abstract

This chapter gives a compact, self-contained tutorial survey of reinforcement learning, a tool that is increasingly finding application in the development of intelligent dynamic systems. Research on reinforcement learning during the past decade has led to the development of a variety of useful algorithms. This chapter surveys the literature and presents the algorithms in a cohesive framework.

Contents

C3.1   Introduction

S Sathiya Keerthi and B Ravindran

Abstract

See the abstract for Chapter C3.

Reinforcement learning (RL), a term borrowed from animal learning literature by Minsky (1954, 1961), refers to a class of learning tasks and algorithms in which the learning system learns an associative mapping, $\pi : X \rightarrow A$ by maximizing a scalar evaluation (reinforcement) of its performance from the environment (user). Compared to *supervised learning*, in which for each x shown the environment provides the learning   B3.1 system with the value of $\pi(x)$, RL is more difficult since it has to work with much less feedback from the environment. If, at some time, given an $x \in X$, the learning system tries an $a \in A$ and the environment immediately returns a scalar reinforcement evaluation of the (x, a) pair (that indicates how far a is from $\pi(x)$) then we are faced with an *immediate* RL task. A more difficult RL task is *delayed* RL, in which the environment only gives a single scalar reinforcement evaluation, collectively for $\{(x_t, a_t)\}$, a sequence of (x, a) pairs occurring in time during the system operation. Delayed RL tasks commonly arise in optimal control of dynamic systems and planning problems of artificial intelligence (AI). In this chapter our main interest is in the solution of delayed RL problems. However, we also study immediate RL problems because methods of solving them play a useful role in the solution of delayed RL problems.

Delayed RL encompasses a diverse collection of ideas having roots in animal learning (Barto 1985, Sutton and Barto 1987), *control theory* (Bertsekas 1989, Kumar 1985), and AI (Dean and Wellman 1991).   F1.9 Delayed RL algorithms were first employed by Samuel (1959, 1967) in his celebrated work on playing checkers. However, it was only much later, after the publication of Barto, Sutton and Anderson's work (Barto *et al* 1983) on a delayed RL algorithm called the *adaptive heuristic critic* and its application to   C2.3.3 the control problem of pole balancing, that research on RL got off to a flying start. Watkins' Q-learning algorithm (Watkins 1989) made another impact on the research. A number of significant ideas have emerged rapidly during the past five years and the field has reached a certain level of maturity. In this chapter we provide a comprehensive tutorial survey of various ideas and methods of delayed RL. To avoid distractions and an unnecessary clutter of notations, we present all the ideas in an intuitive, not-so-rigorous fashion. In preparing this tutorial, we have obtained a lot of guidance from the works of Watkins (1989), Barto *et al* (1990, 1992), Bradtke (1994) and Barto (1992).

To illustrate the key features of a delayed RL task let us consider a simple example.

C3.1.1   Example: navigating a robot

Figure C3.1.1 illustrates a grid world in which a robot navigates. Each blank cell on the grid is called a *state*. Shaded cells represent barriers; these are not states. Let X be the state space, that is, the set of states. The cell marked G is the goal state. The aim is to reach G from any state in the least number of time steps. Navigation is done using four *actions*: $A = \{N, S, E, W\}$, the actions denoting the four possible movements along the coordinate directions.

Rules of transition are defined as follows. Suppose that the robot is in state x and action N is chosen. Then the resulting next state, y, is the state directly to the north of x, *if there is such a state*; otherwise $y = x$. For instance, choosing W at the x shown in figure C3.1.1 will lead to the system staying at x. The goal state is a special case. By definition we will take it that any action taken from the goal state results in a transition back to the goal state. In more general problems, the rules of transition can be stochastic.

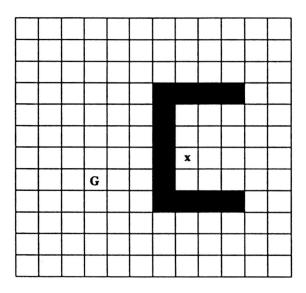

Figure C3.1.1. Navigating in a grid world.

The robot moves at discrete (integer) time points starting from $t = 0$. At a time step t, when the robot is at state x_t, we define an immediate reward† as

$$r(x_t) = \begin{cases} 0 & \text{if } x_t = G \\ -1 & \text{otherwise} . \end{cases}$$

In effect, the robot is penalized for every time step spent at non-goal states. It is simple to verify that maximizing the *total reward* over time,

$$V(x) = \sum_{t=0}^{\infty} r(x_t)$$

is equivalent to achieving minimum time navigation from the starting state, $x_0 = x$. Let $V^\star(x)$ denote the maximum achievable (optimal) value of $V(x)$.

We are interested in finding a feedback policy, $\pi : X \to A$ such that, if we start from any starting state and select actions using π, then we will always reach the goal in the minimum number of time steps.

The usefulness of immediate RL methods in delayed RL can be roughly explained as follows. Typical delayed RL methods maintain $\hat{V}$, an approximation of the optimal function, $V^\star$. If action a is performed at state x and state y results, then $\hat{V}(y)$ can be taken as an (approximate) immediate evaluation of the (x, a) pair—an optimal action at x is one that gives the maximum value of $V^\star(y)$. By solving an immediate RL problem that uses this evaluation function we can obtain a good suboptimal policy for the delayed RL problem. We present relevant immediate RL algorithms in Section C3.2.

Delayed RL problems are much harder to solve than immediate RL problems for the following reason. Suppose, in the example of navigating a robot, performance of a sequence of actions, selected according to some policy, leads the robot to the goal. To improve the policy using the experience, we need to evaluate the goodness of each action performed. But the total reward obtained gives only the cumulative effect of all actions performed. Some scheme must be found to reasonably apportion the cumulative evaluation to the individual actions. This is referred to as the *temporal credit assignment problem*. (In the previous paragraph we have already given a hint of how delayed RL methods do temporal credit assignment.)

C2.3.4   *Dynamic programming* (DP) (Bertsekas 1989, Ross 1983) is a well known tool for solving problems such as navigating a robot. It is an off-line method that requires the availability of a complete model of the environment. But the concerns of delayed RL are very different. To see this clearly let us return to the example of navigating a robot and impose the requirement that the robot has no knowledge of the

† Sometimes r is referred to as the primary reinforcement. In more general situations, r is a function of x_t as well as a_t, the action at time step t.

environment and that the only way of learning is by on-line experience of trying various actions† and thereby visiting many states. Delayed RL algorithms are particularly meant for such situations and have the following general format.

C3.1.2 Delayed reinforcement learning algorithm

Initialize the learning system.
Repeat

(i) *With the system at state x, choose an action a according to an exploration policy and apply it to the system.*
(ii) *The environment returns a reward, r, and also yields the next state, y.*
(iii) *Use the experience, (x, a, r, y) to update the learning system.*
(iv) *Set $x := y$.*

Even when a model of the environment is available, it is often advantageous to avoid an off-line method such as DP and instead use a delayed RL algorithm. This is because in many problems the state space is very large; while a DP algorithm operates with the entire state space, a delayed RL algorithm only operates on parts of the state space that are most relevant to the system operation. When a model is available, delayed RL algorithms can employ a simulation mode of operation instead of on-line operation so as to speed up learning and avoid doing experiments using hardware. In this chapter, we will use the term *real-time operation* to mean that either on-line operation or simulation mode of operation is used.

In most applications, representing functions such as V^* and π exactly is infeasible. A better alternative is to employ parametric function approximators, for example, neural networks. Such approximators must be suitably chosen for use in a delayed RL algorithm. To clarify this, let us take V^*, for instance, and consider a function approximator, $\hat{V}(\cdot; w) : X \to R$, for it. Here R denotes the real line and w denotes the vector of parameters of the approximator that is to be learnt so that $\hat{V}$ approximates V^* well. Usually, at step (iii) of the delayed RL algorithm, the learning system uses the experience to come up with a direction η in which $\hat{V}(x; w)$ has to be changed for improving performance. Given a step size, β, the function approximator must alter w to a new value, w^{new} so that

$$\hat{V}(x; w^{\text{new}}) = \hat{V}(x; w) + \beta\eta. \tag{C3.1.1}$$

For example, in *multilayer perceptrons* (Hertz *et al* 1991, Haykin 1994) w denotes the set of weights and thresholds in the network and their updating can be carried out using *backpropagation* so as to achieve (C3.1.1). In the rest of the chapter we will denote the updating process in (C3.1.1) as   C1.2   C1.2.3

$$\hat{V}(x; w) := \hat{V}(x; w) + \beta\eta \tag{C3.1.2}$$

and refer to it as a *learning rule*.   B3.3

The chapter is organized as follows. Section C3.2 discusses immediate RL. In Section C3.3 we formulate delayed RL problems and mention some basic results. Methods of estimating total reward are discussed in Section C3.4. These methods play an important role in delayed RL algorithms. DP techniques and delayed RL algorithms are presented in Section C3.5. Sections C3.6–C3.8 address various practical issues. We make a few concluding remarks in Section C3.9.

References

Barto A G 1985 Learning by statistical cooperation of self-interested neuron-like computing elements *Human Neurobiol.* **4** 229–56
——1986 Game-theoretic cooperativity in networks of self-interested units *Neural Networks for Computing* ed J S Denker (New York: AIP) pp 41–6
——1992 Reinforcement learning and adaptive critic methods *Handbook of Intelligent Control: Neural, Fuzzy, and Adaptive Approaches* ed D A White and D A Sofge (New York: Van Nostrand Reinhold) pp 469–91
Barto A G, Bradtke S J and Singh S P 1992 Real-time learning and control using asynchronous dynamic programming *Technical Report COINS 91–57* University of Massachusetts, Amherst, MA, USA

† During learning this is usually achieved by using a (stochastic) exploration policy for choosing actions. Typically the exploration policy is chosen to be totally random at the beginning of learning and made to approach an optimal policy as learning nears completion.

Barto A G, Sutton R S and Anderson C W 1983 Neuronlike elements that can solve difficult learning control problems *IEEE Trans. Syst. Man Cybern.* **13** 835–46

Barto A G, Sutton R S and Watkins C J C H 1990 Learning and sequential decision making *Learning and Computational Neuroscience: Foundations of Adaptive Networks* ed M Gabriel and J Moore (Cambridge, MA: MIT Press) pp 539–602

Bertsekas D P 1989 *Dynamic Programming: Deterministic and Stochastic Models* (Englewood Cliffs, NJ: Prentice-Hall)

Bradtke S J 1994 Incremental dynamic programming for online adaptive optimal control *CMPSCI Technical Report* pp 94–62

Dean T L and Wellman M P 1991 *Planning and Control* (San Mateo, CA: Morgan Kaufmann)

Haykin S 1994 *Neural Networks: A Comprehensive Foundation* (New York: Macmillan)

Hertz J A, Krogh A S and Palmer R G 1991 *Introduction to the Theory of Neural Computation* (Redwood City, CA: Addison-Wesley)

Kumar P R 1985 A survey of some results in stochastic adaptive control *SIAM J. Control Opt.* **23** 329–380

Minsky M L 1954 Theory of neural–analog reinforcement systems and application to the brain-model problem *PhD Thesis* Princeton University, Princeton, NJ, USA

——1961 Steps towards artificial intelligence *Proc. Inst. Radio Eng.* **49** 8–30 1963 (Reprinted in *Computers and Thought* ed E A Feigenbaum and J Feldman (New York: McGraw-Hill) pp 406–50)

Ross S 1983 *Introduction to Stochastic Dynamic Programming* (New York: Academic)

Samuel A L 1959 Some studies in machine learning using the game of checkers *IBM J. Res. Develop.* pp 210–29 (Reprinted in 1963 *Computers and Thought* ed E A Feigenbaum and J Feldman (New York: McGraw-Hill))

——1967 Some studies in machine learning using the game of checkers II—recent progress *IBM J. Res. Develop.* pp 601–17

Sutton R S and Barto A G 1987 A temporal-difference model of classical conditioning *Proc. Ninth Ann. Conf. of the Cognitive Science Society* (Erlbaum, Hillsdale, NJ)

Watkins C J C H 1989 Learning from delayed rewards *PhD Thesis* Cambridge University, Cambridge, UK

C3.2   Immediate reinforcement learning

S Sathiya Keerthi and B Ravindran

Abstract

See the abstract for Chapter C3.

Immediate reinforcement learning (RL) refers to the learning of an associative mapping, $\pi : X \to A$ given a reinforcement evaluator. To learn, the learning system interacts in a closed loop with the environment. At each time step, the environment chooses an $x \in X$ and the learning system uses its function approximator, $\hat{\pi}(\cdot; w)$, to select an action: $a = \hat{\pi}(x; w)$. Based on both x and a, the environment returns an evaluation or 'reinforcement', $r(x, a) \in R$. Ideally, the learning system has to adjust w so as to produce the maximum possible r value for each x; in other words, we would like $\hat{\pi}$ to solve the parametric global optimization problem,

$$r(x, \hat{\pi}(x; w)) = r^{\star}(x) \stackrel{\text{def}}{=} \max_{a \in A} r(x, a) \qquad \forall x \in X. \tag{C3.2.1}$$

Supervised learning is a popular paradigm for learning associative mappings (Hertz *et al* 1991, Haykin 1994). In supervised learning, for each x shown the supervisor provides the learning system with the value of $\pi(x)$. Immediate RL and supervised learning differ in the following two important ways.

- In supervised learning, when an x is shown and the supervisor provides $a = \pi(x)$, the learning system forms the directed information, $\eta = a - \hat{\pi}(x; w)$ and uses the learning rule: $\hat{\pi}(x; w) := \hat{\pi}(x; w) + \alpha\eta$, where α is a small (positive) step size. For immediate RL such directed information is not available and so the system has to employ some strategy to obtain such information.
- In supervised learning, the learning system can simply check if $\eta = 0$ and hence decide whether the correct map value has been formed by $\hat{\pi}$ at x. However, in immediate RL, such a conclusion on correctness cannot be made without exploring the values of $r(x, a)$ for all a.

Therefore, immediate RL problems are much more difficult to solve than supervised learning problems.

A number of immediate RL algorithms have been described in the literature. Stochastic learning automata algorithms (Narendra and Thathachar 1989) deal with the special case in which A is a finite set and $r \in [0, 1]$. The associative reward–penalty ($A_{\text{R–P}}$) algorithm (Barto and Anandan 1985, Barto *et al* 1985, Barto and Jordan 1987, Mazzoni *et al* 1990) extends the learning automata ideas to the case where X is a finite set. Williams (1986, 1987) has proposed a class of immediate RL methods and has presented interesting theoretical results. Gullapalli (1990, 1992) has developed algorithms for the general case in which X, A are finite-dimensional real spaces and r is real valued. Here we will discuss only algorithms which are most relevant to, and useful in, delayed RL.

One simple way of solving (C3.2.1) is to take one x at a time, use a global optimization algorithm (e.g. complete enumeration) to explore the A space and obtain the correct a for the given x, and then make the function approximator learn this (x, a) pair. However, such an idea is not used for the following reason. In most situations where immediate RL is used as a tool (e.g. to approximate a policy in delayed RL), the learning system has little control over the choice of x. When, at a given x, the learning system chooses a particular a and sends it to the environment for evaluation, the environment not only sends a reinforcement evaluation but also alters the x value. Immediate RL seeks approaches which are appropriate to these situations.

Let us first consider the case in which A is a finite set: $A = \{a^1, a^2, \ldots, a^m\}$. Let R^m denote the m-dimensional real space. The function approximator, $\hat{\pi}$, is usually formed as a composition of two functions: a function approximator, $g(\cdot; w) : X \to R^m$ and a fixed function, $M : R^m \to A$. The idea behind this setup is as follows. For each given x, $z = g(x; w) \in R^m$ gives a vector of merits of the various a^i values. Let z_k denote the kth component of z. Given the merit vector z, $a = M(z)$ is formed by the max selector,

$$a = a^k \qquad \text{where} \quad z_k = \max_{1 \le i \le m} z_i \,. \qquad (C3.2.2)$$

Let us now come to the issue of learning (i.e. choosing a w). At some stage, let x be the input, z be the merit vector returned by g, and a^k be the action having the largest merit value. The environment returns the reinforcement $r(x, a^k)$. In order to learn we need to evaluate the goodness of z^k (and therefore the goodness of a^k). Obviously, we cannot do this using existing information. We need an estimator, call it $\hat{r}(x; v)$, that provides an estimate of $r^\star(x)$. The difference, $r(x, a^k) - \hat{r}(x; v)$ is a measure of the goodness of a^k. Then, a simple learning rule is

$$g_k(x; w) := g_k(x; w) + \alpha(r(x, a^k) - \hat{r}(x; v)) \qquad (C3.2.3)$$

where α is a small (positive) step size. If $\hat{r}(\cdot; v) \equiv r^\star$ and (C3.2.3) is repeated a number of times for each (x, k) combination, then it should be clear that all nonoptimal a^ks will get large negative merit values while an optimal a^k will retain its initial merit value.

Learning $\hat{r}$ requires that all members of A are evaluated by the environment at each x. Clearly, the max selector (C3.2.2) is not suitable for such exploration. For instance, if at some stage of learning, for some x, g assigns the largest merit to a wrong action, say a^k, and $\hat{r}$ gives, by mistake, a value smaller than $r(x, a^k)$, then no action other than a^k is going to be generated by the learning system at the given x. So we replace (C3.2.2) by a controlled stochastic action selector that generates actions randomly when learning begins and approaches (C3.2.2) as learning is completed. A popular stochastic action selector is based on the Boltzmann distribution,

$$p_i(x) \stackrel{\text{def}}{=} \text{Prob}\{a = a^i | x\} = \frac{\exp(z_i / T)}{\sum_j \exp(z_j / T)} \qquad (C3.2.4)$$

where T is a nonnegative real parameter (temperature) that controls the stochasticity of the action selector. For a given x the expected reinforcement of the action selector is

$$\tilde{r}(x) \stackrel{\text{def}}{=} \text{E}(r(x, a)|x) = \sum_i p_i(x) r(x, a^i) \,. \qquad (C3.2.5)$$

As $T \to 0$ the stochastic action selector approaches the max selector (C3.2.2) and $\tilde{r}(x) \to r^\star(x)$. The ideas here are somewhat similar to those of *simulated annealing*. Therefore we train $\hat{r}$ to approximate $\tilde{r}$ (instead of $r^\star$). This is easy to do because, for any fixed value of T, $\tilde{r}$ can be estimated by the average of the performance of the stochastic action selector over time. A simple learning rule that achieves this is

C1.4.2

$$\hat{r}(x; v) := \hat{r}(x; v) + \beta(r(x, a) - \hat{r}(x; v)) \qquad (C3.2.6)$$

where β is a small (positive) step size.

Two important comments should be made regarding the convergence of learning rules such as (C3.2.6) (we will come across many such learning rules later) which are designed to estimate an expectation by averaging over time.

- Even if $\hat{r}(\cdot; v) \equiv \tilde{r}$, $r(x, a) - \hat{r}(x; v)$ can be nonzero and even large in size. This is because a is only an instance generated by the distribution, $p(x)$. Therefore, to avoid unlearning as $\hat{r}$ comes close to $\tilde{r}$, the step size, β must be controlled properly. The value of β may be chosen to be slightly smaller than 1 when learning begins, and then slowly decreased to 0 as learning progresses.
- For good learning to take place, the sequence of x values at which (C3.2.6) is carried out must be such that it covers all parts of the space X as often as possible. Of course, when the learning system has no control over the choice of x, it can do nothing to achieve such an exploration. To explore, the following is usually done. Learning is done over a number of *trials*. A trial consists of beginning with a random choice of x and operating the system for several time steps. At any one time step, the system is at some x and the learning system chooses an action, a and learns using (C3.2.6). Depending on x, a and the rules of the environment, a new x results and the next time step begins. Usually, when learning is repeated over multiple trials, the X space is thoroughly explored.

Let us now consider the case in which A is continuous, say a finite-dimensional real space. The idea of using merit values is not suitable. It is better to deal directly with a function approximator, $h(\cdot; w)$ from X to A. In order to do exploration a controlled random perturbation η is added to $h(x; w)$ to form $a = \hat{\pi}(x)$. A simple choice is to take η to be a Gaussian with zero mean and having a standard deviation, $\sigma(T)$ that satisfies $\sigma(T) \to 0$ as $T \to 0$. The setting-up and training of the reinforcement estimator, $\hat{r}$ is as in the case when A is discrete. The function approximator h can adopt the following learning rule:

$$h(x; w) := h(x; w) + \alpha(r(x, a) - \hat{r}(x; v))\eta \qquad (C3.2.7)$$

where α is a small (positive) step size. In problems where a bound on r^* is available, this bound can be suitably employed to guide exploration, that is, to choose σ (Gullapalli 1990).

Jordan and Rumelhart (1990) have suggested a method of 'forward models' for continuous action spaces. If r is a known differentiable function, then a simple, deterministic learning law based on gradient ascent can be given to update $\hat{\pi}$:

$$\hat{\pi}(x; w) := \hat{\pi}(x; w) + \alpha \frac{\partial r(x, a)}{\partial a} . \qquad (C3.2.8)$$

If r is not known, Jordan and Rumelhart suggest that it is learned using on-line data, and (C3.2.8) be used using this learned r. If, for a given x, the function $r(x, \cdot)$ has local maxima then the $\hat{\pi}(x)$ obtained using learning rule (C3.2.8) may not converge to $\pi(x)$. Typically, this is not a serious problem. The stochastic approach discussed earlier does not suffer from local maxima problems. However, we should add that, because the deterministic method explores in systematic directions and the stochastic method explores in random directions, the former is expected to be much faster. The comparison is very similar to the comparison of deterministic and stochastic techniques of continuous optimization.

References

Barto A G and Anandan P 1985 Pattern recognizing stocahstic learning automata *IEEE Trans. Syst. Man and Cybern.* **15** 360–75

Barto A G, Anandan P and Anderson C W 1985 Cooperativity in networks of pattern recognizing stochastic learning automata In *Proc. Fourth Yale Workshop on Appl. Adapt. Systems Theory (New Haven, CT)*

Barto A G and Jordan M I 1987 Gradient following without back-propagation in layered networks *Proc. IEEE First Ann. Conf. on Neural Networks (San Diego, CA)* ed M Caudill and C Butler pp II629–36

Gullapalli V 1990 A stochastic reinforcement algorithm for learning real-valued functions *Neural Networks* **3** 671–92

——1992 Reinforcement learning and its application to control *Technical Report COINS 92–10, PhD Thesis* University of Massachusetts, Amherst, MA, USA

Haykin S 1994 *Neural Networks: A Comprehensive Foundation* (New York: Macmillan)

Hertz J A, Krogh A S and Palmer R G 1991 *Introduction to the Theory of Neural Computation* (Redwood City, CA: Addison-Wesley)

Jordan M I and Rumelhart D E 1990 Forward models: supervised learning with a distal teacher *Center for Cognitive Science Occasional Paper # 40* Massachusetts Institute of Technology Cambridge, MA, USA

Mazzoni P, Andersen R A and Jordan M I 1990 A_{R-P} learning applied to a network model of cortical area 7a *Proc. 1990 Int. Joint Conf. on Neural Networks* **2** 373–9

Narendra K and Thathachar M A L 1989 *Learning Automata: an Introduction* (Englewood Cliffs, NJ: Prentice-Hall)

Williams R J 1986 Reinforcement learning in connectionist networks: a mathematical analysis *Technical Report ICS 8605* Institute for Cognitive Science, University of California at San Diego, La Jolla, CA, USA

——1987 Reinforcement-learning connectionist systems *Technical report NU–CCS–87–3* College of Computer Science, Northeastern University, Boston, MA, USA

C3.3 Delayed reinforcement learning

S Sathiya Keerthi and B Ravindran

Abstract

See the abstract for Chapter C3.

Delayed reinforcement learning (RL) concerns the solution of stochastic optimal control problems. In this section we formulate and discuss the basics of such problems. Solution methods for delayed RL will be presented in Sections C3.4 and C3.5. In these three sections we will mainly consider problems in which   C3.4, C3.5 the state and control spaces are finite sets. This is because the main issues and solution methods of delayed RL can be easily explained for such problems. We will deal with continuous state and/or action spaces briefly in Section C3.5.

Consider a discrete-time stochastic dynamic system with a finite set of states, X. Let the system begin its operation at $t = 0$. At time t the *agent (controller)* observes state† x_t and selects (and performs) action a_t from a finite set, $A(x_t)$, of possible actions. Assume that the system is Markovian and stationary, that is,

$$\text{Prob}\{x_{t+1} = y \mid x_0, a_0, x_1, a_1, \ldots, x_t = x, a_t = a\}$$
$$= \text{Prob}\{x_{t+1} = y \mid x_t = x, a_t = a\} \overset{\text{def}}{=} P_{xy}(a).$$

A *policy* is a method adopted by the agent to choose actions. The objective of the decision task is to find a policy that is optimal according to a well defined sense, described below. In general, the action specified by the agent's policy at some time can depend on the entire past history of the system. Here we restrict attention to policies that specify actions based only on the current state of the system. A deterministic policy, π, defines for each $x \in X$ an action $\pi(x) \in A(x)$. A stochastic policy π defines, for each $x \in X$, a probability distribution on the set of feasible actions at x, that is, it gives the values of $\text{Prob}\{\pi(x) = a\}$ for all $a \in A(x)$. For the sake of keeping the notations simple we consider only deterministic policies in this section. All ideas can easily be extended to stochastic policies using appropriate detailed notations.

Let us now precisely define the optimality criterion. While at state x, if the agent performs action a, it receives an immediate *payoff* or *reward*, $r(x, a)$. Given a policy π we define the *value function*, $V^\pi : X \to R$ as follows‡:

$$V^\pi(x) = E\left\{\sum_{t=0}^{\infty} \gamma^t r(x_t, \pi(x_t)) | x_0 = x\right\}.$$

Here future rewards are discounted by a factor $\gamma \in [0, 1)$. The case $\gamma = 1$ is avoided only because it leads to some difficulties associated with the existence of the summation in (C3.3). Of course, these difficulties can be handled by putting appropriate assumptions on the problem solved. But, to avoid unnecessary distraction we do not go into the details; see Bradtke (1994) and Bertsekas and Tsitsiklis (1989).

The expectation in (C3.3) should be understood as

$$V^\pi(x) = \lim_{N \to \infty} E\left\{\sum_{t=0}^{N-1} \gamma^t r(x_t, \pi(x_t)) | x_0 = x\right\} \tag{C3.3.1}$$

---

† If the state is not completely observable then a method that uses the observable states and retains past information has to be used; see Bacharach (1991, 1992), Chrisman (1992), Mozer and Bacharach (1990a, b), Whitehead and Ballard (1990). See Jaakkola *et al* (1995) and Singh *et al* (1994) for a direct treatment of partially observable Markovian decision processes.

‡ Most RL researchers have concerned themselves with the optimization of the expected total discounted reward in (C3.3). See Heger (1994) for a discussion of an alternative objective function: the minimax criterion.

where the probability with which a particular state sequence $\{x_t\}_{t=0}^{N-1}$ occurs is taken in an obvious way using $x_0 = x$ and repeatedly employing π and P. We wish to maximize the value function:

$$V^*(x) = \max_\pi V^\pi(x) \qquad \forall x \,. \tag{C3.3.2}$$

V^* is referred to as the optimal value function. Because $0 \le \gamma < 1$, $V^\pi(x)$ is bounded. Also, since the number of πs is finite $V^*(x)$ exists.

How do we define an optimal policy, π^*? For a given x let $\pi^{x,*}$ denote a policy that achieves the maximum in (C3.3.2). Thus we have a collection of policies, $\{\pi^{x,*} : x \in X\}$. Now π^* is defined by picking only the first action from each of these policies:

$$\pi^*(x) = \pi^{x,*}(x) \qquad x \in X \,.$$

It turns out that π^* achieves the maximum in (C3.3.2) for every $x \in X$. In other words,

$$V^*(x) = V^{\pi^*}(x) \qquad x \in X \,. \tag{C3.3.3}$$

This result is easy to see if one looks at Bellman's optimality equation—an important equation that V^* satisfies:

$$V^*(x) = \max_{a \in A(x)} \left[r(x, a) + \gamma \sum_{y \in X} P_{xy}(a) V^*(y) \right] \,. \tag{C3.3.4}$$

The fact that V^* satisfies (C3.3.4) can be explained as follows. The term within square brackets on the right-hand side is the total reward that one would get if action a is chosen at the first time step and then the system performs optimally in all future time steps. Clearly, this term cannot exceed $V^*(x)$ since that would violate the definition of $V^*(x)$ in (C3.3.2); also, if $a = \pi^{x,*}(x)$ then this term should equal $V^*(x)$. Thus equation (C3.3.4) holds. It also turns out that V^* is the unique function from X to R that satisfies (C3.3.4) for all $x \in X$. This fact, however, requires a nontrivial proof; details can be found in Ross (1983), Bertsekas (1989), Bertsekas and Tsitsiklis (1989).

The above discussion also yields a mechanism for computing π^* if V^* is known:

$$\pi^*(x) = \arg \max_{a \in A(x)} \left[r(x, a) + \gamma \sum_{y \in X} P_{xy}(a) V^*(y) \right] \,.$$

A difficulty with this computation is that the system model, that is, the function $P_{xy}(a)$, must be known. This difficulty can be overcome if, instead of the V-function, we employ another function called the Q-function. Let $\mathcal{U} = \{(x, a) : x \in X, a \in A(x)\}$, the set of feasible (state, action) pairs. For a given policy π, let us define $Q^\pi : \mathcal{U} \to R$ by

$$Q^\pi(x, a) = r(x, a) + \gamma \sum_{y \in X} P_{xy}(a) V^\pi(y) \,. \tag{C3.3.5}$$

Thus $Q^\pi(x, a)$ denotes the total reward obtained by choosing a as the first action and then following π for all future time steps. Let $Q^* = Q^{\pi^*}$. By Bellman's optimality equation and (C3.3.3) we get

$$V^*(x) = \max_{a \in A(x)} [Q^*(x, a)] \,. \tag{C3.3.6}$$

It is also useful to rewrite Bellman's optimality equation using Q^* alone:

$$Q^*(x, a) = r(x, a) + \gamma \sum_{y \in X} P_{xy}(a) \{ \max_{b \in A(y)} Q^*(y, b) \} \,. \tag{C3.3.7}$$

Using Q^* we can compute π^*:

$$\pi^*(x) = \arg \max_{a \in A(x)} [Q^*(x, a)] \,. \tag{C3.3.8}$$

Thus, if Q^* is known then π^* can be computed without using a system model. This advantage of the Q-function over the V-function will play a crucial role in Section C3.5 for deriving a model-free delayed RL algorithm called Q-learning (Watkins 1989).

Let us now consider a few examples that give useful hints for problem formulation. These examples are also commonly mentioned in the RL literature.

C3.3.1 Example: navigating a robot with dynamics

In section C3.1.1 the robot is moved from one cell to another like the way pieces are moved on a chess board. True robot motions, however, involve dynamics; the effects of velocity and acceleration need to be considered. In this example we will include dynamics in a crude way, one that is appropriate to the grid world. Let h_t and v_t denote the horizontal and vertical coordinates of the cell occupied by the robot at time t and $\dot{h}_t$ and $\dot{v}_t$ denote the velocities. The vector $(h_t, v_t, \dot{h}_t, \dot{v}_t)$ denotes the system state at time t; each one of the four components is an integer. The goal state is $x^G = (h^G, v^G, 0, 0)$ where (h^G, v^G) is the coordinate vector of the goal cell G. In other words, the robot has to come to rest at G. Let $\dot{h}_{max}$ and $\dot{v}_{max}$ be limits on velocity magnitudes. Thus, the state space is given by

$$\tilde{X} = \{x = (h, v, \dot{h}, \dot{v}) \mid (h, v) \text{ is a blank cell, } |\dot{h}| \le \dot{h}_{max}, \text{ and } |\dot{v}| \le \dot{v}_{max}\}.$$

We will also include an extra state f called failure state to denote situations where a barrier (shaded) cell is entered or a velocity limit is exceeded. Thus,

$$X = \tilde{X} \cup \{f\}.$$

The accelerations (negative acceleration will mean deceleration) along the horizontal and vertical directions, respectively a^h and a^v, are the actions. To keep h and v as integers let us assume that each of the accelerations takes only even integer values. Let a_{max} be a positive even integer that denotes the limit on the magnitude of accelerations. Thus $a = (a^h, a^v)$ is an admissible action if each of a^h and a^v is an even integer lying in $[-a_{max}, a_{max}]$.

As in section C3.1.1 state transitions are deterministic. They are defined as follows. If barrier cells and velocity limits are not present, then application of action (a^h, a^v) at $x_t = (h_t, v_t, \dot{h}_t, \dot{v}_t)$ will lead to the next state $x'_{t+1} = (h'_{t+1}, v'_{t+1}, \dot{h}'_{t+1}, \dot{v}'_{t+1})$ given by

$$\begin{aligned}
h'_{t+1} &= h_t + \dot{h}_t + a^h/2 & v'_{t+1} &= v_t + \dot{v}_t + a^v/2 \\
\dot{h}'_{t+1} &= \dot{h}_t + a^h & \dot{v}'_{t+1} &= \dot{v}_t + a^v.
\end{aligned}$$

Let C denote the curve in the grid world resulting during the transition from (h_t, v_t) at time t to (h'_{t+1}, v'_{t+1}) at time $(t+1)$, that is, the solution of the differential equations: $\mathrm{d}^2 h/\mathrm{d}\tau^2 = a^h$, $\mathrm{d}^2 v/\mathrm{d}\tau^2 = a^v$, $\tau \in [t, t+1]$, $h(t) = h_t$, $\mathrm{d}h/\mathrm{d}\tau|_\tau = \dot{h}_t$, $v(t) = v_t$, $\mathrm{d}v/\mathrm{d}\tau|_\tau = \dot{v}_t$. If either C cuts across a barrier cell or $(\dot{h}'_{t+1}, \dot{v}'_{t+1})$ is an inadmissible velocity vector, then we say failure has occurred during transition. Thus, state transitions are defined as

$$x_{t+1} = \begin{cases}
f & \text{if } x_t = f \\
f & \text{if failure occurs during transition} \\
x^G & \text{if } x^t = x^G \\
x'_{t+1} & \text{otherwise}.
\end{cases}$$

The primary aim is to avoid failure. Next, among all failure-avoiding trajectories we would like to choose the trajectory which reaches the goal state, $x^G = (h^G, v^G, 0, 0)$, in as few time steps as possible. These aims are met if we define

$$r(x, a) = \begin{cases}
-1 & \text{if } x = f \\
1 & \text{if } x = x^G \\
0 & \text{otherwise}.
\end{cases}$$

The following can be easily checked.

- $V^\star(x) < 0$ iff there does not exist a trajectory starting from x that avoids failure.
- $V^\star(x) = 0$ iff, starting from x, there exists a failure-avoiding trajectory, but there does not exist a trajectory that reaches G.
- $V^\star(x) > 0$ iff, starting from x, there exists a failure-avoiding trajectory that also reaches G; also, an optimal policy $\pi^\star$ leads to the generation of a trajectory that reaches G in the fewest number of steps from x while avoiding failure.

C3.3.2 Example: playing backgammon

Consider a game of backgammon (Magriel 1976) between players A and B. Let us look at the game from A's perspective, assuming that B follows a fixed policy. Now A can make a decision on a move only when the current board pattern, as well as its dice roll, are known. Therefore, a state consists of a (board pattern, dice roll) pair. Each action consists of a set of marker movements. State transition is defined as follows.

- A moves its markers in accordance with the chosen action. This step is deterministic and results in a new board pattern.
- B rolls the dice. This step is stochastic.
- B moves its markers according to its policy. This step can be deterministic or stochastic depending on the type of B's policy.
- A rolls the dice. This step is stochastic.

The set of states that correspond to A's win is the set of goal states, G, to be reached. We can define the reward as: $r(x, a) = 1$ if x is a goal state; and $r(x, a) = 0$ otherwise. If $\gamma = 1$, then for a given policy, say π, the value function $V^\pi(x)$ will denote the probability that A will win from that state.

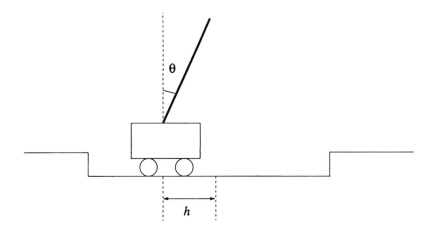

Figure C3.3.1. Pole balancing.

C3.3.3 Example: pole balancing

We now deviate from our problem formulation and present an example that involves continuous state/action spaces. A standard problem for learning controllers is that of balancing an inverted pendulum pivoted on a trolley, a problem similar to that of balancing a stick on one's hand (Barto *et al* 1983). The system comprises a straight horizontal track, like a railway track, with a carriage free to move along it. On the carriage is an axis, perpendicular to the track and pointing out to the side, about which a pendulum is free to turn. The controller's task is to keep the pendulum upright, by alternately pulling and pushing the carriage along the track. Let h and θ be as shown in figure C3.3.1. We say balancing has failed if any one of the following inequalities is violated:

$$h \le h_{\max} \qquad h \ge -h_{\max} \qquad \theta \le \theta_{\max} \qquad \theta \ge -\theta_{\max}$$

where $h_{\max}$ and $\theta_{\max}$ are specified bounds on the magnitudes of h and θ. The aim is to balance without failure for as long a time as possible.

The state of the system is the 4-tuple, $(h, \dot h, \theta, \dot\theta)$, where $\dot h$ and $\dot\theta$ are the time derivatives of h and θ, respectively. The action is the force applied to the carriage. It takes real values in the interval, $[-F_{\max}, F_{\max}]$. To simplify the problem solution, sometimes the action space is taken to be $\{-F_{\max}, F_{\max}\}$ (Michie and Chambers 1968, Barto *et al* 1983, Anderson 1989). A discrete time formulation of the problem is obtained by cutting continuous time (nonnegative real line) into uniform time intervals, each of duration Δ, and taking the applied force to be constant within each interval. (This constant is the action for the time step corresponding to that interval.) The state of the system at the continuous time instant $t\Delta$ is taken to be x_t, the discrete time state at the tth time step. The mechanical dynamics of the system defines state transition, except for one change: once failure occurs, we will assume, for the sake of consistent problem formulation, that the system stays at failure for ever.

As in the example of pole balancing we will take the state space to be $X = \tilde{X} \cup \{f\}$, where

$$\tilde{X} = \{x = (h, \dot{h}, \theta, \dot{\theta}) | -h_{max} \leq h \leq h_{max}, \quad -\theta_{max} \leq \theta \leq \theta_{max}\} \quad \text{(C3.3.9)}$$

and f is the failure state that collectively represents all states not in $\tilde{X}$. Since the aim is to avoid failure, we choose

$$r(x, a) = \begin{cases} -1 & \text{if} \quad x = f, \\ 0 & \text{otherwise.} \end{cases}$$

References

Anderson C W 1989 Learning to control an inverted pendulum using neural networks *IEEE Control Syst. Mag.* 31–7
Bacharach J R 1991 A connectionist learning control architecture for navigation *Advances in Neural Information Processing Systems 3* ed R P Lippman, J E Moody and D S Touretzky (San Mateo, CA: Morgan Kaufmann) pp 457–463
——1992 Connectionist modeling and control of finite state environments *PhD Thesis* University of Massachusetts, Amherst, MA, USA
Barto A G 1992 Reinforcement learning and adaptive critic methods *Handbook of Intelligent Control: Neural, Fuzzy, and Adaptive Approaches* ed D A White and D A Sofge (New York: Van Nostrand Reinhold) pp 469–91
Barto A G, Sutton R S and Anderson C W 1983 Neuronlike elements that can solve difficult learning control problems *IEEE Trans. Syst. Man Cybern.* **13** 835–46
Bertsekas D P 1989 *Dynamic Programming: Deterministic and Stochastic Models* (Englewood Cliffs, NJ: Prentice-Hall)
Bertsekas D P and Tsitsiklis J N 1989 *Parallel and Distributed Computation: Numerical Methods* (Englewood Cliffs, NJ: Prentice-Hall)
Bradtke S J 1994 Incremental dynamic programming for online adaptive optimal control *CMPSCI Technical Report* pp 94–62
Chrisman L 1992 Planning for closed-loop execution using partially observable markovian decision processes *Proc. AAAI*
Heger M 1994 Consideration of risk in reinforcement learning *Proc. 11th Int. Machine Learning Conf. ML-94*
Jaakkola T, Singh S P and Jordan M I 1995 Reinforcement learning algorithm for partially observable Markov decision processes *Advances in Neural Information Processing Systems 7* ed G Tesauro, D Touretzky and T Leen (Cambridge, MA: MIT Press)
Magriel P 1976 *Backgammon* (New York: Times Books)
Michie D and Chambers R A 1968 BOXES: An experiment in adaptive control *Machine Intelligence 2* ed E Dale and D Michie (Oliver and Boyd) pp 137–152
Mozer M C and Bacharach J 1990a Discovering the structure of reactive environment by exploration *Advances in Neural Information Processing 2* ed D S Touretzky (San Mateo, CA: Morgan Kaufmann) pp 439–446
——1990b Discovering the structure of reactive environment by exploration *Neural Comput.* **2** 447–57
Ross S 1983 *Introduction to Stochastic Dynamic Programming* (New York: Academic)
Singh S P 1994 Learning to solve Markovian decision processes *PhD Thesis* Department of Computer Science, University of Massachussetts, Amherst, MA, USA
Singh S P, Jaakkola T and Jordan M I 1994 Learning without state-estimation in partially observable Markov decision processes *Machine Learning*
Watkins C J C H 1989 Learning from delayed rewards *PhD Thesis* Cambridge University, Cambridge, UK
Whitehead S D and Ballard D H 1990 Active perception and reinforcement learning *Neural Comput.* **2** 409–19

C3.4 Methods of estimating V^π and Q^π

S Sathiya Keerthi and B Ravindran

Abstract

See the abstract for Chapter C3.

Delayed reinforcement learning (RL) methods use a knowledge of V^π (Q^π) in two crucial ways: (i) the optimality of π can be checked by seeing if V^π (Q^π) satisfies Bellman's optimality equation; and (ii) if π is not optimal then V^π (Q^π) can be used to improve π. We will elaborate on these details in the next section. In this section we discuss, in some detail, methods of estimating V^π for a given policy, π. (Methods of estimating Q^π are similar and so we will deal with them briefly at the end of the section.) Our aim is to find $\hat{V}(\cdot; v)$, a function approximator that estimates V^π. Much of the material in this section is taken from the works of Watkins (1989), Sutton (1984, 1988) and Jaakkola *et al* (1994).

To avoid clumsiness we employ some simplifying notations. Since π is fixed we will omit the superscript from V^π and so call it V. We will refer to $r(x_t, \pi(x_t))$ simply as r_t. If p is a random variable, we will use p to denote both the random variable as well as an instance of the random variable.

A simple approximation of $V(x)$ is the *n-step truncated return*,

$$V^{[n]}(x) = \sum_{\tau=0}^{n-1} \gamma^\tau r_\tau \qquad \hat{V}(x; v) = E(V^{[n]}(x)). \tag{C3.4.1}$$

(Here it is understood that $x_0 = x$. Thus, throughout this section τ will denote the number of time steps elapsed after the system passed through state x. It is to stress this point that we have used τ instead of t. In a given situation, the use of time—'actual system time' or 'time relative to the occurrence of x'—will be obvious from the context.) If r_{max} is a bound on the size of r then it is easy to verify that

$$\max_x |\hat{V}(x; v) - V(x)| \le \frac{\gamma^n r_{max}}{(1 - \gamma)}. \tag{C3.4.2}$$

Thus, as $n \to \infty$, $\hat{V}(x; v)$ converges to $V(x)$ uniformly in x.

But equation (C3.4.1) suffers from an important drawback. The computation of the expectation requires the complete enumeration of the probability tree of all possible states reachable in n time steps. Since the breadth of this tree may grow very large with n, the computations can become very burdensome. One way of avoiding this problem is to set

$$\hat{V}(x; v) = V^{[n]}(x) \tag{C3.4.3}$$

where $V^{[n]}(x)$ is obtained via either Monte Carlo simulation or experiments on the real system (the latter choice is the only way for systems for which a model is unavailable). The approximation (C3.4.3) suffers from a different drawback. Because the breadth of the probability tree grows with n, the variance of $V^{[n]}(x)$ also grows with n. Thus $\hat{V}(x; v)$ in (C3.4.3) will not be a good approximation of $E(V^{[n]}(x))$ unless it is obtained as an average over a large number of trials. (As already mentioned, a trial consists of starting the system at a random state and then running the system for a number of time steps.) Averaging is achieved if we use a learning rule (similar to (C3.2.6)):

$$\hat{V}(x; v) := \hat{V}(x; v) + \beta \left[V^{[n]}(x) - \hat{V}(x; v) \right] \tag{C3.4.4}$$

where β is a small (positive) step size. Learning can begin with a random choice of v. Eventually, after a number of trials, we expect the $\hat{V}$ resulting from (C3.4.4) to satisfy (C3.4.2).

In the above approach, an approximation of V, $\hat{V}$ is always available. Therefore, an estimate that is *more appropriate than* $V^{[n]}(x)$ is the *corrected n-step truncated return*,

$$V^{(n)}(x) = \sum_{\tau=0}^{n-1} \gamma^\tau r_\tau + \gamma^n \hat{V}(x_n; v) \tag{C3.4.5}$$

where x_n is the state that occurs n time steps after the system passed through state x. Let us do some analysis to justify this statement.

First, consider the ideal learning rule,

$$\hat{V}(x; v) := E(V^{(n)}(x)) \qquad \forall\, x. \tag{C3.4.6}$$

Suppose v gets modified to v_{new} in the process of satisfying (C3.4.6). Then, similarly to (C3.4.2), we can easily derive

$$\max_x |\hat{V}(x; v_{\text{new}}) - V(x)| \le \gamma^n \max_x |\hat{V}(x; v) - V(x)|.$$

Thus, as we go through a number of learning steps we achieve $\hat{V} \to V$. Note that this convergence is achieved even if n is fixed at a small value, say $n = 1$. On the other hand, for a fixed n, the learning rule based on $V^{[n]}$, that is, equation (C3.4.1), is only guaranteed to achieve the bound in (C3.4.2). *Therefore, when a system model is available it is best to choose a small n, say n = 1, and employ (C3.4.6).*

Now suppose that either a model is unavailable or (C3.4.6) is to be avoided because it is expensive. In this case, a suitable learning rule that employs $V^{(n)}$ and uses real-time data is

$$\hat{V}(x; v) := \hat{V}(x; v) + \beta\big[V^{(n)}(x) - \hat{V}(x; v)\big]. \tag{C3.4.7}$$

Which is better: (C3.4.4) or (C3.4.7)? There are two reasons why (C3.4.7) is better.

(i) Suppose $\hat{V}$ is a good estimate of V. Then a small n makes $V^{(n)}$ ideal: $V^{(n)}(x)$ has a mean close to $V(x)$ and it also has a small variance. Small variance means that (C3.4.7) will lead to fast averaging and hence fast convergence of $\hat{V}$ to V. On the other hand n has to be chosen large for $V^{[n]}(x)$ to have a mean close to $V(x)$; but then, $V^{[n]}(x)$ will have a large variance and (C3.4.4) will lead to slow averaging.

(ii) If $\hat{V}$ is not a good estimate of V then both $V^{(n)}$ and $V^{[n]}$ will require a large n for their means to be good. If a large n is used, the difference between $V^{(n)}$ and $V^{[n]}$, that is, $\gamma^n \hat{V}$, is negligible and so both (C3.4.4) and (C3.4.7) will yield similar performance.

The above discussion implies that it is better to employ $V^{(n)}$ than $V^{[n]}$. It is also clear that, when $V^{(n)}$ is used, a suitable value of n has to be chosen dynamically according to the goodness of $\hat{V}$. To aid the manipulation of n, Sutton (1988) suggested a new estimate constructed by geometrically averaging $\{V^{(n)}(x) : n \ge 1\}$:

$$V^\lambda(x) = (1 - \lambda) \sum_{n=1}^{\infty} \lambda^{n-1} V^{(n)}(x). \tag{C3.4.8}$$

Here $(1 - \lambda)$ is a normalizing term. Sutton referred to the learning algorithm that uses V^λ as $TD(\lambda)$. Here TD stands for 'temporal difference'. The use of this name will be justified below. Expanding (C3.4.8) using (C3.4.5) we get

$$\begin{aligned}
V^\lambda(x) &= (1 - \lambda)\big[V^{(1)}(x) + \lambda V^{(2)}(x) + \lambda^2 V^{(3)}(x) + \cdots\big] \\
&= r_0 + \gamma(1 - \lambda)\hat{V}(x_1; v) + \\
&\quad \gamma\lambda\big[r_1 + \gamma(1 - \lambda)\hat{V}(x_2; v) + \\
&\quad \gamma\lambda\big[r_2 + \gamma(1 - \lambda)\hat{V}(x_3; v) + \cdots
\end{aligned} \tag{C3.4.9}$$

Using the fact that $r_0 = r(x, \pi(x))$ the above expression may be rewritten recursively as

$$V^\lambda(x) = r(x, \pi(x)) + \gamma(1 - \lambda)\hat{V}(x_1; v) + \gamma\lambda V^\lambda(x_1) \tag{C3.4.10}$$

where x_1 is the state occurring a time step after x. Putting $\lambda = 0$ gives $V^0 = V^{(1)}$ and putting $\lambda = 1$ gives $V^1 = V$, which is the same as $V^{(\infty)}$. Thus, the range of values obtained using $V^{(n)}$ and varying n from 1 to ∞ is approximately achieved by using V^λ and varying λ from 0 to 1. *A simple idea is to use V^λ instead of $V^{(n)}$, begin the learning process with $\lambda = 1$, and reduce λ towards zero as learning progresses and $\hat{V}$ becomes a better estimate of V.* If λ is properly chosen† then a significant improvement of computational efficiency is usually achieved when compared to simply using $\lambda = 0$ or $\lambda = 1$ (Sutton 1988). In a recent paper, Sutton and Singh (1994) have developed automatic schemes for doing this assuming that no cycles are present in state trajectories.

The definition of V^λ involves all $V^{(n)}$s and so it appears that we have to wait forever to compute it. However, computations involving V^λ can be nicely rearranged and then suitably approximated to yield a practical algorithm *that is suited for doing learning concurrently with real-time system operation.* Consider the learning rule in which we use V^λ instead of $V^{(n)}$:

$$\hat{V}(x; v) := \hat{V}(x; v) + \beta\left[V^\lambda(x) - \hat{V}(x; v)\right]. \tag{C3.4.11}$$

Define the *temporal difference* operator, Δ, by

$$\Delta(x) = r(x, \pi(x)) + \gamma\hat{V}(x_1; v) - \hat{V}(x; v). \tag{C3.4.12}$$

$\Delta(x)$ is the difference of predictions (of $V^\pi(x)$) at two consecutive time steps: $r(x, \pi(x)) + \gamma\hat{V}(x_1; v)$ is a prediction based on information at $\tau = 1$, and $\hat{V}(x; v)$ is a prediction based on information at $\tau = 0$. Hence the name, 'temporal difference'. Note that $\Delta(x)$ can be easily computed using the experience within a time step. A simple rearrangement of the terms in the second line of (C3.4.9) yields

$$V^\lambda(x) - \hat{V}(x; v) = \Delta(x) + (\gamma\lambda)\Delta(x_1) + (\gamma\lambda)^2\Delta(x_2) + \cdots. \tag{C3.4.13}$$

Even equation (C3.4.13) is not in a form suitable for use in (C3.4.11) because it involves future terms, $\Delta(x_1)$, $\Delta(x_2)$, ..., extending to infinite time. One way to handle this problem is to choose a large N, accumulate $\Delta(x)$, $\Delta(x_1)$, ..., $\Delta(x_{N-1})$ in memory, truncate the right-hand side of (C3.4.13) to include only the first N terms, and apply (C3.4.11) at $\tau = N + 1$, that is, $(N + 1)$ time steps after x occurred. However, a simpler and approximate way of achieving (C3.4.13) is to include the effects of the temporal differences as and when they occur in time. Let us say that the system is in state x at time t. When the system goes to state x_1 at time $(t + 1)$, compute $\Delta(x)$ and update $\hat{V}$ according to: $\hat{V}(x; v) := \hat{V}(x; v) + \beta(\gamma\lambda)\Delta(x_1)$. When the system goes to state x_2 at time $(t + 2)$, compute $\Delta(x_1)$ and update $\hat{V}$ according to: $\hat{V}(x; v) := \hat{V}(x; v) + \beta(\gamma\lambda)^2\Delta(x_2)$ and so on. The reason why this is approximate is because $\hat{V}(x; v)$ is continuously altered in this process whereas (C3.4.13) uses the $\hat{V}(x; v)$ existing at time t. However, if β is small and so $\hat{V}(x; v)$ is adapted slowly, the approximate updating method is expected to be close to (C3.4.11).

One way of implementing the above idea is to maintain an *eligibility trace*, $e(x, t)$, for each state visited (Klopf 1972, 1982, 1988, Barto *et al* 1983, Watkins 1989), and use the following learning rule at time t:

$$\hat{V}(x; v) := \hat{V}(x; v) + \beta e(x, t)\Delta(x_t) \qquad \forall x \tag{C3.4.14}$$

where x_t is the system state at time t. The eligibility traces can be adapted according to

$$e(x, t) = \begin{cases} 0 & \text{if } x \text{ has never been visited} \\ \gamma\lambda e(x, t - 1) & \text{if } x_t \neq x \\ 1 + \gamma\lambda e(x, t - 1) & \text{if } x_t = x. \end{cases} \tag{C3.4.15}$$

Two important remarks must be made regarding this implementation scheme.

(i) Whereas the previous learning rules (e.g. (C3.4.4), (C3.4.7) and (C3.4.11)) update $\hat{V}$ only for one x at a time step, (C3.4.14) updates the $\hat{V}$ of all states with positive eligibility trace, at a time step. Rule (C3.4.14) is suitable for neural *hardware implementation*, but not so for implementations on   E1.2 sequential computers. In that case one of the following ideas can be tried.

(a) Keep track of the last k states visited and update $\hat{V}$ for them only. The value of k should depend on λ. If λ is small, k should be small. If $\lambda = 0$ then $k = 1$.

---

† For example, if the underlying dynamic system is deterministic then a value of λ close to 1 is appropriate; on the other hand, if the system is highly stochastic then a value of λ near zero is better.

(b) The following idea is due to Cichosz (1995). Choose a nonnegative integer m (depending on the decay rate $\gamma\lambda$) and truncate the right-hand side of (C3.4.13) to keep only the first $(m+1)$ terms and get

$$\hat{V}(x; v) := \hat{V}(x; v) + \beta\,\delta(x)$$

where

$$\delta(x) = \Delta(x) + (\gamma\lambda)\Delta(x_1) + \cdots + (\gamma\lambda)^m\Delta(x_m)\,.$$

Thus, if x is the state occurring at time step t, $\hat{V}(x; v)$ gets updated at the end of time step $(t+m)$ and, *more importantly, x is the only state for which $\hat{V}$ is updated at time step $(t+m)$*. The recursion,

$$\delta(x_1) = [\delta(x) - \Delta(x)]/(\gamma\lambda) + (\gamma\lambda)^m\Delta(x_{m+1}) \tag{C3.4.16}$$

can be employed so that the δ computation can be done in constant time even if m is large. Cichosz (1995) has suggested (with good justification) another update rule based on truncation which is even better than the idea described above.

(ii) The rule for updating eligibility traces (C3.4.15) assumes that learning takes place in a single trial. If learning is done over multiple trials then all eligibility traces must be reset to zero just before each new trial is begun.

The remark made below equation (C3.2.6) applies also to the learning rules (C3.4.4), (C3.4.7), (C3.4.11) and (C3.4.14). Dayan and Sejnowski (1993) and Jaakkola *et al* (1994) have shown that, if the real time $TD(\lambda)$ learning rule, (C3.4.14) is used, then under appropriate assumptions on the variation of β in time, as $t \to \infty$, $\hat{V}$ converges to V^π with probability one. Practically, learning can be achieved by doing multiple trials and decreasing β towards zero as learning progresses.

Thus far in this section we have assumed that the policy π is deterministic. If π is a stochastic policy then all the ideas of this section still hold with appropriate interpretations: all expectations should include the stochasticity of π, and the $\pi(x)$ used in (C3.4.10), (C3.4.12), and other equations should be taken as instances generated by the stochastic policy.

Let us now come to the estimation of Q^π. Recall from (C3.3.5) that $Q^\pi(x, a)$ denotes the total reward obtained by choosing a as the first action and then following π for all future time steps. Details concerning the extension of Q^π are clearly described in a recent report by Rummery and Niranjan (1994). Let $\hat{Q}(x, a; v)$ be the estimator of $Q^\pi(x, a)$ that is to be learnt concurrently with real-time system operation. Following the same lines of argument as used for the value function, we obtain a learning rule similar to (C3.4.14):

$$\hat{Q}(x, a; v) := \hat{Q}(x, a; v) + \beta e_Q(x, a, t)\Delta_Q(x_t, a_t) \qquad \forall\,(x, a) \tag{C3.4.17}$$

where x_t and a_t are, respectively, the system state and the action chosen at time t;

$$\Delta_Q(x, a) = r(x, a) + \gamma\hat{Q}(x_1, \pi(x_1); v) - \hat{Q}(x, a; v) \tag{C3.4.18}$$

and

$$e_Q(x, a, t) = \begin{cases} 0 & \text{if } (x, a) \text{ has never been visited} \\ \gamma\lambda e_Q(x, a, t-1) & \text{if } (x_t, a_t) \neq (x, a) \\ 1 + \gamma\lambda e_Q(x, a, t-1) & \text{if } (x_t, a_t) = (x, a)\,. \end{cases} \tag{C3.4.19}$$

As with e, all $e_Q(x, a, t)$ must be reset to zero whenever a new trial is begun from a random starting state. If π is a stochastic policy then it is better to replace (C3.4.18) by

$$\Delta_Q(x, a) = r(x, a) + \gamma\tilde{V}(x_1) - \hat{Q}(x, a; v) \tag{C3.4.20}$$

where

$$\tilde{V}(x_1) = \sum_{b\in A(x_1)} \text{Prob}\{\pi(x) = b\}\hat{Q}(x_1, b; v)\,. \tag{C3.4.21}$$

Rummery and Niranjan (1994) suggest the use of (C3.4.18) even if π is stochastic; in that case, the $\pi(x_1)$ in (C3.4.18) corresponds to an instance generated by the stochastic policy at x_1. We feel that, as an estimate of $V^\pi(x_1)$, $\tilde{V}(x_1)$ is better than the term $\hat{Q}(x_1, \pi(x_1); v)$ used in (C3.4.18), and so it fits in better with the definition of Q^π in (C3.3.5). Also, if the the size of $A(x_1)$ is small then the computation of $\tilde{V}(x_1)$ is not much more expensive than that of $\hat{Q}(x_1, \pi(x_1); v)$.

References

Barto A G, Sutton R S and Anderson C W 1983 Neuronlike elements that can solve difficult learning control problems *IEEE Trans. Syst. Man Cybern.* **13** 835–46

Cichosz P 1995 Truncating temporal differences: on the efficient implementation of TD(λ) for reinforcement learning *J. Artif. Int. Res.* **2** 287–318

Dayan P 1993 Improving generalization for temporal difference learning: the successor representation *Neural Comput.* **5** 613–24

Dayan P and Sejnowski T J 1993 TD(λ) converges with probability 1 *Technical Report* CNL The Salk Institute, San Diego, CA, USA

Jaakkola T, Jordan M I and Singh S P 1994 Convergence of stochastic iterative dynamic programming algorithms *Advances in Neural Information Processing Systems 6* ed J D Cowan, G Tesauro and J Alspector (San Mateo, CA: Morgan Kaufmann) pp 703–710

Klopf A H 1972 Brain funtion and adaptive sytems—a heterostatic theory *Technical report AFCRL–72–0164* Air Force Cambridge Research Laboratories, Bedford, MA, USA

——1982 *The Hedonistic Neuron: A Theory of Memory, Learning and Intelligence* (Washington, DC: Hemisphere)

——1988 A neuronal model of classical conditioning *Psychobiology* **16** 85–125

Rummery G A and Niranjan M 1994 Online Q-learning using connectionist systems *Technical Report CUED/F–INFENG/TR 166* University of Cambridge, Cambridge, UK

Sutton R S 1984 Temporal credit assignment in reinforcement learning *PhD Thesis* University of Massachusetts, Amherst, MA, USA

——1988 Learning to predict by the method of temporal differences *Machine Learning* **3** 9–44

Sutton R S and Singh S P 1994 On step-size and bias in TD-learning *Proc. Eighth Yale Workshop on Adaptive and Learning Systems* 91–6 Yale University, USA

Watkins C J C H 1989 Learning from delayed rewards *PhD Thesis* Cambridge University, Cambridge, UK

C3.5 Delayed reinforcement learning methods

S Sathiya Keerthi and B Ravindran

Abstract

See the abstract for Chapter C3.

Dynamic programming (DP) methods (Ross 1983, Bertsekas 1989) are well known classical tools for C2.3.4
solving the stochastic optimal control problem formulated in Section C3.3. Since delayed reinforcement
learning (RL) methods also solve the same problem, how do they differ from DP methods?† The main
differences are as follows.

- Whereas DP methods simply aim to obtain the optimal value function and an optimal policy using
  off-line iterative methods, delayed RL methods aim to *learn the same concurrently with real-time
  system operation* and improve performance over time.

- DP methods deal with the complete state space X in their computations, while delayed RL methods
  operate on $\bar{X}$, the set of states that occur during real-time system operation. In many applications X
  is very large, but $\bar{X}$ is only a small, manageable subset of X. Therefore, in such applications, DP
  methods suffer from the *curse of dimensionality*, but delayed RL methods do not have this problem.
  Also, typically delayed RL methods employ function approximators (for value function, policy etc)
  that generalize well, and so, after learning, they provide near optimal performance even on unseen
  parts of the state space.

- DP methods fundamentally require a system model. On the other hand, the main delayed RL methods
  are model-free; hence they are particularly suited for the on-line learning control of complicated
  systems for which a model is difficult to derive.

- Because delayed RL methods continuously learn in time they are better suited than DP methods for
  adapting to situations in which the system and goals are nonstationary.

Although we have said that delayed RL methods enjoy certain key advantages, we should also add
that DP has been the forerunner from which delayed RL methods obtained clues. In fact, it is correct to
say that delayed RL methods are basically rearrangements of the computational steps of DP methods so
that they can be applied during real-time system operation.

Delayed RL methods can be grouped into two categories: model-based methods and model-free
methods. Model-based methods have direct links with DP. Model-free methods can be viewed as
appropriate modifications of the model-based methods so as to avoid the model requirement. These
methods will be described in detail below.

C3.5.1 Model-based methods

In this section we discuss DP methods and their possible modification to yield delayed RL methods. There
are two popular DP methods: value iteration and policy iteration. Value iteration easily extends to give
a delayed RL method called 'real-time DP'. Policy iteration, though it does not directly yield a delayed
method, forms the basis of an important model-free delayed RL method called actor–critic.

---

† The connection between DP and delayed RL was first established by Werbos (1987, 1989, 1992) and Watkins (1989).

C3.5.1.1   Value iteration

The basic idea in value iteration is to compute $V^*(x)$ as

$$V^*(x) = \lim_{n \to \infty} V_n^*(x) \qquad (C3.5.1)$$

where $V_n^*(x)$ is the optimal value function over a finite horizon of length n, that is, $V_n^*(x)$ is the maximum expected return if the decision task is terminated n steps after starting in state x. For $n = 1$, the maximum expected return is just the maximum of the expected immediate payoff:

$$V_1^*(x) = \max_{a \in A(x)} r(x, a) \qquad \forall\, x \qquad (C3.5.2)$$

Then, the recursion†,

$$V_{n+1}^*(x) = \max_{a \in A(x)} \left[r(x, a) + \gamma \sum_y P_{xy}(a) V_n^*(y) \right] \qquad \forall\, x \qquad (C3.5.3)$$

can be used to compute V_{n+1}^* for $n = 1, 2, \ldots$. (Iterations can be terminated after a large number (N) of iterations, and V_N^* can be taken to be a good approximation of V^*.)

In value iteration, a policy is not involved. But it is easy to attach a suitable policy with a value function as follows. Associated with each value function, $V : X \to R$ is a policy π that is *greedy with respect to V*, that is,

$$\pi(x) = \arg \max_{a \in A(x)} \left[r(x, a) + \gamma \sum_y P_{xy}(a) V(y) \right] \qquad \forall\, x\,. \qquad (C3.5.4)$$

If the state space X has a very large size (e.g. size $= k^d$, where d = number of components of x, k = number of values that each component can take, $d \approx 10$, $k \approx 100$) then value iteration is prohibitively expensive. This difficulty is usually referred to as the *curse of dimensionality*.

In the above, we have assumed that (C3.5.1) is correct. Let us now prove this convergence. It turns out that convergence can be established for a more general algorithm, of which value iteration is a special case. We describe this algorithm as *generalized value iteration*.

Generalized value iteration. Set $n = 1$ and $V_1^* =$ an arbitrary function over states.
Repeat

(i)   *Choose a subset of states, B_n and set*

$$V_{n+1}^*(x) = \begin{cases} \max_{a \in A(x)} \left[r(x, a) + \gamma \sum_y P_{xy}(a) V_n^*(y) \right] & \text{if } x \in B_n \\ V_n^*(x) & \text{otherwise} \end{cases} \qquad (C3.5.5)$$

(ii)   *Reset $n := n + 1$.*

If we choose V_1^* as in (C3.5.2) and take $B_n = X$ for all n, then the above algorithm reduces to value iteration. Later, we will go into other useful cases of generalized value iteration. But first, let us concern ourselves with the issue of convergence. If $x \in B_n$, we will say that the value of state x has been backed up at the nth iteration. Proof of convergence is based on the following result (Bertsekas and Tsitsiklis 1989, Watkins 1989, Barto *et al* 1992).

Local value improvement theorem. Let $M_n = \max_x |V_n^*(x) - V^*(x)|$. Then $\max_{x \in B_n} |V_{n+1}^*(x) - V^*(x)| \le \gamma M_n$.

Proof. Take any $x \in B_n$. Let $a^* = \pi^*(x)$ and $a_n^* = \pi_n^*(x)$, where π_n^* is a policy that is greedy with respect to V_n^*. Then

$$\begin{aligned} V_{n+1}^*(x) &\ge r(x, a^*) + \gamma \sum_y P_{xy}(a^*) V_n^*(y) \\ &\ge r(x, a^*) + \gamma \sum_y P_{xy}(a^*) \left[V^*(y) - M \right] \\ &= V^*(x) - \gamma M_n\,. \end{aligned}$$

---

† One can also view the recursion as doing a fixed-point iteration to solve Bellman's optimality equation (C3.3.4).

Similarly,

$$\begin{aligned}
V_{n+1}^*(x) &= r(x, a_n^*) + \gamma \sum_y P_{xy}(a_n^*) V_n^*(y) \\
&\leq r(x, a_n^*) + \gamma \sum_y P_{xy}(a_n^*) \left[V^*(y) + M \right] \\
&= V^*(x) + \gamma M_n
\end{aligned}$$

and so the theorem is proved.

The theorem implies that $M_{n+1} \leq M_n$ where $M_{n+1} = \max_x |V_{n+1}^*(x) - V^*(x)|$. A little further thought shows that the following is also true. If, at the end of iteration k, K further iterations are done in such a way that the value of each state is backed up at least once in these K iterations, that is, $\cup_{n=k+1}^{k+K} B_n = X$, then we get $M_{k+K} \leq \gamma M_k$. Therefore, *if the value of each state is backed up infinitely often, then (C3.5.1) holds*†. In the case of value iteration, the value of each state is backed up at each iteration and so (C3.5.1) holds.

Generalized value iteration was proposed by Bertsekas (1982, 1989) and developed by Bertsekas and Tsitsiklis (1989) as a suitable method of solving stochastic optimal control problems on multiprocessor systems with communication time delays and without a common clock. If N processors are available, the state space can be partitioned into N sets—one for each processor. The times at which each processor backs up the values of its states can be different for each processor. To back up the values of its states, a processor uses the 'present' values of other states communicated to it by other processors.

Barto *et al* (1992) suggested the use of generalized value iteration as a way of learning during real-time system operation. They called their algorithm *real-time dynamic programming* (RTDP). In generalized value iteration as specialized to RTDP, n denotes system time. At time step n, let us say that the system resides in state x_n. Since V_n^* is available, a_n is chosen to be an action that is greedy with respect to V_n^*, that is, $a_n = \pi_n^*(x_n)$. B_n, the set of states whose values are backed up, is chosen to include x_n and, perhaps, some more states. In order to improve performance in the immediate future, one can do a look-ahead search to some fixed search depth (either exhaustively or by following policy π_n^*) and include these probable future states in B_n. Because the value of x_n is going to undergo change at the present time step, it is a good idea to also include, in B_n, the most likely predecessors of x_n (Moore and Atkeson 1993).

One may ask: since a model of the system is available, why not simply do value iteration or, do generalized value iteration as Bertsekas and Tsitsiklis suggest? In other words, what is the motivation behind RTDP? The answer, which is simple, is something that we have stressed earlier. In most problems (e.g. playing games such as checkers and backgammon) the state space is extremely large, but only a small subset of it actually occurs during usage. Because RTDP works concurrently with actual system operation, it focuses on regions of the state space that are most relevant to the system's behavior. For instance, successful learning was accomplished in the checkers program of Samuel (1959) and in the backgammon program TDgammon of Tesauro (1992) using variations of RTDP. In Barto *et al* (1992), Barto, Bradtke and Singh also use RTDP to make interesting connections and useful extensions to learning real-time search algorithms in artificial intelligence (Korf 1990).

The convergence result mentioned earlier says that the values of all states have to be backed up infinitely often‡ in order to ensure convergence. So it is important to explore the state space suitably in order to improve performance. Barto, Bradtke and Singh have suggested two ways of doing exploration§: (i) adding stochasticity to the policy; and (ii) doing learning cumulatively over multiple trials.

If only an inaccurate system model is available then it can be updated in real time using a system identification technique, such as the maximum likelihood estimation method (Barto *et al* 1992). The current system model can be used to perform the computations in (C3.5.5). Convergence of such adaptive methods has been proved by Gullapalli and Barto (1994).

C3.5.1.2 Policy iteration

Policy iteration operates by maintaining a representation of a policy and its value function, and forming an improved policy using them. Suppose π is a given policy and V^π is known. How can we improve π? An answer will become obvious if we first answer the following simpler question. If μ is another given policy then when is

$$V^\mu(x) \geq V^\pi(x) \qquad \forall x \tag{C3.5.6}$$

† If $\gamma = 1$, then convergence holds under certain assumptions. The analysis required is more sophisticated. See Bertsekas and Tsitsiklis (1989) and Bradtke (1994) for details.
‡ For good practical performance it is sufficient that states that are most relevant to the system's behavior are backed up repeatedly.
§ Thrun (1986) has discussed the importance of exploration and suggested a variety of methods for it.

that is, when is μ uniformly better than π? The following simple theorem (Watkins 1989) gives the answer.

Policy improvement theorem. The policy μ is uniformly better than policy π if

$$Q^{\pi}(x, \mu(x)) \geq V^{\pi}(x) \qquad \forall\, x. \tag{C3.5.7}$$

Proof. To avoid clumsy details let us give a not-so-rigorous proof (Watkins 1989). Starting at x, it is better to follow μ for one step and then to follow π, than it is to follow π right from the beginning. By the same argument, it is better to follow μ for one further step from the state just reached. Repeating the argument we find that it is always better to follow μ than π. See Bellman and Dreyfus (1962) and Ross (1983) for a detailed proof.

Let us now return to our original question: given a policy π and its value function V^{π}, how do we form an improved policy, μ? If we define μ by

$$\mu(x) = \arg\,\max_{a \in A(x)}\, Q^{\pi}(x, a) \qquad \forall\, x \tag{C3.5.8}$$

then (C3.5.7) holds. By the policy improvement, theorem μ is uniformly better than π. This is the main idea behind policy iteration.

Policy iteration. Set π := an arbitrary initial policy and compute V^{π}.
Repeat

(i)   Compute Q^{π} using (C3.3.5).
(ii) Find μ using (C3.5.8) and compute V^{μ}.
(iii) Set: $\pi := \mu$ and $V^{\pi} := V^{\mu}$.

until $V^{\mu} = V^{\pi}$ occurs at step 2.

Nice features of the above algorithm are: (i) it terminates after a finite number of iterations because there are only a finite number of policies; and (ii) when termination occurs we get

$$V^{\pi}(x) = \max_{a} Q^{\pi}(x, a) \qquad \forall x$$

(i.e. V^{π} satisfies Bellman's optimality equation) and so π is an optimal policy. But the algorithm suffers from a serious drawback: it is very expensive because the entire value function associated with a policy has to be recalculated at each iteration (step (ii)). Even though V^{μ} may be close to V^{π}, unfortunately there is no simple shortcut to compute it. In section C3.5.2 we will discuss a well known model-free method called the *actor–critic* method which gives an inexpensive approximate way of implementing policy iteration.

C3.5.2   Model-free methods

Model-free delayed RL methods are derived by making suitable approximations to the computations in value iteration and policy iteration, so as to eliminate the need for a system model. Two important methods result from such approximations: Barto, Sutton and Anderson's actor–critic (Barto *et al* 1983), and Watkins' Q-learning (Watkins 1989). These methods are milestone contributions to the optimal feedback control of dynamic systems.

C3.5.2.1   Actor–critic method

C2.3.3   The *actor–critic* method was proposed by Barto *et al* (1983) (in their popular work on balancing a pole on a moving cart) as a way of combining, on a step-by-step basis, the process of forming the value function with the process of forming a new policy. The method can also be viewed as a practical, approximate way of doing policy iteration: perform one step of an on-line procedure for estimating the value function for a given policy, and at the same time perform one step of an on-line procedure for improving that policy. The actor–critic method—a mathematical analysis of this method has been done by Williams and Baird (1993)—is best derived by combining the ideas of Section C3.2 and Section C3.4 on immediate RL and estimating value function, respectively. Details are as follows.

Actor (π). Let m denote the total number of actions. Maintain an approximator, $g(\cdot; w) : X \to R^{m}$ so that $z = g(x; w)$ is a vector of merits of the various feasible actions at state x. In order to do exploration, choose actions according to a stochastic action selector such as (C3.2.4). (In their original work on pole-balancing, Barto, Sutton and Anderson suggested a different way of including stochasticity.)

Critic (V^{π}). Maintain an approximator, $\hat{V}(\cdot; w) : X \to R$ that estimates the value function (expected total reward) corresponding to the stochastic policy mentioned above. The ideas of Section C3.4 can be used to update $\hat{V}$.

Let us now consider the process of learning the actor. Unlike immediate RL, learning is more complicated here for the following reason. Whereas, in immediate RL the environment immediately provides an evaluation of an action, in delayed RL the effect of an action on the total reward is not immediately available and has to be estimated appropriately. Suppose, at some time step, the system is in state x and the action selector chooses action a^k. For g, the learning rule that parallels (C3.2.3) would be

$$g_k(x; w) := g_k(x; w) + \alpha\left[\rho(x, a^k) - \hat{V}(x; v)\right] \tag{C3.5.9}$$

where $\rho(x; a^k)$ is the expected total reward obtained if a^k is applied to the system at state x and then policy π is followed from the next step onwards. An approximation is

$$\rho(x, a^k) \approx r(x, a^k) + \gamma \sum_y P_{xy}(a^k)\hat{V}(y; v) . \tag{C3.5.10}$$

This estimate is unavailable because we do not have a model. A further approximation is

$$\rho(x, a^k) \approx r(x, a^k) + \gamma \hat{V}(x_1; v) \tag{C3.5.11}$$

where x_1 is the state occurring in the real-time operation when action a^k is applied at state x. Since the right-hand side of (C3.5.11) is an unbiased estimate of the right-hand side of (C3.5.10), using this approximation in the averaging learning rule (C3.5.9) will not lead to errors. Using (C3.5.11) in (C3.5.9) gives

$$g_k(x; w) := g_k(x; w) + \alpha\Delta(x) \tag{C3.5.12}$$

where Δ is as defined in (C3.4.12). The following algorithm results.

Actor–critic trial. Set $t = 0$ and $x =$ a random starting state.
Repeat (for a number of time steps)

(i)   *With the system at state x, choose action a according to (C3.2.4) and apply it to the system. Let x_1 be the resulting next state.*
(ii) *Compute $\Delta(x) = r(x, a) + \gamma \hat{V}(x_1; v) - \hat{V}(x; v)$.*
(iii) *Update $\hat{V}$ using $\hat{V}(x; v) := \hat{V}(x; v) + \beta\Delta(x)$.*
(iv) *Update g_k using (C3.5.12) where k is such that $a = a^k$.*

The above algorithm uses the $TD(0)$ estimate of V^{π}. To speed up learning the $TD(\lambda)$ rule, (C3.4.14) can be employed. Barto *et al* (1983) and others (Gullapalli 1992a, Gullapalli *et al* 1994) use the idea of eligibility traces for updating g also. They give only an intuitive explanation for this usage. Lin (1992) has suggested the accumulation of data until a trial is over, updating $\hat{V}$ using (C3.4.11) for all states visited in the trial, and then updating g using (C3.5.12) for all (state, action) pairs experienced in the trial.

C3.5.2.2   Q-learning

Just as the actor–critic method is a model-free, on-line way of approximately implementing policy iteration, Watkins' Q-learning algorithm (Watkins 1989) is a model-free, on-line way of approximately implementing generalized value iteration. Though the RTDP algorithm does generalized value iteration concurrently with real-time system operation, it requires the system model for doing a crucial operation: the determination of the maximum on the right-hand side of (C3.5.5). Q-learning overcomes this problem elegantly by operating with the Q-function instead of the value function. (Recall, from Section C3.3, the definition of Q-function and the comment on its advantage over value function.)

The aim of Q-learning is to find a function approximator, $\hat{Q}(\cdot, \cdot; v)$ that approximates Q^*, the solution of Bellman's optimality equation (C3.3.7) in on-line mode without employing a model. However, for the sake of developing ideas systematically, let us begin by assuming that a system model is available and consider the modification of the ideas of section C3.5.1 to use the Q-function instead of the value function.

If we think in terms of a function approximator $\hat{V}(x; v)$ for the value function, the basic update rule that is used throughout section C3.5.1 is

$$\hat{V}(x; v) := \max_{a \in A(x)} \left[r(x, a) + \gamma \sum_y P_{xy}(a) \hat{V}(y; v) \right].$$

For the Q-function, the corresponding rule is

$$\hat{Q}(x, a; v) := r(x, a) + \gamma \sum_y P_{xy}(a) \max_{b \in A(y)} \hat{Q}(y, b; v). \tag{C3.5.13}$$

Using this rule, all the ideas of section C3.5.1 can be easily modified to employ the Q-function.

However, our main concern is to derive an algorithm that avoids the use of a system model. A model can be avoided if we: (i) replace the summation term in (C3.5.13) by $\max_{b \in A(x_1)} \hat{Q}(x_1, b; v)$ where x_1 is an instance of the state resulting from the application of action a at state x; and (ii) achieve the effect of the update rule in (C3.5.13) via the 'averaging' learning rule,

$$\hat{Q}(x, a; v) := \hat{Q}(x, a; v) + \beta \left[r(x, a) + \gamma \max_{b \in A(x_1)} \hat{Q}(x_1, b; v) - \hat{Q}(x, a; v) \right]. \tag{C3.5.14}$$

If (C3.5.14) is carried out we say that the Q-value of (x, a) has been backed up. Using (C3.5.14) in on-line mode of system operation we obtain the Q-learning algorithm.

Q-learning trial. Set $t = 0$ and $x = $ a random starting state.
Repeat (for a number of time steps)

(i)   Choose action $a \in A(x)$ and apply it to the system. Let x_1 be the resulting state.
(ii) Update $\hat{Q}$ using (C3.5.15).
(iii) Reset $x := y$.

The remark made below equation (C3.2.6) in Section C3.2 is very appropriate for the learning rule, (C3.5.14). Watkins showed[†] that *if the Q-value of each admissible (x, a) pair is backed up infinitely often, and if the step size β is decreased to zero in a suitable way then as $t \to \infty$, $\hat{Q}$ converges to Q^* with probability one.* Practically, learning can be achieved by: firstly, in step (i), using an appropriate exploration policy that tries all actions[‡]; secondly, doing multiple trials to ensure that all states are frequently visited; and thirdly, decreasing β towards zero as learning progresses.

We now discuss a way of speeding up Q-learning by using the $TD(\lambda)$ estimate of the Q-function, derived in Section C3.4. If $TD(\lambda)$ is to be employed in a Q-learning trial, a fundamental requirement is that the policy used in step (i) of the Q-learning trial and the policy used in the update rule (C3.5.14) should match (note the use of π in (C3.4.18) and (C3.4.21)). Thus $TD(\lambda)$ can be used if we employ the greedy policy

$$\pi(x) = \arg \max_{a \in A(x)} \hat{Q}(x, a; v) \tag{C3.5.15}$$

in step (i)[§] but this leads to a problem: use of the greedy policy will not allow exploration of the action space, and hence poor learning can occur. Rummery and Niranjan (1994) give a nice comparative account of various attempts described in the literature for dealing with this conflict. Here, we only give the details of an approach that Rummery and Niranjan found to be very promising.

Consider the stochastic policy (based on the Boltzmann distribution and Q-values),

$$\text{Prob}\{\pi(x) = a | x\} = \frac{\exp(\hat{Q}(x, a; v)/T)}{\sum_{b \in A(x)} \exp(\hat{Q}(x, b; v)/T)} \qquad a \in A(x) \tag{C3.5.16}$$

where $T \in [0, \infty)$. When $T \to \infty$ all actions have equal probabilities and when $T \to 0$ the stochastic policy tends towards the greedy policy in (C3.5.15). To learn, T is started with a suitably large value

---

† A revised proof was given by Watkins and Dayan (1992). Tsitsiklis (1993) and Jaakkola *et al* (1994) have given other proofs.
‡ Note that step (i) does not put any restriction on choosing a feasible action. So, any stochastic exploration policy that at every x generates each feasible action with positive probability can be used. When learning is complete, the greedy policy $\pi(x) = \arg \max_{a \in A(x)} \hat{Q}(x, a; v)$ should be used for optimal system performance.
§ Although the greedy policy defined by (C3.5.15) keeps changing during a trial, the $TD(\lambda)$ estimate can still be used because $\hat{Q}$ is varied slowly. If more than one action attains the maximum in (C3.5.15) then for convenience we take π to be a stochastic policy that makes all such maximizing actions equally probable.

(depending on the initial size of the Q-values) and is decreased to zero using an annealing rate; at each T thus generated, multiple Q-learning trials are performed. This way, exploration takes place at the initial large T values. The $TD(\lambda)$ learning rule (C3.4.20) estimates expected returns for the policy at each T and, as $T \to 0$, $\hat{Q}$ will converge to Q^*.

An important remark needs to be made regarding the application of Q-learning to RL problems which result from the time-discretization of continuous-time problems. As the discretization time period goes to zero it turns out that the Q-function tends to be independent of action and hence it is unsuitable to use Q-learning for continuous-time problems. For such problems Baird (1993) has suggested the use of an appropriate modification of the Q-function called the advantage function.

C3.5.3 Extension to continuous spaces

Optimal control of dynamic systems typically involves the solution of delayed RL problems having continuous state/action spaces. If the state space is continuous but the action space is discrete then all the delayed RL algorithms discussed earlier can be easily extended, provided appropriate function approximators that generalize a real-time experience at a state to all topologically nearby states are used; see Section C3.6 for a discussion of such approximators. On the other hand, if the action space is continuous, extension of the algorithms is more difficult. The main cause of the difficulty can be easily seen if we try extending RTDP to continuous action spaces: the max operation in (C3.5.5) is nontrivial and difficult if $A(x)$ is continuous. (Therefore, even methods based on value iteration need to maintain a function approximator for actions.) In the rest of this section we will give a brief review of various methods of handling continuous action spaces. Just to make the presentation easy, we will make the following assumptions.

- The system being controlled is deterministic. Let

$$x_{t+1} = f(x_t, a_t) \tag{C3.5.17}$$

describe the transition. (Werbos 1990 describes ways of treating stochastic systems.)
- There are no action constraints, that is, $A(x) =$ an m-dimensional real space for every x.
- All functions involved (r, f, $\hat{V}$, $\hat{Q}$, etc) are continuously differentiable.

Let us first consider model-based methods. Werbos (1990b) has proposed a variety of algorithms. Here we will describe only one important algorithm, the one that Werbos refers to as the *backpropagated adaptive critic*. The algorithm is of the actor–critic type, but it is somewhat different from the actor–critic method of section C3.5.2. There are two function approximators: $\hat{\pi}(\cdot; w)$ for action and $\hat{V}(\cdot; v)$ for critic. The critic is meant to approximate $V^{\hat{\pi}}$; at each time step, it is updated using the $TD(\lambda)$ learning rule (C3.4.14). The actor tries to improve the policy at each time step using the hint provided by the policy improvement theorem in (C3.5.7). To be more specific, let us define

$$Q(x, a) \stackrel{\text{def}}{=} r(x, a) + \gamma \hat{V}(f(x, a); v). \tag{C3.5.18}$$

At time t, when the system is at state x_t, we choose the action $a_t = \hat{\pi}(x_t; w)$ leading to the next state x_{t+1} given by (C3.5.17). Let us assume $\hat{V} = V^{\hat{\pi}}$, so that $V^{\hat{\pi}}(x_t) = Q(x_t, a_t)$ holds. Using the hint from (C3.5.7), we aim to adjust $\hat{\pi}(x_t; w)$ to give a new value a^{new} such that

$$Q(x_t, a^{\text{new}}) > Q(x_t, a_t). \tag{C3.5.19}$$

A simple learning rule that achieves this requirement is

$$\hat{\pi}(x_t; w) := \hat{\pi}(x_t; w) + \alpha \frac{\partial Q(x_t, a)}{\partial a}\bigg|_{a=a_t} \tag{C3.5.20}$$

where α is a small (positive) step size. The partial derivative in (C3.5.20) can be evaluated using

$$\frac{\partial Q(x_t, a)}{\partial a} = \frac{\partial r(x_t, a)}{\partial a} + \gamma \frac{\partial \hat{V}(y; v)}{\partial y}\bigg|_{y=f(x_t, a)} \frac{\partial f(x_t, a)}{\partial a}. \tag{C3.5.21}$$

Let us now come to model-free methods. A simple idea is to adapt a function approximator $\hat{f}$ for the system model function, f, and use $\hat{f}$ instead of f in Werbos' algorithm. On-line experience, that

C2.3.3

is, the combination (x_t, a_t, x_{t+1}), can be used to learn $\hat{f}$. This method was proposed by Brody (1992), actually as a way of overcoming a serious deficiency—this deficiency was also pointed out by Gullapalli (1992b)—associated with an ill-formed model-free method suggested by Jordan and Jacobs (1990). A key difficulty associated with Brody's method is that, until the learning system adapts a good $\hat{f}$, system performance does not improve at all; in fact, at the early stages of learning, the method can perform in a confused way. To overcome this problem Brody suggests that $\hat{f}$ be learnt well, before it is used to train the actor and the critic.

A more direct model-free method can be derived using the ideas of section C3.5.2 and employing a learning rule similar to (C3.2.7) for adapting $\hat{\pi}$. This method was proposed and successfully demonstrated by Gullapalli (Gullapalli 1992a, Gullapalli *et al* 1994). Since Gullapalli's method learns by observing the effect of a randomly chosen perturbation of the policy, it is not as systematic as the policy change in Brody's method.

We now propose a new model-free method that systematically changes the policy similar to what Brody's method and avoids the need for adapting a system model. This is achieved using a function approximator $\hat{Q}(\cdot, \cdot; v)$ for approximating $Q^{\hat{\pi}}$, the Q-function associated with the actor. The $TD(\lambda)$ learning rule in (C3.4.17) can be used for updating $\hat{Q}$. Also, policy improvement can be attempted using the learning rule (similar to (C3.5.20))

$$\hat{\pi}(x_t; w) := \hat{\pi}(x_t; w) + \alpha \frac{\partial \hat{Q}(x_t, a)}{\partial a}\bigg|_{a=a_t}. \tag{C3.5.22}$$

We are currently performing simulations to study the performance of this new method relative to the other two model-free methods mentioned above.

Werbos' algorithm and our Q-learning-based algorithm are deterministic, while Gullapalli's algorithm is stochastic. The deterministic methods are expected to be much faster, whereas the stochastic method has better assurance of convergence to the true solution. The arguments are similar to those mentioned at the end of Section C3.2.

References

Baird III L C 1993 Advantage updating. Wright-Patterson Air Force Base Ohio, USA *Wright Laboratory Technical Report WL-TR-93-1146* (available from the Defence Technical Information Center, Cameron Station, Alexandria, VA 22304-6145, USA)

Barto A G 1992 Reinforcement learning and adaptive critic methods *Handbook of Intelligent Control: Neural, Fuzzy, and Adaptive Approaches* ed D A White and D A Sofge (New York: Van Nostrand Reinhold) pp 469–91

Barto A G, Bradtke S J and Singh S P 1992 Real-time learning and control using asynchronous dynamic programming *Technical Report COINS 91–57* University of Massachusetts, Amherst, MA, USA

Barto A G, Sutton R S and Anderson C W 1983 Neuronlike elements that can solve difficult learning control problems *IEEE Trans. Syst. Man Cybern.* **13** 835–46

Bellman R E and Dreyfus S E 1962 *Applied Dynamic Programming* RAND Corporation

Bertsekas D P 1982 Distributed Dynamic Programming *IEEE Trans. Auto. Control* **27** 610–6

——1989 *Dynamic Programming: Deterministic and Stochastic Models* (Englewood Cliffs, NJ: Prentice-Hall)

Bertsekas D P and Tsitsiklis J N 1989 *Parallel and Distributed Computation: Numerical Methods* (Englewood Cliffs, NJ: Prentice-Hall)

Bradtke S J 1994 Incremental dynamic programming for online adaptive optimal control *CMPSCI Technical Report* pp 94–62

Brody C 1992 Fast learning with predictive forward models *Advances in Neural Information Processing Systems 4* ed J E Moody, S J Hanson and R P Lippmann (San Mateo, CA: Morgan Kaufmann) pp 563–70

Gullapalli V 1992a Reinforcement learning and its application to control *Technical Report COINS 92–10, PhD Thesis* University of Massachusetts, Amherst, MA, USA

——1992b A comparison of supervised and reinforcement learning methods on a reinforcment learning task *Proc. 1991 IEEE Symp. on Intelligent Control (Arlington, VA)* (New York: IEEE Press)

Gullapalli V and Barto A G 1994 Convergence of indirect adaptive asynchronous value iteration algorithms *Advances in Neural Information Processing Systems 6* ed J D Cowan, G Tesauro and J Alspector (San Francisco, CA: Morgan Kaufmann) pp 695–702

Gullapalli V, Franklin J A and Benbrahim H 1994 Acquiring robot skills via reinforcement learning *IEEE Control Syst. Mag.* 13–24

Jaakkola T, Jordan M I and Singh S P 1994 Convergence of stochastic iterative dynamic programming algorithms *Advances in Neural Information Processing Systems 6* ed J D Cowan, G Tesauro and J Alspector (San Mateo, CA: Morgan Kaufmann) pp 703–710

Jordan M I and Jacobs R A 1990 Learning to control an unstable system with forward modeling *Advances in Neural Information Processing Systems 2* ed D S Touretzky (San Mateo, CA: Morgan Kaufmann)

Korf R E 1990 Real-time heuristic search *Aritif. Intell.* **42** 189–211

Lin L J 1992 Self-improving reactive agents based on reinforcement learning, planning and teaching *Machine Learning* **8** 293–321

Moore A W and Atkeson C G 1993 Memory-based reinforcement learning: efficient computation with prioritized sweeping *Advances in Neural Information Processing Systems 5* ed S J Hanson, J D Cowan and C L Giles (San Mateo, CA: Morgan Kaufmann) pp 263–70

Ross S 1983 *Introduction to Stochastic Dynamic Programming* (New York: Academic)

Rummery G A and Niranjan M 1994 Online Q-learning using connectionist systems *Technical Report CUED/F-INFENG/TR 166* University of Cambridge, Cambridge, UK

Samuel A L 1959 Some studies in machine learning using the game of checkers *IBM J. Res. Develop.* pp 210–29 (Reprinted in 1963 *Computers and Thought* ed E A Feigenbaum and J Feldman (New York: McGraw-Hill))

Tesauro G J 1992 Practical issues in temporal difference learning *Machine Learning* **8** 257–78

Thrun S B 1986 Efficient exploration in reinforcement learning *Technical report CMU–CS–92–102* School of Computer Science, Carnegie Mellon University, Pittsburgh, PA, USA

Tsitsiklis J N 1993 Asynchronous stochastic approximation and Q-learning *Technical Report LIDS–P–2172* Laboratory for Information and Decision Systems, MIT, Cambridge, MA, USA

Watkins C J C H 1989 Learning from delayed rewards *PhD Thesis* Cambridge University, Cambridge, UK

Watkins C J C H and Dayan P 1992 Technical note: Q-learning *Machine Learning* **8** 279–92

Werbos P J 1987 Building and understanding adaptive systems: a statistical/numerical approach to factory automation and brain research *IEEE Trans. Syst. Man Cybern.*

——1989 Neural networks for control and system identification *Proc. 28th Conf. on Decision and Control (Tampa, FL)* pp 260–5

——1990 A menu of designs for reinforcement learning over time *Neural Networks for Control* ed W T Miller, R S Sutton and P J Werbos (Cambridge, MA: MIT Press) pp 67–95

——1992 Approximate dynamic programming for real-time control and neural modeling *Handbook of Intelligent Control: Neural, Fuzzy, and Adaptive Approaches* ed D A White and D A Sofge (New York: Van Nostrand-Reinhold) pp 493–525

Williams R J and Baird III L C 1993 Analysis of some incremental variants of policy iteration: first steps toward understanding actor–critic learning systems *Technical Report NU-CCS-93-11* College of Computer Science, Northeastern University, Boston, MA, USA

C3.6 Use of neural and other function approximators in reinforcement learning

S Sathiya Keerthi and B Ravindran

Abstract

See the abstract for Chapter C3.

A variety of function approximators have been employed by researchers to solve reinforcement learning (RL) problems practically. When the input space of the function approximator is finite, the most straightforward method is to use a *lookup table* (Singh 1992a, Moore and Atkeson 1993). Almost all theoretical results on the convergence of RL algorithms assume this representation. The disadvantage of using a lookup table is that if the input space is large then the memory requirement becomes prohibitive. (Buckland and Lawrence (1994) have proposed a new delayed RL method called transition point dynamic programming (DP) which can significantly reduce the memory requirement for problems in which optimal actions change infrequently in time.) Continuous input spaces have to be discretized when using a lookup table. If the discretization is done finely so as to obtain good accuracy we have to face the 'curse of dimensionality'. One way of overcoming this is to do a problem-dependent discretization; see, for example, the 'BOXES' representation used by Barto *et al* (1983) and others (Michie and Chambers 1968, Gullapalli *et al* 1994, Rosen *et al* 1991) to solve the pole balancing problem.

Non-lookup table approaches use parametric function approximation methods. These methods have the advantage of being able to generalize beyond the training data and hence give reasonable performance on unvisited parts of the input space. Among these, neural methods are the most popular. Connectionist methods that have been employed for RL can be classified into four groups: *multilayer perceptrons*; methods based on clustering; *CMAC*; and *recurrent networks*. Multilayer perceptrons have been successfully used by Anderson (1986, 1989) for pole balancing, Lin (1991a, b, c, 1992) for a complex test problem, Tesauro (1992) for backgammon, Thrun (1993) and Millan and Torras (1992) for robot navigation, and others (Boyen 1992, Gullapalli *et al* 1994). On the other hand, Watkins (1989), Chapman (1991), Kaelbling (1990, 1991), and Shepanski and Macy (1987) have reported bad results. A modified form of Platt's *resource allocation network* (Platt 1991), a method based on *radial basis functions*, has been used by Anderson (1993) for pole balancing. Many researchers have used *CMAC* (Albus 1975) for solving RL problems: Watkins (1989) for a test problem; Singh (1991, 1992b, 1992c) and Tham and Prager (1994) for a navigation problem; Lin and Kim (1991) for pole balancing and Sutton (1990, (1991a, 1991b) in his 'Dyna' architecture. Recurrent networks with context information feedback have been used by Bacharach (1991, 1992) and Mozer and Bacharach (1990a, b) in dealing with RL problems with incomplete state information.

A few nonneural methods have also been used for RL. Mahadevan and Connell (1991) have used statistical clustering in association with Q-learning for the automatic programming of a mobile robot. A novel feature of their approach is that the number of clusters is dynamically varied. Chapman and Kaelbling (1991) have used a tree-based clustering approach in combination with a modified Q-learning algorithm for a difficult test problem with a huge input space.

The function approximator has to exercise care to ensure that learning at some input point x does not seriously disturb the function values for $y \neq x$. It is often advantageous to choose a function approximator and employ an update rule in such a way that the function values of x and states 'near' x are modified

C1.2

C1.1.7, B2.3

C1.6.2

similarly while the values of states 'far' from x are left unchanged†. Such a choice usually leads to good generalization, that is, good performance of the learned function approximator even on states that are not visited during learning. In this respect, CMAC and methods based on clustering, such as RBF, statistical clustering and so on, are more suitable than multilayer perceptrons.

The effect of errors introduced by function approximators on the optimal performance of the controller has not been well understood‡. It has been pointed out by Watkins (1989), Bradtke (1993), Bertsekas (1994) and others (Barto 1992) that if function approximation is not done in a careful way, poor learning can result. In the context of Q-learning, Thrun and Schwartz (1993) have shown that errors in function approximation can lead to a systematic overestimation of the Q-function. Linden (1993) points out that in many problems the value function is discontinuous and so using continuous function approximators is inappropriate. But he does not suggest any clear remedies for this problem.

Mance Harmon of Wright-Patterson Air Force Base, Ohio, has pointed out to us the following explanation as to why function approximators used with RL have difficulties. The generalization that takes place when updating the approximation systems can, as a side effect, change the target value. For instance, when the update rule (C3.4.14), which is based on $\Delta(x_t)$, is performed, the resulting change in $\hat{V}$ together with generalization can lead to a sizeable change in $\Delta(x_t)$. We are then, in effect, shooting at a moving target. This is a cause of instability, and the propensity of the weights, in many cases, to grow to infinity. To overcome this problem Baird and Harmon (1993) have suggested a residual gradient approach in which *gradient descent* is performed on the mean square of residuals such as $\Delta(x_t)$. Then one can expect convergence in a way similar to how convergence takes place in the backpropagation algorithm. A similar approach has also been suggested by Werbos (1987).

B5.2.2

Overall, it must be mentioned that much work needs to be done on the use of function approximators for RL, and clear guidelines are yet to emerge.

References

Albus J S 1975 A new approach to manipulator control: the cerebellar model articulation controller (CMAC) *Trans. ASME J. Dyn. Syst. Meas. Control.* **97** 220–7

Anderson C W 1986 Learning and problem solving with multilayer connectionist systems *PhD Thesis* University of Massachusetts, Amherst, MA, USA

——1989 Learning to control an inverted pendulum using neural networks *IEEE Control Syst. Mag.* 31–7

——1993 *Q*-learning with hidden-unit restarting *Advances in Neural Information Processing Systems 5* ed S J Hanson, J D Cowan and C L Giles (San Mateo, CA: Morgan Kaufmann) pp 81–8

Bacharach J R 1991 A connectionist learning control architecture for navigation *Advances in Neural Information Processing Systems 3* ed R P Lippman, J E Moody and D S Touretzky (San Mateo, CA: Morgan Kaufmann) pp 457–463

——1992 Connectionist modeling and control of finite state environments *PhD Thesis* University of Massachusetts, Amherst, MA, USA

Baird III L C and Harmon M E Residual gradient algorithms *Technical Report* Wright-Patterson Air Force Base, Ohio, USA in preparation

Barto A G 1992 Reinforcement learning and adaptive critic methods *Handbook of Intelligent Control: Neural, Fuzzy, and Adaptive Approaches* ed D A White and D A Sofge (New York: Van Nostrand Reinhold) pp 469–91

Barto A G, Sutton R S and Anderson C W 1983 Neuronlike elements that can solve difficult learning control problems *IEEE Trans. Syst. Man Cybern.* **13** 835–46

Bertsekas D P 1989 *Dynamic Programming: Deterministic and Stochastic Models* (Englewood Cliffs, NJ: Prentice-Hall)

——1994 A counter example to temporal-differences learning *Neural Comput.* **7**

Boyen J 1992 Modular neural networks for learning context-dependent game strategies *Masters Thesis* Computer Speech and Language Processing, University of Cambridge, Cambridge, UK

Bradtke S J 1993 Reinforcement learning applied to linear quadratic regulation *Advances in Neural Information Processing Systems 5* ed S J Hanson, J D Cowan and C L Giles (San Mateo, CA: Morgan Kaufmann) pp 295–302

† The criterion for 'nearness' must be chosen properly depending on the problem being solved. For instance, in section C3.3.1 (see figure C3.1.1) two states on opposite sides of the barrier but whose coordinate vectors are near have vastly different optimal 'cost-to-go' values. Hence the function approximator should not generalize the value at one of these states using the value at the other. Dayan (1993) gives a general approach for choosing a suitable 'nearness' criterion so as to improve generalization.

‡ Bertsekas (1989), Singh and Yee (1993) and Williams and Baird (1993) have derived some general theoretical bounds for errors in value function in terms of function approximator error. Tsitsiklis and Van Roy (1994) have derived bounds for errors when feature-based function approximators are used.

Bradtke S J 1993 Reinforcement learning applied to linear quadratic regulation *Advances in Neural Information Processing Systems 5* ed S J Hanson, J D Cowan and C L Giles (San Mateo, CA: Morgan Kaufmann) pp 295–302

Buckland K M and Lawrence P D 1994 Transition point dynamic programming *Advances in Neural Information Processing Systems 6* ed J D Cowan, G Tesauro and J Alspector (San Fransisco, CA: Morgan Kaufmann) pp 639–46

Chapman D 1991 *Vision, Instruction, and Action* (Cambridge, MA: MIT Press)

Chapman D and Kaelbling L P 1991 Input generalization in delayed reinforcement learning: an algorithm and performance comparisions *Proc. 1991 Int. Joint Conf. on Artificial Intelligence*

Dayan P 1993 Improving generalization for temporal difference learning: the successor representation *Neural Comput.* **5** 613–24

Gullapalli V and Barto A G 1994 Convergence of indirect adaptive asynchronous value iteration algorithms *Advances in Neural Information Processing Systems 6* ed J D Cowan, G Tesauro and J Alspector (San Francisco, CA: Morgan Kaufmann) pp 695–702

Gullapalli V, Franklin J A and Benbrahim H 1994 Acquiring robot skills via reinforcement learning *IEEE Control Syst. Mag.* 13–24

Kaelbling L P 1990 Learning in embedded systems *Technical Report TR–90–04 PhD Thesis* Department of Computer Science, Stanford University, Stanford, CA, USA

——1991 *Learning in Embedded Systems* (Cambridge, MA: MIT Press)

Lin L J 1991a Programming robots using reinforcement learning and teaching *Proc. Ninth Nat. Conf. on Artificial Intelligence* (Cambridge, MA: MIT Press) pp 781–6

——1991b Self-improvement based on reinforcement learning planning and teaching *Machine Learning: Proc. Eighth Int. Workshop* ed L A Birnbaum and G C Collins (San Mateo, CA: Morgan Kaufmann) pp 323–7

——1991c Self-improving reactive agents: case studies of reinforcement learning frameworks *From Animals to Animats: Proc. First Int. Conf. on Simulation of Adaptive Behaviour* (Cambridge, MA: MIT Press) pp 297–305

——1992 Self-improving reactive agents based on reinforcement learning, planning and teaching *Machine Learning* **8** 293–321

——1993 Hierarchical learning of robot skills by reinforcement *Proc. 1993 Int. Conf. on Neural Networks* pp 181–6

Lin C S and Kim H 1991 CMAC-based adaptive critic self-learning control *IEEE Trans. Neural Networks* **2** 530–3

Linden A 1993 *On Discontinuous Q-functions in Reinforcement Learning* (available via anonymous ftp from archive.cis.ohio-state.edu in directory /pub/neuroprose)

Mahadevan S and Connell J 1991 Scaling reinforcement learning to robotics by exploiting the subsumption architecture *Machine Learning: Proc. Eighth Int. Workshop* ed L A Birnbaum and G C Collins (San Mateo, CA: Morgan Kaufmann) pp 328–32

Michie D and Chambers R A 1968 BOXES: An experiment in adaptive control *Machine Intelligence 2* ed E Dale and D Michie (Oliver and Boyd) pp 137–152

Millan J D R and Torras C 1992 A reinforcement connectionist approach to robot path finding in non maze-like environments *Machine Learning* **8** 363–95

Moore A W and Atkeson C G 1993 Memory-based reinforcement learning: efficient computation with prioritized sweeping *Advances in Neural Information Processing Systems 5* ed S J Hanson, J D Cowan and C L Giles (San Mateo, CA: Morgan Kaufmann) pp 263–70

Mozer M C and Bacharach J 1990a Discovering the structure of reactive environment by exploration *Advances in Neural Information Processing 2* ed D S Touretzky (San Mateo, CA: Morgan Kaufmann) pp 439–446

——1990b Discovering the structure of reactive environment by exploration *Neural Comput.* **2** 447–57

Platt J C 1991 Learning by combining memorization and gradient descent *Advances in Neural Information Processing Systems 3* ed R P Lippmann, J E Moody and D S Touretzky (San Mateo, CA: Morgan Kaufmann) pp 714–720

Rosen B E, Goodwin J M and Vidal J J 1991 Adaptive range coding *Advances in Neural Information Processing Systems 3* ed R P Lippmann, J E Moody and D S Touretzky (San Mateo, CA: Morgan Kaufmann) pp 486–94

Shepansky J F and Macy S A 1987 Teaching artificial neural systems to drive: manual training techniques for autonomous systems *Proc. First Ann. Int. Conf. on Neural Networks (San Diego, CA)*

Singh S P 1991 Transfer of learning across composition of sequential tasks *Machine Learning: Proc. Eighth Int. Workshop* ed L A Birnbaum and G C Collins (San Mateo, CA: Morgan Kaufmann) pp 348–52

——1992a Reinforcement learning with a hierarchy of abstract models *Proc. Tenth Nat. Conf. on Artificial Intelligence (San Jose, CA)*

——1992b On the efficient learning of multiple sequential tasks *Advances in Neural Information Processing Systems 4* ed J E Moody, S J Hanson and R P Lippmann (San Mateo, CA: Morgan Kaufmann) pp 251–8

——1992c Transfer of learning by composing solutions of elemental sequential tasks *Machine Learning* **8** 323–39

Singh S P and Yee R C 1993 An upper bound on the loss from approximate optimal-value functions *Technical Report* University of Massachusetts, Amherst, MA, USA

Sutton R S 1990 Integrated architecture for learning, planning, and reacting based on approximating dyanmic programming *Proc. Seventh Int. Conf. on Machine Learning* (San Mateo, CA: Morgan Kaufmann) pp 216–24

——1991a Planning by incremental dynamic programming *Machine Learning: Proc. Eighth Int. Workshop* ed L A Birnbaum and G C Collins (San Mateo, CA: Morgan Kaufmann) pp 353–7

——1991b Integrated modeling and control based on reinforcement learning and dynamic programming *Advances in Neural Information Processing Systems 3* ed R P Lippmann, J E Moody and D S Touretzky (San Mateo, CA: Morgan Kaufmann) pp 471–8

Tesauro G J 1992 Practical issues in temporal difference learning *Machine Learning* **8** 257–78

Tham C K and Prager R W 1994 A modular *Q*-learning architecture for manipulator task decomposition *Machine Learning: Proc. Eleventh Int. Conf.* ed W W Cohen and H Hirsh NJ (San Mateo, CA: Morgan Kaufmann) (available via gopher from Dept of Engineering, University of Cambridge, Cambridge, UK)

Thrun S B 1993 Exploration and model building in mobile robot domains *Proc. 1993 Int. Conf. on Neural Networks* (San Francisco: IEEE Press)

Thrun S B and Schwartz A 1993 Issues in using function approximation for reinforcement learning *Proc. Fourth Connectionist Models Summer School* (Hillsdale, NJ: Erlbaum)

Tsitsiklis J N and Van Roy B 1994 Feature-based methods for large scale dynamic programming *Technical Report LIDS-P-2277* Laboratory for Information and Decision Systems, Massachussetts Institute of Technology, Cambridge, MA, USA

Watkins C J C H 1989 Learning from delayed rewards *PhD Thesis* Cambridge University, Cambridge, UK

Werbos P J 1987 Building and understanding adaptive systems: a statistical/numerical approach to factory automation and brain research *IEEE Trans. Syst. Man Cybern.*

Williams R J and Baird 1993 Tight performance bounds on greedy policies based on imperfect value functions *Technical Report NU-CCS-93-14* College of Computer Science, Northeastern University, Boston, MA, USA

C3.7 Modular and hierarchical architectures

S Sathiya Keerthi and B Ravindran

Abstract

See the abstract for Chapter C3.

When applied to problems with large task space or sparse rewards, reinforcement learning (RL) methods are terribly slow to learn. Dividing the problem into simpler subproblems, using a hierarchical control structure, and so on, are ways of overcoming this.

Sequential task decomposition is one such method. This method is useful when a number of complex tasks can be performed making use of a finite number of 'elemental' tasks or skills, say, $T_1, T_2, \cdots, T_n$. The original objective of the controller can then be achieved by temporally concatenating a number of these elemental tasks to form what is called a 'composite' task. For example,

$$C_j = [T(j, 1), T(j, 2), \ldots, T(j, k)] \text{ where } T(j, i) \in \{T_1, T_2, \ldots, T_n\}$$

is a composite task made up of k elemental tasks that have to be performed in the order listed. Reward functions are defined for each of the elemental tasks, making them more abundant than in the original problem definition.

Singh (1992a, b) has proposed an algorithm based on a *modular neural network* (Jacobs *et al* 1991) B2.9 making use of these ideas. In his work the controller is unaware of the decomposition of the task and has to learn both the elemental tasks and the decomposition of the composite tasks simultaneously. Tham and Prager (1994) and Lin (1993) have proposed similar solutions. Mahadevan and Connell (1991) have developed a method based on the *subsumption architecture* (Brooks 1986) where the decomposition of the task is specified by the user beforehand, and the controller learns only the elemental tasks, while Maes and Brooks (1990) have shown that the controller can be made to learn the decomposition also, in a similar framework. All these methods require some external agency to specify the problem decomposition. Can the controller itself learn how the problem is to be decomposed? Though Singh (1992d) has some preliminary results, much work needs to be done here.

Another approach to this problem is to use some form of hierarchical control (Watkins 1989). Here there are different 'levels' of controllers—controllers at different levels may operate at different temporal resolutions—each learning to perform a more abstract task than the level below it and directing the lower-level controllers to achieve its objective. For example, in a ship a navigator decides in what direction to sail so as to reach the port while the helmsman steers the ship in the direction indicated by the navigator. Here the navigator is the higher-level controller and the helmsman the lower-level controller. Since the higher-level controllers have to work on a smaller task space and the lower-level controllers are set simpler tasks, improved performance results.

Examples of such hierarchical architectures are *feudal RL* by Dayan and Hinton (1993) and *hierarchical planning* by Singh (1992a, 1992c). These methods too require an external agency to specify the hierarchy to be used. This is done usually by making use of some 'structure' in the problem.

Training controllers on simpler tasks first, and then training them to perform progressively more complex tasks using these simpler tasks, can also lead to better performance. Here, at any one stage the controller is faced with only a simple learning task. This technique is called *shaping* in animal behavior literature. Gullapalli (1992a) and Singh (1992d) have reported some success in using this idea. Singh shows that the controller can be made to 'discover' a decomposition of the task by itself, using this technique.

C3.7.1 Other techniques

Apart from the ideas mentioned above, various other techniques have been suggested for speeding-up RL. Two novel ideas have been suggested by Lin (1991a, b, c, 1992): *experience playback* and *teaching*. Let us first discuss experience playback. An experience consists of a quadruple (occurring in real-time system operation) (x, a, y, r) where x is a state, a is the action applied at state x, y is the resulting state and r is $r(x, a)$. Past experiences are stored in a finite memory buffer, $\mathcal{P}$. An appropriate strategy can be used to maintain $\mathcal{P}$. At some point in time let π be the 'current' (stochastic) policy. Let

$$\mathcal{E} = \{(x, a, y, r) \in \mathcal{P} \mid \text{Prob}\{\pi(x) = a\} \geq \epsilon\}$$

where ϵ is some chosen tolerance. The learning update rule is applied, not only to the current experience, but also to a chosen subset of $\mathcal{E}$. Experience playback can be especially useful in learning about rare experiences. In teaching, the user provides the learning system with experiences so as to expedite learning.

B3.4   Incorporating domain-specific knowledge also helps in *speeding-up learning*. For example, for a given problem, a 'nominal' controller that gives reasonable performance may be easily available. In that case RL methods can begin with this controller and improve its performance (Singh *et al* 1994). Domain-specific information can also greatly help in choosing state representation and setting up the function approximators (Barto 1992, Millan and Torras 1992).

In many applications an inaccurate system model is available. It turns out to be very inefficient to discard the model and simply employ a model-free method. An efficient approach is to interweave a number of 'planning' steps between every two on-line learning steps. A planning step may be one of the following: a time step of a model-based method such as real-time dynamic programming (RTDP) or a time step of a model-free method for which experience is generated using the available system model. In such an approach, it is also appropriate to adapt the system model using on-line experience. These ideas form the basis of Sutton's *Dyna* architectures (Sutton 1990, 1991) and related methods (Moore and Atkeson 1993, Peng and Williams 1993).

C2.3.4   In this chapter we have given a cohesive overview of existing RL algorithms. Though research has reached a mature level, RL has been successfully demonstrated only on a few practical applications (Gullapalli *et al* 1994, Tesauro 1992, Mahadevan and Connell 1991, Thrun 1993) and clear guidelines for its general applicability do not exist. The connection between *dynamic programming* and RL has nicely bridged control theorists and artificial-intelligence researchers. With contributions from both these groups in the pipeline, more interesting results are forthcoming and it is expected that RL will make a strong impact on the intelligent control of dynamic systems.

References

Barto A G 1992 Reinforcement learning and adaptive critic methods *Handbook of Intelligent Control: Neural, Fuzzy, and Adaptive Approaches* ed D A White and D A Sofge (New York: Van Nostrand Reinhold) pp 469–91

Brooks R A 1986 Achieving artificial intelligence through building robots *Technical Report AI Memo 899* Massachusetts Institute of Technology, Aritificial Intelligence Laboratory, Cambridge, MA, USA

Dayan P and Hinton G E 1993 Feudal reinforcement learning *Advances in Neural Information Processing Systems 5* ed S J Hanson, J D Cowan and C L Giles (San Mateo, CA: Morgan Kaufmann) pp 271–8

Gullapalli V 1992 Reinforcement learning and its application to control *Technical Report COINS 92–10, PhD Thesis* University of Massachusetts, Amherst, MA, USA

Gullapalli V, Franklin J A and Benbrahim H 1994 Acquiring robot skills via reinforcement learning *IEEE Control Syst. Mag.* 13–24

Jacobs R A, Jordan M I, Nowlan S J and Hinton G E 1991 Adaptive mixtures of local experts *Neural Comput.* **3** 79–87

Lin L J 1991a Programming robots using reinforcement learning and teaching *Proc. Ninth Nat. Conf. on Artificial Intelligence* (Cambridge, MA: MIT Press) pp 781–6

——1991b Self-improvement based on reinforcement learning planning and teaching *Machine Learning: Proc. Eighth Int. Workshop* ed L A Birnbaum and G C Collins (San Mateo, CA: Morgan Kaufmann) pp 323–7

——1991c Self-improving reactive agents: case studies of reinforcement learning frameworks *From Animals to Animats: Proc. First Int. Conf. on Simulation of Adaptive Behaviour* (Cambridge, MA: MIT Press) pp 297–305

——1992 Self-improving reactive agents based on reinforcement learning, planning and teaching *Machine Learning* **8** 293–321

——1993 Hierarchical learning of robot skills by reinforcement *Proc. 1993 Int. Conf. on Neural Networks* pp 181–6

Maes P and Brooks R 1990 Learning to coordinate behaviour *Proc. Eighth Nat. Conf. on Artificial Intelligence* (San Mateo, CA: Morgan Kaufmann) pp 796–802

Mahadevan S and Connell J 1991 Scaling reinforcement learning to robotics by exploiting the subsumption architecture *Machine Learning: Proc. Eighth Int. Workshop* ed L A Birnbaum and G C Collins (San Mateo, CA: Morgan Kaufmann) pp 328–32

Millan J D R and Torras C 1992 A reinforcement connectionist approach to robot path finding in non maze-like environments *Machine Learning* **8** 363–95

Moore A W and Atkeson C G 1993 Memory-based reinforcement learning: efficient computation with prioritized sweeping *Advances in Neural Information Processing Systems 5* ed S J Hanson, J D Cowan and C L Giles (San Mateo, CA: Morgan Kaufmann) pp 263–70

Peng J and Williams R J 1993 Efficient learning and planning within the Dyna framework *Proc. 1993 Int. Joint Conf. on Neural Networks* 168–74

Singh S P 1992a Reinforcement learning with a hierarchy of abstract models *Proc. Tenth Nat. Conf. on Artificial Intelligence (San Jose, CA)*

——1992b On the efficient learning of multiple sequential tasks *Advances in Neural Information Processing Systems 4* ed J E Moody, S J Hanson and R P Lippmann (San Mateo, CA: Morgan Kaufmann) pp 251–8

——1992c Scaling Reinforcement learning algorithms by learning variable temporal resolution models *Proc. Ninth Int. Machine Learning Conf.*

——1992d Transfer of learning by composing solutions of elemental sequential tasks *Machine Learning* **8** 323–39

——1994 Learning to solve Markovian decision processes *PhD Thesis* Department of Computer Science, University of Massachussetts, Amherst, MA, USA

Singh S P, Jaakkola T and Jordan M I 1994 Learning without state-estimation in partially observable Markov decision processes *Machine Learning*

Sutton R S 1990 Integrated architecture for learning, planning, and reacting based on approximating dyanmic programming *Proc. Seventh Int. Conf. on Machine Learning* (San Mateo, CA: Morgan Kaufmann) pp 216–24

——1991 Integrated modeling and control based on reinforcement learning and dynamic programming *Advances in Neural Information Processing Systems 3* ed R P Lippmann, J E Moody and D S Touretzky (San Mateo, CA: Morgan Kaufmann) pp 471–8

Tesauro G J 1992 Practical issues in temporal difference learning *Machine Learning* **8** 257–78

Tham C K and Prager R W 1994 A modular Q-learning architecture for manipulator task decomposition *Machine Learning: Proc. Eleventh Int. Conf.* ed W W Cohen and H Hirsh NJ (San Mateo, CA: Morgan Kaufmann) (available via gopher from Dept of Engineering, University of Cambridge, Cambridge, UK)

Thrun S B 1993 Exploration and model building in mobile robot domains *Proc. 1993 Int. Conf. on Neural Networks* (San Francisco: IEEE Press)

Watkins C J C H 1989 Learning from delayed rewards *PhD Thesis* Cambridge University, Cambridge, UK

PART D

HYBRID APPROACHES

PART D

HYBRID APPROACHES

D1

Neuro-fuzzy Systems

Krzysztof J Cios and Witold Pedrycz

Abstract

In this chapter we describe neuro-fuzzy systems which combine the advantages of numerical computations of neural networks with symbolic processing of fuzzy sets. First, we give a brief introduction to fuzzy sets, sufficient to understand the topics covered in the chapter. This includes a discussion of methods for eliciting membership functions. Next, several typical neuro-fuzzy algorithms are discussed and illustrated. The last few sections concentrate on fuzzy neural networks, where basic processing components (fuzzy neurons) and several general architectures are discussed. In particular, it is shown that some topologies of the networks, such as logic processors, can be exploited in a logic-based approximation of functional relationships.

Contents

D1.1 Introduction

Krzysztof J Cios and Witold Pedrycz

Abstract

See the abstract for Chapter D1.

This chapter deals with neuro-fuzzy computing, a hybrid of two diverse concepts: neural networks and fuzzy sets. These two technologies naturally complement each other by addressing various facets of information processing. The most important features can be outlined briefly as follows: neural networks are massively parallel processing structures aimed at purely numerical processing. Fuzzy sets, with their underlying philosophy of looking at collections rather than individual objects, are naturally appropriate for the representation of knowledge at the higher level of information granularity inherent in human problem solving. As such, fuzzy sets naturally constitute a crucial component in the development of neural network theory, especially at the front end of any neural network. They are particularly important when forming a flexible interface to neural networks and placing numerical computational faculties of the networks in certain well-thought-out settings. Before elaborating on the principles guiding this integration, it is worth characterizing the essence of neural networks and fuzzy sets viewed as two key paradigms. The dominant criteria used in this comparison concern knowledge representation, learning capabilities, and learning plasticity.

Owing to a distributed architecture with a vast number of network parameters, neural networks are equipped with significant learning capabilities. These are essentially of a parametric form and aimed at minimizing a given performance index or objective function by modifying the values of the connections. Fuzzy sets are primarily concerned with issues of uncertain knowledge representation. Their learning capabilities are very much limited, if not nonexistent. The domain knowledge is represented explicitly in terms of easily understood linguistic labels that could be perceived at either numeric or symbolic levels. It is also worth concentrating on explicit versus implicit methods of knowledge representation and learning capabilities, and discussing how these facets are handled by fuzzy sets and neural networks.

There are two main approaches towards building neuro-fuzzy architectures depending upon the area of expertise of a designer. On one hand, one can look at incorporating concepts of fuzzy sets into some 'standard' neural networks at the level of their *topologies, learning schemes, interpretation of results*, and so on: see figure D1.1.1. Quite often these activities fall into a category known as object fuzzification, such as fuzzification of neurons and weights. By fuzzification we mean taking a single numerical value and converting it into a collection of numerical values, or a fuzzy set. While the term itself has been widely used in the literature, we are convinced that this wording does not fully reflect the nature of this enhancement and any generalization involving fuzzy sets needs to be analyzed with respect to its computational efficiency. The dual approach involves the use of neural computation viewed as an integral part of enhancing the computational faculties of fuzzy sets. Some examples of this type of interaction concern membership function estimation and fuzzy inference mechanisms implemented as neural networks: refer again to figure D1.1.1. B2.2, B3.3

Finally, we are also faced with neuro-fuzzy systems—a category of systems where both neural networks and fuzzy sets give rise to a totally new concept embracing the essence of neural computation and fuzzy set computing; figure D1.1.1.

Fully acknowledging the variety of the existing approaches, the aim of this chapter is to outline the main trends, study general development techniques, and discuss in depth some algorithms that are representative of the areas already identified.

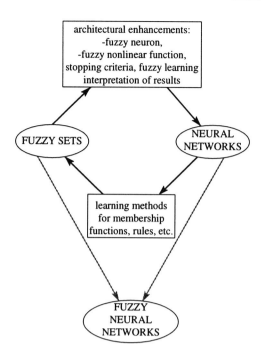

Figure D1.1.1. Different ways of interaction between fuzzy set technology and neural computation.

D1.2 Fuzzy sets and knowledge representation issues

Krzysztof J Cios and Witold Pedrycz

Abstract

See the abstract for Chapter D1.

In this section we are primarily concerned with fuzzy sets viewed as a vehicle for knowledge representation. Our aim is to visualize the essential aspects of fuzzy sets as a tool for explicit knowledge representation capable of handling uncertainty. It is strongly claimed that fuzzy sets and neural networks are complementary with respect to their knowledge representation and learning capabilities or plasticity, making them ideal components for hybridization.

D1.2.1 Sets versus fuzzy sets

In order to introduce the idea of fuzzy sets in more detail, it is worth beginning with the formalism of two-valued logic. In this setting, the notion of a set implies that considering any object, no matter how complex, we are compelled to assign it to one of the two complementary and exhaustive categories specified *a priori*, for instance, *good–bad*, *normal–abnormal* or *odd–even*, etc. Sometimes this discrimination does make sense. In many other situations, this dichotomization tends to be overly restrictive and can easily lead to some serious dilemmas. For example, let us consider natural numbers and define two categories or sets of elements such as odd and even numbers. Within this framework any natural number can be classified without hesitation. On the other hand, in many tasks in engineering, manufacturing, or management, we are faced with classes that are ill defined and do not have clear and well-defined boundaries.

Even within a field of mathematics we encounter some broadly accepted and used notions with gradual rather than abrupt boundaries. We refer to such well known terms as: *sparse* matrix, a linear approximation of a function in a *small* neighborhood of a point x_0, or an *ill*-conditioned matrix, and we accept these notions as conveying useful information. Furthermore, they are not regarded as defects of our everyday language but rather as a beneficial feature indicating our ability to generalize and conceptualize knowledge. Nevertheless, we should stress that these notions are strongly context dependent and by no means can detailed definition be deemed universal.

The key issue of fuzzy sets is one that extends significantly the meaning of a set admitting different grades of belongingness or membership values of an element in a set. This alleviates the previous dichotomization problem by embracing all intermediate conceptual situations arising between total membership and total nonmembership, or truth and falsehood. In the early 1920s Jan Łukasiewicz, a Polish logician, first addressed the problem of the truth of statements being a matter of degree. He introduced multivalued logic which defined a continuum between falsehood and truth, or between zero and one. Many authors, among them Kosko (1993), consider Łukasiewicz to be the father of what later became known as fuzzy logic, a term coined much later by Zadeh (1965).

Formally, a fuzzy set A defined in a universe of discourse $\mathbf{X}$ is described by its membership function viewed as a mapping (Zadeh 1965)

$$A : \mathbf{X} \to [0, 1].$$

The degree of membership $A(x)$ expresses the extent to which x fulfils the category described by A. The condition $A(x) = 1$ identifies elements of $\mathbf{X}$ which are fully compatible with A. The condition $A(x) = 0$ identifies all the elements which definitely do not belong to A. The higher the membership value at x, the higher the adherence of x to A. Any physical experiment whose realization is a matter of energy, like pulling a rubber band, can serve as a useful metaphor for the notion of membership function or membership degree. Usually by discussing a fuzzy set we assume that elements exist with membership grades equal to 1; these sets are called normal.

An intuitive observation that fuzzy sets are generalizations of sets can be formalized in what is usually called the representation theorem (Zadeh 1965, Kandel 1986). Briefly speaking, it states that a fuzzy set can be decomposed, and composed by taking into account elements with membership values not lower than a certain threshold. Let us first introduce the notion of an α-cut. By an α-cut, denoted by A_α, we mean a set of elements of A belonging to it with degrees of membership not less than α.

$$A_\alpha = \{x \in \mathbf{X} \mid A(x) \geq \alpha\} \qquad \alpha \in [0, 1].$$

The representation theorem states that any fuzzy set A can be represented by a union of its α-cuts, namely

$$A = \bigcup_{a \in [0,1]} \alpha A_\alpha.$$

This relationship is also referred to as a resolution identity. It is used quite frequently in situations when a fuzzy set needs to be translated into a collection of sets.

D1.2.2 Membership functions: types and elicitation methods

In many situations it is worth restricting analysis to piecewise linear membership functions. They give rise to a class of triangular and trapezoidal fuzzy numbers or fuzzy sets as shown in figure D1.2.1.

Figure D1.2.1. Examples of triangular and trapezoidal fuzzy numbers.

This characterization of a fuzzy number is sufficient to capture the uncertainty associated with the linguistic term studied. The triangular fuzzy number, denoted by $A(x; \alpha, m, \beta)$ is uniquely characterized by parameters m, α and β, where $\alpha < m < \beta$, see figure D1.2.1(a). The first parameter embodies a modal or typical value. The lower and the upper bounds are denoted by α and β, respectively. For instance, a waiting time W in a queue which typically takes 15 minutes to get service while the bounds are 5 and 29 minutes, respectively, can be described as a triangular fuzzy number $W(t; 5, 15, 29)$. Since no additional information about the waiting time is available, the choice of the linear relationship is fully legitimate. If there is no uncertainty (fuzziness) then $\alpha = m = \beta$ and the fuzzy number reduces to a single quantity regarded as a real number.

A trapezoidal fuzzy number admits an additional degree of freedom that enables us to model a range of equally acceptable typical values. In this class of membership functions the modal value, m, spreads into a closed interval $[n, m]$ as shown in figure D1.2.1(b).

As far as membership function estimation is concerned there are the following essential classes: the first two, described below, elicit the membership functions from experts; the last three estimate membership functions directly from data.

Horizontal approach. Its underlying idea is to gather information about grades of membership of some elements of a universe of discourse in which a fuzzy set is to be defined. The process of elicitation of these membership functions can be stated as follows. Consider a group of N experts. Each of them is asked to answer the following question:

Can x_0 be viewed as compatible with the concept represented by a fuzzy set A?

where x_0 is a fixed element of this universe of discourse and A is a fuzzy set to be determined. The answers are restricted to 'yes' or 'no' statements only. Then, counting the fraction of positive responses, $n(x_0)$, the value of the membership function at x_0 is estimated as

$$A(x_0) = \frac{n(x_0)}{N}.$$

Vertical approach. The main concept behind this method is to fix a certain level of the membership, α, and ask a group of experts to identify a collection of elements in **X** satisfying the concept carried by A to a degree not lower than α. Thus, the essence of the method is to determine α-cuts of the fuzzy set. Once the experimental results are gathered then the fuzzy set is 'reconstructed' by aggregating the estimated α-cuts.

Obviously these two approaches are conceptually simple. The factor of uncertainty reflected by the fuzzy boundaries of A is distributed either vertically, in the sense of the grades of membership, or horizontally, thus being absorbed by the limit points of the α-cuts. The values of α or different elements of the universe of discourse should be selected randomly to avoid any potential bias furnished by the experts.

The evident shortcoming of these two methods lies in the 'local' nature of the experiments. This means that each grade of membership is estimated independently from the rest. Then the results may not fully comply with the general tendency of maintaining a smooth transition from full membership to absolute exclusion. In this situation, a pairwise comparison method introduced by Saaty (1980) can be used to alleviate the inadequacy in the above methods.

The following three methods differ from the two discussed above in that they do not require human experts. Membership functions of any shape, although most often piecewise linear, can be derived directly from a preferably large data set, called training data, collected from the process which is to be described by using fuzzy sets. The three methods are briefly outlined next.

Statistical approach. The assumption is that the membership functions can be initially defined using statistical relationships between the variables of interest. The probability density functions and the corresponding distribution functions can then be estimated from training data on some interval, or range, over which a fuzzy set is to be defined. From the ratios of distribution functions fuzzy membership functions are defined. Details and an example of utilization of the method is described by Cios *et al* (1991).

Machine learning. To define membership functions, usually piecewise linear, the IF... THEN... rules generated by inductive machine learning algorithms are used in the following way. First, the precedent parts of all the rules having the same consequent are aggregated using a generalized fuzzy integration operator. Second, the consequent parts of the same rules are combined to describe a proper linguistic term (membership function) through the use of a generalized fuzzy union operator. Finally, a so-defined membership function can be used directly or converted to, say, a trapezoidal fuzzy number. Details of the method and its utilization on real data can be found in Cios *et al* (1991, 1994).

Neural networks. This method of defining membership functions from numerical data through the use of neural networks is becoming increasingly popular. It takes advantage of division of training examples, performed by neurons/hyperplanes, into those lying on positive/negative sides of a hyperplane, then counting them and taking their ratios to define membership functions. The idea behind the method is explained in Section D1.4 of this chapter, with more details given by Cios and Sztandera (1996).

At this point, it is essential to comment on fuzziness and randomness as two very distinct and somewhat orthogonal facets of uncertainty. In general, randomness deals with the models of statistical inexactness emerging due to the occurrence of random events, while fuzziness concerns situations of modeling of inexactness arising due to perception processes of humans.

D1.2.3 Logical operations on fuzzy sets

The basic operations (logical connectives) can be defined by replacing the characteristic functions of sets by the membership functions of the fuzzy sets. This gives rise to the following expressions:

$$(A \cup B)(x) = \max(A(x), B(x))$$
$$(A \cap B)(x) = \min(A(x), B(x))$$
$$\bar{A}(x) = 1 - A(x)$$

where $x \in \mathbf{X}$ and **X** is a universe of discourse.

Since the grades of membership extend the two-element set of truth values $\{0, 1\}$ into the unit interval $[0, 1]$, it is worth recalling the collection of properties essential for set theory and investigating whether they are satisfied for fuzzy sets.

The De Morgan law of set theory is also preserved in fuzzy sets, namely,

$$\overline{A \cap B} = \bar{A} \cup \bar{B} \qquad \overline{A \cup B} = \bar{A} \cap \bar{B}.$$

The distributivity laws are fulfilled and the properties of absorption and idempotency hold as well. However, the exclusion conditions are not satisfied, that is,

$$A \cup \bar{A} \neq \mathbf{X} \qquad \text{(underlap property)}$$
$$A \cap \bar{A} \neq \emptyset \qquad \text{(overlap property)}.$$

These two properties give rise to a very clear distinction between fuzzy sets and sets.

The semantics of the logical connectives can be expressed in many ways. An example is the product operation, $A(x)B(x)$, studied as a model used for the logic intersection and the probabilistic sum, $A(x) + B(x) - A(x)B(x)$, considered for the union operation. In comparison to the lattice (max and min) operations, the computed degree of membership reflects both values of the membership functions $A(x)$ and $B(x)$. We shall restrict ourselves to a class of binary operations satisfying a collection of the following assumptions:

boundary conditions

$$A \cup \mathbf{X} = \mathbf{X} \qquad A \cap \mathbf{X} = A$$
$$A \cup \emptyset = A \qquad A \cap \emptyset = \emptyset$$

commutativity

$$A \cap B = B \cap A \qquad A \cup B = B \cup A$$

associativity

$$(A \cap B) \cap C = A \cap (B \cap C) \qquad (A \cup B) \cup C = A \cup (B \cup C).$$

Observe that all of the above conditions take on an intuitively clear interpretation: for instance, the boundary conditions indicate that the logical connectives for fuzzy sets coincide with those applied in the two-valued logic. The property of commutativity states that a truth value of a composite expression does not depend on the order in which the predicates have been placed.

By accepting the above conditions, a broad class of models for logical connectives (union and intersection) is formed by triangular norms (Dubois and Prade 1988). The triangular norms (Menger 1942) or t-norms and s-norms originated in the theory of probabilistic metric spaces. By a t-norm we mean a function of two arguments

$$t : [0, 1] \times [0, 1] \rightarrow [0, 1]$$

such that it is

(i) nondecreasing in each argument

$$x \leq y, w \leq z \qquad \text{for} \quad x\, t\, w \leq y\, t\, z$$

(ii) commutative

$$x\, t\, y = y\, t\, x$$

(iii) associative

$$(x\, t\, y)\, t\, z = x\, t\, (y\, t\, z)$$

(iv) satisfies the set of boundary conditions

$$x\, t\, 0 = 0 \qquad x\, t\, 1 = x$$

with $x, y, z, w \in [0, 1]$.

All the properties of the t-norm can be easily identified with the relevant characteristics of the intersection operation (logical AND).

An s-norm is defined as a function of two arguments

$$s : [0, 1] \times [0, 1] \rightarrow [0, 1]$$

such that it:

(i) is a nondecreasing function in each argument
(ii) is commutative
(iii) is associative
(iv) satisfies the boundary conditions

$$x \, s \, 0 = x \qquad x \, s \, 1 = 1 \, .$$

Characteristics (i)–(iv) express the properties of the union operation. An interesting fact is that for each t-norm one can define an associated s-norm such that

$$x \, s \, y = 1 - (1 - x) \, t \, (1 - y) \, .$$

The above relation is simply the De Morgan law found in set theory.

D1.2.4 Frame of cognition: toward a unified data representation

Domain knowledge about a given system can be articulated with the aid of linguistic labels. These are generic pieces of knowledge which are identified by the model developer as being essential in describing and understanding the system. The linguistic labels are represented by fuzzy sets. As demonstrated in Zadeh (1979) they can also be viewed as elastic constraints and identifying regions with the highest degree of compatibility of elements with the given linguistic term. Sometimes the linguistic labels are also referred to as information granules. All the information granules defined in a certain space constitute a frame of cognition of the variable (Pedrycz 1990, 1992). More formally, the family of fuzzy sets $A = \{A_1, A_2, \ldots, A_c\}$ (where $A_i : \mathbf{X} \to [0, 1]$) constitutes a frame of cognition A if the following two properties are satisfied.

(i) A 'covers' the universe $\mathbf{X}$, namely each element of the universe is assigned to at least one granule with a nonzero degree of membership meaning that

$$\forall x \, \exists i \, A_i(x) > 0 \, .$$

This property assures that any piece of information defined in $\mathbf{X}$ is properly represented or described by A_i.

(ii) The elements of A are unimodal fuzzy sets or unimodal membership functions. By stating that, we identify several regions of $\mathbf{X}$, one for each A_i, as highly compatible with the labels.

The frame of cognition can be developed either on a fully experimental basis or in an algorithmic way. In the first instance, the linguistic labels can be specified by studying the problem and recognizing basic relevant information granules as being necessary in describing and handling it. It is the user who provides relevant membership functions for the variables of the system and therefore creates his own individual cognitive perspective. Analogously, the standard methods of membership function estimation, as outlined above, can be utilized directly.

The second approach, which could be helpful when some records of numerical data are available, relies on a suitable utilization of fuzzy clustering techniques. Fuzzy clustering (Bezdek 1981) enables us to discover and conveniently visualize the structure existing in the data set. With its aid the numerical data are structured into a number of clusters according to a predefined similarity measure. The number of clusters is also defined in advance so that they correspond to the linguistic labels constituting the frame of cognition. Fuzzy clustering generates grades of membership of the elements of the data set in the given clusters. The frame of cognition A can be also referred to as a fuzzy partition of $\mathbf{X}$.

Considering the family of the linguistic labels encapsulated in the same frame of cognition, several properties are worth underlining.

Specificity. The frame of cognition $\mathcal{A}$ is more specific than $\mathcal{A}'$ if the elements of $\mathcal{A}$ are more specific than the elements of $\mathcal{A}'$. The specificity of the fuzzy set A can be evaluated using, for example, the specificity measure as discussed in Yager (1980). An example of $\mathcal{A}$ and $\mathcal{A}'$ of different specificity is shown in figure D1.2.2.

Focus of attention. A scope of perception of A_i in frame $\mathcal{A}$ is defined as an α-cut of this fuzzy set. By moving A_i along $\mathbf{X}$ while not changing its membership function we can focus attention on a certain region of $\mathbf{X}$.

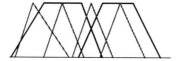

Figure D1.2.2. Two frames of cognition of different specificity levels.

Information hiding. This idea is directly linked with the focus of attention. By modifying the membership function of A being an element of $\mathcal{A}$ we can have the important effect of achieving an equivalence of the elements lying within some regions of $\mathbf{X}$. Consider a trapezoidal fuzzy set A in $\mathbb{R}$ with its 1-cut distributed between a_1 and a_2. All the elements falling within this interval are nondistinguishable: $A(x) = 1$ for x contained in this interval. Thus, the processing module does not distinguish between any two elements in the 1-cut of A, hence the detailed information becomes hidden. By modulating the level of the α-cut we can accomplish an α-information hiding.

There is a question of representing any input datum in the frame of cognition developed in this manner. We shall introduce possibility and necessity measures (Zadeh 1978, Dubois and Prade 1988) as the mechanisms most frequently used to develop this transformation. Let A be one of the elements of the frame of cognition and X constitute an input datum. X and A are defined in the same universe of discourse. The possibility measure, $\mathrm{Poss}(X|A)$,

$$\mathrm{Poss}(X|A) = \sup_{z \in \mathbf{X}}[\min(X(z), A(z))]$$

expresses a degree to which X and A overlap. The necessity measure, $\mathrm{Nec}(X|A)$,

$$\mathrm{Nec}(X|A) = \inf_{z \in \mathbf{X}}[\max((1 - X(z)), A(z))] = \inf_{z \in \mathbf{X}}[\max(\bar{X}(z), A(z))]$$

characterizes an extent to which X is included in A, see figure D1.2.3.

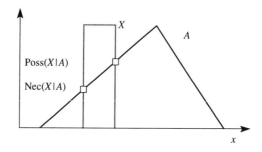

Figure D1.2.3. Calculations of possibility and necessity measures.

Figure D1.2.4 summarizes the performance of these measures for two sets; to discriminate between some of these cases we need to use both measures. Frequently the possibility measure alone might not be sufficient to capture the component of uncertainty residing with X.

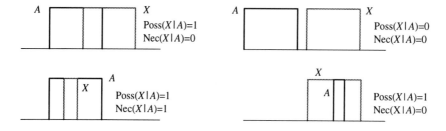

Figure D1.2.4. Possibility and necessity measures for several sets X and A.

For any precise numerical information, $X = \{x_0\}$, these two measures coincide. If X becomes a numerical interval, $X \in \mathbb{R}$, which in fact represents an uncertain input datum, the difference between the possibility and necessity measure is usually different from zero. The following monotonicity property holds:

$$\text{if } X_1 \subset X_2 \text{ then } \text{Poss}(X_1|A) - \text{Nec}(X_1|A) \leq \text{Poss}(X_2|A) - \text{Nec}(X_2|A).$$

This observation may lead us to consider the two measures collectively to quantify uncertainty residing within the input datum. Let us introduce the notation,

$$\lambda = \text{Poss}(X|A)$$

and

$$\mu = 1 - \text{Nec}(X|A).$$

Straightforwardly, for $X = \{x_0\}$, μ becomes a complement of λ, $\mu = 1 - \lambda$, or $\lambda + \mu = 1$. In general, we get either $l + \mu \geq 1$ or $l + \mu \leq 1$. These values depend heavily upon the relative distribution of A and X as well as the form of these fuzzy sets.

References

Bezdek J C 1981 *Pattern Recognition with Fuzzy Objective Function Algorithms* (New York: Plenum)

Dubois D and Prade H 1988 *Possibility Theory—an Approach to Computerized Processing of Uncertainty* (New York: Plenum)

Kandel A 1986 *Fuzzy Mathematical Techniques with Applications* (Reading, MA: Addison-Wesley)

Kosko B 1993 *Fuzzy Thinking* (New York: Hyperion)

Menger K 1942 Statistical metric spaces *Proc. Natl Acad. Sci., USA* **28** 535–7

Pedrycz W 1990 Direct and inverse problem in comparison of fuzzy data *Fuzzy Sets Syst.* **34** 223–36

——1992 Selected issues of frame of knowledge representation realized by means of linguistic labels *Int. J. Int. Syst.* **7** 155–70

Saaty T L 1980 *The Analytic Hierarchy Processes* (New York: McGraw Hill)

Yager R R 1980 On chosing between fuzzy subsets *Kybernetes* **9** 151–4

Zadeh L A 1965 Fuzzy sets *Information Control* **8** 338–53

——1978 Fuzzy sets as a basis for a theory of possibility *Fuzzy Sets Syst.* **1** 3–28

——1979 Fuzzy sets and information granularity *Advances in Fuzzy Set Theory and Applications* ed M M Gupta, R K Ragade and R R Yager (Amsterdam: North-Holland) pp 3–18

D1.3 Neuro-fuzzy algorithms

Krzysztof J Cios and Witold Pedrycz

Abstract

See the abstract for Chapter D1.

Relatively early in neural network research there emerged an interest in analyzing and designing layered, *feedforward* networks augmented by some formalism stemming from the theory of fuzzy sets. One of B2.3 the first approaches was the fuzzification of the binary *McCulloch–Pitts neuron* (Lee and Lee 1975). B1.2 Then, several researchers looked at a typical feedforward neural network architecture and analyzed several combinations of such neurons with fuzzy sets viewed as inputs to the neural network. Similarly, the networks were equipped with connections (weights) viewed as fuzzy sets with triangular membership functions. Interestingly, in all these cases, the outputs of the network were kept numerical. Some representative examples include the work of Yamakawa and Tomoda (1989), O'Hagan (1991), Gupta and Qi (1991), Hayashi *et al* (1992), and Ishibushi *et al* (1992). Commonly, these authors employed fuzzy sets with either triangular or trapezoidal membership functions. The training was accomplished utilizing a standard *delta rule*. In some other cases (Hayashi *et al* 1992) a fuzzified delta rule was used. B3.3.3 The delta rule was also replaced by other algorithms, for instance Requena and Delgado (1992) used a *Boltzmann machine* training. C1.4

D1.3.1 Fuzzy inference schemes and their realizations as neural networks

In the following, we briefly review a certain category of fuzzy inference systems also known as *fuzzy associative memories* (Kosko 1993). This form of memory is often regarded as central to the C1.3, F1.4 implementation of fuzzy-rule-based systems, and, in general, fuzzy systems (Wang and Mendel 1992).

Fuzzy associative memory (FAM) consists of a fuzzifier, fuzzy rule base, fuzzy inference engine, and a defuzzifier. They are static transformations which map input fuzzy sets into output fuzzy sets (Kosko 1993). It carries out a mapping between unit hypercubes. The role of the fuzzifier and defuzzifier is to form a suitable interface between the transformation and the external environment in which modeling is completed. The transformation is based on a set of fuzzy rules, namely rules consisting of fuzzy predicates and reflecting a domain knowledge and usually originating from human experts. This type of knowledge may pertain to some general control policies, linguistic description of systems etc. As will be revealed later on, the knowledge gained from such sources can substantially enhance learning in neural networks by reducing their training time.

The development of a FAM is realized in several steps which are summarized as follows (Kosko 1993). First, we identify the variables of the system and encode them linguistically in terms of fuzzy sets such as *small, medium* and *big*. The second step is to associate these fuzzy sets by constructing rules (if–then statements) of the general form:

$$\text{if } X \text{ is } A \text{ then } Y \text{ is } B$$

where X and Y are system variables, usually referred to as linguistic variables, while fuzzy sets A and B are represented by their corresponding membership functions. Usually each typical application requires from several to many rules of the form given above—their number is implied by the granularity of the fuzzy information captured by the rules. Thus, the rules can be written as:

$$\text{if } X \text{ is } A_k \text{ then } Y \text{ is } B_k.$$

As said before, each rule forms a partial mapping from input space $\mathbf{X}$ into output space $\mathbf{Y}$, which can be written in the form of a fuzzy relation or, more precisely, a Cartesian product of A and B, namely

$$R(x, y) = \min(A(x), B(y))$$

where $x \in \mathbf{X}$, $y \in \mathbf{Y}$ and $A(x)$ and $B(x)$ are grades of membership of x and y in fuzzy sets A and B, respectively.

In the third step we need to decide upon an inference mechanism, used for drawing inferences from a given piece of information and the available rules. The inference mechanism embodies two keys steps (Pedrycz 1993, 1995):

(i) *Aggregation of rules.* This summarization of the rules is almost always done by taking a union of the individual rules. As such, the aggregation of N rules leads to a fuzzy relation of the form

$$R = \bigcup_{k=1}^{N}(A_k \times B_k).$$

(ii) *Producing a fuzzy set from given A and R.* The classic mechanism used here is a max–min operation yielding the expression

$$B = A \circ R$$

namely,

$$B(y) = \sup_{x \in \mathbf{X}}[\min(A(x), R(x, y))]$$

$y \in \mathbf{Y}$. Because of the nature of fuzzy sets no perfect match is required to fire, or activate, a particular rule as is the case when using rules not including linguistic terms.

Finally, although the employed inference strategy will determine the output in a form of a fuzzy set, most of the time a user is interested in a crisp or single value at the output as required in most, if not all, current applications. To achieve that, one needs to use one of several defuzzification techniques. One quite often used is the transformation exploiting a weighted sum of the modal values of the fuzzy sets of conclusion. This gives rise to the expression

$$z = \frac{\sum_{k=1}^{N} \lambda_k b_k^*}{\sum_{k=1}^{N} \lambda_k}$$

where λ_k is the level of activation or possibility measure of the antecedent of the kth rule with

$$\lambda_k = \sup_{x \in \mathbf{X}}[\min(A(x), A_k(x))]$$

where b_k^* is a modal value of B_k, namely

$$B_k(b_k^*) = \max_{y \in \mathbf{Y}} B_k(y).$$

Two features of FAMs are worth emphasizing when analyzing their memorization and recall capabilities. They are very similar to those encountered in correlation-based associative memories:

(i) The learning process is straightforward and instantaneous—in fact FAMs do not require any learning. This could be regarded as an evident advantage but it comes at the expense of a fairly low capacity and potential crosstalk distortions.

(ii) This crosstalk in the memory can be avoided for some carefully selected items to be stored. In particular, if all input items A_k are pairwise-disjoint normal fuzzy sets, $A_k \cap A_l = \emptyset$ for all $k, l = 1, 2, \ldots, N, k \neq 1$, then $B_k = A_k \circ R$, $k = 1, 2, \ldots, N$, meaning a perfect recall.

The functional summary of the FAM system which outlines its main components is shown in figure D1.3.1.

Wang (1992) proved that a fuzzy inference system that is equipped with the max-product composition with scaled Gaussian membership functions is a universal approximator. Let us recall that the main idea of universal approximation states that any continuous function $f : \mathbb{R}^n \to R$, can be approximated using a neural network to any degree of accuracy on a compact subset of $\mathbb{R}$ (Hornik *et al* 1989).

C1.3.2   The above described FAM system is often utilized as part of a so-called *bidirectional associative memory* (BAM). The applications of it can be found in control tasks such as the inverted pendulum (Kosko 1993).

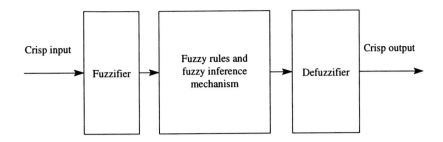

Figure D1.3.1. The architecture of the FAM system.

D1.3.1.1 Fuzzy backpropagation

The fuzzy backpropagation algorithm (Xu *et al* 1992) exploits fuzzy rules for adjusting the activation function and learning rate. By coding the heuristic knowledge about the behavior of the standard backpropagation training Xu *et al* (1992) were able to considerably shorten the time required to train the network, which too often is prohibitive for any real problem.

It should be noted that long training times for backpropagation algorithms arise mainly from keeping both the learning rate and the activation function fixed. Selection of the proper learning rate and 'optimal' activation function in backpropagation algorithms had been studied before (Weir 1991, Silva and Almeida 1990, Rumelhart and McLelland 1986); however, the two parameters were not studied in unison. Rapid minimalization of the training error, e, by proper simultaneous selection of the learning rate, $c(e, t)$, and of the steepness of the activation function, $s(e, t, \text{net}_i)$, where t is time and net_i is the input to the activation function were proposed by Xu *et al* (1992).

As is the most common case, the weights of the network in the backpropagation algorithm are adjusted by using the gradient-descent method according to

$$w_{ji}(t + 1) = w_{ji}(t) - c(e, t) \frac{\partial e}{\partial w_{ji}}$$

where $[w_{ji}]$ represents the weight matrix associated with connections between the neurons and utilizes the following activation function:

$$s(e, t, \text{net}_i) = 1/[1 + \exp(-\sigma(e, t)\text{net}_i)].$$

The activation function, s, is modified by adjusting its steepness factor, $\sigma(e, t)$, as illustrated in figure D1.3.2.

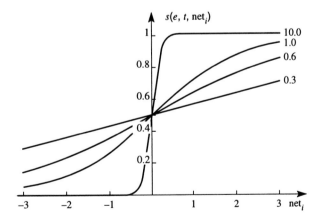

Figure D1.3.2. Activation function for different values of σ.

A set of rules involving linguistic terms (Xu *et al* 1992) used to modify the learning rate $c(e, t)$ is shown in table D1.3.1. The formation of these rules is guided by two straightforward heuristics. First, it is obvious that the learning rate should be *large* when the error is *big*, and *small* when the error is *small*.

Secondly, if training time is *short*, the learning rate should be *large* to promote *faster* learning, and it should be *small* if the training time is *long*, that is, close to a local minimum. Overall, these rules map two input variables with the quantification

$$t = \{\text{short, medium, long}\} \qquad \text{and} \qquad e = \{\text{very small, small, big, very big}\}$$

into the output variable (that is a learning rate)

$$c(e, t) = \{\text{very small, small, large, very large}\}.$$

Table D1.3.1. Rules governing changes of learning rate $c(e, t)$.

Training time	Training error			
	Very small	Small	Big	Very big
Short	Small	Large	Very large	Very large
Medium	Very small	Small	Large	Very large
Long	Very small	Small	Large	Large

These rules can also be expressed in an equivalent 'if–then' format:

rule 1: if $e =$ very small and $t =$ very short then $c(e, t) =$ small
rule 2: if $e =$ very small and $t =$ medium then $c(e, t) =$ very small
$\vdots$
rule 12: if $e =$ very big and $t =$ long then $c(e, t) =$ large.

Similarly, the rules determining the steepness factor $\sigma(e, t)$, as defined in Xu *et al* (1992), are shown in table D1.3.2.

Table D1.3.2. Rules determining steepness factor $\sigma(e, t)$.

Training time	Training error			
	Very small	Small	Big	Very big
Short	Large	Small	Very small	Very small
Medium	Very large	Large	Small	Very small
Long	Very large	Large	Small	Small

The underlying heuristics behind the rules shown in table D1.3.2 can be summarized as follows: if the training time is *short* and the error is *big*, then use a *small* value for the steepness factor so that the activation function becomes flat, and the weights can be quickly adjusted. Second, when the error is *very small* and/or training time is *very long* then the steepness factor should be *large*, so that the activation function becomes almost a step function.

The membership functions for the error, time, steepness factor, and the learning rate are shown in figure D1.3.3.

D1.3.1.2 Fuzzy basis functions

In this section, we shall describe application of the FAM system to the powerful and increasingly popular C1.6.2 *radial basis function* (RBF) network. When the FAM system is incorporated into it, it becomes a fuzzy basis function (FBF) network. We need to briefly introduce radial basis functions first (Moody and Darken 1989), since the FBFs are an augmented version of the RBFs. An RBF is a three-layer network with 'locally-tuned' processing units in the hidden layer. RBF neurons are centered at the training data points, or some subset of them, and each neuron responds only to an input which is close to its center. The

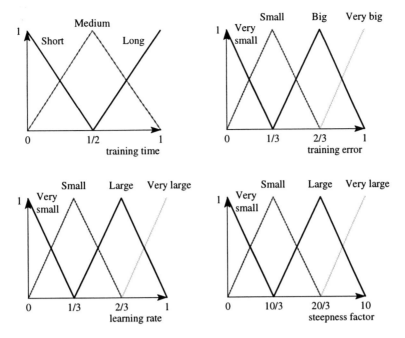

Figure D1.3.3. Membership functions for the linguistic terms used in the above specified rules. Training time and training error are normalized by dividing through by the largest.

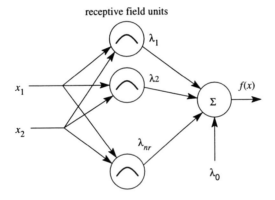

Figure D1.3.4. General RBF network with two inputs.

output layer neurons are linear or use sigmoidal functions and their weights may be obtained by using a supervised learning method, such as a gradient-descent method.

Figure D1.3.4 shows a general RBF network with two inputs and a single linear output. The network performs a mapping $f : \mathbb{R}^n \to \mathbb{R}$ specified by the radial basis function expansion (Chen *et al* 1991):

$$f(x) = \lambda_0 + \sum_{i=1}^{n_r} \lambda_i \, p(\|x - c_i\|)$$

where $x \in \mathbb{R}^n$ is the input vector, $p(\cdot)$ is a function from $\mathbb{R}^n \to \mathbb{R}$ or a radial basis function, $\| \cdot \|$ denotes the Euclidean norm, λ_i are the weights and c_i are the centers, $i = 1, 2, \ldots, n_r$, while n_r is the number of the RBF functions.

One of the most common functions used for $p(\cdot)$ is the Gaussian function

$$p(\|x - c_i\|) = \exp\left(-\frac{\|x - c_i\|}{\sigma_i^2} \right)$$

where σ_i is a constant that determines the width of the ith node; the dimension of vectors c_i is the same as the dimension of the input vectors x.

The centers of the RBF functions, c_i, are usually chosen as elements of the training data points x_i, $i = 1, 2, \ldots, N$. This approach is known as the 'neurons at data points' method (Zahirniak *et al* 1990), and then $n_r = N$. For larger data sets it is not practical to have an RBF center at each data point so other methods are used to reduce the number of RBF centers. Some of them are the random selection of centers, clustering of data points (Zahirniak *et al* 1990), and orthogonal least-squares (OLS) reduction method of Chen *et al* (1991). Jang and Sun (1993) have shown that, under some minor restrictions, RBFs and FAMs are functionally equivalent. Thus, one can apply learning rules of RBFs to fuzzy inference systems, and the learning rules of FAMs to find the number of hidden layers and other parameters of RBFs. Both models are universal approximators if membership functions are scaled Gaussian functions (see also Wang 1992).

In their fuzzy version of the RBF network Wang and Mendel (1992) defined fuzzy basis functions $p(\cdot)$, as follows

$$p_j(x) = \frac{\prod_{i=1}^{n} A_i^j(x_i)}{\sum_{j=1}^{M} \prod_{i=1}^{n} A_i^j(x_i)}$$

where $j = 1, 2, \ldots, M$ is the number of fuzzy if–then rules defined for the system. As can be noticed, the original Gaussian function was replaced by a fuzzy membership function. This was done by multiplying the Gaussian function by a constant (scaling factor), a_i, from the unit interval. The above formula defines fuzzy basis functions for fuzzy systems with singleton fuzzifier, product inference and centroid defuzzifier. The fuzzy Gaussian membership function was defined as

$$A_i^j(x_i) = a_i^j \exp\left[-\frac{1}{2}\left(\frac{x_i - \bar{x}_i^j}{s_i^j}\right)^2\right].$$

These fuzzy basis functions correspond to fuzzy rules of the general form, specified previously as the first part of the FAM system, and they can be determined based only on the 'if' parts of the rules. Note that a more detailed form of a fuzzy rule is

if x_1 is A_1 and x_2 is A_2 and $\ldots$ and x_n is A_n then y is B.

Thus, to calculate the FBF for rule j, or $p_j(x)$, we calculate the product of all membership functions in the 'if' part of the rule j, then we do the same for all M rules, and divide the former through the latter.

FBFs have an interesting property, namely, they seem to combine the Gaussian radial basis functions, which are good for characterizing local properties, with sigmoidal activation functions which have good global characterizing properties (Cybenko 1989). Thus, if fuzzy basis functions are selected using the popular 'neurons at data points' method we achieve high resolution with Gaussian functions, while at the boundaries they look like sigmoidal functions to capture global characteristics of the data.

The FBF expansion can thus be defined in the same manner as for RBF functions, namely by

$$f(x) = \sum_{j=1}^{M} p_j(x)\theta_j$$

where $\theta_j \in R$ are constants or weight parameters. The expansion can be viewed as a linear combination of FBFs, where parameters $p_j(x)$ can be fixed, which allows for an efficient linear estimate of the parameters, in the same manner as in the standard RBF network.

FBFs can be determined in two ways. The first one is to use M fuzzy rules with $M = N$, as described above. The other way is to obtain them from training data and initially position the centers at 'neurons at data points' and require that $a_i^j = 1$ so that the fuzzy Gaussian membership function can achieve unity value at some center $\bar{x}_i^j$. FBFs initial spreads, or their supports, can be determined from

$$\sigma_i^j = [\max(x_i^0(j), j = 1, 2, \ldots, N)) - \min(x_i^0(j), j = 1, 2, \ldots, N)]/n_r$$

where $i = 1, 2, \ldots, n$, $j = 1, 2, \ldots, N$, and n_r is the number of FBFs in the final FBF expansion, $n_r \ll N$. If the number of the training data points is small then this simple method is sufficient for finding a mapping in a reasonably short time. If, on the other hand, this number is large, one can use the standard OLS method (Chen *et al* 1991) to choose the most significant FBFs from the initial FBF determination (Wang and Mendel 1992) before the training is performed.

There exists a multitude of other neuro-fuzzy algorithms which will not be elaborated on here. For a description of fuzzy ART as well as several other algorithms the reader is referred to *IEEE Trans. Neural Networks*, Special Issue on Fuzzy Logic and Neural Networks, September 1992.

References

Chen S, Cowan C F N and Grant P M 1991 Orthogonal least squares learning algorithm for radial basis function networks *IEEE Trans. Neural Networks* **NN-2** 302–9

Cybenko G 1989 Approximation by superpositions of a sigmoidal function *Math. Control. Signals, Systems* **2** 303–14

Gupta M M and Qi J 1991 On fuzzy neuron models *Proc. Int. Joint Conf. Neural Networks (Seattle, WA)* pp 431–6

Hayashi Y, Buckley J J and Czogala E 1992 Direct fuzzification of neural network and fuzzified delta rule *Proc. 2nd Int. Conf. Fuzzy Logic Neural Networks* pp 73–6

Hornik K, Stinchcombe M and White H 1989 Multilayer feedforward networks are universal approximators *Neural Networks* **2** 359–66

Ishibushi H, Fukioka R and Tanaka H 1992 An architecture of neural networks for input vectors of fuzzy numbers *Proc. IEEE Int. Conf. Fuzzy Syst. (San Diego, CA)* pp 1293–300

Jang R and Sun C-T 1993 Functional equivalence between radial basis function networks and fuzzy inference systems *IEEE Trans. Neural Networks* **NN-4** 156–9

Kosko B 1993 *Fuzzy Thinking* (New York: Hyperion)

Lee S C and Lee E T 1975 Fuzzy neural networks *Math. Biosci.* **23** 151–77

Moody J and Darken C 1989 Fast learning networks of locally-tuned processing units *Neural Comput.* **1** 281–94

O'Hagan M 1991 A fuzzy neuron based upon maximum entropy ordered weighted averaging *Uncertainty in Knowledge Bases (Lecture Notes in Computer Science 521)* ed B Bouchon-Meunier, R R Yager and L A Zadeh (New York: Springer) pp 598–609

Pedrycz W 1993 Fuzzy neural networks and neurocomputations *Fuzzy Sets Syst.* **56** 1–28

——1995 *Fuzzy Sets Engineering* (Boca Raton, FL: Chemical Rubber Company)

Requena I and Delgado M 1992 R-FN: a model of fuzzy neuron *Proc. 2nd Int. Conf. Fuzzy Logic Neural Networks (IIZUKA '92)* pp 793–96

Rumelhart D E and McLelland J L 1986 *Parallel Distributed Processing* (Cambridge, MA: MIT Press)

Silva F M and Almeida L B 1990 Acceleration techniques for the back-propagation algorithm *Lecture Notes in Computer Science* **412** 110–9

Wang L-X 1992 Fuzzy systems are universal approximators *Proc. IEEE 1992 Int. Conf. Fuzzy Systems (San Diego, CA)* pp 1163–70

Wang L-X and Mendel J M 1992 Fuzzy basis functions, universal approximation and orthogonal least-squares learning *IEEE Trans. Neural Networks* **3** 807–14

Weir M K 1991 A method for self-determination of adaptive learning rates in back propagation *Neural Networks* **4** 371–9

Xu H Y, Wang G Z and Baird C B 1992 A fuzzy neural networks technique with fast backpropagation learning *Proc. Int. Joint Conf. on Neural Networks, Baltimore, 1992* (IEEE Press) pp I214–9

Yamakawa T and Tomoda S 1989 A fuzzy neuron and its application to pattern recognition *Proc. Third IFSA Congress (Seattle, WA)* pp 30–8

Zahirniak D R, Chapman R, Rogers S K, Suter B W, Kabrisky M and Pyati V 1990 Pattern recognition using radial basis function networks *Sixth Annual Aerospace Application of AI Conf. (Dayton, OH)* pp 249–60

D1.4 Ontogenic neuro-fuzzy F-CID3 algorithm

Krzysztof J Cios and Witold Pedrycz

Abstract

See the abstract for Chapter D1.

The algorithm covered here is a representative of a family of ontogenic (generating their own architecture) algorithms. For details about *supervised ontogenic algorithms* see Section C1.7 of this handbook, and for *unsupervised ontogenic algorithms* see Section C2.4.

    C1.7
    C2.4

    Here we concentrate on the neuro-fuzzy F-CID3 algorithm (Cios and Sztandera 1992, 1996). F-CID3, a hybrid algorithm, may be produced by combining an ontogenic neural network algorithm CID3 (Cios and Liu 1992), which is described in detail in Section G2.12 of this handbook, with fuzzy sets. This ontogenic neuro-fuzzy algorithm generates an initial neural network architecture in the same way as the CID3 algorithm and then defines grades of membership for fuzzy sets associated with hidden layer nodes, where entropy is the first reduced to zero. Then it switches entirely to operations on fuzzy sets. This hybrid approach results in the generation of a simple architecture in a relatively short time.

    G2.12

    A motivation for incorporating fuzzy sets in the CID3 neural network arose from the fact that part of the highly connected architecture generated by the original CID3 algorithm (Cios and Liu 1992) was in a way 'redundant'. That is, after entropy of separation is first decreased to zero (usually at the second hidden layer) there is a need for only one more layer (output) to classify the data correctly. However, the CID3 algorithm generates several additional hidden layers. To illustrate this phenomenon, let us look at the architecture generated by the CID3 algorithm for recognition of the two spirally organized patterns as portrayed in figure D1.4.1. For all layers, subsequent to the second layer, where entropy was for the first time reduced to zero, the entropy was quickly decreasing to zero, see table D1.4.1. For details of how this entropy is calculated see Section G2.12 of this handbook.

    The idea behind adding fuzzy sets was that if one could define fuzzy sets at each node of the second layer and be able to rank them, then this would greatly simplify the network's architecture in terms of both the number of nodes and connections.

Table D1.4.1. Number of nodes neededfor the entropy to decrease to zero; see figure D1.4.1

Hidden layer 1	Hidden layer 2	Hidden layer 3	Hidden layer 4	Hidden layer 5	Output
3 nodes	16 nodes	7 nodes	5 nodes	3 nodes	1 node

    In the F-CID3 algorithm that goal was achieved by introducing a neural fuzzy number tree (Cios and Sztandera 1992, 1996). In a way similar to growing a decision tree in the CID3 algorithm (Cios and Liu 1992), a neural fuzzy number tree, with fuzzy subsets specified at each of its nodes, was incorporated into the F-CID3 algorithm. This resulted in a drastic reduction of the number of nodes and connections in the network, as can be seen in figure D1.4.2. There are actually only two hidden layers with 2 and 11 nodes in them. This example illustrates the advantage of combining neural networks with fuzzy sets.

    In the following we shall study an example explaining in more detail the working of the F-CID3 algorithm after Cios and Liu (1992) and Cios and Sztandera (1992, 1996). The task is to separate the data

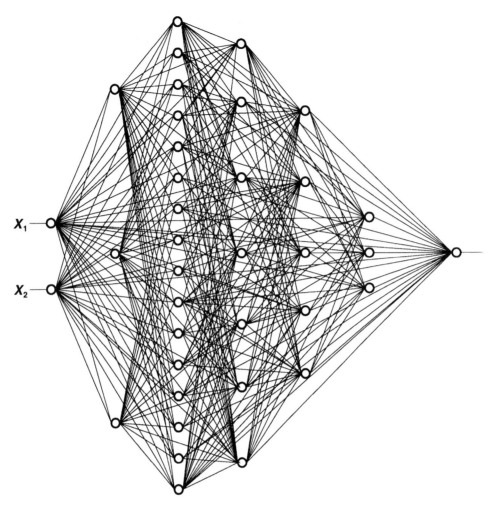

Figure D1.4.1. A fully connected neural network architecture for telling two spirals apart generated by the CID3 algorithm (not all of the connections are shown).

shown in figure D1.4.3 into two categories. The neural fuzzy number tree corresponding to this example is depicted in figure D1.4.4. Connections between the nodes have values equal to the weights of a neural network. The fuzzy sets at each level of the tree correspond to the examples lying on the positive and negative sides of a hyperplane, as generated by the neural network CID3 algorithm; more details of how the hyperplanes are generated can be found in Section G2.12 of this handbook. Each level of the neural fuzzy number tree corresponds to one hyperplane.

The following definition of fuzzy entropy (Kosko 1986, 1992) was utilized in the F-CID3 algorithm:

$$f(F) = \frac{\Sigma \ \mathrm{count}(F \cap \bar{F})}{\Sigma \ \mathrm{count}(F \cup \bar{F})} \tag{D1.4.1}$$

where Σ count (sigma-count) is the scalar cardinality of a fuzzy set. The entropy computations were realized by using, arbitrarily chosen, Dombi's triangular norms (Dombi 1982) giving rise to the following form of union and intersection of fuzzy sets:

$$\frac{1}{1 + [(1/A(x) - 1)^{-\lambda} + (1/B(x) - 1)^{-\lambda}]^{-1/\lambda}} \tag{D1.4.2}$$

$$\frac{1}{1 + [(1/A(x) - 1)^{\lambda} + (1/B(x) - 1)^{\lambda}]^{1/\lambda}} \tag{D1.4.3}$$

where A and B are fuzzy sets defined in the same universe of discourse; the parameter of the triangular norms (λ) assumes the values from $(0, \infty)$.

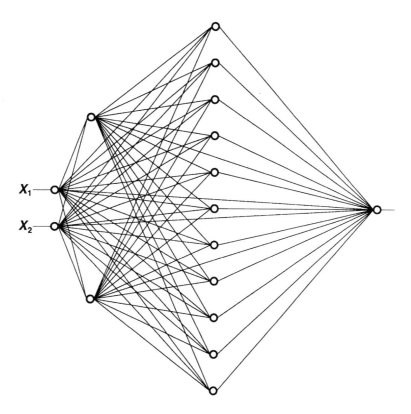

Figure D1.4.2. A neural network architecture for telling two spirals apart generated by the ontogenic neuro-fuzzy F-CID3 algorithm.

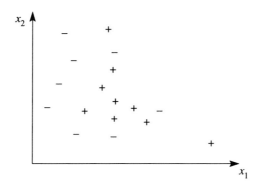

Figure D1.4.3. Simple two-category data.

The F-CID3 algorithm retained the CID3's key feature of generating its own architecture, but then it generated fuzzy sets at each node of the hidden layer, where the entropy was first reduced to zero, based on the numbers of positive ($+$) and negative ($-$) examples on all sides of the hyperplanes. Once fuzzy subsets were defined, it switched to very efficient operations on fuzzy subsets. To explain how the method works, let us introduce here the following notation (Cios and Liu 1992). There are 'N' training examples with N^+ examples belonging to class '$+$', and N^- examples belonging to class '$-$'. A hyperplane divides the examples into two groups: those lying on the positive (1) and negative (0) sides of it. Four possible outcomes are envisaged:

> (i) N_1^+ number of examples from class '$+$' on the side 1;
> (ii) N_0^+ number of examples from class '$+$' on the side 0;
> (iii) N_1^- number of examples from class '$-$' on the side 1;
> (iv) N_0^- number of examples from class '$-$' on the side 0. $\qquad$ (D1.4.4)

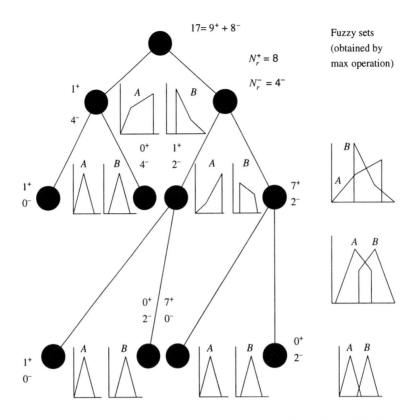

Figure D1.4.4. A neural fuzzy number tree corresponding to figure D1.4.3.

If at a certain level of a decision tree N_r examples are divided by a node r into those belonging to class '+', and class '−' then the values N_{1r}^+ and N_{1r}^- are calculated as follows:

$$N_{1r}^+ = \sum_{i=1}^{N_r} D_i \, \text{out}_i \tag{D1.4.5}$$

$$N_{1r}^- = \sum_{i=1}^{N_r} (1 - D_i) \text{out}_i \tag{D1.4.6}$$

where D_i stands for the desired output, and out_i is a sigmoid function. Then we obtain

$$N_{1r}^+ + N_{1r}^- = \text{out}_1 + \cdots + \text{out}_{Nr} = \sum_i^{N_r} \text{out}_i = \sum_i^{N_r} \left[1 + \exp\left(-\sum_j w_{ij} x_j \right) \right]^{-1}. \tag{D1.4.7}$$

The change in the number of examples, on both the positive and the negative side of a hyperplane, with respect to the weights (Cios and Liu 1992) was given by

$$\Delta N_{1r} = \sum_{i=1}^{N_r} D_i \text{out}_i (1 - \text{out}_i) \sum_j x_j \Delta w_{ij} \tag{D1.4.8}$$

$$\Delta N_{1r}^- = \sum_{i=1}^{N_r} (1 - D_i) \text{out}_i (1 - \text{out}_i) \sum_j x_j \Delta w_{ij}. \tag{D1.4.9}$$

The learning rule used to minimize the fuzzy entropy $f(F)$ (Cios and Sztandera 1992, 1996) was of the form

$$w_{ij} = -\rho \frac{\partial f(F)}{\partial w_{ij}} \tag{D1.4.10}$$

where ρ is a learning rate, and $f(F)$ is a fuzzy entropy function defined in (D1.4.1).

The key point of the F-CID3 algorithm was its definition of the membership function of a fuzzy set F which was specified as follows:

$$F = \left\{ \frac{N_{0r}^-}{N_{0r}}, \frac{N_{0r}^+}{N_{0r}}, \frac{N_{1r}^-}{N_{1r}}, \frac{N_{1r}^+}{N_{1r}} \right\}. \tag{D1.4.11}$$

This fuzzy set quantifies the extent to which a hyperplane separates positive and negative examples. It can be rewritten in the form:

$$F = \{A(m_1), A(m_2), B(m_1), B(m_2)\}. \tag{D1.4.12}$$

Obviously we get

$$\bar{F} = 1 - F. \tag{D1.4.13}$$

The four grades of membership (equations (D1.4.12) and (D1.4.13)) were used (Cios and Sztandera 1992, 1996) in the generalized Dombi operations (with $\lambda = 4$) and calculations of fuzzy entropy. The obtained fuzzy entropy was used to calculate the weights using the learning rule specified in (D1.4.10). In order to increase the chance of finding the global minimum, the learning rule was combined with Cauchy training (Szu and Hartley 1987) in the same manner as in Cios and Liu (1992):

$$W_{k+1} = W_k + (1 - \zeta)\Delta W + \zeta \Delta W_{\text{random}} \tag{D1.4.14}$$

where ζ is a learning rate. By changing the weight by the ΔW_{random} value, the algorithm might escape from local minima.

To show how fuzzy sets for the neural fuzzy number tree were generated, let us again look at the example shown in figure D1.4.3. The corresponding neural fuzzy number tree had two fuzzy subsets, denoted by A and B, defined at its nodes, as shown in figure D1.4.4.

The grades of membership for the fuzzy subsets A and B were initially defined for only two arbitrary points m_1 and m_2 from which the two fuzzy subsets were constructed. These grades of membership were defined, using the mutual dependence of positive and negative examples on both sides of a hyperplane, as follows:

$$A(m_1) = \frac{N_{0r}^-}{N_{0r} + N_{0r}^-} = \frac{N_r - N_r^+ - N_{1r}^+}{N_r - N_{1r}^+ - N_{1r}^-} \qquad A(m_2) = \frac{N_{0r}^+}{N_{0r}^+ + N_{0r}^-} = \frac{N_r - N_r^- - N_{1r}^-}{N_r - N_{1r}^+ - N_{1r}^-}$$

$$B(m_1) = \frac{N_{1r}^-}{N_{1r}^+ + N_{1r}^-} \qquad B(m_2) = \frac{N_{1r}^+}{N_{1r}^+ + N_{1r}^-} \tag{D1.4.15}$$

where fuzzy set A represents a collection of positive and negative examples on the negative side of hyperplane r, while fuzzy set B represents the same on the positive side of the hyperplane.

Fuzzy sets A and B were defined by the following membership functions:

$$A(x) = \begin{cases} \dfrac{x A(m_1)}{m_1} & \text{for } x \leq m_1 \\[2mm] \dfrac{A(m_2)(x - m_1) + A(m_1)(m_2 - x)}{m_2 - m_1} & \text{for } m_1 \leq x \leq m_2 \\[2mm] 0 & \text{for } x > m_2 \end{cases}$$

$$B(x) = \begin{cases} 0 & \text{for } x < m_1 \\[2mm] \dfrac{B(m_2)(x - m_1) + B(m_1)(m_2 - x)}{m_2 - m_1} & \text{for } m_1 \leq x \leq m_2 \\[2mm] \dfrac{B(m_2)(m_1 + m_2 - x)}{m_1} & \text{for } m_2 \leq x \leq m_1 + m_2 \\[2mm] 0 & \text{for } x > m_1 + m_2. \end{cases} \tag{D1.4.16}$$

For fuzzy subsets A and B, specified at some node of a fuzzy neural number tree, the classification rule was based on the following definition. The data samples are fully separated if the following values for the ranking indices are established:

$$x_A = \tfrac{1}{3}(m_1 + m_2) \qquad x_B = \tfrac{2}{3}(m_1 + m_2) \tag{D1.4.17}$$

using the center-of-gravity transformation method. Equation (D1.4.17) corresponds to fuzzy entropy equal to zero. More information about ranking indices can be found in the article by Sztandera and Cios (1993).

Since at the second and third level we have two subsets A and two subsets B, the union operation is used to obtain the resultant single sets for ranking (one A and one B), see figure D1.4.4. Table D1.4.2 shows the grades of membership for fuzzy sets A and B at each level of the neural fuzzy number tree. A neural network architecture corresponding to this tree is depicted in figure D1.4.5. Table D1.4.3 lists grades of membership for fuzzy subsets F and $\bar{F}$ used in calculation of fuzzy entropy.

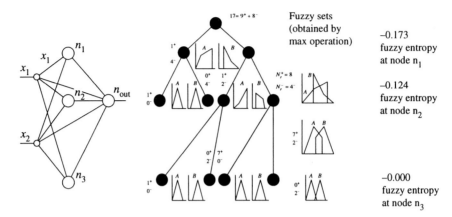

Figure D1.4.5. A neural network architecture (left) and neural fuzzy number tree (middle) corresponding to the hidden layer, and corresponding entropies (right).

Table D1.4.2. Grades of membership for fuzzy subsets A and B at points m_1 and m_2 at each level of the neural fuzzy number tree corresponding to figure D1.4.5.

Level of a tree	$A(m_1)$	$A(m_2)$	$B(m_1)$	$B(m_2)$	Resulting grades (max operation) $A(m_1)$	$A(m_2)$	Resulting grades (max operation) $B(m_1)$	$B(m_2)$
1	4/12	8/12	4/5	1/5	4/12	8/12	4/5	1/5
2	4/4	0/4	0/1	1/1				
	2/9	7/9	2/3	1/3	4/4	7/9	2/3	1/1
3	2/2	0/2	0/1	1/1				
	2/2	0/2	0/7	7/7	1	0	0	1

Table D1.4.3. Grades of membership for fuzzy subsets F and $\bar{F}$ and corresponding fuzzy entropies at each level of the neural fuzzy number tree.

Level of a tree	Grades of membership $F(x)$				Grades of membership $\bar{F}(x)$				Fuzzy entropy $f(F)$
1	4/12	8/12	4/5	1/5	8/12	4/12	1/5	4/5	0.173
2	4/4	0/4	0/1	1/1	0/4	4/4	1/1	0/1	0.124
	2/9	7/9	2/3	1/3	7/9	2/9	1/3	2/3	
3	2/2	0/2	0/1	1/1	0/12	2/2	1/1	0/1	0.000
	2/2	0/2	0/7	7/7	0/2	2/2	7/7	0/7	

To summarize, the F-CID3 algorithm consists of five steps. Step (i) divides the input space into several subspaces; step (ii) counts the number of samples in those subspaces; step (iii) generates membership functions for fuzzy sets using the results obtained at step (ii); step (iv) executes ranking of the fuzzy sets formed in this manner; finally step (v) determines separation of categories based on faithful ranking. For details see the article by Cios and Sztandera (1992).

The F-CID3 algorithm is an example of a host of methods where the neural network technology is used as a 'tool' for generating fuzzy sets.

References

Cios K J and Liu N 1992 A machine learning method for generation of a neural network architecture: a continuous ID3 algorithm *IEEE Trans. Neural Networks* **NN-3** 280–91

Cios K J and Sztandera L M 1992 Continuous ID3 algorithm with fuzzy entropy measures *Proc. 1st Int. Conf. on Fuzzy Systems and Neural Networks (San Diego, CA)* pp 469–76

——1996 Ontogenic neuro-fuzzy algorithm: F-CID3 *Neurocomputing* in press

Dombi J 1982 A general class of fuzzy operators, the De Morgan class of fuzzy operators and fuzziness measures *Fuzzy Sets Syst.* **8** 149–63

Kosko B 1986 Fuzzy entropy and conditioning *Info. Sci.* **40** 165–74

——1992 *Neural Networks and Fuzzy Systems* (Englewood Cliffs, NJ: Prentice Hall)

Sztandera L M and Cios K J 1993 Decision making in a fuzzy environment generated by a neural network architecture *Proc. 5th IFSA World Congress (Seoul)* pp 73–6

Szu H and Hartley R 1987 Simulated annealing *Phys. Lett.* **8A** 157–62

D1.5 Fuzzy neural networks

Krzysztof J Cios and Witold Pedrycz

Abstract

See the abstract for Chapter D1.

D1.5.1 Logic-based neurons

In this section we introduce and study basic properties of neurons developed with the aid of logic operations (fuzzy set connectives) (Pedrycz 1991, 1993, Pedrycz and Rocha 1993). By this class of processing units we mean the neurons whose architecture and computations are directly guided by the mechanisms of fuzzy sets and logic operators (logical connectives). Owing to that, each neuron possesses a straightforward interpretation—a facet not encountered in 'standard' neural networks. From now on, we will be treating the inputs as well as the parameters (connections) of the neurons as the elements in a unit hypercube. According to the general taxonomy outlined in figure D1.5.1, the first class of the neurons consists of aggregative (AND, OR, OR/AND) logic neurons while the second category embraces the neurons aimed at referential processing.

D1.5.1.1 Aggregative logic neurons

The class of aggregative neurons embraces two general types of processing unit such as OR and AND neurons; the subsequent OR/AND neurons emerge as a straightforward combination of the first two. The OR neuron, denoted by $y = \mathrm{OR}(\boldsymbol{x}; \boldsymbol{w})$, realizes a mapping $[0, 1]^n \rightarrow [0, 1]$ that is given in the form,

$$y = \mathrm{OR}[x_1 \text{ AND } w_1, x_2 \text{ AND } w_2, \ldots, x_n \text{ AND } w_n]$$

where $\boldsymbol{w} = [w_1, w_2, \ldots, w_n] \in [0, 1]^n$ is a vector of the connections (weights) of the neuron and $\boldsymbol{x} = [x_1, x_2, \ldots, x_n]$ summarizes its inputs.

The standard implementation of the fuzzy set connectives usually involves triangular norms that means that the OR and AND operators are realized by some s- and t-norms, respectively. This produces the following expression of the neuron:

$$y = \mathop{\mathrm{S}}_{i=1}^{n} [x_i \text{ t } w_i].$$

In the AND neuron, the OR and AND operators are utilized in reverse order: first the inputs interact OR-wise with the connections and those results are finally aggregated through the AND operation. We obtain

$$y = \mathrm{AND}(\boldsymbol{x}; \boldsymbol{w})$$

which, making use of the notation of the triangular norms, reads as

$$y = \mathop{\mathrm{T}}_{i=1}^{n} [x_i \text{ s } w_i].$$

The AND and OR neurons realize 'pure' logic operations on their inputs (membership values). The role of the connections $\boldsymbol{w}$ is to differentiate between the particular levels of impact that the individual inputs could have on the final result of aggregation. Due to the boundary conditions of the triangular norms,

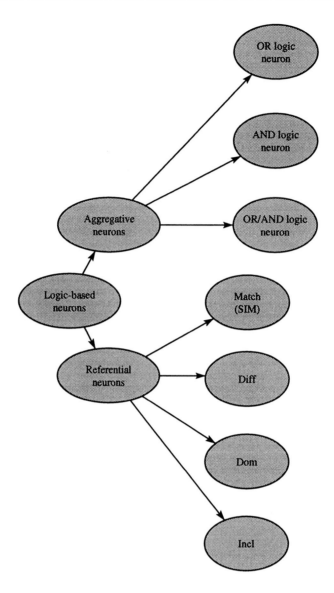

Figure D1.5.1. Classes of fuzzy neurons.

we conclude that the higher values of the connections in the OR neuron emphasize a stronger influence than the corresponding inputs pose on the output of the neuron. The opposite weighting (ranking) effect takes place in the case of the AND neuron: the values of w_i close to 1 make that influence of x_i almost negligible, cf Pedrycz (1993). In limit, the neurons reduce to the straightforward AND and OR operations; then all the connections are set to 0 or 1, namely, $y = \text{AND}(x; 0)$ and $y = \text{OR}(x; 1)$. The specific numerical form of the or or and characteristics conveyed by the logic-based neurons depends upon the triangular norms being utilized in their implementation, figures D1.5.2 and D1.5.3.

As a straightforward generalization of these two neurons, we introduce an OR/AND neuron characterized by some intermediate logical characteristics that could easily be modified according to the specificity of the problem. The OR/AND neuron (Hirota and Pedrycz 1994) is constructed by combining the previously discussed AND and OR neurons into a single two-layer structure as shown in figure D1.5.4.

Considering this structure as a single computational entity, it is easy to notice that the neuron can synthesize a spectrum of intermediate logical characteristics. The response coming from the OR (AND) part of the neuron can be properly balanced by selecting (learning) the relevant values of the connections v_1 and v_2. In limit, when $v_1 = 1$ and $v_2 = 0$, the OR/AND neuron operates like a pure AND neuron. In the second extremal situation for which $v_1 = 0$ and $v_2 = 1$, the structure functions as a pure OR neuron.

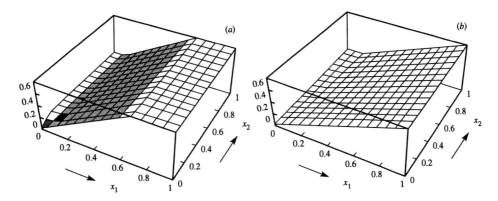

Figure D1.5.2. Three-dimensional characteristics of the OR neuron with $w = [0.7, 0.1]$ for two combinations of the triangular norms (a) t-norm: minimum s-norm: maximum, (b) t-norm: product s-norm: maximum.

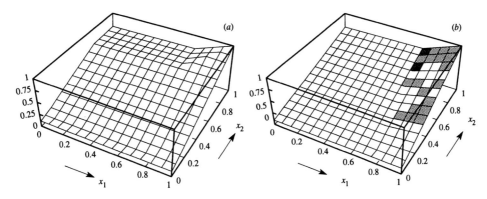

Figure D1.5.3. 3D characteristics of the AND neuron with $w = [0.7, 0.1]$ for two combinations of the triangular norms (a) t-norm: minimum s-norm: maximum, (b) t-norm: product s-norm: maximum.

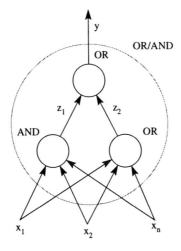

Figure D1.5.4. Architecture of an OR/AND neuron.

We will use the notation

$$y = \text{OR/AND}(x; w, v)$$

to emphasize the intermediate characteristics produced by the neuron. The relevant detailed formulas describing this architecture read accordingly,

$$y = \text{OR}([z_1, z_2]; v)$$

$$z_1 = \text{AND}(x; w_1) \qquad \text{and} \qquad z_2 = \text{OR}(x; w_2)$$

with $v = [v_1, v_2]$, $w_i = [w_{i1}, w_{i2}, \ldots, w_{in}]$, $i = 1, 2$, being the connections of the corresponding neurons. We encapsulate the above expressions into a single formula,

$$y = \text{OR/AND}(x; \text{connections})$$

where now the connections summarize the weights of the network.

D1.5.2 Computational enhancements of fuzzy neurons

We discuss two further enhancements of the fuzzy neurons aimed at increasing their conceptual and computational flexibility.

D1.5.2.1 Representing inhibitory information in fuzzy neurons

The task of representing an inhibitory behavior of some of the inputs of the neurons does not constitute any problem to the 'classic' networks; we simply admit negative connections between the units. Here, as all the numerical manipulation encountered in fuzzy sets is realized within the unit interval, the question of the inhibitory information requires a thorough treatment. Our intention is to maintain the [0,1] style of coding for the sake of preserving the logical nature of the set-theoretic operations utilized in the construction of the neuron. The reader should be aware that an attempt (quite naive and fully unjustifiable, yet encountered in the existing literature) to extend the triangular norms to the $[-1, 1]$ interval and sustain their fundamental properties is not feasible. Being more specific, the well known boundary condition $0 \, t \, x = 0$ is no longer valid; to visualize this put $x = -1$ (this corresponds to the zero boundary condition) and consider the product operation: obviously we get $(-1) \, t \, (-1) \neq -1$.

The intuitively straightforward and convincing solution to the problem is to admit the complemented variables among the inputs of the neuron. Hence, the higher the input x_i, the lower the contribution it provides to the output of the aggregative neuron. In limit, when $x_i = 1$, the impact of this variable is completely eliminated. The detailed numerical form of the inhibitory effect depends upon the t-norm being used to realize this aggregation; refer to some illustrative cases given in figure D1.5.5.

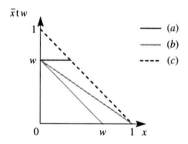

Figure D1.5.5. Inhibitory effect realized by some t-norms: (a) minimum, (b) product, (c) Lukasiewicz connective $(a \, t \, b = \max(0, a + b - 1))$.

This approach is directly motivated by the basic form of minterms and maxterms encountered in the representation of two-valued (Boolean) functions. Remember that these constructions are completely sufficient to represent any two-valued functions of many variables. In our context, a binary (two-valued) OR neuron (in which the entries of w are equal either to 1 or 0) realizes a maxterm. Similarly, the AND neuron with the 0–1 weights realizes a minterm.

As the inhibitory phenomenon described above takes place at the local level of the specific connection (refer again to figure D1.5.5) this does not mean (and does not guarantee) that the inhibitory effect could be always visible at the output of the neuron. The reason for that is the monotonicity of the triangular norms. Bearing this in mind, we can refer to the above scheme as the local mechanism of inhibition.

The mechanism of a global inhibition is realized through some structural enhancements to the neuron as displayed in figure D1.5.6. As shown there, the inhibitory inputs (x_i) are fed into an additional OR (or AND) neuron whose output triggers an inhibitory signal applied to the AND neuron located in the next layer.

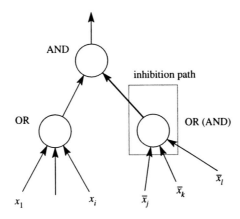

Figure D1.5.6. Realization of a global (structural) inhibition.

The type of aggregative neuron therein depends on the way in which the inhibitory effect needs to be summarized (disjunctive versus conjunctive aggregation). The illustration of these two mechanisms of inhibition is shown in figure D1.5.7 (here $n = 2$; the connections of the neuron are included as well).

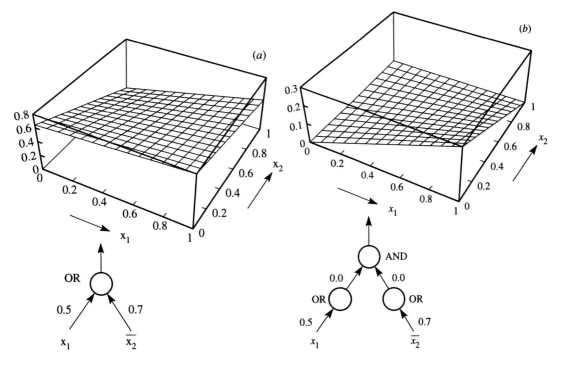

Figure D1.5.7. Two mechanisms of inhibition of logic-based neurons: (*a*) local, (*b*) structural t-norm: minimum s-norm: probabilistic sum.

D1.5.2.2 Nonlinear processing element

Despite the well-defined semantics of the logic-based neurons, the main concern one may raise about their functioning occurs on the numerical side. Once the connections (weights) are set (after learning) each neuron realizes an 'in' (rather than an 'on') mapping, that means that the values of y for all possible inputs cover a subset of the unit interval. More specifically, for the OR neuron the values of the output y are included in the range $[0, S_{i=1}^{n} w_i]$ whereas the accessible range of the output values of the dual (AND) neuron is limited to the interval $[T_{i=1}^{n} w_i, 1]$.

This shortcoming could be alleviated by augmenting the neuron by a nonlinear element placed in series with the previous purely logical component (figure D1.5.8).

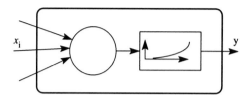

Figure D1.5.8. Fuzzy neuron equipped with a nonlinear processing element.

The neurons obtained in this manner are formalized accordingly,

$$y = \Psi\left(\mathop{S}_{i=1}^{n} (x_i \; t \; w_i) \right) = \Psi(u)$$

$$y = \Psi\left(\mathop{T}_{i=1}^{n} (x_i \; s \; w_i) \right) = \Psi(u)$$

where $\Psi : [0, 1] \rightarrow [0, 1]$ is a nonlinear monotonic mapping. In contrast to the standard nonlinearities discussed commonly in neural computation, we admit both monotonically increasing as well as monotonically decreasing continuous mappings.

A useful two-parametric family of the sigmoidal nonlinearities can be specified in the form

$$y = \frac{1}{1 + \exp[-(u - m)\sigma]}$$

where $u, m \in [0, 1]$, $\sigma \in \mathbb{R}$.

By adjusting the parameters of the function (m and σ), various forms of the nonlinear characteristics of the element can be easily obtained. Especially, the positive or negative values of σ determine either an increasing or decreasing type of the characteristics of the obtained neuron. The other parameter (m) shifts the entire characteristics along the unit interval. The incorporation of this nonlinearity changes the numerical characteristics of the neuron—however—its essential logical behavior is sustained—refer to figures D1.5.9 and D1.5.10 which summarize some of the static input–output relationships encountered there (with the triangular norms set up as the product and probabilistic sum).

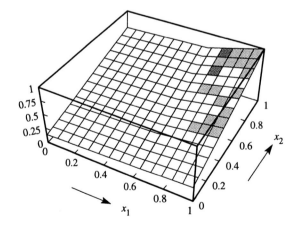

Figure D1.5.9. 3D characteristics of the AND neuron, $w = [0.7, 0.2]$ without a nonlinear element.

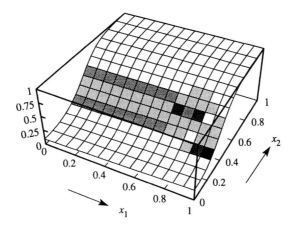

Figure D1.5.10. Three-dimensional characteristics of the AND neuron, $w = [0.7, 0.2]$ with a nonlinear sigmoidal element $m = 0.3$, $\sigma = 15$.

D1.5.3 Logic-based neurons with feedback

The logic neurons studied so far realize a static memoryless nonlinear mapping in which the output depends solely upon the inputs of the neuron. In this form, the neurons are not capable of handling dynamical (memory-based) relationships between the inputs and outputs. This aspect might, however, be essential in a proper description of any dynamical system. Take, for instance, a classification problem in which a decision about a system's failure should be issued while one of the system's sensors provides information about an abnormal (elevated) temperature of an engine. The duration of this phenomenon itself has a primordial impact on expressing the confidence about particular classes (namely, failures). If the elevation of the temperature prolongs, the confidence about the failure rises. On the other hand, some short temporary temperature elevations (spikes) recorded by the sensor might be almost ignored (filtered out) and should not have any impact on the classification decision. To capture properly this dynamical effect about class assignment, one has to equip the standard logic neuron with a certain feedback loop as illustrated in figure D1.5.11.

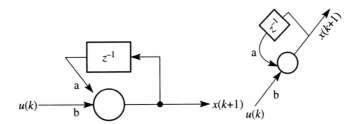

Figure D1.5.11. Logic-based neuron with feedback.

An example of the neuron with feedback can be described as

$$x(k + 1) = [b \text{ OR } u(k)] \text{ AND } [a \text{ OR } x(k)].$$

The dynamics of the neuron are uniquely defined by the strength (a) of a feedback loop that, in fact, determines a speed of evidence accumulation ($x(k)$). The initial condition, $x(0)$, expresses *a priori* confidence associated with this class. After a sufficiently long period of time, $x(k + 1)$ could take on higher values in comparison to the level of the original evidence being present in the input. Figure D1.5.12 summarizes the dynamical behavior of the OR neuron with the positive and negative feedback.

Higher-order dynamical dependencies to be accommodated by the network call for a feedback loop consolidating several pieces of temporal information, for example,

$$x(k + 2) = [b \text{ OR } u(k)] \text{ AND } [a_1 \text{ OR } x(k)] \text{ AND } [a_2 \text{ OR } x(k + 1)].$$

One can also consider the above expressions as examples of fuzzy difference equations.

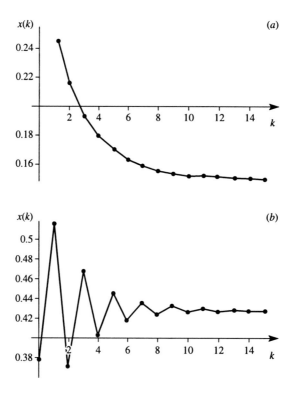

Figure D1.5.12. Dynamics of the neurons with positive or negative feedback displayed in a phase plane: (*a*) positive feedback $x(k+1) = \text{OR}([a, b][x(k), u])$, (*b*) negative feedback $x(k+1) = \text{OR}([a, b][\bar{x}(k), u])$ (t-norm: product, s-norm: probabilistic sum, $a = 0.7$, $b = 0.1$, $u = 0.5$; $x(0) = 0.3$).

References

Hirota H and Pedrycz W 1994 OR/AND neuron in modeling fuzzy set connectives *IEEE Trans. Fuzzy Systems* **2** 151–61

Pedrycz W 1991 Neurocomputations in relational systems *IEEE Trans. on Pattern Analysis and Machine Intelligence* **13** 289–96

——1993 Fuzzy neural networks and neurocomputations *Fuzzy Sets Syst.* **56** 1–28

Pedrycz W and Rocha A F 1993 Fuzzy-set based models of neurons and knowledge-based networks *IEEE Trans. Fuzzy Systems* **FS-1** 254–66

D1.6 Referential logic-based neurons

Krzysztof J Cios and Witold Pedrycz

Abstract

See the abstract for Chapter D1.

In comparison to the AND, OR, and OR/AND neurons realizing logic operations of the aggregative form, the class of neurons now discussed is useful in carrying out referential computations. The main idea behind this neuron is that the input signals are not directly aggregated as took place in the aggregative neuron, but the processing consists of two phases. First, the inputs are analyzed (e.g. compared) with respect to the given reference point. The results of this analysis are subsequently summarized in the aggregative part of the neuron along the lines described earlier. In general, one can describe the reference neuron as

$$y = \text{OR}(\text{REF}(x; \text{reference\_point}), w)$$

(a disjunctive form of aggregation) or

$$y = \text{AND}(\text{REF}(x; \text{reference\_point}), w)$$

(that constitutes a conjunctive form of aggregation). The term REF($\cdot$) stands for the reference operation carried out with respect to the provided point of reference. Figure D1.6.1 underlines more profoundly a composite character of the processing realized by the neuron.

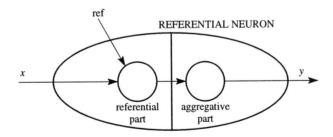

Figure D1.6.1. General two-step processing in referential neurons.

Depending on the form of the reference operation, the functional behavior of the neuron is described accordingly (all the formulas below pertain to the disjunctive form of aggregation).

(i) MATCH neuron:

$$y = \text{MATCH}(x; r, w)$$

or equivalently

$$y = \mathop{S}_{i=1}^{n} [w_i \; t \; (x_i \equiv r_i)]$$

where $r \in [0, 1]^n$ stands for a reference point defined in the unit hypercube. The matching operator is defined as follows (Pedrycz 1990),

$$a \equiv b = \tfrac{1}{2}[(a \; \varphi \; b) \wedge (b \; \varphi \; a) + (\bar{a} \; \varphi \; \bar{b}) \wedge (\bar{b} \; \varphi \; \bar{a})]$$

and

$$a \varphi b = \{c \in [0, 1] \mid a \, t \, c \leq b\}.$$

Quite often the above φ-operator is also referred to as the fuzzy implication.

To emphasize the referential character of this processing carried out by the neuron one can rewrite the expression of the MATCH neuron as

$$y = OR(\boldsymbol{x} \equiv \boldsymbol{r}; \boldsymbol{w}).$$

The use of the OR neuron implies an 'optimistic' (disjunctive) character of the final aggregation. The pessimistic form of this aggregation is produced by using the AND operation.

(ii) Difference neuron. The neuron combines degrees to which x is different from the given reference point $\boldsymbol{g} = [g_1, g_2, \ldots, g_n]$. The output is interpreted as a global level of difference observed between the input x and this reference point,

$$y = DIFFER(\boldsymbol{x}; \boldsymbol{w}, \boldsymbol{g})$$

that is,

$$y = \overset{n}{\underset{i=1}{S}} [w_i \, t \, (x_i | \equiv g_i)]$$

where the difference operator $| \equiv$ is defined as a complement of the equality index introduced before,

$$a| \equiv b = 1 - a \equiv b.$$

As before, the referential character of processing is emphasized by noting that

$$DIFFER(\boldsymbol{x}; \boldsymbol{w}, \boldsymbol{g}) = OR(\boldsymbol{x}| \equiv \boldsymbol{g}; \boldsymbol{w}).$$

(iii) The inclusion neuron summarizes the degrees of inclusion stating the extent to which x is included in the reference point f,

$$y = INCL(\boldsymbol{x}; \boldsymbol{w}, \boldsymbol{f}) \qquad y = \overset{n}{\underset{i=1}{S}} [w_i \, t \, (x_i \rightarrow f_i)].$$

The relationship of inclusion is expressed in the sense of the pseudocomplement operation (implication). The two properties of the φ-operator (already discussed with regard to the MATCH neuron),

$$\text{if } a < b \text{ then } a \varphi b = 1$$
$$\text{if } a > b' > b \text{ then } a \varphi b' \geq a \varphi b$$

where $a, b, b' \in [0, 1]$, assure us that the output of the neuron becomes a monotonic function of the degree of satisfaction of the inclusion property.

(iv) The dominance neuron expresses a relationship dual to that carried out by the inclusion neuron

$$y = DOM(\boldsymbol{x}; \boldsymbol{w}, \boldsymbol{h})$$

where h is a reference point. In other words, the dominance relationship generates the degree to which x dominates h (or, equivalently, h is dominated by x). The coordinate-wise notation of the neuron reads as

$$y = \overset{n}{\underset{i=1}{S}} [w_i \, t \, (h_i \rightarrow x_i)].$$

The referential operations provide a variety of processing elements. The tolerance neuron is a good example of an element exploiting this diversity.

(v) Tolerance neuron. It consists of DOMINANCE and INCLUSION neurons placed in the hidden layer and a single AND neuron in the output layer (figure D1.6.2).

Handbook of Neural Computation release 97/1   © 1997 IOP Publishing Ltd and Oxford University Press

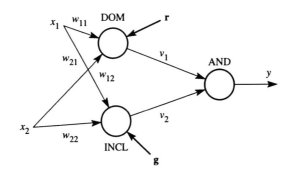

Figure D1.6.2. Architecture of a tolerance neuron.

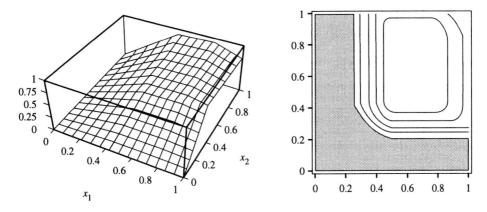

Figure D1.6.3. 2D and 3D characteristics of a tolerance neuron AND neuron: min operator, INCL and DOM neuron: $a \to b = \min(1, b/a) w_{ij} = 0.05$, $v_i = 0.0$.

D1.6.1 Fuzzy threshold neuron

This class of fuzzy neurons constituting a straightforward generalization of threshold computing units (threshold gates), cf the book by Muroga (1971), is formed by a serial composition of the aggregative neuron followed by the inclusion operator which generalizes the two-valued threshold element. More formally, this neuron is defined as

$$y = \mathrm{INCL}(\lambda; \mathrm{OR}(x; w)) = \lambda \to \mathrm{OR}(x; w)$$

where $l \in [0, 1]$ denotes a threshold level. The output values of the OR unit exceeding the threshold are elevated to 1, see figure D1.6.4. In particular, when $\lambda \approx 0$, the neuron behaves very much as an on–off device.

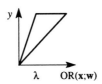

Figure D1.6.4. Characteristics of a single-input threshold neuron; $a \varphi b = \min(1, b/a)$.

References

Muroga S 1971 *Threshold Logic and Its Applications* (New York: Wiley)
Pedrycz W 1990 Direct and inverse problem in comparison of fuzzy data *Fuzzy Sets Syst.* **34** 223–36

D1.7 Classes of fuzzy neural networks

Krzysztof J Cios and Witold Pedrycz

Abstract

See the abstract for Chapter D1.

As we have proposed several clearly distinct types of fuzzy neuron, they could potentially give rise to a tremendous diversity of neuro-fuzzy networks. The variety of some schemes will be exemplified in the next sections. For the time being, we will introduce and study some architectures of pattern classifiers that, due to their functional characteristics, are encountered in many applications forming an essential part of the overall processing structures. Those are logic processors—the networks realizing tasks of logic-oriented approximation—and referential processors—extended logic processors aimed at the mapping of the referential properties between the feature and class membership spaces.

D1.7.1 Approximation of logical relationships: development of the logic processor

An important class of fuzzy neural networks concerns approximation of mappings between the unit hypercubes (from $[0, 1]^n$ to $[0, 1]^m$ or $[0, 1]$, in particular) that are realized in a logic-based format. To fully comprehend the fundamental idea behind this architecture, let us recall some very simple yet powerful concepts emerging from the realm of two-valued systems. The well known Shannon theorem (Schneeweiss 1989) states that any Boolean function $\{0, 1\} \rightarrow \{0, 1\}$ can be represented uniquely as a logical sum (union) of minterms (a so-called SOM representation) or, equivalently, a product of some maxterms (known as a POM representation). From a functional point of view, the minterms can be identified with the AND neurons while the OR neurons can be used to produce the corresponding maxterms. It is also noticeable that the connections of these neurons are restricted to the two-valued set $\{0, 1\}$ thus making these neurons two-valued selectors (on–off units). Considering the representation form of the Boolean functions, two complementary (dual) architectures are envisaged. In the first case, the network includes a single hidden layer that is constructed with the aid of the AND neurons followed by the output layer consisting of the OR neurons (SOM version of the network). The dual type of the network is of the POM type in which the hidden layer has some OR neurons and the output layer is formed by the AND neurons. The generalization of these networks to the continuous case of the input–output variables will be called a logic processor. Analogously to the topologies of the networks sketched so far for the Boolean cases, we will be interested in the two versions of the logic processor (LP), namely its POM and SOM version (figure D1.7.1).

Depending on the value of 'm', we will be referring to a scalar or vector version of the logic processor. Its scalar version, $m = 1$, could be viewed as a generic LP architecture.

Two points are worth making here as they contrast between the logic processors realized in their continuous and two-valued versions.

(i) The logic processor *represents* Boolean data. Assuming that all the input combinations are different, we are talking about a *representation* of the corresponding Boolean function. In this case the POM and SOM versions of the logic processors for the same Boolean function are fully equivalent (with the equivalence regarded at the input–output level).

(ii) The logic processor used for fuzzy (continuous) data *approximates* a certain unknown fuzzy function. The equivalence of the POM and SOM types of the obtained LPs is not guaranteed at all.

When necessary, we will be using a concise notation $N(x, w, v)$ to describe the network with the connections w and v standing between the successive layers.

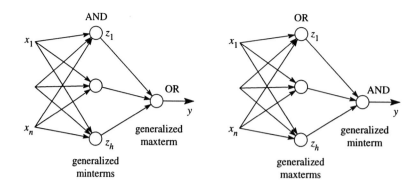

Figure D1.7.1. SOM and POM versions of a logic processor.

D1.7.2 Referential processor

While the role of the logic processor is to implement the logic-based approximation between the unit hypercubes, the essence of the processing developed by a referential processor is concerned with mapping some referential properties between the input and output spaces. Figure D1.7.2 highlights these differences in more detail.

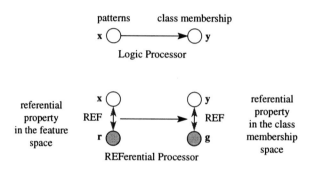

Figure D1.7.2. Logic processing and referential processing.

One among various types of referential computation that is definitely worth discussing deals with analog reasoning. This form of reasoning is oriented toward inferring similarities between some prototypes and current inputs. This form of reasoning has been found useful in pattern recognition, especially when handling relational data. Let us study a reference pattern—class membership pair (r, g) being considered as a given pair of associations, $r \in [0, 1]^n$, $g \in [0, 1]^m$. Qualitatively speaking, the scheme reads as

$$\frac{x \text{ and } r \text{ are } similar}{\qquad r, g \text{ are } associated \qquad}$$
$$y \text{ and } g \text{ are } similar$$

and entails two steps:

(i) determination (quantification) of similarity between y and g
(ii) determination of y based on g and the level of similarity computed at (i).

One could expect, which is intuitively sound, that the more similar the patterns x and r are, the higher the similarity level between this class assignment defined in the membership space (y and g). The architecture of the referential (in particular, analogical) processor in this case is visualized in figure D1.7.3. In fact, the analogical processor dwells upon the logic processor that is now used to transform the referential property of matching expressed in the feature and class membership spaces. Symbolically one can express this function as

$$(\text{matching})_{\text{class membership space}} = LP((\text{matching})_{\text{feature space}}, \text{connections}) .$$

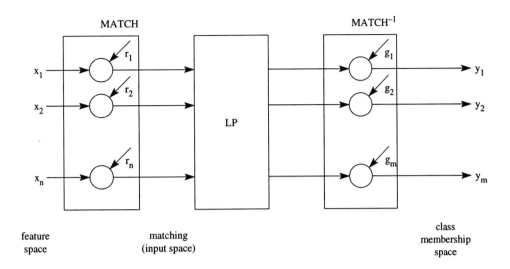

Figure D1.7.3. General architecture of the analogical processor.

In comparison to the plain logic processor, this architecture is augmented by two additional layers. The input layer (MATCH) carries out matching (realized through some matching neurons) while the output layer (marked here symbolically by MATCH^{-1}) is utilized to convert the level of the matching into the objects in the class membership space. From a functional point of view, one can regard the matching (analog) processor as a static input–output structure (figure D1.7.4) with the additional layers of preprocessing and postprocessing.

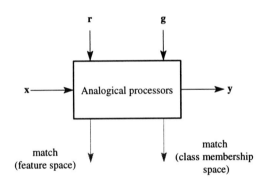

Figure D1.7.4. Analogical classifier—a functional view.

D1.7.3 Learning

The following discussion will be concerned with the supervised mode of learning of fuzzy neural networks. In general, two main tasks are encountered.

- parametric learning
- structural learning.

Most of the existing schemes of learning are preoccupied by parametric learning whose role is to optimize the parameters of the fuzzy neural network. On the other hand, structural learning, being definitely more demanding, is devoted to the optimization of the structure of the network. This could be accomplished in many different ways, for example, by changing the number of layers, adding, replacing, and deleting the individual neurons.

An idea of parametric learning can be portrayed as follows. For a given collection of input–output pairs of pattern–class assignment $(x_1, t_1), \ldots, (x_N, t_N)$, modify the parameters of the network (both the connections and reference points, if included) to minimize the assumed performance index Q (classification error). The general scheme of learning can be qualitatively described as

$$\Delta\_connections = -\alpha \frac{\partial Q}{\partial connections}$$

where α denotes a learning rate, $\alpha \in (0, 1]$. The parameters of the network are adjusted following these increments,

$$new\_connections = connections + \Delta\_connections.$$

The relevant details of the learning scheme can be fully specified once the topology of the network as well as some other details regarding the form of triangular norms have been specified.

D1.7.4 Learning of a single neuron

The standard learning procedure concerning a single logic-based neuron pertains to the parametric modifications of the connections and encompasses a series of iterations aimed at minimizing the following MSE performance index

$$Q = \sum_{k=1}^{N}(t_k - \Psi(\mathrm{OR}(x_k, w)))^2$$

$$Q = \sum_{k=1}^{N}(t_k - \Psi(\mathrm{AND}(x_k, w)))^2$$

where Ψ represents a nonlinear mapping from $[0, 1]$ to $[0, 1]$. In particular one can consider it to be a sigmoid nonlinearity, $\Psi(u) = 1/[1 + \exp(-u)]$.

Two modes of updates of the connections are distinguished:

- on-line learning, the adjustments of the connections are realized after presentation of each individual pattern–class assignment pair of the training data;
- off-line learning, the updates of the connections occur after a complete pass through the training set.

In general, the results of learning (as well as the value of the performance index itself) could differ quite significantly under these two learning modes.

The on-line type of algorithm with the updates (adjustments) worked out on the basis of an individual input–output pair of the training set can be written down as follows:

$$w = w - \alpha \frac{\partial Q}{\partial w}$$

$$\sigma = \sigma - \alpha \frac{\partial Q}{\partial \sigma}$$

$$m = m - \alpha \frac{\partial Q}{\partial m}.$$

Obviously, during all of these modifications, the connections w and the shift parameter m must eventually be clipped to keep them within the unit interval. Denote by z the output of the logical part of the neuron, $z = \mathrm{OR}(w, x)$ (or $z = \mathrm{AND}(w, x)$). Then the above formulas become more detailed,

$$w = w + 2\alpha(t_k - \Psi(z(x_k, w)))z_k(1 - z_k)\sigma \frac{\partial z_k}{\partial w}$$

$$\sigma = \sigma + 2\alpha(t_k - \Psi(z(x_k, w)))z_k(1 - z_k)(z_k - m)$$

$$m = m + 2\alpha(t_k - \Psi(z(x_k, w)))z_k(1 - z_k)(-\sigma).$$

The final computation formulas can be obtained once the appropriate triangular norms have been selected. While most of these detailed computations are to a large extent standard, the calculations of the derivatives for the maximum and minimum operations deserve a special attention. In this framework of learning, the problem was initially addressed by Pedrycz (1991). Briefly speaking, the main issue lies in the piecewise character of these operations. Thus from a formal point of view, the derivative $\partial \max(a, x)/\partial x$ and $\partial \min(a, x)/\partial x$ can be defined for all x but $x = a$. This produces the formulas

$$\frac{\partial \min(a, x)}{\partial x} = \begin{cases} 1 & \text{if} \quad x < a \\ 0 & \text{if} \quad x > a. \end{cases}$$

Similarly

$$\frac{\partial \max(a, x)}{\partial x} = \begin{cases} 1 & \text{if} \quad x > a \\ 0 & \text{if} \quad x < a. \end{cases}$$

Note that neither of them includes the case $x = a$. One can argue that the probability of such a single-point event $\{x = a\}$ is zero and therefore the impact it might have on the learning algorithm is practically negligible. One can eventually slightly modify these definitions by admitting at this critical point the values of the derivatives equal to 1. Nevertheless, the main learning problem is associated with a Boolean (two-valued) character of these derivatives rather than their detailed and specific formulations. The potential, and essentially quite pragmatic aspect of the derivatives defined above is that the learning algorithm could eventually end up being trapped in a nonstationary point. This is primarily caused by an accidental zeroing of all the derivatives that might occur under some configurations of the connections and the learning data. To avoid this highly undesirable phenomenon, several improvements have been proposed.

(i) The above derivative can be viewed as a two-valued predicate (returning either 1 or 0). One can look at the above derivative as a Boolean predicate 'equal to' that returns 1 (true) if and only if the arguments are equal, namely, $\partial \min(a, x)/\partial x = \text{truth}(x < a)$ and $\partial \max(a, x)/\partial x = \text{truth}(x > a)$. This predicate can be relaxed by its multivalued version of 'included in' that yields

$$\frac{\partial \min(a, x)}{\partial x} = \text{INCL}(x, a)$$

and allows for a smooth transition between a full inclusion and complete dominance. For example, the Łukasiewicz implication induces a linear character of the derivative

$$\frac{\partial \min(a, x)}{\partial x} = \begin{cases} 1 & \text{if} \quad x \geq a \\ 1 - x + a & \text{if} \quad x < a. \end{cases}$$

(ii) The modification proposed by Ikoma *et al* (1993) is quite similar to that explained in (i) but now the derivative is defined as a sigmoid-like function.

(iii) The maximum and minimum can be replaced by their smooth, albeit still good, approximations of the original relationships. Feldkamp *et al* (1992) considered a parametric approximation of the minimum and maximum operations,

$$\min_{\delta}(x, a) = \tfrac{1}{2}\left[x + a - [(x - a)^2 + \delta^2]^{1/2} + \delta^2\right]$$

$$\max_{\delta}(x, a) = \tfrac{1}{2}\left[x + a + [(x - a)^2 + \delta^2]^{1/2} - \delta^2\right]$$

where δ is taken as a small real number close to zero, say $\delta = 0.02$. This modification eliminates the edges in the original derivative occurring at $x = a$. More generally, one can look for any parametrized family of the triangular norm that approaches the minimum or maximum at some limit values of its parameters and utilize this representative as a relevant approximation.

While these modifications are conceptually quite different, their final numerical effects of learning, as investigated by Ikoma *et al* (1993) are quite similar.

To come up with a weightless neuron, all its connections w_i have to be kept constant during the learning process, $w_i = w$, $i = 1, 2, \dots, n$. This style of learning leads to the optimization procedure of the form,

$$\min_{w, m, \sigma} Q$$

where the minimization of Q hinges to a significant degree on the parameters of the nonlinear element. The results of the above approximation might usually show higher values of the performance index Q in comparison to those obtained by the previous learning scheme. This phenomenon is quite legitimate considering the lower number of parameters involved in the current optimization. In fact, this behavior reflects a genuine nature of any linguistic modifier as it tends to look at all the variables simultaneously, aggregate them linguistically, and ignore any differences between them. If a substantial discrimination between the variables is necessary, the modifiers (e.g. '*most*' of the variables) might not perform well as this is subsequently reflected by the achieved value of the minimized performance index.

The learning of some other neurons is completed in a similar manner following the general update scheme and including pertinent technical modifications specific to the considered neuron.

D1.7.5 General policies for parametric learning: reductions and expansions

The learning in a fuzzy neural network can vary from case to case and usually depends heavily on the initial information available to the classification problem which can be immediately accommodated in the network. For instance, in many situations it is obvious in advance that some connections need to be weaker or even nonexistent. This allows us to build an initial configuration of the network being very distinct from that of a fully connected network. This initial domain knowledge tangibly enhances the learning procedure eliminating the need to modify all the connections of the network, thus preventing us from proceeding with learning from scratch. On the other hand, if the initial domain knowledge about the problem (network) is not sufficient, then a fully connected structure yielding higher values of its entropy function (Machado and Rocha 1990, Rocha 1992) would be strongly recommended.

In many cases the role of the individual layers is also obvious so that one can project the behavior of the network (and evaluate its learning capabilities) in this way. The following two general strategies of learning are worth pursuing:

(i) *Successive reductions.* One starts with a large and eventually excessive neural network (containing many elements in the hidden layer), analyzes the results of learning and, if possible, resumes the size of the network. These reductions are carried out as far as they do not drastically affect the quality of learning (by slowing it down significantly and/or elevating the values of the minimized performance index). The main advantage of this strategy lies in fast learning. This is achieved due to the 'underconstraint' nature of the successive networks. A deficiency of this approach is that the network constructed in this way can be fairly 'overdistributed'.

(ii) *Successive expansions.* The starting point in this strategy is a small, compact neural network which is afterwards expanded successively, based on the values of the obtained performance index. Excessively high values of the index may suggest further expansions. The network derived in this way could be made compact; nevertheless, under some circumstances, a total computational overhead (many unsuccessfully extended structures of the neural networks) may not be acceptable and could make this approach computationally quite costly.

In addition to the sum of squared errors viewed as a leading indicator of the learning process, the training can be additionally monitored by the entropy function determined at the level of the hidden layer(s). Let us concentrate on a network with a single hidden layer (the same procedure can be immediately applied to the architecture with many hidden layers). The computations of the entropy function proceed accordingly:

(i) The output signals of the neurons situated in the hidden layer first become normalized,

$$p_i = \frac{z_i}{\sum_{j=1}^{h} z_j}$$

where z_i stands for the activation level of the ith neuron in the hidden layer, $i = 1, 2, \ldots, h$. Here p_i is interpreted as a relative normalized frequency (probability) of firing the ith neuron.

(ii) Based on the computed probabilities, the entropy at the level of the hidden layer is next determined in the usual way,

$$H(x) = -\sum_{i=1}^{h} p_i \log p_i = \sum_{i=1}^{h} p_i \log \frac{1}{p_i} = E\left\{\log \frac{1}{p_i}\right\}$$

(where $E\{\cdot\}$ stands for an expectation operator).

The global entropy taken over the available training set of patterns x is obtained by summing the results obtained for the individual patterns,

$$H(X) = \sum_{x \in X} H(x).$$

Too large an increase in the size of the hidden layer could be reflected by significantly lowered values of $H(X)$ pointing out a significant drop in the activities of some neurons after being added to the layer—a visible sign of their underutilization.

In general, the learning in fuzzy neural networks should be made more specific depending upon the architecture of the network. More precisely, the learning formulas need to be calculated from scratch depending upon the topology of the network. As an example, let us consider the network below that realizes a fragment of a qualitative protocol describing a decision problem:

decision d if

$$x_2 \text{ and } x_3 \text{ are close to } 0.5$$

or

$$x_1 \text{ and not } (x_2).$$

The induced fuzzy neural network is shown in figure D1.7.5.

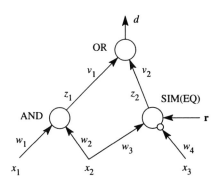

Figure D1.7.5. Fuzzy neural network in mapping qualitative domain knowledge.

We now derive detailed learning formulas for this network. In particular we compute all necessary gradients of the connections of the network,

$$\frac{\partial d}{\partial w_i} = \frac{\partial d}{\partial z_1} \frac{\partial z_1}{\partial w_i} \quad \text{for } i = 1, 2$$

$$\frac{\partial d}{\partial w_i} = \frac{\partial d}{\partial z_2} \frac{\partial z_2}{\partial w_i} \quad \text{for } i = 3, 4$$

$$\frac{\partial d}{\partial v_i} = \frac{\partial \text{OR}([z_1, z_2]; [v_1, v_2])}{\partial v_i} \quad \text{for } i = 1, 2$$

$$\frac{\partial d}{\partial z_i} = \frac{\partial \text{OR}([z_1, z_2]; [v_1, v_2])}{\partial z_i} \quad \text{for } i = 1, 2$$

$$\frac{\partial z_1}{\partial w_i} = \frac{\partial \text{AND}([x_1, x_2]; [w_1, w_2])}{\partial w_i}$$

$$\frac{\partial z_2}{\partial w_i} = \frac{\partial \text{SIM}([x_2, \bar{x}_3]; [w_3, w_4], [r_3, r_4])}{\partial w_i} \quad \text{for } i = 3, 4.$$

Let us discuss the similarity (equality) neuron in more depth. Considering its OR-wise form of aggregation, one gets

$$z_2 = \text{OR}([x_2, x_3] \equiv [r_3, r_4]; [w_3, wr_4]).$$

The logic operations are instantiated accordingly: OR–maximum, AND–product. The similarity operation is induced by the Łukasiewicz implication giving rise to the expression

$$a \equiv b = \begin{cases} 1 - a + b & \text{if } a \geq b \\ 1 - b + a & \text{if } a < b. \end{cases}$$

Thus

$$\frac{\partial z_2}{\partial w_i} = \frac{\partial}{\partial w_i}[(x_i \equiv r_i)w_i \vee (x_j \equiv r_j)w_j]$$

which finally produces the expression,

$$\frac{\partial}{\partial w_i}[(x_i \equiv r_i)w_i \vee (x_j \equiv r_j)w_j] = (x_i \equiv r_i)[(x_j \equiv r_j)w_j \; \varphi \; (x_i \equiv r_i)w_i].$$

References

Feldkamp L A, Puskorius G V, Yuan F and Davis L I Jr 1992 Architecture and training of a hybrid neural-fuzzy system *Proc. 2nd Int. Conf. on Fuzzy Logic and Neural Networks (Iizuka)* pp 131–4

Ikoma N, Pedrycz W and Hirota K 1993 Estimation of fuzzy relational matrix by using probabilistic descent method *Fuzzy Sets Syst.* **57** 335–49

Machado R J and Rocha A F 1990 The combinatorial neural network: a connectionist model for knowledge based systems *Proc. 3rd Int. Conf. on Information Processing and Management of Uncertainty in Knowledge-bases Systems (Paris)* pp 9–11

Pedrycz W 1991 Neurocomputations in relational systems *IEEE Trans. on Pattern Analysis and Machine Intelligence* **13** 289–96

Rocha A F 1992 *Neural Nets: a Theory for Brain and Machine (Lecture Notes in Artificial Intelligence 638)* (Berlin: Springer)

Schneeweiss W G 1989 *Boolean Functions with Engineering Applications* (Berlin: Springer)

D1.8 Induced Boolean and core neural networks

Krzysztof J Cios and Witold Pedrycz

Abstract

See the abstract for Chapter D1.

The elicitation of the structure of the network can be enhanced by pruning some weaker connections of the neurons. Generally, in the OR neuron one eliminates all the connections whose values are below a certain threshold. These connections are set to 0 while the values of the remainder are retained or eventually elevated to 1. The opposite rule holds for the AND neuron: all the connections with the values above the threshold value are set to 1. These threshold levels can be set up arbitrarily or may be subject to optimization.

The optimized way of pruning the connections leads to the approximation of the fuzzy neural network by its Boolean version. Within this procedure all the connections of the network are converted to either 0 or 1. Let $y = N(x, w, v)$ denote the neural network to be approximated, where w, v are collections of the connections between the successive layers. The idea of this approximation is to replace $N(x, w, v)$ by its Boolean counterpart, denoted by $B(x, w_B, v_B)$, in such a way that the results produced by the Boolean network follow as closely as possible those produced by the original network. The quality of the Boolean approximation can be formally characterized by the performance index

$$\sum_{x \in X} \| N(x, w, v, \ldots) - B(x, w_B, v_B, \ldots) \|$$

where $\|\cdot\|$ stands for the distance function. The above sum is taken over a certain collection of the patterns X. The minimization is worked out with respect to the Boolean connections of the network B when approximating the network N over a set of patterns forming X. More precisely, this task pertains to the Boolean approximation of the network carried out with respect to X. Obviously, different forms of X could result in fairly different approximations and, consequently, different Boolean networks induced by the same fuzzy neural network. In particular, one can contemplate two specific families of the inputs:

(i) X is the same as the training data set;
(ii) X covers the entire universe of discourse by including the elements being randomly distributed in the input hypercube.

Obviously, some other options of X might be worth considering (figure D1.8.1).

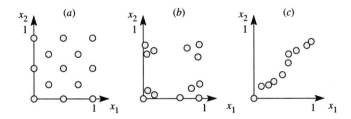

Figure D1.8.1. Training data sets X: (*a*) uniformly distributed in the plane of inputs, (*b*) binary biased (data centered around the vertices of the plane of inputs), (*c*) functionally constrained, $x_2 = g(x_1)$.

The multidimensional optimization task can be reduced by admitting a simplified strategy of building the induced Boolean network. The crux of this simplification is to reduce the dimensionality of the search by selecting a uniform threshold strategy for all the AND and OR neurons. Let us introduce two threshold operations. The first applies to all the OR neurons in the network and replaces their original connections by 0 or 1 depending on their position with respect to the threshold λ,

$$T_\lambda(w) = \begin{cases} 0 & \text{if} \quad w < \lambda \\ 1 & \text{if} \quad w \geq \lambda \end{cases}$$

where $w, \lambda \in [0, 1]$. The second thresholding operation, $T_\mu(w)$, equipped with another threshold value μ, is used for the AND neurons,

$$T_\mu(w) = \begin{cases} 0 & \text{if} \quad w \leq \mu \\ 1 & \text{if} \quad w > \mu. \end{cases}$$

By considering these threshold operations, we arrive at the reduced two-dimensional version of the optimization task,

$$\min_{\lambda, \mu} \sum_{x \in X} \|N(x, w, v, \ldots) - B(x, w, v, \ldots)\|$$

which is computationally definitely much more amenable than the previous one. Another feasible option of network induction retains the most significant ('core') connections of the neurons—hence the resulting architecture will be called the core network. In place of the above transformations we can define less 'drastic' modifications,

$$T_\lambda(w) = \begin{cases} 0 & \text{if } w < \lambda \\ w & \text{if } w \geq \lambda \end{cases}$$

and

$$T_\mu(w) = \begin{cases} w & \text{if } w \leq \mu \\ 1 & \text{if } w > \mu \end{cases}$$

that preserve the values of the connections once they are recognized as being essential in the sense of the assumed criteria. The thresholding operations are illustrated in figure D1.8.2.

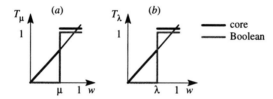

Figure D1.8.2. Core and Boolean thresholding operations (a) AND neuron, (b) OR neuron.

D2

Neural-Evolutionary Systems

V William Porto

Abstract

In this chapter, evolutionary computation is presented as a methodology for solving many current problems encountered in the neural network design process. Several design areas are addressed including alternative training methods, which prevent entrapment in local minima points, automatic selection of optimal neural topologies, and determination of optimal input feature sets. Differences between conventional (i.e. gradient-based learning algorithms, mean-squared-error optimization) and evolutionary computation approaches are discussed along with current application areas and future research directions.

Contents

D2.1 Overview of evolutionary computation as a mechanism for solving neural system design problems

V William Porto

Abstract

See the abstract for Chapter D2.

Although neural networks hold promise for solving a wide variety of problems, they have not yet fulfilled this promise due to limitations in *training*, determination of the most appropriate *topology*, and efficient B3, B2 determination of the best feature set to use as inputs. A number of techniques have been investigated to solve these problems but none has the combination of simplicity, efficiency and algorithmic elegance that is inherent in evolutionary computation (EC). These evolutionary techniques comprise a class of generalized stochastic algorithms which utilize the properties of a parallel and iterative search to solve a variety of optimization and other problems. Evolutionary computation is well suited to solving many of the inherently difficult or time-consuming problems associated with neural networks since most of the difficulties encountered with designing and training neural networks can be expressed as optimization problems.

One of the most common problems encountered in training neural networks is the tendency of the training algorithm to become entrapped in local minima. This leads to suboptimal weight sets which are often insufficient to solve the task at hand. Due to the immense size of the typical search space, an exhaustive search is usually computationally impossible. Gradient methods, such as *error backpropagation*, B6.3.3 are commonly used since these are easy to implement, may be tuned to provide superlinear convergence, and are mathematically tractable given the differentiability of the network nodal transfer functions. But these methods have the serious drawback that when the algorithm converges to a solution, there is no guarantee that this solution is globally optimal. In real-world applications, these algorithms frequently converge to local suboptimal weight sets from which the algorithm cannot escape.

There is also the problem of determining the optimal topology for the application. Much of the research attempting to provide optimal estimates of the number and types of nodes in the topology has been focused on bounding the solutions in a mean squared error (MSE) sense. The notion of nodal redundancy for robustness is often neglected, as is the fact that system performance may be better suited using a different metric for network topology determination.

Finally, if one assumes that the aforementioned problems of network training and topology selection have been surmounted, there still remains the question of optimal input feature selection. Neural networks have been applied to a variety of problems ranging from *pattern recognition* to *signal detection*, yet F1.2, F1.8 very little research has been made into ways to optimally select the most appropriate input features for each application. Typical approaches range the gamut from complex statistical measures to heuristic methodologies, each requiring *a priori* knowledge or specific tuning of the problem at hand.

Fortunately, stochastic evolutionary methods can address not only the weight estimation and topology selection problem, but can also be utilized to help determine the optimal set of input features going into a neural network. Searching and parameter optimization using stochastic methods can provide a comprehensive, self-adaptive solution to parameter estimation problems yet is often overlooked in favor of deterministic, closed form solutions. The most general of these algorithms search the solution space in parallel, and as such are perfectly suited to application and implementation on today's multiprocessor computers.

D2.1.1 Stochastic search

By formal mathematical definition, a stochastic process $X(t, w)$ is a function of two variables, where w is an element of the sampling space and t is a time parameter from the time interval set T (Papoulis 1965). This is typically a real-valued function but can also be complex valued. The terms *random process* and *stochastic process* are often considered synonymous and cover virtually all the theory of probability. In practice, the term stochastic process is generally used when a time parameter is introduced.

Randomness (or noise) in observations of phenomena is often viewed as corruption of the underlying process, and hence, something to be filtered out. From a viewpoint of a deterministic search, this is a common notion, and most optimization algorithms are designed to smooth out any inherent noise processes, either explicitly or implicitly. Algorithms that take advantage of randomness, however, can be effectively utilized to search topologies that contain multiple optima. Certainly, an exhaustive search of the topological parameter space can be computationally impractical, but a number of methodologies exist that selectively use randomness in their search and are not only competitive in convergence speed, but are asymptotically immune to entrapment in suboptimal minima (maxima) points.

D2.1.2 Basic evolutionary computation methodologies and intrinsic differences

Evolutionary computation is based upon simulating the process of evolution in order to iteratively derive better and more appropriate solutions to a variety of problems. As a class of stochastic algorithms, these techniques efficiently utilize randomness as they search through the parameter space for successively better solutions, without the need for explicit derivative information. At the very basis of these algorithms is the presumption that in a statistical sense, phylogenic learning can be encoded in each member of the solution set and is proportional to the fitness of that member. A selection mechanism statistically eliminates suboptimal members of the population. Thus increasingly appropriate solutions can be evolved through competitive selection. Of importance is the parallel nature of these techniques. Other stochastic
C1.4.2 optimization techniques, such as *simulated annealing* and its derivatives, utilize one solution point which is iteratively altered through a mutation process (Metropolis *et al* 1953, Kirkpatrick *et al* 1983, Szu 1986). Evolutionary computation typically utilizes a population of solutions which effect the search for optima points in parallel. The population of solutions can span a large subspace of the parameter set, and efficiently directs the search in the most promising areas of the solution space. A set of parent solutions are iteratively altered to generate offspring solutions. At each iteration, the solutions are scored with respect to their fitness, which may be least-mean squared error, or any other measurable function. The best-scoring solutions are then probabilistically retained to become parents for the next generation of solutions. Selection functions may or may not be *elitist*, that is, the top M scoring solutions are chosen to become the parents for the next generation. A statistical competition may also be used to select among population members. A basic outline of an evolutionary computation algorithm is described below:

D2.1.2.1 Basic evolutionary computation algorithm

$t := 0$;
initialize $P(0) := \{a_1(0), a_2(0), \ldots, a_\mu(0)\}$
evaluate $P(0) : \{\Phi(a_1(0)), \Phi(a_2(0)), \ldots, \Phi(a_\mu(0))\}$
iterate
$\quad\{$
$\qquad$ recombine: $P'(t) := r\Theta_r(P(t))$
$\qquad$ mutate: $P''(t) := m\Theta_m(P(t))$
$\qquad$ evaluate: $P''(0) : \{\Phi(a_1''(t)), \Phi(a_2''(t)), \ldots, \Phi(a_\lambda''(t))\}$
$\qquad$ select: $P(t+1) := s\Theta_s(P''(t) \cup Q)$
$\qquad$ $t := t + 1$;
$\quad\}$

where

$\qquad a$ is an individual member in the population
$\qquad \mu \geq 1$ is the size of the parent population
$\qquad \lambda \geq 1$ is the size of the offspring population

$P(t) := \{bia_1(t), a_2(t), \ldots, a_\mu(t)\}$ is the population at time t

$\Phi : I \to \Re$ is the fitness mapping

$r\Theta_r$ is the recombination operator with controlling parameters Θ_r

$m\Theta_m$ is the mutation operator with controlling parameters Θ_m

is the selection operator $\ni s\Theta_s : (I^\lambda \cup I^{\mu+\lambda}) \to I^\mu$

$Q \in \{\emptyset, P(t)\}$ is a set of individuals additionally accounted for in the selection step, that is, parent solutions.

Three main variations of this basic algorithm, evolutionary programming (EP), evolution strategies (ES) and genetic algorithms (GA), involve differences in the mutation and recombination processes, fitness evaluation, selection mechanism and overall search space representation (Fogel et al 1966, Holland 1975, Goldberg 1989, Koza 1992, Baeck et al 1993, Fogel 1995). Evolutionary programming and evolution strategies are quite similar in their approach to optimization. Evolution strategies can utilize local or global recombination whereas in EP, no recombination is used. Genetic algorithms are among the best known evolutionary algorithms and typically use binary representations. In a GA, an interpretation function mapping between the search space representation and the evaluation space representation is used. Genetic algorithms create new solutions by recombining the representational components of two solution members with a crossover operator. To some degree, mutation operators are also used in GAs. In both EP and ES, however, the representation evaluated by the fitness function is operated upon directly, that is, there is no interpretation function necessary to translate between the search and evaluation spaces. An excellent and more detailed discussion on the similarities and differences between these algorithms can be found in Baeck and Schwefel (1993). It is important to note that the selection process is not necessarily elitist, and thus can permit retention of lower fitness solutions in the next generation of the population. By allowing some percentage of lower scoring solutions, the solution space is often searched more efficiently since higher scoring solutions, while locally optimal, may be far away from the global optimum. Probabilistically, the best solutions are retained, thus convergence is largely monotonic.

Simulated annealing is a generalized Monte Carlo technique with a continuously decreasing variance (Metropolis et al 1953, Kirkpatrick et al 1983). It is a specific case of EP utilizing a single member in the search population with an extrinsic temperature schedule. A semi-local search strategy is used whereby the parametrized representation is mutated according to a specified probability density function. Better scoring solutions are always accepted with probability one, but inferior solutions are also accepted according to the probability used to generate the random process in what is termed as the temperature cooling schedule (Szu 1986). The choice of the probability function determines the convergence rate with Cauchy probabilities and proves considerably faster than Gaussian random processes (Szu 1986).

References

Baeck T, Rudolph G and Schwefel H-P 1993 Evolutionary programming and evolution strategies: similarities and differences *Proc. Second Ann. Conf. on Evolutionary Programming* ed D B Fogel and W Atmar (La Jolla, CA: Evolutionary Programming Society) pp 11–22

Baeck T and Schwefel H-P 1993 An overview of evolutionary algorithms for parameter optimization *Evolutionary Comput.* **1** 1–23

Fogel D B 1995 *Evolutionary Computation* (Piscataway, NJ: IEEE Press) pp 75–84

Fogel L J, Owens A J and Walsh M J 1966 *Artificial Intelligence Through Simulated Evolution* (New York: Wiley) pp 11–26

Goldberg D E 1989 Genetic algorithms *Search, Optimization and Machine Learning* (Reading, MA: Addison-Wesley) pp 1–54

Holland J H 1975 *Adaptation in Natural and Artificial Systems* (Ann Arbor, MI: University of Michigan Press) pp 20–74

Kirkpatrick S, Gelatt C D and Vecchi M P 1983 Optimization by simulated annealing *Science* **220** 671–80

Koza J 1992 *Genetic Programming* (Cambridge, MA: MIT Press) pp 73–7

Metropolis N, Rosenbluth A W, Rosenbluth M N, Teller A H and Teller E 1953 Equation of state calculation by fast computing machines *J. Chem. Phys.* **21** 1087–92

Papoulis A 1965 *Probability, Random Variables, and Stochastic Processes* (New York: McGraw-Hill) p 280

Szu H 1986 Non-Convex Optimization *SPIE* vol 698 *Real-Time Signal Processing* vol IX pp 59–65

D2.2 Evolutionary computation approaches to solving problems in neural computation

V William Porto

Abstract

See the abstract for Chapter D2.

D2.2.1 Training

The number of training algorithms and variations thereof recently published for different neural topologies is exceedingly large. The mathematical basis for the vast majority of these algorithms is to utilize gradient information to adjust the connection weights between nodes in the network. Gradients of the error function are calculated and this information is propagated throughout the topology weights in order to estimate the best set of weights, usually in a least-squared error sense (Werbos 1974, Rumelhart and McClelland 1986, Hecht-Nielsen 1990, Simpson 1990, Haykin 1994, Werbos 1994). A number of assumptions about the local and global error-surface are inherently made when using any of these *gradient-based techniques*. B5.2 Numerous modifications of simple techniques have been made in order to speed up the often exceedingly slow convergence (training) rates. Stochastic training algorithms can provide an attractive alternative by removing many of these assumptions while simultaneously eliminating the calculation of gradients. Thus they are well suited for training in a wide variety of cases, and often perform better overall than the more traditional methods.

D2.2.1.1 Stochastic methods versus traditional gradient methods

A considerable amount of research has been performed in optimization theory in the areas of gradient-based methods, that is, those techniques which utilize derivative information to search for and locate function minima (or equivalently maxima) points. Traditionally, gradient-based techniques have provided the basic foundation for many of the neural network training algorithms (Rumelhart and McClelland 1986, Simpson 1990, Haykin 1994, Werbos 1994). It is important to note that gradient-based methods are not just primarily used in training algorithms for *feedforward networks*, but also in a variety of networks such B2.3 as *Hopfield networks*, *recurrent networks*, *radial basis function networks* and many *self-organizing systems*. B1.3, B2.3, Viewed within the mathematical framework of numerical analysis, gradient-based techniques often provide B1.7.3, C2.1.1 superlinear convergence rates in applications on convex surfaces. First-order (steepest or gradient descent) and second-order (i.e. conjugate gradient, Newton, quasi-Newton) methods have been successfully used to provide solutions to the neural connection weight and bias estimation problem (Kollias and Anastassiou 1988, Kramer and Sangiovanni-Vincentelli 1989, Simpson 1990, Barnard 1992, Saarinen *et al* 1992). While these techniques may prove useful in a number of cases, they often fail due to several interrelated factors. First, by definition, in order to provide guaranteed convergence to a minimum point, first-order gradient techniques must utilize infinitesimally small step sizes (e.g. learning rates) (Luenberger 1973, Scales 1985). Step size determination is most often a balancing act between monotonic convergence and time constraints inherent in the available training apparatus. From a practical standpoint, training should be performed using the largest feasible step size to minimize computational time. Several automated methods for step size determination have been researched with some providing near-optimal step size estimation (Jacobs 1988, Luo 1991, Porto 1993, Haykin 1994). By the Kantorovich inequality, it can be shown that the method of the basic gradient descent algorithm converges linearly to a minimum point with a ratio no greater than $[(\lambda_1 - \lambda_2)/(\lambda_1 + \lambda_2)]$ where λ_1 and λ_2 are the largest and smallest eigenvalues, respectively, of the Hessian of the objective function evaluated at the solution point. However, convergence to a global optimum point is not guaranteed.

Second-order methods attempt to approximate the (inverse) Hessian matrix and utilize a line search for optimal step sizes at each iteration. These methods require the assumption that a reasonably smooth function in N dimensions can be approximated by a quadratic function over a small enough region in the vicinity of an optimal point. In both cases, however, the actual process of iteratively converging on the series of solutions is computationally expensive. For example, convergence of the Davidon–Fletcher–Powell method is inferior to steepest descent with a step size error of only 0.1%, so second-order information does not always provide superior convergence rates (Luenberger 1973, Shanno 1978). It should be noted that problems encountered when the Hessian matrix is indefinite or singular can be addressed by using the method of Gill and Murray, albeit with the added computational cost of solving a nontrivial size set of linear equations (Luenberger 1973, Scales 1985). In practice, quasi-Newton methods work well only on relatively small problems with up to a few hundred weights (Dennis and Schnabel 1983).

One alternative approach to training neural networks is to utilize the numerical solution of ordinary differential equations (ODEs) to estimate interconnection weights (Owens and Filkin 1989). By posing the weight estimation problem as a set of differential equations, ODE solvers can iteratively determine optimal weight sets. These methods, however, are subject to the same prediction-correction errors and, in practice, these too can be quite costly computationally.

Hypothetically, one can find an optimal algorithm for determining step size with the desired gradient-based algorithm. A major problem still remains whereby all of the convergence theorems for these methods prove convergence to an optimum point. There is no guarantee that this is the global optimum point except in the rare case where the function to be minimized is convex. Research has proven convergence to a global optimum point is guaranteed on linearly separable problems when batch mode processing, error-backpropagation learning is used (Gori and Tesi 1992). However, linearly separable problems are easily solved using non-neural network methods such as linear discriminant functions (Fisher 1976, Duda and Hart 1973). In real-world applications, neural network training can, and often does, becomes entrapped in local minima points, generating suboptimal weight estimates (Minsky and Pappert 1988). The most commonly used method to overcome this difficulty is to restart the training process by using a different random starting point. Mathematically, restarting at different initial weight solution 'sample' points is actually an implementation of a simplistic stochastic process.

Stochastic training methods provide an attractive alternative to the traditional methods of training
C1.4 neural networks. In fact, learning in *Boltzman machines* is, by definition, probabilistic and uses simulated annealing for weight adjustments. By their very nature, stochastic search methods, and evolutionary algorithms in particular, are not prone to entrapment in local minima points. Nor are these algorithms subject to step size problems inherent in virtually all of the gradient-based methods. As applied to the weight estimation problem, stochastic methods can be viewed as sampling the solution (weight) space in parallel, retaining those weights which provide the best fitness score. Note that in a stochastic algorithm fitness does not necessarily imply a least-mean squared error criterion. Virtually any metric or combination of metrics can be accommodated. In real-world environments robustness against failure of connections or nodes is often highly important. This robustness can easily be built into the networks during the training phase with stochastic training algorithms.

D2.2.1.2 Case studies

Evolutionary algorithms have been successfully applied to the aforementioned problem of training, that is, estimating the optimal set of weights for neural networks. Numerous approaches have been studied ranging from simple iterative evolution of weights to sophisticated schemes whereby recombination operators exchange weight sets on subtrees in the topology. It is important to note the that these algorithms do not typically utilize gradient information, and hence are often computationally faster due to their simplicity of implementation.

C1.2   Differences between several techniques suitable for training *multilayered perceptrons* (MLPs) and other neural networks were investigated by Porto and Fogel (1992). The computational complexity
C1.4.2 of standard backpropagation (BP), modified (line-search) BP, fast *simulated annealing* (FSA), and evolutionary programming (EP) were compared. In this paper, FSA using a Cauchy probability distribution

for the annealing schedule—the temperature schedule for mutating weights is set inversely proportional to time (number of iterations)—is contrasted with EP. The EP weight optimization is performed with mutation variance inversely proportional to the RMS error of the aggregate input pattern training set. Thus the mutation variance decreases as training converges to more optimal solutions. Computational similarities between the FSA and EP approaches and increased robustness of a parallel search technique such as EP versus the single solution member of an FSA search are shown. A number of tests are performed using underwater mine data using MLPs trained from multiple starting points with each of the aforementioned training techniques in order to ascertain the potential robustness of each to multimodal error surfaces. Results of this research on neural networks with multiple weight set solutions (i.e. local minima points) demonstrate better performance on naive test sets using FSA and EP training methods. These stochastic training methods are proven to be more robust to multimodal error surfaces and hence demonstrate reduced susceptibility to poor performance due to entrapment in local minima points.

The problem of robustness to processing node failure was addressed by Sebald and Fogel (1992). In this paper, adaptation of interconnection weights is performed with the emphasis on performance in the event of node failures. Neural networks are evolved using EP while linearly increasing the probabilistic failure rate of nodes. After training, performance is scored with respect to classification ability given N random failures during the testing of each network. Fault-tolerant networks are demonstrated as often performing poorly when compared against non-fault-tolerant networks if the probability of nodal failure is close to zero, but are shown to exhibit superior performance when failure modes are increased. Evolutionary programming is able to find networks with sufficient redundancy which are capable of dealing with nodal failure.

Using evolutionary computation to evolve network interconnection weights in the presence of hardware weight value limitations and quantization noise was proposed by McDonnell (1992). A modified version of backpropagation is used whereby EP is used for estimating the solutions of bounded and constrained activation functions, and backpropagation is used to refine these solutions. Random competition of the weight sets is used to choose parent networks for each subsequent generation. Results of this research indicate the robustness of this technique and its wide range of applicability to a number of unconstrained, constrained and potentially discontinuous nodal functions.

D2.2.2 Topology selection

Selection of an optimal topology for any given problem is perhaps even more important than optimizing the training technique. It is a well known fact that suboptimal performance of any system can occur by overfitting of data using too many degrees of freedom (network nodes and interconnections) in the model. A balance must be struck between minimizing the number of nodes for generalization in learning and providing sufficient degrees of freedom to fully encode the problem to be learned while retaining robustness to failure. Evolutionary computation is well suited to this optimization problem, and provides for self-adaptive learning of overall topology as well.

D2.2.2.1 Traditional methodology versus self-adaptive approaches

Selection of the most appropriate neural architecture and topology for a specific problem or class of problems is often accomplished by means of heuristic or bounding approaches (Guyon *et al* 1989, Haykin 1994). An eigensystem analysis via a singular value decomposition (SVD) approach has been suggested by Wilson *et al* (1992) to estimate the optimal number of nodes and initial starting weight estimates in a feedforward topology. An SVD is performed on all patterns in the training set with the starting weights initialized using the unitary matrix. The number of nodes in the topology are determined as a function of the sigma matrix in a least-squares sense.

Other analytic and heuristic approaches have also been tried with some success (Sietsma and Dow 1988, Frean 1990, Hecht-Nielsen 1990, Bello 1992) but these are largely based upon probability distribution assumptions, and presence of fully differentiable error functions. In practice, methods which are self-adaptive in determining the optimal topology of the network are the most useful as they are not constrained by *a priori* statistical assumptions. The search space of possible topologies is infinitely large, complex, multimodal, and not necessarily differentiable. Evolutionary computation represents a search methodology which is capable of efficiently searching this complex space.

D2.2.2.2   Case studies

As indicated previously, genetic algorithms (GAs) generate new solutions by recombining representational components of two population members using a function known as crossover. Some degree of mutation is also used but the primary emphasis is on crossover. Specific task environments are characterized as deceptive when the fitness (goodness of fit) is not well correlated with the expected abilities inherent in its representational parts (Goldberg 1989, Whitley 1991). The deception problem is manifested in several ways. Note that identical networks (networks which share identical topologies and common weights when evaluated) need not have the same search representation since the interpretation function may be homomorphic. This leads to offspring solutions which contain repeated components. These offspring networks are often less fit than their parents, a phenomena known as the competing conventions problem (Shaffer *et al* 1992). Second, the crossover operator is often completely incompatible with networks with different topologies. Finally, for any predefined task, a specific topology may have multiple solutions, each with a unique but different distribution of interconnections and weights. Since the computational role of each node is determined by these interconnections, the probability of generating viable offspring solutions is greatly reduced regardless of interpretation function. Fogel (1992) shows GA approaches are indeed prone to these deception phenomena when evolving connectionist networks. Efforts to reduce this susceptibility to deception are studied by Koza and Rice (1991) where they utilize GP techniques which generate neural networks with much more complex representations than traditional GA binary representations. They propose using these alternative representations in an effort to avoid interpretation functions which strongly bias the search for neural network solutions.

The interpretation function which maps the search (representation) space to the evaluation (fitness) space in a GA approach will exceed the complexity of the learning problem (Angeline *et al* 1994). Recent trends have been focused away from binary representations in using GA approaches to solve neural network topology determination problems. Angeline proposes EP for connectionist neural network searches as the representation evaluated by the fitness function is directly manipulated to produce increasingly more appropriate (better) solutions. The generalized acquisition of recurrent links (GNARL) algorithm evolves neural networks using both structural level mutations for topology selection as well as simultaneously evolving the connection weights through mutation. Tests on a food tracking task evolves a number of interesting and highly fit solutions. The GNARL algorithm is demonstrated by simultaneously evolving both the architecture and parameters with very little restriction of the architecture search space on a set of test problems.

C2.1.1   Polani and Uthmann (1993) discuss the use of a GA to improve the topology of *Kohonen feature maps*. In this study, a simple fitness function proportional to the measure of equidistribution of neuron weights is used. Flat network as well as toroidal and Möbius topologies are trained with a set of random input vectors. The GA tests show the existence of networks with nonflat topologies with the ability to be trained to higher quality values than those expected for the optimal flat topology. Given that the optimally trainable topologies may lie distributed over different areas on the topological space, the GA approach is able to find these solutions without *a priori* knowledge and is self-adaptive. Use of this technique could C2.1.1   prove valuable in construction of network topologies for *self-organizing feature maps* where convergence speed or adaptation to a given input space is crucial.

Genetic algorithms are used to evolve both the topology and weights simultaneously as described in a paper by Braun (1993). In weak encoding schemes, genes correspond to more abstract network properties which are useful for efficiently capturing architectural regularities of large networks. However, strong encoding schemes require much less detailed knowledge about the genetic encoding and neural mechanisms. Braun researched a network generator capable of handling large real-world problems. A strong representation scheme is used where every gene of the genotype relates to one connection of the represented network. Once the maximal architecture is specified, potential connections within this architecture are chosen and iteratively mutated and selected. Crossover mutation is performed using distance coefficients to prevent permuted interval representations in order to minimize connection length. This is where crossover alone often proves problematic. Tests on digit recognition, the truck-backer-upper task, and the Nine Men's Morris problem were performed. These experiments concluded that weight transmission from parent to offspring is very important and effectively reduced learning times. Braun also notes that mutation alone is potentially sufficient to get good selection performance.

B1.7.3   The use of evolutionary search to determine the optimal distribution of *radial basis functions* was addressed by Whitehead and Choate (1994). Binary encoding was used in a GA with the evolved networks selected to minimize both the residual error in the function approximation as well as the number of RBF nodes. A set of space filling curves as encoded by the GA are evolved to optimally distribute the RBFs. The weights from the first layer which form linear combinations of the RBFs are trained with a conventional *LMS learning rule*. Convergence is rapid since the total squared error over the training set is a quadratic. C1.1.3 An additional benefit is realized whereby the local response of each RBF can be set to zero beyond a genetically selected radius thus ensuring only a small fraction of the weights need to be modified for each input training exemplar. This methodology strikes a balance between representations which specify all of the weights and require no training, and the other extreme where no weights are specified and full training of each network is required on each pass of the algorithm. Results indicate the superiority of evolving the RBF centers in comparison to k-means clustering techniques. This may possibly be explained by the fact that a large proportion of the evolved centers were observed to lie outside the convex hull of the training data, while the k-means clustering centers remained within this hull.

References

Angeline P, Saunders G and Pollack J 1994 Complete induction of recurrent neural networks *Proc. Third Conf. on Evolutionary Programming* ed A V Sebald and L J Fogel (River Edge, NJ: World Scientific) pp 1–8

Barnard E 1992 Optimization for training neural networks *IEEE Trans. Neural Networks* **3** 232–6

Bello M 1992 Enhanced training algorithms, and integrated training/architecture selection for multilayer perceptron networks *IEEE Trans. Neural Networks* **3** 864–75

Braun H 1993 Evolving neural networks for application oriented problems *Proc. Second Ann. Conf. on Evolutionary Programming* ed D B Fogel and W Atmar (La Jolla, CA: Evolutionary Programming Society) pp 62–71

Dennis J and Schnabel R 1983 *Numerical Methods for Unconstrained Optimization and Nonlinear Equations* (Englewood Cliffs, NJ: Prentice-Hall) pp 5–12

Duda R O and Hart P E 1973 *Pattern Classification and Scene Analysis* (New York: Wiley) pp 130–86

Fisher R A 1976 The use of multiple measurements in taxonomic problems *Machine Recognition of Patterns* (reprinted from 1936 *Annals of Eugenics*) ed A K Agrawala (Piscataway, NJ: IEEE Press) pp 323–32

Fogel D B 1992 *Evolving Artificial Intelligence* PhD dissertation University of California, San Diego, CA

Frean M 1990 The upstart algorithm: a method for constructing and training feedforward neural networks *Neural Comput.* **2** 198–209

Goldberg D E 1989 Genetic algorithms *Search, Optimization and Machine Learning* (Reading, MA: Addison-Wesley) pp 1–54

Gori M and Tesi A 1992 On the problem of local minima in backpropagation *IEEE Trans. Patt. Anal. Mach. Intell.* **14** 76–86

Guyon I, Poujaud I, Personnaz L, Dreyfus G, Denker J and Le Cun Y 1989 Comparing different neural network architectures for classifying handwritten digits *Proc. IEEE Int. Joint Conf. on Neural Networks* vol II (Piscataway, NJ: IEEE) pp 127–32

Haykin S 1994 *Neural Networks, a Comprehensive Foundation* (New York: Macmillan) pp 121–281, 473–584

Hecht-Nielsen R 1990 *Neurocomputing* (Reading, MA: Addison-Wesley) pp 48–218

Jacobs R A 1988 Increased rates of convergence through learning rate adaptation *Neural Networks* **1** 295–307

Kollias S and Anastassiou D 1988 Adaptive training of multilayer neural networks using a least squares estimation technique *Proc. Int. Conf. on Neural Networks* vol I (Piscataway, NJ: IEEE Press) pp 383–9

Koza J and Rice J 1991 Genetic generation of both the weights and architecture for a neural network *IEEE Joint Conf. on Neural Networks* vol II (Seattle, WA: IEEE Press) pp 397–404

Kramer A H and Sangiovanni-Vincentelli A 1989 Efficient parallel learning algorithms for neural networks *Advances in Neural Information Processing Systems 1* ed D S Touretzky (San Mateo, CA: Morgan Kaufmann) pp 40–8

Luenberger D G 1973 *Introduction to Linear and Nonlinear Programming* (Reading, MA: Addison-Wesley) pp 194–201

Luo Z 1991 On the convergence of the LMS algorithm with adaptive learning rate for linear feedforward networks *Neural Comput.* **3** 226–45

McDonnell J M 1992 Training neural networks with weight constraints *Proc. First Ann. Conf. on Evolutionary Programming* (La Jolla, CA: Evolutionary Programming Society) pp 111–9

Minsky M L and Pappert S A 1988 *Perceptrons* expanded edn (Cambridge, MA: MIT Press) pp 255–66

Owens A J and Filkin D L 1989 Efficient training of the back propagation network by solving a system of stiff ordinary differential equations *Proc. Int. Joint Conf. on Neural Networks* vol II (IEEE Press) pp 381–6

Polani D and Uthmann T 1993 Training Kohonen feature maps in different topologies: an analysis using genetic algorithms *Proc. Fifth Int. Conf. on Genetic Algorithms* (San Mateo, CA: Morgan Kaufmann) pp 326–33

Porto V W 1993 A method for optimal step size determination for training neural networks (San Diego, CA: ORINCON Internal Technical Report)

Porto V W and Fogel D B 1992 Alternative methods for training neural networks *Proc. First Ann. Conf. on Evolutionary Programming* (La Jolla, CA: Evolutionary Programming Society) pp 100–10

Rumelhart D E and McClelland J (eds) 1986 *Parallel Distributed Processing: Explorations in the Microstructure of Cognition* vol 1 (Cambridge, MA: MIT Press) pp 318–30

Saarinen S, Bramley R B and Cybenko G 1992 Neural networks, backpropagation, and automatic differentiation *Automatic Differentiation of Algorithms: Theory, Implementation, and Application* ed A Griewank and G F Corliss (Philadelphia, PA: SIAM) pp 31–42

Scales LE 1985 *Introduction to Non-Linear Optimization* (New York: Springer) pp 60–1

Sebald A V and Fogel D B 1992 Design of fault tolerant neural networks for pattern classification *Proc. First Ann. Conf. on Evolutionary Programming* (San Diego, CA: Evolutionary Programming Society) pp 90–9

Shaffer J D, Whitley D and Eshelman L J 1992 Combinations of genetic algorithms and neural networks: a survey of the state of the art *Proc. COGANN-92 International Workshop on Combinations of Genetic Algorithms and Neural Networks* (Baltimore, MD: IEEE Computer Society Press) pp 1–37

Shanno D 1978 Conjugate-gradient methods with inexact searches *Math. Op. Res.* **3**

Sietsma J and Dow R 1988 Neural net pruning—why and how *Proc. Int. Conf. on Neural Networks* **I** (IEEE Press) pp 325–33

Simpson P K 1990 *Artificial Neural Systems* (Elmsford, NY: Pergamon) pp 90–120

Werbos P J 1974 Beyond regression: new tools for prediction and analysis in the behavioral sciences *PhD Thesis* Harvard University

——1994 *The Roots of Backpropagation from Ordered Derivatives to Neural Networks and Political Forecasting* (New York: Wiley) pp 29–81, 256–294

Whitehead B and Choate T 1994 Evolving spacefilling curves to distribute radial basis functions over an input space *IEEE Trans. Neural Networks* **5** pp 15–23

Whitley D 1991 Fundamental principles of deception in genetic search *Foundations of Genetic Algorithms* ed G Rawlins (San Mateo, CA: Morgan Kaufmann) pp 221–41

Wilson E, Umesh S and Tufts D 1992 Resolving the components of transient signals using neural networks and subspace inhibition filter algorithms Proc. Int. Joint Conf. on Neural Networks vol 1 (Baltimore, MD: IEEE) pp 283–8

D2.3 New areas for evolutionary computation research in neural systems

V William Porto

Abstract

See the abstract for Chapter D2.

There are many other areas in which the methodologies of evolutionary computation may be useful in the design and solution of neural network problems. Aside from training and topology selection, EC can be used to select optimal node transfer functions, which are often selected for their mathematical tractability, not their optimality in neural problems. Self-adaptation of input features is another area of current research with great potential. Evolving the optimal set of input features (from a potentially large set of transform functions) can be very useful in refining the preprocessing steps necessary to optimally solve a specific problem.

D2.3.1 Transfer function selection

One recent area of interest is the use of evolutionary computation to optimize the choice of nodal transfer functions. Sigmoidal, Gaussian and other functions are often chosen due to their differentiability, mathematical tractability, and ease of implementation. There exists a virtually unlimited set of alternative transfer functions ranging from polynomial forms and exponentials to discontinuous, nondifferentiable functions. By efficiently evolving the selection of these functions, potentially more robust neural solutions may be found. Simultaneous selection of nodal transfer functions and topology may be the ultimate evolutionary paradigm, as nature has taken this tack in evolving the brains of every living organism.

D2.3.2 Input feature selection

Evolutionary computation is well suited for automatically selecting optimal input features. By iterative self-adaptation of these input features for virtually any neural topology, evolutionary methods can be a more attractive approach than those of principal component analysis and other statistical methods. Efficient, automatic search of this input feature space can significantly reduce the computational requirements of signal preprocessing and feature extraction algorithms.

Brotherton *et al* (1995) devised an algorithm which automatically selects the optimal subset of input features and the neural architecture as well as training the interconnection weights using evolutionary programming. In developing a classifier for electrocardiogram (ECG) waveforms, EP was used to design a hierarchical network consisting of MLPs for the first-layer networks, and *fuzzy min–max networks* for the D1 second output layer. The first-layer networks are trained and outputs fused in the second-layer network. EP is used to select from among several sets of input features. Initial training provided approximately 75% correct classification without including heart rate and phase features in the fusion network. Retraining of the fusion networks was performed with the EP trainer and feature selection mechanism, with the resulting system providing a 95% classification capability. Interestingly, analysis of the final trained network inputs showed the EP feature selection technique had determined that these two scalar input features were not used, but had provided guidance during the training phase.

Chang and Lippmann (1991) examined the use of GAs to determine the input data, storage patterns, and select appropriate features for classifier systems in both *speech* and machine vision problems. Using F1.7

F1.2 an EC approach they found they could reduce the input feature size from 153 features to only 33 features with no performance loss. Their investigations into solving a machine vision *pattern recognition* problem demonstrated the ability of GAs to evolve higher-order features which virtually eliminated pattern classification errors. Finally, in another of their tests with neural pattern classifiers, the number of patterns needed to be stored was reduced by a factor of 8 without significant loss in performance.

The area of feature selection via evolutionary computation will be of increased interest as more and more neural systems are put into the field. Selectively choosing the optimal set of input features can make the difference between a mere idea and a practical implementation.

References

Brotherton T and Simpson P 1995 Dynamic feature set training of neural networks for classification *Proc. Fourth Ann. Conf. on Evolutionary Programming* (Cambridge, MA: MIT Press) pp 79–90

Chang E and Lippmann R 1991 Using a genetic algorithm to improve pattern classification performance *Advances in Neural Information Processing Systems* ed D Touretsky (Palo Alto, CA: Morgan-Kaufmann) pp 797–803

PART E

NEURAL NETWORK IMPLEMENTATIONS

PART E

NEURAL NETWORK IMPLEMENTATIONS

E1

Neural Network Hardware Implementations

Contents

E1.1   Introduction

Timothy S Axelrod

Abstract

A brief overview of neural network hardware implementations, introducing the detailed
discussions that follow.

The main impetus behind the development of neural networks has been the impressive capabilities
of biological systems, and our desire to create systems with similar capabilities, but adapted to other
applications. It was recognized from the outset that biological processing systems owe their abilities both
to a novel *architecture* for computing and to its implementation in *hardware* that has some quite astonishing
properties. The development of neural networks to date has mainly emphasized the architectural aspects,
with implementation largely being performed in software that runs on conventional digital computers. But
it is clear that neural networks will not reach their full potential until we develop hardware that shares more
of the properties of biological *hardware*, while retaining the far superior circuit speeds that characterize
modern computing systems. This section of the Handbook is devoted to the approaches that have been
taken so far to realizing this goal and the possible paths forward from this point.

Any hardware implementation technology must satisfy four basic criteria if it is to be a good foundation
for constructing large-scale neural systems. First, it must allow us to build systems with large numbers of
artificial neurons. Artificial neurons, far more than the biological neurons that they imitate, are extremely
simple computing elements. Although biological systems with desirable properties can be found with
hundreds of neurons, or even fewer, the capabilities we ultimately desire to achieve are mostly found in
systems with millions, or billions, of neurons. Second, it must allow these neurons to make large numbers
of connections to other neurons. Third, the weights associated with each neuron must be changeable, so
that the system can learn and adapt, and yet stably storable for long periods of time. Fourth, all this must
be done in a package that is reasonably small and dissipates a manageable amount of power.

In the sections that follow, the implementation technologies that are currently available are described
in the light of these four criteria. Section E1.2 begins the discussion with a more detailed look at some   E1.2
issues that arise in many hardware implementations, in particular the limited precision available to specify
weights and neural outputs, and the fact that many implementation technologies can represent weights of
only a single sign. Section E1.3 begins a systematic tour of the available implementation technologies   E1.3
with a look at analog integrated circuits. This is followed by an examination of digital integrated circuits
in Section E1.4, and optical techniques in Section E1.5.   E1.4, E1.5

After reading these sections, it will be clear that we do not yet have a hardware technology that is
wholly satisfactory for building large neural systems. Digital and, to a lesser extent, analog integrated
circuits can readily attain interestingly large numbers of neurons, but have major problems when it comes
to interconnecting them sufficiently densely. In large measure this reflects the fact that current integrated
circuits form systems that are basically planar, while biological systems are fully three-dimensional,
exploiting the extra dimension at all scales from microns to centimeters. Optical technologies have much
better prospects for solving the interconnection problem, but they are still in their infancy and do not today
have the capability to implement large numbers of neurons. All implementations currently have difficulties
with economically storing and modifying large numbers of weights. It may well be that the technology
we require can only be built by a manufacturing technology that can fully control the structure of systems
at the molecular level, a capability that is currently unique to biological development, but is unlikely to
remain so for much longer.

E1.2 Neural network adaptations to hardware implementations

Perry D Moerland and Emile Fiesler

Abstract

In order to take advantage of the massive parallelism offered by artificial neural networks, hardware implementations are essential. However, most standard neural network models are not very suitable for implementation in hardware and adaptations are needed. In this section an overview is given of the various issues that are encountered when mapping an ideal neural network model onto a compact and reliable neural network hardware implementation, like quantization, handling nonuniformities and nonideal responses, and restraining computational complexity. Furthermore, a broad range of hardware-friendly learning rules is presented, which allow for simpler and more reliable hardware implementations. The relevance of these neural network adaptations to hardware is illustrated by their application in existing hardware implementations.

E1.2.1 Introduction

Soon after the widespread revival of neural network research in the mid-1980s, it was realized that to fully profit from the massive parallelism inherent in neural network models, hardware implementations are essential. This has led to a large variety of implementations using digital and analog electronics, optics, and hybrid techniques. Even though these implementations are largely different, a common denominator is the mapping of neural network algorithms onto reliable, compact, and fast hardware. Any hardware implementation has to optimize three main constraints: accuracy, space, and processing speed. The design of hardware implementations is governed by a balancing of these criteria. An analog implementation, for example, is very efficient in terms of chip area and processing speed, but this comes at the price of a limited accuracy of the network components. In general, this amounts to a trade-off between the accuracy of the implementation and the reliability of its performance. In this section the influence of the limitations typical for hardware implementations will be outlined. Examples of this phenomenon are the following:

- The quantization of network parameters in digital implementations, specifically its weights, to obtain a far more compact implementation. Its counterpart in analog implementations is a limited accuracy of the network parameters due to system noise.
- Computation in analog hardware, be it electronic or optical, is characterized by the nonuniformity of its components and by the fact that the components are at best approximations of the corresponding mathematical operations in the neural network model.

This section provides a thorough review of the experimental and theoretical research that has been performed on the behavior of existing learning algorithms under the limitations imposed by hardware. Furthermore, training algorithms are discussed that offer an improved performance in the case of limited accuracy and that further simplify the hardware implementation of neural networks.

In section E1.2.2, the effects of a quantization of the network parameters and weight discretization algorithms for various neural network models are reviewed. The different approaches are illustrated with examples from existing neural hardware implementations and several commonly used schemes are discussed in more detail. The influence of hardware nonidealities, such as spatial nonuniformity and nonideal response is outlined in section E1.2.3. Section E1.2.4 contains an overview of *hardware-friendly learning algorithms* which are better suited for hardware implementation and especially for on-chip learning. Finally, in section E1.2.5, a summary and conclusions are presented.

E1.2.2 Quantization effects

The use of very high precision cannot be matched with the goal of developing fast and compact hardware implementations. While in digital implementations a high numerical precision is too area consuming, it is incompatible with the system noise present in analog implementations. Therefore, hardware implementations of neural networks typically use a representation of the network parameters with a limited accuracy. For example, in Philips' L-Neuro 1.0 architecture, which allows the implementation of feedforward networks and on-chip backpropagation training, 16-bit weights are used during the training process and only 4-bit or 8-bit weights are employed during recall (Mauduit *et al* 1992). An example of an analog electronic implementation is Intel's Electrically Trainable Analog Neural Network (ETANN), which can perform an impressive two billion weight multiplications per second. The accuracy of its weights and neurons, however, can be compared with a resolution of only seven bits (Holler *et al* 1989).

Table E1.2.1. Weight discretization in multilayer neural networks: off-chip learning.

Reference	Accuracy (bits)	No of benchmarks Artificial	No of benchmarks Real world	Remarks
Holt and Hwang (1993)	8	1	–	Finite-precision error analysis for the forward retrieving pass
Dündar and Rose (1995)	10	2	–	Statistical model of weight quantization in sigmoidal networks
Piché (1995)	6–10	2	–	Statistical analysis of the effects of weight errors upon an ensemble of multilayer networks

Table E1.2.2. Weight discretization in multilayer neural networks: chip-in-the-loop learning.

Reference	Accuracy (bits)	No of benchmarks Artificial	No of benchmarks Real world	Remarks
Fiesler *et al* (1988) Fiesler *et al* (1990)	2–3	3	–	Forward pass with discrete weights, backward pass with continuous weights
Marchesi *et al* (1993)	3–4	1	1	Power-of-two weights in the forward pass and an adaptive learning rate
Tang and Kwan (1993)	3–4	1	–	Power-of-two weights and adaptive gain of the activation function

Since hardware implementations are characterized by a low numerical precision, it is essential to study the effects of this on the recall and training of the various neural network models. The need for a further reduction of the accuracy, while retaining a satisfactory network performance, has also led to various *weight discretization* algorithms, especially designed for this purpose. Since most research has been C1.2 performed for *multilayer feedforward networks*, these will be discussed separately from the other neural network paradigms. A compact overview of a large variety of results on the effects of limited precision in neural networks can be found in tables E1.2.1 to E1.2.4. These tables list the number of bits that are required for satisfactory (learning) performance and briefly describe the core idea of the algorithms. In order to give an indication of the quality of the experimental evaluation in the cited articles, two columns listing the number of *artificial* and *real-world* benchmarks on which the algorithms have been tested are also included.

E1.2.2.1 Quantization effects in multilayer neural networks

Most methods deal with the various aspects of limited precision calculation in multilayer networks. These approaches can be divided into three categories corresponding to the three different training modes for neural network hardware:

Off-chip learning. In this case the hardware is not involved in the training process, which is performed on a computer using high precision. The weights resulting from the training process are quantized and then downloaded on the chip. Only the forward propagation pass in the recall phase is performed on-chip which makes these quantization effects amenable for mathematical analysis using a statistical model. Some of the results have been summarized in table E1.2.1; these indicate that the accuracy needed in the on-chip forward pass is around 8 bits. Piché (1995) gives a comparison between Heaviside and sigmoidal multilayer networks, showing that the weight precision required in a Heaviside network is much higher and even doubles when a layer is added to the network. An interesting practical example illustrating that low on-chip accuracy is sufficient when mapping a neural network trained with a high precision onto a chip is the application of the analog ANNA chip to high-speed character recognition (Säckinger *et al* 1992). Here, a high precision (32-bit floating point) network is mapped on the ANNA chip which uses a 6-bit weight resolution and a 3-bit resolution for the neuron inputs and outputs. The chip's recognition accuracy is only slightly less than the one obtained with floating-point calculations.

Chip-in-the-loop learning. In this case the neural network hardware is used during training, but only in forward propagation. The calculation of the new weights is done off-chip on a computer, which downloads the updated weights onto the chip after each training iteration. Several learning algorithms have been proposed that take advantage of the fact that in this way the limited precision only plays a role in the forward propagation pass and that floating point calculations can be used in the backward pass (table E1.2.2). One of the first, and perhaps most successful, weight discretization techniques is of the chip-in-the-loop kind (Fiesler *et al* 1988, 1990). It is suitable for feedforward neural networks, easy to implement, and very flexible in that it can handle a large range of discretizations up to the precision of a few bits only (table E1.2.2). The basic idea is to start with a normal neural network with continuous-valued weights. These weights are discretized using a staircase-shaped *multiple-threshold function* and the so-created discrete weights are then used for the forward propagation pass of the learning rule. The errors obtained, which are based on the difference between the obtained network outputs and the desired target outputs, are subsequently used to update the continuous-valued weights during the backward propagation pass. This scheme is repeated until convergence is obtained. This flexible weight discretization method has been successfully used in the development of the Apple Newton (Lyon and Yaeger 1996), and in optical neural networks at Mitsubishi, Japan (Takahashi *et al* 1991) and in Switzerland (Saxena and Fiesler 1995, Moerland *et al* 1996). A similar approach has been applied to design neural networks restricted to single power-of-two weights (see section E1.2.2.3) (Marchesi *et al* 1993, Tang and Kwan 1993).

On-chip learning. Here, the training of the neural network is done entirely on-chip which offers the possibility of continuous training. This means specifically that at least the weight values are represented with only a limited precision. Simulations have shown that the popular backpropagation algorithm (see for example the article by Rumelhart *et al* (1986)) is highly sensitive to the use of limited-precision weights and that training fails when the weight accuracy is lower than 16 bits (first two references in table E1.2.3). This is mainly because the weight updates are often smaller than the quantization step which prevents the weights from changing. In order to reduce the chip area needed for weight storage and to overcome system noise, a further reduction of the number of allowed weight values is desirable. Several weight discretization algorithms have therefore been designed and an extensive list of them and the attainable reduction in required precision is given in table E1.2.3. Some of these weight discretization algorithms have already proven their usefulness in hardware implementations. Battiti's reactive tabu search, for example, has been implemented in the TOTEM processor and successfully applied to a triggering problem in high-energy physics with a weight accuracy as low as 4 bits (Battiti and Tecchiolli 1994). Recently, an analog electronic chip (Kakadu) has been applied successfully to some classification problems by training it with the combined search algorithm and semiparallel weight perturbation algorithms using only a 6-bit weight accuracy (Jabri 1994, Leong and Jabri 1995).

E1.2.2.2 Quantization effects in other neural network models

Also for other neural network models the effects of a coarse quantization of the weight values on recall and learning have been investigated. The small number of weight discretization algorithms proposed can be partly explained from the fact that the required accuracy for successful learning in these models is lower than for gradient descent learning in multilayer networks (table E1.2.4). An interesting example of a hardware implementation is Bellcore's implementation of a Boltzmann machine and mean-field learning, which allows on-chip learning with only 5-bit weights (Alspector 1992). Recently, a weight discretization algorithm for an associative memory with binary $\{-1, +1\}$ weights has been implemented on a digital VLSI chip (Hendrich 1996). The pattern storage capacity that can be obtained with this learning rule is good (0.4 times the number of neurons) and the algorithm is suited for on-chip learning. Verleysen's associative memory training algorithm, that uses the Simplex method to train a network with ternary weights, is best suited for off-chip training (Verleysen *et al* 1989).

Table E1.2.3. Weight discretization in multilayer neural networks: on-chip learning.

Reference	Accuracy (bits)	No of benchmarks		Remarks
		Artificial	Real world	
Asanović (1991)	16	–	1	Coarse weight quantization in the backpropagation algorithm
Holt and Hwang (1993)	14–16	2	–	An error analysis of backpropagation with finite precision
Grossman (1990)	1	1	–	Adaptation of both weights and the internal representation of the neurons
Reyneri and Filippi (1991)	9–10	1	1	Batch backpropagation with a near-optimum learning rate
Xie and Jabri (1992)	10	–	2	Weight perturbation with gain adaptation
Xie and Jabri (1992)	9	–	2	Combination of weight perturbation and a partial random search
Abramson (1991)	2	3	–	A slight modification of the method of Grossman (1990) to train sparsely connected Heaviside networks
Sakaue et al (1993)	8–10	–	2	A weighted error function in the backpropagation algorithm based on an overestimation of the error
Hollis and Paulos (1994)	13	1	–	Weight perturbation with an adaptive gain and learning rate
Jabri (1994)	6	1	1	Semi-parallel weight perturbation algorithms
Simard and Graf (1994)	16	–	1	Backpropagation without multiplication; gradients and states of power-of-two
Battiti and Tecchiolli (1995)	1–8	1	2	Heuristic method for solving combinatorial optimization problems
Dündar and Rose (1995)	10	2	–	Backpropagation with forced weight updates

E1.2.2.3 Some remarks on commonly used schemes

A common point of many weight discretization algorithms is the way in which the effects of having only a limited *weight range* are treated. It has been shown by simulations that as soon as the range of the weights decreases below a certain value, which depends on the problem at hand, the training fails to converge because of the clipping of the weight values (Hoehfeld 1992). This can often be solved by allowing a dynamic rescaling of the weights (and hence the weight range) by adapting the gain β of the activation function. The calculation of an activation value a_j in a multilayer network is namely done as follows:

$$a_j = \phi\left(\beta\left(\sum_i w_{i,j} a_i\right)\right). \tag{E1.2.1}$$

Table E1.2.4. Weight discretization in other neural network models.

Reference	Accuracy (bits)	No of benchmarks		Remarks
		Artificial	Real world	
Self-organizing map, see Kohonen (1989)				
Kohonen (1993)	3–4	–	1	Quantization of input values during recall
Rueping et al (1994)	4	2	1	Power-of-two adaptation factor and quantized weights
Thiran et al (1994)	5	1	–	Uses a conical neighborhood function instead of a rectangular one
Associative memory, see Hopfield (1982)				
Verleysen et al (1989)	2	1	–	A linear programming learning algorithm for associative memories
Johannet et al (1992)	9–11	1	–	Integer arithmetics for learning in associative memory
Hendrich (1996)	1	1	–	Associative memory with binary weights and a good storage capacity
Boltzmann network (Ackley et al 1995)				
Balzer et al (1991)	6–8	2	–	Coarse quantization of the weights during learning
Alspector et al (1992)	5	2	–	Coarse weight quantization for Boltzmann and mean-field learning
Neocognitron (Fukushima 1980)				
White and Elmasry (1992)	3	1	–	Uses power-of-two weights
Cascade topology (Fahlman and Lebiere 1990)				
Hoehfeld and Fahlman (1992)	12	2	1	Coarse weight quantization in the cascade correlation algorithm
Hoehfeld and Fahlman (1992)	6	2	1	Cascade correlation with probabilistic rounding and variable gain
Campbell and Perez Vincente (1995)	1	2	1	A constructive algorithm for Heaviside cascade networks

Thus, a change of the weight range is equivalent to changing the gain β of the activation function. Various strategies have been proposed to perform this gain adaptation, ranging from heuristics based on the average value of the incoming connections to a neuron (Hoehfeld 1992, Xie and Jabri 1992), to approaches that use some form of gradient descent to train the gains (Tang and Kwan 1993, Coggins and Jabri 1994).

In some training algorithms the weight values have been limited to powers-of-two (White and Elmasry 1992, Tang and Kwan 1993, Marchesi et al 1993). The main advantage of this technique is that all costly multiplications can be replaced by easy to implement shift operations. This scheme has also been applied to gradient values, activation values, and learning rates (Hollis and Paulos 1994, Simard and Graf 1994).

Work on limiting the number of weight levels has also been done in the design of Heaviside networks for the computation of boolean functions (majority, parity, comparison, addition) and for the two-spiral problem (Beiu 1996a, 1997). Beiu's concern is to minimize the total number of bits required to represent the weights of a network, since this is a realistic measure of the complexity of VLSI implementations. Moreover, it opens up the possibility of comparing results obtained by learning algorithms with the entropy (*number of bits*) upper bounds of the data set (Beiu 1996b).

Finally, we would like to point out that a comparative benchmarking study of quantization effects on different neural network models and the improvements that can be obtained by weight discretization algorithms has not yet been done. The accuracies listed in table E1.2.1 to E1.2.4 are therefore highly biased by the different benchmarks that were used by the various authors.

E1.2.3 Hardware nonidealities

Both in analog electronic and optical neural network implementations, computation suffers from drawbacks which do not play an important role in digital hardware. Some characteristic examples of such nonidealities inherent to analog computation are the spatial nonuniformity of components and nonideal responses. In this section, examples of these nonidealities are presented, together with their effects on the learning behavior of neural networks.

E1.2.3.1 Component nonuniformity

Variations between the on-chip components, such as multipliers (Cairns and Tarassenko 1994) and the read-out of optical weight matrices (Robinson and Johnson 1992), are inevitable in analog hardware. These nonuniformities are particularly troublesome when the training of the network is done off-chip without taking these component variations into account (Frye et al 1991). It is, however, widely claimed that chip-in-the-loop or on-chip learning can compensate to a considerable extent for these nonuniformities (Card and Schneider 1992). This is also intuitively clear because the use of the analog circuit in the forward pass incorporates the nonuniformities in the learning process. This has been confirmed by experimental results, for example for on-chip learning in backpropagation networks (Cairns and Tarassenko 1994, Dolenko and Card 1995). Their research indicates that backpropagation learning can adapt to the nonuniformity of multiplier gains which are caused by fabrication inaccuracies. The occurrence of additive offsets in the multiplications and especially in weight adaptations do pose serious problems which are not easily overcome by on-chip learning (Dolenko and Card 1995). A possible solution is the use of some dedicated hardware in the weight adaptation circuitry which enables offset-compensation (Annema and Wallinga 1995).

E1.2.3.2 Nonideal response

Computations performed in hardware are approximations of the mathematical operations assumed to be ideal in neural network models. This affects in particular the analog implementation of a linear multiplication and the implementation of a nonlinear activation function like the widely used standard sigmoid. The use of a linear multiplier with a reasonable operating range leads to a large area penalty in VLSI implementations. Therefore, simple nonlinear multipliers are often preferable and are used in both electronic (Lont and Guggenbühl 1992, Hollis and Paulos 1994, Reyneri 1995) and optical implementations (Robinson and Johnson 1992, Neiberg and Casasent 1994). The claims on the learning behavior of a neural network with nonlinear multipliers are rather contradictory. While Cairns and Tarassenko (1994) and Dolenko and Card (1995) find the straightforward use of nonlinear multipliers in simulations of on-chip learning in analog backpropagation networks leads to satisfactory results, Lont and Guggenbühl (1992) find the standard backpropagation algorithm fails to converge with nonlinear synapses. Instead, Lont proposes to incorporate nonlinear multipliers in the formulation of the backpropagation rule, which leads to good results. A disadvantage of this approach is that an accurate model of the on-chip multiplier is needed. This can be alleviated by *chain rule perturbation learning* (Hollis and Paulos 1994), which only performs a forward pass through a multilayer network and hence incorporates the hardware characteristics directly into the training. A solution sometimes applied in optical networks is the use of an additional weight mask which complements and thereby compensates for the nonlinearities in the multiplier (Neiberg and Casasent 1994).

Another problem for analog hardware is the requisite of an activation function that is similar to the standard sigmoid. The incorporation of a model of a sigmoid-like hardware activation function in the training algorithm can compensate for some inaccuracy (Lont and Guggenbühl 1992). This is another example of the opportunism that often plays a role in the design of neural hardware: search for the hidden advantages of apparent drawbacks and try to exploit these instead of trying to approximate the existing mathematical model as closely as possible. Another approach is the use of a simplified activation function, for example the replacement of the Gaussian function in *radial basis networks* by a triangular one (Dogaru et al 1996), leading to a simplified hardware implementation. Additional difficulties arise

C1.6.2

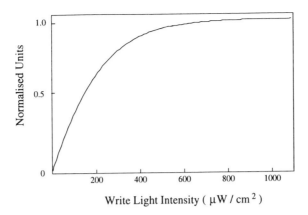

Figure E1.2.1. Response curve of an LCLV.

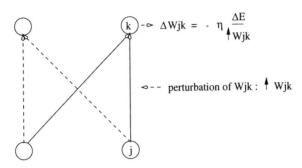

Figure E1.2.2. A schematic of the weight perturbation algorithm.

when the activation functions are implemented by optical hardware, for example in liquid crystal light valves. These optical activation functions are characterized, among other nonidealities, by a gain β that differs greatly from the standard value of one, as can be seen in figure E1.2.1 where a sigmoid with a gain of approximately $1/161$ is depicted (Saxena and Fiesler 1995). While in analog electronics one can try to compensate for a nonstandard gain by including a gain stage, this is not possible in optical implementations. In theory one could add additional optical components whose aim would be a modification of the effective gain, but this would increase the complexity and cost of the system, as well as introducing new side effects. A nice and simple way to solve this problem is by using an adapted backpropagation learning rule that is based on a simple and precise relationship between the gain and two other network parameters (Thimm *et al* 1996), which compensates for a nonstandard gain without any additional hardware, and shows superior results (Moerland *et al* 1995).

E1.2.4 Hardware-friendly learning algorithms

In this section a variety of learning algorithms that are well suited for hardware implementations of neural networks are presented. These *hardware-friendly learning algorithms* (Moerland and Fiesler 1996) can be divided into two classes, namely:

● adaptations of existing neural network learning rules that facilitate their hardware implementation and
● learning algorithms that are by their very conception suitable for hardware implementation.

Here, the emphasis will be on the first of these two classes of hardware-friendly learning algorithms. An example of the second class is *cellular neural networks* which are of special interest for VLSI implementation because of their sparse local connectivity: every unit of the network is a simple analog processor that interacts only with its neighboring units; see the article by Chua and Roska (1993) for a survey. Another example is the class of RAM-based networks which can be easily implemented with standard available components. A recent overview of RAM-based networks and related implementation aspects is given by Austin (1994).

Various hardware-friendlier alternatives have been proposed for several neural network learning rules, especially with the objective to enable *on-chip* learning. The most significant ones are discussed in this section, with an emphasis on hardware-friendly alternatives of the backpropagation algorithm for training multilayer neural networks.

E1.2.4.1 Perturbation algorithms

C1.2.3 The most popular algorithm for the training of multilayer networks is the *backpropagation algorithm* (see for example the book by Rumelhart *et al* (1986)). However, the realization of large backpropagation networks in analog hardware poses serious problems because of the need for separate or bidirectional circuitry for the backward pass of the algorithm. Other problems are the need for an accurate derivative of the activation function and the cascading of multipliers in the backward pass.

The general idea of perturbation algorithms is to obtain a direct estimate of the gradients by a slight random perturbation of some network parameters, using the forward pass of the network to measure the resulting network error. Thus, these on-chip training techniques not only eliminate the complex backward pass but also are likely to be more robust to nonidealities occurring in hardware.

The two main variants of this class of algorithms are *node perturbation* which is based on the
C1.1.4 perturbation of the input value of a neuron, as for example the *madaline-3* rule (Widrow and Lehr 1990), and *weight perturbation*, see for example the article by Jabri and Flower (1992). The basic concepts of weight perturbation (figure E1.2.2) are easily explained by the observation that the gradient descent weight update can be approximated by finite differences ($\uparrow W_{jk}$ denotes the perturbation or change of W_{jk}):

$$\Delta W_{jk} = -\eta \cdot \frac{\partial E}{\partial W_{jk}} \approx -\eta \cdot \frac{\Delta E}{\uparrow W_{jk}}. \tag{E1.2.2}$$

The madaline-3 rule is based on an application of the chain-rule that is standard in the derivation of the backpropagation algorithm (s_k denotes the input to neuron k and $\uparrow s_k$ its perturbation):

$$\Delta W_{jk} = -\eta \cdot \frac{\partial E}{\partial W_{jk}} = -\eta \cdot \frac{\partial E}{\partial s_k} \cdot \frac{\partial s_k}{\partial W_{jk}} \approx -\eta \cdot \frac{\Delta E}{\uparrow s_k} \cdot a_j. \tag{E1.2.3}$$

The main disadvantage of these perturbation algorithms is their sequential nature, as opposed to the weight update calculation in the backpropagation algorithm which can, in principle, be performed in parallel. The main differences between the madaline-3 rule and weight perturbation are the simpler addressing and routing circuitry needed for the latter and the lower computational complexity of the madaline-3 rule. As can be seen in table E1.2.3, weight perturbation also has a good performance with limited precision weights (Xie and Jabri 1992). Moreover, it is more robust against nonidealities occurring in analog hardware: nonuniformity, nonideal circuit response, and noise (Cairns and Tarassenko 1994). The reason for this is that in this algorithm modeling of activation functions and multipliers does not need to be done, since these form an integral part of the training algorithm. It is interesting to note that the derivation of the madaline-3 rule does assume the multiplication to be linear which makes possible the reduction of $\partial s_k / \partial W_{jk}$ to a_j in equation (E1.2.3).

The sequential nature of these simple perturbation algorithms has led to more intricate variants which perform some of the calculations in parallel. A simultaneous perturbation of all weights is a promising alternative (Alspector *et al* 1993, Cauwenberghs 1993), even when for a reliable estimate of the gradient the results of several perturbations should be averaged or a very small and accurate perturbation is required. Other variants use a semiparallel perturbation scheme such as *chain rule* perturbation (Hollis and Paulos 1994), *fan-out* or *fan-in–out* perturbation (Jabri 1994), and *summed weight neuron* perturbation (Flower and Jabri 1993). These semiparallel techniques perturb simultaneously all the weights feeding into or leaving one neuron. An experimental comparison of these perturbation algorithms with an analog multilayer perceptron chip (Kakadu) in-the-loop showed that the semiparallel techniques are best suited for effective learning when the accuracy is low (Jabri 1994). The fan-in–out technique showed the best generalization and training convergence results when the weights and weight updates were quantized to 6 bits.

E1.2.4.2 Local learning algorithms

The implementation of a learning rule can be greatly simplified if it only uses information that is locally available (Palmieri *et al* 1993). This feature minimizes the amount of wiring and communication. Since the backpropagation algorithm is not local, several local learning algorithms have been designed that avoid a global backpropagation of error signals. An example is an anti-Hebbian learning algorithm that is suitable for optical neural networks (Psaltis and Qiao 1993). The weight updates in this algorithm depend only on the input and output of that layer and one global error signal. Although it is not a steepest-descent rule, it is still guaranteed that the weights are updated in the descent direction. Another local learning rule has been developed by Brandt and Lin (1994) which uses only the rates of change of the outgoing weights of a neuron. One of their algorithms is mathematically equivalent to the backpropagation algorithm, but the measurement of the rates of change of the weights could be hard to implement. A promising approach is taken in the Alopex algorithm (Venugopal and Pandya 1991, Unnikrishnan and Venugopal 1994) which is a stochastic algorithm based on the correlation between individual weight changes and changes in the network's error measure. The main advantages of this approach are that the weights can be updated synchronously and that no modeling of the multipliers and activation functions is needed.

E1.2.4.3 Networks with Heaviside functions

The design of a compact digital neural network can be simplified considerably when Heaviside functions are used as activation functions instead of a differentiable sigmoidal activation function. While training algorithms for perceptrons with Heaviside functions abound, training multilayer networks with nondifferentiable Heaviside functions requires the development of new algorithms. One of the earliest examples of such a learning rule is the *madaline-2* rule (Widrow and Lehr 1990), which is closely related to the previously described madaline-3 rule. It is also based on a slight perturbation of the input to a neuron, but in this case the training error is minimized by investigating the effect of an inversion of the activation value of a neuron. If this inversion reduces the Hamming error on the output neurons, the incoming weights of the inverted neuron are adapted with a perceptron training algorithm to reinforce this inversion.

There is also a large variety of *constructive* algorithms which gradually build a Heaviside network by adding neurons and weights (Śmieja 1993). The basis of these algorithms is often formed by a perceptron algorithm that is used to adapt the weights into the freshly added neurons. Recently, some digital and mixed analog/digital architectures have been designed to be suitable for the implementation of a range of these constructive algorithms (Moreno Arostegui 1995).

E1.2.4.4 Robustness

In section E1.2.3 several examples have already been given of the robustness of neural networks to hardware nonidealities. Some research has also been devoted to the robustness of a network to unreliable neurons. This unreliability can consist of sign inversions of hidden neuron values (Judd and Munro 1993) or destruction of hidden neurons (Kerlirzin and Réfrégier 1995). While neural networks trained by standard learning algorithms are not *inherently* fault tolerant, the incorporation of the expected faults in the training phase leads to remarkable improvements. An illustration of this fact is an adaptation of the backpropagation learning rule that uses only a random subset of hidden neurons for each iteration. The trained network is far more robust to the destruction of hidden neurons and shows performance comparable to the noiseless case (Kerlirzin and Réfrégier 1995). This is closely related to the injection of random noise in the weight values during the training of a multilayer neural network, whose effects have been elaborately discussed by Murray and Edwards (1994). It is demonstrated both analytically and experimentally that this synaptic noise improves the network's fault tolerance to weight damage, generalization to unseen patterns, and training time. Similar results have been obtained when injecting additive noise into the weights of recurrent neural networks (Jim *et al* 1994).

E1.2.4.5 Other hardware-friendly neural network models

Although the majority of neural hardware is concerned with the implementation of multilayer networks, because of their wide-ranging applicability, most other popular neural network models have also been implemented in hardware. A few examples of the use of hardware-friendly learning in self-organizing feature maps and recurrent networks are given here.

Self-organizing maps. One of the requisites of a neural network hardware implementation is the effective C2.1.1 use of the processor resources. In general, batch processing is an appropriate alternative to obtain better parallelisation. Kohonen's original algorithm, however, has both an on-line selection of the neuron closest to the input pattern, the *winner neuron*, and an on-line weight update. Two possible variants are to have a batch winner selection combined with either a batch or an on-line weight update. Vassilas *et al* (1995) show the convergence properties of these two variants to be comparable with those of the original algorithm.

C1.4   *Recurrent networks.* Two widely used paradigms for training recurrent networks are *Boltzmann machine* learning and mean field theory learning. The parallelism of a potential hardware implementation is seriously hampered by the required asynchronous update of the neurons. Therefore, in both analog (Pujol *et al* 1994) and optical (Peterson *et al* 1990) implementations, a synchronous neuron update is used. Another C1.4.2 characteristic of the Boltzmann machine is the use of *simulated annealing* to gradually increase the gain of a neuron's activation function. In Bellcore's implementation of a Boltzmann machine this annealing schedule has been replaced by a gradual decrease of additive noise (Alspector 1992), while the main idea of mean field theory learning is to replace the annealing strategy by a deterministic approximation.

E1.2.5   Summary and conclusions

In this section an overview has been given of a variety of adaptations of neural network learning to enable their successful hardware implementation. These problems can be as general as the effects of a quantization of the network parameters or those of the nonidealities of hardware components. Other problems are more specific for a certain neural network model, such as the complications related to the implementation of the backward pass of the standard backpropagation algorithm.

The effects of quantization on a range of neural network models have been outlined, and weight discretization algorithms have been reviewed. These estimations of the required accuracy for well-known learning algorithms and several of the weight discretization algorithms described are already in use in some large-scale hardware implementations. Designers of digital neurocomputers, for example, profit from the fact that the required weight accuracy for backpropagation training is around 16 bits (Mauduit *et al* 1992). An example of a successful implementation of a weight discretization algorithm is Battiti's TOTEM-chip which uses a weight accuracy of 4 bits (Battiti and Tecchiolli 1994).

Compared to the state of the art in digital neural network implementations, the design of analog neural network implementations with nonidealities such as component nonuniformity, nonideal responses, and system noise, is still in a more experimental state. Implementations have therefore been limited to small-scale networks (Leong and Jabri 1995) and it is yet to be shown whether reliable large networks can be realized in practice by analog techniques. An important step towards this goal could be the possibility of on-chip learning, since it has been exemplified that neural network models are remarkably robust to hardware nonidealities when these are incorporated in the training of the network. The development of hardware-friendly learning rules that form an alternative for algorithms which are intricate to implement, like the backpropagation algorithm, is therefore essential. The efficacy of perturbation algorithms illustrates the usefulness of this approach and the first implementations using these training algorithms are emerging (Leong and Jabri 1995).

References

Abramson S, Saad D and Marom E 1993 Training a neural network with ternary weights using the CHIR algorithm *IEEE Trans. on Neural Networks* **4** 997–1000

Ackley D H, Hinton G E and Sejnowski T J 1985 A learning algorithm for Boltzmann machines *Cogn. Sci.* **9** 147–69

Alspector J, Jayakumar A and Luma S 1992 Experimental evaluation of learning in a neural microsystem *Advances in Neural Information Processing Systems (NIPS91)* vol. 4, (San Mateo, CA: Morgan Kaufmann) pp 871–78

Alspector J, Meir R, Yuhas B and Jayakumar A 1993 A parallel gradient descent method for learning in analog VLSI neural networks *Advances in Neural Information Processing Systems (NIPS92)* vol. 5 (San Mateo, CA: Morgan Kaufmann) pp 836–44

Annema A J and Wallinga H 1995 Analog weight adaptation hardware *Neural Processing Lett.* **2** 1–4

Asanović K and Morgan N 1991 Experimental determination of precision requirements for back-propagation training of artificial neural networks *Proc. 2nd Int. Conf. MicroNeuro'91, München, Germany, October 1991* ed U Ramacher, U Rückert and J A Nossek pp 9–15

Austin J 1994 A review of RAM based neural networks *Proc. 4th Int. Conf. on Microelectronics for Neural Networks and Fuzzy Systems, Turin, Italy, September 26–28, 1994* pp 58–66

Balzer W, Takahashi M, Ohta J and Kyuma K 1991 Weight quantization in Boltzmann machines *Neural Networks* **4** 405–9

Battiti R and Tecchiolli G 1994 TOTEM: A digital processor for neural networks and reactive Tabu search *Proc. 4th Int. Conf. on Microelectronics for Neural Networks and Fuzzy Systems, Turin, Italy, September 26–28, 1994* pp 17–25

——1995 Training neural nets with the reactive Tabu search *IEEE Trans. on Neural Networks* **6** 1185–200

Beiu V 1996a Direct synthesis of neural networks *Proc. 5th Int. Conf. on Microelectronics for Neural Networks and Fuzzy Systems, Lausanne, Switzerland, February 12–14, 1996* pp 257–64

——1996b Entropy bounds for classification algorithms *Neural Network World* **6** 497–505

——1997 *VLSI Complexity of Discrete Neural Networks* (New York: Gordon and Breach) in press

Brandt R D and Lin F 1994 Supervised learning in neural networks without explicit error back-propagation *Proc. 32nd Allerton Conf. on Communication, Control, and Computing, Monticello, Illinois, September 28–30, 1994* pp 294–303

Cairns G and Tarassenko L 1994 Learning with analogue VLSI MLPs *Proc. 4th Int. Conf. on Microelectronics for Neural Networks and Fuzzy Systems, Turin, Italy, September 26–28, 1994* pp 67–76

Campbell C and C Perez Vincente 1995 The target switch algorithm: a constructive learning procedure for feed-forward neural networks *Neural Comput.* **7** 1245–64

Card H C and Schneider C R 1992 Analog CMOS neural circuits—*in situ* learning *Int. J. Neural Syst.* **3** 103–24

Cauwenberghs G 1993 A fast stochastic error-descent algorithm for supervised learning and optimization *Advances in Neural Information Processing Systems (NIPS92)*, vol. 5 (San Mateo, CA: Morgan Kaufmann) pp 244–51

Chua L O and Roska T 1993 The CNN paradigm *IEEE Trans. on Circuits and Systems-I: Fundamental Theory and Applications* **40** 147–56

Chua L O and Yang L 1988 Cellular neural networks: theory *IEEE Trans. on Circuits and Systems* **35** 1257–72

Coggins R and Jabri M 1994 Wattle: A trainable gain analogue VLSI neural network *Advances in Neural Information Processing Systems (NIPS93)* vol. 6 (San Mateo, CA: Morgan Kaufmann) pp 874–81

Dogaru R, Murgan A T, Ortmann S and Glesner M 1996 A modified RBF neural network for efficient current-mode VLSI implementation *Proc. 5th Int. Conf. on Microelectronics for Neural Networks and Fuzzy Systems, Lausanne, Switzerland, February 12–14, 1996* pp 265–70

Dolenko B K and Card H C 1995 Tolerance to analog hardware of on-chip learning in backpropagation networks *IEEE Trans. on Neural Networks* **6** 1045–52

G Dündar and Rose K 1995 The effects of quantization on multilayer neural networks *IEEE Trans. on Neural Networks* **6** 1446–51

Fahlman S E and Lebiere C 1990 The cascade-correlation learning architecture *Advances in Neural Information Processing Systems (NIPS89)* vol. 2 (San Mateo, CA: Morgan Kaufmann) pp 524–32

Fiesler E, Choudry A and Caulfield H J 1988 Weight discretization in backward error propagation neural networks *Neural Networks* **1** 380 (special supplement with 'Abstracts 1st Annual (INNS) Meeting')

——1990 A weight discretization paradigm for optical neural networks *Proc. Int. Congr. on Optical Science and Engineering* SPIE vol 1281 (Bellingham, WA: SPIE) pp 164–73

Flower B and Jabri M 1993 Summed weight neuron perturbation: an $O(N)$ improvement over weight perturbation *Advances in Neural Information Processing Systems (NIPS92)* vol. 5 (San Mateo, CA: Morgan Kaufmann) 212–9

Frye R C, Rietman E A, and Wong C C 1991 Back-propagation learning and nonidealities in analog neural network hardware *IEEE Trans. on Neural Networks* **2** 110–17

Fukushima K 1980 Neocognitron: a self-organizing neural network model for a mechanism of pattern recognition unaffected by shift in position *Biol. Cybernet.* **36** 193–202

Grossman T 1990 The CHIR algorithm for feedforward networks with binary weights *Advances in Neural Information Processing Systems (NIPS89)* vol. 2 (San Mateo, CA: Morgan Kaufmann) pp 516–23

Hendrich N 1996 A scalable architecture for binary couplings attractor neural networks *Proc. 5th Int. Conf. on Microelectronics for Neural Networks and Fuzzy Systems, Lausanne, Switzerland, February 12–14* (Los Alamitos, CA: IEEE Computer Society Press) pp 117–124

Hoehfeld M H and Fahlman S 1992 Learning with limited numerical precision using the cascade-correlation algorithm *IEEE Trans. on Neural Networks* **3**

Holler M, Tam S, Castro H and Benson R 1989 An electrically trainable artificial neural network (ETANN) with 10240 'floating gate' synapses *Proc. Int. Joint Conf. on Neural Networks (IJCNN89), Washington, DC* vol. 2, pp 191–6

Hollis P W and Paulos J J 1994 A neural network learning algorithm tailored for VLSI implementation *IEEE Trans. on Neural Networks* **5** 784–91

Holt J L and J-N Hwang 1993 Finite precision error analysis of neural network hardware implementations *IEEE Trans. on Computers* **42** 1380–9

Hopfield J J 1982 Neural networks and physical systems with emergent collective computational abilities *Proc. National Academy of Sciences USA* **79** 2554–8

Jabri 1994 Practical performance and credit assignment efficiency of analog multi-layer perceptron perturbation based training algorithms *SEDAL Technical Report 1-7-94* Systems Engineering and Design Automation Laboratory, Sydney University Electrical Engineering, NSW 2006, Australia

Jabri M and Flower B 1992 Weight perturbation: an optimal architecture and learning technique for analog vlsi feedforward and recurrent multilayer networks *IEEE Trans. on Neural Networks* **3** 154–7

Jim K, Giles C L and Horne B G 1994 Synaptic noise in dynamically-driven recurrent neural networks: convergence and generalization *Technical report UMIACS-TR-94-89 / CS-TR-3322* Institute for Advanced Computer Studies, University of Maryland, College Park, MD 20742, USA

Johannet A, Personnaz L, Dreyfus G, J-D Gascuel and Weinfeld M 1992 Specification and implementation of a digital Hopfield-type associative memory with on-chip training *IEEE Trans. on Neural Networks* **3** 529–39

Judd S and Munro P W 1993 Nets with unreliable hidden nodes learn error-correcting codes *Advances in Neural Information Processing Systems (NIPS92)* vol 5 (San Mateo, CA: Morgan Kaufmann) pp 89–96

Kerlirzin P and Réfrégier P 1995 Theoretical investigation of the robustness of multilayer perceptrons: analysis of the linear case and extension to nonlinear networks *IEEE Trans. on Neural Networks* **6** 560–71

Kohonen T 1989 *Self-Organization and Associative Memory* 3rd edn (Berlin: Springer Verlag)

——1993 Things you haven't heard about the self-organizing map *Proc. 1993 IEEE Int. Conf. on Neural Networks, San Francisco, California, March 28–April 1, 1993* vol. 3, pp 1147–56

Leong P H W and Jabri M A 1995 A low-power VLSI arrhythmia classifier *IEEE Trans. on Neural Networks* **6** 1435–45

Lont J and Guggenbühl W 1992 Analog CMOS implementation of a multilayer perceptron with nonlinear synapses *IEEE Trans. on Neural Networks* **3** 385–92

Lyon R F and Yaeger L S 1996 On-line hand-printing recognition with neural networks *Proc. 5th Int. Conf. on Microelectronics for Neural Networks and Fuzzy Systems, Lausanne, Switzerland, February 12–14, 1996* pp 201–12

Marchesi M, Orlandi G, Piazza F and Uncini A 1993 Fast neural networks without multipliers *IEEE Trans. on Neural Networks* **4** 53–62

Mauduit N, Duranton M, Gobert J and J-A Sirat 1992 Lneuro 1.0: a piece of hardware lego for building neural network systems *IEEE Trans. on Neural Networks* **3** 414–22

Moerland P and Fiesler E 1996 Hardware-friendly learning algorithms for neural networks: an overview *Proc. 5th Int. Conf. on Microelectronics for Neural Networks and Fuzzy Systems, Lausanne, Switzerland, February 12–14, 1996* pp 117–24

Moerland P, Fiesler E and Saxena I 1995 The effects of optical thresholding in backpropagation neural networks *Proc. Int. Conf. on Artificial Neural Networks (ICANN95), Paris, France, October 9–13, 1995* vol. 2, pp 339–43

——1996 Multilayer neural networks for all-optical implementation, in preparation

Moreno Arostegui J M 1995 VLSI architectures for evolutive neural models *PhD Thesis* Technical University of Catalunya, Department of Electronics Engineering, Barcelona, Spain

Murray A F and Edwards P J 1994 Enhanced MLP performance and fault tolerance resulting from synaptic weight noise during training *IEEE Trans. on Neural Networks* **5** 792–802

Neiberg L and Casasent D 1994 High-capacity neural networks on nonideal hardware *Appl. Opt.* **33** 7665–75

Palmieri F, Zhu J and Chang C 1993 Anti-Hebbian learning in topologically constrained linear networks: a tutorial *IEEE Trans. on Neural Networks* **4** 748–61

Peterson C, Redfield S, Keeler J D and Hartman E 1990 An optoelectronic architecture for multilayer learning in a single photorefractive crystal *Neural Comput.* **2** 25–34

Piché S W 1995 The selection of weight accuracies for madalines *IEEE Trans. on Neural Networks* **6** 432–45

Protzel P W, Palumbo D L and Arras M K 1993 Performance and fault-tolerance of neural networks for optimization *IEEE Trans. on Neural Networks* **4** 600–14

Psaltis D and Qiao Y 1993 Adaptive multilayer optical networks *Progress in Optics* vol. 31, ed E Wolf (Amsterdam: Elsevier) ch 4, pp 227–61

Pujol H, Klein O, Belhaire E and Garda P 1994 RA: an analog neurocomputer for the synchronous Boltzmann machine *Proc. 4th Int. Conf. on Microelectronics for Neural Networks and Fuzzy Systems, Turin, Italy, September 26–28, 1994* pp 449–55

Reyneri L M and Filippi E 1991 An analysis on the performance of silicon implementations of backpropagation algorithms for artificial neural networks *IEEE Trans. on Computers* **40** 1380–9

Reyneri L M 1995 A performance analysis of pulse stream neural and fuzzy computing systems *IEEE Trans. on Circuits and Systems-II: Analog and Digital Signal Processing* **42** 642–60

Robinson M G and Johnson K M 1992 Noise analysis of polarization-based optoelectronic connectionist machines *Appl. Opt.* **31** 263–72

Rueping S, Goser K and Rueckert U 1994 A chip for selforganizing feature maps *Proc. 4th Int. Conf. on Microelectronics for Neural Networks and Fuzzy Systems, Turin, Italy, September 26–28, 1994* pp 26–33

Rumelhart D, Hinton G and Williams R 1986 Learning internal representations by error propagation *Parallel Distributed Processing: Explorations in the Microstructure of Cognition* vol. 1: *Foundations* (Cambridge, MA: MIT Press) pp 318–362

Säckinger E, Boser B E, Bromley J, LeCun Y and Jackel L D 1992 Application of the ANNA neural network chip to high-speed character recognition *IEEE Trans. on Neural Networks* **3** 498–505

Sakaue S, Kohda T, Yamamoto H, Maruno S and Shimeki Y 1993 Reduction of required precision bits for backpropagation applied to pattern recognition *IEEE Trans. on Neural Networks* **4** 270–4

Saxena I and Fiesler E 1995 Adaptive multilayer optical neural network with optical thresholding *Opt. Eng.* **34** 2435–40

Simard P Y and Graf H P 1994 Backpropagation without multiplication *Advances in Neural Information Processing Systems (NIPS93)* vol. 6, ed J D Cowan, G Tesauro and J Alspector pp 232–39 (San Mateo CA: Morgan Kaufmann)

Śmieja F J 1993 Neural network constructive algorithms: trading generalization for learning efficiency? *Circuits Syst. Signal Processing* **12** 331–74

Takahashi M, Oita M, Tai S, Kojima K and Kyuma K 1991 A quantized back propagation learning rule and its application to optical neural networks *Opt. Comput. Processing* **1** 175–82

Tang C Z and Kwan H K 1993 Multilayer feedforward neural networks with single power-of-two weights *IEEE Trans. on Signal Processing* **41** 2724–7

Thimm G, Moerland P and Fiesler E 1996 The interchangeability of learning rate and gain in backpropagation neural networks *Neural Comput.* **8** 251–60

Thiran P, Peiris V, Heim P and Hochet B 1994 Quantization effects in digitally behaving circuit implementations of Kohonen networks *IEEE Trans. on Neural Networks* **5** 450–8

Unnikrishnan K P and Venugopal K P 1994 Alopex: a correlation-based learning algorithm for feedforward and recurrent neural networks *Neural Comput.* **6** 469

Vassilas N, Thiran P and Ienne P 1995 How to modify Kohonen's self-organizing feature maps for an efficient digital parallel implementation *Proc. Int. Conf. on Artificial Neural Networks, Cambridge, June 26–28, 1995*

Venugopal K P and Pandya A S 1991 Alopex algorithm for training multilayer neural networks *Proc. Int. Joint Conf. on Neural Networks (IJCNN), Singapore, November, 1991* vol 1 pp 196–201

Verleysen M, Sirletti B, Vandemeulebroecke A and Jespers P G A 1989 A high-storage capacity content-addressable memory and its learning algorithm *IEEE Trans. on Circuits and Systems* **36** 762–6

White B A and Elmasry M I 1992 The digi-neocognitron: a digital neocognitron neural network model for VLSI *IEEE Trans. on Neural Networks* **3** 73–85

Widrow B and Lehr M A 1990 30 years of adaptive neural networks: perceptron, madaline, and backpropagation *Proc. IEEE* **78** 1415–42

Xie Y and Jabri M A 1992 Training limited precision feedforward neural networks *Proc. 3rd Australian Conf. on Neural Networks* pp 68–71

E1.3 Analog VLSI implementation of neural networks

Eric A Vittoz

Abstract

This chapter introduces the motivation for doing signal processing by means of analog VLSI, before discussing the peculiarities and implementation constraints of this approach. The possible modes of operation and the model of the MOS transistor are then recalled before identifying the properties of individual transistors and of their basic combinations to be exploited opportunistically in analog circuits. Some implementations of local and collective operators relevant to neural networks are then discussed. The difficult problems of analog storage of synaptic weights and of communication between cells are addressed in the last part.

E1.3.1 Introduction

With modern scale-down VLSI processes, hardware implementations of traditional signal processing such as filtering are progressively changing from *analog* to *digital* circuits. Figure E1.3.1 represents the E1.4 minimum power consumption P_{min} required to implement one pole of filtering at frequency f by digital and by analog circuits. It shows that digital solutions are more efficient with respect to power consumption when the required signal-to-noise ratio (SNR) exceeds 60 to 80 dB. Indeed, signals represented by codes or by numbers can be regenerated at every step of the process, and noise is limited to the effect of quantization. Power is thus only a weak (logarithmic) function of the signal-to-noise ratio. Qualitatively, the horizontal axis can also represent precision and distortion, whereas the vertical axis also applies to chip area. Furthermore, digital systems are easy to design by mapping algorithms onto silicon in a top-down procedure. Therefore, digital implementations are absolutely needed to meet the requirements of systems aiming at the precise *restitution* of information, later in time after storage or elsewhere in space after transmission.

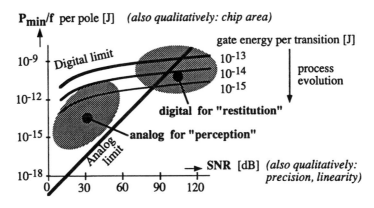

Figure E1.3.1. Minimum power and optimum use of digital and analog processing (Vittoz 1990, 1994a).

Neural networks are intended to carry out tasks of a quite different nature, which are related to the *perception* of an environment or of a situation, on the basis of a large number of data or signals. Here,

---

the need for precision is replaced by that for a massively parallel collective processing of these data, and noise (not to be confused with the quantization noise of digital systems) is not harmful or may even improve the processing speed. As also shown by figure E1.3.1, analog solutions can therefore be expected to be more effective with respect to power consumption and chip area for the implementation of these *perception* systems (Mead 1989, Vittoz 1994b). Analog processing permits continuous time and continuous amplitude. It also allows us to exploit, in an opportunistic manner, all the features offered by the available components, in order to further reduce the area per function. Truly massive parallel computation is then possible in very large structures that can be inspired by biological solutions.

E1.3.2 Characteristics of analog signal processing

Analog processing may first be characterized by the fact that signals are represented not by codes or by numbers, but by physical variables. These may be a voltage, a current, a charge, a frequency, or a time duration. These various modes of signal representation have their respective advantages and drawbacks and can therefore be used in different parts of the same system.

The voltage representation makes it easy to distribute a signal in various parts of a circuit, and to store this signal into a capacitor. Summing voltages requires dedicated active components.

The current representation facilitates the summing of signals. However, it complicates their distribution since replicas must be created which are never exactly equal to the original current. Linear and nonlinear current mode operations are facilitated by powerful techniques such as translinear circuits and pseudoconductance, which will be discussed later.

The charge representation requires time sampling. Packets of charge can be summed or subtracted by means of charge-coupled devices (CCD) or by switched capacitor (SC) techniques (Temes 1987). Multiplication requires a translation into voltage or current representations.

The pulse frequency or time between pulses is used as the dominant mode of signal representation for communication in biological nervous systems. Signals represented in this pseudobinary manner (pulses of fixed amplitude and duration) are easy to regenerate and this representation is therefore preferred for long-distance transfers of information. It is discontinuous in time, but the phase information is kept in asynchronous systems. Alternatively, a representation by pulse duration may facilitate some local operations.

In any case, the value of the physical variable which represents a signal must be related to a reference value. This reference must either be produced internally, with its origin clearly identified to keep it under control with respect to process and ambience variations, or it must be provided from outside. The number of separate references required to implement an operator with k input variables and p output variables may range from none to $k + p$, depending on the signal representations and on the type of operation (Vittoz 1994b). If the operation is time dependent, at least one time reference is needed.

Integrated analog circuits are also characterized by the very poor absolute precision of the available components. The combination of process variations, temperature variations and aging results in a level of absolute precision ranging from 5% in the best case to several hundred per cent for some components. However, all identical components on a chip tend to vary in the same manner, and therefore the relative precision of components is much better, and may be as good as 0.1% for carefully implemented devices.

All circuit architectures should thus be based on *ratios* of matched component values. Matching between components can be improved by respecting the following rules, mostly related to layout (Vittoz 1985):

- Devices to be matched should have the same structure, in order to depend on the same process parameters.

- They should be at the same temperature. This is easily obtained if heat generation on the chip is negligible. Otherwise, matched devices should be placed on the same isothermal curve (by exploiting symmetries), as far as possible from heat sources.

- They should have the same shape and the same size, to minimize the effect of under- or oversizing due to fabrication steps.

- They should be at a minimum distance, in order to eliminate the contribution of spatially correlated fluctuations of parameters.

- Common-centroid shapes should be used to eliminate the effect of constant gradients.

- They should have the same orientation on the chip; this eliminates mismatches due to nonisotropic effects during processing (i.e. shadows) and to nonisotropic physical mechanisms (stress, temperature gradient).
- Devices to be matched should have identical surroundings. Indeed the first and last elements of linear arrays may differ from the other elements; therefore, they should not be used functionally in critical situations. The same is true for the first and last columns and rows in two-dimensional arrays.
- They should have a nonminimum size to provide better averaging on random variations of relevant specific local parameters (mobility, oxide thickness, sheet resistivity, and the like) and on spatially fluctuating width and length.

The relevance and importance of each of these rules depends on the type of device, on the process, and on the required level of precision.

The representation of signals by physical values makes analog circuits very sensitive to physical sources of noise. Most fundamental is the thermally generated noise, which necessitates elevating the signal energy above a minimum level, as shown by figure E1.3.1. However, the noise as well as the limited precision in analog circuits must be clearly distinguished from those due to the limited number of bits in digital circuits: indeed, they are not related to any quantization effect, and thermal noise is truly random. It has been shown that the presence of *random noise* can even be beneficial in improving B3.5.2.8 the learning behavior of analog neural networks (Murray and Edwards 1993). Additional noise, possibly nonrandom, may come from the power supply or from coupling to clock transitions or to other signals generated on the chip.

E1.3.3 Basic components of analog complementary metal oxide semiconductor very large-scale integration

E1.3.3.1 Transistors

The metal oxide semiconductor (MOS) transistor is the most fundamental component of analog VLSI. For most applications related to neural networks, its drain current I_D can be approximated by (Vittoz 1994c)

$$I_D = I_F - I_R \tag{E1.3.1}$$

with

$$I_F = I_S f(V_S, V_G) \qquad \text{(forward component)} \tag{E1.3.2}$$

and

$$I_R = I_S f(V_D, V_G) \qquad \text{(reverse component)} \tag{E1.3.3}$$

where the source voltage V_S, the drain voltage V_D, and the gate voltage V_G are defined with respect to the local substrate (which may be a well), as shown by figure E1.3.2 together with the symbols for p- and n-channel transistors. In practice, the electrode B corresponding to the local substrate is only represented if it is not connected to the positive rail $V+$ for the p-channel, and to the negative rail $V-$ for the n-channel.

The decreasing function $f(V, V_G)$ is shown in the same figure. It can be approximated by

$$f(V, V_G) = \ln^2[1 + e^{(V_P - V)/2U_T}] \qquad \text{with } V = V_S \text{ or } V_D \tag{E1.3.4}$$

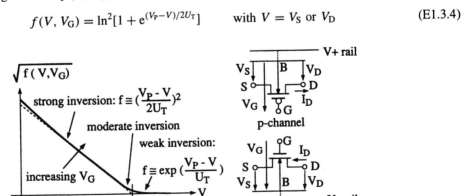

Figure E1.3.2. Modeling and symbols for MOS transistors.

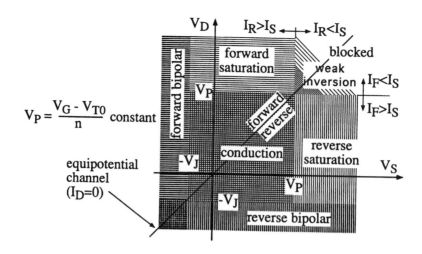

Figure E1.3.3. Modes of operation of a MOS transistor.

where

$$V_P = \frac{V_G - V_{T0}}{n} \tag{E1.3.5}$$

is called the pinchoff voltage and $U_T = kT/q$. In this model, the gate *threshold voltage* V_{T0} is a constant of the process (independent of V_S). It may, however, become dependent on the channel length L and channel width W for very small dimensions (Tsividis 1987). The *slope factor* n is normally smaller than 2 and tends slowly to 1 for very large V_G. The third and last fundamental parameter of the transistor is the *specific current* I_S appearing in (E1.3.2) and (E1.3.3); it is given by

$$I_S = 2n\beta U_T^2. \tag{E1.3.6}$$

It includes the usual transfer parameter

$$\beta = \mu C_{ox} W/L \tag{E1.3.7}$$

where μ is the mobility in the channel, C_{ox} the gate capacitance per unit area, and W/L the effective width-to-length ratio of the gate.

For $V - V_P \ll -U_T$, $f(V, V_G)$ becomes a square law. The corresponding component I_F or I_R given by (E1.3.2) or (E1.3.3) is much larger than I_S, and is said to be in *strong inversion*.

For $V - V_P \gg U_T$, $f(V, V_G)$ becomes exponential. The corresponding component I_F or I_R is much smaller than I_S, and is said to be in *weak inversion*.

The global mode of operation of the transistor depends on the signs of $V_S - V_P$ and $V_D - V_P$ (or on the values of I_F/I_S and I_R/I_S) as represented in figure E1.3.3.

The transistor is in conduction for V_S and V_D smaller than V_P (I_F and I_R larger than I_S); the drain current given by (E1.3.1)–(E1.3.5) can then be approximated by

$$I_D = I_F - I_R = \frac{\beta}{2n}[(V_G - V_{T0} - nV_S)^2 - (V_G - V_{T0} - nV_D)^2] \tag{E1.3.8}$$

where the squared terms are zero for negative arguments. The transistor enters forward (or reverse) saturation when V_D (or V_S) becomes larger than V_P; the second (or respectively the first) term then vanishes and I_D becomes independent of V_D (or V_S).

For V_D and V_S both larger than V_P (I_F and I_R smaller than I_S), the transistor operates globally in weak inversion (also called subthreshold); the drain current given by (E1.3.1)–(E1.3.5) can then be approximated by

$$I_D = I_F - I_R = I_S\, e^{(V_G - V_{T0})/nU_T}(e^{-V_S/U_T} - e^{-V_D/U_T}). \tag{E1.3.9}$$

The drain current I_D in weak inversion also saturates (becomes independent of V_D or V_S) as soon as $|V_D - V_S| \gg U_T$. It decreases exponentially with V_G and $-V_S$ (or $-V_D$) until it becomes so small that the transistor is considered to be blocked.

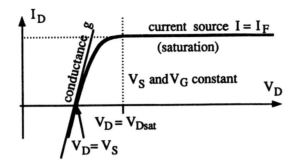

Figure E1.3.4. Generic output characteristics for V_G and V_S constant.

Temperature variations affect the transistor characteristics through the variation of the thermal voltage U_T, and through that of the three parameters V_{T0}, β, and n. The variation of V_{T0} is approximately -2 mV $°C^{-1}$ and dominates the drain current variation in weak inversion. Mobility variations affect the value of β by approximately -0.5% $°C^{-1}$ and tend to dominate in strong inversion. The variation of n of the order of 10^{-3} $°C^{-1}$ is usually negligible.

As also shown by figure E1.3.3, a parasitic bipolar mode appears if the source junction is sufficiently forward biased ($V_S < -V_J$). It is changed into a clean (lateral) bipolar operation if the surface MOS current is blocked by pushing V_P below $-V_J$ (by a very small or negative value of V_G) (Vittoz 1983). This provides a compatible bipolar transistor in which the original source (S), well, and drain (D) terminals become, respectively, the emitter (E), the base (B), and the collector (C). Part of the emitter current of this bipolar transistor flows downwards to the substrate, which plays the role of a second collector.

This vertical grounded collector transistor may also be used independently; the gate and drain structures are then useless and can be removed. The same device can be used as a sensitive photodetector; incident photons create electron–hole pairs which are collected by the base (well) to collector (substrate) junction. If the base is left open, the generated photocurrent is multiplied by the current gain of the transistor.

Coming back to the field effect modes of operation, the generic $I_D(V_D)$ characteristics for V_S and V_G constant are shown in figure E1.3.4, where two qualitatively different behaviors can be identified: voltage-controlled current source and voltage-controlled conductance.

The transistor behaves as a *voltage-controlled current source* $I = I_F$ in forward saturation, that is, for $V_D > V_{Dsat}$ given by (from (E1.3.1)–(E1.3.5))

$$V_{Dsat} = V_S + 3 \text{ to } 4U_T \qquad \text{for } weak\ inversion\ (I \ll I_S) \qquad (E1.3.10)$$

$$V_{Dsat} = V_P = \frac{V_G - V_{T0}}{n} \qquad \text{for } strong\ inversion\ (I \gg I_S). \qquad (E1.3.11)$$

As shown before, the transfer function $I(V_G)$ is exponential in weak inversion and quadratic in strong inversion. Connecting the gate to the drain and imposing a current I provides the inverse functions $V_G(I)$ which are, respectively, *logarithmic* and *square root*.

The small-signal dependence of this current source on the gate voltage can be characterized by the transconductance $g_m = \partial I / \partial V_G$. Equations (E1.3.9) and (E1.3.8) yield

$$g_m = \frac{I}{nU_T} \qquad \text{in } weak\ inversion \qquad (E1.3.12)$$

$$g_m = \frac{(I I_S)^{1/2}}{nU_T} = \left(\frac{2\beta I}{n}\right)^{1/2} = \beta(V_P - V_S) \qquad \text{in } strong\ inversion. \qquad (E1.3.13)$$

Thus, weak inversion provides the maximum possible value of g_m for a given value of current I and the minimum value of saturation voltage V_{Dsat}.

The transistor behaves as a *voltage-controlled conductance* g for $V_D \cong V_S$. It can be easily shown that

$$g = ng_m. \qquad (E1.3.14)$$

E1.3.3.2 Passive components

Capacitors are needed to implement time-dependent processing functions and to store signals or parameters (in particular, synaptic weights). They can be implemented as a sandwich of any pair of conductive layers (metal or polysilicon) separated by a layer of silicon oxide. The thinnest oxide layer available (corresponding to the maximum capacitance per unit area) is often the gate oxide of the MOS transistor. Therefore, a transistor can be used to implement a capacitor of value WLC_{ox}. The source and the drain should be connected to the local substrate and care must be taken to bias the gate sufficiently above threshold V_{T0}, in order to ensure strong inversion of the channel.

Resistors may be needed to convert voltages into currents. The sheet resistivity R_S is usually provided by the polysilicon layer. It does not usually exceed 100 Ω/square, which makes high-value resistors very area consuming.

A transistor operated in conduction provides a sheet resistivity given by (E1.3.13) and (E1.3.14),

$$R_S = \frac{1}{n\mu C_{ox}(V_{Dsat} - V_S)} .$$

(E1.3.15)

This value is inversely proportional to the linear voltage range, as shown by figure E1.3.4. It may reach several 10 kΩ/square for a range 1 V. Several techniques, mostly based on symmetry, have been developed in order to improve the linearity (Tsividis *et al* 1986).

Resistors may also be needed to achieve linear splitting of currents. Such a function may be realized by transistors in the following manner. By defining a *pseudovoltage*

$$V^* = -V_0 f(V, V_G)$$

(E1.3.16)

where V_0 is a scaling voltage of arbitrary value, equation (E1.3.1) may be rewritten as

$$I_D = g^*(V_D^* - V_S^*) .$$

(E1.3.17)

This is the linear Ohm law with a constant *pseudoconductance* of the transistor

$$g^* = I_S/V_0$$

(E1.3.18)

which depends on the width-to-length ratio W/L in I_S. Thus, networks of transistors interconnected by their sources and drains with a common gate voltage behave linearly with respect to currents (Vittoz and Arreguit 1993). The pseudovoltage V^* cannot change sign, but its 0-reference level (pseudoground) is reached as soon as the corresponding real voltage V is large enough to saturate the transistor, which facilitates the extraction of any current flowing to ground. Moreover, if the transistors are maintained in weak inversion, equation (E1.3.9) shows that the dependences on V_G and V_S or V_D can be separated to permit the following alternative definition of V^* and g^*:

$$V^* = -V_0 e^{-V/U_T} \qquad \text{for weak inversion only}$$

(E1.3.19)

$$g^* = \frac{I_S}{V_0} e^{(V_G - V_{T0})/nU_T} \qquad \text{for weak inversion only} .$$

(E1.3.20)

The network of transistors remains linear with respect to currents but the pseudoconductance g^* of each transistor is now *controllable* independently by the value of its gate voltage V_G.

E1.3.3.3 Basic combinations of transistors

The most important and most widespread elementary combinations of transistors are the current mirror and the differential pair illustrated in figure E1.3.5.

The current mirror provides one or more copies of its input current weighted by the ratios of specific currents I_{Si}/I_{S0}. These ratios must be obtained by parallel connections (or even series connections in the same local substrate) of identical unit transistors if precision is needed (Vittoz 1985). Each transistor T_i may also be used as a current-controlled conductance g_i according to (E1.3.12) to (E1.3.14).

The differential pair splits a bias current I_0 into two components I_1 and I_2, the difference of which is an S-shaped function of the input voltage V. The transconductance g_m at the origin is controlled by I_0

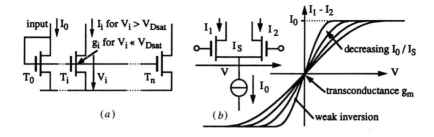

Figure E1.3.5. (a) Current mirror and (b) differential pair.

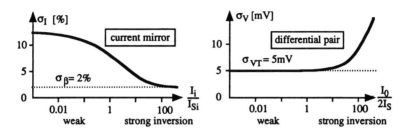

Figure E1.3.6. Effect of mismatch on current mirror and differential pair.

according to (E1.3.12) or (E1.3.13) with $I = I_0/2$. The transfer function in weak inversion obtained from (E1.3.9) is

$$I_1 - I_2 = I_0 \tanh \frac{V}{2nU_T} \tag{E1.3.21}$$

and has the maximum slope for a given value of I_0. The voltage scale is expanded in strong inversion for which saturation is reached at

$$\pm V = 2nU_T (I_0/I_S)^{1/2} = (2nI_0/\beta)^{1/2}. \tag{E1.3.22}$$

The difference $I = I_1 - I_2$ of the two output currents of a differential pair is easily carried out by one or more complementary current mirrors. This results in an operational transconductance amplifier (OTA), the most fundamental block for traditional analog processing (Vittoz 1985).

The characteristics of combinations of transistors are affected by the unavoidable statistical mismatch between devices. For the current mirror, the resulting relative spread of output currents has an RMS value given by (Vittoz 1994c)

$$\sigma_I = \left[\sigma_\beta^2 + \left(\frac{g_{mi}}{I_i} \sigma_{VT} \right)^2 \right]^{1/2} \tag{E1.3.23}$$

where σ_β is the relative mismatch of specific currents (or of transfer parameters β) and σ_{VT} the absolute mismatch of threshold voltages, both in RMS values. In the differential pair, mismatch creates a horizontal offset of the transfer characteristics of RMS value

$$\sigma_V = \left[\sigma_{VT}^2 + \left(\frac{I_0}{2g_m} \sigma_\beta \right)^2 \right]^{1/2}. \tag{E1.3.24}$$

These results are plotted in figure E1.3.6 for typical values for minimum-size devices ($\sigma_T = 5$ mV, $\sigma_\beta = 2\%$, $nU_T = 40$ mV), by using a continuous value of $g_m(I_D)$ computed from (E1.3.2) and (E1.3.4).

It can be seen that the precision of a current mirror is drastically degraded when the transistors operate in weak inversion. If weak inversion is imposed by a very low current level or by the need to minimize the saturation drain voltage V_{Dsat}, much better results are obtained by operating the transistor in the lateral bipolar mode. The offset voltage of a differential pair is limited in weak inversion to the contribution of σ_{VT}.

Another basic combination of transistors is the translinear loop which, in its original form (Gilbert 1975), exploits the exponential transfer characteristics of bipolar transistors to obtain products of currents.

It can also be implemented with a limited precision by MOS transistors operated in weak inversion. A translinear loop made of MOS transistors in a common substrate is illustrated in figure E1.3.7.

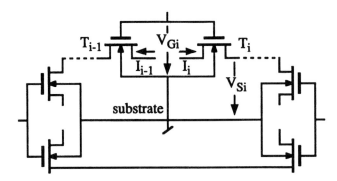

Figure E1.3.7. Translinear loop of MOS transistors in weak inversion (common substrate).

Each transistor in the loop is saturated at a current I_i given by (E1.3.9):

$$I = I_F = I_S \, e^{-V_{T0}/nU_T} \, e^{(V_G/n-V_S)/U_T} \, . \tag{E1.3.25}$$

The loop is made of an even number of transistors connected by their gates and sources, half of them in each direction, clockwise (cw) and counterclockwise (ccw), so that

$$\sum_{cw}(V_{Gi} - V_{Si}) = \sum_{ccw}(V_{Gi} - V_{Si}) \, . \tag{E1.3.26}$$

Now, if the cw and ccw transistors are *alternated*, as in the figure, each gate voltage V_{Gi} is shared by a cw–ccw pair of transistors. It is thus compensated side by side in (E1.3.26) and can therefore be replaced by V_{Gi}/n in this equation. Expressing $V_{Gi}/n - V_{Si}$ as a function of the saturation current I_i according to (E1.3.25) yields

$$\frac{\prod_{cw} I_i}{\prod_{ccw} I_i} = \frac{\prod_{cw} I_{Si}}{\prod_{ccw} I_{Si}} = \lambda \tag{E1.3.27}$$

where the loop factor $\lambda = 1$ for perfectly matched identical transistors.

Circuits exploiting this fundamental result may be built on loops sharing some transistors. The principle can be extended by including any number of transistors connected source to drain in the loop (Andreou and Boahen 1994). These transistors are then no longer saturated and each of them is equivalent to two saturated transistors T_i and T_{i-1} (same gate voltage) connected in parallel to represent the forward and reverse modes. The current flowing through each of these nonsaturated transistors is then $I_i - I_{i-1}$.

If the transistors in the same substrate are not alternated cw–ccw, equation (E1.3.26) is only approximately valid since $n > 1$. Correct translinear operation can be restored by putting each transistor in a separate well connected to its source (all $V_{Si} = 0$), with all transistors saturated.

As shown by figure E1.3.6, currents from MOS transistors operated in weak inversion are not very precise. Furthermore, the slope factor n depends slightly on the gate voltage, and may thus be slightly different from transistor to transistor. Much more precise translinear loops are obtained by using bipolar-operated devices or true bipolar transistors.

E1.3.4 Analog functional blocks

E1.3.4.1 Local operators

Addition/subtraction of signals. The sum of currents is directly provided by applying Kirchhoff's law. If needed, the sign of any current can be changed by a single current mirror.

Summing voltages is usually best obtained by first converting them to currents. If the voltage sources are floating, each conversion may be carried out by the transfer function of a differential pair in an OTA. Several techniques have been developed to increase the range of linearity of the transconductance, mostly for applications in time-continuous filters (Tsividis 1994). Voltage-to-current conversion can be obtained

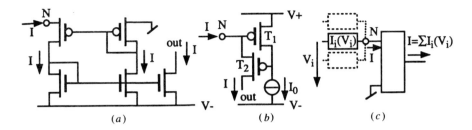

Figure E1.3.8. Current conveyors for weighted sum of voltages.

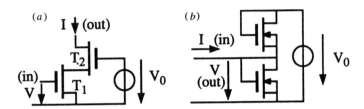

Figure E1.3.9. (a) Linear voltage-to-current and (b) current-to-voltage conversion.

by any resistive element, provided a virtual ground is available to extract the current. This can be achieved by elementary current conveyors such as those shown in figure E1.3.8.

Circuit (a) imposes by symmetry a virtual ground node N at the level of the ground. Version (b) is even simpler, but the virtual ground level is $V_{G1}(I_{F1} = I_0)$ below the positive rail $V+$. Conversion of positive and negative signals can be obtained by adding a bias current at the input of the conveyor and subtracting the same current at the output. The results of several voltage-to-current conversions may be directly added with a single conveyor as shown in part (c) of the figure. The characteristics $I_i(V_i)$ of each dipole may be nonlinear in the general case. It must be a linear conductance for a linear conversion, but this conductance may be different for each input to achieve different weighting before summing. It may be implemented by a transistor operated in conduction according to (E1.3.14); linearity is then maintained if $V_i \ll V_{Dsat}$, and the value may be controlled by the gate voltage. Voltages may be subtracted by converting them by a separate conveyor and by subtracting its output current.

Linear conversion by a transistor may be done without loading the input voltage source by the simple configuration of figure E1.3.9(a).

Transistor T_1 is maintained in conduction by the small drain-to-source voltage imposed by bias V_0 and maintained constant by the large transconductance of follower T_2 (which may alternatively be a bipolar transistor). Then, from the strong-inversion model (E1.3.8):

$$I \cong \frac{\beta_1}{n}(V_0 - V_{T0})(V - V_{T0}) \qquad \text{for} \quad \beta_2 \gg \beta_1 \text{ and } V - V_{T0} \gg V_0 - V_{T0}. \tag{E1.3.28}$$

If needed, the sum/difference of currents may be reconverted into a voltage. A simple and elegant solution shown in figure E1.3.9(b) (Bult and Wallinga 1987) uses two identical transistors operated in saturated strong inversion. The model yields

$$V = \frac{V_0}{2} + \frac{nI}{\beta(V_0 - 2V_{T0})}. \tag{E1.3.29}$$

The standard manner of weighting and adding voltages shown in figure E1.3.10(a) requires a full operational amplifier (low output resistance) and linear resistive elements. Another method is based on switched capacitors as illustrated in part (b) of the same figure. This circuit operates in two clock phases, and the output voltage is only available during the phase shown in the figure. These two classical methods are usually too complicated for applications in neural networks.

Multiplication/division. Multiplication of voltages $(V - V_{T0})$ and $(V_0 - V_{T0})$ is already provided by the voltage-to-current converter of figure E1.3.9(a), and the multiplication of current I_0 by voltage V can be obtained by a single differential pair operated in weak inversion, according to (E1.3.21) with $V \ll 2nU_T$.

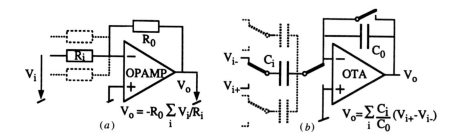

Figure E1.3.10. Classical methods for voltage weighting and summing.

The multiplication of two differential voltages requires three differential pairs operated in strong inversion and configured into a Gilbert multiplier (Gilbert 1968) shown in figure E1.3.11(a).

Application of the model for saturated strong inversion yields

$$I_a - I_b \cong \frac{\beta}{2^{1/2}n}(V_1 - V_2)(V_3 - V_4) \qquad \text{for} \quad (V_1 - V_2) \text{ and } (V_3 - V_4) \ll (2nI_0/\beta)^{1/2}. \qquad \text{(E1.3.30)}$$

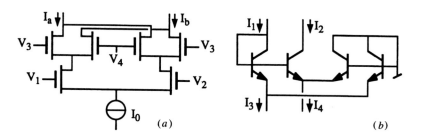

Figure E1.3.11. (a) Voltage and (b) current multipliers.

The multiplication/division of currents is easily obtained by exploiting the translinear principle, as in the example of figure E1.3.11(b) using bipolar transistors. Application of (E1.3.27) with negligible base currents results in

$$I_2 = I_1 I_4/I_3 \qquad \text{for} \quad I_1 \le I_3. \qquad \text{(E1.3.31)}$$

This circuit may be modified to carry out a four-quadrant multiplication (Gilbert 1983).

Vector length calculation. Calculation of the length of a two-dimensional vector (or of the Euclidean distance between two points) can be obtained by the translinear circuit of figure E1.3.12(a) (Gilbert 1981).

This single-quadrant circuit can be extended to n-dimensional calculation by creating n translinear loops $T_{bi} - T_{ai} - T_{ci} - T_0$, each accepting one component I_i of the vector. It can also be implemented by means of MOS transistors if the poor precision to be expected from weak inversion is acceptable. An alternative solution using the square law characteristics of MOS transistors in strong inversion is depicted in figure E1.3.12(b) (Landolt *et al* 1992). It requires $4n + 8$ unit transistors to compute the n-dimensional vector length in all quadrants.

E1.3.4.2 Collective operators

Fast collective processing of a large number of signals is made possible by circuits capable of truly parallel computation, based on simple repetitive cells, that communicate through only one or a few interconnections. Several examples are given here.

Normalization of signals. To permit the comparison of a set of signals in a wide range of levels, it is useful to normalize these signals to obtain a predetermined total value I_{tot}. This operation is performed by the circuit shown in figure E1.3.13 (Gilbert 1984).

All cells have the same voltage difference V between the emitters of their input and output transistors. Therefore, thanks to the exponential transfer characteristics, each cell provides the same current gain I_{out}/I_{in}, whereas the sum of all output currents is imposed as I_{tot}. Alternatively the gain may be imposed

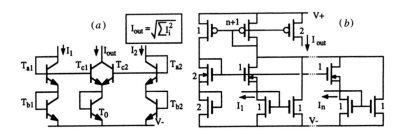

Figure E1.3.12. Circuits for vector length calculation.

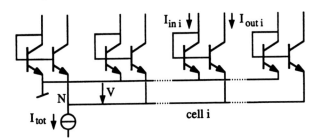

Figure E1.3.13. Normalization of a set of signals.

by imposing the ratio of the sums of input and output currents. The input node of one cell must then be grounded to ensure proper biasing, and the circuit is a generalization of the multiplier of figure E1.3.11(*b*). Bipolars may be replaced by MOS transistors in weak inversion if precision is not required.

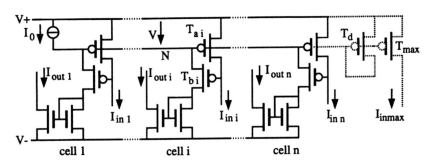

Figure E1.3.14. Winner-take-all circuit and extraction of maximum current.

Winner-take-all. As illustrated in figure E1.3.14, only $4n$ transistors are needed to determine the largest of n currents (Lazzaro *et al* 1988). The voltage V increases until the saturation current I_F of all transistors T_a equals the largest of input currents I_{in}. Therefore, all transistors T_a leave saturation, all transistors T_b are blocked and all output currents I_{out} are zero, except in the winning cell which takes all the bias current I_0 and mirrors it at its output. If the current source I_0 is replaced by diode-connected transistor T_d, then the winner cell delivers $I_{out} = I_{inmax}$. If only the value of the maximum is needed, it can be extracted by an additional transistor T_{max}. The n-channel current mirrors are then no longer needed and the drains of transistors T_b can be directly grounded to $V-$ (Chevroulet *et al* 1995).

Weighted average of signals. The weighted average of a set of voltage sources V_i is easily obtained by connecting each of them to a common node V_{out} through a weighting conductance g_i, as illustrated in figure E1.3.15(*a*). If no current flows out of the common output node, then

$$V_{\text{out}} = \frac{\Sigma(g_i V_i)}{\Sigma g_i}.$$

(E1.3.32)

Each conductance g_i may be replaced by the transconductance g_{mi} of an OTA, as shown in part (*b*) of the same figure (Mead 1989). The input sources are then no longer loaded and the transconductances can be individually controlled by the bias current I_{0i} of the amplifiers.

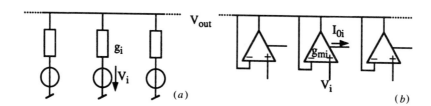

Figure E1.3.15. Weighted averaging of signals.

E1.3.4.3 Resistive diffusion networks

Local averaging, in which the contributions of spatially distant sources are reduced, can be obtained by the resistive diffusion network (Mead 1989) represented in figure E1.3.16 for the two-dimensional case. The contribution to voltage V_j of an input current I_i is maximum locally ($i = j$) and vanishes with distance at a rate approximated by

$$V_j/I_i \approx r^{-1/2} \exp(-r/L) \quad \text{with} \quad L = 1/(RG)^{1/2} \tag{E1.3.33}$$

where r is the distance between nodes i and j measured in grid units. Linearity can be maintained if the real resistors R and $1/G$ are replaced by transistors operated as pseudoconductances, as illustrated by the comparison of subnetworks (b) and (c) in the same figure. The current I_G flowing to the pseudoground can be extracted by an n-channel current mirror. If the transistors are maintained in weak inversion, their gate voltages V_R and V_G can be adjusted independently to control the value of L. Examples of applications are spatial filtering of images and implementation of the weighted excitatory–inhibitory connections that provide the necessary collective behavior in self-organizing feature maps (Vittoz *et al* 1989).

Replacing the p-channel transistors T_G by n-channel devices operated as current sources I_0 results in the nonlinear diffusion network represented in figure E1.3.17 (Heim *et al* 1991). The single input current I_{in} injected at node i is distributed among the I_{in}/I_0 closest current sources I_0, which stay saturated with large node voltage V_j. All sources beyond this limit leave saturation with V_j close to 0.

This circuit can be used to spread an activity in a well-defined area centered at node i.

E1.3.4.4 Storage of synaptic weights

If the synaptic weight values are preestablished by running the learning phase on a computer, they can be stored in a fixed manner as width-to-length ratios of transistors or resistors (Vittoz *et al* 1989). Besides

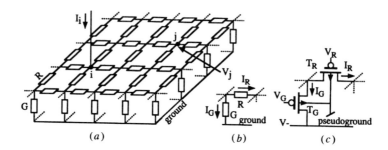

Figure E1.3.16. Two-dimensional resistive diffusion network.

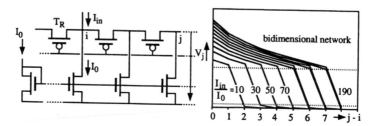

Figure E1.3.17. Nonlinear diffusion network.

preventing any further learning, this solution does not allow the compensation of variations from circuit to circuit.

Short-term storage of adjustable weights can be obtained by sampling and holding a voltage across on-chip capacitors, as illustrated by figure E1.3.18(a). Even if the channel of the switch transistor T_S is completely blocked during the storage phase, the storage time is limited by the leakage of the associated diffusion to substrate junction. This current can be minimized by minimizing the voltage across this junction as in the improved solution of figure E1.3.18(b) (Landolt 1992). To store current I, transistor T_S is switched on. The difference $I' - I$ is then integrated into capacitor C until $I' = I$. This current is then available from output transistor T_O after T_S is blocked. The voltage across the leaking junction at critical node N is limited to a small offset V_{os}. The effect of the charge injected by blocking the transistor can be reduced by various compensation techniques (Wegmann et al 1987). The storage time with integrated capacitors cannot exceed a few minutes; the weight values must thus be maintained by continuous learning, or must be periodically refreshed to predefined analog levels (Vittoz et al 1991) or to a digitally stored value. Digital storage may be local or centralized on the chip, or it may be provided by an external computer. It requires analog-to-digital (AD) and digital-to-analog (DA) conversions with a number of bits compatible with the acceptable amount of quantization.

The best solution for long-term storage of analog weights is to use floating gate transistors. The leakage of charge is drastically reduced by completely wrapping the gate electrode in silicon dioxide, which extends the storage time to several years. Changing the stored value may be achieved in standard complementary MOS (CMOS) technologies by providing a high level of energy to charge carriers, by means of avalanching or UV illumination. Better solutions based on field-aided tunneling usually require special processes adapted for digital E^2PROM, although some solutions are being worked out with standard processes.

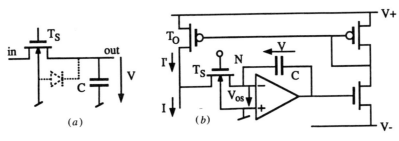

Figure E1.3.18. Storage of synaptic weight as a voltage on a capacitor C.

E1.3.4.5 Neuron circuit

The generic block of a circuit that includes the most important functions of a formal neuron is represented in figure E1.3.19. Not all functions are necessarily implemented in a given circuit, and some neural networks may require completely different types of cell.

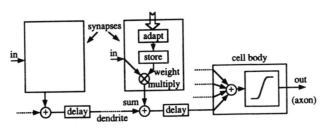

Figure E1.3.19. General form of a neuron.

Each dendrite aggregates and possibly delays the inputs which are weighted by the synapses. The cell body sums up the contributions of all dendrites and produces the output signal to the axon. Its nonlinear transfer function may be obtained by dedicated means, by the natural saturating behavior of the summing circuitry, or by that of the subsequent synaptic circuitry (Coggins et al 1995). An interesting solution creates this nonlinear behavior by the shape of a ramp voltage, common to all neurons, which controls the pulse-width-modulated (PWM) output signal (Murray *et al* 1994); a symmetric ramp avoids synchronous transitions of the neurons.

Since the number of synapses may be very large, each of them must be implemented in a very dense manner by using the rich set of possible basic operators. Efforts to achieve high precision usually result in fairly large synaptic circuits (Heim and Vittoz 1994). The detailed function and implementation of the weight adaptation part of the synapse depends on the type of network and on the learning process.

E1.3.5 Communication of analog data

Collective computation requires a dense network of communication in neural networks, but high connectivity is difficult to achieve in intrinsically two-dimensional VLSI technologies. Some collective actions are possible with just one common wire, as was shown for normalization, averaging, or determination of a spatial maximum. A collective behavior is also possible by limiting the communication to adjacent cells, and by propagating the information from cell to cell. This is the approach used in cellular automata, in cellular neural networks (CNNs) (Chua and Yang 1988a, b), and in resistive diffusion networks.

The standard algorithms for software implementations of neural networks may be modified in order to eliminate long-distance connections between cells. Interconnections may be implemented in a hierarchical manner, or with a density which is a very strong inverse function of the distance, as is the case in biological networks. Communication of signals represented by voltages or by currents is prone to degradation by noise and by crosstalk, and can therefore only be used for short-distance communication. Nature has developed communication schemes based on a pulse frequency representation of activities, which turns out to be very effective for long-distance communication in artificial networks.

On-chip or chip-to-chip communication can exploit the very large speed available in VLSI technologies to implement some form of multiplexing of an analog bus. Standard solutions use a periodic scanning of the array of transmitting cells (Mead and Delbrück 1991). The origin of each sample present on the analog bus is given by its position in the sequence. Yet this solution has several drawbacks. It requires a clock which may interfere with analog signals on the chip; maintaining the timing information content of the signals would require a very high clock frequency, which is usually not feasible. Such a systematic scanning of all cells becomes very inefficient when the overall activity is very sparse. Thus, the ideal scheme would be asynchronous with a priority of access to the common bus based on the level of activity.

These features can be obtained by a particular type of pulse frequency communication using address coding events (Mahowald 1992) as illustrated in figure E1.3.20.

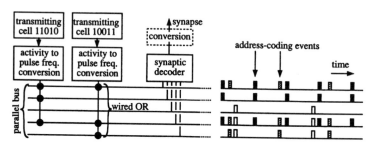

Figure E1.3.20. Communication by address coding pulse events.

The activity to be transmitted from one cell is converted into the frequency of very short pulses. During each pulse, the address of the cell is transmitted on the bus by simple wired OR coding. Each synapse can decode the relevant events and reconvert their frequency into an adequate signal representation.

The pulse duration is as short as possible in order to minimize the collision rate. However, collisions are unavoidable and must be handled in a proper manner. They can be suppressed by using an arbitration scheme (Mahowald 1992); no event is lost if it can be memorized by the transmitting cell until it is given access to the bus. The simplest solution is to let the events access the bus without any arbitration (Mortara and Vittoz 1994, Mortara 1995). The transmitted codes must then be chosen to avoid producing a third valid code by OR-ing the codes of two colliding events. The only possibility is to use codes with equal numbers of 'ones'. Collisions then result in lost events with equal probabilities for all codes.

This corresponds to a transmission attenuation combined with some transmission noise due to the random nature of collisions, both of them increasing with channel occupation. The optimum number of 'ones' in the code is the integer closest to $m/2$ for an m-wire bus, resulting in a redundancy of only about two bits for $10 < m < 20$. One application of this communication scheme is the transfer of the activities of a sending map to a receiving map for further processing (Arreguit and Vittoz 1994). The hardware may then be simplified by transmitting separately the line and column addresses of the origin of each event (Mortara *et al* 1995). Pulse frequency decoding can be effected with just two pulses received (Mortara *et al* 1993), and this conversion may be combined with synaptic weighting and summing. No conversion is necessary if the synapse just weights the pulses and directly accumulates them in the cell body.

An alternative manner of using pulse stream communication is to allocate the channel sequentially to each transmitting cell during one interpulse time (Murray *et al* 1993). Communication is still asynchronous, but not event driven. Moreover, more time is allocated to less active cells, and very low activities must be discarded. This scheme is more efficient to transmit PWM signals (Murray *et al* 1994) for which the pulse width is proportional to the activity.

E1.3.6 Conclusion

Analog VLSI offers a rich medium for the hardware realization of neural networks and should require less power and less chip area than *digital implementations* if a limited precision is acceptable. As a consequence E1.4 of the wide choice of possibilities for signal representation, mode of operation of transistors, and basic computing blocks, no limit exists to the variety of possible implementations of complete systems. However, designing viable analog circuits requires a lot of expertise, since a wide range of problems extending from device physics to system architecture must be addressed. Unlike digital circuits, no real synthesis tools are available, and design is much more time consuming. New ideas, or even new implementations of existing ideas, must be verified not only by computer simulation, but by a physical realization. Testing is also more difficult.

For these reasons, the number of working analog implementations of neural networks is still limited (Vittoz *et al* 1989, Coggins *et al* 1995, Murray *et al* 1994, Graf *et al* 1993, Motishita *et al* 1990, Arima *et al* 1991, Holler *et al* 1989, Graf and Henderson 1990, Van der Spiegel *et al* 1992, Alspector *et al* 1991, Vittoz and Arreguit 1989, Cohen and Andreou 1992, Heim 1993, Masa *et al* 1994) and most of them are still experimental systems waiting for industrial applications. Existing industrial applications are mostly front ends of perception systems that are closely related to neural networks as they exploit massively parallel analog VLSI, but only the most elementary have yet been published (Chevroulet *et al* 1995, Platt and Allen 1995, Venier *et al* 1996, Arreguit *et al* 1996). Most of them are intended for small-size portable microsystems, where power consumption is a major concern.

These existing applications are believed to be only very elementary with respect to what will become feasible in the future with more adequate technological solutions (in particular, for very efficient analog weight storage), with the creation of new dedicated circuit schemes, and with novel architectures. Of special interest is the trend towards deeper exploration of architectures and computational approaches of the brain, as a source of inspiration for analog silicon implementations of very advanced processing systems.

References

Alspector J *et al* 1991 Relaxation networks for large supervised learning problems *Neural Information Processing Systems* vol 3, ed R P Lippmann, J E Moody and D S Touretzky (New York: Morgan Kaufmann) pp 1015–21

Andreou A and Boahen K 1994 Neural information processing II *Analog VLSI Signal and Information Processing* ed M Ismail and T Fiez (New York: McGraw-Hill) pp 358–409

Arima Y *et al* 1991 A 336-neuron 28k-synapse self-learning neural network chip with branch-neuron-unit architecture *ISSCC'91 Dig. Tech. Papers* (Castine, ME: J H Wuorinen) pp 182–4

Arreguit X and Vittoz E 1994 Perception systems implemented in analog VLSI for real-time applications *From Perception to Action Conf.* (Los Alamitos, CA: IEEE Computer Society Press) pp 170–80

Arreguit X *et al* 1996 A CMOS motion detector system for pointing devices *ISSCC'96 (San Francisco, CA) Dig. Tech. Papers* (Castine, ME: J H Wuorinen) pp 98–9

Bult K and Wallinga H 1987 A class of analog CMOS circuits based on the square-law characteristic of an MOS transistor in saturation *IEEE J. Solid-State Circuits* **22** 357–65

Chevroulet M *et al* 1995 A battery-operated optical spot intensity measurement system *Proc. ISSCC'95 Dig. Tech. Papers* (Castine, ME: J H Wuorinen) pp 154–5

Chua L and Yang L 1988a Cellular neural networks: theory *IEEE Trans. Circuits Syst.* **35** 1257–72

——1988b Cellular neural networks: applications *IEEE Trans. Circuits Syst.* **35** 1273–90

Coggins R, Jabri M and Pickard S 1995 A low power network for on-line diagnosis of heart patients *IEEE Micro* **15** 18–25

Cohen M and Andreou A 1992 Current-mode subthreshold MOS implementation of the Hérault–Jutten autoadaptative network *IEEE J. Solid-State Circuits* **27** 714–27

Gilbert B 1968 A precise four-quadrant multiplier with subnanosecond response *IEEE J. Solid-State Circuits* **3** 365–73

——1975 Translinear circuits: a proposed classification *Electron. Lett.* **11** 14

——1981 *Translinear Circuits* private notes

——1983 A four-quadrant analog multiplier/divider with 0.01% distortion *Proc. ISSCC'83 Dig. Tech. Papers* (Coral Gables, FL: Lewis Winner) pp 248–9

——1984 A monolithic 16-channel analog array normalizer *IEEE J. Solid-State Circuits* **19** 956

Graf H P and Henderson D 1990 A reconfigurable CMOS neural network *ISSCC Dig. Tech. Papers* (Castine, ME: J H Wuorinen) pp 144–5

Graf H P, Sackinger E and Jackel L D 1993 Recent developments of electronic neural nets in North America *J. VLSI Signal Processing* **6** 19–31

Heim P 1993 CMOS analog VLSI implementation of a Kohonen map *PhD Dissertation* 1174, EPFL, Lausanne

Heim P *et al* 1991 Generation of learning neighborhood in Kohonen feature maps by means of simple nonlinear network *Electron. Lett.* **27** 275–7

Heim P and Vittoz E 1994 Precise analog synapse for Kohonen feature maps *IEEE J. Solid-State Circuits* **29** 982–5

Holler M *et al* 1989 An electrically trainable artificial neural network (ETANN) *Proc. Int. Joint Conf. on Neural Networks (Washington, DC, 1989)* pp 191–6

Landolt O 1992 An analog CMOS implementation of a Kohonen network with learning capability *3rd Int. Workshop on VLSI for Neural Networks and Artificial Intelligence (Oxford, 1992)*

Landolt O, Vittoz E and Heim P 1992 CMOS selfbiased Euclidean distance computing circuit with high dynamic range *Electron. Lett.* **28**

Lazzaro J *et al* 1988 Winner-take-all network of $0(n)$ complexity advances *Neural Information Processing Systems* (San Mateo, CA: Morgan Kaufman) pp 703–11

Mahowald M 1992 VLSI analogs of neuronal visual processing: a synthesis of form and function *PhD Dissertation* Computation and Neural Systems, California Institute of Technology

Masa P *et al* 1994 A high-speed analog neural processor *IEEE Micro* **14** 40–50

Mead C A 1989 *Analog VLSI and Neural Systems* (Reading, MA: Addison-Wesley)

Mead C A and Delbrück T 1991 Scanners for visualizing activity of analog VLSI circuitry *CNS Memo* 11 California Institute of Technology

Morishita T *et al* 1990 A BiCMOS analog neural network with dynamically updated weights *ISSCC'90 Dig. Tech. Papers* (Castine, ME: J H Wuorinen) pp 142–3

Mortara A 1995 Communication techniques for analog VLSI perceptive systems *PhD Dissertation* 1329, EPFL, Lausanne

Mortara A and Vittoz E 1994 A communication architecture tailored for analog VLSI artificial neural networks: intrinsic performance and limitations *IEEE Trans. Neural Networks* **5** 459–66

Mortara A *et al* 1993 Simple PFM demodulator to be used by analog neural networks which communicate through pulses *Electron. Lett.* **29** 345–6

Mortara A *et al* 1995 A communication scheme for analog VLSI perceptive systems *IEEE J. Solid-State Circuits* **30** 660–9

Murray A and Edwards P 1993 Synaptic weight noise during perceptron training: fault tolerance and training improvements *IEEE Trans. Neural Networks* **4** 722–5

Murray A *et al* 1991 Pulse stream VLSI neural networks mixing analog and digital techniques *IEEE Trans. Neural Networks* **2** 193–204

Murray A *et al* 1994 Pulse stream VLSI neural networks *IEEE Micro* **14** 29–39

Platt J C and Allen T P 1995 A neural network classifier for the I1000 OCR chip *Dig. Conf. on Neural Information Processing Systems (NIPS) (Vail, CO, 1995)* p 60

Temes G C 1987 *Integrated Analog Filters* (New York: IEEE)

Tsividis Y P 1987 *Operation and Modeling of the MOS Transistor* (New York: McGraw-Hill) pp 168–216

——1994 Integrated continuous-time filter design: an overview *IEEE J. Solid-State Circuits* **29** 166–76

Tsividis Y *et al* 1986 Continuous-time MOSFET-C filters in VLSI *IEEE Trans. Circuits Syst.* **33** 125–39

Van der Spiegel J *et al* 1992 An analog neural computer with modular architecture for real-time dynamic computation *IEEE J. Solid-State Circuits* **7** 82–92

Venier P *et al* 1996 Analog CMOS photosensitive array for solar illumination monitoring *ISSCC'96 Dig. Tech. Papers* (Castine, ME: J H Wuorinen) pp 96–7

Vittoz E 1983 MOS transistors operated in the lateral bipolar mode and their applications in CMOS technology *IEEE J. Solid-State Circuits* **18** 273–9

——1985 The design of high-performance analog circuits on digital CMOS chips *IEEE J. Solid-State Circuits* **20** 657–65

——1990 Future trends of analog in the VLSI environment *Proc. ISCAS'90 (New Orleans, LA)* (Piscataway, NJ: IEEE) pp 1372–5

——1994a Low-power design: ways to approach the limits *Proc. ISSCC'94 Dig. Tech. Papers* (Castine, ME: J H Wuorinen) pp 14–8

——1994b Analog VLSI signal processing: why, where, and how *J. VLSI Signal Proc.* **8** 27–44

——1994c Micropower techniques *Design of VLSI Circuits for Telecommunication and Signal Processing* ed J Franca and Y Tsividis (Englewood Cliffs, NJ: Prentice-Hall) pp 53–96

Vittoz E and Arreguit X 1989 CMOS integration of Hérault–Jutten cells for separation of sources *Analog VLSI Implementation of Neural Networks* ed C Mead and M Ismail (Norwell: Kluwer) pp 57–82

——1993 Linear networks based on transistors *Electron. Lett.* **29** 297–9

Vittoz E, Oguey H, Maher M A, Nys O, Dijkstra E and Chevroulet M 1991 Analog storage of adjustable synaptic weights *Introduction to VLSI—Design of Neural Networks* ed U Ramacher (Dordrecht: Kluwer)

Vittoz E *et al* 1989 Analog VLSI implementation of a Kohonen map *Proc. Journées d'Electronique on Artificial Neural Nets, EPFL (Lausanne, 1989)* (Lausanne: Presse Polytechniques Romandes) pp 292–301

Wegmann G *et al* 1987 Charge injection in analog MOS switches *IEEE J. Solid-State Circuits* **22** 1091–7

E1.4 Digital integrated circuit implementations

Valeriu Beiu

Abstract

This section considers some of the alternative approaches towards modeling biological functions by digital circuits. It starts by introducing some circuit complexity issues and arguing that there is considerable computational and physiological justification that shallow threshold gate circuits are computationally more efficient than classical Boolean circuits. We comment on the tradeoff between the depth and the size of a threshold gate circuit, and on how design parameters like fan-in, weights and thresholds influence the overall area and time performances of a digital neural chip. This is followed by briefly discussing the constraints imposed by digital technologies and by detailing several possible classification schemes as well as the performance evaluation of such neurochips and neurocomputers. Lastly, we present many typical and recent examples of implementation and mention the 'VLSI-friendly learning algorithms' as a promising direction of research.

E1.4.1 Introduction

The research on neural networks goes back to the early 1940s (see Section A1.1). The seminal year for the development of the 'science of mind' was 1943 when several articles were published (McCulloch and Pitts 1943, Craik 1943, Rosenbleuth *et al* 1943, 1949, Landahl *et al* 1943).

Almost immediately different approaches to neural network simulation started to be developed. Typical of that era was the development of the first neurocomputer *Snark* (Minsky 1954). It was in fact an electromechanical neurocomputer which was shortly followed by the *Perceptron Mark I* (Rosenblatt 1958). Both were using resistive circuits (motor-driven potentiometers) for implementing the weights. Another successful neurocomputer that used resistive weights was Bernard Widrow's *adaline* and, later, *madaline*. They used a type of electronically adjustable resistor called a memistor. Widrow even founded the first neurocomputer company: the Memistor Corporation. It actually produced neurocomputers during the early and mid-1960s. More details can be found in Nilsson (1965), Anderson and Rosenfeld (1988) and Hecht-Nielsen (1989). The neurocomputer industry was born.

In the last decade the tremendous impetus of VLSI technology has made neurocomputer design a really lively research topic. Hundreds of designs have already been built, and some are available as commercial products. However, we are far from the main objective as can be clearly seen from figure E1.4.1. Here the horizontal axis represents the number of synapses (number of connections), while the vertical axis represents the processing speed in 'connections per second' (CPS). The drawing shows a crude comparison of the computational potential of different neural network hardware 'technologies'. It becomes clear that biological neural networks are far ahead of digital, analog and even future optical implementations of artificial neural networks.

Focusing only on digital implementations, this section will firstly introduce some circuit complexity issues (section E1.4.2) and comment on the constraints imposed by digital technologies (section E1.4.3). Several possible classification schemes for digital implementations of artificial neural networks will also be discussed in the last and most detailed part (section E1.4.4) which will briefly present many different implementations.

A1.1

C1.1.3
C1.1.4

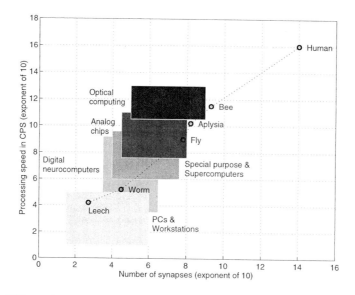

Figure E1.4.1. Different hardware alternatives for implementing artificial neural networks, an enhanced and updated version from Glesner and Pöchmüller (1994) and Iwata (1990).

E1.4.2 Circuit complexity issues

One main line of theoretical research has concentrated on the approximation capabilities of feedforward networks. It was started in 1987 by Hecht-Nielsen (1987) and Lippmann (1987) who were probably the first to point to Kolmogorov's theorem (Kolmogorov 1957), together with Le Cun (1987). The first nonconstructive proof that neural networks are universal approximators was given the following year by Cybenko (1988, 1989) using a continuous activation function. Thus, the fact that neural networks are computationally universal—with more or less restrictive conditions—when modifiable connections are allowed, was established. These results have been further enhanced by Funahashi (1989), Funahashi and Nakamura (1993), Hornik (1991, 1993), Hornik *et al* (1989, 1990), Koiran (1993) and Leshno *et al* (1993). All these results—with the partial exception of Koiran (1993)—were obtained 'provided that sufficiently many hidden units are available'. This means that no claim on the minimality of the resulting network was made, the number of neurons needed to make a satisfactory approximation being in general much larger than the minimum needed.

The other line of research was to find tight bounds, and the problem can be stated as finding the smallest network (i.e., smallest number of neurons) which can realize an arbitrary function given a set of m vectors (examples, or points) in n dimensions. If the function takes as output just 0 or 1, then it is called a dichotomy. This aspect of the smallest network is of great importance when thinking of hardware implementations. The networks considered are feedforward neural networks with threshold activation function. This is probably due to the fact that this line of research was continuing on from the rigorous results already obtained in the literature dealing with threshold logic from the 1960s (Cameron 1969, Cohen and Winder 1969, Cover 1965, Dertouzos 1965, Fischler 1962, Hu 1965, Kautz 1961, Klir 1972, Lewis and Coates 1967, Lupanov 1973, Minnick 1961, Minsky and Papert 1969, Muroga 1959–1979, Muroga *et al* 1961, Neciporuk 1964, Nilsson 1965, Red'kin 1970, Sheng 1969, Winder 1962–1971). The

C1.2 best result was that a *multilayer perceptron* with only one hidden layer having $m - 1$ nodes could compute an arbitrary dichotomy (sufficient condition). The main improvements since then have been as follows:

- Baum (1988b) presented a network with one hidden layer having $\lceil m/n \rceil$ neurons capable of realizing an arbitrary dichotomy on a set of m points in general position in $\mathbb{R}^n$; if the points are on the corners of the n-dimensional hypercube (i.e., binary vectors), $m - 1$ nodes are still needed (the general position condition is now special and strict).

- Huang and Huang (1991) proved a slightly tighter bound: only $\lceil 1 + (m - 2)/n \rceil$ neurons are needed in the hidden layer for realizing an arbitrary dichotomy on a set of m points which satisfy a more relaxed topological assumption as only the points forming a sequence from some subsets are required to be in general position; also the $m - 1$ nodes condition was shown to be the least upper bound needed.

- Arai (1993) recently showed that $m - 1$ hidden neurons are necessary for arbitrary separability (any mapping between input and output for the case of binary-valued units), but improved the bound for the two-category classification problem to $m/3$ (without any condition on the inputs).

A study which somehow tries to unify these two lines of research has been published by Bulsari (1993) who gives practical solutions for one-dimensional cases including an upper bound on the number of nodes in the hidden layer(s). Extensions to the n-dimensional case using three- and four-layer solutions are derived under piecewise constant approximations having constant or variable width partitions and under piecewise linear approximations using ramps instead of sigmoids.

To strengthen such claims, we shall go briefly through some basic circuit complexity results (Papadopoulos and Andronikos 1995, Parberry 1994, Paterson 1992, Pippenger 1987, Roychowdhury *et al* 1991a, b, 1994b, Siu *et al* 1994) and argue that there is considerable computational and physiological justification that shallow (i.e., having relatively few layers) threshold gate circuits are computationally more efficient than classical Boolean circuits. When considering computational complexity, two classes of constraints could be thought of:

- Some arising from the physical constraints (related to the hardware in which the computations are embedded) and including time constants, energy limitations, volumes, geometrical relations and bandwidth capacities.
- Others are logical constraints: (i) computability constraints and (ii) complexity constraints which give upper and/or lower bounds on some specific resource (e.g., size and depth required to compute a given function or class of functions).

The first aspect when comparing Boolean and threshold logic is that they are equivalent in the sense that any Boolean function can be implemented using either logic in a circuit of depth-2 and exponential size (simple counting arguments show that the fraction of functions requiring a circuit of exponential size approaches one as $n \to \infty$ in both cases). Yet, threshold logic is more powerful than Boolean logic as a Boolean gate can compute only one function whereas a threshold gate can compute up to the order of $2^{\alpha n^2}$ functions by varying the weights, with $1/2 \le \alpha \le 1$ (see Muroga 1962 for the lower bound, and Muroga 1971 and Winder 1962, 1963 for the upper bound). An important result which clearly separates threshold and Boolean logic is due to Yao (1985) (see also Håstad 1986 and Smolensky 1987) and states that in order to compute a highly oscillating function like PARITY in a constant depth circuit, at least $\exp[c(n^k)^{1/2}]$ Boolean gates with unbounded fan-in are needed (Furst *et al* 1981, Paturi and Saks 1990). In contrast, a depth-2 threshold gate circuit for PARITY has linear size.

Another interesting aspect is the tradeoff between the depth and the size of a circuit (Beiu 1994, 1997, Beiu and Taylor 1996a, Beiu *et al* 1994c, Siu and Bruck 1990c, Siu *et al* 1991b). There exists a very strong bias in favor of shallow circuits (Judd 1988, 1992) for several reasons. First, for a fixed size, the number of different functions computable by a circuit of small depth is larger than the number of those computed by a deeper circuit. Second, it is obvious that such a circuit is also faster, as having a small(er) depth. Finally, one should notice that biological circuits must be shallow—at least within certain modules like the cortical structures—as the overall response time (e.g., recognizing a known person from a noisy image) of such slow devices (the response time of biological neurons being at least in the 10-ms range due to the refractory period) is known to be in the few hundred millisecond range. Other theoretical results (Abu-Mostafa 1988a, b) also support the shallow architecture of such circuits.

A lot of work has been devoted to finding minimum size and/or minimum constant-depth threshold gate circuits (Hajnal *et al* 1987, Hofmeister *et al* 1991, Razborov 1987, Roychowdhury *et al* 1994a, Siu and Bruck 1990a, Siu *et al* 1990, 1993b, Siu and Roychowdhury 1993, 1994) but little is known about tradeoffs between those two cost functions (Beiu *et al* 1994c, Siu *et al* 1991b), and even less about how design parameters like fan-in, weights and thresholds influence the overall area and time performances of a digital neural chip. Since for the general case only existence exponential bounds are known (Bruck and Smolensky 1992, Siu *et al* 1991b), it is important to isolate classes of functions whose implementations are simpler than that of others (e.g., shallow depth and polynomial size (Rief 1987)). Several of the corner-stone results obtained so far have been gathered in table E1.4.1. Here n is the number of input variables, and the nomenclature commonly in use is (see Amaldi and Mayoraz 1992, Papadopoulos and Andronikos 1995, Parberry 1994, Roychowdhury *et al* 1994b, Siu *et al* 1994, Wegener 1987):

- AC^k represents the circuits of polynomial size with AND and OR unbounded fan-in gates and depth $O(\log^k n)$

- NC^k is the class of Boolean functions with bounded fan-in, and having size n^c (polynomial) and depth $O(\log^k n)$
- TC^0 the family of functions realized by polynomial size threshold gate circuits with unbounded fan-in and constant depth
- $LT_1(\widehat{LT_1})$ is the class of Boolean functions computed by linear threshold gates with real weights (bounded by a polynomial in the number of inputs $|w_i| \leq n^c$ (Bruck 1990))
- $\widehat{LT}_k$ is the class of Boolean functions computed by a polynomial size, depth-k circuit of $\widehat{LT}_1$ gates (Bruck 1990, Siu and Bruck 1990b)
- PT_1 is the class of Boolean functions that can be computed by a single threshold gate in which the number of monomials is bounded by a polynomial in n (Bruck 1990, Bruck and Smolensky 1992)
- PT_k is the class of Boolean functions computed by a polynomial size, depth-k circuit of PT_1 gates
- PL_1 is the class of Boolean functions for which the spectral norm L_1 is bounded by a polynomial in n (Bruck and Smolensky 1989)
- PL_∞ is the class of Boolean functions with the spectral norm L_∞^{-1} bounded by a polynomial in n (Bruck and Smolensky 1989)
- MAJ_1 is the class of Boolean functions computed by linear threshold gates having only ± 1 weights (Mayoraz 1991, Siu and Bruck 1990c)
- MAJ_k is the class of Boolean functions computed by a polynomial size, depth-k circuit of MAJ_1 gates (Albrecht 1992, Mayoraz 1992, Siu and Bruck 1993).

Recently three complexity classes for sigmoid feedforward neural networks have been defined and linked with the (classical) above-mentioned ones:

- NN^k is defined (Shawe-Taylor $et\ al$ 1992) to be the class of functions which can be computed by a family of polynomially sized neural networks with weights and threshold values determined to b bits of precision (accuracy), fan-in equal to Δ and depth h, satisfying $\log \Delta = O[(\log n)^{1/2}]$, $b \log \Delta = O(\log n)$ and $h \log \Delta = O(\log^k n)$
- $NN_{\Delta,\varepsilon}^k$ is defined (Beiu $et\ al$ 1994e, Beiu and Taylor 1996b) to be the class of functions which can be computed by a family of polynomially sized neural networks which satisfies slightly less restrictive conditions for fan-in and accuracy: $\log \Delta = O(\log^{1-\varepsilon} n)$ and $b = O(\log^{1-\varepsilon} n)$
- NN_Δ^k is defined (Beiu $et\ al$ 1994d, Beiu and Taylor 1996b) to be the class of functions which can be computed by a family of polynomially sized neural networks having linear fan-in and logarithmic accuracy ($\Delta = O(n)$ and $b = O(\log n)$).

Still, in many situations one is concerned by the values of a function for just a vanishing small fraction of the 2^n possible inputs. Such functions can also be implemented in poly-size shallow circuits (the size and depth of the circuit can be related to the cardinal of the interesting inputs (Beiu 1996b, Beiu and Taylor 1996a, Beiu $et\ al$ 1994a, Tan and Vandewalle 1992, 1993). Such functions are also appealing from the learning point of view: the relevant inputs being nothing else but the set of training examples (Beiu 1996b, Beiu and Taylor 1995b, Linial $et\ al$ 1989, Takahashi $et\ al$ 1993).

Circuit complexity has certain drawbacks which should be mentioned:

- The extension of the poly-size results to other functions and to the continuous domain is not at all straightforward (Maass $et\ al$ 1991, Siu 1992)
- Even the known bounds (for the computational costs) are sometimes weak
- Time (i.e., delay) is not properly considered
- All complexity results are asymptotic in nature and may not be meaningful for the range of a particular application.

But the scaling of some important parameters with respect to some others represents quite valuable results:

- Area of the chip (wafer) grows like the cube of the fan-in
- Area of the digital chip (wafer) grows exponentially with accuracy.

Furthermore, it was shown recently that the fan-in and the accuracy are linearly dependent parameters. If the number of inputs to one neuron is n, the reduction of the fan-in by decomposition techniques has led to the following results:

- If the fan-in is reduced to (small) constants, the size grows slightly faster than the square of the number of inputs (i.e., $n^2 \log n$) while the depth growth is lower than logarithmic (i.e., $\log n / \log \log n$)

Table E1.4.1. Circuit complexity results.

Author(s)	Result(s)	Remark(s)
Neciporuk (1964)	Lower bound on the *size* of a threshold circuit for 'almost all' n-ary Boolean functions.	$size \geq 2 \cdot (2^n/n)^{1/2}$
Lupanov (1973)	Upper bound for the *size* of a threshold circuit for 'almost all' n-ary Boolean functions.	$size \leq 2 \cdot (2^n/n)^{1/2} \times \{1 + \Omega[(2^n/n)^{1/2}]\}$ $depth = 4$
Yao (1989)	There are Boolean functions for which *depth*-$(k-1)$ threshold gate circuit of unbounded fan-in (i.e., TC^0 circuits) require exponential size *depth-k* Boolean circuits of unbounded fan-in.	Conjecture: $TC^0 \neq NC^1$
Allender (1989)	Any Boolean function computable by a polynomial size constant-*depth* logic circuit with unbounded fan-in (i.e., AC^0) is also computable by a *depth*-3 neural network (threshold gate circuit) of superpolynomial size: $AC^0 \neq TC^0$.	$size = n^{O(\log n)}$ $depth = 3$ *fan-in* unbounded
Immermann and Landau (1989)	Conjecture: $TC^0 = NC^1$	
Bruck (1990)	$LT_1 \subset PT_1 \subset LT_2$	Existence proofs
Siu *et al* (1991a)	$\mathrm{MUL}(x, y) \in \widehat{LT}_4$ $X \bmod p, X^n \bmod p, c^X \bmod p \in \widehat{LT}_2$ $X^n, c^X \in \widehat{LT}_5$ $\mathrm{DIV}(x, y) \in \widehat{LT}_5$ $\mathrm{MUL}(x_1, \ldots, x_n) \in \widehat{LT}_6$	$depth = \mathrm{const}$ $size \leq n^c$ $weights \leq n^c$ *fan-in* unbounded Partly constructive
Siu *et al* (1991b)	Upper bound on the *size* or implementing any Boolean function.	$depth = 3$ $size = O(2^{n/2})$ *fan-in* unbounded Existence proofs
	Lower bound on the *size* or implementing any Boolean function.	$size = \Omega(2^{n/3})$ *fan-in* unbounded Existence proofs
Bruck and Smolensky (1992)	$PL_1 \subset PT_1 \subset PL_\infty$ $PL_1 \subset PT_1 \subset MAJ_2$ $AC^0 \not\subset PL_1$ $AC^0 \not\subset PL_\infty$ $AC^0 \not\subset MAJ_2$	Existence proofs
Siu and Bruck (1992)	$LT_1 \subset \widehat{LT}_3$ $LT_d \subseteq \widehat{LT}_{2d+1}$ $\mathrm{MUL}(x, y) \in \widehat{LT}_4$ $\mathrm{MAX}(x_1, \ldots, x_n) \in \widehat{LT}_3$ $\mathrm{SORT}(x_1, \ldots, x_n) \in \widehat{LT}_4$	$depth = \mathrm{const}$ $size \leq n^c$ $weights \leq n^c$ *fan-in* unbounded Existence proofs
Albrecht (1992)	*Depth*-2 threshold circuits require superpolynomial *fan-in*. Polynomial threshold circuits have more than two layers.	$depth = 2$ $size = (1 \pm \varepsilon) \cdot 2^{n-1}$ $weights \in \{-1, 0, +1\}$ *fan-in* unbounded
Shawe-Taylor *et al* (1992)	$NC^k \subset NN^k \subseteq AC^k$	Partly constructive
Beiu *et al* (1994d)	$NN^k \subset NN_{\Delta,\varepsilon}^k \subset \begin{cases} NC^{k+1} \\ NN_\Delta^k \subset NC^{k+2} \end{cases}$	Constructive proofs (based on binary trees of Boolean gate adders)
Beiu *et al* (1994e)	$NN_{\Delta,\varepsilon}^k \subset NN_\Delta^k \subset \begin{cases} LT_{\log^{k+\varepsilon} n} \\ \widehat{LT}_{\log^{k+1} n} \end{cases}$ $\cap \qquad \cap$ $MAJ_{\log^{k+1} n} \subset MAJ_{\log^{k+2} n}$	Constructive proofs (based on binary trees of threshold gate adders)

Table E1.4.1. Continued.

Author(s)	Result(s)	Remark(s)
Goldmann and Karpinski (1994)	$LT_d \subset MAJ_{d+1}$ (improves on Siu and Bruck 1992 and implies: $$NN_\Delta^k \subset \left\{ \begin{array}{l} LT_{\log^{k+\varepsilon} n} \subset MAJ_{1+\log^{k+\varepsilon} n} \\ \widehat{LT}_{\log^{k+1} n} \end{array} \right) $$	Existence proof. An important aspect is that such a simulation is possible even if the depth d grows with the number of variables n
Beiu and Taylor (1996b)	$$NN_\Delta^k \subset \left\{ \begin{array}{l} NC^{k+1} \\ MAJ_{\log^{k+1} n} \end{array} \right. \subset \widehat{LT}_{\log^{k+1} n}$$	Constructive proofs (based on carry save addition)

- Boolean decomposition can be used for reducing the fan-in, but at the expense of a superpolynomial increase in size $((n^{\log n})^{1/2})$ and a double logarithmic increase in depth $(\log^2 n)$.

Much better results can be achieved for a particular function.

Due to such scaling problems, theoretical results show that we can implement (as digital chips or wafers) only neural networks having (sub) logarithmic accuracy and (sub) linear fan-in (with respect to the number of inputs n). From the practical point of view (the two parameters being dependent) these should be translated to (sub) logarithmic both for accuracy and for fan-in. The main conclusion is that full parallel digital implementations of neural networks (as chips or wafers) are presently limited to artificial neural networks having 10^2–10^3 inputs and about 10^3–10^4 neurons of 10^2–10^3 inputs each. As will be seen later, these values are in good accordance with those from chips and wafers which stick as much as possible to a parallel implementation. Although we do expect that technological advances will push these limits, they cannot be spectacular—at least in the near future.

Such drastic limitations have forced designers to approach the problem from different angles:

- By using time-multiplexing
- By building arrays of (dedicated) chips working together and exploiting as much as possible (in one way or another) the architectural concept of pipe-lining
- By using non-conventional techniques such as: stochastic processing (Gorse and Taylor 1989a, Köllmann *et al* 1996), sparse memory architecture (Aihara *et al* 1996) or spike processing (Jahnke *et al* 1996).

These allow the simulation of far larger neural networks, by mapping them onto the existent (limited) hardware.

E1.4.3 Digital VLSI

Digital neurochips (and, thus, neurocomputers) benefit from the legacy of the most advanced human technology—digital information processing. VLSI technology is the main support for electronic implementations. It has been mature for many years, and allows a large number of processing elements to be mapped onto a small silicon area. That is why it has attracted many researchers (Alla *et al* 1990, Alspector *et al* 1988, Barhen *et al* 1992, Beiu 1989, Beiu and Rosu 1985, Boser *et al* 1992, Del Corso *et al* 1989, Disante *et al* 1989, 1990a, b, Faggin 1991, Fornaciari *et al* 1991b, Holler 1991, Jackel 1992, Mackie *et al* 1988, Personnaz *et al* 1989, Tewksbury and Hornak 1989, Treleaven *et al* 1989, Weinfeld 1990).

The main constraints of VLSI come from the fact that the designer has to implement the processing elements on a two-dimensional limited area and—even more—connect these elements by means of a limited number of available layers. This leads to limited interconnectivity as has been discussed in Akers *et al* (1988), Baker and Hammerstrom (1988), Hammerstrom (1988), Reyneri and Filipi (1991), Szedegy (1989) and Walker *et al* (1989) and limited precision (higher precision requires larger area—both due to storing and processing—leading to fewer neurons per chip (Dembo *et al* 1990, Denker and Wittner 1988, Myhill and Kautz 1961, Obradovic and Parberry 1990, Stevenson *et al* 1990, Walker and Akers 1992)). The shallowness of slow biological neural networks has to be traded off for (somehow) deeper networks made of higher speed elements. Beside these, the power dissipation might impose another severe restriction (especially for wafer scale integration—WSI). The tradeoff is either to reduce the number of neurons per chip (working at high speed) or reduce the clock rate (while having more neurons). Lastly, the number of available pins to get the information on and off the chip is another strong limitation.

From the biological point of view, synapses have to be restricted on precision and range to some small number of levels (Baum 1988a, Baum and Haussler 1989). Lower bounds on the size of the network have been obtained both for the networks with real valued synaptic weights and for the networks where the weights are limited to a finite number of possible values (Siu and Bruck 1990a). These bounds differ only by a logarithmic factor, but to achieve near optimal performance $O(m)$ levels are required (Baum 1988b)—where m is the number of training examples given. A similar logarithmic factor has been proven in Hong (1987), Raghavan (1988) and Sontag (1990) when replacing real weights by integers. Some results concerning the needed number of quantization levels have already been presented in Section E1.2.2 and can be supplemented by many references. For example, Baker and Hammerstrom (1988), Hollis *et al* (1990), Höhfeld (1990), Shoemaker *et al* (1990), Allipi (1991), Asanović and Morgan (1991), Holt and Hwang (1991, 1993), Nigri (1991) and Xie and Jabri (1991) argue that the execution phase needs roughly 8 bits $(6 \ldots 10)$, while learning demands about 16 bits $(14 \ldots 18)$. There are few exceptions: Halgamuge *et al* (1991) being the only pessimistic one claiming that 32 bits are needed, and Reyneri and Filipi (1991) claiming that $20 \ldots 22$ bits are needed *in general*, but explicitly mentioning that this value can be reduced to $14 \ldots 15$ bits or even lower by properly choosing the learning rate (for backpropagation). New weight discretization learning algorithms can go much lower: to just several bits (see Section E1.2.4). This makes them ideal candidates for digital implementations.

Today, the digital VLSI design is still the most important design style. The advantages of the dominant CMOS technology are small feature sizes, lower power consumption and a high signal-to-noise ratio. For neural networks these are supplemented by the following advantages of digital VLSI design styles (see Glesner and Pöchmüller 1994 and Hammerstrom 1995 for more details):

- Simplicity (an important feature for the designer)
- High signal-to-noise ratio (one of the most important advantages over analog designs)
- Circuits are easily cascadable (as compared to analog designs)
- Higher flexibility (digital circuits in general can solve many tasks)
- Reduced fabrication price (certainly of interest for customers)
- Many CAD (computer aided design) systems are available to support a designer's work
- Reliable (as fabrication lines are stable).

Digital VLSI implementations of a neural network are based on several building blocks:

- Summation can easily be realized by adders (many different designs are possible and well-known: combinatorial, serial, dynamic, carry look ahead, manchester, carry select, Wallace tree)
- Multiplication is usually the most area-consuming operation and in many cases a multiplier is time-multiplexed (classical solutions are serial, serial/parallel and fully parallel, each of which differ in speed, accuracy and area)
- Nonlinear transfer function (very different nonlinear activation functions (Das Gupta and Schnitger 1993) can be implemented by using circuits for full calculations, but most digital designs use either a small lookup table (Nigri 1991, Nigri *et al* 1991) or—for even lower area and higher precision—a dedicated circuit for a properly quantized approximation, as can be seen in table E1.4.2 and also in Murtagh and Tsoi (1992), Sammut and Jones (1991)
- Storage elements (are very common—either static or dynamic—from standard RAM cells)
- Random number generators (are normally realized by shift registers with feedback via XOR-gates).

E1.4.4 Different implementations

E1.4.4.1 General comments

As the different number of proposed architectures or fabricated chips, boards and dedicated computers reported in the literature is on the order of hundreds, we cannot mention all of them here. Instead, we shall try to cover important types of architectures by several representation implementations—although certain readers could disagree sometimes with our choice. For a deeper insight the reader is referred to the following books: Eckmiller and von der Malsburg (1988), Eckmiller *et al* (1990), Souček and Souček

Table E1.4.2. Digital implementations of the sigmoid—alternatives to lookup tables.

Author(s)	Result(s)	Remark(s)
Myers and Hutchinson (1989)	Approximation of an A-law sigmoid-like function.	7 segments piecewise. Error $\leq \pm 4.89\%$ for $[-8, 8]$.
Alippi et al (1990a)	Approximations of a classical sigmoid function by sum of 1–5 steps.	Error $\leq \pm 13.1\%$ with 5 steps. Four comparators and several logic gates.
Alippi et al (1990b)	Approximations of a classical sigmoid function.	Sum of 1–5 steps (Alippi et al 1990a).
Pesulima et al (1990)	Approximation of the classical sigmoid by two exponentials.	Digital implementation: LFSR. Error $\leq \pm 2.45\%$ for $[-8, 8]$.
Saucier and Ouali (1990)	Silicon compilation.	Approximation by Taylor series.
Alippi et al (1991a)	Relations between convergence of learning and precision.	Introduces a general class of nonlinear functions.
Alippi and Storti-Gajani (1991)	Approximations of a classical sigmoid function.	Piecewise by the set of points $(\pm n, 1/2^{n+1})$.
Höhfeld and Fahlman (1991)	Probabilistic weight updates (down to 4 bits).	Needed precision for sigmoid 4–6 bits.
Krikelis (1991)	Approximation of the classical sigmoid function.	Piecewise linearization with 3 segments in $[-4, 4]$. Errors $\leq \pm 5.07\%$.
Nigri (1991)	Precision required for backpropagation.	Look-up table for 8 bits; 'exact' only for $[-2, 2]$.
Siggelkow et al (1991)	Analyzes *accuracy* and shows that a problem-dependent synthesis is required.	Piecewise linearization of the sigmoid with 5 segments. No hardware suggested.
Spaanenburg et al (1991)	Bit-serial approximation of binary logarithmic computations (*problem dependent complex parameters*).	Piecewise linearization of the sigmoid with 5 segments (4–6 bits). Errors around $\pm 10\%$. No hardware suggested.
Beiu (1992)	Sum of steps approximation of a particular sigmoid.	Six 'threshold gates' solution with weights $\{-1, 1, 2\}$. Error $\leq \pm 8.1\%$.
Deville (1993)	General method for piecewise linearization. Highest precision $(\leq \pm 1.14\%)$.	*Requires 10 floating-point numbers and 5 multiplications!*
Beiu et al (1993, 1994b)	Piecewise approximation of the classical sigmoid.	Errors $\leq \pm 1.9\%$ using only a shift register and several logic gates.

(1988), Sami (1990), Zornetzer et al (1990), Antognetti and Milutinovic (1991), Ramacher and Rückert (1991), Sanchez-Sinencio and Lau (1992), Hassoun (1993), Przytula and Prasama (1993), Delgado-Frias and Moore (1994) and Glesner and Pöchmüller (1994) together with the references therein. Several overview articles or chapters can also be recommended: Alspector and Allen (1987), Mackie et al (1988), Jackel et al (1987), Jackel (1991), Przytula (1988), DARPA (1989), Del Corso et al (1989), Denker (1986), Goser et al (1989), Personnaz and Dreyfus (1989), Treleaven (1989), Treleaven et al (1989), Schwartz (1990), Burr (1991, 1992), Nordström and Svensson (1991), Graf et al (1991—having many references, 1993), Hirai (1991), Holler (1991), Ienne (1993a, b), Lindsey and Lindblad (1994) and the recent ones—Heemskerk (1995), Hammerstrom (1995) and Morgan (1995). The proceedings of MicroNeuro (International Conference on Microelectronics for Neural Networks) would also prove useful for those readers wishing to find latest details on different implementations or the most recent proposals. Many other conferences on neural networks have special sessions on hardware implementations: NIPS (Neural Information Processing Systems), IJCNN (International Joint Conference on Neural Networks), ICANN (International Conference on Artificial Neural Networks), WCNN (World Congress on Neural Networks), IEEE ICNN (IEEE International Conference on Neural Networks) just to mention some of the most widely known.

One of the difficult problems when discussing dedicated architectures for artificial neural networks is how to classify them. There are many different ways of classifying such architectures, and we shall mention here some which have already been presented and used in the literature.

- A first classification can be made based on the division of computer architectures due to Flynn (1972): single instruction stream, single datastream (SISD); single instruction stream, multiple datastreams (SIMD); multiple instruction streams, single datastream (MISD)—which does not make too much sense; multiple instruction streams, multiple datastreams (MIMD). Most of the architectures proposed for implementing neural networks belong to the SIMD class, and thus the group should be further subdivided into: systolic arrays, processor arrays (linear, mesh, multidimensional) and even pipelined vector processors.

- Another classification has been based on 'how many and how complex' processing elements are (Nordström and Svensson 1991). Computer architectures can be characterized by the level of parallelism which can be: moderately parallel (16 to 256 processors), highly parallel (256 to 4096 processors) or massively parallel (more than 4096 processors). As a coarse measure of the 'complexity' of the processing elements, the bit-length (i.e., the precision) of a processing element has been used.

- A much more simple classification of neurocomputers has been suggested by Heemskerk (1995): those consisting of a conventional computer and an accelerator board; those built from general purpose processors; and those built from dedicated neurochips.

- A completely different classification was suggested by Glesner and Pöchmüller (1994) based on the following three criteria: biological evidence (mimicking biological systems; mimicking on a higher level; or without biological evidence), mapping onto hardware (network-oriented; neuron-oriented; or synapse-oriented) and implementation technology (digital; analog; or mixed).

Only for digital electronic implementations a simple three-class subclassification scheme—somehow similar to that of Heemskerk (1995)—could be the following (Beiu 1994).

- *Dedicated digital neural network chips* (Kung 1989, Kung and Hwang 1988, 1989a), Wawrzynek *et al* (1993) can reach fantastic speeds of up to 1G connections per second. Several examples of such chips are: L-Neuro from Philips (Duranton 1996, Duranton *et al* 1988, Duranton and Maudit 1989, Duranton and Sirat 1989, 1990), X1 and N64000 of Adaptive Solutions (Adaptive Solutions 1991, 1992, Hammerstrom 1990), Ni1000 from Intel (Scofield and Reilly 1991, Holler *et al* 1992), MA16 from Siemens (Ramacher 1990, 1992, Ramacher and Rückert 1991, Ramacher *et al* 1991a, b, 1993), p-RAM from King's College London (Clarkson and Ng 1993, Clarkson *et al* 1989–1993) and Hitachi's WSI (Yasunaga *et al* 1989, 1990) and the 1.5-V chip (Watanabe *et al* 1993), SMA from NTT (Aihara *et al* 1996), NESPINN from the Institute of Microelectronics, Technical University of Berlin (Jahnke *et al* 1996), or SPERT from the International Computer Science Institute, Berkeley (Asanović *et al* 1992, 1993d, Warwzynek 1993, 1996).

- *Special purpose digital coprocessors* (sometimes called neuroaccelerators) are special boards that can be connected to a host computer (PCs and/or workstations) and are used in combination with a neurosimulator program. Such a solution tries to take both advantages: accelerated speed and flexible and user-friendly environment. Well-known are the delta Floating Point Processor from SAIC (DARPA 1989) which can be connected to a PC host, and the ones produced by Hecht-Nielsen Computers (Hecht-Nielsen 1991): ANZA, Balboa. Their speed is in the order of 10M connections per second improving tenfold on a software simulator. Some of them are using conventional RISC microprocessors, some use DSPs or transputers, while others are built with dedicated neurochips.

- *Digital neurocomputers* can be considered the massively data-parallel computers. Several neurocomputers are: WARP (Arnould 1985, Kung and Webb 1985, Annaratone *et al* 1987), CM (Means and Hammerstrom 1991), RAP (Morgan *et al* 1990, Beck 1990), SANDY (Kato *et al* 1990), MUSIC (Gunzinger *et al* 1992, Müller *et al* 1995), MIND (Gamrat *et al* 1991), SNAP (Hecht-Nielsen 1991, Means and Lisenbee 1991), GF-11 (Witbrock and Zagha 1990, Jackson and Hammerstrom 1991), Toshiba (Hirai 1991), MANTRA (Lehmann and Blayo 1991, Lehmann *et al* 1993), SYNAPSE (Ramacher 1992, Ramacher *et al* 1991a, b, 1993, Johnson 1993a), HANNIBAL (Myers *et al* 1993), BACCHUS and PAN IV (Huch *et al* 1990, Pöchmüller and Glesner 1991, Palm and Palm 1991), PANNE (Milosavlevich *et al* 1996), 128 PE RISC (Hiraiwa *et al* 1990), RM-nc256 (Erdogan and Wahab 1992), CNAPS (Adaptive Solutions 1991, 1992, Hammerstrom 1990), Hitachi WSI (Boyd

1990, Yasunaga *et al* 1989–1991), MasPar MP-1 (Grajski *et al* 1990, MasPar 1990a–c, Nickolls 1990), and CNS-1 (Asanović *et al* 1993a)—just to mention only the most well-known.

But even such a subclassification is not very clear cut, as in too many cases there are no borders. For example, many neurocomputers have been assembled based on identical boards built with custom designed neurochips: SNAP uses the HNC 100 NAP chip; MANTRA uses the GENES IV and the GACD1 chips; HANNIBAL uses the HANNIBAL chip; SYNAPSE uses the MA 16 chip; MasPar MP-1 uses the MP-1 chip; CNAPS uses the X1 or the N64000 chip; CNS-1 will use the Torrent and Hydrant chips. That is why we have decided in this section to use a more detailed classification which starts with the first historical neurocomputers and continues through acceleration boards, slice architectures, arrays of DSPs (digital signal processors), arrays of transputers, arrays of RISC processors, SIMD and systolic arrays built of dedicated processing elements and continuing with several other alternatives and ending with some of the latest implementations.

Beside classification and classification criteria, another problem when dealing with neurocomputers and neurochips is their performance evaluation. While the performance of a conventional computer is usually measured by its speed and memory, for neural networks 'measuring the computing performance requires new tools from information theory and computational complexity' (Abu-Mostafa 1989). Although the different solutions presented here will be assessed for size, speed, flexibility and cascadability, great care should be taken especially when considering speed. Hardware approaches are very different, thus making it almost impossible to run the same benchmark on all systems. Even for machines which support backpropagation (which is commonly used as a benchmark), the average number of weight updates per second or CUPS (connection updates per second) reported in publications shows different computational power—even for the same machine! This is due to: different precision of weights; the use of fixed point representation in some cases and the size of the network to be simulated (larger networks may be

C1.2.3   implemented more efficiently). A typical example of two different *backpropagation* implementations on WARP can be found in Pomerleau *et al* (1988). For architectures which do not support learning, the number of synaptic multiplications per second or CPS (connections per second) will be mentioned, but the same caution should be taken due to different word lengths (precision of computation) and network architectures. Normalizing the CPS value by the number of weights leads to CPS per weight or CPSPW, and was suggested as a better way to indicate the processing power of a chip (Holler 1991). Precision can also be included in the processing performance by considering a connection primitive per second (CPPS) which is CPS multiplied by bits of precision and by bits for representing the inputs (van Keulan *et al* 1994). Another reason for taking such speed measurements with a lot of care is that some of the articles report only on a small test chip (and the results reported are extrapolations to a future full-scale chip or to a board of chips and/or neurocomputer), or that only peak values are given.

Finally, for neurochips and neurocomputers which are dedicated to a certain neural architecture (e.g.,
C1.4, C2.1.1   the *Boltzmann machine* (Murray *et al* 1992, 1994); Kohonen's *self-organizing feature maps* (Hochet *et al* 1991, Goser *et al* 1989, Rüping and Rückert 1996, Tryba *et al* 1990, Thiran 1993, Thiran *et al* 1994,
C1.3.4   Thole *et al* 1993); *Hopfield networks* (Blayo and Hurat 1989, Gascuel *et al* 1992, Graf and de Vegvar 1987a, b, Graf *et al* 1987, Savran and Morgül 1991, Sivilotti *et al* 1986, Weinfeld 1989, Yasunaga *et al*
C2.1.3, C1.6.2   1989, 1990); *Neocognitron* (Trotin and Darbel 1993, White and Elmasry 1992); *radial basis functions* and
C1.6.3.1   *restricted coulomb energy* (LeBouquin 1994, Scofield and Reilly 1991)), or for those which are built as
C1.4   *stochastic devices* (Clarkson and Ng 1993, Clarkson *et al* 1993a, b, Köllmann *et al* 1966), it is almost impossible to assess their speed. It should be mentioned that due to such unsurmountable problems there is usually little if any information on benchmarks.

E1.4.4.2 *Typical and recent examples*

We shall firstly mention Mark III and IV from a historical point of view.

- *Mark III* was built at TRW, Inc., during 1984 and 1985 Hecht-Nielsen (1989). The design used eight Motorola M68010-based boards running at 12 MHz, with 512 kbytes of DRAM memory each. The software environment used was called ANSE (Artificial Neural Systems Environment). The original Mark III had a capacity of approximately 8000 processing elements (neurons) and 480 000 connections, and had a speed of 380 000 CPS (large instar network using Grossberg learning).
- *Mark IV* was also built at TRW, Inc., but under funding from the Defense Science Office of the Defense Advanced Research Projects Agency (DARPA). A detailed description is given by Hecht-Nielsen (1989) who, together with Todd Gutschow, was one of the designers. It was capable of

implementing as many as 262 144 processing elements and 5.5 M connections, and had a sustained speed of 5 MCPS, whether or not learning was taking place Kuczewsk *et al* (1988). It had a mass of 200 kg and drew 1.3 kW of power. The basic computing unit was a 16-bit Texas Instruments TMS32020 DSP. The idea was that Mark IV would be a node of a larger neurocomputer (which was never intended to be constructed).

In the meantime most of the neural network simulations have been performed on sequential computers. The performance of such software simulation was roughly between 25 000 and 250 000 CPS in 1989 (DARPA 1989). Fresh results show impressive improvements on computers having just one processing element.

- *IBM 80486/50MHz* exhibits 1.1 MCPS and 0.47 MCUPS (Müller *et al* 1995).
- *Sun* (Sparcstation 10) has 3.0 MCPS and 1.1 MCUPS Müller *et al* (1995).
- *NEC SX-3* (supercomputer) achieves 130 MCUPS (the implementation was presented by Koike from NEC at the Second ERH-NEC Joint Workshop on Supercomputing 1992 Zürich, but no published English reference seems to be available). As NEC SX-3 has 5.9 Gflops it is expected that a similar performance would be obtained on a Cray Y-MP/8 (which has 2.5 Gflops).

Similar results have been reported for Hypercube FPS 20 (Roberts and Wang 1989, Neibur and Brettle 1992) and CM (Deprit 1989, Zhang *et al* 1990). At least one order of magnitude increase can be expected on Fujitsu, Intel Paragon or on the NEC SX-4.

As a first alternative and aimed at increasing the speed of simulations on PCs and workstations, *special acceleration boards* have been developed Williams and Panayotopoulos (1989).

- *Delta Floating Point Processor* from the Science Application International Corporation (SAIC), has separate addition and multiplication parts; it runs at 10 MCPS and 1–2 MCUPS (Souček and Souček 1988, Works 1988).
- SAIC later developed *SIGMA-1* which has a 3.1 M virtual interconnections and has reached 11 MCUPS (Treleaven 1989).
- *ANZA Plus* from Hecht-Nielsen Computers (Hecht-Nielsen 1988) has a 4-stage pipelines Harvard architecture. It can go up to 1 M virtual processing elements, 1.8 MCUPS (Atlas and Suzuki 1989) and 6 MCPS (Treleaven 1989).
- Intel i860 RISC processor is used in the *Myriad MC860* board and in the *Balboa* board from HNC, showing around 7 MCUPS (Hecht-Nielsen 1991).

Many other accelerator boards are mentioned in a tabular form by Lindsey and Lindblad (1994).

One simple way to increase performance even more is to use processors in parallel. A classical design style was used for *slice architectures*, and several representative models are detailed.

- Micro Devices have introduced the NBS (Neural Bit Slice) chip MD1220 (Micro Devices 1989a–c, 1990). The chip has eight processing elements with hard-limit thresholds and eight inputs (Yestrebsky *et al* 1989). The architecture is suited for multiplication of a 1-bit synapse input with a 16-bit weight. The chip only allows for hard-limiting threshold functions. The weights are stored in standard RAM, but only eight external weights per neuron and seven internal weights per neuron are supported. Such a reduced fan-in (maximum 15 synapses per neuron) is quite a drastic limitation. This can be avoided by additional external circuits, but increasing the fan-in decreases the accuracy (as the 16-bit accumulator can overflow). The chip has a processing rate of 55 MIPS which roughly would correspond to 8.9 MCPS.
- A similar chip is the *Neuralogix NLX-420* Neural Processor Slice from Neuralogix (1992), which has 16 processing elements. A common 16-bit input is multiplied by a weight in each processing element in parallel. New weights are read from off-chip. The 16-bit weights and inputs can be user selected as 16 1-bit, 4 4-bit, 2 8-bit or 1 16-bit value(s). The 16 neuron sums are multiplexed through a user-defined piecewise continuous threshold function to produce a 16-bit output. Internal feedback allows for multilayer networks.
- The Philips *L-Neuro 1.0* chip (Duranton and Maudit 1989, Duranton and Sirat 1989, Theeten *et al* 1990, Maudit *et al* 1992) was designed to be easily interfaced to transputers. It also has a 16-bit processing architecture in which the neuron values can be interpreted as 8 2-bit, 4 4-bit, 2 8-bit or 1 16-bit value(s). Unlike the NLX-420, there is a 1 kbyte on-chip cache to store the weights. The chip has 32 inputs and 16 output neurons and only the loop on the input neurons is parallelized (weight parallelism). This chip has on-chip learning with an adjustable learning rate. The transfer function is

computed off-chip. This allows for multiple chips to provide synapse-input products to the neurons and, thus, to build very large networks. An experiment with 16 L-Neuro 1.0 (Maudit *et al* 1991) was able to simulate networks with more than 2000 neurons and reached 19 MCPS and 4.2 MCUPS. The work has been continued: a *L-Neuro 2.0* architecture was reported (Dejean and Caillaud 1994), followed recently (Duranton 1996) by *L-Neuro 2.3* (see the paragraph on the latest implementations).

- *BACCHUS* is another slice architecture which was designed at Darmstadt University of Technology. There have been three successive versions I, II, and III (Huch *et al* 1990, Pöchmüller and Glesner 1991). The neurons perform only a hard-limiting threshold function. The final version was designed as a sea-of-gates in 1.5-μm CMOS (Glesner *et al* 1989, Glesner and Pöchmüller 1991). The chip contains 32 neurons and runs at 32 MCPS (but for 1-bit interconnections!). An associative system PAN IV, based on BACCHUS III chips has been built (Palm and Palm 1991). It has eight BACCHUS III chips (for a total of 256 simple processors) and 2 Mbytes of standard RAM. The system was designed only as a binary correlation matrix memory.

For even higher performances the designers have used SIMD arrays (various one- or two-dimensional systolic architectures (Kung 1988, Kung and Hwang 1988, 1989a, 1989b, Kung and Webb 1985), made of DSPs (digital signal processors), RISC processors, transputers or dedicated chips.

Many neuroprocessors have been built as *arrays of DSPs*.

- One of the first array-processors proposed for neural network simulation was built at IBM Palo Alto Scientific Center (Cruz *et al* 1987). The building block was the *NEP* (Network Emulation Processor) board able to simulate 4000 nodes (neurons) with 16 000 links (weights) and a speed of between 48 000 and 80 000 CUPS. Up to 256 NEPs could be cascaded (through a NEPBUS communication network), thus allowing for networks of 1 million nodes and 4 million links.

- Another DSP neuroprocessor called *SANDY* emerged from Fujitsu Laboratories (Kato *et al* 1990). The DSP used was the Texas Instruments TMS320C30 connected in a SIMD array. SANDY/6 (with 64 processors) was benchmarked on NETtalk (Sejnowski and Rosenburg 1986) at 118 MCUPS and 141 MCPS. SANDY/8 with 256 processors was expected to work at 583 MCUPS (Yoshizawa *et al* 1991).

- The *RAP* (Ring Array Processor) developed at the International Computer Science Institute (ICSI, Berkeley) is an array of between 4 and 40 Texas Instruments TMS320C30 DSPs containing 256 kbytes of fast static RAM and 4 Mbytes of dynamic RAM each (Morgan *et al* 1990, 1992, 1993, Kohn *et al* 1992). These chips are connected via a ring of Xilinx programmable gate arrays, each implementing a simple two register data pipeline and running at the DSP clock speed of 16 MHz. A single board can perform 57 MCPS and 13.2 MUCPS, with a peak performance for a whole system reaching 640 MCPS (tested at 570 MCPS) and 106 MCUPS.

- At the Swiss Federal Institute of Technology in Zürich, a 63-processor system named *MUSIC* (Multiprocessor System with Intelligent Communication) has been developed (Müller *et al* 1992, 1994, 1995). The architecture is similar to that of RAP but differs in the communication interface. Three Motorola 96002 DSPs (32-bit floating-point) are mounted on one board, each one with a Xilinx LCA XC3090 programmable gate array and an Inmos T805 transputer. Up to 21 boards (i.e., 63 processors) fit into a standard 19-inch rack. A global 5-MHz ring connects the nodes and communication can be overlapped with computation. The complete system has achieved 817 MCPS and 330 MCUPS (for a 5000-1575-63 two-layer perceptron), but the peak performance is 1900 MCPS. A fully equipped system consumes 800 W.

- *PANNE* (Parallel Artificial Neural Network Engine) has been designed at the University of Sydney (Milosavlevich *et al* 1996) and exploits the many specialized features of the TMS320C40 DSP chip. One board contains two DSPs together with 32 Mbytes of DRAM and 2 Mbytes of high speed SRAM. These are accessed through a dedicated local bus. Apart from this local bus, each board has a global bus and six programmable unidirectional 8-bit ports specially designed to allow connections of neighboring DSPs at 20 Mbytes per second. The system has up to eight boards and is estimated at 80 MCUPS.

Different solutions have been implemented on *arrays (networks) of transputers* (Ernst *et al* 1990, Murre 1993). Ernoult (1988) reported that a network of 2048 neurons with 921 600 connections running on a 16-transputer system (T800) has reached 0.57 MCUPS. A Megaframe *Hypercluster* from Parsytec (Achen, Germany), having 64 transputers (T800) and implementing backpropagation, runs at 27 MCPS and 9.9 MCUPS (Mühlbein and Wolf 1989). This performance should increase tenfold on the Parsytec's *Gigacluster* which uses T9000 transputers.

Instead of transputers some researchers have used RISC processors and here are some of the neurocomputers built as *arrays of RISC processors.*

- One solution was to design a RISC processor (dedicated for simulating neural networks) and assembling several of them in SIMD arrays. Here we can mention the 16-bit *Neural RISC* developed at University College London (Pacheco and Treleaven 1989, Treleaven *et al* 1989, Treleaven and Rocha 1990). Several neural RISCs have been connected in a linear array. A linear array interconnecting scheme has several advantages: simplified wiring and ease of cascadability. Several arrays are linked by an interconnecting module (Pacheco and Treleaven 1992). This allows for different topologies (rings, meshes, cubes) and is expandable up to a maximum of 65 536 processors. The flexibility is high as the computer is of the MIMD type (multiple instructions multiple data).

- *REMAP*[3] was an experimental neurocomputing project (Bengtsson *et al* 1993, Linde *et al* 1992) with its objective being to develop a parallel reconfigurable SIMD computer using FPGAs. The performance was estimated to be between 100 and 1000 MCUPS.

- Another solution is to use a standard RISC processor. An example is the *128 PE RISC* which uses the Intel 80860 (Hiraiwa *et al* 1990). 128 processors are connected in a two level pipeline array where the horizontal mesh connections serve for information exchange (weights) and vertical meshes share dataflow. For a 256-80-32 network and 5120 training set vectors, the performance is around 1000 MCUPS.

- *BSP400* from Brain Style Processor (Heemskerk *et al* 1991, Heemskerk 1995) used low-cost commercial microprocessors MC68701 (8-bit microprocessor). Due to the low speed of the processor used (1 MHz!) the overall performance reached only 6.4 MCUPS when 400 processors were used.

Because both DSP and RISC processors are too powerful and flexible for the task of simulating neural networks, a better alternative is to use smaller and more specific (less flexible) dedicated processing elements. This can increase the computational power and also maintain a very small cost. The trend has been marked by the use of *SIMD arrays* (Single Instruction Multiple Data) and especially *systolic arrays* (Kung and Hwang 1988) of dedicated chips. Systolic arrays are a class of architecture where the processing elements and the interconnecting scheme can be optimized for solving certain classes of algorithms. Matrix multiplication belongs to this class of algorithms (Leiserson 1982), and it is known that neural network simulation relies heavily on matrix multiplication (Beiu 1989, Kham and Ling 1991, Kung and Hwang 1989b). The SIMD arrays are similar structures, the main difference being that the elementary processing elements have no controllers and that a central controller is in charge of supervising the activity of all the elementary processing elements.

- The *WARP* array was probably the earliest systolic one (Kung and Webb 1985, Arnould 1985, Annaratone *et al* 1987). Although built primarily for image processing, it has also been used for neural network simulation (Pomerleau *et al* 1988). It is a ten (or more) processor programmable systolic array. The system can work either in a systolic mode, or in a local mode (each processor works independently). A performance of 17 MCUPS was obtained on a 10-processor WARP.

- *ARIANE* chip (Gascuel *et al* 1992) is a 64-neuron implementation in a 1.2-μm CMOS of the architecture first proposed by Weinfeld (1989). The chip—having 420 000 transistors in 1 cm^2—implements a fully digital Hopfield-type network, thus continuing on the lines of other Hopfield-type implementations (Sivilotti 1986, Graf *et al* 1986). All operations are performed by a 12-bit adder/subtracter. There are 64 connections per neuron, making it possible to store 4096 weights. The reported speed is 640 MCUPS, but this figure cannot be compared to standard CUPS as the chip does not implement backpropagation. The main drawback is that the chip is not easily cascadable (however, a four chip board has been designed).

- *SNAP* (SIMD Neurocomputer Array Processor) from Hecht-Nielsen Computers, Inc is based on HNC 100 NAP chips (Neural Array Processor). The chip is a one-dimensional systolic array of four arithmetic cells forming a ring (Hecht-Nielsen 1991, Means and Lisenbee 1991) and implementing IEEE 32-bit floating-point arithmetic. Each arithmetic cell contains a 32-bit floating point multiplier, floating point ALU and integer ALU, and runs at 20 MHz with all instructions being executed in one clock cycle. Four NAPs are linked on one SNAP board. SNAP has either 32 (SNAP-32) or 64 (SNAP-64) processors (i.e., either two or four boards). The SNAP-32 performed at 500 MCPS (peak performance being 640 MCPS) and 128 MCUPS. Although the system performs lower than CNAPS (described below), we have to mention that SNAP uses 32-bit floating point arithmetic.

- The *APLYSIE* chip is a two-dimensional systolic array dedicated for Hopfield-type networks (Blayo and Hurat 1989). Since the outputs are only $+1$ and -1, the synaptic multiplication can be performed by an adder/subtracter (like in Weinfeld's 1989 solution). The weights are limited to 8-bit and the partial product is computed by a 16-bit register. The adder/subtracter is of the serial type for minimizing the area, but is also thought for the serial interconnecting scheme used. An advantage of such a solution is its cascadability.

- The *GENES* chip is a generalization of APLYSIE and it was implemented at the Swiss Federal Institute of Technology (Lausanne) as part of the *MANTRA* project (Lehmann and Blayo 1991, Lehmann *et al* 1993, Viredaz *et al* 1992). It is based on the same recurrent systolic array as APLYSIE, but it has been enhanced to simulate several neural network architectures. The first chip of the family was GENES HN8 implementing each synapse as a serial-parallel multiplier. Two versions have been fabricated: 2×2 array of processors and 4×4 array of processors. Weights and inputs are represented on 8 bits. The partial sum is calculated on 24 bits. A full board, GENES SY1, was built as a 9×8 array of GENES HN8 2×2 chips (18×16 synapses) and was able to reach 110 MCPS. A GENES IV chip was later designed as an upgrade of GENES HN8 (Lehmann *et al* 1993, Viredaz *et al* 1992). It has 16-bit inputs and synaptic weights and uses 39 bits for the partial sum. The chip was designed with standard cells in a 1-μm CMOS technology on a 6.2×6.2 mm^2 area. Together with another chip, GACD1 (dedicated to the error computation for delta rule and backpropagation), it was used to build the first MANTRA neurocomputer as a 40×40 array of processing elements. The speed is estimated at 500 MCPS and 160 MCUPS.

- A low-cost high-speed neurocomputer system has recently been proposed (Strey *et al* 1995) and implemented (Avellana *et al* 1996). The system is based on a dedicated AU chip which has been designed so as to *dynamically adapt the internal parallelism to data precision*. It tends to achieve an optimal utilization of the available hardware resources. The AU chip is organized as a pipeline structure where the data path can be adapted dynamically to the encoding of the data values. The chip has been realized in 0.7 μm and has 80 mm^2. Four chips are installed on a board together with: a Motorola DSP96002 (used for the management of the local bus, computation of the sigmoid function, error calculation, winner calculation and convergence check); an FPGA for communication; local weight memories; central memory; and FIFO memory. Several boards can be used together. For 16-bit weights and with only one board the estimated performance is 480 MCPS and 120 MCUPS.

- *TNP* (Toroidal Neural Processor) is a linear systolic neural accelerator engine developed at Loughborough University of Technology (Jones and Sammut 1993, Jones *et al* 1990, 1991). The system is still under development although several prototype chips have been successfully fabricated and tested.

- *HANNIBAL* (Hardware Architecture for Neural Networks Implementing Backpropagation Algorithm Learning) was built at British Telecom. A dedicated HANNIBAL chip contains eight processing elements (Myers *et al* 1991, Orrey *et al* 1991, Naylor *et al* 1993), each one with 9216 bits of local memory (configurable as 512 17-bit words, or 1024 9-bit words). Such a chip allows for high fan-in neurons to be implemented; up to four lower fan-in neurons can be mapped onto one processing element. The neuron activation function is realized by a dedicated approximation for area saving reasons. The chip uses reduced word length (8-bit in the recall phase and 16-bit when learning (Vincent and Myers 1992) and it was fabricated in a 0.7-μm CMOS technology. This has led to 750 000 transistors in a 9×11.5 mm^2 area. The clock frequency is 20 MHz and a single chip can reach 160 MCPS.

- *MM32K* (Glover and Miller 1994) is a SIMD having 32 768 simple processors (bit serial). A custom chip contains 2048 processors. The bit serial architecture allows for the variation of the number of bits (variable precision). The processors are interconnected by a 64×64 full crossbar switch with 512 processors connected to each port of the switch.

- *SYNAPSE 1* and *SYNAPSE X* (Synthesis of Neural Algorithms on a Parallel Systolic Engine) from Siemens (Ramacher 1990, 1992, Ramacher *et al* 1991b, 1993) are dedicated to operation on matrices based on the MA16 chip (Beichter *et al* 1991), which has four systolic chains (of four multipliers and four adders each). The chip runs at 25 MHz and was fabricated in 1.0-μm CMOS. Its 610 000 transistors occupy 187 mm^2. The MA16 alone has 800 MCPS when working on 16-bit weights. SYNAPSE neurocomputer is nothing else but a two-dimensional systolic array of MA16 chips arranged in two rows by four columns. The weights are stored off chip in local memories. Both processor rows are connected to the same weight bus which excludes the operation on different input patterns. The MA16s in a row form a systolic array where input data as well as intermediate results are propagated for obtaining the total weighted sum. Multiple standard 68040s and additional integer ALUs are used as general purpose processors which complement the systolic processor array. The standard configuration has eight MA16s, two MC68040 for control and 128 Mbytes of DRAM. It performs at 5100 MCPS and 133 MCUPS.

- *CNAPS* (Connected Network of Adaptive Processors) is a SIMD array from Adaptive Solutions, Inc (Adaptive Solutions 1991, 1992, Hammerstrom 1990). X1 is a neural network dedicated chip with on-chip learning. It consists of a linear array of elementary processors, each one having a 32-bit adder and a 24-bit multiplier (fixed-point). The structure of an elementary processor is such that it can work with three different weight lengths: 1-bit, 8-bit and 16-bit weights (Hammerstrom 1990, Hammerstom and Nguyen 1991). X1 chips are fully cascadable, allowing the construction of linear arrays having arbitrary many elementary processors. Another chip, the N64000, was produced in 0.8-μm CMOS and 80 elementary processors have been embedded in this design. N64000 is a large chip (one square inch) containing over 11 million transistors (Griffin *et al* 1991) and due to defects in the fabrication process only 64 functioning processing elements are used from one chip (the 16 more being redundant). The same idea will be used at a higher level for the Hitachi's WSI (wafer scale integration) to be discussed later. The maximum fan-in of one neuron is 4096 and there are 256K programmable synapses on the 26.2×27.5 mm^2 chip. The chip alone can perform 1600 MCPS and 256 MCUPS for 8- or 16-bit weights (12 800 MCPS for 1-bit weight). The CNAPS has four N64000 chips running at 20 MHz on one board (256 processing elements). The maximum performance of the system is quite impressive: 5700 MCPS and 1460 MCUPS (Adaptive Solutions 1991, 1992, McCator 1991), but these values are for 8- and 16-bit weights! Hammerstrom and Nguyen (1991) have also compared a Kohonen self-organizing map implemented on the CNAPS: 516 MCPS and 65 MCUPS, with the performance on a SPARC station: 0.11 MCPS and 0.08 MCUPS.

- *MasPar MP-1* is a SIMD computer based on the MP-1 chip (Blank 1990, MasPar 1990). It is a general purpose parallel computer but it exhibits excellent performances when simulating neural networks. The core chip is MP-1 which has 32 processing elements working on 32-bit floating point numbers (each processing element can be viewed as a small RISC processor). MP-1 was fabricated in 1.6-μm CMOS on an area of 11.6×9.5 mm^2 and has 450 000 transistors. The chip works at a moderate clock frequency of only 14 MHz for minimizing the dissipated power. One board uses 32 MP-1 chips, thus having 1024 processing elements which are arranged in a two-dimensional array. The connection scheme is different from others: 16 processing elements are configured as a 4×4 array with an X-net mesh and form a 'processor element cluster'. These clusters are again connected as an X-net mesh of clusters. The processors are connected together from the edges to form a torus. On top of that, a global communication between processing elements is realized by a dedicated 1024×1024 crossbar interconnecting network having three stages for routing. MasPar can have from 1 to 16 boards. The largest configuration has 16 384 processing elements. Grajski *et al* (1990) have simulated neural networks on a MasPar MP-1 with 4096 processing elements (MasPar MP-1 1100). A 900-20-17 backpropagation network obtained 306 MCUPS, but on the largest MasPar MP-1 1200 (16 384 processing elements) performance is expected to be on the order of GCUPS.

Many *other alternatives* have also been presented and we shall shortly enumerate some of them here.

- *WISARD* belongs to the family of weightless neural networks or the RAM model (Aleksander and Morton 1990) and has been used in image recognition. $_{\text{C1.5.4}}$

- The *pRAM* (probabilistic RAM) is a nonlinear stochastic device (Gorse and Taylor 1989a, b, 1990a, $_{\text{C1.5.2}}$ 1991a, c) with neuron-like behavior which—as opposed to the simple RAM model—can implement nonlinear activation functions and can generalize after training (Clarkson *et al* 1993a). It is based on a pulse-coding technique and several chips have been fabricated. The latest digital pRAM has 256 neurons per chip. The 16-bit 'weights' (probabilities) are stored in an external RAM in order to keep the costs at a minimum. Up to 1280 neurons can be interconnected by combining five chips. Learning (Clarkson and Ng 1993, Clarkson *et al* 1991a, b, 1992b, 1993b, c, Gorse and Taylor 1990b, 1991b, Guan *et al* 1992) is performed on-chip. The pRAM uses a 1-μm CMOS gate-array with 39 000 gates. A PC board has been designed and tested. A VMEbus-based neural processor board (using the pRAM-256) has also been recently built (El-Mousa and Clarkson 1996). The current VMEbus version is being used for studying the various different architectures and advantages of hardware-based learning using pRAM artificial neural networks. For this purpose, the board relies heavily on the use of in-system programmable logic devices (ISPLD) to facilitate changing the support hardware logic associated with the actual neural processor without the need to rewrite and/or exchange parts of it.

- Intel has several neural network solutions (Intel 1992a, b). Two commercial chips are dedicated to radial basis functions (Watkins *et al* 1992): the *IBMZISC036* (LeBouquin 1994) and *Ni1000* (Scofield and Reilly 1991) build in cooperation with Nestor. The *ZISC036* (from Zero Instruction Set Computer) contains 36 prototype neurons, where the vectors have 64 8-bit elements and can be assigned to categories from 1 to 16 384 (i.e., the first layer has 36 neurons fully connected by 8-bit weights to the 64 neurons of the second layer). Multiple ZISC036 chips can be easily cascaded to provide additional prototypes, while the distance norm is selectable between city-block (Manhattan) or the largest element difference. The ZISC036 implements a region of influence (ROI) learning algorithm (Verleysen and Cabestany 1994) using signum basis functions with radii of 0 to 16 383. Recall is either according to the ROI identification, or via the nearest-neighbor readout, and takes 4 μs for a 250-K sec-pattern presentation rate.

  The *Ni1000* was developed jointly by Intel and Nestor and contains 1024 prototypes of 256 5-bit elements (i.e., the first layer has 256 neurons, while the second layer is fully connected to the first layer by 5-bit weights and has 1024 neurons). The distance used is the city-block (Manhattan) distance. The third layer has 64 neurons working in a sequential way, but achieving higher precision. All the weights and the threshold are stored on board in a nonvolatile memory, as the chip is implemented in Intel's 0.8-μm EEPROM process. On the same chip a Harvard RISC is used to accelerate learning (Johnson 1993b), and increases the overall number of transistors to 3.7 million. The chip implements two on-chip learning algorithms: restricted coulomb energy or RCE (Reilly *et al* 1982) and probabilistic neural networks or PNN (Specht 1988). Other algorithms can be microcoded. In a pattern processing application the chip can process 40 000 patterns per second (Holler *et al* 1992).

- A *generic neural architecture* was proposed by Vellasco and Treleaven (1992). The idea is to tailor the hardware to the neural network to be simulated. This can increase the performance at the expense of reduced flexibility. The aim of such an approach is to automatically generate application-specific integrated circuits (ASICs). Several chips have been fabricated. Other authors have been working on similar approaches (Disante *et al* 1990b, Fornaciari *et al* 1991a, b), or have tried a mapping onto FPGAs (Beiu and Taylor 1995c, Botros and Abdul-Aziz 1994, Gick *et al* 1993, Nigri *et al* 1991, Nijhuis *et al* 1991, Rossmann *et al* 1996, Rückert *et al* 1991).

C1.4   • Several implementations of the *Boltzmann machine* have also been reported. A high-speed digital one is that of Murray *et al* (1992, 1994). The chip, realized in a 1.2-μm CMOS technology, has 32 neural processors and four weight update processors supporting an arbitrary topology of up to 160 functional neurons. The 9.5×9.8 mm^2 area hosts 400 000 transistors. This includes the 20 480 5-bit weights stored in a dynamic RAM (the activation and temperature memories are static). Although clocked at 125 MHz, the chip dissipates less than 2 W. The theoretical maximum learning rate is 350 MCUPS and the recall rate is typically 1200 patterns per second. An SBus interface board was developed using several reconfigurable Xilinx FPGAs.

- *ArMenX* is a distributed computer architecture (Poulain Maubant *et al* 1996) articulated around a ring of FPGAs acting as routing resources as well as fine grain computing resources (Léonhard *et al* 1995). This allows for a high degree of flexibility. Coarse grain computing relies on transputers and DSPs. Each ArMenX node contains an FPGA (Xilinx 4010) tightly coupled to an Inmos T805 transputer and a Motorola DSP56002, but other processors could be used. The node has 4 Mbytes of transputer RAM and 384 Kbytes of DSP RAM and the FPGA connects to the left and right neighboring nodes. The sustained performance of a node is about 5 MCPS and 1.5 MCUPS, and it is expected that the scale-up will be linear for a 16-node machine: 80 MCPS and 24 MCUPS.

- A solution which uses *on-line arithmetic* has been proposed in Girau and Tisserand (1996) and should be implemented on an FPGA. A redundant number representation allows very fast arithmetic operations, the estimated speed being 5.2 MCUPS per chip.

- The use of *stochastic arithmetic* computing for all arithmetic operations of training and processing backpropagation networks has also been considered (Köllmann *et al* 1996). Arithmetic operations become quite simple. The main problem in this case is the generation of numerous independent random generators. The silicon reported uses a decentralized pseudorandom generator based on the principle of shifting the turn-around code for parities formed on partial stages of a feedback shift register. A 3.5×2.8 mm^2 silicon prototype has been implemented in 1-μm CMOS technology. The prototype delivers a theoretical performance of 400 MCUPS for 12-bit weight length and 15-bit momentum length. It is estimated that a state-of-the-art 0.25-μm process would allow 4K synapses and 64 neurons should fit into 160 mm^2 if standard cells are used; a custom design should increase these values to: 16K synapses and 128 neurons.

Some of the latest implementations are pushing the performances even further and we shall mention here the most promising ones, even if by our classification some of them might also fall in another class.

- The *RM-nc* is a reconfigurable machine for massively parallel-pipelined computations and has been proposed in Erdogan and Wahab (1992). The reconfigurability is not only in the domain of communication and control, but also in the domain of processing elements. A fast floating point sum-of-products circuit using special carry-save multipliers (with built-in on-the-fly shifting capability and extensive pipelining) has been proposed and has to be implemented on FPGAs. The performance of an RM-nc256 machine (with 256 processing FPGAs) has been estimated for NETtalk (203-60-26 network with 13 826 connections) at a speed of 2000 MCUPS. No implementation has yet been reported.

- One interesting development is based on *WSI* (Mann *et al* 1987, Rudnik and Hammerstrom 1988, Tewksbury and Hornak 1989). A first neural network WSI has been developed by Hitachi (Yasunaga *et al* 1989, 1990). This first version was designed only for Hopfield networks without learning. *Hitachi's WSI* has 576 neurons with a fan-in of 64. Weights are represented on 10 bits. If larger fan-in is required, three neurons can be cascaded to increase the fan-in to 190 (this reduces the number of available neurons). A 'small' 5-inch wafer and a 0.8-μ CMOS technology has been used to realize the designed 19 million transistors. The wafer has 64 chips of 12 neurons each; one redundant chip (Zorat 1987) is used to replace faulty neurons from the other chips. Up to 37K synapses are available on chip. For controlling the neurons and the buses there are eight more chips on the wafer. The only way to keep the power to a reasonable 5 W is a quite-slow clock rate: 2.1 MHz, but the actual performance is still around 138 MCPS.

  The same idea has been used by Hitachi (Boyd 1990) to design a WSI for multilayer feedforward networks including the backpropagation algorithm. The weights' accuracy has been increased to 16 bits to cope with the required precision of on-chip learning. One wafer has 144 neurons and eight wafers have been stacked together to form a very small neurocomputer with 1152 neurons (Yasunaga *et al* 1991). The reported speed is 2300 MCUPS. Using a similar architecture and the present day state-of-the-art 0.3-μm CMOS technology it becomes clear that we can expect to have 10 000 neurons WSI in the very near future.

- For portable applications *Hitachi* has also developed a *1.5 V digital chip* with 1 048 576 synapses (Watanabe *et al* 1993). The chip can emulate 1024 fully connected neurons (fan-in of 1024 each) or three layers of 724 neurons. An on-chip DRAM cell array is used to store the 8-bit weights. A 256 parallel circuit for summing product (Baugh parallel multiplier) pushes the processing speed to 1370 MCPS. A scaled-down version of the chip was fabricated using a 0.5-μm CMOS design rule. It allowed an estimation of the full-scale chip: 15.4×18.6 mm^2 and 75 mW.

- The new *L-neuro 2.3* (Duranton 1996) is a fully programmable vectorial processor in a highly parallel chip composed of an array of twelve DSPs which can be used not only for neurocomputing, but also for fuzzy logic applications, real-time image processing and digital signal processing. Beside the twelve DSPs, the chip contains: a RISC processor, a vector-to-scalar unit, a 32-bit scalar unit, an image addressing module and several communication ports. All the DSPs are linked together: by a broadcast bus connecting all DSPs; by two shift chains linking the DSPs as a systolic ring; by fast neighbor-to-neighbor connections existing between adjacent DSPs; and also to an I/O port. All the internal buses are connected together through a programmable crossbar switch. The RISC processor of one chip can be used to control several other L-Neuro chips, allowing an expansion in a hierarchical fashion. The chip was fabricated in 0.6-μm technology and has 1.8 million transistors clocked at 60 MHz. It can implement different learning algorithms such as backpropagation, Kohonen features map, radial basis functions and neural trees (Sirat and Nadal 1990). The peak performance is estimated at 1380 MCUPS and 1925 MCPS but no tests have yet been reported.

- One very interesting approach is the novel *SMA* (Sparse Memory Architecture) neurochip (Aihara *et al* 1996) which uses specific models to reduce neuron calculations. SMA uses two key techniques to achieve extremely high computational speed without an accuracy penalty: 'compressible synapse weight neuron calculation' and 'differential neuron operation'. The compressible synapse weight neuron calculation uses the transfer characteristics of the neuron to stop the calculation for the sum if it is determined that the final sum will be in the saturation region. This also cancels subsequent memory accesses for the synapse weights. The purpose of differential neuron operation is to do calculations only on those inputs whose level has changed. A dedicated processing unit having a 22-bit adder, a 16-bit shifter, an EX-OR gate and two 22-bit registers has been designed. A test chip having 96 processing units has been fabricated in 0.5-μm CMOS and has 16.5×16.7 mm^2. It runs at 30 MHz and dissipates 3.2 W. The chip can store 12 228 16-bit synapse weights and has a peak performance of 30 GCPS (tested at 18 GCPS).

- *SPERT* (from Synthetic Perceptron Testbed) (Asanović *et al* 1992, 1993d, Wawrzynek *et al* 1993, 1996) is a fully programmable single chip neuromicroprocessor which borrowed heavily from the experience gained with RAP (Morgan *et al* 1990, 1992, 1993). It combines a general purpose integer data path with a vector unit of SIMD arrays optimized for neural network computations and with a wide connection to external memory through a single 128 VLIW instruction format. The chip is implemented in 1.2-μm CMOS and runs at 50 MHz. It has been estimated at a peak performance of 350 MCPS and 90 MCUPS. The chip is intended to be a test chip for the future Torrent chip: the basic building block of CNS-1 (see below). Recent developments have led to SPERT-II (Wawrzynek *et al* 1996) which has a vector instruction set architecture (ISA) based on the industry standard MIPS RISR scalar ISA.

- *NESPINN* (Neurocomputer for Spike-Processing Neural Networks) is a mixed SIMD/dataflow neurocomputer (Roth *et al* 1995, Jahnke *et al* 1996). It will allow the simulation of up to 512K neurons with up to 10^4 connections each. NESPINN consists of the spike-event list (the connectivity of sparsely connected networks is performed by the use of lists), two connectivity units containing the network topology (a regular and a nonregular connection unit), a sector unit controlling the processing of sectors and the NESPINN chip. The chip has a control unit and eight processing elements; each processing element has 2 Kbytes of on-chip local memory and an off-chip neuron state memory. The chip has been designed and simulated and will be implemented in 0.5-μm CMOS. It will operate at 50 MHz in either SIMD or dataflow mode. The estimated performance of the system with one NESPINN chip for a model network with 16K neurons of 83 connections each is 10^{11} CUPS.

- *CNS-1* from University of California Berkeley is the acronym from Connectionist Network Supercomputer-1 (Asanović *et al* 1993a–c, 1994) and is currently under development. It is targeted for speech and language processing as well as early and high-level vision and large conceptual knowledge representation studies. The CNS-1 is similar to other massively parallel computers with major differences in the architectural details of the processing nodes and the communication mechanisms. Processing nodes will be connected in a mesh topology and operate independently in a MIMD style. The processor node, named Torrent, includes: an MIPS CPU with a vector coprocessor running at 125 MHz, a Rambus external memory interface, and a network interface. The design is scalable up to 1024 Torrent processing nodes, for a total of up to 2 TeraOps and 32 Gbytes of RAM. The host and other devices will connect to CNS-1 through custom VLSI I/O nodes named Hydrant connected to one edge of the mesh and allowing up to 8 Gbytes of I/O bandwidth. A sketch of the future CNS-1 can be seen in figure E1.4.2. The goal set ahead is to be able to evaluate networks with one million neurons and an average of one thousand connections per unit (i.e., a total of a billion connections) at a rate of 100 times per second, or 10^{11} CPS and 2×10^{10} CUPS.

Several of the implementations presented here have been plotted in figures E1.4.3 (digital neurochips) and E1.4.4 (dedicated neurocomputers and supercomputers). As can be seen from these two figures, some architectural improvements are to be expected from the techniques used in designs like the SMA and the NESPINN, which could reach speed performances similar to the CNS-1.

We shall not end this section before mentioning an interesting alternative that has recently emerged. To cope with the limited accuracy, new learning algorithms with quantized weights have started to appear (see also Section E1.2.4). One might call them 'VLSI-friendly learning algorithms', which was the topic covered in MicroNeuro'94. Such algorithms could be used to map neural networks onto FPGAs or to custom-integrate circuits. The first such learning algorithm (Armstrong and Gecsie 1979, 1991) is in fact synthesizing Boolean functions using adaptive tree networks whose elements—after training and elimination of redundant elements—perform classical (Boolean) logical operations (AND and OR). This

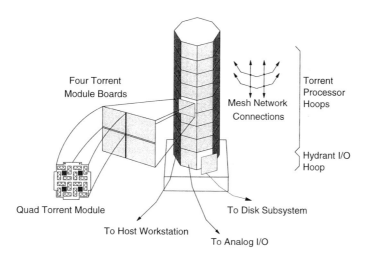

Figure E1.4.2. Connectionist Network Supercomputer CNS-1 (adapted from Asanović *et al* 1993b).

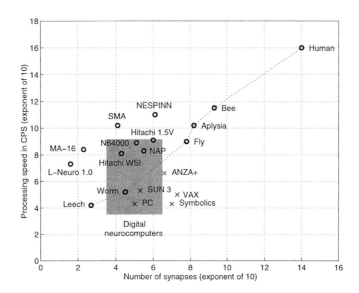

Figure E1.4.3. Different neurochips (circles) and classical computers (crosses) used for implementing artificial neural networks.

line of research has been extended by using a combination of AND and OR gates after an initial layer of threshold gates (Ayestaran and Prager 1993, Bose and Garga 1993). New learning algorithms have been developed by quantizing other learning algorithms (Höhfeld and Fahlman 1991, 1992, Jabri and Flower 1991, Makram-Ebeid *et al* 1989, Pérez *et al* 1992, Sakaue *et al* 1993, Shoemaker *et al* 1990, Thiran *et al* 1991, 1993) or by devising new ones (Fiesler *et al* 1990, Höhfeld and Fahlman 1991, Hollis and Paulos 1994, Hollis *et al* 1991, Mézard and Nadal 1989, Nakayama and Katayama 1991, Oliveira and Sangiovanni-Vincentelli 1994, Walter 1989, Xie and Jabri 1992), a particular class being the one dealing with threshold gates (Beiu *et al* 1994a, Beiu and Taylor 1995a, b, 1996a, Diederich and Opper 1987, Gruau 1993, Krauth and Mézard 1987, Kim and Park 1995, Littlestone 1988, Raghavan 1988, Roy *et al* 1993, Tan and Vandewalle 1992, 1993, Venkatesh 1989). Four overviews have compared and discussed such constructive algorithms (Śmieja 1993, Fiesler 1994, Moerland and Fiesler 1996, Beiu 1996c).

The main conclusion is that a lot of effort and creativity has been used recently to improve digital solutions for implementing artificial neural networks. The many designs proposed over the years make this area a lively topic confirming its huge interest. Fresh proposals together with estimates and/or results already show impressive performances competing with analog chips and reaching towards an area which, not so long ago, was considered accessible only for future optical computing.

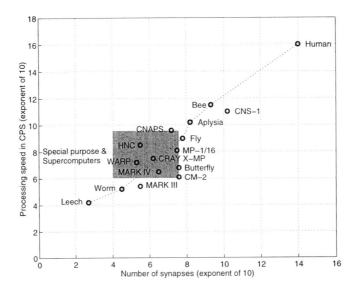

Figure E1.4.4. Different neurocomputers and supercomputers used for implementing artificial neural networks.

References

Abu-Mostafa Y S 1988a Connectivity versus entropy *Proc. Conf. on Neural Information Processing Systems* pp 1–8

——1988b Lower bound for connectivity in local-learning neural networks *J. Complexity* **4** 246–55

——1989 The Vapnik-Chervonenkis dimension information versus complexity in learning *Neural Comput.* **1** 312–7

Adaptive Solutions 1991 *CNAPS Neurocomputing Information Sheet on the CNAPS Neurocomputing system* (Adaptive Solutions Inc, 1400 NW Compton Drive Suite, 340 Beaverton, OR 97006, USA)

——1992 *CNAPS Server Preliminary Data Sheet* (Adaptive Solutions Inc, 1400 NW Compton Drive Suite, 340 Beaverton, OR 97006, USA)

Aihara K, Fujita O and Uchimura K 1996 A digital neural network LSI using sparse memory access architecture *Proc. Int. Conf. on Microelectronics for Neural Networks (1996)* pp 139–48

Akers L A, Walker M R, Ferry D K and Grondin R O 1988 Limited interconnectivity in synthetic neural systems *Neural Computers* eds R Eckmiller and C von der Malsburg (Berlin: Springer) pp 407–16

Albrecht A 1992 On bounded-depth threshold circuits for pattern functions *Proc. Int. Conf. on Artificial Neural Networks (1992)* (Amsterdam: Elsevier) pp 135–38

Aleksander I and Morton H B 1990 An overview of weightless neural nets *Proc. Int. Joint Conf. on Neural Networks (Washington, 1990)* vol II pp 499–502

Alippi C 1991 Weight representation and network complexity reductions *The Digital VLSI Implementation of Neural Nets Research Note RN/91/22* Department of Computer Science University College, London, February

Alippi C, Bonfanti S and Storti-Gajani G 1990a Some simple bounds for approximations of sigmoidal functions *Layered Neural Nets Report No 90-022* (Department of EE, Polytechnic of Milano)

——1990b Approximating sigmoidal functions for VLSI implementation of neural nets *Proc. Int. Conf. on Microelectronics for Neural Networks (1990)* pp 165–170.4

Alippi C and Nigri M 1991 Hardware requirements for digital VLSI implementation of neural networks *Proc. Int. Joint Conf. on Neural Networks (1991)* pp 1873–8

Alippi C and Storti-Gajani G 1991 Simple approximation of sigmoidal functions realistic design of digital neural networks capable of learning *Proc. Int. Symp. on Circuits and Systems (Singapore, 1991)* (Los Alamitos, CA: IEEE Computer Society Press) pp 1505–8

Alla P Y, Dreyfus G, Gascuel J D, Johannet A, Personnaz L, Roman J and Weinfeld M 1990 Silicon integration of learning algorithm and other auto-adaptive properties in a digital feedback neural network *Proc. Int. Conf. on Microelectronics for Neural Networks (1991)* pp 341–6

Allender E 1989 A note on the power of threshold circuits *IEEE Symp. on the Foundation of Computer Science* p 30

Alon N and Bruck J 1991 Explicit construction of depth-2 majority circuits for comparison and addition *Research Report RJ 8300 (75661)* (IBM Almaden, San Jose, CA)

Alspector J and Allen R B 1987 Neuromorphic VLSI Learning System *Advanced Research in VLSI, Proc. 1987 Stanford Conf.* ed P Losleben (Cambridge MA: MIT Press)

Alspector J, Allen R B, Hu V and Satyanaranayana S 1988 Stochastic learning networks and their electronic implementation *Proc. Conf. on Neural Information Processing Systems (1987)* pp 9–21

Amaldi E and Mayoraz E (eds) 1992 *Mathematical Foundations of Artificial Neural Networks (Summer School, Sion, September 1992)* (Swiss Federal Institute of Technology–Lausanne and Kurt Bösch Academic Institute–Sion 1992)

Annaratone M, Arnould E, Gross T, Kung H T, Lam M, Menzilcioglu O and Webb J A 1987 The WARP computer architecture implementation and performance *IEEE Trans. Comput.* **36** 1523–38

Anderson J A and Rosenfeld E 1988 *Neurocomputing: Foundations of Research* (Cambridge, MA: MIT Press)

Antognetti P and Milutinovic V (eds) 1991 *Neural Networks: Concepts Applications and Implementations* vol 2 (Englewood Cliffs, NJ: Prentice Hall)

Arai M 1993 Bounds on the number of hidden units in binary-valued three-layer neural networks *Neural Networks* **6** 855–60

Armstrong W W and Gecsei J 1979 Adaption algorithms for binary tree networks *IEEE Trans. Syst. Man Cybern.* **9** 276–85

Armstrong W W, Dwelly A, Liang J, Lin D and Reynolds S 1991 Some results concerning adaptive logic networks *Technical Report* (Department of Computer Science, University of Alberta, Edmonton, Canada)

Arnould E 1985 A systolic array computer *Proc. IEEE Int. Conf. on Application Specific Signal processing (Tampa, FL, 1985)* pp 232–5

Asanović K, Beck J, Callahan T, Feldman J, Irissou B, Kingsbury B, Kohn P, Lazzaro J, Morgan N, Stoutamire D and Wawrzynek J 1993a *CNS-1 Architecture Specification, Technical Report TR-93-021* (International Computer Science Institute and Unviersity of California, Berkeley)

Asanović K, Beck J, Feldman J, Morgan N and Wawrzynek J 1993b Development of a connectionist network supercomputer *Proc. Int. Conf. on Microelectronics for Neural Networks (1993)* pp 253–62

——1993c Designing a Connectionist Network Supercomputer *Int. J. Neural Systems* **4** 317–26

——1994 A supercomputer for neural computation *Proc. IEEE Int. Conf. on Neural Networks* vol 1 (Los Alamitos, CA: IEEE Computer Society Press) pp 5–9

Asanović K, Beck J, Kingsbury B E D, Kohn P, Morgan N and Wawrzynek J 1992 SPERT: A VLIW/SIMD neuro-processor *Proc. Int. Joint Conf. on Neural Networks (1992)* vol II pp 577–82

Asanović K and Morgan N 1991 Experimental determination of precision requirements for back-propagation training of artificial neural networks *Proc. Int. Conf. on Microelectronics for Neural Networks (1991)* pp 9–15

Asanović K, Morgan N and Wawrzynek J 1993d Using simulations of reduced precision arithmetic to design a neuro-microprocessor *J. VLSI Signal Processing* **6** 3–44

Atlas L E and Suzuki Y 1989 Digital systems for artificial neural networks *IEEE Circuits and Devices Mag.* **5** 20–4

Avellana N, Strey A, Holgado R, Fernández J A, Capillas R and Valderrama E 1996 Design of a low-cost and high-speed neurocomputer system *Proc. Int. Conf. on Microelectronics for Neural Networks (1996)* pp 221–6

Ayestaran H E and Prager R W 1993 *The Logical Gates Growing Network Technical Report 137* (Cambridge University Engineering Department, F-INFENG, July)

Baker T and Hammerstrom D 1988 Modifications to artificial neural network models for digital hardware implementation *Technical Report CS/E 88-035* (Department of Computer Science and Engineering, Oregon Graduate Center)

Barhen J, Toomarian N, Fijany A, Yariv A and Agranat A 1992 New directions in massively parallel neurocomputing *Proc. NeuroNimes '92* pp 543–54

Baum E B 1988a Supervised learning of probability distributions by neural networks *Proc. Conf. on Neural Information Processing Systems (1987)* pp 52–61

——1988b On the capabilities of multilayer perceptrons *J. Complexity* **4** 193–215

Baum E B and Haussler D 1989 What size net gives valid generalization? *Neural Comput.* **1** 151–60

Beck J 1990 The ring array processor (RAP) hardware *Technical Report TR-90-048* (International Computer Science Institute, Berkeley, CA, September)

Beichter J, Bruels N, Meister E, Ramacher U and Klar H 1991 Design of a general-purpose neural signal processor *Proc. Int. Conf. on Microelectronics for Neural Networks (1991)* pp 311–5

Beiu V 1989 From systolic arrays to neural networks *Sci. Ann. Informatics* **35** 375–85

——1994 Neural networks using threshold gates a complexity analysis of their area- and time-efficient VLSI implementations *PhD Dissertation* Katholieke Universiteit, Leuven, Belgium, x-27-151779-3

——1996a Constant fan-in digital neural networks are VLSI-optimal *1st Int. Conf. on Mathematics of Neural Networks and Applications (Oxford, 1995)* (*Ann. Math. Artif. Intell.* to appear)

——1996b Entropy bounds for classification algorithms *Neural Network World* **6** 497–505

——1996c Optimal VLSI implementation of neural networks: VLSI-friendly learning algorithm *Neural Networks and Their Applications* ed T G Taylor (Chichester: Wiley) pp 255–76

——1997 VLSI components *Complexity of Discrete Neural Networks* (New York: Gordon and Breach) accepted for publication

Beiu V, Peperstraete J A and Lauwereins R 1992 Using threshold gates to implement sigmoid nonlinearity *Proc. Int. Conf. on Artificial Neural Networks (1992)* vol II pp 1447–50

Beiu V, Peperstraete J, Vandewalle J and Lauwereins R 1993 Close approximations of sigmoid functions by sum of steps for VLSI implementation of neural networks *Proc. Romanian Symp. on Computer Science (Jassy, Romania 1993)* pp 31–50

——1994a Learning from examples and VLSI implementation of neural networks *Cybernetics and Systems '94, Proc. 12th Euro. Meeting on Cybernetics and Systems Research (Vienna, 1994)* vol II ed R Trappl (Singapore: World Scientific) pp 1767–74

——1994b VLSI Complexity reduction by piece-wise approximations of the sigmoid function *Proc. Euro. Symp. on Artificial Neural Networks (Brussels)* ed M Verleysen (Brussels: De facto) pp 181–6

——1994c Area-time performances of some neural computations *Proc. IMACS Int. Sump. on Signal Processing Robotics and Artificial Neural Networks (Lille, France)* ed P Borne, T Fukuda and S G Tzafestas (Lille: GERF EC) pp 664–8

——1994d On the circuit complexity of feedforward neural networks *Proc. Int. Conf. on Artificial Neural Networks (1994)* pp 521–4

——1994e Placing feedforward neural networks among several circuit complexity classes proceedings *World Congr. on Neural Networks (San Diego, CA, 1994)* vol II (Lawrence Erlbaum Associates/INNS Press) pp 584–9

Beiu V and Rosu I 1985 VLSI implementation of a self-testable real content addressable memory *Proc. 6th Int. Conf. on Control System and Computer Science (Bucharest, Romania, 1985)* vol 2 pp 400–5

Beiu V and Taylor J G 1995a VLSI optimal learning algorithm ed D W Pearson, N C Steele and R F Albrecht *Artificial Neural Nets and Genetic Algorithms, Proc. Int. Conf. on Artificial Neural Networks and Genetic Algorithms (Alès, France, 1995)* (Berlin: Springer) pp 61–4

——1995b Area-efficient constructive learning algorithm *Proc. 10th Int. Conf. on Control Systems and Computer Science (Bucharest, Romania, 1995)* vol 3 pp 293–310

——1995c Optimal mapping of neural networks onto FPGAs–a new constructive algorithm *From Natural to Artificial Neural Computations Lecture Notes in Computer Science* vol 930 eds J Mira and F Sandoval (Berlin: Springer) pp 822–9

——1996a Direct synthesis of neural networks *Proc. Int. Conf. on Microelectronics for Neural Networks (1996)* pp 257–64

——1996b On the circuit complexity of sigmoid feedforward neural networks *Neural Networks* accepted

Bengtsson L, Linde A, Svensson B, Taveniku M and Ehlander A 1993 The REMAP massively parallel computer platform for neural computations *Proc. Int. Conf. on Microelectronics for Neural Networks (1993)* pp 47–62

Blank T 1990 The MasPar MP-1 architecture *Proc. 35th IEEE Computer Society Int. Conf., Spring COMPCON '90 (San Francisco)* pp 20–4

Blayo F and Hurat P 1989 A VLSI systolic array dedicated to Hopfield neural networks *VLSI for Artificial Intelligence* ed J G Delgado-Frias and W R Moore (New York: Kluwer)

Bose N K and Garga A K 1993 Neural network design using Voronoi diagrams *IEEE Trans. Neural Networks* **4** 778–87

Boser B E, Sackinger E, Bromley J, le Cun Y and Jackel L D 1992 Hardware requirements for neural network pattern classifiers *IEEE Micro Mag.* **12** 32–40

Botros N M and Abdul-Aziz M 1994 Hardware implementation of an artificial neural network using field programmable gate arrays (FPGA's) *IEEE Trans. Indust. Electron.* **41** 665–8

Boyd J 1990 Hitachi's neural computer *Electronic World News* 10 December, 6–8

Bruck J 1990 Harmonic analysis of polynomial threshold functions *SIAM J. Discrete Math.* **3** 168–77

Bruck J and Smolensky R 1989 Polynomial threshold functions, AC^0 functions and spectral norms *Research Report RJ 7410 (67387)* (IBM Yorktown Heights, New York)

——1992 Polynomial threshold functions AC^0 functions and spectral norms *SIAM J. Comput.* **21** 33–42

Bulsari A 1993 Some analytical solutions to the general approximation problem for feedforward neural networks *Neural Networks* **6** 991–6

Burr J B 1991 Neural network implementations *Neural Networks Concepts: Applications and Implementations* vol 2 ed P Antognetti and V Milutinovic (Englewood Cliffs, NJ: Prentice Hall)

——1992 Digital neurochip design *Digital Parallel Implementations of Neural Networks* ed K W Przytula and V K Prasanna (Englewood Cliffs, NJ: Prentice Hall)

Cameron S H 1969 An estimate of the complexity requisite in a universal decision network *Bionics Symp. (Wright Airforce Development Division WADD Report 60-600)* pp 197–212

Clarkson T G, Gorse D and Taylor J G 1989 Hardware realisable models of neural processing *Proc. 1st IEE Int. Conf. on Artificial Neural Nets, IEE Publication 313* (London: IEE) pp 242–6

——1990 pRAM automata *Proc. IEEE Int. Workshop on Cellular Neural Networks and Their Applications (Budapest, 1990)* pp 235–43

——1991a Biologically plausible learning in hardware realisable nets *Proc. Int. Conf. on Artificial Neural Networks (1991)* pp 195–9

——1992a From wetware to hardware reverse engineering using probabilistic RAMs *J. Intell. Syst.* **2** 11–30

Clarkson T G, Gorse D, Taylor J G and Ng C K 1992b Learning probabilistic RAM nets using VLSI structures *IEEE Trans. Computer* **41** 1552–61

Clarkson T G, Guan Y, Taylor J G and Gorse D 1993a Generalization in probabilistic RAM nets *IEEE Trans. Neural Networks* **4** 360–3

Clarkson T G and Ng C K 1993 Multiple learning configurations using 4th generation pRAM modules *Proc. Int. Conf. on Microelectronics for Neural Networks (1993)* pp 233–40

Clarkson T G, Ng C K, Gorse D and Taylor J G 1991b A serial update VLSI architecture for the learning probabilistic RAM neuron *Proc. Int. Conf. on Artificial Neural Networks (1991)* pp 1573–6

Clarkson T G, Ng C K and Guan Y 1993b The pRAM An adaptive VLSI chip *IEEE Trans. Neural Networks* **4** 408–12

Cohen S and Winder R O 1969 Threshold gate building blocks *IEE Trans. Computer* **18** 816–23

Cover T M 1965 Geometrical and statistical properties of systems of linear inequalities with applications in pattern recognition *IEEE Trans. Electron. Computer* **14** 326–34

Craik K J W 1943 *The Nature of Explanation* (Cambridge: Cambridge University Press)

Cruz C A, Hanson W A and Tam J Y 1987 Neural network emulation hardware design considerations *Proc. Int. Joint Conf. on Neural Networks (1987)* vol III pp 427–34

Cybenko G 1988 Continuous valued neural networks with two hidden layers are sufficient *Technical Report* (Tufts University)

——1989 Approximations by superpositions of a sigmoidal function *Math. Control Signal Syst.* **2** 303–14

DARPA 1989 DARPA neural network study final report October 1987–February 1988 *Technical Report 840* (Lincoln Laboratory, MIT)

Das Gupta B and Schnitger G 1993 The power of approximating a comparison of activation functions *Conf. on Neural Information Processing Systems (1992)* pp 615–22

Dejean C and Caillaud F 1994 Parallel implementations of neural networks using the L-Neuro 2.0 architecture *Proc. 1994 Int. Conf. on Solid State Devices and Materials (Yokohama Japan)* pp 388–90

DelCorso D, Grosspietch K E and Treleaven P 1989 European approaches to VLSI neural networks *IEEE Micro Mag.* **9**

Delgado-Frias J and Moore W R 1994 *VLSI for Neural Networks and Artificial Intelligence, An Edited Selection of the Papers Presented at the Int. Workshop on VLSI for Neural Networks and Artificial Intelligence (Oxford, 2–4 September, 1992)* (New York: Plenum)

Dembo A, Siu K-Y and Kailath T 1990 Complexity of finite precision neural network classifier *Proc. Conf. on Neural Information Processing Systems (1989)* pp 668–75

Denker J S (ed) 1986 Neural network for computing *Proc. AIP Conf. on Neural Networks for Computing (Snowbird, Utah, 1986)* (New York: American Institute of Physics)

Denker J S and Wittner B S 1988 Network generality training required and precision required *Proc. Conf. on Neural Information Processing Systems (1987)* pp 219–22

Deprit E 1989 Recurrent backpropagation on the connection machine *Neural Networks* **2** 295–314

Dertouzos M L 1965 *Threshold Logic: A Synthesis Approach* (Cambridge, MA: MIT Press)

Deville Y 1993 A Neural implementation of complex activation functions for digital VLSI *Neural Networks Microelectron. J.* **24** 259–62

Diederich S and Opper M 1987 Learning of correlated patterns in spin-glass networks by local learning rules *Phys. Rev. Lett.* **58** 949–52

Disante F, Sami M G, Stefanelli R and Storti-Gajani G 1989 Alternative approaches for mapping neural networks onto silicon *Proc. Int. Workshop on Artificial Neural Networks (Vietri sul Mare, Italy, 1989)* pp 319–28

——1990a A configurable array architecture for WSI implementation of neural networks *Proc. IEEE IPCC* Phoenix AZ March

——1990b A compact and fast silicon implementation for layered neural nets *Proc. Int. Workshop on VLSI for Artificial Intelligence and Neural Networks (Oxford)*

Duranton M 1996 L-Neuro 2.3: a VLSI for image processing by neural networks *Proc. Int. Conf. on Microelectronics for Neural Networks (1996)* pp 157–60

Duranton M, Gobert J and Maudit N 1989 A digital VLSI module for neural networks *Neural Networks from Models to Applications, Proc. nEuro '88 (Paris, June 1988)* (Paris: IDSET) pp 720–4

Duranton M and Maudit N 1989 A general purpose digital architecture for neural network simulation *Proc. IEE Int. Neural Network Conf. (1989)* (London: IEE) pp 62–66

Duranton M and Sirat J A 1989 A general purpose digital neurochip *Proc. Int. Joint Conf. on Neural Networks (Washington, 1989)*

——1990 Learning on VLSI: a general-purpose digital neurochip *Philips J. Res.* **45** 1–17

Eckmiller R, Hartman G and Hauske G (eds) 1990 *Parallel Processing in Neural Systems and Computers* (Amsterdam: North-Holland)

Eckmiller R and von der Malsburg C (eds) 1988 Neural computers *Proc. NATO Advanced Research Workshop on Neural Computers (Neuss, Germany)* (Berlin: Springer)

El-Mousa A H and Clarkson T G 1996 Multi-configurable pRAM based neurocomputer *Neural Network World* **6** 587–96

Erdogan S S and Wahab A 1992 Design of RM-nc a reconfigurable neurcomputer for massively parallel-pipelined computations *Proc. Int. Joint Conf. on Neural Networks (1992)* vol II pp 33–8

Ernoult C 1988 Performance of backpropagation ona parallel transputer-based machine *Proc. Neuro Nimes '88 (Nimes, France)* pp 311–24

Ernst H P, Mokry B and Schreter Z 1990 A transputer based general simulator for connectionist models *Parallel Processing in Neural Systems and Computers* ed G Hartmann and G Hauske (Amsterdam: North-Holland) pp 283–6

Faggin F 1991 Hardware (VLSI) Implementations of Neural Networks. Tutorial 3(b) *Int. Conf. on Artificial Neural Networks 1991)*

Faggin F and Mead C A 1990 VLSI Implementation of neural networks *An Introduction to Neural and Electronic Networks* eds S F Zornetzer, J L Davis and L Clifford (San Diego, CA: Academic Press) pp 275–92

Fiesler E 1994 Comparative bibliography of ontogenetic neural networks *Proc. Int. Conf. on Artificial Neural Networks (1994)* vol I pp 793–6

Fiesler E, Choudry A and Caulfield H J 1990 A universal weight discretization method for multi-layer neural networks *IEEE Trans. Syst. Man Cybern.* accepted (see also Fiesler E, Choudry A and Caulfield H J 1990 A weight discretization paradigm for optical neural networks *Proc. Int. Congr. on Optical Science and Engineering (Bellingham, Washington)* SPIE vol 1281 (SPIE) pp 164–73

Fischler M A 1962 Investigations concerning the theory and synthesis of linearly separable switching functions *PhD Dissertation* Department EE, Stanford University, USA

Flynn M J 1972 Some computer organization and their effectiveness *IEEE Trans. Comput.* **21** pp 948–60

Fornaciari W, Salice F and Storti-Gajani G 1991a Automatic synthesis of digital neural architectures *Proc. Int. Joint Conf. on Neural Networks (1991)* pp 1861–6

——1991b A formal method for automatic synthesis of neural networks *Proc. Int. Conf. on Microelectronics for Neural Networks (1991)* pp 367–80

Funahashi K-I 1989 On the approximate realization of continuous mapping by neural networks *Neural Networks* **2** 183–92

Funahashi K-I and Nakamura Y 1993 Approximation of dynamic systems by continuous time recurrent neural networks *Neural Networks* **6** 801–6

Furst M, Saxe J B and Sipser M 1981 Parity circuits and the polynomial-time hierarchy *Proc. IEEE Symp. on Foundations of Computer Science* **22** 260–70 (also in 1984 *Math. Syst. Theory* **17** 13–27)

Gamrat C, Mougin A, Peretto P and Ulrich O 1991 The architecture of MIND neurocomputers *Proc. MicroNeuro Int. Conf. on Microelectronics for Neural Networks (1991)* pp 463–9

Gascuel J-D, Delaunay E, Montoliu L, Moobed B and Weinfeld M 1992 A custom associative chip used as building block for a software reconfigurable multi-networks simulator *Proc. 3rd Int. Workshop on VLSI for Artificial Intelligence and Neural Networks (Oxford)*

Gick S, Heusinger P and Reuter A 1993 Automatic synthesis of neural networks to programmable hardware *Proc. Int. Conf. on Microelectronics for Neural Networks (1993)* pp 115–20

Girau B and Tisserand A 1996 On-line arithmetic-based reprogrammable hardware Implementation of multilayer perceptron back-propagation *Proc. Int. Conf. on Microelectronics for Neural Networks (1996)* pp 168–75

Glesner M, Huch M, Pöchmüller W and Palm G 1989 Hardware implementations for neural networks *Proc. IFIP Workshop on Parallel Architectures on Silicon (Grenoble, France)* pp 65–79

Glesner M and Pöchmüller W 1991 *Circuit Diagrams and Timing Diagrams of BACCHUS III* (Darmstadt University of Technology, Institute for Microelectronic Systems, Karlstraße 15, D-6100 Darmstadt, Germany)

——1994 *Neurocomputers–An Overview of Neural Networks in VLSI* (London: Chapman and Hall)

Glover M A and Miller W T 1994 A massively-parallel SIMD processor for neural networks and machine vision applications *Proc. Conf. on Neural Information Processing Systems (1993)* pp 843–49

Goldmann J and Karpinski M 1994 Simulating threshold circuits by majority circuits *Technical Report TR-94-030* (International Computer Science Institute, Berkeley, California) (a preliminary version appeared in 1963 *Proc. 25th ACM Symp. on Theory of Computation* (New York: ACM) pp 551–60)

Gorse D and Taylor J G 1989a On the identity and properties of noisy neural and probabilistic RAM nets *Phys. Lett.* A **131** 326–32

——1989b An analysis of noisy RAM and neural nets *Physica* D **34** 90–114

——1990a A general model of stochastic neural processing *Biol. Cybern.* **63** 299–306

——1990b Hardware-Realisable Learning Algorithms *Proc. Int. Conf. on Neural Network (Paris, 1990)* (Dordrecht: Kluwer) pp 821–4

——1991a Universal associative stochastic learning automata *Neural Network World* **1** 192–202

——1991b Learning sequential structure with recurrent pRAM nets *Proc. Int. Joint Conf. on Neural Networks (1991)* vol II pp 37–42

——1991c A continuous input RAM-based stochastic neural model *Neural Networks* **4** 657–65

Goser K, Hilleringmann U, Rückert U and Schumacher K 1989 VLSI Technologies for artificial neural networks *IEEE Micro Mag.* **9** 28–44

Graf H P and de Vegvar P 1987a A CMOS implementation of a neural network model *Advanced Research in VLSI, Proc. Stanford Conf. on Advanced Research on VLSI* ed P Losleben (Cambridge, MA: MIT Press) pp 351–67

——1987b A CMOS associative chip based on neural networks *Proc. IEEE Int. Solid-State Circuits Conf. (New York, 1987)* pp 304, 305 and 437

Graf H P, Hubbard W, Jackel L D and de Vegvar P 1987 A CMOS associative memory chip *Proc. Int. Joint Conf. on Neural Networks (1987)* vol III pp 461–8

Graf H P, Jackel L D, Howard R E, Straughn B, Denker J S, Hubbard W, Tennant D M and Schwartz D 1986 Implementation of a neural network memory with several hundreds of neurons *Neural Network for Computing, Proc. AIP Conf. on Neural Networks for Computing (Snowbird, Utah)* ed J S Denker (New York: American Institute of Physics) pp 182–7

Graf H P, Sackinger E, Boser B and Jackel L D 1991 Recent developments of electronic neural nets in USA and Canada *Proc. Int. Conf. on Microelectronics for Neural Networks (1991)* pp 471–88

Graf H P, Sackinger E and Jackel L D 1993 Recent developments of electronic neural nets in North America *J. VLSI Signal Processing* **6** 19–31

Grajski K A, Chinn G, Chen C, Kuszmaul C and Tomboulian S 1990 *Neural Network Simulation on the MasPar MP-1 Massively Parallel Computer, MasPar information sheet TW007 0690* (MasPar Computer Corporation, 749 North Mary Avenue, Sunnyvale, CA 94086, USA)

Griffin M, Tahara G, Knorpp K, Pinkham P and Riley B 1991 An 11 million transistor neural network execution engine *Proc. IEEE Int. Solid-State Circuits Conf. (San Francisco, CA, 1991)* pp 180–1

Gruau F 1993 Learning and pruning algorithm for genetic boolean neural networks Proc. Euro. Symp. on Artificial Neural Networks (Brussels, 1993) ed M Verleysen (Brussels: de facto) pp 57–63

Guan Y, Clarkson T G, Gorse D and Taylor J G 1992 The application of noisy reward/penalty learning to pyramidal pRAM structures *Proc. Int. Joint Conf. on Neural Networks (1992)* vol III pp 660–5

Gunzinger A, Müller U, Scott W, Bäumle B, Kohler P and Guggenbühl W 1992 Architecture and realization of a multi signal processor system *Proc. Application Specific Array Processors (1992)* ed J Fortes, E Lee and T Meng (Los Alamitos, CA: IEEE Computer Society Press) pp 327–340.2

Hajnal A, Maass W, Pudlák P, Szegedy M and Turán G 1987 Threshold circuits of bounded depth *Proc. IEEE Symp. on Foundations of Computer Science* **28** 99–110 (also in 1993 *J. Computing System Science* **46** 129–54)

Halgamuge S K, Pöchmüller W and Glesner M 1991 Computational hardware requirements for the backpropagation algorithm *Proc. Int. Conf. on Microelectronics for Neural Networks (1991)* pp 47–52

Hammerstrom D 1988 The connectivity analysis of simple associations–or–how many connections do you need *Proc. Conf. on Neural Information Processing Systems (1987)* pp 338–47

——1990 A VLSI Architecture for high-performance low-cost on-chip learning *Proc. Int. Joint Conf. on Neural Networks (1990)* vol II pp 537–43

——1995 Digital VLSI for neural networks *The Handbook of Brain Theory and Neural Networks* ed M A Arbib (Cambridge, MA: MIT Press) pp 304–9

Hammerstrom D and Nguyen N 1991 An implementation of Kohonen's self-organizing map on the adaptive solution neurocomputer *Proc. Int. Conf. on Artificial Neural Networks (1991)* vol I pp 715–20

Hassoun M H (ed) 1993 *Associative Neural Memories Theory and Implementation* (New York: Oxford University Press)

Håstad J 1986 Almost optimal lower bounds for small depth circuits *Proc. ACM Symp. on Theory of Computing (1986)* vol 18 pp 6–20

Heemskerk J N H Neurocomputers for brain-style processing. Design, implementation and application *PhD Thesis* Leiden University, The Netherlands (Chapter 3: 'Overview of Neural Hardware' is available via ftp from: ftp.mrc-apu.cam.ac.uk/pub/nn)

Heemskerk J N H, Murre J M J, Hoekstra J, Kemna L H J G and Hudson P T W 1991 The BSP400: a modular neurocomputer assembled from 400 low-cost microprocessors *Proc. Int. Conf. on Artificial Neural Networks (1991)* vol I pp 709–14

Hecht-Nielsen R 1987 Kolmogorov's mapping neural network existence theorem *Proc. Int. Joint Conf. on Neural Networks (1987)* vol III pp 11–13

——1988 Neurocomputing picking the human brain *IEEE Spectrum* **25** 36–41

——1989 *Neurocomputing* (Reading, MA: Addison Wesley)

——1991 *Computers Information sheet on HNC neural network products* (HNC Inc., 5501 Oberlin Drive, San Diego, CA 92121, USA)

Hirai Y 1991 Hardware implementation of neural networks in Japan *Proc. Int. Conf. on Microelectronics for Neural Networks (1991)* pp 435–46

Hiraiwa A, Kurosu S, Arisawa S and Inoue M 1990 A two level pipeline RISC processor array for ANN *Proc. Int. Joint Conf. on Neural Networks (1990)* vol II pp 137–40

Hochet B, Peiris V, Abdo S and Declercq M 1991 Implementation of a learning Kohonen neuron *IEEE J. Solid-State Circuits* **26** 262–7

Hofmeister T, Hohberg W and Köhling S 1991 Some notes on threshold circuits and multiplication in depth 4 *Info. Processing Lett.* **39** 219–26

Höhfeld M 1990 Fixed point arithmetic in feedforward neural networks *Technical Report FKS3-108* (Siemens AG, Münich)

Höhfeld M and Fahlman S E 1992 Probabilistic rounding in neural network with limited precision *Proc. Int. Conf. on Microelectronics for Neural Networks (1991)* pp 1–8 (also in 1992 *Neurocomputing* **4** 291–9)

——1992 Learning with limited numerical precision using the cascade-correlation algorithm *Technical Report CMU-CS-91-130* (School of Computer Science, Carnegie Mellon) (also in *IEEE Trans. Neural Networks* **3** 602–11)

Holler M A 1991 VLSI implementation of learning and memory systems: a review *Proc. Conf. on Neural Information Processing Systems (1990)* pp 993–1000

Holler M A, Park C, Diamond J, Santoni U, The S C, Glier M, Scofield C L and Núñez L 1992 A high performance adaptive classifier using radial basis functions *Proc. Government Microcircuit Application Conf. (Las Vegas, Nevada)*

Hollis P W, Harper J S and Paulos J J 1990 The effects of precision constraints in a backpropagation learning network *Neural Comput.* **2** 363–73

Hollis P W and Paulos J J 1994 A neural network learning algorithm tailored for VLSI implementation *IEEE Trans. Neural Networks* **5** 784–91

Hollis P W, Paulos J J and D'Costa C J 1991 An optimized learning algorithm for VLSI implementation *Proc. Int. Conf. on Microelectronics for Neural Networks (1991)* pp 121–6

Holt J L and Hwang J-N 1991 Finite precision error analysis of neural network hardware implementations *Technical Report FT-10* (University of Washington, Seattle)

——1993 Finite precision error analysis of neural network hardware implementations *IEEE Trans. Computer* **42** 281–90

Hong J 1987 On connectionist models *Technical Report* (Department of Computer Science, University of Chicago)

Hornik H 1991 Approximation capabilities of multilayer feedforward networks *Neural Network* **4** 251–7

——1993 Some new results on neural network approximation *Neural Network* **6** 1069–72

Hornik K, Stinchcombe M and White H 1989 Multilayer feedforward neural networks are universal approximators *Neural Network* **2** 359–66

——1990 Universal approximation of an unknown mapping and its derivatives using multilayer feedforward networks *Neural Network* **3** 551–60

Hu S 1965 *Threshold Logic* (Berkeley and Los Angeles: University of California Press)

Huang S-C and Huang Y-F 1991 Bounds on number of hidden neurons of multilayer perceptrons in classification and recognition *IEEE Trans. Neural Networks* **2** 47–55

Huch M, Pöchmüller W and Glesner M 1990 Bacchus: a VLSI architecture for a large binary associative memory *Proc. Int. Conf. on Neural Networks (Paris, 1990)* vol II pp 661–4

Hush D R and Horne B G 1993 Progress in supervised neural networks *IEEE Signal Proc. Mag.* **10** 8–39

Ienne P 1993a Quantitative comparison of architectures for digital neuro-computers *Proc. Int. Joint Conf. on Neural Networks (Nagoya, 1993)* pp 1987–1990

——1993b Architectures for neuro-computers: review and performance evaluation *Technical Report no 21/93* (Swiss Federal Institute of Technology, Lausanne)

Immerman N and Landau S 1989 The complexity of integrated multiplication *Proc. Structure in Complexity Theory Symp.* pp 104–111

Intel Corporation 1992a *Intel Neural Network Solutions Order Number 296961-002* (Intel Corporation, 2200 Mission College Boulevard, Mail Stop RN3–17, Santa Clara, CA 95052-8119, USA)

——1992b *Intel Neural Network Products Price and Availability Order Number 296961-002* (Intel Corporation, 2200 Mission College Boulevard, Mail Stop RN3-17, Santa Clara, CA 95052-8119, USA)

Iwata A 1990 Neural devices and networks *Sixth German–Japanese Forum on Information Technology (Berlin, 1990)*

Jabri M A and Flower B 1991 Weight perturbation: an optimal architecture and learning technique for analog VLSI feedforward and recurrent multi-layer networks *SEDAL Technical Report* Department of EE, University of Sydney (1992 *IEEE Trans. Neural Networks* **3** 154–7)

Jackel L D 1991 Practical issues for electronic neural-nets hardware–tutorial notes *Conf. on Neural Information Processing Systems (1991)*

——1992 Neural nets hardware. Tutorial 4 *CompEuro '92 (The Hague, The Netherlands, 1992)*

Jackel L D, Graf H P and Howard R E 1987 Electronic neural network chips *Appl. Opt.* **26** 5077–80

Jackson D and Hammerstrom D 1991 Distributed back propagation networks over the Intel iPSC/860 hypercube *Proc. Int. Joint Conf. on Neural Networks (1991)* vol I pp 569–74

Jahnke A, Roth U and Klar H 1996 A SIMD/DataflowArchitecture for a neurocomputer for spike-processing neural networks (NESPINN) *Proc. Int. Conf. on Microelectronics for Neural Networks (1996)* pp 232–7

Johnson R C (ed) 1993a Siemens shows off its first neural network chip *Cognizer Report* **4** 9–11 (Frontline Strategies, 516 S E Chkalov, Drive Suite 164, Vancouver, WA 98684, USA)

——(ed) 1993b Intel/Nestor Announce delivery of chip to DARPA *Cognizer Report* **4** 17–19 (Frontline Strategies, 516 S E Chkalov, Drive Suite 164, Vancouver, WA 98684, USA)

Jones S R and Sammut K 1993 Learning in systolic neural network engines *Proc. Int. Conf. on Microelectronics for Neural Networks (1993)* pp 175–85

Jones S R, Sammut K and Hunter J 1990 Toroidal neural network processor architecture operation performance *Proc. Int. Conf. on Microelectronics for Neural Networks (1990)* pp 163–9

Jones S R, Sammut K, Nielsen C and Staunstrup J 1991 Toroidal neural network processor architecture and processor granularity *VLSI Design of Neural Networks* ed U Ramacher and U Rückert (New York: Kluwer) pp 229–44

Judd J S 1988 On the complexity of loading shallow neural networks *J. Complexity* **4** 177–92

——1990 *Neural network design and the complexity of learning* (Cambridge, MA: MIT Press)

——1992 Constant-time loading of shallow 1-dimension networks *Proc. Conf. on Neural Information Processing Systems (1991)* pp 863–70

Kato H, Yoshizawa H, Iciki H and Asakawa K 1990 A parallel neurocomputer architecture toward billion connection updates per second *Proc. Int. Joint Conf. on Neural Networks (1990)* vol II pp 47–50

Kautz W 1961 The realization of symmetric switching functions with linear-input logical elements *IRE Trans. Electron. Computer* **10**

Kham E R and Ling N 1991 Systolic Architectures for artificial neural nets *Proc. Int. Joint Conf. on Neural Networks (1991)* vol 1 pp 620–7

Kim J H nad Park S-K 1995 The geometrical learning of binary artificial neural networks *IEEE Trans. Neural Networks* **6** 237–47

Klir G J 1972 *Introduction to the Methodology of Switching Circuits* (New York: Van Nostrand)

Kohn P, Bilmes J, Morgan N and Beck J 1992 Software for ANN training on a ring array processor *Proc. Conf. on Neural Information Processing Systems (1991)* pp 781–8

Koiran P 1993 On the complexity of approximating mappings using feedforward networks *Neural Networks* **6** 649–53

Kolmogorov A N 1957 On the representation of continuous functions of many variables by superposition of continuous functions of one variable and addition *Dokl Akad Nauk SSSR* **114** 679–81 (Engl. transl. 1963 *Math. Soc. Transl.* **28** 55–59

Köllmann K, Reimschneider K-R and Zeidler H C 1996 On-chip backpropagation training using parallel stochastic bit streams *Proc. Int. Conf. on Microelectronics for Neural Networks (1996)* pp 149–56

Krauth W and Mézard M 1987 Learning algorithms with optimal stability in neural networks *J. Phys A: Math. Gen.* **20** L745–52

Krikelis A 1991 A novel massively parallel associative processing architecture for the implementation of artificial neural networks *Proc. Int. Conf. on Acoustics, Speech and Signal Processing (Toronto, 1991)* vol II (Los Alamitos, CA: IEEE Computer Society Press) pp 1057–60

Kuczewsk R, Meyers M and Crawford W 1988 Neurocomputer workstation and processors approaches and applications *Proc. Int. Joint Conf. on Neural Networks (1988)* vol III pp 487–500

Kung S Y 1989 *VLSI Array Processors (Prentice Hall Information and System Sciences Series)* (Englewood Cliffs, NJ: Prentice Hall)

Kung S Y and Hwang J-N 1988 Parallel architectures for artificial neural nets *Proc. Int. Joint Conf. on Neural Networks (1988)* vol II pp 165–72

——1989a Digital VLSI architectures for neural networks *Proc. IEEE Int. Symp. on Circuits and Systems (Portland, Oregon, 1989)* vol I (Los Alamitos, CA: IEEE Computer Society Press) pp 445–8

——1989b A unified systolic architecture for artificial neural networks *J. Parallel Distrib. Comput.* **6** 358–87

Kung H T and Webb J A 1985 Global operations on a systolic array machine *Proc. IEEE Int. Conf. on Computer Design VLSI in Computers (Port Chester, New York, 1985)* pp 165–71

Landahl H D, McCulloch W S and Pitts W 1943 A statistical consequence of the logical calculus of nervous system *Bull. Math. Biophysiology* **5** 135–7

Le Bouquin J-P 1994 IBM Microelectronics ZISC, zero instruction set computer *Proc. World Congr. on Neural Networks (San Diego, CA, 1994) (supplement)*

Le Cun Y 1985 A learning procedure for asymmetric threshold networks *Proc. Cognitiva '85* pp 599–604

——1987 Models connexionistes de l'apprentisage *MSc thesis* Université Pierre et Marie Curie, Paris

Lehmann C and Blayo F 1991 A VLSI Implementation of a generic systolic synaptic building block for Neural Networks *VLSI for Artificial Neural Networks* ed J G Delgado-Frias and W R Moore (New York: Plenum) pp 325–34

Lehmann C, Viredaz M and Blayo F 1993 A generic systolic array building block for neural networks with on-chip learning *IEEE Trans. Neural Networks* **4** 400–7

Leiserson C E 1982 *Area-Efficient VLSI Computation* (Cambridge, MA: MIT Press)

Leshno M, Lin V Y, Pinkus A and Schocken S 1993 Multilayer feedforward neural networks with a nonpolynomial activation function can approximate any function *Neural Networks* **6** 861–7

Lewis P M II and Coates C L 1967 *Threshold Logic* (New York: Wiley)

Léonhard G, Cousin E, Laisne J D, Le Drezen J, Ouvradou G, Poulain Maubant A and Thépaut A 1995 ArMenX: a flexible platform for signal and image processing *Field Programmable Gate Arrays (FPGAs) for Fast Board Development and Reconfigurable Computing* ed J Schewel vol 2607 (SPIE)

Linde A, Nordström T and Taveniku M 1992 Using FPGAs to implement a reconfigurable highly parallel processor *Proc. 2nd Int. Workshop on Field Programmable Logic and Applications (Vienna)*

Lindsey C S and Lindblad T 1994 Review of hardware neural networks: a user's perspective plenary talk given at the *Third Workshop on Neural Networks: From Biology to High Energy Physics (Isola d'Elba, Italy, 1994)* (see also the following two WWW sites: http://www1.cern.ch/NeuralNets/nnwInHep.html and also http://www1.cern.ch/NeuralNets/nnwInHepHard.html)

Linial N, Mansour Y and Nisan N 1989 Constant depth circuits Fourier transforms and learnability *Proc. IEEE Symp. on Foundations of Computer Science* p 30

Lippmann R P 1987 An introduction to computing with neural nets *IEEE ASSP Mag.* **4** 4–22

Littlestone N 1988 Learning quickly when irrelevant attributes abound a new linear-threshold algorithm *Machine Learning* **2** 285–318

Losleben P (ed) 1987 Advanced research in VLSI *Proc. Stanford Conf. on Advanced Research on VLSI* (Cambridge, MA: MIT Press)

Lupanov O B 1973 The synthesis of circuits from threshold elements *Problemy Kibernetiki* **20** 109–40

Maass W, Schnitger G and Sontag E 1991 On the computational power of sigmoid versus Boolean threshold circuits *IEEE Symp. on Foundation of Computer Science (1991)*

Mackie S, Graf H P, Schwartz D B and Denker J S 1988 Microelectronic implementations of connectionist neural networks *Proc. Conf. on Neural Information Processing Systems (1987)* pp 515–23

Makram-Ebeid S, Sirat J-A and Viala J-R 1989 A rationalized error back-propagation learning algorithm *Proc. Int. Joint Conf. on Neural Networks (1989)*

Mann J, Berger B, Raffel J, Soares A and Gilbert S 1987 A generic architecture for wafer-scale nuromorphic systems *Proc. Int. Joint Conf. on Neural Networks (1987)* vol IV pp 485–93

MasPar 1990a MasPar 1100 series computer systems *Information sheet PL003 0490* (MasPar Computer Corporation, 749 North Marry Avenue, Sunnyvale, CA 94086, USA)

——1990b MasPar 1200 series computer systems *Information sheet PL004 0490* (MasPar Computer Corporation, 749 North Marry Avenue, Sunnyvale, CA 94086, USA)

——1990c The MP-1 family data-parallel computer *Information sheet PL006 0490* (MasPar Computer Corporation, 749 North Marry Avenue, Sunnyvale, CA 94086, USA)

Maudit N, Duranton M, Gobert J and Sirat J A 1991 Building up neuromorphic machines with L-Neuro 1.0 *Proc. Int. Joint Conf. on Neural Networks (1991)* pp 602–7

——1992 L-Neuro 1.0: a piece of hardware LEGO for building neural network systems *IEEE Trans. Neural Networks* **3** 414–22

Mayoraz E 1991 On the power of networks of majority functions *Proc. Int. Workshop on Artificial Neural Networks (1991)* (Berlin: Springer) pp 78–85

——1992 Representation of Boolean functions with democratic networks *Internal Report* (ÉPFL, Lausanne)

McCator H 1991 Back propagation Implementation on the Adaptive Solution CNAPS neurocomputer chip *Proc. Conf. on Neural Information Processing Systems (1990)* pp 1028–31

McCulloch W S and Pitts W 1943 A logical calculus of the ideas immanent in nervous activity *Bull. Math. Biophysiol.* **5** 115–33

Means E and Hammerstrom D 1991 Piriform model execution on a neurocomputer *Proc. Int. Joint Conf. on Neural Networks (1991)* vol I pp 575–80

Means R W and Lisenbee L 1991 Extensible linear floating point SIMD neurocomputer array processor *Proc. Int. Joint Conf. on Neural Networks (1991)* vol I pp 587–92

Mézard M and Nadal J-P 1989 Learning in feedforward layered networks the tiling algorithm *J. Phys. A: Math. Gen.* **22** 2191–203

Micro Devices 1989a *Data Sheet MD1220* (Micro Devices 5695B Beggs Road, Orlando, FL 32810-2603, USA)

——1989b Neural bit slice *Data Sheet no DS102300P on circuit MD1200* (Micro Devices 5695B Beggs Road, Orland, FL 32810-2603, USA)

——1989c *Design Manual for the NBS Part No DM102500* (Micro Devices 5695B Beggs Road, Orland, FL 32810-2603, USA)

——1990 Neural bit slice *Data sheet no DS102301 on circuit MD120* (Micro Devices 5695B Beggs Road, Orland, FL 32810-2603, USA)

Milosavlevich I Z, Flower B G and Jabri M A 1996 PANNE: a parallel computing engine for connectionist simulation *Proc. Int. Conf. on Microelectronics for Neural Networks (1996)* pp 363–8

Minnick R C 1961 Linear-Input Logic IRE *Trans. Electron. Comput.* **10** 6–16

Minsky M L 1954 Neural nets and the brain–model problem *PhD Dissertation* (Princeton, NJ: Princeton University Press)

Minsky M L and Papert S A 1969 *Perceptron: An Introduction to Computational Geometry* (Cambridge, MA: MIT Press)

Moerland P D and Fiesler E 1996 hardware-friendly algorithms for neural networks: an overview *Proc. Int. Conf. on Microelectronics for Neural Networks (1996)* pp 117–24

Morgan N 1995 Programmable neurocomputing systems *The Handbook of Brain Theory and Neural Networks* ed M A Arbib (Cambridge, MA: MIT Press) pp 264–8

Morgan N, Beck J, Kohn P and Bilmes J 1993 Neurocomputing on the RAP *Parallel Digital Implementations of Neural Networks* ed K W Przytula and V K Prasanna (Englewood Cliffs, NJ: Prentice Hall)

Morgan N, Beck J, Kohn P, Bilmes J, Allman E and Beer J 1990 The RAP: a ring array processor for layered network calculations *Proc. IEEE Int. Conf. on Application Specific Array Processes* (Los Alamitos, CA: IEEE Computer Society Press) pp 296–308

——1992 The ring array processor (RAP) a multiprocessing peripheral for connectionist applications *J. Parallel Distrib. Comput.* **14** 248–59

Mühlbein H and Wolf K 1989 Neural network simultation on parallel computers *Parallel Computing (1989)* ed D J Evans, G G Joubert and F J Peters (Amsterdam: North-Holland) pp 365–74

Müller U A, Bäumle B, Kohler P, Gunzinger A and Guggenbühl W 1992 Achieving supercomputer performance for neural net simulation with an array of digital signal processors *IEEE Micro Mag.* **12** 55–65

Müller U A, Gunzinger A and Guggenbühl W 1995 Fast neural net simulation with a DSP processor array *IEEE Trans. Neural Networks* **6** 203–13

Müller U A, Kocheisen M and Gunzinger A 1994 High performance neural net simulation on a multiprocessor system with 'intelligent' communication *Proc. Conf. on Neural Information Processing Systems (1993)* pp 888–95

Muroga S 1959 The principle of majority decision logic elements and the complexity of their circuits *Proc. Int. Conf. on Information Processing (Paris)*

——1961 Functional forms of majority decision functions and a necessary and sufficient condition for their realizability In switching circuit theory and logical design *AIEE Special Publication S134* pp 39–46

——1962 Generation of self-dual threshold functions and lower bounds of the number of threshold functions and a maximum weight in switching circuit theory and logical design *AIEE Special Publication S134* pp 170–84

——1971 *Threshold Logic and Its Applications* (New York: Wiley)

——1979 *Logic Design and Switching Theory* (New York: Wiley) ch 5

Muroga S, Toda I and Takasu S 1961 *Theory of Majority Decision Elements Journal* vol 271 (Franklin Institute) pp 376–418

Murray M, Burr J B, Stork D G, Leung M-T, Boonyanit K, Wolff G J and Peterson A M 1992 Deterministic Boltzmann machine VLSI can be scaled using multi-chip modules *Proc. Int. Conf. on Application Specific Array Processors (Berkeley, CA)* (Los Alamitos, CA: IEEE Computer Society Press) pp 206–17

Murray M, Leung M-T, Boonyanit K, Kritayakirana K, Burr J B, Wolff G J, Watanabe T, Schwartz E and Stork D G 1994 Digital Boltzmann VLSI for constraint satisfaction and learning *Proc. Conf. on Neural Information Processing Systems (1993)* pp 896–903

Murre J M J 1993 Transputers and neural networks an analysis of implementation constraints and performance *IEEE Trans. Neural Networks* **4** 284–92

Murtagh P and Tsoi A C 1992 Implementation issues of sigmoid function and its derivative for VLSI digital neural networks *IEE Proc.-E Computer and Digital Techniques* **139** 207–14

Myers D J and Hutchinson R A 1989 Efficient implementation of piecewise linear activation function for digital VLSI neural networks *Electron. Lett.* **25** 1662–3

Myers D J, Vincent J M and Orrey D A 1991 HANNIBAL A VLSI building block for neural networks with on-chip backpropagation learning *Proc. Int. Conf. on Microelectronics for Neural Networks (1991)* pp 171–81

——1993 HANNIBALL A VLSI building block for neural networks with on-chip backpropagation learning *Neurocomputing* **5** 25–37

Myhill J and Kautz W H 1961 On the size of weights required for linear-input switching functions *IRE Trans. Electron. Comput.* **10**

Nakayama K and Katayama H 1991 A low-bit learning algorithm for digital multilayer neural networks applied to pattern recognition *Proc. Int. Joint Conf. on Neural Networks (1991)* pp 1867–72

Naylor D, Jones S, Myers D and Vincent J 1993 Design and application of a real-time neural network based image processing system *Proc. Int. Conf. on Microelectronics for Neural Networks (1993)* pp 137–47

Neciporuk E I 1964 The synthesis of networks from threshold elements *Problemy Kibernetiki II* 49–62 (Engl. transl. 1964 *Automation Express* **7** 35–9 and **7** 27–32

Neibur E and Brettle D 1994 Efficient simulation of biological neural networks on massively parallel supercomputers with hypercube architecture *Proc. Conf. on Neural Information Processing Systems (1993)* pp 904–10

Neuralogix 1992 *NLX420 Data Sheet Neurologix Inc* (800 Charcot Avenue Suite, 112 San Jose, California)

Nickolls J R 1990 The design of the MasPar MP-1: a cost effective massively parallel computer *Proc. 35th IEEE Computer Society Int. Conf. Spring COMPCON '90 (San Francisco, CA)* pp 25–8

Nigri M E 1991 Hardware emulation of back-propagation neural networks *Research Note RN/91/21* (Department of Computer Science, University College London)

Nigri M E, Treleaven P and Vellasco M 1991 Silicon compilation of neural networks *CompEuro '91* ed Pröebster W E and Reiner H (Los Alamitos, CA: IEEE Computer Society Press) pp 541–6

Nijhuis J, Höfflinger B, Neußer S, Siggelkow A and Spaanenburg L 1991 A VLSI implementation of a neural car collision avoidance controller *Proc. Int. Joint Conf. on Neural Networks (1991)* vol 1 pp 493–9

Nilsson N J 1965 *Learning Machines* (New York: McGraw-Hill)

Nordström T and Svensson B 1991 Using and designing massively parallel computers for artificial neural networks *Technical Report TULEA 91:1* (Division of Computer Engineering LuleåUniversity of Technology S-95187 LuleåSweden) (also in *J. Parallel Distrib. Comput.* **14** 260–85)

Obradovic Z and Parberry I 1990 Analog neural networks of limited precision I: computing with multilinear threshold functions *Proc. Conf. on Neural Information Processing Systems (1989)* pp 702–9

Oliveira A L and Sangiovanni-Vincentelli A 1994 Learning complex Boolean functions algorithms and applications *Proc. Conf. on Neural Information Processing Systems (1993)* pp 911–8

Orrey D A, Myers D J and Vincent J M 1991 A high performance digital processor for implementing large artificial neural networks *Proc. IEEE Custom Integrated Circuits Conf. (San Diego, CA)*

Pacheco M and Treleaven P 1989 A VLSI word-slice architecture for neurocomputing *Proc. 1989 Int. Symp. on Computer Architecture and Digital Signal Processing (Hong Kong)* (IEEE)

——1992 Neural-RISC a processor and parallel architecture for neural networks *Proc. Int. Joint Conf. on Neural Networks (1992)* vol II pp 177–82

Palm G and Palm M 1991 Parallel associative networks the PAN-System and the BACCHUS-Chip *Proc. Int. Conf. on Microelectronics for Neural Networks (1991)* pp 411–6

Papadopoulos C V and Andronikos T S 1995 Modelling the complexity of parallel and VLSI computations with Boolean circuits *Microprocess. Microsyst.* **19** 43–50

Parberry I 1994 *Circuit Complexity and Neural Networks* (Cambridge, MA: MIT Press)

Paterson M S (ed) 1992 Boolean function complexity *London Mathematical Society Lecture Notes Series* **169** (Cambridge: Cambridge University Press)

Paturi R and Saks M 1990 On threshold circuits for parity *Proc. IEEE Symp. on Foundation of Computer Science (1990)*

Pérez C J, Carrabina J and Valderrama E 1992 Study of a learning algorithm for neural networks with discrete synaptic couplings *Network* **3** 165–76

Personnaz L and Dreyfus G (eds) 1989 Neural networks from models to applications *Proc. nEuro '88 (Paris, 1988)* (Paris: IDSET)

Personnaz L, Johannet A and Dreyfus G 1989 Problems and trends in integrated neural networks *Connectionism in Perspective* eds R Pfeifer, Z Schreter and F Fogelman-Soulié (Amsterdam: Elsevier)

Pesulima E E, Pandya A S and Shankar R 1990 Digital implementation issues of stochastic neural networks *Proc. Int. Joint Conf. on Neural Networks (1990)* vol II pp 187–90

Pippenger N 1987 The complexity of computations by networks *IBM J. Res. Dev.* **31** 235–43

Pöchmüller W and Glesner M 1991 A cascadable architecture for the realization of large binary associative networks *VLSI for Artificial Intelligence and Neural Networks* ed J G Delgado-Frias and W R Moore (New York: Plenum) pp 265–74

Pomerleau D A, Gusciora G L, Touretzky D S and Kung H T 1988 Neural network simulation at warp speed how we got 17 million connections per second *Proc. Int. Joint Conf. on Neural Networks (1988)* (Los Alamitos, CA: IEEE Computer Society Press) vol II pp 143–50

Poulain Maubant A, Autret Y, Léonhard G, Ouvradoui G and Thépaut A 1996 An efficient handwritten digit recognition method on a flexible parallel architecture *Proc. Int. Conf. on Microelectronics for Neural Networks (1996)* pp 355–62

Przytula K W 1988 A survey of VLSI implementations of artificial neural networks *VLSI Signal Processing III* ed R W Brodersen and H S Moscovitz (New York: IEEE Computer Society Press) pp 221–31

Przytula K W and Prasanna V K 1993 *Parallel Digital Implementations of Neural Networks* (Englewood Cliffs, NJ: Prentice Hall)

Raghavan P 1988 Learning in threshold networks: a computational model and applications *Technical Report RC 13859* (IBM Research July 1988) (also in 1988 *Proc. Workshop on Computational Learning Theory* (Cambridge, MA: Cambridge) pp 19–27

Ramacher U 1990 The VLSI Kernel of neural algorithms *Proc. 1st Int. Workshop on Cellular Neural Networks and their Applications (Budapest, 1990)* pp 185–96

——1992 SNAPSE–a neurocomputer that synthesizes neural algorithms on a parallel systolic engine *J. Parallel Distrib. Comput.* **14** 306–18

Ramacher U, Beichter J and Brüls N 1991a Architecture of a general-purpose neural signal processor *Proc. Int. Joint Conf. on Neural Networks (1991)* vol I pp 443–6

Ramacher U, Raab W, Anlauf J, Hachmann U, Beichter J, Brüls N, Weßeling M and Sicheneder E 1993 Multiprocessor and memory architecture of the neurocomputer SYNAPSE-1 *Proc. Int. Conf. on Microelectronics for Neural Networks (1993)* pp 227–31

Ramacher U, Raab W, Anlauf J, Hachmann U and Weßeling M 1991b SYNAPSE-X a general-purpose neurocomputer *Proc. Int. Conf. on Microelectronics for Neural Networks (1991)* pp 401–9 (also in *Proc. Int. Joint Conf. on Neural Networks (1991)* pp 2168–76

Ramacher U and Rückert U (eds) 1991 *VLSI Design of Neural Networks* (New York: Kluwer)

Razborov A A 1987 Lower bounds for the size of circuits of bounded depth with basis $\{\wedge, \oplus\}$ *Math. Not. Acad. Sci. USSR* **41** 333–8

Red'kin N P 1970 Synthesis of threshold circuits for certain classes of Boolean functions *Kibernetika* **5** 6–9 (Engl. transl. 1970 *Cybernetics* **6** 540–4)

Reilly D L, Cooper L N and Elbaum C 1982 A neural model for category learning *Biol. Cybern.* **45** 35–41

Reyneri L M and Filipi E 1991 An analysis on the performance of silicon implementations of backpropagation algorithms for artificial neural networks *IEEE Trans. Comput.* **40** 1380–9

Rief J H 1987 On threshold circuits and polynomial computations *Proc. 2nd Annual Structure in Complexity Theory Symp.* pp 118–23

Roberts F and Wang S 1989 Implementation of neural networks on a hypercube FPS T20 *Parallel Processing* ed M Cosnard M, M H Barton and M Vanneschi (Amsterdam: North-Holland) pp 189–200

Rosenblatt F 1958 The perceptron a probabilistic model for information storage and organization *Brain Psych. Revue* **62** 386–408

——1961 Principles of neurodynamics *Perceptrons and the Theory of Brain Mechanism* (Washington, DC: Sparton Press)

Rosenbleuth A, Wiener N and Bigelow J 1943 Behaviour, purpose and teleology *Phil. Sci.* **10** 18–24

Rosenbleuth A, Wiener N, Pitts W and Garcia Ramos J 1949 A statistical analysis of synaptic excitation *J. Cell Comput. Physiol.* **34** 173–205

Rossmann M, Hesse B, Goser K, Bühlmeier and Manteuffel G 1996 Implementation of a biologically inspired neuron-model in FPGA *Proc. Int. Conf. on Microelectronics for Neural Networks (1996)* pp 322–9

Roth U, Jahnke A and Klar H 1995 Hardware requirements for spike-processing neural network 1995 *From Natural to Artificial Neural Computations Lecture Notes in Computer Science* vol 930 ed J Mira and F Sandoval (Berlin: Springer) pp 720–7

Roy A, Kim L S and Mukhopadhyay S 1993 A polynomial time algorithm for the construction and training of a class of multilayer perceptrons *Neural Networks* **6** 535–45

Roychowdhury V P, Orlitsky A and Siu K-Y 1994a Lower bounds on threshold and related circuits via communication complexity *IEEE Trans. Info. Theory* **40** 467–74

Roychowdhury V P, Siu K-Y and Orlitsky A (eds) 1994b *Theoretical Advances in Neural Computation and Learning* (Boston: Kluwer)

Roychowdhury V P, Siu K-Y, Orlitsky A and Kailath T 1991a A geometric approach to threshold circuit complexity *Proc. Workshop on Computational Learning Theory COLT (Santa Cruz, CA, 1991)* pp 97–111

——1991b On the circuit complexity of neural networks *Proc. Conf. on Neural Information Processing Systems (1990)* pp 953–59

Rückert U, Kleerbaum C and Goser K 1991 Digital VLSI implementations of an associative memory based on neural networks 1991 *VLSI for Artificial Intelligence and Neural Networks* ed J G Delgado-Frias and W R Moore (New York: Plenum) pp 275–84

Rudnick M and Hammerstrom D 1988 An interconnecting structure for wafer scale neurocomputers 1988 *Connectionist Models Summer School 1988 Proc.* ed D S Touretzky and G Hinton (San Mateo, CA: Morgan Kaufmann)

Rüping S and Rückert U 1996 A scalable processor array for self-organizing feature maps *Proc. Int. Conf. on Microelectronics for Neural Networks (1996)* pp 285–91

Sanchez-Sinencio E and Lau C (eds) 1992 Artificial neural networks *Paradigms Applications and Hardware Implementations* (New York: IEEE Computer Society Press)

Saucier G and Ouali J 1990 Silicon compiler for neuron ASICs *Proc. Int. Joint Conf. on Neural Networks (1990)* vol II pp 557–61

Sakaue S, Kohda T, Yamamoto H, Maruno S and Shimeki Y 1993 Reduction of required precision bits for back-propagation applied to pattern recognition *IEEE Trans. Neural Networks* **4** 270–5

Sami M (ed) 1990 *Workshop on Silicon Architectures for Neural Nets (St Paul de Venice, France)* (Amsterdam: Elsevier)

Sammut K and Jones S R 1991 Implementing non-linear activation functions in neural network emulators *Electron. Lett.* **27** 1037–8

Savran M E and Morgül Ö 1991 On the associative memory design for the Hopfield neural network *Proc. Int. Joint Conf. on Neural Networks (1991)* vol II pp 1166–71

Schwartz T J 1990 *A Neural Chips Survey AI Expert* **5** 34–9

Scofield C L, Reilly D L 1991 Into silicon real time learning in a high density RBF neural network *Proc. Int. Joint Conf. on Neural Networks (1991)* vol I pp 551–6

Sejnowski T J and Rosenberg C R 1986 NETtalk A parallel network that learns to read aloud *Technical Report JHU/EECS-86/01* (Johns Hopkins University, Electrical Engineering and Computer Science, Baltimore)

Shawe-Taylor J S, Anthony M H G and Kern W 1992 Classes of feedforward neural networks and their circuit complexity *Neural Networks* **5** 971–7

Sheng C L 1969 *Threshold Logic* (New York: Academic)

Shoemaker P A, Carlin M J and Shimabukuro R L 1990 Back-Propagation learning with coarse quantization of weight updates *Proc. Int. Joint Conf. on Neural Networks (1990)* vol I pp 573–6

Siggelkow A, Nijhuis J, Neußer S and Spaanenburg L 1991 Influence of hardware characteristics on the performance of a neural system *Proc. Int. Conf. on Artificial Neural Networks (1991)* vol 1 pp 697–702

Singer A 1990a Exploiting the inherent parallelism of artificial neural networks to achieve 1300 million interconnects per second *Proc. INNC '90 (Paris)* pp 656–60

——1990b Implementations of artificial neural networks on the connection machine *Parallel Comput.* **14** 305–15

Sirat J A and Nadal J-P 1990 *Neural trees: a new tool for classification network: computation in neural systems* **1** 423–8

Siu K-Y 1992 On the complexity of neural networks with sigmoid units *Neural Networks for Signal Processing II. Proc. IEEE-SP Workshop on Neural Networks and Signal Processing (1992)* ed S Y Kung, F Fallside, J Aa Sorenson and C A Kamm (Helsingöer, Denmark) (Los Alamitos, CA: IEEE Computer Society Press) pp 23–28

Siu K-Y and Bruck J 1990a On the dynamic range of linear threshold elements *Research Report RJ 7237* (IBM, Yorktown Heights, New York)

——1990b Neural computation of arithmetic functions *Proc. IEEE* **78** 166–75

——1990c On the power of threshold circuits with small weights *Research Report RJ 7773 (71890)* (IBM, Yorktown Heights, New York) (see also *SIAM J. Discrete Math.* **4** 423–35 1991)

——1992 Neural computing with small weights *Proc. Conf. on Neural Information Processing Systems (1991)* pp 944–9

——1993 On the dynamic range of linear threshold elements *SIAM J. Discrete Math.* to appear

Siu K-Y, Bruck J and Kailath T 1991a Depth efficient neural networks for division and related problems *Research Report RJ 7946 (72929)* (IBM, Yorktown Heights, New York) (see also Siu 1993b)

Siu K-Y, Bruck J, Kailath T and Hofmeister T 1993a Depth-efficient neural networks for division and related problems *IEEE Trans. Info. Theory* **39** 946–56

Siu K-Y and Roychowdhury V P 1993 Optimal depth neural networks for multiplication and related problems *Proc. Conf. on Neural Information Processing Systems (1992)* pp 59–64

——1994 On optimal depth threshold circuits for multiplication and related problems *SIAM J. Discrete Math.* **7** 284–92

Siu K-Y, Roychowdhury V and Kailath T 1990 Computing with almost optimal size threshold circuits *Technical Report* (Information System Laboratory, Stanford University) (also in *Proc. IEEE Int. Symp. on Information Theory (Budapest, 1991)*)

——1991b Depth-size tradeoffs for neural computations *IEEE Trans. Comput.* **40** 1402–12

——1993b Computing with almost optimal size neural networks *Proc. Conf. on Neural Information Processing Systems (1992)* pp 19–26

——1994 *Discrete Neural Computation: A Theoretical Foundation* (Englewood Cliffs, NJ: Prentice-Hall)

Sivilotti M A, Emerling M R and Mead C A 1986 VLSI architectures for implementation of neural networks *Neural Networks for Computing* (New York: American Institute of Physics) pp 408–13

Smolensky R 1987 Algebraic methods in the theory of lower bounds for Boolean circuit complexity *Proc. ACM Symp. on Theory of Computing (1987)* vol 19 pp 77–82

Sontag E D 1990 On the recognition capabilities of feedforward nets *Report SYCON* (Rutgers Center for System and Control, 90-03 Department of Mathematics, Rutgers University, New Brunswick, NJ 08903, USA)

Souček B and Souček M 1988 *Neural and Massively Parallel Computers—the Sixth Generation* (New York: Wiley)

Spaanenburg L, Hoefflinger B, Neußer S, Nijhuis J A G and Siggelkow A 1991 A multiplier-less digital neural network *Proc. Int. Conf. on Microelectronics for Neural Networks (1991)* pp 281–9

Specht D F 1988 Probabilistic neural networks for classification, mapping, or associative memory *Proc. Int. Joint Conf. on Neural Networks (1988)* vol I pp 525–32

Stevenson M, Winter R and Widrow B 1990 Sensitivity of feed-forward neural networks to weight errors *IEEE Trans. Neural Networks* **1** 71–80

Strey A, Avellana N, Hogado R, Fernández J A, Capillas R and Valderrama E 1995 A massively parallel neurocomputer with a reconfigurable arithmetical unit 1995 *From Natural to Artificial Neural Computations Lecture Notes in Computer Science* ed J Mira and F Sandoval vol 930 (Berlin: Springer) pp 800–6

Szedegy M 1989 Algebraic methods in lower bounds for computational models with limited communication *PhD Dissertation* University of Chicago

Śmieja F 1993 Neural network constructive algorithm trading generalization for learning efficiency? *Circuits, Syst. Signal Processing* **12** 331–74

Takahashi H, Tomita E and Kawabata T 1993 Separability of internal representations in multilayer perceptrons with application to learning *Neural Networks* **6** 689–703

Tan S and Vandewalle J 1992 Efficient algorithm for the design of multilayer feedforward neural networks *Proc. Int. Joint Conf. on Neural Networks (1992)* vol II pp 190–5

——1993 On the design of feedforward neural networks *Technical Report* (National University of Singapore, Department of EE) (also in *Neurocomputing* **6** 565–82)

Tewksbury S K and Hornak L A 1989 Wafer level system integration: a review *IEEE Circuits and Devices Mag.* **5** 22–30

Theeten J B, Duranton M, Maudit N and Sirat J A 1990 The L-Neuro chip: a digital VLSI with on-chip learning mechanism *Proc. INNC '90 (Paris)* ed B Angeniol and B Widrow (Dordrecht: Kluwer) pp 593–6

Thiran P 1993 Self-organization of a Kohonen network with quantized weights and an arbitrary one-dimensional stimuli distribution *Proc. Euro. Symp. on Artificial Neural Networks (Brussels)* ed M Verleysen (Brussels: de facto) pp 203–8

Thiran P, Peiris V, Heim P and Hochet B 1994 Quantization effects in digitally behaving circuit implementations of Kohonen networks *IEEE Trans. Neural Networks* **5** 450–8

Thole P, Speckmann H and Rosenstiel W 1993 A hardware supported system for Kohonen's self-organizing map *Proc. Int. Conf. on Microelectronics for Neural Networks (1993)* pp 29–34

Treleaven P C 1989 Neurocomputers international *J. Neuro-computing* **1** 4–31

Treleaven P C, Pacheco M and Vellasco M 1989 VLSI architectures for neural networks *IEEE Micro Mag.* **9** 8–27

Treleaven P C and Rocha P V 1990 Towards a general-purpose neurocomputing system *Workshop on Silicon Architectures for Neural Nets (St Paul de Venice, France, 1990)* ed M Sami (Amsterdam: Elsevier)

Trotin A and Darbel N 1993 A neocognitron for digits classification on a VLSI chip *Proc. Int. Conf. on Microelectronics for Neural Networks (1993)* pp 21–8

Tryba V, Speckmann H and Goser K 1990 A digital hardware implementation of a self-organizing feature map as a neural coprocessor to a von Neumann computer *Proc. Int. Conf. on Microelectronics for Neural Networks (1990)* pp 177–86

van Keulan E, Colak S, Withagen H and Hegt H 1994 Neural network hardware performance criteria *Proc. IEEE Conf. on Neural Networks (1994)* vol III (Los Alamitos, CA: IEEE Computer Society Press) pp 1885–8

Vellasco M and Treleaven P C 1992 A VLSI architecture for the automatic generation of neuro-chips *Proc. Int. Joint Conf. on Neural Networks (1992)* vol II pp 171–6

Venkatesh S S 1989 A new linear threshold algorithm for learning binary weights *On-Line Workshop on Neural Network for Computing (Snowbird, Utah, 1989)*

Verleysen M and Cabestany J 1994 Project ESPRIT ELENA *Realisation VLSI de reseaux de neurones VLAGO, ISSN 1243-4835 No 94-1: Les processeurs neuronaux 1994*

Vincent J and Myers D 1992 Weight dithering and wordlength selection for digital backpropagation networks *BT Technology J.* **10** 124–33

Viredaz M A, Lehmann C, Blayo F and Ienne P 1992 MANTRA a multi-model neural network computer *Proc. 3rd Int. Workshop on VLSI for Neural Networks and Artificial Intelligence (Oxford)*

Walker M R and Akers L A 1992 Information-theoretic analysis of finite register effects in neural networks *Proc. Int. Joint Conf. on Neural Networks (1992)* vol II pp 666–71

Walker M R, Haghighi S, Afgan A and Akers L A 1989 Training a limited-interconnect synthetic neural IC *Proc. Conf. on Neural Information Processing Systems (1988)* pp 777–84

Watanabe T, Kimura K, Aoki M, Sakata T and Ito K 1993 A single 1 5-V digital chip for a 10^6-Synapse neural network *IEEE Trans. Neural Networks* **4** 387–93

Watkins S S, Chau P M and Tawel R 1992 Different approaches to implementing a radial basis function neurocomputer *Proc. RNNS/IEEE Symp. on Neuroinformatics and Neurocomputing (Rostov-on-Don, Russia)* pp 1149–55

Wawrzynek J, Asanović K, Kingsbury B, Beck J, Johnson D and Morgan N 1996 SPERT-II: a vector microprocessor system and its applications to large problems in backpropagation training *Proc. Int. Conf. on Microelectronics for Neural Networks (1996)* pp 227–31

Wawrzynek J, Asanović K and Morgan N 1993 The Design of a neuro-microprocessor *IEEE Trans. Neural Networks* **4** 394–9

Wegener I 1987 *The Complexity of Boolean Functions* (Chichester: Wiley)

Weinfeld M 1989 A fully digital integrated CMOS Hopfield network including the learning algorithm *VLSI for Artificial Intelligence* ed Delgado-Frias J G and Moore W R (Boston: Kluwer) pp 169–78

——1990 Integrated artificial neural networks components for higher level architectures with new properties *NATO Advance Workshop on Neurocomputing* ed Fogelman-Soulié F and Hérault J (Berlin: Springer)

White B and Elmasry M 1992 The digi-neocognitron: a digital neocognitron neural network model for VLSI *IEEE Trans. Neural Networks* **3** 73–85

Williams P and Panayotopoulos G 1989 Tools for neural network simulation *Report ANNR04 from ESPRIT project 2092 (ANNIE)*

Winder R O 1962 Threshold logic *PhD Dissertation* Mathematics Department, Princeton University, Princeton, NJ

——1963 Bounds on threshold gate realizability *IRE Trans. Electron. Comput.* **12** 561–4

——1969a *Fundamentals of Threshold Logic* AAT pp 235–318

——1969b The status of threshold logic *RCA Review* **30** 62–84

——1971 Chow parameters in threshold *J. ACM* **18** 265–89

Witbrock M and Zagha M 1990 An implementation of backpropagation learning on GF11 a large SIMD parallel computer *Parallel Comput.* **14** 329–46

Works G 1988 The creation of delta: a new concept in ANS processing *Proc. Int. Joint Conf. on Neural Networks (1988)* vol II pp 159–64

Xie Y and Jabri M A 1991 Analysis of the effect of quantization in multi-layer neural networks using statistical model *SEDAL Technical Report 1991-8-2* (Department of EE, University of Sydney, Australia)

——1992 Training algorithms for limited precision feedforward neural networks *SEDAL Technical Report 1991-8-3* (Department of EE, University of Sydney, Australia) (also in *Proc. Australian Conf. on Neural Networks (Canberra, Australia, 1992)* pp 68–71

Yao A C 1985 Separating the polynomial-time hierarchy by oracles *Proc. IEEE Symp. on Foundations Computer Science (1985)* vol 26 pp 1–10

——1989 On ACC and threshold circuits *Proc. ACM Symp. on Theory of Computing* pp 186–96

Yasunaga M, Masuda N, Asai M, Yamada T, Masaki A and Hirai Y 1989 A wafer scale integration neural network utilizing completely digital circuits *Proc. Int. Joint Conf. on Neural Networks (1989)* vol II pp 213–7

Yasunaga M, Masuda N, Yagyu M, Asai M, Yamada T and Masaki A 1990 Design fabrication and evaluation of a 5-inch wafer scale neural network LSI composed of 576 digital neurons *Proc. Int. Joint Conf. on Neural Networks (1990)* vol II pp 527–35

——1991 A self-learning neural net composed of 1152 digital neurons in wafer-scale LSIs *Proc. Int. Joint Conf. on Neural Networks (1991)* vol III pp 1844–9

Yestrebsky J, Basehore P and Reed J 1989 Neural bit-slice computing element information *Sheet No TP102600* (Micro Devices, 5695B Beggs Road, Orlando, FL, 32810-2603, USA)

Yoshizawa H, Ichiki H K H and Asakawa K 1991 A highly parallel architecture for back-propagation using ring-register data path *Proc. Int. Conf. on Microelectronics for Neural Networks (1991)* pp 325–32

Zhang X, Mckenna M, Mesirov J P and Waltz D L 1990 An Efficient Implementation of the back-propagation algorithm on the connection machine CM-2 *Technical Report RL-89-1* (Thinking Machines Corp., 245 First St., Cambridge, MA 02114, USA) (also in *Proc. Conf. on Neural Information Processing Systems (1989)* pp 801–9)

Zorat A 1987 Construction of a fault-tolerant grid of processors for wafer-scale integration *Circuits, Syst. Signal Processing* **6**

Zornetzer S F, Davis J L and Clifford L (eds) 1990 *An Introduction to Neural and Electronic Networks* (San Diego, CA: Academic)

E1.5 Optical implementations

I Saxena and Paul G Horan

Abstract

An overview of neural network implementations using optics is presented in this section. To begin with, the motivations for using optical hardware are introduced. Some of the core techniques are described, and the suitability of particular neural network algorithms to optical implementation is discussed. For the nonspecialist reader, relevant basic optical physics is briefly reviewed, and the operation of the principal optical devices and techniques in use is described. Following this are a profile of major recent demonstration systems and innovations and some final comments.

E1.5.1 Introduction and overview

E1.5.1.1 Why optics?

To produce machines rivaling the processing capabilities of biological (human) brains with 10^8–10^{10} neurons and 10^{10}–10^{12} interconnections is a daunting task (DARPA 1988, Caulfield *et al* 1989). Following the astonishing success of silicon and the modern computer industry, it is natural to turn to this technology to make artificial 'neural' computers, and this approach is described in Sections E1.3 and E1.4. The E1.3, E1.4 amazing computational power and speed of modern serial processors is well known. However, artificial neural networks (ANNs) are characterized by relatively low levels of computational complexity, but very high degrees of parallelism and interconnectivity. In electronic processors, the information channels are made of conducting material on a two-dimensional surface. Hence, surface area and power dissipation concerns limit very high interconnectivity or massive parallelism from being realized, which are essential for efficient neural network implementations.

Optics offers the promising alternative of exploiting the third dimension by allowing free-space (three-dimensional) interconnections. Noninterference among intersecting optical channels (paths), essentially instantaneous transport over the short distances involved and insensitivity to electromagnetic interference are inherent advantages in choosing optics. Scaling of the number of processing elements without compromising speed appreciably, a decrease in cross-talk problems and a reduction in energy requirements as compared to electronics are further benefits offered by optical systems. Advantageous alternatives in optical technology are expected specifically for very-large-scale dedicated systems tackling such problems as real-time speech and vision processing, which have proved to be beyond the reach of the standard serial processor.

A comparison of the interconnections and the processing speeds that are characteristic of the domains of electronics, optics and biology are illustrated in figure E1.5.1, adapted from earlier reviews by Alspector and a Defense Advanced Research Projects Agency report (DARPA 1988, Alspector 1991). Electronic neural network implementations in VLSI (i.e. two-dimensional structures) are limited by 10 000 to 100 000 weighted interconnections per chip, whereas optical techniques offer the possibility of orders of magnitude improvement of up to 10^{10} interconnections per cubic centimeter in photorefractive crystals (Psaltis *et al* 1988a). Furthermore, power dissipation in optical interconnections is lower (Feldman *et al* 1988) as compared to electrical ones for the same data transmission rates. Optical neural networks form part of a greater effort in optical computing and photonic switching. Progress in this broader area is driven by developments in telecommunications, where optics already plays a major role, and in massively parallel

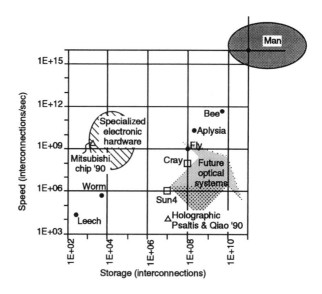

Figure E1.5.1. Computational resources, adapted from Alspector (after DARPA).

processing to relieve the communications bottlenecks. These motivations have sustained research in optical neural networks for over a decade, and as this chapter reveals, encouraging advances have been made during this period.

E1.5.1.2 Functional aspects

Optics can perform many of the functions required in neural computing in a neat and elegant manner. A simple lens can provide fan-in (addition), fan-out, or image inversion. Anamorphic optics, such as cylindrical lenses, can perform two-dimensional ↔ one-dimensional transformations. Arrays of micro-optic components such as prisms can do shift-and-shuffle operations. Holographic and diffractive optical elements, as described later, can perform almost arbitrary input-to-output mappings. Multiplication is performed by passing an optical beam through a semitransparent medium; for example, photographic film. The intensity of the transmitted (or reflected) beam is the product of the original intensity times the medium's transmittance (or reflectance). At the heart of most neural networks is the requirement to carry out a weighted sum of the inputs, which is the equivalent of a vector–matrix multiplication. This is usually followed by some sort of nonlinear transformation or thresholding operation, which is discussed later in this section. This basic process finds a very natural expression in optical hardware.

The most common configuration is the so-called Stanford vector–matrix multiplier (Goodman *et al* 1978). This implements the product of an input vector and an interconnection weight matrix by means of the optical operations discussed above, and is illustrated in figure E1.5.2. A linear array of light sources, each encoding the input value in their intensity, are fanned out vertically by a cylindrical lens. In this way each input is smeared across a column of a two-dimensional array. By adjusting the transmission of each pixel of the two-dimensional array weight matrix, which is typically implemented by a light-modulating device (section E1.5.3), a unique weighted path or interconnection from each source to detector is defined. A second cylindrical lens does fan-in along the horizontal giving the total weighted summation at each detector in the array which corresponds to an element of the resultant product vector.

In contrast to manipulations of interconnection arrays, much of the early interest in optical neural networks arose from the analogy between holography and associative memory. Essentially, a hologram can establish an arbitrary input-to-output mapping. The manner in which multiple holograms are stored in the same volume, and can be recalled by only a partial input, is very reminiscent of the distributed storage mechanism of associative memory. The continuous nature of the holographic media allows potentially astounding amounts of data to be stored per unit volume. It could be argued that holography currently offers the only viable route to matching the information storage and interconnection density of biological neural networks.

The nature of the interconnection weight matrix is one of the most important issues in any implementation. Optical interconnections may be fixed for *recall-only* neural networks, or they may

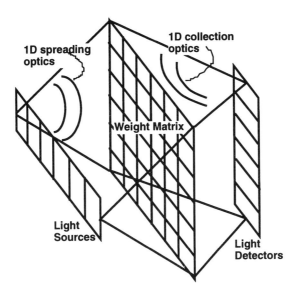

Figure E1.5.2. 'Stanford' optical matrix–vector multiplier.

be adaptive for *learning* networks. Fixed interconnections are determined in advance, usually through simulation, and implemented in some permanent medium such as transparencies or fixed holograms. In *adaptive* optical neural networks the actual weights in hardware are updated or modified during the training process. These adaptive interconnections must be easy to change, must each be independent of the others, and must also be stable. Such adaptive interconnections call for far more complex optical hardware. They can be implemented either by making use of two-dimensional planar devices having spatially variable transmission or reflectivity, known as spatial light modulators (SLMs), or, alternatively, by using adaptive holographic interconnections. These individual technologies are discussed in section E1.5.3. The major   E1.5.3 advantage of adaptive over fixed interconnections is the possibility of the network adapting to local nonuniformities or nonfunctioning units during the training phase, and so overcoming some of the inevitable limitations of real hardware.

An essential function of the basic neuron is a nonlinear transform or thresholding of the weighted summations. Typically this is some type of sigmoidal function. Working with light, there are a number of possible avenues. One is to detect the light intensity corresponding to the weighted sum at each node, and transform it to an electrical signal. Any further processing can then be done with standard electronics. Arrays of optoelectronic optical–electrical–optical devices benefit from VLSI techniques to readily engineer nonlinear and other desirable responses and bear the name 'smart pixels' (see section E1.5.3.2). An alternative or second approach is to do all-optical nonlinear processing. In the presence of suitable materials two light beams can be made to interact in a nonlinear fashion. Initial efforts in the 1970s and early 1980s were therefore directed toward an all-optical computer. However, the noninteracting properties of photons, which make them so suitable for dense interconnectivity, make nonlinear optical operations very difficult. The nonlinear material coefficients involved are so small that nonlinear effects can only be observed at high powers. Despite a concerted research effort, the energies involved in these processes have so far remained prohibitively large for processing with arrays of light beams. Nonlinear optical effects can be used to advantage in interference filters and in phase conjugate mirrors. In using nonlinear interference filters for thresholding, Wang and co-workers (Wang *et al* 1988) found the major shortcomings to be nonuniformity of the threshold and the high optical powers required. A nonlinear gain in phase conjugate mirrors and alternative methods of thresholding have been proposed (Bergeron *et al* 1994). A good example of a third and middle ground between these two approaches is a liquid crystal light valve-based system (Hsu *et al* 1988). In these types of devices (see section E1.5.3.2) which are spatially analog, the incoming light beam causes an electric field to be generated which controls a light-modulating material to allow a nonlinear response approximating a soft threshold. This approach opens up the possibility for multilayer optical neural networks in which recall takes place with light propagation uninterrupted by an electronic processing plane (Saxena and Fiesler 1995, Collings and Xue 1994).

In optical neural networks where information is coded in light intensity, all variables, including the interconnection weight matrix (IWM) elements, have to be positive. Typically, neural network algorithms are based on IWMs containing both negative and positive weights (referred to as bipolar weights). That is, the interconnections are inhibitory as well as excitatory. Three possible solutions are adopted when working with unipolar weights in multilayer optical neural networks. Bipolar weights may be separated into two sets of unipolar weights, which, after being photodetected separately, either (i) *spatially* (Kasama *et al* 1990) or (ii) *temporally* (Yu 1990), are subtracted electronically. This method is typically used in two-layer systems (with no hidden layer), where the second (or final) layer is electronic. However, spatial separation into two sets of unipolar weights requires duplicating hardware and is not optimum use of expensive equipment. Alternatively, bipolar weights can be transformed to unipolar weights by adding a bias term and compensating accordingly (Jang *et al* 1988), or by altering the algorithm to operate with unipolar weights (Shariv and Friesem 1989, Shariv *et al* 1991). Orthogonal polarizations may also be used to encode the weights, given suitable hardware (Kranzdorf *et al* 1989, Ittycheriah *et al* 1990, White *et al* 1988).

It is far easier in optics to add light than to subtract. Although destructive interference of light may be used to give subtraction, this is intrinsically difficult, as it requires mutually coherent light beams, and requires very high physical stability of the components. The most fruitful approach in this area has been to use optoelectronic optical–electrical–optical devices, as discussed in reference to a nonlinear response. The response of the light source or modulator, for example, can be electronically engineered to decrease as light intensity to the detector increases, thus giving an effective inversion, which can be construed as subtraction (Kelly *et al* 1996).

Almost all optical implementations of neural networks to date have been analog in nature. This rests on the inherent simplicity of analog representation, and the belief that a single analog device can perform a nonlinear transform that would require considerable complexity in electronics. This is often at the cost of technologically disparate optical hardware. The use of analog hardware also brings in effective limits on the available numerical resolution, as also discussed in section E1.5.1.2. The number of resolvable gray levels available in an analog system is often limited. Similar problems are experienced in electronic hardware implementations. This is particularly important in adaptive systems, where a far higher resolution is often required to arrive at a suitable set of weights than is needed in operation with the known weights. Thus it is crucial that the requirements of a given algorithm are matched to the available hardware. If hardware requirements are relaxed, such as by using pairs of sources and detectors in a differential format, lower contrast operation and a tolerance to local power nonuniformities is permitted, as values are then expressed as the difference in intensities of two optical beams. As optical systems become more complex it may become necessary to go to some form of signal coding. One possible approach is to use stochastic bit streams, where an analog value is represented by the relative probability of 1s and 0s in a random digital bitstream. Multiplication is done by a bit-wise 'AND-ing' and the accuracy of a value increases with sampling time. Other interesting options for optical nonlinear thresholding also exist (Hands *et al* 1995) with this approach.

E1.5.2 Neural network architectures for optics

This section reviews and compares some neural network algorithms for which optical architectures have been examined. Many issues affect the suitability of an algorithm to optical implementation. The nature and density of the interconnection are of primary importance; whether it is fixed or adaptive, regular or random, local or global. Equally well, the number of neuron layers and the complexity of the individual neuron have to be considered. As must already be clear, the types of calculations that can be done optically are very specific. If an algorithm calls for complex arithmetic then it might be better to consider electronic hardware. Compared to silicon electronics, optical hardware is relatively expensive, so efforts should be made to make maximum use of the hardware. Thus algorithms that have an iterative procedure are attractive. The aim of the system designer should be to tailor the hardware to the algorithm, and vice versa, so as to make maximum use of the unique opportunities offered by optics, without trying to impose unsuitable tasks.

E1.5.2.1 Supervised optical neural networks

C1.3.4 *Hopfield neural networks.* Much attention has been paid to the *Hopfield network*. This was due, in part, to the general interest in the network shown in the early 1980s. But more important is the fact that the network has very natural implementations in optics (Farhat *et al* 1985). The algorithm calls for a single layer, which acts as input and output, is globally and adaptively connected, and operates with a relatively simple learning rule (Hopfield 1982). Thus it involves all essential aspects of a neural network and yet has relatively low complexity. Furthermore, the 'expensive' optical hardware is repeatedly used in an iterative process. For these reasons, as can be seen later in the implementations section, this algorithm has provided a fertile testbed for many of the developments in the field. The algorithm itself has also been adapted to the limitations of optics. For example, a scheme called reversal-input superposing technique (RIST), which yields unipolar, all-positive interconnection weights, has been implemented (Hayasaki *et al* 1994).

The Hopfield network functions as an autoassociative memory. A generalization on this theme is the bidirectional associative memory (BAM) network which is hetero-associative (Kosko 1987). This algorithm is remarkable for the manner in which it immediately evokes the idea of an optical cavity. Information (light) oscillates between two neural planes (mirrors) repeatedly passing through an intervening weight matrix. Although two neural planes are involved, the learning rule is simple and guarantees convergence. Like the Hopfield network, this is an iterative algorithm. As a result a number of optical implementations have been suggested (Guest and TeKolste 1987).

Multilayer neural networks. Multilayer networks, and especially those trained by error backpropagation (BP) and its variations, have proved very popular to the neural network community due to their universal applicability (Hornik *et al* 1989). Hence, in spite of the presence of multiple layers and the requirements for a fairly complex learning rule such as *backpropagation*, which make these algorithms difficult to implement, they are a much desired goal. A promising modular multilayer optical neural network design based on liquid crystal devices has 256 inputs, 256 hidden neurons and 64 outputs (Saxena and Fiesler 1995). Alternatively, the use of the optical hardware is maximized by temporal multiplexing, where a two-layer structure is first configured as input and hidden layer, and the output of this stage is then electronically fed-back to become the input to a reconfigured hidden and output layer (Robinson and Johnson 1992). Methodologies proposed (Wagner and Psaltis 1987) to benefit from the bidirectionality inherent in optical interconnections and implement backpropagating multilayer networks using devices with a derivative-type response have yet to be realized. However, adaptations are required to backpropagation networks in order to implement the nonlinearity by optical thresholding devices (Moerland *et al* 1995).

Meanwhile there are ongoing efforts to search for new training algorithms which obviate the need for backpropagation of errors and/or bipolar weight matrices. See Section E1.2 for an overview on this subject. A result of such effort is the work by Psaltis and Qiao (1990) who implemented a multilayer optical neural network which is a modification of the method based on Kanerva's *sparse distributed memory*. The first IWM has fixed random values and the second IWM is updated by a simple rule. Wagner and Slagle (1993) suggest the use of unsupervised competitive learning between the input and hidden layer followed by (supervised) perceptron learning in the final layer to overcome error backpropagation and yet enable multilayer nonlinearly separable classifier optical neural networks. The anti-Hebbian local learning (or ALL) algorithm is applicable for three-neuron layer neural networks (two IWM layers). As compared to the BP algorithm, it does not require knowledge of the weights of the second IWM in order to determine the weight updates for the first IWM. (In that sense it needs less backward information flow than in BP.) Instead it does require the inputs and outputs at the output layer for determining the first IWM weight updates, and training time seems to be compromised as compared to backpropagation. Other hybrid approaches for special image processing applications, such as a vector feature extracting device (Kuratomi *et al* 1993) are developed which avoid inhibitory synaptic interconnections that are otherwise considered essential for superior pattern recognition capabilities.

Higher-order networks. Higher-order networks involve autocorrelation, to a given degree, of the input, before weighted summation (Psaltis *et al* 1988c). The large increase in the number of weighted terms means that complex, input–output transforms, normally only possible in a multilayer structure, can be accomplished in a single layer. Thus, as there are no hidden layers, simple training rules can be used. In higher-order networks the complexity lies in the interconnection rather than in the processing, making it more attractive for optical implementations. This complexity can also be traded to give a network invariant to a prescribed transformation, e.g. translation (Giles and Maxwell 1987) which can be harnessed to simplify optical implementation (Horan *et al* 1990).

E1.5.2.2 Unsupervised competitive and inhibitory networks

Some problems lend themselves well to finding solutions by unsupervised neural networks, e.g. the traveling salesman problem (Collings *et al* 1990). Competitive networks form the basis for most unsupervised learning algorithms, amongst which are MAXNET, *Kohonen's self-organizing feature maps*, *generalized competitive learning* and *adaptive resonance theory*. Competitive networks are based on finding a maximum (or minimum) among a specified population, which is a nonlocal function, and thus well suited to parallel implementation. This is generally done by some form of mutual inhibition, where the strong (large) will inhibit the weak (small) and thus 'win' the competition. As mentioned, inhibition is difficult to do in an all-optical manner, and best carried out by optoelectronic device arrays, where the mutual inhibition signal may be either optical or electrical.

C2.1.1
C2.2.1

E1.5.3 Hardware

The optical implementation of neural networks is ultimately constrained by our ability to produce, direct, modulate and detect light. The technologies of light sources and detectors are reasonably well developed, especially when compared with the problems of light modulation or optical implementation of the weights (see section E1.5.1.2 and figure E1.5.2). In this section some of the basic physical effects used for light modulation will be outlined, and the manner in which these effects can be used in practical devices described. The use of holography and photo-refractive materials will also be introduced, and a brief mention made of some of the supporting technologies. For a more in-depth discussion of some of the issues in this section see Jahns and Lee (1993).

E1.5.3.1 Materials for light modulation

The weak nonlinear interaction of light with light means that almost all practical SLMs involve (optical)–electrical–optical interactions. SLMs operate on a number of different physical effects, which will be briefly reviewed here, and can be studied in further detail in optics textbooks (Saleh *et al* 1991, Bass 1995).

Electro-optic materials have an electric-field-dependent refractive index. This is an intrinsically fast effect, the speed usually being dictated by the capacitance of the devices. A linear dependence on the field is known as the Pockels effect, while a quadratic dependence is known as the Kerr effect. The induced phase change can be made apparent by including the electro-optic material in an interferometer. Alternatively, the electro-optic effect can be used to vary the coupling between two parallel optical waveguides. The electric-field-induced anisotropy in the material results in a fast and slow optical axis, so that when the modulator is placed between a pair of crossed polarizers, a voltage-dependent transmittance can be observed, as in figure E1.5.3.

Liquid crystal materials consist of ordered elongated organic molecules. These have a natural polarization rotation ability. Under the influence of an applied electric field the orientation of the molecules can be changed to effect a change in polarization. Again, this is made visible as an intensity change using

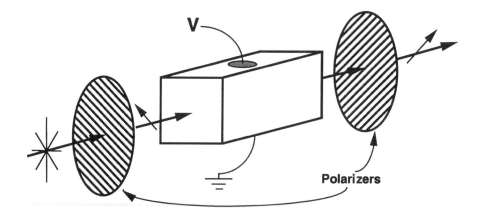

Figure E1.5.3. Electro-optic modulation with crossed polarizers.

external polarizers. Depending on the liquid crystal and the cell construction, continuous analog or bistable operation is possible. Speeds are slower than in crystalline solids, varying from (10^{-3} to 10^{-6}) seconds. Analogous to the electro-optic effect is the magneto-optic or Faraday effect. This effect is characterized by a rotation of the plane of polarization of a light beam in response to an applied magnetic field. Similar to the other polarization modulating materials, an intensity modulator can be made using external polarizers.

The refractive index of a material is also affected by sound, known as the acousto-optic effect. The passage of a sound wave gives rise to a periodic refractive index change. Sound waves can be efficiently generated by electric transducers such as piezoelectric crystals. An incoming light beam, satisfying certain angular conditions, is diffracted from the acousto-optically induced periodic refractive index variation (or grating) in the piezoelectric crystal, and the diffraction efficiency is dependent on the intensity of the sound wave. Thus by varying the sound frequency and intensity, optical modulators, switches or scanners can be made.

Semiconductor materials provide another optical modulation material. The most widely used materials are engineered alloys of III–V semiconductors, such as gallium arsenide. The optical absorption in these materials can vary under the influence of an electric field. However, in materials engineered on the scale of 10s–100s of nm, where quantum mechanical effects come into play, the energy levels of the electrons and holes can be radically altered. This can give rise to an enhanced and qualitatively different change of the absorption in response to an applied electric field (Schmitt-Rink et al 1989). For GaAs a contrast of 2:1 is typically observed. This can be used in a dual rail logic system, where a pair of signals are compared, or alternatively, the contrast can be significantly enhanced by incorporating the material in an optical resonant cavity. The electric field can be efficiently applied by simultaneously incorporating the material in a *pin* diode structure.

There are many other interesting and esoteric optical modulation materials. Physical deflection devices use arrays of micro-machined membranes or cantilevers, which deflect under electrostatic forces (Boysel 1991). Organic materials such as bacterio-rhodopsin have complex optical activation paths. Long-lived quasi-stable states are also important for electron trapping materials.

E1.5.3.2 *Devices: spatial light modulators*

The many different (electro-) optical spatial light modulators that have been proposed and manufactured have been well reviewed (Fisher and Lee 1986, Neff *et al* 1990) and will only be briefly described here. Many SLMs work by the direct application of an electric field. At its simplest this may be no more than a pair of electrodes on either side of a sheet or block of suitable material, making a variable attenuator or switch. Arrays of devices may be individually addressed, but this will only work for one-dimensional or small two-dimensional arrays. As the two-dimensional array size grows, wiring will consume most of the surface area. More commonly, rows of electrodes, vertical and horizontal, address an individual pixel. Only when a voltage is on both wires is a field applied at the point where they cross. Liquid crystal TV screens are typical of this approach. As this involves a serial raster scan of the array, a diode or transistor per pixel may be added to maintain the field at the pixel until the next address cycle. Liquid crystal TVs and modulators have been extensively used in demonstration systems, as these readily available TVs can be easily integrated with PCs and provide cheap and versatile spatial light modulators. The same approach can be used with magneto-optic materials, where currents flowing in the wires generate a local magnetic field (Farhat and Shae 1989).

Many applications call for the direct control of one light beam by another. Optically addressed SLMs (OASLMs) use light to generate an electric field which is then applied to an electro-optic material. Figure E1.5.3 shows a typical electro-optic modulator configuration. A typical OASLM device is an LCLV, using a liquid crystal as the electro-optic material (Bleha *et al* 1978), as shown in figure E1.5.4. Sandwiched between two transparent electrodes is a photoconductive layer (if needed, a light-blocking layer), an insulating mirror and the electro-optic or liquid crystal material. The device is charged up, and acts like a capacitor. Input light to the photoconductor locally discharges the capacitor, thus reducing the field across the modulation material. This can then be read out with a second beam. The OASLM can be used to perform an incoherent-to-coherent light transform, or a change of wavelength. Depending on materials and the applied voltages this device can also perform image inversion and thresholding operations (Armitage and Thackara 1989, Takimoto *et al* 1991). In a related device the photoconductor is replaced by a photoemitter and a microchannel plate. In the MSLM (microchannel SLM) light incident on the photoemitter generates photoelectrons which are accelerated to a microchannel plate which amplifies the

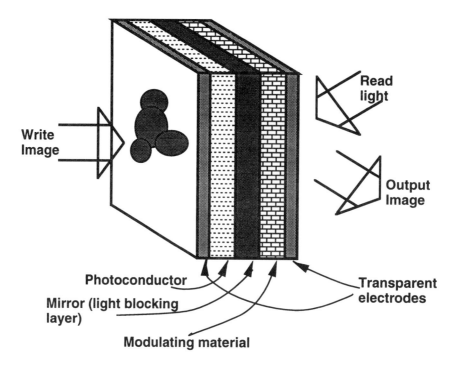

Figure E1.5.4. Basic structure of an OASLM (optically addressed spatial light modulator).

charge which then modulates the electro-optic material (Warde *et al* 1981).

One of the most exciting developments in SLM technology has been the increasing pace of integration of SLMs with electronic circuitry. This allows for local detection and processing of the incoming light signal. This is especially relevant to neural network applications. Such 'smart pixel' processing can include operations such as weight updating, local maxima detection, image processing functions or packet routing. Silicon is the obvious choice for electronic circuitry, but because it is an indirect semiconductor it must be used in a hybrid combination with some optically active materials such as liquid crystals, PLZT (lead lanthanum zirconate titanate) or III–V semiconductors (Wagner and Slagle 1993, Ersen *et al* 1992, Goossen *et al* 1995). The alternative approach is monolithic integration of light sources or modulators with electronics in III–V group semiconductors, such as gallium arsenide alloys. Although the electronic technology is much less well developed in these materials, impressive performance has been demonstrated with integrated arrays of quantum well modulators and field-effect transistors, as can be seen in the technical digest on smart pixels (Smart 1994). Work on integrating LEDs with circuitry is proceeding apace (Grot *et al* 1994).

E1.5.3.3 Techniques: holography

The natural analogies between holography and associative memory are very attractive. Associative memory benefits greatly from holographic techniques of recording images. Since the initial ideas on associative memory appeared, the field has expanded to include real-time holographic storage and fixed interconnections. A conventional photograph records only the amplitude of an optical wavefront at a point in space, and the phase (direction) information is lost. A hologram, on the other hand, stores the wavefront in detail, in that it transforms a specified incoming wave to fully reproduce an image wave in all details of *both* amplitude and phase, giving the now familiar three-dimensional image. Recording phase information is difficult as it is obtained by making an input beam interfere with a mutually coherent beam. The resulting amplitude interference pattern is recorded as the hologram shown in figure E1.5.5(*a*). Illuminating the hologram with the reference beam will recreate the original input beam as shown in figure E1.5.5(*b*) (Caulfield 1979, Saxby 1994). Theoretically, an arbitrary input-to-output mapping can be established, thus acting as an associative memory (Collings 1988). Continuing the analogy, illuminating any one small part of the hologram can replay the full stored image; the hologram operates effectively as a distributed memory. By including a lens in the input beam the Fourier transform of the input may

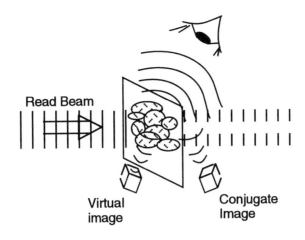

Figure E1.5.5. (*a*) Viewing a hologram.

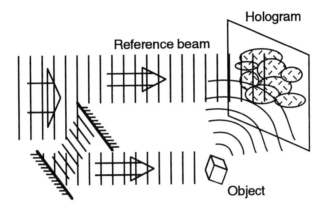

Figure E1.5.5. (*b*) Recording a hologram.

be holographically recorded. If an arbitrary input is presented, the output is the convolution of the input with the stored image. This configuration, shown in figure E1.5.6, is known as a Van der Lugt filter (Van der Lugt 1964). This system may also be configured to yield the correlation of the input and stored data, whereby the filter yields a large response only when the input matches one of the stored memories. If several reference holograms (stored memories) are located in different specific areas of one hologram, then the matching item can be identified. When a hologram is formed in a bulk material, a thick or volume hologram is formed. In this case the reading beam must satisfy certain angular constraints, known as the Bragg condition, before the output beam is reconstructed (Saleh and Teich 1991). Such angular selectivity offers the possibility of very high density data storage, as many holograms can be superimposed in the same volume, but selected individually by choice of angle. Photorefractives are a class of materials which produce a local electric field in response to light, which then modulates the refractive index by the electro-optic effect (see section E1.5.3.4). They provide a means to store rewriteable volume holograms. Methods of ameliorating problems of angular degeneracy and overwriting of volume holograms have been demonstrated (Lee *et al* 1989).

E1.5.3.4 Technology: sources and optics

The above modulation technologies all rely on external optical sources. The field of optical sources is very mature, and laser, laser diode and incoherent sources are available at a range of wavelengths and output radiant powers. The output of conventional cathode ray TVs or electro-luminescent displays may be used directly or transformed to a coherent input using an OASLM. Many experiments have used one- and two-dimensional arrays of light-emitting diodes (LEDs) as input, where the input is intensity encoded on the LED outputs. One-dimensional arrays of edge emitting lasers have been available for some time. Recent developments (Jewel *et al* 1989, 1990) in vertical cavity surface emitting lasers (VCSELs) have

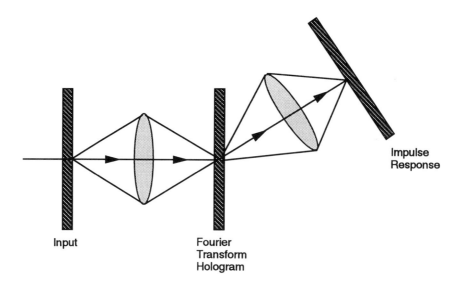

Figure E1.5.6. A Van der Lugt filter.

led to two-dimensional arrays of lasers emitting normal to the array plane being reported (Lear *et al* 1995). These presently use direct electrical matrix-addressing schemes that are not sufficiently fast for parallel interconnections, thereby motivating the investigation of faster optical-addressing (Lee *et al* 1993). A 16×16 VCSEL array of active laser area 4.5×3.7 mm^2 and total chip size 1.1×1.1 cm^2 is a promising step in this direction (Gulden *et al* 1995). Relatively high threshold currents, and hence power consumption, have limited the extent of the arrays. However, recent improvements in processing have seen threshold currents decrease from mA to μA. Further improvements in device characteristics, such as resistivity, are to be expected resulting in VCSELs proving to be very real alternatives to modulation technologies.

The miniaturization and integration of optical sources has been accompanied by similar trends in the field of optics. The range and sophistication of arrays of micro-optical refractive and diffractive components is increasing rapidly (Jahns 1994). Typically, the components are produced using established lithographic techniques, making for cheap and reliable replication. Refractive components can be made by various methods, such as molding, melting, diffusion or micro-machining. Components based on diffraction rely on a complex surface profile, engineered on the scale of the wavelength of light, to produce the required optical field at a specified distance from the piece. In the related field of computer-generated holograms, the hologram of an arbitrary object is calculated using advanced numerical techniques. Again the hologram is manufactured using VLSI technology. Such holographic, or diffractive, optical elements (HOEs or DOEs) now find widespread use in optical interconnection and processing systems. For a general review see Taghizadeh and Turunen (1992) and references therein. They are mainly used for array illumination, fan-in, fan-out and interconnection. Array illuminators, where a single input is fanned out into an array of equal intensity spots, have been fabricated up to 256×256 spots. Two-dimensional arrays of HOE elements can provide almost arbitrary point-to-point interconnection, and fan-out elements can provide neighborhood interconnection.

The problems of how to combine all these components, both active and passive, in a stable, robust yet compact package, are in the process of being solved (Jahns 1994). To date, most optical systems are primarily for demonstration and have been made using large, bulky, individual components, making systems that are measured in meters. For example, a globally interconnected 256×256 network using a 4-f imaging bulk optics system (Collings 1994) is just under a meter long. Special optical elements designed for optical interconnect applications, such as holographic optical elements and lenslet arrays, will yield more compact optical interconnect stages in the near future. Lithographic technology offers device arrays on the micron level; the opto-mechanics must match this scale and level of integration. Recent developments in slot-plate technology offer an intermediate scale of integration, producing stable and relatively cheap systems on the centimeter scale. On a more practical front, impressive progress has been demonstrated in integrating a liquid-crystal-based Van der Lugt correlator onto a PC expansion board to act as a co-processor in a standard PC (Bains 1995). For the future, approaches based on solid optics, where the space between components is a transparent solid such as glass, looks very promising for the next level of miniaturization. Planar optics is a proposal to integrate component arrays on either or both sides of a glass block (Jahns 1994, Prongue and Herzig 1994). Optical interconnections zig-zag between elements, thereby preserving the three-dimensional advantage of free-space propagation while profiting from planar integrated circuit fabrication techniques. Stacked optics is a related idea that uses beam-splitters to maintain normal incidence (Brooke and DeWeerth 1993). Micro-optic components are fabricated in the solid glass substrate, and active components are attached by VLSI techniques, such as flip-chip solder bumps. Developments in opto-mechanics and packaging shall play a key role in the success of future optoelectronic systems.

E1.5.4 Implementations

The optical implementation of neural networks represents an interplay between algorithms, devices and the ingenuity of the researchers. Many different approaches and methodologies have been attempted, and we shall highlight some of the most significant of these. Work in this area can be said to begin with the seminal work of Psaltis and Farhat (1985). In this paper they describe two possible implementations of the *Hopfield model* (Hopfield 1982). One scheme is based on the Stanford vector–matrix multiplier C1.3.4 (Goodman *et al* 1978) with either electrical or optical feedback. An alternative approach using holography and coherent optics is also outlined. These two approaches indicate the main directions research was to undertake in the coming years.

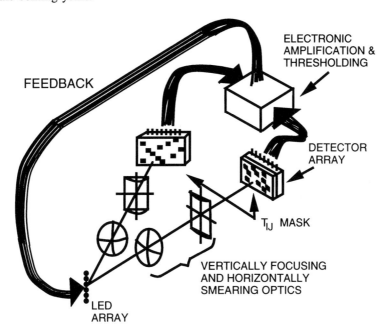

Figure E1.5.7. Hopfield neural network vector–matrix multiplier implementation, Farhat *et al* (1985).

The vector–matrix approach was the first to be exploited. Using an LED array and two-dimensional detector arrays, Farhat *et al* (1985) made a system utilizing electronic thresholding, amplification and feedback, as shown in figure E1.5.7. Anamorphic optics (cylindrical lenses) spread the light from one LED across a row of the detector array. Bipolar weights are implemented on separate weight transparencies, with electronic subtraction of the detected totals. This system illustrates the strengths and weaknesses of optics. Complex interconnection is achieved, but bipolar values are difficult, and amplification and thresholding are done electrically. Although it is suggested that these elements can be replaced by all-optical devices, as discussed earlier, these present certain difficulties.

The basic vector–matrix formalism was to prove very fruitful. Variations on the Hopfield model, (Athale *et al* 1986) and other learning rules such as *Widrow-Hoff* and *Hebbian* rules were implemented B3.3.3, B3.3.1 (Fisher *et al* 1987). Farhat (1987) introduced a scheme for partitioning the vector–matrix processor to implement multilayer networks, whereby the input and output are divided into blocks or sub-matrices, corresponding to input, hidden and output layers, as shown in figure E1.5.8. Thus if it is required that there be no interconnection between layers, the relevant sub-matrix is set to zero. All these implementations

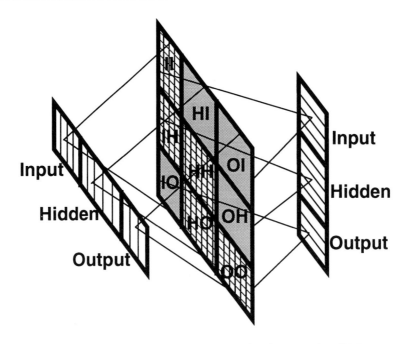

Figure E1.5.8. Multilayer neural network scheme (Farhat 1987).

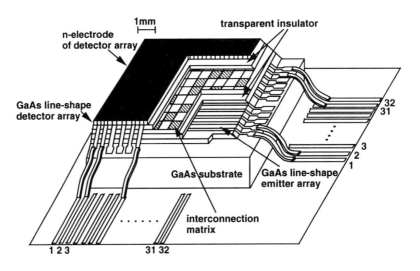

Figure E1.5.9. Optical neuro-chip, (Ohta *et al* 1989).

were based on a one-dimensional input/output and two-dimensional (or 1D–2D–1D) interconnection matrix. Very often it is required to have a two-dimensional–four-dimensional–two-dimensional (i.e. 2D–4D–2D) interconnection. A raster decomposition to the easier 1D–2D–1D format is always possible but not always attractive. Other than the direct implementation of the interconnection with optical fibers between two-dimensional arrays (Ito and Kitayama 1989), bulk holography (discussed below) offers the most obvious alternative. Caulfield (1987) proposed another approach whereby the two-dimensional interconnection matrix is itself composed of sub-matrices for each input–output coupling. An interesting alternative approach to this problem is demonstrated by Lee *et al* where a fixed random interconnection is implemented between the two-dimensional planes using a scatter plate, and local adaptive weightings are made at the inputs and outputs (Lee *et al* 1993).

Most of these implementations rely on anamorphic optics to provide fan-out and fan-in, thus resulting in moderately bulky systems. Athale and Stirk (1989) showed how compact inner product multiplications could be done using mutually orthogonal, finger-shaped one-dimensional arrays of light sources and detectors, with a two-dimensional interconnection matrix sandwiched between. The length of the light source fingers distributes the light to each element of the interconnection matrix, and the detector sums

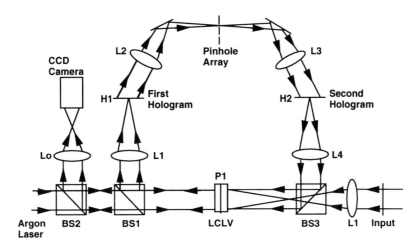

Figure E1.5.10. Hopfield neural network implementation with holographic interconnections (Paek and Psaltis 1987), © 1990 IEEE.

the components. The same ideas are adapted for outer product calculation. This compact scheme was exploited by Ohta *et al* (1989) in their optical neurochip, in which integrated arrays of light sources (LEDs) and photodetectors are combined in a hybrid cross-bar configuration on either side of a fixed interconnection matrix, as shown in figure E1.5.9. The resulting 32-input chip had an area of 0.5 cm^2 and was used to demonstrate operation in a Hopfield network with external electronics. Bipolar weights were implemented on separate chips. In their next generation chip, excitatory and inhibitory connections were implemented on a single chip (Ohta *et al* 1990). The 66 inputs on the 1 cm^2 chip were organized in a three layer (35-29-26) topology to implement a backpropagation character recognition network. To implement dynamic interconnections and so allow for on-line learning Ohta *et al* (1991) introduced variable sensitivity photodiodes. These allow for analog bipolar weights and can exhibit a memory function (Nitta *et al* 1993). Estimated densities of up to 2000 neurons cm^{-2} are possible with this particular technology. The use of fast, bipolar switching in dense array devices, such as multiplexed gray-scale devices, has been demonstrated (Burns *et al* 1994). Variable sensitivity photodiodes have also been used by Rietman *et al* (1991). In particular, they have focused on the nonidealities introduced by the hardware, especially the limitations introduced by finite, quantized weights (Frye *et al* 1991). Work on liquid crystalline, polarization modulation based systems by Robinson and Johnson (1992) have also explored the limits of nonideal hardware. They found nonlinearities in the weight mapping to be important, but were optimistic about the ability of on-line training to overcome nonuniformities. In a detailed analysis of an associative memory algorithm, again based on a vector–matrix processor, Neiberg and Casasent (1994) demonstrate impressive performance on nonideal hardware.

In parallel with developments in vector–matrix machines, the power of holographic interconnection was also being investigated. The holographic implementation of the Hopfield network, first proposed by Psaltis and Farhat, was built (Paek and Psaltis 1987, Hsu *et al* 1990). The system, as built, used two holograms. The input is first correlated with a number of stored memories stored in the first hologram, as shown in figure E1.5.10. Any strong correlation yields a bright spot of light. This spot is then incident on a second hologram and replays the stored image, which becomes the input to another iteration of the cycle. The optical loop is completed by use of a liquid crystal light valve. After a number of iterations the system settles to a state which represents the closest match between the input and the stored images. Variations on this theme were explored by Jang *et al* (1988) and White and Wright (1988). These examples used fixed two-dimensional holograms. Bulk photorefractive holography offers far greater potential storage capacity and the possibility of variable adaptive interconnection (Anderson and Lininger 1987). Furthermore, photorefractives can store their 'memories' for considerable time, giving one of the few viable forms of *optical* memory. Architectures for backpropagation and perceptron networks have been proposed (Psaltis *et al* 1988a, Kitayama *et al* 1989). An improved local learning rule for use in these systems has been proposed by Qiao and Psaltis (1992). Problems of degeneracy between the stored memories reduces the potential storage capacity. However, methods for achieving optimal storage have been found (Psaltis *et al* 1988b). Placing a hologram in an optical resonant cavity gives rise to a rich field of possibilities. The

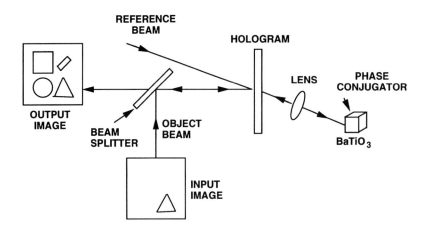

Figure E1.5.11. Soffer *et al*'s associative memory implementation.

stored information can be viewed as modes of the cavity. Using phase conjugate mirrors to form the cavity provides nonlinear gain in self-aligning format (Anderson 1986, Yariv and Kwong 1986). Soffer and Owechko (Soffer *et al* 1986, Owechko *et al* 1987) built a system, as in figure E1.5.11, where the individual holograms were angularly multiplexed and demonstrated recall of a partial input of one of the stored images.

Choosing and adapting algorithms suited to optical implementation has been an on-going activity. The *bidirectional associative memory* (BAM) proposed by Kosko finds a very natural expression in optical terms (Kosko 1987, Guest and TeKolste 1987). As mentioned, higher-order neural networks are attractive for optical implementation in that they trade-off complex interconnection for simple single-layer operation (Giles and Maxwell 1987, Psaltis *et al* 1988c). These have been implemented using liquid crystal (Von Lehmen *et al* 1990, Zhang *et al* 1991) and semiconductor quantum well devices (Jennings *et al* 1994). Inhibition plays a major part in neural network models and, as discussed earlier, is especially difficult to implement optically. Jenkins and Wang (1988) proposed an incoherent optical neuron (ION) based on the response of a liquid crystal light valve. Kawakami *et al* (1989) demonstrated similar inhibitory behavior using the transfer function of a micro-channel plate SLM. This approach has been extended by Wang *et al* (1993) to model early visual processing. In further experiments Kawakami *et al* (1991) built integrated pnpn light source/detector devices having an inhibitory response. It has also been proposed that the inverted response observed with self-linearized SEED devices can be used for lateral inhibition (Horan 1995). Inhibition is also at the heart of most competitive algorithms, which involve finding a maximum. Pattern recognition using Kohonen's self-organizing feature map algorithm has been implemented by Lu *et al* (1990) using LCTVs. Memory enhancement for the number of patterns stored was achieved by introducing forbidden regions for previously learned patterns, to prevent erasure during subsequent learning. Special devices for speeding the implementation of winner-take-all algorithms (Slagle and Wagner 1992) are under development. Investigations by Duvillier *et al* (1994) of a self-organizing architecture based on the Kohonen map model using two LC bistable optically addressed SLMs in a resonator configuration provides a way in which the optical loop can remain closed (uninterrupted by optical to electronic conversion) and allow neural decision, weight updating and spatial ordering of information.

Recently, the increased integration of optics and electronics has led to some very exciting developments. Detecting the maximum among a number of outputs is crucial to many networks, such as MAXNET and many of the unsupervised networks. Radehaus *et al* (1992) have built arrays of pnpn photothyristors, which detect light signals, and only that pixel which receives the largest signal lights up, giving winner-take-all behavior. Great progress has been made with integrating liquid crystal (LC) directly with silicon electronics. In this technology the LC is laid directly on the silicon circuitry, and overlaid by a transparent conductor. Output pads in the silicon circuitry directly modulate the overlying LC (Johnson *et al* 1993). The advanced state of silicon technology allows complex circuits to be easily and quickly developed. This has allowed smart pixel SLMs to be developed for such applications as early vision zero-crossing detection (Jared and Johnson 1991, 1992) or a winner-take-all SLM as part of an unsupervised learning holographic system (Wagner and Slagle 1993). The integration of PLZT on silicon has allowed similarly ambitious hybrid systems to be proposed (Krishnamoorthy *et al* 1992). Most recently, the hybrid integration of III–V materials on Si circuitry using solder bump technology (Goossen *et al* 1995) and epitaxial lift-off (Camperi-Ginstet *et al* 1991) has progressed rapidly, and looks very promising. Monolithic integration in gallium arsenide offers similar opportunities, although the technology is not so well developed. A sigmoidal neuron response has been implemented using LEDs integrated with detectors and electronics (Lin *et al* 1993). A related promising approach is the integration of quantum well modulators with resonant tunneling diodes and transistors (Mehanian *et al* 1991). For the future, though currently more experimental, is persistent spectral hole-burning, which offers the possibility of very high density holographic storage (Maniloff *et al* 1995, Ollikainen 1993). Electron trapping materials combine detection, memory and light source in one material (Jutamulia *et al* 1991), while the complex photochemical cycle of bacteriorhodopsin can reproduce an excitatory and inhibitory response (Takei *et al* 1991) having fast response times only a few picoseconds long.

E1.5.5 Future directions

The future for optical neural networks looks very promising. It is felt by many that optics can be justified only for large scale ($> 10^3$ neuron) systems, as some of the current largest electronic neurocomputers (see Ramacher *et al* 1993, SYNAPSE 93) can achieve the order of a few thousand neurons. The integration of optical technology into computing will, nevertheless, be a gradual evolutionary process. The advantages of optics for point-to-point communication have been illustrated by the dramatic shift from copper to fiber-optic cables for telecommunications. The advantage of optics for communication on a scale greater than centimeters drives interest in using optical interconnections in massively parallel computing; such as for data communication on an optical bus, or for low skew clock distribution. Such passive optical systems seem certain to be increasingly used in parallel computing systems. Many laboratory systems have demonstrated active optical switching, which is redirecting of optical signals as a function of time, but it has yet to make a major impact in the commercial arena. The impressive progress being made in the field of telecom switching makes it most likely that the first applications will be in the field of signal routing, where a switching network is actively configured in response to information carried by the signal packet itself. If established, such technology would be equally useful for reconfiguring the interconnections of a parallel processing system. Active data processing by means of reconfigurable interconnections, as represented by many of the examples discussed here, is a step further.

The course of optoelectronic integration seems set to continue. Realization of the advantages and limitations of both optical and electronic technologies leads to a natural convergence, where both technologies can complement rather than compete. Smart pixel integration offers the possibility of local memory and weight updating. Liquid crystal on silicon device arrays benefit from advances in display technology. Recent developments in GaAs on silicon integration have produced impressive arrays of solid state devices. Other silicon with light source or modulator combinations all have their own attractions. Whether any one technology will *win out*, or individual technologies become specialized to particular applications, remains to be seen.

Most demonstration optical systems built to date have been relatively modest in scale; the challenge, therefore, is to build large-scale optical systems with a clear operational advantage. Potential applications in associative memory and data retrieval would suggest holographic techniques as the most promising, while those requiring more complex learning or control paths might benefit from the flexibility offered by integration with silicon. Continuing progress towards advantageous optical neural networks requires coordinated development on both the algorithm and the hardware front.

Many issues have yet to be explored in the optical implementation of neural networks. The field is still at a relatively immature phase. The full potential of such systems can really only be exploited by large-scale, parallel systems. The advantages of parallel input and output can only be exploited if an individual network layer is integrated into a larger system, perhaps comprising different layers of different functionality, with information flowing in many parallel paths; systems incorporating both feedforward 'bottom up' propositions from the data and 'top down' feedback expectations from the output. Such multilayer systems, and how they can be interfaced to the world and other computing systems, remain as tasks for the future.

Acknowledgements

The authors are grateful to J Hegarty, N Collings, T C B Yu and P D Moerland for their helpful comments.

References

Alspector J 1991 Parallel implementations of neural networks: electronics, optics, biology *Technical Digest, Optical Computing '91*

Anderson D Z 1986 Coherent optical eigenstate memory *Opt. Lett.* **11** 56–8

Anderson D Z and Lininger D M 1987 Dynamic optical interconnects: volume holograms as optical two-port operators *Appl. Opt.* **26** 5031–8

Armitage D and Thackara J I 1989 Photoaddressed liquid-crystal edge-enhancing spatial light modulator *Appl. Opt.* **28** 219–25

Athale R and Stirk C W 1989 Compact architectures for adaptive neural nets *Opt. Eng.* **28** 447–55

Athale R, Szu H H and Friedlander C B 1986 Optical implementations of associative memory with controlled nonlinearity in the correlation domain *Opt. Lett.* **11** 482

Bains S 1995 Miniature optical correlator fits inside a PC *Laser Focus World* pp 17–8

Bass M (ed) 1995 *Handbook of Optics* (New York: McGraw-Hill)

Bergeron A, Arsenault H H, Eustache E and Gingras D 1994 Optoelectronic thresholding module for winner-take-all operations in optical neural networks *Appl. Opt.* **33** 1463–8

Bleha W P, Lipton L T, Wiener-Avnear E, Grinberg J, Reif P G, Casasent D, Brown H B and Markevitch B V 1978 Application of the liquid crystal light valve to real-time optical data processing *Opt. Eng.* **17** 371–84

Boysel R M 1991 A 128 × 128 frame-addressed deformable mirror spatial light modulator *Opt. Eng.* **30** 1422–7

Brooke M A and DeWeerth S P 1993 Merging optics and electronics in neural networks *Opt. and Photonics News* pp 26–9

Burns D C, Underwood I, Murray A F and Vass D G 1994 An optoelectroninc neural network with temporally multiplexed grey-scale weights *MicroNeuro '94* pp 3–7

Camperi-Ginstet C, Hargis M, Jokerst N and Allen M 1991 Alignable epitaxial liftoff of GaAs material with selective deposition using polyimide diaphragms *IEEE Trans. Photonics Technology Lett.* **3** 1123–6

Caulfield H J (ed) 1979 *Handbook of Optical Holography* (New York: Academic)

——1987 Parallel n^4 weighted optical interconnections *Appl. Opt.* **26** 4039–40

Caulfield H J, Kinser J and Rogers S K 1989 Optical neural networks *Proc. IEEE* **77**

Collings N 1988 *Optical pattern recognition using holographic techniques* (Wokingham: Addison-Wesley)

——1994 Design of a useful two-layered neural network *Euro-American Workshop on Optical Pattern Recognition (La Rochelle)* pp 14–17

Collings N, Sumi R, Weible K J, Acklin B and Xue W 1990 The use of optical hardware to find good solutions of the travelling salesman problem. *Proc. SPIE* **1806**

Collings N and Xue W 1994 Liquid-crystal light valves as thresholding elements in neural networks: Basic device requirements *Appl. Opt.* **33** 2829–33

DARPA Neural Network Study 1988 AFCEA International Press, 4400 Fair Lakes Court, Fairfax, Virginia 22033–3899, USA

Duvillier J, Killinger M, Heggarty K, Yao K and de Bougrenet de la Tocnaye J L 1994 All-optical implementation of a self-organizing map: a preliminary approach *Appl. Opt.* **33** 258–66

Ersen A, Krishnakumar S, Ozguz V, Wang J, Fan C, Esener S and Lee S H 1992 Design issues and development of monolithic silicon/lead lanthanum zirconate titanate integration technologies for smart spatial light modulators *Appl. Opt.* **31** 3950–64

Farhat N H 1987 Optoelectronic analogs of self-programming neural networks: architectures and methodologies for implementing fast stochastic learning by simulated annealing *Appl. Opt.* **26** 5093–103

Farhat N H, Psaltis D, Prata A, and Peak E 1985 Optical implementation of the Hopfield model *Appl. Opt.* **24** 1469–75

Farhat N H and Shae Z Y 1989 Scheme for enhancing the frame rate of magnetooptic spatial light modulators *Appl. Opt.* **28** 4792–800

Feldman M R, Esener S C, Guest C C, and Lee S H 1988 Comparison between optical and electrical interconnects based on power and speed considerations *Appl. Opt.* **27** 1742–51

Fisher A D and Lee N J 1986 Current status of two-dimensional spatial light modulator technology *SPIE Optical and Hybrid Computation* ed H H Szu **634** 352

Fisher A D, Lippincott W and Lee J N 1987 Optical implementations of associative networks with versatile adaptive learning capabilities *Appl. Opt.* **26** 5039–54

Frye R C, Reitman E A and Wong C C 1991 Back-propagation learning and nonidealities in analog neural network hardware *IEEE Trans. on Neural Networks* **2** 110–7

Giles C L and Maxwell T 1987 Learning, invariance, and generalization in high-order neural networks *Appl. Opt.* **26** 4972–8

Goodman J W, Dias A R and Woody L M 1978 Fully parallel, high-speed incoherent optical method for performing discrete fourier transforms *Opt. Lett.* 21–3

Goossen K W *et al* 1995 Demonstration of a dense, high-speed optoelectronic technology integrated with silicon cmos via flip-chip bonding and substrate removal in optical computing *Optical Computing, OSA Technical Digest series* **10** 142–4

Grot A C, Psaltis D, Shenoy K V and Constad C G 1994 Large scale integration of LEDs and GaAs circuits fabricated through Mosis *Tech. Digest of the Int. Conf. on Optical Computing, OC '94 (Edinburgh)* pp 3–4

Guest C C and TeKolste R 1987 Designs and devices for optical bidirectional assocative memories *Appl. Opt.* **26** 5055–60

Gulden K H, Ruffieux D, Thelen K, Moser M, Leipold D, Epler J, Schweizer H P, Greger E and Riel P 1995 16 × 16 individually addressable top emitting vcsel array with high uniformity and low threshold voltages *Optics and Information, Topical Meetings Digest Series* **6** p 6.1

Hands M A, Peiffer W, Kirk A and Hall T J 1995 A case study for the implementation of a stochastic bit stream neuron; the choice between electrical and optical interconnects (Washington: IEEE Computer Society Press) pp 63–75

Hayasaki Y, Tohyama I, Yatagai T, Mori M and Ishihara S 1994 Reversal-input superposing technique for all-optical neural networks *Appl. Opt.* **33** 1477–84

Hopfield J J 1982 Neural networks and physical systems with emergent collective computational abilities *Proc. Natl Acad. Sci. USA* **79** 2554–8

Horan P, Uecker D and Arimoto A 1990 Optical implementation of second-order neural network discriminator model *Japan J. Appl. Phys.* **29** 1328–31

Horan P 1995, 1994 *Optical Lateral Inhibition Networks Using Self-Linearised SEED's* pp 403–6 (Bristol: IOP Publishing)

Hornik K, Stinchcombe M and White H 1989 Multilayer feedforward networks are universal approximators *Neural Networks* **2** 359–66

Hsu K, Brady D and Psaltis D 1988 *Neural Information Processing Systems* ed D Z Anderson (New York: IEEE, American Institute of Physics) pp 377–386

Hsu K-Y, Li H-Y, and Psaltis D 1990 Holographic implementation of a fully connected neural network *Proc. IEEE* **78**

Ito F and Kitayama K-I 1989 Optical implementation of the hopfield neural network using multiple fiber nets *Appl. Opt.* **28** 4176–81

Ittycheriah A P, Walkup J F, Krile T F, and Lim S L 1990 Outer product processor using polarization encoding *Appl. Opt.* **29** 275–83

Jahns J 1994 Planar packaging of free-space optical interconnections *Proc. IEEE* **82** 1623–31

Jahns J and Lee S H 1993 *Optical Computing Hardware* (New York: Academic)

Jang J-S, Jung S-W, Lee S-Y and Shin S-Y 1988 Optical implementation of the Hopfield model for two-dimentional associative memory *Opt. Lett.* **13** 248–50

Jared D A and Johnson K M 1991 Optically addressed thresholding very-large-scale-integration/liquid-crystal spatial light modulator *Opt. Lett.* **16** 967–9

——1992 Early vision zero-crossing spatial light modulators *Tech Digest of the IEEE LEOS Summer Topical Meeting on Smart Pixels, (Santa Barbara, CA) IEEE Catalog No 92TH0421-8 ISBN 0-7803-0522-1* Paper MB3

Jenkins B K and Wang C H 1988 Model for an incoherent optical neuron that subtracts *Opt. Lett.* **13** 892–4

Jennings A, Horan P and Hegarty J 1994 Optical neural network with quantum well-devices *Appl. Opt.* **33** 1469–76

Jewell J L, Lee Y H, Scherer A, McCall S L, Olsson N A, Harbison J P and Florez L T 1990 Surface-emitting microlasers for photonic switching and interchip connections *Opt. Eng.* **29** 210–4

Jewell J L , Scherer A, McCall S L, Lee Y H, Walker S J, Harbison J P and Florez L T 1989 Low threshold electrically pumped vertical cavity surface emitting micro-lasers *Electron. Lett.* **25** 1123–4

Johnson K M, McKnight D J and Underwood I 1993 Smart spatial light modulators using liquid crystals on silicon *IEEE J. Quantum Electron.* **29**

Jutamulia S, Storti G M, Lindmayer J and Seiderman W 1991 Use of electron trapping materials in optical signal processing. 2: two-dimensional associative memory *Appl. Opt.* **30** 2879–84

Kasama N, Hayasaki Y, Yatagai T, Mori M and Ishihara S 1990 Experimental demonstration of optical three layer neural network *Japan J. Appl. Phys.* **29** L1565–8

Kawakami W, Kitayama K I, Nakano Y and Ikeda M 1991 Lateral inhibitory action in an optical neural network using an internal-light-coupled optical device array *Opt. Lett.* **16** 1028–30

Kawakami W, Yoshinaga H and Kitayama K-I 1989 Demonstration of an optical inhibitory neural network *Opt. Lett.* **14**

Kelly B, Horan P, Tooley F A P, Taghizadeh M R and Hegarty J 1996 Optical lateral inhibition networks that use self-linearized self-electro-optic-effect devices: theory and experiment *Appl. Opt.* **34** to appear

Kirk A G and Kendall G D *et al* 1991 An optical neural network with reconfigurable holographic interconnection *Optical Memory and Neural Networks* vol 1402 (Bellingham, WA: SPIE)

Kitayama K-I, Yoshinaga H and Hara T 1989 Experiments of learning in optical perceptron-like and multilayer neural networks *Proc. Int. J. Conf. on Neural Networks* vol 2 (San Diego, CA: IEEE and INNS, IEEE TAB Neural Network Committee/SOS Printing) pp 465–71

Kosko B 1987 Adaptive bidirectional associative memories *Appl. Opt.* **26** 4947–60

Kranzdorf M, Bigner B J, Zhang L and Johnson K M 1989 Optical connectionist machine with polarization-based bipolar weight values *Opt. Eng.* **28** 844–8

Krishnamoorthy A V, Yayla G and Esener S C 1992 A scalable optoelectronic neural system using free space optical interconnects *IEEE Trans. Neural Networks* **3** 404–13

Kuratomi Y, Takimoto A, Akiyama K and Ogawa H 1993 Optical neural network using vector-feature extraction *Appl. Opt.* **32** 5750–8

Lear K L, Choquette K D, Schneider R P, Kilcoyne S P and Geib K M 1995 *Electron. Lett.* **31** 208

Lee H-J, Lee S-Y and Shin S-Y 1993 Random interconnections with ground glass for optical TAG *Neural Networks* Optical Society of America, Technical Digest Series, vol 7 pp 104–7

Lee H, Gu X-G and Psaltis D 1989 Volume holographic interconnections with maximal capacity and minimal crosstalk *J. Appl. Phys.* **65** 2191–4

Lin S, Grot A, Luo J and Psaltis D 1993 GaAs optoelectronic neuron arrays *Appl. Opt.* **32** 1275–89

Lu T, Yu T S and Gregory D A 1990 Self-organizing optical neural network for unsupervised learning *Opt. Eng.* **29** 1107–13

Maniloff E, Altner S B, Bernet S, Graf F R, Renn A and Wild U P 1995 Recording of 6000 holograms by the use of spectral hole burning *Appl. Opt.* **34** 4140–48

Mehanian C, Aull B F, and Nichols K B 1991 An optoelectronically implemented neural network for early visual processing *Proc. SPIE* **1469** 275–80

Moerland P, Fiesler E and Saxena I 1995 The effects of optical thresholding in backpropagation neural networks *Proc. Int. Conf. on Artificial Neural Networks (ICANN'95 and NeuroNimes '95)* vol 2, ed F Fogelman-Soulie and P Gallinari (ENNS) pp 339–43

Neff J A, Athale R A and Lee S H 1990 Two-dimensional spatial light modulators: a tutorial *Proc. IEEE* **78** 826–54

Neiberg L and Casasent D 1994 High capacity neural networks on nonideal hardware *Appl. Opt.* **33** 7665–75

Nitta Y, Ohta J, Tai S and Kyuma K 1993 Optical learning neurochip with internal analog memory *Appl. Opt.* **32** 1264–74

Ohta J, Kojima K, Nitta Y, Tai S and Kyuma K 1990 Optical neurochip based on a three-layered feed-forward model *Opt. Lett.* **15** 1362–4

Ohta J, Nitta Y and Kyuma K 1991 Dynamic optical neurochip using variable-sensitivity photodiodes *Opt. Lett.* **16** 744–6

Ohta J, Takahashi M, Nitta Y, Mitsunaga K and Kyuma K 1989 GaAs/AlGaAs optical synaptic interconnection device for neural networks *Opt. Lett.* **14** 844–846

Ollikainen O 1993 Optical implementation of quadratic associative memory by use of persistent spectral hole burning *Appl. Opt.* **32** 1943–7

Owechko Y, Dunning G J, Maron E and Soffer B H 1987 Holographic associative memory with nonlinearities in the correlation domain *Appl. Opt.* **26** 1900–10

Paek E G and Psaltis D 1987 Optical associative memory using fourier transform holograms *Opt. Eng.* **26** 428–33

Prongue D and Herzig H P 1994 Total internal reflection holography for optical interconnections *Opt. Eng.* **33** 636–42

Psaltis D, Brady D and Wagner K 1988a Adaptive optical networks using photorefractive crystals *Appl. Opt.* **27** 1752–9

Psaltis D and Farhat N H 1985 Optical information processing based on an assosiative-memory model of neural nets with thresholding and feedback *Opt. Lett.* **10** 98–100

Psaltis D, Gu X-G and Brady D 1988b Fractal sampling grids for holographic interconnections *Proc. ICO Topical Meeting on Optical Computing (Toulon)* (SPIE) pp 963–70

Psaltis D, Park C H and Hong J 1988c Higher order associative memories and their optical implementations *Neural Networks* **1** 149–63

Psaltis D and Qiao Y 1990 Optical neural networks *Opt. and Photonics News* 17–21

Qiao Y and Psaltis D 1992 Local learning algorithm for optical neural networks *Appl. Opt.* **31** 3285–8

Radehaus C V, Pankove J I, Kuijk M, Heremans P and Borghs G 1992 Maximum detection with a two-dimensional optoelectronic winner-take-all network *Appl. Opt.* **31** 6303–6

Ramacher U, Raab W, Anlauf J, Hachmann U and Wesseling M 1993 SYNAPSE-1—a general purpose neurocomputer *Technical Report, Siemens AG, Corporate Research and Development Division (Munich)*

Rietman E A, Frye R C, and Wong C C 1991 Signal prediction by an optically controlled neural network *Appl. Opt.* **30** 950–7

Robinson M G and Johnson K M 1992 Noise analysis of polarization-based optoelectronic connectionist machines *Appl. Opt.* **31** 263–72

Saleh B E A and Teich M C 1991 *Fundamentals of Photonics* 2nd edn (New York: Wiley)

Saxby G 1994 *Practical Holography* 2nd edn (New York: Prentice-Hall)

Saxena I and Fiesler E 1995 Adaptive multilayer optical neural network with optical thresholding *Opt. Eng.* **34** 2435–40

Schmitt-Rink S, Chemla D S and Miller D A B 1989 Linear and nonlinear properties of semiconductor quantum wells *Advances in physics* **38** 89–188

Shariv I and Friesem A A 1989 All-optical neural network with inhibitory neurons *Opt. Lett.* **14** 485–7

Shariv I, Gila O and Friesem A A 1991 All-optical bipolar neural network with polarization-modulating neurons *Opt. Lett.* **16** 1692–4

Slagle T M and Wagner K 1992 Winner-take-all spatial light modulator *Opt. Lett.* **17** 1164–6

Smart 1994 *Summer topical meeting digest on smart pixels* (New York: IEEE) vol 94 TH 0606-4

Soffer B H, Dunning G J, Owechko Y and Marom E 1986 Associative holographic memory with feedback using phase-conjugate mirrors *Opt. Lett.* **11** 118–20

Taghizadeh M R and Turunen J 1992 Synthetic diffractive elements for optical interconnection *Optical Computing and Processing* **2** 221–42

Takei H, Lewis A, Chen Z and Nebenzahl I 1991 Implementing receptive fields with excitatory and inhibitory optoelectrical responses of bacteriorhodopsin films *Appl. Opt.* **30** 500–9

Takimoto A, Akiyama K, Miyauchi M, Kuratomi Y, Asayama J and Ogawa H 1991 A new optical neuron device for all-optical neural networks *Extended Abstracts of the 1991 Int. Conf. on Solid State Devices and Materials* pp 335–7

Van der Lugt A B 1964 Signal detection by complex spatial filtering *IEEE Trans. Information Theory* **10**

Von Lehmen A, Paek E G, Carrion, L C, Patel J S and Marrakchi A 1990 Optoelectronic chip implementation of a quadratic associative memory *Opt. Lett.* **15** 279–81

Wagner K and Psaltis D 1987 Multilayer optical learning networks *Appl. Opt.* **26** 5061–76

Wagner K and Slagle T 1993 Optical competitive learning with VLSI/liquid-crystal winner-take-all modulators *Appl. Opt.* **32** 1408–35

Wang C-H, Jenkins B K and Wang J-M 1993 Visual cortex operations and their implementation using the incoherent optical neuron model *Appl. Opt.* **32** 1876–87

Wang L, Esch V, Feinleib R, Zhang L, Jin R, Chou H M, Sprague R W, Macleod H A, Khitrova G, Gibbs H M, Wagner K and Psaltis D 1988 Interference filters as nonlinear decision making elements for three-spot pattern recognition and associative memories *Appl. Opt.* **27** 1715–20

Warde C, Weiss A M, Fisher A D and Thackara J I 1981 Optical information processing characteristics of the microchannel spatial light modulator *Appl. Opt.* **22** 2066–74

White H J, Aldridge N B and Lindsay I 1988 Digital and analogue holographic associative memories *Opt. Eng.* **27** 30–7

White H J and Wright W A 1988 Holographic implementations of a Hopfield model with discrete weights *Appl. Opt.* **27** 331–8

Yariv A and Kwong S-K 1986 Associative memories based on message-bearing optical modes in phase-conjugate resonators *Opt. Lett.* **11** 186–8

Yu F T S, Lu T, Yang X, and Gregory D A 1990 Optical neural network with pocket-sized liquid-crystal televisions *Opt. Lett.* **15** 863–5

Zhang L, Robinson M G, and Johnson K M 1991 Optical implementation of a second-order neural network *Opt. Lett.* **16** 45–7

Further reading

1. Abu Mostafa Y S and Psaltis D 1987 Optical neural computers *Scientific American* pp 88–95

2. Yu F T S, 1993 Optical neural networks: architecture, design and models *Progress in Optics* ed E Wolf (Amsterdam: North-Holland) vol 32

3. Psaltis D and Qiao Y 1993 Adaptive multilayer optical networks *Progress in Optics* ed E Wolf (Amsterdam: North Holland) vol 31

4. Saleh B E A and Teich M C 1991 *Fundamentals of Photonics* ch 4, 18, 19, 20 and 21 (New York: Wiley)

5. Neff J A, Athale R A and Lee S H 1990 Two-dimensional spatial light modulators: a tutorial *Proc. IEEE* vol 78 pp 826–54

6. Jahns J and Lee S H (eds) 1994 *Optical Computing Hardware* (San Diego: Academic Press)

7. Krishnamoorthy A V, Yayla G, Esener S and Lee S H 1994 Free-space optoelectronic technology for neural networks *Optical Memory and Neural Networks* **3** 261–89

8. Caulfield H J (ed) 1979 *Handbook of Optical Holography* (New York: Academic Press)

9. Saxby G 1994 *Practical Holography* 2nd edn (New York: Prentice Hall)

10. Collings N 1988 *Optical Pattern Recognition Using Holographic Techniques* (Wokingham: Addison-Wesley)

11. Bass M (ed) 1995 *Handbook of Optics* 2nd edn sponsored by the Optical Society of America (New York: McGraw-Hill).

PART F

APPLICATIONS OF NEURAL COMPUTATION

PART F

APPLICATIONS OF NEURAL COMPUTATION

F1

Neural Network Applications

Contents

F1.1 Introduction

Gary Lawrence Murphy

Neural networks have become a serious contender for real-world computing and industrial control. Neural techniques in nonlinear and associative problems have enabled automation of tasks that previously eluded mechanization, and in the next few years, the growing use of *parallel processing* and *VLSI* will further   E1.1 broaden the scope of neural network applications. Neural solutions, from telephone noise filters to process control, are becoming common in industry and commerce, and this expanding popularity is bringing the topic to new fields of application with intractable problems to solve.

Neural technology has also become a complicated topic, and the scientists and engineers exploring these methods have lacked any clear view of the state of the art within their domain. The sections that follow address the needs of these practitioners by presenting detailed discussions of neural network applications in important problem domains. The case studies in Part G further refine this view to individual solutions, following each case from requirements and design through to training and evaluation.

Several sections in this chapter discuss general solutions applicable to many industrial problems. There are, for example, networks used for *stochastic modeling*, *control issues* and *function optimization*.   F1.9, F1.3 The survey of pattern classification may also find a wide audience. Other sections, such as the discussion on *data compression* and the survey on *speech processing*, are more focused on particular industries but   F1.5, F1.7 contain aspects of interest for similar time-series synthesis and recognition problems in other domains.

These studies offer a general guide and also a catalog of ideas. By showing many domains together, pictures emerge of the roles of neural networks and relationships between topology and task. Taken as a whole, the studies map the range from simple textbook networks to highly customized designs, and from generalized problems to highly specific applications. Any such presentation cannot hope to be complete or current, but from this chapter and the case studies which follow in Part G, neural network practitioners can survey their own or a similar domain and also view the design considerations in other worlds. The terms and objectives may be very different, but this multidisciplinary view may provide seeds for surprising cross-pollinations.

Our computing world was changed forever by the first neural solution of the classical XOR (exclusive OR) problem. Modern possibilities for massively parallel processing and integrated circuit networks have brought the connectionist's machine into the industrial workplace, and new doors for automation and control are now open.

F1.2 Pattern classification

Thierry Denœux

Abstract

Pattern classification consists in assigning entities, described by feature vectors, to pre-defined groups of patterns. When the statistical characteristics of the problem under consideration are perfectly known, minimal error probability can be achieved by means of the Bayes decision rule. In practice, however, a suboptimal classifier has to be constructed from training data. Several neural network approaches to this problem have been proposed. *Nearest-neighbor* models are based on assessing the similarity between the input pattern and a set of reference patterns with known classification. The *regression* approach consists in predicting category from pattern by minimizing a certain error criterion. In the finite sample case, the definition of the structural complexity of these models is shown to have considerable influence on classification error. Finally, a taxonomy of the main neural network and alternative techniques of pattern classification are presented.

F1.2.1 Introduction

In many application domains such as *character recognition, speech understanding, medical diagnosis,* *process fault detection* or *financial decision making,* problems arise that consist of classifying entities, represented by feature vectors, into one of several groups of patterns, or *classes.* A classification system is typically composed of two parts (Duda and Hart 1973, Fukunaga 1990). A *preprocessor* transforms raw data produced by sensors or extracted from computer databases into vectors of *observations* or *features.* Features are defined so as to encode in compact form most of the information needed to discriminate between pattern categories. Feature vectors are then passed to a *classifier* that evaluates the evidence presented and makes a decision regarding the class assignment of the entity under consideration.

G1.3, F1.7, G5
G2.8, G6.3

Ever since the pioneering work of Rosenblatt (1958) and Widrow (Widrow and Lehr 1990), a large part of connectionist research has been devoted to the development and theoretical analysis of pattern classifiers having neural-network-like structure and learning capabilities. In recent years, the development of several new models with previously unequaled performance in real-world applications (Rumelhart *et al* 1986, Kohonen 1987) has generated a wave of interest in connectionism and pattern recognition in general. Although this enthusiasm was first considered with some skepticism by researchers in mainstream statistical pattern recognition (Duin 1994), artificial neural networks are now generally seen as particular types of *statistical pattern classifiers* (Schmidt 1993, Werbos 1991).

B6

In the next section, the basic notation and definitions underlying statistical pattern recognition will first be defined. The main neural network approaches to pattern classification will then be described, with an overview of their asymptotic and small-sample properties. In the last section, a taxonomy of statistical and neural network classifiers will be presented.

F1.2.2 Problem description

We consider a finite number M of populations or classes, $\omega_1, \ldots, \omega_M$. An entity of interest is assumed to belong to one and only one of these populations. Each entity is described by a feature vector $x \in \mathbb{R}^d$ which is seen as a realization of a random vector X. The probability density function of X in class ω_i is denoted by $f_X(x|\omega_i)$. Each entity is generally assumed to be drawn from a mixture of the M populations, in proportions $P(\omega_1), \ldots, P(\omega_M)$, respectively, with $\sum_{i=1}^{M} P(\omega_i) = 1$. The mixture density of X is then

$$f_X(x) = \sum_{i=1}^{M} P(\omega_i) f_X(x|\omega_i) . \tag{F1.2.1}$$

$P(\omega_i)$ can be seen as the prior probability that the entity belongs to ω_i. Having observed feature vector x, the posterior probability $P(\omega_i|x)$ can be computed by applying the Bayes theorem:

$$P(\omega_i|x) = \frac{f_X(x|\omega_i) P(\omega_i)}{f_X(x)} . \tag{F1.2.2}$$

If the class-conditional probability distributions and the priors are all known, then an optimal solution to the classification problem is provided by Bayes decision theory. Let us denote by $A = \{\alpha_1, \ldots, \alpha_a\}$ a finite set of actions; α_i is often interpreted as the decision of allocating x to class ω_i. However, other actions such as ambiguity or distance rejection (Chow 1970, Dubuisson and Masson 1993) can also be considered in the analysis.

If, as a result of observing pattern x, we take action α_i while the entity under consideration belongs to class ω_j, we incur a loss $\lambda(\alpha_i|\omega_j)$. The expected loss $R(\alpha_i|x)$ is

$$R(\alpha_i|x) = \sum_{j=1}^{M} \lambda(\alpha_i|\omega_j) P(\omega_j|x) . \tag{F1.2.3}$$

A *decision rule* is a function $\alpha : \mathbb{R}^d \mapsto A$ that prescribes an action $\alpha(x)$ each time an observation vector x is encountered. The overall risk associated to α is

$$R(\alpha) = \int_{\mathbb{R}^d} R(\alpha(x)|x) f_X(x) \, dx . \tag{F1.2.4}$$

The decision rule that minimizes the risk can be shown to be the *Bayes rule*, which selects for each vector x the action α_i for which $R(\alpha_i|x)$ is minimum.

In the particular case of a zero-one loss function $\lambda(\alpha_i|\omega_j) = 1 - \delta_{ij}$, where δ is the Kronecker symbol, we have

$$R(\alpha_i|x) = 1 - P(\omega_i|x) \tag{F1.2.5}$$

and the overall risk is the average probability of misclassification. Consequently, the Bayes rule consists in this case of selecting the class with the highest posterior probability. This rule has optimal classification performance in the sense that it minimizes the average probability of error.

In practice, however, this rule cannot be applied because the exact posterior probabilities are unknown. However, approximations to that rule can be constructed if a training set $\mathcal{T} = \{(x^{(1)}, t^{(1)}), \ldots, (x^{(\ell)}, t^{(\ell)})\}$ of ℓ patterns with known classification is available; $t^{(i)}$ denotes a vector of M zero-one indicator variables defining the known class of pattern $x^{(i)}$:

$$t_k^{(i)} = 1 \qquad x^{(i)} \in \omega_k \tag{F1.2.6}$$
$$t_k^{(i)} = 0 \qquad x^{(i)} \notin \omega_k . \tag{F1.2.7}$$

The construction of allocation rules based on a limited amount of training data is one of the fundamental problems in statistical pattern recognition and connectionism.

F1.2.3 Neural network classifiers

In the past thirty years, a large number of neural network models have been proposed for performing pattern classification tasks. Although these models are characterized by a variety of architectures and learning rules, most of them can be seen as instances of two main paradigms, the *nearest-neighbor* approach and the *regression* approach, which are summarized in the following sections.

F1.2.3.1   The nearest-neighbor approach

In the nearest-neighbor approach, the most probable classification of an unknown pattern is determined by assessing its similarity with a set of reference vectors or *prototypes* of each class. The pattern is assigned to the class of the nearest prototype. As a consequence, the surface separating the different decision regions is piecewise linear. In such models, learning is essentially a process of prototype formation and adaptation. Two important models in this category are the *restricted Coulomb energy* (RCE) network (Reilly *et al*   C1.6.3.1
1982) and the *learning vector quantization* (LVQ) network (Kohonen 1987).   C1.1.5

In the RCE model, each prototype of a given class is characterized by a weight vector and a receptive field size. The learning algorithm combines two mechanisms of prototype formation and receptive field modification. If input x belonging to class ω_j does not fall into the receptive field of any prototype of that class, then a new prototype of class ω_j is created at the location of x. If x falls inside the receptive field of some prototype of class $c \neq \omega_j$, then the receptive field of that prototype is reduced so as to exclude x. This algorithm has been shown experimentally to be able to resolve class boundaries of arbitrary complexity. However, since no adaptation of prototype vectors is performed, the required number of prototypes may grow very large. Also, the learning process usually becomes unstable in regions where there is a strong overlap between classes. Some improvements to this basic model have been proposed (Reilly *et al* 1982).

The LVQ model introduced by Kohonen (1987, 1990) essentially differs from the previous one in that the number of prototypes is fixed, but their weight vectors are continuously updated in the course of the learning process by a *competitive learning* mechanism. Upon presentation of input vector x of class ω_j, the nearest prototype i is selected. If that prototype belongs to class $c^{(i)}$, its weight vector $p^{(i)}$ is updated as

$$p^{(i)} \leftarrow p^{(i)} + \eta(t)(x - p^{(i)}) \qquad \text{if} \quad c^{(i)} = \omega_j \qquad (\text{F1.2.8})$$

$$p^{(i)} \leftarrow p^{(i)} - \eta(t)(x - p^{(i)}) \qquad \text{if} \quad c^{(i)} \neq \omega_j \qquad (\text{F1.2.9})$$

where $\eta(t)$ is a time-decreasing scalar parameter ($0 < \eta(t) < 1$). After training, the prototype vectors acquire values such that classification using the nearest-neighbor principle approximates the Bayes rule with zero-one costs. Variants of this basic scheme have been proposed by Kohonen (1990) and others (e.g. Poirier and Ferrieux 1991).

Simulations performed with both models (RCE and LVQ) on a simple two-class problem are reported in figure F1.2.1. The LVQ algorithm can be seen to yield a smoother decision boundary with a comparatively smaller number of neurons, as a result of prototype adaptation during training.

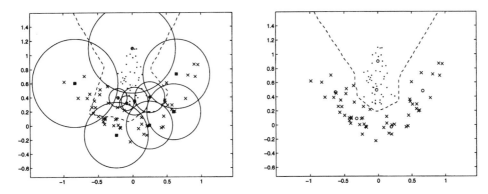

Figure F1.2.1. Prototypes ($\circ$) and decision boundaries ($- - -$) obtained by RCE (left) and LVQ (right) networks in a two-class problem. The receptive fields of RCE prototypes are indicated as circles.

Neural network classifiers based on the nearest-neighbor approach have the advantage of being fast during both training and operation. Experimentally, they are generally found to offer good performance as compared to other, more computationally demanding methods (Kohonen *et al* 1988). Although it is conjectured that the methods relying on competitive learning allow approximation to the Bayes rule for large sample sizes (Kohonen 1990), the determination of the quality of this approximation is a difficult theoretical problem. When classification is performed by considering the nearest neighbor *among samples*, the asymptotic error rate is known to be bounded between the Bayes error and twice the Bayes error (Cover and Hart 1967). This result can be seen as a heuristic justification of the good performance of nearest-neighbor techniques in large-sample problems.

F1.2.3.2 The regression approach

Classification by regression is certainly the most popular approach in the field of artificial neural networks. A regression classifier attempts to predict category from pattern by minimizing a measure of expected error between output and target patterns (Thomas and Mitiche 1994).

More precisely, let us denote the input–output function implemented by a neural network with specified architecture by

$$F : \mathbb{R}^d \times \mathbb{R}^W \mapsto \mathcal{O} \tag{F1.2.10}$$

$$(\boldsymbol{x}, \boldsymbol{w}) \rightarrow F(\boldsymbol{x}, \boldsymbol{w}) \tag{F1.2.11}$$

where $\mathcal{O}$ is the set of possible output values and $\boldsymbol{w}$ is the vector of weights of size W.

C1.2   In the case of *multilayer perceptrons* (MLPs) (Rumelhart *et al* 1986) with one hidden layer and a logistic activation function in the hidden layer, the kth component $F_k(\boldsymbol{x}, \boldsymbol{w})$ of output vector $F(\boldsymbol{x}, \boldsymbol{w})$ is defined as

$$F_k(\boldsymbol{x}, \boldsymbol{w}) = \sum_{j=1}^{N_2} w_{kj}^{(2)} \sigma \left(\sum_{i=1}^{d} w_{ji}^{(1)} x_i + \theta_j^{(1)} \right) + \theta_k^{(2)} \tag{F1.2.12}$$

where $w_{ji}^{(1)}$ is the weight from input unit i to hidden unit j, $\theta_j^{(1)}$ is the bias of hidden unit j, $w_{kj}^{(2)}$ is the weight from hidden unit j to output unit k, $\theta_k^{(2)}$ is the bias of output unit k, N_2 is the size of the hidden layer, and σ is a sigmoid function.

C1.6.2   In the case of *radial basis function* (RBF) networks (Poggio and Girosi 1988, Girosi 1994), the output from hidden unit j is defined as a function of the Euclidean distance between input $\boldsymbol{x}$ and a prototype vector $\boldsymbol{p}^j$. As in the previous model, output units compute a weighted sum of the outputs from the hidden layer. The output $F_k(\boldsymbol{x}, \boldsymbol{w})$ from unit k is given by

$$F_k(\boldsymbol{x}, \boldsymbol{w}) = \sum_{j=1}^{N_2} w_{kj} \exp\left(-\frac{1}{2\sigma_j^2} \|\boldsymbol{x} - \boldsymbol{p}^{(j)}\|^2 \right) \tag{F1.2.13}$$

where w_{kj} is the weight from hidden unit j to output unit k, σ_j is a parameter defining the size of the receptive field of prototype j, and N_2 is defined as above.

An important distinction between MLPs and RBF networks concerns the nature of the internal representation of input patterns. In MLPs, an input signal may activate an arbitrary number of hidden units, resulting in a *distributed* representation. In RBF networks, one input predominantly activates the hidden unit with the closest weight vector, which creates a *local* representation. From this point of view, RBF networks are related to the nearest-neighbor classifiers described in the previous section. A comparison of both models on the same two-class problem as above is shown in figure F1.2.2.

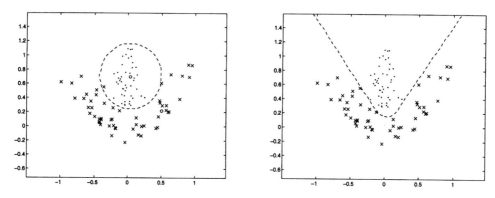

Figure F1.2.2. Decision boundaries (− − −) obtained by an RBF network with three prototype units (left) and by an MLP with two hidden units (right) in a two-class problem.

Both MLPs and RBF networks share the fundamental property of being universal approximators; that is, given enough hidden units, they can approximate any continuous mapping with arbitrary accuracy (Poggio and Girosi 1988, Hornik 1991).

In the MLP and RBF network models, training is performed by optimizing the performance on a training set $\mathcal{T} = \{(\boldsymbol{x}^{(1)}, \boldsymbol{t}^{(1)}), \ldots, (\boldsymbol{x}^{(\ell)}, \boldsymbol{t}^{(\ell)})\}$, using some iterative procedure (Rumelhart *et al* 1986).

Performance is assessed by computing the mean of some error measure between the classifier output and target values. Different output coding schemes and error measures have been proposed. Typically, the desired output for training vector $x^{(i)}$ is taken as $t^{(i)}$, and the error for that pattern is defined as $\|t^{(i)} - F(x^{(i)}, w)\|^2$. The empirical performance on the training set is then

$$J_\ell(w) = \frac{1}{\ell} \sum_{i=1}^{\ell} \|t^{(i)} - F(x^{(i)}, w)\|^2 . \tag{F1.2.14}$$

During training, one seeks a weight vector w_ℓ solution of the problem:

$$\min_w J_\ell(w) . \tag{F1.2.15}$$

However, in most cases, the ultimate goal of learning is in fact to minimize the overall performance for any possible input vector, which can be measured by

$$J(w) = E(\|T - F(X, w)\|^2) \tag{F1.2.16}$$

where X is a random input vector and T is the corresponding random target vector. For large ℓ, $J_\ell(w)$ can be seen as an approximation to $J(w)$, and w_ℓ approximates the solution w^* to

$$\arg \min_w J(w) . \tag{F1.2.17}$$

If the training set is now seen as a realization of a random sample

$$((X^{(1)}, T^{(1)}), \ldots, (X^{(\ell)}, T^{(\ell)})) \tag{F1.2.18}$$

then $J_\ell(w)$ and w_ℓ become realizations of random variables $\hat{J}_\ell(w)$ and $\hat{w}_\ell$, respectively. White (1989) discusses conditions on which the sequence of real-valued random variables $\hat{w}_\ell$ converges, in some strict mathematical sense, to w^*.

So far, we have assumed performance to be assessed by a measure of the distance between desired and obtained output patterns. Intuitively, a classifier whose outputs are close to target values for each x can be expected to have low error probability. As the number ℓ of training vector becomes infinitely large, it is interesting to study the relationships of this approach with the Bayes rule. This has been done by many authors (White 1989, Hampshire and Pearlmutter 1991, Lee *et al* 1991, Thomas and Mitiche 1994). The main result is that w^* minimizes

$$\int_{\mathbb{R}^d} \|E(T|x) - F(x, w)\|^2 \, dx . \tag{F1.2.19}$$

By definition of T, $E(T_j|x) = P(T_j = 1|x) = P(\omega_j|x)$. Consequently, $F_j(w^*, x)$ is a mean-squared approximation to the posterior probability $P(\omega_j|x)$. A classifier trained by minimization of the mean-squared error criterion therefore approximates the Bayes rule asymptotically in ℓ. This result has been extended to other error functions by Hampshire and Pearlmutter (1991) and to more general output coding schemes by Thomas and Mitiche (1994). Note, however, that the quality of this approximation depends on the architecture of the network under consideration, as well as on the training procedure employed, which may not be able to reach a global minimum of the error function.

F1.2.3.3 Small-sample problems

As remarked by Raudys and Jain (1991a), the asymptotic classification error of a regression classifier, assuming a perfect training algorithm, cannot be increased by introducing new hidden units. If the classifier is made more complex, this can only result in a closer approximation to the Bayes classification rule. This remark also applies, to some extent, to nearest-neighbor classifiers, since increasing the number of prototypes results in a closer approximation to the 1-NN classifier, which is known to have near-optimal asymptotic performance.

In practice, however, one is always in a situation where only a finite number of training samples is available. In such a case, numerical simulations reveal the existence of the so-called *peaking phenomenon* (Raudys and Jain 1991a). As the complexity of the classifier increases, classification error initially drops,

then attains a minimum, and then begins to increase. Intuitively, this is due to the fact that inexact estimation of additional parameters increases classification error. At some point, this effect becomes larger than the gain resulting from greater flexibility of the classifier. For that reason, the design of patterns classifiers with optimal complexity is of the utmost importance in practical applications, and the
B2.10   development of *heuristic methods* for automatic determination of a near-optimal number of hidden neurons has been the subject of very intensive research. A variety of techniques have been proposed, which can be categorized as relying on *destructive, constructive* or *direct* strategies. In the *destructive* approach, the complexity of the network is gradually reduced either by penalizing complexity through addition of a bias term to the error function, or by pruning the least relevant units in the course of the training process (Reed 1993). In the *constructive* strategy, a small initial network is gradually expanded until the task is considered to be solved. Examples of such techniques are described in Fahlman and Lebiere (1990, Hirose *et al* (1991), Lengellé and Denœux (1992, 1996), Platt (1991), Lee (1992). The *direct* approach consists in using prior information, acquired through preprocessing or readily available from domain knowledge, to design a neural network that can then be further trained using a standard learning procedure such as
C1.2.3   *backpropagation* (Sethi 1990, Denœux and Lengellé 1993, Karouia *et al* 1995).

In all cases, the classification error of the classifier has to be either estimated, or derived from theoretical considerations. Raudys and Jain (1991b) discuss several methods of error estimation including the resubstitution, hold-out, cross-validation and bootstrap methods. The hold-out method consists in dividing the available data into a training set and a test set used for independent error estimation. This method provides an unbiased error estimate, but it has the disadvantage of preventing the use of all the data for the learning process. The cross-validation and bootstrap methods are more efficient, but also more computationally demanding.

As an alternative, an idea of the generalization performance of a classifier can sometimes be gained
B2.5.2.2   as a result of some kind of theoretical analysis. Recent investigations based on the *Vapnik–Chervonenkis theory* (Vapnik 1982) and the PAC learning model (Valiant 1984) have led to the derivation of bounds for the true and estimated classification errors of minimum empirical error classifiers (Baum and Haussler 1989, Anthony 1994, Kraaijveld 1993). However, these results are based on a worst-case analysis and lead to very pessimistic estimates of the number of samples needed to train a classifier (Raudys 1994). Nevertheless, some of the most recent results are already applicable with approximations as an initial aid to neural network design (Holden and Niranjan 1994). Further improvements are expected from the consideration of specific input distributions and training algorithms in this analysis (Kraaijveld 1993).

F1.2.4 Alternative approaches

Since the 1950s, substantial progress has been achieved in the design of statistical classifiers from empirical data. According to Raudys and Jain (1991b), the number of classification methods already published exceeds two hundred. These methods are described in a number of standard textbooks such as Duda and Hart (1973), Fukunaga (1990) and McLachlan (1992).

A useful taxonomy of classification techniques, including statistical and neural network approaches, has been proposed by Lippmann (1994). Pattern classifiers can be seen as belonging to three main categories. *Probability density function* classifiers estimate class-conditional probability densities separately for each class. They include parametric normal classifiers with different forms of covariance matrices, and nonparametric methods of density estimation such as the Parzen-window approach. *Posterior probability classifiers* estimate the posterior probability of each class, using all the available data simultaneously. Examples of such methods are MLPs and RBF networks, and the voting *k*-NN rule. The third category of classification method includes techniques for directly partitioning the feature space into decision regions, using binary indicator outputs. Examples of such *boundary forming* methods are the nearest-neighbor methods, such as RCE or LVQ networks, and tree-structured classifiers (Breiman *et al* 1984).

A further distinction can be drawn between *model-based* and *data-driven* approaches. In the model-based approach, a particular classifier is chosen among a predefined family of functions, or *model*. Parametric classifiers, MLPs and LVQ classifiers fall in this category. In contrast, the form of data-driven classifiers is not fixed in advance, but determined by the data. This is the case for Parzen-window,
C1.7, C2.4   *k*-NN and tree-structure classifiers, as well as for *ontogenic neural networks* that adapt their structure during the learning process.

This multiplicity of classification techniques obviously poses a serious problem to the practitioner. Many comparative studies have been made to assess the strengths and weaknesses of various methods (e.g. Tsoi and Pearson 1991, Ng and Lippmann 1991, Brown *et al* 1993, Blue *et al* 1994). In general, comparable error rates are achieved by several techniques, provided they are properly tuned. As noted by Ng and Lippmann (1991), the selection of a classifier for a particular task should primarily be guided by practical considerations such as training and classification time, and memory storage requirements. Neural network classifiers usually offer a good compromise between performance and practical applicability.

References

Anthony M 1994 Probabilistic analysis of learning in artificial neural networks: the PAC model and its variants *Technical Report* NC-TR-94-3 Royal Holloway, University of London, Egham, Surrey TW20 0EX, UK

Baum E B and Haussler D 1989 What size net gives valid generalization *Neural Comput.* **1** 151–60

Blue J L, Candela G T, Grother P J, Chellappa R and Wilson C L 1994 Evaluation of pattern classifiers for fingerprint and OCR applications *Patt. Recog.* **27** 485–501

Breiman L, Friedman J H, Olshen R A and Stone C J 1984 *Classification and Regression Trees* (Belmont, CA: Wadsworth)

Brown D E, Corruble V and Pittard C L 1993 A comparison of decision tree classifiers with backpropagation neural networks for multimodal classification problems *Patt. Recog.* **26** 953–61

Chow C K 1970 On optimum recognition error and reject tradeoff *IEEE Trans. Inform. Theory* **16** 41–6

Cover T M and Hart P E 1967 Nearest neighbor pattern classification *IEEE Trans. Inform. Theory* **13** 21–7

Denœux T and Lengellé R 1993 Initializing back-propagation networks with prototypes *Neural Networks* **6** 351–63

Dubuisson B and Masson M 1993 A statistical decision rule with incomplete knowledge about classes *Patt. Recog.* **26** 155–65

Duda R O and Hart P E 1973 *Pattern Classification and Scene Analysis* (New York: Wiley)

Duin R P W 1994 Superlearning and neural network magic *Patt. Recog. Lett.* **15** 215–7

Fahlman S E and Lebiere C 1990 The cascade-correlation learning architecture *Advances in Neural Information Processing Systems 2* ed D S Touretzky (San Mateo, CA: Morgan Kaufmann) pp 524–32

Fukunaga K 1990 *Introduction to Statistical Pattern Recognition* 2nd edn (Berlin: Academic)

Girosi F 1994 Regularization theory, radial basis functions and networks *From Statistics to Neural Networks* ed V Cherkassky, J H Friedman and H Wechsler (Berlin: Springer) pp 166–87

Hampshire J B and Pearlmutter B 1991 Equivalence proof for multilayer perceptron networks and the Bayesian discriminant function *Connectionist Models, Proc. 1990 Summer School* ed D S Touretzky, J L Elman, T J Sejnowski and G E Hinton (San Mateo, CA: Morgan Kaufmann) pp 159–72

Hirose Y, Yamashita K and Hijiya S 1991 Back-propagation algorithm which varies the number of hidden units *Neural Networks* **4** 61–6

Holden S B and Niranjan M 1994 On the practical applicability of VC dimension bounds *Technical Report* CUED/F-INFENG/TR155 Cambridge University Engineering Department, Cambridge CB2 1PZ, UK

Hornik K 1991 Approximation capabilities of multilayer feedforward networks *Neural Networks* **4** 251–7

Karouia M, Lengellé R and Denœux T 1995 Performance analysis of a MLP weight initialization algorithm *Proc. ESANN'95 European Symp. on Artificial Neural Networks* (Brussels: De facto) pp 347–52

Kohonen T 1987 *Self Organisation and Associative Memory* 2nd edn (Berlin: Springer)

——1990 The self-organizing map *Proc. IEEE* **78** 1464–80

Kohonen T, Barna G and Chrisley R 1988 Statistical pattern recognition with neural networks: benchmarking studies *Proc. ICNN'88 Int. Conf. on Neural Networks* vol I (IEEE Computer Society Press) pp 61–8

Kraaijveld M A 1993 Small sample behavior of multi-layer feedforward network classifiers: theoretical and practical aspects *PhD Thesis* Delft University, Delft, The Netherlands

Lee D-S, Srihari S N and Gaborski R 1991 Bayesian and neural network pattern recognition: a theoretical connection and empirical results with handwritten characters *Artificial Neural networks and Statistical Pattern Recognition* ed I K Sethi and A K Jain (Amsterdam: Elsevier) pp 89–108

Lee S 1992 Supervised learning with Gaussian potentials *Neural Networks for Signal Processing* ed B Kosko (Englewood Cliffs, NJ: Prentice-Hall) pp 189–227

Lengellé R and Denœux T 1992 Optimizing multilayer networks layer per layer without back-propagation *Artificial Neural Networks II* ed I Aleksander and J Taylor (Amsterdam: North-Holland) pp 995–8

——1996 Training multilayer perceptrons layer by layer using an objective function for internal representations *Neural Networks* **9** 83–97

Lippmann R P 1994 Neural networks, Bayesian *a posteriori* probabilities, and pattern classification *From Statistics to Neural Networks* ed V Cherkassky, J H Friedman and H Wechsler (Berlin: Springer) pp 83–104

McLachlan G J 1992 *Discriminant Analysis and Statistical Pattern Recognition* (New York: Wiley)

Ng K and Lippmann R P 1991 A comparative study of the practical characteristics of neural networks and conventional pattern classifiers *Advances in Neural Information Processing Systems* vol 3 ed R L Lippman, J E Moody and D S Touretzky (San Mateo, CA: Morgan Kaufmann) pp 970–6

Platt J C 1991 Learning by combining memorization and gradient descent *Neural Information Processing 3* ed R P Lippman, J E Moody and D S Touretzky (San Mateo, CA: Morgan Kaufmann) pp 714–20

Poggio T and Girosi F 1988 A theory of networks for approximation and learning *Technical Report* AI Memo No 1140 MIT

Poirier F and Ferrieux A 1991 DVQ: dynamic vector quantization—an incremental LVQ *Artificial Neural Networks* vol 2 ed T Kohonen, M Mäkisara, O Simula and J Kangas (Amsterdam: Elsevier) pp II-1333–1336

Raudys S J 1994 Why do multilayer perceptrons have favorable small sample properties *Pattern Recognition in Practice IV* ed E S Gelsema and L N Kanal (Amsterdam: Elsevier) pp 287–98

Raudys S J and Jain A K 1991a Small sample problems in designing artificial neural networks *Artificial Neural networks and Statistical Pattern Recognition* ed I K Sethi and A K Jain (Amsterdam: Elsevier) pp 33–50

——1991b Small sample size effects in statistical pattern recognition: recommendations for practitioners *IEEE Trans. Patt. Anal. Machine Int.* **13** 252–64

Reed R 1993 Pruning algorithms: a survey *IEEE Trans. Neural Networks* **4** 740–7

Reilly D L, Cooper L N and Elbaum C 1982 A neural model of category learning *Biol. Cybern.* **45** 35–41

Rosenblatt F 1958 The perceptron: a probabilistic model for information storage and organization in the brain *Psychol. Rev.* **65** 386–408

Rumelhart D E, Hinton G E and Williams R J 1986 Learning internal representations by error propagation *Parallel Distributed Processing* ed D E Rumelhart and J McClelland (Cambridge, MA: MIT Press)

Schmidt W F 1993 Neural pattern classifying systems *PhD Thesis* Delft University, Delft, The Netherlands

Sethi I K 1990 Entropy nets: from decision trees to neural networks *Proc. IEEE* **78** 1605–13

Thomas D S and Mitiche A 1994 Asymptotic optimality of pattern recognition by regression analysis *Neural Networks* **7** 313–20

Tsoi A C and Pearson R A 1991 Comparison of three classification techniques CART C4.5 and multi-layer perceptrons *Advances in Neural Information Processing Systems* vol 3 ed R L Lippman, J E Moody and D S Touretzky (San Mateo, CA: Morgan Kaufmann) pp 963–9

Valiant L G 1984 A theory of the learnable *Commun. ACM* **27** 1134–42

Vapnik V N 1982 *Estimation of Dependences Based on Empirical Data (Springer Series in Statistics)* (Berlin: Springer)

Werbos P J 1974 Beyond regression: new tools for prediction and analysis in the behavioral sciences (Cambridge, MA: Harvard University)

——1991 Links between artificial neural networks and statistical pattern recognition *Artificial Neural networks and Statistical Pattern Recognition* ed I K Sethi and A K Jain (Amsterdam: Elsevier Science) pp 11–31

White H 1989 Learning in artificial neural networks: a statistical perspective *Neural Comput.* **1** 425–64

Widrow B and Lehr M A 1990 30 years of adaptive neural networks: perceptrons, madaline and backpropagation *Proc. IEEE* **78** 1415–42

F1.3 Combinatorial optimization

Soheil Shams

Abstract

Combinatorial optimization problems are frequently encountered in a broad range of disciplines from economics to engineering. These problems are notoriously difficult to solve for many applications, especially when the number of free parameters is large. In many situations, high-performance computers are required to solve optimization problems in practical times. Many of the problems solved by neural networks (e.g. pattern recognition and associative memory) can be viewed as special forms of combinatorial optimization. Due to their inherent parallel computation and simple computational requirements, neural networks have the potential to solve difficult optimization problems while taking full advantage of parallel and simple processing hardware. The methodology for applying neural networks to combinatorial optimization problems will be discussed in this section. A brief summary of alternative optimization techniques is also presented.

F1.3.1 Project overview

Combinatorial optimization problems span a wide spectrum of application areas in such diverse fields as *economics, biology, physics* and *engineering*. Combinatorial optimization involves searching a large G6, G4, G3, G2 space of possible solutions to a problem for a specific solution which is in some sense an 'optimum' solution. Optimality of a specific solution is generally measured by an analytical function of the modifiable parameters of the problem.

One of the most widely studied problems in combinatorial optimization is the traveling salesman problem (TSP) (Lawler *et al* 1985). In this problem, a salesman is to visit M cities while passing through each city exactly once. In addition, the salesman has to end his tour in the city where he started, completing a full circuit. The inter-city distances for all pairs of cities are known *a priori* (usually specified in the form of an $M \times M$ matrix). The objective of the optimization process is to find the shortest tour among all valid tours (those which pass through each city exactly once). The solution space of this problem (number of possible valid tours) can be determined to be $M!/2M$. The rapid growth of the solution space can be explicitly demonstrated through an example. For a seven-city TSP ($M = 7$), the solution space consists of 360 valid paths. Therefore, it is reasonable to calculate the tour length associated with each of the 360 paths and select the shortest one. This solution is called the global optimum solution since we have explicitly determined that among all possible valid tours, no other tour has a shorter traversal length. However, with a slightly larger problem consisting of 20 cities ($M = 20$), the solution space explodes into greater than 6×10^{16} possible valid tours. Even if one could calculate the length of each tour in 100 ns (using a very fast processor) it would require more than 190 years to measure all the possible tour lengths. Obviously such an 'exhaustive search' of the solution space is impractical, especially since in most real-world applications, the number of cities is of the order of several thousand. The TSP belongs to the NP-complete class of problems, meaning that no polynomial time algorithm has yet been found that can find the global optimum solution. Therefore, most approaches to solving the TSP are heuristic techniques which tend to only sample the solution space in search of a 'good' solution. Since the entire solution space is not searched, there is no guarantee that the chosen 'good' solution is the global optimum. However, in many real-world applications, it is common to prefer solutions that are sufficiently optimal (as determined by the system designer) and can be found rapidly over more exhaustive search techniques which can find the global optimum solution but consume valuable computational resources.

Many efficient heuristics have been developed to find good solutions to the TSP (Lawler *et al* 1985, Lin 1965). In this section the TSP will be used to demonstrate how neural networks can be utilized to solve optimization problems. The TSP is a good exemplar of a large class of NP-complete combinatorial optimization problems (e.g. resource allocation and scheduling) since solutions for solving the TSP can also be applied to many of these other problems by noting that an NP-complete problem can be reduced into another NP-complete problem in polynomial time (Cormen *et al* 1990). In addition, the wealth of information accumulated on the analysis of the TSP algorithms makes this problem a good baseline for comparing different optimization techniques. Numerous articles have been published on the application of neural networks to the TSP (e.g. Burke 1994, El Ghaziri 1991, Fritzke and Wilke 1991, Gelenbe *et al* 1994, Hopfield and Tank 1985, Mehta and Fulop 1993, Peterson 1990, Xu and Tsai 1991). In addition, many techniques have been proposed for a systematic application of neural networks to optimization problems (Fang and Li 1990, Gee *et al* 1993, Looi 1992, Peterson and Soderberg 1989, Sun and Fu 1993, Tagliarini *et al* 1991, Zhang and Constantinides 1992). In this section, the motivation as well as the benefits and shortcomings of the fundamental neural network approaches for combinatorial optimization will be described. Also included is a brief overview of alternative stochastic optimization methods, namely D2.1, C1.4.2 *genetic algorithms* (GA) and *simulated annealing* (SA). The reader is cautioned that a general comparison of multiple optimization techniques cannot be performed on a single example problem (such as the TSP). The best method for solving an optimization problem is strongly dependent on the problem itself.

F1.3.2 Neural network approaches

The resurgence of interest in neural networks during the 1980s has been attributed in large part to the work of John Hopfield (Hopfield 1982, Hopfield and Tank 1985). Hopfield popularized the concept that a network of simple bistable (on/off) processing units can collectively perform complex computational C1.3, F1.4 tasks, such as *associative memory* (Hopfield 1982). To see how an ensemble of simple processing units can implement complex computation we can examine many natural phenomena. Optimization processes occur naturally in a variety of systems, from natural selection to particle physics. For example, a particle traveling from point x to point y, in the absence of any external force field, will follow a trajectory of minimum distance. Similarly, a spring pulled or pushed away from its 'optimum' resting configuration will tend to bounce back into its optimum shape after the external force is removed. The optimum configuration of the spring is a global state which is realized by the 'natural' dynamics of the molecules within the spring. These molecules are implementing simple local computation, but the result (returning to the optimum resting configuration) is a global feature of the spring.

The goal of neural network approaches to combinatorial optimization is to formulate the desired objective function being optimized, such that it can be viewed as a 'natural' energy minimization problem. This concept can be interpreted visually by imagining an energy landscape having a corresponding energy level for each possible solution state (see figure F1.3.1), with low energy levels indicating more optimal solutions. The global optimum solution will thus correspond to the lowest point in this energy space. The objective of the optimization algorithm will be to search the energy landscape for the lowest possible minima. This task can be performed using a local search technique by starting at some random state (a random point in the energy landscape) and making small local changes to the solution which tend to lower the energy. In other words, moving the solution in the direction opposite to the energy gradient. B5.2.2 This *gradient descent* approach can find the global optimum solution if the energy landscape is convex (see, for example, figure F1.3.1(*a*)). However, as shown graphically in figure F1.3.1(*b*), in most practical applications, the energy surface is cluttered with many locally optimum solutions that impede attempts to perform a direct gradient descent in energy space. I will describe a number of methods proposed to address the local minima issue later in this section.

F1.3.2.1 Hopfield network

C1.3.4 The general approach to solving constrained optimization problems using the *Hopfield network* involves equating the problem objective function with the neural network energy function which has the form

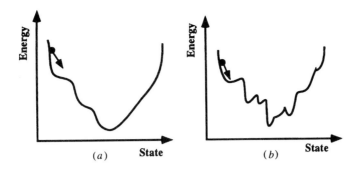

Figure F1.3.1. One-dimensional energy landscape. (*a*) Smooth surface with no local minima. (*b*) Many local minima states are present.

$$E = \frac{1}{2} \sum_{i=1}^{N} \sum_{j=1}^{N} w_{ij} n_i n_j + \sum_{i=1}^{N} \theta_i n_i \tag{F1.3.1}$$

where n_i is one of the components of the state vector, and w_{ij} and θ_i are constants determined based on the objective function. In the neural network representation, a state variable n_i corresponds to the output value of one of the N distinct neurons, w_{ij} corresponds to the synaptic weight value of the connection between neurons i and j, and θ_i corresponds to the bias value of neuron i. The neuron output values can be either discrete binary-valued (Hopfield 1982), or a bounded continuous monotonically increasing function (e.g. sigmoid) (Hopfield 1984). In this section only the more general continuous output-valued neurons will be discussed. The output value of a neuron is determined by the sigmoid function, defined as

$$n_i = g(u_i) = \frac{1}{1 + e^{-\beta u_i}} \tag{F1.3.2}$$

where β controls the gain (or steepness) of the sigmoid and u_i is the total input to neuron i. The dynamics of u_i are designed such that the system converges to the closest local-minima state in energy space, following a direct gradient descent trajectory. In other words, $\mathrm{d}E/\mathrm{d}t \leq 0$. These dynamics are given as

$$\tau \frac{\mathrm{d}u_i}{\mathrm{d}t} = -u_i + \sum_{j=1}^{N} w_{ij} n_j + \theta_i . \tag{F1.3.3}$$

The gain parameter β in (F1.3.2) is analogous to the inverse temperature $(1/T)$ of a thermodynamic system. A method similar to simulated annealing (Kirkpatrick *et al* 1983) (described in section F1.3.3.2 below) can be used to skip over small local minima states and generate a close-to-optimum solution. In this process, the gain parameter starts at a small value (high temperature) and is gradually increased (annealed). In the low-gain configuration, the energy surface is smooth with only a single minimum state, such as the one shown in figure F1.3.1(*a*). As the gain value is increased more of the details of the energy surface are revealed, as shown in figure F1.3.1(*b*).

In constrained optimization problems, such as the TSP, the energy function is generally constructed as the sum of two separate terms,

$$E = E^o + \lambda E^c \tag{F1.3.4}$$

where E^o is the objective term having a global minimum at the optimum point of the objective function, and E^c is the constraint term having a global minimum when all the problem constraints are satisfied. To illustrate this point, let us look at the TSP. The first issue is how to represent a tour using binary state variables. We can use an $N \times N$ array of neurons n_{ia} to indicate a path traversal from city i to city a when $n_{ia} = 1$ (Xu and Tsai 1991). Another more commonly used approach is to define $n_{ia} = 1$ when the salesman visits city i at the ath stop of the tour (Hertz *et al* 1991, Hopfield and Tank 1985, Peterson and Soderberg 1989). The double indices (i and a) on each neuron are used to simplify the formulation of the cost function, and in the implementation of the algorithm the n_{ia} matrix is unfolded into a long vector. In this formulation, the N-city TSP requires N^2 neurons, since both $1 \leq i \leq N$ and $1 \leq a \leq N$.

There are two constraints associated with the TSP. First, the salesman cannot be in two places at the same time. Second, each city is visited only once. These constraints can be stated mathematically as

$$\sum_{a=1}^{N} n_{ia} = 1 \qquad \text{(for each city } i \text{ the tour passes through it only once at some stop } a)$$

and

$$\sum_{i=1}^{N} n_{ia} = 1 \qquad \text{(for each stop } a \text{ the tour passes through only one city } i).$$

Both of these constraints can be satisfied by minimizing the constraint energy term

$$E^c = \left[\sum_{i=1}^{N} \left(1 - \sum_{a=1}^{N} n_{ia} \right)^2 + \sum_{a=1}^{N} \left(1 - \sum_{i=1}^{N} n_{ia} \right)^2 \right]. \tag{F1.3.5}$$

If the cost of travel between all pairs of cities i and j is given *a priori* as d_{ij}, the total tour cost can be calculated as

$$E^o = \frac{1}{2} \sum_{i=1}^{N} \sum_{j=1}^{N} \sum_{a=1}^{N} d_{ij} n_{ia} (n_{j,a-1} + n_{j,a-1}). \tag{F1.3.6}$$

As stated previously, $n_{ia} = 1$ indicates that the salesman is in city i at the ath stop of the tour. At the previous stop $(a - 1)$, the salesman was in city j, as indicated by $n_{j,a-1} = 1$, so the cost d_{ij} is added to the total cost. Similarly, at the next stop, $(a + 1)$, the salesman visits a different city j, as indicated by $n_{j,a+1} = 1$, so the cost d_{ij} is also added to the total. Since the cost of travel between two cities is added twice (once as the previous stop and once as the next stop), the true total cost is determined by dividing the accumulated sum by two, as shown in equation (F1.3.6). It should be noted here that this cost function represents the actual tour cost only when the neuron output values have been saturated to their zero/one limit and the constraint energy term $E^c = 0$.

The total energy function for the TSP can be arrived at by summing (F1.3.5) and (F1.3.6) according to (F1.3.4). The update equation (F1.3.3) can then be used to minimize the energy function iteratively, starting from a random (not necessarily valid) state. The synaptic weight matrix elements $w_{ia\,jb}$ are determined by factoring the quadratic terms $n_{ia}n_{jb}$ of the energy function. Similarly, the bias values θ_{ia} are determined by factoring the linear terms. As noted in Hertz *et al* (1991) the formulation of the TSP presented here requires only N^3 (as opposed to N^4) connections since the problem formulation does not require full interconnectivity of the neurons. The parameter λ in (F1.3.4) is used to weight the cost of satisfying all the problem constraints against producing optimum tours. If λ is very small (close to zero), the problem constraints are ignored in favor of optimizing the objective function. In such a case, a possible outcome is having all neurons converge to zero, thus yielding an invalid solution with a very good objective energy ($E^o = 0$). On the other hand, if the λ parameter is set very large, every valid tour (those satisfying the problem constraints) will represent deep local minimas in the energy landscape. This will cause the network to converge to the closest valid solution in solution space from the starting random state with little or no regard to the cost of the solution (as given by E^o). A method for addressing this issue has been proposed in Simic (1991), where λ is set proportional to the annealing term β. Therefore, the network starts with a small λ value allowing it to move to an area with low-cost solutions, followed by a gradual increase in the λ values. The high λ values at the end of processing will push the network into a valid solution state. A different approach is proposed in Gee *et al* (1993) where all the problem constraints are lumped into a single constraint term. This greatly simplifies the analysis and processing of the network dynamics.

The Hopfield network has demonstrated that a distributed system of simple processing elements can collectively solve optimization problems (Hopfield 1984, Hopfield and Tank 1985). However, the original Hopfield network has very poor scaling characteristics, in that systems using a moderately large number of neurons generate poor quality and/or invalid solutions (Wilson and Pawley 1988, Kunz 1991, Gee 1993). There have been a number of approaches proposed for improving the performance of the Hopfield network (Cuykendall and Reese 1989, Peterson and Soderberg 1989, Simic 1991, Van den Bout and Miller 1989, Yuille 1990). The fundamental problem with the Hopfield network is in its treatment of the problem constraints. As described in the previous section, the Hopfield network attempts to satisfy all the problem constraints by penalizing constraint violations via increasing the system energy ('soft' enforcement). What is truly needed is a method of restricting the search space that the Hopfield network operates within ('hard' enforcement) (Simic 1990). There have been a number of methods proposed that incorporate hard constraints on the energy function (Peterson and Soderberg 1989, Simic 1991, Yuille 1990). In most cases, however, not all of the problem constraints can be implemented as hard constraints, so a mixture of hard and soft constraint enforcement strategies is used (Simic 1990, 1991). Nevertheless, such approaches can lead to a significant improvement in system performance. In addition, the approach proposed by Gee *et al* (1993) forces the network state to fall onto a 'valid subspace' in the solution space during each step of the network evolution. This technique has also demonstrated significant improvement in performance over the generic Hopfield network.

F1.3.2.2 Other neural network approaches

In addition to the Hopfield-style dynamics, neural network models with self-organizing dynamics have also been used to solve combinatorial optimization problems. The *elastic network algorithm* (Durbin $\quad$ C2.1.1 and Willshaw 1987), the *Kohonen self-organizing maps* (SOM) (Kohonen 1987) and the *multiple elastic* $\quad$ C2.1.1 *modules* (MEM) (Shams 1995a), are some examples of such models. It is interesting to note that most of these neural network approaches utilize a common energy minimization framework. For example, the elastic network algorithm can be viewed as a special case of the Hopfield network with certain constraints being strongly enforced (Simic 1990). The strong relations between Kohonen's SOM and the elastic network have also been discussed in Wong (1994). In spite of these fundamental similarities, the self-organizing models exhibit a much superior scaling property than the Hopfield network, allowing these methods to solve 1000-city TSP problems (Angeniol *et al* 1988). The main problem with using self-organizing dynamics for solving combinatorial optimization problems is their explicit use of geometric topology-preserving constraints. These models have been effectively applied to many-dimensional reduction problems (Goodhill and Willshaw 1994, Kohonen 1987, Sirosh and Miikkulainen 1994, von der Malsburg and Willshaw 1977), but their application to general optimization requires a nontrivial transformation of the problem to a geometric representation. Nevertheless, these models are making progress in a wider realm of applications, such as object recognition (Lades *et al* 1993, Sabourin and Mitiche 1993, Shams 1995a, 1995c), target tracking (Shams 1995b) and VLSI optimum cell placement (Hemani and Postula 1990).

F1.3.3 Alternative approaches

The field of combinatorial optimization has a deep and extensive history in applied mathematics. There are a number of well established techniques, such as integer, linear, quadratic and dynamic programming, which have been developed to solve many difficult optimization problems. Algorithms such as the simplex method and the Karmarkar method are widely used in many commercial applications, ranging from airline flight scheduling to aircraft design. Since there are many excellent sources for these methods, I will not discuss them in this section. In this section, I will give a brief discussion and comparison of the somewhat related stochastic search techniques of genetic algorithms and simulated annealing.

F1.3.3.1 Genetic algorithms

As discussed in the introductory section, neural networks attempt to express a given combinatorial optimization problem as a 'natural' energy minimization process. Similarly, *genetic algorithms* (GAs) $\quad$ D2.1 (Holland 1975) are used to express a given optimization problem as an objective of a 'natural' survival-of-the-fittest evolutionary process. A brief overview of GA processing will be presented here.

Genetic algorithms utilize a 'population' of candidate solutions, referred to as chromosomes. This population is typically initialized with randomly generated solutions, which can be viewed as random points on the energy landscape. Three basic operations are performed iteratively on this population, namely reproduction, crossover and mutation. The reproduction operator selects members of the population which are best-fits, as determined by the objective function, to be mated in order to produce the next generation. The crossover operator interchanges a randomly selected portion of a pair of parent chromosomes to generate new offspring. The mutation operator is used to introduce a level of noise into the system by changing a randomly selected point on the chromosome. Function optimization is accomplished through repeated application of these operators to the population.

The main advantage of genetic algorithms over neural networks is that they place no explicit restriction on the form of the objective function being optimized. As discussed earlier, when using neural networks of the Hopfield type, the objective function must be transformed into a quadratic energy function of the form specified by (F1.3.1). Therefore, an objective function such as $f(x) = \sum_{i,j} e^{-x_i} \tan(x_j)$ is difficult to implement using a neural network, whereas it would not cause any difficulties for the GA approach. On the other hand, direct implementation of neural networks, using analog hardware, can offer a fast, low-power, small-size solution which is not possible with the GA technique. In general, the choice of an approach is primarily dependent on the specific problem being solved (Owechko and Shams 1994).

F1.3.3.2 Simulated annealing

C1.4.2   *Simulated annealing* (SA) is yet another method, inspired by 'natural' processes, which is used for combinatorial optimization (Aarts and Korst 1989, Kirkpatrick *et al* 1983). In physics, 'annealing' refers to a thermal process in which a solid material is heated to a molten state and then gradually cooled. If the cooling rate is sufficiently slow the molecules in the material will be aligned in a low-energy-state configuration (for example, in a crystal lattice) at the end of the cooling process. The use of simulated annealing in combinatorial optimization takes advantage of this fact and requires the interpretation of the problem objective function as an energy function. However, as opposed to neural networks where the objective function must be 'transformed' into an energy function of a specific form (as given by equation (F1.3.1)), the SA approach can accommodate a wide range of objective functions. Simulated annealing is a local search strategy, similar to gradient descent with the exception that state transitions causing an increase in system energy are allowed based on a specific 'temperature'-dependent probability. This temperature parameter can be interpreted as a degree of noise injected into the system. The basic SA algorithm can be outlined as follows:

(i) Start from a random initial state X at a high temperature T.
(ii) Create a new state X' by making a small local change to the current state (only change a few of the state parameters).
(iii) Calculate the change in energy $\Delta E = E(X') - E(X)$.
(iv) Accept the state X' as the new state if $\Delta E < 0$, otherwise accept it with a probability $e^{-\Delta E / T}$.
(v) Reduce the temperature T and go back to step (ii).

If T is set constant to zero, the SA algorithm becomes equivalent to the simple gradient-descent method, since the probability $e^{-\infty} = 0$ means no uphill jumps. However, if the initial system temperature is above a critical value, and the cooling schedule is asymptotically slow, the final resting state will be the global optimum solution. Since this approach is too slow for practical applications, a faster cooling schedule is generally employed to find close-to-optimum solutions.

C1.4   The basic stochastic annealing idea of the SA algorithm has been incorporated in many neural network models (e.g. the *Boltzmann machine* (Hinton *et al* 1984, Aarts and Korst 1989)). The main advantage of the SA approach over neural networks is the greater flexibility in the form of the objective function. On the other hand, SA is inherently a serial algorithm which cannot effectively take advantage of parallel processing hardware to improve its throughput.

F1.3.4 Conclusion

As mentioned previously in this section, a general statement on the superiority of one optimization method over another is not possible since each technique offers a unique set of advantages as well as shortcomings. In general, however, neural network models applied to combinatorial optimization problems can offer a unique potential for a direct route to parallel implementations. Due to their inherently parallel structure,
F1.3   neural network techniques are especially suitable for direct hardware implementation, using *analog VLSI* techniques (Mead 1989), or parallel simulations (Shams and Gaudiot 1993, 1995). On the other hand, great care must be taken in the formulation of the problem representation so that the number of possible network states closely corresponds to that of valid solutions. In addition, it is crucial to incorporate as many problem-specific constraints as possible into the structure of the neural network to improve scalability by limiting the search space. Furthermore, network parameters, such as the temperature cooling rate, which are usually set in a rather *ad hoc* manner, can significantly affect the network performance and must therefore be carefully determined. Nevertheless, with continued research in the area of neural computation, many

of these problems can be addressed in the future leading to a general method for efficient processing of difficult optimization problems via distributed computation.

Acknowledgements

I wish to thank Dr Yuri Owechko for helpful discussions and comments on the original draft of this manuscript.

References

Aarts E and J Korst 1989 *Simulated Annealing and Boltzmann Machines: A Stochastic Approach to Combinatorial Optimization and Neural Computing* (New York: Wiley)

Angeniol B, Vaubois G D L C and Texier J-Y L 1988 Self-organizing feature maps and the traveling salesman problem *Neural Networks* **1** 289–93

Burke L I 1994 Neural methods for the traveling salesman problem: insights from operations research *Neural Networks* **7** 681–90

Cormen T H, Leiserson C E and Rivest R L 1990 *Introduction to Algorithms* (Cambridge, MA: MIT)

Cuykendall R and Reese R 1989 Scaling the neural network TSP algorithm *Biol. Cybern.* **60** 365–71

Durbin R and Willshaw D 1987 An analogue approach to the traveling salesman problem using an elastic net method *Nature* **326** 689–91

El Ghaziri H 1991 Solving routing problems by a self-organizing map *Artificial Neural Networks* vol 1, ed T Kohonen, K Mükisara, O Simula and J Kangas (New York: Elsevier)

Fang L and Li T 1990 Design of competition-based neural networks for combinatorial optimization *Int. J. Neural Syst.* **1** 221–35

Fritzke B and Wilke P 1991 FLEXMAP—a neural network for the traveling salesman problem with linear time and space complexity *Proc. Int. Joint Conf. on Neural Networks (Singapore)* vol 2 pp 929–34

Gee A H 1993 Problem solving with optimization networks *Dissertation* Cambridge University

Gee A H, Aiyer S V B and Prager R W 1993 An analytical framework for optimizing neural networks *Neural Networks* **6** 79–97

Gelenbe E, Vassiluda K and Pekergin F 1994 *Elektrik* **2** 1–9

Goodhill G J and Willshaw D J 1994 Elastic net model of ocular dominance: overall stripe pattern and monocular deprivation *Neural Comput.* **6** 615–21

Hemani A and Postula A 1990 Cell placement by self-organization *Neural Networks* **3** 377–83

Hertz J, A Krogh and R G Palmer 1991 *Introduction to the Theory of Neural Computation* (Reading, MA: Addison-Wesley)

Hinton G, Ackley D and Sejnowski T 1984 Boltzmann machines: constraint satisfaction networks that learn *Technical Report CMU-CS-84-119* Carnegie-Mellon University, Department of Computer science

Holland J H 1975 *Adaptation in Natural and Artificial Systems* (Ann Arbor, MI: University of Michigan Press)

Hopfield J J 1982 Neural networks and physical systems with emergent collective computational abilities *Proc. Natl Acad. Sci., USA* **79** 2554–8

——1984 Neurons with graded response have collective computational properties like those of two-state neurons *Proc. Natl Acad. Sci., USA* **81** 3088–92

Hopfield J J and Tank D W 1985 'Neural' computation of decisions in optimization problems *Biol. Cybern.* **52** 141–52

Kirkpatrick S, Gelatt C D Jr and Vecchi M P 1983 Optimization by simulated annealing *Science* **220** 671–80

Kohonen T 1987 *Self-Organization and Associative Memory* 2nd edn *(Springer Series in Information Sciences)* (New York: Springer)

Kunz D 1991 Suboptimum solutions obtained by the Hopfield–Tank neural network algorithm *Biol. Cybern.* **65** 129–33

Lades M, Vorbruggen J C, Buhmann J, Lange J, von der Malsburg C, Wurtz R R and Konen W 1993 Distortion invariant object recognition in the dynamic link architecture *IEEE Trans. Comput.* **42** 300–11

Lawler E L, Lenstra J K, Kan A H, Rinnooy G and Shmoys D B 1985 *The Traveling Salesman Problem: a Guided Tour of Combinatorial Optimization* (New York: Wiley)

Lin S 1965 Computer solutions of the traveling salesman problem *Bell Syst. Tech. J.* **44** 2245–69

Looi C-K 1992 Neural network methods in combinatorial optimization *Comput. Ops Res.* **19** 191–208

Mead C 1989 *Analog VLSI and Neural Systems* (Reading, MA: Addison-Wesley)

Mehta S and Fulop L 1993 An analog neural network to solve the Hamilton cycle problem *Neural Networks* **6** 869–81

Owechko Y and Shams S 1994 Comparison of neural network and genetic algorithms for a resource allocation problem *Proc. Int. Conf. Neural Networks (Orlando, FL)* **7** 4655–60

Peterson C 1990 Parallel distributed approaches to combinatorial optimization: benchmark studies on traveling salesman problem *Neural Comput.* **2** 261–9

Peterson C and Soderberg B 1989 A new method for mapping optimization problems onto neural networks *Int. J. Neural Syst.* **1** 3–22

Sabourin M and Mitiche A 1993 Modeling and classification of shape using a Kohonen associative memory with selective multiresolution *Neural Networks* **6** 275–83

Shams S 1995a Multiple elastic modules for visual pattern recognition *Neural Networks* **8** 1439–56

——1995b A new self-organizing model for angle-only target deghosting and tracking *Proc. World Congress on Neural Networks '95 (Washington, DC)* vol 2, pp 646–51

——1995c Simultaneous recognition of multiple objects using the MEM model *Proc. Int. Workshop on Artificial Neural Networks '95 (Malaga)* ed J Mira and F Sandoval pp 919–25

Shams S and Gaudiot J-L 1993 Parallel implementations of neural networks *Int. J. Artificial Intelligence Tools* **2** 557–81

—— 1995 *IEEE Trans. Neural Networks* **6** 407–21

Simic P D 1990 Statistical mechanics as the underlying theory of 'elastic' and 'neural' optimisations *Network* **1** 89–103

——1991 Constrained nets for graph matching and other quadratic assignment problems *Neural Comput.* **3** 268–81

Sirosh J and Miikkulainen R 1994 Cooperative self-organization of afferent and lateral connections in cortical maps *Biol. Cybern.* **71** 65–78

Sun K T and Fu H C 1993 A hybrid neural network model for solving optimization problems *IEEE Trans. Comput.* **42** 218–27

Tagliarini G A, Christ J F and Page E W 1991 Optimization using neural networks *IEEE Trans. Comput.* **40** 1347–58

Van den Bout D E and Miller T K 1989 Improving the performance of the hopfield-tank neural network through normalization and annealing *Biol. Cybern.* **62** 129–39

von der Malsburg C and Willshaw D J 1977 How to label nerve cells so that they can interconnect in an ordered fashion *Proc. Natl Acad. Sci., USA* **74** 5176–8

Wilson G V and Pawley G S 1988 On the stability of the traveling salesman algorithm of Hopfield and Tank *Biol. Cybern.* **58** 63–70

Wong Y-F 1994 A comparative study of the Kohonen self-organizing map and the elastic net *Computational Learning Theory and Natural Learning Systems* vol 2, ed S J Hanson, T Petsche, M Kearns and R L Rivest

Xu X and Tsai W T 1991 Effective neural algorithms for the traveling salesman problem *Neural Networks* **4** 193–205

Yuille A L 1990 Generalized deformable models, statistical physics, and matching problems *Neural Comput.* **2** 1–24

Zhang S and Constantinides A V 1992 Lagrange programming neural networks *IEEE Trans. Circ. Syst-II: Ana. Dig. Sig. Proc.* **39** 441–52

F1.4 Associative memory

James Austin

Abstract

This section considers how neural networks can be used as associative memory devices. It first describes what an associative memory is, and then moves on to describe associative memories based on feedforward neural networks and associative memories based on recurrent networks. The section also describes associative memory systems based on cognitive models. It also highlights the ability of neural-network-based systems to deal with uncertain data as compared with conventional associative memory systems.

F1.4.1 Introduction

This section describes the use of neural networks as associative memories (AMs). There has been significant interest in AMs for a long time. Thus it is only possible to cover the basic facts about their implementation and operation.

It will be shown that neural-network-based AMs are far more flexible, in terms of their ability to handle noisy and corrupt data, than conventional AMs. In addition, due to the way that they operate, they can operate at very high speeds and can be implemented at very low cost.

The first section starts with a description of what an AM does. The next section describes how conventional non-neural AMs are constructed. The emphasis is on the construction of the memories, as a major factor in the design of AMs is the trade-off between the speed, cost and flexibility of the various implementations. The problems with conventional AMs are then described. The next section describes the difference between neural-network-based AMs and conventional AMs. It then goes on to sample the major neural network architectures in a comparative sense; it does not go into the detailed *operation of* C1.3 *the networks* as this is discussed elsewhere in this handbook. The conclusion then identifies remaining weaknesses of AM design.

F1.4.2 What is an associative memory?

In its simplest form AM is a memory system that allows one data item to be associated to another, so that access to one data item allows access, by association, to the other. Note that the association can be one-to-one as in this example, one-to-many or many-to-many. A particular feature of AMs is the ability to allow or not to allow symmetrical referencing. That is, if the system stores the association of 'A' with 'B' then in the symmetrical case the user can ask both what is associated with A (i.e. B) and what is associated with B (i.e. A). A nonsymmetrical AM would only allow associative recall in one direction.

In the neural network literature, AMs are referred to as being autoassociative or heteroassociative (Kohonen 1977). An auto-AM allows recall of the same item that is input, i.e. a memory may store a picture of a car, so that when a wheel of a car is input to the memory the complete picture of the car is recalled. This type of network is sometimes called a 'clean-up' network as it can be used to remove noise from a corrupted piece of data. A hetero-AM allows recall of an associated item that is different from the input query, i.e. a picture of a wheel may recall the word 'wheel'. Because conventional AM systems do not deal with incomplete or noisy inputs, these terms are not used in the conventional literature on AM.

F1.4.3 Implementing conventional associative memory systems

Conventional memory systems access and store data by reference. This means that every data item has a number associated with it which is commonly known as a memory address. By going to, or referencing, that memory address the value associated with the address can be recalled. That is, if a variable 'A' has to have the value 'B' associated with it, the system allocates a memory cell to store the item 'B'. The address of the location where B is to be found is held in a table next to a reference to A. When an associative access on A is required, the table is consulted and the address is 'de-referenced' to find the associated value or values. In computing systems the table which holds A with the address for B is only used at compile time, after which all references to A are replaced by the actual memory address for B. More complex arrangements are used that allow array indexing and memory management.

It can be seen that a conventional memory is basically nonsymmetrical, and cannot implicitly cope with noise or errors in the item used to access the associated item. Furthermore, because the system must de-reference every time an access is made, it can be slow to perform access. Moreover, storing a new item requires a search of the memory allocation list to find an unused location.

Many databases use AM methods to allow access of a record from a query. A conventional memory system used to do this would require a large list containing every query against the address where the data is stored. This large list is in effect another AM. This approach can be slow, because of the time to search the list. Because of this, complex indexing methods have been developed to allow fast access to the associated memory address from a query; the memory address is then used to access the associated record. Methods such as hashing, superimposed coding and B-trees (Knuth 1973) are used for this. Although applicable to conventional databases, these methods can be slow and cannot easily deal with large input queries (made up of many variables) or noise in the input field (such as that found when AMs are used F1.2 for *pattern recognition*).

A final, important class of AM is the content addressable memory (CAM). These are often confused with AM devices, but are just a means of implementing fast AM systems. A CAM memory is a method of speeding up the recall and storage of associated items. They are heavily used in computers in many cache memory systems (Hayes 1988) because of their speed. Instead of storage by reference, the memories store data by content, i.e. whereas in a conventional system each storage location has a separate address, in a CAM system the item that is stored is the address, thus the concept of content addressed memory. The most simple approach to CAM is to assume all data items are a binary address and store the associated item at that address, i.e. A associated with B would take the ASCII value for A and use that as the address of B. This means that the memory is unsymmetrical and potentially very large, but extremely fast. A more conventional approach to CAM systems is to hold the item to be used as the address in a list which is paired to the associated item. The list is supported by complex hardware that allows any query to be matched to all items in the list simultaneously and therefore very quickly. This allows both a fast memory and one that supports partial matches, as the hardware can allow similarity metrics to be built in. In addition, the memory need only be as big as the number of associated items to be stored. Unfortunately, it is very expensive to implement due to the dedicated hardware. For this reason it is only found in small AM systems, such as cache memories. However, many experimental machines have been built that have explored construction of large AMs (Krikelis and Weems 1994). But the construction of large, cheap and flexible associative memories is still a problem, to which neural networks may offer a solution.

This short review has shown that although there are a wide variety of AM systems, it is very expensive to achieve fast response and an ability to associatively retrieve on noisy and incomplete examples.

F1.4.4 Neural networks for associative memory

B2.3 There are two major classes of neural-network-based AMs. These are based, respectively, on *feedforward networks* and *recurrent networks*. The feedforward AM networks operate by recalling data in one pass through a network where there are no recurrent connections. The recurrent AMs operate by presenting a piece of data and iterating the network until the associated item is recalled.

A further major separation of neural-network-based AMs is those that use distributed representations and those that use local representations. In a local representation each weight in the network is responsible for holding one piece of information on one association. In a distributed AM, each weight is responsible for holding information on one or more associations. It will be shown later that almost all neural networks can be seen as AMs, and that some have only academic interest while others provide very useful and practical capabilities.

While many conventional neural network architectures can be interpreted as AMs, there are a number of neural architectures that have been designed specifically to act as AM systems. These systems arise from work on models of human memory systems. These will be discussed under a separate heading. The major benefits of using neural networks as AMs are the ability of neural networks to cope with noise in the query data and also the speed at which a network can recall data. Peripheral benefits include efficient use of memory and an ability to cluster the data.

F1.4.5 Feedforward associative neural networks

Any feedforward neural network that is capable of classifying data can, in principle, be used to build an AM. There are two ways in which this can be achieved, either through a two-stage or a one-stage process. In the case of a two-stage system they can be separated into systems that are purely neural-network-based and systems that are a hybrid of neural networks and conventional memory systems.

F1.4.5.1 Hybrid two-stage associative memory neural network systems

In a two-stage approach the first stage is a neural network classifier and the second stage is a simple conventional memory system. The front-end neural network is used to associate one half of the association with a pointer, which can be a single neuron output coupled to a memory location that stores a memory address. The second stage in this approach is then a conventional memory which can be accessed to find the associated item. The neural network front-end can be any network with a classification ability, as shown in figure F1.4.1.

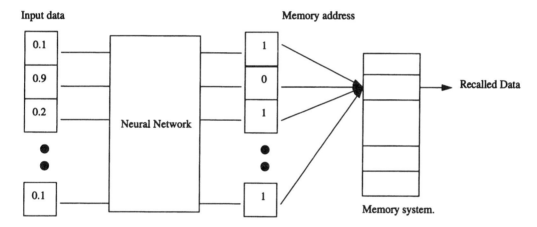

Figure F1.4.1. Simple two-stage neural associative memory. The input data are fed to a network which returns a binary address that is then used to access a conventional random access memory to retrieve data.

One can choose any of the existing networks for this, for example, CMAC, MLN (*multilayer network*), SLN (*single-layer network*), RBF (*radial basis function network*), and SOM (*self-organizing map*). They all allow the ability to perform a partial match on the query to be used to access the association. The great advantage of this approach is the ability to select the network that you feel is most suitable for the problem you have. If a Kohonen network is used as the front-end network, and one or more units are used uniquely for each association, then the network is a nondistributed AM.

C1.2

C1.1, C1.6.2, C2.1.1

To reduce the number of weights used to perform the front-end process, it is possible to let the front-end network output the memory address of the storage cell that contains the associated item. If the memory address contains 32 bits, then the 32-bit address of the associated item can be trained onto the output units of the front-end network. When an associative query is input to the network, the front-end network will directly recall the memory address. This is a faster approach than the first method.

Although the second approach works well when the query is non-noisy, it starts to fail if the input is noisy. This is because in order to form an address to use in the conventional memory a binary number is required. If the query is noisy this can cause output units not to be fully on or off. This then requires a threshold to be set which will decide when an output unit should be on or off. Unfortunately, there is no reliable way of setting this threshold. One method of overcoming this is to use an L-max representation of the memory address (Casasent and Telfer 1992, Austin 1987). In this approach, each memory address is associated with a binary pattern that is unique for each address and contains exactly L bits set to 1. When this is used, the front-end neural network is trained to recall the L-max code when the input query is presented. To recover the code when the input query is noisy, the L highest responding neurons in the output of the front-end network are selected and set to 1, all others are set to 0. Once recovered the code can be passed to a store which recalls the associated memory address. This secondary store can assume that the L-max code is non-noisy and so use conventional AM methods.

When the output of the first-layer neural network is L-max coded or a binary code is used the AM is known to be partly distributed. This arises because the association between the input query and the address is distributed, in that each output unit shares its work between each association. However, the output neural network uses a local representation of the data. As long as the associated items are one-to-one this approach works well. Unfortunately, if the associations are one-to-many or many-to-many, then the system can fail, as any input query will need two addresses to be recalled, which is not possible.

F1.4.5.2 Pure two-stage neural associative memories

To increase the speed and flexibility of the AMs, the second stage of the neural AM can be replaced by a neural network as shown in figure F1.4.2. In effect, the system then becomes one large neural network, but it is initially simpler to view such a system as two neural networks. The second stage can take the 'address' provided by the front-end network and associate this to the associated item. To do this, each output unit in the output network represents a single bit or byte of information in the associated data. The output network is then a distributed memory, making the whole network a distributed AM. As with the front-end network, the back-end network can be any network capable of classifying data. Thus it too can have an ability to perform partial matches.

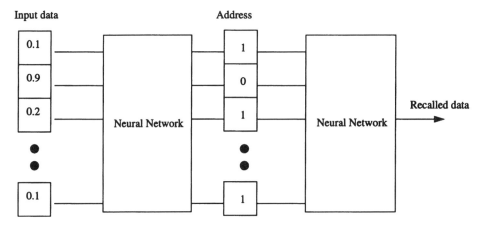

Figure F1.4.2. A two-stage neural associative memory. The input data are fed to a network which returns a binary address that is then fed to a second neural network that is used to retrieve the associated data.

Notable neural network AMs that use this approach are Kanerva's sparse distributed AMs (SDMs) (Kanerva 1988), Austin's advanced distributed AMs (ADAMs) (Austin 1987) and Albus' CMAC (Albus 1975). The first two are considered here.

The SDM uses a simple front-end network that is similar to a Kohonen network and a back-end network that is basically an SLN. The system sets the input network neurons so that they can measure the amount by which they are similar to the input pattern using the Euclidean metric. During training the input data are presented, and a number of the best firing front-end units are selected and set to 1, all others are set to 0. This pattern is then sent to the second layer which is trained, using Hebbian learning, to recall the associated item. Recall works in a similar way. The input is presented and the best firing units in the first layer are selected and set to 1. This is then sent to the second layer which performs the conventional sum of products. The output pattern is then thresholded to recover the associated data. The memory is quite costly to implement, as it requires each input unit to compute a Euclidean distance.

The ADAM system works in a similar way to the SDM. Its first stage is a binary correlation matrix memory (CMM) with an N-tuple preprocessor (Austin 1987) that allows nonlinearly separable patterns to be associated. The system uses L-max encoding of the address from the first network, which it passes to an output network which is also a CMM. The output network allows the assumption that the recovered 'address' is free from any corruption and uses a Wilshaw type threshold to recover the associated data (Wilshaw *et al* 1969). The network has the advantage of simple implementation in dedicated hardware (Kennedy and Austin 1994) and because it uses binary CMM it can learn associations in one presentation of the data. It does so at the cost of reduced generalization ability; AM networks that use MLN as a front-end network do not suffer from this. The ADAM does not suffer the implementation cost of the SDM as it uses CMM and N-tuple networks which, because they are RAM nets (Austin 1994), are simple to implement.

In the SDM and ADAM the back-end network was not a conventional MLN. Although this provides speed of learning it is at the cost of recall reliability. To improve the performance of the networks, an MLN can be used in the second stage. There are a number of advantages of this approach. These relate particularly to the AM's ability to recall associations in noisy situations. If the input data are very noisy the address formed at the output of the front-end network can be so corrupt that even the L-max process can fail. Since the back-end network is a neural network, this can recognize the address even in noisy conditions. Unfortunately, this is at the cost of learning time, which can be very large for large sets of associative data.

The aim of the SDM and ADAM systems is to allow fast training of very large numbers of associative data. AM networks that incorporate MLNs using backpropagation learning work better on noisy data, but are very slow to train. Thus their use is restricted to situations where the number of associations is small and/or where a large amount of corruption in the input queries is expected.

F1.4.5.3 One-stage neural associative memories

The essential feature of the two-stage AM is the careful assignment of the 'address' used to link the two memory systems. It will be obvious to the alert reader that the pure two-layer networks are the same as multilayer networks, apart from their careful selection of the 'address' pattern. In the MLN the 'address' is the hidden-layer pattern which is not chosen by the user but determined through the use of a learning algorithm. As a result, the training times of MLNs using backpropagation can be very long because MLNs have to determine the pattern arrangement on the hidden-layer units (see figure F1.4.3).

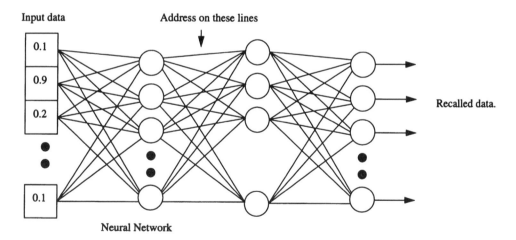

Figure F1.4.3. MLN as an associative memory. This combines the two stages in an MLN. The address would be present in the second hidden layer.

In the two-stage approaches, the hidden-layer pattern (the address) is selected so that all associations can be reliably recalled. The MLN with backpropagation learning is designed to perform generalization over training class data. This allows any pattern that is 'similar' to the training data to be recognized. In the AM case this is the same as training the AM on a set of associations that are many to one (i.e. that many different inputs can elicit the same output). Such an AM would generalize well to example inputs that are somehow similar to the trained associations. This aspect of the two-stage AMs has not been considered in current research.

The final type of one-stage AM is one that has a single layer of units between the input query and the associated data. This type of system can be viewed as a simple perceptron-type network, or as a more complex correlation matrix memory. Because it is equivalent to an SLN the system only works well when the patterns to be associated have orthogonal query inputs (Palm 1980, Kohonen 1977). The networks of this type (Casasent and Telfer 1992) can be made to learn fast, and have been shown to be applicable to a number of problems including rule-based searches (Austin 1996). Many approaches to allow accurate recall from these types of memories have been described, including training through the pseudo-inverse method (Kohonen 1984), through *Hebb-type learning* (Kohonen 1972) and using the *perceptron* and *Widrow–Hoff* learning rules (Widrow *et al* 1987). A number of studies have considered the capacity of such AMs (Nadal and Toulouse 1990, Casasent and Telfer 1992, Willshaw *et al* 1969, Austin and Stonham 1987) as well as studies which have examined the possibility of sparse connectivity (Nadal and Toulouse 1990).

B3.3.1

B3.3.2, B3.3.4

Although one-stage AMs can generalize well, they suffer from slow learning or limited storage ability with respect to memory use.

F1.4.6 Recurrent associative neural networks

Recurrent AMs have been the subject of a great deal of study and mathematical analysis. In particular, the *Hopfield network* has been studied the most and is also attributed to regenerating interest in neural networks at the beginning of the 1980s. This class of network operates through an iterative recall process, where the recall of one pattern through the presentation of another develops through time. The Hopfield neural network (1982) can only be used in the autoassociative mode as it is used for pattern completion. After being trained on a number of patterns, a part of any of the trained patterns can be presented and the network iterated to 'rebuild' a complete pattern of the part presented. The network architecture consists of a single layer of neurons which receive inputs from the outputs of all other neurons. All units have bipolar weights and bipolar inputs are used. To train a new pattern, the pattern is 'clamped' onto the neurons and the network trained by a simple learning rule. The process of recall uses an iterative technique where the output 'evolves' over time. Although the network has interesting properties, it suffers from limited storage ability and can only deal with autoassociative inputs.

C1.3.4

The bidirectional AM (Kosko 1987, 1988) can be seen as an extension to the Hopfield network because it allows the storage and recall of heteroassociated patterns by a recurrent network. It achieves this by using a two-layer architecture where the output of the second layer feeds back to the input of the first layer. The system uses iterative recall, but now the first layer is used to recall the first pattern of the associative pair and the second layer recalls the second pattern. For example, if pattern A were associated with pattern B, the network would recall B through the application of a part of pattern A to the first layer. This will evoke (some of) pattern B which would be sent to the second layer. This would then evoke a more complete version of pattern A, which would then be reapplied to the input. This new input would yield a more complete version of pattern B, and so on. Through this iterative process the network would recall a complete version of both patterns A and B.

These two examples of associative recurrent networks show the interesting property of pattern completion in an AM.

F1.4.7 Associative memories used for automated reasoning

This class of network requires a special mention. These AM systems are particularly designed to support reasoning based on association. The primary motivation for these systems is to understand the operation of the human mind—to generate a model of the human reasoning process. The most notable systems are those of Ajjangadde and Shastri (1991) and Dolan and Smolensky (1989). A review of such systems is presented in Sun (1994).

These systems are motivated by a need to model the properties of human AM. This includes our ability to access associated information rapidly and independent of the number of associations stored. In addition, the memory systems allow associative reasoning, which implicitly allows one associated fact to trigger another. The simplest form of such AM systems is a semantic network, where an association is formed through a single connection between two neurons. The meaning of the association (the data) is attached to the units through a separate system.

F1.4.8 Conclusions

This section has pointed out the advantages and disadvantages of neural associative memories compared to conventional computer associative memories. In particular, it has shown that associative memories based on neural networks are faster and more robust to noisy inputs than conventional associative memories. More details on the specific types of neural networks are given in other sections of the handbook.

References

Ajjangadde V and Shastri L 1991 Rule and variables in neural network *Neural Comput.* **3** 121–34

Albus J S 1975 A new approach to manipulator control: the cerebellar model articulation controller *Trans. ASME J. Dynam. Syst. Measure. Cont.* September 220–7

Austin J A 1987 ADAM: a distributed associative memory for scene analysis *Proc. 1st Int. Conf. on Neural Networks* vol IV ed M Caudill and C Butler (San Diego, CA: IEEE Press) p IV-285

——1994 Review of RAM based neural networks *Proc. Fourth Int. Conf. on Microelectronics for Neural Networks and Fuzzy Systems* (Turin: IEEE Computer Society Press) pp 58–66

——1996 Distributed associative memories for high speed symbolic reasoning *Int. J. Fuzzy Sets Syst.* ed N Kasabov (*Special Issue on Connectionist and Hybrid Connectionist Systems for Approximate Reasoning*) to be published

Austin J and Stonham T J 1987 An associative memory for use in image recognition and occlusion analysis *Image Vision Comput.* **5** 251–61

Casasent D and Telfer B 1992 High capacity pattern recognition associative processors *Neural Networks* **5** 687–98

Dolan C P and Smolensky P 1989 Tensor product production system: a modular architecture and representation *Connect. Sci.* **1** 53–68

Hayes J P 1988 *Computer Architecture and Organisation* (New York: McGraw-Hill)

Hopfield J J 1982 Neural networks and physical systems with emergent collective computational abilities *Proc. Natl Acad. Sci., USA* **79** 2554–8

Kanerva P 1988 *Sparse Distributed Memory* (Cambridge, MA: MIT Press)

Kennedy J and Austin J 1994 A hardware implementation of a binary neural network *MicroNeuro 1994* (Los Alamitos, CA: IEEE Computer Society Press)

Knuth D E 1973 *The Art of Computer Programming* vol 3 (Reading, MA: Addison-Wesley)

Kohonen T 1972 Correlation matrix memories *IEEE Trans. Comput.* **21** 353–9

——1977 *Associative Memories: A System Theoretical Approach* (Berlin: Springer)

——1984 *Self-Organisation and Associative Memory* (Berlin: Springer)

Kosko B 1987 Adaptive bidirectional associative memories *Appl. Opt.* **26** 4947–59

——1988 Adaptive bidirectional associative memories *IEEE Trans. Syst. Man Cybern.* **18** 49–60

Krikelis A and Weems C C (ed) 1994 *IEEE Computer* **27** (Special issue on *Associative Processors*)

Nadal J P and Toulouse G 1990 Information storage in sparsely coded memory nets *Network* **1** 61–74

Palm G 1980 On associative memory *Biol. Cybern.* **36** 19–31

Sun R 1994 *Integrating Rules and Connectionism for Robust Commonsense Reasoning* (New York: Wiley)

Widrow B, Winter R G and Baxter R A 1987 Learning phenomena in layered neural networks *IEEE 1st Int. Conf. on Neural Networks* ed M Caudill and C Butler pp II-411, II-429

Willshaw D J, Buneman O P and Longuet-Higgins H C 1969 Non-holographic associative memory *Nature* **222** 960–2

F1.5 Data compression

Andrea Basso

Abstract

This section is devoted to the applications of neural paradigms on the domain of data and, in particular, still and motion picture compression. Neural networks are inherent adaptive systems, thus they are suitable for handling nonstationaries in image data. Artificial neural networks have been employed with success in at least three approaches to image compression, namely, predictive coding, transform coding, and vector quantization. In predictive coding compression is obtained by exploiting spatial and spatiotemporal redundancy by means of prediction methods, while in transform coding an image or a sequence of images is transformed and the coefficients of the transformation are coded. In vector quantization the data are organized in vectors. The space of vectors which is obtained is then divided into regions and a reproduction vector is calculated for each region. In this section these techniques will be briefly illustrated and for each of them the corresponding neural paradigms will be analyzed. The drawbacks and advantages of the neural techniques with respect to the more traditional ones will be outlined.

F1.5.1 Introduction

The exchange of visual information has become one of the crucial aspects in much of our daily life. The interests in digital television and high-definition digital television (HDTV), video telephone, video conference, and video on demand (VoD) are only some examples of the enormous effort that governments and research communities are making in the direction of video technologies.

Images and image sequences, however, contain a large number of data. Their efficient handling requires the capability of compacting and decompacting such information. Storage technology and broadband network development, even with the recent advances, still require efficient image compression techniques. There are today a number of algorithms developed in the framework of international committees that allow still image compression. The algorithm proposed by the Joint Photographic Expert Group (JPEG) is a well known example of still image coding. As far as image sequence compression is concerned, the algorithm proposed by the Motion Picture Expert Group (MPEG) is largely used in multimedia applications.

Artificial neural networks (ANNs) have been successfully employed in three approaches to image compression: predictive coding, transform coding, and vector quantization.

In the area of predictive coding, compression is obtained by exploiting spatial and spatiotemporal redundancy by means of prediction methods that, on the basis of the values of neighboring pixels in space and/or in time, are capable of predicting the value of the current one. Linear prediction is somewhat limited as an approach because the dependences in image data are not restricted to correlation. Traditional nonlinear prediction methods are in general difficult to design and to handle. The neural network approach is successful in this area because its inherent structure is very well suited to the design and easy implementation of flexible nonlinear predictors.

Another approach to the problem of image compression is the one of *transform coding* in which an image is transformed and the transform coefficients are coded. The goal of such an approach is to represent the image in terms of a smaller set of values. It is possible to minimize the number of transform coefficients representing the image with a distortion not perceivable by the end-user. A common approach is the use of linear block transform coding in which a given image is partitioned into a set of nonoverlapping blocks and each block is transformed and coded separately. Different neural paradigms have been proposed in the area of transform and adaptive transform coding with results comparable to the traditional techniques such as discrete cosine transform (DCT).

F1.6, F1.8
C2.1.1 Vector quantization methods, widely studied in the area of *signal and image processing*, are closely related to certain paradigms of *self-organizing feature maps* (SOFMs). Particular success has been obtained by the frequency-sensitive competitive learning paradigm. Vector quantization based on neural processing has also been applied to image sequences (Manikopoulos and Li 1989).

This section is organized as follows. In section F1.5.2 a general introduction to the image compression problem will be given. Predictive coding, transform coding, and vector quantization methods will be briefly presented. In section F1.5.3 the neural approaches relative to predictive coding will be discussed. In section F1.5.4 the neural approaches relative to transform coding will be presented. In section F1.5.5 neural approaches to the problem of vector quantization will be presented. Section F1.5.6 will give our conclusions.

F1.5.2 Image compression approaches

ANNs have been successfully employed in three approaches to image compression: predictive coding, transform coding, and vector quantization. In this section the basic principles of these approaches will be discussed.

In *predictive coding*, spatial correlation among neighboring pixels of an image is exploited by means of linear or nonlinear prediction techniques. A given set of neighboring pixels is used to predict the value of the current pixel. The compression gain is obtained by the fact that only the prediction error (i.e. the difference between the real value of the pixel and the predicted one) is coded. The better the prediction, the better the coding gain. The traditional algorithm for linear prediction is the delta pulse code modulation (DPCM). Linear prediction is, however, a limited approach because dependences in image data often have a nonlinear nature. On the other hand, classical nonlinear prediction methods are difficult to design. The advantage of neural networks in this area is their inherent structure, which is very well suited to the design and easy implementation of nonlinear predictors.

In another class of algorithms called *transform coders*, an image is mapped into a domain which is significantly different from the intensity domain, and the transform coefficients are coded. This transformation allows us to represent the image in terms of a smaller set of values with minimal distortion. A common approach is the use of linear block transform coding in which a given image is partitioned in a set of nonoverlapping blocks and each block is transformed and coded separately. The optimal linear transformation in terms of minimization of the mean square error (MSE) is the Karhunen–Loève transform (KLT), which also has the property of decorrelating the input data. In practice it is very expensive to realize such a transformation because it is computationally intensive and it is dependent on the input data. In general the discrete cosine transform (DCT) is used. It does not perform a complete data decorrelation, but it is not computationally intensive and it is independent of the input image. The JPEG and MPEG standards are based on the DCT transformation.

Neural paradigms such as the generalized Hebbian algorithm and the principal component extraction algorithm and its adaptive versions have been applied to the problem of transform coding. Furthermore, neural algorithms have been developed which are able to compute the optimal transformation coefficients for a given neural architecture. In this framework, known as *autoassociation* or identical mapping, a network with only one hidden layer is employed. The target pattern is identical to the input pattern, so that the network globally maps the input vector into itself. Because of the smaller number of neurons in the hidden layer, with respect to the number of neurons in the input and output layers, the network is forced to find an intermediate compressed representation of the input data.

In order to represent an image with a finite number of bits, image intensities, transform coefficients, or model parameters must be quantized. Quantization involves assignment of the quantization (reconstruction) levels and decision boundaries. The quantization applied to individual real-valued samples is called *scalar quantization*. Scalar quantization is optimal only if the data source can be regarded as statistically uncorrelated. In practical coding schemes the DCT is used for decorrelating the data. It does not perform a complete data decorrelation and so scalar quantization results in suboptimal performances. Recently, another technique based on quantization of groups of samples organized in vectors, namely *vector quantization*, has been developed. In the first place a set of vectors is generated starting from the data after transformation or directly from the image pixels. The space of vectors obtained is divided into regions and a reproduction vector is calculated for each region. For each data vector, the region in which the vector resides is calculated and the reproduction vector of this region is used in its place. Instead of coding the reproduction vector a symbol which represents it (often called an *index*) is used.

F1.5.3 Neural prediction coding

In traditional approaches the predictor design relies on autoregressive (AR) models. In such models the predicted value of the current pixel is obtained as a weighted sum of the values of the neighboring pixels. The minimization of the mean square error between the predicted and the current pixel value is obtained by computing the weighting coefficients in terms of autocorrelation of the neighboring pixels and cross variance, according to the principles of linear predictive coding (LPC).

Multilayer perceptrons have been used by Dianat *et al* (1991) to implement nonlinear prediction. C1.2 A three-layer perceptron (one input, one hidden, and one output layer) has been employed. The neuron structure used is the traditional weighted sum of the inputs followed by a sigmoidal function. Improvements with respect to the AR model both at reconstruction quality level with more than 4 dB signal-to-noise ratio (SNR), and at compression level with around 0.6 bits per pixel (bpp) gain, have been obtained. High-order nonlinear predictors showing comparable results have been proposed by Manikopoulos (Li and Manikopoulos 1990, Manikopoulos 1992) and Pao (1989).

F1.5.4 Image transform coding based on neural networks

In this section we will review some neural approaches related to principal values decomposition problems and the use of the autoassociative neural network paradigm for image compression.

F1.5.4.1 Principal component extraction based on neural networks

Several researchers in the neural network community have concentrated their efforts on developing new neural methods for optimal linear and nonlinear transformations that are similar to but less computationally expensive than the KLT. The basic neural paradigm employed is *Hebbian learning* (Hebb 1949) which B3.3.1 has been extended by Oja (1989) in order to extract a given number of principal components of the input data in parallel. Sanger (1989) improved the method by simplifying the computational procedure. Practical results show that even if these approaches show an interesting alternative to classical methods for computing the principal components of a set of data, their use is of less help for image compression purposes.

F1.5.4.2 Autoassociation

Multilayer perceptrons (MLPs) have been quite successfully employed in information processing and pattern recognition. The idea of using MLPs for feature space reduction was proposed by Rumelhart and McClelland (1989) and then applied in image compression by Cottrell *et al* (1989). In this framework, known as *autoassociation* or identical mapping, a network with only one hidden layer is employed. The target pattern is identical to the input pattern, so that the network globally maps the input vector into itself. Because of the smaller number of neurons in the hidden layer, with respect to the number of neurons in the input and output layers, the network is forced to find an intermediate compressed representation of the input data. The technique consists of two major steps: (i) *vector formation* in which the pixels of the image are grouped into vectors using a certain criterion and (ii) *dimensionality reduction* obtained by mapping the defined vector space in a lower-dimensional one. There are two important steps in the vector formation process: the vector formation strategy and the choice of the vector size.

The goal of the vector formation strategy is to exploit the dependences among the elements of the input data set. These dependences can be linear (correlations) or nonlinear (all the other dependences). Due to the nonstationarity of natural image sequences, it is difficult to give a precise definition of an optimal vector formation strategy. Furthermore it should be noted that there is a tradeoff between the feature space definition and the type of mapping.

Experiments show that an autoassociative neural network performs a projection of the input data on a lower-dimensional space, but does not decorrelate (i.e. diagonalization of the output covariance matrix)

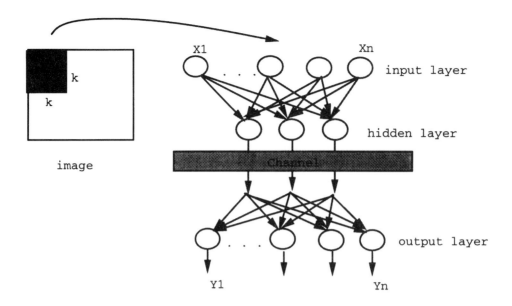

Figure F1.5.1. Three-layer perceptron for image compression.

(Cottrell *et al* 1989, Sicuranza *et al* 1991, Mougeot *et al* 1990). For natural images, a common vector formation strategy consists of forming vectors out of blocks of contiguous image pixels, exploiting in this way the inherent pixel-to-pixel dependence of the images. If the domain is different (subband, transform) the dependences among the data are usually different and, as a consequence, the vector formation strategy is also different (Gersho *et al* 1980, Abut and Luse 1984, Makoul *et al* 1985).

The vector size defines the dimensionality of the input feature space and so it is strictly related to the compression ratio and image quality that can be obtained.

Once the n-dimensional vectors are formed they constitute elements of an n-dimensional space in which a certain norm is supposed to be defined. The particular task of the input and hidden layers of the neural network is to map this n-dimensional space onto a k-dimensional one, with $k < n$, while minimizing the MSE at the output layer. In autoassociative neural networks the teacher vector is identical to the input vector; thus the networks learn the identical mapping. Generally a three-layer perceptron in autoassociative mode is used (figure F1.5.1). An image is sampled in $k \times k$ blocks to form a vector of k^2 elements by row-wise raster scanning. The network performs data compaction of the input since there are fewer neurons in the hidden layer than in the input or output. The numbers of neurons in the input and in the output layers are identical.

In the learning phase, which is unsupervised due to the autoassociative mode, the network is forced to compute a set of good hidden layer weighting factors for representing the input data. The output layer uses this internal representation for the reconstruction of the input pattern. The transmission channel can be assumed present immediately after the hidden layer neurons, with possible modulation.

After the work of Cottrell *et al* (1989) which opened the exploration of a possible utilization of MLPs for image compression, many efforts (Sicuranza *et al* 1991, Basso 1992, Sonehara *et al* 1989) have been made to clarify the behavior of the network and the role of the different parameters involved, such as the number of neurons in the hidden layer and the use of other norms than the L_2 in error computation (Burrascano 1991).

Experimental results (Sicuranza *et al* 1991, Basso 1992) show that in the linear case the learning capabilities and the reconstruction quality of the coded images are higher than in the nonlinear case.

F1.5.5 Vector quantization with neural networks

C1.1.5 In the area of *vector quantization* (VQ) neural networks have been widely studied and compared with vector quantization techniques such as the Linde–Buzo–Gray (LBG) algorithm.

The LBG algorithm is an iterative method based on the Lloyd clustering algorithm (Linde *et al* 1980). Given an initial partitioning of the set of training vectors, the goal of the algorithm is to classify them into clusters and to compute a *centroid*, that is, the representative of the elements of the cluster. The classification process is performed on the basis of a metric defined on the set of training vectors and the nearest-neighbor rule. The centroids are iteratively updated whenever a new element is added to the cluster that they represent. Each iteration goes through all the elements of the training set. The stopping criterion is based on a predefined distortion measure. If the average distortion at the current iteration is under a predefined threshold, the classification process is stopped. For image applications the L_2 norm is used as the metric and MSE as the distortion measure.

The possibility of applying self-organizing maps to the problem of vector quantization is discussed by Kohonen (1990). The self-organizing neural network architecture is based on a two-dimensional lattice formed by interconnected processing elements. Such a lattice receives its input through a relay layer. It is a simple fan-out of the input samples to each of the processing elements of the lattice.

In the learning phase input vectors are presented one at a time, and the weights interconnecting the input signals to the neurons are adaptively updated so that the point density function of the weights tends to approximate the probability density function of the input vector. Once the learning is terminated, the synaptic strengths between the input and the output nodes represent the components of the centroids of the clusters.

Such a neural network paradigm has been widely applied to still and sequence image compression. Nasrabadi and Feng (1988) present results for various coded images using the codebook designed with the self-organizing feature map. The results are compared to images coded by means of codebooks designed with the Linde–Buzo–Gray (LBG) algorithm. Results show that the neural technique gives a performance that is very close to optimal.

One major problem of using the basic self-organization technique is that some neural units may be underutilized. Several variations of the algorithms have been proposed to solve the problem. One of these is the frequency-sensitive competitive learning neural network which is presented by Ahalt *et al* (1989).

This neural paradigm has been successfully employed in the area of image sequence coding. The algorithm has been modified to handle variations in the image content due to scene changes. After an initial training of the codebook, the adaptation resumes in order to respond to scene changes and motion in the image sequence. The main advantage of the technique is the effective adaptation to motion content of scene changes and the ability to adjust the instantaneous bit-rate quickly in order to keep image quality constant.

F1.5.6 Conclusions

The use of neural paradigms for image compression is promising for several reasons. First of all, neural networks—because of their inherent parallel structure—are well suited to the processing of image data. Second, the ability of neural algorithms to compute nonlinear mappings exploiting spatial and spatiotemporal image redundancies shows great potential for new compression methods. Third, neural networks are inherent adaptive systems. This very important feature can be effectively employed for the handling of nonstationarities in image data.

In this section we have shown how different neural paradigms such as multilayer perceptrons, self-organizing feature maps, and autoassociation networks can be efficiently used for image compression purposes with comparable or better results than more classical and well known techniques. Nevertheless, the full potential of the neural approaches will only be exploited completely when massively parallel implementations of such approaches are available. In this sense we share the opinion of Dony and Haykin (1995).

References

Abut H and Luse S A 1984 Vector quantization for sub-band coded waveforms *Proc. IEEE Int. Conf. on Acoustics, Speech, Signal Processing* (New York: IEEE Press)

Ahalt S, Chen P and Krishnamurthy A 1989 Performance analysis of two image vector quantization techniques *Proc. IEEE* **1** 169–75

Basso A 1992 A massively parallel implementation of autoassociative neural networks for image compression 1992 *Artificial Neural Networks, Proc. 1992 Conf. on Artificial Neural Networks (ICANN-92) (Brighton)* vol 2 (Amsterdam: Elsevier) pp 1465–8

Burrascano P 1991 A multilayer perceptron in the Chebyshev norm for image data compression *Proc. IEEE Int. Symp. on Circuits and Systems (Singapore)* vol 3 (New York: IEEE Press) pp 1396–9

Cottrell G W, Munro P and Zipser D 1989 Image compression by back propagation: an example of extensional programming *ICS Report 8702*

Dianat S, Nasrabadi N and Venkataraman S 1991 A non-linear predictor for differential pulse-code encoder (DPCM) using artificial neural networks *Proc. Int. Conf. on Acoustics, Speech and Signal Processing (ICASSP-91) (Toronto)* vol 4 (New York: IEEE Press) pp 2793–6

Dony R and Haykin S 1995 Neural network approaches to image compression *Proc. IEEE* **83** 288–303

Gersho A, Ramstad T and Versvik I 1980 Fully vector-quantized subband coding with adaptive codebook allocation *Proc. IEEE Int. Conf. on Acoustics, Speech, Signal Processing* (New York: IEEE Press) pp 1580–9

Hebb D O 1949 *The Organization of Behavior* (New York: Wiley)

Kohonen T 1990 The self-organizing map *Proc. IEEE* **78** 1464–80

Li J and Manikopoulos C 1990 Nonlinear predictor in image coding with ADPCM *Electron. Lett.* **26** 1357–9

Linde Y, Buzo A and Gray R M 1980 An algorithm for vector quantization design *IEEE Trans. Commun.* **C-28** 84–95

Makoul J, Roucos S and Gish H 1985 Vector quantization in speech coding *Proc. IEEE Int. Conf. on Acoustics, Speech, Signal Processing* (New York: IEEE Press) pp 720–38

Manikopoulos C 1992 Neural network approach to DPCM system design for image coding *Proc. IEE-I* **139** 501–7

Manikopoulos C and Li J 1989 Adaptive image sequence coding with neural network vector quantization *IJCNN-89* p 573

Mougeot M, Azencott R and Angeniol B 1990 Image compression with back propagation: improvement of the visual restoration using different cost functions *Neural Networks* **4** 467–76

Nasrabadi N and Feng Y 1988 Vector quantization of image based upon a neural-network clustering algorithm *SPIE-VCIP* **1001** 207–13

Oja E 1989 Neural networks principal components and subspaces *Int. J. Neural Syst.* **1** 61–8

Pao Y H 1989 *Adaptive Pattern Recognition and Neural Networks* (Reading, MA: Addison-Wesley)

Rumelhart D E and McClelland (eds) 1989 *Parallel Distributed Processing* (Cambridge, MA: MIT Press)

Sanger T D Optimal unsupervised learning in a single-layer linear feedforward neural network *Neural Networks* **2** 459–73

Sicuranza G, Ramponi G and Marsi S 1991 Artificial neural networks for image compression *Electron. Lett.* **26** 477–9

Sonehara N, Kawato M, Miyake S and Nakane K 1989 Image data compression using a neural network model *Proc. Int. Joint Conf. on Neural Networks (IJNN-89) (Washington, DC)* vol 2 (New York: IEE Press) pp 35–41

F1.6 Image processing

John Fulcher

Abstract

This section introduces fundamental image processing concepts and follows with a discussion of how artificial neural network (ANN) techniques can be successfully applied to perform such tasks. This is illustrated by way of representative examples taken from classification, image enhancement, compression and face recognition.

F1.6.1 Introduction

Digital image processing is concerned with processing images via digital computer once they have been captured and converted into a suitable form. Digital image processing offers two main advantages over alternative techniques: precision and flexibility. The former means that image quality is maintained during subsequent processing of the image (by contrast, some photographic and electronic techniques degrade the image quality with each successive process). Flexibility results from working with *digital* versions of the image; once encoded, we are able to perform many and varied operations on them.

Computer (machine) vision systems use information extracted from such images in order to assist in decision making. For example, we may want to develop a biometric recognition system (based on human facial images, iris information, handprints, fingerprints and the like).

Images may originate from scanning a photograph or other hard copy image (via a process similar to that used in photocopiers), from a CCD camera or from some other source, such as X-ray, gamma ray, computer tomography scans, magnetic resonance imaging, or infrared or ultrasound devices.

F1.6.2 Digital image resolution and quantization

Digitizing an image necessarily yields a limited-resolution (quantized) version of the original. The digitized image is stored within the computer as a two-dimensional array of pixels (dots) in either integer or real form. Typical image resolutions range from 1024×1024 down to 64×64 pixels. Obviously the finer the resolution, the more memory (storage) overhead this entails.

We can represent digital images using between 2 and 32 bits per pixel. In the case of a 64×64 8-bit pixel image, this would necessitate 32 kbits (4 kbytes) of storage. Alternatively, a 256×256 16-bit pixel image would require 1 Mbits (12 kbytes) per image! It comes as no surprise then that images are often stored *off*-line (on disks and/or magnetic tape), rather than on-line in the computer's memory (or hard disk for that matter). Moreover, the memory overhead can be reduced if images are *compressed* in some manner prior to storage.

There is a direct relationship between the number of bits per pixel (precision) and the number of discrete quantization levels which can be represented within the computer. For example, 8 bits can discriminate 2^8 or 256 discrete levels in a gray-scale image. Obviously, a greater number of bits is required to represent *color* pixel information. For example, 12 bits could be allocated as three sets of 4 bits (or $2^4 = 16$ discrete intensity levels), one for each primary color—red, green, blue (RGB). By contrast, the same number of bits could represent 2^{12} (4096) discrete gray-scale levels. Of course, if we are dealing with black (0) and white (1) images, this requires only 2-bit precision.

F1.6.3 Image storage and transfer

We have already seen how bit precision is directly related to storage overhead. Entire images can be stored in numerous different formats, some of which are device specific (e.g. Sun raster file, Silicon Graphics, Microsoft BMP or X-Windows Dump formats) or software specific (e.g. MacPaint format on the Apple Macintosh), while others adhere to an interchange standard format (e.g. graphics interchange format—GIF—or tagged image file format—TIFF as used with IBM PC images).

The transfer of images between users, over networks or via magnetic disks and/or tape is usually in compressed form (typically 5–10% of the original, using chain coding, run length encoding, or the JPEG standard, say). Such compression reduces both transfer time (over networks) and disk storage (capacity).

F1.6.4 Typical image processing tasks

The purpose of digital image processing is either to enhance an image in some manner or to extract information from it. Common high-level operations include:

- enlarging, reducing or rotating the image
- deblurring the image (i.e. forming a sharper image)
- smoothing (to decrease graininess, speckle and/or noise)
- improving the contrast (or some other characteristic) prior to display
- removing warps or distortions (e.g. gamma correction, to compensate for a nonlinear camera lens)
- encoding (compressing) the image prior to storage and/or transmission
- segmenting the image into regions and characterizing these in terms of shape, size and orientation (e.g. objects and background)
- edge detection (as a prelude to object recognition)
- feature (or shape) extractions (as a preliminary to pattern recognition, say).

In order to realize such high-level functionality, we require a repertoire of low-level (morphological) operations, typical ones being:

- erosion/skeletonization/thinning (or the stripping away of outer pixels)
- dilation (the addition of a new layer of pixels around an image)
- thresholding (clipping)
- edge/boundary enhancement
- linear (and nonlinear) pixel mappings
- statistical operations
- filtering
- transforms.

An image histogram—such as the one in figure F1.6.1—shows the distribution of pixel values within an image, namely the proportion of pixels which fall within each gray-scale (or color) range. Either standard or specialized pattern analysis statistical techniques can be applied to such histograms (examples of the latter being scattergrams, template matching or nearest-neighbor techniques).

Histogram equalization is a technique used to evenly redistribute concentrated gray-scale (intensity) values throughout the image, so that each range has roughly the same frequency of occurrence. Likewise histogram matching can be used to fit *arbitrary* shapes to an image histogram. For example, contrast enhancement could be achieved by superimposing a gradient on the histogram in order to compensate for uneven background lighting.

Density slicing is a technique used to map pixel intensity ranges to specific (artificial) colors, in order to highlight patterns within an image. Conversely, color pixel information can be converted to gray-scale information, using the weighted mean of the RGB values. Moreover, the distance between adjacent color values can be used as the basis for color thresholding or region growing.

Thresholding can be used, for example, to convert a gray-scale image into a black and white one in which all pixel intensities below the threshold are converted to black, and all those above to white. This could be a useful preprocessing stage in the recognition of printed text or objects.

Linear pixel mappings, in which gray-scale values are translated (mapped) from one range to another, are appropriate when we wish to vary (enhance) the image brightness and/or contrast. A typical example of the latter would be in the production of sharper medical X-ray images. Apart from linear variations,

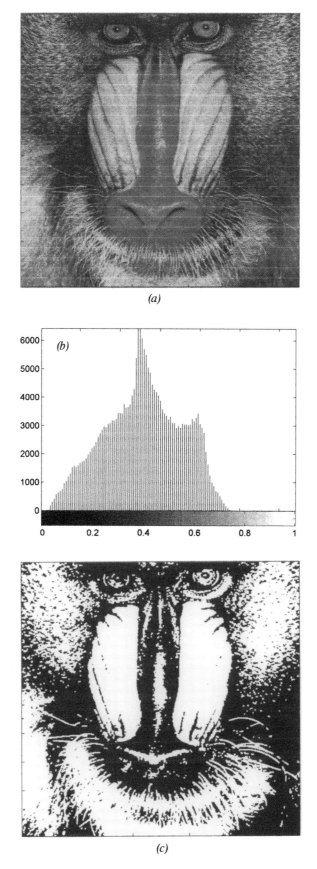

Figure F1.6.1. (*a*) mandrill; (*b*) image intensity histogram and (*c*) dilation using the MatLab image processing toolbox.

we could instead vary the mean and/or standard deviation. Piecewise linear adjustments may also be appropriate, as well as logarithmic or exponential ones.

Now apart from global pixel mappings, *local* pixel mappings may be useful in some cases. For example, statistical differencing could be used to produce an even image contrast. Moreover, the relationship between individual pixels within small regions of an image (such as four- or eight-nearest pixel neighbors) can yield important information for subsequent higher-level processing. Typically such properties are the degree of connectedness and the image geometry (characterized by area, perimeter, length, center of mass, and so forth). It can further lead to the recognition of simple geometric shapes (such as circles, rectangles, triangles and polygons), as well as holes and the degree of convexity (i.e. the number of indentations).

Unwanted noise (introduced during the initial digitization process, transmission of the image via some medium, or by some other process applied to the image) can be reduced but never removed altogether. Noise may be either signal-independent (Gaussian, white or random, such as that produced by vidicon cameras), signal- (or level-) dependent (as with graininess in photographic films), or salt and pepper in appearance (as caused by thresholding a noisy black and white image). A common approach to removing noise is to use smoothing filters, based on the (arithmetic) mean (of a 3×3 pixel neighborhood, say), the weighted mean, mode (most common neighbor), or median (of k nearest neighbors).

Apart from filtering, various transforms also find extensive use in digital image processing. The fast fourier transform (FFT), for instance, can be used to compensate for nonlinearities in an image capture subsystem. The discrete cosine transform (DCT) is used in image coding/compression (and forms part of the JPEG standard). The radon transform is used to compute the projected sum of intensities along a given orientation within an image. This enables principal components (preferred directions) to be extracted from an image, and is used in biomedical imaging. The Hough transform is used to detect straight lines, and can be readily extended to detect circles and other shapes.

Edges occur at regions within an image where the intensity changes rapidly (in other words, the first derivative exceeds some threshold and the second derivative is zero). They represent boundaries between different image regions, and are reflected as abrupt changes of gray-scale within the digitized image. Commonly used edge detection methods include first- (and second-) derivative (gradient) techniques, surface fitting, and local energy methods. A common limitation with all these techniques is their sensitivity to noise and surface texture. This usually means that for practical computer vision systems, we need to restrict the environment and illumination conditions.

Prior to edge detection proper, we may first need to perform edge enhancement in order to simplify this task. A typical edge enhancement technique involves convolving the image with a mask of some description. A Laplacian filter uses the following convolution:

$$\begin{pmatrix} 0 & -1 & 0 \\ -1 & 4 & -1 \\ 0 & -1 & 0 \end{pmatrix}. \tag{F1.6.1}$$

A Sobel filter, by contrast, uses *two* such convolution masks in order to detect horizontal and vertical edges:

$$\begin{pmatrix} -1 & -2 & -1 \\ 0 & 0 & 0 \\ 1 & 2 & 1 \end{pmatrix} \quad \begin{pmatrix} -1 & 0 & 1 \\ -2 & 0 & 2 \\ -1 & 0 & 1 \end{pmatrix}. \tag{F1.6.2}$$

Figure F1.6.1 shows a *de facto* standard (mandrill) image, together with its intensity histogram and the result of invoking the `edges` function within the MatLab image processing toolbox (Thompson and Shure 1993).

F1.6.5 Artificial neural network approaches

The upsurge of interest in the field since the mid-1980s, artificial neural networks (ANNs) have been applied to problems which previously proved difficult if not intractable. It is well known that ANNs perform particularly well at both *pattern classification* and *pattern recognition*. It comes as no surprise then that ANNs have been applied to image processing and computer vision problems (since what we are interested in here is essentially pattern classification/recognition).

F1.2, B6

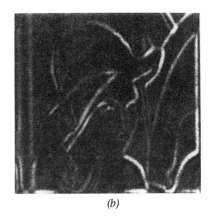

(a)   (b)

Figure F1.6.2. Edge detection using a single-layer ANN and competitive learning: (*a*) original Lena image; (*b*) edges (reprinted from Bhatia *et al* 1991, with permission of IEEE).

(a)   (b)

Figure F1.6.3. Deblurring using a self-organizing neural network: (*a*) original blurred image; (*b*) deblurred image (reprinted from Dhawan and Dufresne 1990, with permission of IEEE).

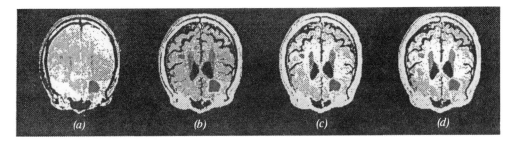

(a)   (b)   (c)   (d)

Figure F1.6.4. MLP applied to magnetic resonance image segmentation: (*a*) number of hidden layer nodes = 1; (*b*) 3; (*c*) 5; (*d*) 10 (reprinted from Ozkan *et al* 1990, with permission of IEEE).

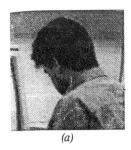
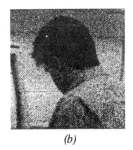
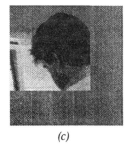

(a)   (b)   (c)

Figure F1.6.5. (*b*) Recall of noisy and (*c*) incomplete images using a Hopfield network (reprinted from Muller *et al* 1993, with permission of IEEE).

(a)   *(b)*

(c)   *(d)*

Figure F1.6.6. Depth recovery and surface reconstruction using a Boltzmann machine: (*a*) original image; (*b*) noisy image; (*c*) initial estimate; (*d*) after 200 annealing iterations (reprinted from Mundkur *et al* 1992, with permission of IEEE).

The number of image processing and computer vision applications to which ANNs have already been applied is considerable and continues to grow. The first *International Joint Neural Networks Conference* included image processing streams, as have all subsequent major ANN conferences. In recent times, we have even witnessed entire specialist conferences on ANN image processing (1st *IEEE International Symposium on Speech, Image Processing and Neural Networks*, Hong Kong, July 1994), as well as the emergence of specialized textbooks (Linggard *et al* 1992, Kulkarni 1994).

Prior to discussing specific image processing applications, we should first make brief mention of significant early developments with *specialized* image processing architectures:

C1.5   •   *WISARD* (Aleksander *et al* 1984) is a network of RAM discriminators, each of which is trained to recognize a different object class. It has been successfully applied to character recognition as well as security applications such as face recognition (see section F1.6.5.3 below).

C2.1.3   •   The *Neocognitron* (Fukushima and Wake 1991) is a hierarchical, feedforward ANN which boasts good generalization ability and enables it to perform deformation-invariant character recognition.

E1.3   •   The *Silicon Retina* (Mead 1989) comprises an array of photoreceptors modeled on the mammalian visual cortex and fabricated in analog VLSI form. It is capable of performing pattern recognition and motion detection operations in real time.

A comprehensive survey of ANN image processing applications is beyond the scope of this section. Instead, we restrict ourselves to a few *representative* examples: image classification/recognition, image compression, and face recognition.

F1.6.5.1 Image classification/recognition

As is common in many application areas, ANNs have commonly been used to replicate standard image processing functions, such as filtering (Pham and Bayro-Corrochano 1992). Other examples include edge detection and deblurring, as illustrated in figures F1.6.2 and F1.6.3, respectively.

Pattern classification is another typical image processing application to which ANNs have been applied. Figure F1.6.4, for example, shows the effect of varying the number of nodes in the hidden layer of an MLP used in the classification of magnetic resonance images. A different medical imaging application—the application of LVQ to ultrasonic image processing—is discussed in Kotropoulus *et al* (1994). Other interesting applications are discussed in Chapter G5 of this handbook.   G5

Autoassociative memories are particularly good at recognizing noisy or incomplete images, providing they have been sufficiently well trained beforehand. The *Hopfield network* of figure F1.6.5 is able to   C1.3.4
produce perfect recall for both noisy (32% Gaussian noise) and incomplete images presented to the network. Figure F1.6.6 shows the application of a *Boltzmann machine* to surface reconstruction in sparse (60% data   C1.4
loss) and noisy (SNR = 8.5 dB) range data.

F1.6.5.2 Image compression

ANN techniques which have been successfully applied to image compression include *multilayer perceptrons*   C1.2
(MLPs), *Kohonen self-organizing maps, competitive learning* and *learning vector quantization*.   C2.1.1, C1.1.5

The dimensionality reduction which results from applying MLPs to the encoder–decoder problem (e.g. 8:3:8), can also be exploited for image compression. For example, Cottrell *et al* (1987) compressed 8×8 pixel images using a 64-16-64 MLP. MLPs were also favored by Sonehara *et al* (1989), Arduni *et al* (1992) and Qui *et al* (1993).

Lu and Shin (1992) used a combination of an MLP classifier together with Kohonen self-organizing maps for the generation of codebook vectors. Thacore *et al* (1988), and Krovi and Pracht (1991), on the other hand, used self-organizing networks exclusively. Dunstone and Andrew (1994) reported success using a specialized 'surface learning network'.

F1.6.5.3 Face recognition

Early attempts at the automatic recognition of human faces focused on distance measures (Goldstein *et al* 1971, Kaya and Kobayashi 1972, Nixon 1985). In more recent times, limited success has been achieved using algebraic extraction methods (or principal component analysis—Kirby and Sirovich 1990, Turk and Pentland 1991) and isodensity lines (Sakaguchi *et al* 1989, Nakamura *et al* 1991). Common limitations with most approaches are the limited number of faces able to be recognized, their inability to operate in real time, and the need for consistent lighting conditions.

ANN approaches to face recognition include standard architectures like the MLP (Cottrell and Fleming 1990, Perry and Carney 1990, Kosugi 1991), as well as specialized networks, such as Von der Malsburg's dynamic link architecture, which uses multiple Kohonen SOMs (Konen *et al* 1994). Crowley (1994) likewise favored a local rather than a global approach to the problem. The WISARD system mentioned earlier has also been applied to face recognition.

F1.6.6 Conclusion

ANNs are being used not only to replicate standard image processing functions, but also as *alternative* approaches to image enhancement, pattern classification, feature extraction, object recognition, computer vision and similar image processing tasks.

Acknowledgements

This work was a direct outcome of the Face Recognition Research Project for Airport Security, sponsored by Société Internationale de Telecommunications Aeronautiques Société Cooperative (SITA). Financial assistance was also provided by the Australian Telecommunications Research Board ATERB through grant no N032/185 and by the Intelligent Systems Research Group within the University of Wollongong.

References

Aleksander I, Thomas W and Bowden P 1984 WISARD, a radical new step forward in image recognition *Sensor Rev.* **4** 120–4

Arduni F, Fioravanti S and Giusto D 1992 Adaptive image coding using multilayer neural networks *Proc. ICASSP* **2** 381–4

Bhatia P, Srinivasan V and Ong S 1991 Single-layer edge detector with competitive unsupervised learning *Proc. Int. Joint Conf. on Neural Networks (Singapore)* vol I pp 634–9

Cottrell G, Munro P and Zipser D 1987 Image compression by backpropagation: an example of extensional programming *ICS Report 8702* University of California, San Diego, CA

Cottrell W and Fleming M 1990 Categorization of faces using unsupervised feature extraction *Proc. Int. Joint Conf. on Neural Networks (San Diego, CA)* vol II pp 65–70

Crowley J 1994 A local feature based human face recognition system *Proc. 2nd Australian and New Zealand Conf. on Intelligent Information Systems (Brisbane)*

Dhawan A and Dufresne T 1990 Low-level image processing and edge enhancement using a self organising neural network *Proc. Int. Joint Conf. on Neural Networks (San Diego, CA)* vol I pp 503–10

Dunstone E and Andrew J 1994 Super-high scale invariant image compression using a surface learning neural network *Proc. IEEE Int. Symp. on Speech, Image Processing and Neural Networks (Hong Kong)* pp 397–400

Fukushima K and Wake N 1991 Handwritten alphanumeric character recognition by the neocognitron *IEEE Trans. Neural Networks* **2** 355–65

Goldstein R, Harmon L and Lesk A 1971 Identification of human faces *Proc. IEEE* **597** 48–60

Kaya Y and Kobayashi K 1972 A basic study on human face recognition *Frontiers of Pattern Recognition* ed S Watanabe (New York: Academic) pp 265–89

Kirby M and Sirovich L 1990 Application of the Karhunen–Loeve procedure for the characterisation of human faces *IEEE Trans. Patt. Anal. Mach. Intell.* **12** 103–8

Konen W, Maurer T and Von der Malsburg C 1994 A fast dynamic link matching algorithm for invariant pattern recognition *Neural Networks* **7** 1019–30

Kosugi M 1991 Human-face identification using mosiac pattern and BPN *Proc. ACNN'91 (Sydney)* vol II pp 111–4

Kotropoulos C, Magnisalis X, Pitas I and Strintzis M 1994 Nonlinear ultrasonic image processing based on signal-adaptive filters and self-organizing neural networks *IEEE Trans. Image Proc.* **3** 65–77

Krovi R and Pracht W 1991 Feasability of self organisation in image compression *Proc. IEEE/ACM Int. Conf. on Developing and Managing Expert System Programs* pp 210–4

Kulkarni A 1994 *Artificial Neural Networks for Image Understanding* (New York: Van Nostrand-Reinhold)

Linggard R, Myers D and Nightingale C (eds) 1992 Neural networks for vision, speech and natural language (London: Chapman and Hall)

Lu C-C and Shin Y-H 1992 A neural network based image compression system *IEEE Trans. Consumer Electronics* **38** 25–9

Mead C 1989 *Analog VLSI and Neural Systems* (Reading, MA: Addison-Wesley)

Muller K-R, Stiefvater T and Lanben H 1993 Associative storage and retrieval of highly correlated natural pattern sets in diluted hopfield networks *Proc. Int. Joint Conf. on Neural Networks (San Francisco, CA)* vol II pp 889–94

Mundkur P, Kapoor S and Desai U 1992 Boltzmann machines for depth recovery using a MRF model *Proc. Int. Joint Conf. on Neural Networks (Baltimore)* vol III pp 260–5

Nakamura O, Mathur S and Minami T 1991 Identification of human faces based on isodensity lines *Patt. Recog.* **24** 263–72

Nixon M 1985 Eye spacing measurement for facial recognition *Proc. SPIE—Applications of Digital Image Processing VIII* **575** 413–6

Ozkan M, Hendrik G, Sprenkels M and Dawant B 1990 Multi-spectral magnetic resonance image segmentation using neural networks *Proc. Int. Joint Conf. on Neural Networks (San Diego, CA)* vol I pp 429–34

Perry J and Carney J 1990 Human face recognition using a multilayer perceptron *Proc. Int. Joint Conf. on Neural Networks (Washington)* vol II pp 413–6

Pham D and Bayro-Corrochano E 1992 Neural computing for noise filtering, edge detection and signature extraction *J. Syst. Eng.* **2** 111–22

Qiu G, Varley M and Terrell T 1993 Image compression by edge pattern learning using multilayer perceptrons *Electron. Lett.* **29** 601–2

Sakaguchi T, Nakamura O and Minami T 1989 Personal identification through facial images using isodensity lines *Proc. SPIE—Visual Communications and Image Processing IV* **1199** 643–54

Sonehara N, Kawato M, Miyake S and Nakane K 1989 Image data compression using a neural network model *Proc. Int. Joint Conf. on Neural Networks* vol II pp 35–41

Thacore S, Pang V, Palaniswami M and Bairaktaris D 1988 Image data compression using a self-organising neural network with adaptive thresholds *Proc. IEEE Int. Joint Conf. on Neural Networks* vol I pp 646–51

Thompson C and Shure L 1993 Image Processing Toolbox for use with Matlab (Natick, MA: The MathWorks Inc)

Turk M and Pentland A 1991 Eigenfaces for recognition *J. Cog. Neurosci.* **3** 71–86

F1.7 Speech processing

Kari Torkkola

Abstract

Speech processing comprises automatic speech recognition, speech synthesis, speech coding, speech enhancement, speaker recognition and verification, language identification, and so on. This section discusses the application of artificial neural networks (ANNs) to these areas. The viewpoint will be that of an engineer; that is, the question we try to answer is, 'How can ANNs be used to solve engineering problems in speech processing?' We will present some conventional approaches to these problems and point out where ANNs could be applicable. As a lot of the ANN effort in speech processing seems to be concentrated around speech recognition, this will also be our focal point. Other areas will be briefly reviewed. Due to the breadth of the field and space limitations, this section can only remain superficial: more of a commented list of bibliographic references.

F1.7.1 Introduction

Speech is a medium for *communication*, and there is always a *language* behind it. While some speech processing applications can be regarded as pure 'signal processing', one usually cannot avoid taking into account that speech is a signal produced by human articulators. Furthermore, it may also be necessary to incorporate knowledge of the language to reach the best solutions. Thus, in addition to engineering, a successful speech processing application might need a combination of speech, hearing, and language sciences.

Many kinds of characteristics of the speech signal can be learned by automated procedures using large databases. Statistical methods (including ANNs) rely on this fact. Some knowledge, especially linguistic, still needs to be obtained through manual coding and some needs to be taken into account implicitly. For instance, knowledge may be incorporated in the structure of a speech recognizer or a speech coder.

Engineering problems in speech communication can roughly be placed in two categories: man-to-man communication and man–machine communication. Examples of the former category include speech coding for transmission and storage, speech enhancement, and speaker separation. Automatic speech recognition (ASR), speech synthesis, speaker identification and verification, and language identification would go in the latter category. Besides speech communication, another super-category is speech analysis, some aspects of which are necessary in every speech processing application. As here we can only touch the surface of these areas, the reader is encouraged to consult O'Shaughnessy (1987), Keller (1994), or Rabiner and Juang (1993) for background in speech processing. Some texts which concentrate on connectionist aspects of speech processing include Morgan and Scofield (1991), Bourlard and Morgan (1994), and Robinson (1993).

Before going through the subcategories in detail, we will take a quick look at some of the generic capabilities of ANNs, and how they match problems in speech processing. It is well known that under some ideal conditions *multilayer perceptrons* (MLP) are universal approximators (Hornik *et al* 1989). There are C1.2 many *function approximation* (or relation approximation) tasks in speech processing, for example, many kinds of mappings in speech synthesis, probability estimation tasks in ASR, mappings for noise reduction, and nonlinear prediction of speech. *Pattern classification* is another prominent capability of ANNs. F1.2

This is needed in ASR, though due to the sequential and dynamic nature of the speech signal, static pattern classifiers are insufficient except for the simplest problems. This has resulted in research in hybrid methods, where ANNs are combined with more traditional sequence processing methods including hidden Markov models (HMM) and dynamic time warping (DTW) (Rabiner and Juang 1993). The vast majority of speech processing applications only take advantage of the two above-mentioned capabilities of ANNs. *Optimization* capabilities of ANNs have been used in some search problems, like in the Viterbi search in F1.5 ASR, and in codebook search for speech coding. *Data compression* capabilities of ANNs have not been used directly in speech coding very often; rather, ANNs appear as components in traditional speech coding methods.

F1.7.2 Speech recognition

Automatic speech recognition is one of the 'grand challenges' in engineering. In essence, the purpose is to find the linguistic content in a spoken message. The problem is made difficult by the *variability* in the speech signal: there are variations in the talking speed, enormous variations between the vocal tract characteristics of different people and consequently in the spectral characteristics of the uttered speech, there are dialectal variations, variations due to the origin of the speaker, and so on.

A speech recognition device typically consists of a feature extraction module, a pattern matching and time warping module which uses an inventory of speech subunit models, and a language processing module, whose purpose is to reduce the search space of the pattern matcher according to a (limited) language to be recognized. Speech recognizers typically tackle the variability by statistical methods, most notably by using hidden Markov models (HMM) (Rabiner and Juang 1993). HMMs provide both the capability to absorb temporal variations and to model the speech variability at the feature vector level. In addition, they are trainable from large databases. Dynamic time warping (DTW) is another related method to match a speech feature vector sequence against a set of models (Rabiner and Juang 1993). Both DTW and the Viterbi algorithm (Rabiner and Juang 1993), which aligns speech with a set of HMMs, are instances of dynamic programming algorithms.

One of the subtasks in speech recognition is *pattern classification* (Lippmann and Gold 1987, Leung and Zue 1988, Makhoul 1991). ANNs are especially amenable to this, because many classification tasks require the construction of complex decision surfaces. In an extension to this classifying function, one can use artificial neural networks to estimate the posterior probabilities for the classes. This property permits the use of a network as a component in a system that incorporates other probabilistic evidence, such as an HMM system (Bourlard *et al* 1992).

In the case of speech, input to the network may be a short-time spectral representation of speech, which may include some context. A problem with this kind of a scheme is that the input, if it includes context, is a fixed-time window without any possibility for alignment or time stretching.

To avoid the problem of fixed-time windows, one can introduce feedback into the network graph. This allows the network to keep information about past inputs for an amount of time that is not fixed *a* C1.2.3 *priori*, but that depends on weights and on the input data. Variations of the *backpropagation algorithm* have been developed to train this kind of *recurrent* ANN (Watrous and Shastri 1987, Robinson 1989).

C1.2.8.3 So-called *time-delay neural networks* (TDNNs) are MLPs that approximate recurrent networks (Waibel *et al* 1989a, 1989b). Instead of a feedback connection from a unit to itself, a fixed number of delayed previous activation values are stored in a shift register, and they are all connected to the units in the above layer (a signal processing analogy is used to approximate an IIR filter by an FIR filter).

In addition to MLPs many other ANN architectures are useful for classification tasks. Two examples C1.1.5 that can be mentioned briefly are *learning vector quantization* (LVQ) (Kohonen *et al* 1988), which is an C1.6.2 algorithm to train a two-layer network for optimum discrimination between pattern classes, and *radial basis function* (RBF) networks (Moody and Darken 1989).

Some ANN architectures are appropriate for producing new kinds of representations of complex C2.1.1 data, like speech data. An example of such a network is the *Kohonen network* which is also called the *self-organizing map* (SOM) (Kohonen 1990, 1995). This kind of network organizes itself automatically by so-called competitive learning, according to the structure of the input data (unsupervised training). Incoming speech can be mapped as the path of best-responding cells of the SOM. Such a mapping can be used as a basis of speech recognition (Kohonen 1988, Torkkola and Kokkonen 1991). In addition, SOMs can also be used to illustrate, analyze and characterize speech (Tognieri *et al* 1992). One application of this is to diagnose phonation disorders (Leinonen *et al* 1992).

F1.7.2.1 How to cope with the time-sequential nature of speech: hybrid recognizers

Although ANNs with delays and recurrent connections can, in theory, model any temporal structure, current architectures are inefficient in capturing some important aspects of temporal structure. Pure ANNs are currently at their best when recognizing short utterances or smaller speech subunits. A suitable ANN architecture to do the time-matching procedure still remains to be found. One attempt in this direction which replaces the Viterbi algorithm by using the optimization capabilities of Hopfield networks is presented in Aiyer and Fallside (1992). However, the mapping ability of ANNs can be used to derive new representations of the speech signal. Combinations of ANNs with other tools that have been proven useful in modeling the temporal structure of speech are thus of interest. Such combinations, in particular with DTW and HMMs, are discussed next.

One possible hybrid configuration is to use ANNs *instead of* vector quantizers. While traditional vector quantizers aim to *represent* speech parameter vectors with minimum distortion, ANNs can be trained to *discriminate* specific features relevant to the task. For example, if the task is phoneme recognition, LVQ-type networks can be trained as frame-level or segment-level phoneme classifiers. These ANNs then provide a stream of information to HMMs consisting of phoneme labels (Iwamida *et al* 1990, Torkkola *et al* 1991), or distance information (Schmidbauer and Tebelskis 1992, Torkkola 1994). Since HMMs can combine the outputs of independent streams, information from other parallel networks computing some other relevant aspects of the task can be integrated (Mäntysalo *et al* 1994, Le Cerf *et al* 1994). The topology-preserving properties of SOMs have also been useful together with discrete observation HMMs (Zhao 1992, Monte *et al* 1992).

Another hybrid approach is the use of multilayer perceptrons (MLPs) (Bourlard *et al* 1992, Bourlard and Morgan 1994), recurrent networks (Robinson 1994), or radial basis function networks (Singer and Lippmann 1992) as *discriminant local probability generators* for HMMs, instead of using, for example, mixtures of Gaussians to generate observation probabilities. The training of HMMs is reduced to training transition probabilities. This kind of hybrid combines several advantages of ANNs and HMMs: HMMs furnish their temporal processing abilities and provide embedded training procedures (thus obviating the need to segment training data), while ANNs are employed for their strong discriminative abilities, and to eliminate the need to formulate assumptions about likely observation probability densities. Further, when an ANN uses context in addition to its current frame, the correlations between consecutive acoustic observations can be taken into account (factors that are ignored in pure HMMs). MLPs can also be used as *local distance generators* for DTW. In this architecture, the discrimination power of MLPs is combined with the time alignment abilities of DTW for word recognition (Sakoe *et al* 1989).

One step further from an ANN–DTW hybrid is to construct *subword models* by ANNs. The idea is to find out which concatenation of these subword models best matches incoming speech using either DTW or Viterbi-related search. So-called multistate TDNNs model the speech as a concatenation of TDNN phone models (Haffner *et al* 1991, Haffner 1992, Tebelskis and Waibel 1993). *Segmental* models classify or model entire segments of speech, instead of short-time observations. It is possible to use MLPs or LVQ as the basis of such models (Leung *et al* 1992, Cheng *et al* 1992, Austin *et al* 1992). This is one way to overcome the HMM assumption of independence between successive observations: to construct models that take a longer duration of speech signal into account. *Predictive* subword models aim at being able to predict the next frame of a particular subword unit (a phone, for example). MLPs can be used as such predictors using input that includes the phonetic context (Levin 1990, Iso and Watanabe 1991, Tebelskis *et al* 1991, Mellouk and Gallinari 1994). However, if a predictor for each speech unit is trained using examples of that particular speech unit only, this approach may result in poor discrimination.

From a theoretical point of view, it has been shown that HMMs are a specific instance of a certain type of recurrent ANN (Bridle 1990). In this case, the forward–backward algorithm is equivalent to backpropagation. Inspired by *error-driven training methods* developed for ANNs, several researchers have applied the same philosophy to HMM training, as well as to the training of ANN–HMM and ANN–DTW hybrids. This involves finding a suitable cost or error function whose derivatives with respect to parameters can be easily computed. Gradient descent can then be used to minimize this function. For example, a cost function based on a maximum mutual information criterion is used to train pure HMMs (Young 1991). Furthermore, it is shown in Driancourt and Gallinari (1992) how different kinds of ANN modules can be combined with DTW and trained using similar principles. This training could even be extended to DTW templates. Similarly, ANN and HMM parameters can be trained within the same framework as presented in Bengio *et al* (1992). Also, the work of Juang and others on error-correcting learning and generalized probabilistic descent (Juang and Katagiri 1992) is close to this spirit although ANNs are not directly involved.

To account for the large variability between speakers, recognition systems intended for *speaker-independent* usage are trained using large databases. However, if such a system will mainly be used by a single individual for some period of time, the existing models of speech subunits within the system can be adapted to the particular speaker to improve the performance (Lee *et al* 1991, Cox 1995). For some ANN-based architectures, adaptation of only a relatively small number of parameters may be sufficient (Schmidbauer and Tebelskis 1992, Hild and Waibel 1993). ANNs can also be used to construct a normalizing mapping for the new speaker (Watrous 1993).

There has not yet been much work in language modeling using ANNs. Currently, the best working language models are statistical (Jelinek *et al* 1992). Their purpose is to predict the next spoken word on the basis of the previous ones. As this is again an approximation task it may come as no surprise that MLPs have also been applied here. Using MLP-based word category prediction, better word recognition scores were reported than by using a standard trigam language model (Nakamura and Shikano 1989).

F1.7.3 Speaker identification and verification

Speaker identification and verification differ in the number of decision alternatives. Identification involves determining the identity of the speaker from a prespecified pool of speakers. Speaker verification entails either accepting or rejecting the claim on the speaker's identity. A good review of the current technology is given in Furui (1994).

As in any pattern recognition problem, feature extraction is followed by similarity comparison to speaker models, either to the whole pool (identification) followed by maximum selection, or only to the model of the claimed speaker (verification) followed by thresholding. Often, the models and the comparison methods are related to speech recognition algorithms. For example, in the text-dependent case the models could be just stored words of each speaker, against which the same word uttered by an unknown speaker is compared using dynamic time warping. Or in the text-independent case the models can be hidden Markov models of phonemes that are concatenated according to prompted text, after which the likelihood of the utterance having been generated by a particular speaker's model can be evaluated. One can also construct a mixture Gaussian model for each speaker (Reynolds 1994). In this case the temporal modeling aspect of speech (as in ASR) can be avoided.

A good overview of ANNs applied to these problems is given in Bennani and Gallinari (1994). A straightforward approach is that of pattern recognition: to train a single discriminative network for the speaker recognition problem (Bennani *et al* 1990). Adding new speakers, however, requires retraining the whole network. The next obvious approach is to model each speaker by an ANN, be it an MLP or RBF network (Oglesby and Mason 1990, 1991, Tsoi *et al* 1994). These two approaches work while the speaker population is small. For a larger number of speakers modular approaches have been proposed (Bennani 1993). Modeling a speaker by a predictive ANN system also allows new speakers to be added easily, as not all of the networks need to be retrained (Hattori 1994).

So far, it is not clear whether ANN-related methods have any edge over traditional ones, because comparisons on the same realistic task have not been performed.

F1.7.4 Language identification

Language identification is a classification problem, in which the difficulty lies in extracting suitable features from an utterance as the basis for this classification. Muthusamy and Cole (1992) describe a system that performs a broad class phonetic segmentation using an MLP, then derives various features from a sequence of the class labels, and performs the final classification using another MLP (Muthusamy and Cole 1992). The system has been developed further by replacing the broad classifier by a phonetic classifier (Berkling *et al* 1994). Making use of linguistic knowledge is essential here to determine what kinds of features to use for the final language classification.

As an example of a non-ANN system, Zissman describes a statistical model, where speakers of each language are modeled as a mixture of Gaussians. This configuration seems to work as well as the ANN-based systems (Zissman 1993). Again, as to which approach is better, there is no answer yet.

F1.7.5 Speech synthesis

The classic approach to speech synthesis from text is synthesis by rule (Klatt 1987). This involves rules to map text into phonemes, phonemes to allophones, and allophones to control parameters of a sound generator, which may be waveform concatenation, a formant synthesizer, or an articulatory model. These mappings are extremely complex, and derivation of good sets of rules or other non-rule-based mappings is a tedious task.

The most famous example in this area is NETTALK (Sejnowski and Rosenberg 1987). An MLP was used to implement a text-to-phoneme mapping. Though the performance fell short of hand-crafted rules (just using a dictionary is even better), NETTALK was a powerful and driving demonstration of the capabilities of ANNs.

It is possible to use ANNs for other mappings, too. For instance, several researchers have used MLPs with or without recurrent connections to generate prosody-related parameters either from syllable, phoneme, or allophone representations. Examples of these parameters include pitch contours (Scordilis and Gowdy 1989, Sagisaka 1990, Traber 1992) and allophone durations (Scordilis and Gowdy 1990, Karjalainen 1991). NETTALK also generated stress-related parameters in addition to phoneme identities. Articulatory analysis/synthesis requires mapping of the speech signal or its spectra onto vocal tract shapes or other geometrical parameters. For this kind of work by ANNs see Rahim *et al* (1991) and Kobayashi *et al* (1991).

Cohen and Bishop (1994) argue that the whole approach of synthesis-by-rule using linguistic notations is incorrect (as demonstrated by the failure of NETTALK). They propose to use self-organizing maps to derive new diphone-based subsymbolic intermediate forms.

Another type of mapping is described by Fels and Hinton (1993, 1995). They implemented a mapping from hand gestures and hand movements to speech synthesizer parameters using several MLPs.

Synthesis of *intelligible* speech is very viable today, but *natural sounding* speech from text seems to require far more use of linguistic knowledge than current synthesizers use. This is basically the same problem as with speech recognition: human knowledge about language is hard to capture and exploit.

F1.7.6 Speech coding

To be successful, low-bit-rate high-quality speech coding must involve taking into account the characteristics of the speech signal, like the fact that the speech signal stays stationary over short periods of time (about 25 ms). As a baseline, we describe the CELP coder (code-excited linear prediction) (Shroeder and Atal 1985). This coder typically involves linear waveform predictors to remove dependencies in the speech signal, both long-term (pitch) and short-term (stationarity). The excitation signal is then vector quantized and transmitted. CELP is 'codebook excited', because the best codebook vector is searched by reconstructing the signal through the predictors. The codebook vector resulting in the best match between the original and the reconstruction, perhaps through a perceptually weighted error criterion, is chosen and its code is transmitted.

Recurrent networks have been used as nonlinear predictors in CELP resulting in better-quality speech at low bit rates (Wu *et al* 1993). TDNNs can also be used for the same purpose (Thyssen *et al* 1994). In this work it was noticed that pitch can be predicted by a nonlinear short-term predictor, instead of a linear long-term predictor. Experiments using hierarchical mixtures of experts as predictors of acoustic vectors are presented in Waterhouse and Robinson (1995).

Vector quantizers can also be replaced by ANNs. It has been shown by Wang and Hanson (1993) and Li *et al* (1994) how a codebook search can be eliminated. They used an MLP to map an input vector directly to a code to be transmitted. A way to use the optimization capabilities of a Hopfield network to perform codebook search resulting in computational savings is presented in Easton and Goodyear (1991). It is also possible to make use of the topological organization in a SOM-trained codebook to reduce the bit rates (Hernández-Gomez and López-Gonzalo 1993).

Switched coders first try to classify the short-time speech frames into a small number of classes, and then use an appropriate coding scheme for each class. Here, the ANNs can be used to classify the frames, for example, as voiced, unvoiced, or silence (Bendiksen and Steiglitz 1990, Cohn 1991, Ghiselli-Crippa and El-Jaroudi 1991).

So far, we have only discussed *source coding*. *Channel coding* entails wrapping the compressed speech bits into a suitable error correction and choosing a modulation scheme. These three steps have traditionally been done separately. Codebook design, error protection, and QAM modulation can be considered jointly using a SOM (Skinnemoen and Perkis 1994). By choosing the QAM modulation scheme corresponding to the topological organization of the SOM codebook, channel errors result in decoded vectors that are similar to the error-free ones. Considerable noise robustness was obtained using this scheme.

F1.7.7 Speech enhancement

Speech enhancement involves separation of the speech signal from unwanted noise, which can be a speech signal of another speaker, or more often, environmental noise such as car noise.

The function approximation capabilities of MLPs have found use in this problem, too. MLPs have been applied directly in the time domain to create a mapping from a noisy signal to a noiseless one (Tamura and Waibel 1988, Tamura and Nakamura 1990). Many researchers have done the same in the feature vector domain, either spectral or cepstral (Sorensen 1991, Barbier and Chollet 1991, Sorensen 1992, Trompf 1992). Training the noise removal network jointly with the speech recognizer improves results over those obtained by training the networks separately (Moon and Hwang 1993, Gao and Haton 1994).

As these methods basically remove *stationary* noise, an adaptive version is presented in Xie and Compernolle (1994), and in Sorensen and Hartmann (1994) an extension to an HMM decomposition of speech and noise using radial basis function networks is presented.

ANNs have also been used in blind separation problems (so-called cocktail party problems) (Jutten and Herault 1991, Burel 1992, Wang *et al* 1995). In Bell and Sejnowski (1995) impressive results are presented with a network that is based on maximizing the information transferred through the nonlinearities of the network.

F1.7.8 Discussion

In most of the reviewed applications ANNs act as nonlinear approximators, and it is not clear whether in this role they offer an advantage over modern statistical tools (Ripley 1993, Sarle 1994). In many cases all that is being done is parameter estimation for nonlinear models. However, there are good reasons for using ANNs, as mentioned in Bourlard and Morgan (1994, p 88) (some of these properties apply, of course, to statistical methods, such as HMMs):

- ANNs can learn
- discriminative training can be used with ANNs
- ANNs have universal approximation capabilities
- ANNs can combine disparate data (e.g. symbolic and real-valued data)
- no strong assumptions about the statistics of the input data need be made
- some ANNs exhibit properties that cannot be replicated using any other methods
- some ANN architectures are amenable to parallel hardware implementations.

Where one or several of these functional properties match the problem domain, it seems that ANNs are capable of adding some desired properties to an existing system. This, in turn, results in improved performance.

How to make use of linguistic knowledge remains a substantial problem in many speech processing fields. This problem is most compelling in research for ASR and natural sounding speech synthesis, and remains a topic for active research, both in the traditional and ANN domains.

References

Aiyer S V B and Fallside F 1992 A Hopfield network implementation of the Viterbi algorithm for hidden Markov models *Technical Report CUED/F-INFENG/TR60* Cambridge University Engineering Department

Austin S, Zavaliagkos G, Makhoul J and Schwartz R 1992 Speech recognition using segmental neural nets *Proc. IEEE Int. Conf. on Acoustics, Speech and Signal Processing (ICASSP92) (San Francisco, CA)* vol I pp 625–8

Barbier L and Chollet G 1991 Robust speech parameters extraxtion for word recognition in noise using neural networks *Proc. IEEE Int. Conf. on Acoustics, Speech and Signal Processing (ICASSP91) (Toronto)* pp 145–8

Bell A J and Sejnowski T J 1995 An information-maximisation approach to blind separation and blind deconvolution *Neural Comput.* **7** 1129–59

Bendiksen A and Steiglitz K 1990 Neural networks for voiced/unvoiced speech classification *Proc. IEEE Int. Conf. on Acoustics, Speech and Signal Processing (ICASSP90) (Albuquerque, NM)* pp 521–4

Bengio Y, de Mori R, Flammia G and Kompe R 1992 Global optimization of a neural network—hidden Markov model hybrids *IEEE Trans. Neural Networks* **3** 252–9

Bennani Y 1993 Probabilistic cooperation of connectionist expert modules: validation on a speaker identification task *Proc. IEEE Int. Conf. on Acoustics, Speech and Signal Processing (ICASSP93) (Minneapolis, MN)* vol I pp 541–4

Bennani Y and Gallinari P 1994 Connectionist approaches for automatic speaker recognition *Proc. ESCA Workshop on Automatic Speaker Recognition Identification Verification (Martigny)* (European Speech Communication Association) pp 95–102

Bennani Y, Soulie F F and Gallinari P 1990 A connectionist approach for automatic speaker identification *Proc. IEEE Int. Conf. on Acoustics, Speech and Signal Processing (ICASSP90) (Albuquerque, NM)* pp 265–8

Berkling K M, Arai T and Barnard E 1994 Analysis of phoneme-based features for language identification *Proc. IEEE Int. Conf. on Acoustics, Speech and Signal Processing (ICASSP94) (Adelaide)* vol I pp 289–92

Bourlard H and Morgan N 1994 *Connectionist Speech Recognition: a Hybrid Approach* (Boston, MA: Kluwer)

Bourlard H, Morgan N and Renals S 1992 Neural nets and hidden Markov models: review and generalizations *Speech Commun.* **11** 237–46

Bridle J S 1990 Alphanets: a recurrent neural network architecture with a hidden Markov model interpretation *Speech Commun.* **9** 83–92

Burel G 1992 Blind separation of sources: a nonlinear neural algorithm *Neural Networks* **5** 937–47

Cheng Y, O'Shaughnessy D, Gupta V, Kenny P, Lenning M, Mermelstein P and Parthasarathy S 1992 Hybrid segmental-LVQ/HMM for large vocabulary speech recognition *Proc. IEEE Int. Conf. on Acoustics, Speech and Signal Processing (ICASSP92) (San Francisco, CA)* vol I pp 593–6

Cohen A D J and Bishop M J 1994 Self-organizing maps in synthetic speech *Proc. World Congress on Neural Networks (San Diego, CA)* vol 4 pp 544–9

Cohn R P 1991 Robust voiced/unvoiced speech classification using a neural net *Proc. IEEE Int. Conf. on Acoustics, Speech and Signal Processing (ICASSP91) (Toronto)* vol 1 pp 437–40

Cox S 1995 Predictive speaker adaptation in speech recognition *Comput. Speech Lang.* **9** 1–17

Driancourt X and Gallinari P 1992 A speech recognizer optimally combining learning vector quantization dynamic programming and multilayer perceptron *Proc. IEEE Int. Conf. on Acoustics, Speech and Signal Processing (ICASSP92) (San Francisco, CA)* vol I pp 609–12

Easton M G and Goodyear C C 1991 A CELP codebook and search technique using a Hopfield net *Proc. IEEE Int. Conf. on Acoustics, Speech and Signal Processing (ICASSP91) (Toronto)* vol 1 pp 685–8

Fels S S and Hinton G E 1993 Glove-talk: a neural network interface between a data-glove and a speech synthesizer *IEEE Trans. Neural Networks* **4** 2–8

——1995 Glovetalk II: mapping hand gestures to speech using neural networks *Advances in Neural Information Processing Systems* vol 7 (Cambridge, MA: MIT Press)

Furui S 1994 An overview of speaker recognition technology *Proc. ESCA Workshop on Automatic Speaker Recognition Identification Verification (Martigny)* (European Speech Communication Association) pp 1–9

Gao Y and Haton J-P 1994 A hierarchical LPNN network for noise reduction and noise degraded speech recognition *Proc. IEEE Int. Conf. on Acoustics, Speech and Signal Processing (ICASSP94) (Adelaide)* vol II pp 89–92

Ghiselli-Crippa T and El-Jaroudi A 1991 A fast neural net training algorithm and its application to voiced–unvoiced–silence classification of speech *Proc. IEEE Int. Conf. on Acoustics, Speech and Signal Processing (ICASSP91) (Toronto)* vol 1 pp 441–4

Haffner P 1992 Connectionist word-level classification in speech recognition *Proc. IEEE Int. Conf. on Acoustics, Speech and Signal Processing (ICASSP92) (San Francisco, CA)* vol I pp 621–4

Haffner P, Franzini M and Waibel A 1991 Integrating time alignment and neural networks for high performance continuous speech recognition *Proc. IEEE Int. Conf. on Acoustics, Speech and Signal Processing (ICASSP91) (Toronto)* vol 1 pp 105–8

Hattori H 1994 Text-independent speaker verification using neural networks *Proc. ESCA Workshop on Automatic Speaker Recognition Identification Verification (Martigny)* (European Speech Communication Association) pp 103–6

Hernández-Gomez L A and López-Gonzalo E 1993 Phonetically driven CELP coding using self-organizing maps *Proc. IEEE Int. Conf. on Acoustics, Speech and Signal Processing (ICASSP93) (Minneapolis, MN)* vol II pp 628–31

Hild H and Waibel A 1993 Multispeaker/speaker-independent architectures for the multistate time delay neural network *Proc. IEEE Int. Conf. on Acoustics, Speech and Signal Processing (ICASSP93) (Minneapolis, MN)* pp 255–8

Hornik K, Stinchcombe M and White H 1989 Multilayer feedforward networks are universal approximators *Neural Networks* **2** 359–66

Iso K and Watanabe T 1991 Large vocabulary speech recognition using neural prediction model *Proc. IEEE Int. Conf. on Acoustics, Speech and Signal Processing (ICASSP91) (Toronto)* vol 1 pp 57–60

Iwamida H, Katagiri S, McDermott E and Tohkura Y 1990 A hybrid speech recognition system using HMMs with an LVQ-trained codebook *Proc. IEEE Int. Conf. on Acoustics, Speech and Signal Processing (ICASSP90) (Albuquerque, NM)* vol 1 pp 489–92

Jelinek F, Mercer R L and Roukos S 1992 Principles of lexical language modelling for speech recognition *Advances in Speech Signal Processing* ed S Furui and M M Sondhi (New York: Dekker)

Juang B H and Katagiri S 1992 Discriminative learning for minimum error classification *IEEE Trans. Acoust. Speech Signal Processing* **40** 3043–54

Jutten C and Herault J 1991 Blind separation of sources part I: an adaptive algorithm based on neuromimetic architecture *Signal Processing* **24** 1–10

Karjalainen M 1991 Neural networks for prosody control in speech synthesis *Artificial Neural Networks* vol 2 *(Proc. Int. Conf. on Artificial Neural Networks (Espoo))* ed T Kohonen, K Mäkisara, O Simula and J Kangas (Amsterdam: North-Holland) pp 1641–4

Keller E (ed) 1994 *Fundamentals of Speech Synthesis and Speech Recognition* (Chichester: Wiley)

Klatt D H 1987 Review of text-to-speech conversion for English *J. Acoust. Soc. Am.* **82** 137–81

Kobayashi T, Yagyu M and Shirai K 1991 Application of neural networks to articulatory motion estimation *Proc. IEEE Int. Conf. on Acoustics, Speech and Signal Processing (ICASSP91) (Toronto)* vol 1 pp 489–92

Kohonen T 1988 The 'neural' phonetic typewriter *IEEE Comput.* **21** 11–22

——1990 The self-organizing map *Proc. IEEE* **78** 1464–80

——1995 *Self-Organizing Maps* (Berlin: Springer)

Kohonen T, Barna G and Chrisley R 1988 Statistical pattern recognition with neural networks: benchmarking studies *Proc. IEEE Int. Conf. on Neural Networks (San Diego, CA)* pp 61–8

Le Cerf P, Ma W and Van Compernolle D 1994 Multilayer perceptrons as labelers for hidden Markov models *IEEE Trans. Speech Audio Processing* **2** 185–93

Lee C H, Lin C H and Juang B H 1991 A study on speaker adaptation of the parameters of continuous density hidden Markov models *IEEE Trans. Signal Processing* **39** 806–14

Leinonen L, Kangas J, Torkkola K and Juvas A 1992 Dysphonia detected by pattern recognition of spectral composition *J. Speech Hear. Res.* **35** 287–95

Leung H, Hetherington I and Zue V 1992 Speech recognition using stochastic segmental neural networks *Proc. IEEE Int. Conf. on Acoustics, Speech and Signal Processing (ICASSP92) (San Francisco, CA)* vol I pp 613–6

Leung H C and Zue V W 1988 Some phonetic recognition experiments using artificial neural nets *Proc. IEEE Int. Conf. on Acoustics, Speech and Signal Processing (ICASSP88) (New York)* pp 422–5

Levin E 1990 Word recognition using hidden control neural architecture *Proc. IEEE Int. Conf. on Acoustics, Speech and Signal Processing (ICASSP90) (Albuquerque, NM)* vol 1 pp 433–6

Li X, Bodruzzaman M and Szu H 1994 Neural network codebook search for digital speech synthesis *Proc. World Congress on Neural Networks (San Diego, CA)* vol 4 pp 512–7

Lippmann R P and Gold B 1987 Neural-net classifiers useful for speech recognition *Proc. 1st Int. Conf. on Neural Networks (San Diego, CA)* vol IV pp 417–25

Makhoul J 1991 Pattern recognition proprerties of neural networks *Neural Networks for Signal Processing (Proc. 1991 IEEE Workshop)* (New York: IEEE) pp 173–87

Mäntysalo J, Torkkola K and Kohonen T 1994 Mapping context dependent acoustic information into context independent form by LVQ *Speech Commun.* **14** 119–30

Mellouk A and Gallinari P 1994 Discriminative training for improved neural prediction systems *Proc. IEEE Int. Conf. on Acoustics, Speech and Signal Processing (ICASSP94) (Adelaide)* vol I pp 233–6

Monte E, Mariño J B and Leida E L 1992 Smoothing hidden Markov Models by means of a self-organizing feature map *Proc. ICSLP'92 Int. Conf. on Spoken Language Processing (Alberta)* vol 1 pp 551–4

Moody J and Darken C 1989 Fast learning in networks of locally tuned processing units *Neural Comput.* **1** 281–94

Moon S and Hwang J-N 1993 Coordinated training of noise removing networks *Proc. IEEE Int. Conf. on Acoustics, Speech and Signal Processing (ICASSP93) (Minneapolis, MN)* vol I pp 573–6

Morgan D P and Scofield C L 1991 *Neural Networks and Speech Processing* (Boston, MA: Kluwer)

Muthusamy Y K and Cole R A 1992 A segment-based automatic language identification system *Advances in Neural Information Processing Systems* vol 4 ed J E Moody S J Hanson and R P Lippmann (San Mateo, CA: Morgan Kaufmann)

Nakamura M and Shikano K 1989 A study of English word category prediction based on neural networks *Proc. IEEE 1989 Int. Conf. on Acoustics, Speech and Signal Processing (ICASSP89) (Glasgow)* pp 731–4

Oglesby J and Mason J S 1990 Optimisation of neural models for speaker identification *Proc. IEEE Int. Conf. on Acoustics, Speech and Signal Processing (ICASSP90) (Albuquerque, NM)* pp 261–4

——1991 Radial basis function networks for speaker recognition *Proc. IEEE Int. Conf. on Acoustics, Speech and Signal Processing (ICASSP91) (Toronto)* vol 1 pp 393–6

O'Shaughnessy D 1987 *Speech Communication* (London: Addison-Wesley)

Rabiner L and Juang B-H 1993 *Fundamentals of Speech Recognition* (Englewood Cliffs, NJ: Prentice Hall)

Rahim M G, Kleijn W B, Schroeter J and Goodyear C C 1991 Acoustic to articulatory parameter mapping using an assembly of neural networks *Proc. IEEE Int. Conf. on Acoustics, Speech and Signal Processing (ICASSP91) (Toronto)* vol 1 pp 485–8

Reynolds D A 1994 Speaker identification and verification using Gaussian mixture speaker models *Proc. ESCA Workshop on Automatic Speaker Recognition Identification Verification (Martigny)* (European Speech Communication Association) pp 27–30

Ripley B D 1993 Statistical aspects of neural networks *Networks and Chaos–Statistical and Probabilistic Aspects* ed O Barndorff-Nielsen, J Jensen and W Kendall (London: Chapman and Hall) pp 40–123

Robinson A J 1989 Dynamic error propagation networks *PhD Thesis* Cambridge University Engineering Department, Trumpington Street, Cambridge, UK

——1994 An application of recurrent nets to phone probability estimation *IEEE Trans. Neural Networks* **5** 298–305

Robinson T 1993 Artificial neural networks: the mole-grips of the speech scientist *Visual Representations of Speech Signals* ed M Cooke, S Beet and M Crawford (New York: Wiley) pp 83–94

Sagisaka Y 1990 On the prediction of global F0 shape for Japanese text-to-speech *Proc. IEEE Int. Conf. on Acoustics, Speech and Signal Processing (ICASSP90) (Albuquerque, NM)* vol 1 pp 325–8

Sakoe H, Isotani R, Yoshida K, Iso K and Watanabe T 1989 Speaker-independent word recognition using dynamic programming neural networks *Proc. IEEE 1989 Int. Conf. on Acoustics, Speech and Signal Processing (ICASSP89) (Glasgow)*

Sarle W S 1994 Neural networks and statistical models *Proc. 19th Ann. SAS Users Group Int. Conf.* (SAS Institute, Cary, NC) pp 1538–50

Schmidbauer O and Tebelskis J 1992 An LVQ based reference model for speaker-adaptive speech recognition *Proc. IEEE Int. Conf. on Acoustics, Speech and Signal Processing (ICASSP92) (San Francisco, CA)* vol I pp 441–4

Schroeder M R and Atal B S 1985 Code excited linear prediction (CELP): High-quality speech at very low bit rates *Proc. IEEE Int. Conf. on Acoustics, Speech and Signal Processing (ICASSP85) (Tampa, FL)* pp 937–40

Scordilis M and Gowdy J N 1989 Neural network-based generation of fundamental frequency contours *Proc. IEEE 1989 Int. Conf. on Acoustics, Speech and Signal Processing (ICASSP89) (Glasgow)*

——1990 Neural network control for a cascade/parallel formant synthesizer *Proc. IEEE Int. Conf. on Acoustics, Speech and Signal Processing (ICASSP90) (Albuquerque, NM)* vol 1 pp 297–300

Sejnowski T and Rosenberg C R 1987 Parallel networks that learn to pronounce english text *Complex Syst.* **1** 145–68

Singer E and Lippmann R P 1992 A speech recognizer using radial basis function neural networks in a HMM framework *Proc. IEEE Int. Conf. on Acoustics, Speech and Signal Processing (ICASSP92) (San Francisco, CA)* vol I pp 629–32

Skinnemoen H and Perkis A 1994 Efficient vector quantisation of LPC parameters for noisy channels *Proc. IEEE Int. Conf. on Acoustics, Speech and Signal Processing (ICASSP94) (Adelaide)* vol I pp 497–500

Sorensen H B D 1991 A cepstral noise reduction multilayer network *Proc. IEEE Int. Conf. on Acoustics, Speech and Signal Processing (ICASSP91) (Toronto)* pp 933–6

——1992 Speech recognition in noise using a self-structuring noise reduction model and hidden control models *Proc. Int. Joint Conf. on Neural Networks (Baltimore, MD)* (New York: IEEE) vol II pp 279–84

Sorensen H B D and Hartmann U 1994 Hybrid model decomposition of speech and noise in a radial basis function neural model framework *Proc. IEEE Int. Conf. on Acoustics, Speech and Signal Processing (ICASSP94) (Adelaide)* vol II pp 657–60

Tamura S and Nakamura M 1990 Improvements to the noise reduction neural network *Proc. IEEE Int. Conf. on Acoustics, Speech and Signal Processing (ICASSP90) (Albuquerque, NM)* pp 825–8

Tamura S and Waibel A 1988 Noise reduction using connectionist models *Proc. IEEE Int. Conf. on Acoustics, Speech and Signal Processing (ICASSP88) (New York)* pp 553–6

Tebelskis J and Waibel A 1993 Performance through consistency: MS-TDNN's for large vocabulary speech recognition *Advances in Neural Information Processing Systems* vol 5 (San Mateo, CA: Morgan Kaufmann) pp 696–703

Tebelskis J, Waibel A, Petek B and Schmidbauer O 1991 Continuous speech recognition using linked predictive neural networks *Proc. IEEE Int. Conf. on Acoustics, Speech and Signal Processing (ICASSP91) (Toronto)* vol 1 pp 61–4

Thyssen J, Nielsen H and Hansen S D 1994 Nonlinear short-term prediction in speech coding *Proc. IEEE Int. Conf. on Acoustics, Speech and Signal Processing (ICASSP94) (Adelaide)* vol I pp 185–8

Tognieri R, Alder M D and Attikiouzel Y 1992 Dimension and structure of the speech space *IEE Proc. I* **139** 123–7

Torkkola K 1994 LVQ as a feature transformation for HMMs *Neural Networks for Signal Processing IV (Proc. 1994 IEEE Workshop (Ermioni))* (New York: IEEE) pp 299–308

Torkkola K, Kangas J, Utela P, Kaski S, Kokkonen M, Kurimo M and Kohonen T 1991 Status report of the Finnish phonetic typewriter project *Proc. Int. Conf. on Artificial Neural Networks (ICANN-91) (Espoo)* (Amsterdam: North-Holland) pp 771–6

Torkkola K and Kokkonen M 1991 Using the topology-preserving properties of SOFMs in speech recognition *Proc. IEEE 1991 Int. Conf. on Acoustics, Speech and Signal Processing (ICASSP91) (Toronto)*

Traber C 1992 F0 generation with a database of natural F0 patterns and with a neural network *Talking Machines: Theories, Models and Designs* ed G Bailly and C Benoit (Amsterdam: North-Holland) pp 287–304

Trompf M 1992 Neural network development for noise reduction in robust speech recognition *Proc. Int. Joint Conf. on Neural Networks (Baltimore, MD)* vol IV (New York: IEEE) pp 722–7

Tsoi A C, Shrimpton D, Watson B and Back A 1994 Application of artificial neural network techniques to speaker verification *Proc. ESCA Workshop on Automatic Speaker Recognition Identification Verification (Martigny)* (European Speech Communication Association) pp 143–52

Waibel A, Hanazawa T, Hinton G, Shikano K and Lang K 1989a A phoneme recognition using time-delay neural networks *IEEE Trans. Acoust. Speech Signal Processing* **37** 328–39

Waibel A, Sawai H and Shikano K 1989b Consonant recognition by modular construction of large phonemic time-delay neural networks *Proc. IEEE Int. Conf. on Acoustics, Speech and Signal Processing (Glasgow)* pp 112–5

Wang L, Karhunen J, Oja E and Vigario R 1995 Blind separation of sources using nonlinear PCA type learning algorithms *Proc. ICNNSP95 (Int. Conf. on Neural Networks and Signal Processing) (Nanjing)*

Wang Z and Hanson J V 1993 Code-excited neural vector quantization *Proc. IEEE Int. Conf. on Acoustics, Speech and Signal Processing (ICASSP93) (Minneapolis, MN)* vol I pp 573–6

Waterhouse S R and Robinson A J 1995 Nonlinear prediction of acoustic vectors using hierarchical mixtures of experts *Advances in Neural Information Processing Systems* vol 7 (San Mateo, CA: Morgan Kaufmann) pp 835–42

Watrous R 1993 Speaker normalization and adaptation using second-order connectionist networks *IEEE Trans. Neural Networks* **4** 21–30

Watrous R and Shastri L 1987 Learning phonetic features using connectionist networks: an experiment in speech recognition *Proc. 1st Int. Conf. on Neural Networks (San Diego, CA)* vol 2 pp 619–27

Wu L Niranjan M and Fallside F 1993 Nonlinear predictive vector quantisation with recurrent neural nets *Neural Networks for Speech Processing III (Proc. 1993 IEEE Workshop) (Linthicum, MD)* (New York: IEEE) pp 372–81

Xie F and Compernolle D V 1994 A family of MLP based nonlinear spectral estimators for noise reduction *Proc. IEEE Int. Conf. on Acoustics Speech and Signal Processing (ICASSP94) (Adelaide)* vol II pp 53–6

Young S 1991 Competitive training: a connectionist approach to discriminative training of hidden Markov models *Proc. IEEE* **138** 61–8

Zhao Z 1992 Integration of neural networks and hidden Markov models for continuous speech recognition *Artificial Neural Networks* ed I Aleksander and J Taylor (Amsterdam: North-Holland) vol I pp 779–82

Zissman M A 1993 Automatic language identification using Gaussian mixture and hidden Markov models *Proc. IEEE Int. Conf. on Acoustics, Speech and Signal Processing (ICASSP93) (Minneapolis, MN)* vol II pp 399–402

F1.8 Signal processing

Shawn P Day

Abstract

Many neural network models possess two significant properties that often allow them
to outperform more conventional techniques in signal processing applications. Their
ability to adapt continuously to new data allows them to track changes in a system
over time, and their ability to learn arbitrary, nonlinear transfer functions permits them
to solve problems that cannot be handled adequately with more conventional adaptive
linear techniques. However, linear methods generally converge to a solution much
faster than neural networks, and they currently have a stronger theoretical foundation
for predicting their behavior. This section shows how neural networks can be used for
channel equalization, signal prediction, and noise canceling tasks.

F1.8.1 Introduction

The area of signal processing encompasses much of what neural networks have been applied to over
the past several years. Speech recognition, control, vision, image processing, pattern classification, data
compression, and time-series analysis can all be viewed as signal processing applications. Most of these
areas are thoroughly covered in other sections of the handbook, so this section will concentrate on the
types of signal processing commonly employed in areas like communication and real-time data analysis.

A key property of neural networks in signal processing applications is their ability to implement
arbitrary nonlinear transfer functions. Many signal processing techniques employed in the past have
been restricted to linear approximations of the desired solution. Conventional approaches that can handle
nonlinear problems have typically been designed using *a priori* information about the problem at hand.
Unfortunately, this type of information is not always available. Neural networks can learn to implement
nonlinear functions without any prior knowledge about the problem domain.

Another key property of neural networks in these applications is their ability to adapt continuously
to incoming data, allowing them to track changes in the system over time. Conventional techniques like
adaptive linear filtering (Widrow and Stearns 1985) can also adapt to new data, but they generally lack
the full representational power of neural network solutions.

The remainder of this section describes how continuous adaptation and the ability to learn nonlinear
transfer functions make some forms of neural networks well suited to the tasks of channel equalization,
signal prediction, and noise canceling.

F1.8.2 Neural network approaches

F1.8.2.1 Channel equalization

Many communication channels have some degree of nonlinear frequency response or nonlinear phase
response, leading to a corrupted version of the transmitted signal when it reaches the receiving end. Some
channels (e.g. mobile radio) can even have a significant time-varying component to their transfer functions.
In addition to the distortions caused by the channel transfer function, noise due to random sources may
also get added to the signal during transmission. It is the goal of adaptive equalization to remove as much
of the noise and distortion as possible to provide a clean signal at the receiving end (Quereshi 1985).

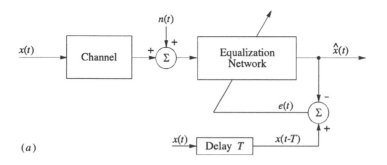

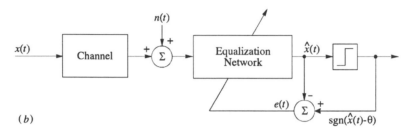

Figure F1.8.1. (a) Channel equalization. (b) Decision feedback equalization. The network in (a) periodically adapts in response to a known training signal, $x(t)$, that is transmitted over the channel. If the channel transfer function does not vary significantly over short time scales, then useful data can be transmitted between training sessions. In (b) the transmitted signal is known to be binary, so a quantized version of the neural network output can be used as the 'truth' value for adaptation (θ is an adaptable threshold for the quantization). This technique works only if the initial network weights cause it to produce the correct output a significant fraction of the time.

Figure F1.8.1(a) shows how a neural network can be trained to compensate for distortion in a communication channel. The channel imparts a distortion, and possibly an additive noise source, $n(t)$, on the input signal, $x(t)$. The neural network takes the corrupted signal and transforms it into a new signal, $\hat{x}(t)$, which is intended to be a close representation of the original signal, $x(t-T)$, with some time lag, T. One of the most common techniques for applying neural networks to this problem is to pass the signal through a series of time-delay elements, and take taps off these elements as the inputs to a conventional

C1.2 *multilayer perceptron* (Chen *et al* 1990). If the delays are each of length d, then the input to the network will be the set of signals $x(t)$, $x(t-d)$, $x(t-2d)$, etc. In the absence of noise, and for certain types of linear channel transfer functions, the network must learn to implement the inverse of the channel transfer function. For most channels, however, the inverse transfer function is not optimal for reducing the error between $\hat{x}(t)$ and $x(t-T)$.

During adaptation, the difference between the output signal, $\hat{x}(t)$, and a delayed version of the input signal, $x(t-T)$, forms an error signal, $e(t)$. For multilayer perceptrons, the backpropagation algorithm can be employed to minimize $e^2(t)$. As the network adapts to reduce the squared error, it learns to reproduce $x(t-T)$ using observations of the corrupted signal up to time t. Obviously, if $x(t-T)$ were continuously available at the receiving end of the channel, there would be no need for the transmission. In practice, $x(t)$ can be a prerecorded signal that is played back periodically at both ends of the channel to permit the network to adapt to slowly changing channel conditions. Useful data can be transmitted between these short periods of adaptation.

For digital signals, figure F1.8.1(b) shows a technique known as 'decision-feedback' equalization (Lucky 1966) which decides whether the output of the network represents a high or a low binary value by comparing it against an adaptable threshold. Based on the decision, the threshold and the parameters of the network adapt to minimize the difference between the actual output of the network and the value of the decision. Of course, the decision can be wrong occasionally, and the network weights will adapt in the wrong direction. However, by starting with an appropriate initial threshold and network weights, most decisions near the beginning of operation will be correct, leading to fewer wrong decisions as the network adapts.

Adaptive linear filters (Widrow and Stearns 1985) often work well in the configurations shown in figure F1.8.1, but their accuracy suffers when the channel distortion is nonlinear. Recently, feedforward neural networks with internal time delays have been applied to channel equalization. These networks can be viewed as a nonlinear generalization of the adaptive linear filter. However, unlike the adaptive linear filter, they can implement arbitrary nonlinear transfer functions, and therefore learn to compensate for nonlinear channel distortions. There are many neural network models with such delays, and some of the most useful for signal processing applications have been described in Waibel *et al* (1989), Wan (1990), Lin *et al* (1992), Day (1993) and Day and Davenport (1993). *Radial basis function networks* have also   C1.6.2 been used for equalization (Chen *et al* 1991, 1993), and more recently, recurrent neural networks have shown promise in this area (Kechriotis *et al* 1994).

Other related areas where neural networks have been successfully applied are in blind deconvolution (e.g. the cancellation of the effects of an unknown filter) and blind separation of sources (Jutten and Herault 1991, Bell and Sejnowski 1995), where a mixture of several signals must be separated into its constituent parts.

F1.8.2.2 Signal prediction

Accurate predictions about future values of a signal can be useful in many engineering applications. Often, the process generating the signal is either completely unknown or far too complex to permit a practical predictive model, leaving only the observed behavior of the system for use in making predictions.

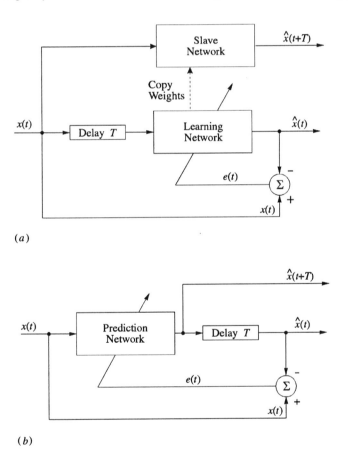

Figure F1.8.2. (*a*) Signal prediction with a slave network. (*b*) Signal prediction with a single network. In (*a*) the learning network adapts to predict $x(t)$ by using a delayed version of $x(t)$ as its input, and the present value of $x(t)$ as its desired output. A slave network, whose weights are continuously copied from the learning network, performs a true prediction by using the present value of $x(t)$ as its input. In (*b*), the delay is placed at the output of the network, and the network adapts to reproduce $x(t)$ at the output of the delay element. A tap taken before the delay element produces a prediction of $x(t)$, T time units into the future. In this diagram, $x(t)$ is the input signal, and $\hat{x}(t)$ is the reproduced signal.

Figure F1.8.2(a) shows one method for training an adaptive network to predict a signal T time units into the future. The same types of neural network topologies and training techniques used for channel equalization can be employed here as well. The delay, T, at the input allows the undelayed signal, $x(t)$, to be used as the target during training. Minimizing the squared error, $e^2(t)$, by adapting the weights causes the network to 'predict' the current value of the input signal based on its values up to time $t - T$. This configuration is often used in time-series prediction tasks. For real-time prediction, a second 'slave' copy of the network can use the undelayed input signal and a copy of the weights from the learning network to provide a true prediction T time units into the future. The weights must be continuously copied from the learning network to the slave network.

Unfortunately, the slave network may require additional hardware, and the continuous weight copying presents a high-bandwidth communication problem between the two networks. Figure F1.8.2(b) shows another technique, where the delay has been moved to the output of the network. As the network adapts to minimize the difference between $\hat{x}(t)$ and $x(t)$, the tap taken off before the fixed delay will provide a true future prediction of $x(t + T)$. To adapt through the time delay, the temporal training techniques described in Wan (1990), Lin *et al* (1992), Day *et al* (1992), and Day and Davenport (1993) can be used.

F1.8.2.3 Noise canceling

Another interesting application of neural networks is in the area of adaptive noise canceling. Adaptive noise cancellation received much attention in the late 1960s (Widrow *et al* 1975), primarily using adaptive linear filters.

Figure F1.8.3 shows the system configuration. Once again, the same types of networks used for channel equalization can be employed in noise canceling applications. The input is a signal, $s(t)$, with uncorrelated additive noise, $n(t)$. The output from the neural network is subtracted from the corrupted signal in an attempt to remove the noise. If the output of the network is identical to the additive noise, then the output from the system will be the noise-free signal.

For the neural network to learn to reproduce the noise, its input must be correlated with the noise in a way that permits its reconstruction. Typically, a localized noise source will generate noise that gets coupled into the communication channel and added to the signal, and that also propagates through another path to a sensor that can be used as a reference noise source. The relationship between the noise added to the signal and the reference noise source may be unknown, and possibly time-varying or nonlinear. It is the task of the adaptive neural network to learn the mapping between the reference noise source and the unwanted additive noise.

If all the signals involved are statistically stationary and have zero means, then minimizing the power in the system output will maximize the output signal-to-noise ratio (Widrow *et al* 1975). Thus, the system output power, $e^2(t)$, can be used as the error signal which the adaptive noise canceler learns to minimize.

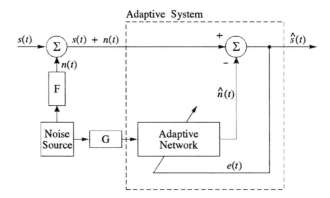

Figure F1.8.3. Adaptive noise cancellation. The input to the adaptive system is a signal, $s(t)$, with additive noise, $n(t)$. A noise source generates $n(t)$ by coupling into the communication channel with an unknown, and possibly time-varying transfer function, $F(t)$. The noise source can also be observed via a sensor with transfer function $G(t)$. The noise picked up by the sensor is correlated with $n(t)$, but the exact relationship is unknown and often time-varying. The network attempts to reproduce $n(t)$ by learning to implement $F(G^{-1}(t))$.

Much of the early work in adaptive noise canceling employed adaptive linear filters, but more general nonlinear neural network models can be more effective when the mapping between the reference noise source and the additive noise is nonlinear. A recent tutorial on the use of neural networks for adaptive noise cancellation (among other applications) was presented in Piché (1994). In addition to adaptive noise cancellation, neural networks can also be used for *active* noise cancellation, where the output of the network is a signal that is out of phase with the original noise. When injected back into the original system via an appropriate transducer, this synthesized noise can cancel out the undesired noise through destructive interference.

F1.8.3 Alternative approaches

This brief introduction to the use of neural networks in signal processing applications has shown how they can be applied to problems in channel equalization, signal prediction, and noise canceling. The ability of multilayer neural networks to learn nonlinear mappings is mediated by the fact that they often take much longer to converge to a solution than conventional linear techniques. In nearly all of the applications described in this section, the neural network can be replaced with an adaptive linear filter (Widrow and Stearns 1985), which will usually converge much faster at the expense of reduced representational capacity. In fact, the techniques presented here were first investigated using adaptive linear filters, and only recently have neural networks been substituted for them in applications where a linear approximation to the solution is not acceptable.

Two early papers describing linear techniques for adaptive channel equalization are Lucky (1966) and Gersho (1969). Much of the subsequent work in this area has been concerned with increasing the speed of convergence of these adaptive filters, using techniques like the Kalman estimation algorithm (Falconer and Ljung 1978) and least-squares lattice algorithms (Satorius and Pack 1981). An excellent tutorial on linear techniques for signal prediction was presented in Makhoul (1975). A more recent investigation of the performance of linear adaptive filters was given in Zeidler (1990), and Widrow and Stearns (1985) is a good introductory text on the subject.

Nonlinear techniques based on polynomial approximations of unknown functions have also been investigated, starting in the early 1960s (Gabor *et al* 1961). Unfortunately, many nonlinear functions require high-order polynomials which do not interpolate well to data points between those encountered during adaptation. Related techniques rely on the Volterra series expansion of the nonlinear function (Biglieri *et al* 1984, Falconer 1978).

Probably the best approach to many signal processing problems is to use an adaptive linear filter whenever possible, because of its fast convergence and the large body of accumulated knowledge that can help predict its behavior. However, when linear techniques cannot provide a sufficiently accurate solution, adaptive nonlinear neural networks offer a promising alternative.

References

Bell A J and Sejnowski T J 1995 An information maximisation approach to blind separation and blind deconvolution, *Technical report* no INC-9501, Institute for Neural Computation, UCSD, San Diego, CA (to appear in *Neural Computation*)

Biglieri E, Gersho A, Gitlin R D and Lim T L 1984 Adaptive cancellation of nonlinear intersymbol interference for voiceband data transmission *IEEE J. Sel. Areas Commun.* **2** 765–77

Chen S, Gibson G J, Cowan C F N and Grant P M 1990 Adaptive equalization of finite nonlinear channels using multilayer perceptrons *Signal Proc.* **20** 107–19

——1991 Reconstruction of binary signals using an adaptive radial-basis-function equalizer *Signal Proc.* **22** 77–93

Chen S, Mulgrew B and Grant P M 1993 A clustering technique for digital communications channel equalization using radial basis function networks *IEEE Trans. Neural Networks* **4** 570–79

Day S P 1993 Dispersive neural networks for adaptive signal processing *PhD Dissertation* University of British Columbia, Vancouver, Canada

Day S P and Davenport M R 1993 Continuous-time temporal back-propagation with adaptable time delays *IEEE Trans. Neural Networks* **4** 348–54

Day S P and Davenport M R and Camporese D S 1992 Dispersive networks for nonlinear adaptive filtering *Neural Networks for Signal Processing II—Proc. 1992 IEEE Workshop* ed S Y Kung, F Fallside, J Aa Sorenson and C A Kamm (Piscataway, NJ: IEEE) pp 540–49

Falconer D D 1978 Adaptive equalization of channel nonlinearities in QAM data transmission systems *Bell Syst. Tech. J.* **57** 2589–611

Falconer D D and Ljung L 1978 Application of fast Kalman estimation to adaptive equalization *IEEE Trans. Commun.* **26** 1439–46

Gabor D, Wilby W P L and Woodcock R 1961 A universal nonlinear filter, predictor and simulator which optimizes itself by a learning process *IEEE Proc.* B **108** 422–38

Gersho A 1969 Adaptive equalization of highly dispersive channels for data transmission *Bell Syst. Tech. J.* **48** 55–70

Jutten C and Herault J 1991 Blind separation of sources part I: an adaptive algorithm based on neuromimetic architecture *Signal Proc.* **24** 1–10

Kechriotis G, Zervas E and Manolakos E S 1994 Using recurrent neural networks for adaptive communication channel equalization *IEEE Trans. Neural Networks* **5** 267–78

Lin D-T, Dayhoff J E and Ligomenides P A 1992 Adaptive time-delay neural network for temporal correlation and prediction *SPIE Intelligent Robots and Computer Vision XI: Biological, Neural Net, and 3-D Methods* **1826** pp 170–81, Boston

Lucky R W 1966 Techniques for adaptive equalization of digital communication systems *Bell Syst. Tech. J.* **45** 255–86

Makhoul J 1975 Linear prediction: a tutorial review *Proc. IEEE* **63** 561–80

Piché S W 1994 Steepest descent algorithms for neural network controllers and filters *IEEE Trans. Neural Networks* **5** 198–212

Quereshi S 1985 Adaptive equalization *Proc. IEEE* **73** 1349–87

Satorius E H and Pack J D 1981 Application of least squares lattice algorithms to adaptive equalization *IEEE Trans. Commun.* **29** 136–42

Waibel A, Hanazawa T, Hinton G and Shikano K and Lang K 1989 Phoneme recognition using time-delay neural networks *IEEE Trans. Acoustics, Speech, Signal Proc.* **37** 328–39

Wan E A 1990 Temporal backpropagation for FIR neural networks *Int. Joint Conf. on Neural Networks I (San Diego, CA)* pp 575–80

Widrow B, Glover J R Jr, McCool J M, Kaunitz J, Williams C S, Hearn R H, Zeidler J R, Dong E Jr and Goodlin R C 1975 Adaptive noise cancelling: principles and applications *Proc. IEEE* **63** 1692–716

Widrow B and Stearns S D 1985 *Adaptive Signal Processing* (Englewood Cliffs, NJ: Prentice-Hall)

Zeidler J R 1990 Performance analysis of LMS adaptive prediction filters *Proc. IEEE* **78** 1780–806

Further reading

1. Haykin S 1994 *Neural Networks: A Comprehensive Foundation* (New York: Maxwell Macmillan)

   A good introduction to the theory and application of neural networks, with several examples relating to signal processing applications.

2. Kosko B 1992 *Neural Networks for Signal Processing* (Englewood Cliffs, NJ: Prentice-Hall)

   Discusses the application of neural networks to signal processing problems such as speech recognition, spectral estimation, and robotics, and discusses the implementation of neural networks using analog VLSI technology and optical devices.

3. Masters T 1994 *Signal and Image Processing with Neural Networks: A C++ Sourcebook* (New York: Wiley)

   Explores signal and image processing techniques using complex-domain neural networks. Includes software on disk.

F1.9 Control

Paul J Werbos†

Abstract

Neurocontrol is a *subset* of the larger field of control theory, which designs systems
for a broad spectrum of applications, ranging from simple regulators (like thermostats
or muscle neurons) to optimal decision-making in complex environments (as in the
brain *as a whole system*). Neurocontrol, like classical control and artificial intelligence,
includes general designs for three basic types of task: cloning, tracking and optimization.
Neural cloning systems copy the input–output behavior of human experts or automatic
controllers. Tracking systems may be regulators, or systems to make a robot arm follow
(track) a desired path in space, etc. Optimization over time may be used to solve
tracking problems, with improved stability, or to solve planning problems which require
real intelligence. This section compares the practical advantages and disadvantages of
a wide variety of control designs, neural and otherwise, ranging from simple regulators
through to designs which begin to provide an explanation of intelligence in brain circuits.

F1.9.1 Overview

Control theory encompasses any system whose outputs control or recommend overt, physical actions,
like movements of motors, muscles or dollars. Neurocontrol—a subset of control theory—offers the
cost, learning and simplicity advantages discussed in Chapter A2, plus specific new capabilities in three   A2
areas—*cloning, tracking* and *optimization*—plus methods to blend multiple capabilities.

Conventional artificial intelligence or fuzzy control 'clones' experts by implementing what the experts
say in a database of rules. Neural networks can imitate what experts *do* as a function of sensor inputs and
past information. Similarly, they can clone the input–output behavior of existing automatic controllers;
this may not improve controller performance, but it may allow a vast reduction in implementation cost,
for example, by permitting the use of high-throughput neural chips in place of large computers.

Conventional adaptive control maintains a desired set point or *tracks* a reference model, using direct or
indirect (i.e. model-based) designs. (For example, a thermostat *tracks* or maintains a desired temperature.)
Neural adaptive control does likewise, but offers: (i) *generalized* nonlinearity and (ii) the ability to *learn*
the parameters of the adaptation process itself, thereby permitting rapid response to changes in familiar
variables such as center of gravity, mass and friction. Many stability theorems exist for conventional and
neural adaptive control, but delays or sign changes over time easily destabilize both; however, designs
based on optimization over time can overcome such instabilities.

If a system can learn to maximize any arbitrary utility function summed over future time in an
arbitrary environment, then logically it should automatically have the ability to 'plan', to solve problems,
etc. The field of neurocontrol includes designs which enhance conventional deterministic optimization
methods, like calculus of variations or model-predictive control. It also includes designs which approximate
dynamic programming and promise truly brain-like capabilities. Critical applications include, among
others, minimizing fuel consumption, pollution or product loss in the chemical process, automotive and
aerospace industries.

† The views presented in this section are those of the author, and are not necessarily those of the National Science Foundation.

F1.9.2 The problem domain

The field of control encompasses a vast and heterogeneous collection of applications, designs and fundamental theory. Some historians claim that the field began in earnest when James Watt developed a very elaborate feedback control mechanism to keep a steam engine within its operating range. The simple thermostat—a feedback mechanism designed to keep temperature close to a desired set-point (a point set by the consumer)—served as a dominant paradigm in the early days of the field. As the field developed, it focused more and more on two fundamental design challenges, which permeate a wide variety of application domains:

- The challenge of *tracking*—making systems settle down into a fixed desired set-point, or into a moving set-point (a desired trajectory or a 'reference model').
- The challenge of *optimization* over time—finding a strategy or policy which maximizes the sum of some utility function over future time (Von Neumann and Morgenstern 1953, Raiffa 1968). Utility functions can be formulated which represent a wide variety of concepts—maximizing profit, minimizing cost, pollution or energy use, maximizing throughput, maximizing satisfaction of particular long-term goals, etc. In principle, the user formulates the utility function (Werbos 1990a); the control system only maximizes it. (In some designs, the system or the control engineer must devise a kind of *secondary* utility function, as will be discussed.)

Furthermore, success in these tasks often depends on one's ability to *model* or *predict* the environment or plant that one is trying to control; therefore, research into 'system identification' (Ljung 1987) and 'system dynamics' (Peterson 1991) has become a large part of the control field.

As the field evolved, it became apparent that engineers and economists were both studying different applications of the same underlying mathematical challenges. Therefore, these groups came together in large conferences and university programs on 'decision and control'. It also became apparent that control was a central issue both in engineering and in biology (Wiener 1961). Wiener's term 'cybernetics' was perhaps a better name for this field than 'control', but the word lost favor in the United States several decades ago because of its popularization and misuse by enthusiasts and consultants who were ignorant of the underlying mathematics. Despite the semantic problems, the field of decision and control began, by 1970, to view itself as a unified approach to all problems involving the design or understanding of systems which output 'control signals'—signals to control or recommend *actions* such as the movement of motors or muscles or levels of investment.

Note that the human brain itself is a 'control system' in this broad sense. The *entire* brain—not just the 'motor centers'—is part of a unified computing system, whose purpose is to calculate control signals—signals to control muscles or glands. (Many would argue that there are other, more spiritual purposes of the brain; however, even that does not invalidate this paradigm, see Levine and Elsberry (1996).) In describing the wiring of this system, Nauta and Feirtag (1986) have shown very concretely how futile and misleading it is to try to separate out the parts of the brain which support motor control and those which do not; they all do.

Circa 1970, the emerging field of artificial intelligence (AI) challenged the existing paradigms of control theory, by suggesting alternative ways to solve control problems, most notably the following:

- To optimize goal-satisfaction over time; formal *task-oriented planning* designs will sometimes work on problems that are too nonlinear and too complex to respond to conventional control techniques (Miller *et al* 1990). Typically, such designs involve complex hierarchies of discrete goals, subgoals, tasks, subtasks and so on (Albus 1991).
- As an alternative to tracking and optimization, one may simply 'clone' a human expert. One may ask a human expert for if–then rules which state how to perform a complex decision or control task.

F1.9.3 Functions performed by neural networks in control

The field of neurocontrol includes generic designs to perform all three fundamental tasks described above—cloning, tracking and optimization over time. These designs are generic in the sense that a *single* computer program could be used, in principle, on a *wide variety* of applications, without changing anything but a few parameters like the number of inputs and outputs; the other differences between applications could be handled by the program itself, as it learns the dynamics of the plant or environment it is trying to control. Thus the underlying program or design is not application specific. (There are, however, a variety of tricks for exploiting whatever application-specific information may be available.)

Neural networks can also be used to perform subordinate tasks—such as *pattern recognition, sensor* F1.2
fusion, diagnostics and *system identification*—within a larger control system; however, in neurocontrol G2.7, G2.9
proper, the actual control signals are output directly from a neural network. (See Werbos (1989) and
Miller *et al* (1990) for the first published definition of neurocontrol; the latter book was the result of the
1988 National Science Foundation conference which essentially created neurocontrol in the United States
as an organized, self-conscious field.) This section will focus mainly on neurocontrol proper.

This definition does not exclude the possibility of using a fixed, nonadaptive postprocessor to provide
a buffer between the decisions of the neural network and the low-level actuators. For example, many
people have used standard neurocontrol designs to output 'actions' which set the parameters of a simple
classical PID controller, which, in turn, controls an industrial plant. This is similar to what the human
nervous system does, in using signals from the brain as inputs to low-level 'spindle cells' and 'gamma
efferents' which provide low-level feedback control of human muscles.

Unlike AI, neurocontrol is logically a subset of control theory. The basic designs now used in
neurocontrol can all be understood completely within the broad framework of control theory. Nevertheless,
there is significant novelty in these designs. For example, classical control theory included only two popular
methods to perform optimization over time in a noisy (stochastic) environment: (i) linear-quadratic (LQ)
methods (Bryson and Ho 1969); (ii) dynamic programming (Howard 1960). Neither was suitable for
solving complex planning problems, because the first required linearity, and the second was computationally
infeasible for problems with many possible states. Neurocontrol contains new methods for approximate
dynamic programming (ADP) which overcome both problems, and provide an alternative to the more rigid
rule-based methods used in AI planning. Useful designs for reinforcement learning—described in Chapter
A2 of this handbook—are a special case of ADP. Simple forms of reinforcement learning, developed in
a neurocontrol context, have been widely popularized and reassimilated into the AI field, largely through
the efforts of Andrew Barto and collaborators (See chapters by Barto in Miller *et al* 1990 and in White
and Sofge 1992). Tesauro at IBM has demonstrated that such designs can be very effective in solving
classical, difficult AI problems such as beating human beings in board games like backgammon.

Complex neurocontrol designs typically do not consist of a single neural network. Typically, they
consist of a higher-level recipe for how to combine several neural networks (and/or non-neural networks)
to perform a higher-level task. Usually, there is at least one module in the design which can be filled in
by *any* supervised learning design. Successful research teams usually begin by implementing very simple
designs, of limited power, in a modular software system. Then, when the simple designs fail on harder
problems, they gradually enhance their software system, and progress to more sophisticated, optimization-
based designs. They usually make it easy to switch the choices of supervised learning methods used in
the various component modules, so as to accommodate different types of applications.

There are some applications in the control field which are even more difficult than the previous
paragraph suggests. For example, consider the problem of balancing three poles, one on top of the other,
like a team of acrobats in a circus. There is probably no neural network system which could learn to
perform this task, starting from zero prior information. Logically, this is an example of the 'local minimum'
problem.

Local minimum problems are far more serious, in practice, in complex decision and control tasks than
in applications like pattern recognition. Random search techniques like *genetic algorithms* can be useful D2.1
in small problems of this sort. But for large problems, the most valuable technique by far is something
which Barto calls 'shaping' (White and Sofge 1992). In shaping, one first adapts an entire neural network
system to solve a simplified version of the task at hand. One then uses the resulting network and weights
as the initial values of a network trained to solve a more realistic version of the task. One may construct
a graded series of tasks, ranging from the easiest through to the most realistic, and adapt a series of neural
systems to solve them. In a similar fashion, one may initialize a neural network with a *fuzzy controller*, D1
and so on (Werbos 1993a). One may use cloning techniques, at an early stage, to stabilize a system, and
then use optimization at a later stage to improve performance while retaining stability. The parallels with
human learning are many. (In practice, shaping requires the use of flexible learning rules, such as the
adaptive learning rate given in Chapter 3 of White and Sofge (1992), to avoid locking in a new network
to the old problem.)

F1.9.4 Neural network approaches

F1.9.4.1 Neural network approaches to cloning experts

Probably the first example of neurocontrol actually working in simulation was the original broom balancer developed by Widrow in the 1960s (Widrow 1987). Widrow's approach has been reinvented many times in the past decade, in part because it seems very obvious to people who know nothing about control theory.

Widrow began by training human students to balance a broom. Then he recorded how the humans did it. At each sampling time, for each student, he recorded two things: (i) what the student *saw* (the state of the broom); (ii) what the student *did* (the correct action). He built a database or 'training set' out of these records. He then trained a simple neural network to learn the mapping from what the student saw to what the student did. This was a straightforward application of supervised learning. This particular work was later refined by Guez and Selinsky (1988).

Most of the people reinventing this approach did not place great emphasis on the human expert. They simply reported that they had trained a neural network to input sensor data and to output the correct control action. Clearly, the performance of this approach depends critically on how one constructs the database containing the 'correct actions'. This must unavoidably come from some other existing controller—either a human expert, or an animal expert or a computer program.

High-quality human operators of chemical plants or high-performance aircraft typically do *not* base their actions solely on sensor data at the current time. Like good automatic controllers, they typically account for things like trends, or experience over multiple time periods, or a sense of how the underlying system parameters are changing. Therefore, one cannot capture their expertise in a static supervised learning exercise. A better approach to cloning is to treat it as a task in *dynamic* modeling or system identification. As McAvoy has said, it is an exercise in 'modeling the human operator' (White and Sofge 1992). The first step in this approach is to collect a *time series* of what the expert sees and what the expert does; then, one may simply apply neuroidentification techniques to build a model of this data—using more difficult and more advanced techniques (White and Sofge 1992, Chapter 10) only if the simpler ones do not perform well enough.

An instructive example of this approach came from the Accurate Automation Corporation (AAC) circa 1992. AAC proposed that *optimizing* neurocontrol could be used to solve the critical efficiency and weight problems in controlling the National Aerospace Plane (NASP), a prototype under design for a future airplane fast enough to reach earth orbit *as* an airplane, at airplane-like costs. Before exploring the neural option, the NASP program office first challenged AAC to prove that it could even stabilize this craft—a highly nontrivial, nonlinear control problem, for which the conventional solution had required a great deal of development work. AAC first built a simple but credible simulation of the vehicle, running on a Silicon Graphics machine, at a slowed-down rate so that humans could stabilize the simulation. AAC recorded the vehicle states and human actions for those few humans able to control the simulation. Then they modeled the human response pattern, using a simple time-delay neural network to perform the neuroidentification. The result—within just a few weeks—was a well-defined algorithm, able to perform at electronic speeds. The resulting neural network was also suitable for use as the initial state of a network to be improved on, later, via optimization designs. Because of this and later successes, AAC is now the prime contractor on the follow-up to NASP (LoFlyte), and is currently flight-testing a physical prototype which they have built.

Another, more proprietary example from the robotics industry is also interesting. In 1994, a major corporation considered using neural networks to replace human workers in a very difficult process which had resisted conventional techniques. They did not know where the real problem was—in the robots themselves, or in the computer programs, or whatever. I proposed that they begin with a kind of 'virtual reality' exercise—equipping human beings with visual displays showing only what the robot would see and dressing them up in data gloves to directly control the robot arms. Naturally, the humans would be permitted to take their time, and would be rewarded if successful. The virtual reality approach would not be of *direct* economic benefit here, because it would not reduce labor costs. However, it would make it possible to test whether the given sensors and actuators might be good enough, in principle. If the exercise were in fact successful, one might then simply 'clone' the successful operators based on data recorded during this exercise. (In other kinds of plants, such as big chemical plants or electric utilities, there is often enough data recorded already to permit cloning without such a special exercise.)

The two-step strategy of cloning followed by improvement does have a crude analogy to what happens in human learning. The phenomenon of *imitation* is amazingly pervasive in early learning by human beings in natural settings. Nevertheless, the phenomenon of imitation in human children is far more complex and subtle than the cloning approaches described above. I would speculate that it involves new, higher-order capabilities which can only be understood at the most advanced level (see Chapter 10 of Werbos 1994).

F1.9.4.2 Neural network approaches to tracking

There are two main approaches to solving tracking problems, both in classical adaptive control and in neurocontrol: the 'direct' approach, and the 'indirect' approach. In the direct approach, one tries to learn the mapping from the location of the plant back to the actuator settings which could move the plant to that location. In the indirect approach, one constructs a model of the plant (e.g. by using neuroidentification) and one then uses optimization techniques to train a neural network to minimize the tracking error.

Robot arm control is the classic paradigm of the direct approach. Suppose that the location of your robot hand is specified by three spatial coordinates—x_1, x_2 and x_3, forming a three-dimensional vector x. Suppose that you control three joint angles in the robot arm—θ_1, θ_2 and θ_3, forming a vector θ. Then we would expect x to be a function f of θ. However, if the function f is a one-to-one invertible function, then θ is also a function f^{-1} of x. Our goal, in tracking, is to calculate the joint angles θ^* which would move the robot hand to some desired location in space, x^*. To solve this problem, we can simply train a neural network to approximate the function f^{-1}. We can do this simply by moving the robot arm around, and recording actual values of x and θ, and training the neural network to learn the mapping from x to θ. Any supervised learning design can be used to learn this mapping.

The first working example of direct neural adaptive control was a physical robot developed by Kuperstein (1988). Kuperstein used a very elaborate, fixed, biologically based preprocessor as his neural network, topped off by a simple adaptive output layer trained by Widrow's *LMS algorithm*. Kuperstein's tracking error was approximately 3%—enough to be interesting scientifically, but not enough to be useful in practice. Miller (Miller *et al* 1990) later used a similar approach, but with a CMAC network augmented by time- delayed inputs. In other words, Miller treated this as a problem in neuroidentification, rather than a problem in static supervised learning. This led to tracking errors of less than 0.1%. Miller produced an impressive video of his robot arm, pushing an unstable cart around a figure-of-eight track with great accuracy. Even after he put a heavy new weight on the cart, it would re-adapt and return to high accuracy within three trips around the track. Similar accuracies have been achieved by a few researchers using static supervised learning, but not with this real-time readaptation capability.

One disadvantage of Miller's approach is that it uses real-time learning to adapt to simple, routine changes like changes in mass. Whenever the mass or the friction change, the network acts as if it is learning a totally new problem, unrelated to anything experienced before. This is similar to the behavior of primitive organisms when confronted with pattern reversals (Bitterman 1965). Werbos (1990b) proposed a different approach: to use a time-lagged recurrent network (TLRN) here. If powerful enough neuroidentification methods were used, then the recurrent nodes *themselves* should learn to detect changes in familiar variables like mass and friction, so long as these variables do, in fact, vary during the training period. This kind of detection—tuned to specific variables and exploiting past experience—should be much more rapid than real-time learning. We could even use this approach to build systems which 'learn offline to be adaptive online'. To my knowledge, no one has applied this approach, as yet, to direct tracking designs; however, Feldkamp of Ford Motor (in Narendra 1994) reports great success with this general approach, plus a few additional features, which he calls 'multistreaming', applied to model-based designs.

An advantage of real-time methods, like Miller's, is the ability to cope with unfamiliar, fundamental structural changes in the plant to be controlled. It is possible to combine real-time learning with TLRNs in an efficient way, but no one has done this yet, to my knowledge (see Chapter 13 of White and Sofge 1992).

Indirect tracking designs are more complicated than direct designs, but also more powerful. There is no need to assume that f is a one-to-one function. Direct designs have been developed which do not become invalid when the number of controls (components of θ) exceed the number of state variables (x); however, they generally waste the additional degrees of freedom. Indirect designs can make good use of such additional controls, especially if they are adapted to minimize a *sum* of tracking error plus some measure of jerkiness or energy consumption (see Kawato in Miller *et al* 1990). In the US, classical adaptive control is dominated by the indirect approach, in part because of the well known work of Narendra (Narendra and Annaswamy 1989). The same is true of neural adaptive control. (See the papers by Narendra in Miller *et al* 1990, in White and Sofge 1992, Narendra 1994.)

B3.3.3

Most of the neural tracking systems in the literature today are indirect systems which fit the following general description. At every time t, there are M sensor inputs $X_1(t), \ldots, X_M(t)$, forming a vector X. The desired set-point or trajectory can be represented, for all practical purposes, as a set of desired values $X_1^*(t), \ldots, X_m^*(t)$ for the first m components of X; they form a vector x (usually $m = M$, but not always). The control signals at time t form a vector $u(t)$. Sometimes the neural system is represented as a time-sampled system (proceeding from time t to $t + 1$ to $t + 2$, etc) and sometimes (as in Narendra's case) it is represented in terms of ordinary differential equations (ODEs). The neural system consists of three components: (i) a function $v(X, x^*)$ representing tracking error—usually just a square error; (ii) a model of the plant—either a neural network or a first-principles model—which predicts changes in X as a function of X and u; (iii) an action network (or 'controller') which inputs $X(t)$, $x^*(t)$ and (in many cases) other information from the model network, and outputs $u(t)$.

In true adaptive control, the model network and the action network are both adapted in real time. The model network is usually adapted by one of the neuroidentification methods described by Narendra. The action network is adapted so as to minimize v in the immediate future; this is done by using some form of backpropagation and adapting the weights in the action network in proportion to the derivatives of v. (Werbos (1994) and White and Sofge (1992) explain these forms of backpropagation, which predate the simplified versions popularized in the 1980s.) This is a straightforward generalization of classical adaptive control, where the model and action networks are usually just matrices (section F1.9.5 discusses exceptions).

When the plant to be controlled is truly linear, or when it stays so close to a desired set-point that it can be treated as linear, then conventional adaptive control can perform just as well as the neural version. The latter tends to stabilize nonlinear plants more effectively, but stability is harder to prove in the nonlinear case. Many stability theorems have been proved both for classical adaptive control and for neural adaptive control; however, all of these theorems involve stringent assumptions which are often violated in practical applications. The problem for practical applications here is not that the mathematics is hard (though it is) or that we need more theorems (though we do). The problem is that all forms of adaptive control can become unstable in practical applications, either when learning rates are too high or when effects like deadtimes or sign reversals exist.

The underlying problem with deadtimes and sign reversals is that actions which reduce tracking error in the *immediate* future (or which have no immediate effect) may actually result in *greater* error over time. We can call this the problem of 'myopia'. Myopia is a central issue in many control problems. For example, consider the problem of deciding how many fish to harvest, so as to maximize long-term profits. The myopic strategy is simply to harvest the largest possible number of fish, using all the boats and networks available, in order to maximize profits in the *immediate* future. However, this strategy could actually wipe out the fish population, and zero out profits in future years. The bioreactor benchmark problem in Miller *et al* (1990) exemplifies this issue; it is an excellent first test for neurocontrol designs. This test has been passed by designs which *explicitly* perform optimization over time (Prokhorov *et al* 1995). When classical adaptive control led to unstable results in the chemical industry, in the 1970s, the industry moved towards model-predictive control—an explicit design for optimization over time—which is now a mainstay of the industry.

Strictly speaking, there is reason to believe that neural adaptive controllers could be devised which could stabilize almost any plant which can, in fact, be stabilized. The challenge lies in finding a loss function $v(X, x, \text{etc})$ which is appropriate for the particular plant. It can be extremely difficult to find good enough loss functions simply by 'guessing'; however, several approximate dynamic programming (ADP) designs can be used to *learn* the optimal function v for specific plants (see Chapter 2 in Pribram 1994). The Wunsch–Prokhorov work can be interpreted in this way. This is a difficult but promising area for future research. Prokhorov and Wunsch (1996) have developed some preliminary stability theorems for a hybrid optimal control scheme, in which an ADP design acts as a kind of supervisor, sending value signals to a lower-level classical linear controller.

In addition to the usual direct and indirect designs, several alternative arrangements have been tried. Probably the most important is the use of a neural network to estimate the current *parameters* of the plant to be controlled, followed by use of a controller—neural or non-neural—which inputs these estimates. Lapedes and Farber, and Farrell (White and Sofge 1992) used this approach previously. Urnes of McDonnell–Douglas is using this approach for a Phase I reconfigurable flight control system, which helps F-15s to recover from 'involuntary configuration changes' (like being hit in combat). This approach has some of the same advantages as 'learning offline to be adaptive online', *if* the controller is properly designed. Offline training also simplifies the process of flight qualification—the rigorous testing process by which new aircraft and aircraft controllers are certified as reliable enough to permit their routine use, with human lives at stake. C Jorgensen of NASA Ames has stated that NASA flight testers have certified a neural network controller (trained offline) which was recently used to land a huge MD-11 aircraft with all of its hydraulic actuators disabled.

Also significant is Kawato's feedback error learning (Miller *et al* 1990), which is really just a way to *blend* a classical feedback controller with a neural network. It is formally equivalent to a particular ADP design (DHP) with the critic network hard-wired in advance. Less interesting are 'model-free' indirect designs which, instead of a model network or matrix, use a kind of correlation matrix, explicitly or implicitly; such designs are not truly model-free, because the correlation matrix (or equivalent) is simply a naive form of plant model.

F1.9.4.3 Neural networks for optimization over time

Optimization over time accounts for a smaller share of the published academic literature on neurocontrol than do cloning or basic tracking designs. However, it probably accounts for the bulk of the dollar value of neurocontrol products actually working in industry. In some cases, optimization over time is used to minimize tracking error *plus* some measure of cost, *accounting* for linkages over time. There is reason to believe that the human brain itself is a member of this family of designs (Pribram 1994, Chapter 31).

There are two major approaches to optimization over time: (i) the *explicit* approach, involving a backpropagation of utility; (ii) an *implicit* approach, based on approximate dynamic programming (ADP).

The explicit approach is similar to indirect adaptive control, discussed in the previous section. The control system usually consists of a model (neural or non-neural), a utility function $U(X)$, and an action network. The main difference is that we pick actions $u(t)$ so as to maximize the *sum* of $U(X(\tau))$ over future times $\tau \geq t$. To do this, we must choose between two forms of backpropagation: (i) backpropagation through time (BTT), a method which I first implemented in 1974 (Werbos 1994) and (ii) a forwards propagation of derivatives. BTT is exact and efficient, like simple backpropagation, but—because it uses calculations which proceed backwards through time through an explicit record of past experience—it is not even remotely plausible as a model of biology. The latter operates in a more real-time mode, but the cost of calculating derivatives is proportional to mN, where N is the number of neurons in the network and m is the *total* number of weights; this, too, is biologically implausible, because the cost rises substantially with the size of the network, and the calculations do not even remotely resemble anything found in the brain.

By 1988, there were already four working examples of explicit optimization based on BTT: Widrow's truck-backer-upper, the simulated robot arm controllers of Kawato and of Jordan, and an official (nonneural) Department of Energy model of the natural gas industry, which the author had previously developed (Miller *et al* 1990). In recent years, Widrow's system has demonstrated ever more interesting capabilities, outperforming human experts both on simulated trucks and on a physical model of a two-trailer truck. Hrycej (1992) of Daimler-Benz and Feldkamp of Ford (in Narendra 1994) have reported many important applications, some of them leading to proprietary products still in the pipeline. McAvoy used this approach in a nonlinear generalization of model predictive control (MPC), for use in the chemical process industries (White and Sofge 1992, Chapter 10). MPC is not a 'real-time' technique, in a formal sense; however, because special-purpose chips can perform calculations very quickly (compared with changes in chemical plants), it can still provide real-time control in a practical sense in these applications. McAvoy's neural network club includes more than twenty large corporate sponsors who have deployed a variety of the techniques he has developed in profit-making applications, albeit on a proprietary basis. Feldkamp and Narendra have also worked with the time-forwards propagation of derivatives, but less so now than in the past, because of the cost issue (and perhaps because of some stability questions).

Explicit optimization methods depend critically on the assumption that the user's model is an exact, deterministic model of the plant to be controlled. Subject to this assumption, they yield *exact* answers, at least for the planning horizon used in the training process. Implicit designs, based on ADP, provide a true real-time capability; however, the solutions they provide are *approximate*. ADP—like dynamic programming itself—is explicitly designed to control stochastic plants, and to use a stochastic plant model (if such a model is available).

The ADP family of designs is far too complex to review thoroughly here. These designs form a kind of ladder, rising up from the simplest but least powerful designs, up to more complex designs like the human brain itself. The simplest reinforcement learning designs work very well on small problems, especially when the choice of actions is small and discrete; however, their learning speed becomes quite slow on larger, more continuous problems. The most powerful designs in operation today are 'brain-like' designs which include at least three components, in addition to the utility function $U(X)$: (i) a Critic network, which provides a kind of 'emotional system', or strategic assessment system; (ii) a Model network, which may be thought of as an 'expectations' system; (iii) an Action network, adapted at least in part by the backpropagation of 'value' signals computed by the critic and backpropagated through the model to the action network. Between late 1993 and late 1994, five groups reported working systems of this sort, including Wunsch and Prokhorov (Prokhorov *et al* 1996), Santiago and I, AAC, Balakrishnan (Balakrishnan and Biega 1995) and Jameson. AAC claims that these designs provide unique capabilities crucial to solving the problems of hypersonic flight, as discussed above. Balakrishnan reports far less error than with the usual methods used on missile interception problems. The other three groups also report substantial improvements in performance, relative to various alternatives, on the bioreactor benchmark problem an autolander benchmark problem and a robot arm simulation. Most of this work was presented at a recent NASA Ames workshop organized by Jorgensen and Pellionisz; the papers are still at press. The underlying principles are described in White and Sofge (1992) and in Pribram (Pribram 1994, Chapter 31). For some additional information see Narendra (1994).

F1.9.5 Nonneural alternatives

The major classical alternatives to these methods have already been discussed. The neurocontrol designs themselves can be applied directly to adapt *nonneural* networks as well, as discussed in Chapter A2 of this handbook. This section will mention only a few additional fine points.

In cloning, the neural network copies what an expert *does*, while the AI approach implements what an expert *says* to do. As an example, consider what would happen if you asked a child how to ride a bicycle; the resulting rules would not be enough to keep the bicycle from falling over. But the child may, nevertheless, know how to ride a bicycle on a nonverbal level. Usually, what an expert *does* will work better than what he *says*; however, when there is a local minimum problem—as discussed above—then fuzzy logic or simpler neural designs may be crucial to providing a good enough starting point for the neural system. When there is very complex reasoning required, then classical AI systems may often be adequate in some applications and far simpler to set up (depending on software availability) than neural networks with similar capability.

In tracking control, there are two techniques often used to keep the classical systems from blowing up when applied to nonlinear systems—gain scheduling and feedback linearization.

In gain scheduling, we try to patch together a nonlinear control rule, by switching back and forth between different linear controllers, designed to operate in different regions of space. Similar improvements in capability can be had with neural networks, by using 'mixture of experts' networks—see, for example, Jordan and Jacobs (Jacobs *et al* 1991), or recent work by Neurodyne (Long 1993), or some proposals I have made for 'syncretism' (Werbos 1993b). (Intuitively, 'syncretism' involves *remembering* observations in real time, adapting a *generalized model* by a combination of ordinary real-time learning and learning from memory, and making predictions based on a *combination* of memory association and a generalized model.) With classical systems, gain scheduling patches together linear domains to try to approximate a smooth surface; however, the same sort of additional complexity allows neural networks to patch together smooth nonlinear surfaces to represent the harder idea of fundamental structural change across different regions of space. See White and Sofge (1992) for a more detailed criticism of gain scheduling.

In feedback linearization, we try to make a plant *behave* as if it were linear, by canceling out simple forms of linearity in restricted parts of a plant model. This process only works on a limited class of plants. Furthermore, Slotine of MIT has shown how neural networks can be useful even in feedback linearization (Sanner and Slotine 1992). Baras and Natel (1995) have developed a more general classical technique to stabilize nonlinear plants, requiring the solution of a nonlinear stochastic optimization problem; adaptive critics an be used in the implementation of this technique.

In optimization over time, I have neglected to mention many methods which are less well known but of serious practical value. Balakrishnan, for example, tests his designs against a variety of methods found (after much investment) to be useful in the missile interception area. The missile interception work has yet to be published, but similar (albeit simpler) work in aircraft control is in the open literature (Balakrishnan and Biega 1995). The explicit methods used most often with BTT are equivalent, in some sense, to the calculus of variations (Bryson and Ho 1969) or to differential dynamic programming (Jacobson and Mayne 1970). True backpropagation simply reduces the cost of calculating derivatives in these applications. The use of a neural network as an action network provides a greater degree of open-loop flexibility than the usual alternatives (a fixed action schedule or a fixed-form policy). DDP is an *explicit* method which nevertheless *does* use stochastic models, in a very interesting way; however, for reasons beyond the scope of this discussion, its convergence rate grows worse than that of well designed ADP systems when the effective planning horizon goes further into the future.

F1.9.6 Preprocessing

Because decision and control are such all-pervasive tasks, drawing on inputs from a multitude of sources, it is not possible here to review all the many forms of preprocessing which can be useful. However, there is one form of preprocessing which is especially crucial—the effort to build up a representation, R, of the true state of the plant or environment to be controlled.

Many neural network papers do not emphasize the difference between the current state of the world, $R(t)$, and the state of the variables observed or sensed by the control system, $X(t)$. However, virtually all of the designs in neurocontrol *implicitly* assume that the controller does in fact 'see' the true state of the world. They assume that there is an approximately one-to-one relationship between states of the world and states of the vector input to the network. As a result, the performance of neurocontrol systems depends critically on obtaining such inputs.

There are three common ways to obtain such inputs: (i) simply obtain more sensor inputs when necessary; (ii) use Kalman filtering (Bryson and Ho 1969) or extended Kalman filtering to calculate an estimated state vector, which is then fed into the network and (iii) use neuroidentification methods (White and Sofge 1992, Chapter 10, Werbos 1994) to adapt a TLRN model of the plant, and then feed in the outputs of the recurrent nodes of the TLRN as additional inputs to the control system. The third is the most brain-like approach.

References

Albus J S 1991 Outline for a theory of Intelligence *IEEE Trans. Syst. Man Cybern.* **3** 473–509

Balakrishnan S N and Biega V 1995 Adaptive critic based neural networks for aircraft control AIAA-95-3297 *Proc. AIAA GNC Conf. (Washington, DC: AIAA)* (a more complete version of this paper is at press in *J. Guidance, Control Dynam.*)

Baras J and Patel N 1995 Information state for robust control of set-valued discrete time systems *IEEE Conf. on Decision and Control* (New York: IEEE Press)

Bitterman M E 1965 The evolution of intelligence *Sci. Am.* January

Bryson A E and Ho Y C 1969 *Applied Optimal Control* (Waltham, MA: Ginn)

Guez A and Selinsky J 1988 A trainable neuromorphic controller *J. Robot. Syst.* **5** 363–88

Howard R 1960 *Dynamic Programming and Markhov Processes* (Cambridge, MA: MIT)

Hrycej T 1992 Model-based training method for neural controllers *Artificial Neural Networks* vol 2 ed I Aleksander and J Taylor (Amsterdam: North-Holland) pp 455–8

Jacobs R A, Jordan M I, Nowlan S J and Hinton G E 1991 Adaptive mixtures of local experts *Neural Comput.* **3** 79–87

Jacobson D and Mayne D 1970 *Differential Dynamic Programming* (New York: Elsevier)

Kuperstein M 1988 Neural model of adaptive hand-eye coordination for single postures *Science* **239** 1308–11

Levine D and Elsberry W (eds) 1996 *Optimality in Biological and Artificial Networks* (Hillsdale, NJ: Erlbaum)

Long T W 1993 A learning controller for decentralized nonlinear systems *American Control Conf.* (New York: IEEE Press)

Ljung L 1987 *System Identification Theory for the User* (Englewood Cliffs, NJ: Prentice-Hall)

Miller G A Galanter E H and Pribram K 1960 *Plans and the Structure of Behavior* (New York: Holt, Rinehart and Winston)

Miller W T, Sutton R and Werbos P (eds) 1990 *Neural Networks for Control* (Cambridge, MA: MIT)

Narendra K and Annaswamy A 1989 *Stable Adaptive Systems* (Englewood Cliffs, NJ: Prentice-Hall)

Narendra K S (ed) 1994 *Proc. Eigth Yale Workshop on Adaptive and Learning Systems* (New Haven, CT: Department of Electrical Engineering, Yale University)

Nauta W and Feirtag M 1986 *Fundamental Neuroanatomy* (San Francisco, CA: Freeman)

Peterson I 1991 Ribbon of chaos *Sci. News* **139** 60–1

Pribram K (ed) 1994 *Origins: Brain and Self-Organization* (Hillsdale, NJ: Erlbaum) (IEEE, 1995)

Prokhorov D, Santiago R and Wunsch D 1995 *Neural Networks* **8** 1367–72

Prokhorov D and Wunsch D 1996 Stability of control with adaptive critic *IEEE Trans. Neural Networks*

Raiffa H 1968 *Decision Analysis: Introductory Lectures on Making Choices Under Uncertainty* (Reading, MA: Addison-Wesley)

Sanner R M and Slotine J J 1992 Gaussian networks for direct adaptive control *IEEE Trans. Neural Networks* **3** 837–63

Von Neumann J and Morgenstern O 1953 *The Theory of Games and Economic Behavior* (Princeton, NJ: Princeton University Press)

Werbos P 1989 Backpropagation and neurocontrol *Proc. Int. Joint Conf. on Neural Networks* (New York: IEEE Press)

——1990a Rational approaches to identifying policy objectives *Energy: Int. J.* **15** 171–85

——1990b Neurocontrol and related techniques *Handbook of Neural Computing Applications* ed A Maren (Orlando, FL: Academic)

——1993a Elastic fuzzy logic: a better fit to neurocontrol and true intelligence *J. Intell. Fuzzy Syst.* **1** 365–77

——1993b Supervised learning: can it escape its local minimum? *Proc. WCNN93* (Hillsdale, NJ: Erlaum)

——1994 *The Roots of Backpropagation: from Ordered Derivatives to Neural Networks and Political Forecasting* (New York: Wiley)

——1995 Optimization methods for brain-like intelligent control *Conf. on Decision and Control* (New York: IEEE Press)

White D A and Sofge D A (eds) 1992 *Handbook of Intelligent Control: Neural, Fuzzy and Adaptive Approaches* (New York: Van Nostrand)

Widrow B 1987 The original adaptive broom balancer *IEEE Conf. on Circuits and Systems* (New York: IEEE Press)

Wiener N 1961 *Cybernetics, or Control and Communications in the Animal and the Machine* 2nd edn (Cambridge, MA: MIT)

PART G

NEURAL NETWORKS IN PRACTICE: CASE STUDIES

PART G

NEURAL NETWORKS IN PRACTICE: CASE STUDIES

G1

Perception and Cognition

Contents

G1.1   Unsupervised segmentation of textured images

Nigel M Allinson and Hu Jun Yin

Abstract

This section describes the use of a hierarchy of neural networks, in association with
nonneural techniques, for the accurate segmentation of natural textured images. The
segmentation is unsupervised as only the total number of different textures is prespecified
for a particular input image and no exemplars of the expected textures are provided. The
system uses a model estimator, based on a Markov random field model, for texture
parameter extraction. Two unsupervised neural networks, a Kohonen self-organizing
map and a local-voting network, are employed to produce a 'clean' segmented image.
A further refinement to improve on the detailed positioning and form of the texture
boundaries is provided by a boundary relaxation algorithm. Typical results and details
of operational parameters are provided.

G1.1.1   Project overview

There are numerous applications where the automatic and accurate *segmentation of images*, based on the identification of textured regions, is important. Remote sensing and microtomed biological samples are just two such applications. Textures can be vaguely defined as the repetition of some elementary, usually small-scale, pattern of pixel intensities. They can conveniently, but not unambiguously, be delineated into two main types: natural (distinguished by their inherent randomness) and artificial (distinguished by their more deterministic periodic appearance). The difficulty in defining texture, and the wide variety in textural properties, means that numerous approaches to texture segmentation have been investigated—each with varying success over a different restricted scope of texture types. Approaches to texture image analysis have progressed along two different main routes: feature-based and model-based approaches. Most earlier work in texture analysis sought to discover useful features that would characterize textures and so establish specific discriminating measures. Later work has concentrated on a model-based approach which seeks a deeper understanding of the interpixel relationships by employing stochastic models. More recently still, multiscale signal processing, using Gabor or wavelet transformations, has received considerable attention (Tuceryan and Jain 1993). One of the most popular statistical feature-based approaches has been based around the spatial grayscale co-occurrence matrix. This texture measure uses the relative frequencies or probabilities of transition from one gray level to another at defined spatial distances. The co-occurrence matrix method has proved successful over a wide range of texture types and applications. A good review of much of this activity is provided by Haralick and Shapiro (1992).

On the other hand, the model-based approach aims to describe textures by a mathematical model from which a set of parameters can be extracted as texture features for description and discrimination. Such models can be used to regenerate the textures and so provide us with a physical (and visual) comparison. The practicality of a model is a combination of its generality and its complexity. Markov random field (MRF) models and the Gibbs distribution (GD), an MRF equivalence, have received considerable attention for the analysis of natural textures. The textures of interest in this project were primarily natural. MRF models have been shown to be powerful descriptors of such statistical textures (Cross and Jain 1983, Lakshmanan and Derin 1989, Manjunath and Chellappa 1991) and these form the basis of the initial stage of information reduction through representing the spatial textures by a small number of parameters. A

F1.6

further requirement of the project was that the segmentation process should be unsupervised. Most texture segmentation systems employ known, explicitly-labeled texture samples (for example, representative samples of 'deciduous woodland' or 'muscular tissue') and the system acts as a conventional classifier using these representative samples as its sought classes. The chief difficulty for unsupervised segmentation, where no representative samples are available, is the obvious lack of *a priori* knowledge. The system is merely presented with an image and given the task of segmenting it into a predefined number of textured regions or some other general criterion. So the model parameters for the different regions are unknown or have to be estimated using the state of the evolving segmentation as the system learns. These initial estimates will be inaccurate and incomplete.

This section, hopefully, not only describes a practical solution to a particular processing task, namely, textured image segmentation, but also indicates how a number of neural networks can successfully cooperate within a single integrated system together with other nonneural pattern processing techniques.

G1.1.2 Design process

The underlying design philosophy can be summarized as follows:

- Unsupervised learning for large data sets

- Reduce network sizes to promote efficient operation

- Natural textured image input

- Little *a priori* knowledge, crude initial parameter estimates

- Detailed segmentation

- Accurate and smooth region boundaries

- Computationally efficient unsupervised network (Kohonen self-organizing map)

- Use texture parameter estimation operation

- Use appropriate estimator (Markov random field model)

- Use large input image window

- Reduce input window size during training

- Second-level network and boundary smoothing (second unsupervised network and boundary relaxation)

G1.1.2.1   *System description*

The general structure of the system is shown in figure G1.1.1. The data input to the parameter estimator are the pixel intensity values from a square region of the input image (the estimator window). The parameter estimator's function is to derive a limited set of parameters that can usefully describe the textured nature of the current estimator window. A variety of estimator types could be used here, but in the current example a Markov random field model estimator is employed. These models describe the statistical interactions between the intensity values of neighboring image pixels. The output of the estimator forms the input to a one-dimensional self-organizing map. The number of neurons in this map is set by the required number of different texture regions in the input image. The winning neuron from this map indicates which texture type the current window is most like and forms the input to the next-level network, which estimates the underlying region texture type. This final network layer is based on the computationally simple local voting network. The effects of the shrinking window and this second network are to reduce noise, and so assist in clear image segmentation. After this first phase of segmentation, the textured regions are essentially well separated but there will be some pixel errors at the texture boundaries. It is not possible to implement any form of single-pixel-based correction as the texture type of a single pixel is a meaningless concept. Boundary improvement is based on the supposition that region boundaries are more likely to be smooth. To achieve this final stage of the segmentation process, a simple relaxation algorithm is employed to smooth out boundary irregularities over small spatial distances. The following paragraphs describe each element of the system in further detail.

Estimator window. In an image consisting of many different texture regions, it is most likely that the individual regions will be concentrated into patches (of very variable spatial extent, but texture only has a meaning when considered over some meaningful spatial region) and that the boundaries between texture

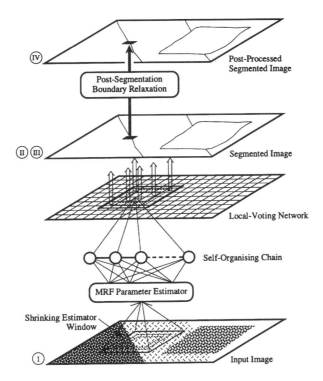

Figure G1.1.1. Schematic of texture segmentation system. The circled Roman numerals refer to the appropriate immediate and final segmented images as shown in figure G1.1.2.

regions will often be smooth. During the initial part of the learning, the system needs to examine a large fraction of the image at one time so that the overall arrangement of textures and their parameters are fed, via the parameter estimator, to the subsequent neural networks. This window shrinks in size during learning in order to reduce the effects of noise and to provide distinct segmentation of the patchlike textured regions (of possibly very small spatial extent). The window is positioned randomly for each learning iteration as this ensures that consecutive inputs to the SOM are uncorrelated.

Parameter estimator. Each image pixel can be assigned a state, $s_{k,l}$ (where k, l are its coordinates), that represents which region it belongs to. The task is to determine the set of states for *all* pixels that possess the largest probability. If Y is the set of all observed pixel intensities and S is the total set of states, then the goal is to determine S so that $\Pr[S|Y]$ is a maximum. This is equivalent to determining the *maximum a posteriori* estimate:

$$p_{Y|S}(Y|S) \cdot \Pr[S] . \tag{G1.1.1}$$

The expression for $\Pr[S]$ can be decomposed into its constituent terms using a Markov random field (MRF). The underlying concept of the MRF, in the context of two-dimensional images, is that the conditional probability for the state $s_{k,l}$, given all other states in the entire image, is the same as the conditional probability of $s_{k,l}$ given only the states in the local neighborhood. In more formal terms, the condition for a Markov process can be expressed as

$$\Pr[s_{k,l}|S'] = \Pr[s_{k,l}|S_{k,l}] \tag{G1.1.2}$$

where $S_{k,l}$ denotes the set of states for pixels within some local neighborhood and S' denotes the set of all states in the image except $s_{k,l}$. An important class of MRF models is known as Gibbsian random fields (GRFs), where the conditional probabilities of equation (G1.1.2) are assumed to be Gaussian in profile. Assuming that all the neighbors of $s_{k,l}$ have the same state as $s_{k,l}$, the conditional probability density of the intensity of this pixel is given by

$$\Pr[s_{k,l}|s_r, r \in N_{s_{k,l}}, L_{s_{k,l}} = l] = \frac{\exp(-U(s_{k,l}|s_r, r \in N_{s_{k,l}}, L_{s_{k,l}} = l))}{Z(l|r \in N_{s_{k,l}})} \tag{G1.1.3}$$

where $N_{s_{k,l}}$ is the symmetric neighborhood of a site $s_{k,l}$ and $L_{s_{k,l}}$ is the corresponding texture label of the site; and

$$U(s_{k,l}|s_r, r \in N_{s_{k,l}}, L_{s_{k,l}} = l) = \frac{1}{2\sigma_l^2}\left(s_{k,l}^2 - 2\sum_{r \in N_{s_{k,l}}} \Theta_{s_{k,l},r}^l s_{k,l} s_r\right) \tag{G1.1.4}$$

where $U(\cdot)$ is the summation of the neighborhood functions over all possible neighborhoods and is termed the Gibbs measure and $Z(\cdot)$ is a normalizing constant termed the partition function. The parameters σ_l and Θ^l are the GRF model parameters of the lth texture class. In practice, only simple functions (such as the linear sum of products) over small neighborhoods are employed. There are several methods which can be employed to estimate these parameters. One of the simplest, certainly in terms of computational effort, is to use the least-squares algorithm.

C2.1.1 *Self-organizing map.* Kohonen's *self-organizing map* (SOM) (Kohonen 1984), when used as a data clustering method, will converge to minimize the mean-square error in its $D2$-dimensional approximation of the $D1$-dimensional input space, at least locally (where $D1 > D2$). For a $D1$-dimensional input space, X, the SOM consists of a $D2$-dimensional array of neurons, Y, so that it forms a $D1 \rightarrow D2$ quantization map. Each of these neurons is connected in parallel to the input by its $D1$ synaptic weights, that is, $w_c(t) = [w_{c1}(t), w_{c2}(t), \ldots, w_{cD1}(t)]^T$, $c \in Y$. The number of neurons is set by the number of predefined texture types expected in the input image, and the dimension of the input weight vector, $D2$, is set by the number of output parameters from the model estimator. The synaptic weights are initially set to random values. At each time step, an input vector, $x(t)$, is applied to the network. The best-matching neuron, termed the *winner*, is found by comparing the inner products $w_c^T(t)x(t)$, $c \in Y$ and identifying the largest. It is often more convenient to normalize the weight vectors to a constant Euclidean norm. Hence the matching condition becomes

$$\arg\min_c \|x(t) - w_c(t)\| \qquad c \in Y \tag{G1.1.5}$$

where $\|\cdot\|$ is the Euclidean norm of the argument vector. The weight vectors of all neurons are updated according to

$$w_c(t+1) = \begin{cases} w_c(t) + \alpha(t)[x(t) - w_c(t)] & c \in N_v(t) \\ w_c(t) & \text{otherwise} \end{cases} \tag{G1.1.6}$$

where $N_v(t)$ is the neighborhood around the winner, v, and $\alpha(t)$ is the scalar adaptation gain. Both $N_v(t)$ and $\alpha(t)$ decrease monotonically during training. There are no theoretical foundations that can be used to determine the selection of these two parameters, and in practice a certain amount of trial and error is required. However, the following general observations can be made (Yin and Allinson 1995).

- Adaptation gain, $\alpha(t)$, must satisfy certain essential conditions in order that the network will converge, namely:

$$0 < \alpha_k(t) < 1 : \sum_{k=0}^{t} \alpha_k(t) \overset{t\to\infty}{\longrightarrow} 1 : \sum_{k=0}^{t} \alpha_k^2(t) \overset{t\to\infty}{\longrightarrow} 0. \tag{G1.1.7}$$

$\alpha(t)$ should start with a value close to unity and then fall gradually to about 0.1 for the first part of the training sequence. This initial phase is often referred to as the *ordering phase* as this is when the global topological ordering of the map occurs. For the fine tuning of the map, or *convergence phase*, $\alpha(t)$ can be kept at a small value—typically 0.01. The detailed nature of reduction of $\alpha(t)$ is not critical.

- The shrinking rate of the neighborhood function, $N_v(t)$, does have a major effect on the final topological order of the map. It usually starts with a radius that encompasses all, or nearly all, of the neurons in the map and then shrinks with time during the *ordering phase* to a radius that includes only one or two neighboring neurons. During the *convergence phase*, the neighborhood can be held constant with a radius of one or none neighbors. For the present application, the ordering of the texture region categories is not critical and so careful consideration of the shrinking rate of $N_v(t)$, or its detailed extent, is unnecessary.

As long as the adaptation gains meet the conditions of (G1.1.7), then the minimum mean-square-error estimate for the parameters of each of the texture regions will be achieved. Though a variety of clustering algorithms could be employed in place of the SOM, the SOM does offer a number of advantages. It is a relatively fast algorithm and the use of a neighborhood function will increase the possibility of escaping from local minima during the learning process. This latter point is particularly important in this application as the initial parameters generated by the estimator are very noisy.

Local voting network. The second-level neural network is a local voting network (LVN). This second network could be another SOM, but for computational ease it is preferable to employ this simpler network. The essential difference between an SOM and an LVN is that the former will converge to the mean of the input distribution, while the latter converges to the time average. When the distribution of the estimator noise is symmetrical, which is always the case, the final result will be equivalent. Every pixel, s, in the image plane possesses a voting neuron, and each of these neurons possesses an N-dimensional weight vector corresponding to the labels for each of the N textured regions, i.e. $[l_s^1(t), l_s^2(t), \dots, l_s^N(t)]$. These weights are set to zero at the commencement of the training sequence. If the winning neuron in the SOM is v, then the LVN weight vectors are updated according to the learning rule

$$\left. \begin{aligned} l_s^v(t+1) &= l_s^v(t) + 1 \\ l_s^k(t+1) &= l_s^k(t) \end{aligned} \right\} \qquad \text{for } s \in \Omega(t) \text{ and } \forall k \neq v \tag{G1.1.8}$$

where $\Omega(t)$ is the estimating window at time t. This updating is restricted to the area within the current *estimating window*. The largest element of the LVN weight vector (*voting number*) for each pixel indicates its texture type.

Boundary relaxation. After both the networks have been trained, the different texture regions will be essentially delineated; but, as can be seen by examining the results given in figure G1.1.2, there will be some segmentation errors. These are more likely at the boundaries between texture regions. It is not possible to use some form of single-pixel-based correction as textures only have a meaning over a patch of pixels. Region boundaries can, of course, take any shape but over a limited spatial extent they can be assumed to be linear or, at least, smooth in profile. A relaxation algorithm can be used to improve the boundary profiles. A small window is moved along the texture boundaries produced by the trained LVN, and a short straight boundary with random orientation is generated within this window. It is accepted or rejected on the basis of the energy change (error) within this small region. The result is then fed forward to the LVN and the resulting voting numbers changed. The local energy, $u_2(x/W, t+1)$ in the window is calculated using (G1.1.8). If the previous local energy was $u_1(x/W, t)$ then this new boundary is accepted or rejected according to the energy change, $\Delta u = u_2 - u_1$, that is

$$P(\text{change}) = \begin{cases} 1 & \text{if } \Delta u < 0 \\ \exp\left(-\dfrac{\Delta u}{T}\right) & \text{if } \Delta u > 0. \end{cases} \tag{G1.1.9}$$

This is an example of a *Metropolis algorithm* (Metropolis *et al* 1953) where T is a parameter similar to a temperature and can be slowly decreased during the learning.

G1.1.3 Training methods

G1.1.3.1 Test data

The complete system has been extensively tested on both synthetic binary and natural textured images. It is essential to employ both types of images during testing as the former possess predictable and specified statistical properties while, of course, the latter does not. It is also an advantage to use natural textures that are commonly available to the research community as this allows some degree of comparison between the results and those published previously. The textured images contained in Brodatz's photographic album (Brodatz 1966), originally intended for artists and graphic designers, have become the *de facto* standard for examining the performance of texture segmentation techniques. The test images are 128×128 pixels and composed (for the tests discussed here) of two different textured regions but with a variety of boundaries (as indicated in figure G1.1.3).

G1.1.3.2 Training schedule

For the purposes of initial testing a second-order MRF model was used, together with LMS estimation of the model parameters. This is about the simplest model possible and so possesses a low computational

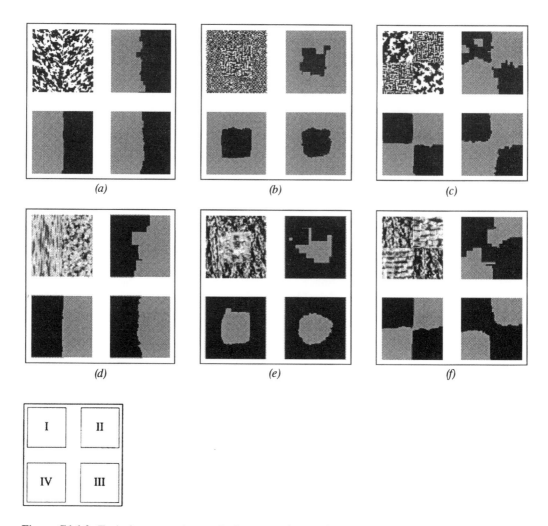

Figure G1.1.2. Typical segmentation results for two-region test images. The four images in each block are (I) input image, (II) output of fully trained SOM, (III) output of fully trained LVN, i.e. completion of first phase of segmentation, and (IV) output after boundary relaxation. Images (*a*), (*b*) and (*c*) are synthetic texture images, and images (*d*), (*e*) and (*f*) are combinations of natural images from Brodatz (1966).

cost. The estimating window, Ω, is set to be large (e.g. 70×70 pixels) at the start of training and then shrinks linearly with training time to a small prespecified size. The size of these windows is a tradeoff between two competing factors. The larger the window size, the more accurate the parameter estimation will be as textures are better represented by a block of pixels rather than, in the limiting case, a single pixel. However, for an accurate resolution of the texture boundaries, that is good segmentation, the opposite is required. The minimum practical size of the estimating window is also a factor of the texture types. For synthetic images, where the homogeneity of even small regions can be maintained, this minimum size is typically 5×5 pixels. For natural textures, local characteristics of the textures can be very variable so a minimum window size greater than about 10×10 pixels is required so that local homogeneity is maintained. During the latter part of the training schedule the LVN will assist in reducing the effects of noise and the large window sizes.

The relaxation window for boundary improvement has been found to provide optimum results with a fixed size between 10×10 and 25×25 pixels. The window may be repeatedly placed randomly on the image or it may be moved in a sequential scan. The number of possible straight-line orientations can be reduced dramatically if *a priori* information concerning the allowable boundary orientations is assumed.

Figure G1.1.3. Typical segmentation boundaries for two-region test images.

G1.1.4 Preprocessing

There is little need to provide any form of input preprocessing other than ensuring that the images have a sufficient grayscale dynamic range. Standard histogram equalization methods can be usefully employed for this.

G1.1.5 Output interpretation

The system differs from many neural network classifiers in that there are no separate training and application phases. The number of required texture regions is prespecified, an input textured image is applied and the system adapts to produce the required segmentation of this image. The resulting regions are not explicitly classified as no class labels are provided. They may, of course, be postclassified manually.

G1.1.6 Development

The system was developed using C and conventional programming support tools in an MS Windows environment. This was appropriate as the eventual use of the system would employ a similar software environment.

G1.1.7 Performance

A typical set of results is shown in figure G1.1.2. Images (*a*), (*b*) and (*c*) are synthetic binary textures, while the images (*d*), (*e*) and (*f*) are natural grayscale textures taken from Brodatz (1966). The following should be read using the identifying key for this figure. The general form of the segmentation is visible if the system's output is taken directly from the output of the SOM layer. However, the region boundaries are ill-formed with occasionally very broken outlines. There are also some instances where small isolated regions are wrongly classified. The system's output after the LVN layer (but without applying the boundary relaxation algorithm) shows a considerable improvement. There are no isolated regions misclassified and the boundaries are much smoother—in some cases, too smooth.

Finally, the results of incorporating the boundary relaxation provide a substantial improvement in boundary placement and profile. For these six test images, the resulting misclassified pixel errors for the complete system are given in table G1.1.1. It should be noted that these test images are small in relation to the estimating and relaxation window sizes. This is a more critical situation than in the practical application of the system for unsupervised texture segmentation. The system has been extensively tested on a very diverse range of textured images with similar excellent results.

Table G1.1.1. Pixel errors for test images of figure G1.1.2.

Boundary relaxation	Image					
	(*a*)	(*b*)	(*c*)	(*d*)	(*e*)	(*f*)
Not used	3.14%	5.78%	9.20%	3.42%	5.44%	8.11%
Used	1.01%	2.37%	2.30%	0.72%	3.25%	2.79%

For two-region texture images, the neighborhood function of the SOM neurons is not important as long as both neurons are active. For multiregion textures, all the neurons must be sufficiently active otherwise one neuron may represent two or more texture regions or two or more neurons may share one region.

Maintaining a large neighborhood for a longer than normal time produces a correct global segmentation. The SOM neighborhood is held constant for the first 1000 iterations, with the amplitude of this function for the nearest and next-nearest neighbors being initially set to 0.6 and 0.2, respectively, and decreasing linearly to zero. The segmentation process can be halted by monitoring the changes in the final layer or simply by setting a sufficiently large number of iterations. A typical iteration cycle for four-region textures is 5000.

System validation was conducted by maintaining a fixed number of SOM neurons and varying the number of distinct texture regions in a series of test images. The validation criterion was to calculate the total mean-square error (MSE) between the actual texture labels and the derived ones on a pixel by pixel basis. Plots of log(MSE) as a function of the region number consistently showed a minimum at the correct match between number of regions and SOM neurons. Further work involved using *real* images (aerial photographs) and excellent results have been obtained.

G1.1.8 Conclusions

The section has shown how a number of techniques, both neural and nonneural, may be integrated to produce a computationally robust system for the unsupervised segmentation of two-dimensional images based on the predefined number of textured regions. It is important to appreciate that a solution of this apparent complexity will often provide a superior solution, from a number of perspectives, than a solution which attempts to use a single network with the minimum of data preprocessing.

It is impossible to state that an approach using neural networks (let alone our own work) for texture segmentation is superior to all other approaches. This difficulty arises from several factors: the inability to produce a formal universal definition of textures (i.e. their inherent variability makes the possibility of a universal segmentation system very unlikely), differing functionality for a delineated range of images, and the absence of agreed standardized test data and procedures. However, neural networks do offer advantages in terms of computational ease (particularly so for SOM and LV networks), flexibility of operation (e.g. in our work, specifying the number of region types is simply a case of modifying the number of SOM neurons), increased probability of escaping local minima during training (the neighborhood function of the SOM assists in this), and the potential for parallel implementation.

References

Brodatz P 1966 *Textures: a Photographic Album for Artists and Designers* (New York: Dover)

Cross G R and Jain A K 1983 Markov random field texture models *IEEE Trans. Patt. Recogn. Mach. Intell.* **5** 25–39

Haralick R M and Shapiro L G 1992 *Computer and Robot Vision* (Reading, MA: Addison-Wesley) ch 7

Kohonen T 1984 *Self-Organisation and Associative Memory* (Berlin: Springer)

Lakshmanan S and Derin H 1989 Simultaneous parameter estimation segmentation of Gibbs random fields using simulated annealing *IEEE Trans. Patt. Recogn. Mach. Intell.* **11** 799–813

Manjunath B S and Chellappa R 1991 Unsupervised texture segmentation using Markov random field models *IEEE Trans. Patt. Recogn. Mach. Intell.* **13** 478–82

Metropolis N, Rosenbluth A W, Rosenbluth M N, Teller A H and Teller E 1953 Equations of state calculations by fast computing machines *J. Chem. Phys.* **21** 1087–91

Tuceryan M and Jain A K 1993 Texture analysis *Handbook of Pattern Recognition* ed C H Chen, L F Pau and P S P Wang (Singapore: World Scientific) ch 2.1

Yin H and Allinson N M 1995 On the convergence and distribution of feature space in self-organizing networks *Neural Comput.* **7** 1178–87

G1.2 Character recognition

John Fulcher

Abstract

Handwritten character recognition is discussed from both a conventional and an artificial neural network (ANN) perspective. Reference is also made to printed character recognition. The importance of segmentation and other forms of preprocessing is emphasized. Several ANN systems are described, from which it is seen that special purpose, hierarchical networks are most suited to this task, with recognition accuracies of around 95% being commonplace.

G1.2.1 Introduction

Interest in optical character recognition (OCR) predates the era of electronic computers (Mori *et al* 1992). In more recent times, much attention has been paid to on-line handwritten rather than printed character recognition (Nouboud and Plamondon 1990, Tappert *et al* 1990). We concern ourselves in this paper with both types of automated character recognition but devote more attention to the former because it is the more difficult of the two problems to solve.

G1.2.1.1 Printed characters

We can consider a printed character as comprising a two-dimensional array (matrix) of pixels (dots) similar to that formed by a dot matrix printer or a computer terminal screen (be it CRT or LCD). Obviously, the more dots (pixels) we use to represent the character, the better resolution we are able to achieve. Moreover, more dots enable different font types and styles to be represented.

Character recognition amounts to the ability to discriminate among a finite number of patterns; for example, 62 for the alphanumeric character set (both upper and lower case). It is a straightforward matter to reformulate these two-dimensional patterns into one-dimensional vectors (63-element vectors if we assume 7×9 dot matrices). Our task then becomes one of distinguishing 62 different vectors. This task can be performed using either conventional *pattern classification* techniques, or artificial neural networks (since ANNs are widely known to excel at pattern classification and recognition tasks).

F1.2

G1.2.1.2 Handwritten characters

Handwritten characters likewise will be captured, either directly or by way of a digitizing tablet, in pixel matrix form. However, in order to cater to the wide variety of handwriting styles, a larger-size (higher-resolution) dot (pixel) matrix will usually be necessary. For example, the handwritten zip codes used in the study of LeCun *et al* (1989) were first stored as 40×60 pixel images, then reduced to 16×16 images prior to presenting them to the ANN for classification—and this was only for the digits 0 through 9, not the entire alphanumeric character set!

It should be emphasized at this point that a *human* classifier cannot classify such handwritten characters with anything like 100% accuracy. Thus, it is unreasonable to expect machine (computer) classifiers to perform perfectly. The best we can hope for (the 'bottom line' as it were) is that the machine (computer, ANN or whatever) will perform at least *as well as* a human classifier.

G1.2.2 Conventional character recognition techniques

Some form of preprocessing will usually be required prior to handwritten character recognition proper. Typical preprocessing operations are segmentation (using a sliding window, say, to isolate characters), smoothing, line thinning, noise filtering, and correction for size and orientation (in other words, normalization).

Template matching is a technique that works well with printed character recognition but is not particularly suited to handwritten character recognition due to the much more extensive shape and size variation encountered with the latter. Curve matching is more appropriate for handwritten characters, especially *elastic* matching techniques, which can be further enhanced by incorporating local affine transformations (Wakahara 1994).

By analyzing the strokes that comprise handwritten characters, we can determine their number, order and direction. Such characteristics can then be stored as stroke codes. More specifically, chain codes (a standard image processing technique for representing objects within images) are a useful method for encoding handwritten characters. Boundaries can be specified in terms of segments of specified length and direction (the latter in terms of 1-of-n nearest pixel neighbors, with n being typically 4, 6 or 8). Such an (efficient) encoding scheme can be likened to Huffman or run length coding. Alternatively, *vector quantization* can be used to encode these strokes.

C1.1.5

Feature extraction is another popular handwritten character recognition technique (e.g. ascenders/descenders, dots, cusps, closures, and so on, but note that the choice of features is somewhat critical). Alternatively, if we represent the character image in algebraic form, we can apply standard image processing transforms, such as the Karhunen–Loeve expansion (Mori *et al* 1992). Conversely, by considering the x- and y-coordinates of the image, we can apply Fourier series techniques. Other standard approaches include decision trees, dictionary lookups and Bayesian classifiers.

Statistical (stochastic) techniques appropriate for handwritten character recognition include k-nearest neighbors (Kovacs *et al* 1993) and hidden Markov models (HMMs), respectively (Kundu *et al* 1989, Chen *et al* 1994, Veltman and Prasad 1994). In the latter approach, the sequence of character segments (symbols) which comprise a word is likened to the intermediate states traversed within the HMM.

G1.2.3 Non-English character recognition

So far we have concentrated on English-language character recognition. There has also been a considerable amount of effort devoted to non-English character recognition, especially Chinese/Japanese (Kanji) characters, and to a lesser extent Arabic. Once again we are faced with problems of scaling, orientation and variability between writers. Furthermore, we often need to deal with larger numbers of more complex characters—this is especially true of the Chinese (Japanese) language, in which over 5000 (3000) characters are encountered in everyday usage. This means we need to be able to discriminate between more *mutually similar* characters.

G1.2.3.1 Chinese/Japanese character recognition

Typical of Chinese character recognition systems is the two-stage one described by Huang (1993)—the first stage involves stroke extraction, achieved using thinning and curve fitting; the second stage performs structural matching, achieved using either deterministic or probabilistic means.

Alternative approaches include dynamic programming (Nouboud and Plamondon 1990), Euclidean distance, graph matching (Chen and Lieh 1990), and local affine transformations (Wakahara 1994). The importance of preprocessing is covered by Chang and Wang (1993).

G1.2.3.2 Arabic character recognition

The Arabic alphabet comprises 29 characters, whose shape depends on their position in a word. Moreover, each character can be formed by between one and four strokes, each of which can occupy one of three zones: on, over or under the center (base) line. These strokes can be augmented by the presence of dots or hazma, which render different meanings to the character.

Stroke segmentation and thinning are common preprocessing stages performed in Arabic character recognition. A variety of techniques can then be employed for classification proper. For example, El-Desouky *et al* (1992) used chain codes and reported up to 94% correct classification in around 3 seconds on an IBM PC. El-Sheikh and Guindi (1988), on the other hand, use feature extraction. They represent isolated characters by their outer contour and stress marks, and store these contours as a series of x, y coordinates. Fourier series coefficients are, in turn, derived from these coordinate series (only three of which are necessary to accurately reconstruct the characters). They reported a 99% classification rate using this approach. A decision tree approach was followed by El-Sheikh and El-Taweel (1990), utilizing an *a priori* classification (into four subsets: isolated, end, beginning and middle characters) in their real-time system, which delivered a recognition rate of 99.6%.

G1.2.4 Artificial neural network character recognition techniques

Despite efforts using *single-layer feedforward networks* for isolated handwritten digit recognition (Knerr *et al* 1992), the most common approach to handwritten character recognition has been with multilayer, feedforward (backpropagation) networks or *multilayer perceptrons* (MLPs). More specifically, specialized hierarchical, layered MLPs have been found to be especially suited to character recognition (Jackel *et al* 1988, LeCun *et al* 1989, Sabourin and Mitchie 1992). We shall describe a couple of representative systems in section G1.2.4.2.

    Work has also been reported using *associative memories*, both the familiar Hopfield type, as well as more specialized ones such as the one based on Chua attractors reported by Baird *et al* (1993). Unsupervised *Kohonen SOM networks* have also been applied to this task (Lisboa 1992), as has Kohonen's supervised *LVQ algorithm* (Idan and Chevallier 1991). Moreover, the Sharp Corporation has recently developed a multilayer ANN which uses LVQ to recognize Japanese Kanji characters (Hammerstrom 1993).

    Apart from research projects, *commercial* neural network based products have also appeared in the marketplace in recent years. Typical of such products are OmniTools and Nestor Reader from Nestor Inc, Providence, Rhode Island.

C1.1

C1.2

C1.3

C2.1.1

C1.1.5

G1.2.4.1 Printed character recognition

Zurada (1992) compared the performance of several MLP configurations in classifying a full set (95) of lowercase letters, uppercase letters, digits and punctuation 7×10 dot matrix (Apple Imagewriter) characters. The output patterns were classified either as 95 discrete pattern classes (characters) or 8-bit ASCII, depending on the target application. Best classification results were obtained using two hidden layers, each comprising 70 neurons (in fact any number between 25 and 80 performed well). Moreover, the eight-output-node networks performed better than the 95-output ones.

    Autoassociative memories are also good candidates for printed character recognition, since they are well known to be able to recall *perfect* characters when presented with noisy or incomplete inputs. Figure G1.2.1 shows the correct recall of a 10×12 pixel '3' corrupted by 25% random noise using a 120-node (14 400 weight) Hopfield network trained to recognize eight different characters, as indicated (Lippmann 1987).

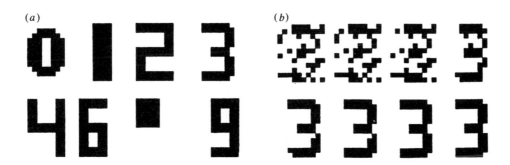

Figure G1.2.1. (*a*) Eight exemplar patterns. (*b*) Output patterns for noisy '3' input. Hopfield network autoassociator for printed character recognition. (Reprinted from Lippmann 1987, with permission of IEEE.)

G1.2.4.2 Handwritten character recognition

ARNIE is an MLP simulator developed for the Apple Macintosh (Cheng 1993). Both the number of hidden layers and the number of nodes in each layer are user selectable via dialog boxes. Handwritten characters can be generated either dynamically using the mouse, or from previously created files with the inbuilt graphical editor. Figure G1.2.2 shows both the main window used for training and recall and a session report window. New character classes are added by selecting the 'new' button. Different training exemplars for each respective class are added using the 'add' button (like the uppercase character 'A' in figure G1.2.2(*a*)).

In the study by LeCun *et al* (1989) of handwritten zip codes, a three-hidden-layer feedforward network with the following configuration was used:

input layer	=	256 neurons	(16 × 16 pixel images)
hidden layer#1	=	768 neurons	(12 × 64)
hidden layer#2	=	192 neurons	(12 × 16)
hidden layer#3	=	30 neurons	
ouptut layer	=	10 neurons	(one for each digit).

Each hidden layer searches for (small-sized) features in the previous layer. For example, each neuron in the first hidden layer is connected to a 5 × 5 window of the 16 × 16 input layer image. This 5 × 5 window is translated by two pixels for neighboring neurons in the hidden layer, which enables this first hidden layer to detect certain predefined features. In a similar manner, the second hidden layer searches for features within a 5 × 5 window of the first hidden layer pattern.

The network of LeCun *et al* was trained using 7291 of the 9298 available handwritten zipcode training

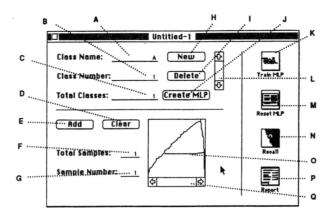

A	Shows the current class name being displayed (eg. as in this document the current class name is "A").
B	Shows the current class number being displayed.
C	Shows the total number of classes in the document.
D	The "**Clear**" button is for deleting a sample from the current class.
E	The "**Add**" button is used to add a new sample into the current class (see adding samples to a class).
F	Shows the total number of samples in the currently shown class.
G	Shows the current sample number that is being displayed.
H	The "**New**" button is used to create a new class; after this button is selected, a dialog box will allow entering of a class name (see inserting a new class into the network).
I	The "**Delete**" button is used to delete the currently shown class, the current class will be removed from the list of classes.
J	The "**Create MLP**" button is for creating a new multilayered perceptron network using backpropagation.
K	The "**Train MLP**" is used to train the multilayered perceptron network that has been created.
L	This scroll bar is used to scroll through all the classes in the document.

Figure G1.2.2. (*a*) ARNIE user interface.

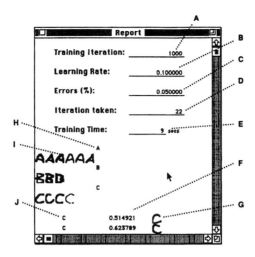

A   Maximum number of training iterations that were set for the training.
B   Learning rate used in the training of the MLP network.
C   Convergence errors.
D   Iterations taken for the training of the MLP network.
E   Time taken to train the MLP network.
F   The highest value given by the MLP network to determine the class
      which the image/pattern belongs to.
G   The image used for recall.

Figure G1.2.2. Continued. (*b*) ARNIE user interface.

exemplars. Following training, the network performed with a recognition accuracy of 95%. However, if there are marginal characters (those which *human* classifiers have trouble distinguishing, and which account for 12.1% of the zipcode database), then this accuracy rises to 99%.

In the system presented by Jackel *et al* (1988) gray-scale character images are initially converted into (40 × 60 pixel) black and white ones and then reduced to 16 × 16 pixels prior to skeletonization. These skeletonized black and white images are then converted into 20 different 16 × 16 feature maps, which are then compressed into 3 × 3 arrays in order to form a 180-element output vector. This vector is then used to classify the handwritten digit into one of 10 categories. The recognition accuracy of this system varied between 95% and 99% when tested on the same zipcode database used by LeCun *et al*.

The *neocognitron* is a special purpose, hierarchical, feedforward network which has a good C2.1.3 generalization ability that enables it to perform deformation-invariant character recognition. However, this ability is heavily reliant on the choice of training set. Figure G1.2.3 shows the neocognitron structure organized in alternate S-cell and C-cell layer pairs (Fukushima and Wake 1991, Fukushima 1992). The number of layer pairs is dictated by the complexity of the patterns to be recognized. Four stages are required for alphanumeric character recognition (35 different characters in this study, the letter 'O' and the number '0' being regarded as the same pattern).

The S-cells perform (preselected) feature extraction, while the C-cells compensate for positional shift (translation) of these features within layers. Each such S- or C-cell layer is divided into subgroups (cell planes) which correspond to the specific features of interest. Connections to S-cells are variable (and reinforced by training), whereas connections are fixed between S- and C-cells. Local features extracted in the earlier stages become gradually integrated into more global features in the later stages. Figure G1.2.4 shows some of the digits used to test this network, which gives an idea of how well the network is able to tolerate deformation, changes in scale, translation and additive noise (the digits on the left being correctly recognized, those on the right not).

Average recognition time was around 3.3 seconds (on a Sun SPARCstation). Training took around 13 minutes, compared with around 3 days in the study by LeCun *et al*. A neocognitron simulator is available via anonymous ftp from unix.hensa.ac.uk (129.12.21.7).

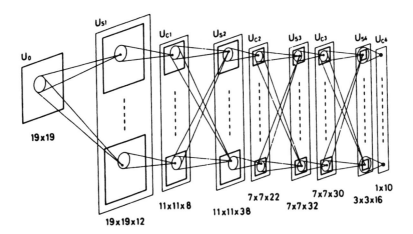

Figure G1.2.3. Neocognitron architecture. (Reprinted from Fukushima *et al* (1983) with permission of IEEE.)

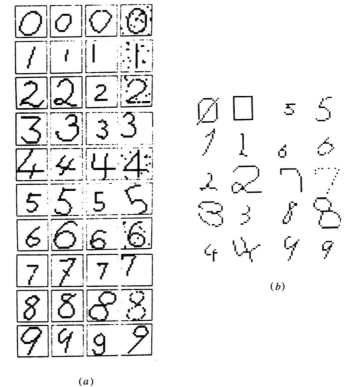

(a)

(b)

Figure G1.2.4. Deformed character test patterns. (Reprinted from Fukushima *et al* (1983) with permission of IEEE.)

G1.2.4.3 Chinese character recognition

ANN approaches to handwritten Chinese character recognition include MLPs (Lua and Gan 1990, Joe *et al* 1990), the neocognitron (Fukushima 1992), and Hopfield networks (Liao *et al* 1993). In the latter network, the rows connect to the test character and the columns to the training exemplar (character). Preprocessing comprises both thinning and the removal of undesirable holes and noise. The stroke extraction process utilizes the *k*-curvature method to locate turning points in the straight-line segments which constitute a curve. The method presented by Liao *et al* (1993) is flexible (able to handle differing numbers of rows and columns), and is able to recognize characters even if the number of strokes in the test and exemplar characters differ. It can also cater to positional translations and rotations. They report a recognition accuracy of around 90%.

G1.2.5 Conclusion

Unconstrained handwritten character recognition remains a difficult problem. Nevertheless, typical state-of-the art systems are capable of emulating human classifiers, with recognition accuracies of around 95% commonly being reported. Some form of preprocessing is usually required, however, to achieve such recognition accuracy (such as partitioning/segmentation into isolated characters, noise filtering, thinning and the like).

Feedforward ANNs are most often employed, with most success being reported for special-purpose hierarchical networks (LeCun *et al* 1989, Fukushima and Wake 1991). For readers interested in investigating handwritten character recognition for themselves, the following CD-ROM databases are recommended: Cedar CD-ROM1 (State University of New York at Buffalo) and the NIST Special Database 3 (US National Institute of Standards and Technology). Samples of the latter are available via anonymous ftp from sequoyah.ncsl.nist.gov.

References

Baird B, Hirsch M and Beckman F 1993 A neural network associative memory for handwritten character recognition using multiple Chua characters IEEE *Trans. Circuits Syst.* **40** 667–74

Chang H-D and Wang J-F 1993 Preclassification for handwritten Chinese character recognition by a peripheral shape coding method *Patt. Recog.* **26** 711–19

Chen L-H and Lieh J-R 1990 Handwritten character recognition using a 2-Layer random graph model by relaxation matching *Patt. Recog.* **23** 1189–205

Chen M-Y, Kundu A and Zhou J 1994 Off-line handwritten word recognition using a hidden Markov model type stochastic network *IEEE Trans. Patt. Anal. Mach. Intell.* **16** 481–93

Cheng G 1993 ARNIE—neural network simulator Preprint 93-3 University of Wollongong, Department Computer Science

El-Desouky A, Salem M, El-Gwad A and Arafat H 1992 A handwritten Arabic character recognition technique for machine reader *Int. J. Mini Microcomput.* **14** 57–61

El-Sheikh T and El-Taweel S 1990 Real-time arabic handwritten character recognition *Patt. Recog.* **23** 1323–32

El-Sheikh T and Guindi R 1988 Computer recognition of Arabic cursive scripts *Patt. Recog.* **21** 293–302

Fukushima K 1992 Character recognition with neural networks *Neurocomput.* 221–33

Fukushima K, Miyake S and Ito T 1983 Neocognitron: a neural network model for a mechanism of visual pattern recognition *IEEE Trans. Syst. Man Cybern.* **13** 826–34

Fukushima K and Wake N 1991 Handwritten alphanumeric character recognition by the neocognitron *IEEE Trans. Neural Networks* **2** 355–65

Hammerstrom D 1993 Neural networks at work *IEEE Spectrum* **30** 26–32

Huang J-S 1993 Optical handwritten Chinese character recognition ed C Chen, L Pau and P Wang *Handbook of Pattern Recognition and Computer Vision* (Singapore: World Scientific) pp 595–624

Idan Y and Chevallier R 1991 Handwritten digits recognition by a supervised kohonen-like learning algorithm *Proc. Int. Joint Conf. on Neural Networks (Singapore) III* pp 2576–81

Jackel L, Graf H, Hubbard W, Denker J and Henderson D 1988 An application of neural net chips: handwritten digit recognition *Proc. Int. Joint Conf. on Neural Networks (San Diego, CA) II* pp 107–15

Joe K, Mori Y and Miyake S 1990 Construction of a large-scale neural network: simulation of handwritten Japanese character recognition on ncube *Concurrency Practice Exp.* **2** 79–107

Knerr S, Personnaz L and Dreyfus G 1992 Handwritten digit recognition by neural networks with single-layer training *IEEE Trans. Neural Networks* **3** 962–8

Kovacs Z, Raggazoni R, Rovatti R and Guerrieri R 1993 Improved handwritten character recognition using second-order information from training set *Electron. Lett.* **29** 1308–10

Kundu A, He Y and Bahl P 1989 Recognition of handwritten word: first and second order hidden Markov model based approach *Patt. Recog.* **22** 283–97

LeCun Y, Boser B, Denker J, Henderson D, Howard R, Hubbard W and Jackel L 1989 Backpropagation applied to handwritten zip code recognition *Neural Comput.* **1** 541–51

Liao H-Y, Huang J-S and Huang S-T 1993 Stroke-based handwritten Chinese character recognition using neural networks *Patt. Recog. Lett.* **14** 833–40

Lippmann R 1987 An introduction to computing with neural nets *IEEE ASSP Magazine* April 4–22

Lisboa P 1992 Single layer perceptron for the recognition of handwritten digits *Int. J. Neural Networks* **3** 17–22

Lua K T and Gan K W 1990 Recognising chinese characters through interactive activation and competition *Patt. Recog.* **23** 1311–21

Mori S, Suen C and Yamamoto K 1992 Historical review of OCR research and development *Proc. IEEE* **80** 1029–58

Nouboud F and Plamondon R 1990 On-line recognition of handprinted characters: survey and beta tests *Patt. Recog.* **23** 1031–44

Sabourin M and Mitchie A 1992 Optical character recognition by a neural network *Neural Networks* **5** 843–52

Tappert C, Suen C and Wakkahara T 1990 The state of the art in on-line handwriting recognition *IEEE Trans. Patt. Anal. Mach. Intell.* **12** 787–808

Veltman S and Prasad R 1994 Hidden Markov models applied to on-line handwritten isolated character recognition *IEEE Trans. Image Proc.* **3** 314–8

Wakahara T 1994 Shape matching using LAT and its application to handwritten numeral recognition *IEEE Trans. Patt. Anal. Mach. Intell.* **16** 618–29

Zurada J 1992 *Introduction to Artificial Neural Systems* (St Paul, MN: West) section 8.2

G1.3 Handwritten character recognition using neural networks

Thomas M Breuel

Abstract

Automating the transcription of handwriting on forms and mail pieces can result in significant cost savings. Neural networks have proven to be a robust and reliable component of handwriting recognition systems. In this paper, we review some of the major concepts and approaches to handwriting recognition using neural networks.

G1.3.1 Introduction

Despite the widespread availability of computers, or perhaps because of it, an ever increasing amount of data is obtained in handwritten form on paper. The offline handwriting recognition problem is the problem of transforming this data into machine readable form, accurately and quickly.

A typical application for computer-based capture of handwritten data is the tax forms issued by the US Internal Revenue Service. Forms contain preprinted instructions and outlines of response fields. Respondents fill alphanumeric information (name, address, income, deductions, etc) into the preprinted response fields. Each of these responses needs to be transcribed into machine readable form and stored in an online database. There exist numerous similar applications in business, banking, health care and government.

Currently, most of these data capture problems are solved manually, sometimes by small secretarial staffs and sometimes by large, offshore service companies. However, while manual data capture is still competitive, computer and pattern recognition technology holds the promise of greatly reducing costs.

The goal of this article is to give the practitioner who wants to build a working isolated handwritten character recognizer a useful starting point, and to give neural network researchers who are looking to isolated handwritten character recognition as a benchmark for new architectures an idea of what has already been accomplished. The paper is based on the experience that the author has gained building neural-network-based handwriting recognition systems for US Census applications (Breuel 1994).

G1.3.2 Issues

We start our discussion assuming that we get an image of an isolated handwritten string consisting of multiple characters. This might, for example, be the contents of a forms field given in response to a question like 'Please state your name in the space provided'.

G1.3.2.1 Segmentation

In order to recognize and transcribe the handwritten input contained in a complete field image (field-level recognition as opposed to character recognition), two problems need to be solved†:

---

† A few systems and approaches attempt to recognize whole words or do not overtly try to isolate individual characters before attempting word recognition; examples of such systems are Martin and Rashid (1992), Martin *et al* (1993), Fukushima (1992), Imagawa and Fukushima (1993), Bunke and Roth (1993), Giloux *et al* (1993), Mulgaonkar *et al* (1994). Such systems fall outside the scope of this paper.

- the image of the string needs to be segmented into images of individual characters
- the individual characters need to be recognized.

There are two major approaches to the segmentation problem.

(i) Have the respondent write isolated characters into preprinted boxes and extract them based on a geometric forms model. Such an approach is easy to implement, fast and makes segmentation errors unlikely. It helps resolve some of the inherent ambiguities in Western handwriting. However, user acceptance for writing long responses in preprinted per-character boxes tends to be low, and preprinted boxes require more space on forms, possibly increasing printing costs.

(ii) Unconstrained fields consist of a single blank space for a complete response (one or more words). Recognition is carried out by generating multiple, possibly overlapping candidate isolated character images. These are then assembled into a complete interpretation of the input (see Burges *et al* 1992, Giloux *et al* 1993 and Breuel 1994 for examples of systems that use such an approach). Unconstrained fields have higher user acceptance, accommodate long responses better, and require less space on the form. But a system for the recognition of responses in unconstrained fields requires more computation and is inherently more susceptible to unresolvable ambiguities.

Both approaches have their place in practice. Preprinted character boxes are best reserved for responses that are difficult to verify and where low-perplexity language models are not available. Examples are bank check amounts, social security numbers, telephone numbers and personal names.

Fields without preprinted character boxes are best used where space is at a premium, high user acceptance is desirable, or where low-perplexity language models or small dictionaries are available. Examples are fields for occupation, city names, street names, race, etc.

Regardless of which of the two approaches we use, we need an isolated character recognizer that takes the output of the segmentation step and recognizes the individual characters.

G1.3.2.2 *Variability*

Isolated character recognition is a difficult problem because inputs exhibit considerable variability. The most common way of compensating for this variability is by using normalization steps. Normalization steps attempt to remove the variability in a model-independent, bottom-up way. Major sources of variability in handwriting recognition and corresponding normalization steps are:

- *noise and dropouts*: scanning hardware adds noise and dropout pixels and may affect image contours
  * morphological filtering to remove salt-and-pepper noise
  * two-dimensional image processing or mathematical morphology to compensate for aliasing and pixelization

- *translation*: characters are not always written in the same position relative to the enclosing box
  * move the centroid of the character image to the origin
  * move the character into a given bounding box

- *size*: characters can be written in many different sizes without changing their meaning
  * force the character image inside a fixed-size bounding box
  * alternatively, normalize the second moments of the character image by scaling

- *writing instrument*: a variety of writing instruments are in use, differing in thickness, color and stroke quality
  * normalize for stroke thickness by skeletonization (thinning)
  * normalize for stroke texture by two-dimensional image processing or mathematical morphology

- *slant*: different writers and the same writer under different conditions will slant their letters differently; that is, their handwriting undergoes a shear along the horizontal axis (note that characters are not, in general, invariant under rotation, namely '6' versus '9')
  * normalize at the character level by forcing the xy moment of the character image to zero by shearing
  * normalize at the field level by identifying likely vertical strokes and forcing them to be vertical by shearing

- *writing style*: we find significant variations in writing style with nationality, region of origin, age, education, profession, physiological state, forms type, size and writing instrument.

A normalization step for translation and size will allow us to use a recognition method that is not invariant under translation and size for the recognition of handwritten characters. Normalization methods are described and referenced in more detail in the literature, both in surveys and evaluations (e.g. Blue *et al* 1994, Wilkinson *et al* 1994, Geist *et al* 1994, Kasturi and O'Gorman 1992, Davis and Lyall 1986, Chatterji 1986) and in specific recognition systems (Caesar *et al* 1993, Grother 1992, Shustorovich 1994, Guillevic and Suen 1993, among many others).

However, normalization is not the only possible approach. Several other approaches have been used in the literature.

- We can base recognition on a feature set that is invariant. This is closely related to applying normalization steps, but the focus is different. Common techniques that yield invariant feature sets are based on Fourier transforms, moments and topological descriptors. For some typical examples see Lu *et al* (1993), Chatterji (1986), Davis and Lyall (1986).
- We can make the recognition method itself invariant. Translation-invariant recognition has been the subject of extensive research in neural networks and computer vision. Such methods tend to be more computationally intensive, but can give lower error rates (Drucker and LeCun 1991, Simard *et al* 1992, Breuel 1993). There are two main reasons for this. First, normalization during preprocessing might throw away important information that is important for recognition (image processing sometimes closes small loops in characters like 'e'), although this can be overcome by passing on both the normalized input along with information about the normalization step, e.g. the centered character image plus the translation. Second, multiple normalization steps can interact in deleterious ways.
- Instead of implementing invariants directly in the recognition algorithms, we treat them as idiosyncratic variations. That is, we simply apply a noninvariant recognition method to a problem that we happen to know has some invariants. We can still take advantage of our knowledge of invariants by increasing the size of the training set through generating transformed (translated, scaled, thickened, etc) instances of characters artificially. Since invariants in character recognition tend to be only approximate and apply over only limited ranges of parameter values, such an approach can work quite well in practice.
- We can also simply note the normalizing transformation (e.g. letter slant) and pass this on to the classifier as an additional feature of the input, alongside the unnormalized input.

G1.3.2.3 Probability estimates

For most applications, it is not sufficient for a recognizer to return a 'hypothesis' of what character an unknown input represents, but also some degree of 'confidence'. This confidence is usually used for two purposes: rejection of characters or fields, and integration of character hypotheses into a complete transcription of a whole field.

Bayesian decision theory (see Berger 1985, Kiefer 1987) has been a useful tool for implementing theoretically well founded methods for rejection and integration. For example, for implementing rejection, Bayesian decision theory tells us that under a zero–one loss function, we should return as a hypothesis the class with the highest posterior probability $P(\omega|I)$ (where ω is the class and I is the input image), and that the probability of error is $1 - P(\omega|I)$. For an example of integration of character hypotheses in a Bayesian framework, see Breuel (1994).

In order to use probabilities for rejection and integration, we need to estimate them. An important result about conditional probabilities and neural networks is that the standard multilayer perceptron (MLP), when trained on a classification problem, actually estimates conditional probabilities (Bourlard and Wellekens 1989). This is a special instance of a number of more general results about using regression methods for probability estimates in classification problems. For a simple MLP, the outputs only approximate posterior probabilities and are not guaranteed to sum to 1, as would be expected for exact posterior probabilities ($\sum_{\omega \in \Omega} P(\omega|x) = 1$, where Ω is the set of all character classes). This can be remedied by using a slightly different network architecture.

Many non-neural-network architectures, as well as some neural network architectures, return 'confidence measures' that are not directly related to posterior probabilities. The most common way of using such confidences for the purposes of rejection or integration is to use a simple binning or thresholding strategy; this second step can be viewed as a simple kind of classifier, yielding a system that is very similar to stacked classifiers (see below).

G1.3.2.4 Input representation

We have already addressed questions of normalization of the two-dimensional input signal to a handwritten character recognizer. What remains is the question of how to transform this two-dimensional signal into a feature vector that can serve as input to a neural network. An annotated bibliography alone of all the different kinds of representations tried for handwritten character recognition would be longer than this whole paper. We limit ourselves here to some general points, some starting points for reading and some simple recommendations.

An important distinction is between *topographic* and *nontopographic* representations. In a topographic representation, the components of the feature vectors correspond in some simple way to the two-dimensional locations in the input image. An example of a topographic representation is a two-dimensional representation of the local orientations found in an input image (perhaps inspired by the existence of topographically organized orientation-selective cells in early biological visual processing); a nontopographic representation would be a moment-based representation.

Some systems attempt to *learn feature extraction* methods, starting only with a raw image, while others rely on sophisticated *hand-coded* feature extraction methods. Two important examples of learning feature extraction methods applied to handwritten character recognition are the Karhunen–Loeve transformation (KLT; e.g. Grother 1992), and the topographic feature maps obtained through weight sharing in the system described by Burges *et al* (1993) and LeCun *et al* (1989).

To get a feel for the kinds of representations in use in working handwritten character recognition systems, the reader should start with several of the available practical comparisons and surveys (e.g. Davis and Lyall 1986, Govindan and Shivaprasad 1990, Kasturi and O'Gorman 1992, Blue *et al* 1994). It appears that seemingly very different representations can give very similar performance. There are several reasons for this. First, the quality of a chosen representation is only one of a large number of factors determining overall system performance; as long as a representation is 'reasonable' and well matched to the classifier used, a system based on it may still outperform other systems. Second, seemingly very different representations may actually be closely related on closer inspection; for example, moment-based, Fourier-based and KLT-based techniques can all be viewed as expansions of an input image in terms of two-dimensional basis functions.

For MLP-based handwritten character recognizers, hand-coded topographic representations of local orientation and the presence of features such as endpoints, junctions, holes and bays are a common and proven choice.

G1.3.3 Neural network architectures

The output of preprocessing is a representation of the unknown character as a high-dimensional feature vector. For a typical topographic representation (Breuel 1994), each two-dimensional feature map is 10×10 pixels in size, and there may be as many as eight such feature maps (four for gradient features, and another four for structural and topological features). Perhaps surprisingly, the high dimensionality of this feature vector does not appear to pose a significant problem for many neural network architectures.

The theory and implementation of the networks described below can be found elsewhere in the handbook. Here, we mention only general properties that specifically relate to isolated character recognition.

G1.3.3.1 Multilayer perceptrons

C1.2
C1.2.3
Perhaps the most widely used neural network architecture is the three-layer *perceptron* with sigmoidal activation function trained using the *backpropagation algorithm*. Performance of such three-layer perceptrons on raw images appears not to be competitive. However, when proper normalization and feature-extraction steps precede the three-layer perceptron, performance becomes competitive. Several successful systems have used such neural network architectures for recognition.

Additional layers can be added before the input layer of a basic three-layer perceptron for isolated character recognition if we wish to carry out feature extraction itself using a neural network architecture;

such a network would then be presented essentially with a raw input image. The system described by Burges *et al* (1993) and LeCun *et al* (1989) takes such an approach. It adds two additional layers before a fully interconnected three-layer perceptron. In order to reduce the number of free parameters and training time, the first two layers in that network use shared or replicated weights.

We can think of these two input layers as performing convolutions of the input image with specific two-dimensional patterns (patterns that are learned automatically using the backpropagation algorithm), followed by a nonlinear (i.e. sigmoidal) transformation of the resulting values. The first two layers of such an architecture can easily perform a computation similar to that of hand-coded feature maps representing the magnitude of directional derivatives. However, features like the detection of 'interior regions' can easily be hand-coded but would be difficult to compute reliably from raw images using the two feedforward perceptron layers at the input of the five-layer perceptron.

G1.3.3.2 Classifier combination

A method of increasing importance is classifier combination, in which the outputs of multiple neural networks and statistical classifiers are combined, generally resulting in a significantly improved overall recognition rates. *Linear combination* of classifiers has been advocated in Perrone (1993) and applied to handwritten character recognition; similar methods have been used in Bayesian statistics (Berger 1985). *Stacking classification* or sequential classifiers refers to combining the outputs from several first-stage classifiers into a single feature vector and using that as input to a second-stage classifier (see Wolpert 1992), often a neural network (linear combination methods can be viewed as a special kind of stacking, with a kind of perceptron as the second classifier). *Boosting* refers to a technique that compares and combines the outputs from several neural networks. For a description and application of the technique to character recognition, see Drucker *et al* (1993, 1994). Examples of classifier combination can also be found in Geist *et al* (1994), Xu *et al* (1992), Loncelle *et al* (1992), Breuel (1993), Ho *et al* (1994), Breuel (1994), Battiti and Colla (1994), and Xu *et al* (1994).

G1.3.3.3 Other neural network techniques

Probabilistic neural networks (PNN; Specht 1990) are similar to kernel density estimators (see Duda and C1.4 Hart 1973). They are noteworthy because they have performed well in tests conducted by NIST (Blue *et al* 1994), and because a free implementation is available from NIST (see the appendix for contact information).

Radial basis functions have been studied extensively from a theoretical point of view (e.g. Powell C1.6.2 1987) and applied successfully to problems in handwritten character recognition (e.g. Lee 1991, Lemarie 1993).

All the neural network techniques mentioned above are inherently not invariant under translation or skew. Some architectures have tried to address invariance directly as part of the network architecture. An interesting example of this is given in Bottou and Vapnik (1992) and Simard *et al* (1992).

Isolated handwritten character recognition has become somewhat of a benchmark for neural networks, perhaps due to the wide availability of large amounts of training data, the challenges involved in coping with input variability and implementing invariances and the high dimensionality of the input. Numerous methods have therefore been applied; a taxonomy can be found in Blue *et al* (1994).

G1.3.4 Alternative approaches

Perhaps the most important and basic non-neural-network method is k-nearest-neighbor classification. It is easy to implement and train, requires no 'parameter tuning', tends to be robust, has asymptotic performance guarantees, and in many cases performs well in practice (see, for example, Weideman *et al* 1989, Lee 1991, Yan 1994). Its main disadvantage is that recognition is slow compared to other approaches. Nearest-neighbor methods are also important because of close ties to several neural network methods, including radial basis functions, multilayer perceptrons (both of which can be interpreted as a kind of 'weighted nearest-neighbor classifier'), and probabilistic neural networks, among others.

Classical and modern statistical approaches are plausible alternatives to neural network models. In fact, there is often considerable similarity between methods independently conceived in the statistical

C1.1.5 community and the neural network community. Examples of statistical methods closely related to neural network techniques are *additive models, vector quantization, mixture models* and *logistic regression*.

Direct geometric matching of geometric character models to images of unknown characters is a technique borrowed from computer vision and achieves high recognition rates (Breuel 1993). Like Drucker and LeCun (1991) and Simard *et al* (1992), this method achieves invariance under translation not through normalization, but by incorporating it directly into the matching step.

F1.7 Hidden Markov models (HMMs) were originally mainly applied to *speech recognition*, but have become popular for offline handwriting recognition as well (e.g. Giloux *et al* 1993). HMMs integrate the segmentation and recognition steps. HMMs can be combined with neural network techniques (e.g. Bourlard and Wellekens 1989).

G1.3.5 Training and evaluation

B3 The general concepts of neural network *training* and *validation* are described elsewhere in the handbook (see also Grother 1993).

Several *standard databases* exist for training and testing handwriting recognition systems (Garris 1992; for information on how to obtain the NIST database of handwriting taken from US Census forms and the CEDAR database of handwriting taken from live US mail pieces, see the appendix). These databases are easily available, and their use should be considered almost a prerequisite for publishing performance results on specific handwriting recognizers.

It must be stressed again that in order to obtain accurate estimates of real-world performance, *a strict separation between training and test sets must be scrupulously maintained*. In fact, even repeated use of a test set solely for evaluating different versions or revisions of a single recognition system can easily lead to an overestimation of the performance of a particular method. For this reason, retaining multiple test sets is prudent.

Rather than modifying or improving the recognition algorithm, simply *using more training data* is often an easy way of improving recognition performance of an isolated character recognizer. In order to determine and extrapolate how well a system might perform when given more training data, the system can be trained on different size subsets of the available training set and the test set error can be plotted against the training set size.

An important property of handwriting is that there is considerable *variation in writing styles*, often correlated with factors like writer origin, age and education. This has several important consequences. Most importantly, the writing style of a training set may be different from the writing style actually encountered in an application. A particularly severe example would be trying to use a system trained on US American digit fields for recognition of ZIP codes written by European writers: even some of the basic letter shapes are different. But even within a particular population, we will encounter some systematic variation. This can even be observed in particular training and test sets. For example, some clusters of consecutive writers found in the NIST databases seem to be sometimes significantly easier and sometimes more difficult to recognize than the database as a whole. This means that when selecting a test set, it should probably be drawn randomly from a database, and it may be useful even to randomize the order of fields in the training set.

The issue of *ground truth* for training and test sets is itself thorny, and which definition we pick depends on the particular use that we put the resulting error rate to. The basic distinction is whether we want to measure the error rate relative to the character that the writer intended to write, or relative to the character that the writer actually wrote. (It should be noted that the 'intention' of the writer can usually be inferred or verified for training and testing purposes from other sources, even if those sources are not available or too costly to take advantage of in the actual application.) The problem with the first approach is that it penalizes recognition systems for errors made by the writer (but, of course, error rates are still comparable between recognizers and between man and machine). The problem with the second approach is that a certain fraction of characters cannot be classified unambiguously even by careful human inspection.

Another important distinction is between measuring *character-level* and *field-level* accuracy. If our training and test sets already consist of isolated, transcribed characters, this is not an important issue. However, if systems can take advantage of field-level information, both measuring and comparing character-level accuracy between systems becomes difficult. As a simple example, in the results reported in Geist *et al* (1994), character-level and field-level errors were not always proportional.

G1.3.6 Recommendations and conclusions

For the practitioner, perhaps the most instructive reference available on handwritten character recognition is Wilkinson *et al* (1994). In it, 57 different handwritten character recognition systems were tested on the same database of handwritten digits. The top ten performers in that comparison all used either some form of MLP or *k*-nearest-neighbor method (MLP was quite popular overall: a little less than a third of the entries used some form of MLP). The reader is referred to the report for detailed analysis and data. MLP-based character recognition was also popular and successful in Geist *et al* (1994), where systems were tested on field-level handwritten word recognition.

Results like these as well as the extensive literature suggest that MLP-based handwritten character recognizers are both proven and practical. Compared to methods like *k*-nearest neighbors, which also have low error rates, they also have practical advantages in terms of throughput and the ability to estimate probabilities well.

For the neural network researcher looking for benchmark tasks, the 'standard' handwritten character recognition task (isolated, truthable characters, large available training and test sets) may not be very rewarding: it appears that current systems perform close to, and in some cases may even surpass, human performance; the best existing systems primarily fail on input that is genuinely ambiguous or was not well represented in their training data. But while 'standard' handwritten character recognition may not leave much room for breakthroughs, demonstrating state-of-the-art performance on it is still an important achievement indicating that a new neural network architecture is at least no worse than existing methods.

However, handwritten character recognition and connected-handwriting recognition still offers a rich field for future research. Some important research topics are:

- achieving robustness to background, dropouts, underlining, manual corrections, etc
- on-the-fly adaptation to new writing styles
- recognition of symbols, diacritics, and non-European writing systems
- better integration of segmentation and recognition in connected handwriting recognition systems
- better integration of linguistic and domain knowledge into the recognition process; automatic acquisition of such knowledge
- word-level recognition.

Progress in these areas will require far more than a simple extension of the classifier paradigm that has proven so successful for isolated handwritten character recognition.

G1.3.7 Databases

Further information on the NIST (National Institute of Standards and Technology) databases for training and testing handwriting recognition systems, as well as the freely available NIST reference character recognition system, can be obtained by contacting:

Standard Reference Data
National Institute of Standards and Technology
221/A323
Gaithersburg, MD 20899
(301) 975-2208
(301) 926-0416 (fax).

The CEDAR (Center for Excellence in Document Analysis and Recognition) database derived from postal addresses can be obtained from:

Jonathan J Hull
Associate Director, CEDAR
226 Bell Hall
State University of New York at Buffalo
Buffalo, NY 14260
716-636-3195 (voice)
716-636-3966 (fax)
hull@cs.buffalo.edu (email).

References

Battiti R and Colla A M 1994 Democracy in neural nets: voting schemes for classification *Neural Networks* **7** 691–707

Berger J O 1985 *Statistical Decision Theory and Bayesian Analysis* (Berlin: Springer)

Blue J L, Candela G T, Grother P J, Chellappa R and Wilson C L 1994 Evaluation of pattern classifiers for fingerprint and OCR applications *Patt. Recog.* **27** 485–501

Bottou L and Vapnik V 1992 Local learning algorithms *Neural Comput.* **4** 888–900

Bourlard H, Wellekens C J 1989 Links between Markov models and multilayer perceptrons ed D S Touretzky *Advances in Neural Information Processing Systems* vol 1 (San Mateo, CA: Morgan Kaufmann) pp 502–10

Breuel T M 1993 Recognition of handprinted digits using optimal bounded error matching *Int. Conf. on Document Analysis and Recognition (ICDAR) (Tsukuba Science City)* (Los Alamitos, CA: IEEE Computer Society Press) pp 493–6

—— 1994 Design and Implementation of a system for the recognition of handwritten responses on US census forms *Document Analysis Systems '94 (DAS '94) (Kaiserslautern)* (Los Alamitos, CA: IEEE Computer Society Press) pp 109–34

Bunke H and Roth M 1993 Off-line recognition of cursive handwriting using hidden Markov models *Actes des Journèes Sur le Tratement d'Image, la Reconnaissance des Formes et Leurs Application* ed R Ingold (Laboratoire d'Informatique de l'Université de Fribourg)

Burges C J C, Matan O, LeCun Y, Denker J S, Jackel L D, Stenard C E, Nohl C R and Ben J I 1992 Shortest path segmentation: a method for training a neural network to recognize character strings *IJCNN Int. Joint Conf. on Neural Networks (Cat. No. 92CH3114-6)* (New York: IEEE Press) pp 165–72

Burges C J C, Ben J I, Denker J S, Lecun Y and Nohl C R 1993 Off line recognition of handwritten postal words using neural networks *Int. J. Patt. Recog. Artif. Int.* **7** 689–704

Caesar T, Gloger J M and Mandler E 1993 Preprocessing and feature extraction for a handwriting recognition system *Int. Conf. on Document Analysis and Recognition (ICDAR) (Tsukuba Science City)* (Los Alamitos, CA: IEEE Computer Society Press) pp 408–11

Chatterji B N 1986 Feature extraction methods for character recognition *IEEE Tech. Rev.* **3** 9–22

Davis R H and Lyall J 1986 Recognition of handwritten characters—a review *Image Vis. Comput.* **4** 208–18

Drucker H, LeCun Y 1991 Double backpropagation increasing generalization performance *IJCNN-91-Seattle: Int. Joint Conf. on Neural Networks (Cat. No. 91CH3049-4)* (New York: IEEE Press) pp 145–50

Drucker H, Schapire R and Simard P 1993 Boosting performance in neural networks *Int. J. Patt. Recog. Artif. Int.* **7** 705–19

Drucker H, Cortes C, Jackel L D, LeCun Y and Vapnik V 1994 Boosting and other ensemble methods *Neural Comput.* **6** 1289–301

Duda R O and Hart P E 1973 *Pattern Classification and Scene Analysis* (New York: Wiley)

Fukushima K 1992 Character recognition with neural networks *Neurocomputing* **4** 221–33

Garris M D 1992 Design and collection of a handwriting sample image database *Social Sci. Comput. Rev.* **10** 196–214

Geist J, Wilkinson R A, Janet S, Grother P J, Hammond B, Larsen N W, Klear R M, Matsko M J, Burges C J C, Creecy R, Hull J J, Vogl T P and Wilson C L 1994 The Second Census Optical Character Recognition Systems *Conf. NISTIR 5452* US Department of Commerce, National Institute of Standards and Technology (NIST), Computer Systems Laboratory, Gaithersburg, MD 20899, USA

Giloux M, Leroux M and Bertille J-M 1993 Strategies for handwritten word recognition using hidden Markov models *Int. Conf. on Document Analysis and Recognition* (Los Alamitos, CA: IEEE Computer Society Press) pp 299–304

Govindan V K and Shivaprasad A P 1990 Character recognition—a review *Patt. Recog.* **23** 671–83

Grother P J 1992 Karhunen Loeve feature extraction for neural handwritten character recognition *Proc. SPIE* **155** 155–66

—— 1993 Cross validation comparison of NIST OCR databases *Proc. SPIE* **296** 296–307

Guillevic D and Suen C Y 1993 Cursive script recognition: a fast reader scheme *Int. Conf. on Document Analysis and Recognition (ICDAR) (Tsukuba Science City)* (Los Alamitos, CA: IEEE Computer Society Press) pp 311–4

Ho T K, Hull J J and Srihari S N 1994 Decision combination in multiple classifier systems *IEEE Trans. Patt. Anal. Mach. Int.* **16** 66–75

Imagawa T and Fukushima K 1993 Character recognition in cursive handwriting with the mechanism of selective attention *Syst. Comput. Japan* **24** 89–97

Kasturi R and O'Gorman L 1992 Document image analysis: a bibliography *Mach. Vision Appl.* **5** 231–43

Kiefer J C 1987 *Introduction to Statistical Inference* (Berlin: Springer)

LeCun Y, Boser B, Denker J S, Henderson D, Howard R E, Hubbard W and Jackel L D 1989 Backpropagation applied to handwritten zip code recognition *Neural Comput.* **1** 541–51

Lee Y 1991 Handwritten digit recognition using *k* nearest-neighbor radial-basis function, and backpropagation neural networks *Neural Comput.* **3** 440–9

Lemarie B 1993 Radial basis function network for handwritten digit recognition *Proc. SPIE* **645** 645–52

Loncelle J, Derycke N and Soulie F F 1992 Cooperation of GBP and LVQ networks for optical character recognition *IJCNN Int. Joint Conf. on Neural Networks (Cat. No. 92CH3114-6)* (New York: IEEE Press) pp 694–9

Lu Y, Schlosser S and Janeczko M 1993 Fourier descriptors and handwritten digit recogition *Mach. Vis. Appl.* **6** 25–34

Martin G L and Rashid M 1992 Recognizing overlapping hand-printed characters by centered-object integrated segmentation and recognition *Advances in Neural Information Processing Systems* vol 4 ed J E Moody, S J Hanson and R P Lippmann (San Mateo, CA: Morgan Kaufmann) pp 504–11

Martin G L, Rashid M and Pittman J A 1993 Integrated segmentation and recognition through exhaustive scans or learned saccadic jumps *Int. J. Patt. Recog. Artif. Int.* **7** 831–47

Mulgaonkar P G, Chen C H and DeCurtins J L 1994 Word recognition in a segmentation-free approach to OCR *Proc. SPIE* **135** 135–41

Perrone M P 1993 Improving regression estimation: averaging methods for variance reduction with extensions to general convex measure optimization *PhD Thesis* Brown University, RI, USA

Powell M J D 1987 Radial basis functions for multivariable interpolation: a review *Algorithms for Approximation* ed J C Mason and M G Cox (Oxford: Clarendon)

Shustorovich A 1994 A subspace projection approach to feature extraction: the two-dimensional Gabor transform for character recognition *Neural Networks* **7** 1295–301

Simard P, LeCun Y, Denker J and Victorri B 1992 An efficient algorithm for learning invariance in adaptive classifiers *Proc. 11th IAPR Int. Conf. on Patt. Recog. Vol II. Conference B: Pattern Recognition Methodology and Systems* (Los Alamitos, CA: IEEE Computer Society Press) pp 651–5

Specht D F 1990 Probabilistic neural networks *Neural Networks* **3** 109–18

Weideman W E, Manry M T and Yau H C 1989 A comparison of a nearest neighbor classifier and a neural network for numeric handprint character recognition *IJCNN: Int. Joint Conf. on Neural Networks (Cat. No. 89CH2765-6)* (New York: IEEE TAB Neural Network Committee) pp 117–20

Wilkinson R A, Geist J, Janet S, Grother P J, Burges C J C, Creecy R, Hammond B, Hull J J, Larsen N J, Vogl T P and Wilson C L 1994 *The First Census Optical Character Recognition Systems Conf. NISTIR 4912* US Department of Commerce, National Institute of Standards and Technology (NIST), Computer Systems Laboratory, Gaithersburg, MD 20899, USA

Wolpert D H 1992 Stacked generalization *Neural Networks* **5** 241–59

Xu L, Krzyzak A and Suen C Y 1992 Methods of combining multiple classifiers and their applications to handwriting recognition *IEEE Trans. Syst., Man Cybern.* **22** 418–35

——1994 Associative switch for combining multiple classifiers *J. Artif. Neural Networks* **1** 77–100

Yan H 1994 Handwritten digit recognition using an optimized nearest neighbor classifier *Patt. Recog. Lett.* **15** 207–11

Further reading

I have tried to compile a list of references that will be useful to the reader interested in implementing their own handwriting recognition system using neural network technology. The reading list is divided by topic into sections. Within each section, papers are listed in the suggested order for reading. References already cited in the paper are usually not repeated here.

The following publications frequently contain papers on handwriting recognition, preprocessing and systems, both using neural networks and 'classical' approaches:

- *IEEE Pattern Recognition and Machine Intelligence* (PAMI)
- *Pattern Recognition* (Journal of the Pattern Recognition Society)
- *Neural Networks*
- *IEEE Transactions on Neural Networks*
- *Pattern Recognition Letters*
- *International Journal of Pattern Recognition.*

Furthermore, there are a number of conferences and conference proceedings that are of interest:

- Neural Networks for Signal Processing (NNSP)
- International Conference on Document Analysis and Recognition (ICDAR)
- International Joint Conference on Neural Networks (IJCNN)
- International Conference on Pattern Recognition (ICPR)
- Neural Information Processing Systems (NIPS)
- Document Analysis Systems (DAS).

General surveys of character recognition systems

1.   Kasturi R and O'Gorman L 1992 Document image analysis: a bibliography *Mach. Vis. Appl.* **5** 231–43

      A comprehensive biliography covering the period from 1986–1991.

2.   Govindan V K and Shivaprasad A P 1990 Character recognition—a review *Patt. Recog.* **23** 671–83

3.   Davis R H and Lyall J 1986 Recognition of handwritten characters—a review *Image Vis. Comput.* **4** 208–18

      Gives a review of classical techniques for handwritten character recognition.

4.   Mori S, Suen C Y and Yamamoto K 1992 Historical review of OCR research and development *Proc. IEEE* **80** 1029–58

      A general survey of OCR, both handwritten and printed.

5.   Tappert C C, Suen C Y and Wakahara T 1990. The state of the art in on-line handwriting recognition *IEEE Trans. Patt. Anal. Machine Int.* **12** 787–808

      A review of online handwriting recognition techniques that has some relevance to offline work as well.

Comparative evaluations

6.   Blue J L, Candela G T, Grother P J, Chellappa R and Wilson C L 1994 Evaluation of pattern classifiers for fingerprint and OCR applications *Patt. Recog.* **27** 485–501

      Evaluation of several approaches by a single group. Also contains a readable survey and taxonomy of preprocessing and recognition methods.

7.   Wilkinson R A, Geist J, Janet S, Grother P J, Burges C J C, Creecy R, Hammond B, Hull J J, Larsen N J, Vogl T P and Wilson C L 1994 The first census optical character recognition systems *Conf. NISTIR 4912* US Department of Commerce, National Institute of Standards and Technology (NIST), Computer Systems Laboratory, Gaithersburg, MD 20899, USA

      A comprehensive evaluation of 57 isolated handwritten character recognition systems on a standard database.

8.   Geist J, Wilkinson R A, Janet S, Grother P J, Hammond B, Larsen N W, Klear R M, Matsko M J, Burges C J C, Creecy R, Hull J J, Vogl T P and Wilson C L 1994 The Second Census Optical Character Recognition Systems *Conf. NISTIR 5452* US Department of Commerce, National Institute of Standards and Technology (NIST), Computer Systems Laboratory, Gaithersburg, MD 20899, USA

      A comprehensive evaluation of 31 field-level handwriting recognition systems on a standard database.

9.   Lee Yuchun 1991 Handwritten digit recognition using k nearest-neighbor, radial-basis function, and backpropagation neural networks *Neural Comput.* **3** 440–9

      Useful comparison between MLP, kNN, and RBF. Suggests that MLP does not have a significant advantage over MLP in terms of error rates, but that it is usually faster at classification time. RBF has some advantages in being able to reject noncharacters.

10.   Baker T and McCartor H 1992 A comparison of neural network classifiers for optical character recognition *Proc. SPIE* **191** 91–202

11.   Weideman W E, Manry M T and Yau H C 1989 A comparison of a nearest neighbor classifier and a neural network for numeric handprint character recognition *Int. Joint Conf. on Neural Networks (Cat. No. 89CH2765-6)* (New York: IEEE TAB Neural Network Committee) pp 117–20

Preprocessing and feature extraction

12.   Chatterji B N 1986 Feature extraction methods for character recognition *IEEE Tech. Rev.* **3** 9–22

      Presents an extensive review of classical preprocessing and feature extraction methods.

13.   Caesar T, Gloger J M and Mandler E 1993 Preprocessing and feature extraction for a handwriting recognition system *Int. Conf. on Document Analysis and Recognition (ICDAR) (Tsukuba Science City)* (Los Alamitos, CA: IEEE Computer Society Press) pp 408–11

14. Lu Y, Schlosser S and Janeczko M 1993 Fourier descriptors and handwritten digit recogition *Mach. Vis. Appl.* **6** 25–34

   Description and evaluation of using different Fourier transformations of character boundaries for digit recognition.

15. Grother P J 1992 Karhunen Loeve feature extraction for neural handwritten character recognition *Proc. SPIE* **155** 155–66

16. Shustorovich A 1994. A subspace projection approach to feature extraction: the two-dimensional Gabor transform for character recognition *Neural Networks* **7** 1295–301

   Describes and compares various feature extraction methods based on integral transformations, including Gabor wavelets, Gaussians, and sine transforms. Uses quality of reconstruction as an evaluation criterion.

17. Guillevic D and Suen C Y 1993 Cursive script recognition: a fast reader scheme *Int. Conf. on Document Analysis and Recognition (ICDAR) (Tsukuba Science City)* (Los Alamitos, CA: IEEE Computer Society Press) pp 311–4

   The paper mainly talks about morphological methods for feature extraction.

18. Roberts A and Yearworth M 1992 A comparison of pre-processing transforms for neural network classification of character images *Int. Conf. on Image Processing and its Applications (Conf. Publ. 354)* (London: IEE) pp 189–92

   Brief paper describing applications of Fourier, sine, cosine, and Hadamard transforms for feature extraction and preprocessing of handwritten digits for MLPs.

19. Jameel A and Koutsougeras C 1993 On features used for handwritten character recognition in a neural network environment *Proc. 1993 IEEE Conf. on Tools with AI (TAI-93)* (Los Alamitos, CA: IEEE Computing Society Press) pp 280–4

20. Kageyu S, Ohnishi N and Sugie N 1991 Augmented multi-layer perceptron for rotation- and scale-invariant hand-written numeral recognition *IEEE Int. Joint Conf. on Neural Networks (Cat. No. 91CH3065-0)* (New York: IEEE Press) pp 54–9

   Uses Fourier transform and complex/log representation to achieve invariance.

21. Perantonis S J and Lisboa P J G 1992 Translation, rotation, and scale invariant pattern recognition by high-order neural networks and moment classifiers *IEEE Trans. Neural Networks* **3** 241–51

22. Awwal A A S and Ahmed F 1993 Complex associative memory neural network model for invariant pattern recognition *Proc. NAECON '93—National Aerospace and Electronics Conf.* (New York: IEEE Press) pp 892–6

   Demonstrates use of Fourier descriptors of character boundary as an invariant representation.

23. Lisboa P J G and Perantonis S J 1991 Invariant pattern recognition using third-order networks and Zernike moments *IEEE Int. Joint Conf. on Neural Networks (Cat. No. 91CH3065-0)* (New York: IEEE Press) 1421–5

   A simple example of designing invariant feature sets for handwriting recognition.

Character segmentation

24. Fenrich R 1992 Segmentation of automatically located handwritten numeric strings *From Pixels to Features III: Frontiers in Handwriting Recognition* (Amsterdam: North-Holland) pp 47–59

   Describes several segmentation methods.

25. Shridhar M and Badreldin A 1987 Context-directed segmentation algorithm for handwritten numeral strings *Image Vis. Comput.* **1** 3–9

26. Holt M J J, Mohammad Beglou M and Datta S 1992 Slant independent letter segmentation for off-line cursive script recognition *From Pixels to Features III: Frontiers in Handwriting Recognition* (Amsterdam: North-Holland) pp 41–6

MLP-based recognizers

27. LeCun Y, Boser B, Denker J S, Henderson D, Howard R E, Hubbard W and Jackel L D 1989 Backpropagation applied to handwritten zip code recognition *Neural Comput.* **1** 541–51

A widely cited reference on handwritten character recognition using MLPs. The network uses multiple hidden layers. The use of weight sharing and local receptive fields in the first two layers means that those layers perform feature extraction functions, similar to handcoded feature extraction methods in other systems.

28. Casey R G and Takahashi H 1992 Experience in segmenting and classifying the NIST data base *From Pixels to Features III: Frontiers in Handwriting Recognition* (Amsterdam: North-Holland) pp 5–16

Describes a successful system for isolated character recognition using MLPs, geometric and topographic features.

29. Lee H Y, Lee Y C and Chen H H 1989 Hand written letter recognition with neural networks *IJCNN: Int. Joint Conf. on Neural Networks (Cat. No. 89CH2765-6)* vol 2 (New York: IEEE TAB Neural Network Committee) pp 618

30. Martin G L and Pittman J A 1991 Recognizing hand-printed letters and digits using backpropagation learning *Neural Comput.* **3** 258–67

31. Hepp D J 1991 An application of backpropagation to the recognition of handwritten digits using morphologically derived features. *Proc. SPIE* **228** 228–33

32. Huang S C, Huang Y F and Jou I C 1991 Analysis of perceptron training algorithms and applications to hand-written character recognition *ICASSP 91 Int. Conf. on Acoustics, Speech and Signal Processing (Cat. No. 91CH2977-7)* (New York: IEEE Press) pp 2153–6

33. Rovner R M, Gillies A M, Ganzberger M J and Hepp D J 1994 Strategies for the automatic interpretation of handwritten addresses *Proc. SPIE* **174** 174–85

34. Wang P S P, Nagendraprasad M V, and Gupta A 1992 A neural net based 'hybrid' approach to handwritten numeral recognition *From Pixels to Features III: Frontiers in Handwriting Recognition* (Amsterdam: North-Holland) pp 145–54

35. Yan Hong 1994 Handwritten digit recognition using an optimized nearest neighbor classifier *Patt. Recog. Lett.* **15** 207–11

Uses MLP/BP for optimizing nearest-neighbor classifier.

36. Blue J L and Grother P J 1992 Training feed-forward neural networks using conjugate gradients *Proc. SPIE* **179** 179–90

37. Jean J S N and Jin Wang 1994 Weight smoothing to improve network generalization *IEEE Trans. Neural Networks* **5** 752–63

Describes one method for modeling local correlations among neighboring pixels in the image of an input character (cf methods like KLT and Fourier transforms, which achieve a similar goal).

38. Lin J T and Inigo R 1991 Hand written zip code recognition by back propagation neural network *IEEE Proc. SOUTHEASTCON '91 (Cat. No. 91CH2998-3)* (New York: IEEE Press) pp 731–5

Uses MLPs with more than three layers and local receptive fields for some units.

39. Sabourin M and Mitiche A 1992 Optical character recognition by a neural network *Neural Networks* **5** 843–52

Printed omnifont OCR. Interesting for its use of hierarchical architectures and momentum term adjustment.

Other neural network recognizers

40. Drucker H and LeCun Y 1991 Double backpropagation increasing generalization performance *IJCNN-91-Seattle: Int. Joint Conf. on Neural Networks (Cat. No.91CH3049-4)* (New York: IEEE Press) pp 145–50

Presents double backpropagation, a method related to notions of invariant recognition.

41. Specht D F 1990 Probabilistic neural networks *Neural Networks* **3** 109–18

PNNs performed best when a number of different methods (kNN, MLP, PNN, and others) were implemented at NIST and applied to the same database. Closely related to classical density estimation methods.

42. Flachs B and Flynn M 1994 Sparse adaptive memory and handwritten digit recognition *Proc. Int. Conf. on Neural Networks (ICNN'94)* (New York: IEEE Press) pp 1098–102

Uses a BP-like algorithm to generate prototypes for a modified *k*-nearest neighbor classifier.

43. Lemarie B 1993 Practical realization of a radial basis function network for handwritten digit recognition *New Trends in Neural Computation. International Workshop on Artificial Neural Networks. IWANN '93 Proc.* ed J Mira and J Cabestany and A Prieto (Berlin: Springer) pp 131–6

44. Knerr S, Personnaz L and Dreyfus G 1991 A new approach to the design of neural network classifiers and its application to the automatic recognition of handwritten digits *Int. Joint Conf. on Neural Networks* (New York: IEEE Press) pp L91–6

Trains and uses pairwise linear discriminants between character classes in the first layer.

45. Bottou L and Vapnik V 1992 Local learning algorithms *Neural Comput.* **4** 888–900

Suggests that, at classification time, the nearest neighbors of an unknown input are used to train a MLP that then performs the final classification step.

46. Hinton G E, Williams C K I and Revow M D 1992 Adaptive elastic models for hand-printed character recognition *Advances in Neural Information Processing Systems* (San Mateo, CA: Morgan Kaufmann) pp 512–9

Elastic models have a long tradition in computer vision and an application to handwritten character recognition seems natural. It remains to be seen whether they can be competitive.

47. Solaiman B and Autret Y 1994 Application of the HLVQ neural network to hand-written digit recognition *Proc. IEEE Workshop on Neural Networks for Signal Processing* ed J Vlontzos, J-N Hwang and E Wilson (New York: IEEE Press) pp 384–93

48. Sabourin M and Mitiche A 1993 Modeling and classification of shape using a Kohonen associative memory with selective multiresolution *Neural Networks* **6** 275–83

Printed omnifont OCR. Users contour based features. Interesting for its use of multiresolution techniques and Kohonen associative memory.

Non-neural-network recognizers

49. Kimura F, Shridhar M and Chen Z 1993 Improvements of a lexicon directed algorithm for recognition of unconstrained handwritten words *Int. Conf. on Document Analysis and Recognition (ICDAR) (Tsukuba Science City)* (Los Alamitos, CA: IEEE Computer Society Press) pp 18–22

Description of a complete system, sources of errors, and how to go about improving performance. These are very important lessons for any recognition system.

50. Kimura F and Shridhar M 1992 Segmentation-recognition algorithm for zip code field recognition *Mach. Vis. Appl.* **5** 199–210

The paper describes important preprocessing and segmentation steps; recognition uses a non-neural classifier.

51. Breuel T M 1993 Recognition of Handprinted Digits using Optimal Bounded Error Matching *Int. Conf. on Document Analysis and Recognition (ICDAR) (Tsukuba Science City)* (Los Alamitos, CA: IEEE Computer Society Press) 493–6

Describes a character recognizer based on computer vision techniques and decision trees with low error rate. Interesting because invariance under translation is part of the matching process itself.

52. Downton A C, Tregidgo R W S, Leedham C G and Hendrawan 1992 Recognition of handwritten British postal addresses *From Pixels to Features III: Frontiers in Handwriting Recognition* (Amsterdam: North-Holland) pp 130–44

Description of a complete system.

Classifier combination

53. Ho T K, Hull J J and Srihari S N 1994 Decision combination in multiple classifier systems *IEEE Trans. Patt. Anal. Mach. Int.* **16** 66–75

    An important paper on classifier combination applied to OCR.

54. Drucker H, Cortes C, Jackel L D, LeCun Y and Vapnik V 1994 Boosting and other ensemble methods *Neural Comput.* **6** 1289–301

    Interesting and well-founded method for classifier combination based on computational learning theory.

55. Xu Lei, Krzyzak A and Suen C Y 1994 Associative switch for combining multiple classifiers *J. Artif. Neural Networks* **1** 77–100

56. Xu L, Krzyzak A and Suen C Y 1992 Methods of combining multiple classifiers and their applications to handwriting recognition *IEEE Trans. Syst., Man Cybern.* **22** 418–35

57. Perrone M P 1993 Improving regression estimation: averaging methods for variance reduction with extensions to general convex measure optimization *PhD Thesis*, Brown University, RI, USA

    Classifier combination using linear methods.

58. Wolpert D H 1992 Stacked generalization *Neural Networks* **5** 241–59

    The paper is about achieving classifier combination by merging the outputs of a multiple classifiers into a new feature vector and using that feature vector as input to a second classification stage.

59. Soulie F F, Viennet E and Lamy B 1993 Multi-modular neural network architectures: applications in optical character and human face recognition *Int. J. Patt. Recog. Artif. Int.* **7** 721–55

60. Battiti R and Colla A M 1994 Democracy in neural nets: voting schemes for classification *Neural Networks* **7** 691–707

61. Ho T K, Hull J J and Srihari S N 1992 On multiple classifier systems for pattern recognition *Proc. 11th IAPR Int. Conf. on Patt. Recog. vol II Conference B: Pattern Recognition Methodology and Systems* vol 2 (Los Alamitos, CA: IEEE Computer Society Press) pp 84–7

    Describes methods for classifier combination based on rankings.

Integrated segmentation and recognition

These systems generate a large number of possibly overlapping character hypotheses, recognize each hypothesis using neural or non-neural classifiers, and find globally optimal interpretations using a Viterbi- or 'best-path' algorithm. Such methods are analogous to the generation and interpretation of phone lattices in speech recognition.

62. Bozinovic R M and Srihari S N 1989 Off-line cursive script word recognition *IEEE Trans. Patt. Anal. Mach. Int.* **5** 265–91

    Describes an early but complete non-neural integrated segmenter and recognizer.

63. Breuel T M 1994 Design and implementation of a system for the recognition of handwritten responses on US census forms *Document Analysis Systems '94 (DAS '94) (Kaiserslautern)* pp 109–34

    Successful integrated segmentation and recognition system using MLPs as a module for character recognition. Discussion of probabilistic foundations of integrated segmentation and recognition methods.

64. Burges C J C, Ben J I, Denker J S, Lecun Y and Nohl C R 1993 Off line recognition of handwritten postal words using neural networks *Int. J. Patt. Recog. Artif. Int.* **7** 689–704

    Describes a system for integrated segmentation and recognition using MLPs for isolated character recognition.

65. Matan O, Burges C J C, LeCun Y and Denker J S 1992 Multi-Digit recognition using a space displacement neural network *Advances in Neural Information Processing Systems* (San Mateo, CA: Morgan Kaufmann) pp 488–95

    An integrated segmentation and recognition method using Viterbi algorithm for selecting optimal segmentation. Note the use of shared weights and avoidance of recomputation of feature values for each segment hypothesis.

66. Mulgaonkar P G, Chen C-H and DeCurtins J L 1994 Word recognition in a segmentation-free approach to OCR *Proc. SPIE* **135** 135–41

Neural network models that integrate segmentation and recognition

Unlike the methods referenced in the previous section, these methods do not attempt to generate isolated character hypotheses, but use a neural network architecture to perform segmentation and recognition simultaneously.

67. Keeler J D, Rumelhart D E and KI Leow W 1992 Integrated segmentation and recognition of hand-printed numerals *Advances in Neural Information Processing Systems* (San Mateo, CA: Morgan Kaufmann) pp 557–63

Interesting idea for how a neural network can integrate recognition and segmentation with possible extensions to visual object recognition.

68. Senior A W and Fallside F 1992 Off-line handwriting recognition by recurrent error propagation networks *Proc. BMVC '92. British Machine Vision Conf.* ed D Hogg and R Boyle (Berlin: Springer) pp 453–61

69. Fukushima K and Imagawa T 1993 Recognition and segmentation of connected characters with selective attention *Neural Networks* **6** 33–41

Integrated segmentation and recognition using neural networks.

70. Martin G L, Rashid M, and Pittman J A 1993 Integrated segmentation and recognition through exhaustive scans or learned saccadic jumps *Int. J. Patt. Recog. Artif. Int.* **7** 831–47

A neural-network-based approach reminiscent of HMMs with contextual modeling.

Pattern recognition

71. Duda R O and Hart P E 1973 *Pattern Classification and Scene Analysis* (New York: Wiley)

A classic reference for pattern recognition techniques. Good introduction to k-nearest neighbor methods, density estimation, Bayesian methods, and clustering. A greatly expanded and updated version is due to be published.

72. Bourlard H and Wellekens C J 1989 Links between Markov models and multilayer perceptrons *Advances in Neural Information Processing Systems* ed D S Touretzky (San Mateo, CA: Morgan Kaufmann) pp 502–10

An important paper on speech recognition using neural networks; describes link between probabilities and network outputs.

73. Kiefer J C 1987 *Introduction to Statistical Inference* (Berlin: Springer)

Readable and useful introduction to Bayesian methods.

74. Berger J O 1985 *Statistical Decision Theory and Bayesian Analysis* (Berlin: Springer)

Comprehensive description of Bayesian methods.

75. Breiman L *et al* 1984 *Classification and Regression Trees* (Belmont, CA: Wadsworth)

Important non-neural pattern recognition method used in some handwriting recognizers.

Other literature

76. Baird H S 1988 Feature identification for hybrid structural/statistical pattern classification *Comput. Vision, Graphics, Image Proc.* **42** 318–33

A paper on omnifont OCR, but the methods and the approach are worth reading about.

77. Edelman S, Flash T and Ullman S 1990 Reading cursive handwriting by alignment of letter prototypes *Int. J. Comput. Vis.* **3** 303–31

Geometric, model-based approach to script recognition inspired by visual object recognition; very different from standard neural network approaches.

78. Bromley J and Denker J S 1993 Improving rejection performance on handwritten digits by training with 'rubbish' *Neural Comput.* **5** 367–70

Presents a useful idea used by a number of recognition systems. Rejection of known bad shapes is particularly important for integrated segmentation and recognition systems.

79. Casey R, Ferguson D, Mohiuddin K and Walach E 1992 Intelligent forms processing system *Mach. Vis. Appl.* **5** 143–55

   Description of system architecture for forms processing. Further information on each module is found in the references.

80. Guyon I, Vapnik V, Boser B, Bottou L and Solla S A 1992. Structural risk minimization for character recognition *Advances in Neural Information Processing Systems* (San Mateo, CA: Morgan Kaufmann) pp 471–9

   Addresses, in the context of character recognition, questions related to how large a network to choose in order to achieve good generalization.

81. Simon J-C 1992 Off-line cursive word recognition *Proc. IEEE* **80** 1150–61

82. Gupta A, Nagendraprasad M V, Liu A, Wang P S P and Ayyadurai S 1993 An integrated architecture for recognition of totally unconstrained handwritten numerals *Int. J. Patt. Recog. Artif. Int.* **7** 757–73

83. Bischoff A and Wang P S P 1992 Handwritten digit recognition using neural networks *Proc. SPIE* **436** 436–47

84. Yanikoglu B A and Sandon P A 1994 Recognizing off-line cursive handwriting *Proc. IEEE Conf. on Comput. Vis. and Patt. Recog.* (Los Alamitos, CA: IEEE Computer Society Press) 397–403

G1.4 Improved speech recognition using learning vector quantization

Kari Torkkola

Abstract

We present a case study where neural networks are used to improve the performance of a hidden-Markov-model-based speech recognition system. The improvement stems from viewing learning vector quantization (LVQ) as a nonlinear feature transformation to enhance phonetic discriminations. *Classwise quantization errors* of LVQ are modeled by *continuous density hidden Markov modeling* (HMM). As decision-making at frame level is suboptimal for speech recognition, more information can be preserved for the HMM stage than in schemes where LVQ is used as classifier. Experiments in both speaker-dependent and speaker-independent phoneme spotting tasks show that significant improvements are attainable over continuous-density HMMs, and over the hybrid of LVQ and discrete HMMs.

G1.4.1 Introduction

Hidden Markov modeling (HMM) is the most successful technique in automatic speech recognition today with well-studied and mature training algorithms (see Rabiner 1989 for a tutorial). These techniques can be roughly divided into two main categories: continuous observation density HMMs (CHMMs) and discrete observation HMMs (DHMMs) with semicontinuous (tied mixture) HMMs in between. Either continuous or discrete, the aim of the models is to learn a faithful *representation* of the feature vector sequence derived from the speech signal, either directly by mixtures of multivariate Gaussian or other distributions, or through vector quantization (VQ). Both the maximum likelihood training algorithms of the HMMs, and, in the discrete case, also the codebook construction algorithms aim at good representations. These learnt representations of various speech units (e.g. phones or words) serve as models against which unknown speech is compared. Usually, the model sequence that most likely could have produced the unknown speech is chosen as the recognition result.

As representation is not necessarily optimal for recognition, enhancement of the discrimination capabilities of HMMs has received some attention. These enhancement methods include, among others, training criteria other than maximum likelihood (Bahl *et al* 1986, Young 1991, Chou *et al* 1992, Kapadia *et al* 1993) and hybrids of some discriminative methods with the HMMs. The latter types of system have recently been dominated by artificial neural networks (ANNs). These hybrids can further be grouped into two main clusters: ANNs as probability estimators for HMMs (Bourlard *et al* 1992, Robinson 1994, Dugast *et al* 1994), or ANNs as codebooks or labelers for DHMMs (Iwamida *et al* 1990, Torkkola *et al* 1991, Mäntysalo *et al* 1994, LeCerf *et al* 1994). (See also Section F1.7 in this handbook for a discussion of ANNs in speech processing). F1.7

This work is concerned with a combination falling into the latter category using *learning vector* C1.1.5 *quantization* (LVQ) as the ANN (Kohonen *et al* 1988, 1992). The goal is to improve an existing CHMM-based phonetic speech recognition stage using LVQ in a novel way as a feature transformation. As such, a phonetic speech recognition system is not a complete application. However, subunit-based recognition is the basic component in every large vocabulary automatic speech recognition system.

G1.4.2 Learning vector quantization codebooks in speech recognition

The role of conventional vector quantization algorithms in speech recognition, such as the Linde–Buzo–Gray algorithm, or the K-means algorithm, is to *represent* speech feature vectors with the smallest possible distortion. This is not the case with the LVQ methods, which try to aim at *discrimination* of pattern classes, whatever they may be (Kohonen *et al* 1988, 1992). In an LVQ network, the weight vectors (also called codebook vectors) directly define the class borders in the feature space according to the nearest-neighbor rule. The learning algorithm of the LVQ modifies the weight vectors adaptively so that the borders between classes will approximate Bayes' decision surfaces. Quantization error (distortion) is of secondary interest.

LVQ has been employed in speech recognition as a substitute to conventional vector quantization. Short time feature vectors act as the basis for classification, most commonly to phoneme classes. DHMMs are then employed to combine local classification decisions by treating the stream of classification labels just as the stream of VQ codebook indices is treated with DHMMs. In phoneme-related tasks, training has been done either by single frames (Kimber *et al* 1990), or by concatenating several frames together to represent some context (Yu *et al* 1990, Iwamida *et al* 1990, Torkkola *et al* 1991, Iwamida *et al* 1991). When the task has been speaker-dependent and phoneme oriented, significant improvements have been observed due to LVQ when compared against ordinary VQ (Iwamida *et al* 1990, Torkkola *et al* 1991).

In addition to enhanced discriminative properties, another advantage of the LVQ over conventional VQ is that the discrete alphabet in phonetically motivated classification, that is, the number of classes, is smaller than the number of codebook vectors in usual VQ (a few tens as opposed to a few hundreds). This results in an order of magnitude smaller number of output probability parameters to be estimated for the DHMMs. Probability smoothing schemes are thus usually unnecessary.

G1.4.3 Extracting more information from the LVQ

As already mentioned, the previous work on LVQ codebooks in speech recognition has concentrated on making use of the *class label of the closest codebook vector* only. To decode this label stream, discrete observation HMMs have been employed.

However, the normal practice with pattern classifiers, extracting only the final decision of the classification (the class label) is desirable only when that really is the final decision stage of the whole task. This is not the case with the LVQ/HMM hybrids in speech recognition; the final decision is made by the Viterbi search at the HMM stage. It is suboptimal to resort to hard classification decisions too early at the frame level.

In addition to the classification label, also the distance to the closest vector in the whole codebook (i.e. the quantization error) was utilized in Mäntysalo *et al* (1994) and modeled by a CHMM. Torkkola (1994b) discards the label, but the classwise quantization errors are preserved. This is the approach we use in this work. Distances between the feature vector and all the codebook vectors need to be computed first, as in normal vector quantization and in LVQ. Now, instead of choosing the closest vector in the whole codebook, the closest vectors are searched within each pattern class, that is, among those codebook vectors that bear the same label (phonetic, in this work). Retaining these minimum distances to each pattern class gives an idea of how close the pattern vector is to all the classes, not just the closest class. A new feature vector is constructed by concatenating these distance values. *Continuous observation density* HMMs can then be applied to model a stream of these vectors. Figure G1.4.1 illustrates the computation of the classwise quantization errors.

G1.4.4 Experiments

G1.4.4.1 Speaker-dependent phoneme spotting with high-quality speech

We will now describe experiments to compare the performance of the LVQ–CHMM hybrid with some established algorithms. A comparison between four architectures is presented in table 1. The architectures are as follows:

(i)   conventional CHMMs modeling a stream of cepstral vectors;
(ii) LVQ–DHMM hybrids, where LVQ produces a stream of best class labels, which is modeled by discrete observation HMMs;
(iii) the new approach proposed in this paper, where LVQ produces a stream of classwise quantization error vectors, which is modeled by CHMMs;

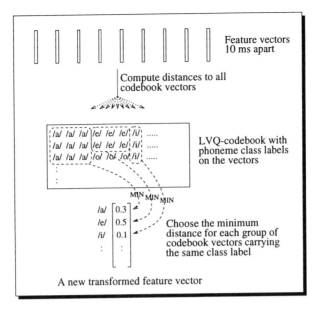

Figure G1.4.1. An illustration of the feature transformation. The stream of resulting new feature vectors is then modeled by HMMs. (From Torkkola 1994a, copyright IEEE Press.)

(iv) parallel use of the classwise quantization error with the cepstral vectors.

Table G1.4.1. Comparison between CHMMs (i) LVQ–DHMM hybrids (ii), and LVQ–CHMM hybrids (iii, iv) in a speaker-dependent task. 'MFCC' refers to 20-component mel-scale cepstral vectors, Δ refers to difference coefficients, 'context' denotes the 220 ms context vector described in Mäntysalo *et al* (1994), 'qerr' refers to classwise quantization errors computed from the LVQ stage, 'mixts/stream' denotes the largest number of Gaussian mixtures used in HMMs to model each stream, after which adding new mixtures did not improve the results significantly. In all cases the HMMs had three emitting states. (From Torkkola 1994a, copyright IEEE Press.)

Combination	LVQ input	HMM input	Streams	Mixts/stream	Covariance	Error rate
(i)	—	MFCC+Δ	2	5,3	Diagonal	3.2
	—	MFCC+Δ	2	2,1	Full	2.9
(ii)	MFCC	LVQ-best label	1	—	—	8.5
	Context	LVQ-best label	1	—	—	4.6
(iii)	Context	LVQ-qerr+Δ	2	7,7	Diagonal	5.4
	Context	LVQ-qerr+Δ	2	2,1	Full	2.5
	MFCC	LVQ-qerr+Δ	2	2,1	Full	4.9
(iv)	Context	MFCC+Δ+LVQ-qerr+Δ	4	2,1,2,1	Full	1.8

The task in this comparison is *speaker-dependent phoneme spotting* in the Finnish language (Mäntysalo *et al* 1994, Torkkola *et al* 1991). The database contains four repetitions of a set of 311 utterances spoken by three male Finnish speakers. Each set consists of 1737 phonemes. In the original Finnish language, there are only 21 different phoneme classes: 8 vowels and 13 consonants. Four additional phonemes have been adopted with loan words from other languages, but none of these were represented in the database. There were thus 22 phonemic classes for the LVQ to differentiate (21 phonemes and silence), which was also the dimensionality of the classwise quantization error vectors. Three of the repetitions were used each time for training, and the remaining one for testing. Four independent runs were made for each speaker by leaving one set at a time for testing. Thus all speaker-dependent recognition results presented in this work are averages of 12 test runs, and based on 20 844 phoneme spotting scores.

Speech analysis conditions were the following: 12.8 kHz sampling rate, pre-emphasis coefficient 0.95, 25.6 ms Hamming window every 10 ms, and 20-component mel-frequency cepstral coefficients (MFCC) computed for every window. Where difference coefficients (Δ) have been used, they were computed from a period of 40 ms. This applies also to the quantization error vectors.

Where context vectors were used as LVQ input, the codebook size was 2000. For the single MFCC vector as input, it was 500. The context vectors had a time span of 220 ms (Mäntysalo *et al* 1994). The LVQ training procedure was exactly the same as described in Mäntysalo *et al* (1994): using only phoneme centers, initialization by K-nearest-neighbor, OLVQ1 for 10 000 iterations with $\alpha(0) = 0.3$, and LVQ1 using $\alpha(0) = 0.2$. In the LVQ1 stage, the number of iterations was 100 times the number of codebook vectors. An example of the output is presented in figure G1.4.2. Phoneme center locations can be clearly distinguished in the stream of quantization errors of the LVQ.

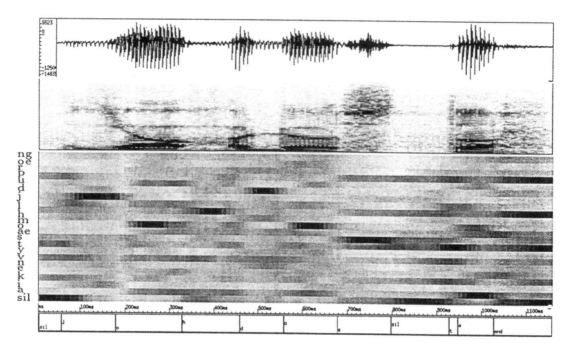

Figure G1.4.2. An example of the output of the LVQ stage. From top to bottom: the speech signal (a Finnish word */johdosta/*), its spectrogram, classwise quantization errors with the phoneme labels on the left (darker shade denotes small values), and the manual labeling of the utterance.(From Torkkola 1994a, copyright IEEE Press.)

Looking at the results in table G1.4.1, it is obvious that exploiting all classwise quantization errors instead of using only the class identity of the closest codevector preserves more information for the latter decision stages. This is reflected in the difference between the LVQ–DHMM hybrid (ii) and the LVQ–CHMM hybrid (iii). We can also see that the classwise quantization errors are interdependent, since modeling them by diagonal covariances for the Gaussians in CHMMs produces poorer results. We obtained our best results by using full covariance matrices.

Although the LVQ–CHMM hybrid (iii) is better than plain CHMMs (i), the difference is barely significant. On the other hand, the LVQ + MFCC–CHMM hybrid (iv) surpasses plain CHMMs by a clear margin. Due to a relatively large number of tests, the confidence limits (99%) are relatively tight: ±0.23% for the best result (1.8%). This enables us to state that the proposed architecture is significantly better than a CHMM system (2.9%), or an LVQ–DHMM hybrid (4.6%) in this task. In the case of plain CHMMs, using second-order derivatives, increasing the number of mixtures, or using diagonal covariance matrices with a larger number of mixtures did not improve our best CHMM result.

In the LVQ–DHMM hybrid we used only the label sequence produced by LVQ, not any other information suggested in Mäntysalo *et al* (1994). Including another LVQ codebook for phoneme center/transition classification, or the whole codebook quantization error, would no doubt improve the performance of the LVQ–DHMM hybrid, as it did for Mäntysalo *et al* (1994).

G1.4.4.2 Speaker-independent phoneme spotting with telephone speech

In this experiment, our aim was to find out whether this LVQ–HMM hybrid is applicable to the speaker-independent case. The database consisted of Swiss-French telephone speech with 56 speakers (about two hours of speech). Half of the speakers were used for training and the other half for testing. The vocabulary also varied across the speakers.

Speech was sampled at 8 kHz, and 12-component mel-scale cepstra were computed each 10 ms. As the input to LVQ, we used slightly narrower context windows whose duration was 140 ms. The LVQ codebook size remained as 2000. 36 *context-independent* HMM phoneme models were used with four emitting states and full covariances throughout this experiment.

Table G1.4.2. Comparison between CHMMs (i) and LVQ–CHMM hybrids (iii, iv) in a speaker-independent task with telephone speech.(From Torkkola 1994a, copyright IEEE Press.)

Combination	HMM input	Streams	Mixts/stream	Error rate
(i)	MFCC + Δ	2	2,1	46.6
(iii)	LVQ-qerr + Δ	2	2,1	44.2
	LVQ-qerr + Δ + $\Delta\Delta$	3	1,1,1	42.6
(iv)	MFCC + Δ + LVQ-qerr + Δ	4	2,1,2,1	42.0
	MFCC + Δ + LVQ-qerr + Δ + $\Delta\Delta$	5	1,1,1,1,1	40.6
	MFCC + Δ + LVQ-qerr + Δ + $\Delta\Delta$	5	3,1,1,3,1	40.0
	MFCC + Δ + (LVQ-qerr) + Δ + $\Delta\Delta$	4	3,1,(1),3,1	38.3

The gain of using LVQ in this (much harder) task is not as dramatic as in the first one, but comparing the baseline CHMM recognizer (the first row of table G1.4.2) to the result on the last row we can see that the improvement is anyway very significant. In addition to Δ coefficients we also tried second-order difference coefficients for the quantization error stream, which turned out to be advantageous. In addition, it seems that the actual quantization errors are less important than their difference and second-order difference coefficients. In the result of the last row of table G1.4.2 we only retained the Δ and $\Delta\Delta$ streams, and the results improved.

G1.4.5 Discussion

We have reviewed ways of employing LVQ-based codebooks with HMMs in speech recognition. We have demonstrated that modeling classwise quantization errors of LVQ by continuous-density hidden Markov models leads to a significant improvement over the mainstream HMM techniques. The resulting system could well serve as a phonetic recognition engine in a large-vocabulary continuous-speech recognition system. How the technique would work with *context-dependent* phone models remains a topic for further research.

It should be noted that, throughout these comparisons, exactly the same training conditions and algorithms (embedded Baum–Welch training) have been used. A phoneme bigram model was used in both experiments as the language model. The basic phoneme model structure was also been the same throughout each experiment. Both comparisons are thus actually made between different input representations, and not, for example, between different HMM software packages. In all of the experiments we used a public domain software package LVQ_PAK (Kohonen *et al* 1992), and a commercial package HTK (Young 1992) for the HMMs. A discrete-observation version of the HTK was written for DHMM experiments.

On the downside, one should mention increased computational cost, as we are in fact combining the computationally intensive parts of both vector quantization and CHMMs.

References

Bahl L, Brown P, de Souza P and Mercer R 1986 Maximum mutual information estimation of hidden Markov model parameters for speech recognition *Proc. IEEE Int. Conf. on Acoustics, Speech and Signal Processing (Tokyo, Japan)* (Piscataway, NJ: IEEE Press) pp 49–52

Bourlard H, Morgan N and Renals S 1992 Neural nets and hidden Markov models: review and generalizations *Speech Commun.* **11** 237–46

Chou W, Juang B and Lee C 1992 Segmental GPD training of HMM based speech recognizer *Proc. IEEE Int. Conf. on Acoustics, Speech and Signal Processing (San Francisco, CA)* vol I (Piscataway, NJ: IEEE Press) pp 473–6

Dugast C, Devillers L and Aubert X 1994 Combining TDNN and HMM in a hybrid system for improved continuous-speech recognition *IEEE Trans. Speech Audio Proc.* **2** 217–23

Iwamida H, Katagiri S and McDermott E 1991 Speaker independent large vocabulary word recognition using an LVQ/HMM hybrid algorithm *Proc. IEEE Int. Conf. on Acoustics, Speech and Signal Processing (Toronto)* vol 1 (Piscataway, NJ: IEEE Press) pp 553–6

Iwamida H, Katagiri S, McDermott E and Tohkura Y 1990 A hybrid speech recognition system using HMMs with an LVQ-trained codebook *Proc IEEE Int. Conf. on Acoustics, Speech and Signal Processing (Albuquerque, NM)* vol 1 (Piscataway, NJ: IEEE Press) pp 489–92

Kapadia S, Valtchev V and Young S J 1993 MMI training for continuous phoneme recognition on the TIMIT database *Proc. IEEE Int. Conf. on Acoustics, Speech and Signal Processing (Minneapolis, MN)* vol II (Piscataway, NJ: IEEE Press) pp 491–4

Kimber D G, Bush M A and Tajchman G N 1990 Speaker-independent vowel classification using hidden Markov models and LVQ2 *Proc. IEEE Int. Conf. on Acoustics, Speech and Signal Processing (Albuquerque, NM)* vol 1 (Piscataway, NJ: IEEE Press) pp 497–500

Kohonen T, Barna G and Chrisley R 1988 Statistical pattern recognition with neural networks: Benchmarking studies *Proc. IEEE Int. Conf. on Neural Networks (San Diego)* (Piscataway, NJ: IEEE Press) pp 61–8

Kohonen T, Kangas J, Laaksonen J and Torkkola K 1992 LVQ_PAK: A program package for the correct application of Learning Vector Quantization algorithms *Proc. Int. IEEE Joint Conf. on Neural Networks (Baltimore)* vol I (Piscataway, NJ: IEEE Press) pp 725–30

Le Cerf P, Ma W and Van Compernolle D 1994 Multilayer perceptrons as labelers for hidden Markov models *IEEE Trans. Speech and Audio Proc.* **2** 185–93

Mäntysalo J, Torkkola K and Kohonen T 1994 Mapping context dependent acoustic information into context independent form by LVQ *Speech Commun.* **14** 119–30

Rabiner L R 1989 A tutorial on Hidden Markov Models and selected applications in speech recognition *Proc. IEEE* **77** 257–86

Robinson A J 1994 An application of recurrent nets to phone probability estimation *IEEE Trans. Neural Networks* **5** 298–305

Torkkola K 1994a LVQ as a feature transformation for HMMs *Neural Networks for Signal Processing IV, Proc 1994 IEEE-SP Workshop (Ermioini, Greece, 6–8 September 1994)* (Piscataway, NJ: IEEE Press) pp 299–308

——1994b New ways to use LVQ-codebooks together with hidden Markov models *Proc. IEEE Int. Conf. on Acoustics, Speech and Signal Processing (Adelaide)* (Piscataway, NJ: IEEE Press) pp 401–4

Torkkola K, Kangas J, Utela P, Kaski S, Kokkonen M, Kurimo M and Kohonen T 1991 Status report of the Finnish phonetic typewriter project *Proc. Int. Conf. on Artificial Neural Networks (Espoo, Finland)* (North-Holland) pp 771–6

Young S 1991 Competitive training: a connectionist approach to discriminative training of hidden Markov models *Proc. IEE* **138** 61–8

——1992 *HTK: Hidden Markov model toolkit V1.4 – Reference Manual* (Cambridge: Cambridge University Engineering Department)

Yu G, Russel W, Schwartz R and Makhoul J 1990 Discriminant analysis and supervised vector quantization for continuous speech recognition *Proc. IEEE Int. Conf. on Acoustics, Speech and Signal Processing (Albuquerque, NM)* vol 1 (Piscataway, NJ: IEEE Press) pp 685–88

G1.5 Neural networks for alphabet recognition

Mark Fanty, Etienne Barnard and Ron Cole

Abstract

In a system which performs name retrieval from spellings, neural networks were used
for three components: a pitch tracker, a broad phonetic classifier used to find letters and
letter–internal segment boundaries, and a letter classifier. The broad phonetic classifier
uses spectral features in a fixed window around each 3 ms frame of the utterance. Its
output is a score for each of several broad phonetic classes (e.g. voiced stop, iy—the
vowel in E). A Viterbi search finds the most likely letter segment sequence based on
these scores. This defines the letter boundaries and internal segmentation of each letter.
The letter classifier uses carefully selected features from the whole letter based on our
knowledge of the acoustic differences between the letters. The features are anchored by
segment boundaries because they are especially useful in fine phonetic distinctions (e.g.
'B' versus 'D'). A neural network classifier is trained using these features extracted from
letters spoken by 120 different speakers. We achieved 96% accuracy on an independent
test set from 30 new speakers. When searching a database of 50 000 names, we achieved
95% first-choice name retrieval for 1020 spelled names. This section describes all three
neural networks briefly, but focuses on the broad phonetic and letter classifiers.

G1.5.1 Project overview

We have used neural networks for *classification of speech data* extensively since 1989. They are the major F1.7
tool for virtually all speech recognition research at the Center for Spoken Language Understanding (Fanty
et al 1993, Cole *et al* 1994, Muthusamy *et al* 1994).

For several years, our major focus was alphabet recognition. Alphabets are interesting scientifically
because they require fine phonetic distinctions, such as M/N, B/D, B/V, S/F, G/J. They are interesting
commercially because spelling may be the only practical way to communicate with an automatic spoken
language system in many situations, such as when callers are giving their names.

Figure G1.5.1 is a schematic of the system. The complete system performs name retrieval from a
database based on the spoken spelling of the name. For example, the user might spell 'S M I T H E
R S' and thereby retrieve the telephone extension for the employee Smithers. Speech is captured and
digitized (via a microphone, telephone, or the like). For the pitch tracker, the signal is filtered to remove
high-frequency components and peaks in the waveform are classified as to whether they begin a pitch
period or not. These pitch marks are used to create features for the other classifiers.

Spectral analysis of the waveform is used to provide features for the broad phonetic classifier and
letter classifier. In early versions of the system, we used FFT but switched to PLP (Hermansky 1990)
because it is more compact and yields results as good as or better than FFT (Fanty *et al* 1991). A number
of other features are derived from the waveform, such as the zero-crossing rate and peak-to-peak amplitude.

Before a letter can be classified, it must be located in the signal and its internal phonetic boundaries
must be defined. To this end, every 3 ms frame of the signal is phonetically classified using a mixture of
22 broad and fine phonetic classes. After every frame receives a score for the 22 broad phonetic classes,
a Viterbi search finds the optimal alignment of those scores with the pronunciation models of the letters.
Every letter consists of some sequence of phonemes. For example, a T has some number of 'ptk' frames

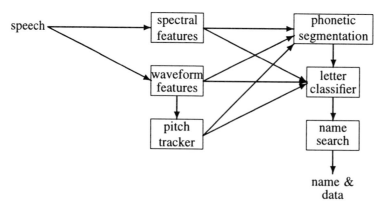

Figure G1.5.1. Schematic of the name retrieval system.

followed by some number of 'iy' frames. Because we do not do fine classification of the stop, this is also the model for a P.

For each letter, the segment boundaries are used to compute features based on pitch, spectral analysis, zero crossing, and peak-to-peak amplitude. These features are the inputs of a neural network which classifies the letter. The output of this net is a score for each letter of the alphabet. These scores are used to search the database of names. The name with the highest score, and any associated data, is retrieved.

G1.5.2 Design process

Our guiding philosophy is to study speech in order to discover which acoustic features provide the best evidence for correct classification. We then build a system which measures those features and uses them as input to a neural network classifier. We believe this is the optimal compromise between knowledge intensive, rule-based approaches and simple automated learning approaches. We have found, as have many others, that purely rule-based approaches are too fragile. There are too many interactions and special cases to capture in hand-crafted rules. Purely automatic approaches, which use simple frame-based input features, have had a good deal of success in speech recognition. However, we believe that by using knowledge to

B4.4 *preprocess* the input before classification we can create an input space in which the class boundaries are much easier to learn and thereby enhance the performance. In practice, all speech recognition systems do this to some extent, for example, by using spectral coefficients instead of the raw speech samples. The letter classifier described here goes much further.

All of the networks used have the same topology: all input units are connected with all hidden units. All hidden units are connected with all output units. Hidden and output units take a weighted sum of their inputs and apply a sigmoid function to produce their output. There is no bias. Rather, there is an 'extra' input which has the constant value 1.0.

G1.5.3 Pitch tracker

Pitch tracking is achieved through a network which classifies each peak in the waveform as to whether it begins a pitch period (Barnard *et al* 1991). The waveform is low-pass filtered at 700 Hz and each positive peak is classified using information about it and the preceding and following four peaks. For each of the nine peaks, the following information is provided: (i) the amplitude, (ii) the time difference between the peak and the candidate peak, (iii) a measure of the similarity of the peak and the candidate peak (point-by-point correlation), (iv) the width of the peak, and (v) the negative amplitude or most negative value preceding the peak. The network was trained on the TIMIT database, and agrees with expert labelers about 98% of the time, compared to 99% agreement among human labelers. It performs well on our high-bandwidth data without retraining. The pitch tracker does not perform as well for telephone speech because the waveform characteristics are so different; pitch is not yet a feature in the telephone systems.

G1.5.4 Frame-based phonetic classification

The recorded speech is divided into letters and broad phonetic segments within the letters by first phonetically classifying every 3 ms frame. The neural network which does this classification has 22 outputs. Some correspond to phonemes, such as 'eh', the vowel in M, N, F and S; others are collections of phonemes, such as the output we call 'ptk' which includes the voiceless stops in P, T and K. The motivation was to have enough discrimination to separate letters and find their internal boundaries, but to avoid making difficult phonetic distinctions at this stage. We found that by forcing the network to choose between easily confused classes the score for each of these classes might be low in some cases, allowing some other class to win and produce an incorrect segmentation. Research showed that grouping acoustically similar classes into broad categories produced more accurate classification of these segments (Roginski 1991).

The inputs for the frame-based phonetic classifier consist of the spectrum of the frame to be classified, as well as frames before and after it so the classifier is provided with information about the acoustic context. The total window size is 170 ms. In addition to the spectrum in this window, there are waveform features representing the zero-crossing rate and peak-to-peak amplitude, and a binary feature indicating the presence of consistent pitch peaks.

Training data were created by hand-labeling the phonetic boundaries in recorded letters. Frames of these letters were selected as training data. Limited resources precluded using every frame of training data. The short frame size means adjacent frames are very similar in any case, and largely redundant. Also, some categories are much more prevalent than others. We found it worked best to sample these less densely so that approximately the same number of each category are in the training set†.

Our first strategy was to pick frames at random with the desired density, but this resulted in very poor performance near certain boundaries, especially those following background silence. The reason for this was the sampling strategy. Because there is so much silence surrounding each letter, only a very small number of training data were selected near speech boundaries. Thus, the classifier was misled by the presence of speech in the look-ahead features. The solution to this problem was to pick extra frames near boundaries (Roginski 1991). The more general lesson is that if there are sparse but especially important regions of the input space, random sampling of training data is not enough.

G1.5.5 Letter classification

After segmenting the signal, features must be extracted based on segment boundaries. Every letter in the English alphabet has a single sonorant segment except W, which is treated as a special case. In F, this is the 'eh'; in M this is both the 'eh' and the 'm'. This segment always exists, and provides the temporal anchor for most of the feature measurements. The previous consonant is the stop or fricative (e.g. B or C) before the sonorant. If there is no stop or fricative (e.g. E), the 200 ms interval before the sonorant is treated as a single segment for feature extraction. After dozens of experiments, we arrived at the following feature set:

- FFT coefficients from the consonant preceding the sonorant. The consonant is divided into thirds temporally; from each third, 32 averaged values are extracted linearly from 0 to 8 kHz. All FFT inputs are normalized locally so that the largest value from a given time slice becomes 1.0 and the smallest becomes 0.0 (96 values).
- FFT coefficients from the sonorant. From each seventh of the sonorant, 32 averaged values are extracted linearly from 0 to 4 kHz (224 values).
- FFT coefficients following the sonorant. At the point of maximum zero-crossing rate in the 200 ms after the sonorant, 32 values are extracted linearly from 0 to 8 kHz—for F, S, X and H (32 values).
- FFT coefficients from the second and fifth frames of the sonorant—32 values from each frame extracted linearly from 0 to 4 kHz. These are not averaged over time, and will reflect formant movement at the sonorant onset (64 values).

† Bourlard and Morgan (1994) have explained why this should be so. It can be shown that when all the data are used, the classifier estimates the probability of each phoneme given the data from a frame. These are to be blended with the pronunciation models to estimate the probability of an observed sequence of frames according to each model. Under a Markov assumption, the probability of the sequence is the product over frames of the probability of each frame given the phoneme in the model. These are the Bayes' rule inverses of the classifier outputs. Given a uniform prior distribution over frames, the Bayes inversion can be accomplished (up to a constant factor) simply by dividing the classifier outputs by the unconditional probabilities of their corresponding classes. It is easy to show that this is equivalent to subsampling the training data to equalize the unconditional class probabilities they represent.

- FFT coefficients from the location in the center of the sonorant with the largest spectral difference (amount of frame-to-frame change in spectrum) extracted linearly from 0 to 4 kHz. This samples the formant locations at the vowel–nasal boundary in case the letter is M or N (32 values).

- Zero-crossing rate in 11 18 ms segments (198 ms) before the sonorant, in 11 equal-length segments during the sonorant and in 11 18 ms segments after the sonorant. This provides an absolute time scale before and after the sonorant which could help overcome segmentation errors (33 values).

- Amplitude from before, during, and after the sonorant represented the same way as zero crossing (33 values).

- Low-pass-filtered amplitude represented the same way as amplitude (33 values).

- Spectral difference represented like zero crossing and amplitude except the maximum value for each segment is used instead of the average, to avoid smoothing the peaks which occur at boundaries (33 values).

- Inside the sonorant, the spectral center of mass from 0 to 1000 Hz, measured in 10 equal segments (ten values).

- Inside the sonorant, the spectral center of mass from 1500 to 3500 Hz, measured in 10 equal segments (ten values).

- Median pitch, the median distance between pitch peaks in the center of the sonorant. This correlates with vocal tract length, which determines characteristic formant locations (one value).

- Duration of the sonorant (one value).

- Duration of the consonant before the sonorant (one value).

- High-resolution representation of the amplitude at the sonorant onset: five values from 12 ms before the onset to 30 ms after the onset (five values).

- Abruptness of onset of the consonant before the sonorant, measured as the largest two-frame jump in amplitude in the 30 ms around the beginning of the consonant (one value).

- The category of the segment before the sonorant: closure, fricative, or stop (three values).

- The largest spectral difference value from 100 ms before the sonorant onset to 21 ms after, normalized to accentuate the difference between B and V (one value).

- The number of consistent pitch peaks in the previous consonant (one value).

- The number of consistent pitch peaks before the previous consonant (one value).

- The presence of the segment sequence closure fricative after the sonorant (an indicator of X or H) (one binary value).

Later networks, including those used for telephone speech, replaced the 32 FFT coefficients with eight PLP coefficients, considerably reducing the size of the network (Cole *et al* 1992). Each of these features was designed with some discrimination(s) in mind. Care must be taken that all features have reasonable values for all letters, even those for which the feature makes no sense. Thus, letters with no preceding consonant still need values for features which represent attributes of the consonant. These are extracted from a fixed interval of silence preceding the letter.

All inputs to our network were normalized: mapped to the interval [0.0, 1.0]. We attempted to normalize so that the entire range was well utilized. In some instances, the normalization was keyed to particular distinctions. For example, the center of mass in the spectrum from 0 to 1000 Hz was normalized so that E was low and A was high. Other vowels, such as O, would have values 'off the scale' and would map to 1.0, but the feature was added specifically for E/A distinctions.

G1.5.6 Results

All results are for letters spoken with pauses between them. Using high-quality speech (16 kHz sampling; Sennheiser noise-canceling microphone) in the ISOLET spoken letter corpus (Cole *et al* 1990), we achieved 96% correct on 30 spoken alphabets from 30 different speakers after training on alphabets from 120 different speakers. We did a separate study for the especially difficult E-set (B, C, D, E, G, P, T, V, Z) in which we added a few more specialized features. For just this set, we achieved 95% correct (Fanty and Cole 1990). Name retrieval performance for 1020 test callers was 95% when searching a list of the 50 000 most common names (Cole *et al* 1991).

When we retrained the system using telephone speech and the CSLU Whitepages corpus (Cole *et al* 1992), performance on the alphabet dropped to 89.1% (compared to 93% for human listeners). Bandwidth limitations of the telephone channel make recognition of some letters especially difficult (e.g. S and F).

G1.5.7 Development

All neural networks were trained using OPT, which is based on conjugate gradient optimization (Barnard and Cole 1991). OPT is an implementation of *backpropagation* which eliminates the learning rate and momentum parameters. It is available as part of the OGI speech tools. For more information, ftp speech.cse.ogi.edu:pub/tools/ANNOUNCE or see our WWW home page: http://www.cse.ogi.edu/CSLU.

C1.2.3

For information about acquiring OGI speech corpora, send email to Mike Noel: noel@cse.ogi.edu.

G1.5.8 Comparison with traditional methods

The most popular traditional approaches to speech recognition are based on dynamic time warping (DTW) and hidden Markov models (HMMs). Since English alphabet recognition has been a popular task domain in computer speech recognition for a number of years, several results using these approaches have been published. Unfortunately, many different corpora have been used to test the various methods, so direct comparison is generally not possible.

Early work, reviewed by Cole *et al* (1984), used template matching with DTW to perform frame-by-frame matching of an input utterance to reference templates. This approach produced speaker-dependent recognition rates of 60% to 80% on the alphabet and alphadigit (letters and digits) vocabularies.

A substantial improvement in recognition accuracy was demonstrated in the FEATURE system, which combined knowledge-based feature measurements and multivariate classifiers in a hierarchical decision tree (Cole *et al* 1983). FEATURE performed speaker-independent recognition of spoken English letters at 89.5% accuracy, a substantial improvement over speaker-dependent, DTW-based approaches.

More recently, increased recognition accuracy has been obtained by applying HMMs to the letter recognition problem. HMMs construct a more detailed statistical model of the speech sounds to be recognized. A first-order Markov model is typically used to describe the temporal evolution of the speech signal, and mixtures of multivariate Gaussians describe the density function of the observed acoustic signals in each of the states of the Markov model.

Brown (1987) obtained 92% correct classification of the E-set on a multispeaker task using 100 speakers. (For a multispeaker task, as opposed to a speaker-independent task, the same speakers occur in the training and test sets.) Researchers at AT&T Bell Labs have applied HMM-based approaches to multispeaker recognition of a 39-word vocabulary of letters, digits, and the words 'stop,' 'error', and 'repeat'. Using whole-word continuous density HMMs, the best result obtained for 100 speakers was 89.5% (Rabiner and Wilpon 1987). The addition of acoustic segmentation and probabilistic word modeling resulted in multispeaker recognition rates of 92 to 93% (Euler *et al* 1990).

In principle, neural network systems that classify segments of speech rather than single frames should be able to outperform HMMs, since they are not constrained by the Markovian assumption (namely that the differences between successive frames in the same state are uncorrelated with one another). However, this theoretical advantage has not yet been translated to substantial improvements in performance for neural networks—possibly since the technology based on HMMs is more mature. It is thus widely believed (see e.g. Bourlard and Morgan 1994) that it is possible to obtain comparable results with HMM-based systems and systems based on neural networks, and the results on alphabet classification seem to confirm this trend. Currently, the main advantage of neural networks seems to be that they can achieve this level of performance with much lower system complexity (Cohen *et al* 1993), which is extremely important for real-time implementations.

Our work on alphabet recognition, and that of Waibel and his colleagues at CMU (Hild and Waibel 1993, Bregler *et al* 1993) is consistent with that view.

G1.5.9 Conclusions

The use of neural networks in the name retrieval system has been a great success. The results on high-quality speech are the best reported in the literature, and the telephone implementation has been implemented on a DSP board, licensed commercially, and ported to other languages. It was used in the spring of 1995 by the Bureau of the Census in a technology pretrial for automated name entry.

Acknowledgements

The authors wish to thank Richard Rohwer for his assistance. This research was supported by the National Science Foundation, US West, Apple Computer, the Office of Naval Research and Digital Equipment Corporation.

References

Barnard E B and Cole R A 1991 A neural-net training program based on conjugate-gradient optimization *Oregon Graduate Institute of Science and Technology Technical Report* CSE 89-014

Barnard E B, Cole R A, Vea M and Alleva F 1991 Pitch detection with a neural net classifier *IEEE Trans.* **39** 298–307

Bourlard H A and Morgan N 1994 *Connectionist Speech Recognition* (Boston, MA: Kluwer)

Bregler C, Hild H, Manke S and Waibel A 1993 Improving connected letter recognition by lipreading *IEEE Proc. Int. Conf. on Acoustics, Speech, and Signal Processing (Minneapolis, MN)* pp I-557–560

Bourlard H A and Morgan N 1994 *Connectionist Speech Recognition* (Boston, MA: Kluwer)

Brown P F 1987 The acoustic-modeling problem in automatic speech recognition *PhD Thesis* Carnegie Mellon University, Department of Computer Science

Cohen M, Franco H, Morgan N, Rumelhart D and Abrash V 1993 Context-dependent multiple distribution phonetic modeling with MLPs *Advances in Neural Information Processing Systems 5* ed S J Hanson, J D Cowan and L C Giles (San Mateo, CA: Morgan Kaufmann) pp 649–57

Cole R A, Fanty M, Gopalakrishnan M and Janssen R D T 1991 Speaker-independent name retrieval from spellings using a database of 50 000 names *Proc. Int. Conf. on Acoustics, Speech and Signal Processing* pp 325–8

Cole R A, Fanty M and Roginski K 1992 A telephone speech database of spelled and spoken names *Proc. Int. Conf. on Spoken Language Processing* pp 891–3

Cole R A, Muthusamy Y K and Fanty M 1990 The ISOLET spoken letter database *Oregon Graduate Institute of Science and Technology Technical Report* CSE 90-004

Cole R A, Novick D, Fanty M, Vermeulen P, Sutton S, Burnett D and Schalkwyk J 1994 A prototype voice response questionnaire for the US census *Proc. IEEE Int. Conf. on Acoustics, Speech and Signal Processing* pp 683–6

Cole R A, Stern R M and Lasry M J 1984 Performing fine phonetic distinctions: templates vs. features *Invariance and Variability of Speech Processes* ed J S Perkell and D H Klatt (Lawrence Erlbaum) pp 325–45

Cole R A, Stern R M, Phillips M S, Brill S M, Pilant A and Specker P 1983 Feature-based speaker-independent recognition of isolated English letters *Proc. IEEE Int. Conf. on Acoustics, Speech and Signal Processing* pp 731–4

Euler S A, Juang B H, Lee C H and Soong F K 1990 Statistical segmentation and word modeling techniques in isolate word recognition *Proc. IEEE Int Conf. on Acoustics, Speech and Signal Processing* pp 745–8

Fanty M and Cole R A 1990 Speaker-independent English alphabet recognition: experiments with the E-set *Proc. 1990 Int. Conf. on Spoken Language Processing* pp 1361–4

Fanty M, Cole R and Slaney M 1991 A comparison of DFT, PLP, and cochleagram for alphabet recognition *Proc. 25th Asilomar Conf. on Signals, Systems and Computers* pp 326–9

Fanty M, Schmid P and Cole R A 1993 City name recognition over the telephone *Proc. IEEE Int. Conf. on Acoustics, Speech and Signal Processing* pp I-549–552

Hermansky H 1990 Perceptual linear predictive (PLP) analysis of speech *J. Acoust. Soc. Am.* **87** 1738–52

Hild H and Waibel A 1993 Speaker-independent connected letter recognition with a multi-state time delay neural network *Proc. Eurospeech '93* pp 1481–5

Muthusamy Y, Barnard E and Cole R 1994 Reviewing automatic language identification *IEEE Signal Proc. Mag.* **11** 33–41

Rabiner L and Wilpon J 1987 Some performance benchmarks for isolated word, speech recognition systems *Computer Speech and Language* pp 343–57

Roginski K 1991 A neural network phonetic classifier for telephone spoken letter recognition *Masters Thesis* Oregon Graduate Institute of Science and Technology, Department of Computer Science and Engineering

G1.6 A neural network for image understanding

Heggere S Ranganath, Govindaraj Kuntimad and John L Johnson

Abstract

The pulse-coupled neuron, which is significantly different from the conventional artificial neuron, is a result of recent research conducted on the visual cortex of cats and monkeys. Pulse-coupled neural networks (PCNNs) are modeled to capture the essence of recent understanding of image interpretation processes in biological neural systems. Our study indicates that the PCNN is capable of image smoothing, image segmentation and feature extraction. The PCNN reduces noise in digital images better than traditional smoothing techniques. As an image segmenter the PCNN performs well even when the intensity varies significantly within regions, and adjacent regions have overlapping intensity ranges. This article describes the theory, design and implementation of an image segmentation/detection system based on the PCNN.

G1.6.1 Project overview

The primary objective of the project was to design, implement and study the pulse-coupled neural network, in order to determine its potential for use in real-time image understanding applications. The task was accomplished by developing an automatic target-detection system shown in figure G1.6.1. The system consists of five functional modules: a smoothing module, segmentation module, evaluation module, detection module and knowledge base. The smoothing and segmentation modules are implemented as pulse-coupled neural networks in which there is a neuron corresponding to each pixel in the image. The smoothing network reduces the random noise present in the input image. The segmentation network partitions the image into several regions. The evaluation module checks each segment and, if satisfied with the overall segmentation result, forwards the segmented image to the detection module. Otherwise, neurons corresponding to regions of no interest are disabled, the network parameters are modified, and the image is resegmented. The detection module, with the help of the knowledge base, identifies all subsets of regions that can potentially form a target for further processing by the recognition system.

G1.6.2 Design process

The development of a PCNN-based target-detection system was motivated by the recent results published by Eckhorn *et al* (1990) and Johnson (1994). Eckhorn has modeled the cat visual cortex by using a network of pulse-coupled neurons. His simulations show that the PCNN has *associative memory* embedded in its architecture and is capable of filling in the missing spatial and temporal information. Johnson has studied the effects of linking in a single-layer laterally connected network. The primary objective of his research is to map 2D spatial distributions to 1D temporal patterns called *time series*.   C1.3, F1.4

Eckhorn's neuron was modified to suit the needs of image understanding applications (figure G1.6.2). The neuron, N_k, consists of a linking receptive field comprising several leaky integrators and a spike generator. A leaky integrator is a first-order linear time-invariant system with an exponentially decaying output. Its impulse response is given by

$$I_l(t) = \begin{cases} V_l \exp(-t/\tau_l) & t \geq 0 \\ 0 & t < 0 \end{cases} \qquad (G1.6.1)$$

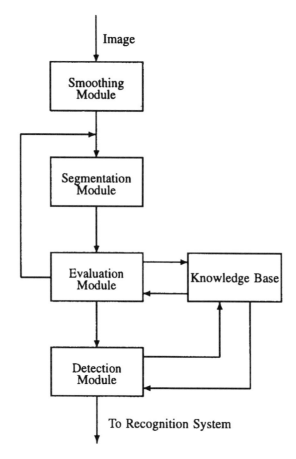

Figure G1.6.1. PCNN-based automatic target recognition system. Smoothing and segmentation modules utilize PCNN-based algorithms.

where V_l is the amplification factor and τ_l is the decay time constant of the leaky integrator. Therefore, the output of a leaky integrator is the convolution of its impulse response and its input.

The linking receptive field consists of Q linking leaky integrators ($LLI_{k1}, \ldots, LLI_{kQ}$). The input to LLI_{kj} is $Y_n(t)$, the output of the linking neuron N_n. Each leaky integrator LLI_{kj} is connected to a different linking neuron. The network linking input to N_k, $L_k(t)$, is the weighted sum of the outputs of all the leaky integrators in the neuron's linking receptive field. The weighting factor between the neurons N_k and N_n is inversely proportional to the square of the distance between them.

The feeding input X_k, the intensity of the associated pixel in the image, is modulated by the linking input to yield the neuron's internal activity, $U_k(t)$:

$$U_k(t) = X_k[1 + \beta_k L_k(t)] \tag{G1.6.2}$$

where β_k is a positive constant referred to as the linking coefficient.

The spike generator consists of a step-function generator and a threshold generator. The output of the step-function generator is the output of neuron $Y_k(t)$. When the internal activity exceeds the threshold value, $Y_k(t)$ goes to 1; otherwise $Y_k(t)$ is zero. The output of the threshold generator, $\theta_k(t)$, is set to a predetermined value, V_θ, whenever $Y_k(t)$ goes to 1. When $Y_k(t)$ is zero, $\theta_k(t)$ decays exponentially with decay time constant τ_θ. We have chosen V_θ to be much greater than the highest possible value of $U_k(t)$. Therefore, when $Y_k(t)$ goes high, the threshold value exceeds the internal activity of the neuron instantaneously, forcing $Y_k(t)$ to zero. Thus, ideally, the neuron N_k generates an impulse on $Y_k(t)$ every time $U_k(t)$ exceeds $\theta_k(t)$. If the charging of the threshold generator is not instantaneous, $Y_k(t)$ will be a pulse of finite width.

C1.1   The PCNN used in image understanding applications is a *single-layer network* in which there is a neuron corresponding to every pixel in the image. The value of the linking coefficient is the same for all neurons ($\beta_k = \beta$, for all k). A neuron N_k in the network receives linking input from every neuron in the

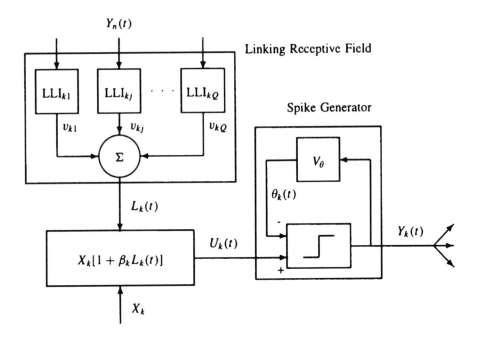

Figure G1.6.2. Image processing neuron.

circular neighborhood of radius r centered about N_k. In an unlinked network (linking coefficient β is zero) the firing rate of N_k is determined by its primary input X_k. If β is greater than zero the neuron's primary input is modulated by the linking input from its neighbors. As a result, if $U_k(t)$ exceeds $\theta_k(t)$, N_k will fire synchronously with its neighbors. Now, we say that N_k is *captured* by its neighbors. This property of the PCNN where a group of firing neurons capture other neighboring neurons in a similar environment forms the foundation for image smoothing, image segmentation and feature extraction algorithms.

G1.6.2.1 Image smoothing

Image smoothing is the process of removing or reducing the random noise present in images (Gonzalez and Wintz 1977). The image to be smoothed is applied as an input to a weakly linked PCNN. The intensity of a noisy pixel is significantly different from the intensities of its surrounding pixels. Therefore, most neurons corresponding to noisy pixels do not capture neighboring neurons or get captured by the neighboring neurons. Image smoothing is accomplished by adjusting the intensity of each pixel based on the neuron firing pattern in its neighborhood. If a neuron fires sooner than a majority of its neighbors fire, its intensity is adjusted downwards. If a neuron fires after the majority of its neighbors have fired, its intensity is adjusted upwards. If a neuron fires with the majority of its neighbors no change is made. After completing the firing cycle (all neurons fire exactly once) the network may be reset by forcing the threshold values of all neurons to zero and the smoothing process may be repeated.

G1.6.2.2 Image segmentation

Image segmentation is the process of partitioning an image into its component regions. Consider an image consisting of two regions R and B. Spatially connected object pixels form R and background pixels form B. Perfect segmentation is possible if there exist a linking radius r and linking coefficient β which force all neurons belonging to a region to pulse together periodically. Of course, different regions need to pulse at different rates.

Let $[I_1, I_2]$ and $[I_3, I_4]$ be the intensity ranges of background and object pixels, respectively. When $I_3 < I_2$ thresholding techniques do not produce a perfect result. On the other hand, PCNN can produce perfect segmentation if: (i) object neurons of intensity I_4 pulsing at T_R capture all the remaining object neurons and (ii) background neurons of intensity I_2 pulsing at T_B capture all the remaining background neurons. We have shown that perfect segmentation occurs when the following inequalities are satisfied

(Ranganath and Kuntimad 1994):

$$I_3[1 + \beta L_{\min r}(T_R)] \geq I_4 \tag{G1.6.3}$$

$$I_2[1 + \beta L_{\max b}(T_R)] < I_4 \tag{G1.6.4}$$

$$I_1[1 + \beta L_{\min b}(T_B)] \geq I_2. \tag{G1.6.5}$$

In the above inequalities $L_{\min r}(T_R)$ is the minimum of all the linking inputs received by object neurons at T_R, $L_{\max b}(T_R)$ is the maximum of all the linking inputs received by background neurons at T_R and $L_{\min b}(T_B)$ is the minimum of all the linking inputs received by background neurons at T_B. The values of $L_{\min r}$, $L_{\max r}$ and $L_{\min b}$ increase as r increases and the rates of increase are determined by the object-background boundary geometry. When perfect segmentation is not possible, the challenge is to find optimal parameters (β^*, r^*) which minimize the error.

G1.6.2.3 Feature extraction and target detection

Assume that an image is connected to primary inputs of a pulse-coupled neural network. Initially, at time $t = 0$, all neurons pulse together. The subsequent pulsing of a neuron is determined by the neuron's primary input, linking input and network parameters. The plot of the number of neurons firing as a function of time is referred to as the *time series*. Johnson is of the opinion that if the network is allowed to pulse, it will eventually converge producing a periodic time series. He further believes that the resulting time series characterizes the image, and when normalized, is invariant to rotation, translation and scaling. He has demonstrated this property of the PCNN through simulation using simple test images (Johnson 1994). However, there is no mathematical proof for the existence of periodic time series.

Also, by making slight modifications to the current version of the PCNN, it is possible to obtain geometric features such as the size, centroid and radius of gyration of each pulsing region. The features so obtained may be utilized for target detection or recognition in the postprocessing stage.

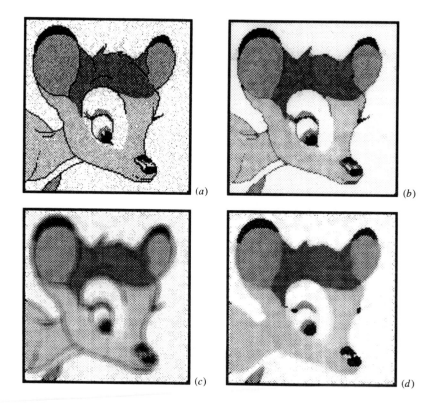

Figure G1.6.3. An example of image smoothing: (*a*) noisy image; (*b*) result of PCNN processing; (*c*) result of neighborhood averaging; (*d*) result of median filtering.

G1.6.3 Development

To the best of our knowledge there are no PCNN software or hardware packages available commercially. Therefore, we implemented all modules of the PCNN-based target-detection system in C language on a SUN workstation running a UNIX operating system.

G1.6.4 Comparison with traditional methods

The ability of PCNN to smooth images was compared with that of neighborhood averaging and median filtering methods. The neighborhood averaging method blurred edges. The median filtering method often suffered from the problem of edge erosion and dilation. However, simulation results indicated that PCNN smoothed images without blurring, eroding or dilating edges. These facts are well illustrated in figure G1.6.3.

A number of artificial and real images were segmented using the PCNN and several known segmentation methods such as optimal thresholding, region growing and probabilistic relaxation. The performance of the PCNN was far superior to that of optimal thresholding and region-growing techniques. The PCNN was better than the relaxation technique in segmenting regions with fuzzy edges.

We also verified the perfect segmentation property of the PCNN through simulation. Figure G1.6.4(a) shows a 64×64 image consisting of two regions. The intensity ranges of the background and object are [100, 175] and [150, 250], respectively. For $r = 1$, it can be shown that $L_{\min r} = 2, L_{\min b} = 2$ and $L_{\max b} = 1$. Perfect segmentation is possible when the value of β is in the range [3/8, 3/7]. The image was segmented using PCNN with $\beta = 0.4$. The regions of the image as determined by the synchronous firing of neurons are shown in figures G1.6.4(b) and G1.6.4(c).

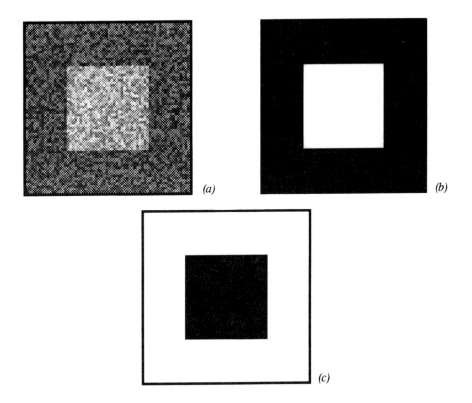

Figure G1.6.4. An example of perfect segmentation: (a) noisy image; (b) segmented region; (c) segmented background.

The PCNN-based target-detection system, shown in figure G1.6.1, was implemented in C on a SUN workstation. The 128×128 input image shown in figure G1.6.5(a) was smoothed using PCNN. The smoothed image was iteratively segmented until satisfactory results were obtained, as determined by the evaluation module. A low value of β (0.1) was chosen initially to ensure over-segmentation. For each segment, size, radius of gyration and centroid were computed. Any segment whose attributes violated the size and elongation constraints specified in the knowledge base was eliminated from further consideration. This was accomplished by disabling the neurons corresponding to the eliminated segment. If the number of remaining segments was greater than a predetermined number (18), the value of β was increased by 0.05, and the remaining image was resegmented. After successful segmentation, all groups of spatially connected segments which satisfied the constraints in the knowledge base were detected as potential targets for further processing by a target recognition system. A graph-theoretic approach was used to reduce computation during grouping. For the image shown in figure G1.6.5(a) the system detected one candidate group, shown in figure G1.6.5(b), as a potential target.

 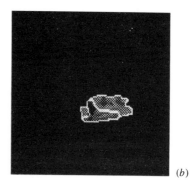

(a) (b)

Figure G1.6.5. Image detection system: (a) noisy image containing tank; (b) detected tank.

G1.6.5 Summary and conclusions

Single-layered, laterally linked pulse-coupled neural networks, being fairly insensitive to noise and local intensity variations in digital images, are highly effective for image smoothing, segmentation and feature extraction applications.

A comparison of the PCNN with Markov random fields and relaxation methods is relevant. The smoothing method based on the Markov random fields modifies the intensity of each pixel in the image, but the PCNN selects and modifies a small group of pixels which do not pulse with their neighbors. Also, the PCNN-based smoothing yields much sharper edges than the Markov random fields method. The segmentation performance of the relaxation method is comparable to that of the PCNN. However, unlike the relaxation method, the PCNN does not require prior knowledge of the number of regions in the image. Simulation has shown that the PCNN segments boundary pixels more accurately than the relaxation method.

E1.5 Several *optical devices* capable of implementing the linking receptive field efficiently have been developed. For example, the optically addressed spatial light modulator developed by SY Technology Inc. facilitates large-scale electro-optical implementation of the PCNN. As a result, the PCNN has potential in real-time image processing.

Acknowledgement

This project, led by Ranganath and Johnson, was funded in part by the US Army Missile Command and supported by three graduate students.

References

Eckhorn R, Reitboeck H J, Arndt M and Dicke P 1990 Feature linking via synchronization among distributed assemblies: simulations of results from cat visual cortex *Neural Comput.* **2** 293–307
Gonzalez R C and Wintz P 1977 *Digital Image Processing* (Reading, MA: Addison-Wesley)
Johnson J L 1994 Pulse-coupled neural networks *Adaptive Computing: Mathematics, Electronics and Optics* vol CR55 (Proceedings of a conference held on 4–5 April 1994 Orlando, FL)
Ranganath H S and Kuntimad G 1994 Image segmentation using pulse coupled neural networks *Proc. IEEE Int. Conf. on Neural Networks (Orlando, FL)* vol II, pp 1285–90

G1.7 The application of neural networks to image segmentation and way-point identification

James Austin

Abstract

The analysis of images of the ground taken from aircraft and satellites is of immense importance. This case study describes work using the ADAM neural network aimed at finding way-points, roads, towns and rural areas in infrared line scan images of the ground. The network is capable of on-line, single-pass training and simple computer implementation, making it particularly applicable in image processing tasks. In addition, no image preprocessing is needed, and images may be very large, which makes the approach particularly simple to adopt.

G1.7.1 Project overview

This project was aimed at finding way-points, roads, towns and rural areas (segmentation) in infrared line scan (IRLS) images of the ground taken from aircraft. The problem of finding way-points is vital for airborne vehicles (i.e. landmarks that are used to guide a vehicle) and can be useful in many other image processing tasks. In this work we were particularly interested in a system that could be trained rapidly on new features and way-points. The more conventional network architectures, based on gradient descent learning, could not be used in the problem because they could not be rapidly trained on the large and complex data sets used in the work. The use of the *advanced distributed associative memory* (ADAM)   C1.5.8 (Austin 1987) combined the fast feature-recognition ability of the network with the large capacity of the network so that a large number of way-points could be stored. Section G1.7.2 describes the feature-recognition system that could find roads and segment town areas in maps from rural areas. Section G1.7.3 builds on this and describes how way-points can be trained into the network. Due to space limitations, the neural network methods used are not described in detail, but can be found in other sections of the handbook and in the papers cited.

G1.7.2 First stage processing: feature recognition

G1.7.2.1 Design process

The IRLS problem involved the use of an ADAM to recognize a small feature in an image; from this a label could be recalled to indicate the type of feature found. To perform feature recognition a small square window is scanned over the image in regular steps. At each step the image data in the window are passed to the inputs of the neural network. The network is tested and the classification of the data in the window outputs from the network. This can be passed to the second stage of the network which outputs a label used to indicate the type of feature found, so that the user can assess the quality of the recognition performance. Figure G1.7.1 shows a typical output showing how the network can recognize the roads in the image in figure G1.7.4.

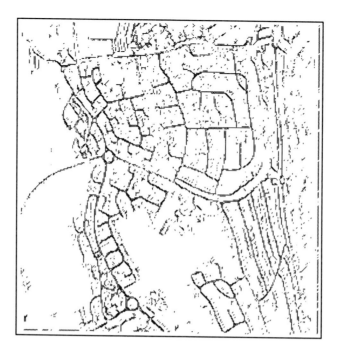

Figure G1.7.1. The result of finding roads using ADAM.

G1.7.2.2 Network architecture

The approach used the ADAM network with ranked gray-scale preprocessing. This network is described fully in Austin and Stonham (1987).

G1.7.2.3 Training methods

Prior to the recognition phase the ADAM network is trained on image data. This is achieved by the following process. A training image was selected from areas which contain typical data of the class of interest. The data were separated into a training set and a validation set. In the road example, we selected the training set shown in figure G1.7.2. There were four classes of road representing four road orientations. The fifth class was the nonroad class. For each class, an idealized version of the feature was to be recalled from the network, so that once the feature was recognized a 'clean' image of the data could be built up. The training process was as follows.

(i)   Set the initial parameters of the neural network. These can be set to N-tuple size 4, number of N-tuples to about 100, and ranks set to 2.
(ii) Place the sampling window at the first position to be trained.
(iii) Train the example into the network.
(iv) Evaluate the performance of the network (see section G1.7.2.4).
(v)   Is the current performance better by an amount T than that achieved so far? If it is, place the training window at the next point to be trained and go to step (iii). If the performance is not better than an amount T, stop training.

This process was extremely fast, taking less than one second to train on a UNIX-based 68040 workstation. Because of this, the search for the set of optimal parameters for the network is very quick. In this problem the typical ranges of values that would be examined were: a range of N-tuple sizes between 4 and 8, the number of tuples from 10 to 200 and rank sizes from 2 to the size of the N-tuple. These apply for the image size of 8×8 pixels used in this problem.

G1.7.2.4 Validation process

This was done by recording the average recognition success over all window positions in the validation set. The recognition success is taken as the average response of the class bits that are on (see Austin and Stonham 1987). In this case the validation set consisted of a complete scan of the example image.

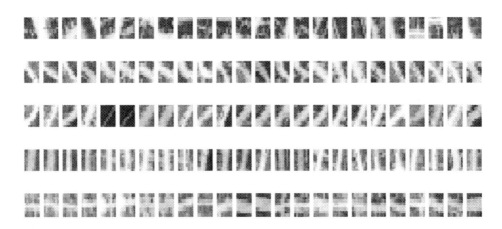

Figure G1.7.2. The road training examples used for finding roads.

G1.7.2.5 Preprocessing

A particular feature of the ADAM network is its ability to use data that have not been extensively preprocessed. Because the network is capable of training large amounts of pattern data without long training times, the image data can be fed directly to the network. Had an MLP network been used, this would have necessitated the use of preprocessing to reduce the size of the image that is fed to the network.

G1.7.2.6 Output interpretation

The interpretation of the output was shown in figure G1.7.1. At each point in the input image, the ADAM memory recalled the 'ideal' image for that feature. This was placed into the output image at the same point as the image feature was found in the input image. The example given here is covered in detail in Austin and Buckle (1995).

G1.7.3 The segmentation of towns and rural areas

The same approach was used to segment images into town areas and rural areas. The results of this work are shown in figure G1.7.3. The white areas indicate urban areas, the black areas are rural regions. Details of this are also given in Austin and Buckle (1995).

G1.7.3.1 Finding a way-point feature in a large image

This work used the network's ability to recognize image segments to find known locations in unknown images (a way-point). This built on the ADAM network's ability to store many thousands of image segments for subsequent matching, allowing it to be used effectively in a vehicle guidance system. The work examined methods needed to achieve accurate way-point finding. The capacity of the memory has been examined in Austin and Stonham (1987). An example of a way-point to be found is given in figure G1.7.4. Because the image feature is small, and not particularly salient, it could match in many positions. To allow such small features to be accurately found the method used a multi-image resolution approach to provide context for the recognition process.

G1.7.3.2 Training the network

To solve the problem five ADAM networks were used. Each ADAM network received an input image at a different resolution, as shown in figure G1.7.5. These images were gathered by first selecting a feature that was to be trained, then taking a window of pixels centered on the feature of size 21×21 pixels for window 5, 42×42 for window 4, 84×84 for window 3, 168×168 for window 2, and 336×336 for window 1. Window 1 is the 'high-resolution' window, window 5 is the 'low-resolution' window. All these images were then reduced in resolution to match the size of the highest-resolution image (i.e. 21×21 pixels) that is, by 2, 4, 8 and 16, respectively. This ensures that all ADAMs have the same input image size. Each image was then trained into an ADAM network specific to each resolution. If other examples of the same

Figure G1.7.3. The segmentation of images into town and rural areas.

Figure G1.7.4. Way-points in an IRLS image.

feature in different images were available (different days or time of year) these could be trained in the same way. If the feature is to be recognized invariant to rotation, the procedure is repeated, using rotated image data. Further way-points (features) are trained in the same way.

G1.7.3.3 Preprocessing

Again, no preprocessing was used other than the ranked gray-scale N-tuple process.

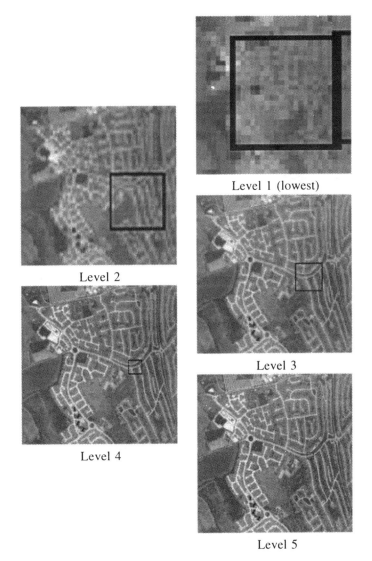

Level 1 (lowest)

Level 2

Level 3

Level 4

Level 5

Figure G1.7.5. Way-point 'A' trained into the networks.

G1.7.3.4 Finding the feature in another image

The five networks were used to find the way-point position in an unknown image. The initial image to be searched was reduced in resolution by 16, to match the resolution of window 5 in the trained data. The ADAM trained on this resolution was then scanned over the image. At each point the class and the confidence of the class were recorded (this is obtained from the class pattern). After a complete scan the position where the best response was obtained was recorded (position p, q). Because the image is small, this operation is fast.

Next, a scan area around the position p, q was defined, equal to about twice the image area covered by window 4, in this case 336×336 pixels. This area was scanned by the ADAM trained on resolution 4, and the point at which the best response was found is noted. The scan area around this point is defined as before. The process is repeated for the remaining three ADAM networks. The final position of the feature will identify the way-point.

If two equally likely way-points were found during the scanning process, the selection of way-points will be random (the method could have been altered to allow both way-points to be recorded). If the image is expected to be very noisy or the way-points are unreliable, many way-points can be trained in a local area. The system will have a better chance of finding at least one of these. An example of the system finding way-point 'B' is shown in figure G1.7.6.

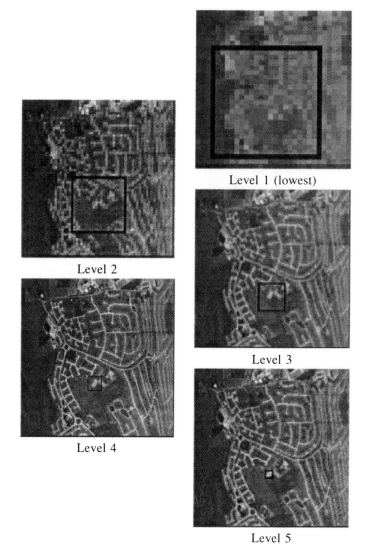

Figure G1.7.6. The result of testing the hierarchical ADAM with the memory trained to recognize position 'B' shown in figure G1.7.4.

G1.7.3.5 Comparison with traditional methods

There are many methods available for segmenting images. The approach described here has the unique ability of fast training coupled to the absence of any image preprocessing. This not only makes the system simple to use, but provides recognition at very high speed. However, this must be traded against the quality of the final result. If your problem requires high accuracy then you should select a method that uses a more robust (and slower) training method, or you should resort to a careful selection and preprocessing of the image. We have recently constructed a dedicated parallel processor that can perform the operations given above hundreds of times faster than conventional workstations (Austin *et al* 1995)

G1.7.4 Conclusions

The work described here has shown how way-points and image segmentation can be achieved very easily using the ADAM network. Because the network trains rapidly, the optimal parameters used by the network can be easily found.

Acknowledgement

This work was partly supported by a grant GR/F36330 from the Department of Trade and Industry and by UK Science and Engineering Research Council.

References

Austin J 1987 ADAM: A Distributed Associative Memory For Scene Analysis *Proc. First Int. Conf. on Neural Networks (San Diego, CA)* vol IV ed M Caudill and C Butler (New York: IEEE) p 285

Austin J and Buckle S 1995 Segmentation and matching in infrared airborne images using a binary neural network *Neural Networks* ed J Taylor (Waller) pp 95–118

Austin J, Kennedy J, Buckle S, Moulds A and Pack R 1995 The cellular neural network associative processor, C-NNAP *IEEE Monograph on Associative Computers* (to be published)

Austin J and Stonham T J 1987 An associative memory for use in image recognition and occlusion analysis *Image and Vision Computing* **5** 251–61

G2

Engineering

Contents

G2.1 Control of a vehicle active suspension model using adaptive logic networks

William W Armstrong and Monroe M Thomas

Abstract

Adaptive logic networks (ALNs) were used to control a nonlinear mechanical model of a vehicle active suspension system. An ALN consists of a tree of logic gates having linear threshold units (simple perceptrons) at its leaves. ALNs learned to predict future states based on relationships among values of variables recorded during operation. Piecewise linear functions were extracted from trained ALNs, and executed with the aid of a decision tree so that only a small number of linear pieces had to be evaluated to compute any output value. A 486DX2-66 PC was able to produce a control output in 250 μs, much faster than was required to control the test system in real time. The results are applicable to a broad range of real-time control problems.

G2.1.1 Project overview

An active suspension system may be used to improve the ride and handling qualities of a vehicle. Actuator struts exert forces controlling the position and orientation of the vehicle body with respect to the road. Because terrain and the characteristics of a vehicle vary, it is desirable to use an adaptive controller for the task. Several approaches to using neural networks in *control*, including their application to active suspension systems, are discussed by Hampo and Marko (1992). The dynamic analysis in a paper by Sunwoo and Cheok (1991) served as a useful starting point for the present study. F1.9

The vehicle used was a Bombardier Iltis 1/4 ton truck at the Defence Research Establishment Suffield (DRES) in Alberta, Canada. It was equipped with two on-board 386DX-33 computers, one dedicated to processing ultrasound measurements of the height of the front bumper above ground (preview), and one dedicated to controlling the active suspension system.

The study was carried out under the direction of the authors working for Dendronic Decisions Limited. The Department of Computing Science, University of Alberta, constructed some of the electronics, and mechanical tests were done at the Department of Mechanical Engineering. A special ALN accelerator board was designed and constructed by the Alberta Microelectronic Centre in Edmonton. Testing used a 1/4-vehicle mechanical model consisting of a mass (approximately 90 kg) supported by a coil spring and an actuator strut (see figure G2.1.1). The simulated road disturbance was provided by a modified MTS vibration tester able to generate about 5 cm of motion at 5 Hz.

G2.1.2 Design process

A radically different approach to control was adopted with this project. The approach taken was to develop two predictors for the dynamic state of the 1/4-vehicle model 10 ms into the future. One predictor was for the future vertical displacement of the sprung mass (represented by its *difference* from the present displacement), and one predictor for the future velocity (represented by the *difference* of two displacements at 5 and 10 ms into the future). The use of neural networks allows one to depart significantly from the paradigm of differential equations, and to use representations capturing the desired information but which are possibly difficult to analyze mathematically.

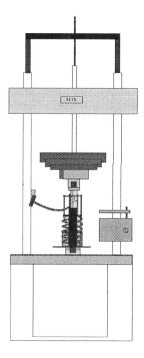

Figure G2.1.1. 1/4-vehicle model of the active suspension system.

Extensive computer simulations were first carried out. It was assumed that the goal of control was to put the sprung mass into a desired state of displacement and vertical velocity one time step into the future. Inversion of predictors of these quantities gave two values for the strut force u that would, if used separately, give the correct vertical displacement (force u_1) or vertical velocity (force u_2), respectively. A strategy of combining the two calculated forces by taking their average worked about as well as taking an optimal convex combination. The optimal mixture for a linear problem is given by

$$u_{\text{opt}} = \frac{1}{\frac{1}{\omega^2}\left(\frac{B}{m} - \frac{2}{T}\right)^2 + 1} \left[\frac{1}{\omega^2}\left(\frac{B}{m} - \frac{2}{T}\right) u_2 + u_1 \right]$$

where m is the mass, B is the damping coefficient, T is the prediction time, and ω is the natural circular frequency. Since, for the system under test, the masses, stiffnesses and damping factors were not measured, the use of an average of the two calculated control forces seemed appropriate.

G2.1.2.1 Inputs and outputs

The following inputs were available as raw 12-bit integers from the NI-DAC data acquisition card:

(i)   Vertical position of the mass.

(ii) Extension of the actuator strut.

(iii) Downward force exerted on the actuator strut.

(iv) Vertical position of the 'road'.

(v)   10 ms preview of the vertical position of the 'road'.

The preview signal of the 'road' was not used in most trials, since the ultrasound preview on the vehicle was felt to be unreliable over some types of terrain. The output expected from the control software was a 12-bit integer representing the desired downward force of the actuator strut. This output value from the control software was filtered through a nonlinearity before being passed to the NI-DAC card to guarantee a nonlinear system. Wooden blocks attached to the coils of the spring surrounding the actuator performed the same function.

G2.1.2.2 Applying an artificial neural network

In applying a neural network to any real-time problem, speed of execution is an issue. In the present case, 5 ms was allowed for computing each control output. The *multilayer perceptron* (MLP) with differentiable C1.2
squashing functions is a widely used type of neural network; however, for real-time use it may require special hardware to accelerate evaluation. Inversion of MLPs is a difficult problem, requiring training of a separate MLP. MLPs are also difficult to test for conformity to a specification, since in high-dimensional problem domains testing MLPs other than by statistical methods becomes mathematically intractable. A different technique was called for, and a suitable form of *adaptive logic network* was developed during C1.8
the course of the work on this project which had the following advantages compared to the usual MLPs:

(i) ALNs use hard limiters instead of sigmoids in the first hidden layer, and logic gates AND or OR in the remaining hidden layers and the output layer. This allows fast training and evaluation through use of efficient flow of control in the simulation, whereby large parts of an expression can simply be omitted from most computations without changing the result. This is often referred to as *lazy evaluation*, and is to be contrasted with the usual *massive parallelism* approach to neural networks.

(ii) ALNs represent *relations* among the inputs and outputs of the system. That is, the *same* ALN can be used for representing a function and *its inverse* without any retraining.

(iii) ALNs permit the use of *a priori* knowledge in the form of convexity and slope constraints on the learned function. This can enhance learning performance when training data are sparse, and specifying convexity diminishes the influence of noise on the results of training.

(iv) A piecewise linear function can be extracted from an ALN in the form of a decision tree which partitions the input space so that, for any input, only a small number of linear expressions have to be computed and the resulting values combined by maximum and minimum operations. Besides the potential for great speed of evaluation, this also makes it possible to *understand* what the system is doing and *check* it for safety.

G2.1.3 Training methods

ALNs were trained by having the system undergo simulated road motions in the form of a sum of sinusoids while the actuator strut was driven with random forces within the limits of safety. (Perturbations around the output of an existing control system would have been chosen if random inputs had been dangerous.) Based on the sampled data, ALNs were trained to learn the relationships among past and present values of sprung mass displacement, as well as values 10 ms into the future (velocities were represented by differences, so differential equations were not an exact model of the process).

A priori knowledge of the monotonicities of the future quantities with respect to strut force was used to constrain the result of training. For example, an increase in downward strut force will cause an increase in vertical mass displacement (over a short period of time), all other things being equal. This ensures that the predictor for vertical displacement can be inverted to yield a value used in control.

G2.1.4 Output interpretation

ALNs produce logical values as outputs; they only *represent* functions from reals to reals, they do not compute them directly. It is possible to extract from the trained ALN a function which combines linear functions by maximum and minimum operations to compute a piecewise linear output. To greatly increase speed of computation, the input space is partitioned by a decision tree, whereby in each block of the partition only a few of the linear pieces have to be evaluated. This does not cause the function to change; it is simply an efficient way of evaluating it.

Decision trees computing inverse functions were extracted from two trained ALNs predicting cab vertical displacement and cab vertical velocity, respectively. The decision trees partitioned the input space so that sometimes only one linear piece had to be evaluated, and, in general, at most six linear pieces had to be evaluated, corresponding to the dimension of the space of ALN inputs. The two values obtained from the decision trees were averaged to give the final control signal sent to the NI-DAC card which then developed the analog control force signal for the hydraulic actuator. No calibrations were performed to assign any meaningful physical units to the input or output values of the ALN.

G2.1.5 Development of platform and hardware tools

The ALNs and decision trees used for the project were trained, tested, and evaluated using proprietary software for MS Windows developed by Dendronic Decisions Limited. Software embodying similar training algorithms, Atree 3.0, is now commercially available, and a free Educational Kit is available from CompuServe (GO DENDRONIC) or from ftp.cs.ualberta.ca in pub/atree/atree3/atree3ek.exe. Data acquisition software for MS Windows was supplied by National Instruments with their data acquisition hardware. The remainder of the hardware used in the project was (see figure G2.1.2):

Actuator strut: TRW Active Control System Inc, Model #P30XC46A08.
Strut LVDT: Moog Inc, Model #BA04-039-1 (integrated).
Strut valve: Moog Inc, Model #E773-004 (integrated).
Strut load cell: Eaton Model #039564 (integrated).
Servo amplifier card: Moog Inc, Model #F122-202.
Control computer: 486DX2/66.
NI-DAC card: National Instruments AT-MIO-16F-5.
MTS machine: MTS vibration tester, modified to use larger pump for greater motion capability.
ALN accelerator board: Alberta Microelectronic Centre designed card with Altera FLEX EPF81188 FPGA.

The objective of the accelerator board was to rapidly compute the index of the region of the input space where an input vector lies. Such hardware may be useful, perhaps combined with DSP or D/A chips, in very demanding applications.

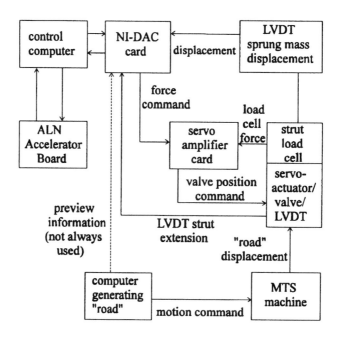

Figure G2.1.2. Schematic of the hardware test system.

G2.1.6 Conclusions

In many trials, the sprung mass could be held motionless to within about 2 mm, eliminating all but about 4% of the simulated road disturbance at low frequency (7 Hz). It was also possible to control the sprung mass to execute a slow sinusoidal motion independent of the disturbance of the ground. Experiments were successful both with and without the simulated ultrasound preview. The system showed no signs of instability during ALN control, even when a system trained at low frequency was tested at high frequency (29 Hz) or when the sprung mass was roughly doubled by someone putting his weight on it. An update rate of 200 commands per second was specified, but the 486DX2-66 PC was shown to be capable of computing 4000 commands per second. This rendered the accelerator board unnecessary. The feasibility of using ALNs to derive piecewise linear functions for real-time control of a vehicle active suspension system has been demonstrated. The use of *a priori* knowledge in training, and the relational approach used by ALNs, allow predictors of the future dynamic state to be easily inverted to provide control outputs.

Acknowledgements

This research was conducted for DRES, under SSC contract W7702-2-R328/01XSG. The authors wish to acknowledge the guidance and support of the late Alan McCormac, which led to successful adaptation algorithms for logic trees combined with linear threshold elements. We also wish to thank Nicole Armstrong, John Evans, Bernie Faulkner, Doug Hanna, Rene Leiva, Debi Owens, Harold Peacock, Steve Sutphen, Roger Toogood and Hong Zhang.

References

Hampo R and Marko K 1992 *Neural Network Architectures for Active Suspension Control* vol 2 (Seattle, WA: IJCNN) pp 765–70
Sunwoo M and Cheok K C 1991 Investigation of adaptive control approaches for vehicle active suspension systems *American Control Conf.* vol 2, pp 1542–7

G2.2   ATM network control by neural network

Atsushi Hiramatsu

Abstract

Application of neural networks to asynchronous transfer mode (ATM) network traffic control has been proposed. ATM is a key technology for building a broadband integrated services digital network (B-ISDN), which gives users universal high-speed communication channels for all kinds of communications services. One important issue in ATM networks is the design of an efficient traffic control architecture that guarantees quality of service (QoS) for all network users. However, in future networks for multimedia communication, it will be impossible to model user activity and to exhaustively analyze the network traffic dynamics by mathematical calculations and computer simulations. A neural network is an important technology for deriving an unknown nonlinear function for estimating QoS from the network status by real-time training. The advantage of this method is that the QoS can be accurately estimated without detailed user action models or knowledge about the switching system architecture. This section gives an overview of neural network applications in ATM traffic control, and describes ATM connection admission control as a typical example.

G2.2.1   Project overview

G2.2.1.1   ATM network

Asynchronous transfer mode (ATM) is a high-speed packet switching technology for the broadband integrated services digital network (B-ISDN), in which various kinds of communications services such as voice, video and data are transferred over common high-speed links. Users send all kinds of information as a series of 53-byte packets called 'cells' (consisting of a 5-byte header plus 48 bytes of user data). The cell format and various interfaces and protocols have been standardized at ITU-TS and the ATM Forum. For details about ATM architecture and applications, see Kawarasaki and Jabbari (1991), Newman (1992), Write *et al* (1992) and Armbrüster and Wimmer (1992).

Figure G2.2.1 shows a typical image of an ATM network, where two nodes are connected by a high-speed link. Typically the capacity of the link is 150 or 600 Mbps. Audio, video and data terminals are represented as A_i, V_i and D_i, respectively. Here terminals A_1, V_2 and D_1 are sending cells to A_4, V_3 and D_2. An ATM node (node 1) multiplexes cells from the terminals into an ATM link through output buffers. When another ATM node (node 2) receives these cells, it distributes them to the proper terminals according to the header information.

G2.2.1.2   ATM traffic control and the problem

In the ATM network, users can send cells with arbitrary timing on demand; an output buffer in node 1 is used to keep cells temporarily when many cells arrive at one time at a node. This means that, when cells are sent from many users at a time, the cell waiting time (cell delay) in the output buffer will increase and some of the cells may even be discarded (cell loss) when the buffers are full. Users usually set certain values for the cell delay and cell loss probability to keep their communications services at a proper level

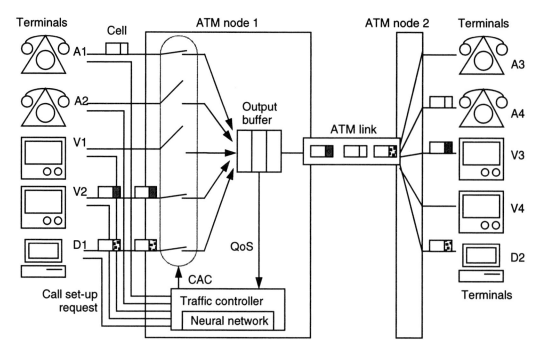

CAC: Call admission control, QoS: Quality of service

Figure G2.2.1. ATM traffic control using a neural network.

of quality. Thus an ATM node must control the cell traffic to guarantee users' QoS parameters such as cell delay time and cell loss probability.

Various traffic control functions have been proposed for the ATM network (for example, see Eckberg 1992): they can be categorized into the following three classes by control period:

(i)   cell transfer level: priority control, cell discard (μs order);
(ii) call admission level: call admission control, routing (ms order);
(iii) network level: link capacity assignment, network design (more than several minutes).

The biggest problem in traffic controller design is the diversity of traffic characteristics in a multimedia network. Each user has different cell generation characteristics, and each communications service has different QoS requirements, like cell delay and cell loss probability. This diversity complicates the traffic control system to the extent that it is almost impossible to design an effective and efficient traffic control system based on traditional mathematical calculations and computer simulations employed in the conventional communications networks. Also, the traffic controller should have the flexibility to handle the traffic for new communications services which may be installed at any time in the future.

G2.2.1.3 Neural network applications in the ATM network

Neural networks are thought to have many potential applications in ATM traffic control, because a neural network can extract an unknown nonlinear curve from a set of examples (Takahashi and Hiramatsu 1990). The following gives some examples in this field.

Call admission control (CAC) is one of the main topics in ATM traffic control using neural networks, where the neural network is used to determine whether the network can accept a new user connection under a certain required QoS (described later). This method was first proposed in Hiramatsu (1990), and then further studied in Tran-Gia and Gropp (1992) and Hiramatsu (1994b). It was also combined with other traffic control functions like link capacity control in Hiramatsu (1991). Hiramatsu (1994a) gives a review of this topic.

Okuda *et al* (1994) are trying to estimate QoS from a traffic descriptor declared by users in CAC, but it is difficult to describe cell generation exactly with a small set of parameters, so there is always a difference between the declared traffic and the actual traffic. Tarraf *et al* (1993, 1994) are trying use a neural network to derive the characteristic parameters from the history of the number of arrived cells.

Other related topics are cell rate control, like priority control (Schwartz 1994, Chen and Leslie 1991). Examples of neural-network-based switching network control can also be seen in Brown (1989, 1994). Traffic prediction is another topic in this field.

G2.2.1.4 Call admission control

The inputs to and outputs from a neural network depend on the application, and the neural network capability and training method also vary. In the following part of this section, a neural network application for call to admission control is described in detail as an example.

Connection admission means that an ATM node allows a user to start a new call only when the node judges that the QoS does not degrade the QoS required by all of the users already connected. In figure G2.2.1, when terminal A_2 starts to talk with terminal A_3, A_2 has to send a call set-up request to a traffic controller of ATM node 1. The call set-up request declares information like the destination terminal ID, the required QoS values, and a 'traffic descriptor' to define the cell generation characteristics of the call. The ATM node 1 estimates the post-connection QoS from the information, and if the estimated value does not violate the QoS requirements for all users, it allows A_2 to start the call. Otherwise it rejects the request. When the QoS estimation is accurate, the guaranteed QoS for all users is maintained.

G2.2.2 Design process

G2.2.2.1 Motivation for a neural network solution

Accurate QoS estimation requires a nonlinear curve mapping the network status to the QoS value, but the curve is usually unknown and hard to obtain by calculation or exhaustive computer simulation. The motivation for using a neural network is the expectation of extracting the nonlinear curve by *learning* and the extrapolation of the curve by *generalization*. B3.5

Theoretically, accurate QoS estimation requires all detailed information about the statistical characteristics of cell arrival to the buffer, which means that the accurate cell generation characteristics of each user should be known. However, users do not send cells at constant intervals, but send them with arbitrary timing, so many parameters like the average cell generation interval, the minimum interval and the variance of the interval are required in order to define the cell generation characteristics. The ATM Forum standards propose a set of four parameters as a 'traffic descriptor'.

Accurate QoS estimation is, however, still difficult for the following reasons.

(i)   Users have to declare the traffic descriptor before they start their call. There is always a big difference between the declared and actual values.

(ii) Since many users share a buffer at a time, the total number of parameters to be considered in the QoS estimation is very large.

G2.2.2.2 General description of the neural network function

To solve the problems in the QoS estimation described above, various ways of using a neural network have been proposed.

The simplest way is to use a neural network as a compact nonlinear curve generator. First, computer simulations are done to evaluate the QoS for a set of preselected buffer statuses. Then a neural network is trained with the simulation results to teach it the relationship between the buffer status and the QoS. It is expected that the neural network will not only memorize the relationship for the training status, but also extrapolate the relationship to the unexperienced buffer status, since it is impossible to simulate the possible situations exhaustively. After the training, the neural network is installed in a traffic controller and used to estimate the QoS when a traffic controller receives a call for request setup.

Another approach is real-time training. A neural network trained as above is trained again with the QoS observed in a running ATM node to adjust for the difference between the simulations and the actual situation. This process eliminates the error in the traffic descriptor declared by a user as well as the error in the simulation model.

G2.2.2.3 Black-box description

The input to the neural network is a set of parameters that define the output buffer capability and buffer status; the output is the QoS or the judgement of whether or not the estimated QoS exceeds the required value.

The buffer capability is represented by parameters like buffer size and link capacity, which are usually constant but may be changed by the network operator. The buffer status is represented by the combination of traffic descriptors declared by all users. When the number of users multiplexed is large, however, it is inconvenient to feed all of the traffic descriptors to the neural network. Thus the traffic descriptors are categorized into groups, and the buffer status is usually represented by the set of the numbers of users in the group. The simple way to define the groups is categorization by the communication services used. In the case shown in figure G2.2.1, for example, the buffer status is represented by $(n_A, n_V, n_D) = (1, 1, 1)$, where n_A, n_V and n_D represent the number of users using audio, video and data communication services at that time. When terminal A_2 sends a call setup request for audio communication and the traffic controller estimates the postconnection QoS, the input to the neural network will be $(2, 1, 1)$.

The history of the buffer status may be useful for estimating QoS much more accurately, since the buffer status is always changing and the observed status usually has some delay relative to the actual status (Hiramatsu 1990); the performance difference between these two inputs is analyzed in Tran-Gia and Gropp (1992). The following section considers only the present buffer status.

G2.2.2.4 Requirements and constraints

The number of inputs of the neural network is mainly determined by the number of traffic descriptor categories, and at this moment it is thought that the number of categories is less than 100. The number of outputs is the same as the number of QoS parameters considered.

The requirement for the forward calculation speed is determined by the number of call setup requests arriving at a node in one second. The typical target is 1 ms so as to handle 1000 requests per second. The required training speed is rather slow, because the target curve does not change quickly.

There is another constraint in this application: 'safe-side control'. Suppose that the traffic controller overestimates the QoS and accepts more calls than it can actually accept while maintaining the QoS. In this case, all users multiplexed into the same buffer experience QoS worse than required, and none of the users can continue their communications. On the other hand, if the traffic controller underestimates the QoS, more calls are rejected than necessary. In this case, all users can continue their communications, although the link utilization is low. This is much better than the acceptance of too many calls. The safe-side control policy requires the neural network to estimate the QoS to be worse than the actual value when it makes errors. This is especially important for public communications networks.

G2.2.2.5 Topology

C1.2   For QoS estimation in CAC, only the conventional three-layer *backpropagation neural network* with
C1.6.2 sigmoid function has been reported. If the number of inputs is small, the *radial basis function* (RBF) neural network is useful.

G2.2.3 Training methods

C1.2.3 The training algorithm used is conventional *backpropagation*. Usually a neural network is trained in two steps: the first step is training with data from computer simulations to determine the initial weights for the second step. The second step is real-time training to improve the estimation accuracy by adjusting the initial weights according to the training data observed from output buffers in a running ATM node.

G2.2.3.1 Training data

It is easy to prepare the training data for the first step. We can choose any buffer status and simulation period. It is easy to get a very accurate QoS value by a very long-period simulation, but the QoS data obtained may be very different from an actual situation.

It is not easy to obtain the training data for the second training step, because buffer status, like the number of connections, is always changing and cannot be set to an arbitrary status from outside. It is impossible to monitor the average QoS over a long period for a constant status. Also the 'safe-side control policy' requires that the number of observations at the worse QoS should be at a minimum.

G2.2.3.2   Pattern table method

In the real-time training, the QoS observation period is set very short. A typical observation period is 100 ms. With such a short observation period, the QoS data actually observed are distributed widely around the average. The cell loss probability typically ranges from 10^{-10} to 10^{-0}. For example, an observed data set for the average cell loss probability of 10^{-5} may consist of 99 zeros and just one 0.001. Thus a neural network must estimate the average from many data observed. This is the reason why a large memory called a 'pattern table' is used to store the data observed. When a new datum is observed, an entry in the table is randomly chosen and an old datum in the entry is overwritten by the new one.

Here it should be noticed that both good and bad QoS data are needed to train a neural network but that observed QoS that is worse than required is very rare when the QoS is well controlled. Then it is proposed to use separate pattern tables for good and bad QoS data so as to prevent rare bad QoS data from being replaced with the more frequent good QoS data (Hiramatsu 1990). For example, suppose that the target cell loss probability is 10^{-4}. The observed buffer status is stored in the pattern table for bad QoS when the cell loss probability observed exceeds 10^{-4} and in the table for good QoS otherwise. Another advantage of this method is that the training ratio of bad QoS data can easily be weighted by the table selection rate in the training phase.

G2.2.3.3   Virtual output buffer method

Even with the pattern table method, it is very hard for a neural network to estimate QoS accurately, because the target cell loss probability is very small, the data contain much noise, and the number of bad QoS data is small. The 'virtual output buffer method' is a novel approach to estimating a very small cell loss probability from virtual cell loss data (Hiramatsu 1994b). A virtual buffer is a set of counters that simulates an imaginary cell buffering process, where the cell arrival at the virtual buffer is entirely the same as the actual probability, but the capability to transfer cells is less than the actual. Thus the cell loss probability at each virtual buffer is much larger than the actual probability, and the accuracy in the observed data is high. By extrapolating QoS data from virtual buffers with smaller capability, we can accurately estimate QoS for the actual buffer. For example, when an actual output buffer is connected to a 150 Mbps link, four virtual output buffers connected to 30, 60, 90, and 120 Mbps links, respectively, are simulated in parallel with the actual, and the virtual cell loss probabilities for these virtual buffers being observed. We can estimate the cell loss probability for the 150 Mbps buffer by extrapolating these virtual cell loss probabilities.

G2.2.4   Preprocessing and output interpretation

Suppose that the output from a neural network is the cell loss probability for the buffer status fed to the neural network. Each time a new call setup request arrives at a node, the QoS after accepting the call is estimated using a neural network. The cell loss probability range, from 10^{-10} to 10^{-0}, is too wide for a sigmoid function to handle linearly. Therefore the logarithm of the cell loss probability is used in both training and estimation phases to map the probability to a 0–1 range. Also, a smoothing technique is used to obtain the average of the logarithm of the cell loss probability (Hiramatsu 1994b).

G2.2.5   Comparison

G2.2.5.1   Performance

The advantages of the neural-network-based approach are summarized as follows.

(i)   A neural network is compact hardware for representing a nonlinear curve. The neural network VLSI will make the hardware much more compact and reduce the calculation time.

(ii)   The traffic controller designer is not required to know the detail characteristics of users and services. The real-time training absorbs the difference between the actual traffic and the model used to determine the initial weights.

(iii) It is easy to install a new service on the traffic controller, because it can be accommodated simply by adding the number of calls for the new service as an input to the neural network.

QoS control using neural networks has shown good performance in preliminary computer simulations. However, practical comparison between the conventional CAC and CAC using neural networks is still in progress, since practical ATM network services have just started.

G2.2.6 Conclusions

In ATM traffic control, the conventional mathematical calculations and computer simulations do not work effectively in the controller design process because of the diversity and the ambiguity of traffic characteristics of the users and services. It is thought that there are many potential applications of neural networks in ATM traffic control other than the call admission control described in this section. The neural-network-based approach will be particularly welcome in private ATM networks, where users have a very strong interest in the efficiency of their network. For example, when a high-speed link is leased between two remote private networks, the gateways at the two ends should have a traffic control function to efficiently use the calls on the link. There is a big difference between a private network and a public network, where safe control is the first priority. The ATM network service has just started, and the performance of QoS estimation using neural networks will be tested in a real network. It can be said that the expectations for neural networks are very large.

References

Armbrüster H and Wimmer K 1992 Broadband multimedia applications using ATM Networks: high-performance computing, high-capacity storage, and high-speed communication *IEEE J. Sel. Areas Commun.* **10** 1382–96

Brown T X 1989 Neural networks for switching *IEEE Commun. Mag.* **27** 72–81

——1994 Neural networks for switching *Neural Networks in Telecommunications* ed B Yuhas and N Ansari (Boston, MA: Kluwer) pp 11–36

Chen X and Leslie I M 1991 A neural network approach towards adaptive congestion control in broadband ATM networks *IEEE Global Telecommunications Conf. (Phoenix)* pp 115–9

Eckberg A E 1992 B-ISDN/ATM traffic and congestion control *IEEE Network* September 1992 pp 28–37

Hiramatsu A 1990 ATM communications network control by neural networks *IEEE Trans. Neural Network* **1** 122–30

——1991 Integration of aTM call admission control and link capacity control by distributed neural networks *IEEE J. Sel. Area Commun.* **9** 1131–8

——1994a ATM traffic control using neural networks *Neural Networks in Telecommunications* ed B Yuhas and N Ansari (Boston, MA: Kluwer) pp 63–89

——1994b ATM call admission control using a neural network trained with a virtual output buffer method *Int. Conf. on Neural Networks 94 (Orlando)* pp 3611–6

Kawarasaki M and Jabbari B 1991 B-ISDN architecture and protocol *IEEE J. Sel. Areas Commun.* **9** 1405–15

Newman P 1992 ATM technology for corporate networks *IEEE Commun. Mag.* **30** 90–101

Okuda T, Anthony M and Tadokoro Y 1994 A neural approach to performance evaluation for teletraffic system *IEEE Int. Conf. on Communications (New Orleans)* pp 774–8

Schwartz D B 1994 Learning from rare events: dynamic cell scheduling for ATM networks *Neural Networks in Telecommunications* ed B Yuhas and N Ansari (Boston, MA: Kluwer) pp 91–108

Takahashi T and Hiramatsu A 1990 Integrated ATM traffic control by distributed neural networks *Int. Switching Symp. 90 (Stockholm)* vol III pp 59–65

Tarraf A A, Habib I W and Saadawi T N 1993 Characterization of packetized voice traffic in ATM networks using neural networks *IEEE Global Telecommunications Conf. 93 (Houston)* pp 996–1000

——1994 A novel neural network traffic enforcement mechanism for ATM networks *IEEE Int. Conf. on Communications 94 (New Orleans)* pp 779–83

Tran-Gia P and Gropp O 1992 Performance of a neural net used as admission controller in ATM systems *IEEE Global Telecommunications Conf. '92 (Orlando)* pp 1303–9

Wright D J, Wright M, Verbiest W, Shimasaki N and Prycker M D 1992 (eds) B-ISDN applications and economics *IEEE J. Sel. Areas Commun.* **10**

G2.3 Neural networks to configure maps for a satellite communication network

Nirwan Ansari

Abstract

This article reports and summarizes a neural-network-based approach that was used to dynamically configure maps of a satellite communication network, and was incorporated with a state-dependent routing scheme to manage the network traffic. We modified Kohonen's self-organization paradigm to automate the map configuration task. The modified algorithm consisted of three phases: (i) pattern recognition for selecting an exemplar map which most resembled the input traffic, (ii) a learning phase for fine tuning the chosen exemplar map, and (iii) a decision-maker for replacing the original exemplar map. The intrinsic properties of the proposed unsupervised learning allowed efficient tracking of the random traffic. The proposed traffic management scheme was effective in reducing the block rate of the network, as demonstrated through simulations.

G2.3.1 Project overview

The objective of traffic management is to best meet the communication requirement of users under the constraint of a fixed network capacity. That is, the probability of having users blocked from accessing the network should be at a minimum. A satellite communication network is flexible—as opposed to the rigidity of a terrestrial communication network—and can be configured to various maps. To take advantage of the flexibility of a satellite communication network, neural networks can be applied naturally to configure a satellite communication network adaptively. We (Ansari and Chen 1990, 1991) first reported the initial results of configuring maps of a satellite communication network using self-organization learning; subsequent results, in combination with a routing module, were reported later (Ansari and Liu 1991, 1995). Further improvements using a different approach and concept by means of annealing procedures (Balasekar and Ansari 1993, Arulambalam and Ansari 1994, Ansari *et al* 1996) were also investigated. The traffic management scheme employed in this project, shown in figure G2.3.1, consisted of two components: a map configuration module and a routing module. An assignment of channels to links of a satellite communication network constitutes a map. The map configuration task was achieved with a modified *Kohonen self-organization learning algorithm*, while simple state-dependent routing was used for the routing module. C2.1.1

G2.3.2 Design process

Reducing operational costs and improving quality of service while maximizing network utilization of a communication network are the major goals and reasons cited in justifying a traffic management scheme. The map configuration task, also known as demand assignment (Pritchard *et al* 1993), allows the network to assign/allocate a varying number of channels to each link of the network according to the changing traffic loads of the satellite network. The primitive approach still employed in many systems involves a network operator manually assigning channels to each link of the network via a monitoring console. Most of the systems deployed today use a fixed set of 'canned' or stored plans to achieve the task. Aiming at reducing the operational costs and improving network efficiency, we proposed self-organization learning to automate the map configuration task.

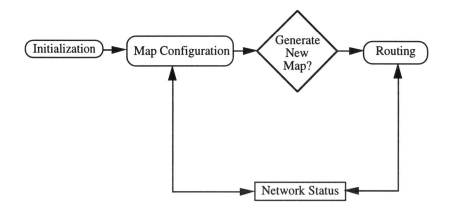

Figure G2.3.1. The proposed traffic management scheme.

G2.3.2.1 Why self-organization?

Kohonen's self-organization (Kohonen 1989), summarized below in figure G2.3.2, is an unsupervised learning scheme known for clustering, topographical mapping, and learning familiarity. These desirable characteristics, with a slight modification to the algorithm, enable tracking and monitoring of the random and non-coincident nature of the telephone or data traffic. Note that this type of learning involves selection of a winner (analogous to pattern recognition) and modification of the parameters associated with the winner. The array shown in the figure is simply a graphical representation of the output of the neural network. It is a feature mapping (clustering) from the input vector to the output array. If the input to this neural network is the traffic load of the satellite network and the winner is one of the pre-planned maps, this neural paradigm essentially conducts a pattern recognition task, selecting the map that best meets the input traffic load, and tunes the selected winner to better fit the input.

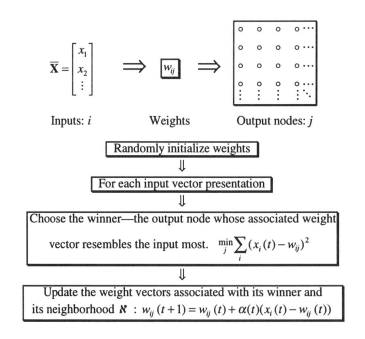

Figure G2.3.2. Kohonen's self-organization.

G2.3.2.2 The map configuration module

Assume that the original N exemplar maps are given. The exemplar maps have functions similar to the 'canned' plans deployed in most of today's systems to reduce the block rate when employed at appropriate schedules. However, as opposed to the canned plans, these maps are dynamically configured to better fit the underlying traffic, and thus better results are expected. The state of the network is also assumed to be updated constantly from the common signaling channel (CSC) of the satellite system, and quantified as $R_i(t)$, which denotes the number of channels in link i required by users at time t. The capacity of the satellite communication network, which is fixed, is denoted by T. Note that the notations N, L, i and j are used within the context of this section (G2.3). Similarly to Kohonen's self-organization, the map configuration task is carried out in three stages, as shown in figure G2.3.3. The algorithm can thus be

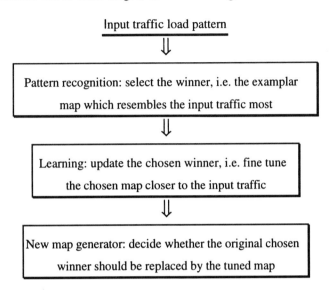

Figure G2.3.3. The map configuration module.

summarized according to the three stages (Ansari and Liu 1995) depicted in the figure:

Pattern recognition (first stage). Compute the distance (metric) between the input data and each of the exemplar maps:

$$D_j(t) = \sum_{i=1}^{L} \frac{|D_{ij}(t)|}{R_i(t)} \qquad j = 1, 2, \ldots, N$$

where $D_{ij}(t) = R_i(t) - C_{ij}(t)$, $(i = 1, 2, \ldots, L$ and $j = 1, 2, \ldots, N)$. The total number of links in the network is denoted by L, $C_{ij}(t)$ is the total number of channels assigned to link i of the jth exemplar map at time t, and $D_{ij}(t)$ indicates the busyness (load) of link i of the network if the jth exemplar map is used. $D_{ij}(t) > 0$ implies that link i of the network is overloaded when map j is used. $D_{ij}(t) < 0$ means that link i of map j provides more than enough channels required by the users in link i. Thus $D_j(t)$ indicates the resemblance between the input data $R_i(t)$ and exemplar map j.

Select the best exemplar map, $j\left(j : \min_j\{D_j(t)\}\right)$, that most resembles the input traffic load.

Learning (second stage). Modify the chosen exemplar map to better meet the network demands according to the following updating rule:

$$C'_{ij}(t+1) = C'_{ij}(t) + \beta(n(j))D'_{ij}(t) \qquad (i = 1, 2, \ldots, L \text{ and } j = 1, 2, \ldots, N)$$

where $D'_{ij}(t) = R_i(t) - C'_{ij}(t)$, $C'_{ij}(t)$ is the number of channels assigned to link i of map j being modified, $n(j)$ is the number of occurrences of exemplar map j being selected, and $\beta(n(j))$ is a gain factor, $0 < \beta(n(j)) < 1$.

The 'prime' is used to distinguish the map which is being modified from the original exemplar map chosen in the first stage. The function $\beta(n(j))$ is monotonically decreasing with respect to $n(j)$:

$$\beta(n(j)) = k_1^{-(n(j)+k_2)}$$

where k_1 and k_2 are nonnegative constants. When a new exemplar map j is generated by the third stage, $n(j)$ is reset to 0.

New map generator (third stage). Compute a convergence factor:

$$E(j) = \max_i |\beta(n(j))D'_{ij}(t)| \qquad (i = 1, 2, \ldots, L).$$

A new map is generated according to:

$$\text{If } E \geq r, \qquad C_{ij}(t+1) = C_{ij}(t) \qquad (i = 1, 2, \ldots, L)$$

$$\text{else,} \qquad C_{ij}(t+1) = \frac{T}{B}C'_{ij}(t)$$

where $B = \sum_{i=1}^{L} C'_{ij}(t)$ is the total number of channels assigned to the modified map j at time (iteration) t, and r is the threshold set to define whether a new 'channel assignment' is made.

It is readily seen that $E(j)$ is getting smaller when the jth map is selected more often, indicating that the modified map j is approaching the network requirement more closely. A new map is more likely to be generated. If the jth map has been selected often enough (i.e. $\beta(n(j))$ is small enough to make E less than r), the self-organization model generates a new map j to replace the original chosen exemplar map. Since the value of $\beta(n(j))$ also depends on the other two parameters, k_1 and k_2, the rate of generating a new map is also affected by these two parameters. A larger k_2 causes a faster rate of map change. On the other hand, with a fixed k_2 and r, a larger k_1 requires the system to adapt a few more iterations before a new map may be generated. These parameters, k_1, k_2 and r, were determined empirically.

G2.3.2.3 The routing module

We adopted a simple state-dependent routing scheme and restricted to two-link routing; that is, each routing could consist of up to two links:

(i)   Route an A–B call over:

    (a)   The direct route from A to B, if the A–B link has an available channel.

    (b)   An alternate two link path, A–M–B, if $C_{AM} + C_{MB} < C_{Ai} + C_{iB}$ and the A–M link and the M–B link have available channels, where i is any other third node in alternate routes and C_{AB} is the cost of using the A–B link.

(ii) Block the call in all other cases.

Here, we referred to an A–B call as a call originating from A and destined for B, and an A–B link as the link between node A and node B. The cost of using a particular path was computed based on its path length and busyness (Ansari and Liu 1995).

G2.3.3 Simulation and comparison results

The viability of the proposed concept was demonstrated through various simulations. A network consisting of 10 links with a capacity of 1000 channels and different traffic loads was simulated. The network was characterized by seven traffic conditions corresponding to seven different time slots. Corresponding to each time slot, the traffic of each link was defined by a specific arrival rate (calls/time unit) and a specific service time (time units). The traffic condition of the network not only varied from time slot to time slot, but also in links. Calls were generated according to a Poisson process. All simulations were written in C and run on a Sun Sparc II workstation. The performance was evaluated and quantified by the average block rate at time t (Ansari and Liu 1991, 1995), defined as:

$$\frac{1}{L} \frac{\sum_{i=1}^{L} u\left(R_i(t) - C_{ij}(t)\right)}{\sum_{i=1}^{L} R_i(t)}$$

where $u(\cdot)$ is a unit-step function, and L is the number of links in the network. Without any traffic management, the block rate of the raw traffic is shown in figure G2.3.4. Since all channels were available initially, no call was blocked, and thus the block rate was zero until the traffic started to peak at $t = 120$. The peak block rate in the graph occurred in the period between $t = 480$ and $t = 720$, which corresponded to the peak traffic condition (higher arrival rates and longer service times). The block rate was tapering off at the end because the traffic in the last time slot was low.

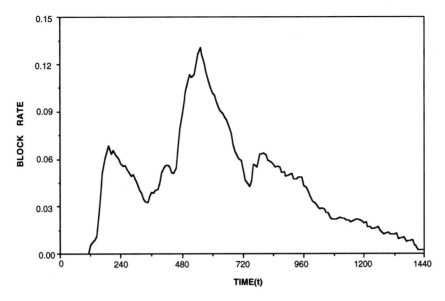

Figure G2.3.4. The block rate of the raw traffic. (Reprinted from Ansari and Liu (1995), with permission of Elsevier Science BV, Amsterdam.)

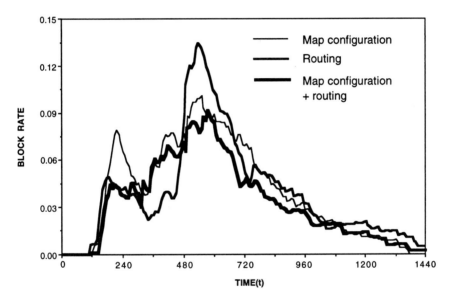

Figure G2.3.5. Block rates of the network managed by the map configuration module alone, the routing module alone, and the proposed scheme, respectively.

Instead of comparing the scheme with the demand assignment approach using canned plans, which was quite subjective, the proposed scheme was compared with those using the self-organization mechanism without routing and the 'status quo' approach with the dynamic routing mechanism only. Figure G2.3.5 shows the average block rate of the network managed by the map configuration module without routing, the routing module only, and the proposed scheme combining the map configuration module and the routing module, respectively. Compared with the raw traffic, the overall block rate using the map configuration module without routing had been reduced, and in particular, the block rate at peak traffic was reduced from 0.13 to 0.1. As seen from the figure, the routing algorithm sacrificed the communication facility efficiency by routing calls through various paths, and thus the block rate was slightly higher than that of the raw traffic at the peak load. The combined scheme provided the best compromise performance at different load conditions. It also inherited the merits of the two aspects of traffic management: demand assignment and routing.

G2.3.4 Conclusions

We may conclude from the simulation results that self-organization learning was successfully applied to automate and dynamically modify the demand assignment of a satellite network, thus reducing the blocking rate, especially in cases where the traffic in various links varied significantly in the transition from one time slot to another. In combination with a state-dependent routing algorithm, the best compromise performance was achieved compared to schemes with either demand assignment alone or routing only. Many other neural network applications to telecommunications can be found in Yuhas and Ansari (1994). The reader is also referred to Sections G2.2 and G2.4 of this handbook.

References

Ansari N and Chen Y 1990 A neural network model to configure maps for a satellite communication network *Proc. IEEE GLOBECOM '90 (San Diego, CA)* pp 1042–6

——1991 Configuring maps for a satellite communication network by self-organization *J. Neural Network Comput.* **2.4** 11–7

Ansari N, Arulambalam A and Balasekar S 1996 Traffic management of a satellite communication network using stochastic optimization *IEEE Trans. Neural Network* (to appear)

Ansari N and Liu D 1991 The performance evaluation of a new neural network based traffic management scheme for a satellite communication network *Proc. IEEE GLOBECOM '91 (Phoenix, AZ)* pp 110–4 (also published in *Neurocomput.* 1995 **8** 263–82)

Arulambalam A and Ansari N 1994 Traffic management of a satellite communication network using mean field annealing *Proc. IEEE Int. Conf. on Neural Networks (Orlando, FL)* pp 1777–82

Balasekar S and Ansari N 1993 Adaptive map configuration and dynamic routing to optimize the performance of a satellite communication network *Proc. IEEE GLOBECOM '93 (Houston, TX)* pp 986–90

Kohonen T 1989 *Self-Organization and Associative Memory* 3rd edn (Berlin: Springer)

Pritchard W L, Suyderhoud H G and Nelson R A 1993 *Satellite Communication Systems Engineering* 2nd edn (Englewood Cliffs, NJ: Prentice-Hall)

Yuhas B and Ansari N 1994 *Neural Networks in Telecommunications* (Boston, MA: Kluwer)

G2.4 Neural network controller for a high-speed packet switch

M Mehmet Ali and Huu Tri Nguyen

Abstract

A neural network input access scheme in a high-speed packet switch for broadband ISDN is presented. In this switch each input port maintains a separate queue for each of the outputs, thus there are n^2 queues in an ($n \times n$) switch. Using synchronous operation at most one packet per input and output port will be transferred in any slot in such a way as to maximize the throughput of the switch. The high transmission rates in broadband ISDN, with slot durations of the order of microseconds, demand that the choice of the packets be performed in real time. A conventional solution of this optimization problem cannot meet timing constraints. We propose a recurrent neural network maximizing the throughput of the switch, and determine the corresponding energy function, its optimized parameters, and the connection matrix. The energy function has linear cost terms with excellent convergence properties. The neural network has null programming complexity that avoids readjusting the parameters before presenting new inputs. From the hardware implementation point of view, because the neural network has $O(n^2)$ neurons and $O(n^3)$ connections, the sparse connection matrix will help in implementation. Finally, comparing simulation results with analytically derived upper and lower bounds we show close to optimal throughput.

G2.4.1 Project overview

As explained in Asatani (1988), the asynchronous transfer mode (ATM) has been accepted as the transfer mode for the broadband integrated services digital networks (B-ISDN). According to this concept, all information is digitized and formed into small packets (referred to as cells) and transmitted over a synchronous network. Since the implementation of the ATM requires a fast packet switch, a number of switching fabrics have been studied which have the characteristics of being modular, easily expandable to a large number of inputs–outputs, and nonblocking, meaning no contention within the switch.

As has been described in Hluichyj and Karol (1989) and Hui and Arthurs (1988) several queuing mechanisms have been proposed for these switches, such as input and output queuing. We considered such a queuing discipline with multiple queues at each input port and synchronous operation where each input will have a separate queue for each output. Thus in the switching fabric of size ($n \times n$) there were n queues at each input port, one per output, or a total of n^2 input queues (figure G2.4.1) each operating according to the FCFS discipline. During each slot at most one packet per input port would be transferred and the packets chosen in such a way as to maximize the throughput of the switch. The choice of packets to be transmitted during a slot has to be performed in real time. Because of the high transmission rates in broadband ISDN, with slot durations of the order of microseconds, this choice is too time critical for conventional solutions; in this work, we propose a neural network for the implementation of this queuing discipline.

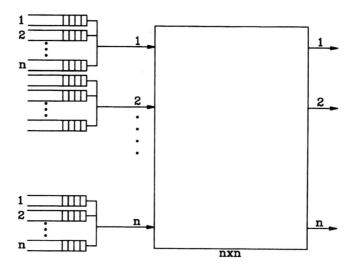

Figure G2.4.1. Architecture of the input access scheme in an $(n \times n)$ switching fabric. (A single queue per output at each input.)

G2.4.2 Design process

This design uses a total of n^2 input queues, each operating according to the FCFS discipline. At the beginning of each slot, the status of all input queues (busy or idle) is used to determine the packets to be transferred by the switch. Let $q_{l,i}$ denote the number of packets in the queue from input l to output i, then the status of the input queues may be given by a matrix $\mathbf{V}$ defined as

$$
\mathbf{V} = \begin{bmatrix}
a_{1,1} & a_{1,2} & \cdot & \cdot & \cdot & a_{1,n} \\
a_{2,1} & a_{2,2} & \cdot & \cdot & \cdot & a_{2,n} \\
\cdot & \cdot & & \cdot & & \cdot \\
a_{l,1} & a_{l,2} & \cdot & a_{l,i} & \cdot & a_{l,n} \\
\cdot & \cdot & & \cdot & & \cdot \\
a_{n,1} & a_{n,2} & \cdot & \cdot & \cdot & a_{n,n}
\end{bmatrix}
$$

where

$$
a_{l,i} = \begin{cases} 0 & \text{if } q_{l,i} = 0 \\ 1 & \text{if } q_{l,i} \geq 1. \end{cases}
$$

During each slot the input access scheme will choose, if possible, a single packet per input port to be transmitted to an output. Since an output may receive from only one input port at a time, no two packets may be chosen simultaneously for the same output. In matrix $\mathbf{V}$, any row or column with at least a single nonzero element is a candidate for the selection of a packet. Thus the objective was to maximize the number of packets chosen per slot under the above switching constraints. The neural network was designed to solve this optimization problem.

G2.4.2.1 Motivation for a neural solution

Neural networks can achieve high computation rates through massive parallelism. As shown in Hopfield (1982, 1984) and Hopfield and Tank (1985) they provide good solutions to difficult optimization problems. Those applications that require fast, good solutions reliably (but not necessarily the best solutions) are considered very good candidates for neural networks. Certainly, choice of the packets under the studied queuing discipline fell into this domain. This problem also belongs to a class of optimization problems that are best suited for neural network applications. As argued in Protzel (1990), Brandt *et al* (1988) and Moopenn *et al* (1990) optimization problems which can be formulated with linear (as opposed to quadratic) cost terms in the energy function are most congruous for a neural network approach. Furthermore, the neural network corresponding to this problem has null programming complexity. This means that the neural network parameters do not need to be modified each time before presenting new inputs, thus avoiding a significant overhead.

The choice of the packets during a slot is a type of discrete optimization problem, known as assignment problems, and as described in Stenoy (1989) there exist conventional algorithms for solving these problems; such as the branch-and-bound technique and network flow algorithms. However, the computational time complexity of these algorithms is third-order polynomial, $O(n^3)$. Table G2.4.1 presents times required with different polynomial complexity functions to compute the solutions for different problem sizes, n, from Garey and Johnson (1979). It is assumed that size 1 takes 1 μs to execute, which is quite reasonable. As may be seen, the computation times of the $O(n^3)$ algorithm for problem sizes of $n = 40$ and $n = 60$ are 64 and 216 ms, respectively. These will be unacceptable for B-ISDN since it has slot durations of the order of microseconds.

Table G2.4.1. Comparison of several polynomial time complexity functions. (We assume that the problem of size 1 takes 1 μs to compute.)

Time complexity function	Problem size (n)					
	10	20	30	40	50	60
n	0.01 ms	0.02 ms	0.03 ms	0.04 ms	0.05 ms	0.06 ms
n^2	0.1 ms	0.4 ms	0.9 ms	1.6 ms	2.5 ms	3.6 ms
n^3	1 ms	8 ms	27 ms	64 ms	125 ms	216 ms

On the other hand, the computation time of neural networks is not expected to grow rapidly with problem size, as has been argued in Takeda and Goodman (1986), because the larger the problem size, the more neurons participate in solving the problem and the higher the parallelism. Finally, as Kosko (1992) shows, *opto-electronic implementations* promise very large neural networks (10^5 neurons, with 10^{10} interconnections) and exceedingly high bandwidths (100 MHz). As a result of these advantages, our input access scheme is implemented with neural networks.

E1.5

G2.4.2.2 General description of the neural function

The recurrent type of neural networks from *Hopfield* (1982, 1984) were considered for implementation of this input access mechanism. In this network, neurons are arranged in a matrix where each neuron is identified by a set of double indices l and i indicating its row and column number, respectively. The input–output voltage relationship of a neuron on row l and column i is given as $a_{l,i} = g(u_{l,i})$. There is a feedback path among pairs of neurons, designated as w_{l_i,m_j} and referred to as a connection matrix. Further, there is an external bias $\theta_{l,i}$, supplied to each neuron. The differential equation describing the dynamics of a neuron is given by

C1.3.4

$$\frac{du_{l,i}}{dt} = -\frac{u_{l,i}}{\tau} + \sum_{m=1}^{n}\sum_{j=1}^{n} w_{l_i,m_j} a_{m,j} + \theta_{l,i} \qquad (G2.4.1)$$

where $\tau = RC$ is the time constant of the neuron. It has been shown that the quadratic energy function defined as

$$E = -\frac{1}{2}\sum_{l=1}^{n}\sum_{i=1}^{n}\sum_{m=1}^{n}\sum_{j=1}^{n} w_{l_i,m_j} a_{l,i} a_{m,j} - \sum_{l=1}^{n}\sum_{i=1}^{n} a_{l,i}\theta_{l,i} \qquad (G2.4.2)$$

for a neural network operates in the interior of the n-dimensional hypercube defined by $a_{l,i} = 0$ or 1. Further, it has been proven that the local minima of this energy function occur only on the 2^n corners of this hypercube.

In expressing an optimization problem through neural networks, we construct an energy function in the form of the above equation, where local mimima correspond to the solutions of the particular problem. Then from the correspondence between the constructed energy function and equation (G2.4.2), the values of the neural network's parameters are determined.

G2.4.2.3 Topology

At the beginning of each slot, the status of all input queues (busy or idle), would be fed to a neural network which would determine the packets to be transferred. The status of each input queue would be given as an initial condition to a neuron; thus for an $(n \times n)$ switching fabric n^2 neurons were needed. The initial output voltages of the neurons would reflect the status of the input queues, or the matrix **V** defined in section G2.4.2. Then the neural network would choose the maximum number of packets to be transferred with no more than one per row and column. Let us define x and y as the number of rows and columns with at least a single nonzero element, then the maximum number of nonzero elements that may be chosen is bounded by $\min(x, y)$. The energy function of the switching problem is given by

$$E = \frac{A}{2} \sum_{l=1}^{n} \sum_{i=1}^{n} \sum_{\substack{j=1 \\ j \neq i}}^{n} a_{l,i} a_{l,j} + \frac{B}{2} \sum_{l=1}^{n} \sum_{i=1}^{n} \sum_{\substack{m=1 \\ m \neq l}}^{n} a_{l,i} a_{m,i}$$

$$+ \frac{C}{2} \left[\min(x, y) - \sum_{l=1}^{n} \sum_{i=1}^{n} a_{l,i} \right]^2 + \frac{D}{2} \sum_{l=1}^{n} \sum_{i=1}^{n} a_{l,i}(1 - a_{l,i}) \qquad \text{(G2.4.3)}$$

where A, B, C and D are positive parameters.

In the above, the first (second) term is minimized when a solution has at most a single nonzero element per row (column) and makes sure that at most a single packet per input (output) is chosen. Thus any solution satisfying this requirement is a feasible solution and meets the physical constraints of the switching fabric. The third term is minimized when the number of nonzero elements in the solution is maximized. Further, it makes sure that the trivial solution of all zero rows and columns, which is permissible by the first two terms, is not chosen. Finally, the fourth term was added to ensure that the final solution consists of only binary values. The comparison of equation (G2.4.3) with (G2.4.2) gives the elements of the connection matrix and external biases as

$$w_{l_i, m_j} = -A\delta_{l,m}(1 - \delta_{i,j}) - B\delta_{i,j}(1 - \delta_{l,m}) + D\delta_{l,m}\delta_{i,j} - C$$

$$\theta_{l,i} = C \min(x, y) - \frac{D}{2} \qquad \text{(G2.4.4)}$$

where

$$\delta_{l,i} = \begin{cases} 1 & \text{if } l = i \\ 0 & \text{otherwise} . \end{cases}$$

The drawbacks of this neural network were that the biases were input dependent, and the connection matrix was fully connected, i.e. $(n^4 - n^2)$ connections. Therefore, we tried the same energy function with a linear third term,

$$E = \frac{A}{2} \sum_{l=1}^{n} \sum_{i=1}^{n} \sum_{\substack{j=1 \\ j \neq i}}^{n} a_{l,i} a_{l,j} + \frac{B}{2} \sum_{l=1}^{n} \sum_{i=1}^{n} \sum_{\substack{m=1 \\ m \neq l}}^{n} a_{l,i} a_{m,i}$$

$$+ \frac{C}{2} \left[n - \sum_{l=1}^{n} \sum_{i=1}^{n} a_{l,i} \right] \qquad \text{(G2.4.5)}$$

$$w_{l_i, m_j} = -A\delta_{l,m}(1 - \delta_{i,j}) - B\delta_{i,j}(1 - \delta_{l,m})$$

$$\theta_{l,i} = \frac{C}{2} . \qquad \text{(G2.4.6)}$$

This resulted in the desired sparse connection matrix, with only $(2n^3 - n^2)$ nonzero connections. However, now the third term could be negative if the number of nonzero elements in the initial input matrix was greater than n. As the neural network searched for a solution, the value of this term would be increasing while the first two terms would be decreasing. Fortunately, the conversion of a nonzero input element to zero resulted in a greater change in the first two terms than the third term:

$$\sum_{\substack{i=1 \\ i \neq j}}^{n} a_{l,i}, \sum_{\substack{m=1 \\ m \neq l}}^{n} a_{m,j} > 1.$$

Further, we chose $A, B > C$ to ensure a monotonically decreasing energy function which is needed for convergence. The analog circuit of this neural network is shown in figure G2.4.2 for a (4×4) switch. As may be seen, only connections among the neurons in the same row and column are nonzero, reflecting the fact that only a single element from every row and column is chosen.

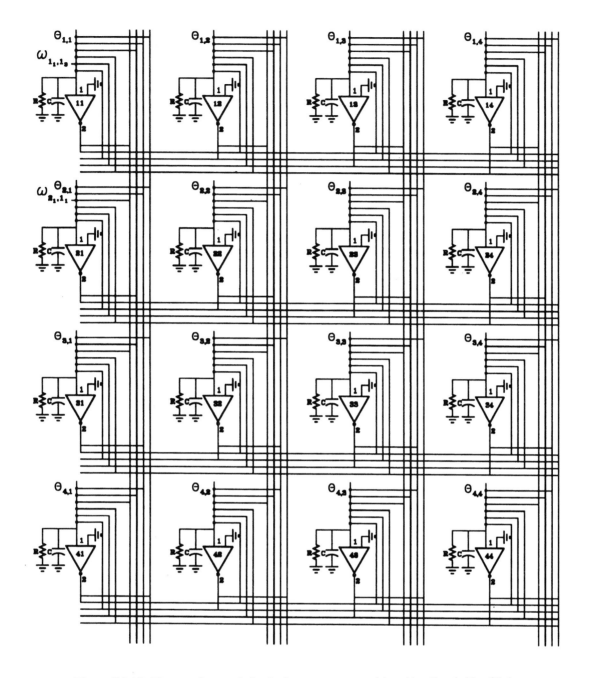

Figure G2.4.2. The neural network for the input access control in a (4×4) switching fabric.

G2.4.2.4 Performance features of the chosen topology

The above energy function has a number of nice properties. First, it has linear (as opposed to quadratic) cost terms in the energy function. Also, this type of neural network has been found to exhibit excellent convergence properties. In addition, there may be a large number of optimal solutions, each of which forms a valid basin of attraction, also known as a local energy attractor, and each of these local attractors is a candidate for the final stable state of the neurons. Thus for our application, the neural network searched for one among the many valid solutions—'getting the optimal solution' was not a concern for us.

From a hardware point of view, the complexity of our neural network approach is $O(n^2)$ neurons and $O(n^3)$ connections and the sparse connection matrix will be helpful in the implementation as a fundamental limitation of the VLSI neural networks is the large number of connections. Since, in general, the number of connections increases quadratically with that of neurons, the silicon area is mainly occupied by the connections.

One salient feature of the proposed neural architecture is that it has a configuration which is problem-independent, in the sense that neither the connection matrix, w_{l_i,m_j}, nor the input biases $\theta_{l,i}$, depend on the problem data. Thus, there is no need to readjust the weights every time a new set of inputs is presented to the neural network. The data were fed into the neural network as an initial state condition, rather than being stored in the weights or the input biases. As explained earlier, our neural network has a null programming complexity.

G2.4.3 Performance

The performance of this neural network was studied through simulation. In this section, we present these results as well as determine the optimized values of the neural network parameters. The simulation used the simultaneous solution of n^2 first-order differential equations describing the dynamics of neurons, given by equation (G2.4.1) and was done by examining and updating the output voltages of neurons, concurrently, at the intervals δt. Let $u_{l,i}^{(\alpha)}$ and $a_{l,i}^{(\alpha)}$ denote the input and output voltages, respectively, of neuron (l, i) at the end of the αth interval. Substitution of equation (G2.4.6) into (G2.4.1) resulted in the following differential equation describing a neuron during the αth interval:

$$\frac{du_{l,i}^{(\alpha)}}{dt} = -\frac{u_{l,i}^{(\alpha)}}{\tau} - A \sum_{\substack{j=1 \\ j\neq l}}^{n} a_{l,j}^{(\alpha)} - B \sum_{\substack{m=1 \\ m\neq i}}^{n} a_{m,i}^{(\alpha)} + \frac{C}{2}. \tag{G2.4.7}$$

For the input–output characteristic of a neuron, we made the following assumption:

$$a_{l,i}^{(\alpha)} = \frac{1}{2}\left[1 + \tanh\left[\frac{u_{l,i}^{(\alpha)}}{h_0}\right]\right] \tag{G2.4.8}$$

where h_0 is the gain-width parameter of the amplifier.

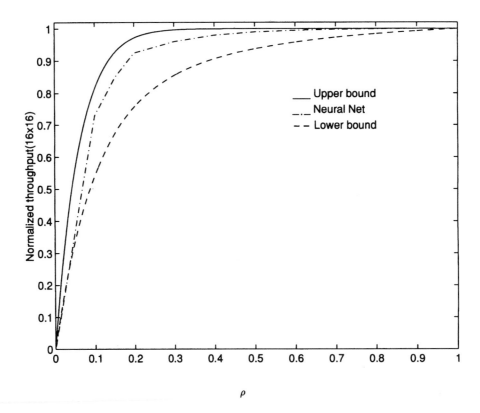

Figure G2.4.3. Neural network simulation results and theoretical bounds of the normalized throughput as a function of the probability, ρ, that an input queue for an output is busy (for a switch size $(n \times n)$).

From Wilson and Pawley (1988), the input voltages of neurons were updated using the following rule at each step:

$$u_{l,i}^{(\alpha+1)} = u_{l,i}^{(\alpha)} + \left[\frac{du_{l,i}^{(\alpha)}}{dt}\right]\delta t\,. \tag{G2.4.9}$$

In the above equations τ was set to 1 without any loss of generality and δt was chosen to be 10^{-4}, smaller values of δt do not improve the results but increase the simulation run time. The output voltages of the neurons may be updated simultaneously by using equations (G2.4.7)–(G2.4.9). During this process the value of the energy function dropped down monotonically. At each update, the new values of the neuron output voltages were compared to the previous ones. If no two consecutive values differed by more than a threshold of 10^{-6}, then the system was assumed to have reached a stable state and the simulation was stopped. At the stable state, if the output voltage of a neuron was greater than 0.5, then the neuron was ON ($a_{l,i} = 1$) and otherwise the neuron was OFF ($a_{l,i} = 0$). The initial input voltages $u_{i,j}^{(0)}$ were given values of ($+h_0$) or ($-h_0$) depending on whether the corresponding elements in the input matrix V had values of 1 or 0, respectively. Following a number of trial runs, the following parameter values were found to give accurate results: $A = 100$, $B = 100$ and $C = 40$.

A typical example for a (4×4) switch is shown in table G2.4.2 where the elements in the rectangle are the ones chosen by the neural network. The input row 1 and columns 2 and 3 each have a single non-zero element and the optimal solution should contain these elements, then for the last row and column there is a single choice, their common element. As may be seen, the neural network solution is indeed this.

Table G2.4.2. An example input/output matrix.

	1	2	3	4
1	1	0	0	0
2	1	1	0	0
3	0	0	1	1
4	1	0	0	1

In a large number of simulations for a switch size of 16×16 with random matrices as inputs, each element of the random input matrices was chosen as an independent identically distributed Bernoulli random variable with parameter ρ. Then in a similar way random biases were chosen. For every value of ρ, 200 independent runs were made and their averages were determined. Figure G2.4.3 shows the throughput per input port as a function of ρ from the simulation results together with the theoretical upper and lower bounds for the same switch size from Mehmet Ali and Youseffi (1991). The simulation results fall in between the bounds, but closer to the upper bound. Unfortunately, due to long run times we could not obtain any simulation results for larger switch sizes.

References

Asatani K 1988 Network node interface for new synchronous digital network-concept and standardization *Globecom'88* 4.5.1–7

Brandt R D *et al* 1980 Alternative networks for solving the traveling salesman problem and the list-matching problem *Int. Joint Conf. on Neural Networks* vol II pp 333–40

Garey M R and Johnson D S 1979 *Computers and Intractability: A Guide to the Theory of NP-Completeness* (San Francisco, CA: Freeman)

Hluichyj M G and Karol M J 1988 Queuing in high performance packet switching *IEEE J. Select. Areas Commun.* **6** 1587–97

Hopfield J J 1982 Neural networks and physical systems with emergent collective computational abilities *Proc. Natl Acad. Sci.* **79** 2554–8

——1984 Neurons with graded response have collective computational properties like those of two-state neurons *Proc. Natl Acad. Sci.* **81** 2554–8

Hopfield J J and Tank D W 1985 'Neural' computation of decisions in optimization problems *Biol. Cybern.* **52** 141–52

Hui J and Arthurs E 1987 A broadband packet switch for integrated transport *IEEE J. Select. Areas Commun.* **5** 1264–73

Kosko B 1992 *Neural Networks for Signal Processing* (Englewood Cliffs, NJ: Prentice-Hall)

Mehmet Ali M and Youseffi M 1991 The performance analysis of an input access scheme in a high-speed packet switch *Infocom* 454–61

Moopenn A, Duong T and Thakoor A P 1988 Digital–analog hybrid synapse chips for electronic neural networks *Advances in Neural Information Processing Systems* vol 2 ed D S Touretzky (Morgan Kaufmann) pp 769–76

Protzel P P 1990 Comparative performance measure for neural networks solving optimization problems *Int. Joint Conf. on Neural Networks* vol II pp 523–6

Stenoy G V 1989 *Linear Programming Methods and Applications* (New York: Wiley)

Takeda M and Goodman J W 1986 Neural networks for computation: number representations and programming complexity *Appl. Opt.* **25** 3033–45

Wilson G Y and Pawley G S 1988 On the stability of the travelling salesman problem algorithm of Hopfield and Tank *Biol. Cybern.* **58** 63–70

G2.5 Neural networks for optimal robot trajectory planning

Dan Simon

Abstract

This case study discusses the interpolation of minimum-jerk robot joint trajectories through an arbitrary number of knots using a hard-wired neural network. Minimum-jerk joint trajectories are desirable for their similarity to human joint movements and their amenability to accurate tracking. The resultant trajectories are numerical functions of time. The interpolation problem is formulated as a constrained quadratic minimization problem over a continuous joint angle domain and a discrete time domain. Time is discretized according to the robot controller rate. The outputs of the neural network specify the joint angles (one neuron for each discrete value of time) and the Lagrange multipliers (one neuron for each trajectory constraint). An annealing method is used to prevent the network from getting stuck in a local minimum. We show via simulation that this trajectory planning method can be used to improve the performance of other trajectory optimization schemes.

G2.5.1 Project overview

G2.5.1.1 Robot trajectory planning

The industrial robot is a highly nonlinear, coupled multivariable system with nonlinear constraints. For this reason, robot control algorithms are often divided into two stages: *path planning* and *path tracking* (Craig 1989). Path planning is often done without much consideration for the robot dynamics, and with simplified constraints. This reduces the computational expense of the path planning algorithm. The output of the path planning algorithm is then input to a path tracking algorithm.

There are algorithms for the robot control problem which do not separate path planning and path tracking. These algorithms take source and destination Cartesian points as inputs, and determine optimal joint torques. Shiller and Dubowsky (1989) provide a concise review of such algorithms. While such methods are attractive in that they provide optimal solutions to some robot control problems, they result in impractically complicated algorithms and a large computational expense. A simpler approach to the robot control problem is to generate a suboptimal joint trajectory, and then track the trajectory with a controller. This approach ignores most of the dynamics of the robot. So the resultant trajectories do not take full advantage of the robot's capabilities, but are computationally much easier to obtain. In this approach, a number of knot points are chosen along the desired Cartesian path. The number of knots chosen is a tradeoff between exactness and computational expense. The Cartesian knots are then mapped into joint knots using inverse kinematics. Finally, for each robot joint, an analytic interpolating curve is fit to the joint knots. Some of the initial and final derivatives of the curve are constrained to zero so as to ensure that the robot begins and ends its motion smoothly. 'Smoothness' is a concept which combines the ideas of derivative continuity and derivative magnitudes.

The most popular type of interpolation is algebraic splines (Lin and Chang 1983, Lin *et al* 1983, Thompson and Patel 1987). Higher-order splines result in continuity of higher-order derivatives, which reduces wear and tear on the robot (Craig 1989) but this is at the expense of large oscillations of the trajectory. Trigonometric splines can be used to provide a less oscillatory interpolating curve (Simon and Isik 1993).

G2.5.1.2 Motivation for a neural solution

Consider a sequence of knots through which an interpolating curve is required to pass. A human could create an interpolating curve, but in a different way than a computer algorithm would. Computer algorithms can calculate analytic functions which pass through given knots. A human can draw a smooth curve through a given set of knots, but without performing any mathematical calculations. In contrast with the computer algorithm, the interpolating curve drawn by the human would not be an analytic function of time. In addition, the human would not satisfy the constraints exactly, but only approximately. For example, if the human was requested to maintain a zero slope at the endpoints, the resulting slope would not be zero, but would be very small. Such a result would be satisfactory for most robot path planning applications. These facts indicate that an artificial neural network may be able to do well at interpolation.

E1.2   Of course, artificial neural networks are still quite far from any biological neural networks. Further motivation for seeking a neural solution to the robot trajectory optimization problem is obtained from the possibility of *implementation in parallel hardware*. This would give the advantage of quick solutions to large problems which would not otherwise be practical using more conventional optimization methods.

The robot path planning problem can be viewed as an optimization problem: given a desired set of knots and endpoint constraints, find the 'best' interpolating curve such that the knot errors and endpoint derivatives are not too 'large'. Several researchers have solved continuous optimization problems using neural networks (Zhao and Mendel 1988, Jeffrey and Rosner 1986a, b, Jang *et al* 1988). Platt and Barr (1988) formulate a neural network which can calculate a minimum of a general function subject to inequality or equality constraints. Their network has the important property of local stability for the problem considered in this section. Due to its stability and generality, this is the network which is used to determine a minimum-jerk robot joint path through a given set of knots.

In order to plan an optimal robot trajectory, the measure of optimality must be defined. Human arm movements satisfy some optimality criterion, and this would seem to be a desirable criterion to adopt when planning trajectories for robot arms. Flash and Hogan (1985) suggest that human arm movements minimize a measure of Cartesian jerk, while Flanagan and Ostry (1990) present evidence that a function of joint jerk is minimized. Uno *et al* (1989) and Kawato *et al* (1990) argue that the objective function is a measure of the derivative of the joint torques, and propose a neural network to learn such a trajectory. In this section, a joint jerk objective function is used. While this choice ignores the dynamics of the robot, it reduces the error of the path tracker (Kyriakopoulos and Saridis 1988) and thus is suitable for robotics applications.

G2.5.2 Design process

G2.5.2.1 Topology

Platt and Barr (1988) formulate a neural network which can be used for constrained minimization. Their algorithm, along with some straightforward extensions, is summarized in the following paragraphs.

Consider the following constrained minimization problem:

$$\min f(x) \text{ subject to } g(x) = 0 \tag{G2.5.1}$$

where $f(\cdot)$ is a scalar functional, x is an n-vector of independent variables, and $g(\cdot)$ is a vector-valued function mapping $\mathcal{R}^n \to \mathcal{R}^m$.

Lagrange multipliers can be used to convert the constrained problem of (G2.5.1) into the following unconstrained problem:

$$\min[f(x) + \lambda^{T} g(x)] \tag{G2.5.2}$$

where λ is an m-vector of Lagrange multipliers associated with the constraints $g(\cdot)$. A necessary condition for the solution of (G2.5.2) is

$$\frac{\partial f}{\partial x} + \lambda^{T} \frac{\partial g}{\partial x} = 0. \tag{G2.5.3}$$

Now consider a neural network with dynamics of the form

$$
\begin{aligned}
\dot{x}_i &= -\frac{\partial f}{\partial x_i} - \sum_{\alpha=1}^{m} (\lambda_\alpha + c_\alpha g_\alpha) \frac{\partial g_\alpha}{\partial x_i} \quad (i = 1, \ldots, n) \\
\dot{\lambda}_j &= c_j g_j \quad (j = 1, \ldots, m)
\end{aligned}
\tag{G2.5.4}
$$

where c is an m-vector of constants. Assume that the constraints $g(\cdot)$ of the original problem (G2.5.1) are linear functions of x. Then differentiating $\dot{x}_i$ in (G2.5.4) gives

$$
\ddot{x}_i + \sum_{j=1}^{n} \frac{\partial^2 f}{\partial x_i \, \partial x_j} \dot{x}_j + \sum_{\alpha=1}^{m} \left[c_\alpha \left(g_\alpha + \sum_{j=1}^{n} \frac{\partial g_\alpha}{\partial x_j} \dot{x}_j \right) \frac{\partial g_\alpha}{\partial x_i} \right] = 0 .
\tag{G2.5.5}
$$

Now consider the candidate Lyapunov energy function

$$
E = \tfrac{1}{2} \sum_{i=1}^{n} (\dot{x}_i)^2 + \tfrac{1}{2} \sum_{\alpha=1}^{m} c_\alpha g_\alpha^2 .
\tag{G2.5.6}
$$

The derivative of this energy function is a quadratic function

$$
\dot{E} = -\dot{x}^{\mathrm{T}} A \dot{x} .
\tag{G2.5.7}
$$

It has been shown in the literature (Platt and Barr 1988, Arrow *et al* 1958) that there exists a finite vector c such that matrix A is positive definite at the constrained minima of (G2.5.1). If A is continuous, then it is positive definite in some region surrounding each constrained minimum. Therefore, if the dynamic system defined by (G2.5.4) begins in that region and remains in that region, the system will settle into the zero-energy state where

$$
\dot{x} = 0
\tag{G2.5.8}
$$

$$
g(x) = 0 .
\tag{G2.5.9}
$$

Now $g(x) = 0$ implies that the original constraints are satisfied, and $\dot{x} = 0$ implies (G2.5.4) that

$$
\frac{\partial f}{\partial x_i} + \sum_{\alpha=1}^{m} (\lambda_\alpha + c_\alpha g_\alpha) \frac{\partial g_\alpha}{\partial x_i} = 0
\tag{G2.5.10}
$$

which satisfies the necessary conditions for a local minimum of the original constrained problem (G2.5.3).

To sum up, equation (G2.5.4), with an appropriately chosen c, converges to a solution of the original constrained minimization problem of (G2.5.1). Equation (G2.5.4) is in the form of first-order differential equations, which implies that it could be implemented in parallel hardware to yield a very quick solution.

G2.5.2.2 Development details

When interpolating the path of a robot joint between a set of joint space knots, it is desirable to obtain as smooth a solution as possible. This results in an appearance of coordination (Flanagan and Ostry 1990), reduces wear on the robot joints and prevents the excitation of resonances (Craig 1989), and improves the accuracy of the path tracker (Kyriakopoulos and Saridis 1988). Therefore, in robot trajectory generation, the interpolation problem for each joint can be stated as follows.

Given a set of L knots for a robot joint, determine a function $\theta(t)$ which

- is as 'smooth' as possible
- has 'small' errors at the knots
- has 'small' derivatives at the endpoints.

Smoothness can be defined as the integral of the square of the jerk of the position trajectory (Flanagan and Ostry 1990). In order for the robot joint to start and stop its motion in a smooth manner, the first three derivatives at the endpoints should be small. If the path length is T s, and the desired knot angles are $\theta(t_j) = \phi_j$ $(j = 1, \ldots, L)$, then the optimization problem for each joint can be written as

$$\min \int_0^T [\theta'''(t)]^2 \, dt \qquad \text{(G2.5.11)}$$

$$\begin{aligned}
\text{subject to} \quad \theta(t_j) &= \phi_j \qquad (j = 1, \ldots, L) \\
\theta'(0) &= 0 \\
\theta'(T) &= 0 \\
\theta''(0) &= 0 \\
\theta''(T) &= 0 \\
\theta'''(0) &= 0 \\
\theta'''(T) &= 0.
\end{aligned}$$

If the L knots are equally spaced in time, then the knot times t_i satisfy

$$t_i = (i - 1)T/(L - 1) \qquad (i = 1, \ldots, L). \qquad \text{(G2.5.12)}$$

The joint trajectory at the endpoints is exactly constrained. That is, the joint angles at $t = 0$ and $t = T$ are fixed constants. But the joint angles at the interior knot times are not truly equality constraints; the interior knot angles are more like centers of tolerance *near* which the joint trajectory is required to pass. Also, the first three endpoint derivatives do not need to be exactly zero. As long as they are very small, the robot motion will begin and end smoothly. Therefore, the constraints $\theta(t_1) = \phi_1$ and $\theta(t_L) = \phi_L$ can be considered 'hard' constraints, while the remaining $(L + 4)$ constraints in (G2.5.11) can be considered 'soft' constraints.

Since the joint trajectory is input to the path tracker at discrete values of time, the trajectory does not need to be a continuous function of time. It can be a discrete set of joint angles, defined only at times kh $(k = 0, 1, \ldots, N)$ where h is the sample period of the path tracker (typically on the order of 0.01 s), and Nh is the length of the trajectory.

The angle θ_i is input to the path tracker every h s, starting at $t = 0$ and ending at $t = T$. There are exactly M discrete times per knot, so each knot angle is separated from its neighboring knots by Mh s. Thus, the path length T satisfies

$$T = M(L - 1)h. \qquad \text{(G2.5.13)}$$

Also, from $t = 0$ to $t = T$, there are exactly $N + 1$ discrete time steps. Thus, the number of discrete time steps satisfies

$$N + 1 = M(L - 1) + 1. \qquad \text{(G2.5.14)}$$

These relationships are depicted graphically in figure G2.5.1.

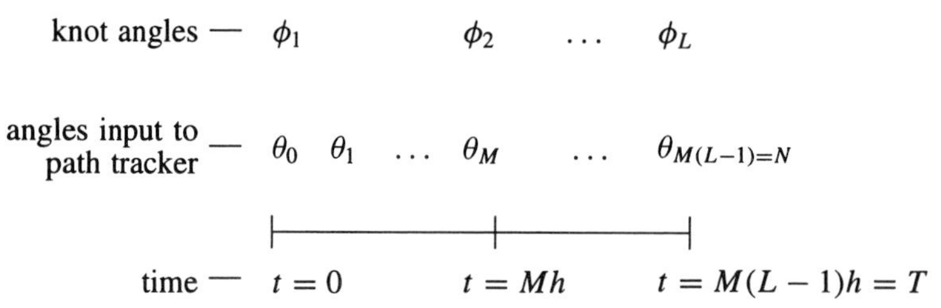

knot angles — ϕ_1 $\qquad \phi_2 \quad \cdots \quad \phi_L$

angles input to path tracker — $\theta_0 \ \theta_1 \ \cdots \ \theta_M \quad \cdots \quad \theta_{M(L-1)=N}$

time — $t = 0 \qquad t = Mh \qquad t = M(L-1)h = T$

Figure G2.5.1. Relationships between network variables.

So the optimization problem of (G2.5.11) can be discretized (using the the trapezoidal integration rule) into the following problem:

$$\min \left[\tfrac{1}{2}(\theta_0''')^2 + \sum_{i=1}^{N-1} (\theta_i''')^2 + \tfrac{1}{2}(\theta_N''')^2 \right] \qquad \text{(G2.5.15)}$$

$$\text{subject to} \quad \theta_{M(j-1)} = \phi_j \quad (j = 1, \ldots, L)$$
$$\theta'_0 = 0$$
$$\theta'_N = 0$$
$$\theta''_0 = 0$$
$$\theta''_N = 0$$
$$\theta'''_0 = 0$$
$$\theta'''_N = 0$$

where $\theta_0 = \phi_1$ and $\theta_{M(L-1)} = \phi_L$ are hard constraints, and the rest of the constraints are soft.

Since the values of θ_0 and θ_N are hard constraints, they can be considered constants. Then the independent variables of the optimization problem are θ_i $(i = 1, \ldots, N - 1)$. Note that since we are constraining θ'''_0 and θ'''_N to zero, they can be omitted from the objective function of (G2.5.15). Then, using finite-difference expressions for the first three derivatives of $\theta(t)$, the optimization problem of (G2.5.15) can be converted into the equivalent problem

$$\min \sum_{i=1}^{N-1} (-\theta_{i-2} + 2\theta_{i-1} - 2\theta_{i+1} + \theta_{i+2})^2 \qquad (G2.5.16)$$

$$\text{subject to} \quad \theta_{M(j-1)} = \phi_j \quad (j = 2, \ldots, L - 1)$$
$$\theta_1 = \phi_1$$
$$\theta_2 = \phi_1$$
$$\theta_{N-2} = \phi_L$$
$$\theta_{N-1} = \phi_L$$

where we have defined $\theta_{-1} \equiv \theta_0$ and $\theta_{N+1} \equiv \theta_N$. Now (G2.5.16) can be written as

$$\min(\boldsymbol{\theta}^{\mathsf{T}} \mathbf{A} \boldsymbol{\theta} + \boldsymbol{b}^{\mathsf{T}} \boldsymbol{\theta}) \text{ subject to } g(\boldsymbol{\theta}) = 0 \qquad (G2.5.17)$$

where $\boldsymbol{\theta} = [\theta_1 \ldots \theta_{N-1}]^{\mathsf{T}}$, $g(\boldsymbol{\theta})$ is the $(L + 2)$-element constraint vector defined by (G2.5.16), and $\mathbf{A}$ and $\boldsymbol{b}$ are, respectively, an $(n - 1) \times (n - 1)$ matrix and an $(n - 1)$-vector. Matrix $\mathbf{A}$ is a positive semidefinite matrix of bandwidth 4 (Golub and Van Loan 1989) whose diagonal and first through fourth upper and lower diagonals are given as follows:

$$\text{diagonal} = (5 \quad 9 \quad 10 \quad 10 \quad \ldots \quad 10 \quad 10 \quad 9 \quad 5)$$
$$\text{first upper and lower diagonal} = (-2 \quad -4 \quad -4 \quad \ldots \quad -4 \quad -4 \quad -2)$$
$$\text{second upper and lower diagonal} = (-4 \quad -4 \quad \ldots \quad -4 \quad -4) \qquad (G2.5.18)$$
$$\text{third upper and lower diagonal} = (4 \quad 4 \quad \ldots \quad 4 \quad 4)$$
$$\text{fourth upper and lower diagonal} = (-1 \quad -1 \quad \ldots \quad -1 \quad -1).$$

Vector $\boldsymbol{b}$ is given by

$$\boldsymbol{b} = (-4\phi_1 \quad -4\phi_1 \quad 6\phi_1 \quad -2\phi_1 \quad 0 \quad 0 \quad \ldots \quad 0 \quad 0 \quad -2\phi_L \quad 6\phi_L \quad -4\phi_L \quad -4\phi_L)^{\mathsf{T}}. \qquad (G2.5.19)$$

According to the results given by (G2.5.4), (G2.5.17) is solved by the dynamic system

$$\dot{\boldsymbol{\theta}} = -2A\boldsymbol{\theta} - \boldsymbol{b} - \frac{\partial g}{\partial \boldsymbol{\theta}}(\boldsymbol{\lambda} + c \circ g)$$
$$\dot{\boldsymbol{\lambda}} = c \circ g \qquad (G2.5.20)$$

where $c \circ g$ is the $(L + 2)$-vector Hadamard product of c and g whose ith element is given by $c_i g_i$. The element in the ith row and jth column of $\partial g / \partial \boldsymbol{\theta}$ is given by $\partial g_j / \partial \theta_i$.

If matrix $\mathbf{A}$ were positive definite, we could set c equal to the zero vector and still be guaranteed convergence. However, if $\mathbf{A}$ is only positive semidefinite, we need to use a nonzero c. Even if $\mathbf{A}$ is positive definite, a nonzero c will improve the convergence properties of the neural network.

Note that the neural net may converge to a local minimum rather than a global minimum. Some sort of *simulated annealing* technique can be used in conjunction with the network (Jeffrey and Rosner 1986a, b). This idea results in the long computational time characteristic of annealing, but it also enables the network to find the best solution among many local minima.

C1.4.2

The annealing-type method which is suggested is as follows. Once the network converges to a local minimum, the network state is perturbed in a random direction and by a random magnitude. Then the network dynamics are reactivated, and another local minimum is found. During this process, the algorithm keeps track of the best solution. After a predetermined number of local minima are found, the algorithm terminates and the solution with the lowest energy is accepted as the best solution.

G2.5.3 Comparison with other methods of robot trajectory planning

Two methods were used to generate minimum-jerk robot joint trajectories: a minimum-jerk trigonometric spline method was used by Simon and Isik (1993), and the neural network proposed above was used. The trigonometric spline method is analytical and was coded using MATLAB on a Sun-4 workstation. The neural network is a numerical method and was simulated on a Sun-4 workstation in the C programming language. The neural net dynamics were integrated using a basic fourth-order Runge–Kutta method with an integration step size of 5 ms.

Six multiple-knot, 35-second joint trajectories were calculated using the trigonometric spline method and the simulated neural network. Each joint trajectory has eight evenly spaced knots, corresponding to the examples given in previous work (Lin *et al* 1983, Thompson and Patel 1987).

Plots of the six neural-network-based trajectories which pass through the six sets of knots are given by Simon (1993). The trajectory corresponding to joint 2 (a typical example) is reproduced here in figure G2.5.2. The initial state of the neural nets consisted of the minimum-jerk trigonometric trajectories (Simon and Isik 1993), λ was initialized to the zero vector, and c was a vector in which each element was 1. Note from figure G2.5.2 that the neural-network-based trajectory does not pass exactly through the knots. The neural network trajectories have small nonzero derivatives at the endpoints. The trigonometric splines have zero velocity, acceleration and jerk at the endpoints, and pass exactly through the knots.

Table G2.5.1 shows the decrease of the jerk objective function due to the evolution of the network dynamics. It is seen that the use of the neural network for this typical example gives an average improvement of almost 20% in the objective function. Although we cannot quantify the result of this decrease at this point in time, we can state that two results are a corresponding decrease in the error of the path tracker, and robot arm movement which appears more smooth and coordinated.

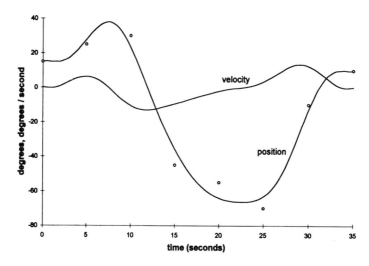

Figure G2.5.2. Minimum-jerk trajectory for joint 2.

G2.5.4 Conclusions

Minimum-jerk joint trajectories have the properties of similarity to human joint movements (Flanagan and Ostry 1990) and amenability to tracking (Kyriakopoulos and Saridis 1988). This makes them attractive choices for robotics applications in spite of the fact that the dynamics are not taken into account.

In this section, the minimum-jerk joint trajectory formulation problem is posed as a constrained quadratic optimization problem. A hard-wired neural network is proposed to solve the problem numerically.

Table G2.5.1. Jerk objective function values.

Joint	Minimum jerk		Decrease (%)
	Trigonometric	Neural net	
1	127	106	16.5
2	44	28	36.3
3	558	462	17.2
4	765	662	13.5
5	252	206	18.3
6	38	33	13.2
Averages	297	250	19.2

The network may converge to a local minimum rather than the global minimum. The solution obtained by the network depends on the initial state of the network. An annealing-type technique is used in conjunction with the network to climb out of local minima and find the best among many solutions. This prevents the algorithm from being appropriate for real-time use, but significantly improves the quality of the final solution. The simulation results presented verify that the network can be successfully applied to robot trajectory generation.

The neural-network-generated trajectories pass near but not exactly through the specified knots. If it is important that the trajectory pass exactly through the knots, this method may not be suitable for joint interpolation. While this section has dealt specifically with minimum-jerk joint trajectories, there are no theoretical limitations to applying this method to other objective functions. More specifically, minimum-energy or minimum-torque-change trajectories could be generated with the network discussed in this section.

Acknowledgements

Much of this section has been adapted from the article by Simon (1993) where additional details can be found, and the permission of the publisher is gratefully acknowledged.

References

Arrow K, Hurwicz L and Uzawa H 1958 *Studies in Linear and Nonlinear Programming* (Stanford, CA: Stanford University Press)

Cohen M and Grossberg S Absolute stability of global pattern formation and parallel memory storage by competitive neural networks *IEEE Trans. Systems, Man, Cybern.* **13** 815–26

Craig J 1989 *Introduction to Robotics* (Reading, MA: Addison-Wesley)

Flanagan J and Ostry D 1990 Trajectories of human multi-joint arm movements: evidence of joint level planning *Experimental Robotics I, Ist Int. Symp.* ed V Hayward and O Khatib (New York: Springer)

Flash T and Hogan N 1985 The coordination of arm movements: an experimentally confirmed mathematical model *J. Neurosci.* **5** 1688–703

Golub G and Van Loan C 1989 *Matrix Computations* 2nd edn (Baltimore, MD: Johns Hopkins University Press)

Jang J *et al* 1988 An optimization network for matrix inversion *Neural Information Processing Systems* ed D Anderson (New York: American Institute of Physics) pp 397–401

Jeffrey W and Rosner R 1986a Optimization algorithms: simulated annealing and neural network processing *Astrophys. J.* **310** 473–81

——1986b Neural network processing as a tool for function optimization *Neural Networks for Computing* ed J Denker (New York: American Institute of Physics) pp 241–6

Kawato M *et al* 1990 Trajectory formation of arm movement by cascade neural network model based on minimum torque-change criterion *Biol. Cybern.* **62** 275–88

Kyriakopoulos K and Saridis G 1988 Minimum jerk path generation *IEEE Int. Conf. on Robotics and Automation* vol 1, pp 364–9

Lin C and Chang P 1983 Joint trajectories of mechanical manipulators for Cartesian path approximation *IEEE Trans. Syst. Man Cybern.* **13** 1094–102

Lin C, Chang P and Luh J 1983 Formulation and optimization of cubic polynomial joint trajectories for industrial robots *IEEE Trans. Automatic Control* **28** 1066–73

Platt J and Barr A 1988 Constrained differential optimization *Neural Information Processing Systems* ed D Anderson (New York: American Institute of Physics) pp 612–21

Shih L 1984 On the elliptic path of an end-effector for an anthropomorphic manipulator *Int. J. Robot. Res.* **3** 51–7

Shiller Z and Dubowsky S 1989 Robot path planning with obstacles, actuator, gripper, and payload constraints *Int. J. Robot. Res.* **8** 3–18

Simon D 1993 The application of neural networks to optimal robot trajectory planning *Robot. Autonomous Syst.* **11** 23–34

Simon D and Isik C 1993 A trigonometric trajectory generator for robotic arms *Int. J. Control* **57** 505–17

Thompson S and Patel R 1987 Formulation of joint trajectories for industrial robots using B-splines *IEEE Trans. Indust. Electron.* **34** 192–9

Uno Y *et al* 1989 Formation and control of optimal trajectory in human multijoint arm movement *Biol. Cybern.* **61** 89–101

Zhao X and Mendel J 1988 An artifical neural minimum-variance estimator *IEEE Conf. on Neural Networks* vol 2, pp 499–506

G2.6   Radial basis function network in design and manufacturing of ceramics

Krzysztof J Cios, George Y Baaklini, Laszlo Berke and Alex Vary

Abstract

This case study has two goals. One is to show the application of the radial basis function (RBF) neural network in aiding in all aspects of design and manufacturing of advanced ceramics, where it is desirable to find which of the many processing variables contribute most to the desired properties of the material. The second goal of the chapter is to compare the RBF network results with those obtained by using fuzzy sets on the same data collected at the NASA Lewis Research Center. To set the RBF hidden layer centers and to train the output layer weights the nodes at data points and the gradient descent methods were used, respectively. The RBF network predicted strength with an average error of less than 12% and density with an average error of less than 2%, and demonstrated a potential for accelerating the development and processing of emerging ceramic materials.

G2.6.1   Project overview

In this case study our intent is to show how *RBF networks* could be used in the design and fabrication of ceramics. RBF networks were utilized to identify trends indicating which input variable contributed most to the increase of a desired output parameter, say strength. Such identification could potentially speed up the process of designing a new material. Although human designers could easily notice such trends for a few variables, it becomes difficult to do so for a large number of variables.   C1.6.2

This case study is based on our previous work (Cios *et al* 1994a, b) in which we utilized the data originally collected by Sanders and Baaklini (1986). Silicon nitride ceramics were chosen for our study since it is an important material for heat engine applications due to its high operating temperature, reduced weight, resistance to oxidation, thermal shock resistance, and good high-temperature strength (Klima and Baaklini 1984). Their scatter in strength and low toughness are generally attributed to discrete defects such as voids, inclusions, and cracks introduced during processing (Sanders and Baaklini 1986). Current cost-effective fabrication procedures also frequently produce ceramics containing bulk density variations and microstructural anomalies that can adversely affect performance (Klima and Baaklini 1984).

Scatter in mechanical properties of ceramics is a great drawback from a design/reliability standpoint. This scatter is attributed to defects and inhomogeneities occurring during processing of silicon nitride powder compositions and during part fabrication. From the research work on silicon nitride composition at the NASA Lewis Research Center it was evident that density gradients were strongly dependent upon sintering conditions (Sanders and Baaklini 1986, Klima and Baaklini 1984). The results of an investigation of one silicon nitride composition involving sintering trials of several batches of material were described by Sanders and Baaklini (1986), and these particular data were utilized to show that RBF neural networks were a useful tool which could provide much needed information to advanced materials designers.

Sanders and Baaklini (1986) were concerned with the problem of designing a silicon nitride ceramic with the goal of achieving fully dense material that possesses high strength with the lowest amount of scatter. In the process of manufacturing they tried to optimize several variables such as milling time, sintering temperature, sintering time, nitrogen pressure and setter contact. In addition, they investigated the effects of sintering and temperature variations and whether wet powder sieving was superior to dry sieving. They were also trying to optimize the manufacturing process by using sound engineering judgment coupled with trial and error methodology.

From the data collected at the NASA Lewis Research Center we selected three input variables, namely the milling time of the silicon nitride powder, the sintering time, and the nitrogen pressure employed during sintering of the modulus of rupture (MOR) test bars. From the output variables we selected flexural strength and density. Only the above-mentioned variables were used since there were not enough training pairs (outputs associated with inputs) for processing variables such as temperature and sieving. In our investigation we concentrated on determining how effectively an RBF neural network can be trained to predict the resultant strength and density of a batch of MOR bars.

G2.6.2 Data used

RBFs were trained using the data from 273 silicon nitride modulus of rupture bars (MOR) that were tested at room temperature and 135 MOR bars that were tested at 1370°C. For the room temperature, 18 different combinations of milling time, sintering time, and nitrogen pressure yielded the composition strengths and densities listed in table G2.6.1. Also listed in table G2.6.1 are the strengths and densities for nine combinations at 1370°C.

In order to determine the validity of the network predictions for the previously untried compositions, it was necessary to test the RBF network using known test vectors and then calculate the error of the predictions. Of particular interest was the ability of the network to predict the output values for batch number 6Y25, as this batch number represented the optimum combination for the processing variables from the available data set. Batch 6Y25 was considered optimal because although the average value (of

Table G2.6.1. Strength and density at room temperature for different processing and sintering conditions.

Batch No	No of specimen	Milling time (h)	Sintering time (h)	Nitrogen pressure (MPa)	Actual strength (MPa)	Actual density (g cm^{-3})
6Y1B	30	24	1	2.5	556	3.12
6Y2B	30	24	1	2.5	532	3.18
6Y11	15	100	1	2.5	490	3.23
6Y12	15	300	1	2.5	579	3.25
6Y13	15	100	1	2.5	684	3.24
6Y14	14	300	1	2.5	746	3.24
6Y15, 6Y16	19	24	2	5	664	3.22
6Y17	10	100	2	5	646	3.23
6Y18	10	100	1.5	5	608	3.21
6Y19	10	100	1.5	5	570	3.22
6Y20	10	100	2	5	650	3.22
6Y23	15	100	1.25	5	631	3.24
6Y24A	15	100	1.25	3.5	586	3.26
6Y24B	15	100	2	3.5	619	3.26
6Y25	10	300	2	5	714	3.28
6Y26A	15	100	1	3.5	479	3.20
6Y26B	15	100	1	5	503	3.18
6Y28	10	100	2	5	671	3.21
1370°C						
6Y9B	29	24	1	2.5	382	3.12
6Y11	13	100	1	2.5	445	3.23
6Y12	14	300	1	2.5	417	3.25
6Y13	15	100	1	2.5	405	3.24
6Y14	14	300	1	2.5	424	3.24
6Y15, 6Y16	20	24	2	5	402	3.22
6Y17	10	100	2	5	441	3.23
6Y18	10	100	1.5	5	460	3.21
6Y25	10	300	2	5	467	3.28

ten specimens, table G2.6.1) of strength is 714 MPa, it is accompanied by low scatter (not shown in table G2.6.1). Batch number 6Y14 had higher (746 MPa) average strength but it was accompanied by much higher scatter.

Batch number 6Y25 was first removed from the data sets. The data sets were then pseudorandomly divided into a ratio of 70% training to 30% testing. Batch number 6Y25 was then inserted into the test data set. This was repeated five times in order to have five different pairs of training and test data sets. This entire process was then repeated using a ratio of 60% training to 40% testing. The 60% proportion of training data was used in order to give an indication as to how much processing information was required to make accurate predictions.

Next, a training data set consisting of all the batch numbers (100%) except 6Y25 was created. Batch number 6Y25 was placed in the test data set as the sole vector. Finally, all the batch numbers were placed in a training data set and the test data set was constructed using vectors for which the outputs were not known in order to demonstrate the capability of the RBF network in material process optimization. This gave a total of 12 pairs of training and test data sets for the room-temperature-tested materials, and another 12 for materials tested at 1370°C.

Third, several new combinations of the three input parameters were used to determine whether a material having equal or higher values of flexural strength and density, close to the optimal (6Y25) value, could be obtained. Thus, a training data set consisting of all the batch numbers (100%) except 6Y25 was created. Batch number 6Y25 was then placed in the training data set and we made predictions for different combinations of the input vectors not tried in previous experiments (Cios *et al* 1994a).

G2.6.3 Radial basis functions

For details of RBF networks the reader is referred to section C1.6.2 of the handbook and Cios *et al* (1994a). C1.6.2
Here we only very briefly summarize the main ideas of the radial basis function (RBF). It is a three-layer network with locally tuned processing units in the hidden layer. RBF neurons are centered at the training data points, or some subset of them, and each neuron only responds to an input which is close to its center. The output layer neurons are linear or sigmoidal functions and their weights may be obtained by using a supervised learning method, such as a gradient descent method.

Figure G2.6.1 shows a general RBF network with n inputs and one linear output. This network performs a mapping $f : \mathbb{R}^n \to \mathbb{R}$ given by the following equation:

$$f(x) = \lambda_0 + \sum_{i=1}^{n_r} \lambda_i \phi(\|x - c_i\|)$$

where $x \in \mathbb{R}^n$ is the input vector, $\phi(\cdot)$ is a function from $\mathbb{R}^n \to \mathbb{R}$, $\|$ denotes the Euclidean norm, λ_i $(0 \leq i \leq n_r)$ are the weights of the output node, c_i $(0 \leq i \leq n_r)$ are the RBF centers, and n_r is the number of the RBF centers.

One of the most common functions used for $\phi(\cdot)$ is the Gaussian function:

$$\phi(\|x - c_i\|) = \exp\left(-\frac{\|x - c_i\|^2}{\sigma_1^2}\right)$$

where σ_1 is a constant which determines the width of the ith node. This function has a maximum value of 1 when $\|x - c_i\|$ is 0, and drops off to 0 as $\|x - c_i\|$ approaches infinity. The centers of the RBF functions, c_i, are usually chosen from the training data points x_i $(1 \leq i \leq N)$. This method is known as the neurons at data points method.

G2.6.4 Results of the radial basis function

The RBF networks were trained using several training data sets described above. The neurons at data points method was used to set up the hidden layer, and the gradient descent method was used to train the output layer neurons which use the sigmoidal function. The RBF networks consisted of three input neurons and two output neurons which corresponds to the number of input and output variables, respectively. The number of neurons in the hidden layer depended on the number of training vectors.

Table G2.6.2 shows the detailed results for the 70% training and 30% test data set, for one of the combinations, at room temperature. The overall results for five combinations, and for 6Y25, are shown

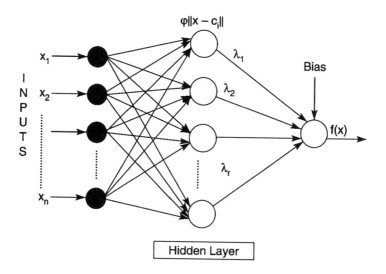

Figure G2.6.1. Radial basis function network with single output.

in Table G2.6.3 for 70% training, at room temperature. Table G2.6.4 shows the same for 60% training. Table G2.6.5 shows predictions made for selected, not tried, combinations of milling and sintering values that resulted in strengths and densities similar to that of the optimum batch 6Y25. Table G2.6.6 and table G2.6.7 show the overall results obtained for 1370°C.

Relatively large errors occurred in several cases. In table G2.6.2, the error of 29.84% on the predicted strength can be explained by the fact that the training vector from batch 6Y14 biased the results of 6Y12 and this was totally due to a sintering variable that was not included as an input feature. In table G2.6.6, the 16.45% error can be attributed to the absence of training vectors with 300 h grinding time.

The information in table G2.6.5 and table G2.6.7 suggested that there might be other combinations of sintering and processing variables that would produce results almost as good as that obtained for 6Y25, but more efficiently. For example, in table G2.6.5, using a milling time of 250 h, a sintering time of 1.5 h, and a nitrogen pressure of 3 MPa, the RBF network predicted that a strength of 709 MPa could have been obtained. This was only slightly less than the 6Y25 value of 712 MPa, but with a reduction in milling time of 50 h.

Table G2.6.2. Predicted room-temperature strength with 70% training.

Batch No	Actual strength (MPa)	Predicted strength (MPa)	% error	Actual density (g cm^{-3})	Predicted density (g cm^{-3})	% error
6Y2B	556	544	2.26	3.18	3.17	0.46
6Y12	579	752	29.84	3.25	3.24	0.27
6Y17	646	660	2.13	3.23	3.21	0.49
6Y18	608	616	1.37	3.21	3.24	0.91
6Y24A	586	507	13.51	3.26	3.23	0.88
6Y25	714	681	4.85	3.28	3.21	2.28
		Average error	8.95			0.88

Table G2.6.3. Overall results for room temperature strength and density with 70% training.

Strength—average % error for all test vectors (and 6Y25)	Density—average % error for all test vectors (and 6Y25)
10.54 (10.17)	0.98 (2.50)

Table G2.6.4. Overall results for room temperature strength and density with 60% training.

Strength—average % error for all test vectors (and 6Y25)	Density—average % error for all test vectors (and 6Y25)
11.34 (5.96)	1.03 (1.74)

Table G2.6.5. Prediction of input variables for highest strength and density for 100% plus 6Y25 training data.

Milling time (h)	Sintering time (h)	Nitrogen pressure (MPa)	Predicted strength (MPa)	Predicted density (g cm^{-3})
150	1.5	3	692	3.28
175	1.5	3	700	3.28
200	1.5	3	706	3.28
200	1.75	4	689	3.27
250	1.5	3	709	3.28
250	1.5	4	705	3.28
250	1.75	4	705	3.28
300	1.5	4	711	3.28
300	1.75	4	713	3.28
300	2	5	712	3.28

Table G2.6.6. Overall results for 1370°C, 70% training.

	Strength—average % error for all test vectors	Strength—% error for 6Y25	Density—average % error for all test vectors	Density—% error for 6Y25
	8.77	5.80	0.83	1.28
	7.61	11.88	1.50	2.71
	7.22	11.17	1.69	1.14
	10.36	16.45	1.62	2.34
	6.69	3.80	1.52	2.82
Combined average error	8.21	9.82	1.43	2.06

Table G2.6.7. Prediction of selected processing and sintering variables for optimum density and strength at 1370°C with 100% plus 6Y25 training.

Milling time (h)	Sintering time (h)	Nitrogen pressure (MPa)	Predicted strength (MPa)	Predicted density (g cm^{-3})
150	1.5	4	466	3.24
175	1.5	4	469	3.25
200	1.5	4	470	3.26
200	1.5	5	471	3.25
200	1.75	5	471	3.27
250	2.0	5	467	3.27
300	1.5	4	468	3.27
300	1.5	5	470	3.26
300	1.75	5	471	3.27
300	2.0	5	467	3.27

Similarly, table G2.6.7 indicated that a slightly higher value than optimal for 6Y25 of 471 MPa could be achieved with milling time of 200 h, sintering time of 1.5 h, and nitrogen pressure of 5 MPa, which is a 100 h saving in milling time over 6Y25. A word of caution here: these predictions need to be confirmed by manufacturing of ceramics using the same input parameters.

Using even the smaller training data set of 60% did not increase the prediction errors in a significant way. This suggested a potential for speeding up the optimization of ceramics processing by using RBF neural networks.

G2.6.5 Comparison of radial basis function results with those obtained using fuzzy sets

G2.6.5.1 Basics of fuzzy sets

Fuzzy sets allow us to deal with phenomena that are vague or too ill defined to be analyzed by conventional mathematical tools. For more information on fuzzy sets the reader is referred to Chapter D1 of this handbook, Cios *et al* (1991), and Cios *et al* (1994b). Definitions essential for subsequent explanation of the method used follow. Let $\mathbb{R}$ be the set of real numbers and U be the conventional (crisp) set. Let u be a generic element of U. A fuzzy subset A of U is defined by a membership function $\mu : U \rightarrow [0, 1]$. The fuzzy subset A of U can be expressed as (Klir and Folger 1988):

$$A = \{\mu(u) \mid u; u \in U, \mu_A(u) \in [0, 1]\}$$

where μ_A is referred to as the grade of membership of u in A.

The support of A is the set of elements in U whose memberships in fuzzy subset A, $\mu_A(u)$, are positive:

$$\text{Supp}(A) = \{u \mid u \in U, \mu_A(u) > 0\}.$$

Aggregation of fuzzy sets is an operation by which several fuzzy sets are combined into a single set. In general, any aggregation operation is defined by the function

$$h : [0, 1]^n \rightarrow [0, 1]$$

for some $n \geq 2$. When applied to n fuzzy sets defined on U, h produces an aggregate fuzzy set A by operating on the grades of membership of each element of U in the sets being aggregated.

From the several classes of averaging operations we chose generalized means defined as follows:

$$h_a(a_1, a_2, \ldots, a_n) = \left(\frac{a_1^\alpha + a_2^\alpha + \ldots + a_n^\alpha}{n} \right)^{1/\alpha}$$

where $\alpha \in \mathbb{R}$ ($\alpha = 0$) is a parameter by which different means are distinguished; $\alpha = 2$ was used.

The data shown in table G2.6.1 were used to define fuzzy sets for each batch for both input and output variables. The input fuzzy sets were defined for three support values: nitrogen pressure (p), sintering time (st), and milling time (mt), while the output fuzzy sets had support of two elements: flexural strength (s) and density (d). The grades of memberships were normalized elementwise, and the normalization was repeated for every step of prediction. The resulting membership grades were combined by means of generalized mean operation. After that, a dissimilarity measure (Cios *et al* 1994b) was used to calculate the difference between the actual and generalized fuzzy sets of input parameters. Next, the k-fraction of the measure, where $k \in (0, 1)$, was either added to or subtracted from the generalized grades of memberships of the output parameters. The graphical explanation of the method is shown in figure G2.6.2, for the 6Y12 test batch. The generalized input fuzzy set consisted of grades of membership obtained by generalized mean operation performed on normalized values of input parameters: mt, st, and p. The dissimilarity measure was then used to calculate the sum of the elementwise differences between grades of membership of actual and generalized input fuzzy sets. The k-fraction of the measure was then added to the grades of membership of the generalized output fuzzy set. The generalized output fuzzy set was obtained by generalized mean operation performed on normalized values of output parameters: s and d. Addition of the k-fraction of the dissimilarity measure results in the predicted fuzzy set. The latter was then compared with the actual grades of membership obtained by normalization of the values of the 6Y12 batch output thus yielding a measure of error for strength and density.

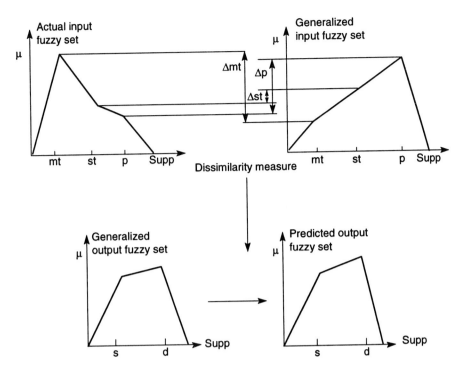

Figure G2.6.2. Explanation of the fuzzy prediction method.

Table G2.6.8. Overall results for strength and density for room temperature.

Strength—average % error for all test vectors (and 6Y25)	Density—average % error for all test vectors (and 6Y25)
5.7 (4.4)	2.4 (0)

G2.6.5.2   Results of fuzzy sets

The method described above for fuzzy sets was used to predict, for randomly chosen values of input variables, the values for output variables, namely flexural strength and density of batch samples at room temperature. The overall results are shown in table G2.6.8. Since the errors were reasonably small, predictions were made for selected new combinations of processing and sintering variables. Table G2.6.9 shows the results. We could notice that the resultant strengths and densities were lower than those for the optimum batch (6Y25), which can be explained by the fact that fuzzy sets are bounded by the values [0, 1].

G2.6.6   Discussion

If, in the process of designing new ceramics, the designers were to use RBF networks in order to notice the correlations between the input and output variables, it might greatly shorten the fabrication cycle. We have shown (Cios *et al* 1994a, b) that this is true for even a small number of input variables. If a larger number of input variables could be used, that would certainly improve the reliability of predictions and their accuracy.

Comparison of results obtained by using fuzzy sets with those obtained by using RBF neural networks indicates that both were successful in modeling relationships existing between the processing variables and output variables. This is shown graphically in figure G2.6.3. As could be seen, there were small differences in terms of errors. When we tried to predict the untried combinations of input variables which might yield higher values for strength and density, the results were again only slightly different.

Table G2.6.9. Prediction of input variables for highest strength and density for 100% plus 6Y25 training data.

Milling time (h)	Sintering time (h)	Nitrogen pressure (MPa)	Predicted strength (MPa)	Predicted density (g cm^{-3})
150	1.5	3	596	3.15
175	1.5	3	604	3.18
200	1.5	3	611	3.21
200	1.75	4	634	3.28
250	1.5	3	619	3.25
250	1.5	4	634	3.28
250	1.75	4	649	3.28
300	1.5	4	649	3.28
300	1.75	4	656	3.28
300	2	5	686	3.28

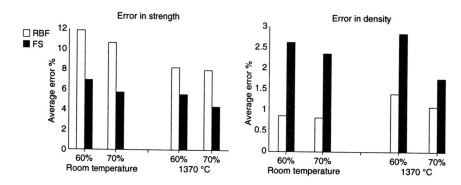

Figure G2.6.3. Average errors in predicting strength and density using 60% and 70% of data for training.

G2.6.7 Conclusions

The radial basis function network was found to be applicable for learning silicon nitride processing and consequently for predicting strength and density using three processing variables as input features. Predicting strength and density values for the 30% or 40% of the modulus of rupture batches subsets which were not used for training was successful with an average error of less than 12% for strength and 2% for density, for both room and high temperatures. Predicting strength for the optimum batch was successful when the training set reflected a reduced gradient and less biased regions. Predicting bulk density of ceramics was more successful than predicting strength. This may be explained by noting that bulk density was more directly related to milling time, sintering time, and pressure, whereas the flexural strength was additionally dependent on pore morphology, on microstructure, and on the presence of failure causing defects. Our work (Cios *et al* 1994a) showed that RBF neural networks had a great potential for accelerating improvements in ceramic material processing.

We have shown that RBFs, if they were part of the design process, could help in optimizing the process of fabricating ceramics with high strength, accompanied by low scatter. We concentrated on three input variables and two output parameters. The available data set was divided into training and test parts. The former was used for training RBF neural networks and defining fuzzy sets, and the second to validate them on the test part as to how accurately they can predict the strength and density of new 'unknown' inputs.

Then, we showed that it was possible to indicate combinations of input variables, other than those tried, which resulted in at least as strong material as the one from the known training data (6Y25), but more optimal in terms of either shorter milling and sintering times, or lower pressure. RBF networks may not necessarily yield the optimal solution, but in many situations, a robustly obtained 'acceptable' solution is preferred to an optimal solution which may take a lot of time to compute.

The obtained results indicated that RBF networks could be a powerful tool for both process modeling and process control. They can speed the development and fabrication of emerging ceramic materials by capturing imprecise relationships between the input variables and output parameters. In turn, these learned relationships can be used for predicting strength and density for new combinations of the input variables. The reliability of our predictions was validated by calculating the errors on the test data encompassing 30% or 40% of available data. The maximum combined error, between the two methods, was less than or equal to 5.7% for strength and 0.98% for density. The latter clearly shows that by using a hybrid neuro-fuzzy approach one could achieve even better results.

References

Cios K J, Baaklini G Y, Vary A and Tjia R E 1994a Radial basis function learns ceramic processing and predicts related strength and density *J. Testing Evaluation* **22** 343–50

Cios K J, Baaklini G Y, Vary A and Sztandera L M 1994b Fuzzy sets in the prediction of flexural strength and density of silicon nitride ceramics *Mater. Evaluation* **52** 600–6

Cios K J, Shin I and Goodenday L S 1991 Using fuzzy sets to diagnose artery coronary stenosis *IEEE Comput. Mag., special issue on Computer-Based Medical Systems* **24** 57–63

Klima S J and Baaklini G Y, 1984 Ultrasonic characterization of structural ceramics *NASA CP-2383*

Klir G J and Folger T A 1988 *Fuzzy Sets, Uncertainty and Information* (Englewood Cliffs, NJ: Prentice-Hall)

Sanders W A and Baaklini G Y 1986 Correlation of processing and sintering variables with the strength and radiography of silicon nitride *Ceram. Eng. Sci. Proc.* **7** 839–60

G2.7   Adaptive control of a negative ion source

Stanley K Brown, William C Mead, P Stuart Bowling and Roger D Jones

Abstract

We describe a project in which we developed an automated adaptive controller based on the CNLS artificial neural network and evaluated its applicability for the tuning and control of a small-angle negative ion source on the discharge test stand at Los Alamos. The controller processes information obtained from the beam current waveform to determine beam quality. The controller begins by making a sparse scan of the four-dimensional operating surface. The independent variables of this surface are the anode and cathode temperatures, the hydrogen flow rate, and the arc voltage. The dependent variable is a figure of merit that is composed of terms representing the magnitude of the beam current, the stability of operation, and the quietness of the beam. Once the sparse scan is finished, the neural network formulates a model from which it predicts the best operating point. The controller takes the ion source to that operating point for a reality check. The operating data are compared with the predicted data to determine the validity of the model. As real data are fed in, the model of the operating surface is updated until the neural network model agrees with reality. The controller then uses a gradient ascent to optimize the operation of the ion source. Initial tests of the controller indicate that it is remarkably capable. It has optimized the operation of the ion source on six different occasions bringing the beam to excellent quality and stability.

G2.7.1   Project overview

The design of this ion source evolved at Los Alamos from an initial Russian design. Its internal processes are so complicated that no one has been able to model them. Consequently, control has always required human operators. This project was to develop a model of the operation of the ion source using experimental data and a neural network. Once the model was developed it would be used to optimally control the ion source in normal operation. All of this was accomplished.

G2.7.2   Design process

G2.7.2.1   Motivation for a neural solution

Several attempts at control of this ion source using classical linear techniques as well as some using statistical pattern recognition techniques were only partially successful. Several characteristics combine to present a difficult control problem:

- large multidimensioned control space
- complex relationships between diagnostics and required control actions
- non-linear responses
- multiple operating modes with strong history dependence
- substantial drifts within operating and maintenance cycles
- relatively long settling times.

The multidimensional nature of the source translates to a large complicated control surface that is time consuming to map. Because of this and the aforementioned characteristics, especially the nonlinear responses, conventional control theory methodology will not work. A more sophisticated controller based on a neural network has characteristics that allow it to deal adequately with these issues.

G2.7.2.2 General description of the ion source

A cut-away schematic of the ion source is shown in figure G2.7.1. The important parts of the ion source that should be noted are the small (0.37 mm) dimension of the gap between the anode and the cathode where a 600 V arc occurs, the emission slit where the ion beam emerges as a result of a large (35 000 V) electric pulse being applied to the source, the small hole in the center of the anode where the hydrogen is introduced into the arc region, and the presence of a fairly substantial magnetic field that is used to trap electrons in the arc region. These electrons interact with the hydrogen to produce a plasma. Interaction of hydrogen molecules with the plasma and with the walls of the arc region form the negative hydrogen ions that will make up the ion beam. The ions inside the plasma can be extracted from the source through the emission slit by applying a large electric pulse.

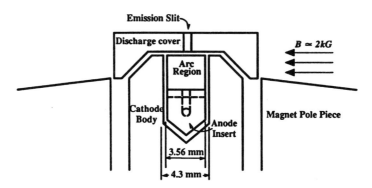

Figure G2.7.1. Cutaway schematic through the center of the negative hydrogen ion source.

The operation of the source is dependent on a proper temperature for both the anode and the cathode, the proper amount of hydrogen introduced into the arc region, and the proper voltage applied between the anode and cathode to produce an arc. These four variables, the control variables, make up the four independent variables of the control function. The dependent variable is a figure of merit that includes the quietness of the produced beam, the reproducibility of the beam from pulse to pulse as the arc voltage is pulsed, and the total amount of current in the beam produced. This figure of merit is an expression of what human operators indicated to us were the important things they looked for when they operated this device.

G2.7.2.3 General description of the neural network controller

The controller that uses the neural network performs process identification, retains a history of both training and operating data, controls the ion source during identification, tuning, and operation, and keeps the operator fully informed of the status of the process as it proceeds. The neural network module and the control/operator interface module were set up to reside on two different Sun workstations on our local area network. Communication was accomplished through files that were passed between the two processes.

The various computational blocks showing their relationships can be seen in figure G2.7.2. As with most process control, the control system is the heart of it. Requests are made of the control system to take a control variable to a new setpoint, to read a data channel, to set a controller to manual, and so on. In our experiment, the control system contained setpoint control, PID control, alarm enunciators, data logging, and all data readouts. The neural network controller was able to communicate with the control system using the same connections as the operator interface. When it decided a change should be made in one of the four control variables, it sent a request to the control system and the control system carried out the task. The neural network controller block provided for sequencing through the plant identification phase, maintaining the database of training and operating data, requesting training of the neural network

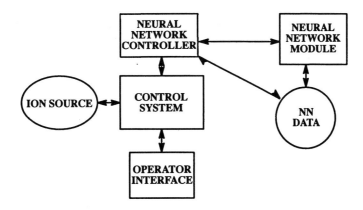

Figure G2.7.2. Block diagram of the controller and control system.

module when required, and reading ion source data from the control system as well as sending changes to the setpoints for the control variables.

G2.7.2.4 Requirements of the neural network controller

We noted early in the study that changing one of the variables, cathode temperature for example, caused changes in other variables. We forced independence of variables by connecting PID (proportional/integral/differential) controllers to the two temperatures and the hydrogen gas flow. A PID controller will control the cathode temperature, for example, by changing the corresponding voltage on the cathode heater to maintain the temperature on the setpoint. We now felt confident that if the model generated by the neural network required a change in cathode temperature, the cathode temperature alone would change.

G2.7.2.5 Mathematical description of the neural network

The neural network we used was developed at Los Alamos in the Center for Non-Linear Science and named the *connectionist normalized local spline* (CNLS) neural network. It is based on a modified *radial* basis function.   C1.6.2

Consider the identity:

$$g(x) = \frac{\sum_{j=1}^{N} g(x)\rho_j(x)}{\sum_{k=1}^{N} \rho_k(x)} \tag{G2.7.1}$$

where $g(x)$ is the unknown multivariable function that represents the output variable we are attempting to control. The modified radial basis function is represented by $\rho_j(x)$ which is defined by

$$\rho_j(x) = \beta_j \, e^{[-\beta_j|x-x_j|^2]}$$

where β_j is related to the inverse of the width and the center of the function is at x_j. Note that the vector notation x indicates a vector of all of the independent variables (dimensions), in our case the anode and cathode temperatures, the arc voltage, and the hydrogen gas flow. The $g(x)$ we are trying to approximate is a figure of merit based on the quality of the beam current waveform and will be discussed in a subsequent section. Expanding $g(x)$ in (G2.7.1) as a Taylor series yields

$$g(x) \simeq \phi(x) = \sum_{j=1}^{N} [f_j + (x - x_j) \cdot d_j] \frac{\rho_j(x)}{\sum_k \rho_k(x)} . \tag{G2.7.2}$$

The reason that $g(x)$ is approximated by $\phi(x)$ is that all of the terms of order greater than one have been dropped from the Taylor expansion. Note that this is a mathematical approximation rather than a heuristic approximation. In equation (G2.7.2), f_j is the zero-order term and d_j is the gradient term. Written this way, we see that these terms can also be regarded as adaptive weights and since they are linear the 'training' is very fast. We also will not get caught in a local minimum. The widths and centers show up in the exponent and consequently are difficult to train. In our work, choosing widths to extend over the full space (i.e. setting the β equal to 1) and setting the centers randomly throughout the space seemed to work well. The iterative fitting algorithms become:

$$f_j^{p+1} = f_j^p + \alpha[g(x_p) - \phi(x_p)] \frac{\rho_j(x) \sum_k \rho_k(x_p)}{\sum_k [\rho_k^2(x_p) + \beta_k(x - x_k)^2 \rho_k^2(x_p)]} \qquad \text{(G2.7.3)}$$

and

$$d_j^{p+1} = d_j^p + \alpha[g(x_p) - \phi(x_p)] \frac{\beta_j(x_p - x_j)\rho_j(x) \sum_k \rho_k(x_p)}{\sum_k [\rho_k^2(x_p) + \beta_k(x - x_k)^2 \rho_k^2(x_p)]} \qquad \text{(G2.7.4)}$$

where α is a 'learning' rate. We set this to 0.1.

In practice, one uses the training data—in our case all the vectors of independent variables—and applies equation (G2.7.2) to obtain corresponding approximate values of the dependent variable. The f_j and d_j are found by applying (G2.7.3) and (G2.7.4) over and over for each value calculated from (G2.7.2) until the difference from one superscript p to the next is minimized in a least-squares sense.

G2.7.2.6 Performance features

Although one wants to react to changes in operation within a control cycle (200 ms for the ion source) in this case we realized that the long settling times for temperatures would allow us to perform whatever calculations were required between the times when the controller could make decisions. The control system that we were interfaced with provided us with an environment that removed the problems of interacting directly with the device hardware. Even reading operating data was a matter of issuing a call for data to the network through subroutine calls. Our only problem was working out a semaphore-like method to indicate when the neural network calculations were finished. There were no real performance constraints.

G2.7.3 Preprocessing of data

We derived the figure of merit from measurements of the beam current waveform. A Faraday cup intercepted the entire beam and the beam current waveform was recorded by the EPICS control hardware. Once it was transferred to the workstation, we calculated a figure of merit by summing three terms extracted from a time window in the middle of the pulse. The first term, which is positive, is the average of the beam current in the time window. The remaining two terms were negative, being formulated as penalties for adverse beam qualities. Beam noise was evaluated by taking the difference between the actual and a low-pass-filtered version of the waveform. We evaluated the pulse-to-pulse variation by computing the rms variation of the beam current integral. These three terms were combined with coefficients that made the controller about equally sensitive to each factor for typical source operation conditions.

G2.7.4 Training methods

Before the ion source could generate a beam the anode and cathode temperatures had to be brought to the proper point. Following this step, the hydrogen valve was opened and the gas was pulsed into the arc region. This was followed shortly afterwards by pulsing of the arc power supply. Once an arc was established, the extraction pulser was slowly ramped to operating level. These steps normally produced a beam, although it was not optimized. An automated recipe was implemented for this sequence. We performed sparse scans of the operating space to initialize the controller after major ion source maintenance. At other times when the machine had been shut down and idle for a short period, e.g. overnight, we were able to use operating data from preceding runs to initiate the optimization.

The first problem was how to cover the space without generating so many data that training and operating would take excessively long. To address this, we first decided to bound each of the variables with a maximum and minimum value determined from operators' experience. Second, each of the variables was 'discretized'. Discretization was probably critical to the success of the project. We divided the ranges of each of the variables into seven discrete parts, numbered from zero to six. The values of our independent variables could only take on the values associated with the zero to six on their scale. This arrangement still contained 2041 (7^4) different coordinates on the control map. Consequently, for training, we also chose to scan the control surface sparsely.

We began the identification phase by generating a set of 36 fairly evenly spaced points on the control surface. This was done by using discrete points 1, 3, and 5 for anode temperature and arc voltage and 1 and 5 for cathode temperature and hydrogen flow rate. We did this by holding three of the variables constant while the fourth was stepped through its discrete points. The next variable was stepped and the first was then backed to its starting point. Since the changes to the arc voltage and the hydrogen flow were effectively instantaneous, a typical identification scan required no more than two hours.

A trial model of the process was built by the neural network using the identification data. Next, the controller asked the network to predict the coordinates for the best figure of merit. The controller then adjusted the control variables to those coordinates. After the ion source was allowed to settle, waveforms of the beam current were taken and a new figure of merit was calculated. This was compared with the prediction. If the difference between the two was greater than a preset convergence criterion, this new observed figure of merit, along with its independent variables, was added to the database and a new model was generated. This iteration cycle from (i) model training to (ii) predicting a new operating point to (iii) data acquisition was repeated until adequate convergence between the measure and predicted figures of merit was obtained. The computation time required to train the network with a set of data was much less than 30 s on a Sun 3 workstation, giving us enough time to generate models from networks configured with centers spread in three different random patterns. We used the network that produced the best result at each step.

Once the network had produced a model that agreed with the data obtained from the ion source (rms error over the whole data set of less than 0.05) the control variables were adjusted to provide the largest figure of merit. Then, to optimize the operation the control variables were adjusted again, using a gradient ascent approach until the figure of merit was maximized. To remove problems of long-term drifts the control variables were occasionally dithered and the maximum figure of merit was again sought by gradient ascent.

G2.7.5 Interpretation of the network output

By using an ordered set of four numbers to indicate the coordinates of each of the independent variables (anode temperature, cathode temperature, arc voltage supply, hydrogen gas flow rate) we can identify an operating point on the control surface. Thus, the set (6352) indicates that the anode temperature is at its highest operating point of discretized values, the cathode temperature is at its median value, the arc voltage is at the second-highest value, and the hydrogen flow rate is two from the lowest. Using this method, on the six different runs we found that the best operating points were at (6641), (6633), (4464), (6133), (6134), and (6160). With so few runs it is difficult to see any trends in these operating points. These six points seem not to be related. However, each of these points was a good operating point from the standpoint of the figure of merit and provided a very acceptable beam. The figures of merit were 0.692, 0.691, 0.691, 0.696, 0.694, and 0.682, respectively. Between the fifth and sixth runs, the ion source was disassembled, cleaned, and reassembled. One might have expected, therefore, that the first and sixth runs would have had figures of merit that would have been much closer to each other.

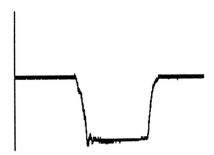

Figure G2.7.3. Single waveform of the beam current on the Faraday cup.

Figure G2.7.3 shows a single waveform of the beam current. For comparison purposes figure G2.7.4 shows a suboptimal waveform with some noise on the beam current. This waveform would be graded down due to this noise. Figure G2.7.5 shows the six optimized waveforms from six different days plotted on top of one another.

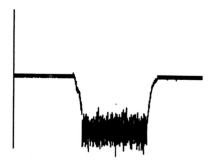

Figure G2.7.4. Single waveform from a suboptimally tuned ion source.

Figure G2.7.5. Six optimized waveforms overplotted.

It is easy to see from this figure that the waveforms produced by the controller optimizing the ion source are quite similar. In fact, there is a less than 10% deviation in these six waveforms. At the end of these six runs the ion source was preempted for ion source development work.

G2.7.6 Development environment

G2.7.6.1 Description of the real-time control system

The real-time control system we used is a software toolkit known as EPICS (experimental physics and industrial control system). The toolkit contains drivers for the various hardware interface modules in use, a graphical operator interface builder, a graphical database builder, and a set of library routines that allow application type programs (C, Fortran, and the like) to interface directly with the control system to obtain data or issue commands, independent of where on the network the device exists. A graphical operator interface window is automatically connected to the control and monitor points that are defined when it is built. The actual interface to the ion source is contained in a VME crate and is controlled by a Motorola 68020. The operating system it uses is the VxWorks system. The crate is connected to an ethernet backbone to which the Sun workstations are also connected. This provides a modular method of connecting various pieces of large experiments together and providing for operator monitoring and supervisory control.

G2.7.6.2 Description of the user interface environment

Several operator interface windows were built for this project using the EPICS graphical interface builder. These provided for operator setpoint changes as well as operator monitoring facilities. These could be tiled on and off the operator workstation.

Another type of user interface was built for the neural network controller that uses the X-window library. This controller performed the actual identification and optimization. Using this interface, the controller is started and stopped. The various phases that the controller moves through are monitored. It was also set up to run in simulation mode. This turned out to be very beneficial when it came to ensuring that the controller was running properly prior to attempting control of the ion source.

G2.7.7 Conclusions

The neural network controller that was built to tune and optimize the operation of a negative hydrogen ion source at Los Alamos has proved itself to be remarkably capable. We have been able to quantitatively map the operating space and detect and compensate for both small and large drifts in source operation. It has tuned the ion source from first arc to a good stable beam in an average time of 2.5 hours on six different days. It has also shown itself capable of recovering from faults (usually a system crash) quickly and with little difficulty. Once it has optimized ion source operation it has maintained good beam quality for 5 hours with no apparent limit. Further experimental effort might provide some indication of why, on subsequent runs, the operation does not end up at the same or a neighboring spot in the operating space. Further work would also undoubtedly uncover many questions that would benefit from further research.

Acknowledgement

This project was funded from internal laboratory research and development funds. The authors are very grateful for that support.

Further reading

1. Hiskes J R, Karo A and Gardner M 1976 Mechanism for negative-ion production in the surface-plasma negative-hydrogen-ion source *J. Appl. Phys.* **47** 3888–95

   A fairly detailed description of the processes that occur in the surface plasma source that produce the charged species. It also contains a description of the ion source itself.

2. Jones R D *et al* 1990 Nonlinear adaptive networks: a little theory, a few applications *Los Alamos National Laboratory Report* LA-UR-91-273

   This text contains much of the theory behind the formulation of the CNLS net along with its relationship to other networks. Some interesting applications are also discussed.

G2.8   Dynamic process modeling and fault prediction using artificial neural networks

Barry Lennox and Gary A Montague

Abstract

This case study presents two practical applications where artificial neural networks (ANNs) have been used to solve difficult process engineering problems. Firstly, ANNs are shown to provide a more accurate process model of a vitrification process than was possible using linear techniques. In the second application ANNs are applied in a novel way in which the residuals of the models are monitored in order to detect the imminent failure of a vessel used in the vitrification process.

G2.8.1   Introduction

Two applications of ANNs are demonstrated in this case study using real process data. Firstly, in section G2.8.3, the methodology followed to develop an accurate ANN model of a vitrification process is demonstrated. Vitrification is the process which encapsulates highly active liquid waste in glass to provide a safe and convenient method of storage. The second application, detailed in section G2.8.4, again employs ANNs, but this time they are applied in a novel way in which they are used to capture nonlinear system characteristics and then recalled to provide a means of detecting the imminent failure of a vessel used in the same vitrification process.

The following section provides a detailed description of the process which has been studied in this work.

G2.8.2   Process description

The system under investigation in this work concerns a vitrification process operated by British Nuclear Fuels Limited at their Sellafield site in Cumbria. This process encapsulates highly active liquid waste obtained in the reprocessing of spent nuclear fuel elements in glass to form solid blocks of waste for safe and convenient storage.

The process is a two-stage semicontinuous operation. The liquid waste is initially fed continuously into the first stage of the process, known as the calciner. The calciner is a long, cylindrical vessel which is rotated inside a heated furnace. As the liquid waste flows down this vessel it is successively evaporated, dried, and partially denitrated. The resulting dry powder, known as calcine, is then discharged under gravity into the second stage, the melter vessel. This vessel is elliptically shaped and heated by electrical induction coils. After every 10 minute period the wall temperatures in the melter are compared to a high and low preset temperature limit. The power supplied to the induction coil is then adjusted accordingly using a PLC controller.

Glass frit is also fed continuously into the melter vessel, in which it forms a molten mixture with the calcine. The level in the melter rises steadily until a certain point when the contents of the vessel are discharged into a product storage container positioned below the melter. This container is then sealed, cleaned, and moved to the vitrification product store. The operation of discharging the melter contents is known as 'pouring'.

Heat transfer mechanisms in the melter are complex and highly dependent upon the contents of the vessel. During pouring, molten waste will be drained from the vessel resulting in an increase in the heat transfer from the vessel walls to the melt. This causes a sudden increase in the melter power requirement. Unfortunately, the response of the control system to the increased power requirement is slow and therefore the melter wall temperature falls sharply.

This large thermal disturbance exerts significant thermal stresses on the walls of the melter vessel. These periodic stresses are thought to have been responsible for a small number of these vessels fracturing before their full life expectancy had been reached. These fractures resulted in increased downtime costs as well as extra costs incurred in the disposal of the radioactive vessel itself.

In an attempt to reduce these costs BNFL is investigating techniques for improving the present control strategy for this process by utilizing both linear and nonlinear models.

The objectives of this study were firstly to attempt to develop an accurate model of the process and secondly to provide a technique which could allow the detection of imminent vessel failure. The next section in this paper describes, in detail, the procedure which was followed to develop a model capable of predicting the wall temperature of the melter vessel.

G2.8.3 Model development

G2.8.3.1 Process data

The raw data supplied for this modeling exercise were the temperature of the melter, the power supplied to the induction coil, and the level of waste in the vessel. Previous studies by BNFL had shown that the vessel temperature was dependent upon the power supply and also the level of waste in the vessel. These measurements were supplied from the *histories* of three melter vessels.

G2.8.3.2 Modeling results using artificial neural networks

The objective of this study was to develop an ANN model of the process which could then be compared to the prediction accuracy of a linear model.

The term *artificial neural network* encompasses a massive range of model structures and architectures (Lippman 1987). The choice of architecture is very much problem specific; however, for the mapping of nonlinear systems a layered architecture is used ordinarily. This form of architecture is commonly referred to as a *feedforward network* and comprises an input layer, which introduces the input variables into the network, the output layer, from which the network outputs are obtained, and one or more hidden layers located in between the input and output layers.

B2.3

Since the melter vessel is clearly a dynamic system, a dynamic element must also be incorporated into the neural network. This is typically achieved by utilizing a time series of input variables in the same manner as used in linear modeling. This technique can, however, lead to large numbers of input variables and network weights, which in turn leads to very long and inefficient training times. A more elegant approach to introducing dynamics into the ANN model is to pass the output of the input and hidden layer nodes through a first-order low-pass filter. A discrete time representation of these filters transforms the neuron output as follows:

$$y^f(t) = \Omega y^f(t-1) + (1-\Omega)y(t).$$

The values of the filter time constants Ω are not known *a priori* and must therefore be determined along with the network weights, when the network is trained.

Once the network architecture and topology have been selected the network is trained on actual process data. The aim of the training procedure is to reduce the sum of the squared difference between predictions of the network and the desired output over the training data set. The training algorithm used in this study was the Levenberg–Marquardt search direction method. This was used in preference to the more commonly used *backpropagation* training algorithm because it has been shown that this algorithm can significantly reduce training times (Demuth and Beale 1994).

C1.2.3

When training an ANN it is possible, due to the efficiency of the training algorithm, to minimize prediction errors greatly and hence fit training data sets with extreme accuracy. This can occur to such a degree that the ANN model will begin to fit secondary system characteristics such as noise and measurement errors. A network trained to such accuracy will be too specific to the training data set and will generalize poorly when applied to other plant data. In order to prevent this occurrence some form of model validation is typically employed during the training procedure.

In this study, the prediction accuracy of the network was measured over a validating data set at periodic intervals during the training procedure. The training procedure was terminated when the prediction accuracy of the model over the validating data set began to increase. At this point the network weights were stored. The network was then finally tested by measuring the prediction accuracy over a testing data set. It is important that within the training data set all the system characteristics are represented and that the testing and validating data sets contain a similar quality of data as used to train the model and at least half the quantity of data.

To summarize, the network architecture used to model this process was a feedforward network with dynamic processing in the nodes. The inputs into the model were the power supplied to the induction coil and the level of waste in the melter vessel and the output of the model was the temperature of the melter. The best ANN modeling results were obtained using five nodes in a single hidden layer. The prediction errors over the training, validating, and testing data sets obtained using this model were 8.8 °C, 9.5 °C, and 9.9 °C, respectively.

These figures compare with 11.4 °C, 9.7 °C, and 11.3 °C obtained using a simple linear autoregressive model with exogeneous signal (ARX). Figure G2.8.1 compares the actual wall temperature of the vessel with that predicted using both the ARX and the ANN models. It is evident from this graph and also from the error statistics that the neural network is slightly better able to model this system than the linear technique. Other linear modeling techniques, such as autoregressive with moving average and exogeneous models, were also investigated and found to be outperformed by the ARX model.

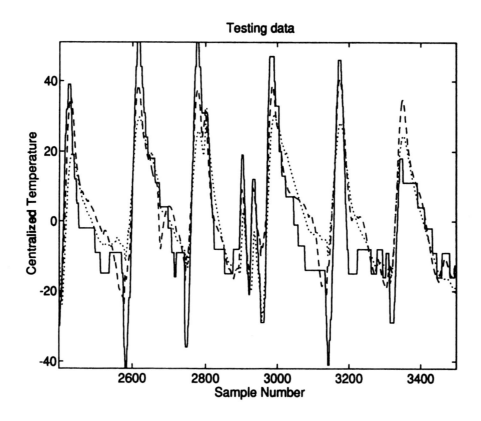

Figure G2.8.1. Comparison of linear and ANN model predictions.

The objectives of this work were not only to develop an accurate model of the process but also to investigate whether it was possible to detect signs of imminent vessel failure by analyzing the process data. This next section details the development of a condition monitoring system for just such a purpose.

G2.8.4 Application of neural networks to condition monitoring

As described earlier in this case study, as a result of the thermal stresses the melter vessels are subjected to, a small number of vessels failed before their full life expectancy had been reached. These unexpected melter failures incur large disposal and handling costs for BNFL and it is therefore desirable for a system to be developed which could predict when these failures may occur.

It was postulated that by studying the thermal characteristics of the melter an indication of the present condition of the melter vessel could be obtained. As a melter aged it was expected that due to the distortion of the melter vessel the thermal characteristics of the melter would possibly change. This distortion should bring the vessel walls closer to the induction heating coils, thereby improving heat generation and hence similar temperatures are achieved with apparently less power. It was therefore believed that if an accurate temperature prediction model could be generated to model the early thermal characteristics of the vessel, then as the thermal characteristics changed the prediction accuracy of the model would begin to deteriorate until the point of melter failure. This model or, rather, the prediction error obtained using this model, could be used as an indication of forthcoming melter failure.

The previous section illustrated the suitability of ANNs for modeling the melter process. It identified a feedforward neural network with localized dynamics as the most suitable network architecture with which to model the process. Therefore, this is the model which was used throughout this condition monitoring study.

To investigate the relationship between the age of the melter and the melter's thermal characteristics, ANN models were trained using data collected from the early stage of a melter life; the melter used for this initial work was known as melter 2. The prediction accuracy of these models was then tested on data collected later in the vessel's life. The prediction errors produced by these models were determined for each individual pour in the melter's life. These error statistics were then analyzed to see whether there were any signs of the vessel aging present in the prediction errors of the model.

The methodology used in this work was to train the ANN on a series of 12 'pours' collected at the start of the melter life. This model was then tested using the following six pours and finally the RMS errors obtained using this model were monitored over the life of the vessel. Initially, this hypothesis was tested on a melter vessel which actually failed before its full life expectancy was reached. It was found that by plotting the average of the last five melter batch RMS errors calculated throughout the life of the melter vessel, as shown in figure G2.8.2, it became clear that there was a trend in the error profile towards the end of the melter lifetime. The performance of the ANN model deteriorates as the point of vessel failure is approached. Investigations on two more melter vessels confirmed that signs of melter failure could be detected using this methodology.

In summary, it would appear from the investigation of three melter vessels that signs of imminent melter failure are visible in the error trends of the temperature prediction. It is also evident that this vessel failure seems to occur when the RMS error profile reaches approximately twice the error obtained over the training and testing data sets.

G2.8.5 Conclusions

This contribution has shown that the thermal characteristics of the melter vessel used in the vitrification process can be modeled successfully by utilizing the techniques of artificial neural networks. Prediction errors were found to be lower when using ANNs to model the process rather than a linear ARX model. This is due to the ability of ANNs to capture the system nonlinearities present in this process.

This contribution has also described a novel condition monitoring method which was devised for the melter vessel. This procedure involved training an ANN model on the early thermal characteristics of the melter vessel and then monitoring the prediction error produced by this model later in the lifetime of the melter. The prediction accuracy of this model was found to deteriorate significantly towards the end of the life of two melter vessels, clearly indicating a change in thermal characteristics.

The potential for using ANNs as a condition monitoring tool for the melter process has been illustrated. Further development work is now required in the form of testing the developed condition monitoring procedure on data collected from other melter vessels.

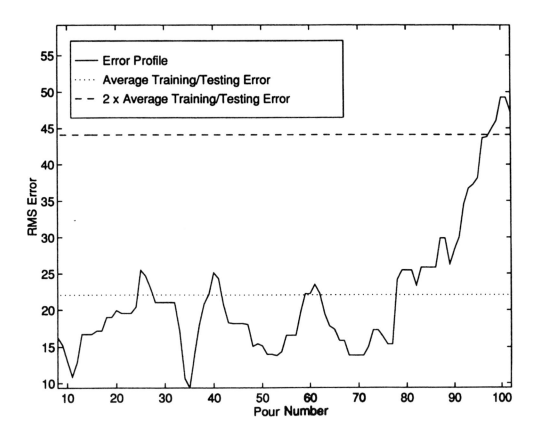

Figure G2.8.2. RMS error profile for melter 2.

Acknowledgements

The authors would like to acknowledge the financial assistance of the Department of Trade and Industry and the industrial members of the *NeuroControl* Club. The authors are also grateful to the contribution made by the Department of Chemical and Process Engineering at Newcastle University and to Bill Harper and Craig Haughin at BNFL.

References

Demuth H and Beale M 1994 *Neural Network Toolbox for MATLAB* (MathWorks)
Lippman R P 1987 An introduction to computing with neural nets *IEEE ASSP Mag.* 4–22

G2.9   Neural modeling of a polymerization reactor

Gordon Lightbody and George W Irwin

Abstract

Model predictive control techniques such as generalized predictive control (GPC) (Clarke *et al* 1987) and dynamic matrix control (DMC) (Rovlak and Corlis 1990) have proven successful when applied to the control of industrial processes. It has been demonstrated that such linear predictive control techniques can be improved by including nonlinear system models (Morningred *et al* 1990). In particular, both GPC (Montague *et al* 1991) and DMC (Hernandez and Arkun 1990) have been extended by utilizing a nonlinear neural predictive model of the process. This industrial case study focuses on the application of neural modeling to improve the control of a polymerization reactor. The industrial system is introduced, highlighting the problems of accurate polymer viscosity control, based on a delayed measurement. This work presents a nonlinear predictor developed around the multilayer perceptron that can be used to remove this measurement delay. Finally, a platform is proposed to allow for the on-line implementation of neural-network-based predictive controllers.

G2.9.1   Project overview

The polymerization reactor is essentially a continuously stirred tank reactor, into which a number of constituent ingredients are fed. The contents of this reactor are continuously stirred, using a variable-speed drive, for which both measurements of speed and torque are available. On-line measurements are also provided for all the flow rates and the viscosity of the polymer product. It is the objective of the control system to regulate the product viscosity, keeping it constant and immune from disturbances, particularly those due to feed-rate changes. Two catalysts are added to the reactor, a compound C_A which promotes polymerization and a compound C_B which acts to inhibit polymerization. Hence by increasing the ratio of the flow rate of C_A to that of C_B, the probability of longer polymer chains is higher and hence the viscosity is increased. For this plant the flow rate of catalyst C_A and the flow rates of all the other constituent compounds along with the speed of the variable-speed drive are set for specific feed rates, with the flow rate of the inhibitor catalyst C_B manipulated to regulate the viscosity. A cascaded PID control structure is used here, with the faster inner loop operating from the motor torque measurement (which essentially is a measure of viscosity) and the outer loop utilizing the slower and more accurate measurement of viscosity provided by an on-line viscometer. The present viscosity control system is as shown in figure G2.9.1.

From a detailed analysis of plant data within the Matlab package, the reactor could be represented by two separate subsystems (figure G2.9.1). The first subsystem, S1, represents the process within the reactor itself and how the torque depends on the flow rate of catalyst C_B and on the various disturbances affecting the plant. The second subsystem, S2, represents the change in viscosity between the reactor and the viscometer and the transformation of torque to viscosity. As such, all the disturbances affect the first subsystem and are reflected by fluctuations in the torque which are then passed into the second subsystem to cause fluctuations in the viscosity.

It was determined that there was present in S2 a significant pure time delay which was recognized as being primarily responsible for the problems in providing accurate control for this system. In order to

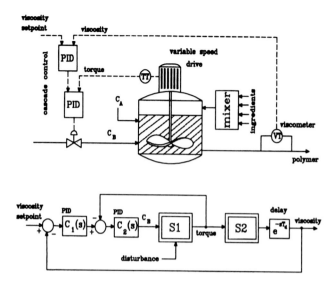

Figure G2.9.1. The polymer viscosity control structure.

obtain a value for this time delay the linear ARX model structure of (G2.9.1) was assumed to model the dynamic relationship between the torque signal and the measured viscosity. Here, $y(k)$ and $u(k)$ represent the viscosity and torque, respectively, with $e(k)$ being the prediction error:

$$A(z^{-1})y(k) = z^{-d}B(z^{-1})u(k) + e(k). \tag{G2.9.1}$$

The A and B polynomials in the delay operator are assumed to be both of fixed order m. The data are split into a modeling and a test set. The dead time d, which resulted in the optimum generalization results over the test data set, was then assumed to be the best estimate of the actual system dead time. In this particular problem, with a sample time of one minute, and with the order $m = 6$, a system dead time of three minutes resulted.

G2.9.2 Predictor design process

G2.9.2.1 Neural predictive modeling

C1.2 Due to the nonlinear nature of the plant, it was proposed that a *multilayer perceptron* (MLP) be trained to predict polymer viscosity from past torque and viscosity data and hence remove the three minute time delay. Data were collected from the distributed control system (DCS) and then analyzed using Matlab. The viscosity and torque data were conditioned using a third-order low-pass Butterworth filter, then normalized so that both torque and viscosity sequences were constrained to the range $[-1.0, 1.0]$. These normalized, filtered data were then decimated by a factor of six to yield a sample time of one minute. The model structure of (G2.9.2) was proposed, with a multilayer perceptron utilized to form the nonlinear function. Here $T(k)$ and $v(k)$ represent the normalized torque and viscosity measurements:

$$\hat{v}(k+3|k) = \hat{f}(\phi(k), \psi(k))$$
$$\phi(k) = [v(k), v(k-1), \ldots, v(k-n+1)]^{\mathrm{T}} \tag{G2.9.2}$$
$$\psi(k) = [T(k), T(k-1), \ldots, T(k-m+1)]^{\mathrm{T}}.$$

G2.9.2.2 Training algorithms

The training of the multilayer perceptron network to approximate this function represents a nonlinear
C1.2.3 optimization problem. Steepest-descent-based algorithms, such as *backpropagation*, have been shown
B3.4.4.4 to be restrictively slow and subject to local minima. Second-order techniques, such as the *Levenberg–*
B3.4.4.4 *Marquardt* method (Ruanno *et al* 1991) or the *Broyden–Fletcher–Goldfharb–Shanno* (BFGS) method (Battiti and Masulli 1990), have been found to provide a significant acceleration of the training process

over backpropagation. In this work the memoryless version of the BFGS algorithm was used. This is a batch method in which the cost is as determined in (G2.9.3), where $w(k)$ represents the present weight vector for the network, N_{T1} is the size of the training set with $\hat{v}(i)$ and $v(i)$ the estimated and actual viscosities, respectively:

$$J(w(k)) = \frac{1}{2N_{T1}} \sum_{i=1}^{N_{T1}} (\hat{v}(i) - v(i))^2. \tag{G2.9.3}$$

The weight update equation, utilizing the memoryless BFGS algorithm, is as summarized below, where the gradient of the cost is determined at each instant using batch backpropagation.

$$
\begin{aligned}
g(k) &\triangleq \frac{\partial J(w(k))}{\partial w(k)} \\
p(k) &= g(k) - g(k-1) \\
s(k) &= w(k) - w(k-1) \\
r(k) &= -g(k) + \lambda_1 p(k) + \lambda_2 s(k) \\
\lambda_1 &= \frac{s(k)^\mathrm{T} g(k)}{s(k)^\mathrm{T} p(k)} \qquad \lambda_2 = \frac{p(k)^\mathrm{T} g(k)}{s(k)^\mathrm{T} p(k)} - \lambda_1 \left(1 + \frac{p(k)^\mathrm{T} p(k)}{s(k)^\mathrm{T} p(k)}\right) \\
w(k+1) &= w(k) + \eta(k) r(k).
\end{aligned}
\tag{G2.9.4}
$$

The step size $\eta(k)$ is chosen on each iteration using an efficient single-line search technique. In extended tests for a wide variety of nonlinear approximation example problems this training algorithm was found to be typically 20 times faster than standard batch backpropagation and was less subject to the problems of local minima. Likewise, this algorithm was found to provide consistently comparable performance to the *Fletcher–Reeves conjugate gradient technique* with optimal reset value (Irwin *et al* 1994). The choice of this reset value is not straightforward and has been found to greatly affect the performance of conjugate gradient algorithms. As such, the use of the memoryless version of the BFGS algorithm is to be recommended, due to its speed and its ease of use, as it requires no reset value and the gain choice is automatic.

B3.4.4.4

A parallel version of the memoryless BFGS algorithm was then devised, taking advantage of the concurrency present in the training set. To improve the processing efficiency, a novel parallel single-line search routine was developed. This training algorithm was mapped efficiently onto a pipeline of six T800 transputers mounted on a Niche platform connected to a Sun 4/330 server. It was found that this parallel algorithm could typically reduce the training time of the multilayer perceptron to 1 per cent of that achieved using standard batch backpropagation (Lightbody and Irwin 1992).

G2.9.2.3 Neural predictive modeling of viscosity

From the data available, a range of training and test sets was generated, corresponding to a number of possible model orders. It was assumed that the number of hidden units was fixed as ten hyperbolic tangent nodes. A model structure was selected with orders $n = 6$ and $m = 3$, that provided the lowest generalization cost. For this model structure, the number of hidden units was selected in a similar manner by training a wide range of networks and selecting the one that best balanced generalization performance against network size. Using this technique a network with an MLP(9:14:1) structure was decided upon, with a linear output neuron. Figure G2.9.2 shows the response of the resultant neural model over the training set.

When the network was applied to the test data, it was found that although the high- and middle-frequency dynamics were accurately reproduced, there was a low-frequency or DC offset. Figure G2.9.3 shows the response of the neural predictor over the test set. This was compensated for by utilizing the present output of the plant, $v(k)$, and the predicted estimate of the viscosity, $\hat{v}(k|k-3)$, to generate a correction term $d(k)$, as in (G2.9.5):

$$d(k) = v(k) - \hat{v}(k|k-3). \tag{G2.9.5}$$

This correction $d(k)$ was filtered to increase immunity to noise and to ensure that it reflected unmodeled low-frequency errors. The correction term can then be expressed as in (G2.9.6), where $T(z^{-1})$ represents a suitable filter:

$$\hat{v}_\psi(k+3|k) = \hat{v}(k+3|k) + T(z^{-1})d(k). \tag{G2.9.6}$$

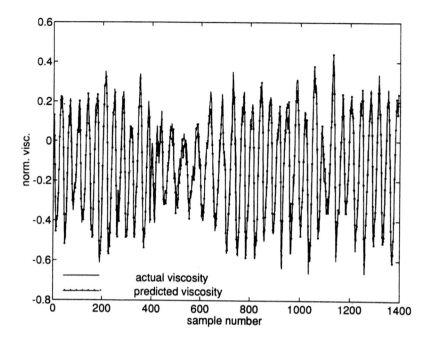

Figure G2.9.2. The response over the training set.

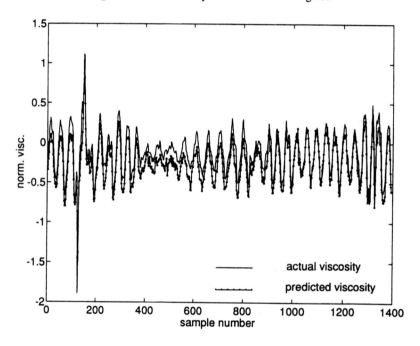

Figure G2.9.3. Response of the neural predictor over the test set.

The complete neural predictive estimator is given in figure G2.9.4, including low-pass Butterworth filters and normalization at the inputs, tapped delay lines to provide the past window of data, and a multilayer perceptron with structure MLP(9:14:1), to provide the nonlinear function, and a correction filter to remove DC and low-frequency offsets. The filter $T(z^{-1})$ was chosen to be the simple first-order filter of (G2.9.7):

$$T(z^{-1}) = \frac{0.1}{1 - 0.9z^{-1}}. \qquad (G2.9.7)$$

When applied to the data of the test set this corrected predictor provided excellent results, predicting accurately over the measurement delay as shown in figure G2.9.5.

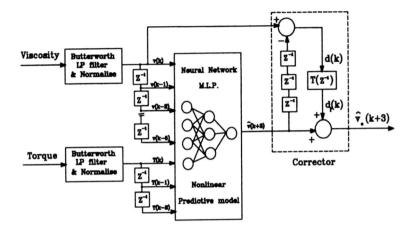

Figure G2.9.4. The complete neural Smith predictor for viscosity.

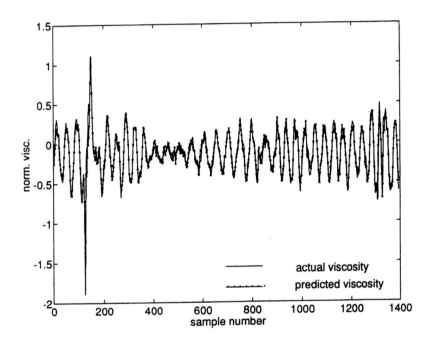

Figure G2.9.5. The response of the corrected predictor over the test set.

G2.9.2.4 On-line implementation of a neural viscosity predictor

Many distributed control systems (DCSs) do not have the capability to allow the implementation of sophisticated algorithms, such as neural network models. To facilitate the development of on-line predictive models and for on-line training of neural networks, a hardware platform was developed. This was based on a personal computer, running Lab-Windows software and connected to the DCS system using a data acquisition board. This software offered a powerful environment for the development of neural models and also allowed for control in software of the data acquisition board. The structure is as described in figure G2.9.6.

In this manner, it was not necessary either to add or to break connections between the DCS and the plant. Separate channels were set up in software within the DCS, so that key measurements could be copied onto these channels and hence accessed, via the interface, by the software on the development platform. Similarly, outputs from the development platform could appear both on the screen of the personal computer and via the extra DCS channels, set up as inputs, on the operating console. In this way, they would be treated and logged as if they were process variables.

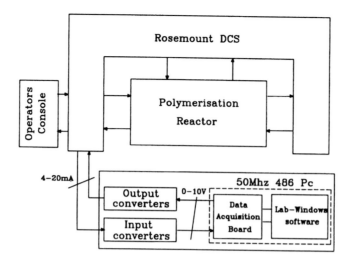

Figure G2.9.6. The development platform structure.

G2.9.3 Conclusions

This work has presented an industrial polymerization reactor as a suitable case study to demonstrate the potential of neural modeling. The viscosity control structure was discussed, highlighting the problems introduced by the measurement delay at the viscometer. A neural network was trained to predict over this three minute measurement delay, using past torque and viscosity measurements. The importance of a correction filter was demonstrated for the removal of errors caused by the presence of low-frequency unmodeled system dynamics. Finally, this predictor was implemented on-line, interfaced to the plant DCS using a commercial data acquisition board.

Acknowledgement

The financial support of du Pont (UK) PLC and the Industrial Research and Technology Unit (IRTU), is gratefully acknowledged.

References

Battiti R and Masulli F 1990 BFGS optimisation for faster and automated supervised learning *Proc. Int. Neural Net. Conf.* vol 2, pp 757–60

Clarke D W, Mohtadi C and Tuffs P S 1987 Generalised predictive control—Part 1. The basic algorithm *Automatica* **23** 137–48

Hernandez H and Arkun Y 1990 Neural network modelling and an extended DMC algorithm to control nonlinear systems *Proc. ACC* vol 2, pp 2454–9

Irwin G W, Lightbody G and McLoone S F 1994 Offline training of feedforward neural networks *Proc. Irish DSP and Control Conf., IDSPCC'94 (Dublin)*

Lightbody G and Irwin G W 1992 A parallel algorithm for training neural network based nonlinear models *Proc. 2nd IFAC Symp. on Algorithms and Architectures for Real-Time Control (S Korea)*

Montague G A, Willis M J, Tham M T and Morris A J 1991 Artificial neural network based control *Proc. IEE Int. Conf. on Control* vol 1, pp 266–71

Morningred J D, Paden B E, Seborg D E and Mellichamp D A 1990 An adaptive nonlinear predictive controller *Proc. ACC* vol 2, pp 1614–9

Rovlak J A and Corlis R 1990 Dynamic matrix based control of fossil power plants *Proc. Int. Joint Power Gen. Conf. (Boston, MA)*

Ruanno A E B, Fleming P J and Jones D I 1991 A connectionist approach to PID auto-tuning *Proc. IEE Int. Conf. on Control* vol 2, pp 762–8

G2.10 Adaptive noise canceling with nonlinear filters

Wolfgang Knecht

Abstract

Standard adaptive noise canceling uses linear filters to minimize the mean-squared difference between the filter output and the desired signal. For non-Gaussian signals, however, nonlinear filters can further reduce the mean-squared difference, thereby improving signal-to-noise ratio at the noise canceler output. This work investigates a two-microphone beamformer for suppressing directional background noise—an important task in, for example, radar, seismic or hearing aid applications. The beamformer includes an adaptive noise canceler with a nonlinear filter. Two nonlinear filters are examined: the Volterra filter (a specific sigma-pi neuron) and the multilayer perceptron. In the case of a single noise source emitting an independent, identically distributed (IID) random process, optimum linear and nonlinear performance limits are known for uniformly distributed noise. These limits were compared to the actual performance of the two nonlinear filters adapted off-line. The third-order Volterra filter and the perceptron with 20 hidden neurons performed equally well. For on-line adaptation, convergence speed and steady-state performance were scrutinized. In these experiments, the RLS-adapted Volterra filter outperformed the perceptron adapted with on-line backpropagation.

G2.10.1 Introduction

The Bayes conditional mean is the optimum filter for the mean-squared error (MSE) criterion. Generally, the optimum filter output is a nonlinear function of the observed data. An important exception exists: when the observed data and the data to be estimated are jointly Gaussian then the Bayes filter is linear (Papoulis 1991). In the following, we note several equations from Bayes estimation theory which are relevant to our application.

Suppose we measure a random data vector

$$\boldsymbol{X}(k) = (X(k), X(k-1), \dots, X(k-M))^{\mathrm{T}} \qquad \text{(G2.10.1)}$$

at time k. The data vector $\boldsymbol{X}(k)$ consists of successive components of the stochastic process $X(\cdot)$. The number M is called the filter length. The task is to find an estimate $\hat{S}(k)$ of a random variable $S(k)$ based on the data vector $\boldsymbol{X}(k)$ such that

$$\mathrm{MSE} = E[(S(k) - \hat{S}(k))^2] \qquad \text{(G2.10.2)}$$

is minimized. The symbol $E[\cdot]$ denotes the expectation operator. To simplify notation, the time argument k is omitted in the following discussion. If the conditional probability density function $p_{S|\boldsymbol{X}}(\cdot|\boldsymbol{x})$ is known, the optimum (Bayes) filter estimates S from a given data vector $\boldsymbol{X} = \boldsymbol{x}$ by

$$\hat{s}_B(\boldsymbol{x}) = E[S|\boldsymbol{X} = \boldsymbol{x}] = \int\limits_{-\infty}^{+\infty} s \, p_{S|\boldsymbol{X}}(s|\boldsymbol{x}) \, \mathrm{d}s. \qquad \text{(G2.10.3)}$$

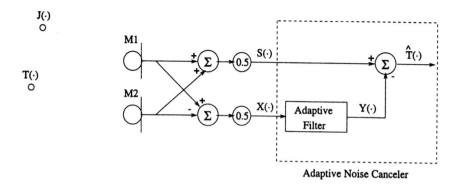

Figure G2.10.1. Two-microphone beamformer for suppressing directional background noise.

The Bayes estimator yields the minimum mean-squared error (MMSE) defined by

$$\text{MMSE} = \int_{-\infty}^{+\infty} ds \int_{R^{N+1}} dx \, (s - \hat{s}_B(x))^2 \, p_{S,X}(s, x) \tag{G2.10.4}$$

where $p_{S,X}(\cdot, \cdot)$ is the joint probability density function.

Adaptive noise canceling and adaptive beamforming will now be viewed within the framework of Bayes estimation theory. Consider the adaptive beamformer (Griffiths and Jim 1982) with microphones M1 and M2 in figure G2.10.1. The target source emitting the signal $T(\cdot)$ is equidistant from M1 and M2. An off-axis jammer signal $J(\cdot)$ impinges on the microphones with a time delay Δ between M1 and M2. We model both signals $T(\cdot)$ and $J(\cdot)$ as stochastic processes. The scaled difference between the two microphone signals $X(\cdot) = \frac{1}{2}(J(\cdot) - J(\cdot - \Delta))$ contains no target components and is the reference input to the adaptive noise canceler (Widrow and Stearns 1985). The scaled sum of the signals $S(\cdot) = T(\cdot) + [(J(\cdot) + J(\cdot - \Delta)]/2$ is the primary input to the noise canceler. Assuming that $T(\cdot)$ and $J(\cdot)$ are independent, the beamformer produces a target estimate $\hat{T}(\cdot)$ by minimizing its output power $E[(S(k) - Y(k))^2]$ for all k, where $Y(\cdot)$ is the output of the adaptive filter. The signal $\hat{T}(\cdot)$ is called the 'minimum variance distortionless estimate' of the target signal because the beamformer attenuates the interference without affecting the target. It must be emphasized, however, that this holds true only when no target components exist in the reference channel. A misalignment of the target location, or a microphone gain mismatch, will violate this condition and, consequently, the system will partially cancel the target.

G2.10.2 Nonlinear filtering

The conditional probability density required for the calculation of the Bayes filter (G2.10.3) is usually not available for real signals. Consequently, the (unknown) Bayes filter function must be approximated. This work employs the *perceptron* and the *Volterra filter* as the adaptive filter of the beamformer in figure G2.10.1. Recently, these two filter architectures have been chosen quite often for approximating nonlinear functions. Both filters will attempt to approximate the Bayes filter. For non-Gaussian signals, the Bayes filter is generally nonlinear so that a nonlinear filter architecture is required to reach the MMSE.

The perceptron filter is a simplified version of the *time-delay neural network* (TDNN) proposed by Waibel *et al* (1989). The Volterra filter can be considered as a single sigma-pi or higher-order neuron employing the identity function to its weighted and summed inputs. Replacing the activation function of the perceptron filter by a polynomial leads to the Volterra filter for which optimum weights can be calculated (Knecht 1994).

G2.10.2.1 *The perceptron filter*

We use a fully interlayer-connected perceptron with one hidden layer. It has one output neuron whose activation function is the identity function. For the input vector $X(k)$, the output of the perceptron filter at time k is

$$Y(k) = \theta_3 + \sum_{i=1}^{N_2} w_{2i,3} \tanh[w_{1,2i}^{T} X(k) + \theta_{2,i}] \tag{G2.10.5}$$

where the weight vector $w_{1,2_i}$ contains the weights from the input layer $l = 1$ to the ith neuron in the hidden layer $l = 2$ and the superscript T denotes the matrix transpose. The weights connecting the hidden layer to the output are denoted by $w_{2,3}$. The biases of the hidden neurons are designated by $\theta_{2,i}$ and the bias of the output neuron is θ_3. Finally, the total number of hidden neurons is N_2.

In the off-line experiment, the perceptron filter was adapted with the Levenberg–Marquardt algorithm. This technique has been shown to be more efficient than backpropagation with adaptive learning rate or conjugate gradient backpropagation provided that the total number of weights is limited to a few hundred (Hagan and Menhaj 1994). In the on-line experiment, we adapted the perceptron filter with standard backpropagation without momentum and with a fixed learning rate. Note that both off-line and on-line algorithms cannot guarantee finding the global minimum of the mean-squared error surface in weight space.

G2.10.2.2 The Volterra filter

The polynomial or Volterra filter is one of the most popular nonlinear filter realizations. It has been used in various applications including channel equalization, echo or noise cancelation, and distortion analysis in semiconductor devices. For tutorials on this filter and its training, see Mathews (1991) and Sicuranza (1992) which also list references for these applications.

For the input vector $X(k)$, the Pth-order Volterra filter of length M yields the output

$$Y(k) = h_0 + \sum_{p=1}^{P}\sum_{n_1=0}^{M}\sum_{n_2=n_1}^{M}\cdots\sum_{n_p=n_{p-1}}^{M} h(n_1,\ldots,n_p)X(k-n_1)\cdots X(k-n_p) \qquad (G2.10.6)$$

with $n_0 = 0$. The kernels or weights are denoted by h_0 and $h(n_1,\ldots,n_p)$. The filter output depends linearly on the weights such that the mean-squared error surface in weight space is a hyper-paraboloid with a single minimum. This fact has a very useful consequence, that is, adaptive Volterra filters can be described by linear adaptive filter theory.

The off-line calculation of the optimum Volterra weights for a given order P is as follows. We rewrite (G2.10.6) as

$$Y(k) = h^T X_e(k) \qquad (G2.10.7)$$

where

$$h^T = [h_0, h_1(0),\ldots,h_1(M), h_2(0,0),\ldots,h_P(M,\ldots,M)] \qquad (G2.10.8)$$
$$X_e(k) = [1, X(k),\ldots,X(k-M), X(k)^2,\ldots,X(k-M)^P]^T. \qquad (G2.10.9)$$

Analogously to linear filter theory, the optimum weights solve the 'extended' Wiener–Hopf equations:

$$E[X_e X_e^T]h = E[X_e S]. \qquad (G2.10.10)$$

In section G2.10.3, the expectation $E[\cdot]$ was approximated by averages over time assuming ergodicity. The Volterra filter was adapted on-line with the standard LMS and RLS algorithms described in Mathews (1991).

G2.10.3 Experiments and results

For all experiments in this section, we chose the jammer delay $\Delta = 1$ between microphones. The jammer signal $J(\cdot)$ consisted of independent, identically distributed (i.i.d.) samples with a uniform probability density function. For this particular jammer, we calculated the Bayes filter and the corresponding MMSE according to (G2.10.3) and (G2.10.4). The Bayes filter achieves MMSE $= 12/(N+3)(N+4)$, while the best linear filter reaches only the suboptimal MSE $= 2/(N+2)$. The derivations of these formulas can be found in Knecht (1995). They are not included here because they are quite complex and would not contribute to the understanding of the main concepts of this section. Note that all mean-squared errors are normalized to the variance of the primary signal which is one half of the jammer variance.

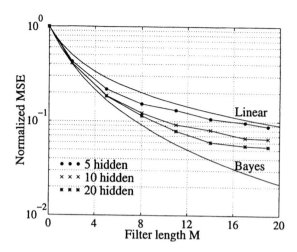

Figure G2.10.2. Normalized jammer power (MSE) at the beamformer output for perceptron filters with 5, 10 and 20 hidden neurons versus the filter length $M = N_1 - 1$.

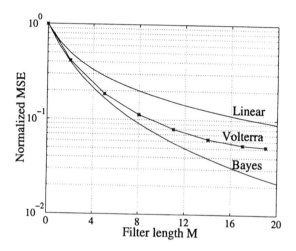

Figure G2.10.3. Normalized jammer power (MSE) at the beamformer output for the third-order Volterra filter versus the filter length $M = N_1 - 1$.

G2.10.3.1 Off-line experiment

We examined the ideal situation where the target signal remains unaffected by the beamformer. Because no target components exist in the reference channel, the optimum linear and nonlinear filters do not depend on the target. Therefore, the target signal was set to zero.

The weights and biases of the perceptron filter were initialized with a simple and effective method described in Nguyen and Widrow (1990). The MATLAB™ routine TRAINLM implements the Levenberg–Marquardt algorithm and was used in this experiment to adapt the weights and biases. The training set consisted of 6000 input vectors $X(1), \ldots, X(6000)$. For each number of hidden neurons N_2 and for each number of input neurons N_1, the filter was adapted over 80 epochs. After training, a test set of 100 000 input vectors was filtered by the perceptron and the test MSE was determined by averaging the squared beamformer output samples. The results are summarized in figure G2.10.2.

The optimum weights of the third-order ($P = 3$) Volterra filter were calculated according to (G2.10.10) with 6000 extended input vectors $X_e(1), \ldots, X_e(6000)$. Note that for i.i.d. jammers with symmetric probability density functions, the Volterra weights belonging to even order components of X_e vanish. Hence, the third-order Volterra filter was used without second-order terms. As for the perceptron, a test set of 100 000 input vectors was processed by the beamformer with the fixed optimum Volterra weights. The normalized test MSE is depicted in figure G2.10.3.

Table G2.10.1. Maximum learning rates, steady-state on-line and off-line normalized MSEs of the linear and various nonlinear filters. For the perceptron, the entries represent 'quasi' steady-state MSEs (see text). The off-line results were taken from the off-line experiment in the previous section. The filter length was $M = 8$.

Filter	Maximum learning rate	Steady state MSE	Offline MSE
Linear FIR, LMS	0.02	0.2068	0.2000
Perceptron, $N_2 = 5$	0.01	0.1866	0.1519
Perceptron, $N_2 = 20$	0.005	0.1962	0.1136
Volterra, LMS	0.01	[a]	0.1129
Linear FIR, RLS	—	0.2050	0.2000
Volterra, RLS	—	0.1131	0.1129

[a] The LMS-adapted Volterra filter did not converge within the sample index interval [10 000, 30 000].

G2.10.3.2 On-line experiment

In a beamforming hearing aid, for example, the adaptive filter must converge sufficiently fast to adapt to the changing environment and to compensate for head movements. The experiments in this section compare the convergence speed and steady state performance of the on-line adapted perceptron and Volterra filter. Although the test involved only one particular jammer (uniform i.i.d. noise) at filter length $M = 8$, the results reflect a typical filter behavior which was also observed in other simulations with different filter lengths and signals.

The delay between the microphones was again set to $\Delta = 1$. The target signal was female speech (one sentence) sampled at 8 kHz and the input target-to-jammer ratio was zero dB. The perceptron was adapted with on-line backpropagation and the third-order (without second-order terms) Volterra weights were adjusted with the standard LMS and RLS algorithms. Because the jammer was stationary, the RLS forgetting factor λ was set to unity.

When the target is present, the learning rate in the backpropagation (or LMS) algorithm must stay below a certain limit to avoid target cancelation. Note that this limit is generally not identical with the maximum learning rate which would render the adaptive filter unstable. For linear filters, this form of target cancelation is discussed in more detail in Widrow and Stearns (1985).

The maximum learning rates for the linear finite-impulse-response (FIR) filter, for the perceptron with 5 and 20 hidden neurons and for the LMS-adapted Volterra filter were determined as follows. The beamformer was run with a series of different learning rates. For each learning rate, we listened to the *filter output* (not the beamformer output) and chose the maximum rate for which the target signal was not audible in the output. The maximum learning rates are shown in table G2.10.1. Using these learning rates ensured an undistorted target signal at the beamformer output. Larger learning rates would have allowed a faster adaptation at the expense of some target distortion.

Because the beamforming did not affect the target, it was set to zero in the subsequent experiments. The beamformer was run ten times employing ten different sets of initial weights and ten different uniformly-distributed jammer signals for each filter in table G2.10.1. For the perceptron, the weights were initialized again with Nguyen and Widrow's method. The linear FIR and the Volterra coefficients were chosen from a normal distribution with zero mean and one-quarter variance. Figure G2.10.4 depicts the ensemble-averaged learning curves as a function of the sample index. The steady-state normalized MSEs in table G2.10.1 were estimated from these curves by *time-averaging* the instantaneous squared errors from sample index 10 000 to index 30 000. For the RLS algorithm, the averages were calculated between the indices 1000 and 20 000.

The perceptron learning curves in figure G2.10.4 seem to have reached the steady state after about 10 000 iterations. In a test simulation of 60 000 iterations, however, the MSE decreased further. For example, between the indices 40 000 and 60 000, the MSE of the perceptron with $N_2 = 20$ declined to 0.1843. Because the perceptron error decayed very slowly over many iterations, the entries in table G2.10.1 are called 'quasi' steady-state MSEs. With a sampling rate of 8 kHz, the perceptron required more than one second to reach a quasi-stationary state. It is striking that the perceptron did not perform significantly better than the linear FIR filter.

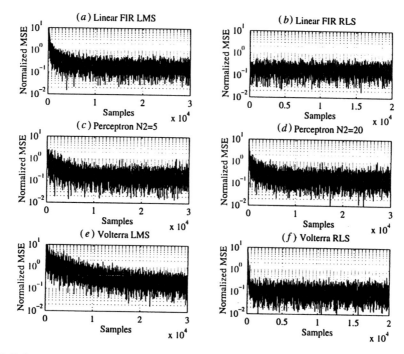

Figure G2.10.4. Ensemble averaged learning curves for various on-line adaptive linear and nonlinear filters. Note the different abscissa scaling for the RLS-adapted filters.

The LMS-adapted Volterra filter converged extremely slowly, i.e. the MSE (measured in blocks of 20 000 samples) still decreased after 100 000 iterations. The RLS Volterra filter converged after approximately 1000 iterations with a steady-state MSE close to its optimum value. Simultaneously, the computational burden of this algorithm is the highest of all algorithms in this section. For the linear filter, the RLS algorithm requires $O(M^2)$ operations per iteration. The third-order Volterra filter has $O(M^3)$ coefficients and thus, it requires $O(M^6)$ operations per iteration. Backpropagation with N_2 hidden neurons entails $O(N_2 M + 2N_2)$ operations per iteration.

G2.10.4 Summary

The results of this work can be summarized as follows.

- For small filter lengths $N < 20$, the perceptron with 20 hidden neurons and the third-order Volterra filter could approximate the optimum Bayes filter.
- For $N > 20$, the memory requirements of the Levenberg–Marquardt routine TRAINLM exceeded our computer capacity. A similar effect was observed for the Volterra filter.
- Backpropagation could not adapt the perceptron appropriately in the on-line experiment. The Volterra filter could be adjusted fast enough with the RLS algorithm and attained a satisfactory steady-state MSE. The computational load of the third-order Volterra RLS, however, was prohibitive.

References

Griffiths L J and Jim C W 1982 An alternative approach to linearly constrained adaptive beamforming *IEEE Trans. Antenn. Propag.* **30** 27–34

Hagan T H and Menhaj M B 1994 Training feedforward networks with the Marquardt algorithm *IEEE Trans. Neural Networks* **5** 989–93

Knecht W G 1994 Nonlinear noise filtering and beamforming using the perceptron and its Volterra approximation *IEEE Trans. Speech Audio Proc.* **2** 55–62

—— 1995 On nonlinear filtering for noise reduction using a sensor array *PhD Thesis* Swiss Federal Institute of Technology, Zurich

Mathews V J 1991 Adaptive polynomial filters *IEEE Signal Processing Magazine* vol 8, no 3, pp 10–26

Nguyen D and Widrow B 1990 Improving the learning speed of 2-layer neural networks by choosing initial values of the adaptive weights *Int. Joint Conf. on Neural Networks* (IEEE Publishing) vol III pp 21–6

Papoulis A 1991 *Probability, Random Variables and Stochastic Processes* (New York: McGraw-Hill)

Sicuranza G L 1992 Quadratic filters for signal processing *Proc. IEEE* **80** 1263–85

Waibel A, Hanazawa T, Hinton G, Shikano K and Lang K J 1989 Phoneme recognition using time-delay neural networks *IEEE Trans. Acoustics Speech and Signal Processing* **37** 328–39

Widrow B and Stearns S D 1985 *Adaptive Signal Processing* (Englewood Cliffs, NJ: Prentice-Hall)

G2.11 A concise application demonstrator for pulsed neural VLSI

Alan F Murray and Geoffrey B Jackson

Abstract

Current research at the University of Edinburgh has developed pulse-stream neural systems to operate on the boundary between the analog sensory environment and that of conventional digital processors. The issues of where, how and why pulse stream neural hardware should be applied are examined in this section. We present here a chip, EPSILON II (Edinburgh Pulse Stream Implemenation of a Learning Oriented Network) and a processor card incorporating it that have been designed to bring pulse stream neural hardware to bear on real applications. As an example, an autonomous mobile robot is described.

G2.11.1 Introduction

Applications of analog neural hardware have been few and slow to emerge despite the success of neural networks in many diverse applications areas. For example, in the DARPA (Defence Advanced Research Projects Agency) neural networks study of 1988, of the 77 neural network applications investigated, none of the field-tested systems (Widrow 1988) used dedicated neural network hardware. The situation has not changed dramatically in the subsequent five years. While this handbook shows that there is an increase in 'real' use of neural networks, it is our view that the reasons for the dearth of hardware demonstration systems can be summarized as follows:

- Most neural applications will be served optimally by fast, generic digital computers.
- Dedicated digital neural accelerators have a limited lifetime as 'the fastest' neural networks, since standard computers are developing so rapidly.
- Analog neural VLSI is a niche technology, optimally applied at the interface between the real world and higher-level digital processing.

This attitude has some profound implications with respect to the size, nature, and constraints we place on new hardware neural designs. After several years of research into hardware neural network implementation, we are now concentrating on the areas in which *analog neural network technology* has an 'edge' over well established *digital technology*.

E1.3
E1.4

Clearly, neural network technology must compete with more conventional digital techniques in solving real-world problems, and neural networks must concentrate on areas where their advantages are most prominent and their disadvantages (the inability to interrogate a solution fully, for example) are least problematic.

Within the pulse stream neural network research at the University of Edinburgh, the EPSILON chip's areas of strength can be summarized as:

- analog or digital inputs, digital outputs
- compact, low power
- modest size
- scaleable and cascadeable design.

This list points naturally and strongly to problems on the boundary of the real, analog world and digital processing, such as preprocessing/interpretation of analog sensor data. Here a modest neural network can act as an *intelligent analog-to-digital converter* presenting preprocessed information to its host. It is our conclusion that this is an area where analog neural networks will make the most significant impact. We are now engaged in a two-pronged approach, whereby development of technology to improve the performance of pulse stream neural network chips is occurring concurrently with a search for and development of applications to which this technology can be applied.

The key requirements of this technological development are that devices must:

- work directly with analog signals
- provide a moderate size network to process data for further digital processing
- have the potential for a fully integrated solution.

The next subsection describes the EPSILON II chip (specifically, the features of the chip that have been developed to make the hardware more amenable to use in real applications) and examines the system-level considerations and the specifics of the EPSILON processor card (EPC), a flexible environment for applications and chip-level development. Finally, the nature of appropriate applications is discussed and a demonstration application of an autonomous mobile robot is presented.

G2.11.2 The EPSILON II chip

The EPSILON II chip has been designed around the requirements of an application-based system. It follows from an earlier generation of pulse stream neural network chips, the EPSILON chip (Murray *et al* 1992).

The EPSILON II chip represents neural states as a pulse-encoded signal. These pulse-encoded signals have digital signal levels which make them highly immune to noise and ideal for inter- and intrachip communication, facilitating efficient cascading of chips to form larger systems. The EPSILON II chip can take as inputs either pulse-encoded signals or analog voltage levels, thus facilitating the fusing of analog and digital data in one system. Internally the chip is analog in nature allowing the synaptic multiplication function to be carried out in compact and efficient analog cells (Jackson *et al* 1994).

Table G2.11.1. EPSILON II specifications.

No of state input pins	32
Input modes	Analog, PW or PF
Input mode programmability	Bit programmable
No of state outputs	32 pinned out
Output modes	PW or PF
Digital recovery of analog I/P	Yes—PW encoded
No of synapses	1024
Additional *autobias* synapses	4 per output neuron
No of weight load channels	1
Weight load time	2.3 ms
Weight storage	Dynamic
Programmable activity voltage	Yes
Maximum speed (cps)	102.4 Mcps
Technology	ES2 1.5 μm CMOS
Die size	6.9 mm $\times$ 7 mm
Packaging	120-pin PGA
Maximum power dissipation	320 mW

Table G2.11.1 shows the principal specifications of the EPSILON II chip which is based around a 32×32 synaptic matrix allowing efficient interfacing to digital systems. A plot of the layout of the chip (figure G2.11.1(*a*)) shows the structure of and the signal flow within the chip. Several features of the device have been developed specifically for applications-based usage. The first of these is a programmable input mode. This allows each of the network inputs to be programmed as either a direct analog input or a digital pulse-encoded input. We believe that this is vital for application-based usage where it is often necessary to *fuse* real-world analog data with historical or control data generated digitally. The second major feature is a pulse recovery mode. This allows conversion of any analog input into a digital value for direct use by the host system. Such a facility is necessary if learning is to be done with the system in operation using, for example, the backpropagation algorithm as input state values are needed for learning.

Other concurrent work in the neural group in Edinburgh seeks to make future chips more 'application friendly' by using amorphous silicon for nonvolatile weight storage (Holmes *et al* 1993) and developing on-chip learning circuits to render chips more autonomous (Woodburn *et al* 1994).

An example of the characteristics of the EPSILON II device is shown in figure G2.11.1(*b*). This plot shows the characteristics of an individual synapse/neuron on the chip as a plot of output pulse width against the input range for various weight values. This characteristic represents a significant improvement over the earlier EPSILON pulse stream neural network chip (Murray *et al* 1992). This improvement arises from careful layout and architecture changes while still using the same basic circuits.

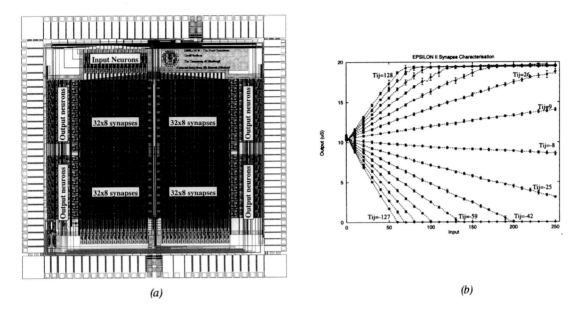

(a)   (b)

Figure G2.11.1. EPSILON II layout and synapse characteristics.

G2.11.3 The EPSILON processor card

The need to embed the EPSILON chip in a processor card is driven by several considerations. Firstly, working with pulse-encoded signals requires substantial processing to interface directly to digital systems. If the neural processor is to be transparent to the host system and is not to become a substantial processing overhead, then all pulse support operations must be carried out independently of the host system. Secondly, to respond to further chip-level advances and allow rapid prototyping of new applications as they emerge, a certain amount of flexibility is needed in the system. It is with these points in mind that the design of the flexible EPSILON processor card (EPC) was undertaken.

G2.11.3.1 Design specification

The EPC has been designed to meet the following specifications. The card must:

- operate on a conventional digital bus system
- be transparent to the host processor, that is, carry out all the necessary pulse encoding and decoding
- carry out the refresh operations of the dynamic weights stored on the EPSILON chip
- generate the ramp waveforms necessary for pulse width coding
- support the operation of multiple EPCs
- allow direct input of analog signals.

As all data used and generated by the chip are effectively of 8-bit resolution, the STE bus, an industry standard 8-bit bus, was chosen for the bus system. This is also cost effective and allows the use of readily available support cards such as processors, DSP cards, and analog and digital signal conditioning cards.

To allow the transparency of operation the card must perform a variety of functions. A block diagram indicating these functions is shown in figure G2.11.2.

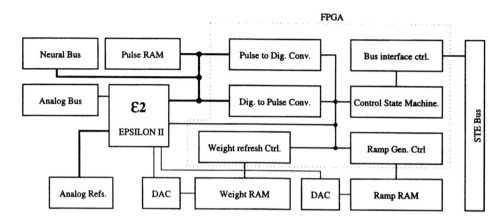

Figure G2.11.2. EPSILON processor card.

A substantial amount of digital processing is required by the card, especially in the pulse conversion circuitry. To conform to the *Eurocard* standard size of the STE specification an FPGA device is used to 'absorb' most of the digital logic. A twin mother/daughter board design is also used to isolate sensitive analog circuitry from the digital logic. The use of the FPGA makes the card extremely versatile as it is now easily reconfigurable to adapt to specialist applications. The dotted box of figure G2.11.2 shows functions implemented by the FPGA device. An onboard EPROM can hold multiple FPGA configurations such that the board can be reconfigured 'on the fly'. All EPSILON support functions, such as ramp generation, weight refresh, pulse conversion, and interface control are carried out on the card. Also, the use of the FPGA means that new ideas are easily tested as all digital signal paths go via this device. Thus a card with new functionality can be designed without the need to design a new PCB.

G2.11.3.2 Specialist buses

The digital pulse bus is buffered under control of the FPGA to the neural bus along with two control signals. Handshaking between EPCs is done over these lines to allow the transfer of pulse stream data between processors. This implies that larger networks can be implemented with little or no increase in computation time or overhead. A separate analog bus is included to bring analog inputs directly onto the chip.

G2.11.3.3 Future extensions

As all control and pulse stream signals are generated by the FPGA, the EPC stands ready to accept the next generation of the EPSILON chipset. By judicious chip design, chips incorporating on-chip learning or nonvolatile analog storage currently being developed at Edinburgh (see Murray *et al* 1994) will readily plug into the EPC for evaluation in a stable environment.

G2.11.4 Applications

The overriding reason for the development of the EPC is to allow the easy development of hardware neural network applications. We have already indicated that we believe that this form of neural technology will find its niche where its advantages of direct sensor interface, compactness, and cost-effectiveness are of prime importance. As a good and intrinsically interesting example of this genre of application, we have chosen autonomous mobile robotic control as a first test for EPSILON II. The object of this demonstrator is not to advance the state of the art in robotics. Rather, it is to demonstrate analog neural VLSI in an appropriate and stimulating context.

G2.11.4.1 'Instinct rule' robot

The 'instinct rule' robotic control philosophy is based on a software-controlled exemplar from the University's Department of Artificial Intelligence (Nehmzow 1992) (see figure G2.11.3). The robot incorporates an EPC which interfaces all the analog sensor signals and provides the programmable neural link between sensor/input space and the motor drive actuators.

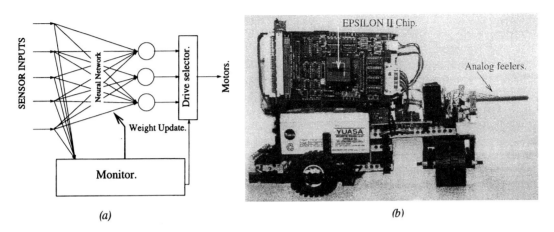

Figure G2.11.3. (*a*) Controller architecture. (*b*) 'Instinct rule' robot.

The controller architecture is shown in figure G2.11.3. The neural network implemented on the EPC is the *plastic* element that determines the mapping between sensory data and motor actions. The majority of the monitor section is currently implemented on a host processor and monitors the performance of the neural network by regularly evaluating a set of *instinct rules*. These rules are simple behavior-based axioms. For example, we use two rules to promote simple obstacle avoidance competence in the robot, as listed in column one of table G2.11.2.

Table G2.11.2. Instinct rules.

Simple obstacle avoidance	Wall following
1. Keep crash sensors inactive.	1. Keep crash sensors inactive.
2. Move forward.	2. Keep side sensors active.
	3. Move forward.

If an instinct rule is violated the drive selector then chooses the next strongest output (motor action) from the neural network. This action is then performed to see if it relieves the violation. If it does, it is used as a target to train the neural network. If it does not, the next strongest action is tried. Using this scheme the robot can be initialized with random weights (i.e. no mapping between sensors and motor control) and within a few epochs obtains basic obstacle avoidance competence.

It is a relatively easy matter to promote more complex behavior with the addition of other rules. For example, to achieve a wall-following behavior a third rule is introduced as shown in column two of table G2.11.2. Navigational tasks can be accomplished with the addition of a rule to '*maximize the navigational signal*'. An example of this is a light sensor mounted on the robot producing a behavior to move towards a light source. Equally, a signal from a more complex, higher-level, navigational system could be used. Thus the instinct rule controller handles basic obstacle avoidance competence and motor/sensory interface tasks leaving other resources free for intensive navigational tasks.

G2.11.5 Conclusions

This case study has discussed the use of pulse stream neural networks in practical applications. We have presented new results from a novel analog neural chip, EPSILON II, and offered reasoned opinions regarding the optimal use of neural analog VLSI.

To aid the development of practical application the EPSILON II chip and the EPSILON processor card have been designed. These resources have been designed to process data on the boundary between the analog real world and the digital world of conventional computing. The analog VLSI nature of the neural hardware makes it extremely versatile for this type of purpose. Reasons for this include:

(i)   Direct interfacing to analog signals.
(ii) The ability to fuse direct analog sensor data with digital sensor data processed elsewhere in the system.
(iii) Distributed processing. Several EPCs may be embedded in a system to allow multiple networks and/or multilayer networks.
(iv) Speed. Guaranteed calculation times (as per table G2.11.1). The speed of software solutions is not so readily defined or achievable in a compact unit. This has implications for real-time applications.
(v)   The EPC represents a flexible system-level development environment.
(vi) The EPC requires very little computational overhead from the host system and can operate independently if needed.
(vii) The flexibility of the EPC with major digital functions carried out in programmable logic means that it is easily reconfigured for new applications or improved chip technology.

In conclusion, we believe that the immediate future of neural analog VLSI is in small applications-based systems that interface directly with the real world. We see this as the niche area where VLSI neural networks can compete most effectively with conventional digital systems. The EPSILON II chip and processor card are now of a form that can prototype real-world applications in the analog domain rapidly and efficiently. The example of the instinct rule robot readily demonstrates this.

References

Caudell M and Butler C 1990 *Naturally Intelligent Systems* (Cambridge, MA: MIT Press)
Holmes A J *et al* 1993 Use of α-Si:H memory devices for non-volatile weight storage in artificial neural networks *15th Int. Conf. on Amorphous Semiconductors*
Jackson G, Hamilton A and Murray A F 1994 Pulse stream VLSI neural systems: into robotics *Proc. ISCAS'94* vol 6 (New York: IEEE) pp 375–8
Maren A, Harston C and Pap R 1990 *Handbook of Neural Computing Applications* (San Diego, CA: Academic)
Murray A F, Baxter D J, Churcher S, Hamilton A, Reekie H M and Tarassenko L 1992 The Edinburgh pulse stream implementation of a learning-oriented network (EPSILON) chip *Neural Information Processing Systems (NIPS) Conf.*
Murray A F, Churcher S, Hamilton A, Holmes A J, Jackson G B and Woodburn R 1994 Applications of pulsed neural VLSI *IEEE MICRO*
Nehmzow U 1992 Experiments in competence acquisition for autonomous mobile robots *PhD Thesis* University of Edinburgh
Widrow B 1988 *DARPA Neural Network Study* (AFCEA)
Woodburn R, Reekie H M and Murray A F 1994 Pulse stream circuits for on-chip learning in analogue VLSI neural networks *Proc. ISCAS'94* vol 4 (New York: IEEE) pp 103–6

G2.12 Ontogenic CID3 algorithm for recognition of defects in glass ribbon

Krzysztof J Cios

Abstract

This case study describes an ontogenic CID3 algorithm and its application to recognition of defects in a floating glass ribbon. The structure of this case study is as follows. First, the CID3 algorithm is described in sufficient detail to give the reader the feeling of how ontogenic algorithms generate their architectures. Second, a step-by-step application of the algorithm to a problem of distinguishing true defects (bubbles, stones and tin drops) from surface anomalies (water droplets and water spots) is provided. The second step also includes a description of the preprocessing steps crucial for achieving high accuracy of recognition. Finally, the ontogenic CID3 algorithm results are compared with those obtained by RBF and backpropagation algorithms on the same data.

G2.12.1 Motivation

A commercial system for detection of defects in manufactured glass was unable to distinguish between actual defects and the glass surface anomalies, usually caused by airborne debris. These anomalies were detected by a commercial system as defects and the section of glass containing them must have been discarded thus resulting in the loss of usable glass. It was estimated that 2–3% of net glass production was lost due to this problem. If it were possible to distinguish between permanent defects and correctable surface anomalies the company could recover a significant portion of glass production that was normally discarded. We believed that neural network analysis of defect images in the float glass ribbon could achieve that goal.

G2.12.1.1 The ontogenic CID3 algorithm

The *continuous ID3* (CID3) algorithm (Cios and Liu 1992) utilizes inductive machine learning to specify   D1.4 conversion of a decision tree into a hidden layer of a neural network. The algorithm is representative of a host of ontogenic algorithms which are very similar to one another. One of the first ones was the tiling algorithm of Nadal (1989); the cascade-correlation algorithm of Fahlman and Lebiere (1990) was a variation of it. CID3 is similar to the algorithm of Bischel and Seitz (1989) although it was developed from a very different perspective. It was based on the machine learning ID3 algorithm (Quinlan 1983, 1990). The CID3 algorithm creates a hidden layer in a manner similar to the ID3's generation of a decision tree. In a learning process new hidden layers are being added to the network until a learning task becomes linearly separable at the output layer. By combining machine learning algorithms with neural networks, the CID3 algorithm not only generates a feedforward neural network architecture but also enables translation of the knowledge embedded in the connections and weights into decision rules. Another machine learning algorithm is described in Cios and Liu (1995a, b).

In order to explain the main ideas of the CID3 algorithm, let us briefly introduce the ID3 algorithm. The latter generates decision rules from a set of training examples. Each example is represented by a list of features. The idea is to examine training examples and find the minimum number of original features that suffice in determining class memberships. ID3 uses information theory to select features which give the

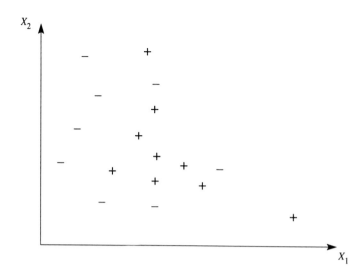

Figure G2.12.1. Seventeen training examples belonging to two classes.

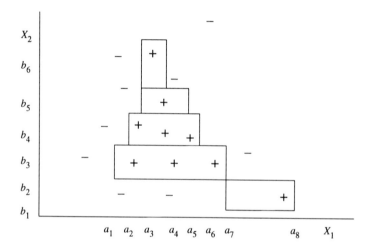

Figure G2.12.2. Five decision regions covering nine positive training examples.

greatest information gain, or decrease of entropy. Entropy is defined as $-p \log_2 p$, where probability p is determined from the frequency of occurrence. To generate decision rules that correctly classify training examples, a *feature test* is performed by first selecting a feature, and then dividing examples into subclasses using the selected feature. Next, information entropy is calculated to determine how significant the feature is. The ID3 algorithm requires features to have discrete values.

The drawback of knowledge representation based on a *feature test* is that the correlations between features are ignored. Also, it is not easy to detect features yielding the minimum entropy when training examples are represented by continuous data. Let us repeat here after Cios and Liu (1992) an example of distinguishing nine positive examples from eight negative examples, shown in figure G2.12.1.

When the ID3 algorithm is applied to this problem, the thresholds are represented as vertical and horizontal lines in the two-dimensional space. Figure G2.12.2 shows five rectangular decision regions to cover the nine positive examples.

However, the decision region covering the same positive training examples can be formed by using C1.1.3 hyperplanes defined by *adalines* (Widrow *et al* 1988). The output of an adaline, which defines a hyperplane $\sum w_i x_i + x_0 = 0$ is given by

$$\text{output} = \begin{cases} 1, & \sum w_i x_i + w_0 > 0 \\ 0, & \sum w_i w_i + w_0 \leq 0 \end{cases}.$$

Thus the *feature test* performed by ID3 can be treated as a special case of an adaline with its hyperplane parallel to an axis; that is, the weight vector is a base vector. Thus, the decision region covering nine positive training examples can be described by three hyperplanes, figure G2.12.3, where arrows indicate positive sides of the hyperplanes.

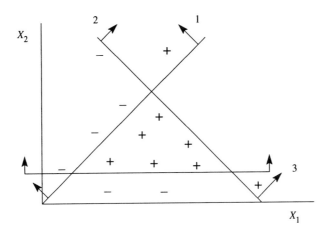

Figure G2.12.3. Decision region specified by hyperplanes.

To describe a decision region in terms of decision rules, Feature$_i$ in a decision rule will correspond to a hyp$_i$, $i = 1, 2, 3$. If an example is on the positive side of a hyperplane then Feature$_i = 1$; otherwise Feature$_i = 0$, which is an abbreviation of a statement: $\{\forall x = (x_1, x_2) : \sum w_j x_j + w_0 < 0\}$. Thus, a decision rule can be simply specified as

IF Feature$_1 = 0$ and Feature$_2 = 0$ and Feature$_3 = 1$ THEN class $=$ positive.

IF Feature$_2 = 1$ and Feature$_3 = 0$ THEN class $=$ positive.

IF Feature$_1 = 1$ and Feature$_2 = 1$ THEN class $=$ positive.

The above example will be used to illustrate conversion of a decision tree into a hidden layer. First, if an example is tested on the positive side of hyp$_1$, then that example will be classified along edge 1, as shown in figure G2.12.4, otherwise that example will be classified along edge 0. Starting at the root node a, the training examples are divided, by adaline #1, into two nodes, b and c. The corresponding entropy of 0.861 is shown. At the second level of the decision tree, the examples from nodes b and c are tested against hyp$_2$ (adaline #2). The second hyperplane is obtained with minimum entropy of 0.567. The training examples on the positive side of the second hyperplane will be classified along edge 1 to a node descending from their parent node. Those on the negative side will be classified along edge 0. Now, class memberships of training examples in nodes d and e are already correct so only one more (third) hyperplane (adaline #3) is needed to divide the examples at nodes f and g. As a result, a hidden layer with three nodes is generated, as shown on the left-hand side of figure G2.12.4. The directional vector of a hyperplane is taken as the weight vector of an adaline. For hyp$_1$, the weights w_1 and w_2 are the connection strengths of inputs x_1 and x_2 to adaline #1 (node #1).

In order to derive CID3's learning rule let us introduce the following notation after Cios and Liu (1992). There are N training examples, N^+ examples belonging to class '+', and N^- examples belonging to class '−'. A hyperplane divides the examples as lying either on its positive (1) or negative (0) side, with four possible outcomes:

- N_1^+ denotes the number of examples from class + on side 1;
- N_0^+ denotes the number of examples from class + on side 0;
- N_1^- denotes the number of examples from class − on side 1;
- N_0^- denotes the number of examples from class − on side 0.

The following relations hold:

$$N = N^+ + N^- = N_1^+ + N_1^- + N_0^+ + N_0^- \qquad \text{(G2.12.1a)}$$

$$N_0^+ = N^+ - N_1^+ \qquad \text{(G2.12.1b)}$$

$$N_0^- = N_r^+ - N_{1r}^+. \qquad \text{(G2.12.1c)}$$

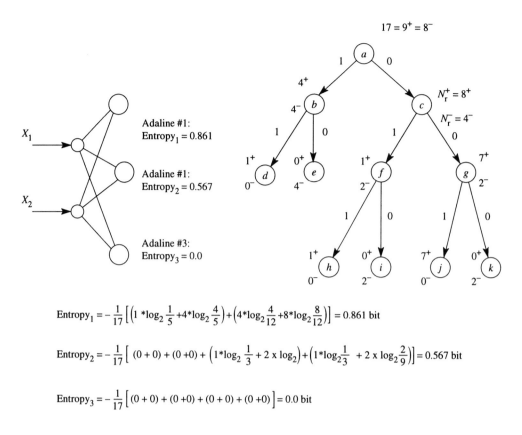

$$\text{Entropy}_1 = -\frac{1}{17}\left[\left(1*\log_2\frac{1}{5}+4*\log_2\frac{4}{5}\right)+\left(4*\log_2\frac{4}{12}+8*\log_2\frac{8}{12}\right)\right] = 0.861 \text{ bit}$$

$$\text{Entropy}_2 = -\frac{1}{17}\left[(0+0)+(0+0)+\left(1*\log_2\frac{1}{3}+2 \times \log_2\right)+\left(1*\log_2\frac{1}{3}+2\times\log_2\frac{2}{9}\right)\right] = 0.567 \text{ bit}$$

$$\text{Entropy}_3 = -\frac{1}{17}\left[(0+0)+(0+0)+(0+0)+(0+0)\right] = 0.0 \text{ bit}$$

Figure G2.12.4. Hidden layer corresponding to a decision tree and entropies calculated using (G2.12.3).

At a certain level of a decision tree we assume that N_r examples are divided by node r into N_r^+, belonging to class $+$, and N_r^-, belonging to class $-$. Relations analogous to (1) follow:

$$N_r = N_r^+ + N_r^- = N_{1r}^+ + N_{1r}^- + N_{0r}^+ + N_{0r}^- \tag{G2.12.2a}$$

$$N_{0r}^+ = N_r^+ - N_{1r}^+ \tag{G2.12.2b}$$

$$N_{0r}^- = N_r^- - N_{1r}^-. \tag{G2.12.2c}$$

The information entropy at level L of a decision tree is an average of entropies of all R nodes in this layer:

$$\begin{aligned}
E &= -\sum_{r=1}^{R}\frac{N_r}{N}\text{entropy}(L,r) \\
&= -\frac{1}{N}\sum_{r=1}^{R}\left[N_{1r}^+\log_2\frac{N_{1r}^+}{N_{1r}^+ + N_{1r}^-} + N_{1r}^-\log_2\frac{N_{1r}^-}{N_{1r}^+ + N_{1r}^-}\right. \\
&\quad\left. + (N_r^+ - N_{1r}^+)\log_2\frac{N_r^+ - N_{1r}^+}{N_r - N_{1r}^+ - N_{1r}^-} + (N_r^- - N_{1r}^-)\log_2\frac{N_r^- - N_{1r}^-}{N_r - N_{1r}^+ - N_{1r}^-}\right]. \tag{G2.12.3}
\end{aligned}$$

This formula is obtained by employing the mutual dependency of positive and negative examples given by (G2.12.2b) and (G2.12.2c).

The values of N_{1r}^+ and N_{1r}^- can be calculated as follows:

$$N_{1r}^+ = \sum_{i=1}^{N_r} D_i\text{out}_i = \sum_{i=1}^{N_r} D_i\left[1+\exp\left(-\sum_j w_{ij}x_j\right)\right]^{-1} \tag{G2.12.4}$$

$$N_{1r}^- = \sum_{i=1}^{N_r}(1-D_i)\text{out}_i = \sum_{i=1}^{N_r}(1-D_i)\left[1+\exp\left(-\sum_j w_{ij}x_j\right)\right]^{-1} \tag{G2.12.5}$$

where D_i stands for the desired output of a training example, and out_i is a sigmoid function of inputs to a node:

$$\text{out}_i = f\left(-\sum_{j=1}^{\text{dim}} w_{ij}x_j\right) = \left[1 + \exp\left(-\sum_{j=1}^{\text{dim}} w_{ij}x_j\right)\right]^{-1}. \tag{G2.12.6}$$

The partial derivatives of information entropy with respect to the number of examples on the positive and negative sides of a hyperplane are

$$\frac{\partial E}{\partial N_{1r}^+} = -\frac{1}{N}\sum_{r=1}^{R}\left[\log_2 \frac{N_{1r}^+}{N_{1r}^+ + N_{1r}^-} - \log_2 \frac{N_r^+ - N_{1r}^+}{N_r - N_{1r}^+ - N_{1r}^-}\right] \tag{G2.12.7}$$

$$\frac{\partial E}{\partial N_{1r}^-} = -\frac{1}{N}\sum_{r=1}^{R}\left[\log_2 \frac{N_{1r}^-}{N_{1r}^+ + N_{1r}^-} - \log_2 \frac{N_r^- - N_{1r}^-}{N_r - N_{1r}^+ - N_{1r}^-}\right]. \tag{G2.12.8}$$

Thus, the change in information entropy is stated as

$$\Delta E = \sum_{r=1}^{R}\left[\frac{\partial E}{\partial N_{1r}^+}\Delta N_{1r}^+ + \frac{\partial E}{\partial N_{1r}^-}\Delta N_{1r}^-\right]. \tag{G2.12.9}$$

Although the values for N_{1r}^+ and N_{1r}^- may not come out as integers from (G2.12.4) and (G2.12.5), the analytic approximations make it possible to calculate the partial derivatives, (G2.12.7) and (G2.12.8), representing the relation between the change in the number of examples on both sides of a hyperplane with respect to the weights:

$$\Delta N_{1r}^+ = \sum_{i=1}^{N_r} D_i \text{out}_i (1 - \text{out}_i) \sum_{j=1}^{\text{dim}} x_j \Delta w_{ij} \tag{G2.12.10}$$

$$\Delta N_{1r}^- = \sum_{i=1}^{N_r} (1 - D_i)\text{out}_i (1 - \text{out}_i) \sum_{j=1}^{\text{dim}} x_j \Delta w_{ij}. \tag{G2.12.11}$$

Relations (G2.12.10) and (G2.12.11) make it possible to define a learning rule which minimizes the entropy function:

$$\begin{aligned}
\Delta w_{ij} &= -\rho\frac{\partial E}{\partial w_{ij}} \\
&= -\rho\sum_{r=1}^{R}\left(\frac{\partial E}{\partial N_{1r}^+}\sum_{i=1}^{N_r} D_i \text{out}_i (1 - \text{out}_i)x_j + \frac{\partial E}{\partial N_{1r}^-}\sum_{i=1}^{N_r}(1 - D_i)\text{out}_i (1 - \text{out}_i)x_j\right) \\
&= -\rho\sum_{r=1}^{R}\sum_{i=1}^{N_r}(\text{out}_i (1 - \text{out}_i)x_j\left(D_i\left(\frac{\partial E}{\partial N_{1r}^+} - \frac{\partial E}{\partial N_{1r}^-}\right) + \frac{\partial E}{\partial N_{1r}^-}\right) \tag{G2.12.12}
\end{aligned}$$

where ρ is a learning rate. Thus, the learning process for adjusting the weights can be stated in vector form as follows:

$$W_{k+1} = W_k + \Delta W. \tag{G2.12.13}$$

When the rule specified by equation (G2.12.13) is used the learning process might converge to a local minimum. The gradient method does not guarantee constant information gain while generating a hidden layer. In order to increase the chance of finding the global minimum the learning rule was combined (Cios and Liu 1992) with *Cauchy training*. The Cauchy training method (Szu and Hartley 1987) uses C1.4.2 statistically determined steps to converge to a global minimum

$$\Delta w = T(t)\tan[\pi P(\Delta W \le \Delta w) - \pi/2] \tag{G2.12.14}$$

To calculate the size of this weight change, a random number is selected from a uniform distribution over $[0, 1]$, and substituted for Cauchy distribution $P(\cdot)$. Artificial temperature $T(t)$ changes, from initial high value T_0 down to zero, with time t, according to $T(t) = T_0/(1+t)$. To determine whether to accept the weight change, Boltzmann distribution was used (Cios and Liu 1992). The probability of the error, err, was calculated using equation (G2.12.15), where k is the Boltzmann constant.

$$P(\text{err}) = \exp\left(\frac{-\text{err}}{kt}\right) \tag{G2.12.15}$$

The final learning rule of the CID3 algorithm Cios and Liu (1992) is stated in equation (G2.12.16), where the random weight vector ΔW_{random} is calculated from (G2.12.14) and η is a control parameter, $0 \le \eta \le 1$.

$$W_{k+1} = W_k + (1 - \eta)\Delta W + \eta \Delta W_{\text{random}} \tag{G2.12.16}$$

The random weight change W_{random} in (G2.12.16) enables the algorithm to escape from local minima and hopefully achieve the global minimum, which would ensure that a hidden layer will be created with the smallest possible number of nodes.

Pseudocode for the CID3 algorithm follows:

(i) For a given problem with N training examples, follow the notations given in (G2.12.1a)–(G2.12.1c) and (G2.12.2a)–(G2.12.2c). Start with a random initial weight vector W_0.

(ii) Utilize learning rule (G2.12.13) and search for a hyperplane that minimizes the following entropy function:

$$\min_{w_L^*} E = -\sum_{r=1}^{R} \frac{N_r}{N} \text{entropy}(L, r).$$

(iii) If the minimized entropy is not zero, but smaller than the previous value, add a node to the current layer and return to step (ii). Otherwise, go to step (iv).

(iv) If a hidden layer consists of more than one node, generate a new layer that utilizes inputs from both the original training data and the outputs from all previously generated layers, and go to step (ii). If the hidden layer consists of only one node, then the problem is reduced to a linearly separable one; stop.

The CID3 algorithm generates a multilayer network with a single node at the output. To solve multiple-category classification problems one can easily build a network consisting of many such subnetworks.

After a hidden layer is generated the outputs from all the generated hidden layers, together with the original inputs, are used to generate a new hidden layer. For instance, if a hidden layer with three adalines were generated the dimension of an input vector to the second hidden layer would be five and could be specified as follows:

$$[x_1, x_2, 1, 0, 1]$$

where the last three values are the outputs from the first hidden layer. The usage of the information from both the original training data and the outputs from the previously generated hidden layers allows a learning process to converge faster because of the increase of the dimensionality of training data (Nilsson 1990).

The connections between nonadjacent layers are called shortcuts. The use of shortcuts plays a vital role in the convergence of the algorithm. The learning which uses the knowledge from both original training examples and the outputs from hidden layers is actually a generalization process. The process of adding new hidden layers can be seen as a process of knowledge refinement. A single decision rule is specified at the end of learning and it gives the most general description of all training examples.

Analysis of the complexity of the CID3 algorithm shows that in contrast to backpropagation, where correct classification of training examples is achieved only at the output layer, training examples are correctly recognized by CID3 at a hidden layer for which the information entropy is for the first time reduced to zero.

For a description of the ontogenic neuro-fuzzy CID3 (F-CID3) algorithm, which after reducing entropy to zero switches to efficient operations on fuzzy sets (Cios and Sztandera 1992, 1996), see also Section D1.4 of this handbook. In it we describe how all the subsequent layers can be eliminated.

In the next section we shall describe a problem, mentioned already at the beginning of the section, that the CID3 algorithm has already been applied to.

G2.12.2 Definition of defects in glass ribbon

A commercial laser imaging system was used to obtain gray-scale images of a number of true defects and surface anomalies (Cios *et al* 1991a, b) in glass ribbon. The basic types of defects were defined as follows.

True defects. Permanent structures that degraded the homogeneity and optical quality of the glass. They were divided into:

- bubble—round or elongated gaseous inclusions within the glass; sometimes open at top or bottom surface;
- stone—a variety of crystalline or amorphous inclusions within the glass; might be opaque or slightly translucent;
- tin drop—a depression on the surface caused by a drop of molten tin adhering to the glass surface during forming; the solidified tin drop remained in the depression.

Surface anomalies. Nonrejectable, temporary marks or spots on the glass surface. They were divided into:

- water droplet—a more or less hemispherical drop of liquid water; might occur on either surface;
- water spot—mineral residue from a dried drop of water; again, might occur on either surface.

G2.12.2.1 Data acquisition

Samples of glass with defects were collected and the defect categories determined by a factory expert. Due to their transitory nature surface anomalies, such as water droplets and water spots, were recreated in the laboratory.

Images of the defects were then obtained using an imaging system at a resolution of 133 pixels per inch horizontally and 40 lines per inch vertically, with gray-scale of eight bits per pixel. These images were placed in a database along with information on the imaged defects including: size, type, sample number and so on. The sizes of the images obtained by the imaging system varied in proportion to the size of the actual defect. Images ranged from 30×20 pixels to 250×200 pixels in size.

G2.12.2.2 Data processing

In order to use the defect images as input for neural networks the following preprocessing steps were performed.

(i)   The first line of an image was used as a baseline to normalize the image. The intensity values of the first line were subtracted from each line of the image, thus zeroing out the effects of normal glass. This step also compensated for anomalies in image illumination.

(ii) The image was smoothed using a standard low-pass filtering technique.

(iii) The region of interest was found by cutting out a rectangular region around the defect in order to eliminate parts of the image that depicted normal (nondefective) glass. The normal glass was distinguished as being near zero in value.

(iv) The large number of pixels in the defect images prohibited direct neural network analysis of raw image data. In addition, the major problem in applying neural networks to real life problems, like this one, was that the dimension of the input data must be the same. Therefore, the image data needed to be reduced and the number of features (pixels) normalized. The following four methods were used to accomplish this goal.

Two-feature data. This method involved finding the defect width and the maximum intensity for each line of the defect. Then the number of lines was normalized to 30. Each defect having fewer than 30 image lines was expanded by duplicating lines, while a defect having greater than 30 lines was compressed by omitting lines. For example, for a large defect having 90 horizontal lines, the reduction to 30 lines was done by calculating for the first row line—the average of the first 3 original lines, and so on. Data created using this reduction technique resulted in a 60 element input vector.

Three-feature data. This technique was identical to the two-feature data except that the position of the maximum intensity was also included for each line. This resulted in a 90-element input vector.

Image reduction. This method was not only the most interesting but also, as we shall see later, resulted in best recognition results. It reduced two-dimensional defect images into 10×10 pixels $\times$ amplitude (intensity), three-dimensional images. The image was scaled down to a 10 × 10 array using the same scaling factor for the length and width. This scaling factor was such that the larger of the length or width would just fit into the 10 × 10 array. The amplitude values were not scaled. In terms of a three-dimensional object this had the effect of scaling the length and width but keeping the height, or intensity of the image, unchanged. The original data were thus reduced to an input vector of length 100, each element corresponding to an intensity value (amplitude).

FFT image reduction. A fast Fourier transform (FFT) was performed on each line of the 10 × 10 reduced image. An FFT algorithm for an arbitrary number of samples per period was used (Brigham 1974). Ten points in the time domain resulted in 10 points in the frequency domain. Again, the image data were reduced to a 100 element input vector.

G2.12.2.3 Preparation of training and learning data

Neural networks which learn in a supervised mode, and only those were studied, require a number of known input/output examples for training. Thus the available data were divided into training and testing data sets using the standard in machine learning ratio of 7/3. That is, 70% of the collected examples were used as training data, with the remaining 30% used as testing data.

After preprocessing, it was found that amplitude of some of the defect images was so small that it was impossible to distinguish them from noise. Thus, although a larger number of samples was collected, the neural network analysis was performed on 293 usable samples, 88 of which were chosen as test samples. Breakdown of measurements into training and testing data was as follows.

Table G2.12.1. Number of training and testing samples.

	True defects	Surface anomalies
Training	121	84
Testing	52	36

As described above, four different types of preprocessed data were obtained. For the neural network using two-feature data each vector in the training file consisted of 61 elements; the first 60 were the inputs and the last was the desired output (1 for true defects and 0 for surface anomalies). Likewise, for the three-feature data, 10 × 10 image, and 10 × 10 FFT image, the training files consisted of 91, 101 and 101 element vectors, respectively.

G2.12.3 Results

The goodness of the four kinds of input data for recognition purposes was tested by analyzing the accuracy of the classification results obtained by the CID3 algorithm. After training the CID3 algorithm with 205 samples, it was applied to the test data of 88 samples. The predicted outputs were compared with the desired output and the results of classification were as follows.

As can be seen, the best results were achieved by using the 10 × 10 image data. That result warrants a comment. As all practitioners know very well, any successful application of a neural network depends more on careful preparation, or preprocessing, and proper choice of training data than on a particular algorithm used. All would work on 'good' data and none would work on 'bad' (difficult) data. The more time one spends on studying the process which generated the data and on data preprocessing (often over 50% of the entire effort), the better the results.

When a dimension of the input data varies from sample to sample, like in this application problem, some clever schemes have to be used to keep the input dimension constant, which is an input requirement of any neural network. The biggest lesson to be learned from this study was that by transforming the original two-dimensional defect data, a collection of signals each having a single spike representing, for example, a stone, into a three-dimensional image, was that only the 10 × 10 image data representation made it possible to distinguish between true defects and surface anomalies with acceptably high accuracy. Without using that transformation of the data there would be no success. The defect data application

Table G2.12.2. Results of the CID3 algorithm for different kinds of input data.

	CID3 algorithm results
2-feature data	
true defects	44/52[†]
surface anomalies	30/36
3-feature data	
true defects	45/52
surface anomalies	30/36
10×10 image data	
true defects	50/52
surface anomalies	35/36
10×10 FFT image data	
true defects	46/52
surface anomalies	34/36

[†] (correct recognition)/(total number of test examples).

Table G2.12.3. Comparison of results of different algorithms and their architectures.

Method	True defects	Surface anomalies	Total
CID3 (100:7:6:6:1)	50/52 (96.15%)	35/36 (97.22%)	85/88 (96.59%)
RBF neurons at data points (100:205:2)	47/52 (90.38%)	34/36 (94.44%)	81/88 (92.04%)
RBF neurons at cluster centers, $R < 75$, (100:89:2)	45/52 (86.54%)	35/36 (97.22%)	80/88 (90.90%)
RBF neurons at cluster centers, $R < 100$, (100:54:2)	49/52 (94.23%)	32/36 (88.89%)	81/88 (90.90%)
Backpropagation (100:20:1)	51/52 (98.07%)	34/36 (94.44%)	85/88 (96.59%)

Table G2.12.4. Comparison of training times.

Method	Normalized CPU time
CID3	161
RBF—neurons at data points	333
RBF—neurons at cluster centers, $R < 75$	18
RBF—neurons at cluster centers, $R < 100$	10
Backpropagation	615

clearly showed the importance of data preparation, or preprocessing. It simply could not be overstated in any real application.

After performing the above analysis the next step was to compare the results achieved by the CID3 algorithm with a powerful *radial basis function* (RBF) network on the same 10×10 image data. RBFs C1.6.2 were used with two different methods of selecting the RBF centers: 'neurons at data points' and 'neurons at clusters' centers' (Zahirniak *et al* 1990). The latter method was tested with two different radii, shown in the table below. For comparison, a popular *backpropagation* network was also run on the data. The C1.2 table below shows the architecture, in parenthesis below the name of a method, for each network used.

The normalized CPU times required to train these networks were also calculated and were as shown in table G2.12.4.

G2.12.4 Discussion

The classification results indicate that the CID3 algorithm gave almost a 97% correct recognition rate. Using the RBF network with the neurons at data points method, in which all 205 training examples were used, the recognition rate was 92%. When training vectors which were close together (within a radius R) were clustered to reduce the number of training examples to 89 ($R < 75$) and 54 ($R < 100$) the resulting recognition rate was almost 91% for both cases.

The time required to train the networks varied greatly. An RBF network using 205 neurons in the hidden layer required a training time twice as long as that of the CID3 algorithm, but almost half of that required by backpropagation. However, when clustering was performed to reduce the number of training vectors in the RBF networks, the training time dropped considerably, at the cost of accuracy.

The CID3 algorithm did not require the network architecture to be *a priori* specified. Based on the information entropy function, the algorithm added the necessary number of layers and nodes to correctly recognize all the input–output pairs in the training data. The RBF network using the neurons at data points method also had its architecture determined by the size of the data set. With backpropagation, the number of hidden layers and the number of nodes in each layer had to be guessed.

As a result, the CID3 algorithm might be useful in situations where the neural networks are to be generated automatically, and in real time, while backpropagation networks could not be used. There might also be situations where there is a time constraint on the training time, like in many control problems. Then the choice of the CID3 algorithm would be appropriate.

G2.12.5 Conclusions

The goal of the case study described above was to determine whether it was possible to distinguish between true defects and surface anomalies by using ontogenic neural networks. The results using the CID3 algorithm show that the correct recognition rate, depending on the input data (2 and 3 features, 10×10 image and FFT), was in the range of 84% to 97%.

As far as data preprocessing techniques were concerned, the best results of classification were obtained using the 10×10 reduced image data. The results show that in spite of the drastic reduction of the original image, from (in an extreme case) 250×200 pixels to 10×10 pixels, the reduced image retained most of the key original features. The 10×10 matrix containing the reduced image was well-filled, as opposed to the matrix containing the FFT image, which was sparse with most of the information clustered about the center. This was probably why FFT was not as good as the 10×10 image.

Acknowledgement

This research was partially supported by the National Science Foundation, grant no DDM-901533.

References

Bischel M and Seitz P 1989 Minimum class entropy: a maximum information approach to layered networks *Neural Networks* **2** 133–41

Brigham E O 1974 *The Fast Fourier Transform* (Englewood Cliffs, NJ: Prentice-Hall)

Cios K J, Langenderfer R A, Tjia R and Liu N 1991a Recognition of defects in glass ribbons using neural networks *Proc. 1991 NSF Design and Manufacturing Systems Conf.* (Dearborn, MI: SME Publishing) 203–200

Cios K J and Liu N 1992 A machine learning method for generation of neural network architecture: a continuous ID3 algorithm *IEEE Trans. Neural Networks* **2** 280–91

——1995a An algorithm which learns multiple covers via integer linear programming, part I–the CLILP2 algorithm *Kybernetes* **24**(2) 29–50

——1995b An algorithm which learns multiple covers via inter linear programming, part II–experimental results and conclusions *Kybernetes* **24**(3) 28–40

Cios K J and Sztandera L 1992 Continuous ID3 with fuzzy entropy measures *First IEEE Int. Conf. on Fuzzy Systems (San Diego, CA)* (New York: IEEE Press) pp 469–76

Cios K J and Sztandera L 1996 Ontogenic neuro-fuzzy algorithm: F-CID3 *Neurocomputing* in press

Cios K J, Tjia R, Liu N and Langenderfer R A 1991b 'Study of continuous ID3 and radical basis function algorithms for the recognition of glass defects' *Proc. Int. Joint. Conf. on Neural Networks (Seattle, WA)* vol 1 149–54

Fahlman S and Lebiere C 1990 The cascade-correlation learning architecture *Technical Report CMU-CS-90-100* Carnegie Mellon University

Nadal J P 1989 New algorithms for feedforward networks *Neural Networks and SPIN Glasses* ed Theumann and Koberle (Singapore: World Scientific) pp 80–8

Nilsson N J 1990 *The Mathematical Foundations of Learning Machines* (Los Altos, CA: Morgan Kaufmann)

Quinlan J R 1983 Learning efficient classification procedures and their application to chess end-games *Machine Learning: An Artificial Intelligence Approach* vol I ed R S Michalski, J G Carbonnell and T M Mitchell (Palo Alto, CA: Tioga) pp 463–82

——1990 Probabilistic decision trees *Machine Learning: An Artificial Intelligence Approach* vol III ed Y K Kodratoff and R S Michalski (Los Altos, CA: Morgan Kaufmann) pp 140–52

Szu H and Hartley R 1987 First simulated annealing *Phys. Lett.* A **122** 157–62

Widrow B, Winter R G and Baxter R A 1988 Layered neural nets for pattern recognition *IEEE Trans. Acoust., Speech, Signal Process.* **36** 1109–18

Zahirniak D R, Chapman R, Rogers S K, Suter B W, Kabrisky M and Pyati V 1990 Pattern recognition using radial basis function networks *Sixth Annual Aerospace Appl. of AI Conference (Dayton, OH)* pp 249–60

G3

Physical Sciences

Contents

G3.1 Neural networks for control of telescope adaptive optics

T K Barrett and D G Sandler

Abstract

We report on the use of artificial neural networks to estimate phase distortion in astronomical telescopes using focused images of a stellar source or an artificial laser guide star. The method was first developed as a means of measuring distortion induced by atmospheric turbulence and controlling an adaptive optics system for compensation of this atmospheric aberration. The method was then extended for use as a means of estimating static aberrations in the Hubble Space Telescope. We have tested the neural network aberration estimates against wavefront measurements of a Hartmann sensor, one of the traditional means of aberration measurement in adaptive optics systems, and have found good agreement. We have also compared the neural network with traditional high-resolution phase-retrieval methods with good agreement. The neural network approach offers a simple inexpensive way to implement adaptive optics in astronomical telescopes. It can also provide a quick and easy diagnostic tool for astronomical telescopes by providing estimates of static aberrations without any modification or disassembly of the telescope.

G3.1.1 Project overview

During the last five years we have investigated the application of artificial neural networks to the task of estimating aberrations in astronomical telescopes. The majority of our effort has been directed towards the development of neural networks suitable for use in optical systems designed to compensate in real time for the effects of aberrations induced by atmospheric turbulence in large monolithic astronomical telescopes (Sandler *et al* 1991a, 1991b). We have, however, also extended the method for use as an off-line (non-real-time) tool for estimating the static aberration in the Hubble Space Telescope (Barrett and Sandler 1993) and other authors have used the neural network method for controlling atmospheric compensation systems in astronomical array telescopes (Angel *et al* 1990, Wizinowich *et al* 1992, Lloyd-Hart *et al* 1992).

The objective of the networks which we developed was to use intensity images formed from aberrated wavefronts to determine the actual phase aberrations of that wavefront. This is a form of phase-retrieval or phase-recovery problem which has been studied by other authors who have used iterative techniques to obtain solutions (Fienup 1982, 1987) or have used a linearized curvature sensing technique based on intensity measurements taken in two out-of-focus planes (Roddier 1988). Specifically, the inputs to our neural networks were pixelized intensity measurements of two point-spread functions (PSF) taken at two different image planes near the best focus of the optical system. The light source for the PSF is either a natural guide star, or an artificial laser guide star created by scattering of a laser beacon from particles high in the atmosphere or resonant excitation of sodium atoms in the mesosphere (Gardner 1989). The output of the networks was an estimate of phase aberration in terms of coefficients for orthogonal fitting polynomials.

In general, the development team involved in designing and building a complete adaptive optics system can be quite large; including optical scientists, physicists, and mechanical, software and electrical engineers. However, the development of the neural network portion of such a system can be achieved with a much smaller group. In our work, the actual development and training of the neural network has been accomplished by ourselves, with significant support in the form of suggestions and analysis from our coauthors and sponsors.

Ground-based imaging of objects in space is hampered by the blurring effects caused by non-uniformities in the index of refraction of the Earth's atmosphere. These fluctuations in the index of refraction are stirred and randomized by atmospheric turbulence and, as a result, an optical wavefront passing through the atmosphere becomes aberrated in a random way. Images formed from the light are distorted, blurred and often scintillated. To further complicate the problem, the turbulence and the prevailing wind convect the fluctuations across the field of view of a telescope, tending to induce rapid changes in the magnitude and shape of the atmospheric distortion. Adaptive optics systems attempt to measure the distorted wavefront reaching a telescope and manipulate a specialized optical component designed to compensate or flatten the wavefront. The measurements need to be made quickly in order to keep up with the changing atmospheric aberrations and the typical optical component used for compensation consists of a mirror with a deformable surface (Ealey and Wellman 1994). Figure G3.1.1 is a block diagram which illustrates how the neural network fits into a generic adaptive optics system. The neural network receives intensity data from two image planes and estimates the residual aberration remaining after the incoming light reflects from the surface of the deformable mirror. This results in a closed-loop-type system in which the neural network is always trying to estimate and correct for the error between the mirror surface and the wavefront surface. The error is due to inaccuracies in the computation of actuator positions and by the changing atmospheric distortion. Each loop, the network's estimate of aberration is passed to a postprocessor which converts the estimate of residual aberration into electronic commands which drive the deformable mirror's surface to a new shape.

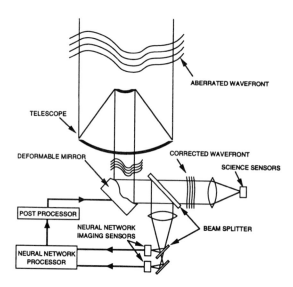

Figure G3.1.1. Simplified schematic diagram showing where the neural network fits into a generic adaptive optics system. The beam paths are indicated by full lines, with light traveling down through the atmosphere and off the primary mirror. Electronic information travels from the neural network imaging sensors to the neural network processor and then to the postprocessor and deformable mirror.

The neural networks we developed to estimate static aberrations in telescopes worked in basically the same manner as those suitable for the adaptive optics systems. However, these networks were tuned to measure the low spatial frequency aberrations in telescopes which are caused by faulty or misaligned optical components. This made the neural network method useful to NASA's Jet Propulsion Laboratory (JPL) during its effort to determine the exact aberration in the primary mirror of the Hubble Space Telescope. Our neural network estimates of the aberration were combined with estimates obtained with several other methods, to produce a prediction of the Hubble aberration which could be regarded with a high degree of confidence (Barrett and Sandler 1993).

G3.1.2 Design process

As an optical scientist gains experience with imaging systems he or she often begins to develop the ability to recognize the presence of certain optical aberrations by the characteristic shapes and features which the phase distortion induces in the PSF. This process is analogous to the learning which occurs in many neural network methods. The ability to learn a mapping or correlation from features in observed data to a quantitative description of the data is a property of neural networks which has been exploited for many applications, and motivated us to use this method for the phase-retrieval problem. The ability of the optical scientist is qualitative at best and often fails when the aberration is complicated. The neural network, however, can be trained to accurately recognize complicated aberrations even in the presence of high spatial frequency distortion which scintillates the image (Sandler *et al* 1991a).

For each of our optical phase-recovery applications we employed neural networks consisting of *multilayer perceptrons* (Rosenblatt 1962). Each network consisted of an input layer, a single hidden layer C1.2 and an output layer. Adjacent layers were fully connected. The transfer functions of the input and output layer were linear and the transfer function of the hidden layer was a sigmoid. The number of input and output nodes for each network was related to the desired resolution of the predicted aberration in terms of spatial frequency across the telescope aperture. The higher the resolution the greater the number of nodes required. For any given application the number of nodes in the hidden layer of each network was determined empirically by testing networks with increasing numbers of hidden nodes until the performance of the network no longer increased with increasing nodes. A typical network would have 128 input nodes, 64 hidden nodes and 18 output nodes.

Several considerations affected our choice of neural network architecture. First, phase recovery from stellar image data is a nonlinear problem since the angular distribution of intensity in the PSF, $I(\vec{\theta})$, of an astronomical telescope may be approximated by the following nonlinear equation (Goodman 1968):

$$I(\vec{\theta}) \propto \left| \int w(r) \exp[i\phi(r)] \exp\left[\frac{2\pi i}{\lambda}\vec{\theta} \cdot r\right] dr \right|^2 . \tag{G3.1.1}$$

In (G3.1.1), $\phi(r)$ is the phase distortion of the system projected onto the entrance pupil, r is a position vector in the plane of the pupil, $w(r)$ is the pupil function and λ is the wavelength. In order to learn the required nonlinear input–output mapping, at least one of our layers needed a nonlinear transfer function. Secondly, supervised training algorithms, such as backpropagation, have so far proved superior to unsupervised or self-organizing networks in learning complex functional relationships. Also, after training, this architecture can be implemented quite efficiently using digital hardware and is therefore appropriate for a real-time control system.

Examination of (G3.1.1) reveals one final factor influencing our design for the neural network. Notice that for an even pupil function ($w(r) = w(-r)$) the PSF intensity is the same for the two phase distortions $\phi(r)$ and $-\phi(-r)$. Thus, any single PSF may be caused by a pair of related but different phase distortions. Without added information the neural network processing architecture cannot resolve the ambiguity of the mapping from image data to phase distortion. One possible solution, and the one which we utilized, consists of using data from two images obtained at two distinct image planes slightly out of focus. The added information in the second image plane breaks the ambiguity of the problem making the mapping from input intensity data to phase distortion unique (Gonsalves 1982, Paxman and Fienup 1988).

G3.1.3 Preprocessing

Specific applications of the phase-recovery neural network required different *preprocessing procedures*. In B4.4 general, only one preprocessing step was required for all applications. In every case, we normalized each input intensity image by the magnitude of the brightest pixel within that image.

Other preprocessing depended upon the individual circumstances of the application. For instance, the preprocessing for the Hubble data consisted of centroiding each PSF image, subtracting off the background pedestal intensity introduced by the camera electronics, binning adjacent pixels of the image to decrease the resolution of the image and the number of inputs into the neural network and, as discussed before, normalizing the resultant image to the brightest pixel.

Our experience has shown that the neural network does not require high-resolution input data to recover low spatial frequency distortion. We have found (Sandler *et al* 1991b) that the lowest 18 spatial distortion modes can be recovered from input pixels with angular resolution three times larger than diffraction-limited. Therefore, we typically used pixels much larger than is common for imaging. The ability to use large input pixels increased the signal-to-noise ratio of the input data. It also reduced the number of inputs to the neural network allowing the size and computational complexity of the network to be reduced; shortening the training time required, and increasing the throughput of the system.

G3.1.4 Training methods

C1.2.3 The network was trained by adjusting the synaptic weights using the externally supervised *backpropagation algorithm* (Rumelhart 1986). In principle, the training data can be generated either with numerical simulations or direct optical measurements, but for all the applications which we investigated it was more practical to use the former method.

Our experience has shown that our neural networks typically converged to a small residual error after a few 100 000 training iterations. Generally, this was sufficient for the real-time adaptive optics applications. However, when estimating static aberrations an extremely high degree of accuracy was desired, and approximately 1 000 000 training iterations were used. Since the generation of training data requires the computation of a two-dimensional FFT, and was therefore quite time consuming, the generation of 100 000 to 1 000 000 sets of training data was prohibitive. Instead, a smaller number of training patterns were generated and passed through the network several times. There is a limit to the minimum size of the training set though. Care must be taken to include enough independent realizations of distortion to ensure that the network learns a general functional input–output mapping for the phase-retrieval problem as opposed to 'memorizing' the mapping for only a few patterns. By testing the neural network with data not in the original training set, we have found that approximately 4000–6000 individual input patterns are all that are required to ensure that no memorization occurs and that a general mapping is learned for the phase-retrieval problem.

G3.1.5 Output interpretation

The neural network may be trained to estimate phase aberration in terms of a variety of representations. For example, we have generated networks which determine average phase and wavefront slopes over small subapertures of the entrance pupil, or alternatively, we have trained networks which determine the phase aberration with respect to orthogonal functions defined over the entrance pupil. The former representation can sometimes be useful, but for problems such as the recovery of the low spatial frequency static aberrations of the Hubble Space Telescope, the latter representation has conspicuous advantages since most low-order optical aberrations may be described with only a few orthogonal functions. Therefore, we chose to train most of our neural networks to estimate phase distortion in terms of a finite number of the well known Zernike polynomials (Born and Wolf 1970). In this configuration each output node of the network produced a coefficient, Z_i, such that the phase aberration was approximated by

$$\phi(r) = \sum_i^N Z_i P_i(r).$$
(G3.1.2)

In (G3.1.2), $P_i(r)$ is a radial Zernike polynomial orthogonal over a circular entrance pupil.

G3.1.6 Development

Our neural network development was accomplished on a PC compatible 386 computer hosting a general purpose digital signal processing (DSP) board. The DSP board was manufactured by Atlanta Signal Processing and contained a single Texas Instruments TMSC3020C30 DSP microprocessor running at 33 MHz. All source code was written in C and was developed by ourselves. The software ran under the DOS operating system with a minimal real-time kernel running on the DSP board. The development tools required were minimal and consisted only of a C compiler and debugger for the PC/DOS environment and the standard Texas Instruments C compiler and assembler for the TMSC3020C30.

The performance of the network was quite good despite the minimal nature of the software and hardware required to develop the system. For a typical phase-recovery network as described here the process of generating 4000–6000 training data sets and then training the network could be accomplished in a 36 hour period.

Although our inherently nonparallel development system was not the most efficient for the real-time implementation of the neural network, we tested the throughput rate of a phase-retrieval neural network on our system in order to estimate the rate at which a single modest processor could measure phase distortion. The network was designed to estimate eight Zernike coefficients per cycle and could complete one cycle in 122 μs. An adaptive optics system compensating the lowest eight Zernike modes at a few hundred Hertz is sufficient for a modest telescope at a good astronomical site with relatively small atmospheric aberration, so even our modest computing power is sufficient for this case. A more sophisticated processor could easily implement a network designed for sites with more atmospheric turbulence or a larger telescope, where it is necessary to estimate more coefficients at similar or faster rates.

G3.1.7 Comparison with traditional methods

Conventional methods of estimating wavefront aberration in adaptive optics systems measure local slopes of the wavefront over subapertures within the larger telescope aperture (Hardy *et al* 1977). A linear least-squares algorithm is then used to reconstruct the phase profile from the slope data. Conventional sensors require complicated beam-train optics, which tend to add extra distortion to the wavefront, lead to photon losses, and introduce uncommon and therefore hard to measure optical aberrations between the sensor and viewing camera. The neural network approach eliminates these difficulties. It operates directly on the quantity of primary interest for astronomical imaging, namely the point spread function (PSF) of the system. The optical requirements are much simpler and allow the PSF sensor to be located near the telescope aperture and the astronomical viewing camera, minimizing the uncommon optics. The approach is also flexible because the network can be optimized for a variety of different conditions.

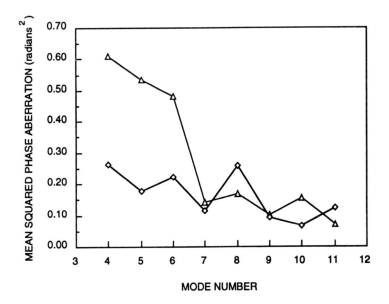

Figure G3.1.2. Experimental statistics comparing Hartmann sensor measurements of atmospheric turbulence with neural network estimates. ($\triangle$), average squared phase aberration per Zernike mode measured by a Hartmann sensor. ($\lozenge$), average squared difference between a Hartmann sensor measurement and a neural network estimate per Zernike mode. Mode 4 is focus, 5 and 6 are astigmatisms, 7 and 8 are coma, and 11 is third-order spherical aberration. (Reprinted with permission from *Nature* vol. 351, 23 May 1991, page 302. Copyright 1991 Macmillan Magazines Limited.)

With the assistance of our collaborator R Q Fugate, director of the Starfire Optical Range (SOR) at the Phillips Laboratory located on Kirtland Airforce Base, we were able to test the performance of a neural network system by making simultaneous measurements of atmospheric aberrations with a Hartmann sensor and a neural network (Sandler *et al* 1991b). The measurements were obtained with the 1.5 m telescope located at SOR. Figure G3.1.2 shows the mean-squared magnitude of the phase aberration per Zernike mode as reconstructed by the Hartmann sensor and the mean-square difference between the reconstructed phase and the neural network estimates of phase for modes 4 (focus) through 11 (spherical aberration). Note that the difference between the network's predictions and the Hartmann sensor reconstructions for

---

modes 4–7 inclusive differ by less than $\lambda/14$ rms. These data indicate that if used in an adaptive optics system, the network would reduce the mean-square wavefront error from 1.77 rad^2 to at most 0.78 rad^2, corresponding to an increase in effective image resolution by a factor of approximately 3 (Fried 1966, Angel *et al* 1990). For Zernike modes ≥ 8 there is less power in the turbulent aberration spectrum, making it difficult to compare results. The measurement uncertainties introduced by Hartmann-sensor noise, unshared optical paths, alignment errors and aberrations in the static beam train are of the same order as the phase distortions induced by the atmosphere.

The Hubble Aberration Recovery Project (HARP) mentioned earlier provided us with a chance to compare the neural network method with traditional iterative techniques for phase retrieval. The iterative nature of the traditional algorithms and their computational complexity make these methods unsuitable for real-time systems, but they can be used to produce very accurate estimates of static telescope aberrations. During the HARP effort the neural network was tested on simulated images produced by the Space Science Telescope Institute. The network performed quite well and was able to estimate the aberration to within 0.3%. On real Hubble Space Telescope data the neural network estimates of aberration agreed with the average of the estimates of other algorithms to within the 5% scatter found in the estimates made by all the investigators.

G3.1.8 Conclusions

We have demonstrated that a simple optical sensor with a neural network processor can measure low-order aberrations created by atmospheric turbulence. We have also proven the method as a simple and quick means of estimating the static aberration in an astronomical telescope. Good agreement between the neural network method and more conventional methods of estimating optical wavefront distortion show that the neural network can be an effective tool for both adaptive optics and testing of large optics. The quick throughput of the technique along with the ease with which it may be implemented make it an attractive means of checking and adding supplementary data even when other, more traditional algorithms, are used.

References

Angel J R P, Wizinowich P, Lloyd-Hart M and Sandler D G 1990 Adaptive optics for array telescopes using neural-network techniques *Nature* **348** 221

Barrett T K and Sandler D G 1993 Artificial neural network for the determination of the Hubble Space Telescope aberration from stellar images *Appl. Opt.* **32** 1720–7

Born M and Wolf E 1970 *Principles of Optics* (New York: Pergamon) pp 464–6

Ealey M A and Wellman J A 1994 Xinetics low cost deformable mirrors with actuator replacement cartridges *Adaptive Optics in Astronomy (Proc. SPIE* **2201**) ed M A Ealey and F Merkle pp 680–7

Fienup J R 1982 Phase retrieval algorithms: a comparison *Appl. Opt.* **21** 2758

——1987 Reconstruction of a complex-valued object from the modulus of its Fourier transform using a support constraint *J. Opt. Soc. Am.* A **4** 118

Fried D L 1966 Optical resolution through a randomly inhomogeneous medium for very long and very short exposures *J. Opt. Soc. Am.* **56** 1372–9

Gardner C S 1989 Sodium resonance fluorescence lidar applications in atmospheric science and astronomy *Proc. IEEE* **77** 408–18

Gonsalves F A 1982 Phase retrieval and diversity in adaptive optics *Opt. Eng.* **21** 829–32

Goodman J W 1968 *Introduction to Fourier Optics* (San Francisco, CA: McGraw-Hill) pp 57–76

Hardy J W, Lefebvre J E and Koliopoulous C L 1977 Realtime atmospheric compensation *J. Opt. Soc. Am.* **67** 360–9

Lloyd-Hart M, Wizinowich P, McLeod B, Wittman D, Colucci D, Dekany R, McCarthy D, Angel J R P and Sandler D G 1992 First results of an on-line adaptive optics system with atmospheric wavefront sensing by an artificial neural network *Astrophys. J. Lett.* **390** L41–4

Paxman R G and Fienup J R 1988 Optical misalignment sensing and image reconstruction using phase diversity *J. Opt. Soc. Am.* A **5** 914

Roddier F 1988 Curvature sensing and compensation: a new concept in adaptive optics *Appl. Opt.* **27** 1223–5

Rosenblatt F 1962 *Principles of Neurodynamics* (Washington, DC: Spartan)

Rumelhart D E, Hinton G E and Williams R J 1986 *Parallel Distributed Processing: Explorations in the Microstructure of Cognition* vol 1 (Massachusetts: MIT Press) pp 318–62

Sandler D G, Barrett T K and Fugate R Q 1991a Recovery of atmospheric phase distortion from stellar images using an artificial neural network *Active and Adaptive Optical Components (Proc. SPIE* **1543**) ed M A Ealey pp 491–9

Sandler D G, Barrett T K, Palmer D A, Fugate R Q and Wild W J 1991b Use of a neural network to control an adaptive optics system for an astronomical telescope *Nature* **351** 300–2

Wizinowich P, Lloyd-Hart M, McLeod B, Colucci D, Dekany R, Wittman D, Angel J R P, McCarthy D, Hulburd W G and Sandler D G 1991 Neural network adaptive optics for the multiple-mirror telescope *Active and Adaptive Optical Components (Proc. SPIE* pp **1542**) ed M A Ealey pp 148–58

G3.2 Neural multigrid for disordered systems: lattice gauge theory as an example

Martin Bäker, Gerhard Mack and Marcus Speh

Abstract

Multigrid relaxation algorithms for discretized partial differential equations require learning steps when disorder is present. They have to determine the interpolation operators from coarse to fine grids (disordered 'wavelets'). The matrix elements of these operators are considered as connection strengths of a neural net. Learning by backward propagation is too slow. An efficient alternative algorithm is presented. It is based on the multiscale philosophy where objects on larger scales are built from objects of smaller scales. Applications include gauge-covariant propagators in lattice gauge theory, fissures in materials, and so on.

G3.2.1 Project overview

G3.2.1.1 Scope of application: lattice gauge theory as a special case

The multigrid method is an extremely efficient method for solving discretized partial differential equations, especially linear ones such as the Laplace equation or Maxwell's equations (Brandt 1984). It fails, however, in the disordered case, that is, when there is no approximate translational invariance. In this case, the interpolation operators from coarse to fine grids cannot be guessed *a priori*. These operators must be able to approximate the poorly converging ('smooth') parts of the error. They can be regarded as wavelets. (By 'wavelets' we mean a set of localized objects out of which every function can be generated. As the problem is not translationally invariant, the usual notion of wavelets is not appropriate here. To be more specific, the operators correspond to the scaling functions of multiresolution analysis.) We use a neural network design to compute them.

There are many potential applications: propagation of fissures in materials, low-lying states and their localization properties in continuous spin glasses, growth of snowflakes, and gauge-covariant propagators in lattice gauge theory. In the last case, the disorder is in the gauge field, see below. In hybrid Monte Carlo simulations of lattice gauge theories with dynamical fermions (Montvay and Münster 1994) the computation of the Dirac propagator is the most time consuming step.

G3.2.1.2 Differential equations and lattice gauge theory

Consider (real or) complex vector-valued functions f, ξ, ϕ on a d-dimensional hypercubic lattice Λ_0 of lattice spacing $a_0 = 1$. The value of ξ at site $z \in \Lambda_0$ is denoted by $\xi(z)$, etc.

Given a linear operator L, we consider the inhomogeneous linear equation and the associated eigenvalue problem for the associated positive operator D, $D = L$ if $L > 0$; $D = L^*L$ otherwise:

$$L\xi = f \tag{G3.2.1}$$

$$D\phi_n = \epsilon_n \phi_n. \tag{G3.2.2}$$

We are interested in sparse matrices L which come from discretizing partial differential equations, especially elliptic ones. To be more specific, we consider lattice gauge theory as an example.

SU(2) lattice gauge theory. We define a link as a pair $b = (w, z)$ of nearest-neighbor sites; $-b = (z, w)$ is the link in the opposite direction. A lattice gauge field assigns an $SU(2)$-matrix $U(b)$ to every link b of the lattice, with $U(-b) = U(b)^{-1}$. $SU(2)$-matrices are unitary complex 2×2 matrices of determinant 1. These matrices are distributed randomly with a Boltzmannian probability distribution $\propto \exp(-\beta S_W(U))$, completely analogous to a thermodynamical problem. S_W is the standard Wilson action of lattice gauge theory (Creutz *et al* 1983):

$$S_W(U) = \sum_p \mathrm{Tr}(1 - U(\partial p)) \qquad \text{with} \quad U(\partial p) = U(b_4)U(b_3)U(b_2)U(b_1)$$

for an elementary square p of the lattice with links $b_1, \ldots, b_4$ at its boundary. Note that variables $U(z, w)$ at different links are correlated.

The $SU(2)$ matrices act on the lattice functions f, ξ, ϕ, which therefore have to be two-component complex vectors. The matrices are used as parallel transporters—whenever two vectors $\xi(z_1)$, $\xi(z_2)$ have to be compared (to calculate their difference), a path $C = b_n \circ \cdots \circ b_2 \circ b_1$ leading from z_1 to z_2 has to be chosen. The vector $\xi(z_1)$ is transported along the path using the matrix $U(C) = U(b_n)U(b_{n-1}) \cdots U(b_1)$. The result of this transport is path-dependent. The equations of lattice gauge theory are, for instance, gauge-covariant. They involve discretized versions of covariant differential operators in the continuum. These discretized versions are obtained from their noncovariant relatives by including a parallel transporter into finite differences between nearest neighbors: $\xi(z) - \xi(w)$ gets replaced by $\xi(z) - U(z, w)\xi(w)$. Standard discretized differential operators have to be changed accordingly. The negative covariant Laplacian $-\Delta$ is a positive operator defined by

$$-\Delta\xi(z) = a_0^{-2} \sum_{w \text{ n.n. } z} [\xi(z) - U(z, w)\xi(w)].$$

Summation is over all nearest neighbors w of z.

G3.2.1.3 Criticality and the multiscale principle

Consider the inhomogeneous equation (G3.2.1) with

$$L = -\Delta + (\delta m^2 - \epsilon^\Delta)\mathbb{1} > 0. \tag{G3.2.3}$$

ϵ^Δ is the lowest eigenvalue of $-\Delta$ so that the lowest eigenvalue of L is δm^2.

The problem is ill-posed when there is an eigenvalue of zero. When the lowest eigenvalue δm^2 is very close to zero the problem is called critical and traditional local relaxation algorithms and the conjugate gradient algorithm suffer from critical slowing down—the time needed for the solution of the equation grows when δm^2 decreases because the convergence of this algorithm is determined by the condition number, the quotient between the largest and the smallest eigenvalue. Local algorithms are not able to address the parts of the error corresponding to low eigenmodes. In ordered systems this is due to the fact that these modes are the smoothest, i.e. they do not change appreciably on a small length scale of order a_0.

The multiscale approach consists in using nonlocal updating steps of the form

$$\delta\xi(z) = c_x w_x(z) \tag{G3.2.4}$$

(or $\delta\xi(z) = c_x L^* w_x(z)$ if $D = L^*L$). Herein w_x is an appropriate set of functions (called interpolation operators or wavelets), having supports $[x]$ of diameter of order 2^k lattice spacings, $k = 1, 2, 3, \ldots$ These functions have to be able to approximate the low eigenmodes of D, which are not affected by the relaxation. In a multigrid method we define a sequence of lattices Λ_k, $k = 1, 2, 3, \ldots, N$, of increasing lattice spacing $a_k = 2^k a_0$ and we label these functions by the sites $x \in \Lambda_k$ of the kth layer. After doing some relaxation steps to eliminate the high-frequency parts of the error, the equation is transported to the coarser layers of the multigrid where the appropriate weights c_x are determined, usually by performing a relaxation on these layers.

G3.2.1.4 Features of disordered systems: localized states

The covariant Laplace operator (G3.2.3) contains disorder through the randomly distributed gauge fields. As stated above, the low-lying modes have to be approximated well by the functions w_x. Frequently, low-lying states in disordered systems show localization properties. An example is shown in figure G3.2.1.

It shows the ground state of the gauge-covariant Laplacian in a fairly disordered $SU(2)$ lattice gauge field on a two-dimensional square lattice with periodic boundary conditions. The figure is for a particular gauge-field configuration from the ensemble with $\beta = 1$.

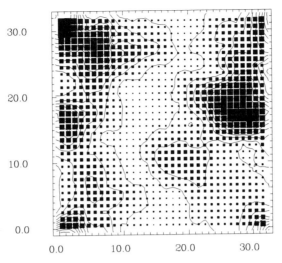

Figure G3.2.1. Lowest mode of the two-dimensional covariant Laplace operator in an $SU(2)$-gauge field at $\beta = 1$ as a 3D plot.

G3.2.2 Design

G3.2.2.1 Motivation for a neural network solution

Necessity for the computation of accurate wavelets for disordered systems. In standard local relaxation algorithms, the slow-to-converge modes are the smooth modes. The iterative solution of the inhomogeneous equation (G3.2.1) will converge quickly only if all smooth modes ϕ_n can be well approximated by a superposition of just a few wavelets. This motivates the prescription to make the wavelets as smooth as is consistent with their support properties. In disordered systems the differential operator D defines an appropriate notion of smoothness. The smoothest functions are those obtained by superposition of the lowest eigenmodes of D. Smooth functions are not known *a priori*, instead they have to be calculated by the neural network.

Let $[x] \subset \Lambda_0$ be a hypercube of sidelength of the order of $a_k = 2^k a_0$ which is determined by a site $x \in \Lambda_k$. $[x]$ is called a block. Demand that w_x is restricted to this block

$$w_x(z) = 0 \qquad \text{for} \quad z \notin [x]. \tag{G3.2.5}$$

Extremalization of $\langle w_x, Dw_x \rangle = \sum_z \bar{w}_x(z)Dw_x(z)$ subject to (G3.2.5) and to the normalization constraint $\langle w_x, w_x \rangle = 1$ is equivalent to finding the lowest eigenmode

$$D^{D,[x]} w_x(z) = w_x(z)\epsilon(x) \qquad z \in [x] \tag{G3.2.6}$$

of the eigenvalue problem with Dirichlet boundary conditions on the boundary of $[x]$. $w_x(z)$ are matrices, as they are used to transport vector-valued functions between the layers of the network. In our example, f, ξ etc are two-component vectors, and each eigenvalue problem has two degenerate two-component vector-valued functions as solutions. They are combined into one 2×2 matrix. In other problems, an appropriate number of nearly degenerate vector-valued eigenvectors are to be combined into a matrix.

For large hypercubes $[x]$ this eigenvalue problem looks just as hard as the original one. The iteratively smoothing unigrid algorithm (ISU) (Bäker *et al* 1992, Bäker 1995a, b) is designed to solve it. This algorithm can be considered as a neural net—the coefficients $w_x(z)$ are naturally identified with connection strengths between nodes x, z in a neural network whose nodes are the sites of the layers $\Lambda_0, \ldots, \Lambda_N$ of the multigrid; their iterative determination amounts to a learning process. In contrast to standard neural networks these connection strengths are matrices rather than numbers because they map vectors (the field on one layer of the grid) on other vectors (the field on another layer).

Example of localized states and its decomposition into wavelets. In figure G3.2.2 we show the modulus $\|w_x(z)\|^2$ of the solutions of the eigenvalue problem equation (G3.2.6) for the problem whose ground state was shown in figure G3.2.1. The solution for the four largest overlapping blocks [x] is shown; the eigenvalues $\epsilon(x)$ are also indicated. One clearly sees how this furnishes a decomposition of the ground state into separate patterns. The patterns in different blocks [x] have slightly different eigenvalues. The contribution to the ground state of those patterns with slightly larger eigenvalues appear to be significantly suppressed. The example shows that the determination of the wavelets is really a problem in cognition. One determines constituent parts of objects. Here, the objects are the low-lying modes of D.

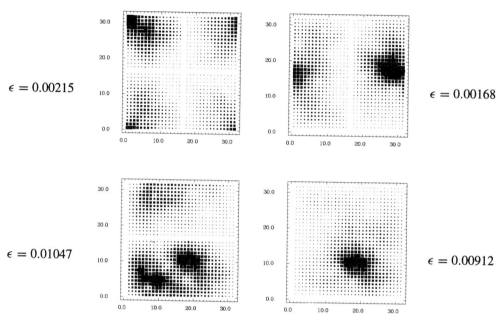

$\epsilon = 0.00215$

$\epsilon = 0.00168$

$\epsilon = 0.01047$

$\epsilon = 0.00912$

Figure G3.2.2. Solutions of the eigenvalue problem (G3.2.6) for the same gauge-field configuration as in figure G3.2.1. Comparing with this figure, it can be clearly seen that the modes with the lowest eigenvalues contribute most to the eigenmode with periodic boundary conditions.

Black box description. During the learning phase, the algorithm does not need the input of test patterns. It uses the given connection strengths on layer Λ_0 (the problem operator) to generate the connection strengths $w_x(z)$ and $D^k(x_1, x_2)$, see below. As a byproduct one obtains the ground state $\phi_0(z)$. It comes out as the strength of the connection from the single node x in the last layer Λ_N to node z of the input/output layer Λ_0. The algorithm can be generalized to yield several lowest-lying eigenmodes of D (Bäker 1995b).

Afterwards, the right-hand side $f(z)$ is given as an input pattern to node $z \in \Lambda_0$. After the computation, node z furnishes the result $\xi(z)$. This step can be repeated for arbitrarily many right-hand sides without any need to compute the connection strengths anew.

G3.2.2.2 Topology

Neural multigrid, implementation of wavelets as connection strengths. The topology of the neural network in the simplest case of a three-grid is shown in figure G3.2.3. The bottom layer Λ_0 is the input/output layer. The top layer consists of a single node. The connections whose strength determines the wavelets are shown in black. They are determined by a learning process. The input connections which define the problem (i.e. D) are the dotted lines between nodes of Λ_0. The other dotted lines are auxiliary connections. The nonhorizontal ones are computed anew in every iteration step of the learning process, each as a solution of a quadratic equation. The dotted horizontal lines stand for connections which define coarse-grained relatives D^j of the basic operator D on scale a_j. They are determined by D and by the wavelets.

In general, there are N layers and the wavelets $w_x(z)$ connect sites x of these layers Λ^j to sites z in the overlapping blocks [x] of sidelength $2^{j+1} - 1$ lattice spacings inside the fundamental layer Λ_0.

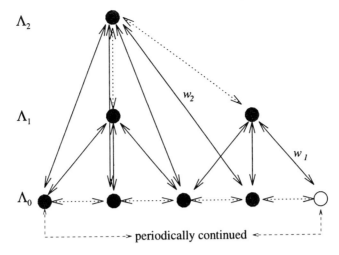

Figure G3.2.3. Topology of the neural network for the case of a one-dimensional three-grid. The fundamental grid consists of four points, the intermediate of two. Full lines denote the connections of the network $w_x^k(z)$, dotted lines are auxiliary connections used for the updates and the connections on the input/output layer are given by L.

All neural connections are bidirectional because they are used to transport functions back and forth between the layers of the network. If $w_x(z)$ is the connection strength from x to z then the adjoint matrix $w_x(z)^*$ gives the connection strength from z to x.

G3.2.2.3 Learning

Necessity to deviate from textbook learning rules. The layers of the neural network correspond to the layers of the multigrid. Their number increases logarithmically with the lattice size. We are interested in very large lattices, that is, many layers. The absence of critical slowing down means that the convergence rate should not increase much faster than the lattice volume. However, learning by a standard *backpropagation* C1.2.3 *algorithm* deteriorates quickly with the number of layers. Tests confirmed that it is totally useless for our purpose.

From scale to scale. For clarity we write $w_x^k(z)$ in place of $w_x(z)$ for the strength of the connection from node x in layer Λ^k to $z \in \Lambda_0$. These wavelets (or interpolation operators) are matrices to be determined as solutions of the eigenvalue equations (G3.2.6).

These solutions are determined recursively for $k = 0, 1, 2, \ldots$, using the wavelets (connection strengths) w_y^j for $j < k$ which were determined previously. This is the crucial point of the learning process—the larger the blocks become, the harder it is to determine the eigenvectors on the blocks. Only by using all the information about slowly-converging modes (smooth wavelets) already gained on the smaller scales are we enabled to solve the problem on the larger scale in a reasonable amount of time. On the larger scales there are fewer functions smooth on this scale and therefore we need fewer wavelets, that is, fewer grid points.

We will now describe the learning process in greater detail: let the effective operators D^k be defined by

$$D^k(x_1, x_2) = \sum_{z \in \Lambda_0} w_{x_1}^* D w_{x_2}(z).$$

Only the diagonal part $x_1 = x_2$ is needed. The eigenvalue problem (G3.2.6) is equivalent to the extremality condition

$$\operatorname{tr} D^k(x, x) = \operatorname{extr}$$

subject to the constraint $\sum_z w_x^k(z)^* w_x^k(z) = 1$. It is solved by an iterative procedure as follows.

(i) Layer Λ_0:

$$w_x^0(z) = 1 \delta_{x,z} \quad (x \in \Lambda_0).$$

(ii) Layer Λ^k, $k > 0$: start with $w_x^k(z) = \mathbb{1}\delta_{\hat{x},z}$, where $\hat{x}$ is the central site of hypercube $[x]$. Use the already known wavelets w_y^j for $j < k$ to perform updatings of the following form. Sweep through all $j < k$, $y \in \Lambda_j$ and update the connection strengths $w_x^k(z)$ for all $z \in [y]$ by

$$\delta w_x^k(z) = w_y^j(z)c$$

with a matrix $c \equiv c(y, x)$ which is determined from the extremality condition $\operatorname{tr} D^k(x, x) = \operatorname{extr}$ subject to the constraint.

$c(y, x)$ are the auxiliary connection strengths mentioned above. They can be determined by the Lagrange multiplier method in terms of the solution of a quadratic equation (Meyer 1987). The neurons y have to perform two tasks. They must add up inputs $\sum_z w_y(z)\eta(z)$ linearly, and they must solve the quadratic equations to determine $c(y, x)$.

An alternative approach is also possible—the connection strengths are directly calculated as a solution to the eigenvalue equation (G3.2.6) via inverse iteration (Press *et al* 1989). This is done in the standard unigrid manner: first we relax the equation on the fundamental layer, afterwards it is transported to the next coarser layer, relaxed there to smoothen the error on this scale, and so forth going through all layers $j < k$. Finally, the error will be smooth on layer Λ_k and there seems to be a problem because the connection strengths to this layer are not yet known—we are trying to compute them. However, the error now has exactly the shape we are looking for, namely that of the lowest mode on this block; therefore a simple rescaling suffices to fulfill the normalization condition. This latter implementation is the one we actually used for the calculations.

G3.2.3 Performance

G3.2.3.1 Critical Laplace equation in an external non-Abelian gauge field.

In the example, about six sweeps through all the layers $j < k$, $y \in \Lambda_j$ sufficed to determine the wavelets $w_x^k(z)$ sufficiently accurately, irrespective of k, i.e. irrespective of the size of the support $[x]$. The larger the lattice, the larger k can be. In total, this gives a computational workload for computing the connection strengths $w_x(z)$ which goes like $V \ln^2 V$ with the volume V of the lattice. Afterwards, the iteration of the inhomogeneous equation by updates (G3.2.4) converged with asymptotic convergence time (time needed to reduce the error by a factor e) of about one V-cycle sweep through the multigrid irrespective of the lattice size and irrespective of how critical the problem is, that is, of how small δm^2 is (see figure G3.2.4). In a V-cycle sweep, each site of the multigrid is visited twice. The total computational workload goes with the volume like $V \ln V$. For large lattices, the network takes longer (by a factor of order $\ln V$) to learn how to do the job of solving the inhomogeneous equation than to finally do it.

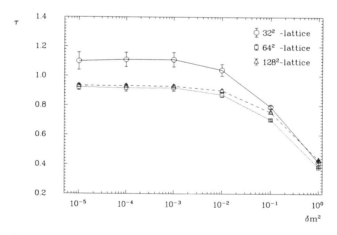

Figure G3.2.4. Performance of the ISU algorithm for the critical Laplace equation in an $SU(2)$-field. The inverse asymptotic convergence time τ is shown for different grid sizes as a function of the critical parameter δm^2. All configurations were equilibrated at $\beta = 1.0$. Lines are drawn only to guide the eye.

G3.2.4 Generalization to general problem solving strategies

The solution of equations like (G3.2.1) and (G3.2.2) can be viewed as an extremalization task, $\langle L\xi - f, L\xi - f \rangle = \min$. Inspection of the algorithm reveals that two basic pieces of structure are made use of to solve such a task.

(i) Composability of the connections. Given a connection with some strength $c(y, x)$ from node x to node y and a connection of strength $w_y(z)$ from node y to node z, a connection from x to z is specified.
(ii) Linearity is used to add strengths of connections between the same two nodes.

The second of these requirements could be relaxed, although this is fairly complicated and cannot be explained here. Moreover, here we used *a priori* chosen block shapes (hypercubes). This can be relaxed and replaced by an optimizing strategy for block shapes. Taking all of this for granted, one sees that the algorithm appears to be capable of generalization to a general multiscale strategy for solving optimization problems of general complex adaptive systems. A general framework was developed in Mack (1994, 1995a, b).

References

Bäker M 1995a Localization in two-dimensional lattice gauge theory and a new multigrid method *Int. J. Mod. Phys.* C **6** 85

——1995b A multiscale view of propagators in gauge fields *PhD Thesis* Hamburg, DESY-95-134

Bäker M, Kalkreuter T, Mack G and Speh M 1992 Neural multigrid for gauge theories and other disordered systems *Proc. Physics Computing '92 (Prague)* ed R A de Groot and J Nadrchal (River Edge, NJ: World Scientific)

Brandt A 1984 *Multigrid Techniques: 1984 Guide with Applications to Fluid Dynamics* GMD–Studie Nr. 85, Bonn

Creutz M, Jacobs L and Rebbi C 1983 *Phys. Rev.* **95** 201

Mack G 1994 Gauge theory of things alive and universal dynamics DESY-94-184 *Preprint*; also from the Los Alamos electronic bulletin board hep-lat@ftp.scri.fsu.edu 9411059

——1995a Gauge theory of things alive *Nucl. Phys. (Proc. Suppl.)* B **42** 923

——1995b Gauge theory of things alive: Universal dynamics as a tool in parallel computing *Prog. Theor. Phys. (Suppl.)* in press

Meyer A 1987 *Modern Algorithms for Large Sparse Eigenvalue Problems* (Berlin: Academic)

Montvay I and Münster G 1994 *Quantum Fields on a Lattice (Cambridge Monographs on Mathematical Physics)* (Cambridge: Cambridge University Press)

Press W H, Flannery B P, Teukolsky S A and Vetterling W A 1989 *Numerical Recipes* (Cambridge: Cambridge University Press)

Further reading

1. Briggs W L 1987 *A Multigrid Tutorial* (Philadelphia, PA: SIAM)

   An excellent introduction into the multigrid method.

2. Creutz M 1983 *Quarks, Gluons, and Lattices* (Cambridge: Cambridge University Press)

   A short but thorough introduction to lattice gauge theory. For more recent results see 1994 Montvay and Münster (references).

3. Farge M 1992 *Ann. Rev. Fluid Mech.* **24** 395

   This article explains the basics of wavelet transforms and explores some of their applications to fluid dynamics.

4. Hackbusch W 1985 *Multigrid Methods and Applications (Springer Series in Computational Mathematics 4)*

   Another multigrid introduction, broader-ranged and more mathematical than the book by Briggs.

G3.3 Characterization of chaotic signals using fast learning neural networks

Shawn D Pethel and Charles M Bowden

Abstract

The characterization of nonlinear and chaotic systems has become increasingly important in many areas of science and engineering (Campbell and Rose 1983). Features such as broadband power spectra and a lack of long-term predictability often make chaotic phenomena difficult to distinguish from purely random processes. In characterizing data, the most basic question to ask is whether or not the data is deterministic, and if so, what dimensionality? To this end, we show that neural networks can be used to detect determinism and to estimate dimensionality. Furthermore, we show that neural networks are capable of detecting multiple processes with different dimensionality in the same data set. Model-generated chaotic time series from the Mackey–Glass systems (Raisband 1990) are used to measure performance and robustness. The procedure is applied to the analysis of experimental results of spontaneously generated Brillouin signals from intense laser-field-excited single-model fibers (Harrison *et al* 1990).

G3.3.1 Background

A neural network trained on a time series of a single dynamical variable, say $x(t)$, can become a functional realization of the time series, and more profoundly, a global characterization of the chaotic attractor. It does this by the process of embedding. Using the embedding theorem of Takens (1981) with an embedding time, τ, between samples, a delay coordinate map may be constructed in the form:

$$f[x(t), \ldots, x(t+n\tau)] = x[t + (n+1)\tau].$$ (G3.3.1)

The argument of f is a delay coordinate vector which constitutes a point in a reconstructed phase space of embedding dimension d_e. The allowable range of embedding dimensions is governed by the Hausdorf–Besechovitch fractal dimension d_f (Raisband 1990) such that $d_f + 1 \le d_e \le 2\,d_f + 1$ and is sufficient to completely unfold the chaotic attractor in a subspace of the full phase space associated with the dynamical system. The embedding time τ is chosen to be small compared to the mean orbital period of the system and can be taken as e^{-1} of the peak of the correlation function for the variable $x(t)$, or alternatively as the first minimum of the average mutual information (Abarbanel *et al* 1994). Characterization of a dynamically chaotic system from a time series of a single dynamical variable is contingent upon the determination of the function f, equation (G3.3.1). There are several well known techniques which have been developed to model f from a single time series (Abarbanel 1993). These methods involve fitting data points in a reconstructed phase space using polynomials (Farmer and Sidorowich 1987), *radial basis functions* C1.6.2 (Casdagli 1989), or neural networks (Lapedes and Farber 1987, Albano *et al* 1992). Only neural networks offer a global approximation, i.e. the data need not be partitioned into small regions to be fitted separately. Global fitting is superior to local fitting in that it avoids discrepancies between neighborhoods and provides a smoother fit in the presence of noise (Casdagli 1989). A neural network trained accurately on a window of a time series becomes a functional approximation, in the form of (G3.3.1), of that time series. We show that the neural network can also answer basic questions about determinism and dimensionality in a data set. Of primary importance is the ability to distinguish pure noise from chaos. Both are aperiodic and broadband in frequency. Methods of calculating dimensionality, such as box-counting or correlation (Grassberger and Procaccia 1993), can be fooled by pure noise. In addition, a prohibitively large amount of data is required to make an estimate of dimensionality using these methods. We demonstrate that a neural network can make estimates of dimensionality using much smaller data sets and can distinguish between chaos and noise.

G3.3.2 Architecture

C1.2   The topology is best represented by considering the equation for a *single hidden layer neural network*,

$$G(I_i A_{ij}) B_{jk} = \Theta_k \tag{G3.3.2}$$

where $i = 1, \ldots, n$; $j = 1, \ldots, m$; and $k = 1, \ldots, q$. Here, we take A and B to be weight matrices, I and Θ are input and output vectors, respectively, and G is a threshold function, $[G(x)]_{ij} = \frac{1}{2}(1 + \tanh(x_{ij}))$. Input nodes are represented by I_i, hidden layer nodes by $[G(I_i A_{ij})]_j$ and output nodes by Θ_k. We also define a sequence of 'patterns' labeled by $\mathcal{P} = 1, \ldots, p$, where $\mathcal{P}$ refers to the pattern number, and Φ_{pk} is a matrix of correct or target outputs associated with an input matrix of patterns I_{pi}. In our case, I_{pi} is a matrix whose rows are delay coordinate vectors. Since equation (G3.3.1) has a scalar output, $k = 1$ and Θ_{pk} is a column vector of outputs of (G3.3.1) associated with the delay coordinate vectors I_{pi}. For a given set of training patterns I_{pi}, the process of learning is done by comparing the actual output, Θ_{pk} with the ideal target output Φ_{pk} and adjusting the weight matrices A and B such that the cost function

$$E \equiv \sum_p \|\Theta_p - \Phi_p\|^2 \tag{G3.3.3}$$

is minimized.

Training a neural network to model chaotic systems is extremely time consuming using conventional C1.2.3 methods based upon steepest-descent procedures such as *backpropagation* (Rumelhard and McClelland 1987). A common feature of steepest-descent algorithms is an asymptotic approach to a global minimum. This feature makes it computationally difficult to obtain the high-accuracy fits that are mandatory when building a predictive model. For this reason, backpropagation typically requires the use of a supercomputer. Further complications arise from the presence of local minima.

G3.3.3 Training

We have introduced a new training procedure and applied it to the analysis of nonlinear dynamical systems that achieve the high-accuracy, global approximation needed in modeling chaotic systems, while using computational resources such as a PC or workstation (Pethel *et al* 1993a, to be published). We note from (G3.3.2) that multilayer, feedforward neural networks are essentially several linear maps, separated by a simple and, in our case, invertible nonlinear function. Least-squares solutions for linear systems suffer none of the problems mentioned above and are well known through the Moore–Penrose generalized inverse formalism (Penrose 1955, Rao and Mitra 1971). We take advantage of the mostly linear structure of multilayer neural networks by using linear algebraic techniques to produce training. We call our method generalized inverse learning (GIL). Training neural networks using GIL makes the global modeling of chaotic systems practical.

The principal concept of GIL is based upon the application of a Moore–Penrose generalized inverse of a matrix M (Rao and Mitra 1971), defined as

$$I_L(M) = (M^T M)^{-1} M^T \tag{G3.3.4a}$$

where M^T is the transpose of M for the left generalized inverse, and as

$$I_R(M) = M^T (M M^T)^{-1} \tag{G3.3.4b}$$

for the right generalized inverse. When used to calculate the solution to systems of linear equations, the generalized inverse provides the solution that minimizes the mean square error (Rao and Mitra 1971). For a given set of target elements Φ_{pk} for the output corresponding to a set, I_{pi}, of input patterns, we drop the subscripts for convenience and define the zeroth-order hidden-layer output as $C_0 \equiv G(IA_0)$, where A_0 is a random matrix. Thus,

$$C_0 B = \Phi \tag{G3.3.5}$$

constitutes a linear system of equations for which a first-order, least-squares solution of B can be written as B_1, where

$$B_1 = \left(C_0^T C_0\right)^{-1} C_0^T \Phi \tag{G3.3.6}$$

with an associated root mean square error, (equation (G3.3.3)). A reduction in the error E is possible by modifying the weight matrix A_0. This is done by first calculating a new hidden-layer output, C_1, from which A_1 can be determined. Substituting B_1 into (G3.3.5), we calculate $C_1 = \Phi B_1^T (B_1 B_1^T)^{-1}$. Using $C_1 = G(IA_1)$, a new weight matrix can be calculated,

$$A_1 = \left(I^T I\right)^{-1} I^T G^{-1}(C_1) \tag{G3.3.7}$$

where the left generalized inverse of I has been used in (G3.3.7). Thus we have defined an iterative algorithm—generalized inverse learning (GIL)—for the calculation of weight matrices of multilayer neural networks. The algorithm can be written as follows:

$$B_{n+1} = I_L[G(IA_n)]\Phi \tag{G3.3.8}$$

$$A_{n+1} = I_L(I)G^{-1}[\Phi I_R(B_{n+1})]. \tag{G3.3.9}$$

In practice, we find convergence to a high-accuracy solution to be extremely fast—usually one or two iterations. GIL is generalizable to any number of hidden layers, but for most applications we find one hidden layer to be sufficient. Shepanski (1988) has developed optimal estimation theory (OET) for training single hidden-layer neural networks. OET is equivalent to GIL for the single hidden-layer case in which the hidden and input layers are the same size. An equivalent learning algorithm was developed independently by Biegler-König and Bärmann (1993) and tested using a simple nonlinear mapping example.

G3.3.4 Examples

We have used a variety of model equations to demonstrate that the neural network approach, together with GIL, can be a powerful new tool in the characterization and analysis of time series information from chaotic dynamical systems (Pethel et al to be published, 1993a). The neural network trained on an arbitrary chaotic time series using GIL becomes a functional realization of the entire series and thus forms a global approximation to the chaotic attractor. Using the ability of neural networks to form accurate functional approximations with small sets of data, we have demonstrated how data window extension, for stationary time data, can provide short-term prediction as well as long-term statistical properties (Pethel et al 1993b). The introduction of the fast and accurate training method, GIL, renders this powerful new method practical. Here we apply this new method to distinguish noise from chaos in an experimental data signal output. We have used GIL and the functional realization property of the neural network equation (G3.3.1) to show that the training error as a function of the embedding dimension undergoes a dramatic reduction above the fractal dimension of the chaotic signal, in strong contrast to the smooth response for a stochastic signal (Pethel et al 1993a). The algorithm was shown to train, with high accuracy, on low-dimensional chaotic signals, whereas the response of the training algorithm to stochastic signals is a least-squares coarse graining (Pethel et al 1993a). As a test, we trained a neural net using GIL on data generated from numerical integration of the Mackey–Glass delay differential equation (Raisband 1990) for three different sets of parameters in the region of chaotic dynamical evolution,

$$\frac{\mathrm{d}x}{\mathrm{d}t} = \frac{ax(t-s)}{1 + [x(t-s)]^{10}} - bx(t) \tag{G3.3.10}$$

where $a = 0.2$ and $b = 0.1$, with $s = 18, 30$ and 50. The dimensionality of the chaotic attractor increases with the parameter s. The neural net was trained using GIL with 30 hidden-layer nodes and 500 data points. The training error ε versus embedding dimension d_e is displayed in figure G3.3.2 for the Mackey–Glass system as well as for white noise. The dramatic dips in training error for the Mackey–Glass system indicate determinism as well as the increasing dimensionality with the parameter s. There are no significant dips in training error for the white noise.

This method was recently applied to the analysis of stimulated Brillouin scattering under CW laser pump conditions, involving a single Stokes and pump signal in a single-mode optical fiber (Pethel *et al* 1993a), as shown in figure G3.3.1. The Stokes signal data generated from a standard model was used to correlate the training performance of GIL with statistical and dynamical characteristics of the system determined by other calculational means. This procedure was applied to the temporal Stokes signal, using parameters which represent recent experiments (Harrison *et al* 1990, Gaeta and Boyd 1991) to show that the signal is largely the result of noise-generated phase waves (Englund and Bowden 1990, 1992) in the nonlinear strong-pump regime, whereas in the linear regime the signal is simply an amplification of the stochastic initiation process.

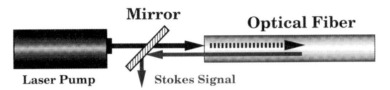

Figure G3.3.1. Experimental setup used by Harrison *et al* (1990) to measure Stokes output in a single-mode optical fiber under CW laser pump conditions.

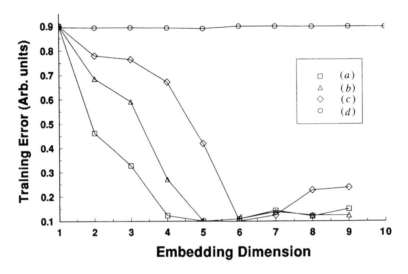

Figure G3.3.2. Training error versus embedding dimension for Mackey–Glass $a = 0.2$, $b = 0.1$, and (*a*) $s = 18$, (*b*) $s = 30$, (*c*) $s = 50$ and (*d*) for white noise.

Here, we demonstrate confirmation of these results by applying the procedure to experimental data (Harrison *et al* 1994). Subsequent to their initial experiments (Harrison *et al* 1990, Gaeta and Boyd 1991), where great care was taken to ensure the absence of feedback, the experiments were repeated with less than 3% reflectivity from the pump output end of the fiber (Harrison *et al* 1994). We report here, for the first time, the results of our method using GIL applied to 500 data points of the experimental time trace (Harrison *et al* 1994). The topology used here is a feedforward neural network with a single hidden layer of 30 nodes and a single output node (see equation (G3.3.2)). The number of input nodes is commensurate with the prescribed state space dimensionality d_s. Figure G3.3.3 shows the training error using GIL, ε, versus the prescribed state space dimensionality, d_s, for two separate experimental conditions. The open circles in figure G3.3.3 show the result of the response of the neural network training for the Stokes time trace for the condition without any reflectivity. The error ε as a function of d_s is approximately invariant, indicating a high-dimensional system characteristic of a stochastic process. For a Gaussian process with zero mean, the average of the training error over d_s is an index of the variance of the noise. In strong contrast are the results for the response to training using Stokes signal data observed with approximately 3% reflectivity at the end of the fiber. This is exhibited by the open squares in figure G3.3.2, which clearly indicate two strong dips, one at dimension $d = 3$ and another at dimension $d = 6$. This indicates the coexistence of two distinct, weakly coupled, chaotic dynamical processes, one process at low embedding

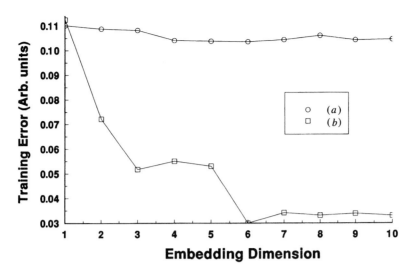

Figure G3.3.3. Training error versus embedding dimension for experimental Stokes data with (*a*) no reflectivity at the fiber end, and (*b*) with 3% reflectivity at the fiber end.

dimension ($d_e \approx 3$) and another process of higher dimensionality ($d_e = 6$). This is unexpected from the standard model, but subsequent recent analysis of the polarization sensitivity of the fiber used in the experiment revealed a significant qualitative difference in the Stokes signal for orthogonal polarizations for the incident pump field. The results have led to further study of the polarization properties of the single-mode fibers used in the experiments (Harrison and Lu, private communication).

G3.3.5 Conclusion

The accuracy and speed of GIL combine to facilitate a powerful new method to distinguish chaos from noise, as well as global characterization of chaotic attractors (Pethel *et al* to be published). In addition to providing an estimate of the fractal dimension (Raisband 1990) of arbitrary chaotic time signals, the procedures described allow the detection and characterization of multiple processes with different dimensionality.

References

Abarbanel H D J 1993 *Rev. Mod. Phys.* **65** 1331
Abarbanel H D J, Carroll T A, Pecora L M, Sidorowich J L and Tsimring L S 1994 *Phys. Rev.* E **49** 1840
Albano A M, Passamonte A, Hediger T and Farrell M E 1992 *Physica* **58D** 1
Biegler-König F and Bärmann F 1993 *Neural Networks* **6** 127
Campbell P and Rose H 1983 *Order in Chaos* (Amsterdam: North-Holland)
Casdagli M 1989 *Physica* **35D** 335
Englund J C and Bowden C M 1990 *Phys. Rev.* A **42** 2870
——1992 *Phys. Rev.* A **46** 578
Farmer J D and Sidorowich J L 1987 *Phys. Rev. Lett.* **59** 845
Gaeta A L and Boyd R W 1991 *Phys. Rev. Lett.* **44** 3205
Grassberger P and Procaccia 1993 *J. Phys. Rev. Lett.* **50** 346
Harrison R G and Lu W Private communication
Harrison R G, Ripley P M and Lu W 1994 *Phys. Rev.* A **49** R24
Harrison R G, Uppal J S, Johnstone A J and Moloney J V 1990 *Phys. Rev. Lett.* **65** 167
Lapedes A and Farber R 1987 *Technical Report* LA-UR-87-2662 Los Alamos National Laboratory
Penrose R 1955 *Proc. Camb. Phil. Soc.* **51** 406
Pethel S D, Bowden C M and Scalora M 1993a Chaos in optics *SPIE* **2039** 129
Pethel S D, Bowden C M and Sung C C 1993b *US Army Missile Technical Report* TR-RD-WS-93-5 Redstone Arsenal, AL
—— 1996 Global characterization of chaotic attractors: a novel, high-speed neural network approach (to be published)
Raisband S 1990 *Chaotic Dynamics of Nonlinear Systems* (New York: Wiley)
Rao C R and Mitra S K 1971 *Generalized Inverse of Matrices and Its Applications* (New York: Wiley)

Rumelhart D E and McClelland J L (ed) 1987 *Parallel Distributed Processing* vol 2 (Cambridge, MA: MIT Press)

Shepanski J F 1988 *Proc. IEEE Int. Conf. on Neural Networks* A I-464

Takens F 1981 Dynamical systems and turbulence *Lecture Notes in Mathematics* vol 898 ed D Rand and L S Young (Berlin: Springer) p 366

G4

Biology and Biochemistry

Contents

G4.1 A neural network for prediction of protein secondary structure

Burkhard Rost

Abstract

Currently, the prediction of a three-dimensional protein structure from a protein sequence poses insurmountable difficulties. As an intermediate step, a much simpler task has been pursued extensively: predicting one-dimensional strings of secondary structure. Here, a composite neural network is described which predicts three secondary-structure states (helix, strand, loop). The network system comprises two levels of feedforward networks (one hidden layer each) and a final jury decision over differently trained networks. Training is done by an adaptive-like backpropagation. An important key feature of the system is that the input is not only the sequence of one protein but the profile of a set of sequences from proteins which have the same three-dimensional structure. The combination of the problem-specific topology and the preprocessing of the input improve prediction accuracy from 62% to 72%. Furthermore, the specific topology and training procedure successfully correct for shortcomings of both simpler neural network and classical methods. Over the last few years, the network system has been the best automatic predictor in a very competitive area of research.

G4.1.1 Introduction to protein structure prediction

G4.1.1.1 Protein folding

Proteins are formed by joining amino acids into a long stretched chain, the protein sequence. They differ in length (from 30 to 30 000 amino acids) and in the arrangement of the amino acids (called residues, when joined in proteins). In water, the chain folds into a unique three-dimensional structure. The main driving force for folding is the need to pack residues for which a contact with water is energetically unfavorable (hydrophobic residues) into the interior of the molecule. This is only possible if the protein forms regular patterns of a macroscopic substructure called secondary structure (figure G4.1.1); for an introduction see Brändén and Tooze (1991).

G4.1.1.2 Sequence–structure gap

Today the sequence is known for more than 40 000 proteins (Bairoch and Boeckmann 1992), but the three-dimensional structures for only 3000 have been determined by crystallography (Bernstein *et al* 1977). Large-scale gene sequencing projects increase this sequence–structure gap further (Oliver *et al* 1992).

G4.1.1.3 Protein structure prediction

Protein three-dimensional structure determines protein function. It is well established that the three-dimensional structure is uniquely determined by the sequence (Anfinsen 1973). Thus, in principle, three-dimensional structure could be predicted from first principles. Unfortunately, the CPU time required is many orders of magnitude beyond today's scope (van Gunsteren 1993, Yun-yu *et al* 1993). However, it is of practical importance to know the three-dimensional structure, for example, for rational drug design.

G4.1.1.4 Protein structure prediction by alignment

The evolutionary pressure conserves protein function. Thus, protein structure is more conserved than sequence. Evolution has created pairs of proteins which have similar structure but only 25% identical residues (Sander and Schneider 1991). Therefore, three-dimensional structure can be predicted accurately by homology if a protein with sufficient sequence identity and known three-dimensional structure is found in the databank. Homology modeling reduces the sequence–structure gap by about 10 000 proteins (Sander and Schneider 1993, Rost and Sander 1994d).

G4.1.1.5 Drastic simplification of the prediction problem

If homology modeling is not applicable, that is, for about 30 000 of the known sequences, the prediction problem has to be simplified. An extreme simplification is the prediction of one-dimensional strings of secondary-structure assignment (figure G4.1.1). One tool that has been applied to various aspects of the protein structure prediction problem is the artificial neural network (ANN) (McGregor et al 1989, Bengio and Pouliot 1990, Bohr et al 1990, Bossa and Pascarella 1990, Holbrook et al 1990, Kneller et al 1990, Petersen et al 1990, Brunak 1991, Friedrichs et al 1991, Hirst and Sternberg 1991, Böhm et al 1992, Ferrán and Ferrara 1992b, Ferrán and Ferrara 1992a, Frishman and Argos 1992, Goldstein et al 1992a,

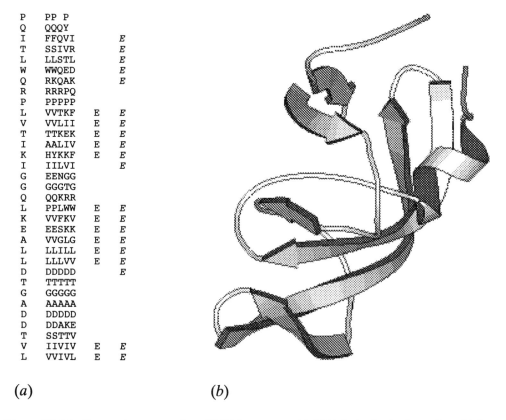

P	PP P		
Q	QQQY		
I	FFQVI		E
T	SSIVR		E
L	LLSTL		E
W	WWQED		E
Q	RKQAK		E
R	RRRPQ		
P	PPPPP		
L	VVTKF	E	E
V	VVLII	E	E
T	TTKEK	E	E
I	AALIV	E	E
K	HYKKF	E	E
I	IILVI		E
G	EENGG		
G	GGGTG		
Q	QQKRR		
L	PPLWW	E	E
K	VVFKV	E	E
E	EESKK	E	E
A	VVGLG	E	E
L	LLILL	E	E
L	LLLVV	E	E
D	DDDDD		E
T	TTTTT		
G	GGGGG		
A	AAAAA		
D	DDDDD		
D	DDAKE		
T	SSTTV		
V	IIVIV	E	E
L	VVIVL	E	E

(a)   (b)

Figure G4.1.1. Structural representation of HIV-1 protease with PDB (a databank of proteins with known three-dimensional structure) code 1HHP (Bernstein et al 1977)) in one and three dimensions. (a) Amino acids for the first 33 residues (one letter code, first column); alignment of five proteins with the same three-dimensional structure as HIV-1 protease (second column); secondary structure computed from three-dimensional structure using the program DSSP (dictionary of secondary structures of proteins, a program that computes secondary-structure segments from three-dimensional coordinates, Kabsch and Sander 1983a), H; strand = E, rest = blank (third column); and a typical prediction by the neural network program (Rost and Sander 1994b) for secondary structure (in italics, fourth column). (b) The protein chain in three-dimensions is plotted schematically as a ribbon. Strands are indicated by arrows; the short helix is on the right towards the end of the protein. Graph by Christos Ouzounis (European Molecular Biology Laboratory) using the program MOLSCRIPT (Kraulis 1991).

1992b, Hayward and Collins 1992, Muskal and Kim 1992, Pancoska *et al* 1992, Xin *et al* 1992, Andrade *et al* 1993, Dubchak *et al* 1993, Fariselli *et al* 1993, Ferrán and Pflugfelder 1993, Maclin and Shavlik 1993, Metfessel *et al* 1993, Presnell and Cohen 1993, Rost and Sander 1993c, Rost and Sander 1993a, Sasagawa and Tajima 1993, Tchoumatchenko *et al* 1993, Dombi and Lawrence 1994, Radomski *et al* 1994, Rost and Sander 1994a, 1994c, Tolstrup *et al* 1994).

G4.1.2 Design process

G4.1.2.1 Motivation for a neural network solution

Even the simplified task of predicting secondary structure is a difficult problem. Thus, secondary-structure prediction became a playground to apply any fancy new pattern classification techniques, for example, neural networks (Bohr *et al* 1988, Qian and Sejnowski 1988, Holley and Karplus 1989). The hope was that neural networks could use higher-order correlation in the data. However, this failed—neural networks with and without a hidden layer were equally accurate (Holley and Karplus 1989). The motivation to try again was twofold: first, evolutionary records provide a rich resource of structural information which should contain higher orders of correlation; and second, some disadvantages of both neural network and non-neural network predictions should be correctable by alternatives to *backpropagation training* (Stolorz *et al* 1992) or composite neural networks.   C1.2.3

G4.1.2.2 General description of the neural function

The task is to classify residues from a protein into three secondary-structure types. A window of a adjacent residues is taken from a protein sequence and input to the network. The output consists of three units for the secondary structure of the residue in the center of the input window. The window is shifted through the whole protein, such that a protein with R residues provides R classification examples.

G4.1.2.3 Topology

Helices extend over at least four residues; the average length of a helix is typically some ten residues. A simple neural network as described in the previous paragraph does not capture the correlation between secondary-structure states of adjacent residues. Thus, for example, the average length of a predicted helix is about four instead of ten residues. Correlations between adjacent residues can be introduced by using a second level of structure-to-structure neural network (figure G4.1.2). Such a second level of neural network improves overall prediction accuracy only marginally (Qian and Sejnowski 1988), but the average length of predicted secondary-structure segments is more similar to observed averages than for the first-level sequence-to-structure neural network (Rost and Sander 1992, 1993b, 1994b). A further difficulty with a simple neural network is that different training procedures result in different predictions. Which one to take? A simple solution is to compute an arithmetic average over differently trained networks (jury decision or committee machine, Hansen and Salamon 1990). Such a third level improves overall accuracy and tends to combine the advantages of differently trained networks.

G4.1.3 Training methods

G4.1.3.1 Balanced training

Neural networks trained by backpropagation (Rumelhart *et al* 1986) in an on-line mode (updated for each training pattern) typically result in a three-state accuracy of around 62% Rost and Sander 1993b, Rost *et al* 1993). The accuracy is very unbalanced between the three secondary-structure types (helix 56%, strand 41%, loop 76%). This reflects the typical distribution of secondary structure in the data set: 32% helix, 21% strand, 47% loop (Rost and Sander 1992, Rost and Sander 1994a). A simple way to balance the prediction and thus to more accurately predict the most abundant class of strand is an adaptive-like training: instead of choosing the training samples at random from all examples, now at each time step an example is chosen at random from each of the three classes (helix, strand, loop):

$$\Delta W_{ij}\,(t+1) = -\varepsilon \frac{\partial E_{\text{sum3}}(t)}{\partial W_{ij}} + \alpha \Delta W_{ij}(t-1)$$

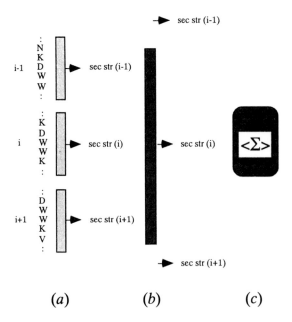

Figure G4.1.2. Three-level system for prediction of secondary structure. (*a*) First level, sequence-to-structure network: a window of $a = 13$ adjacent residues is shifted through all proteins. For each window the task of the network is to predict the secondary–structure state of the central residue (D, W, W). Neural network: unidirectional connections. Number of units (see figure G4.1.4): $N_1 = 536$, $N_2 = 15$, $N_3 = 3$. (*b*) Second level, structure-to-structure network: a window of $a = 17$ adjacent residues is shifted through all proteins. Again the task is to predict the secondary structure for the central residue. But now the input are the output values (i.e. the predictions) of the first-level network (as shown, the second level predicts the secondary structure for W at position i). Neural network: unidirectional corrections. Number of units (see figure G4.1.3): $N_1 = 627$, $N_2 = 15$, $N_3 = 3$. (*c*) Third level, jury decision: the output from differently trained networks (figure G4.1.4) for the same sequence position is summed. The secondary-structure prediction for residue W at sequence position i is assigned to the unit with the maximal sum.

with the learning rate ε (set to 0.05), the momentum term α (set to 0.2), the algorithmic time t, and error E_{sum3}:

$$E_{\text{sum3}} = \sum_{\mu=1}^{3} \sum_{k=1}^{3} \left(a_k^\mu - d_k^\mu\right)^2 \tag{G4.1.1}$$

where a_k^μ is the value of output unit k (helix, $k = 1$; strand, $k = 2$; loop, $k = 3$) for pattern μ, and d_k^μ the desired value for unit k (e.g. for $k = 1$ and $\mu = 1$, i.e. the first output unit of the helix example; $d = 1$ if the central residue of pattern μ in helix, and $= 0$ otherwise). The three patterns μ are chosen such that, for example, $\mu = 1$ represents a helix; $\mu = 2$ a strand, and $\mu = 3$ a loop. Training is stopped when the accuracy has reached 76%. This empirical value reflects a flat curve for overtraining; that is, stopping at values of 76–85% resulted in only marginal differences in terms of generalization). Such a training results in a more balanced prediction accuracy (helix 59%, strand 58%, loop 61%).

G4.1.3.2 Training and testing set

B3.5 To evaluate the *generalization performance*, multifold cross-validation experiments have to be performed: the data set containing 126 proteins is split into seven partitions $108 + 18$ proteins. The 108 are used for training, the 18 for testing. This is repeated seven times (i.e. seven neural networks are trained independently) until each protein has been used once for testing. Two problem-specific constraints are imposed on the data set. First, sequence similarity between any two proteins used has to be lower than 25% (Sander and Schneider 1991), as above 25% sequence identity homology modeling is applicable and is clearly superior to any *ab initio* prediction; Rost *et al* (1994b). Second, the size of the set should be sufficiently large as prediction accuracy differs between proteins (Rost and Sander 1993a, Rost *et al* 1993). Sets are taken from PDB, the databank of known three-dimensional structures (Bernstein *et al*

1977). Currently, there are more than 200 unique proteins of known three-dimensional structure with more than 60 000 residues (i.e. patterns) in total (Hobohm and Sander 1994). Secondary structure can be compiled automatically from three-dimensional structure and is stored in databases such as DSSP (Kabsch and Sander 1983a) or HSSP (a database of the homology-derived structures of proteins, Sander and Schneider 1993).

G4.1.4 Input preprocessing

G4.1.4.1 Input coding, single sequences

Each residue is coded by 20 input units for 20 different amino acids. Binary coding (19 units = 0; one unit = 1) is as good as or better than any alternative coding scheme (Cherkauer and Shavlik 1993, Rost 1993, Rost and Sander 1993b, Maza 1994). To allow the first and last residues of a protein to be used as the central residue in a window, an additional 21st input unit is used as a spacer.

G4.1.4.2 Input coding, multiple alignment profiles

The elaborated neural network system described so far is still limited to a performance accuracy of about 65%. The input information is not sufficient. As stated above, naturally evolved proteins can exchange about 75% of their residues without changing the three-dimensional structure. Such evolutionary information is highly specific for three-dimensional structure (figure G4.1.3) and can thus be used for prediction (Dickerson et al 1976, Maxfield and Scheraga 1979, Zvelebil et al 1987). Profiles of evolutionary exchanges are taken from HSSP, a database of homology-derived predictions (Sander and Schneider 1993).

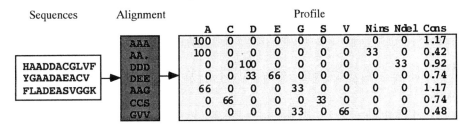

Input local in sequence

Sequences	Alignment		A	C	D	E	G	S	V	Nins	Ndel	Cons
	AAA		100	0	0	0	0	0	0	0	0	1.17
	AA.		100	0	0	0	0	0	0	33	0	0.42
HAADDACGLVF	DDD		0	0	100	0	0	0	0	0	33	0.92
YGAADAEACV	DEE		0	0	33	66	0	0	0	0	0	0.74
FLADEASVGGK	AAG		66	0	0	0	33	0	0	0	0	1.17
	CCS		0	66	0	0	0	33	0	0	0	0.74
	GVV		0	0	0	0	33	0	66	0	0	0.48

Input global in sequence

Amino acid content in whole protein	=	20 units
Length of protein	=	4 units
Distance of window to beginning and end of protein	=	2 × 4 units

Figure G4.1.3. Preprocessing input data. First, a protein is taken from PDB (Bernstein et al 1977), then proteins with similar sequence are searched in SWISSPROT (a databank of known protein sequences, Bairoch and Boeckmann 1992). For naturally evolved proteins it is possible to select proteins of homologous three-dimensional structure purely on the basis of sequence identity (Sander and Schneider 1991). Homologues (three here) are aligned with the alignment program MAXHOM (Sander and Schneider 1991). At each residue position the occurrence (percentage) of each amino acid (given in one-letter code) is compiled along with the number of insertions (Nins) and deletions (Ndel) necessary to render an optimal alignment. Such a profile is fed as input into the neural network, instead of just the sequence of the first protein. Acids E and D are mutually more similar in terms of their biochemical properties than E and C. The conservation weight (Cons) reflects the degree of similarity of the residues found at a particular position of the alignment (Rost and Sander 1993b). In addition to the information locally available from, for example, 13 adjacent residues, global information can be compiled, such as the content of each amino acid in the whole protein, the length of the protein, or the distance of the window from the beginning and end of the protein.

G4.1.4.3   Further preprocessing of input

Alignments of homologous proteins contain further details (figure G4.1.3). First, the more insertions and deletions necessary to render an optimal alignment the more likely this region occurs in a loop. Second, consecutive stretches of high conservation of physicochemical properties of exchanged amino acids often indicate the presence of either a helix or a strand. Third, the amino acid composition of the whole protein is specific for certain types of proteins (e.g. all-helical proteins). Information about the protein class (e.g. all-helical) can improve prediction accuracy further (Kneller *et al* 1990); however, in practice this marginal gain is lost by the inaccuracy in predicting the class (Rost and Sander 1993c).

G4.1.5   Output interpretation

G4.1.5.1   Jury decision over various neural networks

The final output of the composite neural network is an arithmetic average over 12 second-level structure-to-structure neural networks (figure G4.1.2) which differ both in the training method and the input preprocessing (figure G4.1.4).

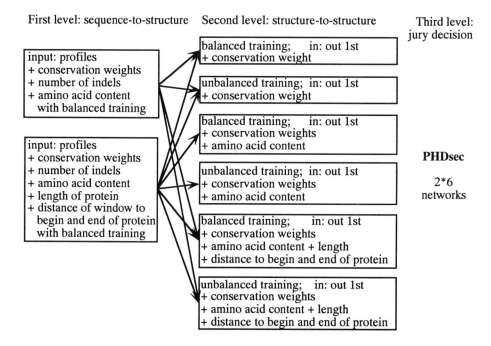

Figure G4.1.4. Generating different networks for jury decision. The final prediction of the composite neural network as an arithmetic average (jury decision) over 12 different neural networks. The neural networks differ in training procedure (unbalanced and balanced training (see section G4.1.3) and different preprocessing of the evolutionary information (see section G4.1.4), both in the first- and second-level neural networks (figure G4.1.2).

G4.1.5.2   Output to prediction

The final prediction is derived by a winner-take-all decision, that is, the unit with the largest sum after the jury decision is chosen as the neural network prediction. An additional filtering is applied: helices shorter than three and strands shorter than two residues are elongated or interpreted as loops, depending on the strength of the prediction. The final composite neural network using evolutionary information as input—dubbed PHDsec, a profile neural network system from Heidelberg, Germany, for prediction of secondary structure—has an expected overall accuracy greater than 72% (Rost and Sander 1994b).

G4.1.5.3 Reliability index

The strength of the prediction correlates with prediction accuracy. An empirically reasonable index for the reliability of the prediction is

$$RI = INT\{10 \times (a_{\max} - a_{\text{next}}\}$$ (G4.1.2)

where $a_{\max}$ is the output value of the output unit with highest value and a_{next} that of the unit with the next highest value. The factor 10 normalizes RI to integer values from 0 to 9.

G4.1.6 Comparison with traditional methods

G4.1.6.1 Neural network versus traditional predictions of secondary structure

Prediction accuracy, direct comparison from literature. Predictions of neural networks have been reported to yield a three-state prediction accuracy of better than 66% (Zhang *et al* 1992). This is comparable to non-neural network methods (Biou *et al* 1988, Munson *et al* 1994) as shown in table G4.1.1. Predictions using multiple alignment information as input are, in general, significantly more accurate than those using single sequences only (table G4.1.2). For most methods the comparisons are problematic, as results are based on different evaluation sets, and most data sets used were too small or contained proteins of significant pairwise sequence identity (see table G4.1.3). For example, a simple neural network, if evaluated on 126 unique proteins, scores at some 62% accuracy (Rost and Sander 1993b), and at greater than 64% if evaluated on 15 proteins with homologies to the training set (Qian and Sejnowski 1988). For an appropriate comparison the accuracy has to be evaluated on identical, sufficiently large, and unique data sets.
Prediction accuracy, identical data sets. Laborious comparisons based on identical data sets have revealed two results. First, the composite neural network PHDsec is clearly superior to any other prediction method published so far. Second, comparisons have to be based on identical data sets; for example, for a 'favorable' data set (such as used by Levin *et al* 1994) prediction accuracy PHDsec had an accuracy of about 75% (see also the comparison between Biou *et al* 1988 in table G4.1.2 and in table G4.1.3).

G4.1.6.2 Specific improvements of the network system PHDsec

Improvements on the network side. The composite neural network improves performance in three ways (Rost and Sander 1994b). First, balanced training (see section G4.1.3) yields more accurate strand   G4.1.3
predictions than most traditional methods (exception Gascuel and Golmard 1988). Second, the second-level structure-to-structure neural network (figure G4.1.2) results in more protein-like predictions than most published traditional methods. Third, the final jury average (see section G4.1.5) improves overall accuracy   G4.1.5
by about one to two percentage points, and finds a compromise between unbalanced (overall more accurate) and balanced (strands more accurate) neural networks. The latter improvement is comparable to classical 'joint prediction methods' (Biou *et al* 1988, Nishikawa and Noguchi 1991, Viswanadhan *et al* 1991).
Improvements by using biological information. Using only profiles as input improves prediction accuracy by more than five percentage points (table G4.1.2). The composite neural network successfully uses further important input information. For all steps of adding relevant input information, the composite neural network has, so far, outperformed traditional methods (table G4.1.2).

G4.1.6.3 Practical impact of the neural network system PHDsec

How good is the prediction for a protein of unknown three-dimensional structure? Prediction accuracy varies with the protein, thus the expected prediction accuracy of PHDsec is $72 \pm 9\%$ (one standard deviation). This implies that users cannot deduce from the prediction whether it is 45% or 95% correct. Here, the definition of a reliability index (equation (G4.1.2)) proves to be of immense practical importance as it correlates with prediction accuracy; that is, residues predicted with higher reliability are on average predicted more accurately. Comparable indices exist for traditional methods but the composite neural network is significantly more accurate: half the residues are predicted at an expected accuracy of 88% (Rost and Sander 1994b).
How can the neural network predictions be obtained? Predictions from the composite neural network system PHDsec are available via a fully automatic prediction service (Rost *et al* 1994a). The user sends a sequence or an alignment and the prediction is returned. (Send the word 'help' by electronic mail

---

Table G4.1.1. Secondary structure prediction accuracy (from the literature). Methods are abbreviated as in the reference list ('Rost and Sander 1993—reference' is a simple neural network used as reference point for the performance on a large unique data set). All methods given use single sequences as input. Abbreviations used: 'accuracy', percentage of correctly predicted residues in three states; 'number of proteins', number of proteins used for evaluation; 'unique set', a set allowing for pairwise sequence identity greater than 25% is dubbed 'not unique'. For more recent methods more than 100 proteins is a sufficiently large data set. KS, Kabsch and Sander (1983b); subKS, subset of KS; QS, Qian and Sejnowski (1988) (unfortunately this completely inadequate set allowing for pairwise identities greater than 50% is widely used); subQS, subset of QS; RS, globular proteins of Rost and Sander (1993b).

Method	Accuracy	Number of proteins	Unique set?
Non-neural network predictions			
Asai et al (1993)	66.0	120	?
Biou et al (1988)	65.5	62[KS]	yes
Garratt et al (1991)	61.0	53[subKS]	yes
Gascuel and Golmard (1988)	58.7	62[KS]	yes
Geourjon and Deléage (1994)	69.0	239	no
King and Sternberg (1990)	60.0	18	yes
Leng et al (1994)	68.2	74	no
Munson et al (1994)	65.9	67	no
Nishikawa and Noguchi (1991)	64.8	27	yes
Salzberg and Cost (1992)	65.1	128	no
Viswanadhan et al (1991)	64.0	45	?
Yi and Lander (1993)	68.0	110	no
Neural network predictions			
Fariselli et al (1993)	64.0	62	yes
Fogelman-Soulié and Mejía (1990)	58.8	62[KS]	yes
Holley and Karplus (1989)	63.2	14[subQS]	yes
Kneller et al (1990)	65.0	105[QS]	no
Maclin and Shavlik (1993)	63.4	106[QS]	no
Qian and Sejnowski (1988)	64.3	14[subQS]	no
Rost and Sander (1994) reference	62.1	126[RS]	yes
Sasagawa and Tajima (1993)	60.1	29	yes
Stolorz et al (1992)	64.4	14[subQS]	no
Zhang et al (1992)	63.1	107	no
Zhang et al (1992)	66.4	107	no

Table G4.1.2. Prediction accuracy for alignment-based methods (from the literature): all methods given use multiple alignments as input and are evaluated on unique data sets. Only the PHDx methods use neural networks. The following abbreviations indicate different stages of input preprocessing (section G4.1.4): PHD0, alignment profiles; PHD1, PHD0+ conservation weight; PHD2, PHD1 + insertions and deletions; PHDsec, PHD2 + amino acid content. The following data sets are labeled to indicate identical sets: LPAG, Levin et al (1993); RS, Rost and Sander (1993b); and superRS, a super set of RS = RS + RS2 (Rost and Sander 1994b). Further abbreviations used as in table G4.1.1.

Method	Accuracy	Number of proteins
Rost and Sander (1994) reference	62.1	126[RS]
Boscott et al (1993)	64.0	31
Levin et al (1993)	68.5	60[LPAG]
Rost and Sander (1994)—PHD0	69.7	126[RS]
Rost and Sander (1994)—PHD1	70.8	126[RS]
Rost and Sander (1994)—PHD2	71.41	26[RS]
Rost and Sander (1994)—PHDsec	72.1	250[superRS]
Wako and Blundell (1994)	69.0	13
Zvelebil et al (1987)	66.1	11

Table G4.1.3. PHDsec versus other methods evaluated on identical data sets: abbreviations used as in table G4.1.1 and table G4.1.2. For comparison results on set RS are given.

Method	Accuracy	Number of proteins
Chou and Fasman (1974)	49	62[KS]
Gascuel and Golmard (1988)	58.7	62[KS]
Rost and Sander (1994)—PHD1	72.5	62[KS]
Rost and Sander (1994)—PHD1	70.81	26[RS]
Levin et al 1993	68.5	60[LPAG]
Rost and Sander (1994)—PHD2	74.8	60[LPAG]
Rost and Sander (1994)—PHD2	71.4	126[RS]
Gibrat et al (1987)	58.9	124[RS2]
Biou et al (1988)	60.9	124[RS2]
Rost and Sander (1994)—PHDsec	72.5	124[RS2]
Rost and Sander (1994)—PHDsec	71.6	126[RS]

to the internet address 'PredictProtein@EMBL-Heidelberg.de', or use the WWW site 'http://www.embl-heidelberg.de/predictprotein/predictprotein.html'.) Both improved prediction accuracy and rigorous testing procedures have led to about 100 prediction requests per day.

G4.1.7 Conclusions

Neural networks can easily be tailored to the problem. The three improvements on the network side (see above) illustrate that a deeper understanding of the stochastic behavior of the 'black-box pattern classifier neural network' can be used to avoid problem specific disadvantages of a simple neural network.

Highest gain from preprocessing input data by biological expertise. It is not enough to tailor the composite network system to the problem. Instead, the most significant improvement of the prediction accuracy stems from the incorporation of biological knowledge (evolutionary information).

Composite system superior to any other prediction method. Often neural networks are shown to be the second-best solution of a problem. The composite Neural network described here, today, is clearly better than any other prediction method. Further improvements of the method appear possible. Thus, the neural network for secondary-structure prediction is likely to remain one of the best tools in a very competitive field of research.

Appropriate evaluation and availability of methods is the key to applications. Most methods developed in the field of 'biocomputing' rely upon time-consuming literature searches (step 1), appropriate testing procedures (step 2) and making the program available (step 3). However, theoretical tools for the prediction of protein structure can influence research in molecular biology only if these simplifications are avoided.

Perspectives for the future? The goal is to predict protein three-dimensional structure. The explosion of protein databases may bring this goal in reach in the near future. Neural networks have a fair chance to be part of a hybrid system that will first predict three-dimensional structure. But even if one heads for less ambitious projects, there are many problems for which sufficiently tested, available neural network solutions would be highly welcomed by experimentalists.

References

Andrade M A, Chacón P, Merelo J J and Morán F 1993 Evaluation of secondary structure of proteins from UV circular dichroism spectra using an unsupervised learning neural network *Protein Eng.* **6** 383–90

Anfinsen C B 1973 Principles that govern the folding of protein chains *Science* **181** 223–30

Bairoch A and Boeckmann B 1992 The SWISS-PROT protein sequence data bank *Nucleic Acids Res.* **20** 2019–22

Bengio Y and Pouliot Y 1990 Efficient recognition of immunglobulin domains from amino acid sequences using a neural network *Comput. Appl. Biol. Sci.* **6** 319–24

Bernstein F C, Koetzle T F, Williams G J B Meyer E F Brice M D Rodgers J R Kennard O Shimanouchi T and Tasumi M 1977 The protein data bank: a computer based archival file for macromolecular structures *J. Mol. Biol.* **112** 535–42

Biou V, Gibrat J F, Levin J M, Robson B and Garnier J 1988 Secondary structure prediction: combination of three different methods *Protein Eng.* **2** 185–91

Böhm G, Muhr R and Jaenicke R 1992 Quantitative analysis of protein far UV circular dichroism spectra by neural networks *Prot. Eng.* **5** 191–5

Bohr H, Bohr J, Brunak S, Cotterill R M J, Lautrup B, Nørskov L Olsen O H and Petersen S B 1988 Protein secondary structure and homology by neural networks *FEBS Lett.* **241** 223–8

Bohr H, Bohr J, Brunak S, Fredholm H, Lautrup B and Petersen S B 1990 A novel aroach to prediction of the 3-dimensional structures of protein backbones by neural networks *FEBS Lett.* **261** 43–6

Bossa F and Pascarella S 1990 PRONET: a microcomputer program for predicting the secondary structure of proteins with a neural network *Comput. Appl. Biol. Sci.* **5** 319–20

Brändén C and Tooze J 1991 *Introduction to Protein Structure* (New York: Garland)

Brunak S 1991 Non-linearities in training sets identified by inspecting the order in which neural networks learn *Neural Networks From Biology to High Energy Physics* ed O Benhar, C Bosio P Del Giudice and E Tabet (Italy: Elba) pp 277–88

Cherkauer K J and Shavlik J W 1993 Protein Structure Prediction: Selecting Salient Features from Large Candidate Pools *Proc. First Int. Conf. on Intelligent Systems for Molecular Biology* (Bethseda, MD: AAAI Press) in press

Dickerson R E, Timkovich R and Almassy R J 1976 The cytochrome fold and the evolution of bacterial energy metabolism *J. Mol. Biol.* **100** 473–91

Dombi G W and Lawrence J 1994 Analysis of protein transmembrane helical regions by a neural network *Prot. Sci.* **3** 557–66

Dubchak I, Holbrook S R and Kim S-H 1993 Prediction of protein folding class from amino acid composition *Prot.: Struct. Func. Gen.* **16** 79–91

Fariselli P, Compiani M and Casadio R 1993 Predicting secondary structures of membrane proteins with neural networks *Europ. Biophys. J.* **22** 41–51

Ferrán E and Ferrara P 1992a Clustering proteins into families using artificial neural networks *Comput. Appl. Biol. Sci.* **8** 39–44

——1992b A neural network dynamics that resembles protein evolution *Physica* **185A** 395–401

Ferrán E A and Pflugfelder B 1993 A hybrid method to cluster protein sequences based on statistics and artificial neural networks *Comput. Appl. Biol. Sci.* **9** 671–80

Friedrichs, M S Goldstein R A and Wolynes P G 1991 generalized protein tertiary structure recognition using associative memory Hamiltonians *J. Mol. Biol.* **222** 1013–34

Frishman D and Argos P 1992 Recognition of distantly related protein sequences using conserved motifs and neural networks *J. Mol. Biol.* **228** 951–62

Gascuel O and Golmard J L 1988 A simple method for predicting the secondary structure of globular proteins: implications and accuracy *Comput. Appl. Biol. Sci.* **4** 357–65

Goldstein R A, Luthey-Schulten Z A and Wolynes P G 1992a Optimal protein-folding codes from spin-glass theory *Proc. Natl Acad. Sci.* **89** 4918–22

——1992b Protein tertiary structure recognition using optimized Hamiltonians with local interactions *Proc. Natl Acad. Sci.* **89** 9029–33

Hansen L K and Salamon P 1990 Neural Network Ensembles *IEEE Trans. Patt. Anal. Machine Intell.* **12** 993–1001

Hayward S and Collins J F 1992 Limits on a-helix prediction with neural network models *Proteins* **14** 372–81

Hirst J D and Sternberg M J E 1991 Prediction of ATP-binding motifs a comparison of a perceptron-type neural network and a consensus sequence method *Prot. Eng.* **4** 615–23

Hobohm U and Sander C 1994 Enlarged representative set of protein structures *Prot. Sci.* **3** 522–4

Holbrook S R, Muskal S M and Kim S-H 1990 Predicting surface exposure of amino acids from protein sequence *Prot. Eng.* **3** 659–65

Holley H L and Karplus M 1989 Protein secondary structure prediction with a neural network *Proc. Natl Acad. Sci.* **86** 152–6

Kabsch W and Sander C 1983a Dictionary of protein secondary structure: pattern recognition of hydrogen bonded and geometrical features *Biopolymers* **22** 2577–637

——1983b how good are predictions of protein secondary structure? *FEBS Lett.* **155** 179–82

Kneller D G, Cohen F E and Langridge R 1990 Improvements in Protein Secondary Structure Prediction by an Enhanced Neural Network *J. Mol. Biol.* **214** 171–82

Kraulis P 1991 MOLSCRIPT: a program to produce both detailed and schematic plots of protein structures *J. Appl. Crystallogr.* **24** 946–50

Levin J M, Pascarella S Argos P and Garnier J 1993 Quantification of secondary structure prediction improvement using multiple alignments *Prot. Eng.* **6** 849–54

Maclin R and Shavlik J W 1993 Using knowledge-based neural networks to improve algorithms: refining the Chou-fasman algorithm for protein folding *Machine Learning* **11** 195–215

Maxfield F R and Scheraga H A 1979 Improvements in the prediction of protein topography by reduction of statistical errors *Biochemistry* **18** 697–704

Maza M d l 1994 Generate, test, and explain: synthesizing regularity exposing attributes in large protein databases *27th Hawaii Int. Conf. on System Sciences* ed L Hunter (Wailea, Hawaii: IEEE Society Press) pp 123–32

McGregor M J, Flores T P and Sternberg M J E 1989 Prediction of -turns in proteins using neural networks *Prot. Eng.* **2** 521–6

Metfessel B A, Saurugger P N Connelly D P and Rich S S 1993 Cross-validation of protein structural class prediction using statistical clustering and neural networks *Prot. Sci.* **2** 1171–82

Munson P J, Di Francesco V and Porrelli R 1994 Prediction of protein secondary structure using linear and quadratic logistic models with penalized maximum likelihood estimation *27th Hawaii Int. Conf. on System Sciences* ed L Hunter (Wailea, HI: IEEE Computer Society Press) pp 375–84

Muskal S M and Kim S-H 1992 Predicting protein secondary structure content. A tandem neural network approach *J. Mol. Biol.* **225** 713–27

Nishikawa K and Noguchi T 1991 Predicting protein secondary structure based on amino acid sequence *Meth. Enz.* **202** 31–44

Oliver S *et al* 1992 The complete DNA sequence of yeast chromosome III *Nature* **357** 38–46

Pancoska P, Blazek M and Keiderling T A 1992 Relationships between secondary structure fractions for globular proteins. Neural network analyses of crystallographic Data sets *Biochemistry* **31** 10 250–7

Petersen S B, Bohr H, Bohr J, Brunak S, Coterill R M J, Fredholm H and Lautrup B 1990 Training neural networks to analyse biological sequences *TIBTECH* **8** 304–8

Presnell S R and Cohen F E 1993 Artificial Neural Networks for Pattern Recognition in Biochemical Sequences *Ann. Rev. Biophys. Biomol. Struct.* **22** 283–98

Qian N and Sejnowski T J 1988 Predicting the secondary structure of globular proteins using neural network models *J. Mol. Biol.* **202** 865–84

Radomski J P, van Halbeek H and Meyer B 1994 Neural network-based recognitioin of oligosaccharide 1H-NMR spectra *Nature Struct. Biol.* **1** 217–8

Rost B 1993 Neural networks and evolution—advanced prediction of protein secondary structure *Doctoral Thesis* Department of Physics and Astronomy, University of Heidelberg, Germany

Rost B and Sander C 1992 Exercising Multi-layered Networks on Protein Secondary Structure *Neural Networks: From Biology to High Energy Physics* ed O Benhar, S Brunak, P DelGiudice and M Grandolfo (Italy: Elba) *Int. J. Neural Systems* 209–20

——1993a Improved prediction of protein secondary structure by use of sequence profiles and neural networks *Proc. Natl Acad. Sci.* **90** 7558–62

——1993b Prediction of protein secondary structure at better than 70% accuracy *J. Mol. Biol.* **232** 584–99

——1993c Secondary structure prediction of all-helical proteins in two states *Prot. Eng.* **6** 831–6

——1994a 1D secondary structure prediction through evolutionary profiles *Prot. Struct. Distance Analysis* ed H Bohr and S Brunak (Amsterdam, Oxford, Washington: IOS Press) pp 257–76

——1994b Combining evolutionary information and neural networks to predict protein secondary structure *Proteins* **19** 55–72

——1994c Conservation and prediction of solvent accessibility in protein families *Proteins* **20** 216–26

——1994c

Rost B, Sander C and Schneider R 1993 Progress in protein structure prediction? *Trends in Biochem. Sci.* **18** 120–3

——1994a PHD—an automatic server for protein secondary structure prediction *Comput. Appl. Biol. Sci.* **10** 53–60

——1994b Redefining the goals of protein secondary structure prediction *J. Mol. Biol.* **235** 13–26

Rumelhart D E, Hinton G E and Williams R J 1986 Learning representations by back-propagating error *Nature* **323** 533–6

Sander C and Schneider R 1991 Database of homology-derived structures and the structurally meaning of sequence alignment *Proteins* **9** 56–68

——1993 The HSSP data base of protein structure-sequence alignment *Nucleic Acids Res.* **21** 3105–9

Sasagawa F and Tajima K 1993 Prediction of protein secondary structures by a neural network *Comput. Appl. Biol. Sci.* **9** 147–52

Stolorz P, Lapedes A and Xia Y 1992 Predicting protein secondary structure using neural net and statistical methods *J. Mol. Biol.* **225** 363–77

Tchoumatchenko I, Vissotsky F and Ganascia J-G 1993 *How to Make Explicit A Neural Network Trained to Predict Proteins Secondary Structure* ACASA, LAFORIA-CNRS, Universitè Paris VI, 4 Place Jussieu, 75 252 Paris, CEDEX 05, France

Tolstrup N, Toftgård J, Engelbrecht J and Brunak S 1994 Neural network model of the genetic code is strongly correlated to the GES scale of amino acid transfer free energies *J. Mol. Biol.* submitted

van Gunsteren W F 1993 Molecular dynamics studies of proteins *Current Opinion in Struct. Biol.* **3** 167–74

Viswanadhan V N, Denckla B and Weinstein J N 1991 New Joint Prediction Algorithm (Q7-JASEP) Improves the Prediction of Protein Secondary Structure *Biochemistry* **30** 11 164–72

Xin Y, Carmeli T T, Liebman M N and Wilcox G L 1992 Use of the backpropagation neural network algorithm for prediction of protein folding patterns *Second Int. Conf. on Bioinformatics, Supercomputing and Complex Genome Analysis* ed H A Lim, J W Fickett, C R Cantor and R J Robbins (St Petersburg Beach, FL: World Scientific) pp 360–76

Yun-yu S, Mark A E, Cun-xin W, Fuhua H, Berendsen H J and van Gunsteren W F 1993 Can the stability of protein mutants be predicted by free energy calculations? *Prot. Eng.* **6** 289–95

Zhang X, Mesirov J P and Waltz D L 1992 Hybrid system for protein secondary structure prediction *J. Mol. Biol.* **225** 1049–63

Zvelebil M J, Barton G J, Taylor W R and Sternberg M J E 1987 Prediction of protein secondary structure and active sites using alignment of homologous sequences *J. Mol. Biol.* **195** 957–61

G4.2 Neural networks for identification of protein coding regions in genomic DNA sequences

E E Snyder and Gary D Stormo

Abstract

We have developed a system which uses neural networks and dynamic programming (DP) to identify protein coding regions in genomic DNA sequences. Nine scores are calculated on all subintervals of the sequence which evaluate the likelihood that the subinterval belongs to one of four classes; first, last or internal exon or intron. These scores are weighted by a neural network and used as input to a DP algorithm. DP is used to find the highest scoring combination of introns and exons subject to a few simple constraints on gene structure. The neural network weights are optimized by training on input vectors which measure the difference between the predicted optimal solution by DP and the biologically correct solution. The system is trained by maximizing the difference between the correct parse and a sample of incorrect parses. On a test set of genomic sequences from GenBank, we obtained correlation coefficients for exon nucleotide prediction as high as 0.94. This is superior to the results obtained by purely rule-based systems.

G4.2.1 Project overview

The DNA molecule is the storage media of the genetic information in every living thing. At its most fundamental level, this media consists of a linear arrangement of nucleotide base pairs which are the rungs of the DNA double-helical ladder. At each position, there are four possible bases which can be symbolized as A, C, G, or T. In the human being, there are about 3×10^9 base pairs (bp) per haploid genome. There are estimated to be some 50 000 genes, most of which code for a single protein. Assuming an average protein consists of 300 amino acids, coded for by three base pairs each or a total of about 1000 bp of DNA, it is clear that only a small fraction ($< 2\%$) of the genome codes for protein. With rapid advances in DNA sequencing technology and the initiation of projects such as the Human Genome Initiative, the ultimate goal of which is to sequence the entire human genome, the problem of identifying coding regions in uncharacterized DNA sequences is of central importance.

In addition to being a small fraction of the total DNA, the identification of coding regions in higher organisms is complicated by the presence of intervening sequences or *introns* which can separate the coding region of a gene into several parts. These parts are called *exons*. There are additional constraints which dictate how exons can be joined together to form a continuous reading frame from which the encoded protein can be translated. These constraints are illustrated in figure G4.2.1.

We have developed a computer program called GeneParser which addresses both of these problems simultaneously†. There are a number of tests which can be used to evaluate the likelihood that a sequence interval belongs to the class exon, intron or neither. These tests are applied to all subintervals in a sequence. Separate neural networks are used to weight these tests to yield a composite score which reflects the likelihood that the interval belongs to a particular class. The weighted scores are the input to

† This work was done as part of the doctoral research of E E Snyder in the laboratory of G D Stormo at the University of Colorado, Boulder, USA. This work was supported by DOE grant ER61606.

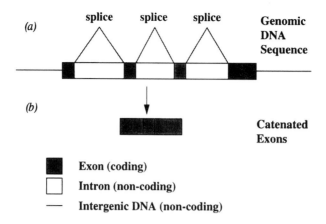

Figure G4.2.1. Eukaryotic gene structure. Part (*a*) shows the arrangement of coding sequences in genomic DNA. Exons which contain protein-coding DNA are separated from one another by intervening sequences called introns which are non-coding. After transcription into RNA, these introns are spliced out. This yields a messenger RNA (mRNA) shown in (*b*) in which the exons are joined together, allowing the gene's protein product to be translated. The successful prediction of gene structure requires both identifying the gene in genomic DNA and the correct prediction of its intron–exon structure.

a dynamic programming (DP) algorithm which finds the highest scoring combination of introns and exons subject to the constraints of eukaryotic gene structure. Figure G4.2.2 illustrates the flow of information in GeneParser.

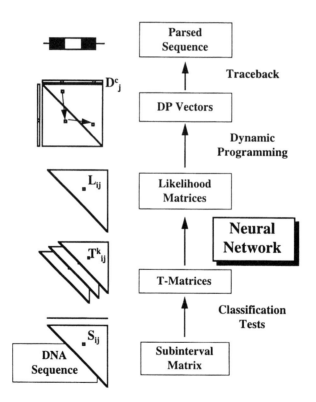

Figure G4.2.2. Information flow in GeneParser. Each operation is shown with its associated data. The DNA sequence is represented by the string of characters, S, of length N. All $N^2/2$ subintervals of S are scored ($S_{ij}, i < j$) for the c classification statistics. This gives rise to t T-matrices, one for each test. For each of the c interval types, a network-weighted score is calculated, L_{ij}^c, which represents the likelihood that interval S_{ij} belongs to class c. This information serves as input to the dynamic programming algorithm which parses the sequence into the c sequence types.

G4.2.2 Design process

G4.2.2.1 Motivation

Our motivation for using a neural approach to solve this problem was threefold. First, it was clear from the outset that the properties which distinguish coding from non-coding DNA are at best only poorly understood. Thus, we expected that the methods available for coding sequence identification may be insufficient to yield an exact solution. For example, mRNA splicing can occur using different factors depending on the mRNA substrate or the tissue in which it is expressed. Optimization techniques such as the simplex method for solving linear inequalities were eliminated in favor of neural network methods which exhibit more graceful failure when confronted with contradictory training data. Our experience with using the simplex method on a similar problem involving protein secondary structure prediction (Batra 1993) had shown that training sets quickly evolved to which no exact solution existed.

Error tolerance was the second property of neural networks which made them attractive in this project. Previous gene identification methods suffer severe degradation in performance when confronted with test data containing even small numbers of sequencing errors (0.5% indels, 0.5% substitutions errors). Because the cost of sequencing increases dramatically as the required accuracy increases, it was very desirable to build an error-tolerant system from the beginning.

Finally, we hoped to exploit the scalability of neural networks to deal with more complex relationships between classification statistics. Our initial development used only a simple network with one layer of weights and no hidden units. We hoped that increasing the complexity of the network might increase its predictive power.

G4.2.2.2 Dynamic programming

To provide background for the following sections, a brief introduction to the application of DP for sequence parsing will be presented here. A more detailed description can be found in Snyder and Stormo (1993, 1995), Snyder (1994). Given a DNA sequence s, let all subintervals in s be represented as elements of the matrix $\mathbf{S}$ such that the sequence starting at s_i and ending at s_j is represented by the element S_{ij}. We postulate a function $L^c(S_{ij})$ which calculates the log-likelihood that the interval S_{ij} belongs to sequence class c (i.e. is either a first, internal or last exon or intron). The score of a solution is defined as the sum of the L-matrix values of the intervals which compose it. A valid solution is one which meets the following constraints on gene structure: introns and exons must be adjacent, alternating and nonoverlapping; first and last exons, if present, must be the extreme left ($5'-$) and extreme right ($3'-$) exons, respectively, in the solution. The space of valid solutions can be searched for the optimum by evaluating the following recursion over all c and on $j : 1 < j < N$ when N is the length of sequence S:

$$D_j^c = \max \left\{ \begin{array}{l} \max_{x \in \mathcal{N}} \left[\max_{k:1 \to j-m^c} \left[L_{kj}^c + D_{k-1}^x \right] \right] \\ 0 \end{array} \right. \tag{G4.2.1}$$

and $D_0^c = 0$. Thus, D_j^c is the score of the best solution ending in an interval of type c which ends at position j. $\mathcal{N}$ is the set of valid transitions between sequence types. To find the end of the optimum parse of the entire sequence, D is scanned for the highest value. Knowing the position and sequence type, the parse which led to that score can be derived.

G4.2.2.3 Network design

The neural networks in GeneParser are simple *feedforward classifiers*, serving as approximations to the likelihood function L^c. Each network takes as input an array of floating-point numbers which describe the interval with respect to one of the four sequence classes. Each network returns a scalar, the magnitude of which is proportional to the log-likelihood that the interval belongs to that particular class.   B2.3

Several network topologies were evaluated. The first network consisted of a single layer of input units connected to a single sigmoidal output unit. This corresponds to the network shown in figure G4.2.3. A variety of multilayered networks were also evaluated. Figure G4.2.4 shows one such design.

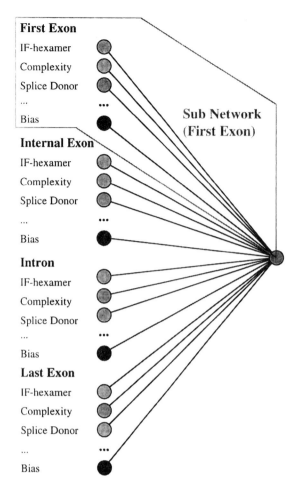

Figure G4.2.3. A simple linear network used to predict gene structure. Each sequence type (first, internal, last exon or intron) is assigned to one subnetwork. These statistics are the values of the gray input units. Given the outputs of the classification statistics for an interval of a particular sequence type, the likelihood that the interval belongs to that sequence type can be calculated using the appropriate subnetwork. For each subnetwork, there is a bias unit (shown in black), the value of which is clamped to unity. The network is trained as a whole to maximize the difference between correct and incorrect gene parsings as described in the text. The values of the input units are calculated as the sum of the intervals of each type in the correct solution less the sum of the values of the intervals in an incorrect solution. The bias units represent the difference between the number of intervals of each type between the correct solution and the incorrect solution.

G4.2.3 Training methods

Figure G4.2.5 illustrates the basic training procedure. The neural network in GeneParser is initialized with random weights. The program is asked to predict the structure of all the genes in the training set based on these weights. Each solution is compared to the correct solution and a single training vector is calculated from each target-predicted pair. These vectors are used to train the delta network described below. After training, the four subnetworks are extracted from the delta network and used to update the weights in GeneParser. The cycle is repeated until performance reaches a plateau. Generalization performance is tested using the weights that performed best on the training data.

Because the number of possible training vectors is so large (exponential in terms of the length of the training sequences), we adopted an 'exploratory learning' approach to training vector collection. Random weights are used only in the first pass of GeneParser through the training sequences. Following that, training vectors are recruited using the weights which give the best parsing based on the data acquired up to that training cycle. As the training progresses, the predicted solutions are closer to the actual solutions and thus the magnitude of the training vectors decreases with training iteration.

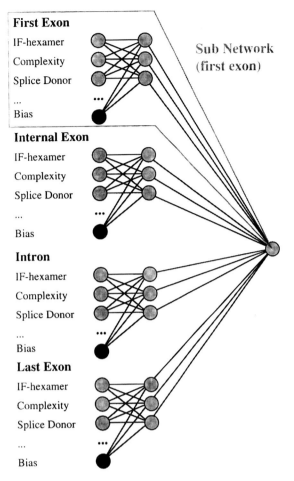

Figure G4.2.4. A multilayered network. Like the linear network, the multilayered network is divided into parts which represent the four sequence classes. Each unit in the hidden layer is connected to all input units within its respective subnetwork.

G4.2.3.1 Error propagation through dynamic programming

Each subnetwork calculates a score based on the properties of a single sequence interval. We considered training each network separately on randomly chosen sequence intervals from many different genes, assigning a target of 1.0 to members of the class, a target of 0.0 to nonmembers. Training would yield weights optimized to identify members of a particular class, leaving DP to implement the structural constraints. This approach was tried with only marginal success (data not shown). We cite two possible reasons for this failure. First, it is known that different genes can have exons and introns with very different statistical properties. It is probably unreasonable to expect these features to be recognized without reference to the background in which they occur. Second, picking a negative population of subintervals at random is not a realistic simulation. The biological constraints on gene structure make certain choices incompatible with others. Indeed, the whole notion of considering exons and introns in isolation seems absurd in the larger context of mRNA splicing. Since exons *define* the locations of introns (and vice versa), it is best to model the system as a whole.

To this end, we sought to train the neural network in the context of DP. An approach which alleviates these two major problems involves training the neural network on complete solutions instead of single intervals. Let $D^{\mu+}$ be the score of a correct (+) solution for sequence μ and $D^{\mu-}$ be the score of an incorrect (−) solution. A perfect set of network weights would make

$$D^{\mu+} > D^{\mu-} \tag{G4.2.2}$$

for all μ for all possible $D^{\mu-}$. Subtracting $D^{\mu-}$ from $D^{\mu+}$ yields the inequality

$$D^{\mu+} - D^{\mu-} > 0 \,. \tag{G4.2.3}$$

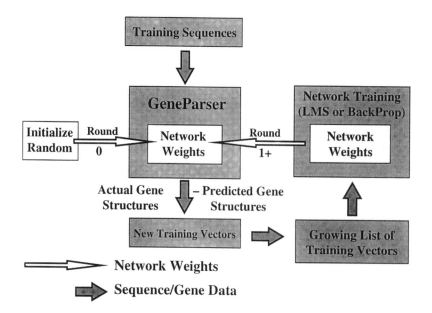

Figure G4.2.5. Training cycle. The neural network in GeneParser is initialized with random weights and used to predict the structure of the genes in the training set. The predictions are compared to the known structures, generating the first set of training vectors. These vectors are used to train the network. The weights are subsequently copied into the GeneParser network. GeneParser makes predictions again on the training set and the cycle is repeated. Each time, the newly calculated training vectors are added to the list of those previously used in training. Each pass through this cycle is referred to as one 'training iteration'.

At this point, it is useful to introduce a notation which makes the classification statistics and their weights explicit:

$$D = \sum_{c \in \{f,e,i,l\}} \left[\sum_{j=1}^{N^c} \sum_{k=1}^{P^c} T_{c,j,k} w_{c,k} + N^c B^c \right] \tag{G4.2.4}$$

where $T_{j,k}^c$ is the score for classification statistic k for the jth interval of type c. The term w_k^c is the corresponding weight for that statistic and B^c is a bias term. P^c is the number of classification statistics used for sequence type c and N^c is the number of intervals of type c in the solution. A neural network is used to find weights which satisfy the following inequality:

$$D^{\mu+} - D^{\mu-} = \sum_{c \in \{f,e,i,l\}} \left[\sum_{k=1}^{P^c} \left[w_k^c \underbrace{\left(\sum_{j=1}^{N^{c+}} T_{j,k}^{c,\mu+} - \sum_{j=1}^{N^{c-}} T_{j,k}^{c,\mu-} \right)}_{\Delta T} \right] + \overbrace{(N^{c,\mu+} - N^{c,\mu-})}^{\Delta N} B^c \right]. \tag{G4.2.5}$$

When written in this form, one can see a simple network implementation to solve this inequality. The inputs are simply $T^+ - T^-$ for each statistic for each sequence type (ΔT) and the difference between the number of each sequence type in the actual and predicted solutions (ΔN). This network design is referred to as the delta network because the network is trained on the difference between the actual solution and an incorrect solution for a particular sequence. If the right-hand side of equation (G4.2.5) is passed through a squashing function such as the symmetrical sigmoid

$$g(x) = \frac{1}{1 + e^{-x}} - \frac{1}{2} \tag{G4.2.6}$$

then training to a target of 0.5 will maximize the difference between correct and incorrect solutions.

G4.2.3.2 Training and test sets

The training set used for GeneParser was based on the collection of human genes used in the development of the program GeneID (Guigó *et al* 1992). These loci are genomic DNA sequences for which the sequence of the mRNA have been independently determined. Thus, there is experimental evidence to confirm the sequence of the gene product and thus the structure of the gene contained within. In addition, loci containing examples of alternative splicing (for which there is not a unique gene product), have been culled from the set. The test data were taken from the test sets for the programs GeneID (Guigó *et al* 1992) and GRAIL (Uberbacher and Mural 1991) with several examples of alternative splicing removed.

There are several properties of this data set which are noteworthy and typical of human DNA sequences. First, the number of coding nucleotides (i.e. nucleotides that are in exons of any type) is small compared to the total length of the sequences. Second, there are large differences between loci in base composition (G + C content). These differences are much larger than would be expected of a random distribution. There are also large variations in the number and size of introns and exons in different loci. These properties combine to make human gene identification a particularly difficult signal recognition problem.

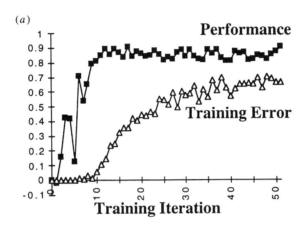

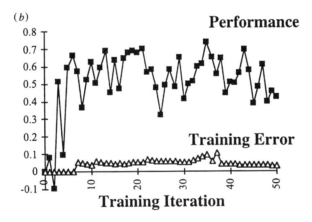

Figure G4.2.6. Learning curves for (*a*) single and (*b*) multilayered networks. The GeneParser-network performance (full squares) is measured as the correlation coefficient for predicting exonic nucleotides. The training error (open triangles) is the fraction of the training set that is not correctly assigned following the network training session.

G4.2.3.3 *Performance*

The single-layered architecture proved to be the best in terms of both speed and accuracy. Figure G4.2.6(*a*) shows a typical learning curve plotting predictive accuracy as a function of training iteration. Starting with random weights, the correlation coefficient for prediction of exon nucleotides in the training set is approximately zero. As training progresses, the performance increases until a plateau is reached after 10 to 15 training iterations. Performance on test data mirrors that on the training data, generalization being 90% to 95% that of the training data. In every instance, the beginning of the plateau phase coincides with the change in slope of the residual training error. This measure is the fraction of training vectors which cannot be correctly classified following the neural network training procedure. Typically, the network is trained until a bail-out criterion is reached (99% of vectors correctly classified) or the maximum number of training epochs is reached.

Figure G4.2.6(*b*) shows a learning curve for a network with six hidden units per sequence class (24 hidden units total). In practice, the more complex network architectures have proven unsatisfactory due to increased training times. More complex networks increase the run time for each sequence considerably. In addition, the increase in the number of free network parameters results in a corresponding increase in the quantity of training data required to obtain good generalization performance. These factors taken together have limited our ability to train and evaluate multilayer networks.

The performance of GeneParser has been measured and compared to other gene identification programs including rule-based and other neural network approaches. These results have been presented elsewhere (Snyder and Stormo 1994, 1995). In summary, GeneParser performs at least as well as other methods and often significantly better when an exhaustive search of the solution space is advantageous. Such cases include the ability to predict very short exons and to correctly parse a sequence in the presence of sequencing errors.

G4.2.4 Conclusions

We have found GeneParser a useful tool for the identification of coding regions in genomic DNA sequences. In addition to being an accurate and sensitive gene identification tool on the benchmark data sets, the neural network architecture allows it to evolve rapidly in a production environment. The system can be retrained to take advantage of new statistics or optimized for the identification of specific sequence targets. Finally, optimization for error tolerance gives the promise of reduced costs by decreasing the coverage required to accurately identify genes in large-scale shotgun sequencing projects.

References

Batra S 1993 A new algorithm for protein structure prediction: using neural nets with dynamic programming *Master's Thesis* Department of Computer Science, University of Colorado, Boulder, CO, USA

Guigó R, Knudsen S, Drake N and Smith T 1992 *J. Mol. Biol.* **226** 141–57

Snyder E E 1994 Identification of protein coding regions in genomic DNA *PhD Thesis* University of Colorado, Boulder, CO 80309-0347

Snyder E E and Stormo G D 1993 *Nucl. Acids Res.* **21** 607–13

——1994 *Nucleic Acid and Protein Sequence Analysis: A Practical Approach* 2nd edn (Oxford: IRL Press) at press

——1995 Identification of protein coding regions in genomic DNA *J. Mol. Biol.* **248** 1–18

Uberbacher E C and Mural R J 1991 *Proc. Natl Acad. Sci., USA* **88** 11 261–5

G4.3 A neural network classifier for chromosome analysis

Jim Graham

Abstract

Analysis of chromosomes is an important and time-consuming task in the diagnosis of inherited or acquired genetic abnormality. Machine vision systems can contribute to the visual inspection of microscope images and the assignment of chromosomes to 24 classes is a critical stage in this analysis. A multilayer perceptron classifier has been developed for use in an automated chromosome analysis system. The inputs to the classifier are chromosome size, centromere position and a representation of the banding pattern measured from microscope images of dividing cells. The outputs are likelihoods of class membership. Optimum performance was obtained by factoring the classifier into two networks, one using size and centromere position alone to provide a first assignment into seven groups, followed by a second step in which the banding information was incorporated to give a final classification. The network is trained by backpropagation and considerable advantage is obtained by using a strategy of gain reduction using both total error and classification accuracy as network monitoring parameters. Classifier performance was tested on fairly large sets of chromosome measurements covering a representative range of data quality. Overall classification accuracy was found to equal or exceed that of a well developed statistical classifier applied to the same data.

G4.3.1 Introduction

In a normal human cell there are 46 chromosomes which, at an appropriate stage of cell division (metaphase), can be observed as separate objects using high-resolution light microscopy. Appropriately stained they show a series of bands along their length and a characteristic constriction called the centromere. Figure G4.3.1(*a*) shows a typical metaphase cell, stained to produce the most commonly used banding appearance (G-banding). Chromosome analysis, which involves visual examination of these cells, is routinely undertaken in hospital laboratories, for example, for diagnosis of inherited or acquired genetic abnormality or monitoring of cancer treatment.

This visual analysis, known as karyotyping, involves counting the chromosomes and examining them for structural abnormalities. To determine the significance of both numerical and structural abnormality it is necessary to classify the chromosomes into 24 groups on the basis of their relative size, the pattern of bands and the centromere position (see figure G4.3.1). Twenty-two of these groups normally contain two homologous (structurally identical) chromosomes. The other two groups contain the sex chromosomes X and Y. In the case of a normal male cell, the X and Y groups contain one chromosome each; in a female cell there is a homologous pair of X chromosomes and the Y group is empty.

The time-consuming nature of chromosome analysis has resulted in considerable interest in the development of automated systems based on machine vision. A number of such systems are now in routine use in many hospitals (e.g. Graham 1987, Graham and Pycock 1987, for a review see Lundsteen and Martin 1989). The processing stages in analyzing the microscope images are illustrated in figure G4.3.2. Chromosomes are isolated from the images, measurements are made of chromosome size, shape and

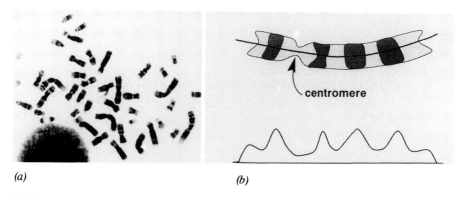

(a)   (b)

Figure G4.3.1. Chromosomes and chromosome features. (*a*) A cell at metaphase. The individual chromosomes show the banding pattern (G banding) produced by staining. (*b*) Schematic drawing of a chromosome showing the position of the centromere. The density profile (below) is formed by projecting the density onto the curved centerline.

banding pattern, these measurements are used in a classifier to assign the chromosome to appropriate groups and the information is displayed to the user, usually in the form of a karyogram in which the chromosomes are arranged in a tabular array of their classes (see Graham and Piper 1994). The chromosome classification performance of these systems depends on the type of material used, but at best the misclassification rate is 6–18% (Piper and Granum 1989) which compares poorly with visual classification by a cytotechnician (Lundsteen and Granum 1976). All automated systems in clinical use operate interactively, allowing an expert operator to correct machine errors in image segmentation, feature extraction and classification, resulting in useful performance (Graham and Piper 1994). However, there is clear scope for improvement in automatic classification. The objective of this study was to investigate the use of a neural network in the classification module.

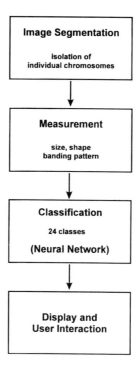

Figure G4.3.2. Block diagram of an automated chromosome analysis system. Classification of chromosomes follows segmentation and measurement modules, and is implemented in this study as a neural network. The display and interaction module permits correction of errors in machine analysis and diagnostic decision making.

G4.3.2 Design process

G4.3.2.1 Design constraints

An important issue for automatic classification is the representation of the banding pattern. Several different classifiers have been reported using statistical or syntactic approaches (e.g. Granlund 1976, Granum 1982, Groen *et al* 1989, Thomason and Granum 1986). Each of these involves the extraction of a number of intuitively defined features, usually associated with the chromosome's density profile. The density profile is a one-dimensional pattern obtained by projecting the chromosome's density onto its center line (figure G4.3.1(*b*)), and reflects the largely linear organization of the chromosome structure. The processing involved in extracting features from the profiles involves the risk of losing information, a risk which may be eliminated by using the density profile itself as the banding representation. This type of one-dimensional pattern is a natural form of input for an artificial neural network. The potential advantage of neural network classifiers lies in their flexibility; they can be readily retrained for classification of new types of data. This property is likely to be useful for chromosome classification as specimen preparation techniques in routine use evolve very rapidly, resulting in changes in chromosome appearance. In particular, there is an increasing clinical requirement to use higher-resolution banding for diagnostic purposes, resulting in routine examination of longer (prometaphase) chromosomes. This will result in the need for greater adaptability in automated karyotyping systems.

Figure G4.3.2 indicates that the classification module is easily isolated from the rest of the system. The outputs of the classifier are the probabilities of membership of each of the 24 classes corresponding to the inputs for each chromosome. The inputs are the chromosome size, the centromeric index and the banding profile.

Size. This may be measured either as the length of the chromosome or its area; the two measures are very highly correlated. In the datasets used in this study, the length was used.

Centromeric index. The centromere divides the chromosome into long and short 'arms' (figure G4.3.1(*b*)). The centromeric index (CI) is the ratio of the length of the short arm to that of the whole chromosome, and gives a measure of shape.

Banding profile. The number of samples representing the banding profile can vary between 10 and 140 depending on the class of the chromosome and the state of contraction of the cell in which it occurred. The classification module requires a consistent input vector and all banding patterns must therefore be represented by the same number of samples. Considerable experimentation (Jennings and Graham 1993, Errington and Graham 1993) gave the result that a constant number of samples could be used to represent the profile, irrespective of the original chromosome length, and that this number could be quite small (as low as 15 samples for all profiles) with very little loss of classification accuracy. The use of a uniform number of samples meant that the profiles of long chromosomes had to be subsampled by local averaging, and the short chromosomes oversampled by interpolation.

The principal requirement of the classifier module is classification accuracy. The overall system performance is closely dependent on presenting the clinical user with a classification of the chromosomes in a cell which requires minimal interactive correction. Statistical classifiers give (barely) acceptable performance and it would be desirable to improve on this using a neural network classifier, although similar performance would be acceptable in view of the potential benefits in adaptability.

G4.3.2.2 Network topology

In this application we have a classification problem using continuous-valued inputs, where the classes are well defined and expert classification of the training data is available. It is a clear case for a *multilayer* C1.2 *perceptron* (MLP). A preliminary study (Jennings and Graham 1993) compared the suitability of the MLP topology with the *Kohonen self-organizing map*, and confirmed the expected result that significantly better C2.1.1 classification was obtained using the supervised training regime of the MLP. Optimum network parameters (starting gain, momentum, number of hidden nodes) were determined empirically (Errington and Graham 1993).

In principle, it is possible to classify chromosomes on the basis of the banding pattern alone. However, the size and centromeric index are extremely powerful classification features, and must be included for the most accurate results. These features might be used as inputs to the network in addition to the banding features as shown in figure G4.3.3(*a*). It is known, however, that size and centromeric index can classify chromosomes into seven groups in the absence of banding information (the 'Denver' classification, Denver Conference 1960). An alternative form of input was therefore investigated, in which these two features were processed by a preclassifier, also an MLP, and trained to produce outputs corresponding to the 'Denver' classes. The seven outputs of the preclassifier were then used along with the banding features as inputs to the main classifier (figure G4.3.3(b)). The main classifier consisted of a network with 15 input nodes for banding features, plus the nodes necessary for the size and centromeric index features, 100 hidden nodes and 24 output nodes (one for each class), as illustrated in figure G4.3.3. The classification results in the three sets of chromosome data (see below) are given in table G4.3.3. It is clear that preprocessing the centromeric index and size features gave a considerable advantage.

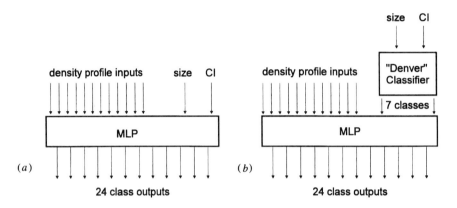

Figure G4.3.3. Two possible configurations for including size and centromeric index features in the input vector. (a) The two features are simply additional features along with the banding profile samples. (b) The features are processed to produce seven values corresponding to the probability of membership of the 'Denver' groups. The banding profile then provides information to refine the classification to 24 classes. In either case there are 24 outputs corresponding to the membership likelihoods of each of the classes.

G4.3.3 Training methods

The network was trained and tested using three data sets of annotated measurements from G-banded chromosomes. The characteristics of these data sets are summarized in table G4.3.1. The data in the Copenhagen set were obtained by densitometry of photographic negatives of selected cells of good appearance. The other two data sets were digitized directly from microscope images of routine material. The preparation techniques in chorionic villus sampling results in poor visual quality of the chromosome images in the Philadelphia set. The three data sets give a reasonably large number of data for network training and testing covering a range of quality representative of that found in a real implementation.

Table G4.3.1. Summary of the data sets of chromosome measurements.

Data set	Tissue of origin	Data acquisition method	Number of chromosomes	'Quality' of chromosome images
Copenhagen	Peripheral blood	Densitometry	8106	High
Edinburgh	Peripheral blood	TV camera	5469	Medium
Philadelphia	Chorionic villus	Linear CCD array	5817	Low

C1.2.3   The training algorithm employed was the classical *backpropagation* method (Rummelhart *et al* 1986), using a strategy of progressive reduction in gain (learning rate) during the training. Two measures were used to monitor performance: total network error and classification accuracy on the training data. These measures are not identical due to the fact that the classification result is determined only by the highest output, but they are both useful measures of performance. During training, the gain was halved if the total network error had increased by more than 10%, or the classification performance had not improved over

© 1997 IOP Publishing Ltd and Oxford University Press

the previous presentation of the training data. Training was halted when the value of gain dropped below 10^{-4}. The gain reduction strategy proved extremely valuable in this application. Table G4.3.2 shows the misclassification rates on training data after convergence of networks trained to classify banding features alone in the preliminary study (Jennings and Graham 1993). There is a clear advantage in using gain reduction and in using two performance characteristics to monitor the network.

Table G4.3.2. The effect on classification performance of gain reduction during training, monitored using total network error and accuracy of classification of the training data.

Training strategy	No gain reduction	Gain reduction (network error only)	Gain reduction (network error and classification accuracy)
Misclassification rate	53 (%)	12 (%)	4 (%)

In the classification experiments, the network was trained using approximately half of each data set, the remainder being used for 'unseen' testing. The roles of the training and test sets were then reversed, and the classification rate obtained as the average of the two unseen tests. In all classification experiments the initial gain value used was 0.1 and the momentum value 0.7.

G4.3.4 Preprocessing

As noted above, the banding profiles were represented by 15 sample values, obtained by averaging or interpolation from the 'raw' profiles. The relative sizes and overall densities of chromosomes in a cell are fairly consistent; however, absolute lengths and densities can vary between cells. Length and density measures were therefore normalized to a constant value for each cell before classification.

The size and CI features were preprocessed using an MLP with two inputs, seven outputs and a hidden layer of 14 nodes (see figure G4.3.3(b)).

G4.3.5 Output interpretation

The network output is a vector of 24 class assignment values for each chromosome, approximating the Bayesian probabilities of the chromosome belonging to each class. The class to which the chromosome is assigned is that with the highest output. Classification results are shown in table G4.3.3. It is worth noting here that the classification of chromosomes is constrained by the fact that (in a normal cell) each class contains exactly two chromosomes (or one in the case of the sex chromosomes in a male cell). Application of this constraint can significantly improve the classification accuracy over 'context-free' classification of individual chromosomes (Tso *et al* 1991). Network approaches can give good results in applying constraints (Errington 1994), but consideration of these methods is beyond the scope of this chapter which is restricted to considering the classification of isolated chromosomes.

Table G4.3.3. Classification performance of two MLP configurations compared with that of a parametric statistical classifier (Piper and Granum 1989).

Classifier	Data set		
	Copenhagen	Edinburgh	Philadelphia
MLP, banding, length and centromeric index	6.9%	18.6%	24.6%
MLP, 'Denver' preclassifier	5.8%	17.0%	22.5%
Parametric classifier	6.5%	18.3%	22.8%
Significance of MLP improvement	2% level	5% level	not significant

G4.3.6 Development

As we were required to carry out a number of experimental investigations using the network, and to arrive at a configuration which could be incorporated with other software modules, we implemented our own network simulators. They were programmed in Pascal and ran on UNIX workstations.

G4.3.7 Comparison with traditional methods

A feature of developing a neural network classifier for chromosome analysis is the possibility of comparing a network solution to classical statistical methods. There have been a number of approaches to chromosome classification, but the most successful prior to this study was that of Granum (1982), subsequently greatly refined by Piper (Piper and Granum 1989). This method extracts banding features using 'weighted density distributions'; essentially, the banding profile is multiplied by a number of intuitively defined weighting functions, approximating a set of basis functions for the banding pattern. The features extracted from the density profiles in this way are combined with length and CI features, and classified using a parametric classifier. Table G4.3.3 compares the best network performance with the statistical method of Piper and Granum (1989) in performing context-free classification of individual chromosomes. The network performance is significantly better for the Copenhagen and Edinburgh data sets and identical for the Philadelphia data set.

The results show that a network classifier can give higher classification accuracy than a classical technique. While the improvement is statistically significant, however, it is not overwhelming. The classification performance of both types of classifier is good for Copenhagen data, probably acceptable for data of routine quality, such as is found in the Edinburgh set, and inadequate in the case of the poor-quality Philadelphia data. The development costs of the network classifier are arguably appreciably smaller, since the time from proposing the concept to arriving at a final configuration was considerably shorter and involved less manpower than was the case for the conventional classifier. From an implementation point of view, the network classifier is likely to be more adaptable. Our experience is that the best network parameters (topology, gain, momentum) are stable in the face of wide variation in the quality of data. It seems likely then that a single 'hard-wired' network would be adequate for any implementation, requiring only a mechanical training process to adapt to the properties of the chromosome data in a new installation. Training 'on the fly' could be applied to account for slow changes in chromosome appearance arising from changes in the nature of the preparation techniques, etc.

G4.3.8 Conclusions

Chromosome classification is an important element in automated cytogenetic analysis. The classification problem in this case is far from trivial; there are few applications where there is a requirement to assign objects to as many as 24 classes. We have constructed a chromosome classifier using a multilayer perceptron network whose performance equals or betters that of a well developed classifier using traditional statistical methods. The form of the network is standard, with the exception that known properties of the classification features allowed the network to be 'factored' into two steps to achieve optimum classification performance. Equivalent performance can be obtained with a single network composed of many more nodes (Errington 1994).

In this study we have had the luxury, not afforded to many network implementations, that data sets have been available with fairly large quantities of expertly classified real-world examples. The data were made available within the Concerted Action of Automated Cytogenetics Groups supported by the European Community (project no II.1.1.13). An interesting feature of this application is that we have been able to make a direct comparison with a statistical classifier applied to the same data.

References

Denver Conference 1960 A proposed standard system of nomenclature of human mitotic chromosomes *Lancet* **1** 1063–5

Errington P A 1994 Application of neural network models to chromosome classification *PhD Thesis* University of Manchester

Errington P A and Graham J 1993 Application of artificial neural networks to chromosome classification *Cytometry* **14** 627–39

Graham J 1987 Automation of routine clinical chromosome analysis I, Karyotyping by machine *Anal. Quantit. Cyt. Hist.* **9** 383–90

Graham J and Piper J 1994 Automatic karyotype analysis *Chromosome Analysis Protocols* ed J R Gosden (Totowa, NJ: Humana) pp 141–85

Graham J and Pycock D 1987 Automation of routine clinical chromosome analysis II, Metaphase finding *Anal. Quantit. Cyt. Hist.* **9** 391–7

Granlund G H 1976 Identification of human chromosomes using integrated density profiles *IEEE Trans. Biomed. Eng.* **23** 183–92

Granum E 1982 Application of statistical and syntactical methods of analysis to classification of chromosome data *Pattern Recognition Theory and Application* ed J Kittler, K S Fu and L F Pau, NATO ASI (Dordrecht: Reidel) pp 373–98

Groen F C A, tenKate T K, Smeulders A W M and Young I T 1989 Human chromosome classification based on local band descriptors *Patt. Recog. Lett.* **9** 211–22

Jennings A M and Graham J 1993 A neural network approach to automatic chromosome classification *Phys. Med. Biol.* **38** 959–70

Lundsteen C and Granum E 1976 Visual classification of banded human chromosomes I, Karyotyping compared with classification of isolated chromosomes *Am. J. Human Genet.* **40** 87–97

Lundsteen C and Martin A O 1989 On the selection of systems for automated cytogenetic analysis *Am. J. Med. Genet.* **32** 72–80

Piper J and Granum E 1989 On fully automatic measurement for banded chromosome classification *Cytometry* **10** 242–55

Rummelhart D E, Hinton G E and Williams R J 1986 Learning internal representations by error propagation *Parallel Distributed Processing: Explorations in the Microstructures of Cognition* vol 1 *Foundations* ed D E Rummelhart and J L McCelland (Cambridge, MA: MIT Press) pp 318–62

Thomason M G and Granum E 1986 Dynamically programmed inference of Markov networks from finite sets of sample strings *IEEE Trans.* **8** 491–501

Tso M K S, Kleinschmidt P, Mitterreiter I and Graham J 1991 An efficient transportation algorithm for automatic chromosome karyotyping *Patt. Recog. Lett.* **12** 117–26

G4.4 A neural network for recognizing distantly related protein sequences

Dmitrij Frishman and Patrick Argos

Abstract

A sensitive technique for protein sequence motif recognition based on neural networks has been developed by Frishman and Argos, and by Vogt *et al*. It involves three major steps. (i) At each alignment position of a set of N matched sequences, a set of N aligned oligopeptides is specified with preselected window length. N neural networks are subsequently and successively trained on $N - 1$ amino acid spans after eliminating each ith oligopeptide. A test for recognition of each of the ith spans is performed. The average neural network recognition over N such trials is used as a measure of conservation for the particular windowed region of the multiple alignment. This process is repeated for all possible spans of given length in the multiple alignment. (ii) The M most conserved regions, delineated by significance thresholds, are regarded as motifs and the oligopeptides within each are used to train extensively M individual neural networks. (iii) The M networks are then applied in a search for related primary structures in a large databank of known protein sequences. The oligopeptide spans in the database sequence with strongest neural net output for each of the M networks are saved and then scored according to the output signals and the proper combination which follows the expected N- to C-terminal sequence order. The motifs found from the database search with highest similarity scores can then be used to retrain the M neural nets which can be subsequently utilized for further searches in the databank, thus providing even greater sensitivity to recognize distant familial proteins. This technique was successfully applied to the integrase, DNA-polymerase and immunoglobulin families.

G4.4.1 Project overview

Comparison and alignment of protein amino acid sequences can provide important biological information (compare Argos 1990) which can substantially reduce experimental effort. The degree of sequence variability in different parts of the protein molecule is determined by complex functional and structural constraints. The most conserved subsequence regions (motifs or patterns) can often be delineated from several aligned protein sequences of a given molecular type, especially if the proteins are distantly related. The most conserved amino acids within the motifs are often the most important functionally; they may form receptor and nucleic acid binding regions or active sites for enzymes. These regions are also very useful for identifying very distant members in a molecular family as their conservation is required to maintain function. Their collection can, in turn, shed further light on protein structure/function relationships.

The objective of the present algorithm was to act as an automatic and sensitive procedure to delineate motifs in multiply aligned sequences and then to use these patterns in a search for other distantly related primary structures. The latter problem is of particular importance for the human genome project which is expected to produce massive quantities of sequence data. The total number of different molecular families is expected to be of the same order of magnitude as the number of genes contained in the bacterial chromosome, but the number of sequences determined will be several orders of magnitude greater. This vast quantity of data can be handled easily if each sequence can be quickly and sensitively assigned to its molecular family from its characteristic sequence patterns.

G4.4.2 Design process

G4.4.2.1 Motivation for a neural solution

When very similar sequences are considered with only a limited number of amino acid substitutions, the problem of defining a protein pattern becomes trivial as all possible exchanges can be enumerated. For more diverged sequences, with many and distributed residue exchanges, derivation of subsequence patterns or motifs becomes a more sophisticated task in pattern recognition (see figure G4.4.1 for an example of a motif). One of the powerful techniques for dealing with poorly determined and noisy patterns are artificial neural networks, which can extract essential features from a set of variable, imperfect objects (see, for example, Wasserman 1989).

```
P2    H A L R H S F A T H F M I N G
186   H V L R H T F A S H F M M N G
P22   H D L R H T W A S W L V Q A G
P1    H S A R V G A A R D M A R A G
λ     H E L R S L S A - R L Y E K Q
φ     H D M R R T I A T N L S E L G
P4    H G F R T M A R G A L G E S G
      +   ~ +  ·  ·    - ~    ~    ·   -
```

Figure G4.4.1. Example of a conserved region in the integrase protein family (modified from Argos *et al* 1986). A short region of aligned integrase sequences from bacteriophages P2, 186, P22, Pl, λ, φ80 and P4 are shown. Amino acids are in one letter code. Symbols +, −, ~ and · denote positions with different degrees of amino acid conservation, from high to low, respectively.

G4.4.2.2 General description of the algorithm

Protein sequence pattern recognition based on neural networks requires a database of sequences known to belong to a family; i.e. the mapping between sequences in the training set and patterns is known in advance. The trained network is used to search for additional representatives of the same pattern in a large database of known primary structures. Here a more difficult task is addressed. From a multiple alignment of protein sequences, the relevant subsequence motifs are first delineated. This is achieved by training individual neural networks for every possible set of matched oligopeptides with given length over the entire multiple alignment. The subsequence regions that are best identified by their associated neural networks are defined as the conserved motif regions in the overall alignment. An entire sequence database can subsequently be searched by submitting every database oligopeptide of given length to the motif networks and the network outputs recorded. Those sequences with sufficient positive response from the individual pattern networks taken in the proper sequential order from the N- to C-termini are then identified as distantly related members of the original family. The motif networks can then be retrained and made more sensitive in the light of the newly found subsequences. A database search can then be re-initiated in an attempt to discover even more distant family members. This process can be repeated as often as necessary.

G4.4.2.3 Data, preprocessing and neural network topology

It is assumed that a set of aligned sequences is available. Such techniques as CLUSTAL (Higgins and Sharp 1989) or PILEUP in the GCG program suite (Genetics Computer Group 1991) can be utilized for such purposes. Figure G4.4.2 illustrates the neural network architecture. In each position k of a multiple alignment of total length L over N protein sequences ($k = 1, L - c + 1$), all N alignable oligopeptides of chosen but constant length c are used as positive (observed) examples of a possible consensus pattern. Negative examples of the pattern are randomly generated oligopeptides with amino acid composition corresponding to that of the N alignable peptides of length c starting at position k of the multiple alignment. A 20-bit binary code is used to represent each of the 20 amino acid types such that only one bit is assigned unity and the rest null values. A different position with value 1 is chosen for each of the residue types. This coding scheme for protein sequences was first proposed by Qian and

© 1997 IOP Publishing Ltd and Oxford University Press

Sejnowski (1988) and Bohr *et al* (1988). There is no special code for gaps. For the sake of simplicity, gaps are substituted by randomly generated amino acids.

The neural networks used consisted of one input, one hidden and one output level. The output level consists of only one neuron, the state of which is compared to the desired response for each particular presentation (1 for a positive output and 0 for a negative one). *Backpropagation procedures* were used for network training (Rumelhart *et al* 1986, Wasserman 1989, White 1989).

C1.2.3

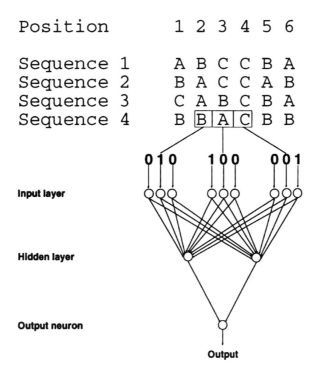

Figure G4.4.2. Illustration of the neural network architecture for calculating the profile of conservation. The figure depicts a multiple alignment of four sequences of length 6 consisting of a three-letter alphabet (A, B and C). A tripeptide segment of the alignment acts as input to a neural network. In this case, the window length c is 3 and the start alignment position k for the fourth oligopeptide is 2. The input layer receives three bits of information representing each of the three symbols A, B and C. The input layer would then consist of nine neurons with binary input values. Outputs from all these neurons act as inputs to each neuron of the hidden layer; the hidden neuron outputs are, in turn, inputs to the single output neuron. In reality, each amino acid is represented by a 20-bit vector such that the number of units in the input layer is $20 \times n$ where n is the number of amino acids in the oligopeptide. The number of units used in the hidden layer for protein sequences was 10.

G4.4.3 Training and recognition methods

The method used for protein pattern recognition consists of three main procedures (figure G4.4.3). Every possible span of alignment positions with given window length is scanned with a neural network. For each aligned oligopeptide in a particular alignment span, a neural network is trained over the remaining peptides and its response for the given peptide recorded. These responses are then averaged for all oligopeptides in the alignment span. A plot of the mean response versus the overall alignment position number of the start point of the span under consideration represents a profile of sequence conservation for a particular protein family. Peaks on this curve correspond to the most conserved regions of the primary structures. In the second step several individual networks selected from the first procedure are intensively trained to recognize only the most conserved regions or motifs of the alignment where all oligopeptides in the corresponding amino acid positions are used as a training set. In the third procedure these resulting networks are applied sequentially to all sequences and all possible oligopeptides in a large protein sequence database, and the best hits are determined. The newly discovered motifs can be used to retrain and further sensitize the networks, subsequently applied to a second search of the database with resultant recognition

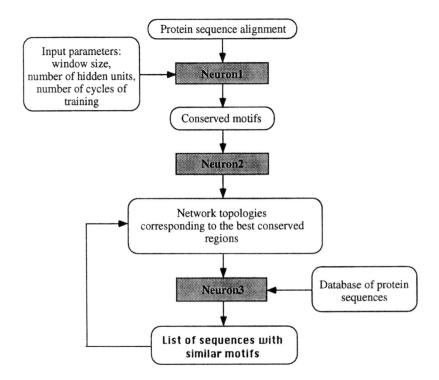

Figure G4.4.3. Flowchart for the procedure to recognize protein sequence motifs. Three consecutive steps of the analysis are implemented as computer programs NEURON1, NEURON2 and NEURON3 (see the text for details).

of even more distantly related sequences. The parameters and thresholds of the analysis may be modified and the analysis repeated until some optimal decision boundaries are achieved (e.g. the number of false positives minimized). In the following sections, the three major steps are described in detail.

G4.4.3.1 Search for unknown protein patterns

To make the training process robust, we adopted a jackknife procedure similar to that described by Hirst and Sternberg (1991). An ith peptide ($i = 1, N$) is taken from the subalignment and an ith network is trained on the set of $N - 1$ remaining peptides used as positive examples and $N - 1$ randomly generated peptides acting as negative examples. The training is repeated N_{cycl} times ($N_{cycl} = 60$) for each of the ith oligopeptides such that the total number of input presentations to the network associated with alignment position k is $2 \times (N - 1) \times N_{cycl}$. The number of times each of the $N - 1$ peptides is presented to the neural network differs according to the similarity of the oligopeptides associated with position k (Sibbald and Argos 1990) such that subsequences with high similarity are not allowed to bias the training. After training of the ith network the removed peptide is presented for recognition and the output of the network $REC(i, k)$ (which lies in the range 0–1) is stored. This procedure is repeated for all N peptides of the subalignment associated with the start position k of the overall alignment. To build a numerical curve characterizing the regions or motifs with primary structure conservation along the protein alignment, the average recognition of all N networks was taken as the measure of conservation in each position of the alignment k:

$$\overline{REC(k)} = \frac{\sum_{i=1}^{N} REC(i, k)}{N}.$$

Each network is trained until the fractional change of the error becomes very small.

In order to derive the most conserved motifs from the resultant plot, it is necessary to define some cutoff level such that, if $\overline{REC(k)}$ is greater than this threshold, then the alignment span, which is c amino acids in length and begins at position k, is declared well conserved. It was found from several protein examples that 12-residue spans with a mean recognition peak value above 0.7 constituted significant motifs. This implies that 70% of the N subsequences associated with start site k will be recognized given that $REC(i, k) = 1.0$ or 0.0 represents, respectively, complete or no recognition.

The number of units in the hidden layer has little effect on the results at this step of the analysis provided that it is not less than 10. Since the optimal window size was found to be in the range 10–15 for the protein examples, 12 was selected as representative.

G4.4.3.2 Generation of final topologies for search neural networks

The M most conserved regions in the multiple alignment were used as input to train several individual neural networks and to generate final sets of weights. Randomly generated peptides were used as negative examples. As the jackknife procedure is not used in this step, many more cycles of training (120–150) were required to reach the same level of recognition accuracy.

Although use of an ensemble of networks based on variable length motifs would certainly improve sensitivity for recognizing distantly related sequences in a full database search, the computer processing time is prohibitive. However, sensitivity can be improved considerably by increasing the number of units in the hidden layer which are optimally more than 10 times greater (100–150) than the number used during the profile calculation. Further increase of the number of hidden units did not improve the results in the protein sequence examples tested here.

G4.4.3.3 Large database searches

The resulting networks were used in an attempt to find distant members of a protein family in a large database of known protein sequences. Release no 21 of the SWISS-PROT database (Bairoch and Böckmann 1991) consisting of over 23 000 individual sequences was searched in the protein examples considered. All oligopeptides of each database sequence are presented to all M networks and the R best recognitions (BESTREC(p, q), $q = 1, R$) for each pth ($p = 1, M$) network as well as the starting sequence positions of these peptides of length c (POS(p, q)) were stored. It is also possible to specify the maximal number NSUB of subunits or domains, each with the M motifs, which are expected in the proteins belonging to the family under question. Then the number of best recognized peptides for each of the M networks can be $R^* = R \times$ NSUB. All possible combinations (NCHAIN) of successive peptides, taken from the recognition set of each of the M networks, are considered. It must be emphasized that only those combinations are allowed that contain the motifs in the proper N- to C-terminal order as they appear in the multiple sequence alignment. For each combination, a score is calculated:

$$\text{SCORE}(i) = \left(\sum_{p=1}^{M} \text{BESTREC}(p, q)^2 \right)^{1/2}, i = 1, \text{NCHAIN}$$

where q for each network p is in the range $1 \leq q \leq R^*$, POS$(p, q) + c <$ POS$(p + 1, q)$, and BESTREC(p, q) is the qth output value for pth network. The largest SCORE(i) among all possible paths is stored as the final score for the database sequence under consideration.

If new motifs in distantly related family members are discovered, then they can be used as additional inputs to retrain the networks of step 2 and then a database search re-initiated as in step 3. Alternatively, to conserve computing effort, the sequences associated with the highest scores from the initial step 3 process (e.g. the first 1000 or so) can be searched after retraining. Obviously the process can be iteratively repeated as appropriate.

G4.4.4 Interpretation of output and comparison with traditional methods

A sliding neural network over each stretch of a multiple alignment in conjunction with a jackknife procedure found conserved motifs in integrases (Argos *et al* 1986, Abremski and Hoess 1992), DNA-directed DNA polymerases (Ito and Braithwaite 1991) and proteins sharing the immunoglobulin fold (Williams and Barclay 1988). For example, the profile of conservation calculated from the sequence alignment of DNA-polymerases clearly reveals the location of the four catalytic and DNA-binding motifs (figure G4.4.4). These four motifs were utilized in a search for other members of the DNA-polymerase family in SWISS-PROT.

The resolving power of the database search with the program NEURON3 is illustrated by figure G4.4.5. In scanning the protein sequence database with neural networks trained to recognize conserved integrase motifs, 24 out of 25 sequences with scores three standard deviations above the average were members of

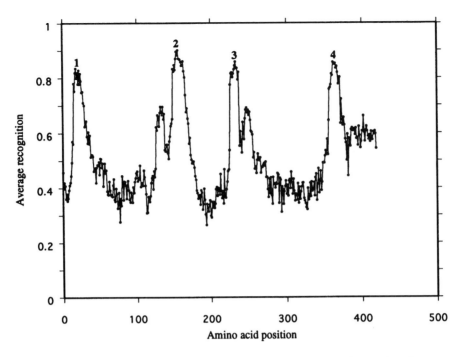

Figure G4.4.4. Profile of conservation of the partial alignment of DNA-dependent DNA polymerases. Four peaks (average recognition above 0.7) show the location of the catalytic and DNA-binding motifs.

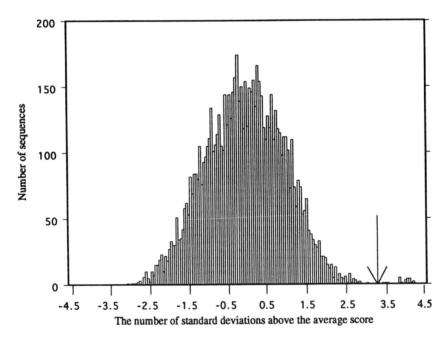

Figure G4.4.5. Statistical distribution of similarity scores for all database sequences after a search for distant members of the integrase family with the program NEURON3. The arrow indicates the border between the 25 highest scoring sequences (24 of which are integrases) and other sequences of the database.

the integrase family. In this and other examples studied, the neural network motif technique performed more sensitively than the present most successful and widely used profile analysis method in detecting distantly related familial sequence members (Gribskov *et al* 1987, 1990). The profile approach relies on the frequency of appearance of each amino acid type at all alignment positions. For example, utilization of the PROFILESEARCH routine (Genetics Computer Group 1991) to detect proteins with an immunoglobulin fold in a large database from an initial set of 32 aligned familial sequences yielded 66 false positives in the top 400 best hits while the neural network motif search had only 12 errors. In the top 600 hits, the profile technique recognized only 18 of 49 immunoglobulin-like molecular types (T-cell receptors, histocompatibility antigens, proto-oncogene tyrosine kinase, etc) while the neural network motifs pointed to 37 members, over twofold more. Furthermore, retraining the networks with the first 250 best hits of the first search resulted in only three missed immunoglobulin family types and no false positives in the 400 best hits of the second search.

The methodology described here is intended for sensitive sequence comparisons where little overall similarity is detectable except for a few conserved regions. For alignments of closely related sequences, the motif/neural network procedure has no advantages over profile analysis or other comparable search techniques based purely on sequence statistics. In these cases the conservation is distributed over practically the entire sequences and it is not possible to distinguish conserved regions. For very distantly related proteins, conserved segments 'float' atop the background noise of a multiple alignment. In such cases, searching a large database with neural networks trained to recognize only motifs results in better recognition of distant sequences as compared with profile-like algorithms which are vulnerable to difficulties in correctly aligning largely dissimilar structures, reliant on the constrained size of insertions and deletions, sensitive to selected gap penalty values required to find the optimal alignment, and which yield alignment assessments based considerably on nonconserved regions in the distant sequences. The motif/neural network method is independent of these factors providing the multiple alignment procedures or the researcher can at least recognize the conserved subsequences. This is not the first indication that neural networks can perform more sensitively in sequence analysis than statistical methods (to which profile-like techniques belong). Lapedes *et al* (1990) investigated the effectiveness of various neural network, machine learning and information theory techniques in DNA sequence pattern searches and found that neural networks provided the highest accuracy. The effectiveness of the network/motif method described here lies in its ability to delineate the motif regions automatically, the sensitivity of the neural networks, the proper weighting of the input subsequences, and the reliance only on motif segments in database searches avoiding problems associated with insertions/deletions and noisy assessments of significance.

The motif search technique has not only been implemented on a single processor computer (Frishman and Argos 1992) but also on a DEC massively parallel machine (Vogt *et al* 1994) referred to as a MASPAR computer. The algorithm is particularly amenable for the multiprocessor environment (4096 in the MASPAR) since motif searches can be performed on individual sequences simultaneously. The 12-hour processing time required on a VAX 9000 mainframe to search about 30 000 sequences was reduced to 0.5 hours on the MASPAR.

References

Abremski K E and Hoess R H 1992 Evidence for a second conserved arginine residue in the integrase family of recombination proteins *Prot. Eng.* **5** 87–91

Argos P 1990 Computer analysis of protein structure *Methods Enzymol.* **182** 751–76

Argos P, Landy A, Abremski K, Haggard-Ljungquist E, Hoess R H, Khan M L, Kalionis B, Narayana S V L, Pierson L S III, Sternberg N and Leong J M 1986 The integrase family of site-specific recombinases: regional similarities and global diversity *EMBO J.* **5** 433–40

Bairoch A and Böckmann B 1991 The SWISS-PROT protein sequence data bank *Nucl. Acids Res.* **19** 2247–9

Bohr H, Bohr J, Brunak S, Cotterill R M J, Lautrup B, Noorskov L, Olsen O H and Petersen S B 1988 Protein secondary structure and homology by neural network *FEBS Lett.* **241** 223–8

Frishman D I and Argos P 1992 Recognition of distantly related protein sequences using conserved motifs and neural networks *J. Mol. Biol.* **228** 951–62

Genetics Computer Group 1991 *Program Manual for the GCG Package, Version 7* April 1991, 575 Science Drive, Madison, Wisconsin, USA 53711

Gribskov M, Lüthy R and Eisenberg D 1990 Profile analysis *Meth. Enzymol.* **183** 146–59

Gribskov M, McLachlan A D and Eisenberg D 1987 Profile analysis: detection of distantly related proteins *Proc. Natl Acad. Sci., USA* **84** 4355–8

Higgins D G and Sharp P M 1989 Fast and sensitive multiple sequence alignments on a microcomputer *Comput. Appl. Biosci.* **5** 151–3

Hirst J D and Sternberg M J 1991 Prediction of ATP-binding motifs: a comparison of a perceptron-type neural network and a consensus sequence method *Prot. Eng.* **4** 615–23

Ito J and Braithwaite D K 1991 Compilation and alignment of DNA polymerase sequences *Nucl. Acids Res.* **19** 4045–57

Lapedes A, Barnes C, Burks C, Farber R and Sirotkin K 1990 Application of neural networks and other machine learning algorithms to DNA sequence analysis *Computers and DNA, SFI Studies in the Sciences of Complexity* vol Vll ed G Bell and T Marr (New York: Addison-Wesley) pp 157–81

Qian N and Sejnowski T J 1988 Predicting the secondary structure of globular proteins using neural network models *J. Mol. Biol.* **202** 865–84

Rumelhart D E, Hinton G E and Williams R J 1986 Learning internal representations by error propagation *Parallel Distributed Processing. Explorations in the Microstructure of Cognition Vol 1: Foundations* ed D E Rumelhart and J L McLelland (Cambridge, MA: MIT Press) pp 318–62

Sibbald P R and Argos P 1990 Weighting aligned protein or nucleic acid sequences to correct for unequal representation *J. Mol. Biol.* **216** 813–8

Vogt G, Frishman D and Argos P 1994 A parallel processor implemementation of an algorithm to delineate distantly related protein sequences with conserved motifs and neural networks *Information Systems and Data Analysis Proc. 17th Annual Conf. of the Gesellschaft für Klassification* ed H H Bock, W Lenski and M M Richter (Berlin: Springer) pp 397–408

Wasserman P D 1989 *Neural Computing. Theory and Practice* (New York: Van Nostrand Reinhold)

White H 1989 Learning in artificial neural networks: a statistical perspective *Neural Comput.* **1** 425–64

Williams A F and Barclay A N 1988 The immunoglobulin superfamily-domains for cell surface recognition *Ann. Rev. Immunol.* **6** 381–405

G5

Medicine

Contents

G5.1 Adaptive logic networks in rehabilitation of persons with incomplete spinal cord injury

Aleksandar Kostov, William W Armstrong, Monroe M Thomas and Richard B Stein

Abstract

Persons with spinal cord injury are generally at least partially paralyzed and are often unable to walk. Some are able to use manually controlled electrical stimulation to act upon nerves or muscles to cause movement of a paralyzed leg so functional walking is achieved. They use crutches or a mobile walker for support, and control stimulation by pressing a switch, usually installed on the walking aid. Machine learning techniques are now making it possible to automate this control. Supervised training can be based on samples of correct stimulation given by the user (e.g. the subject or a resercher), accompanied by data from sensors indicating the state of the person's body and its relation to the ground during walking. A major issue is generalization: whether the result of training can still be used for automatic control after the passage of time or in somewhat different circumstances. As it becomes possible to increase the number and variety of sensors used and to easily implant more numerous stimulation channels, the need is increasing for fast and powerful learning systems to automatically develop effective and safe control algorithms. In the present study, adaptive logic networks were used to develop an experimental walking prosthesis. Successful generalization has been observed up to several days after training.

G5.1.1 Project overview

Today it is possible to apply advanced mechanical, electronic and computing technology to problems of rehabilitation of persons with spinal cord injury (SCI). One of the major thrusts has been in the area of functional electrical stimulation (FES) to cause paralyzed limbs to move and thereby restore a measure of walking capability (Stein *et al* 1992). FES can enable the person to walk reasonably long distances and enter places where a wheelchair does not fit but other means of mechanical support can be used. He or she thereby enjoys a more independent life, with the concomitant benefit of a better blood supply to the paralyzed extremities. The most common and the most reliable method to control stimulation is with hand switches, but this is not appropriate for incomplete quadriplegics or stroke victims who lack adequate hand function. Another problem is that operating a hand switch requires repetitive voluntary action, which can introduce delays and variability. Automatic control of FES is therefore desirable or necessary for some persons. This system for automatic control of FES for locomotion is designed for subjects who have one leg paralyzed after an incomplete SCI, and who have some remaining capability in the other leg. It was developed at the University of Alberta by a team of researchers from a variety of areas under the leadership of Richard B Stein (neuroscience). Much of the work was done by Aleksandar Kostov (biomedical and rehabilitation engineering) to prepare a PhD dissertation in neuroscience. *Adaptive logic network* (ALN) C1.8 software was specially designed and implemented by William W Armstrong (computing science) and Monroe M Thomas (software development). The system was integrated around a desktop PC. Thus, the subjects were electronically linked to the computer for the experiments. During a period of training, an

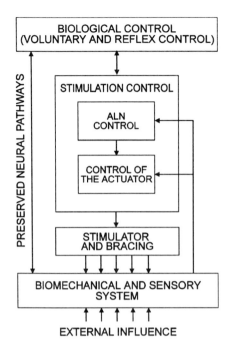

Figure G5.1.1. Control of FES-assisted walking after spinal cord injury in a human–machine system. Two-level machine control takes its inputs from traditional or natural sensory sources and sends its decisions to the assistive system employing FES and mechanical bracing.

artificial neural system learns to copy the stimulation control skill of the physiotherapist or subject. After satisfactory training, the adaptive logic network can take over control of stimulation. It is assumed that the capabilities of the still intact neural pathways are sufficient to enable the subject to move so as to initiate and terminate the swing phase of the stimulated leg. The block diagram in figure G5.1.1 illustrates the hierarchical structure of the automatic stimulation controller. It operates together with control via the preserved neural pathways and uses sensory feedback information in the control loop.

G5.1.2 Design process

G5.1.2.1 Background

After healing from the injury and surgical procedures has occurred, FES-assisted walking is gradually introduced into the rehabilitation program of selected SCI subjects. First, the subject becomes familiar with basic FES principles and learns how an appropriate FES system operates. Then the subject learns how to operate the switch or switches to start and stop stimulation to do simple exercises with an appropriate mechanical aid (parallel bars, harness, frame, four-point walker). Finally, gait training extends the walking distance from a few steps between parallel bars to as many steps as the subject finds comfortable using a mobile mechanical walking aid (a metal frame on wheels). A therapist will begin controlling the walk, but the subject is encouraged to take over as soon as possible.

Taking a step, which is an automatic process for people having normal voluntary control over their extremities, is a very complex process for someone whose extremities are paralyzed. For example, a subject with one leg completely disabled and the other partially disabled has to perform more than ten distinct actions to ensure that body posture and walking aid position result in a safe movement. The two most hazardous phases are the shifts of weight to and from the disabled leg, and need the greatest attention during walking, no matter how walking is controlled. During these phases it is not always obvious to the subject which leg is in charge of supporting most of the body weight.

Despite its many advantages, manual control of FES has a few disadvantages. Even when it becomes a routine motor action presenting little or no cognitive difficulty to the subject, manual switching still requires constant checking to ensure a safe physical movement. Locomotion can be improved by stabilizing the stance phase and reducing its duration, which can be achieved by means of electromechanical sensors and automatic control.

The goal in designing an automatic control system for FES-assisted walking is to preserve or even improve the reliability and safety of a manual system, and to bring more functionality and more efficiency to the disabled gait. A major task in automating control of walking for stroke or incomplete SCI subjects is automatic recognition of the intention to take a step with a disabled leg and to provide the required control signals to the stimulator.

Basic research in neurophysiology suggests a hierarchical structure of natural motor control in vertebrates (Prochazka 1993). This scheme is roughly analogous to the proposed automatic control structure for FES-assisted movement. The external control of FES should consist of at least two major parts: an upper (coordination) level controller should make decisions about the movements to be performed to accomplish a certain task, and a lower (actuator) level controller should initiate actions required to perform a particular movement (see figure G5.1.1). Of course, there is always a third level in the movement control hierarchy—voluntary control. In the present human–machine system, the subject's control of less impaired body parts is assumed to be adequate for this purpose.

One of the first uses of automatic control in the case of foot-drop used a heel switch, which activated a single channel of stimulation to assist in the swing phase whenever the heel came off the ground (Liberson et al 1961). This simple system does not work reliably in subjects where contractions or spasticity prevent a good heel contact with sufficient weight bearing, or in subjects who suffer from clonus (rapid involuntary contraction and relaxation of a muscle) which can cause the heel to lift and touch the ground several times during the stance phase. A rule-based system using hand-crafted threshold logic applied to the signal from a force sensor installed under the toe of the normal leg has been proposed as an alternative method to detect the subject's intention to take a step. The duration of stimulation was either preset or determined by means of another force sensor installed under the toe of the stimulated leg (Kostov et al 1994).

The current study investigated an approach to automatic FES switching based on machine learning of the switching actions of a skilled subject or physiotherapist. It is intended for persons already trained to step periodically by manual switching. This method of cloning a human skill from sensory and output control signals was proposed by Kirkwood and Andrews (1989) and our research team (Stein et al 1992). Feedback information describing the state of the body is derived from force sensors installed in the subject's shoe insoles, though it could also be recorded from biological sensory paths (Popovic et al 1993). An adaptive logic network learns the control signal in a supervised mode based on the manual control signal. If training succeeds, the result can then be used to transform input sensory signals into output control signals for stimulation. Automatic control can then be used in conjunction with manual control to enable the subject to concentrate on other functions during walking, such as shifting the body weight from one leg to the other, avoiding obstacles, moving assistive devices and carrying objects. Manual control or the person's remaining capabilities after an incomplete SCI may be used for safety override functions and to initiate and terminate walking.

During a preliminary feasibility study, ALNs were evaluated offline. The ability of ALNs to learn to generate control signals based on manually controlled stimulation was demonstrated. In addition, it was demonstrated that the quality of ALN learning depends on the number of sensory feedback channels, and that the use of more sensory inputs can reduce errors. To introduce the time dimension into the learning and prediction process, previous sensory signal samples, differences of sensor values and measured time delays were also used. An important feature for control was introduced: early prediction of stimulation events, which provides feedback to the subject about impending stimulation. ALNs were successful in predicting stimulation events up to two seconds in advance (Kostov 1995).

G5.1.2.2 Motivation for a neural network solution

The traditional way to design a rule-based or finite-state control system for FES-assisted locomotion is to apply expert knowledge in generating rules linking sensory feedback information to system actions. This is very labor-intensive and such expertise is in short supply, so it would be feasible to use on a large scale only if the same rules could be applied to many subjects. However, even with very similar physical injuries to the spinal cord, injured persons have functional disabilities that are very specific to the individual. Furthermore, any set of rules must change as the subject advances through rehabilitation. These factors were our main motivation for using machine learning in a system which can generate control functions automatically.

G5.1.2.3 General description of a neural network function

An ALN, which is a feedforward network computing a logical output (see Section C1.8 of this handbook) is used for supervised training based on manual control signals. The goal is to provide initiation of the stimulus and control of its duration based on data coming from force sensors installed in the subject's shoes. The ALN was used to learn the *relationship* between sensor data and the manual on/off signals. It actually represented a real-valued function whose output expresses the level of 'confidence' in 'on' and 'off' as the right decision. For control purposes, that real value was thresholded to produce an on/off decision. Any 'on' decision was then sent to the restriction rule checker (see section G5.1.2.10) which could allow the actual stimulation to take place or not.

C1.8

G5.1.2.4 Requirements and constraints

A practical automatic FES control system is subject to constraints on size, weight, reliability, power consumption and cost. It must permit upgrades when technology advances. The cost factor suggests using inexpensive off-the-shelf components, while the need for real-time control means that a very efficient computational approach is required. Safety is a primary concern of the system's design. The stimulus control function should have a simple form so that a limited number of test samples is sufficient to ensure that the system will not give an unexpected stimulus, which could cause the person to fall. In order to make it extremely unlikely that stimuli are delivered when that is counterindicated by *a priori* knowledge, restriction rules are used to postprocess ALN decisions.

G5.1.2.5 Topology

The form of the control function was assumed to be convex-up, a simple shape that does not allow for any spikes (unless the function *is* just a single spike) thus the topology could be reduced to just one AND node and several LTUs. Larger ALNs were tried but did not give significantly different results, so the simplest system that worked was chosen. Generally, a convex function will not be appropriate, and a more elaborate network topology will be required.

G5.1.2.6 Comparison to other methods

Inductive learning (IL) was also tested for control of FES (Kostov *et al* 1995b). It was used to measure the relative importance of sensors, and to eliminate all but the most useful ones. IL was then evaluated for cloning the control rules for walking of a subject with *complete* spinal cord injury. It was demonstrated that IL is capable of cloning the skill of skilled subjects in controlling two-channel stimulation for FES-assisted walking. ALN and IL techniques were compared on six subjects (Kostov *et al* 1995a). It was demonstrated that, although IL generates its decision trees faster and with lower error on a training set, the ALNs have better generalization. A practical implication of this result is that IL may be better suited for use in control systems where the training set represents the domain very well. It is obvious that training sets acquired during walking of subjects with SCI cannot represent all possible situations, because some high-risk situations that could be valuable for training could give rise to possible injuries (e.g. instability leading to a fall). Both ALN and IL techniques give better results if previous samples are used as inputs together with current ones. Also, both techniques were capable of predicting future stimulation events.

G5.1.2.7 Sources

The ALN system used was specially built in the form of a Windows-based DLL (dynamic link library) to permit interfacing to other parts of the data acquisition and control system. It was based on the Atree 3.0 software of Dendronic Decisions Limited, though Atree 2.7 was used in early trials. An Atree 3.0 ALN Educational Kit is available online that has all of the features of the Atree 3.0 software but is limited to two-input functions (Armstrong and Thomas 1995).

G5.1.2.8 The training set

Three force sensors (Interlink Electronics Inc) installed in insoles were put into each shoe. This set of sensors was chosen on the basis of the following criteria: ability to represent biomechanical measurements useful for describing the state of walking, accuracy, reliability, production of fairly reproducible values (a high signal-to-noise ratio), easy donning and doffing and noninterference with other functions, low cost, easy availability and low power consumption. For the training set, vectors were used consisting of quantized sensor signals together with derived quantities, including values of earlier signal samples and differences. From 1000 to 1500 contiguous data samples were selected, which constituted part of a single walking trial.

G5.1.2.9 Preprocessing

The original sensor signals were amplified and filtered to remove noise before calculation of the derived quantities. During the early development process, ALNs with fixed thresholds and adaptive nodes (Atree 2.7) were used (see Section C1.8 of this handbook) which required a reversible encoding from C1.8 quantized real numbers to Boolean vectors. This was done using random-walk or thermometer encoding. In later experiments, ALNs with LTUs were used, eliminating the need for an encoding step.

G5.1.2.10 Output interpretation

The ALNs were trained to produce a value which was thresholded to obtain a logical signal indicating whether the stimulation was to be on or off. Before stimulation, the ALN-derived decision was checked by the restriction rules. For example, one rule prevented restimulation until a certain time had elapsed. The alternative of separate control of initiation and duration by two ALNs was tried, but was not found useful.

G5.1.3 Development platform and tools

The control system was developed on a desktop IBM-compatible 486DX-50 computer having a multifunctional I/O board (National Instruments Inc). A compatible software development platform (LabView for Windows) was used to integrate signal acquisition, preprocessing, ALN-LabView interfacing, ALN training, output interpretation and stimulator control. ALN-related functions were embodied in special DLLs invoked by the LabView program. Microsoft Visual C++ was used to develop the DLLs.

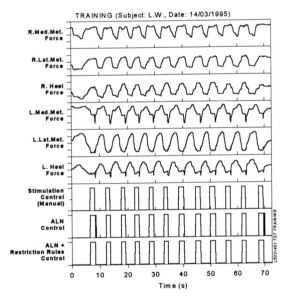

Figure G5.1.2. ALN training and its evaluation: an example of signals from force sensors and manual stimulation recorded during manually controlled FES-assisted walking. ALN + Restriction Rules control is the result of training shown upon replay. Excellent agreement must still be checked for generalization.

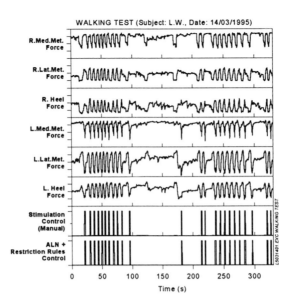

Figure G5.1.3. ALN generalization on a test set: a manually controlled walking sequence is used to test generalization of trained ALNs. An example shows good performance not only during straight-line walking, but also during turning, a process not presented to the ALNs during training.

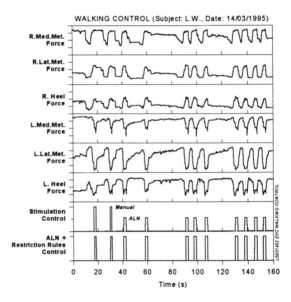

Figure G5.1.4. ALN real-time control of FES-assisted walking: the subject stood up from the chair, took two manually controlled steps with the stimulated leg (represented by two high pulses in the seventh trace) and then walked under automatic ALN control (low pulses in the seventh trace).

G5.1.4 Experimental procedure and results

The results reported below are taken from Kostov (1995). Training data were accumulated during a walking session from the sensors and the switching actions. The subject stood up from the wheelchair supporting herself by a four-point wheeled walker and proceeded to walk using manual control of the stimulation. The walking distance per trial was between 10–12 m with a 180° turn at the half-way point.

The data were then preprocessed according to the procedure described above and analyzed using ALN learning, a process requiring about thirty seconds to finish on a 486DX-50 PC. Figure G5.1.2 shows six signal traces of the force sensors in the shoes, the stimulation control signal produced manually (trace seven), the automatic control signal produced by the ALN decision tree (trace eight) and the signal produced by ALNs plus the restriction rules (trace nine), all evaluated on the same data.

If the output of the ALN decision tree did not contain any functional errors (extra or missing stimuli) when tested on the training data, the tree was tested on new data which were not used during training. Again, if there were no functional errors in predicted output control signals, a similar test was repeated, but this time during real-time, manually controlled walking (figure G5.1.3). The subject still controlled the stimulation manually, but this time she heard a buzzing sound whenever the decision was that stimulation should be on.

After this test was passed without any functional errors, the ALN decision tree was applied in real-time control of stimulation for FES-assisted walking. The subject, after standing up from the wheelchair, took one or more manually controlled steps to check if the whole system was connected and turned on. Then the ALN control was switched on and put in parallel with the manual control, which remained active as a functional override (figure G5.1.4).

G5.1.5 Discussion

The primary target of this work was the design of a coordination level controller for a neuroprosthetic device to control FES for walking in subjects with SCI. To prepare for automatic generation of control rules, manually controlled FES-assisted walking of subjects with incomplete spinal cord injury was studied. Manually controlled stimulation for walking is important in the rehabilitation of SCI subjects as it provides a way for the subject to learn how muscles react to different stimulation conditions. Manual control also remains the backup control system for stimulation during the development of more sophisticated control systems. Various sensors were evaluated for use as sources of sensory feedback information. It was concluded that an affordable array of force sensors built into the subjects' shoe insoles can provide a reliable and reproducible source of feedback information for design of control rules.

Results obtained so far demonstrate the capability of ALNs to control FES-assisted walking successfully. It was also demonstrated that generalization is satisfactory up to several days later (Kostov 1995). Although the 180° turn was excluded from the training set, the subject was able to do the turn under automatic control too. This result implies that an ALN-based control system might be quite robust, and frequent retrainings of the ALNs for calibration may not be necessary. It remains to be seen how fast the walking pattern changes, requiring new ALN training or retraining of the existing ALNs. In case ALNs can generalize over long periods of time, an integrated control system (ICS) can be built consisting of two parts: an FES control fitting station and the FES controller itself. The FES controller can be miniaturized and built into a portable neuroprosthetic device. The control functions can be learned in the laboratory or at home using an FES control fitting station, which can be based on a small notebook computer with data acquisition capability. After the control algorithm is produced, it can be downloaded to the portable FES controller, which can then be used independently.

G5.1.6 Conclusions

ALNs were evaluated for cloning the manual skill of a skilled subject in controlling one channel of stimulation for FES-assisted walking. The ability of ALNs to generate control functions from training based on manually controlled stimulation was demonstrated. After ALN training, the result was tested on the training set, on a new test set, and in real-time walking, whereby stimuli were initiated by the subject and the ALN automatic stimulation was indicated by a buzzer. After these tests were passed without any functional errors, the ALN was used in real-time control of stimulation for FES-assisted walking. ALN control was used in parallel with the manual control, which remained active as a functional override. The subject, after standing up from the wheelchair, took one or more manually controlled steps to check the system; then ALN control was switched on and the subject walked under ALN control. ALN control has been demonstrated to be very robust allowing for the passage of several days between training and test and allowing use in circumstances not presented in training, such as turns.

References

Armstrong W W and Thomas M M 1995 *Atree 3.0 ALN Educational Kit for Windows* from ftp.cs.ualberta.ca in pub/atree/atree3/atree3ek.exe (binary mode, 900 kilobytes).
——1996 Adaptive logic networks *Handbook of Neural Computation* (New York: Oxford University Press) section C1.8

Kirkwood C A and Andrews B J 1989 Finite-state control of FES systems: application of AI inductive learning techniques *Proc. 11th IEEE–EMBS Conf. (Seattle, WA)* (Piscataway, NJ: IEEE Engineering in Medicine and Biology Society) pp 1020–1

Kostov A 1995 Machine learning techniques for the control of FES-assisted locomotion after spinal cord injury *PhD Thesis* Department of Neuroscience, University of Alberta, Edmonton, Alberta, Canada

Kostov A, Andrews B J, Popovic D B, Stein R B and Armstrong W W 1995a Machine learning in control of functional electrical stimulation systems for locomotion *IEEE Trans. Biomed. Eng.* **42** 541–51

Kostov A, Andrews B J and Stein R B 1995b Inductive machine learning in control of FES-assisted gait after spinal cord injury *Proc. 5th Vienna Int. Workshop on Functional Electrical Stimulation (Vienna)* (Sendai: Sendai FES Research Project) pp 59–62

Kostov A, Stein R B, Armstrong W W and Thomas M M 1992 Evaluation of adaptive logic networks for control of walking in paralyzed patients *Proc. 14th IEEE-EMBS Conf. (Paris)* vol 4 (Piscataway, NJ: IEEE Engineering in Medicine and Biology Society) pp 1332–4

Kostov A, Stein R B, Popovic D B and Armstrong W W 1994 Improved methods for control of FES for locomotion, *Proc. IFAC Symp. Modeling and Control in Biomedical Systems (Galveston, TX)* (Galveston, TX: International Federation of Automatic Control) pp 422–7

Liberson W T, Holmquest H J, Scott D and Dow M 1961 Functional electrotherapy, stimulation of the peroneal nerve synchronized with the swing phase of the gait of hemiplegic patients *Arch. Phys. Med. Rehabil.* **42** 101–5

Popovic D B, Stein R B, Jovanovic K L, Dai R, Kostov A and Armstrong W W 1993 Sensory nerve recording for closed-loop control to restore motor functions *IEEE Trans. Biomed. Eng.* **40** 1024–31

Prochazka A 1993 Comparison of natural and artificial control of movement *IEEE Trans. Rehabil. Eng.* **1** 7–17

Stein R B, Kostov A, Belanger M, Armstrong W W and Popovic D B 1992 Methods to control functional electrical stimulation *Proc. First Int. Symp. FES (Sendai)* (Vienna: Department of Biomedical Engineering and Physics, University of Vienna) pp 135–40

Further reading

1. Stein R B, Peckham H P and Popovic D (eds) 1992 *Neural Prostheses: Replacing Motor Function After Disease or Disability* (New York: Oxford University Press)

2. Tomovic R, Popovic D and Stein R B 1995 *Nonanalytical Methods for Motor Control* (Singapore: World Scientific)

G5.2 Neural networks for diagnosis of myocardial disease

Hiroshi Fujita

Abstract

A neural network approach to computer-aided diagnostic systems for coronary artery diseases is described as one of the case studies in cardiac nuclear medicine. Recently, we have been developing a computerized system by using artificial neural networks, called 'BULLsNET', which can aid the physician in the detection and classification of coronary artery diseases in ^{201}Tl myocardial SPECT bull's-eye images. Three-layer feedforward neural networks with a backpropagation algorithm were employed, in which whole or partial images were fed into the input layer. The BULLsNET system, which includes two major neural-network-based elements for the analysis of 'EXTENT' and 'SEVERITY' bull's-eye images, was trained using pairs of training input images and the desired output data ('correct' diagnosis). The system classified the input image data into eight cases, that is, one normal case and seven different types of abnormal cases. The results showed that the recognition performance of the system was comparable to that of a two-year RI-experienced physician. Our study suggests that the neural network approach is useful for developing a computer-aided diagnostic system for coronary artery diseases in myocardial SPECT bull's-eye images.

G5.2.1 Project overview

The nuclear imaging technique is one of the most effective methods of examination for the diagnosis of myocardial disease. However, visual interpretation of nuclear images is subject to substantial variability even by experienced observers. Thallium-201 (^{201}Tl) myocardial SPECT (single-photon emission computed tomography) imaging (Fischer 1990) has been reported to offer major improvements over planar imaging and to be a sensitive and specific examination for the diagnosis of coronary artery disease. However, to overcome the difficulties of interpretation of the myocardial SPECT images, a polar map display, called a bull's-eye image, has been developed to characterize the three-dimensional images of the left ventricle in two dimensions (Garcia *et al* 1985). Even with this technique, many problems have been indicated. Also, the number of experienced physicians or radiologists in this field is substantially limited. The development of a computer-aided diagnostic system or expert system, therefore, is considered to be helpful for the diagnosis of bull's-eye images.

We have been developing a computerized system, which can aid the physician's diagnosis in the detection and classification of coronary artery diseases in ^{201}Tl SPECT bull's-eye images, by employing several artificial neural networks for different tasks. One of the advantages of the neural network approach is its powerful ability to analyze the physician's complicated decision-making or pattern-recognizing process in diagnosis without any need to write a special computer program. As a pilot study, we investigated the applicability of the neural network technique in developing the computerized system for the diagnosis of coronary artery diseases only when the bull's-eye 'EXTENT' images were used for the analysis (Fujita *et al* 1992a), and also studied the effects of image processing and neuro parameters on the system performance (Shinoda *et al* 1993). We also developed an improved system, in which

'EXTENT' and 'SEVERITY' images were used for the analysis with composite neural networks, and reported the results of the system performance comparing it with physicians' recognition rates (Fujita *et al* 1992b, 1993a, 1994, Katafuchi *et al* 1993). The overall flow of our system, called a 'BULLsNET', is shown in figure G5.2.1. Here we present our recent work from these studies, all of which were done as cooperative works with coworkers at the Department of Radiology, National Cardiovascular Center and at the Biomedical Research Center, Osaka University Medical School (Suita, Osaka, Japan).

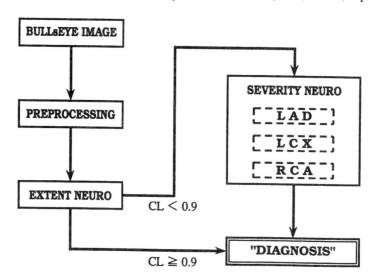

Figure G5.2.1. Processing elements of the BULLsNET system in which two major neural-network-based elements for EXTENT and SEVERITY images are included (Fujita *et al* 1993a, 1994).

G5.2.2 Database

Thirty-six planar images of a 64×64 matrix with 64 gray levels were obtained with a gamma camera (Shimadzu LFOV dual head) and these data were transferred to a data processing system (Shimadzu SCINTIPAC-2400) at the Department of Radiology, National Cardiovascular Center. This system produces three different types of bull's-eye images, that is, 'PIXEL CT', 'EXTENT' and 'SEVERITY' images, which, respectively, represent the original bull's-eye image, the image simply showing the extent of the diseased area relative to the averaged normal case (in two colors), and the image showing the severity of the disease within the extent area (in several colors). In our study, we used both EXTENT and SEVERITY images. Actually, when physicians interpret the bull's-eye images, they first look at the EXTENT image, and then at the SEVERITY image carefully.

Coronary artery territories in the bull's-eye display are illustrated in figure G5.2.2, where the regions of three main coronary arteries, left anterior descending coronary artery (LAD), left circumflex coronary artery (LCX), and right coronary artery (RCA), are segmented (Garcia *et al* 1985). It should be noted that this figure shows approximate territories and many variations, overlaps and exceptions in each territory can exist, preventing the design of a simple artificial intelligence rule-based expert system. The coronary artery diseases can therefore be classified into seven different types due to the existence of single-, double- and triple-vessel diseases. A total of 74 bull's-eye images were collected. Because we selected the cases that had also been examined by coronary angiography (CA), in which a coronary artery of more than 75% stenosis was diagnosed as 'diseased' according to the criteria of the American Heart Association (AHA), these CA results were employed as a gold standard or 'correct diagnosis' in this study.

G5.2.3 Neural network software employed

At an initial stage, we employed a personal 'neuro-computer' system (Neuro-07, NEC), which consists of a personal computer (PC-9801 VX21, NEC) a neuro-engine board (PC-98XL-02, NEC) and a neuro-software package ('Michi-Zane', NEC). The neural network software was written in the C language and B2.3 was based upon a *feedforward layered model* with an input layer, one to three middle or hidden layer(s), and an output layer. Lately, a SUN-type workstation has been employed.

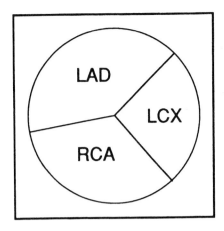

Figure G5.2.2. Coronary artery territories in the bull's-eye image (Fujita *et al* 1993a, 1994).

G5.2.4 Preprocessing

A *preprocessing* of the image data was required due to the limited memory capacity of the neuro-engine B4.4
board. This was also important to save computation time. The effects of the matrix size of the EXTENT
images on the system performance were investigated (Shinoda *et al* 1993); a 16 × 16 matrix image was
judged to be enough by considering the recognition rate, training time and data volume. Therefore, all of
the bull's-eye images studied were compressed to produce the images of 16 × 16 matrices by averaging the
neighboring pixel values and also to produce binary gray-level images for the EXTENT and six gray-level
images for the SEVERITY. As an example, preprocessed images are shown in figure G5.2.3, which is a
case of LAD + LCX double-vessel disease.

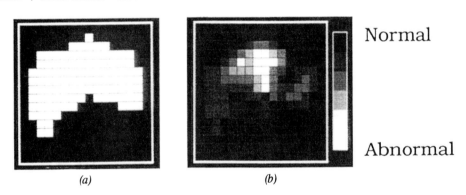

Figure G5.2.3. Preprocessed bull's-eye images in the case of LAD + LCX double-vessel disease. (*a*)
EXTENT image of 16 × 16 matrix size with binary gray levels. (*b*) SEVERITY image of 16 × 16 matrix
size with six gray levels (Fujita *et al* 1993a, 1994).

G5.2.5 Network structure and training method

As shown in figure G5.2.1, the BULLsNET system includes two major image-analysis parts, 'EXTENT
neuro' and 'SEVERITY neuro', and the latter consists of three neural networks for separately analyzing
three artery regions in the bull's-eye image. The architecture of each neuro is shown in figure G5.2.4. The
number of input units in the EXTENT neuro was 256 because the whole compressed image was fed into
the input layer. On the other hand, the ones for the networks in the SEVERITY neuro were 61, 41 and 57
for LAD, LCX and RCA regions, respectively. The number of neurons in the output layer in the EXTENT
neuro was fixed at eight units, corresponding to the eight different types of diagnoses including normal.
The SEVERITY neuro had two units corresponding to normal or abnormal. Three output results from
each network in the SEVERITY neuro were combined to determine the final diagnosis (eight outputs).
The neural network was trained using pairs of training input images (compressed images) and the desired

output data (the 'correct diagnosis' based on the gold standard). The numbers of training iterations for the EXTENT and SEVERITY neuros were 200 and 150 for each region with 100 and 50 units in the hidden layer, respectively. We varied combinations of image data for the training (58 cases) and testing (16 cases) processes in the neural network and made three different combinations, cases A, B and C, in which all images were chosen at random from a database of 74 images.

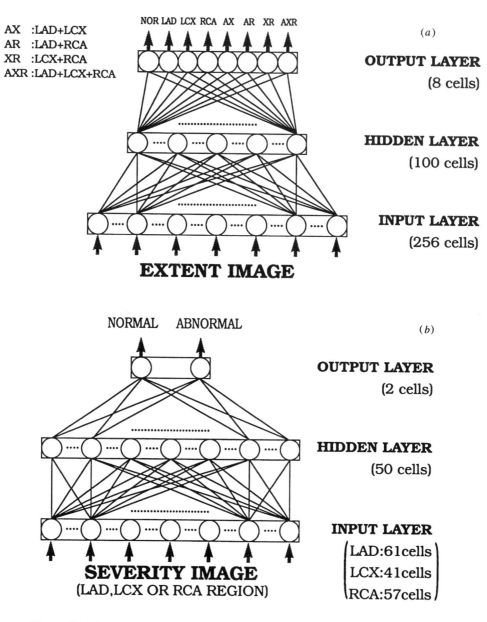

Figure G5.2.4. Architecture of (*a*) the EXTENT neuro and (*b*) the SEVERITY neuro.

G5.2.6 Output interpretation

In the case where the confidence level (CL) of the 'extent neuro' was lower than 0.9, the 'severity neuro' was performed (figure G5.2.1), in which each part of the vessel regions based upon the territories in figure G5.2.2 was examined by LAD, LCX and RCA neural networks, then the output result from the severity neuro was used as a diagnosis. The CL was determined from the weight values in the output layer of the network. On the other hand, in the case where the confidence level was equal to or larger than 0.9, the output result from the extent neuro was simply used as a diagnosis. The percentage of the cases where the confidence level of the extent neuro was smaller than 0.9, that is, the severity neuro was necessary for analysis, was approximately 35%.

G5.2.7 Performance

In the case where only the extent neuro was performed, the following results were obtained. The recognition rates (percentage of correct recognition) determined by the neural network for image data never used in the training process are listed in table G5.2.1, together with those by one resident (I) and two physicians (II and III) for comparison. This table demonstrates that the recognition rate depends on the image combinations, which can be explained from two different viewpoints. One is that image data used for the training may be insufficient to recognize image data for testing. The other is that the inclination of image data in terms of their category in the training process and the degree of difficulty in diagnosis in the recognition process may cause variances in the recognition rate. These effects may be decreased by increasing the image data for training as well as for recognition. By comparing the averaged results, the performance of the neural network is better than that of the resident, and comparable to that of the two-year experienced physician, but worse than that of the ten-year experienced physician. The pure computation time for training was approximately 23 minutes in the case of a personal-computer-based procedure; however, the time for training is not so important, because the user at the hospital may simply utilize the results obtained from the training process. On the other hand, the recognition of one image data in the testing process, including the preprocessing procedure, was performed in 'real time'.

Table G5.2.1. Recognition rates for three different combinations of image data and their average for three observers and the BULLsNET system, only when the extent neuro was employed (Fujita *et al* 1993a, 1994). NN: neural network, I: three-month RI-experienced resident, II: two-year RI-experienced physician, III: ten-year RI-experienced physician

	Case A	Case B	Case C	Average
NN	69%	75%	88%	77%
I	56%	81%	69%	69%
II	75%	75%	88%	79%
III	75%	88%	88%	83%

It is worthwhile including the SEVERITY image for analysis, because it can help to differentiate lesions from artifacts. Actually, in the case of the physicians, we observed that the recognition rate with both EXTENT and SEVERITY images results in a 6–10% higher rate relative to that with only EXTENT images. The recognition rate determined by the neural networks using both images when the confidence level from the extent neuro is lower than 0.9 was 85%. It is considered to be comparable to that of the two-year experienced physician.

G5.2.8 Summary

The approach of using artificial neural networks for a computer-aided diagnostic system of coronary artery disease in stress SPECT examinations appears to show considerable promise. The recognition performance of our present system (BULLsNET) is comparable to that of the two-year RI-experienced physician. However, in order to improve our system, it is required to increase the number of image data for training and testing processes. Moreover, we are now extending our system to redistribution (rest) bull's-eye images so as to interpret ischemia and infarction (Fujita *et al* 1993b). Finally, other clinical information, such as sex, temperature and electrocardiogram data, have to be included in the overall analysis.

Acknowledgements

The author would like to thank all his coworkers, Mr T Katafuchi, Professor T Nishimura, Dr T Uehara, Dr Y Ishida, Mr H Iida, Mr M Horio, Mr M Shinoda, Mr T Hara and Mr Y Torisu.

References

Fischer K C 1990 Qualitative SPECT thallium imaging: technical considerations and clinical applications *Nuclear Cardiovascular Imaging: Current Clinical Practice* ed M J Guiberteau (New York: Churchill Livingstone) pp 133–66

Fujita H, Katafuchi T, Shinoda M, Uehara T, Hara T and Nishimura T 1993a Neural network approach for the computer-aided diagnosis of coronary artery diseases in myocardial SPECT bull's-eye images *Proc. Int. Symp. CAR'93 Computer Assisted Radiology* ed H U Lemke, K Inamura, C C Jaffe and R Felix (Berlin: Springer) pp 606–11

—— 1994 Neural network approach for the computer-aided diagnosis of coronary artery diseases in myocardial SPECT bull's-eye images *Radiol. Diagnost.* **35** 15–8

Fujita H, Katafuchi T, Shinoda M, Uehara T, Ishida Y and Nishimura T 1993b Computer-aided diagnostic system for interpretation of myocardial SPECT bull's-eye images *Radiology* **189**(P) 237 (abstract)

Fujita H, Katafuchi T, Uehara T and Nishimura T 1992a Application of artificial neural network to computer-aided diagnosis of coronary artery disease in myocardial SPECT bull's-eye images *J. Nucl. Med.* **33** 272–6

—— 1992b Neural network approach for the computer-aided diagnosis of coronary artery diseases in nuclear medicine *Proc. Int. Joint Conf. Neural Networks '92* (Baltimore, OH) vol III, pp 215–20

Garcia E V, Train K V, Maddahi J, Prigent F, Friedman J, Areeda J, Waxman A and Berman D S 1985 Quantification of rotational thallium-201 myocardial tomography *J. Nucl. Med.* **26** 17–26

Katafuchi T, Fujita H, Uehara T and Nishimura T 1993 Development of a computer-aided diagnostic system for cardiac nuclear medicine using multi-neural networks *Trans. Inst. Electron., Info. Commun. Eng.* **J76-D-II** 2436-9 (in Japanese, with figure captions in English)

Shinoda M, Fujita H, Katafuchi T, Uehara T and Nishimura T 1993 Development of a computer-aided diagnostic system for myocardial SPECT images: effects of image processing and neuro parameters *Med. Imag. Info. Sci.* **10** 38–45 (in Japanese, with abstract and figure captions in English)

G5.3 Neural networks for intracardiac electrogram recognition

Marwan A Jabri

Abstract

Implantable cardioverter defibrillators are life-saving devices for people with heart disease. They sense the electrical activity of the heart through leads attached to its tissue. The sensed signals are called intracardiac electrograms and their interpretation is in many instances still a challenging pattern recognition task. This is especially the case because the defibrillators are battery powered, and most conventional recognition techniques are computationally intensive. We present here neural network techniques for electrogram recognition and describe their application to the detection of two rhythms that cannot be recognized by present day defibrillators. The implementation of such networks in micropower very large-scale integration is also described. A method for resolving the problem of morphology changes due to tissue growth is addressed by a method in which the neural network continuously learns using patterns that are automatically labeled.

G5.3.1 Introduction

Cardiac arrest is responsible for the death of about half a million people in the USA alone every year. The automated detection of abnormal heart conditions has considerably improved over the last two decades thanks to advances in many aspects of pattern recognition and integrated circuit technologies.

Heart diseases are reflected in cardiac electrical activities. This is illustrated in figure G5.3.1 where electrical signals are shown and are related to the region of the heart where they could be observed. The electrical activity represents the contraction and relaxation of the heart muscle and can be observed in a near-field scheme where electrodes are attached to the actual heart tissue (intracardiac electrograms or ICEG), or in a far-field scheme where electrodes are attached to the surface of the body (external electrocardiograms or ECG).

ECG recognition is performed by Holter monitors, ambulatory systems and coronary care units. ICEG recognition is performed by implantable pacemakers and cardioverter defibrillators. Because of the difference in the sensing distance, far-field (ECG) and near-field (ICEG) observation of the heart activity yield different signal morphologies. Hence, signal processing and recognition techniques developed for ECG may not necessarily be applicable to ICEG and vice versa.

In figure G5.3.2 we show examples of the ICEG for the normal sinus rhythm (NSR) and four common arrhythmia, supraventricular tachycardia (SVT), ventricular tachycardia (VT), ventricular fibrillation (VF) and sinus bradycardia (SRB).

In general, there are over 17 arrhythmia of interest to cardiologists, and they are grouped under four classes defined by the type of therapy they require:

- NSR: this is the normal operation of the heart and no therapy is required
- SVT: present single-channel intracardiac cardioverter defibrillators (ICDs) cannot detect this and so deliver no therapy; however, experiments have shown that pacing of the atrium can terminate SVT
- VT: generally VT is treated with pacing and if not successful then eventually shocking
- VF: VF is usually treated with shock therapy.

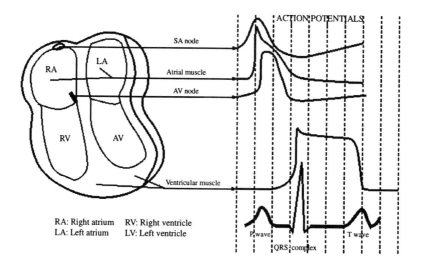

Figure G5.3.1. Diagram of the heart and corresponding electrical activities.

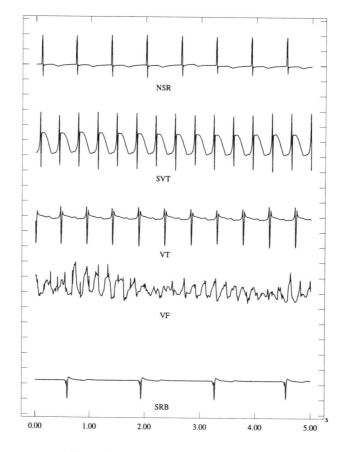

Figure G5.3.2. Examples of ICEG signals.

Note we have listed above the tachycardia-based groups only. The bradycardia-based group which corresponds to arrhythmia with heart rates slower than 60 bpm (beats per minute) is not considered in this section. An 'implantable pacemaker' commonly refers to a device implanted in patients with bradycardia conditions whereas intracardiac cardioverter defibrillators (ICDs) are devices implanted in patients with tachycardia conditions.

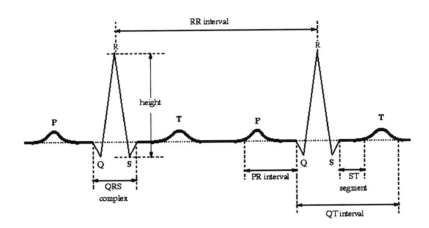

Figure G5.3.3. QRS complex.

Each heart beat in the ECG or ICEG trace is labeled as a QRS complex as shown in figure G5.3.3. The R point corresponds to the peak of the beat. RR is a measure of the interval between two beats and is used to compute the heart beat rate. Most automated ECG and ICEG interpretation systems rely on the beat rate to detect arrhythmia. Some arrhythmia, however, cannot be reliably detected using the heart beat rate alone and other features, such as the signal morphology (e.g. shape of the sensed signal), need to be used for reliable diagnosis.

Morphology analysis is mainly used in ECG recognition, in particular in ambulatory monitoring systems and CCUs. ICD devices rarely use morphology analysis because of its high computational requirements. ICDs are battery operated and because battery replacement is costly, morphology recognition tends to be avoided.

The present section is mainly concerned with morphology recognition techniques for ICDs. We discuss, in particular, the application of *multilayer perceptrons* to the recognition of dangerous arrhythmia C1.2 by the means of morphology analysis. Although we consider only the case of detecting a type of VT, the technology described can be applied to other arrhythmia detection problems which necessitate morphology recognition.

G5.3.2 Neural computing for intracardiac electrogram classification

ICDs monitor the heart's electrical activity through leads attached to its internal surface. There are two types of ICD: single chamber and dual chamber. In a single-chamber ICD, a single lead is attached to the heart's right ventricular apex (RVA). In a dual-chamber ICD, an additional lead is attached to the heart's high-right atrium (HRA). Single-chamber ICDs are aimed at recognizing the NSR, VT and VF arrhythmia. They do that mainly by detecting the QRS complex, computing the RR interval and making use of pattern classifiers to recognize the heart condition. Arrhythmias like SVT are impossible to detect using a single-chamber ICD because atrial and ventricular information is required for reliable detection.

Figure G5.3.4 shows a schematic diagram illustrating the inputs and outputs of an arrhythmia classifier in single- and dual-chamber ICD schemes. Here we describe the signal flow in figure G5.3.4 for the case of a single-chamber ICD. The flow for a dual-chamber ICD is similar. QRS detection is performed on the RVA signal providing events for the RR interval and timing feature extractor. Timing features and RVA samples are passed into the classifier for detection of arrhythmia. The classifier outputs the arrhythmia class to an X out of Y filter which is used to filter out spurious classifications which may be due to noise, QRS detection failures, or misclassifications. The X out of Y produces a decision that is based on a 'majority' (at least X) vote over a number (Y) of classifications. The therapy logic block assigns the therapy that corresponds to the recognized arrhythmia class.

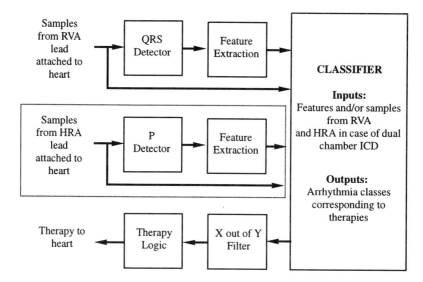

Figure G5.3.4. Inputs and outputs of an ICEG classifier in single- and dual-chamber schemes. The processing enclosed in the dotted box is only used in a dual-chamber ICD.

As stated earlier, the fundamental features used by present ICDs are timing-based, that is the heart rate as computed using the RR interval. Some ICDs do perform some limited forms of morphology analysis. Present ICDs (single and dual chamber) cannot be used to classify several types of arrhythmia. For instance, patients with ventricular tachycardia with one-to-one retrograde conduction (VT 1:1) may develop arrhythmia with heart rates close to their fast NSR rates (or sinus tachycardia, ST) when they are exercising vigorously. In these cases, it is impossible to properly diagnose their conditions on the basis of the heart rate alone and a more elaborate morphology analysis is required (Leong and Jabri 1992). These patients cannot presently take advantage of an ICD solution to their disease and have to rely on other forms of medication.

The research described in this article shows that neural computing can provide an effective morphology analysis which can be implemented in ultra-low-power microelectronics to provide a solution to problems such as the ST/VT 1:1 recognition. Before we describe the neural-computing-based morphology analysis, we present the database used in the research as well as the preprocessing applied to it.

G5.3.3 Training and evaluation data

The data used in the studies were collected from electrophysiological studies (EPS) performed in Australian and British hospitals. EPS is performed by introducing temporary probes into the internal surface of the patient's heart, and artificially inducing arrhythmia through these probes. Once induced, the arrhythmia can then be monitored through the same probes. Our database includes over 150 EPS sessions from different patients. For each patient, data from at least the RVA and HRA leads is available. All data have been classified and labeled by cardiologists. Data are stored as digitized wave forms at a 250 Hz sampling rate. Although cardiologists in some circumstances, and on the basis of the RVA signal alone, may label the data differently, the availability of the signals from the other lead and the history of the patient provide sufficient information for highly reliable labeling.

G5.3.4 Data preprocessing

Most ICDs perform some form of bandpass filtering, with lower cutoff frequencies of a few hertz and a higher cutoff frequency of about 45 Hz. The low-pass filtering is aimed at eliminating rapid baseline 'wandering' of the sensed signal and the high-pass filtering is aimed at eliminating noise and any external interferences. Our classification system makes use of the RVA signal which has already been filtered before storage into our database.

As indicated earlier, the section of the electrical signal associated with each heart beat is termed the QRS complex (see figure G5.3.3). In the last several decades, there have been many implementations of QRS detectors (see Friesen *et al* 1990 for a recent review). The QRS detection algorithm used in our experiments is based on an exponentially decaying threshold and is proprietary to our commercial collaborator.

G5.3.5 VT 1:1/ST morphology classification

The ST/VT 1:1 morphology recognition neural network is implemented using a simple multilayer perceptron (MLP) as shown in figure G5.3.5. The input to the MLP is a window of RVA samples centered around the R peak as detected by the QRS detection algorithm. As our data were sampled at 250 Hz (4 ms) and QRS complexes are typically about 30 to 40 ms long, a window size equivalent to 80 ms was chosen by skipping every second RVA sample. The MLP has a single output which indicates whether the input morphology belongs to the VT 1:1 or ST class.

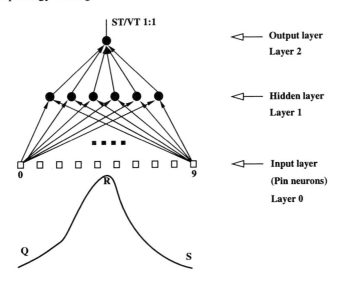

Figure G5.3.5. The morphology recognition MLP. It has ten inputs, six hidden units and one output. The ten inputs are the QRS samples and its output indicates whether the morphology is that of an ST or VT 1:1.

The MLP was implemented using micropower complementary metal oxide semiconductor (CMOS) technology. The actual chip, called *Snake*, is described in Coggins *et al* (1995). We briefly review the multilayer perceptron architecture here.

The synapses are implemented as multiplying analog-to-digital converters with the weights represented as 6-bit signed numbers. An unusual aspect of the network is that its synapses operate as nonlinear multipliers. The outputs of the synapses are differential currents which are summed at the input of the neurons. Neurons are implemented as current-to-voltage converters operating mainly in their linear regions. Hence, the nonlinearities of the network are implemented in the synapses and not the neurons as is usually the case with multilayer perceptrons, but without any degradation in the nonlinear classification capabilities of the MLP (Coggins *et al* 1995).

The MLP chip was interfaced to a commercial ICD. The defibrillator provided its filtered version of the RVA signal as well as the QRS event detection. The RVA samples provided by the defibrillator are provided to the MLP chip which has a built-in analog shift register. These samples are stored on the chip and are shifted every time there is a new sample to be stored. The analog shift register is 10 samples long and provides the MLP with its 10 inputs.

The MLP chip is trained in an in-loop fashion. The response of the chip (its outputs in response to an input pattern) are provided to a personal computer which orchestrates the training. The training algorithm used in the experiments is called summed-weight neuron perturbation (Flower and Jabri 1993) which is a semiparallel version of the weight perturbation algorithm described in Jabri and Flower (1992).

The training of the MLP chip has proven to be a challenging task because:

(i)   the QRS detection is not perfect, and

(ii)   the inputs to the MLP are the outputs of the analog shifted samples.

Nevertheless, the two issues above did not affect the training and generalization of the MLP chip.

The experimental setup (MLP chip and ICD) was used on the data of all relevant patients in our database (seven patients had VT 1:1). We show the training and generalization performance of the chip in tables G5.3.1 and G5.3.2, respectively.

Table G5.3.1. Training performance of the *Snake* chip on seven patients with ICD in-loop.

Patient	Training iterations	% correct ST	% correct VT
p45	56	100	100
p55	200+	100	87.5
p651	200+	87.5	100
p76	46	100	100
p81a2	200+	100	100
p81	140	100	100
p862	14	100	100

The power consumption of the *Snake* chip, assuming 120 bpm heart rate and 3 V supply, was around 186 nW. The ultra-low power consumption and good performance make possible the inclusion of a *Snake*-like device in ICDs enabling their use for VT 1:1 patients.

Table G5.3.2. Classification performance of the *Snake* network on seven patients with ICD in-loop.

Patient	No of complexes ST	No of complexes VT	% correct ST	% correct VT
p45	440	61	100	98.3
p55	94	57	100	95
p651	67	146	77.6	99.3
p76	166	65	91	99.3
p81a2	61	96	97	93
p81	61	99	97	100
p862	28	80	96	99

G5.3.6   Tissue growth, patient dependence and integrated learning

The morphology recognition scheme described above may suffer from morphology changes due to the growth of tissue on the ICD lead tips. The growth has the effect of changing the sensing characteristics which lead to variations in the sensed signal morphology. This means that a neural network targeted to classify morphology has to be either insensitive to tissue-growth-based variations, would require the patient's morphology classifier to be regularly adjusted, or has to be capable of adapting to them.

Making a neural network insensitive to morphology changes due to tissue growth is a difficult if not impossible task. Regular tuning of the patient's morphology classifier is possible but is not as economical as making the classifier adapt to morphology changes.

The method we will present below, for the training and adaptation of the morphology classifier, not only allows a network to adapt to morphology changes, but also simplifies the initial training of a morphology classifier to fit the requirements of a patient. Because the training of a snake-like chip takes a matter of tens of minutes, the easiest (but not necessarily the most economical) approach would be to train the network in an EPS session. However, a scheme where the network could learn and adapt with no morphology labeling supervision would be desirable. Two obstacles need to be overcome to achieve this:

(i)   integrated on-chip learning has to be implemented and has to be economical from a power consumption and hardware overheads point of view, and

(ii) supervisory signals have to be derived, somehow, to replace the labeling of the morphology which has been done so far by a cardiologist in EPS sessions.

On-chip learning has been demonstrated recently in our laboratory (Flower *et al* 1995) and is no longer a serious obstacle or challenge. As for the supervisory signals, it can be resolved as described in the next section.

G5.3.7 A scheme for the automatic labeling of morphology for supervised training

Although the morphologies of ICEG and ECG signals are different, the heart beat rate is the same (assuming reliable QRS detection) whether sensed internally or externally. The beat rate can be used to automatically label morphologies that are 'definitely' NSR/ST or VT 1:1. This is better illustrated by figure G5.3.6. The distributions of NSR and VT 1:1, as functions of the RR interval, are shown here in an abstract fashion to illustrate the existence of what we call the TN region. The TN region is the 'gray' region where the heart beat rate alone cannot reliably determine an arrhythmia. If we define the 'high confidence decision' regions of these distributions as being those where the heart beat rate can definitely determine an NSR or VT 1:1, then we can use the rate to indicate whether the corresponding RVA samples (morphology) are those of NSR or VT 1:1. That is, we can label the RVA QRS samples as being NSR or VT 1:1 by measuring the heart beat and if it is within the 'high confidence decision' regions, the signal morphology can be labeled and used for supervised learning. Note that our scheme is different from the approach where the ICD would apply a VT 1:1 therapy whenever the heart beat rate is outside a safe NSR region. Such an approach leads to excessive use of valuable battery energy, is uncomfortable to the patient and can induce an arrhythmia.

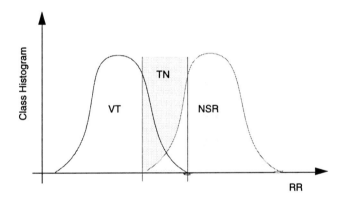

Figure G5.3.6. Distributions of NSR/ST and VT 1:1 with respect to the RR interval. Note that the TN region is where the heart beat rate cannot confidently determine the condition of the heart. Outside of the TN region we can confidently classify the condition to be NSR or VT 1:1.

We have simulated our proposed scheme using the data of the seven VT 1:1 patients in our database. The simulation system consisted of two modules, a timing-based classifier and an MLP similar to that implemented by the *Snake* chip. The timing-based classifier provides an enabling signal for the training of the MLP every time that a high confidence region of the heart beat rate is met. Of course, the MLP need not be trained at every enabling signal. The rate at which QRS samples are considered for training could be programmable.

The results of our simulated system show that the MLP can be trained in an on-line fashion, every time there is a high-confidence timing-based decision. The on-line aspect of the training is essential and is performed once through the data of a particular patient (the data were split into training and testing sets). A summary of the performance of the simulated system on the test sets which includes data from the TN and non-TN regions is shown in table G5.3.3. Note that the number of test patterns is different from those used for the testing of the *Snake* chip as the number of training patterns in the present experiment is larger. This also explains why some patients used in the present simulations are different from those shown in tables G5.3.1 and G5.3.2. The network tends to make more 'false positives' than it does 'false negatives' which is desirable for a life-saving device. Also note that the simulated system described here is implemented in floating-point arithmetic. When mapped to an architecture such as that of *Snake*, some marginal degradation is expected.

Table G5.3.3. Summary of classification performance for the automatically labeled and on-line-trained morphology classifier.

Patient	No. of Complexes		% Correct	
	ST	VT	ST	VT
p25	7	99	100	100
p45	428	49	79.2	100
p55	80	45	100	100
p650	8	80	62.5	82.5
p76	87	53	100	100
p81	30	80	100	100
p862	25	68	100	97.1

G5.3.8 Conclusions

In this article we have described neural computing techniques for ICEG morphology classification. The research shows that multilayer perceptrons implemented in ultra-low-power microelectronics provide solutions to ICEG pattern recognition problems that have not been solved using conventional techniques because of power constraints. We have also described a method which can take advantage of integrated on-chip learning to provide adaptation of the neural network to patient morphology. This adaptation can be used in the initial implantation stage to train the neural network on the patient morphology, and at later stages to adapt to patients' morphology variations due to tissue growth on the ICD's lead tips.

The neural computing techniques described in this section can be applied to other ICEG pattern recognition problems. In particular, low-power pattern analysis of the P-wave can be of assistance in better detection of other arrythmia and will be the subject of future investigations.

Acknowledgements

A part of the work presented in this article was funded by Telectronics Pacing Systems Ltd and the Australian Federal Government under a GIRD project led by the author. Other team members who have contributed to the project are Z Chi, R Coggins, B Flower, P Leong, S Pickard and E Tinker. A Chan has assisted the author with some of the experiments.

References

Coggins R, Jabri M, Flower B and Pickard S 1995 A hybrid analog and digital VLSI neural network for intracardiac morphology classification *IEEE J. Solid State Circuits* **30** 542–50

Flower B and Jabri M 1993 Summed weight neuron perturbation: an O(*N*) improvement over weight perturbation NIPS5 **5** pp 212–9 (San Mateo, CA: Morgan Kauffmann)

Flower B, Jabri M and Pickard S 1995 An analogue on-chip supervised learning implementation of an artificial neural network *IEEE Trans. Neural Networks* re-submitted

Friesen G, Jannett T, Jadallah M, Yates S, Quint S and Nagle H 1990 A comparison of the noise sensitivity of nine QRS detection algorithms *IEEE Trans. Biomed. Eng.* **BE-37** 85–98

Jabri M and Flower B **1992** Weight perturbation: an optimal architecture and learning technique for analog VLSI feedforward and recurrent multilayer networks *IEEE Trans. Neural Networks* **NN-3** 154–7

Leong P and Jabri M 1992 MATIC—An intracardiac tachycardia classification system *Pacing Clin. Electrophys.* **15** 1317–31

G5.4 A neural network to predict lifespan and new metastases in patients with renal cell cancer

Craig Niederberger, Susan E Pursell and Richard M Golden

Abstract

The natural history of patients with renal cell cancer is bizarre: many patients succumb soon after diagnosis, while others live for decades. The lack of an accurate model to predict lifespan and the occurrence of new metastases has hampered the proper selection of therapy. In this project, a neural network programming environment (neUROn) was designed so that compiled neural networks could be tailored to specific medical/urological applications. Using neUROn, neural networks were built for data sets containing lifespan and disease progression outcomes for renal cell cancer patients. After these networks were trained, the Wilks' generalized likelihood ratio test was used to determine which input variables were significant to the network's prediction. An inspection of the results of this statistical test yielded information relevant to the current clinical treatment of renal cell cancer.

G5.4.1 Project overview

For centuries, physicians and medical researchers have attempted to make sense of cancer outcomes by assigning a set of carefully chosen heuristic rules to patient features, a system known as 'staging'. For example, in the current 'TNM' system of staging kidney cancer, a tumor smaller than 2.5 cm in diameter limited to the kidney is termed 'stage T1' (de Kernion 1986, Williams 1987). A cancer larger than 2.5 cm limited to the kidney is labeled 'stage T2'. A tumor invading the adrenal, renal vein, vena cava, or tissue outside the kidney without spreading beyond the fatty capsule surrounding the kidney known as Gerota's fascia is termed 'stage T3'. Tumor extending beyond Gerota's fascia is 'stage T4'. Frequently these rules are posited at international conferences where epidemiologists and cancer specialists present expert opinions. Unfortunately, many cancers do not behave according to a logical progression of stages. Many kidney and prostate cancers 'jump' stages to significantly more aggressive tumors, while others remain quiescent in one stage for years (de Kernion 1986, Williams 1987). If a computational system could be built that accurately modeled cancer outcomes from raw clinical features, such a system would be of invaluable assistance to physicians counseling patients and, by altering features and predicting future outcomes, planning therapeutic strategies. We thus chose to investigate neural computation as an outcome modeling system for renal cancer.

Data were collected from patients entering treatment for renal cancer at a large public hospital in Chicago, and entered into a database. On completion of data entry, it was known whether 341 patients were alive or deceased, and whether or not a patient developed a new metastasis, or new tumor at a site remote from the kidney, in 232. Features tracked in the database were patient ethnicity, gender, date of birth, date of diagnosis, whether or not a nephrectomy was performed, date of surgery, presence of lung or bone metastases at diagnosis (separate features), histologic cell type of tumor, tumor size, chosen therapy, and date of follow-up. In addition, T, N and M stage were also entered into the database, thus allowing both derived and raw data to be tracked simultaneously. Outcomes recorded were whether the patient was alive or deceased at follow-up, and if new metastases were noted.

G5.4.2 Design

We built a neural programming environment to generate neural networks to model urological data analysis problems. We refer to the environment as *neUROn* for *N*eural computational *E*nvironment for *URO*logical *N*umericals. We designed neUROn, shown schematically in figure G5.4.1, to be a general purpose neural programming environment in C rather than a single compiled program. In the environment, network architecture features are coded in preprocessor directives specified by a single header file, *neural_net.h*. In this way, users of neUROn define preprocessor variables in *neural_net.h* and generate machine code tailored to a specific medical data set, thus reducing computing cost.

NeUROn's programs include:

- *big_lotto.c*, which randomizes the initial data set into training and test sets, and maintains the proportion of outcome data types between sets to ensure representative test sets
- *randomize.c*, which randomizes initial connection weights and biases at each network node
- *prepare.c*, which normalizes the input and output values of testing and training data sets
- *train.c*, the training engine, using files generated by *randomize.c* and *prepare.c* to produce files containing network trained weights for each node and
- *predict.c* and
- *test.c*, which use the trained weights for each node to predict outcomes from either an individual input sample or a file containing multiple samples, respectively. *Test.c* also calculates classification accuracy of the network in training and test sets.

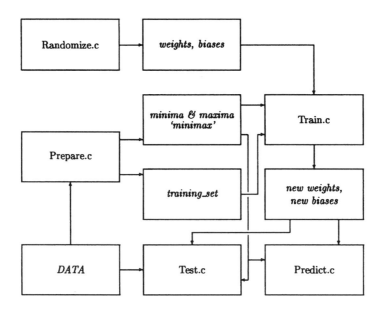

Figure G5.4.1. NeUROn: Neural computational Environment for UROlogical Numericals.

Data were encoded in the input layer as shown in table G5.4.1. Values were either encoded with Q or $Q + 1$ nodes, where Q = number of representational raw data values, and the $(Q + 1)$th node signifies whether or not its companion values are present in the database. Categorical variables were encoded with Q = number of categories. For example, ethnicity was encoded in the input layer as African–American = [0001], Caucasian = [0010], Hispanic = [0100], and other = [1000].

Two neural networks were built: one which classified whether a patient was alive at follow-up, and one which classified if new metastases developed. These targets were encoded as binary. In the network which modeled mortality, 0 represented a patient who was deceased, and 1 represented a patient who was alive at follow-up. The network which modeled new metastases was encoded with 0 if no new metastases were noted, and 1 if new metastases developed.

The two networks implemented in neUROn for the renal cancer data sets are characterized as follows. The topology was *fully interlayer connected* with 1 input, 1 hidden, and 1 output layer. Bias nodes were included on both input and hidden layers. The activation function was *sigmoidal* and the learning rule was *backpropagation*, with the exception that at the output node the error function was selected to be the

B2.5
B3.2.4
C1.2.3

Table G5.4.1. Input-layer preprocessing for the renal cancer data set.

Number input nodes	Variable	Value(s)
4	Ethnicity	Categorical
1	Gender	Binary
1	Diagnosis date available (Yes, No)	Binary
1	Age = Date of diagnosis − Date of birth	Numerical
1	T stage available (Yes, No)	Binary
1	T stage	Numerical
1	N stage available (Yes, No)	Binary
1	N stage	Numerical
1	M stage available (Yes, No)	Binary
1	M stage	Numerical
1	Nephrectomy (yes, no)	Binary
1	Nephrectomy date available (yes, no)	Binary
1	Date of surgery − Date of birth	Numerical
1	Lung metastases information available (yes, no)	Binary
1	Lung metastases (yes, no)	Binary
1	Bone metastases information available (yes, no)	Binary
1	Bone metastases (yes, no)	Binary
10	Histologic subtype	Categorical
1	Tumor size (cm)	Numerical
7	Treatment choice	Categorical

cross-entropy error function since the targets were binary-valued:

$$\eta(\mathbf{W}) = -(1/M) \sum_{i=1}^{M} \log[t^i o^i + (1 - t^i)o^i] \tag{G5.4.1}$$

where M is the number of training stimuli, t^i is the desired activation for stimulus i, and o^i is the neural network's output activation level given that stimulus i has been presented (Baum and Wilczek 1988). Input values were normalized to $[-0.9 \rightarrow +0.9]$, and all initial weights were randomized to $[-0.5 \rightarrow +0.5]$. The learning rate was initially set to 0.05, and increased as the network neared a local minimum during training. Network training was terminated if the reduction in error between iterations was less than 1×10^{-9} or if the network error increased over a window chosen to be 6000 iterations. The number of hidden nodes was initially set to 10, and overlearning was noted by the divergence of training and test set classification errors. The number of hidden nodes was then reduced until training and test set classification error curves were nondivergent, which occurred at six hidden nodes for both the network which modeled mortality as well as the network which modeled new metastases.

Classification accuracy (CA) was defined as

$$\text{CA} = \left[\frac{C}{C + I} \right] \times 100 \tag{G5.4.2}$$

where C is the number of correct network classifications in the data set and I the number of incorrect classifications. The n_1/n_2 cross-validation method was used, so that *the network was not trained using data sequestered in the test set.* The classification accuracy of the neural network which modeled the development of new metastases was 92.5% in the training set and 84.5% in the test set. Classification accuracy in the training set was 90.3% for the network which modeled mortality, and 71.4% in the test set.

G5.4.3 Statistical analysis of network behavior

Golden has noted that, on completion of network training, Wilks' generalized likelihood ratio test may be used to determine if its final error is statistically different than another network of dissimilar topology (Golden to appear, Wilks 1938). By removing input nodes to alter the topology of the network, the contribution of individual input features to the network's model may be studied. This capability is of particular interest to medical researchers who desire to 'open the black box' and dissect the importance of specific clinical parameters.

Use of Wilks' generalized likelihood ratio test begins with the training of a network on a particular data set, and recording the network error $\eta_1(W^1)$. One or more input node(s) corresponding to a specific feature are then removed from the network by setting the r weights in this first network connected to those input nodes to zero. The network is then retrained on the same data set and the network error $\eta_2(W^2)$ is recorded. The procedure requires that both error estimates are associated with strict local minima of their respective error surfaces and the same strict local minimum of the 'true' error function. The error for network $\eta_2(W^2)$ should be greater than or equal to the network $\eta_1(W^1)$ since the second model has fewer free parameters. The question one wishes to test is whether the increase in error is statistically significant (i.e. if the r weights in the original network were really equal to zero).

Using Wilks' classical generalized likelihood ratio test, the null hypothesis that the two networks are equally effective (aside from sampling error) in classification can be rejected if:

$$2M[\eta_2(W^2) - \eta_1(W^1)] > \kappa_\alpha \qquad (G5.4.3)$$

where κ_α is a constant with the property that a chi-squared random variable with r degrees of freedom exceeds κ_α with probability α (Wilks 1938).

NeUROn was programmed so that variables could be specified in a file *class_define* by the position of their corresponding input nodes. For example, the first three variables in the renal cancer data set, ethnicity (four nodes), gender (one node), and age (two nodes) were specified by

$$1 - 4, \quad 5, \quad 6 - 7, \ldots .$$

In this way, groups of nodes corresponding to one variable are held to zero simultaneously to generate subnetworks for comparison to the full network using Wilks' generalized likelihood ratio test. NeUROn was programmed to retrain subnetworks automatically with combinations of variables removed from the full network by holding their corresponding input node(s) to zero; for the renal cancer network trained on the new metastases data set, the variables were removed singly. The resulting p-values for each variable are shown in table G5.4.2.

Table G5.4.2. Wilks' generalized likelihood ratio test p-values for individual variables removed to produce feature-deficient subnetworks.

Variable removed	p-value
Ethnicity	1.000
Gender	0.009†
Age	< 0.001†
T stage	0.004†
N stage	0.007†
M stage	0.428
Nephrectomy	1.000
Surgery Date	1.000
Lung metastases	0.807
Bone metastases	1.000
Histologic subtype	< 0.001†
Tumor size	0.739
Treatment Choice	1.000

†$p < 0.05$

As shown in table G5.4.2, patient gender, age, T stage, N stage and histologic type were all found to be significant features in predicting the development of new metastases. Interestingly, the presence of lung or bone metastases *did not* predict the development of new metastases. This observation supports the currently controversial practice of surgically removing a single metastasis, for one metastasis does not absolutely predict future metastases.

G5.4.4 Comparison with discriminant function analysis

Network performance was compared to the Bayes' classifiers linear-discriminant function analysis (LDFA) and quadratic-discriminant function analysis (QDFA) (James 1985, Duda and Hart 1973). Each divides

M-dimensional decision hyperspace with a single $(M-1)$-dimensional hyperplane in the 2-class case. Linear-discriminant function analysis can be considered to be a special case of quadratic-discriminant function analysis in which covariance is equal among classes. Classification accuracy of the network in comparison to discriminant function analysis applied to the data set which recorded the development of new metastases is shown in table G5.4.3. Comparison to discriminant function analysis in classifying patient mortality is detailed in table G5.4.4. In both cases, the neural network outperformed linear- and quadratic-function analysis, with the classification accuracies for discriminant function analysis applied to the mortality data set no better than chance.

Table G5.4.3. Classification accuracies of the neural network, linear- and quadratic-discriminant function analysis in modeling new metastases in the renal cancer data set.

Data set	LDFA	QDFA	Neural network
Training	68.4%	69.0%	92.5%
Test	67.2%	69.0%	84.5%

Table G5.4.4. Classification accuracies of the neural network, linear- and quadratic-discriminant function analysis in modeling mortality in the renal cancer data set.

Data set	LDFA	QDFA	Neural network
Training	40.1%	39.3%	90.3%
Test	40.5%	39.3%	71.4%

G5.4.5 Discussion

Physicians commonly encounter classification tasks. Although most physicians and medical researchers encounter statistics only once during training, learning to design studies and employ tests of discrimination such as analysis of variance, the most common problem encountered in the practice of medicine is classification. Diagnosis, choice of therapy and outcome prediction are all classification tasks. Tumor staging systems were devised to serve as algorithmic systems to model the latter task, outcome prediction, in cancer. Unfortunately, simple decision trees are insufficient to accurately model many types of tumors. Predicting tumor behavior in individual patients with renal cancer is, to date, an intractable modeling problem (de Kernion 1986, Williams 1987).

We chose to investigate neural computation as a modeling system for renal cancer outcomes. The two outcomes tracked in our database were patient mortality, i.e. whether or not patients were alive at follow-up, and the development of new metastases. In both cases, the trained neural network outperformed linear- and quadratic-discriminant function analysis. We do not know if other Bayesian modeling systems would necessarily perform more poorly than the neural computational system. In fact, we are actively investigating many types of classifiers to find the most accurate model. At present, the neural computational approach yields the most accurate classifier in our renal cancer data set.

Although the neural network's performance in modeling new metastases yielded an 84.5% classification accuracy in the test set, its performance in modeling mortality was lower at 71.4%. We expect this is due to critical missing features. Patients may die of many causes, such as cardiac events, that are not related directly to the variables that we tracked in our database.

Simply building an accurate classifier is not enough for medical researchers who need to know which features are important to the model. The use of Wilks' generalized likelihood ratio test allows such a dissection of the neural computational 'black box'.

Finally, medical classifiers are only useful if actually used by physicians. Many physicians have limited experience with computational systems, requiring highly 'user-friendly' interfaces. We have chosen to investigate the World Wide Web as a front-end for neUROn. Via a set of PERL scripts which allows the use of forms to submit input vectors to compiled and trained neural networks, World Wide Web browsers may efficiently access our trained networks for use in classifying remote patient data. At the time of writing, neUROn trained networks may be accessed at http://godot.urol.uic.edu.

Acknowledgements

The authors would like to acknowledge the substantial contributions of the members of the neUROn team: Luke Cho, Patrick Guinan, Joe Jovero, Vinod Kutty, Dolores Lamb, Larry Lipshultz, Lawrence Ross, Sue Ting, and Yuan Qin.

References

Baum E B and Wilczek F 1988 Supervised learning of probability distributions by neural networks *Neural Information Processing Systems* ed D Z Anderson (New York: American Institute of Physics) pp 52–61

de Kernion J B 1986 Renal tumors *Campbell's Urology* ed P C Walsh, R F Gittes, A D Perlmutter and T A Stamey (Philadelphia, PA: Saunders) pp 1294–342

Duda R O and Hart P E 1973 *Pattern Classification and Scene Analysis* (New York: Wiley) pp 17–20

Golden R M *Fundamentals of Neurocomputer Analysis and Design* (Boston, MA: MIT Press) to appear

James M 1985 *Classification Algorithms* (London: Collins) pp 15–29

Wilks S S 1938 The large sample distribution of the likelihood ratio for testing composite hypotheses *Ann. Math. Stat.* **9** 60–2

Williams R D 1987 Renal, perirenal, and ureteral neoplasms *Adult and Pediatric Urology* ed J Y Gillenwater, J T Grayhack, S S Howards and J W Duckett (Chicago, IL: Year Book Medical Publishers, Inc) pp 513–54

G5.5 Hopfield neural networks for the optimum segmentation of medical images

Riccardo Poli and Guido Valli

Abstract

In this section we present a general-purpose neural architecture for segmenting two-dimensional and three-dimensional medical images. The architecture is based on a continuous Hopfield neural network including one or more sets of two-dimensional layers of neurons with local connections. This architecture can be specialized to perform the segmentation of two-dimensional images, the multiscale segmentation of two-dimensional images and the segmentation of three-dimensional images by simply changing the number of such sets and/or the size of the component layers. By changing synaptic weights the architecture can adapt to the differences existing between tomographic and radiographic images. The segmentation produced by this architecture is optimum with respect to a 'goodness' criterion which establishes the tradeoff between sensitivity and robustness. The section describes the derivation of the architecture and some experimental results obtained with synthetic and real medical images.

G5.5.1 Introduction

The general objective of the *segmentation of medical images* is to find regions which represent single anatomical structures. The availability of such regions not only makes tasks such as interactive visualization and automatic measurement of clinical parameters directly feasible, but is also the starting point for using more sophisticated computer vision techniques and performing higher-level tasks such as three-dimensional shape comparison and recognition (Poli *et al* 1994). F1.6, G1.7

Unfortunately, due to the presence of image noise, masking structures, biological shape variability, tissue inhomogeneity, imaging-chain anisotropy and variability, etc, the segmentation of medical images is a very hard problem. Therefore, to obtain reliable segmentation algorithms researchers have almost invariably been obliged to exploit as much *a priori* information as possible.

Knowledge of statistical properties of the gray levels of the image is a kind of *a priori* information that has been extensively exploited in the case of magnetic resonance (MR) and computed tomography (CT) images (see, for example, Raya 1990, Lei and Sewchand 1992, Gerig *et al* 1992, Amartur *et al* 1992, Özkan *et al* 1993). Despite the differences existing among these methods, they share the idea of considering each pixel as a separate entity to be classified, thus neglecting the spatial correlation between measurements due to cohesion of matter.

Spatial correlation is considered as more important in other methods, such as those based on mathematical morphology operators (Higgins *et al* 1990, Klingler *et al* 1988, Thomas *et al* 1991, Joliot and Mazoyer 1993), on rule-based expert systems (Catros and Mischeler 1988, Manos *et al* 1993, Li *et al* 1993), on special-purpose computer vision techniques (Raman *et al* 1993, Coppini *et al* 1993, Deklerck *et al* 1993) or on neural networks trained with the *backpropagation algorithm* (Silverman and Noetzel 1990, C1.2.3 Toulson and Boyce 1992, Coppini *et al* 1993). However, in addition to the spatial correlation between measurements all these methods exploit another kind of *a priori* information: the anatomical knowledge about which structures are present in the image, where they usually are, what they usually look like, etc.

Whereas on one hand this information considerably improves the robustness of segmentation algorithms, on the other hand it drastically reduces their generality and their applicability to different kinds of images or anatomical districts. Therefore, anatomical-information-based methods do not seem good candidates to build general purpose segmentation systems for medical images.

To overcome these problems and build a general-purpose segmentation system for medical images, we adopted a different approach inspired by biological vision.

G5.5.2 Approach and objectives

Vision is ruled by principles, such as perceptual grouping, selection and discrimination, which mostly depend on regularities of nature such as cohesiveness of matter or existence of bounding surfaces (Marr 1982, Reuman and Hoffman 1986). As these properties are also valid for the anatomical structures present in medical images, they can be exploited to build segmentation systems for such images. If no other source of information is used, the resulting segmentation algorithms are independent of the imaging modality, of the scanning parameters, of the imaged district, and so on and therefore can be used for general-purpose medical-image segmentation.

Regularities of nature can be exploited in a very simple way by using grouping or discrimination criteria based, for example, on the idea that pixels which are close to each other and have similar gray levels have a high probability of representing the same object and therefore should be grouped together. However, even if the strategy is simple, in order to design a general-purpose segmentation algorithm for medical images a number of requirements must be met which can make the actual implementation of the strategy quite complex. Let us analyze these requirements.

- The segmentation algorithm should be maximally sensitive to small structures or to structures with a low contrast (possible lesions or tumors in early stages).
- The algorithm should be maximally robust with respect to the noise, texture and slow intensity changes typically present in medical images.
- The algorithm should be able to adapt to the differences existing among the processes of generation of images obtained from different imaging devices. Therefore, it should be able to process not only two-dimensional tomographic images but also three-dimensional and x-ray projective ones.
- A segmentation algorithm to be integrated in more complex analysis systems should be able to perform multiscale segmentation as, in many applications, segmentation is analyzed by multiple modules requiring different levels of detail.
- An algorithm to be used with imaging devices (e.g. cine-CT scanners) which can produce hundreds of images per patient should be suitable for parallel, high-speed implementation.

The first two requirements counteract each other and, therefore, any segmentation algorithm can only produce results that represent a tradeoff between them. In order to achieve optimum compromises it is first necessary to define a quantitative criterion of goodness of segmentation which takes sensitivity and robustness into account, and then to optimize it for any specific image. Therefore, the problem of medical image segmentation can be seen as a problem of combinatorial optimization.

Unfortunately, for any given image the space of possible solutions to this optimization problem is huge and conventional optimization techniques tend to fail on it. Therefore, following recent approaches in the field of natural scene segmentation (Darrell *et al* 1990, Reed 1992, Wang *et al* 1992) we decided to solve it by using an architecture based on continuous Hopfield neural networks (Hopfield 1984), a computational paradigm which can effectively search huge solution spaces.

C1.3.4   *Hopfield networks* can be seen as dynamical systems which tend to relax into states which minimize the following energy function

$$E_{\text{net}} \approx -\frac{1}{2} \sum_{i=1}^{N} \sum_{j=1}^{N} T_{ij} v_i v_j - \sum_{i=1}^{N} i_i v_i \qquad (\text{G5.5.1})$$

where v_i is the output of neuron i, i_i is its external input and T_{ij} is the weight of the connection from neuron j to neuron i. Thanks to this minimum-seeking behavior, Hopfield networks can be used to solve optimization problems (Hopfield and Tank 1985, 1986). The basic strategy is as follows: (i) to preprocess, when needed, the input data, (ii) to find a binary representation for the solutions of the problem so that they can be mapped into the stable states of the neurons of a Hopfield network, (iii) to define a quadratic (symmetric) energy function whose minimization leads to an optimum solution of the problem and then calculate weights and external inputs, (iv) to initialize and let the network relax into a stable state to then be mapped back into a solution for the original problem.

In the following we describe how these steps, applied to the problem of medical-image segmentation, lead to an architecture that not only provides the optimum sensitivity/robustness tradeoff but also meets the other requirements listed above.

G5.5.3 Segmentation of tomographic images

In this case the input data of the segmentation algorithm is a two-dimensional tomographic image denoted with the symbol $I(x, y)$. Normally these data need no preprocessing and, therefore, the first step for solving the segmentation problem is finding a binary representation for its solutions.

G5.5.3.1 Binary representation

We adopted a representation suggested by the analogy of the segmentation process with that of coloring geographic maps (Bilbro *et al* 1987). This analogy indicates that, in order to represent the regions ('states') obtained from the segmentation of an image, only a reduced number of labels ('colors') are needed, as long as different labels are associated to connected regions ('bordering states'). Therefore, as shown in figure G5.5.1, a segmentation can be represented with a small set of two-dimensional layers of neurons (each layer represents a different label).

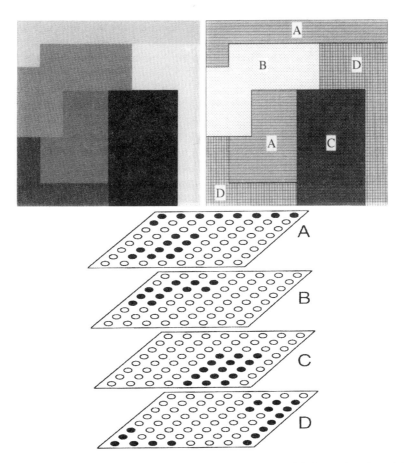

Figure G5.5.1. Synthetic 8 × 8 image (top left), a possible labeling with four colors (top right), and the related binary representation with four layers of neurons (bottom). (Active neurons are represented as filled circles.)

G5.5.3.2 Energy function

The next step is the definition of a quadratic energy function E_{net} whose minimization gives an optimal solution to the segmentation problem. We adopted an energy function partially inspired by the one suggested in Hopfield and Tank (1985, 1986) for the solution of the traveling-salesman problem and the one proposed in Bilbro *et al* (1987) for the segmentation of signals with simulated annealing. As the fixed points of Hopfield networks tend to be the vertices of the hypercube $[0, 1]^N$, we were able to design E_{net} on the hypothesis of binary neurons, i.e. $v_{xyc} \in \{0, 1\}$.

E_{net} includes two parts: the *syntax energy* E_{syntax} which enforces the syntactic correctness of the solutions (i.e. prevents the network from settling into nonbinary states or states which cannot be mapped back to solutions of the segmentation problem), and the *semantics energy* $E_{semantics}$ which is our criterion of goodness of segmentation. The two parts are added so that $E_{net} = E_{syntax} + E_{goodness}$.

Syntax energy. The syntactic correctness of the solutions requires that *one and only one neuron is active among the neurons which represent a given pixel*, i.e. $\exists! c : v_{xyc} = 1$. This constraint can be enforced by including in E_{syntax} terms such as $\sum_{c_1} \sum_{c_2 \neq c_1} v_{xyc_1} v_{xyc_2}$ and $\left(\sum_c v_{xyc} - 1 \right)^2$ (the latter prevent the network from settling in the nonvalid null solution $v_{xyc} = 0$, $c = 1, 2, \ldots$). By summing these terms for all the pixels in the image we obtain

$$E_{syntax} = \frac{K_1}{2} \sum_x \sum_y \sum_c \sum_{\hat{c} \neq c} v_{xyc} v_{xy\hat{c}} + \frac{K_2}{2} \sum_x \sum_y \left(\sum_c v_{xyc} - 1 \right)^2 \tag{G5.5.2}$$

where K_1 and K_2 are constant values.

Semantics energy. The goal of the semantics energy is that of driving the network towards segmentations that represent an optimum compromise between sensitivity and robustness. Therefore the semantic energy includes two terms, the *sensitivity energy* $E_{sensitivity}$ and the *robustness energy* $E_{robustness}$, which are summed up to give $E_{semantics} = E_{sensitivity} + E_{robustness}$.

Sensitivity energy. The sensitivity energy should force the network to perform a segmentation revealing any transition between different tissues; that is, any change in the image gray levels. In order to obtain this effect, $E_{sensitivity}$ must include terms which increase when neighboring pixels lying across a boundary have the same label. We used terms such as $\sum_c v_{xyc} v_{\hat{x}\hat{y}c} [dI(x, y)]/[dn(x, y, \hat{x}, \hat{y})]$ where $n(x, y, \hat{x}, \hat{y}) = [(\hat{x}, \hat{y}) - (x, y)]/[\|(\hat{x}, \hat{y}) - (x, y)\|]$, and (x, y) and $(\hat{x}, \hat{y})$ are neighboring pixels. These terms must be present for all pixels lying in a neighborhood B^{xy} which does not contain pixels too close to or too far from (x, y). (We adopted the simplest neighborhood satisfying these requirements: $B^{xy} = \{(\hat{x}, \hat{y}) \mid 2 \leq [(\hat{x} - x)^2 + (\hat{y} - y)^2]^{1/2} \leq 2(2)^{1/2}\}$.) Thus, the complete expression of the sensitivity energy is

$$E_{sensitivity} = \frac{K_4}{2} \sum_x \sum_y \sum_{(\hat{x}, \hat{y}) \in B^{xy}} \sum_c v_{xyc} \, v_{\hat{x}\hat{y}c} \frac{dI(x, y)}{dn(x, y, \hat{x}, \hat{y})} \tag{G5.5.3}$$

where K_4 is a constant value.

Robustness energy. The aim of $E_{robustness}$ is to reduce the effects of noise and texture. Since noise and texture tend to produce very small regions, $E_{robustness}$ should favor the construction of large regions which have a high probability of representing single anatomical structures. This can be obtained using the constraint: *pixels which are close to each other should have the same label.* The constraint can be implemented using terms of the form $-\sum_c v_{xyc} v_{\hat{x}\hat{y}c}$, for all the pixels $(\hat{x}, \hat{y})$ in a 4-connected neighborhood N^{xy} of any given pixel (x, y). The total robustness energy becomes

$$E_{robustness} = -\frac{K_5}{2} \sum_x \sum_y \sum_{(\hat{x}, \hat{y}) \in N^{xy}} \sum_c v_{xyc} \, v_{\hat{x}\hat{y}c} \tag{G5.5.4}$$

where K_5 is a constant value.

Once E_{net} is defined, the weights and the external inputs of the network can be computed easily (for example by comparing the expression of E_{net} with the left-hand side of equation (G5.5.1)).

G5.5.3.3 Network initialization

Hopfield networks can be simulated by simply integrating numerically their motion equation until a stable state is reached. However, before doing that the state of the network has to be initialized. As the standard random initialization method in the present case gives poor segmentation results, we adopted the strategy suggested in Chen *et al* (1991) which consists of initializing the network in an area of state space where a good solution is present. In this way the network has only to improve on the solution instead of looking for it in the whole state space. As an initial solution we used the segmentation produced by the following algorithm:

(i)   Let $I_{\max} = \max\limits_{x,y} I(x, y)$ and $I_{\min} = \min\limits_{x,y} I(x, y)$.

(ii) For each pixel (x, y) do:

    (a) let $\hat{c}$ be the nearest integer which is less than or equal to $[I(x, y) - I_{\min}]/(I_{\max} - I_{\min}) + 1$.

    (b) For each color $c = 1, \ldots$ do:

$$\text{let } v_{xyc} = \begin{cases} \alpha & \text{if } c = \hat{c} - 1 \text{ or } c = \hat{c} + 1 \\ 1 - 2\alpha & \text{if } c = \hat{c} \\ 0 & \text{otherwise}. \end{cases}$$

G5.5.3.4 Extensions to three-dimensional and multiscale segmentation

The extension of the method to the segmentation of three-dimensional images can easily be obtained by introducing three-dimensional neighborhoods and three-dimensional image derivatives as well as by adding an extra sum in equations (G5.5.2), (G5.5.3) and (G5.5.4).

The extension to multiscale segmentation requires a preprocessing step as segmentation has to be performed simultaneously on multiple, smoothed and decimated versions of the original image. Such images, denoted with the symbol $I(x, y, s)$, are built recursively from one another according to the formula

$$\begin{cases} I(x, y, 1) = I(x, y) \\ I(x, y, s + 1) = \frac{1}{4} \sum\limits_{i=0}^{1} \sum\limits_{j=0}^{1} I(2x + i, 2y + j, s) \qquad \text{for } s = 1, 2, \ldots. \end{cases}$$

After preprocessing, the various components of the energy function can be separately defined for each scale and summed. However, in order for the segmentation performed at a given scale to influence and to be influenced by the segmentation being performed at other scales, additional energetic terms such as $-v_{xycs}v_{(x/2)(y/2)c(s+1)}$ and $-v_{xycs}v_{(2x+i)(2y+j)c(s-1)}$ (for $i = 0, 1$ $j = 0, 1$) are needed.

Derivation of weights and inputs, initialization and relaxation are performed as in the case of two-dimensional segmentation.

G5.5.4 Segmentation of x-ray images

The general criteria of goodness of segmentation introduced in the previous sections are valid also for projective x-ray images. However, the peculiarities of the physical process of generation of this kind of image imposes a few changes.

G5.5.4.1 Preprocessing

The approximate linearization of the image generation process is a *preprocessing step* needed for x-ray B4.4
image segmentation. This is obtained by performing an appropriate logarithmic transformation of the gray levels of the original image after which we can express

$$I(x, y) = \int_0^{d(x,y)} \mu(x, y, z) \, dz$$

where $\mu(x, y, z)$ is the linear absorption coefficient of the tissue at coordinates (x, y, z) and $d(x, y)$ the thickness of the body in (x, y). As any anatomical district contains a discrete number of structures of interest, if we denote with $d_i(x, y)$ the thickness of the ith structure and with μ_i the absorption coefficient of such a structure, we can rewrite

$$I(x, y) = \sum_{i=1}^{N} \mu_i d_i(x, y). \tag{G5.5.5}$$

G5.5.4.2 Binary representation

Anatomical structures which are overlaid or inside one another are represented by the same pixels in an x-ray image and, therefore, regions are no longer constrained to form a tessellation of the image but can overlap.

To represent in binary form a segmentation with overlapping regions we adopted a set of two-dimensional layers of neurons like those used for the segmentation of tomographic images, with the important difference that each layer does not represent a different 'color' but a different anatomical structure.

G5.5.4.3 Energy function

Syntax energy. The syntactic correctness of solutions does not require any more that one and only one neuron inside the set of neurons which represent a given pixel be active, as this would mean that a pixel cannot represent more than one anatomical structure. However, syntax requires that, in stable states, each neuron of the network be completely excited ($v_{xyc} = 1$) or inhibited ($v_{xyc} = 0$). To obtain this effect we used a term of the form $v_{xyc}(1 - v_{xyc})$ for each neuron. As a result:

$$E_{\text{syntax}} = \frac{K_1}{2} \sum_x \sum_y \sum_c v_{xyc}(1 - v_{xyc})$$

where K_1 is a constant value.

Sensitivity energy. The function of $E_{\text{sensitivity}}$ is maximizing the consistency of segmentation with respect to the image gray levels expressed by equation (G5.5.5). Unfortunately, to obtain a quadratic $E_{\text{sensitivity}}$ we had to add the hypothesis (only approximately valid) that the thickness of the structures shown in the x-ray image is constant; that is, $d_i(x, y) = d_i$. On this hypothesis we can define the quantity $D_i = \mu_i d_i$ (estimated on the basis of the typical density and thickness of the structures of interest) and express $I(x, y) = \sum_c v_{xyc} D_c$. To force the network to settle into solutions (approximately) consistent with this equation we defined

$$E_{\text{sensitivity}} = \frac{K_2}{2} \sum_x \sum_y \left(\sum_c v_{xyc} D_c - I(x, y) \right)^2$$

K_2 being a proper constant value.

Robustness energy. The robustness energy for x-ray image segmentation includes the same terms as in equation (G5.5.4). Unfortunately, in this case these terms alone can induce the diffusion of the activation of the neurons representing a given structure outside the boundaries of that structure. This happens because $E_{\text{sensitivity}}$ does not include any terms which force the neurons of a region to change their state in proximity of the boundaries of the structure represented by that region. This can be overcome by also including the constraint: *if a structure is not present in a given pixel, it is also not present nearby*. The resulting robustness energy turns out to be

$$E_{\text{robustness}} = -\frac{K_3}{2} \sum_x \sum_y \sum_{(\hat{x},\hat{y}) \in \mathcal{N}^{xy}} \sum_c v_{xyc}\, v_{\hat{x}\hat{y}c} - \frac{K_4}{2} \sum_x \sum_y \sum_{(\hat{x},\hat{y}) \in \mathcal{N}^{xy}} \sum_c (1 - v_{xyc})(1 - v_{\hat{x}\hat{y}c})$$

where K_3 and K_4 are constant values.

Weights and external inputs can be easily obtained in the standard way. In order to ensure the convergence of the network to good solutions, we initialized it to a point of state space which represents a good segmentation. The initialization algorithm is similar to that used for tomographic images.

G5.5.5 Experimental results

The networks described in the previous sections have been tested both on synthetic images and on real tomographic and x-ray ones. Synthetic images were generated by simulating the operation of a real tomographic device on an ellipsoidal organ surrounded by a homogeneous tissue. In order to test the robustness of the method, in addition to the blurring caused by the finite thickness of the slices (partial-volume effect) Gaussian white noise with zero mean and increasing standard deviation σ was included in the images. The resulting images were segmented using both the single-scale and multiscale networks described in the previous sections and then compared with the exact segmentation obtained manually with images in which noise and partial-volume effect were absent. Table G5.5.1 shows the average errors obtained in these experiments for several different values of σ and for 1–4 interacting scales.

Noise σ	Scales			
	1	2	3	4
0	1.05	1.20	1.49	1.61
5	1.17	1.68	1.81	1.81
10	1.32	1.93	2.05	1.95
20	1.46	1.95	1.98	1.93
40	16.38	4.59	4.17	4.13
80	52.88	51.81	53.71	55.59

Table G5.5.1. Segmentation of synthetic tomograms: wrong assignments (per cent) versus noise standard deviation and number of interacting scales.

The table reveals that, in the presence of noise with relatively small standard deviation, there are no advantages in using multiscale segmentation. Actually, for $\sigma = 0$–20, using 2–4 scales produces 0.15% to 0.73% more wrong assignments than in the single-scale case. However, in the presence of noise of higher intensity ($\sigma = 40$) multiscale segmentation is much more reliable than a single-scale one. Results are not satisfactory only when noise standard deviation is extremely high ($\sigma = 80$).

The accuracy shown by the method in the experiments with synthetic images has been confirmed by numerous experiments with real tomograms. For example, figure G5.5.2 illustrates how, in segmenting an MR image of the thorax, the network has correctly identified most of the anatomical districts of clinical interest (e.g. lungs, subcutaneous fat, muscular tissue, right atrium, right ventricle, backbone and pulmonary artery). Another example is represented by figure G5.5.3 which shows an MR slice of the head along with the multiscale segmentation produced by the network. Segmentation has been performed jointly at three different scales: 128×128, 64×64 and 32×32. At the lowest resolution there are only eight regions, the largest five of which represent the most significant anatomical structures: white matter, gray matter, cerebrospinal fluid (CSF) in the ventricles, fat with bone, and background. These regions can be easily recognized and used to guide a complete interpretation of the image. At 64×64 resolution the boundaries of white matter, gray matter and CSF become more complex and new regions are present to represent the difference between fat and bone and between thin and thick areas of the ventricles. Maximum accuracy is reached at the highest resolution where, despite noise and texture the most important structures are still represented by a single or a small number of large regions.

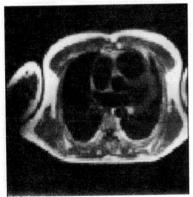

Figure G5.5.2. Segmentation of an MR image of the thorax.

The method has also been tested on x-ray images. For example, figure G5.5.4 (left) shows a cine-angiographic x-ray image of the left ventricle of the heart. The largest structures inside the circular area representing the borders of the image intensifier are: the left ventricle with the descending aorta (center), the diaphragm muscle (lower left) and a metallic filter (upper right). To perform the segmentation of this kind

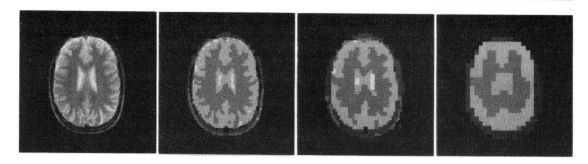

Figure G5.5.3. multiscale segmentation of an MR image of the head.

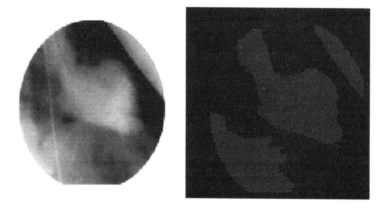

Figure G5.5.4. Segmentation of a cine-angiographic image of the left ventricle.

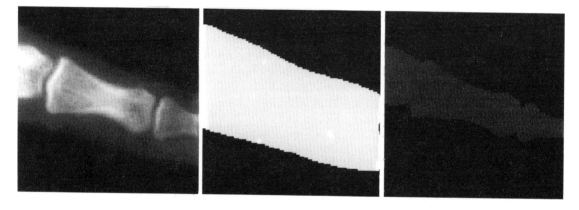

Figure G5.5.5. Segmentation of a radiogram of a tract of a finger.

of images we utilized three layers of neurons: one to represent the image intensifier, one for the background (soft tissues with a low density) and one for the structures just mentioned (they have approximately the same value of D_c). Figure G5.5.4 (right) shows the activation of this last layer. Diaphragm muscle, left ventricle with aorta and metallic filter have been correctly represented as disjunct regions. Another example is given in figure G5.5.5 which illustrates the segmentation of a radiogram of a finger. Although, in this case, the network has not been capable of splitting the bone part of the finger into its anatomical components because of the very limited inter-bone space, the important discrimination between soft tissue and bone is correct, even where bone and soft tissue overlap.

G5.5.6 Conclusion

In this section we have described a neural architecture for the segmentation of medical images. With simple topology and parameter changes the architecture can be adapted to perform the two-dimensional, three-dimensional and multiscale segmentation of tomographic and x-ray images. Thanks to its broad applicability, to the robustness and sensitivity shown in the experiments and to its implementability with fine-grained parallel hardware, this architecture seems to meet the requirements to be considered a possible general-purpose solution to the problem of medical image segmentation.

Acknowledgements

This work has been partially supported by the Italian Ministry for University and Scientific and Technological Research (MURST).

References

Amartur S C, Piraino D and Takefuji Y 1992 Optimization neural networks for the segmentation of magnetic resonance images *IEEE Trans. Med. Imag.* **11** 215–20

Bilbro G L, White M and Snyder W 1987 Image segmentation with neurocomputers *Neural Computers* ed R Eckmiller and C v d Malsburg (Berlin: Springer)

Catros J Y and Mischeler D 1988 An artificial intelligence approach for medical picture analysis *Patt. Recog. Lett.* **8** 123–30

Chen C T, Tsao E C K and Lin W C 1991 Medical image segmentation by a constraint satisfaction neural network *IEEE Trans. Nucl. Sci.* **38** 678–86

Coppini G, Demi M, Poli R and Valli G 1993 An artificial vision system for X-ray images of human coronary trees *IEEE Trans. Patt. Anal. Mach. Int.* **15** 156–62

Coppini G, Poli R, Rucci M and Valli G 1992 A neural network architecture for understanding 3D scenes in medical imaging *Comput. Biomed. Res.* **25** 569–85

Darrell T, Sclaroff S and Pentland A 1990 Segmentation by minimal description *IEEE Int. Conf. Computer Vision III (Osaka)* (Osaka: IEEE Press) pp 112–6

Deklerck R, Cornelis J and Bister M 1993 Segmentation of medical images *Image Vis. Comput.* **11** 486–503

Gerig G, Martin J, Kikinis R, Kubler O, Shenton M and Jolesz F A 1992 Unsupervised tissue type segmentation of 3D dual-echo MR head data *Image Vis. Comput.* **10** 349–60

Higgins W E, Chung N and Ritman E L 1990 Extraction of left-ventricular chamber from 3D CT images of the heart *IEEE Trans. Med. Imag.* **9** 384–95

Hopfield J J 1984 Neurons with graded response have collective computational properties like those of two-state neurons *Proc. Natl Acad. Sci.* **81** 3088–92

Hopfield J J and Tank D W 1985 'Neural' computation of decisions in optimization problems *Biolog. Cybern.* **52** 141–52

—— 1986 Computing with neural circuits: a model *Science* **233** 625–33

Joliot M and Mazoyer B M 1993 Three-dimensional segmentation and interpolation of magnetic resonance brain image *IEEE Trans. Med. Imag.* **12** 269–77

Klingler J W Jr, Vaughan C L, Franker T D Jr and Andrews L T 1988 Segmentation of echocardiographic images using mathematical morphology *IEEE Trans. Biomed. Eng.* **35** 925–35

Lei T and Sewchand W 1992 Statistical approach to X-ray CT imaging and its applications in image analysis—Part II: a new stochastic model-based image segmentation technique for X-ray CT image *IEEE Trans. Med. Imag.* **11** 62–9

Li C, Goldgof D B and Hall L O 1993 Knowledge-based classification and tissue labeling of MR images of human brain *IEEE Trans. Med. Imag.* **12** 740–50

Manos G, Cairns A Y, Ricketts I W and Sinclair D 1993 Automatic segmentation of hand–wrist radiographs *Image Vis. Comput.* **11** 100–11

Marr D 1982 *Vision* (New York: Freeman)

Özkan M, Dawant B M and Maciunas R J 1993 Neural-network-based segmentation of multimodal medical images: a comparative and prospective study *IEEE Trans. Med. Imag.* **12** 534–44

Poli R, Coppini G and Valli G 1994 Recovery of 3D closed surfaces from sparse data *Comput. Vis. Graphics Image Proc.: Image Understanding* **60** 1–25

Raman S V, Sakar S and Boyer K L 1993 Hypothesizing structures in edge-focused cerebral magnetic resonance images using graph-theoretic cycle enumeration *Computer Vision, Graphics, and Image Processing: Image Understanding* **57** 81–98

Raya S P 1990 Low-level segmentation of 3D magnetic resonance brain images—a rule-based system *IEEE Trans. Med. Imag.* **9** 327–37

Reed T R 1992 Region growing using neural networks ed H Wechsler *Neural Networks for Perception* vol 1 (San Diego, CA: Academic) pp 386–97

Reuman S R and Hoffman D D 1986 Regularities of nature: the interpretation of visual motion *From Pixels to Predicates* ed Alex P Pentland (Norwood, NJ: Ablex) pp 201–26

Silverman R H and Noetzel A S 1990 Image processing and pattern recognition in ultrasonograms by backpropagation *Neural Networks* **3** 593–603

Thomas J G, Petersx R A II and Jeanty P 1991 Automatic segmentation of ultrasound images using morphological operators *IEEE Trans. Med. Imag.* **10** 180–86

Toulson D L and Boyce J F 1992 Segmentation of MR images using neural nets *Image and Vision Computing* **10** 324–8

Wang T, Zhuang X and Xing X 1992 Robust segmentation of noisy images using a neural network model *Image and Vision Computing* **10** 233–40

G5.6 A neural network for the evaluation of hemodynamic variables

Tom Pike and Robert A Mustard

Abstract

A standard feedforward backpropagation network was used to perform automated integrity evaluation of arterial pressure waveforms. Our goal was to automatically and reliably read hemodynamic variables directly from patients with our existing lab equipment (Mustard *et al* 1990). The most difficult part turned out to be validating the signals were not corrupted (i.e. suitable for measurement).

G5.6.1 Introduction

Our goal for this study was to automate the collection of arterial pressure data. Whenever a human observer takes a measurement, there is unconscious effort put forward to check the measurement's validity. In a medical environment this effort is often critical since incorrect information could lead to incorrect decisions regarding patient care. When measuring hemodynamic (blood in motion) variables, medical staff automatically compare the measured value with a normal range to ensure the values are reasonable. Another verifying tool is the real-time trace of the variable being measured. In our case an arterial pressure waveform shows the blood pressure as it changes over time. This allows full inspection of the variable over a short time period. Poor catheter positioning and patient movement are common occurrences that can affect signal shape and arterial pressure measurements. A quick glance at the waveform trace is usually enough to verify it is free from external artifacts. If external artifacts occur, measurements should be rejected until the signal returns to a normal state.

Removing the human element from measuring arterial pressure causes a problem. The validation step is not the trivial task that it seems. Checking that the measurement is within normal parameters is not the problem. The huge variability of peak shapes that occur from patient to patient and even within a single patient make it difficult to validate the waveform through pattern matching or screening through characteristics based on shape. Whether the measurement was for an intensive care unit alarm or direct unsupervised support in an animal model, a way is needed to validate signal integrity.

G5.6.1.1 Motivation

A neural approach looked promising since much success had been reported in similar, seemingly more complex, signal processing problems (e.g. *speech recognition*). The only alternative was an exhaustive F1.7 tinkering with statistical methods based on waveform parameters. This would have to be redone for each new signal we wished to study. We hoped the knowledge gained from this initial work could be used towards other types of signal data.

G5.6.1.2 Classifier

The role of the network was to determine the integrity of the input signal. Each peak input would be classified as either clean, contaminated or damped. By definition a clean peak would be suitable for measuring hemodynamic variables. A contaminated peak contained some local shape-distorting phenomenon making measurement inaccurate. Damped peaks are normal peaks with dull features representing some global undesirable signal damping.

G5.6.1.3 Black box description (diagram)

The system consisted of an input module that fed 30 second waveform signal segments into our neural network module. If the output from our neural network module indicated a clean signal, measurements would be recorded and passed to subsequent modules in the experiment. A network output indicating a dirty signal would cause the system to disregard the current waveform. A network output of 'damped' could sound an alarm to alert a technician of equipment malfunction.

G5.6.1.4 Requirements and constraints

Two conditions had to be met: speed and accuracy. We required real-time performance since the network would be monitoring the patient continuously for days at a time. This posed little problem since even a slow personal computer inspects signals using our method more quickly than a human operator. The more difficult constraint involved comparable accuracy with a human expert.

G5.6.1.5 Topology

C1.2 The final network used was a three-layer (one hidden layer) *backpropagation* network. The input layer consisted of 70 neurons. Each neuron in the input layer was unidirectionally connected to each neuron in the hidden layer. The hidden layer contained 20 neurons unidirectionally connected to three output neurons.

G5.6.1.6 Other topologies investigated

Many combinations of the following three-layer network dimensions were tried:

- input neurons: 10, 20, 30, 40, 45, 55, 60, 65, 70, 75, 80, 85, 90, 100
- hidden neurons: 3, 5, 6, 7, 8, 9, 10, 12, 15, 17, 18, 19, 20, 21, 22, 23, 25, 30
- output neurons: 1, 2, 3, 5.

On some network architectures we had two extra output neurons (five in total). They represented the segmentation errors occasionally made by our preprocessing algorithm. We hoped this might convey additional information on which the network could generalize. We thought this might also counterbalance the negative effect of having improperly segmented peaks. The network was able to detect these errors with some success but it did not increase the overall accuracy of the network.

A few four-layer configurations were tried. Using straight backpropagation the connection strengths between the second and third layers were found to grow very large in magnitude compared to the input to second-layer connections. Training never produced acceptable performance when this occurred. An alternative approach was tried. Initially the network was treated as a three-layer network. After reasonably good performance was achieved, the third hidden layer was inserted, inheriting the output connections and starting with random second- to third-layer connections. Training continued as a four-layer backpropagation network. Performance approached that of the three-layer network but progress seemed to be slower and more erratic.

G5.6.1.7 Sources

Our main resource was Rumelhart and McClelland (1986). In addition we used notes from a graduate course on neural computing offered by Professor Geoffrey Hinton at the University of Toronto. All software used was custom-written. A special-purpose database engine was constructed for keeping track of recorded signals, segmenting and labeling peaks, and creating training and testing sets. For network training we designed and implemented a flexible backpropagation system.

G5.6.1.8 Performance features of topology

A feedforward three-layer backpropagation network has two major advantages in applications with a large number of inputs. Firstly, it has a simple training method. Secondly, the training method is fast enough to be used in software. Additionally, backpropagation networks are quite fast executing in software.

G5.6.2   Methods

G5.6.2.1   Training sets

The recorded signal was segmented into peaks (Ellis 1985, Hinton 1986, Klee *et al* 1974, Mustard *et al* 1990). Each peak was manually labeled as overdamped, clean or dirty by the authors. Clinical information about the patient as well as the entire 15 second signal tracing was used to provide the 'correct' label for each peak. This represented the 'gold standard' with which the network was trained and tested. Overdamped indicates inaccuracy due to an obstructed catheter or other problem with the signal collection equipment. Clean denotes an acceptable shape and dirty denotes an irregular or corrupted shape. A corrupted shape may result from patient movement, catheter slippage and so on. Once a large number of patient tracings had been recorded and labeled, two groups were formed with the first 19 patients in one group and the next 19 patients in the second group. Selected peaks from the two groups were used to train two neural networks using backpropagation (Rumelhart *et al* 1986). Then each network was tested on the entire patient group to which it had not been exposed. Our original experiment design was flawed in one important aspect. The peaks recorded in the first 10 patients were any patient and included a large percentage of clean peaks. As we collected, the trend was to record from a more diverse set of signal types, particularly very corrupted signals. This tends to give the first network less experience than the second network. This can be seen in the large false negative error represented in the table. Later we separated the groups into odd and even patient numbers. The performances of these networks were much closer in accuracy.

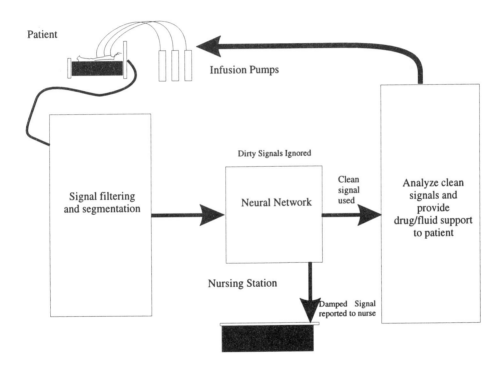

Figure G5.6.1. Data flow diagram.

G5.6.2.2   Preprocessing

The network architecture consisted of a 70-element input array fully connected to a 20-element hidden layer which was then fully connected to a 3-element output layer. The hidden units and the output units all had thresholds learned in the backpropagation step (Rumelhart *et al* 1986). The mapping from an arterial pressure waveform to a decision on its peak's validity is as follows. The waveform is recorded as positive integers at 100 Hz. The signal is then segmented into peaks using the zero crossing algorithm as previously described (Burger 1980, Pike and Mustard 1992). Each peak is then analyzed individually.

The number of points representing the peak is reduced by a factor $1/n$. n is an integer value defined by

$$n = \text{ceil}[(\# \text{ of points in peak})/50] + 1$$

and has a minimum (and typical) value of 2. The number of points in the peak is reduced by replacing each n consecutive values with their average. The calculated averages are then normalized to a range of 0.05 to 0.95 (approximately corresponding to the 40 mmHg to 300 mmHg range). The normalized values are then centered in a 50-element array and empty array positions are set to 0.05. This array representation is used as input to the neural network. Along with this representation of the individual peak, 20 input neurons are used to represent the average and standard deviation of the peaks within the 15 second recording currently being analyzed. All peaks within this recording are transformed into their 'array representation'. All the array representations are split into 10 segments of 5 elements each. The average and standard deviation are calculated for the 10 segments for all peaks in the waveform. The 50 element peak representation, the 10 element average shape representation and the 10 element shape variability representation are used as input to the network.

G5.6.2.3 Training method

The input neurons simply take on the data input as their activations. The equations governing all the remaining neurons are as follows:

$$O_{pj} = \frac{1}{1 + e^{-(\sum_i w_{pij} O_{(p-1)i} + \theta_{pj})}}$$
$$\delta_{pj} = (t_j - O_{pj}) O_{pj} (1 - O_{pj})$$
$$\theta_{pj} = O_{pj}(1 - O_{pj}) \sum_k \delta_{(p+1)k} w_{pkj}$$
$$w_{pji}(n+1) = \alpha(\delta_{pj} O_{pi}) + \beta w_{pji}(n)$$

where O_{pj} = output activation for the pth row and the jth column; w_{pij} = weight connecting the neuron in the $(p-1)$th row and the ith column with the neuron in the pth row and the jth column; δ_{pj} = error signal for O_{pj}; δ_{pj} = threshold for O_{pj}; α = the learning parameter; β = the momentum parameter.

The training was carried out on a 16 MHz 80386 computer system as follows. The backpropagation step (learning step) was initiated after every 10 cases (peaks). The alpha and momentum learning parameters were both fixed at 0.1 until 10 000 epochs (1 epoch = 10 cases). At this point the alpha parameter was set to 0.02 until training was concluded at approximately 15 000 epochs. The sequence of cases was skewed to increase the exposure of problematic peaks. The procedure was as follows. After every 10 passes of the entire training set, each case is used to train the network. In the remaining passes, the network is only trained on cases where the network's output neurons' activations (in the range of 0.0 to 1.0) are incorrect by 0.3 or more. Towards the end of the training cycle the network was evaluated on peaks in its training set. The network is saved if it beats the current best network. When training stops, the current best network is used.

G5.6.2.4 Output

On being exposed to a peak, the network flags, via its output neurons, one of three conditions: acceptable, shape error, or overdamped. The acceptable condition signals the peak is of the correct shape and can be considered (locally) uncorrupted. Shape error implies peak shape is incorrect and should be considered corrupted. The overdamped condition implies shape features are dull and could indicate non-local corruption.

G5.6.2.5 Development

Development of the software took a considerable time—on the order of seven months. When the project began there was little in the way of commercial software for neural computation. What did exist was very inflexible and quite expensive. In addition it was of great benefit to have the neural network simulator tied directly to the signal database.

G5.6.2.6 Comparison with traditional methods

Originally we considered hand-coding statistical methods as being a viable approach. While creating the training data, we were greatly surprised at the difficulty we had classifying a large subset of the peaks. On reflection it would have been an extremely difficult job to embed the huge varieties and interdependent characteristics using traditional coding techniques. Each additional rule or characteristic added to an expert system or fuzzy logic algorithm would have to be balanced with the previously established programming. This makes development difficult and maintenance almost unworkable. In contrast, as other neural network approaches appear they can be tried using the now available commercial network simulators. Of course finding someone specializing in neural networks may, for some time, remain a problem. It would probably be just as difficult as recruiting experienced expert system programmers or fuzzy mathematicians.

G5.6.3 Results

Two basic errors can be made by the network. A false positive error incorrectly indicates a corrupted arterial pressure signal is valid and measurable. The measured hemodynamic variables will be invalid since the signal is corrupted; a very serious error if treatment is based directly on the measurement. A false negative error rejects peaks that would yield accurate parameters, and either decreases availability of derived parameters or increases the number of signals that must be visually inspected. The accuracy of our networks is shown in table G5.6.1. Note that the 'testing' data sets are derived from patients not used in the 'training' data. We found that by allowing the network to learn for longer or shorter periods we could adjust the ratio of false positive errors to false negative errors. Running the learning procedure longer typically reduced the false positive error rate at the expense of a greater false negative error rate. With experimentation, a trade-off can be reached between accuracy and the number of cases that have to be visually inspected. Table G5.6.1 contains results for our initial experiment. Subsequently with an improved segmentation algorithm and better subdivision of training/testing set patients we achieved the results in table G5.6.2.

Table G5.6.1. Published error rates for the two networks.

		False positive	False negative
Network 1			
Group 1	(training)	0.008816	0.031370
Group 2	(testing)	0.012307	0.198583
Network 2			
Group 1	(testing)	0.032258	0.054401
Group 2	(training)	0.004662	0.041488

Table G5.6.2. Subsequent error rates with improved segmentation and balanced training sets.

		False positive	False negative
Network 1			
Group 1	(training)	0.010425	0.047360
Group 2	(testing)	0.022101	0.126603
Network 2			
Group 1	(testing)	0.030570	0.098210
Group 2	(training)	0.008376	0.039841

G5.6.4 Conclusions

We found neural networks well suited to evaluating peak integrity. We only realized how subtle and fuzzy this problem is through manually labeling our data. Some of the mistakes made by the network were difficult to explain (a perfect shape rejected). But many were in the gray zone and difficult even for us to classify as clean or dirty. While researching our experiment we found no literature examining continuous hemodynamic variables at high resolution. Except for the studies involving sleep, continuous high resolution monitoring of disease mechanisms remains a vast unexamined field.

References

Burger D 1980 Analysis of electrophysiological signals: a comparative study of two algorithms *Computers and Biological Research* **13** 73

Ellis M 1985 Interpretation of beat-to-beat blood pressure values in the presence of ventilatory changes *J. Clincal Monitoring* **1**

Hinton G E 1986 Learning distributed representations of concepts, Proc. Eighth Ann. Conf. of the Cognitive Science Society (Amhearst, MA)

Klee G, Ackerman E and Leonard A 1974 Computer detection of distortion in arterial pressure signals *IEEE Trans. Biomed. Eng.* January

Korten J B, Haddad G G 1989 Respiratory waveform pattern recognition using digital techniques *Computers in Biology and Medicine* **19**

Marshall R J 1986 The determination of peaks in biological waveforms *Computers and Biological Research* **19** 319

Mustard R, Cosolo A, Fisher J, Pike T, Shouten D and Swanson H 1990 PC-based system for collection and analysis of physiological data *Computers in Biology and Medicine* **20** 2

Pike T and Mustard R 1992 Automatic recognition of corrupted arterial waveforms using neural network techniques *Computer in Biology and Medicine* **22** 3

Rumelhart D E, Hinton G E and Williams R J 1986 Learning internal representations by error propagation *Parallel Distributed Processing: Explorations in the Microstructures of Cognition* vol 1 ed D E Rumelhart and J L McClelland (Cambridge, MA: MIT Press) pp 318–62

Rumelhart D E and McClelland J L (eds) 1986 *Parallel Distributed Processing: Exploration in the Microstructures of Cognition* (Cambridge, MA: MIT Press)

G6

Economics, Finance, and Business

Contents

G6.1 Application of self-organizing maps to the analysis of economic situations

F Blayo

Abstract

The Kohonen map is a reduction dimension method which can be used for representation of high-dimensional problems. In this case study, we use the Kohonen map for the analysis of economic situations, and we make a comparison with a classical data analysis method: principal component analysis.

G6.1.1 Project overview

Simple observation of a phenomenon cannot be compared to knowledge of a phenomenon, for example, the deep understanding of the relationships (structure, causality, and such) between all the elements involved. The observations provide a set of data, which constitute an image of the phenomenon. The analysis of this image, and its synthesis, constructs our understanding of the phenomenon, transforming the pure data into information. This transformation cannot be easily achieved on multiple and complex data. It requires a reduction of the complexity, using suitable techniques. The different techniques (linear regression, canonic analysis, discriminant analysis) constitute the field of data analysis. Typically, economic data involving a large number of high-dimensional samples are quite difficult to represent and require expertise to extract relevant information to be given to managers. Classical data analysis methods are very efficient in many cases, but generally apply a linear transformation on original data. In this project, we have tried to use a neural method to perform an extraction of characteristics from a set of economic data in order to discover possible relationships between countries described by six economic values. We have also performed a comparison with the principal component analysis (PCA) method which is one of the most classical dimension reduction methods.

G6.1.2 Design process

The most important reason to apply a neuronal solution is the possibility of easily performing a nonlinear dimension reduction on available data. The *self-organizing map*, proposed by Kohonen (1982), is able to perform such a dimension reduction, providing a possible bi-dimensional representation of high-dimensional data. For this application, we have chosen to use an 8×8 map, trained with a finite set of 52 examples. The original version of the algorithm has been used, without any improvement. The design process started with the development of the algorithm, in a classical C language. The input/output code was only developed for the purpose of visualization. The general architecture of the network is given in figure G6.1.1.

C2.1.1

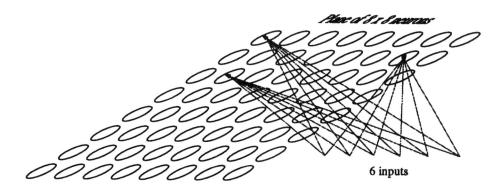

Figure G6.1.1. General architecture of the network. Only three neuron connections are shown.

G6.1.3 Training method

The self-organization algorithm performs a projection onto a subspace spanned by a discrete lattice of formal neurons. The map establishes a correspondence between input data and the neurons of the lattice, such that the topological relationships among the inputs are reflected as faithfully as possible in the arrangement of the corresponding neurons of the lattice. This provides a nonlinearly flattened and bi-dimensional version of the input space (Ritter 1988). The algorithms consist of two steps: for an input vector x, find the neuron whose activity is maximum. Then, in a defined subset of neurons around this maximum $\mathcal{V}(i, l(t))$, the weight vectors are moved in the direction of the input vector x according to the equation

$$
\begin{aligned}
W_i(t+1) &= W_i(t) + \alpha(t)(x(t) - W_i(t)) \qquad i \in \mathcal{V}(i, l(t)) \\
W_i(t+1) &= W_i(t) \qquad i \notin \mathcal{V}(i, l(t)).
\end{aligned}
\tag{G6.1.1}
$$

In equation (G6.1.1), the function $l(t)$ controls the width of the neighborhood, and $\alpha(t)$ controls the amplitude of the weight modification. These two functions are decreasing over the time t. Numerous iterations of these steps build an organized network, where the weights are ordered and quantify the input space. After the convergence of the algorithm, each country, represented by a vector $x(k)$, is presented to the network, and the unit whose activity is maximum is labeled with the name of the country. A two-dimensional representation is obtained, where relationships built by this data analysis appear clearly.

G6.1.4 The training set

The data of the training set are values of economic variables which characterize the state of 52 countries for the year 1984. The gross internal product of a country (GIP) per inhabitant concerns the countries with a planned economy. The infant mortality rate is the number of infants who died before the age of one year, compared to the number of living infants born during the year. The illiteracy ratio is the fraction of illiterate people older than 15 years, except for some countries for which the ratio is estimated compared to people older than 10 years. The school attendance index is the ratio of the education registration for people between 11 and 17 years old. In the example chosen, 52 countries are taken into account. Each one is represented by a six component vector $x = [x_1, x_2, x_3, x_4, x_5, x_6]$. The components represent, respectively, the annual economic growth, the infant mortality, the illiteracy ratio, the school attendance index, the GIP, and the GIP annual increase. Typical vectors are shown in table G6.1.1.

The entire set of data is then represented by a matrix, with 52 lines and 6 columns. The learning process is done on a square network, with 8×8 neurons.

G6.1.5 Preprocessing

For this application, the input data are normalized. This means that, for each variable, the mean value is
B4.4 equal to zero, and the variance is equal to one. This *preprocessing* is important to give an equal importance to the variables. It is also necessary for a correct comparison with the normalized principal component analysis.

Table G6.1.1. Some values associated with each country. All the values are percentages (%) except the gross internal product (GIP).

Country	Annual increase	Infant mortality	Illiteracy ratio	School attendance	GIP	Annual GIP increase
Canada	1.0	1.0	0.9	93.0	9857	3.0
France	0.4	0.9	1.2	86.0	11326	0.5
Mali	2.8	15.2	86.5	16.7	0190	1.5
South Africa	2.9	8.9	50.0	19.0	2690	−2.9

G6.1.6 Output interpretation

After the adaptation phase, the weights are fixed. Each six-dimensional example, among the 52 available, is presented to the network and the winning unit is labeled with the name of the corresponding country. After presentation of all the 52 examples, a map is obtained (see figure G6.1.2). It reflects a certain order which is representative of the similarities and the differences between the countries.

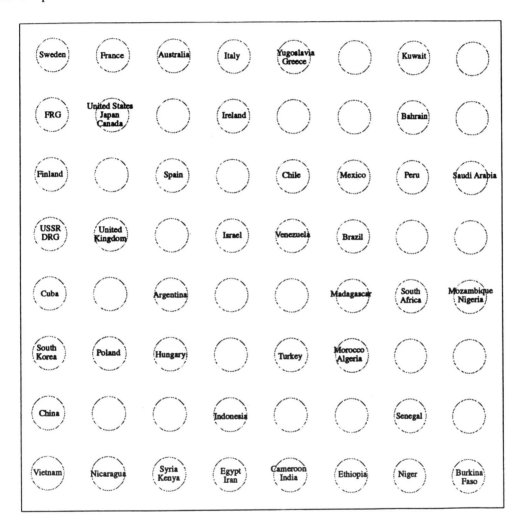

Figure G6.1.2. Organization of an 8 × 8 map of neurons, with the six-dimensional examples.

As we can see in figure G6.1.2, the countries are clustered in a way that emphasizes the socio-economic similarities and differences. Opposite regions correspond to countries with strongly different economic situations. The main clusters correspond to the most industrialized countries, the oil producing countries, the former communist countries, the South American countries and the African countries. Generally speaking, it is easy to see in this example that economic information is extracted from the direct measures. But it is also important to make an economic interpretation of this representation. This task is relevant for an expert in economics whose main task is to define the meaning of the two axes.

G6.1.7 Comparison with traditional methods

The principal component analysis (PCA) method is designed to draw high-dimensional vectors in a lower subspace (generally two dimensions). The main constraint in obtaining the representation is to maintain as much information as possible in the transformation. The transformation performed by the Kohonen algorithm is strongly similar, but essential differences exist and can be valuable in some applications (Blayo 1991).

First of all, the projection obtained with the PCA is continuous, as shown in figure G6.1.3. This is not the case with the Kohonen algorithm. Only discrete locations of the countries are available because there is only a discrete lattice of neurons (8 × 8 in this example).

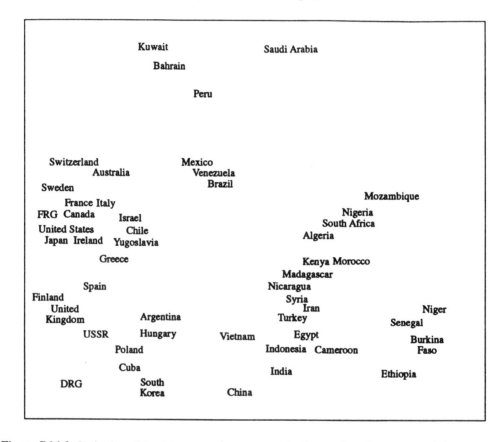

Figure G6.1.3. Projection of the data onto a plane spanned by the two first eigenvectors of the covariance matrix.

The PCA is a linear method. It performs an orthogonal projection on a plane spanned by eigenvectors of the covariance matrix. As we can see in figure G6.1.3, the PCA makes clusters which are significant for economists. The distances between countries, and also between clusters, can contain some information on the relative distances of the countries in the six-dimensional space. This is not always true, and adequate statistical tests can confirm the representation obtained.

The Kohonen algorithm realizes a projection on a surface spanned by the network topology. This is completely defined by the relation of order between the neurons: a simple one-dimensional relation or a bi-dimensional one, square or hexagonal. Other relations (three or higher) can be considered but they are not necessarily useful for representation purposes.

From a computational point of view, the PCA requires the inversion of a covariance matrix. This is a global operation, which can be very costly when applied to large matrices. The Kohonen algorithm is sequential, and does not require any global information. Only local modification occurs, and after a small number of iterations, the global order between data appears in the map.

G6.1.8 Conclusion

From a statistical point of view, the method of self-organizing maps is an original dimension reduction method. It has no real statistical analog, and thus can be very useful for specific applications where a linear method fails. Current research is developing in this direction (Demartines and Hérault 1993), and will reinforce the cross-fertilization of the statistical and neural network fields.

References

Blayo F 1991 Data analysis: how to compare Kohonen neural network to other techniques *Artificial Neural Networks* (*Lecture Notes in Computer Science 540*) ed A Prieto (Berlin: Springer)

Demartines P and Hérault J 1993 Representation of non-linear data structures through a fast VQP neural network *Proc. Int. Conf. NeuroNîmes93* pp 411–24

Kohonen T 1988 *Self-Organization and Associative Memory* 2nd edn ed T S Huang and M R Schroeder (Berlin: Springer)

Ritter H 1988 Kohonen's self-organizing maps: exploring their computational capabilities *Proc. Int. Conf. on Neural Networks* **1** 109–16

G6.2 Forecasting customer response with neural networks

David Bounds and Duncan Ross

Abstract

This case study looks at the application of neural computing to commercial problems. It highlights an area where neural computing has been shown to provide a direct commercial advantage for a company, and indicates why neural networks were the preferred approach. Elements of preprocessing, preparatory work, and network design are discussed.

G6.2.1 Introduction

Since the early 1980s, developments in neural computing have given neural networks the capability to solve complex 'real world' problems. However, it is only more recently that the benefits of neural computing have been applied to commercial and business areas.

This is perhaps surprising given the amount of money and effort that has been put into information technology systems in the last decade, often for relatively small returns. An article in the *Financial Times* made the dangers of this position very clear: 'US service companies spent at least $750 billion on communication systems, computer hardware and software during the 1980s. During this period, their annual productivity gain was a mere 0.7%' (*Financial Times* 1994).

The existence of large corporate databases, built up over this period, provides an ideal opportunity for using neural networks to gain a business benefit, and many companies are now routinely using the technology.

The application of neural computing to corporate data analysis has been given further impetus in the United Kingdom through the Department of Trade and Industry's *Neural Computing: Learning Solutions* program. This two-year-long, 5.7 million campaign drew to a close in 1995, and has enabled many UK companies to use neural computing solutions within their businesses. One of the cornerstones of the campaign has been the work done by six applications demonstrator clubs, each of which has produced examples of practical ways in which neural computing can give an organization a business advantage. Recognition Systems has run one of these clubs, the NeuroData Club, in conjunction with Logica plc. NeuroData has successfully investigated the application of neural computing to corporate data analysis— creating applications in the fields of customer response, database completion, and sales forecasting.

This case study looks at the application of neural networks to customer response and customer targeting—predicting whether a customer will respond to a particular offer, and providing a strategy for maximizing the profit from a customer database. Customer targeting is of great importance to companies. As the amount of money spent on direct marketing continues to rise, the potential savings produced by accurate customer targeting also grow. This article shows that neural networks provide a considerable benefit over conventional approaches for this type of problem.

Although this case study looks in detail at a direct marketing problem, it is worth emphasizing that the benefits of neural computing have been successfully demonstrated in many fields—their success is due to their being an efficient and flexible approach to model building, not to a particular type of problem.

G6.2.2 Project overview

A typical response rate for a mailing campaign, where an offer is posted directly to a potential customer, is in the region of one to two per cent. If people who want to respond to the offer can be identified in advance, then people who will not respond can be removed from the mailing. This has a double benefit: the cost of the campaign is reduced, and people who are not interested in this product (but who *may* be interested in other products from the company) are not annoyed by junk mail.

The problem described in this case study was provided by a company that had a large customer database and approached its customers by direct mail with offers for new products. As a result, they had built up a history of individual customers and how each had responded to certain campaigns. Associated with each customer were a range of parameters, examples of which are listed in table G6.2.1. In addition, each customer had a parameter that indicated whether they had been a respondent or a nonrespondent to a previous campaign.

Table G6.2.1.

Parameter	Description
AGE	The customer's age
PREMIUM	The premium paid by the customer
MOSAIC	Geodemographic classification
SEX	The customer's sex
TV_REGN	The customer's TV region

G6.2.2.1 Visualization

The complexity of the problem can be demonstrated by visualizing the data. This has often proved to be a useful first step in determining the best approach for a particular problem.

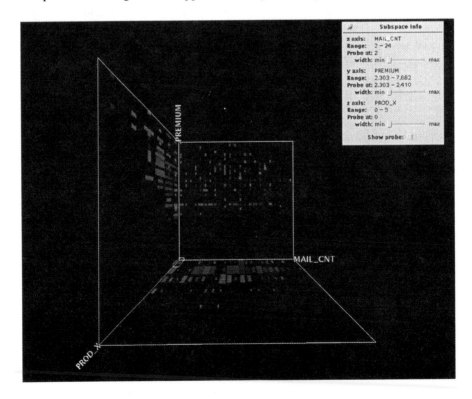

Figure G6.2.1. The 3D+ tool.

'The purpose of computing is insight, not numbers', wrote Richard Hamming (1962). Visualization is concerned with exploring data and information in such a way as to gain understanding and insight into the data. The goal of visualization is to promote a deeper level of understanding of the data under investigation and to generate new insight into the underlying processes.

Visualization is inherently application-dependent and many techniques only make sense within a particular context. An important point to note is that data that are fed into a visualization tool are typically sampled from some underlying phenomenon. It is this underlying phenomenon that we are aiming to visualize and hence understand—not the data themselves. This distinction is fundamental.

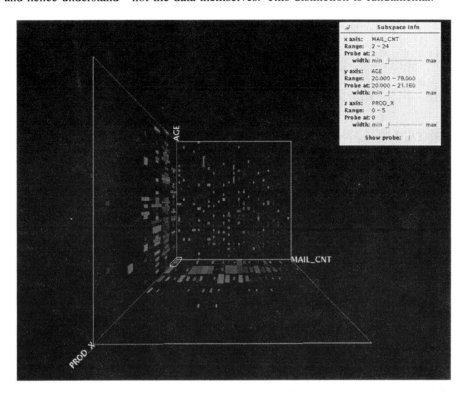

Figure G6.2.2. The 3D+ tool showing the distribution of age.

The initial visualization for this study was done using a tool known as the 3D+ tool. This allows input data distributions to be examined for any three input variables at a time. The data distribution across additional input variables can also be examined for subranges of the first three inputs. Such functionality allows the user to become more familiar with trends and clusters in their data. In addition, the response of the customer can be overlaid on the tool display as a color map. This allows conclusions to be drawn about the nature of the data. The use of the 3D+ tool has been described more fully elsewhere (Bounds and Barrett 1995).

A three-dimensional room (two walls and a floor) is displayed on the screen. Three input variables are displayed on the x, y and z axes with the response being overlaid as color on each data point. Each individual wall or floor has two variables plotted on it as a scatter plot. The variables that are chosen to be displayed on the graph are chosen automatically according to the amount of variance they contain. The first three variables that are used are those that contain the most variance (figure G6.2.1).

This tool is referred to as 3D+, as more than three dimensions can be explored. This is done by the user drawing a cube or probe around an area of points and inspecting them further in relation to the next three most variant variables, thus establishing which variables affect the response rate the most. For each subspace, a window containing subspace information is displayed, which gives the names of the variables being displayed, the axis on which each variable is displayed and the ranges of each variable. The window also contains the ranges that the probe is currently covering.

Rules can be generated from these tools by inspecting the distribution of the data points on the walls and floor and by looking at the colors with which the points have been overlaid. One rule that can be formulated is that the older the clients are, the fewer times they have been sent mail, and the more policies of type x they hold, then the more likely they are to respond.

Other useful information can also be extracted from these plots. For example, from the plot in figure G6.2.2 it can be seen that there is an even distribution of age: clients of all ages exist within the database; it is not biased towards any age range.

The results from the data exploration allow new insight into the database, the data distribution and the nature of the customers currently on file. They even allow simple rules to be derived that may improve the ability to target customers more accurately, for example:

(i)   The older a customer is the more likely they are to respond.

(ii) The higher the premium they are paying, or the richer they are, the more likely they are to respond.

(iii) The more policies of type x they hold, and in general the more policies of any type they hold, the more likely they are to respond.

(iv) The fewer times the customer has been sent mail, the more likely they are to respond.

However, these rules demonstrate one of the limitations that conventional approaches to targeting customers suffer from, and that neural computing can overcome—they are a linear approximation of a nonlinear problem. As can be seen from the 3D+ plots the problem is highly non-linear, and a non-linear technique is required to gain the maximum benefit from these data.

G6.2.3   Preparatory work

Before a neural solution can successfully be deployed it is essential that there is a large enough pool of historical data relating to the problem of training (model building) and testing data sets to be constructed. It is also vital that the data have been prepared in such a way that the neural networks can make maximum use of the information that they contain. Data fields that contained unreliable information and those that were set entirely to one value were removed from the database before model building began. Fields were examined for inconsistencies, and records that were thought to contain too many errors were dropped.

B4.4   Another *preprocessing* technique that can improve the results produced by modeling the problem is that of weighting input data fields. By applying specific functions to individual fields data distributions can be made more uniform, and extreme values can be limited. The data available then had to be split sensibly between a training set (used to build the model) and a test set that would be used to verify the success of the model.

G6.2.4   Neural network design

C1.6.2   The neural network model chosen for this application was the *radial basis function network*. The radial basis
C1.2   function is a supervised neural network that differs from the more commonly used *multilayer perceptron* in that it will produce a solution in one pass of the data, rather than through an iterative process.

Radial basis functions build classifications from ellipses and hyperellipses that partition the input data space. These hyperellipses are defined by radial functions ϕ of the type

$$\phi(||x - y||)$$

where $||\cdots||$ is a distance measure between an input pattern x and a center y that is positioned in the input data space. These centers are defined by the weights associated with the inputs to the nodes in the hidden layer of the radial basis function.

The function f in k-dimensional space that partitions the space is composed of elements f_k, where

$$f_k = \sum_{j=1}^{m} \lambda_{jk} \phi(||x - y_j||).$$

Since this equation requires only the solution for the linear coefficients λ, the technique is rapid, and requires only one pass of the data to produce a solution.

G6.2.4.1 Training data

A frequently used rule of thumb is that the number of records needed to train a system is equal to at least ten times the number of weights in the system. For this application, after data encoding there were approximately 50 input fields for the model. Taking the above rule of thumb this means that approximately 5000 records would be required to train the model.

However, it is important that the training data set has an equal number of respondents and nonrespondents in it. If the set is biased towards nonrespondents it will weaken the model's ability to identify respondents correctly. Unfortunately, this is the case in the real world where the response rate of direct marketing campaigns is about 1%. So if 2500 respondents are required for the training set, then we need to have historical data for approximately 250 000 customers to be able to build a model successfully. It is rare for this number of data to be available, and so other techniques need to be adopted in order to reduce the number of training data required.

One approach which has been found to be effective in reducing the quantity of training data needed is partitioning tasks. To do this, the problem is broken into a number of smaller subproblems, each of which uses only a subset of the input fields to learn the problem. In this problem the input data were split into three subproblems, each of which related a group of input features with the customer's likelihood to respond to a campaign. The groups of input features were:

- personal information
- policy information
- mailing history information.

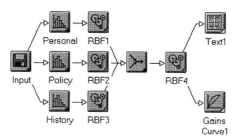

Figure G6.2.3. The topology.

Neural models were built to learn the correlations between these groups and the likelihood of responding. To prevent correlations between these groups from being lost, the outputs from these models are used as the input to another neural network (figure G6.2.3). It is the output from this final network which gives the likelihood that any given customer will respond to a mailing. In the figure, the arrows indicate the flow of data, the icons personal, policy, and history allow data fields outside these groups to be blocked, and the icons RBF1 to RBF4 are the radial basis functions that model the problem.

G6.2.5 Outputs from the neural networks

When the neural networks had been trained, and experimentation had been undertaken to find the best parameters for the neural models used, the test data were passed through the application to establish the success of the models. The test data set had not been seen during model building, but contained customer records where the outcome of a previous campaign (whether or not the customer had responded) was known. When each customer record was passed through the application a value ranging between 0 and 1 was produced, indicating the probability of that customer responding to the mailing. The records in the test data set were then ranked, with the most likely to respond placed first, and the least likely to respond placed last.

The most important measure of success for the company involved is the financial advantage that can be gained from using a neural network model. To evaluate this a gains chart was used and linked to a simple cost-benefit analysis.

G6.2.5.1 Gains charts

The gains curve is useful because it enables a direct calculation of the impact that an application will have if used in a direct marketing campaign. By comparing the results produced by a model to those that would be achieved if customers were mailed at random, it is possible to evaluate the improvements that have been achieved. The test data are scored using the predicted outputs from the application. The scored data are then ranked and compared with the targets. A plot is then produced which shows the proportion of the total database against the number of customers that would have responded if that percentage of the database had been mailed (figure G6.2.4).

If the customers were mailed by random selection from the database, rather than using the model, the equivalent plot would be a straight diagonal line. The difference between the two lines if a certain proportion of the database is mailed is the *gain*. The gains curve shown in figure G6.2.4 clearly demonstrates the benefits that the model provides. However, the company was also interested in how this related to financial benefits. This was done using a cost-benefit analysis (figure G6.2.5).

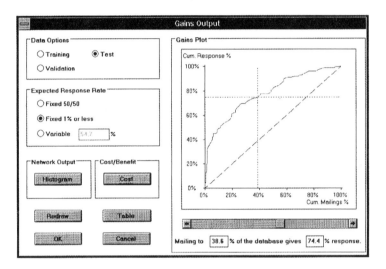

Figure G6.2.4. A gains curve.

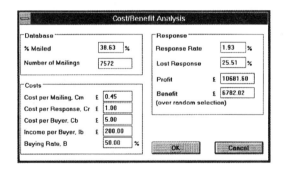

Figure G6.2.5. A typical cost-benefit analysis.

The cost-benefit analysis simply calculates the number of respondents that would be achieved by mailing a proportion of the ranked database, and associates this with the costs of sending each piece of mail. By factoring in the benefit to the company for each purchase, the costs of handling each response and each purchase, and the known buying rate for respondents, a net benefit can be calculated for running the campaign. This is inevitably only an approximation, but has nevertheless proved effective as a means of quickly establishing the relative value of different strategies.

From analysis of the gains curve and cost-benefit analysis a strategy can be chosen for a forthcoming mailing campaign. Two potential strategies that are often deployed are *conservative* and *cherry picking* strategies. In a conservative strategy the aim is to reduce the size of a mail shot, while retaining almost all of the respondents. For example, you could mail 80% of your database hoping to reach 95% of the actual respondents—thus saving one fifth of your mailing costs. In a cherry picking strategy the aim is to send offers only to those few customers who have a significant likelihood of responding to the offer. In this scenario the size of the mail shot is reduced dramatically, but at the cost of losing more of the potential respondents. The strategy used in each case depends on the success of the model and the financial requirements of the company.

G6.2.6 Conclusions

Neural computing has been shown to be a successful way of increasing the value of corporate data. By applying the unique pattern learning capabilities of neural networks a significant advantage can be gained when compared to previous methods of data analysis. Although this article deals with one particular use of the technology, many other applications have been successfully developed in business areas that include:

- predicting customer attrition in the insurance industry
- database completion
- segmentation of databases
- demand forecasting
- fraud detection.

Those companies and organizations that are willing to make use of neural computing are gaining a significant advantage over their competitors throughout the world.

References

Bounds D and Barrett P 1995 Neural networks and data visualization *Neural Networks* ed J G Taylor (Oxford: Waller)
Financial Times 19 July 1994
Hamming R W 1962 *Numerical Methods for Scientists and Engineers* (New York: McGraw-Hill)

G6.3 Neural networks for financial applications

Magali E Azema-Barac and A N Refenes

Abstract

Modeling of financial systems using neural network techniques has attracted a great
deal of attention in the past few years. Neural networks, because of their inductive
nature, can infer complex nonlinear relationships between input and output variables,
and thus bypass the step of theory formulation. This paper reviews the state of the art in
financial modeling using neural networks and describes applications in key areas, such
as foreign exchange and fixed income. It shows that with careful network design, the
backpropagation learning procedure is an effective way of training neural networks for
time-series prediction.

G6.3.1 Introduction

Modeling and prediction of financial systems has traditionally attracted a lot of attention. The basic
methodology has been statistical, enabling a limited number of determinants of any given asset price to be
analyzed at the same time (Ross and Ross 1990, Peters 1991). Because of their inductive nature, neural
networks can infer complex nonlinear relationships between input and output variables. Neural networks
have thus been applied to a number of financial applications and have demonstrated better performance
than conventional approaches (Hoptroff 1993, Diamond *et al* 1993).

In this paper, we review financial modeling using neural networks and describe applications in two
key financial areas: foreign exchange and fixed income. The foreign exchange application deals with
univariate time-series prediction. The bond application deals with multivariate time series.

G6.3.2 Finance and neural networks

G6.3.2.1 Modeling financial systems using neural networks

The development of systems for modeling and predicting financial indicators has traditionally received a
great deal of attention, but success in both long-term and short-term forecasting has been somewhat limited
(Burns 1986). Three main reasons can be identified. Firstly, classical statistical techniques have been used,
and these techniques enable only a limited number of determinants of any given asset price to be analyzed
at the same time. The financial markets, however, operate on a large number of factors at any one time.
Secondly, the relationship between an asset price and its determinants changes over time. These changes
can be abrupt: for example, in the currency markets a rise in interest rate can strengthen a currency one
month and weaken the same currency the next month. Neural networks can, in principle, deal with the
problem of structural instability. Thirdly, many of the rules which govern asset price are qualitative or at
best fuzzy, requiring judgement and hence by definition are not susceptible to purely quantitative analysis.

Because of their inductive nature, dynamical systems such as neural networks can infer complex
nonlinear relationships between input and output variables. For example, neural networks can be used to
determine the structural relationship between a given asset (e.g. bond price) and potential determinants
(e.g. government interest rate, inflation).

Typical applications of neural networks in financial modeling include time-series prediction; for
example, forecasting foreign exchange rates and classification such as stock ranking (Refenes *et al* 1992).

The development of successful applications of neural networks in finance involves two areas: financial engineering and neural network engineering.

Knowledge of the financial application is required in order to achieve good generalization performance. For example, when analyzing the structural relationship between an asset and its determinants, one needs to know the potential economic variables and/or indicators, their significance and correlation. In the case of time-series predictions, it is necessary to be aware of the econometric methods for preprocessing and normalizing data sets.

Concerning neural networks, awareness of the interrelations between 'network engineering' parameters and network performance metrics is essential for successful application development. The next section outlines neural network performance (e.g. convergence, generalization and stability) and control (e.g. activation function, cost function) parameters.

G6.3.2.2 Backpropagation performance and control parameters

This section outlines the performance and control parameters associated with the design and use of *backpropagation neural networks*. The reader is assumed to be familiar with the backpropagation algorithm.

C1.2.3

The backpropagation neural network is generally believed to be an effective learning procedure when the mapping from input to output contains both regularities and exceptions (LeCun 1989) and it is, in principle, capable of solving virtually any nonlinear classification problem. There are three main problems and thus metrics to evaluate the performance of a backpropagation network in nontrivial applications:

(i) *Convergence* concerns the learning process, and whether or not this process is capable of learning the classification defined in the data set, under what conditions it does so, and what are the computational requirements for convergence. Fixed-topology networks prove convergence by showing that in the limit, as training time tends to infinity, the error minimized by the gradient descent method will tend to zero.

B3.5 (ii) *Generalization* measures the ability of a network to recognize patterns outside the training set. Frequently, an analogy is made between learning and curve fitting. There are two problems in curve fitting: finding the *order* of the polynomial, and finding the *coefficients* of the polynomial (once the order has been established). For example, given a certain data set on a second order polynomial, $ax^2 + bx + c$, the values for a, b, c are computed normally by minimizing the sum of the squared differences between required and predicted $f(x_i)$ for x_i in the training set. Once both the order and the coefficients have been computed, the value of $f(x_i)$ can be calculated for any x_i including those not present in the training data set. Choosing orders lower than is appropriate leads to a poor approximation even for the points in the data set.

On the other hand, choosing a higher order implies fitting a high-degree polynomial to the low-order data. Furthermore, in practice the high-order terms do not end up with a zero coefficient. Typically, this leads to a perfect fit for the points in the data set but very bad $f(x_i)$ values for x_i out of the training data, i.e. the system generalizes poorly.

By analogy, a backpropagation network with a structure (network topology and layer size) simpler than necessary cannot give good approximations even to patterns in the training set. On the other hand, a network with a structure more complicated than necessary *overfits*, that is, it gives a good fit for the training set but performs poorly on unseen patterns.

(iii) *Stability* concerns the consistency of the results produced by neural networks when varying the values of the parameters that influence their performance. Neural networks are known to produce wide variations in their predictive features (Gorman and Sejnowski 1988). That is, small changes in network design, learning times, initial conditions, and so on might produce large changes in network behavior. For the types of application considered here, it is important to identify intervals of values for these parameters which give statistically stable results, and to demonstrate that these results persist across various training and test sets.

Controlling and thus improving the performances of a neural network is done using four main control parameters.

B3.2.4 (i) *Activation function.* This parameter controls the choice of activation function for the neuron/unit. The activation function is nonlinear, such as a hard limiter, or a sigmoid. The simple hard limiter functions produce values of either 0 or 1 depending on whether the total input of a unit exceeds a certain threshold value. Sigmoid functions are the most widely used (Refenes and Alipi 1991) in all types of learning. They are more complex and differentiable. There are two types of sigmoid functions: *asymmetric* and *symmetric* (e.g. scaled hyperbolic tangent). It has been shown that symmetric sigmoid functions are capable of improving the speed of convergence over the commonly used asymmetric sigmoid (Refenes and Alipi 1991).

(ii) *Cost function.* The choice of the cost function is believed to play an important role in determining the convergence and generalization characteristics of neural networks (Hinton 1987). The most commonly used function is the family of quadratics, e.g. the least-mean-square error. Several researchers suggested changing the cost function from the quadratic measure (e.g. Fahlman and Lebiere 1990), but the exact relationship between cost function and performance measures is somewhat undefined and is currently the subject of intensive research. In the applications described in the next section it is the standard quadratic cost function that is used.

(iii) *Network architecture.* The architecture of a neural network is determined by the topology of the units and the connections between them. The network's topology is the main parameter controlling the generalization capability. Theoretical studies (Denker 1987) have shown that the likelihood of correct generalization depends on the size of the hypothesis space (i.e. the total number of architectures considered), the size of the solution space (i.e. the number of architectures producing good generalization) and the number of training examples. In our applications, multiple architectures are tested.

(iv) *Gradient descent/ascent.* The most important parameter for controlling the gradient descent is λ, the learning rate. Several researchers have experimented with additional parameters, such as the momentum terms, second derivative, etc, but the learning rate is the parameter controlling both the *speed of convergence* and the *stability*. In principle, there are two approaches to adjusting the learning rate. The simplest one is to use the same learning rate for the whole network and thus experiment to optimize both convergence speed and stability. The second one is to use one learning rate for each weight, and thus use a heuristic rule to adapt each learning rate (Refenes and Azema-Barac 1993). In our applications we use one learning rate for the whole set of connections. We also use a momentum term (Fahlman and Lebiere 1990).

G6.3.3 Neural networks applied to foreign exchange markets

G6.3.3.1 Application environment

Univariate time-series prediction is a core component of many financial modeling systems (Denker 1987). The system described here is designed and trained to predict the exchange rate between the US$ and the DM. Non-model-based techniques such as neural networks rely heavily on the identification of strong empirical regularities in a system which is often contaminated by noise. A common method for identifying such regularities is *windowing* (Refenes and Azema-Barac 1993). That is, two windows W^i and W^0 of fixed sizes n and m are used to analyze the data set. For a given window size the assumption is that the sequence of values $W_0^i, \ldots, W_n^i$ is somehow related to the following sequence: $W_0^o, \ldots, W_n^o$, and that this relationship, although unknown, is defined entirely within the data set. Various methods can then be used to correlate the two sets of values. In the case of neural networks $W^i \rightarrow W^o$ is used as a training vector. Both windows are shifted along the time series using a fixed step size s. The choice of window and step sizes is critical to the ability of any prediction system to identify regularities and thus approximate the hidden relationship accurately. For our simulations, the parameter values were $n = 12$, $m = 1$ and $s = 4$ (Refenes and Azema-Barac 1993).

G6.3.3.2 Neural network system

The architecture of the neural network system at the input/output level is determined by the application sizes, n and m described above. The internal topology of the network is more difficult to determine *a priori*. In this application, a single hidden-layer fully connected backpropagation network was used. This type of network was used principally because of its proven capability in various fields (Gorman and Sejnowski 1988). The neural networks used in this application correspond to a $(12, x, 1)$ fully connected backpropagation network with a learning rate λ set at 0.6 and a momentum term equal to 0.25. A large number of experiments have been done while varying the number of hidden units, x, in order to achieve *stability*.

G6.3.3.3 Training and test sets

The training and test sets consist of currency exchange data for the period 1988–9, with on-hourly updates for a complete year, that is, 260 trading days. The first 200 items of the data set were used for training, while the remaining 60 data items were used for testing the network performance, and, in particular, its generalization capability. With the windowing mechanism described earlier, the resulting training test consists of overlapping snapshots of the time series, each of 12 hours length, moving along the curve at an interval of four hours. The overall size of the training set is therefore equal to 8236 training vectors. This overall size allowed us to conduct extensive tests on learning speed and generalization performance (Refenes and Azema-Barac 1993).

Furthermore, in order to thoroughly test the generalization performance of the network, two types of forecasting were tested: *single-step* and *multi-step* prediction.

- *Multi-step prediction* allows the network to be tested for long-term forecasting which aims to identify general trends and major turning points in a currency exchange rate. In a multi-step prediction, the neural network uses a set of current values to predict the value of the exchange rate over a fixed period, i.e. the prediction at time t is fed back to the network for forecasting at time $(t + 1)$.

- *Single-step prediction* allows the network to predict the exchange rate only one step ahead of time. This serves two purposes. Firstly, it is a good mechanism for evaluating the adaptability and robustness of the prediction system by showing that even when its prediction is wrong, it is not dramatically wrong and the network can use the actual value to correct itself for the next single-step prediction. Secondly, it can act as an alarm generator and would allow traders to buy or sell in advance of a price increase or decrease, respectively.

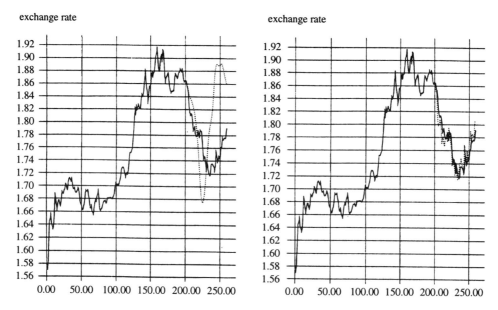

Figure G6.3.1. (*a*) *Multi-step prediction*—the full curve shows the whole time series while the dotted curve shows the forecasted exchange rate produced by the neural network from days 200 to 260 and using only forecasted values. (*b*) *Single-step prediction*—the full curve shows the whole time series while the dotted curve shows the forecasted exchange rate produced by the neural network for days 200 to 260.

G6.3.3.4 Results

As shown in figure G6.3.1(*a*), the results for the multi-step prediction for the general trend in the exchange rate is very accurate. The network predicted a sharp fall and then a rise in the exchange rate. For the first 30 days, it is very accurate both in terms of trends, and also in terms of absolute values. The network predicted a turning point at approximately the time it took place, and estimated quite accurately the pace of the recovery.

Figure G6.3.1(*b*) displays the result for the single-step prediction in which the input values are the values of the observed time series. The prediction is quite accurate in that it follows the actual prices closely. When the network makes a mistake with respect to predicting a turning point, it is still capable of adjusting itself as soon as the actual price is made available at the next step. This type of performance is often cited as indicative of robustness; it is, however, of little use in practical terms.

G6.3.4 Neural networks applied to the bond markets

This application deals with the prediction of a set of bond returns on a month-to-month basis. Each variable in the system is, in fact, a (lead) time series which is treated in the same way as the time series in the previous application, i.e. using windowing.

G6.3.4.1 Application environment

The aim of this application is to perform quantitative asset allocation between bond markets and US cash to achieve returns in dollars significantly in excess of any industry benchmark (e.g. the *JP Morgan* bond index). Assets are allocated in seven markets (United States, Japan, United Kingdom, Germany, Canada, France and Australia) chosen on the basis of capitalization.

Each market is modeled on an individual basis using local (e.g. interest rates) and global (e.g. oil prices) parameters. The system is developed in two stages. In the first stage, each local market is modeled with the aim of producing a local portfolio (i.e. local market and USA cash) which out-performs a standard local benchmark (50% in the market and 50% in cash). In the second stage the results for the individual markets are integrated in the global portfolio (seven markets).

G6.3.4.2 Neural network system

The architecture of the neural network and portfolio system are shown in figure G6.3.2. For each market the dollar-adjusted bond return is predicted one month ahead. The predicted returns are then passed through a portfolio management system which imposes constraints on the allocation to minimize the risk.

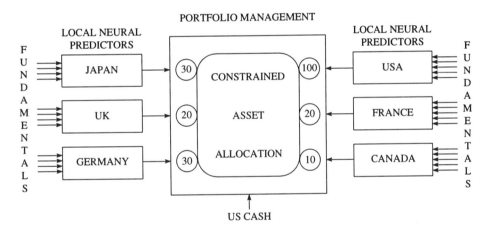

Figure G6.3.2. Neural network and portfolio system.

Each local neural network corresponds to a two-hidden-layer backpropagation network. Each uses past bond-return time series and also market-related parameters, for example, oil prices to predict the next month's bond return. Each local network typically uses between four and eight inputs†. It should be noted that the neural networks described here have been extensively validated with various values for network control parameters, and in particular network architecture. Stability of the results has been achieved for networks with two hidden layers and trained for 10 000 to 20 000 iterations.

G6.3.4.3 Training and test sets

The data are returns derived from government bond yields of the longest maturity for each market, using a fixed elasticity factor as given in table G6.3.1.

† These leading indicators are proprietary to Econstat Ltd.

Table G6.3.1. Elasticity factors.

	Maturity	Elasticity
USA	30	8.25
Japan	10	6.78
UK	15	6.49
Germany	10	5.90
Canada	10	7.12
France	10	6.61
Australia	10	5.26

The training data used for each local market are bond returns from 1974–1988, updated monthly. A typical training vector t_i has the following format:

$$t_i = V_0^0 \ldots V_5^0, \, V_0^1 \ldots V_5^1, \, \ldots, \, V_0^n \ldots V_5^n \to y \qquad (G6.3.1)$$

where 0 through n are fundamental or technical leading indicators, each containing up to six items denoting the rate of change of that indicator over the past six months. The right-hand side, y, is the target or test value, that is, the bond return at time $(t + 1)$. In this application and in addition to the training and test data sets, there is a cross-validation data set composed of 10% of the whole data set. These data are excluded from both the training data set and test data set. The cross-validation data set is used for stopping the training prematurely; this allows the network to avoid overfitting and leads to better generalization performance.

G6.3.4.4 Results

The results obtained by the neural-network-based system are compared to a global benchmark calculated according to the proportion of the global market capitalization represented by each market: United States 42%, Japan 23%, United Kingdom 13%, Germany 11%, Canada 3% France 4% and Australia 4% (this benchmark is not dissimilar from the *JP Morgan* index). When comparing the cumulative return of the neural-based system versus the global benchmark, the neural portfolio outperforms the benchmark by a factor of 3.6 (Diamond *et al* 1993). But more important is the consistency of the outperformance, that is, in fund management one is willing to trade off some outperformance for short-term consistency—the neural system never under-performs the benchmark for two consecutive months. Figure G6.3.3 displays the relative outperformance of the neural-network-based system versus the global benchmark and shows that the neural system has outperformed the benchmark consistently.

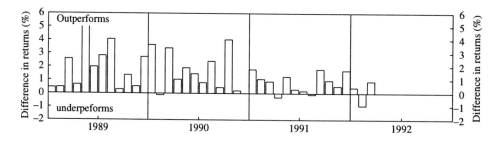

Figure G6.3.3. Neural portfolio relative outperformance versus global benchmark.

G6.3.5 Conclusion

We have reviewed financial modeling using neural networks and described two applications in key areas of forecasting, that is, foreign exchange and asset allocation. We have shown that simple neural learning procedures such as the backpropagation algorithm outperform traditional approaches. In foreign exchange, the backpropagation was applied to the prediction of univariate time series. The resulting neural network is able to predict the general trend and turning point. In the bond markets, backpropagation was applied to the prediction of multivariate time series. The resulting neural-based portfolio consistently outperforms a traditional benchmark in the field: the *JP Morgan* index.

References

Burns T 1986 The interpretation and use of economic predictions *Proc. R. Soc.* A 103–25

Denker J 1987 Large automatic learning. Rule extraction and generalization *Complex Systems* **1**

Diamond C, Shadbolt J, Azema-Barac M and Refenes A 1993 Neural network system for tactical asset allocation in the global bond market *IEEE 3rd Int. Conf. Neural Networks*

Fahlman S and Lebiere C 1990 The cascade correlation learning architecture *cmu-cs-90-100* Carnegie Mellon University

Gorman R and Sejnowski T P 1988 Analysis of hidden units in a layered network trained to classify sonar targets *Neural Networks* **1**

Hinton G 1987 *Connectionist Learning Procedures* Carnegie Mellon University

Hoptroff A 1993 The principles and practice of time series forecasting and business modeling using neural nets *Neural Comput. Appl.* 25–32

LeCun Y 1989 Generalization and network design strategies *Technical Report CRG-TR-89-4* University of Toronto

Peters E 1991 *Chaos and Order in the Capital Markets* (New York: Wiley)

Refenes A N, Zapranis A and Azema-Barac M E 1992 Stock ranking using neural networks *Proc. ICNN (San Fransisco, CA)*

Refenes A N and Alipi C 1991 Histological image understanding by error backpropagation *Microprocess. Microprog.* **32** 18–35

Refenes A N and Azema-Barac M 1993 Currency exchange rate prediction and neural network design strategies *Neural Comput. Appl.* 46–58

Ross R L and Ross F 1990 An empirical investigation of the arbitrage pricing theory *J. Finance* December

G6.4 Valuations of residential properties using a neural network

Gary Grudnitski

Abstract

With the advent of large computerized databases, computational techniques are being relied on more frequently to estimate residential property values. As an alternative to the most commonly used computational technique of multiple regression, this application describes how a neural network was applied to estimate the selling price of single-family residential properties in one area of a large California city. For the holdout sample of 100 properties, the average absolute difference between the actual selling price and the estimated selling price generated by the neural network was 9.48%. In terms of comparative accuracy, the network was able to achieve, on average, more accurate valuations of properties than the multiple regression model in the holdout sample. The network also produced more accurate valuations than the multiple regression model for 57 out of the 100 residential properties in the holdout sample.

G6.4.1 Design process

Accurate, economical and justifiable valuation of residential property is of great importance to mortgage holders who wish to value their portfolios, to prospective lenders who are contemplating the issuance of new mortgages, and to local government authorities who must know the worth of their tax base. As large computerized databases become increasingly more common, computational techniques, especially multiple regression, are being relied on more frequently to assess residential property values.

Residential property, like many other commodities, can be viewed as bundles of attributes. A problem in valuing residential property exists, however, because the prices of a property's individual components are both unobservable and devoid of an implicit market. Empirically, the choice of pricing equations that value a property's individual components often appears to be dictated by the nature of the available data and the tendency of those providing the estimates to fixate on 'goodness of fit' criteria. On one hand, this is understandable because pricing equations for residential property represent, in reduced form, an interaction between both supply and demand, and thus make the specification of an exact functional form difficult. On the other hand, however, housing price estimates that critically depend on the functional form chosen can be negatively impacted by this imprecision in the specification of pricing equations.

In an attempt to mitigate the negative effects on estimates of property values due to imprecision in the specification of the valuation equation, what follows is a description of how a standard *backpropagation* C1.2 *neural network* (Rumelhart and McLelland 1986) is applied to estimate the selling price of single-family residences. To measure the relative performance of the network, prices produced are compared to estimations generated by a multiple regression model.

Table G6.4.1. An example of data downloaded from the MLS describing a sales transaction.

```
PT1 SINGLE FAMILY-DETACHED 08/28/93 03:49 PM
LP: 189,000 STATUS: SOLD MT: 82 LD: 01/12/93 XD: 06/12/93 REF# 69
SP: 186,000 OLP: 189,000 FIN:  OMD: 05/27/93 LNO: 93 6000909
AD: 13131 OLD WEST DR ZIP: 92129 APN: 3151703700
MC: 30F3 XST: TED WMS PRWY COM: RP NCD: CRM YB: 1987 ZN: NONE
BR: 3 OPBR: BATH: 2.5 ESF: 1638 SSF: ASSES TRM:
LR : 11X17 FP : F PTO: SLAB HOF: 101 TLB: 0
DR : 10X10 TV : C EXT: STUCO HFP: MONTHLY TD1: 0
FAM: 12X14 R/O: R/O ELEC RF : CNSHK HFI: GTC IN1: 0.0 AS1:
KIT: 11X10 DW : DISHWASH SWR: SEWER OF : 0 LT1:
MBR: 13X20 MW : MICRO BI SPA: NONE OFP: NONE KNO TD2: 0
BR2: 11X10 TC : IRR: SPRINKLE TOF: NONE KNO IN2: 0.0 AS2:
BR3: 11X11 HT : FAG FLR: SLAB LDY: GAR LT2:
BR4: 0 WH : ALU: NONE KNO LSZ: 8500 AST: NONE KNOWN
BR5: 0 SEC: EQPT OWN GUEST: NONE ACS: 0.00 BF : NONE KNOWN
XRM: 0 VU : NK AGEREST: NONE LSF: 0 EQP: D,E,F,G,K
STY: 2 STO PL: YES CL: CFA PKG: 2G
REMARKS: THIS PLAN 3 CAMBRIDGE HAS IT ALL! MINT CONDITION WITH NEW BERBER CARPET
NEW WINDOW TREATMENTS, NEW FLOORING IN BATHROOMS SEC SYS, 2 PATIOS, PATIO COVER,
BUILT-IN GAS BRICK BBQ, SOFT WTR SYS, REFINISHED KITCHEN CABINETS, LANDSCAPED WITH
AUTO SPRINKLERS. SHOWS TERRIFIC! GATE CODE * 0289
```

G6.4.1.1 Description

Source data representing the sale of a residential property were obtained from the San Diego Board of Realtors' multiple listing service (MLS). For this application, data on single-family homes sold during 1992–93 in Rancho Penasquitos, a northern suburb of San Diego County, California, were electronically downloaded. A typical entry for one of these properties is shown in table G6.4.1.

From the downloaded MLS residential sales data, a parser, written in C, extracted the following nine descriptors for each property (these descriptors are shown in bold in table G6.4.1): SP is the actual selling price, YB is the age of the structure in years, derived by subtracting the year the house was built from 1992, BR is the number of bedrooms, BATH is the number of bathrooms in increments of 1/4 baths, ESF is estimated total square footage of the house, LSZ is the lot size measured in square feet, STY is the number of stories, PL/SPA indicates if a pool or spa existed (0 otherwise) and PKG is the number of car-garages.

For the sample, descriptive statistics for the continuous variables are presented in table G6.4.2. In addition, for the PL/SPA variable, 31% of the houses in the sample had either a pool or spa. Data from the parser were then passed to an Excel spreadsheet. Using the spreadsheet, each of the values of the variables was normalized according to equation (G6.4.1) and output to the neural network software.

$$i_{\text{norm}} = (i - \text{min})/\text{range} \tag{G6.4.1}$$

where i_{norm} is the vector of normalized values of the variable, i is the vector of original values of the variable, min is the minimum original value of the variable, and range is the range of the original values of the variable.

G6.4.1.2 Topology

The topology of the network to estimate the selling price of a house is depicted in figure G6.4.1. This standard backpropagation network consisted of an input layer of eight neurons, a hidden layer of N neurons, and an output layer of a single neuron. The eight neurons in the input layer of the network captured the attributes believed to determine a property's value. The single neuron in the output layer represented the network's determination of the selling price of a house. Values estimated by the network fell within a range of 0 to 1 to achieve comparability to the previously transformed (also according to equation (G6.4.1)) actual selling prices of these houses.

Table G6.4.2. Descriptive statistics of the continuous variables in the sample.

Variable abbreviation	Variable definition	Overall mean (std. deviation)	Minimum (maximum)
SP	Selling price ($)	214 112 (32 997)	150 000 (365 000)
YB	Age (yrs)	9.19 (5.32)	0 (22)
BR	Number of bedrooms	3.72 (0.67)	2 (6)
BATH	Number of bathrooms	2.51 (0.41)	2 (4)
ESF	Total square footage	1991 (413)	1100 (3009)
LSZ	Lot size (sq. ft.)	8246 (4464)	3746 (44 866)
STY	Number of stories	1.75 (0.43)	1 (2)
PKG	Number of car-garages	2.16 (0.37)	1 (3)

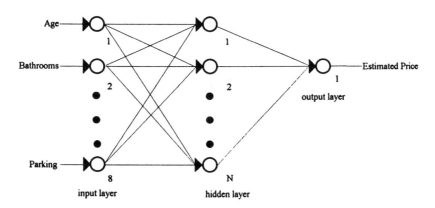

Figure G6.4.1. Topology of the neural network.

G6.4.2 Training methods

The data set was randomly divided into three subsets. The first subset of the data, made up of 119 properties, was used to train the network. The second subset of the data, called the training-test set, consisted of 30 properties. It was used to check the ability of the supposedly trained neural network to *generalize* (i.e. to prevent overtraining), and to select the optimal number of hidden-layer neurons (Masters B3.5 1993, p 183). The third subset of the data consisted of 100 properties, and was used to assess the ability of the network to estimate property values accurately.

The neural network software was written in C for a personal computer and is available as shareware from Roy W Dobbins (Eberhart and Dobbins 1990). The network was run on a 33MHz 486DX. With random starting weights between ±5.0, and a learning coefficient and momentum factor of 0.1 and 0.6, respectively, networks employing a logistic activation function and having from two to four neurons in their hidden layer were trained. Figure G6.4.2 graphs the average absolute error—(i.e. (estimated selling price − actual selling price)/actual selling price)—of the training set against the average absolute error of the training-test set for from two to four hidden-layer neurons at 2000, 4000, 6000, 8000 and 10 000 training iterations.

Figure G6.4.2 indicates for this training and training-test sample the superiority of a network with two neurons in its hidden layer. Specifically, contrast the plot of the error of the network with two neurons in its hidden layer to the plot of the error of the network with three neurons in its hidden layer. While the error of the network with two neurons in its hidden layer moves consistently down and to the left as

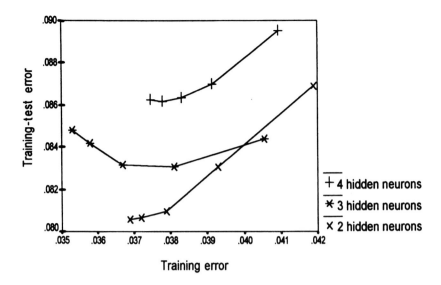

Figure G6.4.2. Average absolute error for the training set and training-test set when the number of hidden-layer neurons is varied from 2 to 4.

the number of iterations increases from 2000 to 10 000, the plot of the training-test error for the network with three neurons in its hidden layer initially declines from 0.0844 at 2000 iterations to 0.0831 at 4000 iterations, but then begins to rise fairly uniformly to 0.0848 at 10 000 iterations.

G6.4.3   Output interpretation

In terms of overall estimation of the selling price of the 100 properties in the test sample, the trained network with two neurons in its hidden layer resulted in an average absolute error of 9.48%. The smallest and largest individual absolute errors in estimating the selling price of the test sample residential properties were 0.3% and 38.7%, respectively. Figure G6.4.3 graphs the absolute error of the network's prediction, ordered by the size of the absolute error, for the test sample of 100 properties. It shows that 28% of the determinations were in error by less than 5%, 65% of the determinations were in error by less than 10%, and 12% of the determinations of the network were in error by more than 20%.

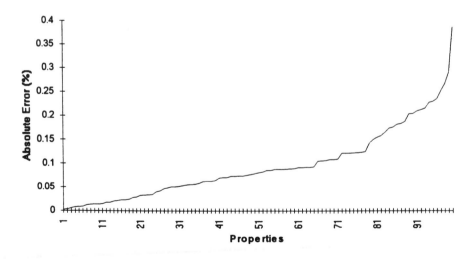

Figure G6.4.3. Absolute error for the 100 test-set properties.

G6.4.4 Comparison with multiple regression

A linear multiple regression model was derived based on the 119 properties in the training sample. The regression coefficients and their corresponding t values are given in table G6.4.3.

Table G6.4.3. Statistics for the multiple regression model.

Variable abbreviation	Variable definition	Coefficient (std. error)	t value (Prob > $|t|$)
	Intercept	132 422 (125 042)	11.00 (0.0001)
YB	Age (yrs)	−1769 (208)	−8.51 (0.0001)
BR	Number of bedrooms	2878 (1964)	1.47 (0.1458)
BATH	Number of bathrooms	−1789 (5227)	−0.34 (0.7328)
ESF	Total square footage	36 (4)	8.95 (0.0001)
LSZ	Lot size (sq. ft.)	0.73 (0.29)	2.56 (0.0118)
STY	Number of stories	−1337 (3349)	−0.40 (0.6909)
PL/SPA	Existence of a pool or spa	5502.23 (3788.47)	1.45 (0.1505)
PKG	Number of car-garages	3281 (4238)	0.77 (0.4405)

In terms of statistical performance, the multiple regression model had an adjusted R-squared of 0.689 and an F value of 33.7. In terms of estimation performance, the multiple regression model resulted in an average absolute error of 11.6% in estimating the selling price of the test sample properties. Thirty-six per cent of the determinations of the multiple regression model were in error by less than 5%, 54% of the determinations were in error by less than 10%, and 9% of the determinations were in error by more than 20%. Further, for 57 out of 100 test sample properties, the absolute error of the multiple regression model exceeded that of the network.

G6.4.5 Conclusion

While for this sample of residential properties the network produced more accurate overall estimates of selling prices than the multiple regression model, the network's average absolute error was still relatively high and some of its errors were unacceptably large. These weaknesses are likely to be attributable to two sources. First and most importantly, a number of potentially significant variables have been omitted from the pricing equation. These include view characteristics of the property such as canyon, mountain, and ocean; specific neighborhood location parameters, such as those that might be obtained by reference to the Thomas Guide 0.25 square-mile grid identifier; and other physical attributes of a house such as the existence of air conditioning, the type of roof, and the presence of a security system.

A second factor that contributed to the size of the network error was the source data. The source data describing a property were supplied by the listing agent and are subject to buyer verification. Although these agents attempt to describe the property as completely as possible, frequently the data were incomplete or erroneous.

References

Eberhart R C and Dobbins R W (eds) 1990 *Neural Network PC Tools* (San Diego, CA: Academic)
Masters T 1993 *Practical Neural Network Recipes in C++* (San Diego, CA: Academic)
Rumelhart D E and McLelland J L 1986 *Parallel Distributed Processing* vol 1 (Cambridge, MA: MIT Press)

G7

Computer Science

Contents

G7.1 Neural networks and human–computer interaction

Alan J Dix and Janet E Finlay

Abstract

There has been much interest over several years in the use of neural networks within human–computer interaction. However, this promise has led to surprisingly few published results. This article reviews those applications which have been addressed by neural networks or similar techniques. It also describes the use of the ADAM neural network for task recognition from traces of user interaction with a bibliographic database. This achieved high accuracy rates in training and in some on-line use. However, there were significant problems with its use. These problems are of interest not just for this system, but for any which is attempting to analyze trace data. The two main problems were due to the continuous sequential data and the presence of literal input (personal names, file names, dates and so on). Those systems which have achieved success in this area have not used neural techniques, but instead more traditional (although often *ad hoc*) methods. However, it is expected that recurrent networks may be suitable but probably only within a hybrid approach.

G7.1.1 Context

The use of neural networks in human–computer interaction (HCI) is largely pragmatic. They are used if they do their job well. The applications to which they are suited are also tackled by other statistical and machine learning techniques. It would be nice to report that the choice between these techniques is based on sound principles, but in fact the choice is usually based on familiarity with a particular technique. So, when considering those applications within HCI it is better to consider neural networks under the wider banner of *pattern recognition techniques*.

B6

There has been considerable interest in the application of neural networks and pattern recognition within HCI. There have now been several well-attended workshops dedicated to the theme, the results of two of which have been collected in a book (Beale and Finlay 1992). However, despite the apparent interest there are relatively few published articles on actual neural network applications (although there are many on more traditional artificial intelligence techniques). This may be because few researchers have skills in both areas and thus do not achieve their desired results.

G7.1.1.1 Applications in human–computer interaction

In common with other domains, applications of neural networks in HCI can be divided into those which require only a behavioral or black-box method and those which care about the manner in which the solution is represented (and perhaps even derived). Also, HCI applications differ in the extent to which the network must mimic human behavior—the network either satisfies a purely *computational* role or else must be to some extent *anthropomorphic*.

First consider purely computational uses, that is where there is no requirement that the behavior is in any way human. Some will be pure black-box applications. One example of this is the use of real-time electrocardiogram data by British Aerospace to detect whether pilots are becoming drowsy.This is basically

F1.7, G1.3   a matter of signal processing (see Section F1.8). Other applications include *speech, handwriting* and *gesture recognition*. Of special note is the use of gesture recognition among the disabled. Normally recognition systems have to be very accurate to be acceptable. However, where normal verbal communication is very difficult, even relatively low recognition rates may be acceptable.

In other cases the system must give some explanation of its behavior—the traditional problem of expert systems. An example of this is Query-by-Browsing (QbB) an intelligent database front-end developed by one of the authors (Dix and Patrick 1994). From examples of required records, the system generates a database query. Although the process of reasoning does not have to resemble that of a human expert, it is important that the query is in a form comprehensible to the user so that it can be verified. For this reason the present version of QbB uses decision tree induction rather than a neural network to perform pattern matching. Similar uses include the vetting of credit and job applications. In both cases explanations may be required for both legal and ethical reasons (Dix 1992).

Now consider the anthropomorphic uses. These uses include various forms of task analysis and recognition (described below) and various forms of simulation where the network takes the place of the user in the testing of software. An example of the latter is in the evaluation of the readability of computer and paper form layouts. In this case human-like behavior is sufficient, so long as the system gives similar responses to humans (especially if it can pinpoint the problem parts of the layout) it needs no further explanation. However, other researchers require that the network models more faithfully the process of human reasoning. For example, McGrew (1992) uses the interconnection weights of a parallel distributed processing (PDP) network to generate a task analysis graph. Also, Booth (1992) models the way misunderstandings give rise to errors in HCI. An important part of this analysis is an understanding of the way different areas of knowledge are used during (incorrect) reasoning.

G7.1.1.2   Trace analysis and task recognition

An important class of HCI applications are those based on trace analysis, that is, where a record of the user's interaction with a system is analyzed to recognize or uncover patterns. The data for this process may be collected automatically, often by keystroke or event logs, or may be generated as the result of observation. This raw data can be recorded for later analysis or used on-line to guide the system during interaction.

The off-line data can be used to aid task analysis. Task analysis involves the identification of patterns of behavior used to accomplish particular goals. Self-organizing networks can be used to find repeated patterns of behavior which can then be examined by the human analyst as possible task sequences. A particular task may often be accomplished by several sequences of actions and so the use of a network does not replace the human analyst. However, hand analysis is very tedious as the logs are often very long and repetitive and so this is an application where the automatic tools truly augment human skills.

On-line data can be used in various ways.

- To identify a particular user (Stacey *et al* 1992). This may be used to recall user-specific preferences or for security purposes.
- To classify the user, for example as novice/expert (Finlay 1990), in order to adapt the interaction to suit the user's knowledge and ability.
- To learn repeated sequences of actions so that they can be offered as potential macros (Hassell and Harrison 1994, Crow and Smith 1992) or as a predictive accelerator (Cypher 1991, Schlimmer and Hermens 1993).
- To recognize known task sequences (which may themselves be the result of human or automatic task analysis). Uses of this include driving a user model during computer-based learning and offering context-sensitive help.

The system we will describe in the rest of this article addresses the last of these, automatic task identification.

G7.1.1.3   Whose error?

Throughout this article we talk about various user errors, but in most software such errors are inevitable because of the design of the system. Hence, the error is most often not so much the users but the designers. However, to constantly use phrases and language to emphasize this important point would detract from the rest of the description. Hence whenever we talk of the user's error please bear in mind that this is a gross simplification.

G7.1.2 System description

G7.1.2.1 Problem domain

We now describe a system designed to recognize tasks in a menu-driven bibliographic database program called REF. More detailed descriptions of this work can be found elsewhere (Finlay 1990, Finlay and Beale 1992, Finlay and Harrison 1990). The program supported a fixed number of tasks and was therefore a very constrained environment in which to examine the issues of task recognition. However, it was a far from trivial domain. The sequence of user commands to accomplish a task varied from 3 to 16 and so the neural network had to cater for time-series data with variable length patterns. The trace was complicated by the fact that some user actions involved typing literal inputs: names, titles, dates etc. For the purposes of event logging such literals were reduced to a single user action. However, this was based on the system's idea of whether the user was entering literal input rather than menu commands. Of course, if the user and system got out of sync—a major incident—the logging did not accurately reflect the user's understanding of the interaction.

G7.1.2.2 System overview

Users interacted with the bibliographic database on an IBM-compatible PC. The event trace was transmitted along a serial link to a SUN workstation which performed the network calculations. In order to deal with the time-series data, the trace was windowed on two or six characters (although two sounds small, many tasks were easily identified by their two initial events). The windowed data was then n-tupled and passed through an ADAM (*Advanced Distributed Associative Memory*) array. The output was thresholded to give   c1.5.8
a task code and associated confidence. Finally this task code was displayed on the experimenters terminal. For example, in figure G7.1.1, the input trace 'SsM#eM' is passed through the ADAM array giving an output of $\langle 8, 5, 6, 2, 8, 0, 6, 2 \rangle$, this is thresholded at a level of 6 to yield $\langle 1, 0, 1, 0, 1, 0, 1, 0 \rangle$. Finally this binary pattern is recognized as representing the 'exit' task, but is obviously not an exact match and gets a confidence rating of 70%.

G7.1.2.3 Input format

Both the event logs and the training set included both the user's actions and some system responses. The system responses were also coded as single characters. Since the selection of menu options in REF was case-insensitive all the user's commands were translated into lowercase and uppercase letters and digits were used to code the system responses.

An example trace is 'MsSn?2'. This translates to: (M) system shows main menu, (s) user types 's', (S) system shows select sub-menu, (n) user types 'n', (?) user types a name to find, (2) system responds that there are two or more matching records.

Of course, in the user's event log such sequences are appended one after the other. Also, whereas this trace represents correct activity, event logs may also include various forms of user error.

G7.1.2.4 Training set

The REF system has 11 main task types (e.g. selecting a set of references, altering existing references, exiting the program). A complete description of the system was produced in CSP (Hoare 1985) and this was used to enumerate all possible correct task sequences. This gave rise to 529 traces which were used for training (including the example above). As these traces varied in length they were padded to a fixed size. In subsequent experiments traces of some known common problems were added to the training set.

G7.1.2.5 Topology

The neural component in the system was the ADAM binary associative network (Austin 1987). This was chosen mainly because of its speed in learning and recall. This consists of an n-tupling stage followed by a form of *Willshaw network*. The output if the network was n-point thresholded yielding a class code and   c1.5.4

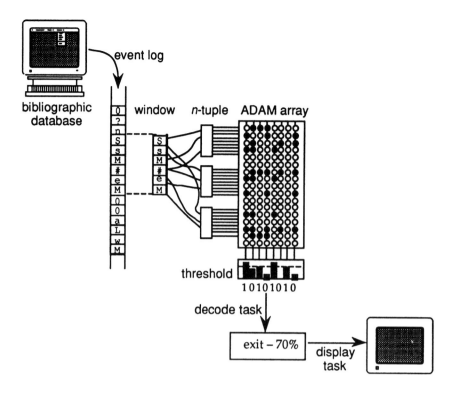

Figure G7.1.1. System components.

a confidence measure.

G7.1.2.6 Preprocessing

As described earlier, the user's entry of literal input was reduced to a single character in the trace, also the sequence of characters was reduced to a finite length by using a moving window. The ADAM network requires binary input and two representations were used, one using the normal ASCII coding and one using one bit for each possible character. The former led to an input size of 2×8 or 6×8 depending on the window size. The latter was much bigger and was expected to give a better performance because of the sparser representation. However, there was no measurable difference in performance, possibly because the latter representation effectively duplicated the job of the n-tupling.

G7.1.3 Evaluation

G7.1.3.1 Results

The system showed very high accuracy and generalization on the training set. With 50% of the full set of tasks used for training the recognition on the complete set was perfect, and stayed high even when only 10% of the examples were used in training.

However, when used on actual traces the picture was more complicated. When the small window size of 2 was used, the accuracy was around 99%. However, this dropped to 65% when the larger window of 6 was used. Apparently the problems with variable length patterns were getting in the way with the larger window. The smaller window did not have this problem (it was smaller than the shortest task sequence). However, it is likely that it was simply recognizing the user's initial main menu choice—acceptable when the user does it right, but not much help when the user and system are out of sync.

G7.1.3.2 Comparison with traditional methods

D1.4, G2.12 The results using ADAM were compared with those obtained using a variant of *ID3* (Quinlan 1979), a machine learning algorithm which builds decision trees by induction. When tested on the training set it too obtained 100% accuracy using 50% of the full set of tasks, and was highly accurate, but slightly worse than ADAM on smaller training sets. When used on the actual logs of user interaction, its accuracy was substantially lower than ADAM, although following the same pattern attaining 85% accuracy with a window size of 2 but only 35% on a window size of 6.

G7.1.3.3 General problems

This application highlights several problems which must be tackled if neural techniques are to be applied within the human–computer interface.

Matching varying length subsequences within an event trace was clearly a substantial problem. There are various issues connected with this. The segmentation problem is well known in other fields, for example in separating words within continuous speech. Recall how the accuracy rate for recognizing the training set (which was already segmented) was high, but that it fell off dramatically when faced with continuous user traces. For some kinds of task recognition it is possible to use information from the state of the computer dialogue, for example, when the REF system was at the main menu. However, as we saw, an important class of interface errors occur when this does not concur with the user's notion of the task state.

Assuming we have segmented the trace, we still face problems due to the omission of required actions or where the task sequence is split by irrelevant or erroneous actions. This is similar to the problems of spelling correction. Windowing techniques are very fragile in the face of changes in the relative position of parts of a sequence.

These problems are also faced by systems (as discussed earlier) which look for repeated sequences in the user input and other sequence-based problems. To our knowledge none of these use neural techniques, but instead rely on symbolic artificial intelligence techniques (Cypher 1991), inductively-built finite-state machines (Schlimmer and Hermens 1993), hidden Markov models (Hanlon and Boyle 1992) or special purpose algorithms (Crow and Smith 1992). However, it seems likely that recurrent neural networks could also be used for this purpose. Indeed, many representations of user interaction can be transformed into some form of finite-state representation which could be used to train recurrent networks such that the network's internal representation matches that of the analyst (Dix *et al* 1992).

In fact, the sequences we dealt with were not as difficult as they could be. The REF system was an old-fashioned DOS application, with only a single thread of control. Consider a windowed application. These typically allow the user to perform several tasks in parallel, even within one window. From the recognition system's point of view, these appear rather like insertion errors. A typical insertion error is caused by the user accidentally typing an extra character which breaks the original pattern in two. In the case of multiple windows the user may begin to perform a task in one window, then swap to another and perform one or more tasks in the second window, and finally return to the first window to complete the initial task. Just like a mis-typing, the original task is broken in two. However, in contrast to simple insertion errors caused by mis-typing, the breaks in a windowed application may often be substantial. Neither is it sufficient to regard each window or application separately; part of the power of windowed systems is that tasks involve interaction with multiple applications.

The other major problem we discussed was literal inputs, such as the typing of author names to search for in the bibliography. These cause three problems.

First, they act as variable-length insertions in the trace. The method used in our system to code them works only when the user and system are in agreement. If the user starts to type a name when the system is expecting further menu choices, then the trace will record the full name. At just the moment when the user is confused and needs help, we find that the network is equally confused!

Second, the values of the literal inputs matter. Although the particular value is typically unimportant it is often important whether the same name is used several times. For example, in an operating system consider the following commands.

```
copy onefile.txt another.txt
delete onefile.txt
```

It is very important that the two commands in this sequence refer to the same file. The Query-by-Browsing system mentioned earlier uses variable matching techniques, but this is in the context of inductively learned decision trees where it is easier to add symbolic constraints (Dix and Patrick 1994).

Third, the insertions resulting from literal input often have a completely different syntactic form to that of the rest of the interaction. This can make it easy to detect and so segment, although the exact start of the insertion may be less clear. However, this suggests that the pattern recognizer needs to have, either explicitly or implicitly, several modes. A similar problem arises when dealing with multiple applications in a windowed system. As with literal input it is no good relying on the system's interpretation of where input belongs—an important error is precisely when the user mistakenly inputs data to the wrong window.

G7.1.4 Conclusions

For this application, the ADAM neural network performed better than an alternative machine learning algorithm. However, there were fundamental problems that arose which need to be tackled by anyone wishing to apply neural networks to on-line or off-line trace analysis. The nature of these suggests that a hybrid rather than pure neural approach will be required.

References

Austin J 1987 ADAM: a distributed associative memory for scene analysis *Proc. First Int. Conf. on Neural Networks (San Diego)* IEEE

Beale R and Finlay J (eds) 1992 *Neural Networks and Pattern Recognition in Human–Computer Interaction* (Chichester: Ellis-Horwood)

Booth P A 1992 Modelling misunderstandings using artificial neural networks *Neural Networks and Pattern Recognition in Human–Computer Interaction* ed R Beale and J Finlay (Chichester: Ellis-Horwood) pp 301–19

Crow D and Smith B 1992 DB_Habits: comparing minimal knowledge and knowledge-based approaches to pattern recognition in the domain of user computer interactions. *Neural Networks and Pattern Recognition in Human–Computer Interaction* ed R Beale and J Finlay (Chichester: Ellis-Horwood) pp 39–63

Cypher A 1991 Eager: programming repetitive tasks by example. *Proc. CHI'91* (New Orleans, LA: ACM Press)

Dix A 1992 Human issues in the use of pattern recognition techniques *Neural Networks and Pattern Recognition in Human Computer Interaction* ed R Beale and J Finlay (Chichester: Ellis-Horwood) pp 429–51

Dix A, Finlay J and Beale R 1992 Analysis of user behaviour as time series *Proc. HCI'92: People and Computers VII* (Cambridge: Cambridge University Press) pp 429–44

Dix A and Patrick A 1994 Query By Browsing *Proc. IDS'94: The 2nd International Workshop on User Interfaces to Databases (Lancaster)* (Berlin: Springer) pp 236–48

Finlay J 1990 Modelling users by classification *D. Phil. Thesis* University of York

Finlay J and Beale R 1992 Pattern recognition and classification in dynamic and static user modelling *Neural Networks and Pattern Recognition in Human–Computer Interaction* ed R Beale and J Finlay (Chichester: Ellis-Horwood) pp 65–89

Finlay J E and Harrison M D 1990 Pattern recognition and interaction models *Human–Computer Interaction INTERACT90* (Amsterdam: North-Holland) pp 149–54

Hanlon S J and Boyle R D 1992 Syntactic knowledge in word level text recognition *Neural Networks and Pattern Recognition in Human–Computer Interaction* ed R Beale and J Finlay (Chichester: Ellis-Horwood) pp 173–93

Hassell J and Harrison M 1994 Generalisation and the adaptive interface. *Proc. HCI'94: People and Computers IX, (Glasgow)* (Cambridge: Cambridge University Press) pp 223–38

Hoare C A R 1985 *Communicating Sequential Processes* (Englewood Cliffs, NJ: Prentice-Hall)

McGrew J K 1992 Task analysis, neural nets, and very rapid prototyping *Neural Networks and Pattern Recognition in Human–Computer Interaction* ed R Beale and J Finlay (Chichester: Ellis-Horwood) pp 91–102

Quinlan J R 1979 Discovering rules by induction from large collections of examples. *Expert Systems in the Micro-Electronic Age* ed D Michie (Edinburgh: Edinburgh University Press) pp 168–201

Schlimmer J C and Hermens L A 1993 Software agents: completing patterns and constructing user interfaces. *J. Artif. Intell. Res.* **1** 61–89

Stacey D, Calvert D and Carey T 1992 Artificial neural networks for analysing user interactions *Neural Networks and Pattern Recognition in Human–Computer Interaction* ed R Beale and J Finlay (Chichester: Ellis-Horwood) pp 103–13

Further reading

1. Beale R and Finlay J (eds) 1992 *Neural Networks and Pattern Recognition in Human–Computer Interaction* (Chichester: Ellis-Horwood)

    A collection of papers from two workshops held in the US and UK, covering both neural networks and related pattern recognition techniques.

2. Finlay J and Beale R 1993 Neural networks and pattern recognition in human–computer interaction *SIGCHI Bulletin* **25** 25–35

3. Finlay J E and Dix A J 1994 Pattern recognition in human–computer interaction a viable approach? *SIGCHI Bulletin* **26** 23–7

4. Reports on the CHI91 and CHI94 workshops of the same names.

    There is also a moderated mailing list; interested parties should send a request to be added to prhci@zeus.hud.ac.uk

G8

Arts and Humanities

Contents

G8.1 Distinguishing literary styles using neural networks

Robert A J Matthews and Thomas V N Merriam

Abstract

Scholars in the humanities have long argued over the authorship of works ranging from Elizabethan histories to religious texts. Most of the debate has been essentially subjective and qualitative. However, the advent of computer technology has led to the development of *stylometry*—the quantitative analysis of literary style based on, for example, frequency of words used by different authors. With their ability to cope with both nonlinear and noisy data sets, neural networks are well suited to the stylometric problem. Here we show that they out-perform linear methods of identifying authors, and illustrate their power with studies of disputed works from the era of William Shakespeare.

G8.1.1 Project overview

Scholarship thrives on debates, and there is no shortage of debates in the study of historical and literary texts. Among the most intriguing are those centering on the authorship of such texts. Is it possible to identify the correct authors of each of *The Federalist Papers*, written pseudonymously in 1787–8 to persuade voters to ratify the US Constitution? Were the divine Mormon texts actually the work of Joseph Smith, founder of the Mormon Church? Did Shakespeare always write masterpieces in isolation, or were some the result of collaboration with contemporaries?

Until recently, the evidence in such debates has been primarily in the form of scholarly opinion, founded on extensive knowledge of both the texts and their putative authors. However, such an approach is inevitably subjective and may (indeed, has) been subject to changing fashion. Nevertheless, questions of authorship are objective ones: ultimately, they do admit a single, correct answer.

Attempts to apply objective and quantitative methods to questions of authorship have their origins in the 19th Century (see Matthews and Merriam 1994a for a popular-level historical review). Only relatively recently, however, with the advent of powerful computer technology, have these methods found much application. Collectively, they belong to a branch of statistical analysis known as *stylometry*. This is founded on the premise that one can extract quantitative features—usually certain word frequencies in texts—which discriminate between the style of one author and another. As such, these 'stylometric discriminators' can be used to cast light on the authorship of a disputed work. The procedure, in principle at least, is simple: extract suitable discriminators for each author from their undisputed texts, and then see which author's discriminators best fit the stylometric data for the disputed works.

Inevitably, capturing literary style is not so simple. First, humans are not automatons, and stylometric discriminators are typically statistically 'noisy'. Second, language involves complex interactions between its components, and attempts to capture its essence by simple discriminators involves dimensionality reduction and feature extraction, an inevitably nonlinear process.

Despite this, stylometry has traditionally tended to rely on parametric linear methods. Neural networks, in contrast, are well known for their abilities to classify data in the face of both nonlinearities and noise. This raises the possibility that the power of conventional stylometry may be boosted by using discriminators as the inputs to neural networks. Here we show that this is indeed possible: after extracting suitable discriminators, neural networks can be trained to recognize the characteristic features of an author's style. When tested on undisputed texts to which the neural network has not previously been exposed, it typically out-performs conventional linear stylometric methods. This suggests that neural networks may be a valuable new source of evidence in literary disputes.

G8.1.2 Design process

The basic approach in any stylometric study is to extract discriminators from large amounts of undisputed text from the likely authors, and then see which provides the best fit to the stylometric 'signature' of the disputed text. In what follows, we use the discriminators as inputs to a neural network.

The design of a stylometric neural network (SNN) then consists of two stages:

(i) determination of suitable stylometric discriminators capable of distinguishing between author A and author B;
(ii) determination of a suitable SNN topology, such that it is complex enough to capture the stylometric features of each author, but not so complex that it cannot be fully trained.

G8.1.2.1 Determination of discriminators

A wide range of stylometric measures has been investigated as potential discriminators of literary style. However, the relative frequencies of so-called 'function words'—conjunctions, prepositions and the definite and indefinite articles—have often been found to be sufficiently different between authors to constitute discriminators (Matthews and Merriam 1994b). For a specific authorship dispute, the relative frequencies (i.e. raw number of occurrences in a sample text, divided by sample length—typically 1–2000 words) of a wide variety of such words can be calculated for many extracts from undisputed texts by authors A and B. Those function words showing the largest separation in relative frequency between A and B (typically 2–3 standard deviations) can then be taken as potential discriminators.

The task of performing this analysis has been greatly eased by the publication of machine-readable versions of many important literary works, such as the entire corpus of Shakespeare by the Oxford University Computing Service.

G8.1.2.2 Determination of suitable topology

C1.2 Our first investigations of neural computation in stylometry centered on the *multilayer perceptron* (MLP), the most widely used form of neural network. Other techniques can be, and have been, used (Lowe and Matthews 1995); the design considerations that follow will, however, apply to most neural approaches.

The classical MLP consists of three layers: an input layer of N_1 neurons and an output layer of N_3 neurons, the two being linked by a hidden layer of N_2 neurons. For a stylometric MLP, N_1 will be the number of function word discriminators used to classify a text as the work of one of various authors. In general, the more discriminators that are used, the better the reliability of the classification. However, a limit is set on the number of discriminators by the fact that if an MLP has too many inputs relative to the amount of training data, it loses its ability to generalize to new data. (Essentially, there are too many unknowns for the data to support.) A useful rule of thumb for setting the topology of a stylometric MLP follows from the requirement of having sufficient undisputed text to both extract discriminators and train the MLP, while still having sufficient text left over to test the MLP (Matthews and Merriam 1994b)

$$N_1 + N_3 < 10^{-4}C$$

where C is the total amount, in words, of undisputed text for each author. Thus for a binary authorship dispute concerning plays of the Elizabethan era, which are typically around 20 000 words long, the number of input neurons should be less than around $2P - 2$, where P is the number of plays available for each author. For Shakespeare and many of his contemporaries, $P > 5$, so $N_1 \sim 8$. In practice, we have found $N_1 = 5$ sufficient to achieve excellent classification results (Matthews and Merriam 1993, Merriam and Matthews 1994).

The size of the hidden layer is set by the competing requirements of capturing as many features in the data as possible while ensuring that the MLP can generalize to new data. The optimal hidden layer size can be found by training with different values of N_2 and seeing which gives the best results during cross-validation (i.e. the use of part of the training data for testing purposes). For plays by Shakespeare and his contemporaries, our experience has been that $N_2 = 3$ neurons gives very acceptable results. This leads to the topology for an SNN shown in figure G8.1.1.

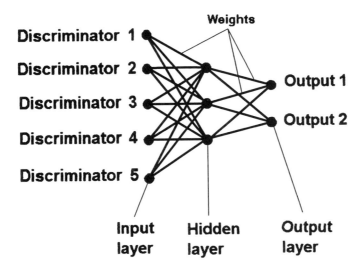

Figure G8.1.1. Typical topology for a stylometric neural network (SNN).

G8.1.3 Training of a specific stylometric neural network

We now consider the training, testing and application of an SNN suitable for investigating texts associated with Shakespeare and John Fletcher (1579–1625), Shakespeare's successor as chief dramatist to the King's Men company. Specifically, we address questions concerning the plays *Henry VIII*, *The Two Noble Kinsmen*, *The Double Falsehood* and *The London Prodigal*. All have been associated with Shakespeare and Fletcher at some time; the central question concerns the balance of contribution of each playwright. Current scholarly opinion on *Henry VIII* and *The Two Noble Kinsmen* is that both may contain some contribution from Fletcher, but nevertheless are sufficiently Shakespearean to merit inclusion in any collection of the Bard's work. In contrast, the evidence for Shakespeare's involvement in *The Double Falsehood* and *The London Prodigal* has generally been deemed insufficient for either to merit inclusion in his works, with Fletcher being seen by some scholars as a substantial, if not principal, contributor to each.

G8.1.3.1 *The training set*

The first task is to gather sufficient undisputed works by both playwrights for training the SNN. This has to be done with care: there are, for example, two different versions of *King Lear*, three of *Hamlet* and six of *Richard III*. Fortunately, there is general agreement among scholars as to which works constitute essentially undisputed 'core canon' works by Shakespeare and Fletcher, and these can be used for training purposes. For Shakespeare we took *Antony and Cleopatra*, *As You Like It*, *Henry IV Part 1*, *Henry V*, *Julius Caesar*, *Love's Labour's Lost*, *A Midsummer Night's Dream*, *Richard III*, *Twelfth Night*, and *The Winter's Tale*. Collectively, these give a representative sample of Shakespeare's work on different themes throughout his career. Similarly, for Fletcher we took: *Bonduca*, *The Chances*, *Demetrius and Enanthe*, *The Island Princess*, *The Loyal Subject* and *The Woman's Prize*.

G8.1.3.2 *Preprocessing inputs*

In order to train the SNN to associate undisputed texts with their correct author, function-word frequencies capable of discriminating between Shakespeare and Fletcher must be extracted from core canon plays. Research by Horton (1987) suggests that the function-word frequency ratios

$$are/N \quad in/N \quad no/N \quad of/N \quad the/N$$

(where N is the total number of words in a sample) can act as suitable discriminators for a Shakespeare–Fletcher SNN; these formed the inputs for our SNN. We then extracted 100 sets of these five discriminators from the core canon plays of each dramatist, with each set of five being preprocessed to give zero mean and unit standard deviation; this ensures that each discriminator contributes equally in the training process.

G8.1.3.3 Output interpretation

The target pattern for training purposes was an output pattern of $(1, 0)$ for Shakespeare and $(0, 1)$ for Fletcher. For ease of interpretation, these were then converted to a so-called Shakespearean characteristics measure (SCM), defined as

$$\text{SCM} = O_S/(O_S + O_F)$$

where O_S and O_F are the values of the outputs from the Shakespeare and Fletcher nodes of the SNN, respectively. Thus the stronger the Shakespeare output signal relative to the Fletcher signal, the higher the SCM. Strongly Fletcherian classifications, on the other hand, give SCM closer to zero, and those on the borderline ($O_S = O_F$) give SCM = 0.5.

G8.1.3.4 Development

Given the small size of the SNN topology, the development platform can be very modest. The Shakespeare–Fletcher SNN was trained on a 286/12 MHz PC running the MS-DOS version of NetBuilder software provided by Recognition Systems Ltd of 140 Church Lane, Marple, Stockport, SK6 7LA, Cheshire, UK.

G8.1.4 Comparison with traditional methods

During training we achieved cross-validation accuracy of 96%, with the 4% misclassified text being equally divided between the two dramatists. This is considerably better than the results achieved by linear stylometric methods: for example, an optimum linear transformation using the Horton ratios gives a 9% misclassification rate, three-quarters of which comprises Fletcher samples wrongly ascribed to Shakespeare. The greater success of the SNN reflects its ability to cope with both the noise and nonlinearity in the data set.

The power of the SNN is, however, most impressive when it is applied to core canon texts to which it has not previously been exposed. This gives a measure of its ability to generalize to new data. Table G8.1.1 shows the results for the eight remaining core canon plays of Shakespeare and two of Fletcher.

Table G8.1.1. SNN results for core canon Shakespeare and Fletcher.

Dramatist	Play	SCM	Verdict
Shakespeare	*Much Ado*	0.71	Shakespeare
Shakespeare	*All's Well*	0.92	Shakespeare
Shakespeare	*Comedy of Errors*	0.91	Shakespeare
Shakespeare	*Coriolanus*	0.98	Shakespeare
Shakespeare	*King John*	0.91	Shakespeare
Shakespeare	*Merchant of Venice*	0.97	Shakespeare
Shakespeare	*Richard II*	0.92	Shakespeare
Shakespeare	*Romeo and Juliet*	0.87	Shakespeare
Fletcher	*Valentinian*	0.30	Fletcher
Fletcher	*Monsieur Thomas*	0.29	Fletcher

As can be seen, the SNN correctly classified every one of the ten remaining core canon plays on which it was tested. This impressive success rate is somewhat higher than that obtained during cross-validation, a reflection of the fact that entire plays are now being used, which are less noisy than the samples used for cross-validation.

Of course, the most interesting results come from the application of the trained and tested SNN to disputed works. This permits comparison between the conclusions of the SNN and the subjective assessments of conventional literary scholarship. For each of the four disputed plays in the Shakespeare–Fletcher debate, discriminator values were extracted from the entire play and from its individual acts, and these were then used as inputs to the SNN. The results are shown in table G8.1.2.

Table G8.1.2. SNN results for disputed plays.

Play	SCM	SNN verdict
Henry VIII		
As whole play	0.94	Shakespeare
Act I	0.98	Shakespeare
Act II	0.85	Shakespeare
Act III	0.97	Shakespeare
Act IV	1.00	Shakespeare
Act V	0.57	Shakespeare
Two Noble Kinsmen		
As whole play	0.65	Shakespeare
Act I	0.93	Shakespeare
Act II	0.30	Fletcher
Act III	0.32	Fletcher
Act IV	0.60	Shakespeare
Act V	0.91	Shakespeare
Double Falsehood		
As whole play	0.37	Fletcher
Act I	0.60	Shakespeare
Act II	0.87	Shakespeare
Act III	0.29	Fletcher
Act IV	0.73	Shakespeare
Act V	0.29	Fletcher
London Prodigal		
As whole play	0.30	Fletcher
Act I	0.89	Shakespeare
Act II	0.29	Fletcher
Act III	0.34	Fletcher
Act IV	0.28	Fletcher
Act V	0.30	Fletcher

The SNN results for the plays taken in their entirety support the qualitative opinions of contemporary scholars, that is, that *Henry VIII* and *The Two Noble Kinsmen* merit inclusion in the Shakespeare canon, while The *Double Falsehood* and *The London Prodigal* do not. More interesting, however, are the results for individual acts. While the SNN gives strong Shakespearean classifications for Acts I to IV of *Henry VIII*, the low SCM for Act V supports claims that this was largely the work of Fletcher (Hoy 1956). Similarly, the SNN classifies Acts I and V of *The Two Noble Kinsmen* as Shakespearean, Acts II and III to Fletcher, and Act IV as borderline. This detailed breakdown is again in broad agreement with contemporary scholarship (Proudfoot 1970). Taken as an entire play *The Double Falsehood* emerges as predominantly Fletcherian in style, agreeing with contemporary scholarship summed up by Metz (1989). Similar remarks apply to the SNN findings with *The London Prodigal*: we find an overall Fletcherian attribution, but with some Shakespearean influence, especially in Act I (cf Hope 1994).

G8.1.5 Conclusions

The results presented here suggest that neural networks can make a valuable contribution to stylometry. With their ability to deal with both noisy and nonlinear data sets, they amplify the power of standard stylometric discriminators.

The principal limitation on the use of SNNs appears to be the demand for sufficient undisputed texts on which to train and test the networks. (This would seem to rule out the use of SNNs in forensic applications, such as the analysis of alleged confessions.) Nevertheless, the recent growth in the number of machine-readable texts of important authors gives plenty of scope for further research. We have ourselves used SNNs to study the influence of Marlowe on Shakespeare's early career, finding evidence that the young Bard leaned heavily on works by his gifted contemporary (Merriam and Matthews 1994). Tweedie, Singh and Holmes have applied a multilayer perceptron SNN to *The Federalist Papers* (Tweedie *et al* 1995), while investigations using other network techniques such as the *radial basis function* are also being started (Lowe and Matthews 1995). Experience to date on all these projects suggests that neural networks provide a valuable new source of evidential weight on which literary scholars may draw.

C1.6.2

References

Hope J 1994 *The Authorship of Shakespeare's Plays: A Socio-linguistic Study* (Cambridge: Cambridge University Press) p 115

Horton T B 1987 The effectiveness of the stylometry of function words in discriminating between Shakespeare and Fletcher *Doctoral Thesis* University of Edinburgh

Hoy C 1956 The shares of Fletcher and his collaborators in the Beaumont and Fletcher canon (VII) *Studies in Bibliography* **15** 129–46

Lowe D and Matthews R A J 1995 Shakespeare vs Fletcher: A stylometric analysis by radial basis function *Computers and the Humanities* **29** 449–61

Matthews R A J and Merriam T V N 1993 Neural computation in stylometry I: an application to the works of Shakespeare and Fletcher *Literary and Linguistic Computing* **8** 203–9

—— 1994a A Bard by any other name *New Scientist* 22 January 23–7

—— 1994b Using neural networks to cast light on literary mysteries *Applications and Innovations in Expert Systems II (Proc. Expert Systems 94; 14th Ann. Conf. British Computer Society Special Interest Group on Expert Systems, Cambridge)* ed R Milne and A Montgomery (Oxford: SGES Publications) pp 237–47

Merriam T V N and Matthews R A J 1994 Neural computation in stylometry II: an application to the works of Shakespeare and Marlowe *Literary and Linguistic Computing* **9** 1–6

Metz G H (ed) 1989 *Sources of Four Plays Ascribed to Shakespeare* (Columbia: University of Missouri Press)

Proudfoot G R (ed) 1970 The Two Noble Kinsmen (London: Edward Arnold)

Tweedie F J, Singh S and Holmes D I 1995 Neural network applications in stylometry: The Federalist Papers *Computers in the Humanities* submitted

G8.2   Neural networks for archaeological provenancing

John Fulcher

Abstract

Artificial neural networks (ANNs) are applied to the problem of classifying obsidian rock samples taken from the West New Britain region of Papua New Guinea. Multilyer perceptrons, self-organizing maps and learning vector quantization are found to be the most appropriate models for this task. A somewhat surprising result is that ANNs are able to yield good results (at least comparable with a human expert) with very few training exemplars.

G8.2.1   Introduction

Provenancing is the study of ancient artifacts, in order to determine their time and place of origin. In doing so, we also hopefully learn something of the culture of the people of that era. An associated study—archaeometry—is the mathematical analysis of archaeological artifacts and data.

In the present study we concern ourselves with obsidian artifacts collected from the Talasea (northern) region of West New Britain, Papua New Guinea. Preprocessing is performed on data samples gathered from several sites in the region, by way of proton-induced x-ray emission (PIXE) analysis. ANNs are then used to classify these samples in terms of their sites of origin.

G8.2.2   Obsidian samples

Obsidian is a glass-like substance produced by rhyolitic flow in errupting volcanos. It is found in several locations around the world, including both sides of the Pacific, namely Papua New Guinea (Torrence *et al* 1992), Oregon (Nelson *et al* 1975, Hughes 1986, Godfrey-Smith *et al* 1993) and Ontario (Godfrey-Smith and Haywood 1984). Obsidian possesses excellent flaking properties, and is readily split into thin slices, which in turn can be used to fabricate knives, axes and other implements. Indeed, it has been traded and used for toolmaking since prehistoric times. In modern times some forms are regarded by many cultures as semiprecious gemstones.

Obsidian has been quarried by the indigenous people of Papua New Guinea for around 20 000 years. By undertaking provenancing studies, we hope to gain insight into the trading practices of these people, from prehistoric times onwards.

The color, translucency and texture of obsidian varies considerably depending on the site from which it is collected. For example, the rhyolitic flows from the Kutau and Bao regions of West New Britain are usually banded (and therefore not as a rule translucent) and range in color from gray or gray green to black. In contrast, obsidian from the Mount Baki and Garala Island regions is usually deep black and translucent, whereas samples from the Mount Hamilton region invariably contain high concentrations of small white phenocrysts. Moreover, Talasea obsidian has been unearthed at archaeological sites over an 8000 km area of the western Pacific, extending from Sabah in the west to Fiji in the east (Torrence *et al* 1992).

It is not possible to identify different obsidian samples solely on the basis of such broad characteristics; further detailed analysis is required, and this is where electron and x-ray techniques come into play (Nelson *et al* 1975, Barton and Krinsley 1987). Proton-induced x-ray emission (PIXE) is used in the present study.

PIXE quantifies 14 elements and 7 oxides, as indicated in table G8.2.1. Normalized major and trace element data is used in order to remove any small, correlated parameter variations, thereby improving variable independence (two detectors are typically used, one normalized to iron, and the other to sodium). The use of ratios also leads to lower dimensionality of independent variables. Moreover, elemental ratios are commonly used in statistical cluster analysis (provided they are ratios of *independent* variables).

Table G8.2.1. Element and oxide data resulting from PIXE analysis.

Major elements	Na	Al	Si	K	Ca	Fe			
Trace elements	F	Ti	Mn	Rb	Sr	Y	Zr	Nb	
Normalized ratios	Al/Na	F/Na	Mn/Fe	K/Fe	Ca/Fe	Rb/Fe	Y/Fe	Zr/Fe	Nb/Fe
Oxides	Na_2O	Al_2O_3	SiO_2	K_2O					
Sum of oxides	CaO	Ti_2O	Fe_2O_3	Oxide sum					

G8.2.3 Neural network classification

Traditional statistical approaches to obsidian classification include dendograms and cluster analysis. In practice, such techniques are used as aids for experts in order to perform manual classification of the data. The motivation for using artificial neural networks to perform obsidian rock classification automatically was twofold:

(i) could ANNs perform comparable classification to that currently performed by experts, and

(ii) would ANNs be able to arrive at any meaningful classifications, given the small number of training data available?

The first of these will be dealt with in section G8.2.4. As regards (ii), there has been some work done on minimum training sets for ANNs. Hepner *et al* (1990) were concerned with the classification of satellite image data. They found that even with minimal training sets, the performance of ANNs matched that of conventional techniques, and moreover was far superior in terms of generalizability.

As a result of their study, Eaton and Oliver (1992) derived an empirical formula in which the learning rate is reduced in proportion to the size of the data set (we shall be making reference to this finding again in section G8.2.3.1). Yan (1992) has suggested that by 'judicious' selection of training data, 'protoypes' can be produced which effectively average the relevant features of each sample class. This approach appears to work well with nearest-neighbor techniques, such as *Kohonen's self-organizing map* (SOM). The use of thresholds was proposed by Tom and Tenorio (1991), in order to lower the incidence of misclassifications. They further found that increasing the size of the training set increases the likelihood of correct recognition (of short speech utterances, in their case).

C2.1.1

The data we had available for the present study were gathered from the eight different sites in the West New Britain area of papua New Guinea indicated in figure G8.2.1 (Potter *et al* 1994). There were a total of 200 training exemplars, 122 of which were obsidian rock samples (which had previously been classified manually); the remainder were artifacts. These artifacts (knives and other such tools) were known to have come from essentially the same source. However, since this source remained unknown to us, these artifact data were removed to avoid the likelihood of cross-training.

Of the remaining sources, two did not yield sufficient numbers of training exemplars, and so were not used. The six sites we used in the end were: Kutau, Gulu, Garala, Baki, Hamilton and Mopir. In the present study, we only have a few (between four and seven) training exemplars from each source, but each sample contains a considerable amount of information.

The obsidian data was preprocessed using PIXE analysis, in order to provide element and oxide (ratio) information. The data had been manually classified by an expert (but not perfectly—some samples were only classified as coming from a particular source in terms of probability, not certainty). This manual classification was the benchmark (yardstick) against which our ANN approach was to be appraised.

We have certain *a priori* knowledge about the data at our disposal. For example, we know that all sources (sites) are close geographically, and that the obsidian samples have similar physical characteristics. As a result, the data can be grouped into four distinct classes. However, apart from this characteristic, the data are well spread.

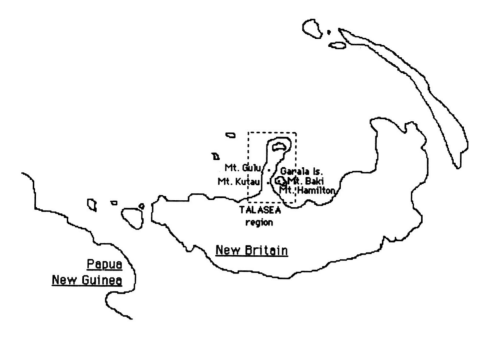

Figure G8.2.1. Obsidian sources in West New Britain, Papua New Guinea.

G8.2.3.1 Multilayer perceptron

Our starting point for this study was the familiar multilayer feedforward network—*multilayer perceptron* c1.2
(MLP) or backpropagation network. We began training MLPs using the public domain PlaNet X-
Windows ANN simulator (Myata 1991) but soon switched to a commercial ANN simulator—NeuralWorks
Professional-II+ (NeuralWare 1993). This latter software simulator had been previously adjudged to be
one of the best available from our experiences with other ANN projects (Fulcher 1994). Furthermore,
it could support many more ANN models than PlaNet, which we required for the present study (see
sections G8.2.3.2 and G8.2.3.3 below).

The MLP configuration was as follows:

> Input layer = 31 neurons (all 31 PIXE characteristics)
>
> Hidden layer = 8 neurons
>
> Output layer = 6 neurons (one for each of the 6 sources).

A preliminary investigation used five MLPs, one for each of the five rows in table G8.2.1. Initial results
were disappointing, however. The networks confused samples from Garala and Baki (but these had also
been misclassified on occasion by the human expert). Of more concern was the confusion between Gulu
and Hamilton samples. This prompted the use of *un*supervised networks, in an attempt to arrive at
independent classifications (see section G8.2.3.2 below).

The next step involved grouping tables together, to determine whether higher dimensionality would
improve classification. The training set was normalized to seven samples only from each site. A new
MLP, with a learning rate of 0.15 and a momentum term of 0.8, was trained on this combined data set,
and yielded 9% misclassification error (most of which could be attributed to samples from either Garala
or Baki).

Further exemplars were removed from the training set and placed in the test set. The MLP was
randomized then retrained using only four samples from each site. As can be seen from figure G8.2.1
(row 2), the effect on classification error was only marginal (but only for MLP). This is a rather surprising
result, given that we had *so few* training exemplars with which to work.

At this juncture we compare our results with the empirical formula developed by Eaton and Oliver
(1992) for optimum learning rate:

$$\eta = 1.5/\left(n_1^2 + n_2^2 + \cdots\right)^{1/2}$$

Table G8.2.2. Effect of sample size on misclassification.

ANN model	% incorrect (7 samples)	% incorrect (4 samples)
MLP	9.0	10.5
SOM	13.1	25.4
LVQ	16.4	39.3

where n_1 is the number of patterns in class 1. In our case, for six distinct classes of four training patterns each, this yields a learning rate of 0.1531 (compared to the value of 0.15 used in the present study, and which had been independently arrived at). The momentum was slightly lower than that used by Eaton—0.8, instead of 0.9. We conclude that our results are consistent with Eaton's assertion that the learning rate should be reduced as the number of exemplars in the training set is decreased.

G8.2.3.2 Self-organizing map

As with MLPs, the starting point for using self-organizing maps (SOMs) was separate networks for each of the five major tables. Once again, misclassification of Garala and Baki samples accounted for the majority of errors, which were higher than those obtained using MLPs. This is not a surprising result in a way, since no one table contains sufficient information to identify adequately the site from which the sample came. Moreover, since SOMs form boundaries within the training data, outlying samples from each class will run the risk of being misclassified.

The real value of SOMs in the present study was to examine overlap between different classes. Actually, a modification of Kohonen's original SOM—SOM with classification—was used here (as implemented within Neural Works Professional-II+). We were able to verify that Gulu and Hamilton samples could indeed be discriminated.

The fundamental behavior of an SOM network is to perform dimensionality reduction of the training data onto a two-dimensional feature map. The Mexican hat function is used to group neighborhoods of neurons into classes. In the case of sparsely distributed data, subgroups will be formed, which together constitute wider neighborhoods (or classes). Different classes will be defined by similar distributed neighborhoods. Thus it is no surprise that SOMs are able to distinguish Gulu and Hamilton classes, given the uneven distribution of training samples.

SOMs did not perform as well as MLPs when the size of the training set was reduced, however. Retraining the SOM using four samples per class instead of seven saw the misclassification error almost double (rising from from 13% to 25%—see table G8.2.2).

G8.2.3.3 Learning vector quantization

C1.1.5 *Learning vector quantization* (LVQ) was even more sensitive to reducing the number of training samples from seven to four, with the overall error increasing from 16% to 39% (table G8.2.2). Samples from Hamilton, Kutau and Mopir are still able to be correctly discriminated; the misclassification occurs between Baki, Garala and Gulu (and more especially Baki and Gulu with four training samples).

G8.2.4 Results

Table G8.2.2 summarizes the performance of the three ANNs used in the present study as a function of the number of training exemplars. We conclude that all three networks yield acceptable performance with seven training samples per class (comparable, at least, with manual classification by a human expert). However, when the number drops to four per class, only MLPs yield acceptable performance. The characteristics of each ANN are summarized in table G8.2.3.

The addition of thresholding reduces the number of misclassified samples, as indicated in table G8.2.4 (four samples per network). The term 'decision rate' refers to the proportion of time the output is greater than the threshold.

Table G8.2.3. ANN characteristics.

ANN model	Summation	Transfer function	Learning rule	Convergence threshold
MLP				0.01
Input layer	Linear	Linear	—	
Hidden layer	Linear	Sigmoid	Cumulative	
Output layer	Linear	Sigmoid	Widrow–Hoff	
SOM				0.0
Input layer	Linear	Linear	—	
Kohonon layer	SOM	Linear	SOM	
SOM with categorization				0.0
Input layer	←	Direct transfer	→	
Kohonen layer	SOM	Linear	SOM	
Output layer	Linear	Linear	Widrow–Hoff	
LVQ				0.01
Input layer	Linear	Linear	—	
Kohonen layer	LVQ	Linear	LVQ1 for 5000 epochs,	thence LVQ2
Output layer	Linear	Linear	—	

Table G8.2.4. ANN performance with thresholding.

ANN model	% incorrect (threshold = 0)	% incorrect (threshold = 0.6)	Decision rate
MLP	10.5	7.4	90.98
SOM	25.4	12.7	82.25
LVQ	39.3	39.7	100.00

In the case of SOM, for example, the percentage of *correctly* classified samples with thresholding increases from 74.6% to 87.3%, but at the expense of lowering the decision rate from 100% to 82.25%. These results confirm the earlier findings of Tom and Tenorio (1991) that the incidence of misclassifications can be reduced by using thresholds. The surprising general finding of this study was that meaningful classifications were obtained using such small training sets (Potter 1993).

G8.2.5 Conclusion

The obvious next step is to repeat the above study using much larger training sets. Apart from this obvious extension, another promising avenue for future research would be to use some form of hybrid ANN, in which an unsupervised network such as an SOM is used to form broad classifications. MLPs would then be used to provide finer discrimination using these preclassifications as part of their training. In doing so, we would be aiming to remove the need for manual (pre)classification using the human expert.

Acknowledgements

This work was made possible by financial assistance from the Australian Telecommunications and Electronics Research Board (grant no 32/185), the Advanced Telecommunications and Intelligent Software Research Programs within the University of Wollongong, as well as the Société d' Informatique et Télécommunications Aéronautiques (who funded the Face Recognition Project for Airport Security at the University of Wollongong). Thanks are also due to Michael Potter who trained the ANNs, Roger Bird (our 'human expert') and Eric Clayton, of the Australian Nuclear Science and Technology Organization, Lucas Heights, who provided the (preprocessed) obsidian data.

References

Barton J and Krinsley D 1987 Obsidian provenance determination by backscattered electron imaging *Nature* **326** 585–7

Eaton H and Oliver T 1992 Learning coefficient dependence on training set size *Neural Networks* **5** 283–8

Fulcher J 1994 A comparison of commercial ANN simulators *Computer Standards and Interfaces* **16** 241–51

Godfrey-Smith D and Haywood N 1984 Obsidian sources in Ontario prehistory *Ontario Archaeology* **41** 29–35

Godfrey-Smith D, Kronfeld J, Strull A and D'Auria J 1993 Obsidian provenancing and magmatic fractionation in central Oregon *Geoarchaeology* **8** 385–94

Hepner G, Logan T, Ritter N and Bryant N 1990 Artificial neural network classification using a minimal training set: comparison to conventional supervised classification *Photogramm. Eng. Remote Sens.* **56**

Hughes R 1986 Energy dispersive x-ray fluorescence analysis of obsidian from Dog Hill and Burns Butte, Oregon *Northwest Sci.* **60** 73–80

Myata Y 1991 *PlaNet Neural Network Simulator* University of Colorado, Boulder, CO

Nelson D, D'Auria J and Bennett R 1975 Characterisation of Pacific northwest coast obsidian by x-ray fluorescence analysis *Archaeometry* **17** 85–97

Neural Ware Inc 1993 *Neural Computing and Reference Guide*

Potter M 1993 Minimal training sets for neural networks *B. Comput. Sci. (Honours) Thesis* University of Wollongong, Department of Computer Science

Potter M, Fulcher J, Bird R and Clayton E 1994 Training artificial neural networks for obsidian provenancing studies *Proc. Australian Conf. Archaeometry (Armidale)*

Tom M and Tenorio M 1991 Short utterance recognition using a neural network with minimum training *Neural Networks* **4** 711–22

Torrence R, Specht J, Fullagar R and Bird R 1992 From Pleistocene to present: obsidian sources in west new Britain, Papua New Guinea *Records Australian Museum (Supplement)* **15** 83–98

Yan H 1992 Building a robust nearest neighbour classifier containing only a small number of prototypes *Int. J. Neural Syst.* **3** 361–9

PART H

THE NEURAL NETWORK RESEARCH COMMUNITY

PART H

THE NEURAL NETWORK RESEARCH COMMUNITY

H1

Future Research in Neural Computation

Contents

H1.1 Mathematical theories of neural networks

Shun-ichi Amari

Abstract

The brain is an enormously complex system having a rich structure and flexible information processing ability. It is a highly parallel, distributed and modifiable system different from the modern computer architecture. It is important to understand the system-theoretic aspects of the brain, such as how information is represented in the brain and what algorithms the brain uses to solve specific tasks of mental activities. The brain should have realized principles of information processing other than those of modern computers through a long history of evolution. Such principles should be analyzed mathematically by using abstract and idealized models of neural networks. The present section remarks on historical efforts and recent trends in mathematical approaches to (i) multilayer networks, (ii) recurrent networks and (iii) information geometry.

H1.1.1 Multilayer perceptrons

Rosenblatt (1961) proposed simple and multilayer *perceptrons* in the late 1950s and early 1960s and proved C1.1, C1.2
the convergence theorem of simple perceptrons (actually single neurons), having opened a new paradigm
in neural learning. Widrow (1966) used analog linear neurons (*adaline*) and proposed the gradient descent C1.1.3
learning rule, the so-called *delta rule*. However, their methods could not be applied directly to multilayer B3.3.3
networks. It was a very old but still not well known paper (Amari 1967) in the late 1960s that proposed
the stochastic descent learning rule for multilayer perceptrons including hidden units. This idea is called
the generalized delta rule which has been rediscovered many times and is now implemented by the *error* C1.2.3
backpropagation algorithm (Rumelhart *et al* 1986). There is research on modification and acceleration of
the method as well as on its application in various fields. Recently, structural learning has been paid much
attention and new learning algorithms have been proposed on the basis of statistical ideas (Jordan and
Jacobs 1994, Amari 1995). In addition to the learning algorithm, learning performances and capacities of
feedforward networks should be elucidated.

A network is trained by examples, which represent the structure to be learnt. Amari (1967) studied
the dynamical process of on-line learning, showing how fast the parameters converge to the desired target
and how large the fluctuating error around the optimal value is. The dynamics of on-line learning is now
an active area revived by a new statistical–physical method (Heskes and Kappen 1991, Sompolinski *et
al* 1995). When the number of available examples is limited, the criterion of minimizing the training
error does not necessarily imply *minimization of the generalization error*. There are a number of ideas B3.5.2
which overcome this difficulty. They are, for example, early stopping by cross-validation, introduction of
regularization terms, model selection by statistical and information-theoretic methods (see, for example,
Amari and Murata 1993, Murata *et al* 1994, Opper and Haussler 1991, Moody 1992, Watkin *et al* 1993,
Amari *et al* 1996).

Concerning the capacity, it is known that a one-layer perceptron can realize only a very limited class
of functions. However, when a network has one additional hidden layer, it has the universal property
that any continuous function can be approximated by it sufficiently well provided the number of hidden
neurons is sufficiently large. This is good but not so surprising. The problem is how well a given function
can be approximated as the number of hidden units increases. A function can be approximated by many

analytical methods, for example, the Taylor series expansions, Fourier expansions, spline functions and so on. It is known that these expansions are not free of the curse of dimensionality: in order to attain an approximation of error ε for a function $f(x)$ of an n-dimensional input x, the number of modifiable parameters increases in the order of $(1/\varepsilon^{1/2})^n$. This is intractable if $n = 100$. The surprising fact was revealed recently by Jones (1992) and then Barron (1993), that a neural network has an ability of function approximation which is free of the curse of dimensionality, that is, the number of required modifiable parameters does not increase exponentially as the input dimensions increase. Neural networks research opens a new approximation scheme of functions.

H1.1.2 Neurodynamics in recurrent networks

Neural networks of recurrent connections have been studied intensively for a long time. Behavior of such a network is represented by the dynamics of state transition; differential equations in the continuous case and difference state update equations in the discrete time case. Macroscopic behavior of networks of random recurrent connections have been analyzed mathematically since the early 1970s (Amari 1972a, Harth *et al* 1970, Wilson and Cowan 1972). Theoretical foundations of such dynamics were studied (Amari *et al* 1977, Rozoner 1969). The autocorrelation associative memory model was studied in the early 1970s by a network of recurrent connections (Nakano 1972, Anderson 1972, Amari 1972b). It was Hopfield (1982) who introduced the asynchronous state transition to the model and the concept of the energy function by

C1.3.4   the spin-glass analogy. Hence, a recurrent network is sometimes called the *Hopfield network*. Dynamics of recalling processes were proposed by Amari and Maginu (1988) and then by many others (Okada 1996, Coolen and Sherrington 1993). The network can memorize and recall temporal pattern sequences when the asymmetric connections are permitted (Amari 1972b).

A recurrent neural network of symmetric connections has a Lyapunov function (energy function) so that it shows no oscillatory or chaotic behavior (Cohen and Grossberg 1983). Such a network is called an attractor neural network because of this property. Much richer dynamical behaviors also emerge in neural fields (Wilson and Cowan 1973, Amari 1977). Self-organization of neural fields has the ability to generate topological maps, as was proposed by Willshaw and von der Malsburg (1976). Kohonen (1982) proposed powerful algorithms of formation of self-organizing topological maps. Takeuchi and Amari (1979) studied dynamical stability and instability of such maps, showing spatial instability of topological maps which generates patch or columnar structures (see also Ritter and Schulten 1988). Much attention has been paid recently to temporal encoding of information and chaotic behaviors.

H1.1.3 Information geometry of manifolds of neural networks

We have so far treated mostly the information processing ability of various types of neural networks. However, it is important to search for the geometry of the set of all neural networks of a fixed architecture. Let $w = (w_1, \ldots, w_p)$ be the set of structural or modifiable parameters (connection weights) of networks. Then, the set is regarded as a manifold and called the manifold of neural networks or, in short, the neural manifold, where w is a coordinate system to specify each network in the set.

It is important to study the intrinsic geometry of the neural manifold. For example, we consider a

C1.2   *feedforward network* (multilayer perceptron) where the input–output relation is written as $z = f(x; w)$ where w summarizes all the modifiable parameters. Let S be the set of all the smooth functions $S = \{\varphi(x)\}$. Then, the set of functions $M = \{f(x; w)\}$ realizable by neural networks corresponds to the neural manifold. It is embedded in S as a curved submanifold. Given a function $\varphi(x)$, it is important to find w_0 such that φ is optimally approximated by $f(x; w_0)$. The capacity of M shows how well a function is approximated. On the other hand, learning is the problem of finding w_0 from examples. When M is curved in S to fill most parts of S, the capacity is large. However, this curvature produces many local minima and learning capability decreases. All of these are related to the intrinsic geometry of M. When the behavior of neurons is stochastic or noise-contaminated, the output z of the network is specified by the conditional distribution $p(z|x; w)$ conditioned on input x. In this case, M is the set of all the conditional probability distributions parametrized by w. Information geometry (Amari 1985) gives an intrinsic structure to the manifold of probability distributions. It is a Riemannian manifold with a dual pair of affine connections and serves as a fundamental basis of statistics. Information geometry provides a geometrical insight for analyzing neural manifolds (Amari 1991, Amari *et al* 1992).

C1.4   *Stochastic neural networks* provide nonlinear modeling of multivariate data. This is related to nonlinear multivariate statistical analysis so that neural modeling is a recent hot topic in statistics. On the other hand, statistical techniques give neural network researchers powerful methods of analysis. They are, for example, projection-pursuit, EM algorithm, asymptotic theories, learning curves, Bayesian priors and overtraining.

All of these are related to information geometry. The geometrical foundation of the EM algorithm and dual minimization procedures are given by Amari (1995). Information geometry will grow as an indispensable method of mathematical theories of neural networks.

References

Amari S 1967 Theory of adaptive pattern classifiers *IEEE Trans. Electr. Comp.* **16** 299–307

——1972a Characteristics of random nets of analog neuron-like elements *IEEE Trans. Syst. Man Cybern.* **2** 643–57

——1972b Learning patterns and pattern sequences by self-organizing nets of threshold elements *IEEE Trans. Comp.* **21** 1197–206

——1977 Dynamics of pattern formation in lateral-inhibition type neural fields *Biol. Cybern.* **27** 77–87

——1985 *Differential-Geometrical Methods in Statistics* (New York: Springer)

——1991 Dualistic geometry of the manifold of higher-order neurons *Neural Networks* **4** 443–51

——1995 Information geometry of EM and em algorithms for neural networks *Neural Networks* **8** 1379–408

Amari S, Kurata K and Nagaoka H 1992 Information geometry of Boltzmann machines *IEEE Trans. Neural Networks* **3** 260–77

Amari S and Maginu K 1988 Statistical neurodynamics of associative memory *Neural Networks* **1** 63–73

Amari S and Murata N 1993 Statistical theory of learning curves under entropic loss criterion *Neural Computation* **5** 140–53

Amari S, Murata N, Müller K R, Finke M and Yang H 1996 Asymptotic statistical theory of overtraining and cross-validation *IEEE Trans. Neural Networks* to appear

Amari S, Yoshida K and Kanatani K 1977 A mathematical foundation for statistical neurodynamics *SIAM J. Appl. Math.* **33** 95–126

Anderson J A 1972 A simple neural network generating interactive memory *Math. Biosci.* **14** 197–220

Barron A R 1993 Universal approximation bounds for superpositions of a sigmoidal function *IEEE Trans. Inf. Theory* **39** 930–945

Cohen M A and Grossberg S 1983 Absolute stability of global pattern formation and parallel memory storage by competitive neural networks *IEEE Trans. Syst. Man Cybern.* **13** 815–25

Coolen A C C and Sherrington D 1993 Dynamics of fully connected attractor neural networks near saturation *Phys. Rev. Lett.* **71** 3886–9

Harth E M, Csermely T J, Beek B and Lindsay R D 1970 Brain functions and neural dynamics *J. Theor. Biol.* **26** 93–120

Heskes T M and Kappen B 1991 Learning processes in neural networks *Phys. Rev.* A **440** 2718–26

Hopfield J J 1982 Neural networks and physical systems with emergent collective computational abilities *Proc. Natl Acad. Sci.* **79** 2445–2458

Jones L K 1992 A simple lemma on greedy approximation in Hilbert space and convergence rates for projection pursuit regression and neural network training *The Annals of Statistics* **20** 608–613

Jordan M I and Jacobs R A 1994 Hierarchical mixtures of experts and the EM-algorithm *Neural Comput.* **6** 181–214

Kohonen T 1982 Self-organized formation of topologically correct feature maps *Biol. Cybern.* **43** 59–69

Moody J E 1992 The effective number of parameters: An analysis of generalization and regularization in nonlinear systems *Advances in Neural Information Processing Systems* ed J E Moody, J Hanson and J Kangas (Amsterdam: Elsevier) pp 847–54

Murata N, Yoshizawa S and Amari S 1994 Network information criterion—Determining the number of hidden units for an artificial neural network model *IEEE Trans. Neural Networks* **5** 865–872

Nakano K 1972 Association—A model of associative memory *IEEE Trans. Syst. Man Cybern.* **2** 381–8

Okada M 1996 Notions of associative memory and sparse coding *Neural Networks* **9** to appear

Opper M and Haussler D 1991 Calculation of the learning curve of Bayes optimal classfication algorithm for learning a perceptron with noise *Proc. 4th Ann. Workshop on Computational Learning Theory* (San Mateo, CA: Morgan Kaufmann) pp 75–87

Ritter H and Schulten K 1988 Convergence properties of Kohonen's topology conserving maps: fluctuation, stability and dimension selection *Biol. Cybern.* **60** 59–71

Rosenblatt F 1961 *Principles of Neurodynamics* (New York: Spartan)

Rozonoer L I 1969 Random logical nets I *Automat. Telemekh.* **5** 137–47

Rumelhart D, Hinton G E and Williams R J 1986 Learning internal representation by error propagation *Parallel Distributed Processing* vol 1 *Foundations* ed Rumelhart D and McClelland J L (Cambridge, MA: MIT Press) pp 318–62

Sompolinski H, Barkai N and Seung H S 1995 On-line learning dichotomies: Algorithms and learning curves *Neural Networks: The Statistical Mechanics Perspectives* ed J H Oh Ch Kwon and S Cho (Singapore: World Scientific) pp 105–30

Takeuchi A and Amari S 1979 Formation of topographic maps and columnar microstructure *Biol. Cybern.* **35** 63–72

Watkin T L H, Rau A and Biehl M 1993 The statistical mechanics of learning a rule *Rev. Mod. Phys.* **65** 499–556

Widrow B 1966 *A Statistical Theory of Adaptation* (Oxford: Pergamon)

Willshaw D J and von der Malsburg C 1976 How patterned neural connections can be set up by self-organization *Proc. R. Soc.* B **194** 431–45

Wilson H R and Cowan J D 1972 Excitatory and inhibitory interactions in localized populations of model neurons *Biophys. J.* **12** 1–24

——1973 Stationary states and transients in neural populations *J. Theor. Biol.* **40** 77-106

H1.2 Neural networks: natural, artificial, hybrid

H John Caulfield

Abstract

A personal reflection on the future of neural network research.

The direction of my future neural network research and that of many of my colleagues is to achieve mammalian functionality: the olfactory capabilities of a dog, the visual world of a primate, the brilliant intellect and intuition of a human. These are easily expressed, easily understood goals. Achieving them would be of immense practical value. In addition, reflexively, it will help us understand what nature is doing. Humans habitually project their technology onto nature. At one time or other the brain has been viewed as a fluidic system, a telephone switchboard, a digital computer, a hologram, a set of attractor neural networks and arrays of pulse-coupled neural networks. Some of these (pumps, switchboards, holograms) are obviously wrong. Yet even they had enough truth to have been useful. The neural models are surely more nearly correct, so they can teach us more. Let us look at the areas just noted in a little more detail.

The olfactory system seems simple, but we know it is not. Real mammalian olfactory neural networks involve chaotic attractors in a way yet to be sorted out in detail. But we do know the functions carried out. Each sniff identifies and notes the strength of one of N possible chemical components. For the next sniff, that component is suppressed and the process is repeated. Thus a feature vector is generated sequentially. The order as well as the magnitudes are encoded, so this is a syntactic description. This, in turn, can be recognized by a more conventional neural network. Thus the power of syntactic pattern recognition is available with the simplicity of statistical pattern recognition. My own version of this is optical. It is a NOSE (neural optical sequencing engine), which, when coupled to available chemical sensor arrays, can be helpful in detecting such things as drugs, chemical agents and explosives.

The mammalian visual system is very complex. The optics (cornea and lens) are followed by an exuberantly complex sensor–preprocessor called the retina. In every meaningful sense, the retina is part of the brain. It is parallel and pipelined through multiple layers to present the various visual layers in the brain with a cleaner edge-enhanced, bandwidth-compressed scene description than an ordinary detector array on the retina would. We are developing an optical multilayer artificial retina which exhibits the same functionality as our RETINA (retinally inspired architecture). RETINA should simplify the task for subsequent processors, for example, for machine vision. RETINA hooked up to a lens system could produce signals which might be directly input to the optic nerve or to V1 to give sight to the blind or create hypervision (better-than-normal vision).

The most exciting venture, to me, is to emulate the functionality of the human brain. Humans have language. They have intention, emotion, meaning. Computers, on the other hand, function like very large look up tables. They can be viewed as symbol manipulators. But they intend nothing, feel nothing, mean nothing. My greatest current interest is to use the pulse-coupled neural network (PCNN) to answer questions such as the following:

- How can a computer mean something? How can it intend anything?
- Can there be a 'language of the mind'? How do we learn language?
- How do instincts get inherited? Do birds have song genes, beavers have dam genes and primates have snake genes? Or is there an inherited neural basis for behavior? How would that work?

- What is 'attention'? What would an attention organ do? How would it connect with the rest of the visual system?
- What is 'subconscious thought'? Can and/or should we create a computer subconscious?
- What is 'consciousness'? Is consciousness, attention to attention? Do mystics attend to consciousness?

In each case, the PCNN appears to offer unexpected and plausible answers which should allow artificial systems to exhibit those features (albeit crudely, compared to a human). What are my future directions? The answer is: to achieve biological functionality without directly imitating biological methodologies. Imitation is too difficult to understand or implement. Similar functionality is within our grasp. My approaches will be different from yours. My goals and vision, however, are ones I commend strongly to others.

What I hope the reader has not missed is the fact that I am not modeling the mammalian nervous system. Rather, I am trying to follow the procedure shown in figure H1.2.1.

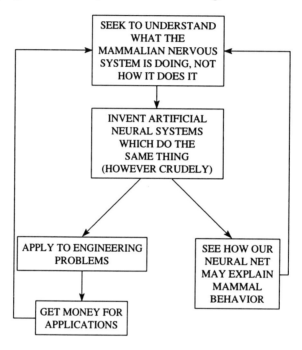

Figure H1.2.1. A flowchart describing my research.

H1.3   The future of neural networks

J G Taylor

Abstract

A personal reflection on the future of neural network research.

It is difficult to predict the future of a subject experiencing such a fast rate of growth as is now occurring in neural networks. Their use is expanding into an increasing number of applied areas—finance, business, industrial process control, energy control, human resource management such as personnel selection, and many areas in which the fast pattern recognition powers of the systems are able to handle difficult template matching problems, and where the template itself is not well defined. Solutions to problems such as credit card fraud detection (accomplished by noting the special temporal pattern of each credit card use, or by voice or other identification approaches to secure recognition) are difficult to achieve at the present success levels and speed with any other method. However, neural networks are not only going to provide ever better applications to hard problems in business and industry but also to help uncover some of the higher powers of the cognitive processes of the human brain. It is through exploration of these higher processes that there is a chance that really hard problems about intelligence, as it actually is in humans, may be solved and moved onto artificial systems (with expected improvements?).

It is clear that neural networks are now part of the toolkit of the adaptive information processor. The applications mentioned above are only a small part of the many that are now being investigated with ever greater power and success. At the same time it is also clear that the use of hybrid techniques can make a neural approach much more effective. Thus the combination of the use of constraint satisfaction (such as arc consistency) and a neural (relaxation) network allows a neural network solution to a hard optimization problem (the radio links frequency assignment problem) to be far more effective than if the neural approach were used on its own (Bouju *et al* 1995). Ever more use of hybridization can be expected and multihybrid solutions are already well developed for some problems. As the understanding of the nature of neural systems grows, then I expect a similar deepening of the understanding of how, when, why, and where to hybridize. Thus it may be appropriate to determine a good solution to a problem by an expert system or an exact solution technique, but then to develop a neural system based on the exact solution which will be more robust to external perturbations or small changes in the parameters of the problem. Having said that I expect neural applications to become ever more 'thick on the ground' over the next few years, I have also said that I assume that there will be a corresponding deepening of the nature of neural network theory. This is a strong trend already, with the advance of the statistical community to our aid (Cherkassky 1995)—and learning something in the process—as well as the use of information theory and related techniques to give a deeper understanding of the nature of optimal learning algorithms (Amari 1991). I will also not forget the enormous insights which have come to neural networks from the use of statistical mechanics techniques (Amit 1989); this is still giving an increased understanding of the nature of learning laws and the manner in which sudden changes in the weight space usually observed in training arise from phase changes of the corresponding statistical mechanical system. There is also an increasing insight from dynamical systems, in which the above phase change corresponds to a bifurcation of the dynamical system in the space of weights, of one of the classical and well-classified sorts. Thus the nature of temporary minima is explained in these terms. Furthermore, the avoidance of local minima is also slowly being achieved, using techniques of *simulated annealing* or by tunneling methods which   C1.4.2
deform the total energy surface itself so as to make the problem simple, and then change it back smoothly so as to avoid falling into the basins of attraction of the local minima as they start to grow and become putatively dangerous.

All of these techniques are sometimes said to be producing 'intelligent' systems. However, the intelligence involved is still remote from that of ourselves. To predict further it is necessary to give a definition of intelligence itself. I favor that which states that intelligence is the ability to manipulate internal (neural) representations so as to aid the achievement of a goal of importance to the system. Such a process clearly occurs in the human brain, and it is my claim that we are now in the process of beginning to understand what the neural underpinning might be for such processing. There are already several simple models of frontal lobe processing to solve simple tasks, such as the Wisconsin card sorting task (Levine and Prueitt 1989) or the recency task (Monchi and Taylor 1995). It now appears feasible to attempt to model the global structure of the frontal system, in terms of the great anatomical loops, of the ventrolateral, dorsolateral, orbitofrontal, eye fields, and motor/supplementary cortices discussed by Alexander *et al* (1986). Such architectures present a specific and finite problem to neural modelers to solve as follows: how does the observed architecture enable executive function to be achieved, and intelligence to thereby be supported?

The answers to this might not be so far away. The ACTION network has been suggested (Alavi and Taylor 1995) as having the abilities that are crucial: to carry representations of inputs over periods of time in a working or active memory, to be able to make choices between different schemata by some form of lateral inhibition, and to allow for the learning of temporal sequences of actions, so developing schemata and higher-order chunking. The interactions between the various frontal loops given by Alexander *et al* (1986) are clearly where the most important features of executive and intelligent function might lie. There is no reason why such processes cannot be amenable to analysis by both dynamical systems theory and simulation. Both are part of the neural networks toolkit presently available, and they are being improved all the time.

This leads us to one of the mysteries of mankind—the nature of consciousness and the mind. There are increasing numbers of groups now working seriously on that question. Especially with the advent of noninvasive techniques it is possible to contemplate constructing large-scale simulations of the brain at a global level (Taylor 1995). It may be possible in this manner to build an ever more precise artificial laboratory to allow for improved understanding of the brain and mind. The next century will surely lead to the computing power necessary to crack the problem; how far after the year 2000 we will need to go to properly understand the principles of mind is not clear to me now, but it may not be so far away as all that.

References

Alavi F and Taylor J G 1995 A basis for long-range inhibition across cortex *Lateral Interactions in the Cortex* ed J Sirosh, R Mikkulainen and Y Choe, sited at http:/www.cs.utexas.edu/users/nn/lateral_interactions_book/cover.html

Alexander G E, DeLong M R and Strick P L 1986 Parallel organization of functionally segregated circuits linking basal ganglia and cortex *Ann. Rev. Neurosci.* **9** 357–81

Amari S-I 1991 Dualistic geometry of the manifold of higher order neurons *Neural Networks* **4** 443

Amit D 1989 *Models of Brain Function* (Cambridge: Cambridge University Press)

Bouju A, Boyce J F, Dimitropolous C H D, vom Scheidt G, Taylor J G, Likas A, Papageorgiu G and Stafylopatis A 1995 Intelligent search for the radio links frequency assignment problem *Proc. Int. Conf. on Digital Signal Processing DSP95* (New York: IEEE Press)

Cherkassky V 1995 Neural network and statistical methods for function estimation *WCNN95 Short Course*

Levine D S and Prueitt P S 1989 Modelling some effects of frontal lobe damage: novelty and perseveration *Neural Networks* **2** 103–16

Monchi O and Taylor J G 1995 A model of the prefrontal loop that includes the basal ganglia in solving the recency task *World Congress on Neural Networks WCNN95* vol III (Erlbaum) pp 48–51

Taylor J G 1995 Modules for the mind of psyche. Invited Talk *World Congress on Neural Networks WCNN95* vol II (Erlbaum) pp 967–72

H1.4 Directions for future research in neural networks

James A Anderson

Abstract

Neural network hardware is sometimes claimed to be inspired by the design of the brain, that is, it is *neuromorphic*. However, the operations of the resulting systems only occasionally act *psychomorphic*, that is, working like the mind. In this article we point out some places where neural network technology must be significantly extended if it is to act more like minds. (1) There is little understanding of intermediate level organization above the level of single units and below the level of the entire system. (2) The theoretical formulation of neural network learning needs to advance beyond 1920s behaviorism. (3) Flexibility of operation and control of the direction of a computation are probably more important to behavior than retrieval accuracy. (4) Neural networks are almost always special purpose devices. Successful system performance lies in the details of the architecture and the data representation.

H1.4.1 Introduction

When neural networks regained popularity in the mid-1980s, a term that was sometimes used to describe systems containing them was 'neuromorphic'. 'Brain-like computing' was another way of saying about the same thing. When one of these terms was used in engineering the implication was that the artificial devices being built were following at least some of the design principles of the mammalian brain.

To those of us professionally concerned with behavior, a parallel set of names might be proposed: 'psychomorphic' systems and 'mind-like computing'. Artificial intelligence, as classically defined, is describable by these names, though when AI first developed in the 1950s and 60s it deliberately paid little attention to the substantial amount known about the facts of human behavior, believing that sheer cleverness was capable of overcoming ignorance. The field of neural networks may be making the same mistake.

A major conceptual problem in the future of neural networks is that, even if neural networks are in some vague architectural sense neuromorphic, they are rarely psychomorphic. Even though there is a large body of lawful, regular and reproducible experimental results in the behavioral sciences, these ideas have rarely had much influence in the neural network community, outside of a small number of researchers who specifically try to model human cognition. Let me state several reasons for this neglect.

H1.4.2 Missing levels of organization: neuroscience

Neural network models are built from elementary computing units. The largest neural network simulations used in practice contain perhaps a few thousand units. The human brain contains billions of neurons. Current neural network models have a severe problem using or even acknowledging the intermediate levels of organization that must exist in this numerical gap in scale between the properties of single units and the coordinated activity of the whole brain.

Consider a large business organization like IBM. We can follow an individual employee during the course of a day. Or we can follow the health of the company as a whole by looking at the annual report. It would be difficult to infer from either of these sources of information the presence of workgroups, departments and divisions, that is, groups of employees and groups of groups of employees, where in fact most of the work of the company is organized and performed. Similarly, government has complex and essential intermediate-level structures, for example, in rough order of size, neighborhood, city, county, state and federal.

Experimental techniques in neuroscience currently allow us to look at single-unit recordings for the behavior of single neurons and gross electrical activity (EEG, evoked responses, imaging) for overall activation levels, roughly the lowest and highest levels of neural organization. As many have pointed out, there are several orders of magnitude of grouping that must exist, have been conjectured to exist, are felt to be important, but about which almost nothing is known. For a large functional neuromorphic system, there is surely much more important structure present than is currently assumed and the details of this additional structure will strongly affect the overall behavior of the system.

H1.4.3 Missing levels of organization: cognitive science

The most commonly used formulations of network learning are limited and often misleading from the point of view of a psychomorphic system. Neural network theory has been strongly influenced, for better B6 and worse, by the mathematics of *classical pattern recognition*. Typically, pattern recognition assumes that sensors have provided a set of input data connected to a classification, say a set of pixels corresponding to the written letter 'A'. A network is presented with a number of examples of the classification in a training set and the weights in the network are adjusted by various learning algorithms so as to make it classify more accurately in the future.

It can be shown—see the many examples in this book—that properly designed neural networks can do this operation effectively enough for many useful applications. However, a psychomorphic engineer might ask if this is all that we want to do. This structure, with an input pattern transformed in the network to an output pattern, reproduces in form classical stimulus–response (S–R) learning from psychology. S–R learning was proposed by the behaviorists in the 1920s and 30s as the only true basis of a scientific psychology. Essentially, we can solve the problem of animal behavior when we make lists of externally observable stimuli, the associated observed responses, and assume the brain is there to make links between them. No hidden mental processes need be invoked.

Clearly there is some truth behind this analysis. Association has been known to be a primary mechanism of learning since Aristotle. Even Aristotle, however, was quite aware, and it has been amply confirmed by work in psychology and cognitive science over the past decades, that such a limited definition of association cannot explain many aspects of behavior. It is therefore distressing to see neural network theorists deliberately, or even worse, unconsciously, reproduce a severely limited and inadequate view of mental operation.

H1.4.4 Controllability, accuracy and flexibility

Focus on the formation of accurate associations has distracted attention from a number of other important requirements for a psychomorphic system. Controllability, flexibility and teachability are at least as important in human cognition as accuracy in retrieval, probably more so. For example, consider the pixel pattern that a letter recognizer classifies as a letter 'A'. Depending on the context this pattern can be labeled as a capital 'A', a grade in a college class, an indefinite article in English, and so on.

The switch between one possible association of the pattern and another is extremely rapid. For example, in a psychological experiment, an 'A' can be first associated with pressing a button on the left. Time to respond to the presentation of the 'A' will become faster with repeated presentations, even though responses have been error-free since the beginning of the experiment. Suppose a verbal instruction now tells the subject to respond to an 'A' by pressing the button on the right. Suddenly the subject is making a different response. The responses may be a little slower at first, but performance is still error-free. This flexibility is common and so trivial that we hardly even think about how difficult it must be to get a neural network to completely and correctly shift its input–output relationships in a matter of milliseconds. My guess is that the need for this flexibility places much more stringent constraints on possible neuromorphic architectures than accurate learned association. The psychomorphic system can constantly and quickly reprogram itself.

This example also suggests the importance of 'teachability' for network operation. Somehow presentation of properly structured inputs can speed up learning by orders of magnitude. In this case, the inputs causing a change in association were not even examples of the association but verbal instructions recombining past learning. Learning in school would be a painful and slow process if it were purely associative. Learning does not proceed by a random pairwise accretion of facts in knowledge space. Something much more complex is occurring, involving the formation of mental structures, use of interlocked concepts and detailed mental models and the presentation of specific factual examples which are explained by a teacher.

The time course of real learning is often strikingly unlike the time course of simple neural network learning. Neural network learning typically starts with a *tabula rasa*, learns the first associations quickly and accurately, and then gets slower and less accurate as it learns more and more.

Real learning often starts slowly—for example, learning the times tables in grade school—and then accelerates, so college mathematics courses provide an immense amount of information very rapidly, once the foundations are built. As William James commented, '... the more other facts a fact is associated with in the mind, the better possession of it our memory retains .... Let a man early in life set himself the task of verifying such a theory as that of evolution and facts will soon cluster and cling to him like grapes to their stems. Their relations to the theory will hold them fast ...' (James 1892/1984). The point here is that real memory has strong high-level structure that uses simple association as an elementary mechanism. Past information can aid in the learning and retrieval of later information.

One of the best critiques of simple neural networks is in the well known paper by Jerry Fodor and Zenon Pylyshyn (1988), who, among other points, observed that simple association is such an inefficient way to build an information processing and retrieval system that an engineer would be strongly advised to use something else if the system was to be in any way useful.

An obvious and practical task for future research is to take today's relatively well understood simple neural network systems and try to combine them in such a way as to reproduce at least a little of the flexibility and controllability observed in human memory.

H1.4.5 Generality versus specificity

Because the history of the field is tied to pattern recognition and computer science, there is a tendency to believe that neural networks form general computing systems in the sense that Turing machines form universal computers. There is absolutely no reason to believe that this is true. The biological nervous system is concerned with specificity and not generality: specific sensory systems, specialized structures, specific kinds of computation.

Although we like to think the human brain is very general, when mental operations are looked at in detail striking limitations appear. For example, the simple logic operation XOR, the *bête noire* of neural networks, can be incorporated into a puzzle. This puzzle can be solved by humans, though often with some difficulty. The same logical structure when instantiated in a different problem often does not generalize. There is a substantial body of research on this observation in cognitive science.

Successful computation in neural networks is dependent on details of the *data representation*, that is, B4 on how the pattern of input and output unit activation relates to the world. Neural networks are extremely sensitive to representations. In a real sense, the data representation is the mechanism by which networks are programmed. The choice of a good data representation is of far more value toward the solution of a problem than is the choice of the learning rule or network.

For various reasons, including the fact that neural structures tend to be noisy, and that small errors can propagate and amplify, it is not possible to have psychomorphic computers perform in sequence the very large number of accurate elementary computational steps that characterize operation of digital computers. A small sequence of computational operations combined with an effective input and output neuromorphic data representation comprises the entire psychomorphic computation. John von Neumann pointed out this essential characteristic of neural computation in 1958.

The biological brain contains true marvels of data representation, using details of neuroanatomy and neurophysiology to respond to useful properties of the world. However, data representations tend to be very problem specific. The more that is known about a given problem, the less general adaptability is needed. Learning requires ignorance; if everything is known, nothing need be learned. Learning and adaptation are dangerous for an animal because they involve rewiring the nervous system and should be used only when necessary. It has been suggested that normal learning is one end of a continuum with pathology lying at the other. Here, perhaps more than in many fields, God is in the details.

I suppose the point of this discussion is that our field, the field presented in this handbook, knows only a little about the earliest stages of intelligent system design. The outlines of intermediate-level network organization and the rules, if there are any, for designing data representations for specific problems remain to be discovered. It is not even clear what is the best way to analyze complex intelligent systems; proper analysis may start with traditional statistics and its extensions to pattern recognition but is unlikely to end that way.

The most important future developments for both intelligent machines and for the understanding of our own mental processes may arise when the constraints and the abilities seen at the highest levels of cognitive function can be connected with low- and intermediate-level neural network architectures.

References

Fodor J A and Pylyshyn Z W 1988 Connectionism and cognitive architecture: a critical analysis *Cognition* **28** 2–72
James W 1892/1984 *Psychology: Briefer Course* (Cambridge, MA: Harvard University Press) pp 257–9
von Neumann J 1958 *The Computer and the Brain* (New Haven, CT: Yale University Press) pp 75–82

LIST OF CONTRIBUTORS

List of Contributors

Igor Aleksander (C1.5)

Professor of Neural Systems Engineering,
Imperial College of Science, Technology and Medicine,
London,
United Kingdom
e-mail: i.aleksander@ic.ac.uk

Nigel M Allinson (G1.1)

Professor of Electronic Systems Engineering,
University of Manchester Institute of Science and
Technology,
United Kingdom
e-mail: allinson@umist.ac.uk

Luis B Almeida (C1.2)

Professor of Signal Processing and Neural Networks,
Instituto Superior Tecnico,
Technical University of Lisbon,
Portugal
e-mail: lba@inesc.pt

Shun-ichi Amari (H1.1)

Director of the Brain Information Processing Group,
RIKEN (Institute of Physical and Chemical Research),
Saitama,
Japan
e-mail: amari@zoo.riken.go.jp

James A Anderson (Foreword, H1.4)

Professor of Cognitive and Linguistic Sciences,
Brown University,
Providence,
Rhode Island,
USA
e-mail: james_anderson@brown.edu

Nirwan Ansari (G2.3)

Associate Professor of Electrical and Computer
Engineering,
New Jersey Institute of Technology,
Newark,
USA
e-mail: ang@hertz.njit.edu

Michael A Arbib (A1.2, B1)

Professor of Computer Science and Neurobiology,
University of Southern California,
Los Angeles,
USA
e-mail: arbib@pollux.usc.edu

Patrick Argos (G4.4)

Professor and Senior Research Group Leader in
Biocomputing,
European Molecular Biology Laboratory,
Heidelberg,
Germany
e-mail: argos@mailserver.embl-heidelberg.de

William W Armstrong (C1.8, G2.1, G5.1)

Professor of Computing Science,
University of Alberta;
and President of Dendronic Decisions Limited,
Edmonton,
Alberta,
Canada
e-mail: arms@cs.ualberta.ca

James Austin (F1.4, G1.7)

British Aerospace Senior Lecturer in Computer Science,
and Director of the Advanced Computer Architecture
Group,
University of York,
United Kingdom
e-mail: austin@minster.york.ac.uk

Timothy S Axelrod (E1.1)

Senior Fellow,
Mount Stromlo Observatory,
Canberra,
Australia
e-mail: tsa@mso.anu.edu.au

Magali E Azema-Barac (G6.3)

Quantitative Researcher,
U S West Inc,
Englewood,
Colorado,
USA
e-mail: mazemab@uswest.com

George Y Baaklini (G2.6)

Nondestructive Evaluation Group Leader,
Structural Integrity Branch,
NASA Lewis Research Center,
Cleveland,
Ohio,
USA
e-mail: baaklini#y#-george@lims-al.lerc.nasa.gov

Martin Bäker (G3.2)

Research Assistant,
Institut für Theoretische Physik,
Universität Hamburg,
Germany
e-mail: baeker@x4u2.desy.de

Etienne Barnard (G1.5)

Associate Professor of Computer Science and Electrical
Engineering,
Oregon Graduate Institute of Science and Technology,
Beaverton,
USA
e-mail: barnard@cse.ogi.edu

T K Barrett (G3.1)

Senior Scientist,
ThermoTrex Corporation,
San Diego,
California,
USA
e-mail: tbarrett@crash.cts.com

Andrea Basso (F1.5)

Senior Researcher,
École Politechnique Fédéreli de Lausanne (EPFL),
Switzerland
e-mail: basso@tcom.epfl.ch

Russell Beale (Preface, B5.1)

Lecturer in Computer Science,
University of Birmingham,
United Kingdom
e-mail: r.beale@cs.bham.ac.uk

Valeriu Beiu (E1.4)

Senior Lecturer in Computer Science,
Bucharest Polytechnic University,
Romania;
and Postdoctoral Fellow,
Los Alamos National Laboratory,
New Mexico,
USA
e-mail: beiu@mth.kcl.ac.uk

Laszlo Berke (G2.6)

Senior Staff Scientist,
NASA Lewis Research Center,
Cleveland,
Ohio,
USA
e-mail: berke#m#-laszlo@lims-a1.lerc.nasa.gov

Christopher M Bishop (B6)

Professor of Neural Computing,
Neural Computing Research Group,
Aston University,
Birmingham,
United Kingdom
e-mail: c.m.bishop@aston.ac.uk

F Blayo (G6.1)

Consultant;
and Director of PREFIGURE,
Lyon,
France;
and Lecturer in Neural Networks,
Swiss Federal Institute of Technology,
Lausanne,
Switzerland
e-mail: fblayo@babel.asi.fr

David Bounds (G6.2)

Professor of Computer Science and Applied Mathematics,
Aston University;
and Recognition Systems Ltd,
Birmingham,
United Kingdom
e-mail: boundsd@aston.ac.uk

P Stuart Bowling (G2.7)

Technical Staff Member,
Los Alamos National Laboratory,
New Mexico,
USA
e-mail: psb@lanl.gov

Charles M Bowden (G3.3)

Senior Research Scientist,
US Army Missile Command,
Redstone Arsenal,
Alabama,
USA;
and Adjunct Professor of Physics and Optical
* Science,*
University of Alabama,
Huntsville,
USA
e-mail: fybt01a@prodigy.com

Thomas M Breuel (G1.3)

IBM Almaden Research Center,
San Jose,
California,
USA
e-mail: tmb@almaden.ibm.com

Stanley K Brown (G2.7)

Technical Staff Member,
Los Alamos National Laboratory,
New Mexico,
USA
e-mail: skbrown@lanl.gov

Masud Cader (C1.4)

CSIS,
Department of Computer Science,
Washington, DC,
USA
e-mail: mcader@worldbank.org

Gail A Carpenter (C2.2.1)

Professor of Cognitive and Neural Systems;
and Professor of Mathematics,
Boston University,
Massachusetts,
USA
e-mail: gail@cns.bu.edu

H John Caulfield (H1.2)

University Eminent Scholar,
Alabama A&M University,
Normal,
USA
e-mail: caulfield@caos.aamu.edu

Krzysztof J Cios (C1.7, D1, G2.6, G2.12)

Professor of Electrical Engineering and Computer Science,
University of Toledo,
Ohio,
USA
e-mail: fac1765@uoft01.utoledo.edu

Ron Cole (G1.5)

*Director of the Center for Spoken Language
Understanding;
and Professor of Computer Science and Engineering,
Oregon Graduate Institute of Science and Technology,
Beaverton,
USA*
e-mail: cole@cse.ogi.edu

Shawn P Day (F1.8)

*Senior Scientist,
Synaptics Inc,
San Jose,
California,
USA*
e-mail: shawn@synaptics.com

Massimo de Francesco (B2.9)

*University of Geneva,
Switzerland*
e-mail: massimo@cui.unige.ch

Thierry Denœux (F1.2)

*Enseignant-Chercheur en Génie Informatique,
Université de Technologie de Compiégne,
France*
e-mail: tdenoeux@hds.univ-compiegne.fr

Alan J Dix (G7.1)

*Reader in Software Technology,
University of Huddersfield,
United Kingdom*
e-mail: alan@zeus.hud.ac.uk

Mark Fanty (G1.5)

*Assistant Professor of Computer Science,
Oregon Graduate Institute of Science and Technology,
Beaverton,
USA*
e-mail: fanty@cse.ogi.edu

Emile Fiesler (Preface, B2.1–B2.8, C1.7, E1.2)

*Research Director,
Institut Dalle Molle d'Intelligence Artificielle Perceptive
(IDIAP),
Martigny,
Switzerland*
e-mail: efiesler@idiap.ch

Janet E Finlay (G7.1)

*Senior Lecturer in Information Systems,
University of Huddersfield,
United Kingdom*
e-mail: j.e.finlay@hud.ac.uk

Dmitrij Frishman (G4.4)

*Postdoctoral Fellow,
European Molecular Biology Laboratory,
Heidelberg,
Germany*
e-mail: frishman@mailserver.embl-heidelberg.de

Bernd Fritzke (C2.4)

*Postdoctoral Researcher in Systems Biophysics,
Institute for Neural Computation,
Ruhr-Universität Bochum,
Germany*
e-mail: fritzke@neuroinformatik.ruhr-uni-bochum.de

Hiroshi Fujita (G5.2)

*Professor of Computer Engineering,
Gifu University,
Japan*
e-mail: fujita@fjt.info.gifu-u.ac.jp

John Fulcher (F1.6, G1.2, G8.2)

*Senior Lecturer in Computer Science,
University of Wollongong,
New South Wales,
Australia*
e-mail: john@cs.uow.edu.au

George M Georgiou (C1.1)

*Associate Professor of Computer Science,
California State University,
San Bernadino,
USA*
e-mail: georgiou@.csci.csusb.edu

Richard M Golden (G5.4)

*Assistant Professor of Psychology,
University of Texas at Dallas,
Richardson,
Texas,
USA*
e-mail: golden@utdallas.edu

Jim Graham (G4.3)

*Senior Lecturer in Medical Biophysics,
University of Manchester,
United Kingdom*
e-mail: jim.graham@man.ac.uk

Stephen Grossberg (C2.2.1, C2.2.3)

*Chairman and Wang Professor of Cognitive and
Neural Systems;
Director of Center for Adaptive Systems;
and Professor of Mathematics, Psychology,
and Biomedical Engineering,
Boston University,
Massachusetts,
USA*
e-mail: steve@cns.bu.edu

Gary Grudnitski (G6.4)

*Professor of Accountancy,
San Diego State University,
California,
USA*
e-mail: gary.grudnitski@sdsu.edu

Mohamad H Hassoun (C1.3)

*Professor of Electrical and Computer Engineering,
Wayne State University,
Detroit,
Michigan,
USA*
e-mail: hassoun@brain.eng.wayne.edu

Atsushi Hiramatsu (G2.2)

Senior Research Engineer,
NTT Network Service Systems Laboratories,
Tokyo,
Japan
e-mail: hiramatsu@csl.ntt.jp

Paul G Horan (E1.5)

Senior Research Scientist,
Hitachi Dublin Laboratory,
Ireland
e-mail: paul.horan@hdl.ie

Peggy Israel Doerschuk (C2.2.2)

Assistant Professor of Computer Science,
Lamar University,
Beaumont,
Texas,
USA
e-mail: doerschupi@hal.lamar.edu

George W Irwin (G2.9)

Professor of Control Engineering,
The Queen's University of Belfast,
United Kingdom
e-mail: g.irwin@ee.qub.ac.uk

Marwan A Jabri (G5.3)

Professor of Adaptive Systems;
and Director of the Systems Engineering and Design
*   Automation Laboratory,*
University of Sydney,
New South Wales,
Australia
e-mail: marwan@sedal.usyd.edu.au

Geoffrey B Jackson (G2.11)

Design Engineer,
Information Storage Devices,
San Jose,
California,
USA
e-mail: gjackson@isd.com

Thomas O Jackson (B4)

Research Manager,
High Integrity Systems Engineering Group,
University of York,
United Kingdom
e-mail: tom@minster.york.ac.uk

John L Johnson (G1.6)

Research Physicist,
US Army Missile Command,
Redstone Arsenal,
Alabama,
USA
e-mail: jjohn@ssdd.redstone.army.mil

Roger D Jones (G2.7)

Director of Basic Technologies,
Center for Adaptive Systems Applications,
Los Alamos,
New Mexico,
USA
e-mail: rdj@lacasa.com

Christian Jutten (C1.6)

Professor of Electrical Engineering,
University Joseph Fourier;
and Director of the Image Processing and Pattern
*   Recognition Laboratory (LTIRF),*
National Polytechnic Institute of Grenoble (INPG),
France
e-mail: chris@tirf.inpg.fr

S Sathiya Keerthi (C3)

Associate Professor of Computer Science and Automation,
Indian Institute of Science,
Bangalore,
India
e-mail: ssk@csa.iisc.ernet.in

Wolfgang Knecht (G2.10)

Doctor of Technical Sciences,
Research and Development Department,
Phonak AG,
Staefa,
Switzerland
e-mail: phonak@dial-switch.ch

Aleksandar Kostov (G5.1)

Research Assistant Professor,
Faculty of Rehabilitation Medicine,
University of Alberta,
Edmonton,
Canada
e-mail: aleks.kostov@ualberta.ca

Cris Koutsougeras (C2.3)

Associate Professor of Computer Science,
Tulane University,
New Orleans,
Louisiana,
USA
e-mail: ck@cs.tulane.edu

Govindaraj Kuntimad (G1.6)

Engineering Specialist,
Rockwell International,
Huntsville,
Alabama,
USA
e-mail: gkuntima@rdyne.rockwell.com

Barry Lennox (G2.8)

Research Associate in Chemical Engineering,
University of Newcastle-upon-Tyne,
United Kingdom
e-mail: barry.lennox@ncl.ac.uk

Gordon Lightbody (G2.9)

Lecturer in Control Engineering,
The Queen's University of Belfast,
United Kingdom
e-mail: g.lightbody@ee.qub.ac.uk

Alexander Linden (B5.2)

Staff Scientist,
General Electric Corporate Research and Development
  Center,
Niskayuna,
New York,
USA
e-mail: alexander.linden@crd.ge.com

Stephen P Luttrell (B5.3)

Senior Principal Research Scientist in Pattern and
  Information Processing,
Defence Research Agency,
Worcestershire,
United Kingdom
e-mail: luttrell@signal.dra.hmg.gb

Gerhard Mack (G3.2)

Professor of Physics,
University of Hamburg,
Germany
e-mail: mack@x4u2.desy.de

Robert A J Matthews (G8.1)

Visiting Research Fellow,
Aston University,
Birmingham,
United Kingdom
e-mail: 100265.3005@compuserve.com

William C Mead (G2.7)

President,
Adaptive Network Solutions Inc,
Los Alamos,
New Mexico,
USA
e-mail: wcm@ansr.com

M Mehmet Ali (G2.4)

Associate Professor of Electrical and Computer
  Engineering,
Concordia University,
Montreal,
Quebec,
Canada
e-mail: mustafa@ece.concordia.ca

Thomas V N Merriam (G8.1)

Independent Scholar,
Basingstoke,
United Kingdom

Perry D Moerland (E1.2)

Researcher,
Institut Dalle Molle d'Intelligence Artificielle Perceptive
  (IDIAP),
Martigny,
Switzerland
e-mail: perry.moerland@idiap.ch

Gary A Montague (G2.8)

Reader in Process Control,
University of Newcastle-upon-Tyne,
United Kingdom
e-mail: gary.montague@ncl.ac.uk

Helen B Morton (C1.5)

Lecturer in Psychology,
Brunel University,
Middlesex,
United Kingdom
e-mail: helen.morton@brunel.ac.uk

Gary Lawrence Murphy (F1.1)

Director of Communications Research,
TeleDynamics Telepresence and Control Systems,
Sauble Beach,
Ontario,
Canada
e-mail: garym@maya.sos.on.ca

Alan F Murray (G2.11)

Professor of Neural Electronics,
University of Edinburgh,
United Kingdom
e-mail: a.f.murray@ee.ed.ac.uk

Robert A Mustard (G5.6)

Assistant Professor,
Department of Surgery,
University of Toronto,
Ontario,
Canada

Huu Tri Nguyen (G2.4)

Systems Engineer,
CAE Electronics Ltd,
Montreal,
Quebec,
Canada

Craig Niederberger (G5.4)

Assistant Professor of Urology,
Obstetrics-Gynecology and Genetics;
Chief of the Division of Andrology;
and Director of Urologic Research,
University of Illinois at Chicago,
USA
e-mail: craign@uic.edu

James L Noyes (B3)

Professor of Computer Science,
Wittenberg University,
Springfield,
Ohio,
USA
e-mail: noyes@wittenberg.edu

Witold Pedrycz (D1)

Professor of Computer Engineering and Computer Science,
University of Manitoba,
Winnipeg,
Canada
e-mail: pedrycz@ee.umanitoba.ca

Shawn D Pethel (G3.3)

Electronics Engineer,
US Army Missile Command,
Redstone Arsenal,
Alabama,
USA
e-mail: sdpethel@ssdd.redstone.army.mil

Tom Pike (G5.6)

Software Engineer,
University of Toronto,
Ontario,
Canada

Riccardo Poli (G5.5)

Lecturer in Artificial Intelligence,
University of Birmingham,
United Kingdom
e-mail: r.poli@cs.bham.ac.uk

V William Porto (D2)

Senior Staff Scientist,
Natural Selection Inc,
La Jolla,
California,
USA
e-mail: bporto@natural-selection.com

Susan E Pursell (G5.4)

Resident,
Department of Urology,
University of Illinois at Chicago,
USA

Heggere S Ranganath (G1.6)

Associate Professor of Computer Science,
University of Alabama,
Huntsville,
USA
e-mail: ranganat@cs.uah.edu

B Ravindran (C3)

Research Scholar,
Department of Computer Science and Automation,
Indian Institute of Science,
Bangalore,
India
e-mail: ravi@bheeshma.csa.iisc.ernet.in

A N Refenes (G6.3)

Associate Professor;
and Director of the Neuroforecasting Unit,
London Business School,
United Kingdom
e-mail: pnr@lbs.co.uk

Duncan Ross (G6.2)

Recognition Systems Ltd,
Stockport,
United Kingdom

Burkhard Rost (G4.1)

Physicist,
European Molecular Biology Laboratory,
Heidelberg,
Germany
e-mail: rost@embl-heidelberg.de

D G Sandler (G3.1)

Chief Scientist,
ThermoTrex Corporation,
San Diego,
California,
USA
e-mail: dsandler@crash.cts.com

I Saxena (E1.5)

Institut Dalle Molle d'Intelligence Artificielle Perceptive
(IDIAP),
Martigny,
Switzerland
e-mail: isaxena@idiap.ch

Soheil Shams (F1.3)

Senior Research Staff Member,
Hughes Research Laboratories,
Malibu,
California,
USA
e-mail: shams@maxwell.hrl.hac.com

Dan Simon (G2.5)

Senior Test Engineer,
TRW Vehicle Safety Systems,
Mesa,
Arizona,
USA
e-mail: d.simon@ieee.org

E E Snyder (G4.2)

Biocomputational Scientist,
Sequana Therapeutics Inc,
La Jolla,
California,
USA
e-mail: eesnyder@sequana.com

Marcus Speh (G3.2)

Director,
Knowledge Management Services,
Andersen Consulting,
London,
United Kingdom
e-mail: marcus.speh@ac.com

Richard B Stein (G5.1)

Professor of Physiology and Neuroscience,
University of Alberta,
Edmonton,
Canada
e-mail: richard.stein@ualberta.ca

Maxwell B Stinchcombe (B2.10)

Associate Professor of Economics,
University of Texas at Austin,
USA
e-mail: maxwell@mundo.eco.utexas.edu

Gary D Stormo (G4.2)

University of Colorado,
Department of MCD Biology,
Boulder,
USA
e-mail: stormo@beagle.colorado.edu

Harold Szu (C1.4)

Alfred and Helen Lamson Professor of Computer Science;
and Director of the Center for Advanced Computer Studies,
University of Southwestern Louisiana,
Lafayette,
USA
e-mail: hszu@cacs.usl.edu

J G Taylor (A1.1, H1.3)

Director of the Centre for Neural Networks;
and Professor of Mathematics,
King's College,
London,
United Kingdom
e-mail: udah057@bay.cc.kcl.ac.uk

Monroe M Thomas (C1.8, G2.1, G5.1)

Vice President of Dendronic Decisions Ltd,
Edmonton,
Alberta,
Canada
e-mail: mmt@msn.com

Kari Torkkola (F1.7, G1.4)

Principal Staff Scientist,
Motorola Phoenix Corporate Research Laboratories,
Tempe,
Arizona,
USA
e-mail: a540aa@email.mot.com

Guido Valli (G5.5)

Associate Professor of Bioengineering,
University of Florence,
Italy
e-mail: valli@cobra.ing.unifi.it

Alex Vary (G2.6)

Deputy Branch Chief, Retired,
Structural Integrity Branch,
NASA Lewis Research Center,
Cleveland,
Ohio,
USA

Michel Verleysen (C2.1)

Research Fellow in Microelectronics and Neural Networks,
National Fund for Scientific Research,
Université Catholique de Louvain,
Belgium
e-mail: verleysen@dice.ucl.ac.be

Eric A Vittoz (E1.3)

Senior Vice President and Head of Bio-inspired Systems,
Centre Suisse d'Electronique et de Microtechnique SA,
Neuchâtel,
Switzerland;
and Professor of Electrical Engineering,
École Politechnique Fédéreli de Lausanne (EPFL),
Switzerland
e-mail: vittoz@csemne.ch

Paul B Watta (C1.3)

Assistant Professor of Electrical and Computer
* Engineering,*
Wayne State University,
Detroit,
Michigan,
USA
e-mail: watta@brain.eng.wayne.edu

Paul J Werbos (A2, F1.9)

Program Director for Neuroengineering,
National Science Foundation,
Arlington,
Virginia,
USA
e-mail: pwerbos@nsf.gov

Hu Jun Yin (G1.1)

Research Fellow,
Department of Electrical Engineering and Electronics,
University of Manchester Institute of Science and
* Technology,*
United Kingdom
e-mail: yin@umist.ac.uk

INDEXES

Index